THE BOUNDARIES OF CONSCIOUSNESS: NEUROBIOLOGY AND NEUROPATHOLOGY

THE BOUNDARIES OF CONSCIOUSNESS: NEUROBIOLOGY AND NEUROPATHOLOGY

EDITED BY

STEVEN LAUREYS

Department of Neurology and Cyclotron Research Center, University of Liège, Liege, Belgium

This volume is an official title of the

Association for the Scientific Study of Conciousness ASSC

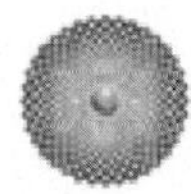

ELSEVIER

AMSTERDAM – BOSTON – HEIDELBERG – LONDON – NEW YORK – OXFORD
PARIS – SAN DIEGO – SAN FRANCISCO – SINGAPORE – SYDNEY – TOKYO

Elsevier
Radarweg 29, PO Box 211, 1000 AE Amsterdam, The Netherlands
The Boulevard, Langford Lane, Kidlington, Oxford OX5 1GB, UK

First edition 2005 (Hardbound version)
First Paperback edition: 2006

Library of Congress Cataloging-in-Publication Data
A catalog record for this book is available from the Library of Congress

British Library Cataloguing in Publication Data

The boundaries of conciousness: neurobiology and
neuropathology. – (Progress in brain research; v. 150)
1. Consciousness 2. Nervous system – Diseases 3. Brain –
Diseases 4. Neurobiology
I. Laureys, Steven
612.8′2

ISBN-10: 0444518517

ISBN-13: 978-0-444-51851-4 (**Hardbound version** of: *Progress in Brain Research Volume 150*)
ISBN-10: 0-444-51851-7 (**Hardbound version** of: *Progress in Brain Research Volume 150*)
ISBN-13: 978-0-444-52876-6 (**Paperback version**; this volume)
ISBN-10: 0-444-52876-8 (**Paperback version**, this volume)
ISSN: 0079-6123 (Series)

Printed and bound in The Netherlands

06 07 08 09 10 10 9 8 7 6 5 4 3 2 1

List of Contributors

S. Adam, Department of Neuropsychology, University of Liège, B32 Sart Tilman, B-4000 Liege, Belgium

J.H. Adams, 31 Burnhead Road, Glasgow G43 2SU, UK

G. Albouy, Cyclotron Research Centre B30, University of Liege - Sart Tilman, 4000 Liege, Belgium

I. Aleksander, Department of Electrical and Electronic Engineering, Imperial College, London SW7 2BT, UK

M.T. Alkire, University of California at Irvine, Medical Center, Department of Anesthesiology *and* the Center for Neurobiology of Learning and Memory, 101 City Drive South, Bldg. 43, Route 81-A, Orange, CA 92868, USA

K. Andrews, Institute of Complex Neuro-disability, Royal Hospital for Neuro-disability, London, UK

B.J. Baars, The Neurosciences Institute, 10640 John Jay Hopkins Drive, San Diego, CA 92121, USA

S.B. Backman, Department of Anesthesia, McGill University, Royal Victoria Hospital, 687 Pine Avenue West, Suite S5.05, Montreal, QC H3A 1A1, Canada

J.L. Bernat, Neurology Section, Dartmouth-Hitchcock Medical Center, 1 Medical Center Drive, Lebanon, NH 03756, USA

J. Berré, Intensive Care Medicine, Hôpital Erasme, Université Libre de Bruxelles, Route de Lennik 808, 1070 Brussels, Belgium

E.L. Berry, MRC Cognition and Brain Sciences Unit, 15 Chaucer Road, Cambridge CB2 2EF, UK

O. Blanke, Laboratory of Cognitive Neuroscience, Ecole Polytechnique Fédérale de Lausanne (EPFL), Swiss Federal Institute of Technology, 1015 Lausanne, Switzerland

H. Blumenfeld, Departments of Neurology, Neurobiology and Neurosurgery, Yale University School of Medicine, 333 Cedar Street, New Haven, CT 06520-8018, USA

M. Boly, Cyclotron Research Centre B30, University of Liege - Sart Tilman, 4000 Liege, Belgium

M. Bullinger, Centre of Psychosocial Medicine, Institute and Policlinics of Medical Psychology, University Clinic Hamburg-Eppendorf, Martinistr. 52, S35, 20246 Hamburg, Germany

S. Bünning, Laboratory of Cognitive Neuroscience, Ecole Polytechnique Fédérale de Lausanne (EPFL), 1015 Lausanne, Switzerland

A. Cleeremans, Cognitive Science Research Unit, Université Libre de Bruxelles CP 191, Avenue F.-D. Roosevelt 50, 1050 Brussels, Belgium

M.R. Coleman, Wolfson Brain Imaging Centre and the Cambridge Coma Study Group, Addenbrooke's Hospital, University of Cambridge, Hills Road, Cambridge CB2 2QQ, UK

F. Collette, Department of Neuropsychology, University of Liege, B32 Sart Tilman, B-4000 Liege, Belgium

T.F.T. Collins, Department of Pharmacology, Mansfield Road, University of Oxford, Oxford OX1 3QT, UK

F. Damas, Intensive Care Medicine, Centre Hospitalier Régional de la Citadelle, Boulevard du 12e de Ligne 1, 4000 Liege, Belgium

T. Dang-Vu, Cyclotron Research Centre B30, University of Liege - Sart Tilman, 4000 Liege, Belgium

M.H. Davis, MRC Cognition and Brain Sciences Unit, 15 Chaucer Road, Cambridge CB2 2EF, UK

M. Desseilles, Cyclotron Research Centre B30, University of Liege - Sart Tilman, 4000 Liege, Belgium

A. Destrebecqz, LEAD, Université de Bourgogne, Pôle AAFE - Esplanade Erasme, BP 26513, 21065 Dijon Cedex, France
M.-E. Faymonville, Pain Clinic and Department of Anesthesiology, University Hospital of Liege, Sart Tilman B33, Liege, Belgium
J.J. Fins, Division of Medical Ethics, Weill Medical College of Cornell University, 525 East 68th Street, F-173, New York, NY 10021, USA
P. Fiset, Department of Anesthesia, McGill University, Royal Victoria Hospital, 687 Pine Avenue West, Suite S5.05, Montreal, QC H3A 1A1, Canada
C.C. French, Anomalistic Psychology Research Unit, Department of Psychology, Goldsmiths College, University of London, New Cross, London SE14 6NW, UK
S. Ghorbel, Neurorehabilitation Medicine, Hôpital Caremeau, CHU Nîmes, 30029 Nimes Cedex, France
J.T. Giacino, New Jersey Neuroscience Institute, 65 James Street, Edison, NJ 08818, USA
H. Gill-Thwaites, Occupational Therapy Department, Royal Hospital for Neuro-disability, London, UK
S. Goldman, Biomedical PET Unit, Hôpital Erasme, Université Libre de Bruxelles, Route de Lennik 808, 1070 Brussels, Belgium
D.I. Graham, Academic Unit of Neuropathology, Institute of Neurological Sciences, Southern General Hospital, South Glasgow University Hospitals NHS Trust, Glasgow G51 4TF, UK
S.A. Greenfield, Department of Pharmacology, Mansfield Road, University of Oxford, Oxford OX1 3QT, UK
J.-M. Guérit, Clinique Edith Cavell, rue Edith Cavell 32, B-1180 Brussels, Belgium
J. Hirsch, Departments of Radiology and Psychology, Center for Neurobiology and Behavior, Columbia University, New York, NY 10032, USA
B. Jennett, Institute of Neurological Sciences, University of Glasgow, Glasgow, Scotland, UK.
E.R. John, Brain Research Laboratories, Department of Psychiatry, NYU School of Medicine, 550 First Avenue, New York, NY 10016, USA
I.S. Johnsrude, MRC Cognition and Brain Sciences Unit, 15 Chaucer Road, Cambridge CB2 2EF, UK
E. Kalbe, Max-Planck Institute for Neurological Research at Cologne, Gleueler Street 50, D-50921 Cologne, Germany
T.T.J. Kircher, Department of Psychiatry and Psychotherapy, University of Aachen, Pauwelsstrasse 30, D-52074 Aachen, Germany
T.W. Kjaer, Department of Clinical Neurophysiology, Copenhagen University Hospital, DK-2100 Copenhagen, Denmark
B. Kotchoubey, Institute of Medical Psychology and Behavioral Neurobiology, Eberhard-Karls-University of Tubingen, Gartenstraat 29, 72074 Tubingen, Germany
A. Kübler, Institute of Medical Psychology and Behavioral Neurobiology, University of Tübingen, Gartenstrasse 29, 72074 Tubingen, Germany
R. Kupers, Center for Functionally Integrative Neuroscience (CFIN), Aarhus University and Aarhus University Hospital, Aarhus, Denmark
M. Lamy, Anesthesiology, Reanimation and Pain Clinic, CHU University Hospital, Sart Tilman B33, 4000 Liege, Belgium
S. Laureys, Centre de Recherches du Cyclotron, Department of Neurology, University of Liege, Sart Tilman-B30, 4000 Liege, Belgium
H.C. Lou, Department of Functionally Integrative Neuroscience, Aarhus University Hospital, Aarhus C, DK-8000, Denmark
S. Majerus, Department of Cognitive Sciences, University of Liege, Liege, Belgium
P. Maquet, Cyclotron Research Centre B30, University of Liege - Sart Tilman, 4000 Liege, Belgium
A. Maudoux, Cyclotron Research Centre B30, University of Liege - Sart Tilman, 4000 Liege, Belgium
W.L. Maxwell, Department of Human Anatomy, I.B.L.S. University of Glasgow, Glasgow G12 8QQ, UK

D.K. Menon, Wolfson Brain Imaging Centre and the Cambridge Coma Study Group, Addenbrooke's Hospital, University of Cambridge, Hills Road, Cambridge CB2 2QQ, UK

J. Miller, University of California at Irvine, Medical Center, Department of Anesthesiology *and* the Center for Neurobiology of Learning and Memory, 101 City Drive South, Bldg. 43, Route 81-A, Orange, CA 92868, USA

G. Moonen, Neurology Department and Cyclotron Research Center, University of Liege, Sart Tillman B30, 4000 Liege, Belgium

E. Myin, Centre for Philosophical Psychology, Department of Philosophy, University of Antwerp, Rodestraat 14, Room R110, 2000 Antwerp, Belgium *and* Centre for Logic and Philosophy of Science, Department of Philosophy, Vrije Universiteit Brussels, Pleinlaan 2, 1050 Brussels, Belgium

L. Naccache, Fédération de Neurophysiologie Clinique, Hôpital de la Pitié-Salpêtrière, 47-83 Boulevard de l'Hôpital, 75013 Paris, France

N. Neumann, Institute of Medical Psychology and Behavioral Neurobiology, University of Tubingen, Gartenstraat 29, 72074 Tubingen, Germany

A. Noë, Department of Philosophy, University of California, Berkeley, CA 92720-2390, USA

M. Nowak, Department of Clinical Physiology, Bispebjerg Hospital *and* Cyclotron and Positron Emission Tomography Center, Rigshospitalet, Copenhagen University Hospital, DK-2100 Copenhagen, Denmark

J.K. O'Regan, Laboratoire de Psychologie Expérimentale, Institut de Psychologie, Centre Universitaire de Boulogne, 71 avenue Edouard Vaillant, 92774 Boulogne-Billancourt Cedex, France

A.M. Owen, MRC Cognition and Brain Sciences Unit, 15 Chaucer Road, Cambridge CB2 2EF, UK

K.-H. Pantke, German Association Locked in Syndrome, Evangelischen Krankenhaus Köningin Elisabeth Herzberge GmbH (Lehrkrankenhaus der Charité), Haus 30, Herzbergstrasse 79, 10365 Berlin, Germany

P. Peigneux, Cyclotron Research Centre B30, University of Liege - Sart Tilman, 4000 Liege, Belgium

F. Pellas, Neurorehabilitation Medicine, Hôpital Caremeau, CHU Nîmes, 30029 Nimes Cedex, France

D. Perani, Vita-Salute San Raffaele University, Via Olgettina 60, 20132 Milan, Italy

F. Perrin, Neurosciences et Systèmes Sensoriels, Unité Mixte de Recherche 5020, Université Claude Bernard Lyon I - CNRS, 69007 Lyon, France

C. Petersen, Centre of Psychosocial Medicine, Institute and Policlinics of Medical Psychology, University Clinic Hamburg-Eppendorf, Martinistr 52, S35, 20246 Hamburg, Germany

J.D. Pickard, Wolfson Brain Imaging Centre and the Cambridge Coma Study Group, Addenbrooke's Hospital, University of Cambridge, Hills Road, Cambridge CB2 2QQ, UK

G. Plourde, Department of Anesthesia, McGill University, Royal Victoria Hospital, 687 Pine Avenue West, Suite S5.05, Montreal, QC H3A 1A1, Canada

M.I. Posner, Department of Psychology, 1227 University of Oregon, Eugene, OR 97403-1227, USA

U. Ribary, Department of Physiology and Neuroscience, NYU School of Medicine, New York, NY 10016, USA

J.M. Rodd, Department of Psychology, University College London, London WC1 H OAP, UK

P. Ruby, Cyclotron Research Centre and Department of Neurology, University of Liege, B35 Sart Tilman, B4000 Liege, Belgium

E. Salmon, Cyclotron Research Centre and Department of Neurology, University of Liege, B35 Sart Tilman, B4000 Liege, Belgium

N.D. Schiff, Department of Neurology and Neuroscience, Weill Medical College of Cornell University, 1300 York Avenue Room F610, New York, NY 10021, USA

C. Schnakers, Neurology Department and Cyclotron Research Center, University of Liege, Sart Tilman B30, 4000 Liege, Belgium

V. Sterpenich, Cyclotron Research Centre B30, University of Liege - Sart Tilman, 4000 Liege, Belgium

R. Thienel, Department of Psychiatry and Psychotherapy, University of Aachen, Pauwelsstrasse 30, D-52074 Aachen, Germany

G. Tononi, Department of Psychiatry, University of Wisconsin at Madison, 6001 Research Park Boulevard, Madison, WI 53719, USA

P. Van Eeckhout, Department of Speech Therapy, Hospital Pitié Salpétrière, Paris and French Association Locked in Syndrome (ALIS), 225 Bd Jean-Jaures, MBE 182, 92100 Boulogne-Billancourt, France

J.-L. Vincent, Department of Intensive Care, Erasme University Hospital, Route de Lennik 808, B-1070 Brusssels, Belgium

B.A. Vogt, Cingulum Neurosciences Institute, 4435 Stephanie Drive, Manlius, NY 13104, USA

P. Vuilleumier, Laboratory for Behavioral Neurology and Imaging of Cognition, Clinic of Neurology and Department of Neurosciences, University Medical Center, 1 rue Michel-Servet, 1211 Geneva 4, Switzerland

N.S. Ward, Wellcome Department of Imaging Neuroscience, Institute of Neurology, University College London, 12 Queen Square, London WC1N 3BG, UK

A. Zeman, Department of Clinical Neurosciences, Western General Hospital, University of Edinburgh, Edinburgh EH4 2XU, UK

Foreword

Consciousness and Cautiousness

There was a time, not so long ago, when the word "consciousness" was not welcome in scientific papers. This attitude, often based on a laudable sense of cautiousness but sometimes bordering on 'taboo', was nicely captured by George Miller:

> "consciousness is a word worn smooth by a million tongues.... Maybe we should ban the word for a decade or two until we can develop more precise terms for the several uses which "consciousness" now obscures." (in *Psychology, The Science of Mental Life*, 1962).

Fortunately, neither Miller himself nor many other great scientists including Michael Posner, Larry Weiskrantz, Tim Shallice, or Dan Schacter heeded this advice. In the 1970s and 1980s, consciousness research saw a slow but persistent progress, owing to the development of many paradigms as diverse as the split-brain, neglect and blindsight conditions, sleep and anesthesia, the implicit–explicit dissociation, the attention-orienting "Posner task", or subliminal priming experiments. A steady flow of papers on consciousness kept irrigating the field, but the C word itself was rarely used — only the attentive reader could see that quite a lot of the research bore directly on problems related to consciousness.

Not until the late 1980s did we see an explosion of work directly aimed at exploring the nature of consciousness. I will not attempt to identify all of the causes for this major change, as this is a task for historians of science. In my opinion, however, the movement started under the impulse of highly visible publications by several neuroscientists and psychologists. Jean-Pierre Changeux's *Neuronal Man* (1983), Gerald Edelman's *Remembered Present* (1989), Bernard Baars's *Cognitive Theory of Consciousness* (1989), and Crick and Koch's *Towards a Neurobiological Theory of Consciousness* (1990) all paved the way toward the present state of affairs. Consciousness has now become an amazingly dynamical and almost dizzying field of research, ripe with many interesting discoveries, and a constant buzzing, blooming confusion of articles and books at the rate of hundreds per year.

This situation is justified, to some extent, by the breadth of the scientific database that suddenly appears as relevant to consciousness research. Contributions in the field exhibit an extreme diversity that cuts across the traditional boundaries of scientific disciplines. Whoever aims at understanding this domain must be ready to jump from philosophy of mind to studies of coma, from minute psychological designs in artificial sequence learning to mathematical definitions of information integration, from anatomical studies of anterior cingulate to the complex methods of primate electrophysiology.

It is in this context that the present volume may be particularly valuable. The distinguished panel of philosophers, modelers, psychologists, physicians, and neuroscientists assembled here provides an accessible, yet in-depth perspective on many of those fields of research. Rarely has such a diversity of points of view been made available in a single volume. Browsing through them provides an exciting window into the forefront of consciousness research, as well as the associated philosophical, ethical, and clinical issues.

Yet has the pendulum swung too far? Shouldn't we be a little bit more cautious in our broad use of the word "consciousness"? Many scientists — who also tend to be referees — still think that we would be better off using only the technical terms of our trade (implicit/explicit, overt/covert, awake/asleep, masked/unmasked), and that appealing to the over-encompassing concept of consciousness does not add much. In many circles, it is still considered as the wisest and most cautious approach to carefully avoid any mention of the term "consciousness". In the field of subliminal priming, for instance, it remains customary to read papers in which awareness of the primes was not even measured.

I personally believe that black-outing consciousness is a major error. If some of our theories are even partly correct, the absence or presence of consciousness corresponds to a major change in the activity of the nervous system (Dehaene and Changeux, 2004; Dehaene and Naccache, 2001). Indeed, radical metabolic changes occur in a highly distributed thalamo-cortical system when normal subjects or comatose patients fall in and out of awareness (Laureys, Owen and Schiff, 2004). Such changes in state also imply changes in access to consciousness: the brain activations evoked by the same stimulus can differ importantly as a function of the state during which they occur (Portas et al., 2000). To take another example, recent experiments using masked priming suggest that the results can change dramatically when one sorts the subjects into two groups, those that perceived some of the primes and those that did not (Kouider and Dupoux, 2001). This is hardly surprising, because conscious access makes available a diversity of processes of immediate interest to any psychologist: access to episodic memory, verbal report, consumption of attentional resources. Indeed, consciousness may permeate almost all fields of human research: the "strategic biases" that have been the curse of psychological experimentation for decades in fact reflect the operation of a conscious brain that has its autonomy and cannot be enslaved by the experimenter's instructions.

I am therefore persuaded that, with the hindsight of time, our past neglect of consciousness will be considered as a major error, one that delayed for decades the progress of our discipline. Still, we have to sort the wheat from the chaff. Cautiousness remains deeply needed as we begin to converge on a set of methodologies for studying consciousness, and of theories with which to assess the results. The present volume, by confronting many points of view in the field, is likely to serve as an important landmark on the way to this ultimate goal.

Suggested Readings:

Dehaene, S. and Changeux, J. P. (2004). Neural mechanisms for access to consciousness. In: M. Gazzaniga (Ed.), The Cognitive Neurosciences (3rd ed.) vol. 82, Norton, New York, pp. 1145–1157.

Dehaene, S. and Naccache, L. (2001) Towards a cognitive neuroscience of consciousness: Basic evidence and a workspace framework. Cognition, 79: 1–37.

Kouider, S. and Dupoux, E. (2001) A functional disconnection between spoken and visual word recognition: Evidence from unconscious priming. Cognition, 82(1): B35–B49.

Laureys, S., Owen, A.M. and Schiff, N.D. (2004) Brain function in coma, vegetative state, and related disorders. Lancet Neurol., 3(9): 537–546.

Portas, C.M., Krakow, K., Allen, P., Josephs, O., Armony, J.L. and Frith, C.D. (2000) Auditory processing across the sleep-wake cycle: Simultaneous EEG and fMRI monitoring in humans. Neuron, 28(3): 991–999.

Stanislas Dehaene
President-elect ASSC
Onsay, France

Foreword

There are probably too many books on consciousness. The topic inspires the imagination of many scholars but truly only a few neurologists. I have barely finished one or two books and I admit more books laid down. The shortcomings of these works are blindly obvious — the material does not really apply to my clinical practice.

This book is different. Laureys and contributors have taken up the task to link biocognitive theories, functional neuroimaging, and clinical descriptions of known comatose states, and the result is impressive. The authors have mined this complex and disputatious field and its entire consolidated work is very readable. True, there are chapters that would make many clinical neurologists cringe, but there is nothing wrong with that. There are several superb clinical chapters that should be required reading for anyone treating patients in altered states of consciousness or is involved with long-term care of patients in a vegetative state. Outstanding chapters are the chapter on brain death (one of my odd hobbies too) and an important chapter on the minimally conscious state that contains a serious attempt to define its legitimate use. Finally, there is a nice chapter on neuropathology of the vegetative state by Graham and associates. Professor Jennett provides an important clinical chapter on vegetative state. The legal problems surrounding withdrawal of support in a patient with a vegetative state continue to surface once in a while, often with family members blaming each other, causing unnecessary tensions. I was particularly interested in the chapter on near-death experience in cardiac arrest survivors, particularly because I thought there would be an easy physiological explanation for this phenomenon. The book provides a good basis for further studies.

Many clinicians have been laying the basis for the understanding of mechanisms of unconsciousness, notably Plum, Posner, and Ropper. Others have defined death by neurological criteria (Schwab and Adams). The role of the clinician — in these modern times often a neurointensivist — is important and new observations are constantly published that help us understand how abnormal consciousness may occur. MRI has been a godsend and now helps to understand why some patients do not awaken after meningitis (e.g. infarcts in pons or thalami), general anesthesia (e.g. multiple territorial cerebral infarctions) and may reassure when serial studies are normal (e.g. status epilepticus). Dr. Laureys' book resists easy simplification of the complex issue of impaired consciousness and comatose states. It is probably one of the most ambitious monograph in this field.

Eelco F.M. Wijdicks
Mayo Clinic College of Medicine
Rochester, USA

Preface

A genuine glimpse into what consciousness is would be ***the*** *scientific achievement, before which all past achievements would pale. But at present, psychology is in the condition of physics before Galileo.*

William James, 1899

This volume has a bold ambition. It aims to confront the latest theoretical, philosophical, and computational insights in the scientific study of human consciousness with the most recent behavioral, neuroimaging, electrophysiological, pharmacological, and neuropathological data on brain function in states of impaired consciousness such as brain death, coma, vegetative state, minimally conscious state, locked-in syndrome, near-death experiences, out-of-body experiences, hysteria, general anesthesia, dementia, epilepsy, sleep, hypnosis, and schizophrenia.

Fig. 1. Lost 'consciousnologist'. The challenging search for the neural correlate of consciousness (NCC) in the brain, the "most complicated material object in the known universe". Reproduced with permission–copyright C. Laureys.

The interest of this is threefold. First, the exploration of brain function in diseases of consciousness represents a unique lesional approach to the scientific study of consciousness and adds to the worldwide effort to identify the "neural correlate of consciousness" (NCC) (Fig. 1). Second, patients with altered states of consciousness continue to represent a major clinical problem in terms of clinical assessment, treatment, and daily management. Third, new scientific insights in this field have major ethical and social implications, which are the topic of the last part of this book.

There are many recent monographs on consciousness (e.g., Metzinger, 2000; Zeman, 2003; Cleeremans, 2003; Edelman, 2004; Koch, 2004 to name but a few) and this one certainly is not the best one around. However, this book's contributions demonstrate a rarely encountered fertile mixture that cuts across traditional boundaries of scientific and medical disciplines and confronted distinguished philosophers, modelers, physiologists, pharmacologists, psychologists, physicians, and neuroscientists from all over the world. Even if still far away from unravelling all the mysteries of consciousness and the NCC, I believe many of the 40 chapters in this book offer a genuine forefront glimpse on the solution to the millennia-old mind–body conundrum.

"Consciousness is the guarantor of all we hold to be human and precious", writes Gerald Edelman (2004). I hope this book will not only be of some guidance to the ever increasing number of "consciousnologists", but, more importantly, might ultimately improve our care and understanding of patients suffering from disorders of consciousness.

This book grew out of the Eighth Conference of the Association for the Scientific Study of Consciousness (ASSC; www.assc.caltech.edu) held on June 26–28, 2004, and its satellite symposium on 'Coma and Impaired Consciousness', which took place on June 24, 2004, both at the University of Antwerp, Belgium. I had the honor to act as the principal organizer for the latter and co-organizer of the former, and it is thrilling to present here a selection of the many excellent contributions that were made there. I am very much indebted to Professor Erik Myin (Philosophy, University of Antwerp), chair of ASSC8.

References

Cleeremans, A. (2003) The Unity of Consciousness: Binding, Integration, and Dissociation. Oxford University Press, Oxford.

Edelman, G.M. (2004) Wider Than the Sky: The Phenomenal Gift of Consciousness. Yale University Press, New Haven and London.

Koch, C. (2004) The Quest for Consciousness: A Neurobiological Approach. Roberts & Company Publishers, Englewood, Colorado.

Metzinger, T. (2000) The Neural Correlates of Consciousness: Empirical and Conceptual Questions. Bradford Books, MIT Press, Cambridge, MA.

Zeman, A. (2003) Consciousness: A User's Guide. Yale University Press, London.

Steven Laureys
Liège, March 2005

Acknowledgments

I thank Erik Myin (Philosophy, University of Antwerp), main organizer and chair of ASSC8, and the other members of the ASSC scientific program committee: Tim Bayne (Philosophy, Macquarie University), Axel Cleeremans (Sciences Cognitives, Université Libre de Bruxelles), Susana Martinez-Conde (Barrow Neurological Institute, Phoenix, Arizona), Jean Petitot (Mathematics, School for Advanced Studies in Social Sciences, Paris), Geraint Rees (Cognitive Neuroscience, University College London), and Patrick Wilken (Biology, California Institute of Technology), and members of the local organizing committee: Joachim Leilich (chair) and Peter Reynaert (Philosophy, Universiteit Antwerpen), Xavier Seron (Neuropsychology, Université Catholique de Louvain), Petra Stoerig (Institute of Physiological Psychology, Heinrich-Heine-University Düsseldorf), and Jean Paul Van Bendegem (Philosophy, Vrije Universiteit Brussel). Local and administrative arrangements were assured by Lars De Nul, Livia Verbrugge, Myriam Segers, and Johan Veldeman (Universiteit Antwerpen) and Annette Konings, Annick Claes, and Lucienne Arena (Université de Liège).

Second, I wish to thank all those who have supported me and who have financially secured the enterprise. In the first place, Professor André Luxen, director of the Cyclotron Research Center and Professor Gustave Moonen, Head of the Department of Neurology and Dean of the Faculty of Medicine of the University of Liège, and Joseph Dial, executive director of the Mind Science Foundation (www.mindscience.org), and main sponsor of the satellite symposium on "Coma and Impaired Consciousness" (http://assc.caltech.edu/assc8/satellite.html).

For their past and present financial aid, I am grateful to the 'Fonds National de la Recherche Scientifique' (FNRS), the University of Liège, the 'Centre Hospitalier Universitaire Sart Tilman', the 'Fondation Médicale Reine Elisabeth', the 'Fondation Léon Fredericq' and the 'Université Libre de Bruxelles'. I also acknowledge the Belgian Neurological Society (Professor Jean Schoenen).

The functional imaging projects in coma and related disorders would not have been possible without the professional and passionate support from Marie-Elisabeth Faymonville and Maurice Lamy (Pain Clinic and Anesthesiology – Reanimation). Other inestimable main actors are Philippe Peigneux, Steve Majerus, Caroline Schnakers, Fabienne Collette (Neuropsychology), Mélanie Boly, Eric Salmon, Gaetan Garraux (Neurology), Christian Degueldre, Christophe Phillips, Guy Del Fiore, Joel Aerts, Evelyne Balteau (Cyclotron Research Center), Pierre Damas, and Bernard Lambermont (Intensive Care Medicine) from the University of Liège; Serge Goldman (PET Unit), Jacques Berré (Intensive Care), and Axel Cleeremans (Cognitive Sciences) from the Université Libre de Bruxelles, and Fabien Perrin (Electrophysiology) from the University of Lyon, France (please apologize for the many I have not mentioned here). I also thank Bernard Baars (San Diego, CA), Brent Vogt (Syracuse, NY) and Ron Kupers (Aarhus, Denmark) for the many stimulating debates.

Long time before this project started, Professor Guy Ebinger (Vrije Universiteit Brussel) aroused my interest for neuroscience. I owe him my clinical formation as a neurologist and my first steps in fundamental research, carried out at the laboratory for Pharmaceutical Chemistry and Drug Analysis. I thank him for sending me to Professor George Franck (University of Liège) and Pierre Maquet (Research Director, FNRS), my ever-inspiring invaluable mentor.

Contents

See Colour Plate Section in the back of this book

S. Laureys (Ed.)
Progress in Brain Research, Vol. 150
ISSN 0079-6123

CHAPTER 1

What in the world is consciousness?

Adam Zeman*

Department of Clinical Neurosciences, Western General Hospital, University of Edinburgh, Edinburgh EH4 2XU, UK

Abstract: The concept of consciousness is multifaceted, and steeped in cultural and intellectual history. This paper explores its complexities by way of a series of contrasts: (i) states of consciousness, such as wakefulness and sleep, are contrasted with awareness, a term that picks out the contents of consciousness, which range across all our psychological capacities; (ii) consciousness is contrasted to self-consciousness, which is itself a complex term embracing self-detection, self-monitoring, self-recognition theory of mind and self-knowledge; (iii) "narrow" and "broad" senses of consciousness are contrasted, the former requiring mature human awareness capable of guiding action and self-report, the latter involving the much broader capacity to acquire and exploit knowledge; (iv) an "inner" conception of consciousness, by which awareness is essentially private and beyond the reach of scientific scrutiny, is contrasted with an "outer" conception which allows that consciousness is intrinsically linked with capacities for intelligent behavior; and finally (v) "easy" and "hard" questions of consciousness are distinguished, the former involving the underlying neurobiology of wakefulness and awareness, and the latter the allegedly more mysterious process by which biological processes generate experience — the question of whether this final distinction is valid is a focus of current debate. Varied interests converge on the study of consciousness, from the sciences and the humanities, creating scope for interdisciplinary misunderstandings, but also for a fruitful dialog.

Introduction

The current surge of excitement about the science of consciousness, attested by this volume, flows from several sources — the techniques of functional imaging are enabling us to *see* something of what happens in the brain during experience, and during its absence in states such as coma (Laureys et al., 2001); the idea that much of the brain's activity proceeds *without* giving rise to awareness is encouraging a "contrastive analysis" of the neural substrates of conscious and unconscious processes (Baars, 2002); progress in the design of intelligent machines is allowing us to glimpse what it might take to create an artificial consciousness (see Aleksander elsewhere in this volume). Interestingly, these scientific developments are also bolstering our confidence in the value and veracity of first-person testimony, as some of the most intriguing but elusive phenomena of experience — imagery (Ishai et al., 2000), hallucinations (Ffytche et al., 1998), shivers down the spine (Blood and Zatorre, 2001; Griffiths et al., 2004) — are shown to have informative, distinctive neural correlates. Against this intellectual background, everyone interested in consciousness is bound to look for ways of resolving the tension between the first- and third-person views of our lives, and ways of understanding how experience can be at once real, functional, and rooted in our physical existence (Zeman, 2001, 2002).

This endeavor is of course an ancient one. The contemporary "problem of consciousness" flags up our version of a time-honored quest for a fully satisfying explanation of the relationship between mind and body. How does what passes through our minds relate to the events occurring in our brains? How do the anatomy and physiology of

*Corresponding author. Tel.: +44-131-537-1167;
Fax: +44-131-537-1106; E-mail: adam.zeman@ed.ac.uk

DOI: 10.1016/S0079-6123(05)50001-3

the hundred billion neurons inside our heads generate awareness — what the poet Louis MacNiece described as "the sluice of hearing and seeing," the taste of honey, a needle's prick, the scent of burning rubber, the bite of a cold wind, the vivid procession of moments of experience that composes our conscious lives?

The belief that the key to a solution lies in a fuller understanding of the brain is also ancient, as revealed by this famous, and astonishingly prescient, passage from Hippocrates" text "On the Sacred Disease"(Jones, 1923):

> Men ought to know that from the brain, and from the brain only, arise our pleasures, joys, laughters and jests, as well as our sorrows, pains, griefs and tears. Through it...we think, see, hear, and distinguish the ugly from the beautiful, the bad from the good, the pleasant from the unpleasant...sleeplessness, inopportune mistakes, aimless anxieties, absent-mindedness, and acts that are contrary to habit. These things that we suffer all come from the brain...Madness comes from its moistness.

But progress in understanding *how* experience "arises" from the brain has been disappointingly slow over the ensuing two and half millennia. E.O. Wilson, in 1998, could still identify the problem as a central issue in science (Wilson, 1998):

> ...the master unsolved problem of biology: how the hundred million nerve cells of the brain work together to create consciousness...

To make matters worse, there have always been skeptics who have doubted that we will ever find what we are seeking in the brain. Leibniz voiced such doubts in *Monadology*, which invites us to imagine walking into the midst of an artificial brain (Leibniz, 1714):

> Perception and that which depend on it are inexplicable by mechanical causes, that is by figures and motions. And supposing there were a machine so constructed as to think, feel and have perception, we could conceive of it as enlarged and yet preserving the same proportions, so that we might enter into it as into a mill. And this granted, we should only find on visiting it, pieces which push against one another, but never anything by which to explain perception.

Such pessimism may seem unwarranted, even curmudgeonly, set against the exciting recent work I have referred to. Yet this skepticism is a recurring theme. It has at least two possible sources. The first is the thought that our experience, unlike the brain from which it "arises", is private, invisible, immaterial, ghostly, and inaccessible to the methods of science. If so, no amount of peering and poking in the brain will allow us to find what we are seeking: in this case an orthodox scientific explanation is simply not available for *consciousness*, although it may be for *behavior*, and for events in the brain. This thought suggests that we have been seeking the wrong kind of explanation for the phenomenon of interest. But there is another possibility: that we have been looking for the right kind of explanation for the wrong kind of thing. Perhaps experience is not what we took it to be. This line of thought is currently popular. It suggests that our concept of consciousness has misled us into hunting for a theory of an illusion.

How could this happen? We need to acknowledge, as Dan Dennett has repeatedly reminded us (Dennett, 1991), that "consciousness" is not a straightforward scientific term. On the contrary, it belongs to a network of associated concepts that has evolved over centuries under the pressure of a host of cultural influences — scientific, artistic, philosophical, and religious. The network includes other terms such as "mind," "self" (Berrios and Markova, 2003), "soul" and "spirit." Making a scientific sense of these, and all their connotations, is a tall order. Figure 1 shows some of the beliefs of 250 undergraduates at the University of Edinburgh on aspects of the relationship of mind to brain (data submitted for publication), underlining the distance that already exists between the beliefs of most neuroscientists about consciousness and those which are held by a wider but well-educated public.

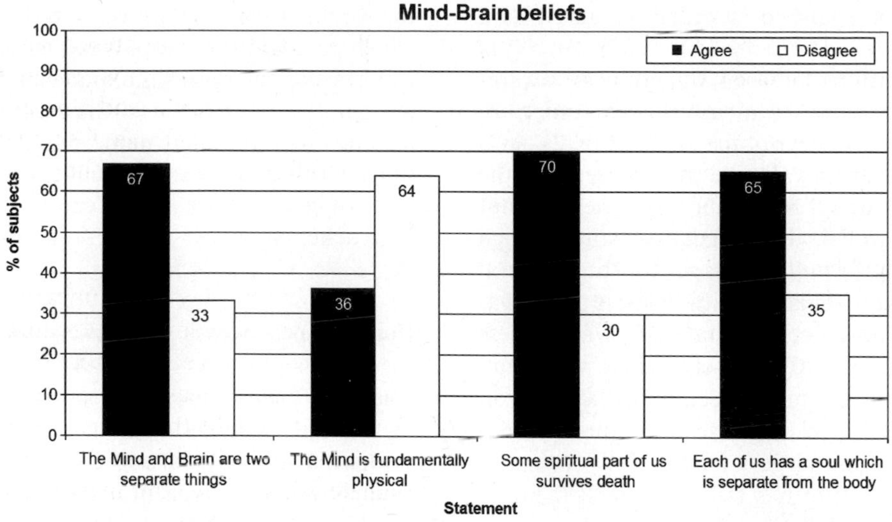

Fig. 1. The results of a survey of 250 students from several disciplines at the University of Edinburgh on attitudes to mind and brain (Liew et al., submitted).

All of us may need to reconceive the scientific target of the quest for "consciousness" in the light of the evidence emerging from the quest. This process will take time.

The complexity and nuances of the concept provide a reason for opening this volume with some introductory thoughts on what we mean by consciousness. I will not be able to resolve its complexities here but can try, at least, to clarify the concept, by way of a series of contrasts designed to highlight its multiple dimensions of ambiguity.

Wakefulness versus awareness

Consciousness has two key senses in colloquial English: wakefulness and awareness. Wakefulness is a *state* of consciousness, distinguished from other states such as sleep and coma. These states admit of degrees: we can be wide awake or half awake, lightly or deeply anesthetized. We are normally confident of our ability to judge an individual's state of consciousness, in this first sense, with the help of objective criteria, like those of the Glasgow Coma Scale (Teasdale and Jennett, 1974): having one's eyes open is generally an indication of wakefulness — being able to converse pretty much settles the matter.

We usually assume that anyone who is awake will also be aware — in other words, not merely conscious but conscious *of* something. Objective criteria are still helpful in ascertaining the presence of consciousness in this second sense — anyone who can obey your instructions and tell you the date is presumably aware — but it has a much stronger connotation of subjectivity than the first sense: as we all know, it is often hard to be sure about what is passing through another person's mind on the basis of their behavior.

The general properties of awareness — the contents of consciousness — have been much discussed. There is a consensus about the following properties: the contents of consciousness are relatively stable for short periods of a few hundred milliseconds, but characteristically changeable over longer ones; they have a narrow focus at a given moment, but over time our awareness can range across the spectrum of our psychological capacities, allowing us to be aware of sensations, percepts, thoughts, memories, emotions, desires, and intentions (our current awareness often combines elements from several of these psychological domains); our awareness is personal, allowing us a distinctive, limited perspective on the world; it is fundamental to the value we place on our lives: keeping people alive once their capacity for awareness has been permanently extinguished is widely regarded as a wasted effort.

The relationships between wakefulness, awareness, and their behavioral indices are more involved than they appear at first sight (Fig. 2). As a rule, while we are awake we are aware. But the phenomena of wakefulness and awareness do not always run parallel. The vegetative state, which results from profound damage to the cerebral hemispheres and thalami, with relative preservation of the brainstem, is a state of "wakefulness without awareness." Conversely, when we dream, we are asleep yet aware. Nor can we always rely on behavioral criteria to diagnose consciousness: patients paralyzed for surgery may be fully aware — if the anesthetic drug has failed to reach them — but completely unable to manifest their awareness; patients "locked in" by a brain stem stroke may appear unconscious until someone recognizes their ability to communicate by movements of their eyes or eyelids (see Laureys et al., elsewhere in this volume).

This is not the place to embark on an account of the science that bears on these two key senses of consciousness, except to point out that a great deal of contemporary neuroscience is relevant. The biological basis of conscious states has been elucidated by work, exploring their electrical correlates in the EEG, and their control by the structures in the brain stem, thalamus, and basal forebrain, which regulate the sleep–wake cycle. The biological basis of awareness is a major target of research in cognitive neuroscience: the study of vision, for instance, has revealed numerous detailed correlations between aspects of visual experience and features of neural processing (Zeman, 2001, 2002) (see Naccache, elsewhere in this volume).

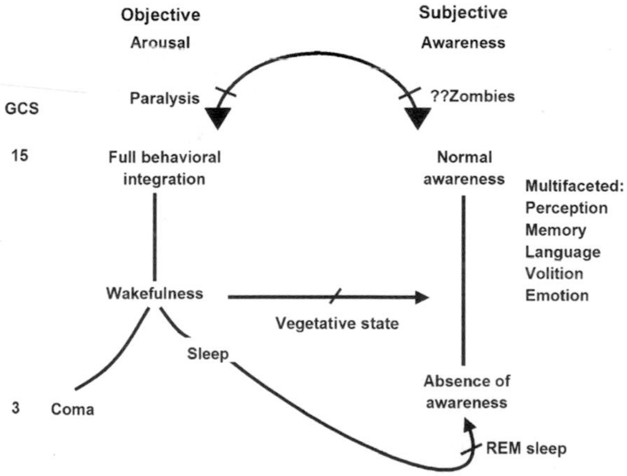

Fig. 2. Interrelationships between arousal and awareness. See text for explanation. "Zombies" are theoretical beings discussed by philosophers who display the signs of consciousness, yet lack experience: the plausibility of such beings is controversial.

Consciousness versus self-consciousness

The term "self-consciousness" is sometimes used in medical contexts as if its meaning were self-evident. I doubt this: self-consciousness is a peculiarly complex idea, not too surprisingly, as it combines two others — "self" and "conscious" — each of which is multifaceted (Berrios and Markova, 2003). I shall try to tease apart its principle strands.

The distinction between "self" and "other" is biologically crucial: there are many activities that we need to direct toward other objects in the world — like eating them — which would be disastrous if we directed toward ourselves. We should expect to find strategies for drawing this distinction in the simplest organisms. But "self-consciousness" implies more than an ability to behave differently toward self and other: it requires a representation of self and other. A variety of different kinds of representation fall out of the senses I shall discuss (and number for ease of reference).

The *colloquial* sense of self-consciousness (i) — a proneness to embarrassment in the presence of others — is rather sophisticated, as it implies the subject's awareness that the awareness of others is directed on her. A second sense (ii), *self-consciousness as self-detection* refers to a family of forms of self-consciousness that are probably present in many animals. This family includes awareness of stimuli that directly impinge on the body (the ant walking up your arm); of proprioceptive information about bodily position that contributes substantially to our body image; of information about actions that we are about to perform or are performing, giving rise to a sense of agency; of information about bodily state (hunger, thirst, etc.); and of emotions, like fear or affection, which signal the state of our relationship to objects and to people around us, and without which we are liable to lose the sense of our own reality or that of the world, as in "depersonalization" and "derealization."

A third sense, (iii) — *self-consciousness as self-monitoring* extends self-detection in time into past and future, and in range, to encompass more plainly cognitive abilities. It refers to the ability to recall the actions we have recently performed (Beninger et al., 1974), and to our ability to predict our chances of success in tasks which challenge memory (Hampton, 2001) or perception (Smith et al., 2003): we undoubtedly possess these abilities, and ingenious experiments in comparative psychology (Beninger et al., 1974; Hampton, 2001; Smith et al., 2003) suggest that many other animals have them too.

Senses (ii) and (iii) require representation of various activities and capacities of self rather than any unified concept. Senses (iv) and (v), discussed below, draw closer to what we normally have in mind, I suspect, when we speak of self-awareness. The fourth sense, (iv) — *self-consciousness as self-recognition* alludes to our ability to recognize our own bodies as our own, for example in mirrors (i.e., mirror self-recognition). Human children develop this ability at around 18 months (Parker et al., 1994). Gordon Gallup discovered in the 1970s that if apes are given experience with a mirror they will soon realize that they are looking at themselves, while their monkey cousins apparently fail to grasp this fact despite extensive exposure (Gallop, Jr., 1970). Recent evidence suggests that dolphins can also acquire mirror self-recognition (Reiss and Marino, 2001); there are claims for pigeons (Epstein et al., 1981) and magpies (Stoerig, P., personal communication).

Between the ages of 18 months and around 5 years, human children take a further major intellectual stride: they come to appreciate that as well as being objects, that can be inspected in mirrors, they are also subjects, of experience — they possess, in other words, not only bodies, but also minds (Parker et al., 1994). The awareness of ourselves as subjects of experience opens up a world of new possibilities for understanding our own behavior and the behavior of others in terms of desires and beliefs, and for implanting and manipulating these (Baron-Cohen, 1995; Frith and Frith, 1999). It has been described as the acquisition of a "theory of mind." Once we realize that others, like ourselves, have a limited, personal perspective on the world, we can choose to inform, misinform, and influence them in a host of entertaining and profitable ways. This "awareness of awareness" is the penultimate sense (v) of self-consciousness. The degree to which animals other than man possess this awareness is debated.

Finally, we use "self-consciousness" to refer to our self-knowledge in its broadest sense (vi) — one's knowledge of oneself as the hero of a personal narrative, deeply conditioned by one's circumstances and cultural background: contrast our self-knowledge at 6, 16 and 60 years or the conceptions of self in a medieval monk and a contemporary scientist. The capacity to relive our past in the form of "mental time-travel" constitutes the "autonoetic awareness," which Endel Tulving has identified as the distinctively human intellectual achievement (Tulving, 1985). Self-depiction is a central focus of art, another distinctive human activity.

What is the relationship between consciousness and self-consciousness? The implication of the last few paragraphs is that it would be unwise to give a general answer to this question: one needs first to specify the senses of consciousness and self-consciousness one has in mind. Two contrasting, controversial, but influential ideas about this relationship deserve a mention here, although space constraint prevents me from pursuing them further: first, the notion that some form of prereflective self-awareness or "ipseity" is presupposed by even simple forms of perceptual awareness (when I look at the glass I experience "my seeing the glass"); and second, the idea that awareness is summoned into being by self-awareness, specifically by the acquisition of a theory of mind, the idea at the core of some "top-down" theories of consciousness (Humphrey, 1978).

Consciousness narrow versus consciousness broad

The third distinction can be introduced via the etymology of "consciousness." The Latin "conscientia," the root of our "consciousness," was formed by the combination of "cum," meaning "with," and "scio," meaning "know" (Lewis, 1960). In its narrow, but strong, sense, conscientia means knowledge shared with another, often guilty knowledge of the kind one might share with a fellow conspirator. By metaphorical extension it could refer to "knowledge shared with oneself," but alongside this "strong" usage, conscientia was sometimes used in a broad, or weak, sense, in which it meant, simply, knowledge.

The tension between these two senses lives on, I suspect, in our current use of "consciousness." We sometimes reserve the word for mental states the contents of which we can "share" fully with ourselves. For example, I am currently conscious of the book lying on my desk: I could describe it to you, read its cover, reach out and pick it up. In other words, I can report my visual awareness of the book and use it to guide a range of actions. This is full-blooded awareness, consciousness "narrow" and "strong." Yet we are often inclined to attribute consciousness to creatures who lack the full repertoire of "conscious" behavior: alinguistic animals, prelinguistic infants, and dysphasic adults for example. We are even tempted to attribute consciousness in a certain sense in cases where its absence in the strong sense is precisely the focus of our interest. For example, consider experiments in which subject's improving performance in a psychological task reveals that they must be learning a rule (or a "grammar") governing the task, yet they are unable to explain how they are improving their performance, unable to articulate the rule (Berns et al., 1997) (see Destrebecqz and Peigneux, elsewhere in this volume). There is an inclination to say that "in some sense" they must have become conscious of the rule even though, in another sense, they are learning it unconsciously. The first sense here is consciousness "broad" or "weak," consciousness as knowledge.

At the "broad" or "weak" extreme of this spectrum lies a range of processes that we normally regard as quintessentially "unconscious": for example, blindsight (Weiskrantz, 1998) and its analogs blind touch and blind smell (Sobel et al., 1999), priming by neglected stimuli (Rees et al., 2000), priming by masked stimuli (Dehaene et al., 1998), the processing of "unseen" visual stimuli in experiments using binocular rivalry (Kanwisher, 2000) or binocular extinction (Moutoussis and Zeki, 2002) or in states of impaired awareness, like the vegetative and minimally conscious states (Laureys et al., 2001) (see Schiff, Giacino, and Owen et al., elsewhere in this volume). Examples like these create an opportunity for neuroscientists to contrast the features of "conscious" and "unconscious" neural processing (Baars, 2002). These contrasts are in their early days but so far indicate that several plausible candidates can influence the chances that a given stream of neural processing will give rise to awareness: the amplitude and duration of the associated activity, its degree of synchronization (see Ribary, elsewhere in this volume), its site (for example, cortical or subcortical) and its neural "reach" or connectivity.

The most popular current model of conscious processing proposes that it occurs when individually unconscious modules of cognitive function — conceived of in either psychological or anatomical terms — join forces and communicate. In Baars' global workspace theory (Baars, 2002) and Dehaene's neuronal workspace model (Dehaene and Naccache, 2003), information becomes conscious when it is broadcast widely through the brain, allowing forms of cognitive performance that are otherwise unattainable (see Baars, elsewhere in this volume). Dehaene suggests, for example, that these include the bridging of delays, inhibition of habitual responses, the planning, evaluation and monitoring of novel strategies, and higher level semantics. Whether awareness really depends on a qualitatively distinct style of neural processing or rather is a straightforward function of variables such as amplitude and site is a major empirical question for the next decade.

Consciousness inner versus consciousness outer

Some readers will have a nagging sense that the distinction between broad and narrow senses of consciousness in the last section misses the real point of interest: it fails to capture the essential difference between conscious and unconscious processes. Whether information is conscious, you may want to say, does not depend at all upon whether we can report or act on it. It simply depends on whether it gives rise to "an experience" — whether there is something it feels like to be conscious of the information. And for any neural process, however one discovers this, there either is or there is not an associated experience. This notion, that consciousness is a determinate, private, invisible, and crucially important internal process or event, is the "inner" concept of consciousness.

I suspect that this notion is the dominant current conception of consciousness, which is widely shared by scientists as well as non-scientists. It suggests the following kind of picture of brain–mind interaction: certain types of process in the brain "generate" consciousness, rather as, in the tale of the *Arabian Nights*, rubbing Aladdin's magic lamp conjures up the genie (Fig. 3). The product of the process, consciousness, is, indeed, rather like the genie: magical and evanescent (although consciousness is even less visible than the genie). We tend to think that this picture of consciousness is simply obvious, necessitated by a range of basic facts about experience and the brain: for example, dreams, hallucinations, mental imagery, and brain stimulation experiments teach us that certain kinds of brain activities are all that is needed to produce experience. The resulting experience looks nothing like the brain activity that causes it and must therefore be different from it in kind, a stream of autonomous mental events, which arises from the physical ones in the brain.

This widespread conception of consciousness implies that that awareness is a deeply private matter, inaccessible to observation by third parties (Zeman, 2004). On this view, awareness casts an "inner light" on a private performance: in a patient just regaining awareness we imagine the light casting a faltering glimmer, which grows steadier and stronger as a richer awareness returns. We sometimes imagine a similar process of illumination at the phylogenetic dawning of awareness, when animals with simple

Fig. 3. Our standard "inner" conception of consciousness makes its generation by the brain appear miraculous (copyright Cliff Laureys).

nervous systems first became conscious. We wonder whether a similar light might one day come to shine in artificial brains. But, bright or dim, the light is either on or off: awareness is present or absent, and only the subject of awareness knows for sure. The light of awareness is invisible to all but its possessor.

This inner concept creates some real difficulties for the "science of consciousness." It implies that its chief focus of interest lies beyond the reach of scientific observation. If so, we may never be able to integrate consciousness fully into scientific theory. If we cannot cash the language of consciousness in the currency of neuroscience, we will neither be able to provide a deeply satisfying explanation of how matter creates mind, nor to give an intuitive account of how mind, that is, our desires and intentions influences matter. The best we can hope for are correlations identifying the bare mysterious fact that certain neural processes give rise to certain kinds of experiences, which by themselves play no part in the explanation of behavior. This inner view creates an explanatory gap that seems unbridgeable. Things, in brief, look bad for the science of consciousness.

Given the gravity of this impasse, we have a reason to backtrack — to consider the possibility that we might have been mistaken about the nature of consciousness, and perhaps even about its contents. This journey back through our assumptions is bound to be a tough one, as it tends to strike us as being obvious that our experience is as it seems to be and obvious also, in the light of what we know from neuroscience, that experience is generated by the brain. So, if we backtrack, does any alternative offer itself?

One alternative emerges quite naturally when we ask ourselves how we establish that others are conscious. We do so by interacting with them, by engaging with their behavior. When we do this, when we communicate or dance or play with other people, we are left with no genuine doubt about their consciousness. This prompts the thought that experience might not arise from the brain in the way we normally envisage: perhaps, rather than viewing it as a mysterious emanation from the brain, we should think of it as a sophisticated form of interaction with the world, an elaborate process of exploration. This is the "outer" conception of consciousness. If there is something to it, our current efforts to account for consciousness may be excessively "neurocentric" (Zimmer, 2004): perhaps, we need to broaden the horizon of our explanation, to consider the mind as "embodied, embedded, and extended" (Broks, 2003) — embodied in the wider frame of our biological being, embedded in the culture in which it has developed, and extended in space and time through which our transactions with the physical world proceed.

To many of us this is an alien, disturbing view of consciousness, a theory that snuffs out its essence. Yet there are grounds for doubting that our grasp of the contents and nature of experience is as firm as we usually take it to be. The evidence from work on change blindness (O'Regan and Noe, 2001) and inattentional blindness suggests that our "internal representation" of our visual surroundings is less rich than we normally take it to be, shaking our certainty about the contents of our experience; evidence from sensory substitution experiments (O'Regan and Noe, 2001) suggests that the "visual" properties of visual experience may be conferred by the manner in which visual behavior explores the world rather than by the "specific nerve energy" of the normal visual pathways, undermining our usual understanding of how experience arises from the brain (see O'Regan et al., elsewhere in this volume).

Neither the inner nor the outer concept of consciousness alone seems to be fully equal to our needs, yet it is not clear how we should reconcile them. Which of these concepts you prefer, the picture of consciousness "inside the head" or the picture of consciousness "at large" in the world, will determine your view of the final distinction I shall draw, between the "hard" and the "easy" questions of consciousness.

Hard versus easy questions of consciousness

If you are drawn to the inner notion of consciousness, to the picture of experience as an entirely private, invisible mental process, you will have the sense that there is a gulf, an apparently unbridgeable gulf, between all that we can learn about the brain from science and the subjective essence of awareness; whereas if you are attracted by the idea that it may be possible to reconstrue the concept of awareness in terms of complex interactions with the world, you will be more optimistic about the prospects for a science of consciousness.

The distinction between the "hard" and "easy" questions of consciousness was drawn by the philosopher David Chalmers to highlight the difference between explaining the neural events that mediate conscious behavior and the (allegedly) more mysterious process by which these events give rise to awareness (Chalmers 1996) — "how the water of the physical brain is turned into the wine of experience" in the words of another philosopher, Colin McGinn (1991). Given the inner notion of consciousness this distinction becomes an important one: science promises to solve the easy not the hard question. Given the outer view, the distinction collapses: explain how the brain facilitates conscious behavior, and consciousness is explained.

Conclusion

I have tried to give a guide, in this chapter, to the ambiguous concept of consciousness. I began with the uncontroversial distinction between its two key colloquial senses, the relatively objective notion of wakefulness and the more subjective concept of awareness. We made a detour via self-consciousness, teasing apart six related but separable senses of the term. Returning to the main topic of consciousness, we identified a spectrum of senses that extends between consciousness "narrow" and consciousness "broad" — consciousness in its strong sense of knowledge shared with oneself, available for report and the control of voluntary action, and consciousness in its broad sense of "knowledge" pure and simple. We then examined the intuition that what really matters to us when we are deciding whether a person or a state is conscious is simply the occurrence of "experience," not the possibility of report or deliberate action: on this "inner" conception of consciousness, experience is conceived as an immaterial, invisible, and private process. I contrasted this to an "outer" conception which regards "consciousness" as shorthand for intelligent behavior. Finally, we saw that those who are wedded to the inner conception are likely to see the science of consciousness as confronted by an insurmountable challenge — to solve the "hard" problem of consciousness, bridging the explanatory gap between the physical and the mental — while for those who accept the outer conception there is no worrying gap between the two.

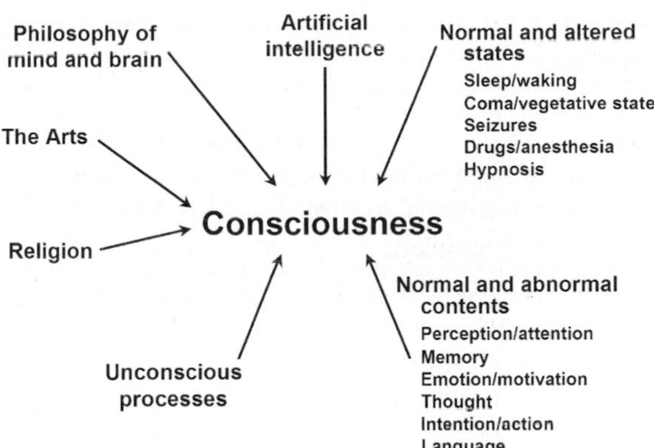

Fig. 4. Sources of evidence for the study of consciousness.

The topic of consciousness is so rich because it lies at the intersection of several intellectual domains (Fig. 4), including some, in the Humanities, which focus on the experiences of subjects, and others, in the Sciences, which highlight processes in objects. The study of the arts, of religion, and of philosophy, all have their contribution to make alongside the study of the brain and its biology. This richness creates a risk of confusion and cross-purposes alongside the possibility of cross-fertilization. We should keep this rich but potentially confusing context in the back of our minds as we explore the fascinating work on disorders of consciousness surveyed in the following chapters.

References

Baars, B.J. (2002) The conscious access hypothesis: origins and recent evidence. Trends Cogn. Sci., 6: 47–52.

Baron-Cohen, S. (1995) Mindblindness. MIT Press, Cambridge, MA.

Beninger, R.J., Kendall, S.B. and Vanderwolf, C.H. (1974) The ability of rats to discriminate their own behaviour. Can. J. Psychol., 28: 79–91.

Berns, G.S., Cohen, J.D. and Mintun, M.A. (1997) Brain regions responsive to novelty in the absence of awareness. Science, 276(5316): 1272–1275.

Berrios, G.E. and Markova, I.S. (2003) The self and psychiatry: a conceptual history. In: Kircher T. and David A. (Eds.), The Self in Neuroscience and Psychiatry. Cambridge University Press, Cambridge.

Blood, A.J. and Zatorre, R.J. (2001) Intensely pleasurable responses to music correlate with activity in brain regions implicated in reward and emotion. Proc. Natl. Acad. Sci. USA, 98(20): 11818–11823.

Broks, P. (2003) Into the Silent Land. Atlantic Books, London.

Chalmers, D.J. (1996) The Conscious Mind. Oxford University Press, Oxford.

Dehaene, S. and Naccache, L. (2003) Towards a cognitive neuroscience of consciousness: basic evidence and workspace framework. Cognition, 79: 1–37.

Dehaene, S., Naccache, L., Le Clec, H.G., Koechlin, E., Mueller, M. and Dehaene-Lambertz, G. (1998) Imaging unconscious semantic priming. Nature, 395: 595–600.

Dennett, D.C. (1991) Consciousness Explained. Little Brown, Boston, MA.

Epstein, R., Lanza, R.P. and Skinner, B.F. (1981) "Self-awareness" in the pigeon. Science, 212: 695–696.

Ffytche, D.H., Howard, R.J., Brammer, M.J., David, A., Woodruff, P. and Williams, S. (1998) The anatomy of conscious vision: an fMRI study of visual hallucinations. Nat. Neurosci., 1(8): 738–742.

Frith, C.D. and Frith, U. (1999) Interacting minds — a biological basis. Science, 286(5445): 1692–1695.

Gallop Jr., G.G. (1970) Chimpanzees: self-recognition. Science, 167(914): 86–87.

Griffiths, T.D., Warren, J.D., Dean, J.L. and Howard, D. (2004) "When the feeling's gone": a selective loss of musical emotion. J. Neurol Neurosurg. Psychiatry, 75(2): 344–345.

Hampton, R.R. (2001) Rhesus monkeys know when they remember. Proc. Natl. Acad. Sci. USA, 98(9): 5359–5362.

Humphrey, N. (1978) Nature's psychologists. New Scientist, 78: 900–903.

Ishai, A., Ungerleider, L. and Haxby, J.V. (2000) Distributed neural systems for the generation of visual images. Neuron, 28: 979–990.

Kanwisher, N. (2000). Neural correlates of changes in perceptual awareness in the absence of changes in the stimulus Towards a science of consciousness, Abstr. No. 164.

Laureys, S., Berre, J. and Goldman, S. (2001) Cerebral function in coma, vegetative state, minimally conscious state, locked-in syndrome and brain death. In: Yearbook of Intensive Care and Emergency Medicine. Springer, Berlin, pp. 386–396.

Leibniz, G. (1714). Monadology.

Lewis, C.S. (1960) Studies in Words. Cambridge University Press, Lansridge.

McGinn, C. (1991) The Problem of Consciousness. Blackwell, Basil.

Moutoussis, K. and Zeki, S. (2002) The relationship between cortical activation and perception investigated with invisible stimuli. Proc. Natl. Acad. Sci USA, 99: 9527–9532.

O'Regan, J.K. and Noe, A. (2001) A sensorimotor account of vision and visual consciousness. Behav. Brain Sci., 24(5): 939–973.

Parker, S.T., Mitchell, R.W. and Boccia, M.L. (1994) Self-awareness in animals and humans. Cambridge University Press, Cambridge.

Rees, G., Wojciulik, E., Clarke, K., Husain, M., Frith, C. and Driver, J. (2000) Unconscious activation of visual cortex in the damaged right hemisphere of a parietal patient with extinction. Brain, 123(Pt 8): 1624–1633.

Reiss, D. and Marino, L. (2001) Mirror self-recognition in the bottlenose dolphin: a case of cognitive convergence. Proc. Natl. Acad. Sci. USA, 98(10): 5937–5942.

Smith, J.D., Shields, W.E. and Washburn, D.A. (2003) The comparative psychology of uncertainty monitoring and metacognition. Behav. Brain Sci., 26(3): 317–339.

Sobel, N., Prabhakaran, V., Hartley, C.A., Desmond, J.E., Glover, G.H., Sullivan, E.V. and Gabrieli, J.D. (1999) Blind smell: brain activation induced by an undetected air-borne chemical. Brain, 122(Pt 2): 209–217.

Teasdale, G. and Jennett, B. (1974) Assessment of coma and impaired consciousness. A practical scale. Lancet, 2(7872): 81–84.

Tulving, E. (1985) Memory and consciousness. Can. Psychol.: 1–12.

Jones, W.H.S. (1923) Hippocrates. William Heinemann and Harvard University Press, London.

Weiskrantz, L. (1998) Blindsight — a Case Study and Implications (Second edition). Clarendon Press, Oxford.

Wilson, E.O. (1998) Consilience — the Unity of Knowledge. Little, Brown and Company, London.

Zeman, A. (2001) Consciousness. Brain, 124: 1263–1289.

Zeman, A. (2004). In: Heywood C.A., Milner D.A. and Blakemore C. (Eds.), Theories of Visual Awareness. Elsevier, Amsterdam, pp. 321–329.

Zeman, A. (2002) Consciousness: a User's Guide. Yale University Press, New Haven.

Zimmer, C. (2004) Soul Made Flesh. Free Press, New York.

S. Laureys (Ed.)
Progress in Brain Research, Vol. 150
ISSN 0079-6123

CHAPTER 2

A neuroscientific approach to consciousness

Susan A. Greenfield* and Toby F.T. Collins

Department of Pharmacology, Mansfield Road, University of Oxford, Oxford, OX1 3QT, UK

Abstract: For a neuroscientist, consciousness currently defies any formal operational definition. However, the phenomenon is distinct from self-consciousness: after all, one can "let oneself go," when experiencing extreme emotion, but still be accessing a sentient, subjective, conscious state. This raw, basic subjective state does not appear to be an exclusive property of the human brain. There is no obvious qualitative transformation in either the anatomy or the physiology of the central nervous system of human or non-human animals, no phylogenetic Rubicon in the animal kingdom. Similarly, there is no clear ontogenetic line that is crossed as the brain grows in the womb, no single event or change in brain physiology, and certainly not at birth, when consciousness might be generated in an all-or-none fashion. A more plausible, and scientific, view of consciousness might be therefore that it is not a different property of the brain, some magic bullet, but that it is a consequence of a quantitative increase in the complexity of the human brain: consciousness will grow as brains grow. Hence, consciousness is most likely to be a continuously variable property of the brain, in both phylogenetic and ontogenetic terms. Here, we describe how modern techniques may be utilized to determine the physiological basis of consciousness.

Introduction

The only way in which the physical brain can accommodate the ebb and flow of a continuously variable conscious state, would be at the intermediate level between macro brain regions and individual neurons: the level of highly transient assemblies of brain cells that wax and wane in size, from one moment to the next. As yet, this rate of turnover of assemblies and their size at any one time has not been quantifiable. Perhaps in the future, however, a unit of measure could be formulated that could determine an "assemblage," i.e., a quantifiable measure reflecting the combined temporal and spatial dynamics of constantly changing neuronal assemblies. Such assemblies, where 10 million neurons can synchronize in activity over merely 230 ms, have been established for over a decade (Grinvald et al., 1994). It has previously been suggested that these transient, three-dimensional configurations of large-scale assemblies throughout the brain, which need not respect conventional anatomical boundaries, will correlate with different degrees of consciousness at any one moment (Greenfield, 1997, 2000a, b). These assemblies will vary in size from one moment to the next, according to (i) the strength of the trigger (a hub of neuronal circuitry, analogous to a stone in a puddle) that initiates their transient synchrony, and also (ii) according to the ease with which the neurons will be synchronized, in turn dependent on (iii) the availability of facilitating "modulatory" chemicals that, through synthesis/breakdown and phosphorylation/dephosphorylation of cellular proteins, function as the physical correlate of "mood". These chemicals will originate from widespread sources in the brain, and also from the rest of the body, including the immune system. It is this two-way iteration of chemicals, of which the assembly size is a mere index, that can be viewed as "consciousness" (Fig. 1).

*Corresponding author. Tel.: +44 1865 271852;
Fax: +44 1865 271853; E-mail: susan.greenfield@pharm.ox.ac.uk

DOI: 10.1016/S0079-6123(05)50002-5

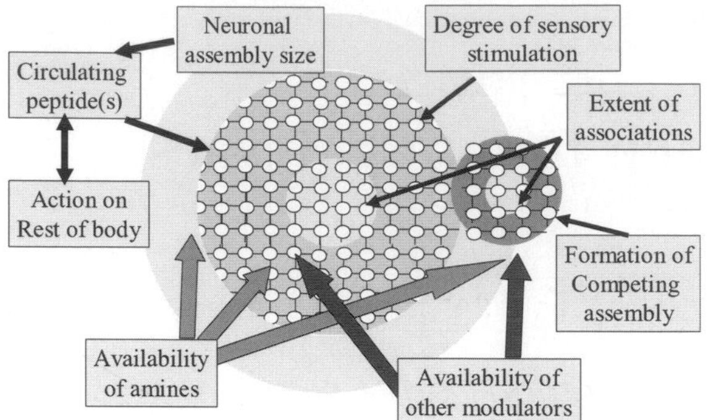

Fig. 1. A possible mechanism for the generation of consciousness. The two sets of concentric circles represent two transient assemblies of tens of millions of brain cells: the largest assembly will dominate at any one moment in the brain, and determine that moment of consciousness. The degree to which cells are recruited, and hence the degree of consciousness, will be determined by a variety of factors, such as vigor of sensory inputs, pre-existing connections ("associations"), and degree of competition ("distractions"), as shown by the smaller assemblies starting to form. Signature chemicals, such as peptides, will be released from the transient assembly. The type, number, and concentration of these chemicals will thus represent the salient assembly in the brain, and convey that information to the rest of the body, via the circulation. In turn, chemicals released from the immune system and the vital organs will modify the working assembly of neurons, as will other chemicals such as hormones and the "amines" that are released in relation to arousal. Consciousness is therefore dependent on the whole body working cohesively (see Greenfield, 2000b).

Imaging of these transient assemblies would provide a net index of consciousness, a much sought after, and as yet elusive, "correlate of consciousness" that is both necessary and sufficient (Greenfield, 2000a). It is indeed possible to generate potentially falsifiable hypotheses as to how different factors, in turn expressible simultaneously in both phenomenological and neuroscientific terminologies, will result in neuronal assemblies of varying size and hence varying degrees of consciousness. In neuroscientific terms, let us therefore assume that a subjective state is a transient neuronal assembly of varying degree, recruited by the activation of a hard-wired hub, or network of neurons. Such a hub, once activated by some input, could trigger a temporary formation of a much larger assembly of brain cells, rather like a highly durable stone that can elicit far more extensive, yet highly transient, ripples in a pond. In such a way, different mental states could be correlated with different sizes of neuronal assembly, and as such provide a bridge between phenomenology and physiology (Table 1).

The appeal of the predictions made in Table 1 is that they are testable, i.e., that different parameters, such as amine availability, age, and stimulation, could be differentially varied, and the ensuing, corresponding assembly size, noted. The primary problem, however, is with the limitations of the techniques currently available.

The neuroscience portfolio: brain imaging

The most obvious means of monitoring the formation of transient neuronal assemblies in the conscious brain is by imaging technology. However, within the range of methodologies available, no particular approach is ideal. First, there is functional magnetic resonance imaging (fMRI). This is a very powerful non-invasive method of visualizing regions of activity within the intact brain of human or animal patients (Hirsch, this volume). Areas of neuronal activity have a higher demand for oxygen, which is carried by the hemoglobin molecule. A powerful magnet is able to detect the changes in the magnetic field that are generated by a hemoglobin molecule when it releases oxygen. The great advantage of this technique is that subjects do not need to be injected with tracer substances, as with positron emission tomography (PET), and since they are not required to be under anesthetic, can be fully cooperative and can perform necessary tasks and receive instruction. The spatial resolution is high and there is little or no long-term exposure damage reported. On the other hand, the time delay in an area of neuronal activity and its detection can be in the region of 6–10 s. Although this can be mathematically corrected, this temporal delay makes it difficult to dissect the sub-second activity of the brain, but instead is used to detect gross regions of activation.

To combat the problem of time resolution, an imaging technique based on changes in brain conductivity has been developed for use on human volunteers, called functional electrical impedance

Table 1. Net assembly size as a neural correlate of consciousness

Neuronal connectivity	Trigger	Amine modulators	Assembly turnover	Assembly size	Physiology
Very extensive	Strong	High	Low	Large	Pain
Extensive	Strong	Low	Low	Large	Meditation
Extensive	Weak	Low	Low	Small	Dreaming
Sparse	Strong	High	High	Small	Childhood
Extensive	Strong	Medium	Low	Large	Abstract thought
Very extensive	Strong	Low	Very low	Large	Depression
Sparse	Strong	Medium	Low	Small	Alzheimer's Disease
Very extensive	Strong	High	Low	Large	Anxiety
Extensive	Strong	High	High	Small	Schizophrenia
Extensive	Strong	High	High	Small	Fear
Age/species/mind	*+ Stimulus/ + strength/ significance*	*+ Thrill*	*+ Distraction*	*Degree of consciousness*	*Phenomenology*

Note: This scheme links the physiological characteristics/features in the brain (e.g., assembly size, turnover, amine modulators, etc.) with phenomenological events (e.g., degree of consciousness, distraction, etc.). The model of neuronal assembly size can serve as a "Rosetta Stone" for extrapolating physiological states (e.g., assembly size) with phenomenological ones (e.g., pain, meditation, etc., right hand column) and vice versa. Four independent parameters, describable in both physiological and phenomenological terms, can be differentially manipulated to yield assembly sizes and, in turn, the relating degrees of consciousness and vice versa. For more details of underlying rationale see Greenfield (2000a).

tomography (fEIT) (Towers et al., 2000). fEIT is at an early stage of development, but relies on an array of electrodes (currently 16) placed externally around the head of a patient, either human or animal. Two of the electrodes provide an electric current pulse and the other 14 electrodes within the array act as receivers and the impedance generated is then calculated. This impedance varies depending on the activity of the brain regions that lie between the electrodes. The switching between electrodes as pulse generators or receivers happens on a microsecond resolution and this technique already offers sub-second imaging of brain activity in response to sensory stimuli.

Like fMRI, fEIT is completely non-invasive and does not require any anesthesia. However, perhaps the greatest advantage of this technique is that although the spatial resolution is comparable to that of fMRI, the temporal resolution is far greater and can show activity on a sub-second scale. Moreover, in clear contrast to the massive magnets required for fMRI, the equipment is very compact and is fully portable. It can be used in an operating theater and can be used to assess the depth of anesthesia of a patient in surgery or, like fMRI, can be used to assess brain activity to command tasks. Anxiety, often a further hazard with fMRI, is reduced by using the equipment in comfortable surroundings. There is no time limit for recording, from minutes to hours and there is no use of injected chemicals, making it safe and cheap. On the other hand, this technique is still at an early stage of development, although early trials look very promising.

An imaging method that is not only well established, but gives a time resolution commensurate with the time scale of neuronal events, is based on voltage-sensitive dyes. Molecules have been developed that have a natural fluorescence, i.e., they absorb light of certain wavelengths and release light of longer wavelengths. These molecules are incorporated into the cellular membrane of neurons and undergo a structural change when the membrane is electrically activated by an action potential. With this structural change there is a shift in the wavelength of emitted light and it is this shift of emitted fluorescence that is measured. It is therefore possible to record the electrical activity of regions of brain tissue, under different conditions, and determine some of the physiological and pharmacological mechanisms that exist.

Hence, the great advantage of this technique is that with the fast acquisition rate of a charge-coupled device (CCD) camera we can measure the activity of small or gross regions of tissue on a sub-millisecond time scale. The dyes do not appear to have any deleterious effect on the tissue. Recording can be performed under a variety of different conditions and on a variety of preparations. However, the big drawback in this case, is that this technique cannot be used on humans. The dyes are best used on in vitro slices of tissue and so this is a fully invasive procedure. Although in vivo studies can be performed, it is essential that the subject is fully anaesthetized with sufficient analgesia that makes the study of conscious awareness very difficult.

In vivo studies with high time resolution can also be performed using other fluorescent probes. Recent years have seen a large increase in the use of these chemicals, especially in the study of calcium signaling and cellular processing. The probes change their emitted fluorescence under certain conditions, for example when binding to calcium. It is possible to gain a temporal resolution in the region of milliseconds and the spatial resolution, with modern confocal or multiphoton microscopes, of tens of micrometers.

To improve further on the spatial and temporal resolutions it is possible to observe protein–protein interactions using FRET (fluorescence resonance energy transfer) producing nanometer-scale resolution. With multiphoton microscopy it is possible to visualize dynamics at the subcellular level within a living animal. But the big disadvantage is that these techniques are restricted to animal models as they require a degree of invasive surgery or genetic modification when certain fluorescent protein expression is desired (e.g., green fluorescent protein — GFP).

Luminescent proteins offer an even better alternative to their fluorescent counterparts. They are involved in the phenomenon of light emission in living systems. Included are the enzymatic and non-enzymatic types of system with or without the presence of oxygen or cofactors, for example aequorin, the photoprotein isolated from the bioluminescent jellyfish *Aequorea*. It emits visible light by an intramolecular reaction when a trace amount of a specific ion is bound. The probes, in contrast to fluorescent dyes, have a very wide dynamic range and excellent signal to noise ratio. Since all the electrical activity of the brain culminates in changes in intracellular calcium in different parts of the neuron, the ability to measure intraneuronal calcium changes in different brain regions, give a real functional readout of neuronal activity.

This technique offers a highly sensitive method of measuring intracellular ion levels with good temporal and spatial resolutions. An additional advantage is that the costs of the necessary measuring equipment is relatively low. Aequorin may be targeted to specific cell types and even to distinct organelles and subcellular compartments by genetic techniques. Then again, the technique is restricted to animal models since they require either surgery/injections or transgenic techniques. Although transgenic zebrafish expressing aequorin are available, more work is needed to generate transgenic mice expressing aequorin.

The problem has been that transient assemblies of neuronal activity that could be interpreted as "moments of consciousness," are on a sub-second level, whereas the responses measured in standard imaging approaches (fMRI) take some 7–10 s to develop. The net, final assembly would probably be too fast and too small to detect with the spatial and temporal limitations of non-invasive imaging. In order to appreciate precisely the mechanisms that are at work, on a time scale commensurate with true events, we need to develop a means of extrapolating from studies that give high time resolution, but are not possible in humans. Hence, if we are to monitor the transient formation of neuronal assemblies, ultimately, in the human brain, the best way forward currently, would be to devise a methodology whereby the speed, durability, and sensitivity of neuronal connectivity changes can be evaluated in the animal brain and extrapolated to the human brain. A three-stage process might circumvent the problem to some extent, whereby brain slice preparations are compared with human in vivo imaging, by referral to an interim "Rosetta Stone," i.e., in vivo animal imaging, using optical probes (Fig. 2).

The neuroscience portfolio: beyond imaging

Although imaging is clearly essential for monitoring the formation of neuronal assemblies in time

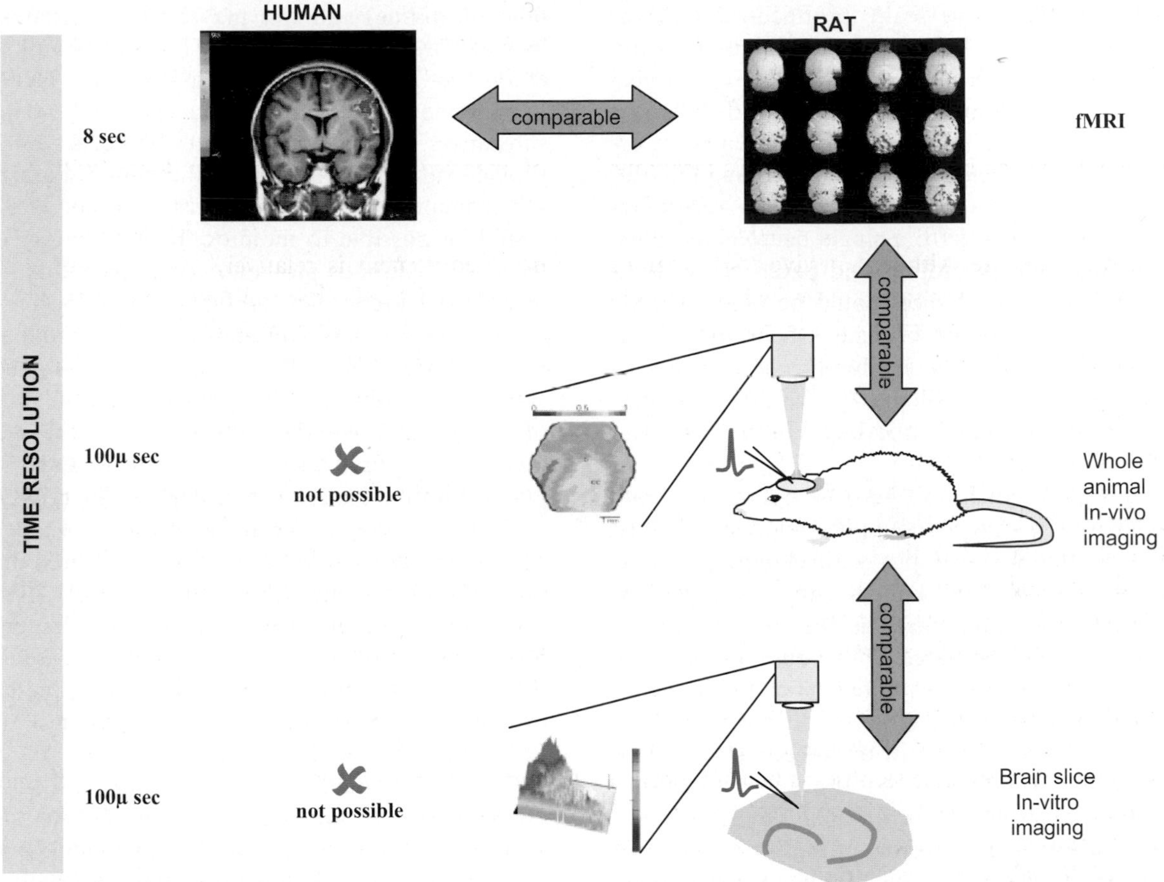

Fig. 2. The different approaches of this study and how they inter-relate to compensate for the respective shortcomings of each. The limits of fMRI are primarily temporal, but by comparing human studies with those of lower mammals we can bridge the gap with invasive optical imaging techniques. Optical imaging provides the necessary temporal and spatial resolutions needed to investigate the characteristics of neural assemblies and the in vivo data can be compared with the commonly used in vitro brain slice preparation. See Plate 2.2 in Colour Plate Section.

and space, it tells us little as to the physiological mechanisms that enable neurons to behave in this way. The nature of conscious experience may reflect the extent of neuronal assembly activation across different interconnected regions of the brain. Each brain area mediates different aspects of cognitive processing, and so the question remains as to how the activity across such dispersed neuronal assemblies is precisely synchronized to generate a single unified percept. This "binding" problem cannot be solved readily by imaging techniques alone, but might be solved by recording background oscillations in cortical activity, which can be synchronized across many different brain regions, thereby providing discrete temporal windows in which neuronal assemblies can be corralled and then disbanded. These cortical oscillations can be recorded from the scalp of human subjects, and so it would be possible to explore whether the coherence and frequency of these brain rhythms can be correlated with the subjective experience of humans.

Synchronous oscillations in neuronal activity can be recorded in the extracellular electrical fields, generated by current flow in this excitable tissue. In humans, these electrical fields would be measured by electrodes on the scalp, producing an electro encephalogram (EEG; John, this volume). In animal models, both in vitro and in vivo, neuronal oscillations could be recorded using field electrodes placed within the brain regions themselves. In order to explore the mechanisms underlying the neuronal oscillations in animal models, recordings could also be made from single neurons using extracellular unit recordings and patch–clamp techniques. Neocortical slices could be used to study how fast oscillations become synchronized over long distances. As well as monitoring neuronal assemblies electrophysiologically, we can also characterize them neurochemically (Whittington et al., 1995; Fisahn et al., 1998).

One major synchronizing system is the forebrain cholinergic system that supplies many cortical areas and the hippocampus with cholinergic afferents. It has been known since the earliest neurochemical investigations that cortical acetylcholine levels correlated with conscious arousal, for example release was greater during wakefulness and dreaming and lower during light sleep and when unconscious in deep-sleep stages (Jouvet, 1965). Cholinergic basal forebrain neurons oscillate autosynchronously at a slow frequency, even when isolated from other neurons (see Khateb et al., 1992). This ability to generate activity spontaneously, in the absence of sensory input, may be an essential property of a system that, through recruiting and synchronizing neuronal activity of large areas of the cortex, is postulated to be involved in consciousness. Electrical stimulation of the forebrain cholinergic system in experimental animals led to the activation of the EEG and caused the phasic large amplitude, slow "delta" oscillations (1–5 Hz) to switch to a "theta" rhythm and to tonic low amplitude, fast (20–40 Hz) gamma oscillations (Metherate et al., 1992). The forebrain cholinergic system is responsible, via the septo-hippocampal pathway, for hippocampal theta activity. This activity is seen in experimental animals during rapid eye movement (REM) sleep, when they explore a new environment, or when learning a behavioral task. Gamma oscillations are superimposed on the theta activity and are synchronous between spatially separated areas. Such synchrony is disrupted by anesthetics (and morphine) and this may be the mechanism by which they disrupt cognitive function (Faulkner et al., 1998) or indeed consciousness itself. Hence, measurement of cholinergic agents and related substances in the hippocampus or cortex might provide valuable neurochemical readout of assembly dynamics. For example, using proteomics it would be possible to monitor the modulatory effects of metabotropic muscarinic receptors, of which there are several subtypes, on inhibition of adenylyl cyclase (AC), excitation of phospholipase C, and activation of K^+ channels. In turn, such activation would alter the phosphorylation states of intracellular proteins, ion channels, enzymes, and transcription factors, from millisecond to long-term time scales. The central cholinergic system could play a role in the reactivation of a gene (Zif268), where activity is markedly increased during wakefulness and REM sleep (Jouvet, 1965). Interestingly, when these animals later entered REM sleep, Zif268 expression is then increased to the high levels normally seen during wakefulness in extra-hippocampal areas, e.g., cortex and amygdala (Ribeiro et al., 2002).

Genetically modified mice, which lack the GluR-A (glutamate receptor subunit A, also known as GluR1) subunit of the glutamate α-amino-3-hydroxy-5-methyl-4-isoxazolepropionic acid (AMPA) receptor do not show normal hippocampal long-term potentiation (LTP) (Zamanillo et al., 1999). It has been recently shown that these mice can nonetheless learn spatial memory tasks that depend on the integrity of the hippocampus, so long as the spatial goals that the animals must find remain in constant positions from day to day. However, the same mice are quite incapable of learning spatial tasks in which the animal's response must flexibly vary from trial to trial, signaled by some other item of information that the animal must remember at the time it makes its spatial choice (Reisel et al., 2002; Schmitt et al., 2003). These mice thus seem to lack a mechanism that normally allows spatially or temporally separated pieces of information to be integrated into coherent, moment-to-moment cognitive states. By comparing them to their wild-type normal siblings, it could be possible to establish the gene and protein expressions associated with the ability to assemble coherent cognitions from spatiotemporally separated underlying elements. It seems therefore that neuronal assemblies could be monitored by a variety of imaging, electrophysiogical, and neurochemical techniques — ideally a combination of all three approaches. The critical issue then is: what will they be used to measure?

A paradigm for evaluating neuronal assemblies as indices of depth of consciousness: pain

One of the most empirically tractable paradigms for exploring neuronal correlates of consciousness is pain. Acute pain is an unpleasant sensation that serves as an alerting mechanism to warn the organism of the potential for injury. There are many good animal models of acute pain and these have led to the development of many effective analgesics, ranging from non-steroidal anti-inflammatory drugs (NSAIDs), to opioids and local anesthetics. Moreover, painful stimulation and its alleviation can be readily standardized within an experimental protocol; on the other hand, the subjective sensation of pain is highly variable and dependent on a host of other factors within the brain/body. Indeed, a distinction is traditionally made between "nociception," i.e., the physiological process of pain signals, as opposed to the subjective experience of pain: the former does not require an ongoing conscious state, while the latter does. It has previously been suggested that these diverse factors, both cognitive and chemical, contribute to the net size of a prevailing neuronal assembly, and in turn the size of that assembly will be the final common factor in determining the degree of subjective pain. Thus it follows, the greater the assembly, the greater the pain (Table 2). However, as yet there is no quantifiable measure for the depth of analgesia.

If the net assembly size does indeed correlate with a degree of pain, then different anti-pain strategies should each reduce, albeit in different ways and to different extents, the net size of transient neuronal assemblies forming in the brain during the experience of pain. It would be interesting to identify subsequently any common factors in such strategies and attempt to interpret them in neuroscientific terms (e.g., high turnover of neuronal assemblies; Greenfield, 2000a, b), as characterized by imaging (invasive and non-invasive), as well as by electrophysiological, biochemical, molecular biological, and morphological markers. Indeed, a third step would then be to test whether different strategies with nonetheless similar neuroscientific correlates, are indeed interchangeable in the alleviation of pain. How well can the techniques outlined in the preceding section, monitor pain and its alleviation?

Table 2. Degree of pain is correlated with size of neuron assembly

Pain is expressed in terms of associations, e.g. pricking, stabbing, burning
Pain thresholds to the same stimulus change throughout the day (and as an aside vary according to point in menstrual cycle in women = hormonal influence?)
Under some conditions, the greater the anxiety, the greater the pain
Phantom limb pain due to stimulation of neuron "matrix," i.e., assembly
Absent in dreams
Morphine (analgesic) gives dreamlike (small assembly state) euphoria
Morphine disrupts assembly formation (Whittington et al., 1998)
Schizophrenia, high pain threshold
Depression, changes in pain threshold
Diverse anesthetics explicable as common final factor: reducing assembly size

Note: Diverse findings suggesting that the greater the pain, the greater the net size of neuronal assembly. For references, see Greenfield (2000a) unless stated otherwise.

Pain and its alleviation, monitored by the neuroscience portfolio

Non-invasive brain imaging in human subjects

To date, fMRI has revealed key brain areas (e.g., cingulate, insula, somatosensory, and motor cortices, the thalamus and basal ganglia) activated during the experience of pain, as shown in Fig. 3, (Peyron et al., 2000; Tracey et al., 2002a). This technique has also highlighted additional regions, such as the hippocampus (Ploghaus et al., 2001), which will contribute to states of anxiety that in turn exacerbate pain sensation. By the same token, other areas, such as the periaqueductal gray (PAG),

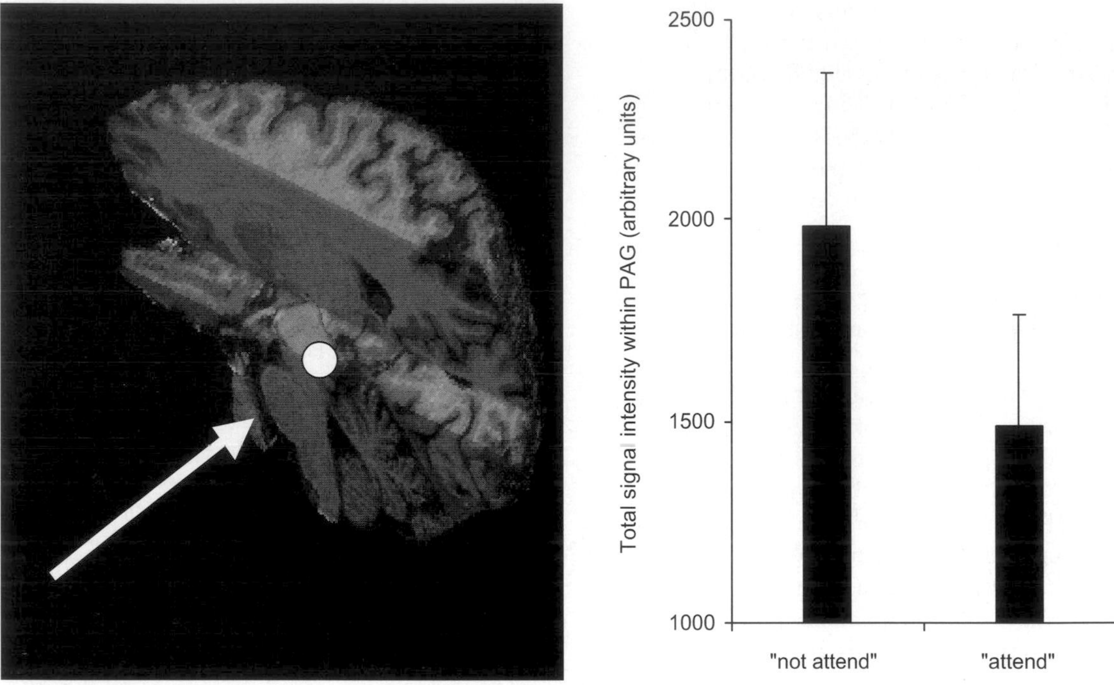

Fig. 3. PAG matter activation in response to noxious thermal stimulation. The right panel shows that activation in this region is increased when subjects do not attend to the noxious thermal stimulus and that this increased activation is concomitant with a decreased perceived pain intensity.

if activated, contribute to pain alleviation, seen at the behavioral level as the shift in attention used by some as an effective non-invasive strategy (see above). This mindset of distraction is characterized not just by an activated PAG, but by a decreased activity in the circuitry otherwise related to pain (Tracey et al., 2002b; Kupers et al., this volume).

Invasive (optical) imaging in non-human brains

To date, our laboratory has succeeded in completing a pilot set of experiments to examine the effects of anesthetics on the spatiotemporal patterns of neuronal assembly formation in the cortex and hippocampus in vitro, using the mammalian brain slice model. Indeed, we found that in response to electrical paired-pulse stimulation, two anesthetics (thiopental and propofol) reduced the spatial spread and diminished the amplitude of neuronal assemblies, despite the two chemicals having diverse chemical structures. It has been reported previously that these two anesthetics induce the release of an amino acid neurotransmitter, γ-aminobutyric acid (GABA), which leads to the activation of chloride channels. Interestingly, another compound used to treat neuropathic pain, gabapentin, is very similar in structure to GABA; however, it is not an anesthetic and it does not produce the same effects, following repeated stimulation, as do the two anesthetics (Fig. 4). These data suggest that loss of awareness with anesthesia does not depend on any one transmitter, such as GABA, but on some higher-order property of neuronal organization, such as macro assembly formations, and most importantly, their duration. Indeed, given that anesthesia itself is well known to be graded in stages (i.e., not all or none) (Rang and Dale, 1991), it would be appropriate for its neural correlate to be correspondingly analogue too (Fiset, this volume, Alkire, this volume). A current monitor for depth of anesthesia relies on a number of collated parameters from a patients EEG trace, which provides a numerical output, the Bispectral index (BIS) (Pomfrett, 1999). If a measure of neuronal "assemblage" (see the

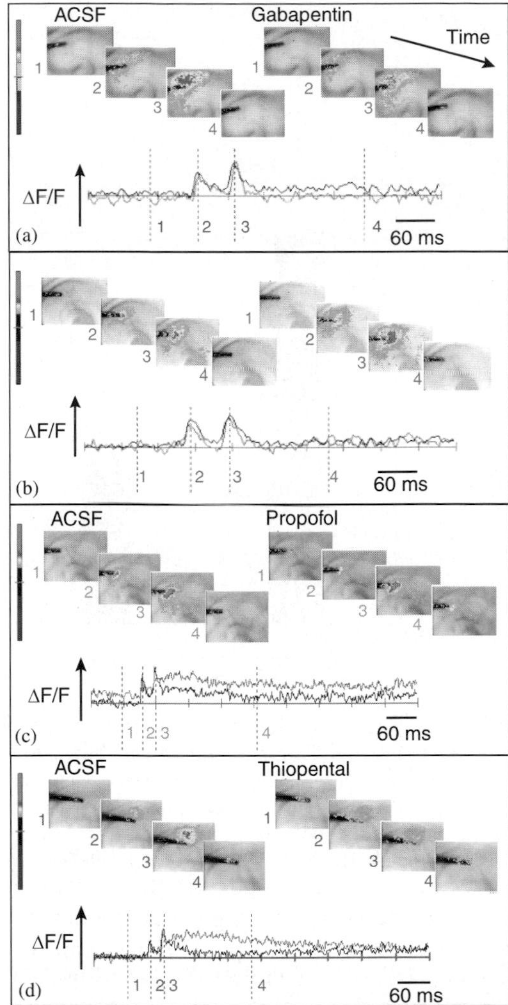

Fig. 4. Analgesic and anesthetic effects on hippocampal activity. Optical measurements of analgesic and anesthetic action in the brain. Voltage-sensitive dye imaging of neuronal population activity in in vitro hippocampal slices of rat brain in response to paired-pulse electrical stimulation along the Schaffer Collaterals. The left panels show the responses under control conditions (in artificial cerebrospinal fluid, ACSF) and the right panels show treatment conditions. Both analgesics and anesthetics are linked with the actions of the inhibitory GABAergic system and so two compounds from each class were tested (analgesics: gabapentin and morphine) and (anesthetics: propofol and thiopental). Warmer colors show a higher degree of depolarizing excitation. The analgesic treatments do not significantly differ from their controls (a) and (b) but the anesthetics (c) and (d) cause a prolonged period of excitation that lasts for up to 500 ms after the second stimulating pulse. The optical recordings are also represented by traces below; black, control; red, treatment (Collins and Greenfield, *submitted*). See Plate 2.4 in Colour Plate Section.

Introduction section) can be developed in the future it would be very interesting to compare the correlation between the two indices.

Electrophysiology in human subjects

A complementary means of comparing functional states in humans with the neural mechanisms underlying those states is, as described earlier, with electrophysiology. Data are already accumulating that characterize the synchronous oscillations of brain cells within a working assembly (Traub et al., 1998), recorded with arrays of extracellular electrodes. Just as the in vivo animal preparation serves as a bridge for comparing imaging data from the high time resolution of brain slices with the less experimentally tractable cognitive states of whole human brains, so the same rationale could be used for studying the significance of the oscillatory activity of neurons within an assembly. Often, particularly after strokes or amputations, severe pain is experienced that is not generated by peripheral pain receptors, but by spontaneous oscillatory activity of the whole central nervous system (CNS) pain matrix; and it is notoriously difficult to treat. Stimulation of the periventricular/PAG region can dampen down these oscillations and with it the patients' experience of pain (Fig. 5).

Neurochemical readout from brain to body

It has been shown in a neuroma model of chronic pain (sciatic nerve transection) that autotomy, an indicator of spontaneous pain, was reduced by depleting cortical noradrenaline with the neurotoxin *N*-ethyl-*N*-(2-chloroethyl)-2-bromobenzylamine hydrochloride (DSP-4; Al-Adawi et al., 2002). At the spinal level, the cyclic adenosine monophosphate (cAMP) signaling cascade is also important in mediating chronic pain. In rats with excitotoxic lesions of the spinal cord, excessive grooming was associated with increased serine phosphorylation of the NR1 (NMDA (*N*-methyl-D-aspartate) receptor subunit) of spinal tissue (Caudle et al., 2003). Similar results were reported by Zou et al. (2002) after intradermal injection of capsaicin. In a model of mechanical hyperalgesia, spinal blockade of adenylate cyclase prevented the expected increase in phosphorylated

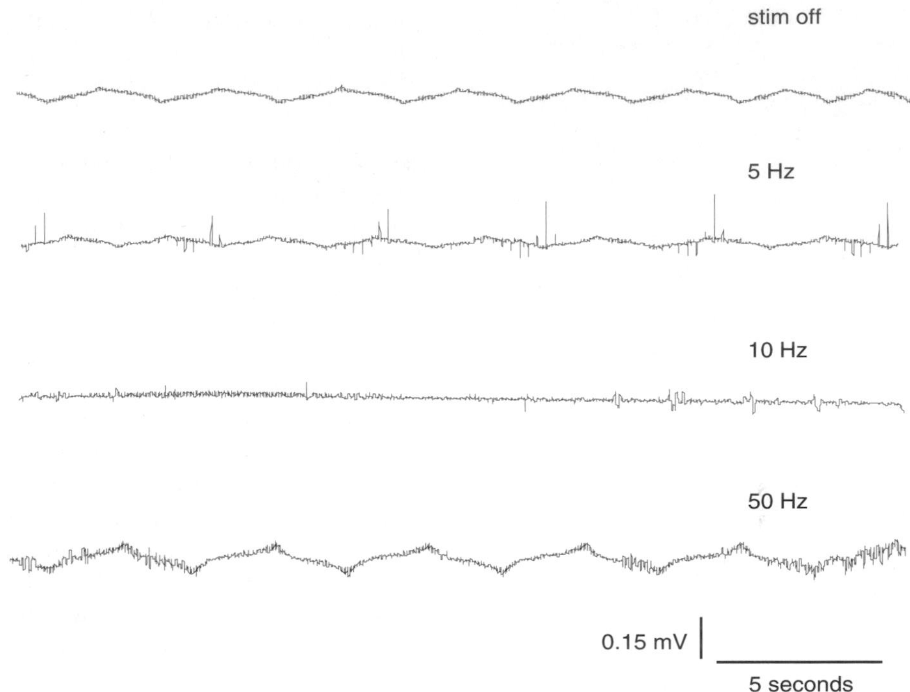

Fig. 5. Measuring thalamic potentials in response to deep brain stimulation. Often, particularly after strokes or amputations, severe pain is experienced that is not generated by peripheral pain receptors, but by spontaneous oscillatory activity of the whole CNS pain matrix; and it is notoriously difficult to treat. Stimulation of the periventricular/PAG region can dampen down these oscillations and with it the patients' experience of pain. With no stimulation in the periventricular/PAG region, the oscillations are evident in the thalamus (top trace). With specific stimulation paradigms, the thalamic oscillations can be reduced (5 Hz, second trace), and completely abolished (10 Hz, third trace) but return again when the frequency is too high (50 Hz, bottom trace) (data taken from Nandi et al., 2003).

cAMP-response-element-binding protein (p-CREB; Hoeger-Bement and Sluka, 2003).

Therefore, it is to be expected that many of the signaling mechanisms that have been implicated in cognitive processing have also been implicated in pain processes because they utilize the same neurotransmitters. The ability to remember painful events, such as electric shock, is important to survival and therefore pain necessarily involves associative memory. Indeed, the meaning of the word "pain" is learned by experience. However, experiments with Calmodulin-dependent protein kinase IV (CaMKIV) knockout mice have shown that it is possible to dissociate memory of the shock from the behavioral response to the shock (Wei et al., 2002b).

Lesions of the anterior cingulate cortex reduce chronic pain in humans and produce antinociceptive effects in animals (for references see Wei et al., 2002a). The anterior cingulate cortex contains high levels of two ACs, AC1 and AC8, that couple NMDA receptor activation to cAMP signaling pathways. Transgenic mice lacking these enzymes did not show allodynia (i.e., pain from stimuli that are not normally painful) in response to intraplantar injection of Freund's adjuvant, a model of inflammatory pain but this could be rescued by focal injections of forskolin (Wei et al., 2002a). In a proteomic study of chronic pain, intraplantar injection of Freund's adjuvant increased the concentration of five proteins in rat cerebrospinal fluid (CSF) (Gineste et al., 2003). Fear conditioning and synaptic potentiation in the anterior cingulate and amygdala was significantly reduced in CaMKIV knockouts (Wei et al., 2002b).

There have been a number of proteomics analyses of brain proteins in animals, for example changes in protein expression during varying sleep conditions (Pavlides and Ribeiro, 2003; Krueger et al., 2003; see Figs. 13 and 14) and analysis of postsynaptic density in rat brain (Li et al., 2004), but no one as yet seems to have combined proteomic technology with changes in brain states resulting from experience/the environment. This lacuna is somewhat surprising, given the wealth of literature on changes in specific proteins as a result of relatively subtle changes in the environment, for example: synaptogenesis and memory on apoE4 (apolipoprotein E4; Levi et al., 2003); enriched housing on synaptophysin and neurotrophic factors (Koo et al., 2003); wheel running on nerve growth factor (Dahlqvist et al., 2003); exposure to toys on glutamic acid decarboxylase (GAD) activity (Frick et al., 2003). In general, we also need to study neurochemical readout, not just from the brain per se, but between the brain and the rest of the body.

It is well established that feedback from the vital organs and, indeed, from the immune and endocrine systems all strongly influence conscious experience (Martin, 1997). Subjective states surely then must entail integration of the three control systems of the body: the immune, endocrine and, nervous systems. "Somatic" markers (Parvizi et al., 2000), such as peptides (Pert, 1997) for example in blood (Greenfield, 2000b, 2003), would have impartial, bioactive access to all three systems. Peptides such as the encephalins might therefore serve as additional parameters for influencing and reflecting baseline states of consciousness during pain and its alleviation, as measured by analytical high-pressure liquid chromatography (HPLC) on peptide-containing fractions of plasma for identification by matrix-assisted laser desorption–ionization–time-of-flight (MALDI–TOF) mass spectrometry. In this way, the paradigm of acute pain might serve as an example, albeit the prototype protocol, for testing the hypothesis that transient neuronal assemblies could be a valuable neuronal correlate of consciousness.

Conclusions

Neuroscience offers a wide portfolio of techniques that nonetheless, on their own, do not inspire any new insights into how the brain generates consciousness. On the other hand, once a hypothesis is developed, then this portfolio can be used to falsify or prove the validity of that model. We suggest that the concept of variable degrees of consciousness as reflected in varying size of neuronal assemblies, is indeed tractable to this type of empirical scrutiny and that the methodologies are all currently possible. The paradigm of pain and its alleviation, as described here, would be particularly appropriate, but should not exclude similar studies on other mental states, for example depression and its alleviation with selective serotonin uptake inhibitors.

If the general concept was indeed validated, then transient neuronal assembly size could serve as the much sought after neural correlate of consciousness. However, it would be important to bear in mind that a transient neuronal assembly would be at best an *index* of degree of consciousness, not consciousness itself. Beyond the type of study outlined here, it would be the issue of how that index, an assembly of neurons, releasing signature substances to the rest of the brain and body, would in turn contribute to the holistic state of the conscious experience. By the same token, experiments characterizing assembly formation and its functional correlates would not in and of themselves, provide any insight beyond correlation, into the causal relationship between physical and phenomenological descriptions. Nonetheless, the contribution of neuroscience to an understanding of consciousness would be of value, not just in enabling the elimination or espousing of hypotheses, but more immediately in prompting new strategies for the alleviation of specific mental disorders.

Abbreviations

AC	adenylyl cyclase
ACSF	artificial cerebrospinal fluid
AMPA	α-amino-3-hydroxy-5-methyl-4-isoxazolepropionic acid
APO E4	apolipoprotein E4 — (increased levels in Alzheimer's disease)
BIS	bispectral index
CaMKIV	calmodulin-dependent protein kinase IV
cAMP	cyclic adenosine monophosphate

CCD	charge-coupled device
CNS	central nervous system
CSF	cerebrospinal fluid
DSP-4	*N*-ethyl-*N*-(2-chloroethyl)-2-bromobenzylamine hydrochloride
EEG	electroencephalogram
fEIT	functional electrical impedance tomography
fMRI	functional magnetic resonance imaging
FRET	fluorescence resonance energy transfer
GABA	γ-aminobutyric acid
GAD	glutamic acid decarboxylase
GFP	green fluorescent protein
GluR-A	glutamate receptor subunit A
HPLC	high-pressure liquid chromatography
LTP	long-term potentiation
MALDI–TOF	matrix-assisted laser desorption–ionization–time-of-flight
NMDA	*N*-methyl-D-aspartate
NR1	NMDA receptor subunit 1
NSAIDs	non-steroidal anti-inflammatory drugs
PAG	periaqueductal gray
p-CREB	phosphorylated cAMP-response-element-binding protein
PET	positron emission tomography
REM	rapid eye movement

Acknowledgments

The authors thank all those who contributed to the preparation of this chapter: Dr. Stephanie Cragg, Mrs. Vivienne Pearson, and to those who provided data including Professor John Stein, Dr. Irene Tracey, Dr. Chris Pomfrett, Mr. Tipu Aziz, Professor Brain Anderton, Dr. John Stephenson, and Dr. Pauline Rudd. Work by S.A.G. and T.F.T.C. was supported by Pfizer and the Mind Science Foundation and we also thank the Templeton Foundation for inspiring the collaborative aspect of this work.

References

Al-Adawi, S., Dawe, G.S., Bonner, A., Stephenson, J.D. and Zarei, M. (2002) Central noradrenergic blockade prevents autotomy in rat: Implication for pharmacological prevention of postdenervation pain syndrome. Brain Res. Bull, 57: 581–586.

Caudle, R.M., Perez, F.M., King, C., Yu, C.G. and Yezierski, R.P. (2003) *N*-methyl-D-aspartate receptor subunit expression and phosphorylation following excitotoxic spinal cord injury in rats. Neurosci. Lett, 349: 37–40.

Dahlqvist, P., Ronnback, A., Risedal, A., Nergardh, R., Johansson, I.M., Seckl, J.R., Johansson, B.B. and Olsson, T. (2003) Effects of postischemic environment on transcription factor and serotonin receptor expression after permanent focal cortical ischemia in rats. Neuroscience, 119(3): 643–652.

Faulkner, H.J., Traub, R.D. and Whittington, M.A. (1998) Disruption of synchronous gamma oscillations in the rat hippocampal slice: A common mechanism of anaesthetic drug action. Br. J. Pharmacol, 125: 483–492.

Fisahn, A., Pike, F.G., Buhl, E.H. and Paulsen, O. (1998) Cholinergic induction of network oscillations at 40 Hz in the hippocampus in vitro. Nature, 394: 186–189.

Frick, K.M., Stearns, N.A., Pan, J.Y. and Berger-Sweeney, J. (2003) Effects of environmental enrichment on spatial memory and neurochemistry in middle-aged mice. Learn Mem, 10(3): 187–198.

Gineste, C., Ho, L., Pompl, P., Bianchi, M. and Pasinetti, G.M. (2003) High-throughput proteomics and protein biomarker discovery in an experimental model of inflammatory hyperalgesia: Effects of nimesulide. Drugs, 63(1): 23–29.

Greenfield, S.A. (1997) The human brain: A guided tour. London, Weidenfeld & Nicolson.

Greenfield, S.A. (2000a) The private life of the brain. London, Penguin.

Greenfield, S.A. (2000b) Brain story — unlocking our inner world of emotions, memories, ideas and desires. London, BBC Worldwide.

Greenfield, S.A. (2003) Tomorrow's people: How 21st Century technology is changing the way we think and feel. London, Penguin (Allen Lane).

Grinvald, A., Lieke, E.E., Frostig, R.D. and Hildesheim, R. (1994) Cortical point-spread function and long-range lateral interactions revealed by real-time optical imaging of Macaque monkey primary visual cortex. J. Neurosci, 14: 2545–2568.

Hoeger-Bement, M.K. and Sluka, K.A. (2003) Phosphorylation of CREB and mechanical hyperalgesia is reversed by blockade of the cAMP pathway in a time-dependent manner after repeated intramuscular acid injections. J. Neurosci, 23: 5437–5445.

Jouvet, M. (1965) Paradoxical sleep — a study of its nature and mechanisms. Prog. Brain Res, 18: 20–62.

Khateb, A., Muhlethaler, M., Alonso, A., Serafin, M., Mainville, L. and Jones, B.E. (1992) Cholinergic nucleus basalis neurons display the capacity for rhythmic bursting activity mediated by low-threshold calcium spikes. Neuroscience, 51: 489–494.

Koo, J.W., Park, C.H., Choi, S.H., Kim, N.J., Kim, H.S., Choe, J.C. and Suh, Y.H. (2003) The postnatal environment can counteract prenatal effects on cognitive ability, cell proliferation, and synaptic protein expression. FASEB J, 17(11): 1556–1558.

Krueger, J.M., Obál, F., Harding, J.W., Wright, J.W. and Churchill, L. (2003) Sleep modulation of the expression of plasticity markers. In: Maquet P., Smith C. and Stickgold R. (Eds.), Sleep and brain plasticity. Oxford University Press, Oxford.

Levi, O., Jongen-Relo, A.L., Feldon, J., Roses, A.D. and Michaelson, D.M. (2003) ApoE4 impairs hippocampal plasticity isoform-specifically and blocks the environmental stimulation of synaptogenesis and memory. Neurobiol. Dis, 13(3): 273–282.

Li, K.W., Hornshaw, M.P., Van Der Schors, R.C., Watson, R., Tate, S., Casetta, B., Jimenez, C.R., Gouwenberg, Y., Gundelfinger, E.D., Smalla, K.H. and Smit, A.B. (2004) Proteomics analysis of rat brain postsynaptic density. Implications of the diverse protein functional groups for the integration of synaptic physiology. J. Biol. Chem, 279(2): 987–1002.

Martin, P. (1997) The sickening mind: brain, behaviour, immunity and disease. HarperCollins, New York.

Metherate, R., Cox, C.L. and Ashe, J.H. (1992) Cellular bases of neocortical activation: Modulation of neural oscillations by the nucleus basalis and endogenous acetylcholine. J. Neurosci, 12: 4701–4711.

Nandi, D., Aziz, T.Z., Carter, H. and Stein, J.F. (2003) Thalamic field potentials in chronic central pain treated by peri-ventricular gray stimulation — a series of eight cases. Pain, 101: 97–107.

Parvizi, J., Van Hoesen, G.W. and Damasio, A. (2000) Selective pathological changes of the periaqueductal gray matter in Alzheimer's disease. Ann. Neurol, 48(3): 344–353.

Pavlides, C. and Ribeiro, S. (2003) Recent evidence of memory processing in sleep. In: Maquet P., Smith C. and Stickgold R. (Eds.), Sleep and brain plasticity. Oxford University Press, Oxford.

Pert, C.B. (1997) Molecules of emotion: Why you feel the way you feel. Scribner, New York.

Peyron, R., Laurent, B. and Garcia-Larrea, L. (2000) Functional imaging of brain responses to pain. A review and meta-analysis. Neurophysiol. Clin, 30(5): 263–288.

Ploghaus, A., Narain, C., Beckmann, C.F., Clare, S., Bantick, S., Wise, R., Matthews, P.M., Rawlins, J.N. and Tracey, I. (2001) Exacerbation of pain by anxiety is associated with activity in a hippocampal network. J. Neurosci, 21(24): 9896–9903.

Pomfrett, C.J.D. (1999) Heart rate variability, BIS and "depth of anaesthesia". Br. J. Anaesth, 82(5): 659–662.

Rang, H.P. and Dale, M.M. (1991) Pharmacology (2nd ed.). Churchill Livingstone, Edinburgh.

Reisel, D., Bannerman, D.M., Schmitt, W.B., Deacon, R.M.J., Flint, J., Borchardt, T., Seeburg, P.H. and Rawlins, J.N.P. (2002) Spatial memory dissociations in mice lacking GluR1. Nat. Neurosci, 5: 868–873.

Ribeiro, S., Mello, C.V., Velho, T., Gardner, T.J., Jarvis, E.D. and Pavlides, C. (2002) Induction of hippocampal long-term potentiation during waking leads to increased extrahippocampal zif-268 expression during ensuing rapid-eye-movement sleep. J. Neurosci, 22(24): 10914–10923.

Schmitt, W.B., Deacon, R.M.J., Seeburgh, P.H., Rawlins, J.N.P. and Bannerman, D.M. (2003) A within subjects, within task demonstration of intact spatial reference memory and impaired spatial working memory in GluRA deficient mice. J. Neurosci, 23: 3953–3959.

Towers, C.M., McCann, H., Wang, M., Beatty, P.C., Pomfrett, C.J. and Beck, M.S. (2000) 3D simulation of EIT for monitoring impedance variations within the human head. Physiol. Meas, 21(1): 119–124.

Tracey, I., Becerra, L., Chang, I., Breiter, H., Jenkins, L., Borsook, D. and Gonzalez, R.G. (2002a) Noxious hot and cold stimulation produce common patterns of brain activation in humans: A functional magnetic resonance imaging study. Neurosci. Lett, 288(2): 159–162.

Tracey, I., Ploghaus, A., Gati, J.S., Clare, S., Smith, S., Menon, R.S. and Matthews, P.M. (2002b) Imaging attentional modulation of pain in the periaqueductal gray in humans. J. Neurosci, 22(7): 2748–2752.

Traub, R.D., Spruston, N., Soltesz, I., Konnerth, A., Whittington, M.A. and Jefferys, J.G.R. (1998) Gamma-frequency oscillations: A neuronal population phenomenon, regulated by synaptic and intrinsic cellular processes, and inducing synaptic plasticity. Prog. Neurobiol, 55: 563–575.

Wei, F., Qiu, C.S., Kim, S.J., Muglia, L., Maas, J.W., Pineda, V.V., Xu, H.M., Chen, Z.F., Storm, D.R., Muglia, L.J. and Zhuo, M. (2002a) Genetic elimination of behavioral sensitization in mice lacking calmodulin-stimulated adenylyl cyclases. Neuron, 36: 713–726.

Wei, F., Qiu, C.S., Liauw, J., Robinson, D.A., Ho, N., Chatila, T. and Zhuo, M. (2002b) Calcium calmodulin-dependent protein kinase IV is required for fear memory. Nat. Neurosci, 5: 573–579.

Whittington, M.A., Traub, R.D. and Jefferys, J.G.R. (1995) Synchronized oscillations in interneuron networks driven by metabotropic glutamate receptor activation. Nature, 373: 612–615.

Whittington, M.A., Traub, R.D., Faulkner, H.J., Jefferys, J.G.R. and Chettiar, K. (1998) Morphine disrupts long-range synchrony of gamma oscillations in hippocampal slices. Proc. Natl. Acad. Sci. USA, 95(10): 5807–5811.

Zamanillo, D., Sprengel, R., Hvalby, O., Jensen, V., Burnashev, N., Rozov, A., Kaiser, K.M., Koster, H.J., Borchardt, T., Worley, P., Lubke, J., Frotscher, M., Kelly, P.H., Sommer, B., Andersen, P., Seeburg, P.H. and Sakmann, B. (1999) Importance of AMPA receptors for hippocampal synaptic plasticity but not for spatial learning. Science, 284: 1805–1811.

Zou, X., Lin, Q. and Willis, W.D. (2002) Role of protein kinase A in phosphorylation of NMDA receptor 1 subunits in dorsal horn and spinothalamic tract neurons after intradermal injection of capsaicin in rats. Neuroscience, 115: 775–786.

S. Laureys (Ed.)
Progress in Brain Research, Vol. 150
ISSN 0079-6123

CHAPTER 3

Functional neuroimaging during altered states of consciousness: how and what do we measure?

Joy Hirsch*

Departments of Radiology and Psychology, Center for Neurobiology and Behavior, Columbia University, New York, NY 10032, USA

Abstract: The emergence of functional neuroimaging has extended the doctrine of functional specificity of the brain beyond the primary stages of perception, language, and motor systems to high-level cognitive, personality, and affective systems. This chapter applies functional magnetic resonance imaging to another high-level realm of cognition and neurology to characterize cortical function in patients with disorders of consciousness. At first pass, this objective appears paradoxical because conventional investigations of a cognitive process require experimental manipulation. For example, to map the location of language-sensitive cortex, a language-related task is performed according to a temporal sequence that alternates the task with rest (no-task) periods. Application of this approach to the study of consciousness would require that levels of consciousness be similarly varied, this is an unlikely technique. Alternatively, another strategy is presented here where the focus is on functional brain activity elicited during various passive stimulations of patients who are minimally conscious. Comparisons between patients with altered states of consciousness due to brain injury and healthy subjects may be employed to infer readiness and potential to sustain awareness. As if a behavioral microscope, fMRI enables a view of occluded neural processes to inform medical practitioners about the health of the neurocircuity-mediating cognitive processes. An underlying point of view is that assessment of recovery potential can be enhanced by neuroimaging techniques that reveal the status of residual systems specialized for essential cognitive and volitional tasks for each patient. Thus, development of imaging techniques that assess the functional status of individual unresponsive patients is a primary goal. The structural integrity of injured brains is often compromised depending on the specific traumatic event, and, therefore, images cannot be grouped across patients, as is the standard practice for investigations of cognitive systems in healthy volunteers. This chapter addresses these challenges and discusses technique adaptations associated with passive stimulation, paradigm selection, and individual patient assessments, where there is "zero tolerance for error," and confidence in the results must meet the highest standards of care. Similar adaptations have been previously developed for the purpose of personalized planning for neurosurgical procedures by mapping the locations of essential functional systems such as language, perception, and sensory–motor functions for each individual patient. Rather than addressing the question of "how does the brain do consciousness" with these techniques, this chapter presents methods for assessment of neurocognitive health in specific patients with disorders of consciousness.

*Corresponding author. Tel.: +1 (212) 342-0291;
Fax: +1 (212) 342-0855; E-mail: jh2155@columbia.edu

DOI: 10.1016/S0079-6123(05)50003-7

Lesions, deficits, and historical milestones leading to the neurobiology of cognition

The organizational principle that drives neuroimaging research is that specific brain areas are involved in specific aspects of behavior such as action, perception, memory, cognition, affect, and consciousness. Francis Crick refers to this as "The Astonishing Hypothesis" (1994):

> "You, your joys, and your sorrows, your memories and your ambitions, your sense of personal identity and free will, are in fact no more than the behavior of a vast assembly of nerve cells and their associated molecules."

Accordingly, the biological principles that underlie cognition fundamentally link the structure and function of the brain.

Active cortical areas associated with cognitive processes in healthy human volunteers are observable using conventional neuroimaging methods, and these techniques have stimulated a renewed focus on the physiological bases of cognition. In particular, the development of non-invasive, functional imaging techniques offers an unprecedented multi-scalar view of the complexities of the intact working human brain including isolated regions of interest and global large-scale systems of interconnected regions. Associations between specific patterns of activity with specific mental processes, such as seeing, hearing, feeling, moving, talking, and thinking provide a basis to undertake a neural foundation for cognition.

Conventional definitions of cognition do not directly address the biological underpinnings of mental events. For example, *Dorland's Illustrated Medical Dictionary* (1988) defines cognition as

> "operations of the mind by which we become aware of objects of thought or perception; it includes all aspects of perceiving, thinking and remembering."

The American Heritage Dictionary (2000) offers a similar definition for cognition as

> "the mental process of knowing, including aspects such as awareness, perception, reasoning and judgment, and that which comes to be known, as through perception, reasoning, or intuition, and knowledge."

However, in his seminal book, *Cognitive Psychology*, Ulrich Neisser (1967) defines cognition as "*all processes* by which the sensory input is transformed, reduced, elaborated, stored, recovered, and used." This definition could be interpreted to encompass biological processes. Here lies an essential advantage of neuroimaging, which is to link models of cognition and awareness to biological processes.

Within this context, the goal of this chapter is to explore the link between brain biology and disorders of consciousness as revealed by neuroimaging. The approach is limited to the medical (rather than the philosophical) aspects of consciousness. In these instances, an observable lesion is present and the functional deficit of loss of consciousness follows. The objectives of this chapter include first the presentation of methodologies to assess the functional consequences of the brain damage leading to deficits of consciousness. In particular, the focus is on patients diagnosed as minimally conscious, a recently defined diagnostic category where severe injury to the brain results in unequivocal, but intermittent, behavioral evidence of awareness of self or the environment (Giacino et al., 2002; Giacino, this volume). The second objective of this chapter is to explore new approaches to guide and inform possible new treatment options for this population of patients for whom few current options for rehabilitation and aggressive therapy exists. The former constitutes the primary objective of this chapter, whereas the latter is a future direction and "wishful thinking."

Medical reports of associations between specific brain injuries and functional deficits provided the initial basis for the linkage between specific brain areas and behavior. As early as 1841, Broca reported language production deficits in patients with specific damage to the left frontal lobe, and in 1874 Wernicke reported deficits in language comprehension in patients following specific damage to the left temporal lobe. Thus, Broca's and Wernicke's areas became established as regions of cortex associated with aspects of speech production and comprehension, respectively, and, the notion that "regions of the mind" correspond to "regions of the brain" became embedded in the foundation of scientific thought. This provided a context to understand well-established observations, such as a severed segment of the optic nerve *always* results in visual field loss, and a severed primary motor projection pathway *always* results in a contralateral plegia. These well-established medical facts are the origin of our understanding of functional specificity. Nonetheless, not all reported functional deficits at the time contributed to understanding the brain-to-mind relationships. For example, in 1861, Phineas Gage recovered from an injury resulting from a dynamite tampering rod propelled through the frontal lobe of his brain with a completely transformed personality and "moral compass." Although his physician, John Harlow, reported this as evidence that personality and even morality might be functions of the brain, those ideas were not accepted (see Damasio et al., 1994). However, the recent emergence of functional imaging technologies has elaborated and developed this notion by enabling investigations on healthy volunteers and hypothesis-based research that probes the complex and less obvious systems of high-level cognition, perception, and volition.

Nearly a century after Broca and Wernicke, Penfield (1975) pioneered the experimental technique of direct cortical stimulation during neurosurgical procedures to directly test the emerging views of functional specificity. His observations confirmed the functional specializations of the speech-related areas, and demonstrated topographical maps associated with sensory and motor functions. His reports of cortical stimulations that elicited memories, tastes, and other mental events enhanced appreciation of the precise link between brain structure and function widely accepted within the mainstream of clinical neurology and neurosurgery (Fig. 1).

Coupling of hemodynamic and neural processes

One of the key principles that links brain function and mental events is the relationship between neural activity and blood flow. In 1881, Angelo Mosso, a physiologist, studied a patient who had survived an injury to the skull. Owing to the nature of the injury it was possible to observe blood flow-related pulsations to the left frontal lobe which occurred during certain cognitive events. Mosso concluded that blood flow within the brain was coupled to mental events, and Roy and Sherrington subsequently proposed a specific mechanism to couple blood flow and neural activity based on direct measurements on dogs. More recently using $H_2^{15}O$ as a tracer of blood flow in the human brain, Raichle et al. (1983) confirmed this fundamental relationship between blood flow and local neural activity. This seminal physiology provided the basis for imaging active cortical tissue during execution of a task, and the early positron emission tomography (PET) studies provided the "proof of principle" that imaging of cognitive process was possible.

The first neuroimaging study of cognitive processes relating to language (Fox and Raichle, 1984; Peterson et al., 1989) determined that different cortical patterns of activation were associated with four separate word tasks: passively viewing words, listening to words, speaking words, and generating words. These early PET investigations firmly established the notion that activity associated with language was observable in the living human brain by tracing the coupled hemodynamic variations within the locally active neural areas. However, owing to the risks associated with injections of radioactive tracers, the limitations to the number of times a subject can be studied, the relatively coarse spatial resolution, the need for registration to high-resolution structural images, and the relatively few PET facilities available for research, the imaging of cortical activity associated with cognitive processes has advanced rapidly using a newer, non-invasive, higher resolution and more available technique, i.e., functional magnetic resonance imaging (fMRI).

Magnetic resonance imaging (MRI) and visualization of brain structure

Starting with the microscope, imaging technologies have guided the mainstream of basic translational

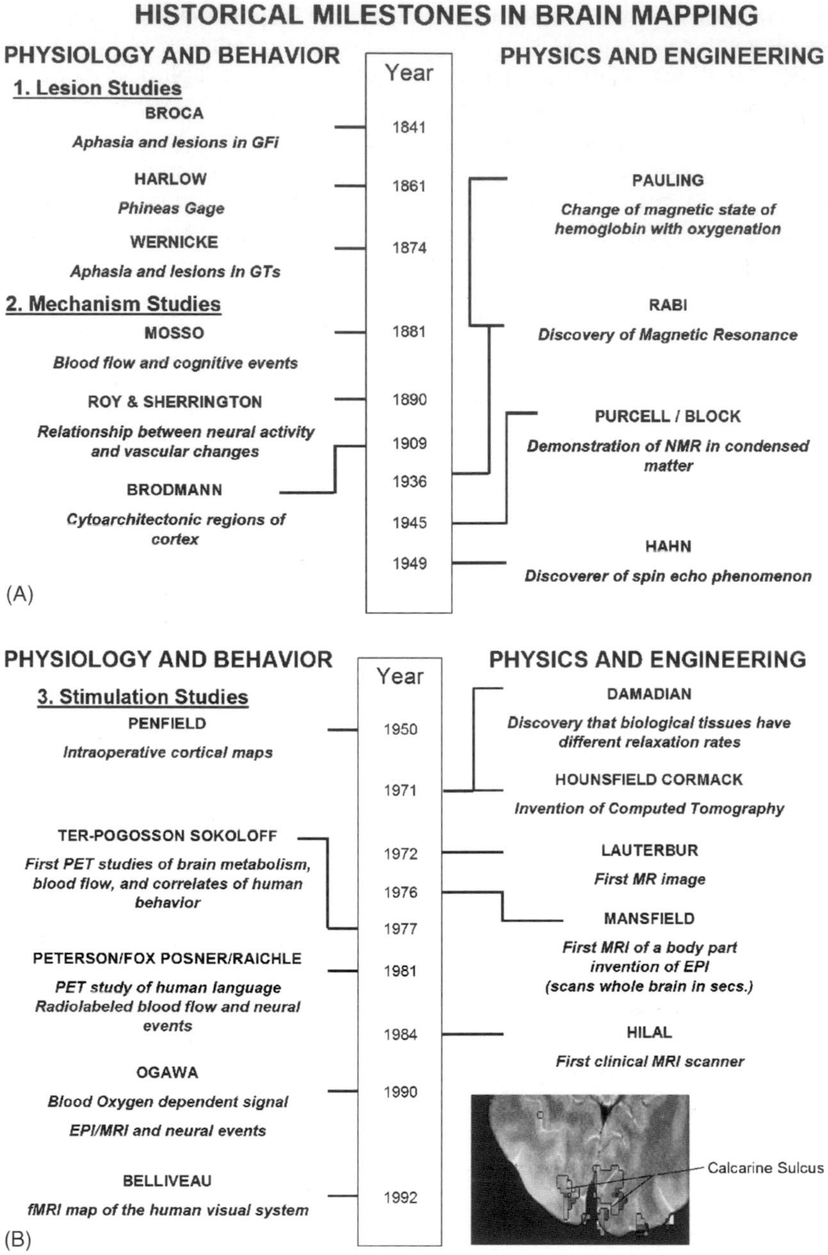

HISTORICAL MILESTONES IN BRAIN MAPPING

PHYSIOLOGY AND BEHAVIOR | **Year** | **PHYSICS AND ENGINEERING**

1. Lesion Studies

BROCA
Aphasia and lesions in GFi — 1841

HARLOW
Phineas Gage — 1861

WERNICKE
Aphasia and lesions in GTs — 1874

2. Mechanism Studies

MOSSO
Blood flow and cognitive events — 1881

ROY & SHERRINGTON
Relationship between neural activity and vascular changes — 1890

BRODMANN
Cytoarchitectonic regions of cortex — 1909

PAULING
Change of magnetic state of hemoglobin with oxygenation — 1936

RABI
Discovery of Magnetic Resonance — 1936

PURCELL / BLOCK
Demonstration of NMR in condensed matter — 1945

HAHN
Discoverer of spin echo phenomenon — 1949

(A)

PHYSIOLOGY AND BEHAVIOR | **Year** | **PHYSICS AND ENGINEERING**

3. Stimulation Studies

PENFIELD
Intraoperative cortical maps — 1950

DAMADIAN
Discovery that biological tissues have different relaxation rates — 1971

HOUNSFIELD CORMACK
Invention of Computed Tomography — 1971

LAUTERBUR
First MR image — 1972

MANSFIELD
First MRI of a body part invention of EPI (scans whole brain in secs.) — 1976

TER-POGOSSON SOKOLOFF
First PET studies of brain metabolism, blood flow, and correlates of human behavior — 1977

PETERSON/FOX POSNER/RAICHLE
PET study of human language Radiolabeled blood flow and neural events — 1981

HILAL
First clinical MRI scanner — 1984

OGAWA
Blood Oxygen dependent signal EPI/MRI and neural events — 1990

BELLIVEAU
fMRI map of the human visual system — 1992

Calcarine Sulcus

(B)

Fig. 1. Historical milestones in human brain mapping. The timeline marks milestone discoveries in physiology and behavior originating with lesion studies and continuing to studies of mechanism, direct cortical stimulation, and hypothesis-based neuroimaging studies in parallel with physics and engineering advances that lead to the development of fMRI and the ability to image active human cortical tissue corresponding to specific cognitive function. The inset illustrates the fundamental "structure to function" relationship between the occipital cortex (calcarine sulcus) and human vision. Cortical activity (assessed by means of fMRI) is observed during viewing of full-field 8 Hz on–off stimulation. The proximity of the activity patterns and the calcarine sulcus (primary visual cortex) is consistent with the hypothesis of functional specificity for this simple stimulation.

research in life sciences by revealing structures not visible to the biological eye including the cell, organelles, molecules, and even atoms. These structures could then be related to physiological mechanisms and subsequently to therapy. The microscope, however, with all its electronic and computational developments, is not suited for the imaging of living structures occluded beneath the surface of the body. However, this "occlusion" problem has been solved with the development of magnetic resonance imaging, where internal structures within the living body are resolved at sub millimeter scales for routine medical and translational research purposes.

Functional magnetic resonance imaging and visualization of brain function

The most recent advances in MRI extend imaging of brain structures to imaging of active neural tissue within the cortex. The chain of discoveries which led to the generation of images that show working brains using MR include Michael Faraday's discovery that blood has magnetic properties, and Linus Pauling's discovery in 1936 that the magnetic properties of hemoglobin change with the state of oxygenation (see National Academy of Sciences, 2001) These discoveries became relevant to functional neuroimaging after the development of high-speed MRI of brain structure, and after it was known that active neural tissue based on the coupling of blood flow was detectable. The fundamental breakthrough of the Blood Oxygen Level Dependent (BOLD) signal was made by Seiji Ogawa, et al. (1990), following the observation that the MR signal originating from the occipital lobe (the area of the brain specialized for visual processing) in rats was higher when the room lights were on than when the lights were off. Ogawa reasoned that oxygenated hemoglobin associated with blood flow coupled to neural activity was related to the increased amplitude of the MR signal. The basic physiological and physics elements are summarized in Table 1.

John Belliveau et al. (1991) and colleagues replicated Ogawa's visual stimulation studies in humans using echo planar MRI demonstrating the potential to reveal not only brain structure, but also brain function using the endogenous signal based on the coupling of hemodynamics and neural dynamics. This signal was based on blood flow and blood volume, and reflected neural events underlying these variations. For example, in the Belliveau experiment a full-field flashing light was viewed by the subject revealing the hemodynamic effects of neural activity along the calcurine sulcus, the location of primary visual cortex. A replication of that experiment is illustrated in Fig. 1b.

Figure 2 illustrates a BOLD signal (right panel) originating from a single voxel located in the left hemisphere, inferior frontal gyrus (indicated by the arrows). The yellow cluster indicates locations of multiple contiguous voxels where the MR signal during an object-naming task was significantly elevated relative to the baseline (rest) period. The active region is well established as an area specialized for language production (Broca's area), and provides a classical example of "functional specialization" as observed by fMRI.

Neuroimaging of individual brains: identification of functionally specialized cortical areas

One of the distinctions between basic research and clinical applications is the focus on the individual patient in clinical applications and the goal of a complete medical summary that is unique and tailored for a specific patient. This is in contrast to the normed and grouped evidence for a conserved feature relating to a general mechanism that is often the goal of basic science (Friston et al., 1999). Although the methods for group studies are well established, the new emphasis on clinical applications calls for establishment of highly reliable methods to accurately describe a single patient. One example of functional mapping that is well developed to serve clinical objectives is the mapping of critical functions to inform neurosurgical procedures by locating the centers of critical functions. These well-established methods guide our efforts to access functional systems in minimally conscious patients (MCS) and are introduced in the following section.

Table 1. The blood oxygen level dependent (BOLD) signal

Physiology	Physics
Neural activation is associated with an increase in blood flow. O_2 extraction is relatively unchanged Result Reduction in the proportion of deoxyhemoglobin in the local vasculature	Deoxyhemoglobin is paramagnetic and distorts the local magnetic field causing signal loss Result Less distortion of the magnetic field results in local signal increase

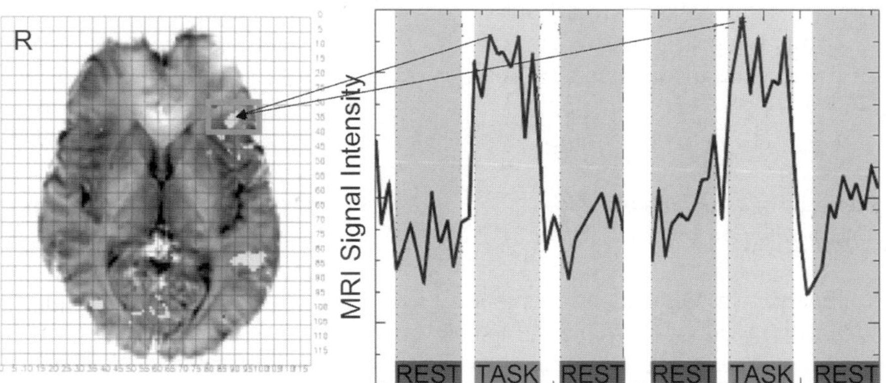

Fig. 2. Cortical activity associated with naming visual objects. Signals illustrate the BOLD changes in MR susceptibility observed in response to naming visually presented objects in a healthy volunteer. Signals originate from a single voxel (1.5 × 1.5 × 4.5 mm) on two separate runs. Each run lasted 2 min 24 sec, during which 36 images were acquired, including 10 images for each of three epochs: initial resting baseline (*blue bar*), task (naming) (*pink bar*), and final resting baseline (*blue bar*). All voxels in the brain for which the statistical criteria were met (the average amplitude of the signal during the activity epoch was statistically different from the baseline signal) are indicated by either a *yellow*, *orange*, or *red* color superimposed on the $T2^*$-weighted image at the voxel address, and signify decreasing levels of statistical confidence. Arrows point to the source voxel, which is centered within a cluster of similar (*yellow*) voxels and located in the left hemisphere of the brain in the inferior frontal gyrus, putative Broca's area. This slice of brain also shows activity in left superior temporal gyrus, putative Wernicke's area, and in the right lateral occipital extra striate area (putative object processing area), and along the medial and posterior occipital regions (putative primary visual cortex). See Plate 3.2 in Colour Plate Section.

Methods for neuroimaging of cognitive processes in individual brains

Functional maps for individual subjects aim to identify critical functional specializations specific to that subject or patient. For example, in the case of functional mapping prior to a surgical procedure, the goal is to identify regions of the individual patient's brain that are employed for functions, such as motor movements, tactile sensation, language, vision, and audition that might be at risk because of the location of the surgery. The presence of a space-occupying lesion or long-term seizure-genic conditions can modify the foci of functional brain tissue, and normal assumptions of functional specificity do not necessarily apply. In these cases, functional brain maps are acquired at the highest possible resolution to locate the functionally eloquent cortex, and this information is integrated into the appropriate treatment plan for the patient.

Because the preservation of function during a brain tumor resection for an individual patient is an essential goal of neurosurgery, various conventional intraoperative and preoperative brain-mapping techniques are routinely employed for this purpose (Burgess, 1995; George et al., 1995; Pujol et al., 1998). These techniques aim to identify cortical areas involved in sensory, motor, and language functions and have become standard practice. They include intraoperative electrophysiology with motor and language mapping, preoperative Wada tests, and visual field examinations. However, the added risk, time, and expense of multiple mapping procedures favors noninvasive, preoperative procedures. MRI maps of sensory and motor functions, either alone or in combination with other neuronavigation techniques, have become widely accepted as effective in directing brain tumor resection procedures away from cortical regions with residual function (George et al., 1995; Latchaw et al., 1995; Puce et al., 1995; Puce, 1995; Yousry et al., 1995; Atlas et al., 1996; Mueller et al., 1996; Stapleton et al., 1997; Lee et al., 1998; Schulder et al., 1998; Nimsky et al., 1999).

Similarly, functional mapping of language areas has also been accomplished by a range of tasks and procedures including object naming and verb generation (Herholz et al., 1997), production of the names of animals starting with a given letter (Latchaw et al., 1995), word generation in alphabetical order (Hinke et al., 1993), or auditory noun presentations with a required category response (Binder et al., 1996). Assessment of cortical activity associated with visual stimulation has been accomplished with intermittent binocular photic stimulation (Fried et al., 1995, Latchaw et al., 1995; Kollias et al., 1998), as well as with various projected pattern stimuli (Tootell et al., 1995). Although all of these tasks for sensory, motor, language, and visual functions may be individually effective, there are strong arguments in favor of an integrated and standardized battery of tasks to facilitate validation and enhancement.

One such battery of fMRI tasks targets cortical regions associated with tactile, motor, language, and visual-sensitive cortical areas considered most critical for surgical decisions (Hirsch et al., 2001). In this battery, all functions are repeated using both "active" (volitional) and "passive" (receptive) modes to assure that it is applicable to patients with a wide range of symptoms and abilities to comply with task directions. This feature also facilitates use in MCS patients where passive stimulation is required. Any subset of these tasks may be selected for specific clinical objectives while retaining the advantages of the standardized procedures with validations based on responses of both healthy volunteers and patients. Thus, objectives of mapping cognitive functions in unresponsive patients requiring passive stimulation techniques are served by these prior validations. These advantages are essential for mapping residual functions in MCS patients.

A multi-function task battery

The specific tasks selected for this task battery are nearly universally applicable and employ simple stimuli and procedures. The tasks consist of four separate procedures (Fig. 3) including:

(1) Passive tactile stimulation of a hand (either the dominant hand or the hand relevant to the hemisphere of surgical interest) is performed with a mildly abrasive plastic surface that is gently rubbed on the palm and fingers. Simultaneously, the patient views a reversing checkerboard pattern (8 Hz). This visual stimulation also aids the patient with head stabilization.
(2) Active hand movement (finger–thumb tapping) using either the same hand as in the passive tactile stimulation or both hands during repeat of the simultaneous visual stimulation (reversing checkerboard).
(3) Picture naming by internal (silent) speech is performed in response to visually displayed, black-and-white, line-drawings presented at 4 sec intervals. These drawings are selected from an appropriate range of the Boston Naming test (Kaplan et al., 1983).
(4) Listening to recordings of spoken words (names of objects) that are presented

Task Battery for Cortical Mapping of Sensory, Motor, Language, Auditory and Vision-Related Functions

1.	**SENSORY** Touch/hand	+	**VISION** Reversing Checkerboard
2.	**MOTOR** Finger/Thumb tapping	+	**VISION** Reversing Checkerboard
3.	**LANGUAGE/active** Picture Naming	+	**VISION** Pictures
4.	**LANGUAGE/passive** Listening to Words	+	**AUDITION** Spoken words

Fig. 3. A summary of each of the functions mapped by the four conditions in the standardized fMRI task battery for neurosurgical planning.

through headphones is designed to reduce scanner noise. A visual "cross-hair" remains present throughout the run to help prevent head movement.

The aims of these four conditions include: localization of primary and secondary visual areas, localization of sensory and motor cortices, and by inference, prediction of the location of the central sulcus, and localization of language-related activity, and by inference, prediction of the locations of Broca's and Wernicke's areas and the dominant hemisphere for speech. The language areas are redundantly targeted by expressive (active) and receptive (passive) language tasks, and by visual and auditory modalities.

In the case of this test battery, development was based upon a total of 63 healthy volunteers (24 females and 39 males), who participated in the evaluation of the specific set of tasks targeted to identify brain regions that are most likely to be surgical regions of interest for (1) primary brain tumor, (2) brain metastasis, (3) seizure disorder, or (4) cerebral vascular malformation (Hirsch et al., 2000). A total of 125 patients also participated. These patients were surgical candidates and presented with surgical regions of interest that included sensorimotor ($n = 63$), language ($n = 56$), or visual ($n = 6$) functions. The importance of valid imaging techniques for assessment of cortical responses in patients who give no bedside evidence of responsiveness motivates inclusion of the validation procedures here.

Accuracy of brain maps: comparison with intraoperative electrophysiology

In the same study, accuracy of the fMRI observations was assessed for these 125 surgical patients. Comparison between conventional and MRI mapping procedures were made when both were included in the treatment plan by comparison of the fMRI maps with conventional intraoperative mapping (Fandino et al., 1999). This method of comparison also serves validity assessments of the same technique when applied to situations where surgical confirmations are not made. Both fMRI preoperative maps and intraoperative electrophysiology were performed in 16 cases. Intraoperative recording of somatosensory evoked potentials (SSEPs) were performed to localize the central sulcus (Dinner et al., 1987), and successful intraoperative recordings were obtained in 15 cases. Direct cortical stimulation (Ojemann et al., 1989) was performed in 11 of these cases with successful stimulations in 9. The areas of electrophysiological response were referenced to axial images with the use of an intraoperative frameless-based stereotactic navigation device and compared to the preoperative fMRI images. In each case where a successful SSEP or a direct stimulation occurred, the surgeon judged the correspondence as consistent with the functional maps (Hirsch et al., 2001).

The fMRI maps revealed precentral gyrus activity in 16/16 (100%) cases and postcentral gyrus in 13/16 (81%) cases. However, the combined maps revealed the location of the central sulcus in all cases. When both methods (fMRI and electrophysiology) reported the central sulcus, the locations concurred in 100% of the cases for the SSEPs (15/15) and also 100% for the direct cortical stimulation (9/9), as determined within the spatial accuracy of both methods, and are in accord with similar investigations (Puce et al., 1995; Cedzich et al., 1996). Thus, functional mapping based on voxed-by-voxed threshold criteria (see below) is relatable to conventional mapping techniques and serves to validate the methodology.

Comparison of fMRI, Wada, and intraoperative language mapping

The identification of the dominant hemisphere for language determined by Wada testing (Wada and Rasmussen, 1960) was consistent with fMRI results in 100% of cases ($n = 16$) with both examinations in a double-blind study, and is consistent with findings of previous investigations (Hirsch et al., 2000). In a subsequent cohort of five patients, this integrated battery of tasks was applied prior to intraoperative language mapping with consistent findings between the two methods (Ruge et al., 1999). Since language functions are highly relevant to emergence from MCS, these tools to passively assess language systems are considered essential. In particular, the passive listening task in this battery was 100% effective in stimulating putative Wernicke's area and 93% effective in stimulating putative Broca's area in 45 healthy individuals who performed this task without volition.

In our clinical experience (nearly 1500 cases to date), this fMRI task battery serves both pre- and intraoperative objectives. On the preoperative side, the fMRI maps have contributed to our estimates of the risk/benefit ratio and to the decision whether or not to offer surgery to the patient, although these decisions are based on the entire medical situation taken together, and not on any single assessment. Communication between the surgeon and the patient is also facilitated by images that summarize the relevant structure and function issues. On the intraoperative side, the fMRI results also serve to direct the intraoperative electrophysiology, and thereby contribute to the efficiency of the intraoperative procedures. Although the accuracy and sensitivity of the maps are evaluated in each case where the corresponding intraoperative tasks are managed, the overall validity of the methodology for individual patients is well established and thus serves as a foundation to assess functional systems in unresponsive patients.

Considerations that differentiate single subject mapping from group investigations

As noted above, functional mapping for clinical applications is focused on the individual patient rather than sample or population average responses. The objective to contribute to the plan for a single patient imposes additional burdens of proof and confidence. The relevance of these considerations for functional mapping of patients with deficits of consciousness justifies inclusion here.

Zero tolerance for error

In contrast to population-based studies where grouped results are taken as evidence and error bars are considered to be variations largely due to individual differences, clinical tests focus on the individual for diagnosis, treatment, and follow up. Thus, the burden for accuracy for the "*n* of one" case is 100%, and dictates methodological adaptations to meet this standard. These adaptations include an extraordinary standard for high image quality as well as clarity, accuracy, and precision for interpretation.

The "zero tolerance for error" standard is further complicated by the special circumstances of many patients. Some key factors include functional deficits that challenge execution of the task, high levels of anxiety leading to claustrophobia, inability to remember or perform instructions, excessive head movements, probability of a seizure or other sudden event that interrupts a scan, the effects of therapeutic drugs, and susceptibility artifacts often resulting from a previous surgical bed, implant, or a vascular abnormality. The methodological adaptations developed to accommodate these special circumstances are discussed in the following text. They include standardized paradigms and tasks that map most relevant functions, employ short-imaging runs, high-resolution grids, and least number-of-assumptions for data analysis. All of these adaptations are relevant to the even more challenging task of imaging unresponsive patients with disorders of consciousness where passive (rather than volitional) responses are required.

Most relevant functions and task selection

Task selection is one of the key aspects of functional imaging because the task defines the interpretations of the functional map. In the case of mapping for surgical planning, a high priority is placed on "most-feared morbidity" including aphasia and paraplegia. Thus, tasks that identify motor and sensory-sensitive cortices and, therefore, the central sulcus landmark on the targeted hemisphere as well as tasks that reliably engage essential language areas are essential. Tasks that utilize features already in use in conventional intraoperative mapping including hand-related sensory/motor tasks and object-naming tasks are also an advantage. Active and passive (listening) modes of object naming exhibits activity in both visual and auditory systems in addition to the language systems, which also serve to enrich the maps by verifying the status and location of these eloquent areas even though they are often not the primary focus of the surgical plan.

Although a specific task elicits a specific brain map, interpretation of the function that each of the areas contributes usually requires careful followup attention and control experiments. For example, in the language listening task, it is possible that the activity patterns employ specific regions involved in attention, languages, audition, imagery, associations, emotion, and memory. Even if each of the repetitions of the task elicit common responses, this diverse group of putative functions cannot be disambiguated. Thus, interpretation of the factors that actually elicit the BOLD signal remains a pivotal issue.

In the case of patients with disorders of consciousness, the tasks must be limited to passive stimulations, and the most relevant questions center around inferences regarding cognition and awareness. The challenge is to illicit evidence that sheds light on the question of internal cognitive processes and potential for recovery. Options for testing with the advantage of prior validation procedures (mentioned above) are tasks that employ listening to spoken language and tactile stimulation of the hands. The tactile stimulation can serve as a procedural control, whereas the passive listening offers possible information regarding the status of language-related cognitive systems. However, the extent to which the BOLD response in MCS can be interpreted similarly to the BOLD response in healthy volunteers must be examined carefully.

Procedure adaptations: short runs

Under conditions where constant monitoring and assessments are necessary, anxiety and claustrophobia are frequent issues for patients, and imaging runs may need to be repeated due to confusion regarding instructions, failure to comply, or other unexpected events. In these instances, the shorter the imaging epoch, the better. The tradeoff, of course, is accuracy and statistical confidence, and both are optimized by increased numbers of acquisitions. One technique that balances these procedural concerns is sometimes referred to as a "double pass" strategy. Short runs are performed for each task consisting of an initial and ending baseline epoch (minimum of 10 acquisitions each) and a central "activity" epoch. Both runs are identical block designs and if not successfully implemented can be repeated without compromising the other (Fig. 4). The optimal analysis rule is that two good runs must be acquired for standardized quality assurance. However, in the case where only one run is possible, a map can be produced that offers meaningful results at a reduced level of certainty.

Analysis adaptation: confidence

The short-run, double-pass method employs an analysis strategy based on a combination of statistical signal-to-noise models and physiological repeatability. Basically, voxel-by-voxel "t" maps are generated separately for each run, and the final report requires that all reported voxels meet that statistical criteria on both runs. This "conjunction" strategy tends to filter out highly significant voxels based solely on signal-to-noise factors that are due to spurious noise events, and adds the criteria that the event is replicable Nichols et al., 2005. Although signals due to spurious sources occurring sequentially at the same place are possible, the likelihood is greatly reduced relative to a single occurrence. Probabilities of false-positive events using this conjunction criteria can be determined empirically, based on a resting brain or phantom. Probabilities of a false-positive result using this technique are conventionally adjusted to be in the range of $p<0.0005$ to <0.0001 (Hirsch et al., 1995a, b, 2000, 2001; Kim et al., 1997; Ruge et al., 1999; Victor et al., 2000; Wang et al., 2003). These empirically determined probabilities are consistent with calculated expected probabilities using the Bonferoni correction for multiple determinations, assuming an appropriate cluster size and our high resolution of 1.5×1.5 mm (Fig. 5). Because of the extensive validation and ease of implementation, the short-run, double-pass method is well suited for imaging patients who are not responsive or cooperative.

Imaging adaptations: field strength, and resolution

Although high-field scanners promise advantages in sensitivity, the increased susceptibility to artifacts within the field of view in some cases favors

Run 1

Time (min)	Image	TASK
2:24	0	
2:20	1	
2:16	2	
2:12	3	
2:08	4	
2:04	5	Rest
2:00	6	
1:56	7	
1:52	8	
1:48	9	
1:44	10	
1:40	11	
1:36	12	
1:32	13	
1:28	14	
1:24	15	
1:20	16	
1:16	17	
1:12	18	Activity
1:08	19	
1:04	20	
1:00	21	
0:56	22	
0:52	23	
0:48	24	
0:44	25	
0:40	26	
0:36	27	
0:32	28	
0:28	29	
0:24	30	Rest
0:20	31	
0:16	32	
0:12	33	
0:08	34	
0:04	35	
0:00	36	

Run 2

Time (min)	Image	TASK
2:24	0	
2:20	1	
2:16	2	
2:12	3	
2:08	4	
2:04	5	Rest
2:00	6	
1:56	7	
1:52	8	
1:48	9	
1:44	10	
1:40	11	
1:36	12	
1:32	13	
1:28	14	
1:24	15	
1:20	16	
1:16	17	
1:12	18	Activity
1:08	19	
1:04	20	
1:00	21	
0:56	22	
0:52	23	
0:48	24	
0:44	25	
0:40	26	
0:36	27	
0:32	28	
0:28	29	
0:24	30	Rest
0:20	31	
0:16	32	
0:12	33	
0:08	34	
0:04	35	
0:00	36	

Fig. 4. Double-pass short-block paradigm. Time series for the two short runs, double-pass method of image acquisition. The total run time is 2 min and 24 sec for each run, and each line illustrates the 4-sec bin boundary separating whole brain acquisition (right columns) into image acquisitions (column 2). The stimulation prescription (column 3) indicates the stimulation cycle where images 0–12 are acquired during resting baseline (with visual fixation), images 13–24 are acquired during the task performance (Fig. 3), and images 25–36 are acquired during the final resting period as in acquisitions 0–12.

the 1.5 T field strength scanners for patients with implants, surgical beds, and vascular abnormalities for functional maps. These postsurgical conditions often apply to patients with head trauma, thus favoring the conventional strength scanners. Nonetheless, high-quality descriptions of functional systems on individual patients requires high-resolution acquisitions for the functional images. In basic science applications where images from multiple subjects are registered together, the potential advantages to high-resolution acquisitions are often not realized due to the multiple registrations and normalizations. However, for a single individual the alignments are restricted to motion correction, whose shifting facilitates high-resolution acquisitions Woods et al., 1993. An example of an optimal set of high-resolution imaging parameters includes a field of view of 19×19 cm and an array size of 128×128. The in-plane voxel size with these parameters is 1.5×1.5 mm, which assures a high standard of spatial accuracy. Most scanners can acquire 20–40 slices for these parameters allowing slice thickness to vary from approximately 2.0–4.0 mm under these conditions. Another advantage to the high-resolution T2* acquisition is the reduced need to display the functional maps on an alternative (such as T1) acquisition for anatomical detail. This avoids distortions due to image registrations and preserves the best description of structure and function for the individual patient. In patients with head injury and altered cortical topography, these advantages have particular relevance.

Functional mapping during sedation

Following the development of the basic standard battery of tasks for neurosurgical planning, we were challenged again by the need to provide similar maps for patients who were sedated during the imaging procedure (Chapman et al., 1995). The population for which this was generally necessary

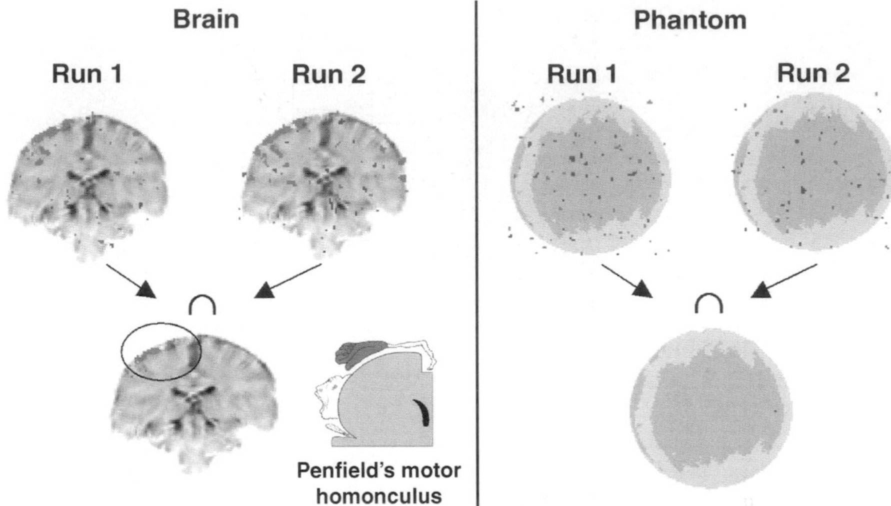

Fig. 5. Double-pass method. Illustrations of voxels that meet statistical criteria at each stage of statistical decision for the brain (coronal slice) of a subject (left panel) and for the copper sulfate phantom simulating the brain (right panel). The stages of the analysis are outlined. Each average signal must be higher during the stimulation epoch than the initial baseline (stage 1) and the recovery baseline (stage 2) on two independent test-runs (runs 1 and 2). The task performed was finger-thumb tapping using the left hand. As would be expected, the right hemisphere of the precentral gyrus (motor strip) is activated (circled activity). The location of activity within the motor strip matches the known location of the hand in the motor strip in the Homonculus, as shown in the schematic on the right. The conjunction-across-2 runs is based on the principle that real signals can be differentiated from the noise by the probability of a repeat occurrence in the same place through multiple acquisitions. This double-pass acquisitions simulation (described above) is shown for a copper sulfate phantom control. Activity passing all conjunction criteria includes only three "active" voxels. This procedure is repeated many times, using both copper sulfate phantoms and also human brains under resting conditions, to empirically determine that the average rate of a false-positive signal is less than 1 part in 10,000, i.e., on average there is one voxel observed by chance for each brain slice. See Plate 3.5 in Colour Plate Section.

included children under 6 years of age. In the face of specific medical needs, the passive tasks (listening, tactile, and visual photic stimulation) were adapted for use under sedation with propofol (Hirsch et al., 1997; Souweidane et al., 1999). One of the adaptations was the use of a familiar speaker for the passive listening, and meaningful narratives were recorded for the purpose of the functional maps. As illustrated for the case of a 15-month-old sedated girl (Fig. 6), this technique yields a system of regions consistent with known language-related areas. In the case of this patient, the language-dominant hemisphere was on the right side which is consistent with the absence of morbidity following a complete and total resection on the left hemisphere. These studies confirm that "consciousness" is not a prerequisite to confirming the presence of an active language-related system in the cortex.

Neuroimaging of cortical areas specialized for high-level cognitive tasks

In contrast to sensory/motor, and visual systems, current understanding of the mechanisms of higher cognitive functions is not closely linked to a specified neurophysiological substrate. Posner and Raichle (1994) note that

> "While there has been general agreement that operations performed by the sensory and motor systems are localized, there has been much more dispute about higher-level cognitive processes. It is still undetermined whether higher-level processes have defined locations, and if so, where these locations would be."

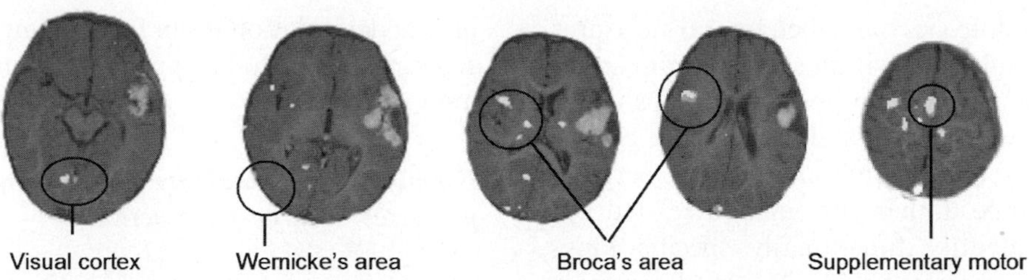

Fig. 6. Sedated child "listening" to mother's narrative. Fifteen-month-old girl with an aggressive tumor in the left hemisphere (hyperintense region). In-plane resolution was 1.5×1.5 mm and the double-pass paradigm and analysis (see text) was employed. The patient was sedated with propofol and maintained at the lightest level that prevented movement of the head. A recorded passage of the mother talking to the patient was played through earphones worn by the patient while scanning. The circles indicate putative Wernicke's area, Broca's area, and supplementary motor area, consistent with the expected language network in preparation for the emerging language function. See Plate 3.6 in Colour Plate Section.

The emergence of neuroimaging, however, provides a new opportunity to test hypotheses about the underlying networks that sustain cognition in healthy individuals without reliance on lesions or disease processes (Cabeza and Nyberg, 2000). Determinations of the anatomy and topography of cortical areas specialized for cognitive tasks informs models of attention, working memory, and executive processes as well as consciousness.

As discussed above the functional neural anatomy of cognitive processes is revealed by comparing the BOLD response elicited by various experimental and control tasks, and is typically characterized by a voxel-by-voxel statistical comparison of the signal amplitude during the activity with the signal amplitude during a baseline resting or control event. The basic assumption is that neuronal activity is increased in a functionally specialized area during the execution of a task that employs that specialization. Locations of active areas, cluster sizes, and dynamical properties of the signal can be compared across cognitive conditions. Tasks are designed to either include or exclude the cognitive component of interest, and the signals elicited from these tasks are compared (Price et al., 1997).

Neural systems and cognitive processes

Although differentiation of brain areas by functional specificity is a well-established principle of cortical organization, recent approaches to study human cognition have focused on the integration of groups of specialized areas into long-range units that collectively serve as the comprehensive neural substrate for specific cognitive tasks. According to Mesulam (1998):

> "A central feature in the organization of the large-scale network is the absence of one-to-one correspondences among anatomical sites, neural computations and complex behaviors."

Thus, an individual cognitive or behavioral domain is subserved by several interconnected sites, leading to a distributed neural system with functional specificity. An early empirical foundation for this emerging view is found in the work of Mishkin and Ungerleider (1982), who described ventral and dorsal pathway segmentation during visual tasks that required either object identification or object localization, respectively. More recently, direct interactions between multiple brain regions that participate in a variety of complex functions and behaviors have been proposed as evidence for this systems model. For example, covariations between BOLD responses in separate cortical areas during complex attention tasks (Buchel and Friston, 1997) have been described using a statistical approach called structural equation modeling to determine whether the covariances between areas are due to direct or indirect interactions. Models of fear, anxiety, and cognitive control have been similarly studied using this technique and other more recent and emerging approaches to determine various interactions between psychophysical and behavioral measures and signal modulations within specific brain areas. These analyses identify the involved groups of areas associated with a task, and characterize changes in regional activity and interactions between regions over time. Other current and related approaches to identify functionally specific long-range systems associated with high-level cognitive and attentive processes include covariation of regional activities with either other areas or other data like reaction time or individual assessments of behavior like personality inventories.

Functional neuroanatomy of language: a large-scale network

The fundamental relevance of the language system to recovery from MCS motivates focus on the underlying network for language. Models of the neural correlates for elementary language processes often include left hemisphere regions involved in a variety of language functions, including Broca's and Wernicke's areas and are generally consistent with a network model. This network is easily demonstrated using an object-naming task and multiple modalities including auditory, visual, and tactile stimuli (Hirsch et al., 2001). A cross-modality conjunction technique isolates effects that are observable in all cases and, therefore, not dependent upon sensory processes (Price et al., 1997). Results are consistent with the view that the task of naming objects elicits activity from a set of areas within a neurocognitive system specialized for language-related functions. The colored circles on the normalized "glass brain" (Fig. 7) represent average locations of activity centroids from a large group of healthy subjects on the standard atlas brain (x-, y-, z-coordinates) as indicated in Table 1 (Talairach and Tournoux, 1988). There are five regions in this neurocognitive system (all located within the left hemisphere) including putative Broca's area (inferior frontal gyrus, BA 44 and 45), putative Wernicke's area (superior temporal gyrus, BA 22), and medial frontal gyrus (BA 6). Thus, these results are consistent with the view that the functional specialization for this elementary language task involves a system of language-related areas rather than a single area. This simple "active" task is clearly similar to the results of the "passive" task under sedation (even for a young child, above), suggesting that both approaches stimulate a common system.

Functional neuroanatomy of attention and volitional processes: a large-scale network

Like language, the ability to attend is involved in a wide range of cognitive tasks. Functional imaging studies by Mesulam and others suggest that spatial attention is also mediated by a large-scale distributed network of interconnected cortical areas within the posterior parietal cortex, the region of frontal eye fields, and the cingulate cortex. Kim et al. (1999), used a conjunction analysis to compare activity associated with two different types of visuospatial attention shifts: one based on spatial priming and the other based on spatial expectancy to test the hypothesis of a fixed area network for both tasks. The activation foci observed for the two tasks were nearly overlapping indicating that both were subserved by a common network of cortical and subcortical areas. The main findings were consistent with a model of spatial attention that is associated with a fixed large-scale distributed network specialized to coordinate multiple aspects of attention. Alternative hypotheses which predict that task variations are associated with an increase in the number of involved areas were rejected. However, an observed rightward bias for the spatial priming task suggested that activation within the system showed variations specific to the attributes of the attentional task.

Neuroimaging and deficits of consciousness

Guided by the "Astonishing Hypothesis" of Francis Crick (1994), we assume that the biological underpinnings of consciousness are observable and are represented by specific neurocorrelates. However, the question is, how and which ones? Although still in a nascent stage, the study of consciousness is an "emerging" area of research (Zeman, 2001, 2002, this volume). According to Antonio Damasio (1998)

> "The fact that consciousness is a private, first-person phenomenon makes

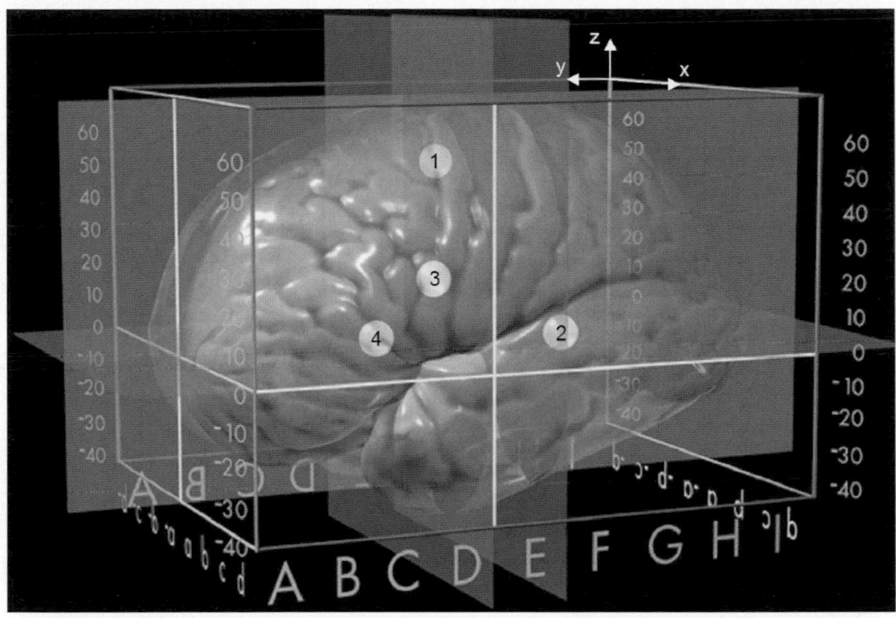

Common Name	Anatomical Region	Brodmann Area	Center of mass x	y	z
① Supplementary Motor	Medial Frontal Gyrus (GFd)	6	9	-6	53
② Wernicke's Area	Superior Temporal Gyrus (GTs)	22	57	-26	9
③ Broca's Area	Inferior Frontal Gyrus (Gfi)	44	49	10	25
④ Broca's Area	Inferior Frontal Gyrus (Gfi)	45	40	25	8

Fig. 7. A fixed large-scale network for object naming. The network of areas which subserve object naming as determined by conjunction across three sensory modalities consists of medial frontal gyrus (GFd, BA 6), superior temporal gyrus (GTs, BA 22; putative Wernicke's area), and inferior frontal gyrus (Gfi, BA 44,45; putative Broca's area), in the left hemisphere. These areas are portrayed as colored circles in a three-dimensional glass brain based upon the Talairach and Tournoux Human Brain Atlas and stereotactic coordinate system. The table appearing below contains the average group coordinates (*x*, *y*, *z*) of the included regions (from Hirsch et al., 2001). See Plate 3.7 in Colour Plate Section.

> it more difficult to study than other cognitive phenomena that, although being equally private, also have characteristic behavioral signatures. Nonetheless, by combining cognitive and neurobiological methods, it is possible to approach consciousness, to describe its cognitive nature, its behavioral correlates, its possible evolutionary origin and functional role; last but not least, it is possible to investigate its neuroanatomical and neurophysiological underpinnings."

As in the neuroimaging of related cognitive processes such as language, attention, working memory, and executive control, an investigation of consciousness is largely dependent upon the development of appropriate paradigms and relevant theory. A major obstacle in the application of neuroimaging techniques to the investigation of consciousness is the inability to establish a task to vary the state of consciousness. That is, consciousness is not started and stopped in synchrony with a particular imaging sequence.

To circumvent this obstacle, we employed a variation of the passive language tasks developed for neurosurgical procedures in both conscious and sedated patients, as described above. The neuroimaging of patients with disorders of consciousness who are exposed to stimulation intended to provoke neural circuitry underlying cognition is a novel and developing area of investigation. Techniques employed by our group are based on the methodology, described earlier in this chapter, originating from clinical needs to map critical functions in single patients (Hirsch et al., 2000).

An illustrative case

A 33-year-old, right-handed male without prior history of neurological disorders suffered severe head trauma secondary to a blow in the right frontal region with a blunt object leading to bilateral subdural hematoma and brainstem compression injury (Hirsch et al., 2003; Schiff et al., 2005). Bedside examinations were consistent with the diagnosis of MCS (Giacino et al., 2002), and the functional imaging study was performed 24 months after the injury. Neurological examination at the time of the study revealed oculocephalic responses with intact visual tracking and saccades to both stimuli and to commands, marked increased motor tone bilaterally, and frontal release signs. The highest level of behavior observed in this patient was his ability to inconsistently follow complex commands including go, no-go, countermanding tasks, and occasional verbalization. Right frontal lobe encephalomalacia and paramedian thalamic infarction were present on the structural MRI and on the functional (T2*) scans. Resting FDG-PET demonstrated 40.6% of normal regional cerebral metabolic rate.

Imaging methods were as described previously for neurosurgical patients, and employed passive auditory stimulation by presenting a prerecorded narrative of a family member recalling past experiences that she had shared with the patient. The narrative was also presented as reversed speech including the same content but without linguistic meaning. Passive tactile stimulation was also employed. Single epoch runs, each lasting 2 min and 24 sec and the double-pass short-block paradigm was employed (Fig. 4). Single-voxel confidence levels were set at $p < 0.0001$ corrected for false positives (Fig. 2). Voxel size was $1.5 \times 1.5 \times 4.5$ mm and 21 continuous slices covered the whole brain. The scanner was a 1.5 T GE equipped for EPI with a standard fMRI sequence (TR = 4000, TE = 60, flip = 60).

The three panels in Fig. 8 show the results of activity associated with the forward speech excluding any areas associated with the backward speech in order to evaluate the hypothesis that the subjects were responsive to the "meaning" of the narrative and not just the sound. The three panels show the results of 10 healthy normal subjects performing a similar task (left panel); a 29-year-old male imaged as a healthy volunteer at the same time as the patient and listened to the same narrative (middle panel); and the MCS patient (right panel). Most notable is the similarity between the patient and both the group and individual results. In particular, activity observed in the left temporal (GTs) and inferior frontal gyres (GFi) are consistent with Wernicke's and Broca's areas, respectively. Of notable interest is the additional activity in the occipital region (calcarine sulcus, SCa, and middle occipital gyres, GOm) of the patient suggestive of mental imagery and/top-down information processing in response to the narrative. Although not shown owing to space and summary constraints, other acquired slices were consistent with the findings as illustrated in these slices. Further, the passive tactile stimulation was indistinguishable in all three cases (see Schiff et al., 2005).

A notable comparison

The neural activity associated with this stimulation is not only remarkably robust for the patient, it is basically indistinguishable from the normals. Together with the inconsistent evidence for receptive and expressive language evident in the bedside examination, and the low resting cerebral metabolic rate, these imaging findings were unexpected. There was, however, one notable difference between this patient and the group of normals. When the stimulating narrative was in time-revered order (lacking linguistic content but containing speech-like qualities), it elicited robust responses from the normals, whereas the MCS patient response was markedly reduced (Schiff et al., 2005). This failure of the time-reversed narratives to elicit robust activity may reflect a failure to provoke a global change in neuronal firing patterns based on stimulus salience. One preliminary interpretation sug-

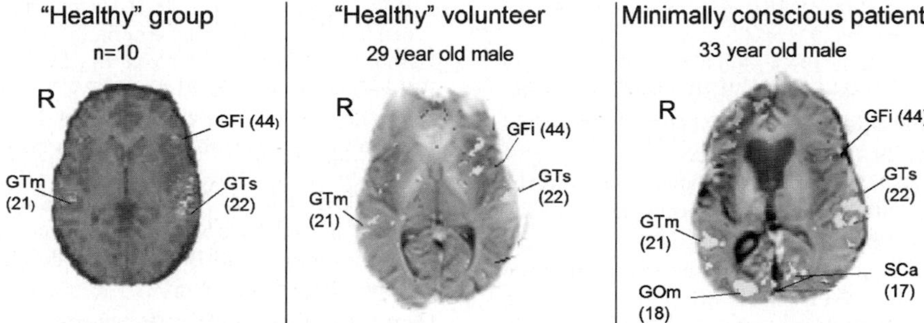

Fig. 8. Passive listening to narratives presented as normal speech (forward) excluding responses elicited when the same speech was presented backwards. Results from the healthy volunteers are shown on one slice of an average brain (left panel), results from an age- and sex-matched healthy volunteer acquired at the same session and using the same narratives as the patient in an MCS are shown on a comparable slice (middle panel), and results from the MCS patient (see text) are shown at a similar slice level (right panel). Note the structural abnormalities in the anterior right hemisphere secondary to the injury responsible for the patient's altered state of consciousness. All three brains show activity in the left inferior frontal gyrus (GFi), Brodmann's area (BA) 44, putative Broca's area; left hemisphere temporal gyrus (GTs), BA 22, putative Wernicke's area; and middle temporal gyrus (GTm), BA21, bilaterally, putative auditory-cortex. Additionally, the MCS subject shows activity in the extrastriate visual cortex (middle occipital gyrus, GOm, BA18), and in the primary visual cortex (calcarine sulcus, SCa, BA17). The levels of statistical stringency and analysis procedures were identical for the patient and comparison healthy volunteer, suggesting that the responses elicited by the patient were (1) more robust as they activated a large volume of cortex (larger number of significant voxels) and (2) more distributed as the global pattern of activity included visual system responses. This was unexpected since there was no visual stimulation (only auditory), and the patient had his eyes closed. We cannot rule out the possibility that this patient was "visualizing" some representation of the narrative which was personalized for him. See Plate 3.8 in Colour Plate Section.

gests that for his patient MCS may be related to a deficit of "baseline" responsiveness consistent with known effects associated with compression of the thalamus and brainstem that occurred during the acute phase of the injury.

What does this mean?

These findings and other similar cases (Schiff et al., 2005) raise a number of complex questions including "Does this patient have an inner cognitive life that is not reflected by his bed-side responses?" If so, what can we do about it? Are any of these findings evidence for "neural preparedness" for emergence? What ethical issues do these findings raise (Fins, 2003, this volume)? What investigations need to be done to lead us to informed and aggressive therapies? Can neuroimaging contribute to assessments that will guide and inform therapies for individual patients? The content of this paper suggests that imaging paradigms established to investigate the neural underpinnings of cognition in normals also provide foundations for emerging investigations of the neural underpinnings of consciousness (Hirsch, 2005). Both bring neural science closer to the goal of understanding the biological underpinnings of the mind, and perhaps closer to the goal of treating patients with disorders of consciousness.

References

Dorland's Illustrated Medical Dictionary (1988). 27th ed. W.B. Saunders Co. (Harcourt Brace Jovanovich Inc.), Philadelphia, PA.

The American Heritage Dictionary of the English Language (2000). 4th ed. Houghton Mifflin Co., Boston.

Atlas, S.W., Howard, R.S., Maldjian, J., Alsop, D., Detre, J.A., Listerud, J., D'Esposito, M., Judy, K.D., Zager, E. and Stecker, M. (1996) Functional magnetic resonance imaging of regional brain activity in patients with intracerebral gliomas: findings and implications for clinical management. Neurosurgery, 38(2): 329–338.

Belliveau, J.W., Kennedy, D.N., McKinstry, R.C., Buchbinder, B.R., Weisskoff, R.M., Cohen, M.S., Vevea, J.M., Brady, T.J. and Rosen, B.R. (1991) Functional mapping of the human visual cortex by magnetic resonance imaging. Science, 254: 716–719.

Binder, J.R., Swanson, S.J., Hammeke, T.A., Morris, G.L., Mueller, W.M., Fischer, M., Benbadis, S., Frost, J.A., Rao, S.M. and Haughton, V.M. (1996) Determination of language dominance using functional MRI: A comparison with the Wada test. Neurology, 46: 978–984.

Buchel, C. and Friston, K.J. (1997) Modulation of connectivity in visual pathways by attention: cortical interactions evaluated with structural equation modeling and fMRI. Cerebral Cortex, 7(8): 768–778.

Burgess, R.C. (Ed.). (1995) Editorial: functional localization by neurophysiologic and neuroimaging techniques. J. Clin. Neurophysiol., 12: 405.

Cabeza, R. and Nyberg, L. (2000) Imaging cognition II: an empirical review of 275 PET and fMRI studies. J. Cog. Neurosci., 12(1): 1–47.

Cedzich, C., Taniguchi, M., Schafer, S. and Schramm, J. (1996) Somatosensory evoked potential phase reversal and direct motor cortex stimulation during surgery in and around the central region. Neurosurgery, 38: 962–970.

Chapman, P.H., Buchbinder, B.R., Cosgrove, G.R. and Jiang, H.J. (1995) Functional magnetic resonance imaging for cortical mapping in pediatric neurosurgery. Ped. Neurosurg., 23: 122–126.

Crick, F. (1994) The Astonishing Hypothesis: The Scientific Search for the Soul. Charles Scribner's Sons, New York.

Damasio, H., Grabowski, T., Frank, R., Galaburda, A.M. and Damasio, A.R. (1994) The return of Phineas Gage: clues about the brain from the skull of a famous patient. Science, 264: 1102–1105.

Damasio, A.R. (1998) Investigating the biology of consciousness. Phil. Trans. R. Soc. Lond. B, 353: 1879–1882.

Dinner, D.S., Luders, H., Lesser, R.P. and Morris, H.H. (1987) Cortical generators of somatosensory evoked potentials to median nerve stimulation. Neurology, 37: 1141–1145.

Fandino, J., Kollias, S., Wieser, G., Valavanis, A. and Yonekawa, Y. (1999) Intraoperative validation of functional magnetic resonance imaging and cortical reorganization patterns in patients with brain tumors involving the primary motor cortex. J. Neurosurg., 91: 238–250.

Fins, J.J. (2003) Constructing an ethical stereotaxy for severe brain injury: balancing risks, benefits and access. Nat. Rev. Neurosci., 4: 323–327.

Fox, P.T. and Raichle, M.E. (1984) Stimulus rate dependence of regional cerebral blood flow in human striate cortex demonstrated by positron emission tomography. J. Neurophys., 51: 1109–1120.

Fried, I., Nenov, V.I., Ojemann, S.G. and Woods, R.P. (1995) Functional MR and PET imaging of rolandic and visual cortices for neurosurgical planning. J. Neurosurg., 83: 854–861.

Friston, K.J., Holmes, A.P., Price, C.J., Büchel, C. and Worsley, K.J. (1999) Multisubject fMRI studies and conjunction analysis. Neuroimage, 10: 385–396.

George, J.S., Aine, C.J., Mosher, J.C., Schmidt, M.D., Ranken, D.M., Schlitt, H.A., Wood, C.C., Lewine, J.D., Sanders, J.A. and Belliveau, J.W. (1995) Mapping function in the human brain with magneto encephalography, anatomical magnetic resonance imaging, and functional magnetic resonance imaging. J. Clin. Neurophysiol., 12: 406–429.

Giacino, J.T., Ashwal, S., Childs, N., Cranford, R., Jennett, B., Katz, D.I., Kelly, J.P., Rosenberg, J.H., Cohyte, J., Zafonte, R.D. and Zasler, N.D. (2002) The minimally conscious state: definition and diagnostic criteria. Neurology, 58: 349–353.

Herholz, K., Reulen, H., von Stockhausen, H., Thiel, A., Ilmberger, J., Kessler, J., Eisner, W., Yousry, T.A. and Heiss, W. (1997) Preoperative activation and intraoperative stimulation of language-related areas in patients with glioma. Neurosurgery, 41(6): 1253–1262.

Hinke, R.M., Hu, X., Stillman, A.E., Kim, S.G., Merkle, H., Salmi, R. and Ugurbil, K. (1993) Functional magnetic resonance imaging of Broca's area during internal speech. Neuro. Report, 4: 675–678.

Hirsch, J., DeLaPaz, R.L., Relkin, N.R., Victor, J., Kim, K., Li, T., Rubin, N. and Shapley, R. (1995a) Illusory contours activate specific regions in human visual cortex: evidence from functional magnetic resonance imaging. Proc. Natl. Acad. Sci., USA, 92: 6469–6473.

Hirsch, J., DeLaPaz, R.L., Relkin, N.R., Victor, J., Kim, K., Li, T., Rubin, N. and Shapley, R. (1995b) Illusory contours activate specific regions in human visual cortex: evidence from functional magnetic resonance imaging. Proc. Natl. Acad. Sci. USA, 92: 6469–6473.

Hirsch, J., Ruge, M.I., Kim, K.H.S., Correa, D.D., Victor, J.D., Relkin, N.R., Labar, D.R., Krol, G., Bilsky, M.H., Souweidane, M.M., DeAngelis, L.M. and Gutin, P.H. (2000) An integrated fMRI procedure for preoperative mapping of cortical areas associated with tactile, motor, language and visual functions. Neurosurgery, 47(3): 711–722.

Hirsch, J., Rodriguez-Moreno, D. and Kim, K.H.S. (2001) Interconnected large-scale systems for three fundamental cognitive tasks revealed by functional MRI. J. Cog. Neurosci., 13(3): 1–16.

Hirsch J., Rodriguez-Moreno, O., Petrovich, N., Giacino, J.T., Plum, F. and Schiff, N.D. (2003). fMRI reveals intact cognitive systems in two minimally conscious patients. 9th International Conference on Functional Mapping of the Human Brain, New York, NeuroImage, 19(2): CD-Rom.

Hirsch, J., Kim, K.H.S., Souweidane, M., McDowall, R., Ruge, M., Correa, D. and Krol, G. (1997) fMRI reveals a developing language system in a 15-month sedated infant. Soc. Neurosci., 23(2): 2227 abstract.

Hirsch, J. (2005) Raising Commission. The Journal of Clinical Investigation, 115: 1102–1103.

Kaplan, E.F., Goodglass, H. and Weintraub, S. (1983) The Boston Naming Test (2nd ed.). Lea and Febiger, Philadelphia.

Kim, K.H.S., Relkin, N.R., Lee, K-M. and Hirsch, J. (1997) Distinct cortical areas associated with native and second languages. Nature, 388: 171–174.

Kim, Y-H., Gitelman, D.R., Nobre, A.C., Parrish, T.B., LaBar, K.S. and Mesulam, M.-M. (1999) The large-scale neural network for spatial attention displays multifunctional overlap but differential asymmetry. NeuroImage, 9: 269–277.

Kollias, S.S., Landau, K., Khan, N., Golay, X., Bernays, R., Yonekawa, Y. and Valavanis, A. (1998) Functional evaluation using magnetic resonance imaging of the visual cortex in patients with retrochiasmatic lesions. Neurosurgery, 89: 780–790.

Latchaw, R.E., Xiaoping, H.U., Ugurbil, K., Hall, W.A., Madison, M.T. and Heros, R.C. (1995) Functional magnetic resonance imaging as a management tool for cerebral arteriovenous malformations. Neurosurgery, 37(4): 619–625.

Lee, C.C., Jack Jr, C.R. and Riederer, S.J. (1998) Mapping of the central sulcus with functional MR: active versus passive activation tasks. Neuroradiology., 19: 847–852.

Mesulam, M.-M. (1998) From sensation to cognition. Brain, 121: 1013–1052.

Mishkin, M. and Ungerleider, L.G. (1982) Contribution of striate inputs to the visuospatial functions of parieto-preoccipital cortex in monkeys. Behav. Brain Res., 6(1): 57–77.

Mueller, W.M., Yetkin, F.Z., Hammeke, T.A., Morris, G.L., Swanson, S.J., Reichert, K., Cox and Haughton, V.M. (1996) Functional magnetic resonance imaging mapping of the motor cortex in patients with cerebral tumors. Neurosurgery., 39(3): 515–521.

National Academy of Sciences (2001). Beyond discovery. A lifesaving window on the mind and body. *The Development of Magnetic Resonance* Imaging. www.beyonddiscovery.org.

Neisser, U. (1967) Cognitive Psychology. Appleton, New York.

Nichols, T., Brett, M., Anderson, J., Wager, T. and Paline, J-P. (2005) Valid conjunction influence with the minimum statistic. Neuro. Image, 25: 653–660.

Nimsky, C., Ganslandt, O., Kober, H., Moller, M., Ulmer, S., Tomandl, B. and Fahlbusch, R. (1999) Integration of functional magnetic resonance imaging supported by magnetoencephalography in functional neuronavigation. Neurosurgery, 44(6): 1249–1255.

Ogawa, S., Lee, T.-M., Nayak, A.S. and Glynn, P. (1990) Oxygenation-sensitive contrast in magnetic resonance image of rodent brain at high magnetic fields. Magn. Reson. Med., 14: 68–78.

Ojemann, G., Ojemann, J., Lettich, E. and Berger, M. (1989) Cortical language localization in left, dominant hemisphere. J. Neurosurg., 71: 316–326.

Penfield, W. (1975) The Mystery of the Mind. Princeton University Press, Princeton, NJ.

Peterson, S.E., Fox, P.T., Posner, M.I., Mintun, M. and Raichle, M.E. (1989) Postiron emission tomographic studies of the processing of single words. J. Cog. Neurosci., 1(2): 153–170.

Posner, M. I., Raichle, M. E. (1994). Chapter 5: interpreting words. In *Images of Mind.* Scientific American Library, New York, (pp. 105–130).

Price, C.J., Moore, C.J. and Friston, K.J. (1997) Subtractions, conjunctions, and interactions in experimental design of activation studies. Human Brain Mapping, 5: 264–272.

Puce, A. (1995) Comparative assessment of sensorimotor function using functional magnetic resonance imaging and electrophysiological methods. J. Clin. Neurophys., 12: 450–459.

Puce, A., Constable, T., Luby, M.L., Eng, M., McCarthy, G., Nobre, A.C., Spencer, D.D., Gore, J.C. and Allison, T. (1995) Functional magnetic resonance imaging of sensory and motor cortex: comparison with electrophysiological localization. J. Neurosurg., 83: 262–270.

Pujol, J., Conesa, G., Deus, J., Lopez-Obarrio, L., Isamat, F. and Capdevila, A. (1998) Clinical application of functional magnetic resonance imaging in presurgical identification of the central sulcus. J. Neurosurg., 88: 863–869.

Raichle, M.E., Martin, W.R.W., Herscovitch, P., Mintun, M.A. and Markham, J. (1983) Brain blood flow measured with intravenous H215O II. Implementation and validation. J. Nuc. Med., 24: 790–798.

Ruge, M.I., Victor, J.D., Hosain, S., Correa, D.D., Relkin, N.R., Tabar, V., Brennan, C., Gutin, P.H. and Hirsch, J. (1999) Concordance between functional magnetic resonance imaging and intraoperative language mapping. J. Stereotactic Funct. Neurosurg., 72: 95–102.

Schiff, N. D., Rodriguez-Moreno. D., Kamal, A., Kim, K. H. S., Giacino, J. T., Plum, F. and Hirsch, J. (2005) fMRI reveals operational cognitive systems in two minimally conscious patients, Neurology, 64: 514–523.

Schulder, M., Maldijian, J.A., Liu, W.C., Holodny, A.I., Kalnin, A.T., Mun, I.K. and Carmel, P.W. (1998) Functional image-guided surgery of intracranial tumors located in or near the sensorimotor cortex. Neurosurcry, 89: 412–418.

Souweidane, M.M., Kim, K.H.S., McDowall, R., Ruge, M.I., Lis, E., Krol, G. and Hirsch, J. (1999) Brain mapping in sedated infants and young children with passive-functional magnetic resonance imaging. Ped. Neurosurg., 30: 86–91.

Stapleton, S.R., Kiriakopoulos, E., Mikulis, D., Drake, L.M., Hoffman, H.J., Humphreys, R., Hwang, P., Otsubo, H., Holowka, S., Logan, W. and Rutka, J.T. (1997) Combined utility of functional MRI cortical mapping and frameless stereotaxy in the resection of lesions in eloquent areas of brain in children. Ped. Neurosurg., 26: 68–82.

Talairach, J. and Tournoux, P. (1988) Co-Planar Stereotaxic Atlas of the Human Brain. Thieme, New York.

Tootell, R.B., Reppas, J.B., Kwong, K.K., Malach, R., Born, R.T., Brady, T.J., Rosen, B.R. and Belliveau, J.W. (1995) Functional analysis of human MT and related visual cortical areas using magnetic resonance imaging. J. Neurosci., 15: 3215–3230.

Victor, J.D., Apkarian, P., Hirsch, J., Packard, M., Conte, M.M., Relkin, N.R., Kim, H.S. and Shapley, R.M. (2000) Visual function and brain organization in non-decussating retinal-fugal fiber syndrome. Cerebral Cortex, 10: 2–22.

Wada, J. and Rasmussen, T. (1960) Intracarotid injection of sodium amytal for the lateralization of cerebral speech dominance. J. Neurosurg., 17: 266–282.

Wang, Y., Sereno, J.A., Jongman, A. and Hirsch, A. (2003) fMRI evidence for cortical modification during learning of Mandarin lexical tone. J. Cog. Neurosci., 15(7): 1019–1027.

Woods, R.P., Mazziotta, J.C. and Cherry, S.R. (1993) MRI-PET registration with automated algorithm. J. Comput. Assist. Tomogr., 17: 536–546.

Yousry, T.A., Schmid, U.D., Jassoy, A.G., Schmidt, D., Eisener, W.E., Reulen, H.J., Reiser, M.F. and Lissner, J. (1995) Topography of the cortical motor hand area: prospective study with functional MR imaging and direct motor mapping at surgery. Radiology, 195: 23–29.

Zeman, A. (2001). Consciousness, Brain, 124: 1263–1289.

Zeman, A. (2002) Consciousness, a User's Guide. Yale University Press, New Haven.

S. Laureys (Ed.)
Progress in Brain Research, Vol. 150
ISSN 0079-6123

CHAPTER 4

Global workspace theory of consciousness: toward a cognitive neuroscience of human experience

Bernard J. Baars*

The Neurosciences Institute, 10640 John Jay Hopkins Dv., San Diego, CA 92121, USA

Abstract: Global workspace (GW) theory emerged from the cognitive architecture tradition in cognitive science. Newell and co-workers were the first to show the utility of a GW or "blackboard" architecture in a distributed set of knowledge sources, which could cooperatively solve problems that no single constituent could solve alone. The empirical connection with conscious cognition was made by Baars (1988, 2002). GW theory generates explicit predictions for conscious aspects of perception, emotion, motivation, learning, working memory, voluntary control, and self systems in the brain. It has similarities to biological theories such as Neural Darwinism and dynamical theories of brain functioning. Functional brain imaging now shows that conscious cognition is distinctively associated with wide spread of cortical activity, notably toward frontoparietal and medial temporal regions. Unconscious comparison conditions tend to activate only local regions, such as visual projection areas. Frontoparietal hypometabolism is also implicated in unconscious states, including deep sleep, coma, vegetative states, epileptic loss of consciousness, and general anesthesia. These findings are consistent with the GW hypothesis, which is now favored by a number of scientists and philosophers.

Introduction

Shortly after 1900, behaviorists attempted to purge science of mentalistic concepts like consciousness, attention, memory, imagery, and voluntary control. "Consciousness," wrote John B. Watson, "is nothing but the soul of theology." But as the facts accumulated over the 20th century, all the traditional ideas of James (1890) and others were found to be necessary. They were reintroduced with more testable definitions. Memory came back in the 1960s; mental imagery in the 1970s; selective attention over the last half century; and consciousness last of all, in the last decade or so.

It is broadly true that what we are conscious of, we can report with accuracy. Conscious brain events are therefore assessed by way of reportability. We now know of numerous brain events that are reportable and comparable ones that are not. This fact invites experimental testing: why are we conscious of *these words at this moment*, while a few seconds later they have faded, but can still be called to mind? Why is activity in visual occipito-temporal lobe neurons reportable, while visually evoked activity in parietal regions is not? Why does the thalamocortical system support conscious experiences, while the comparably large cerebellum and basal ganglia do not? How is waking consciousness impaired after brain damage? These are all testable questions. The empirical key is to treat consciousness as a controlled variable.

A growing literature now compares the brain effects of conscious and unconscious stimulation. Precise experimental comparisons allow us to ask what conscious access does "as such." Many

*Corresponding author. Tel.: +1925-283-2601; Fax: +1925-283-2673; E-mail: baars@nsi.edu

DOI: 10.1016/S0079-6123(05)50004-9

techniques are used for this purpose. In visual backward masking, a target picture is immediately followed by a scrambled image that does not block the optical input, but renders it unconscious (Dehaene et al., 2001). Binocular rivalry has been used for the same reason: it shows that when two competing optical streams enter the two eyes, only one consistent interpretation can be consciously perceived at any given moment (Leopold and Logothetis, 1999). Most recently, several studies have demonstrated inattentional blindness, in which paying attention to one visual flow (e.g., a bouncing basketball) blocks conscious access to another one at the very center of visual gaze (e.g., a man walking by in a gorilla suit) (Simons and Chabris, 1999). These studies generally show that unconscious stimuli still evoke local feature activity in sensory cortex. But what is the use of making something conscious if even unconscious stimuli are identified by the brain? More than a score of studies have shown that although unconscious visual words activate known word-processing regions of visual cortex, the same stimuli, when conscious, trigger widespread additional activity in frontoparietal regions (e.g., Dehaene et al., 2001).

A rich literature has arisen comparing conscious and unconscious brain events in sleep and waking, general anesthesia, epileptic states of absence, very specific damage to visual cortex, spared implicit function after brain damage, attentional control (also see Posner, this volume), visual imagery, inner speech, memory recall, and more (Crick and Koch, 2003). In state comparisons, significant progress has been made in understanding epileptic loss of consciousness (Blumenfeldt and Taylor, 2003; Blumenfeld, this volume), general anesthesia (Fiset et al., 2001; John et al., 2001; Alkire and Fiset et al., this volume) and sleep[1] (Steriade, 2001; Maquet, this volume).

[1]At the level of cortical neurons, bursting rates do not change in deep sleep (Steriade, 2001). Rather, neurons pause together at <4 Hz between bursts. Synchronous pausing could disrupt the cumulative high-frequency interactions needed for waking functions such as perceptual continuity, immediate memory, sentence planning, motor control, and self-monitoring. It is conceivable that other unconscious states display similar neuronal mechanisms.

The global access hypothesis

The idea that consciousness has an integrative function has a long history. Global workspace (GW) theory is a cognitive architecture with an explicit role for consciousness. Such architectures have been studied in cognitive science, and have practical applications in organizing large, parallel collections of specialized processors, broadly comparable to the brain (Newell, 1994). In recent years, GW theory has been found increasingly useful by neuroscientists. The theory suggests a *fleeting memory capacity that enables access between brain functions that are otherwise separate.* This makes sense in a brain that is viewed as a massive parallel set of specialized processors. In such a system, coordination and control may take place by way of a central information exchange, allowing some processors — such as sensory systems in the brain — to distribute information to the system as a whole. This solution works in large-scale computer architectures, which show typical "limited capacity" behavior when information flows by way of a GW. A sizeable body of evidence suggests that consciousness is the primary agent of such a global access function in humans and other mammals (Baars, 1988, 1997, 2002). The "conscious access hypothesis" therefore implies that conscious cognition provides a gateway to numerous capacities in the brain (Fig. 1). A number of testable predictions follow from this general hypothesis (Table 1).

A theater metaphor and brain hypotheses

GW theory may be thought of as a theater of mental functioning. Consciousness in this metaphor resembles a bright spot on the stage of immediate memory, directed there by a spotlight of attention under executive guidance. Only the bright spot is conscious, while the rest of the theater is dark and unconscious. This approach leads to specific neural hypotheses. For sensory consciousness the bright spot on stage is likely to require the corresponding sensory projection areas of the cortex. Sensory consciousness in different modalities may be mutually inhibitory, within

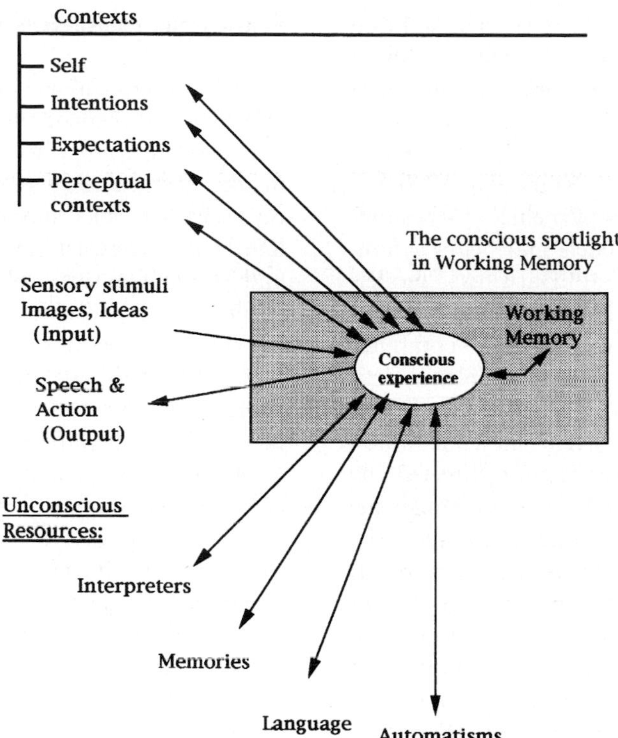

Fig. 1. A schematic diagram of GW theory, viewed metaphorically as a theater of mind. Conscious contents correspond to the bright spot on the stage of working memory. Once conscious, they activate many unconscious regions of the brain, including interpreters, memories, language capacities, and automatisms. In brain terms, those would be involved in certain cortical regions, hippocampus, and basal ganglia, which are believed not to directly support conscious experiences. However, conscious cognitions themselves are always shaped by unconscious contexts. Executive functions (self) may be considered as one set of such contexts (Adapted from Baars, 1997).

Table 1. Theoretical claims: brain capacities enabled by conscious events

1. Conscious perception enables access to widespread brain sources; unconscious sensory processing is much more limited
2. Conscious perception, inner speech, and visual imagery enable working memory functions; there is no evidence for unconscious access to working memory
3. Conscious events enable almost all kinds of learning: episodic and explicit learning, but also implicit and skill learning
4. Conscious perceptual feedback enables voluntary control over motor functions, and perhaps over any neuronal population and even single neurons
5. Conscious contents can evoke selective mechanisms (attention) and be evoked by it
6. Consciousness enables access to the "observing self" — executive interpreters, involving parietal and prefrontal cortex

approximately 100 ms time cycles (Baars and Franklin, 2003). Sensory cortex can be activated internally as well as externally, resulting in the "internal senses" of conscious inner speech and imagery. Once a conscious sensory content is established, it is distributed widely to a decentralized "audience" of expert networks sitting in the darkened theater, presumably using corticocortical and corticothalamic fibers. The transfer of information from conscious visual episodes to the (unconscious) hippocampal system is a clear example of such distribution of conscious information in the brain (Moscovitch, 1995). This is the primary functional role of consciousness: to allow a theater architecture to operate in the brain, in order to integrate, provide access, and coordinate the functioning of very large numbers of specialized networks that otherwise operate autonomously. All the elements of GW theory have reasonable brain interpretations, allowing us to generate a set of specific, testable brain hypotheses about consciousness and its many roles in the brain. Some of these ideas have now received considerable empirical support (Baars, 2002; Baars et al., 2003).

The theory has been implemented in computational and neural net models and bears a family resemblance to Neural Darwinist models (Edelman, 2003). Franklin and colleagues have implemented GW theory in large-scale computer agents, to test its functionality in complex practical tasks (Franklin, 2001). IDA (for "intelligent distributed agent"), the current implementation of the extended GW architecture directed by Franklin, is designed to handle a very complex artificial intelligence task normally handled by trained human beings (also see Aleksander on machine consciousness in this volume). The particular domain in this case is interaction between U.S. Navy personnel experts and sailors who move from job to job. IDA negotiates with sailors via e-mail, and is able to combine numerous regulations, sailors' preferences, time, location and travel considerations into human-level performance. While it has components roughly corresponding to human perception, memory, and action control, the heart of the system is a GW architecture that allows the content or meanings of the messages to be widely distributed, so that specialized programs called "codelets" can respond with solutions to centrally posed problems. Franklin writes that "The fleshed out global workspace theory is yielding hopefully testable hypotheses about human cognition. The architectures and mechanisms that underlie consciousness and intelligence in humans can be expected to yield information agents that learn continuously, adapt readily to dynamic environments, and behave flexibly and intelligently when faced with novel and unexpected situations." (http://csrg.cs.memphis.edu/). Similar architectures have been applied to difficult problems like speech recognition. While such autonomous agent simulations do not prove that GW architectures exist in the brain, they give an existence proof of their functionality. It is worth noting that few integrative theories of mind or brain show functional utility in applied settings.

Sensory consciousness as a test case

Visual consciousness has been studied in depth, and there is accepted evidence that visual features that become conscious are identified by the brain in the ventral stream of visual cortex. There, feature-sensitive cells support visual experiences of light, color, contrast, motion, retinal size, location, and object identity; small lesions can selectively abolish those conscious properties without affecting other aspects of conscious vision (Zeki, 2001; Naccache, in this volume).

However, to recollect the experience of a human face, we need the hippocampal system. To respond to it emotionally, neurons in amygdala may be activated. But hippocampus and amygdala do not seem to support conscious contents directly (Moscovitch, 2001). Thus, the ventral visual stream, which is needed for specific conscious contents, seems to influence regions that are not.

Dehaene and colleagues have shown that backward-masked visual words evoked brain activity confined to the well-known visual word recognition areas of cortex (Dehaene et al., 2001). Identical conscious words triggered higher levels of activity in these areas, but more importantly, they also evoked far more widely distributed activity in parietal and prefrontal cortex. That result has now been replicated more than a dozen times, using different brain imaging techniques and different methods for comparing conscious and unconscious input. Such methods have included binocular rivalry (Sheinberg and Logothetis, 1997), inattentional blindness (Rees et al., 1999), neglect and its extinction (Rees et al., 2002), and different sense modalities, such as audition (Portas et al., 2000), pain perception (Rosen et al., 1996), and sensorimotor tasks (Haier et al., 1992; Raichle et al., 1994). In all cases, conscious sensory input evoked wider and more intense brain activity than identical unconscious input.

Complementary findings come from studies of unconscious states. In deep sleep, auditory stimulation activates only primary auditory cortex (Portas et al., 2000). In vegetative states following brain injury, stimuli that are ordinarily loud or painful activate only the primary sensory cortices (Laureys et al., 2000, 2002). Waking consciousness is apparently needed for widespread of input-driven activation to occur. These findings support the general notion that conscious stimuli mobilize large areas of cortex, presumably to distribute information about the input.

Inner speech, imagery, and working memory

Both auditory and visual consciousness can be activated endogenously. Inner speech is a particularly important source of conscious auditory-phonemic events, and visual imagery is useful for spatial memory and problem-solving. The areas of the left hemisphere involved in outer speech are now known to be involved in inner speech as well (Paulesu et al., 1993). Likewise, mental imagery is known to involve visual cortex (Kosslyn et al., 2001). Internally generated somatosensory imagery may reflect emotional and motivational processes, including feelings of psychological pain, pleasure, hope, fear, sadness, etc. (Damasio, 2003). Such internal sensations may communicate to other parts of the brain via global distribution or activation.

Prefrontal executive systems may sometimes control motor activities by evoking motivational imagery, broadcast from the visual cortex, to activate relevant parts of motor cortex. Parts of the brain that play a role in emotion may also be triggered by global distribution of conscious contents from sensory cortices and insular cortex. For example, the amygdala appears necessary to recognize visual facial expressions of fear and anger. Thus, many cortical regions work together to transform goals and emotions into actions (Baars, 1988).

The attentional spotlight

The sensory "bright spot" of consciousness involves a selective attention system, the ability of the theater spotlight to shine on different actors on the stage. Like other behaviors like breathing and smiling, attention operates under dual control, voluntary, and involuntary. Voluntary attentional selection requires frontal executive cortex, while automatic selection is influenced by many areas, including the brain stem, pain systems, insular cortex, and emotional centers like the amygdala and peri-aqueductal grey (Panksepp, 1998). Presumably, these automatic attentional systems that allow significant stimuli to "break through" into consciousness, as when a subject's name is sounded in an otherwise unconscious auditory source.

Context and the first-person perspective

When we step from a tossing sailboat onto solid ground, the horizon can be seen to wobble. On an airplane flight at night passengers can see the cabin tilting on approach to landing, although they are receiving no optical cues about the direction of the plane. In those cases unconscious vestibular signals shape conscious vision. There are numerous examples in which unconscious brain activities can shape conscious ones, and vice versa. These unconscious influences on conscious events are called "contexts" in GW theory (Fig. 1). Any conscious sensory event requires the interaction of sensory analyzers and contextual systems. In vision, sensory contents seem to be produced by the ventral visual pathway, while contextual systems in the dorsal pathway define a spatial domain within which the sensory event is defined. Parietal cortex is known to include allocentric and egocentric spatial maps, which are not themselves objects of consciousness, but which are required to shape every conscious visual event. There is a difference between the disorders of content systems like the visual ventral stream, compared to damaged context systems. In the case of ventral stream lesions, the subject can generally notice a missing part of normal experience; but for damage to context, the brain basis of expectations is itself damaged, so that one no longer knows what to expect, and hence what is missing. This may be why parietal neglect is so often accompanied by a striking loss of knowledge about one's body space (Bisiach and Geminiani, 1991). Patients suffering from right parietal neglect can have disturbing alien experiences of their own bodies, especially of the left arm and leg. Such patients sometimes believe that their left leg belongs to someone else, often a relative, and can desperately try to throw it out of bed. Thus, parietal regions seem to shape contextually both the experience of the visual world and of one's own body. Notice that neglect patients still experience their alien limbs as conscious visual objects (a ventral stream function); they are just disowned. Such specific loss of contextual body information is not accompanied by a loss of general intelligence or knowledge.

Vogeley and Fink (2003) suggest that parietal cortex is involved in the first-person perspective, the viewpoint of the observing self. When subjects are asked to adopt the visual perspective of another person, parietal cortex became differentially active.

Self-systems

Activation by of visual object regions by the sight of a coffee cup may not be enough to generate subjective consciousness of the cup. The activated visual information may need to be conveyed to executive or self-systems, which serve to maintain constancy of an inner framework across perceptual situations. When we walk from room to room in a building, we must maintain a complex and multileveled organization that can be viewed in GW theory as a higher-level context. Major goals, for example, do not change when we walk from room to room, but conscious perceptual experiences do. Gazzaniga (1996) has found a number of conditions under which split-brain patients encounter conflict between right and left hemisphere executive and perceptual functions. He has proposed the existence of a "narrative self" in the left frontal cortex, based on split-brain patients who are clearly using speech output in the left hemisphere to talk to themselves, sometimes trying to force the right hemisphere to obey its commands. When that proves impossible, the left hemisphere will often rationalize the sequence of events so as to repair its understanding of the interhemispheric conflict. Analogous repairs of reality are observed in other forms of brain damage, such as neglect. They also commonly occur whenever humans are confronted with major, unexpected life changes. The left-hemisphere narrative interpreter may be considered as a higher-level context system that maintains expectations and intentions across many specific situations. Although the inner narrative itself is conscious, it is shaped by unconscious contextual influences.

If we consider Gazzaniga's narrative interpreter of the dominant hemisphere to be one kind of self-system in the brain, it must receive its own flow of sensory input. Visual input from one-half of the field may be integrated in one visual hemicortex, as described above, under retinotopic control from area V1. But once it comes together in late visual cortex (presumably in inferotemporal object regions), it needs to be conveyed to frontal areas on the dominant hemisphere, in order to inform the narrative interpreter of the current state of perceptual affairs. The left prefrontal self system then applies a host of criteria to the input, such as "did I intend this result? Is it consistent with my current and long-term goals? If not, can I reinterpret it to make sense in my running account of reality?" It is possible that the right hemisphere has a parallel system that does not speak but that may be better able to deal with anomalies via irony, jokes, and other emotionally useful strategies. The evidence appears to be good that the isolated right prefrontal cortex can understand such figurative uses of language, while the left does not. Full consciousness may not exist without the participation of such prefrontal self systems.

Relevance to waking, sleeping, coma, and general anesthesia

Metabolic activity in the conscious resting state is not uniformly distributed. Raichle et al. (2001) reported that mesiofrontal and medial parietal areas, encompassing precuneus and adjacent posterior cingulate cortex, can be posited as a tonically active region of the brain that may continuously gather information about the world around, and possibly within, us. It would appear to be a default activity of the brain. Mazoyer et al. (2001) also found high prefrontal metabolism during rest. We will see that these regions show markedly lower metabolism in unconscious states.

Laureys (1999a, b, 2000) and Baars et al. (2003) list the following features of four unconscious states, that are causally very different from each other: deep sleep, coma/vegetative states, epileptic loss of consciousness, and general anesthesia under various agents. Surprisingly, despite their very different mechanisms they share major common features. These include: (i) widely synchronized slow waveforms that take the place of the fast and flexible interactions needed for conscious functions; (ii) frontoparietal hypometabolism; (iii) widely blocked functional connectivity, both corticocortical and thalamocortical; and (iv) behavioral unconsciousness, including unresponsiveness to normally conscious stimuli. Fig. 2 shows marked hypofunction in the four unconscious states compared with conscious controls, precisely where we might expect it: in frontoparietal regions.

In a related study, John and co-workers showed marked quantitative electroencephalogram (EEG)[2] changes between conscious, anesthetic, and post-anesthetic (conscious) states (John et al., 2000). At loss of consciousness, gamma power decreased while lower frequency bands increased in power, especially in frontal leads. Loss of consciousness was accompanied by a significant drop in coherence between homologous areas of the two hemispheres, and between posterior and anterior regions of each hemisphere. However, there was hypersynchronous activity within anterior regions. The same basic changes occurred across all six anesthetics,[3] and reversed when patients regained consciousness (see John, in this volume).

From the viewpoint of globalist theories, the most readily interpretable finding is the coherence drop in the gamma range after anesthetic loss of consciousness. It suggests a loss of coordination between frontal and posterior cortex, and between homologous regions of the two hemispheres. The authors also suggest that the anteriorization of low frequencies "must exert a profound inhibitory influence on cooperative processes within (frontal) neuronal populations. This functional system then becomes de-differentiated and disorganized" (p. 180). Finally, the decoupling of the posterior cortex with anterior regions suggests "a blockade of perception" (p. 180). These phenomena appear to be consistent with the GW notion that widespread activation of nonsensory regions is required for sensory consciousness.

The role of frontoparietal regions in conscious contents and states

Could it be that brain regions that underlie the contextual functions of Fig. 1 involve frontal and parietal regions? In everyday language, the "observing self" may be disabled when those regions are dysfunctional and long-range functional connectivity is impaired. Frontoparietal association areas have many functions, but several lines of evidence suggest that they could have a special relationship with consciousness, even though they do not support the sensory contents of conscious experience directly. (i) Conscious stimulation in the waking state leads to frontoparietal activation, but unconscious input does not; (ii) in unconscious states, sensory stimulation activates only sensory cortex, but not frontoparietal regions; (iii) the conscious resting state shows high frontoparietal metabolism compared with outward-directed cognitive tasks; and (iv) four causally very different unconscious states show marked functional decrements in the same areas. Although alternative hypotheses must be considered, it seems reasonable to suggest that "self" systems supported by these regions could be disabled in unconscious states. From the viewpoint of the narrative observer, this would be experienced as subjective loss of access to the conscious world. Unconscious states might not necessarily block the objects of consciousness; rather, the observing subject might not be at home.

Conclusion

GW theory suggests that consciousness enables multiple networks to cooperate and compete in solving problems, such as retrieval of specific items from immediate memory. Conscious contents may

[2]Although the spike-wave EEG of epileptic seizures appears different from the delta waves of deep sleep and general anesthesia, it is also synchronized, slow, and high in amplitude. The source and distribution of spike-wave activity varies in different seizure types. However, the more widespread the spike-wave pattern, the more consciousness is likely to be impaired (Blumenfeldt and Taylor, 2003). This is again marked in frontoparietal regions.

[3]There is a debate whether ketamine at relatively low doses should be considered an anesthetic. All anesthetic agents in this study were used at dosages sufficient to provide surgical-level loss of consciousness.

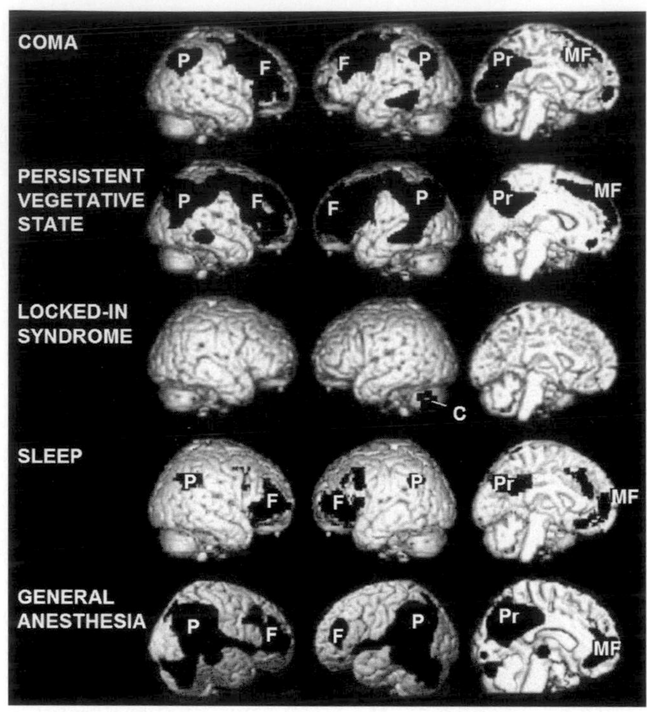

Fig. 2. Neural activity in four types of unconscious states, subtracted from conscious controls. Positron emission tomography scans showing regional decreases in metabolism or blood flow when unconscious states are compared with resting consciousness. Coma, persistent vegetative state, sleep, and general anesthesia all show regional decreases in frontoparietal association cortices. Column 1: the right lateral aspect of the brain; column 2: the left lateral aspect; column 3: a medial view of the left hemisphere. Abbreviations: F, prefrontal; MF, mesiofrontal; P, posterior parietal cortex; Pr, posterior cingulate/precuneus (from Baars et al., 2003).

correspond to brain processes that work much like brief memories whose contents activate widespread regions in the brain. Physiologically such interactions seem to involve multiple high-frequency oscillatory rhythms. The overall function of consciousness is to provide widespread access, which in turn may serve coordination and control. Consciousness is the gateway to the brain.

Acknowledgments

I gratefully acknowledge support from the Neurosciences Institute and the Neurosciences Research Foundation (10640 John Jay Hopkins Drive, San Diego, CA 94549, USA). I thank Steven Laureys, Gerald M. Edelman, Stan Franklin, Bjorn Merker, Anil Seth, Douglas Nitz, E. Roy John, and Walter Freeman for helpful discussions.

References

Baars, B.J. (1988) A cognitive theory of consciousness. Cambridge University Press, New York.

Baars, B.J. (1997) In the theater of consciousness: The workspace of the mind. Oxford, New York.

Baars, B.J. (2002) The conscious access hypothesis: origins and recent evidence. Trends Cogn. Sci., 6: 47–52.

Baars, B. J., Ramsoy, T., Laureys, S. (2003). Trends in Neurosci., 26, 671–675.

Baars, B.J. and Franklin, S. (2003) How conscious experience and working memory interact. Trends Cogn. Sci., 7: 166–172.

Bisiach, E. and Geminiani, G. (1991) Anosognosia related to hemiplegia and hemianopia. In: Prigatano G.P. and Schacter D.L. (Eds.), Awareness of Deficit After Brain Injury: Clinical and Theoretical Issues. Oxford University Press.

Blumenfeldt, H. and Taylor, J. (2003) Why do seizures cause loss of consciousness? Neuroscientist, 9: 1–10.

Crick, F. and Koch, C. (2003) A framework for consciousness. Nat. Neurosci., 6: 119–126.

Damasio, A. (2003) The feeling of knowing. Basic Books, New York.

Dehaene, S., et al. (2001) Cerebral mechanisms of word masking and unconscious repetition priming. Natl. Neurosci., 4: 752–758.

Edelman, G.M. (2003) Naturalizing consciousness: a theoretical framework. Proc. Natl. Acad. Sci. USA, 100: 5520–5534.

Fiset, P., Paus, T., Daloze, T., Plourde, G., Meuret, P., Bonhomme, V., et al. (1999) Brain mechanisms of propofol-induced loss of consciousness in humans: a positron emission tomographic study. J. Neurosci., 19(13): 5506–5513.

Frackowiak, et al. (2004) Human brain function (2nd ed.). Elsevier, London.

Franklin, S. (2000) Deliberation and voluntary action in 'conscious' software agents. Neural Network World, 10: 505–521.

Franklin, S. (2001) Automating human information agents. In: Chen Z. and Jain L.C. (Eds.), Practical Applications of Intelligent Agents. Springer-Verlag, Berlin.

Freeman, W.J. (2003) The wave packet: an action potential for the 21st century. J. Integr. Neurosci., 2(1): 3–30.

Haier, R.J., et al. (1992) Regional glucose metabolic changes after learning a complex visuospatial motor task: a positron-emission tomographic study. Brain Res., 570: 134–148.

James, W. (1890) The principles of psychology. Holt, New York.

John, E.R., Prichep, L.S., Kox, W., Valdes-Sosa, P., Bosch-Bayard, J., Aubert, E., et al. (2001) Invariant reversible QEEG effects of anesthetics. Conscious Cogn., 10(2): 165–183.

Kosslyn, S.M., Ganis, G. and Thompson, W.L. (2001) Neural foundations of imagery. Natl. Rev. Neurosci., 2(9): 635–642.

Laureys, S., et al. (2000) Restoration of thalamocortical connectivity after recovery from persistent vegetative state. Lancet, 355: 1916.

Laureys, S., et al. (1999a) Cerebral metabolism during vegetative state and after recovery to consciousness. J. Neurol. Neurosurg. Ps., 67: 121.

Laureys, S., et al. (1999b) Impaired functional connectivity in vegetative state: preliminary investigation using PET. Neuroimage, 9: 377–382.

Laureys, S., et al. (2000) Auditory processing in the vegetative state. Brain, 123: 1589–1601.

Laureys, S., et al. (2002) Cortical processing of noxious somato–sensory stimuli in the persistent vegetative state. Neuroimage, 17: 732–741.

Mazoyer, B., et al. (2001) Cortical networks for working memory and executive functions sustain the conscious resting state in man. Brain Res. Bull., 54: 287–298.

Moscovitch, M. (1995) Recovered consciousness: a hypothesis concerning modularity and episodic memory. J. Clin. Exp. Neuropsychol., 17(2): 276–290.

Newell, A. (1994) Unified theories of cognition: The William James lectures. Harvard University Press, Cambridge, MA.

Panksepp, J. (1998) Affective Neuroscience. Oxford University Press, New York.

Paulesu, E., Frith, D. and Frackowiak, R.S.J. (1993) The neural correlates of the verbal component of working memory. Nature, 362: 342–345.

Portas, C.M., et al. (2000) Auditory processing across the sleep–wake cycle: simultaneous EEG and fMRI monitoring in humans. Neuron, 28: 991–999.

Raichle, M.E., et al. (1994) Practice-related changes in human brain functional anatomy during non-motor learning. Cereb. Cortex, 4: 8–26.

Raichle, M.E., et al. (2001) A default mode of brain function. Proc. Natl. Acad. Sci. USA, 98: 676–682.

Rees, G., Russell, C., Frith, C.D. and Driver, J. (1999) Inattentional blindness versus inattentional amnesia for fixated but ignored words. Science, 286(5449): 2504–2507.

Rees, G., Wojciulik, E., Clarke, K., Husain, M., Frith, C. and Driver, J. (2002) Neural correlates of conscious and unconscious vision in parietal extinction. Neurocase, 8(5): 387–393.

Rosen, S.D., Paulesu, E., Nihoyannopoulos, P., Tousoulis, D., Frackowiak, R.S., Frith, C.D., et al. (1996) Silent ischemia as a central problem: regional brain activation compared in silent and painful myocardial ischemia. Ann. Int. Med., 124(11): 939–949.

Sheinberg, D.L. and Logothetis, N.K. (1997) The role of temporal cortical areas in perceptual organization. Proc. Natl. Acad. Sci. USA, 94(7): 3408–3413.

Simons, D.J. and Chabris, C.F. (1999) Gorillas in our midst: sustained inattentional blindness for dynamic events. Perception, 28(9): 1059–1074.

Steriade, M. (2001) Active neocortical processes during quiescent sleep. Arch. Ital. Biol., 139: 37–51.

Tononi, G. and Edelman, G. (1998) Consciousness and complexity. Science, 282: 1846–1851.

Vogeley, K. and Fink, G.R. (2003) Neural correlates of the first-person perspective. Trends Cogn. Sci., 7: 38–42.

Zeki, S. (2001) Localization and globalization in conscious vision. Annu. Rev. Neurosci., 24: 57–86.

S. Laureys (Ed.)
Progress in Brain Research, Vol. 150
ISSN 0079-6123

CHAPTER 5

Skill, corporality and alerting capacity in an account of sensory consciousness

J. Kevin O'Regan[1], Erik Myin[2,3,*] and Alva Noë[4]

[1]*Laboratoire de Psychologie Expérimentale, Institut de Psychologie, Centre Universitaire de Boulogne, 71, avenue Edouard Vaillant, 92774 Boulogne-Billancourt Cedex, France*
[2]*Department of Philosophy, Centre for Philosophical Psychology, University of Antwerp, Rodestraat 14, room R110, 2000 Antwerp, Belgium*
[3]*Department of Philosophy, Centre for Logic and Philosophy of Science, Vrije Universiteit Brussel, Pleinlaan 2, 1050 Brussells, Belgium*
[4]*Department of Philosophy, University of California, Berkeley, CA 94720-2390, USA*

Abstract: We suggest that within a skill-based, sensorimotor approach to sensory consciousness, two measurable properties of perceivers' interaction with the environment, "corporality" and "alerting capacity", explain why sensory stimulation is experienced as having a "sensory feel", unlike thoughts or memories. We propose that the notions of "corporality" and "alerting capacity" make possible the construction of a "phenomenality plot", which charts in a principled way the degree to which conscious phenomena are experienced as having a sensory quality.

Introduction

Although knowledge is rapidly accumulating concerning the neurobiological mechanisms involved in consciousness (cf. Rees et al., 2002 for an overview), there still remains the problem of how to capture the "qualitative" aspects with a scientific approach. There would seem to be an unbridgeable "explanatory gap" (Levine, 1983) between what it is like to have a sensory experience, and the neural correlates or physical mechanisms involved.

The purpose of this paper is to show how a step can be made toward bridging this gap. We purposefully leave aside many interesting problems of consciousness, such as self-awareness, the distinction between awake and unconscious states, being aware of facts, etc., and concentrate on the question of the nature of sensation. The fact that contrary to other mental phenomena, sensations have a distinctive qualitative character or sensory "feel" lies at the heart of the explanatory gap problem. Indeed philosopher Ned Block has noted that being conscious of something involves two aspects. First, it involves having "conscious access" to that thing, in the sense that one can make use of that thing in one's decisions, judgments, rational behavior and linguistic utterances (Block, 1995, 2005). This "access consciousness" is amenable to scientific explanation, since it can be formulated in functional terms. On the other hand, being conscious of something also involves a second "phenomenal" aspect, which corresponds to the enigmatic "what it's like" to experience that thing. It is not clear how this "phenomenal consciousness" could be approached scientifically.

Our approach to this question will be to suggest that there is a way of thinking about sensations that is different from the usually accepted way. A first aspect of this new way of thinking involves taking a counterintuitive stance at first sight, namely that sensation consists in the exercise of

*Corresponding author. E-mail: emyin@vub.ac.be; Erik.Myin@ua.ac.be

DOI: 10.1016/S0079-6123(05)50005-0

an exploratory skill (cf. O'Regan and Noë, 2001a; Myin and O'Regan, 2002; see Torrance, 2002, for further references to skill theories). Taking the skill approach allows a first problem about the experiential quality of sensation to be addressed, namely why the experienced qualities of different sensations differ the way they do.

Second, when skill theories are supplemented by two concepts, which we refer to as "corporality" and "alerting capacity", then a second, more profound problem about the experienced quality of sensations can be addressed, namely why they have an experienced sensory quality at all.

We have organized our paper in a main body in which the concepts crucial to our approach are introduced and described, and three "application" sections, in which they are put to use in the context of more specific issues, namely intra- and intermodal differences, dreaming and imagery, and change blindness. In a final section, we consider the issue of whether our approach really constitutes an explanation of phenomenal sensory consciousness.

Sensation as a skill: explaining intra- and intermodal sensory differences

The basic tenet of the skill theory from which we take our start is that having a sensation is a matter of the perceiver knowing that he is currently exercising his implicit knowledge of the way his bodily actions influence incoming sensory information (O'Regan and Noë, 2001a).

An illustration is provided by the sensation of softness that one might experience in holding a sponge (Myin, 2003). Having the sensation of softness consists in being aware that one can exercise certain practical skills with respect to the sponge: one can, for example, press it, and it will yield under the pressure. The experience of softness of the sponge is characterized by a variety of such possible patterns of interaction with the sponge, and the laws that describe these sensorimotor interactions we call, following MacKay (1962), laws of sensorimotor contingency (O'Regan and Noë, 2001a). When a perceiver knows, in an implicit, practical way, that at a given moment he is exercising the sensorimotor contingencies associated with softness, then he is in the process of experiencing the sensation of softness.

Note that in this account, the softness of the sponge is not communicated by any particular softness detectors in the fingertips, nor is it characterized by some intrinsic quality provided by the neural processes involved, but rather it derives from implicit, practical knowledge about how sensory input from the sponge currently might change as a function of manipulation with the fingers.

This approach to sensation has a tremendous advantage. It avoids a fundamental problem that is encountered by any approach that assumes that sensation is generated by a neural mechanism: namely the problem why one particular neural process (whatever its neural specification) should give rise to one specific sensation (and not to another one). In addition, the skill-based sensorimotor description of experiencing softness in terms of an exploratory finding out that the object yields when one presses "fits" the experience of softness in a way a description in terms of a correlated neural process cannot. Thus, for example, while under a "neural correlate" explanation it is always possible to imagine the presumed neural process for softness to be paired with the sensation of hardness (i.e., nothing of the specifics of the neural description seems to forbid this), it would seem impossible to imagine one is going through the exploratory pattern of softness, yet experiencing hardness.

Application 1 on intra- and intermodal differences in sensory quality (see below) describes how the sensorimotor way of thinking can be applied to perceptual sensations in general, even to cases like color perception where no active exploration appears necessary. Just as the difference between hard and soft can be accounted for in terms of the different exploratory strategies required to sense hard and soft objects, the differences between red and blue, for example, can be accounted for in terms of the different exploratory strategies involved in exploring red and blue surfaces.

Another, related question can also be dealt with by this approach, namely the question of the differences between the sensory qualities of the different sensory modalities. As suggested in Application 1, the difference, for example, between hearing and seeing is accounted for in terms of the different laws of sensorimotor contingency that characterize hearing and seeing. Again, under this approach, no appeal is necessary to special, as yet unexplained intrinsic properties of neural mechanisms.

The sensorimotor theory and its explanation of intra- and intermodal sensory differences, as just reviewed, has previously been treated in a number of papers (O'Regan and Noë, 2001a, b, c; Myin and O'Regan, 2002; Noë, 2002a, b; Noë, 2004). We now come to the main purpose of this chapter, which is to address a more profound question, namely the question of why sensations have a sensory experiential quality at all.

Corporality and alerting capacity: explaining sensory presence

What is special about sensory experience that makes it different from other mental phenomena, like conscious thought or memory? In particular, consider the difference between actually feeling a terrible pain and merely imagining or thinking that you are feeling one. Or consider actually feeling softness or seeing red, compared to thinking that you are feeling softness or seeing red (see Application 2 for a discussion of dreams, imagery and hallucinations).

Theorists have tried to describe and capture such differences in various ways. Hume, for example, opposed (perceptual) sensations and "ideas" (recollections of sensations and thoughts), in terms of "vivacity" and "force" (Hume, 1777/1975). Husserl proposed the notion of an object being experienced as "being present in the flesh" (having "Leibhaftigkeit") as an essential ingredient for truly perceptual experience (Husserl, 1907/1973; Merleau-Ponty, 1945; cf. Pacherie, 1999) for similar use of the notion "presence". In contemporary descriptions of perceptual consciousness, such a distinction is often made in terms of "qualia", those special qualitative or phenomenal properties that characterize sensory states, but not cognitive states (Levine, 1983; Dennett, 1988).

While these notions seem descriptively adequate, we propose they should and can be complemented with an explanatory story that accounts for why sensory experience differs in these respects from other conscious mental phenomena. Our claim is that, within a skill-based, sensorimotor theory, the notions of corporality and alerting capacity provide precisely this missing explanatory addition. Corporality and alerting capacity are complementary aspects of an observer's interaction with the environment: corporality concerns the way actions affect incoming sensory information, and, conversely, alerting capacity concerns the way incoming sensory information potentially affects the attentional control of behavior.

Again we wish to claim that corporality and alerting capacity are not merely descriptive, but actually possible first steps toward explanations. We will return to this distinction later.

Corporality or "bodiliness"

We define corporality as the extent to which activation in a neural channel systematically depends on movements of the body (in previous publications we used the term "bodiliness" (O'Regan and Noë, 2001b; Myin and O'Regan, 2002; O'Regan et al., 2004). Sensory input from sensory receptors like the retina, the cochlea, and mechanoreceptors in the skin possesses corporality, because any body motion will generally create changes in the way sensory organs are positioned in space, thereby causing changes in the incoming sensory signals. Proprioceptive input from muscles also possesses corporality, because there is proprioceptive input when muscle movements produce body movements.

Note that we intend the term corporality to apply to any neural channels in the brain whatsoever, but because of the way it is defined, with the exception of muscle commands themselves and proprioception, only neural activation that corresponds to sensory input from the outside environment will generally have corporality. For example, neural channels in the autonomic nervous system that measure parameters such as the heartbeat or digestive functions, because they are not very systematically affected by movements, will have little corporality even though they may carry sensory information. Note also that memory processes or thinking have no corporality, because body movements do not affect them in any systematic way.

We shall see below that corporality is an important factor that explains the extent to which a sensory experience will appear to an observer as being truly sensory, rather than non-sensory, like a thought, or a memory. In Philipona et al. (2003) it is shown mathematically how this notion can be used by an organism to determine the extent of its own body and the fact that it is embedded in a three-dimensional physical world in which the group-theoretic laws of Euclidean translations and rotations apply.

Alerting capacity or "grabbiness"

We define the alerting capacity of sensory input as the extent to which that input can cause automatic orienting behaviors that peremptorily capture the organism's cognitive processing resources. Alerting capacity could also be called: capacity to provoke exogenous attentional capture, but this would be more cumbersome. In previous papers, we have also used the term "grabbiness" (O'Regan and Noë, 2001b; Myin and O'Regan, 2002; O'Regan et al., 2004).

Pain channels, for example, have alerting capacity, because not only can they cause immediate, automatic and uncontrollable withdrawal reactions, but they also can cause cognitive processing to be modified and attentional resources to be attributed to the source of the pain. Retinal, cochlear and tactile sensory channels have alerting capacity, since not only can abrupt changes in incoming signals cause orienting reflexes, but the organism's normal cognitive functioning will be modified to be centered upon the sudden events. For example, a sudden noise not only can cause the organism to turn toward the source of the noise, but the noise will also additionally, peremptorily, modify the course of the organism's cognitive activity so that if it is human, it now takes account of the noise in current judgments, planning, and linguistic utterances. Autonomic pathways do not have alerting capacity, because sudden changes in their activation do not affect cognitive processing. For example, while sudden changes in vestibular signals cause the organism to adjust its posture and blood pressure automatically, these adjustments themselves do not generally interfere in the organism's cognitive processing (interference occurs only indirectly, when, for example, the organism falls to the ground and must interact in a new way with its environment). Like corporality, we take alerting capacity to be an objectively measurable parameter of the activation in a sensory pathway.

Using corporality and alerting capacity to explain "sensory presence"

We now consider how the notions of corporality and alerting capacity can contribute to understanding what provides sensory experiences with their particular sensory quality, and more precisely, what makes for the difference between truly sensory and other experiences.

To see our notions at work, consider the difference between seeing an object in full view, seeing an object partially hidden by an occluding object, being aware of an object behind one's back, and thinking, remembering or knowing about an object. It is clear that these different cases provide different degrees of sensory "presence" (Merleau-Ponty, 1945; O'Regan and Noë, 2001a; Noë, 2002b). Our claim is that these different degrees of sensory presence precisely reflect different degrees in corporality and alerting capacity.

Thus, when an object is in full view, it comes with the fullest intensity of sensory presence. But it is precisely in this case that observer motion will immediately affect the incoming sensory stimulation. Also, any change that occurs in the object, such as a movement, a shape, color, or lightness change, will immediately summon the observer's attention. This is because low-level transient-detection mechanisms exist in the visual system that peremptorily cause an attention shift to a sudden stimulus change. In terms of the concepts we defined above, this means that an object in full view has both high corporality and high alerting capacity.

Contrast this with just knowing that an object is somewhere, but out of view. While knowledge about an object in another room might certainly be conscious, it lacks real sensory presence. Clearly, in this case, there is no corporality, since the stimulus changes caused by bodily movements do not concern that object. Similarly, there is no alerting capacity, as the changes that the object might undergo do not immediately summon the perceiver's attention.

An object that is only partially in view because of an occluding object or an object known to be behind one's back provides borderline cases. For example, the occluded part might be said to still have some presence (Merleau-Ponty, 1945; Gregory, 1990; O'Regan and Noë, 2001a; Noë and O'Regan, 2002; Noë, 2002b) because it has a degree of corporality, as we can easily bring it into view by a slight movement. The "boundary extension" phenomenon of Intraub and Richardson (1989), according to which observers overestimate what can be seen of a partially occluded object, is coherent with this view. Amodal completion may be an example where one has an intermediate kind of "almost-visual" feeling of presence of a shape behind an occluder. Application 3 gives examples of "change blindness", showing that when alerting capacity is interfered with, the experience of perception ceases.

These examples show that the differing degrees of what one might call "sensory presence" (perhaps Hume's "vividness" or Husserl's "Leibhaftigkeit") can be accounted for plausibly in terms of the physically measurable notions of corporality and alerting capacity.

The "sensory phenomenality plot"

The exercise of contrasting sensations with other mental phenomena can be systematized in a "sensory phenomenality plot" (Fig. 1).

By plotting the degree of corporality and alerting capacity for different mental phenomena, such a figure reveals that those states that possess both corporality and alerting capacity correspond precisely to cases that provide true sensory experiences. (But note, importantly: we consider that our plot only charts the degree to which mental phenomena have sensory or perceptual quality, and not consciousness *per se*. In particular, when we claim that thought has no sensory quality, we are not saying that thought is not conscious–more on this in section "Consciousness".

Thus, vision, touch, hearing, and smell are the prototypical sensory states and indeed have high

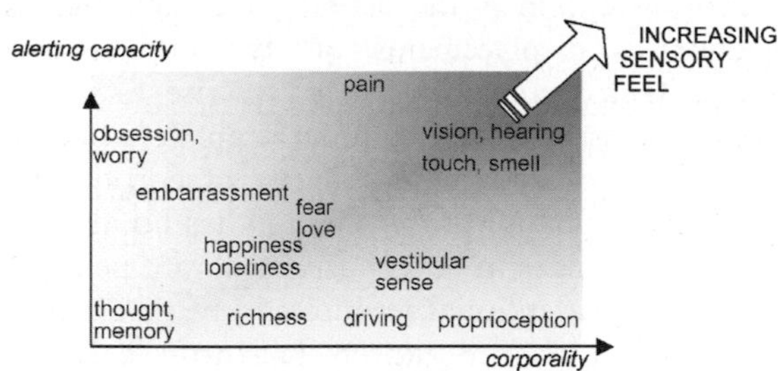

Fig. 1. A sensory phenomenality plot.

corporality and high alerting capacity, as mentioned above in the definition of these terms. High corporality derives from the fact that changes in head or limb positions have an immediate effect on visual, auditory or tactile sensory input (smell is less clear, but sniffing, blocking the nose, and moving the head do affect olfactory stimulation; Steriade, 2001). High alerting capacity is provided by the fact that sudden changes in visual, tactile, auditory, or olfactory stimulation provoke immediate orienting behaviors that peremptorily modify cognitive processing.

What characterizes pain is its particularly large amount of alerting capacity. Here it is virtually impossible to prevent oneself from attentively focusing on the noxious stimulation. Pain also has corporality, but to a lesser extent. Moving one's body can generally modify the pain (one can remove one's finger from the fire; rub the aching limb and change the incoming sensations), but there are cases like headaches or toothaches, which are more problematic. Headaches and toothaches are characterized by the fact that associated sensory input changes only moderately as a function of things that one can do such as press on the head or chew with one's teeth. This lack of an ability to easily modulate the sensory stimulation by body motions, i.e., a reduced corporality, could possibly correspond to a particular aspect of pain, such as headaches, which distinguishes them from vision, touch, hearing, and smell, namely that they have an interior quality, often not clearly localized.

We have plotted thinking and recalling from memory at the other extreme, because they have neither corporality or alerting capacity, as we have pointed out above.

Proprioception is the neural input that signals mechanical displacements of the muscles and joints. Motor commands that give rise to movements necessarily produce proprioceptive input, so proprioception has a high degree of corporality. On the other hand, proprioception has no alerting capacity: changes in body position do not peremptorily cause attentional resources to be diverted to them. We therefore expect that proprioception should not appear to have an experienced sensory quality. Indeed it is true that though we generally know where our limbs are, this position sense does not have a sensory nature.

The vestibular system detects the position and motion of the head, and so vestibular inputs have corporality. However, they have no alerting capacity. This is because although sudden changes in body orientation immediately result in re-adjusting reactions, these do not *per se* interfere with current cognitive processing. Coherent with our expectations, therefore, the vestibular sense is not perceived as corresponding to an experience. We know we are standing vertical, but we do not have the experience of this in the same sense as we have the experience of hearing a bell or seeing a red patch.

Speculatively, we suggest our plot also can track phenomena intermediate between sensory and mental states. Richness is one of the several examples very tentatively included as points in Fig. 1. The feeling of being rich is a case where there is a limited form of corporality (there are things one can do when one is rich, like getting money from the bank teller, buying an expensive car, but this is nothing like the immediate and intimate link that action has on visual perception, for example), and little alerting capacity (there is no warning signal when one's bank account goes empty). As a consequence, the feeling of being rich is somewhat, though not entirely, sensory.

Application 1: intra- and intermodal differences in sensory quality

One important aspect of sensory experience concerns the differences and the similarities between sensations of a same modality. Why, for example, is the sensation of red different from the sensation of blue? It seems that any account in terms of different neural processes correlated with red and blue immediately encounters an insurmountable problem: why should this particular neural process, say (whatever its specification in neural terms), provide the red sensation, rather than the blue sensation?

In the preceding sections, it was claimed, with reference to the example of softness, that an account in terms of sensorimotor contingencies sidesteps such difficulties. This same approach can now be applied to color. The incoming sensory data concerning a fixated patch of color depend on eye position. Because of non-uniformities in macular pigment and retinal cone distributions, eye movements provoke different patterns of change in sensory input, depending on which colors are being fixated. Such sensorimotor contingencies are part of what constitute the sensations of the different colors. Another type of sensorimotor contingency associated with colors depends on body motions. Consider the light reflected from a colored piece of paper. Depending on where the observer is positioned with respect to ambient illumination, the paper can, for example, reflect more bluish sky light, more yellowish sunlight, or more reddish lamplight. Such laws of change constitute another type of sensorimotor contingency that constitute the sensations of different colors. The fact that color sensation can indeed depend on body motions has been suggested by Broackes (1992) and further philosophical work on color from a related perspective is reported in Myin (2001); cf. also Pettit (2003). A mathematical approach applied to the idea that the differences between color sensations are determined by differences in sensorimotor laws has recently been used to quantitatively predict the structure of human color categories (Philipona and O'Regan, submitted).

Research by Ivo Kohler (1951) provides empirical confirmation for this application of the sensorimotor approach to color. Kohler's subjects wore goggles in which one side of the field was tinted one color (e.g., yellow) and the other another color (e.g., blue). Within a period of some days the subjects came to see colors as normal again. The sensorimotor theory would indeed predict such an adaptation to the new sensorimotor contingencies associated with each color. Kohler's experiments have been criticized (e.g., McCollough, 1965), but recent further work using half-field tinted spectacles (see Fig. 2) shows that adaptation of this kind is indeed possible (O'Regan et al., 2001; Bompas and O'Regan, in press).

A second important aspect of sensory experience concerns intermodal differences in sensory quality: the fact that hearing involves a different quality as compared with seeing, which has a different quality as compared with tactile sensation.

We propose to again apply the idea that sensation involves the exercising of sensorimotor contingencies: differences between modalities come from the different skills that are exercised. The difference between hearing and seeing amounts to the fact that among other things, one is seeing if, when one blinks, there is a large change in sensory input; one is hearing if nothing happens when one blinks, but, there is a left/right difference when one turns one's head, etc. Some other modality-specific sensorimotor contingencies are specified in Table 1.

In addition to providing a more principled account of sensory modality, the sensorimotor approach leads to an interesting prediction. According to this approach, the quality of a sensory modality does not derive from the particular sensory input channel or neural circuitry involved in that modality, but from the laws of sensorimotor contingency that are implicated. It should, therefore, be possible to obtain a visual experience from auditory or tactile input, provided the sensorimotor laws that are being obeyed are the laws of vision (and provided the brain has the computing resources to extract those laws).

Fig. 2. Half-field tinted spectacles worn by A. Bompas. See Plate 5.2 in Colour Plate Section.

The phenomenon of sensory substitution is coherent with this view. Sensory substitution has been experimented with since Bach-y-Rita (1967) constructed a device to allow blind people to "see" via tactile stimulation provided by a matrix of vibrators connected to a video camera. Today there is renewed interest in this field, and a number of new devices are being tested with the purpose of substituting different senses: visual-to-tongue (see Fig. 3, from Sampaio et al., 2001); visual-to-auditory (Veraart et al., 1992); auditory-to-visual (Meijer, 1992); and auditory-to-tactile (Richardson and Frost, 1977). One particularly interesting finding is that the testimonials of users of such devices at least sometimes come framed in terms of a transfer of modalities. For example, a blind woman wearing a visual-to-auditory substitution device will explicitly describe herself as seeing through it (cf. the presentation by Pat Fletcher at the Tucson 2002 Consciousness Conference, available on http://www.seeingwithsound.com/tucson2002.html). Sensory substitution devices are still in their infancy. In particular, no systematic effort has been undertaken up to now to analyze the laws of sensorimotor contingency that they provide. From the view point of the sensorimotor approach, it will be the similarity in the sensorimotor laws which such devices recreate, that determines the degree to which users will really feel they are having sensations in the modality being substituted.

Related phenomena which also support the idea that the experience associated with a sensory modality is not wired into the neural hardware, but is rather a question of sensorimotor contingencies, comes from the experiment of Botvinick and Cohen (1998), where the "feel" of being touched can be transferred from one's own body to a rubber replica lying on the table in front of one (see Fig. 4; also related work on the body image in tool use: Iriki et al., 1996; Farne and Ladavas, 2000;

Table 1. Some sensorimotor contingencies associated with seeing and hearing

Action	Seeing	Hearing
Blink	Big change	No change
Move eyes	Translating flowfield	No change
Turn head	Some changes in flow	Left/right ear phase and amplitude difference
Move forward	Expanding flowfield	Increased amplitude in both ears

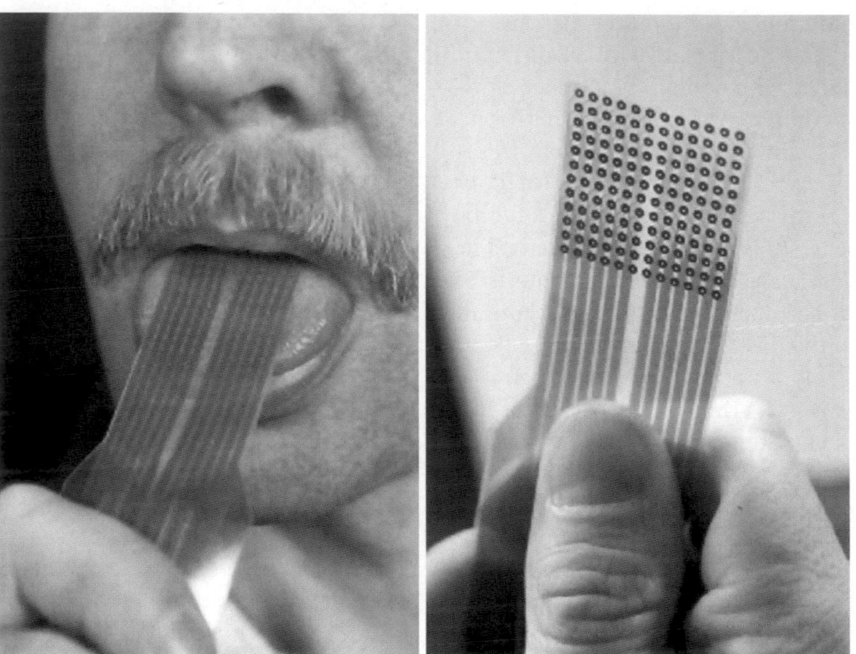

Fig. 3. Tongue stimulation device. This device, connected to a video camera, creates a 12 × 12 sensory pattern on the tongue (from Sampaio et al., 2001) (Photo courtesy of Paul Bach-y-Rita).

Yamamoto and Kitazawa, 2001). The finding of the Sur group (Roe et al., 1990), according to which ferrets can see with their auditory cortex can also be interpreted within the context of the present theory (Hurley and Noë, 2003).

Application 2: dreaming and mental imagery

Dreams are characterized by the fact that while people are dreaming they seem to assume that they are having the same full-blown perceptual experiences that they have in real life. Clearly, however dreams do not involve corporality or alerting capacity in the normal fashion, since there is no sensory input at all.

On the other hand, it is also clear that it is precisely corporality that ultimately allows people to realize that they are actually dreaming — the classic way of knowing that you are dreaming is to try to switch on the light: this kind of "reality-checking" is nothing more than testing for corporality — checking that your actions produce the normal sensory changes expected when you are having real sensory experiences.

It is important to note however that what counts in giving the particular "sensory" feel of sensation is not the actual sensory input itself, but the knowledge that the sensory input possesses corporality and alerting capacity. This means that an observer can have a sensation even though he is, at a given moment, doing nothing at all, and even

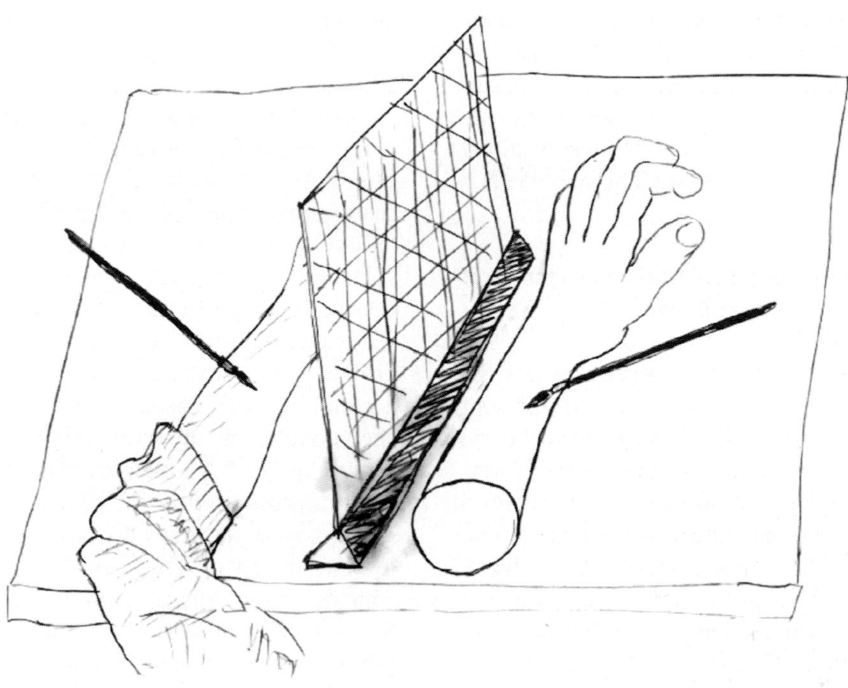

Fig. 4. Illustration of the experiment of Botvinick and Cohen (1998). The subject's arm is placed behind a screen. The subject only sees a rubber arm replica placed in front of him. The experimenter simultaneously stimulates the replica and the arm with a brush. After a few minutes the subject has the impression that the rubber arm is his own arm.

though he is receiving no sensory input at all. It suffices for this that he be in the same mental state that he would usually be in when he has implicit knowledge that the sensorimotor contingencies associated with a sensation are currently applicable.

We can therefore understand how it might happen that a person would have experience of reality without sensory input, and therefore no corporality and alerting capacity. The person merely has to be in a state where he thinks (in point of fact incorrectly) that if he were to move, then those changes would occur that normally occur when he moves. He just has to implicitly think (incorrectly) that were there to be a sudden event, his attention would be automatically attracted to it.

Dreaming therefore poses no problem for the sensorimotor approach that we are proposing. Indeed the approach actually makes it easier to envisage brain mechanisms that enable convincing sensory experiences without any sensory input, since the sensation of richness and presence and "ongoingness" can be produced in the absence of sensory input merely by the brain implicitly "supposing" (in point of fact incorrectly) that if the eyes were to move, say, they would encounter more detail. This state of "supposing that one can get more detail" would be a much easier state to generate than having to actually recreate all the detail somewhere in the brain. In dreaming, furthermore, the state would be particularly easy to maintain because what characterizes dreaming would seem to be a lack of attention to the absence of disconfirming evidence, which is quite unsurprising, since one is asleep. This lowering of epistemic standards implies that, while dreaming, one is easily led into thinking one is perceiving, while — if only one were to pay attention — it would be obvious that one is not. Thus one can remain convinced for the whole duration of one's dream that one is experiencing reality. A whole series of different bizarre dream events may be taken at face value.

Similar remarks apply to mental imagery. As for dreams, mental imagery would correspond to a kind of perceptual action without an actual stimulus and without "going through" the motions — it would involve having implicit expectancies without these being actually fulfilled by worldly responses (for a detailed account of mental imagery along roughly "sensorimotor" lines, see Thomas, 1999).

Application 3: spatial and temporal completeness of the visual world — "change blindness"

When one looks out upon the world, one has the impression of seeing a rich, continuously present visual panorama. Under the sensorimotor theory, however, the richness and continuity of this sensation are not due to the activation of a neural representation of the outside world in the brain. On the contrary, the "ongoingness" and richness of the sensation derive from implicit knowledge of the many different things one can do (but need not do) with one's eyes, and the sensory effects that result from doing them. Having the impression of seeing a whole scene comes, not from every bit of the scene being present in the mind, but from every bit of the scene being immediately available for handling by the slightest flick of thc eye. In terms of the core concepts of this paper: the "feeling of seeing everything" comes from exercise of implicitly knowing one is in a relation with the visually perceived part of the environment which has a high degree of both corporality (moving the body causes changes in sensory input coming from the visual field) and alerting capacity (if something suddenly changes inside the visual field, attention will immediately be drawn to it).

But now a curious prediction can be made. Only one aspect of the scene can be "handled" at any one moment. The vast majority of the scene, although perceived as present, is not actually being "handled". If such currently "unhandled" scene areas were to be surreptitiously replaced, such changes should go unnoticed. Under normal circumstances, the alerting capacity of visual input ensures that any change made in a scene will provoke an eye movement to the locus of the change. This is because low-level movement detectors are hard-wired into the visual system and detect any sudden change in local contours. Attention is peremptorily focused on the change, and visual "handling" is the immediate result. But if the alerting capacity could be inactivated, then we predict that it should indeed be possible to make big changes without this being noticed.

An extensive current literature on "change blindness" confirms this prediction (for a review see Simons, 2000). By inserting a blank screen or "flicker", or else an eye movement, a blink, "mud-splashes" (see Fig. 5), or a film cut between successive images in a sequence of images or movie sequence, the local transients that would normally grab attention and cause perceptual "handling" of a changing scene aspect are drowned out. Under such conditions, observers remain unaware of very large changes. Another method of obviating the usual alerting action of local changes is to make them so slow that they are not detected by the low-level transient detectors in the visual system (see Fig. 6, from Auvray and O'Regan, 2003; also Simons et al., 2000). Demonstrations of change blindness phenomena can be found on the web sites: http://nivea.psycho.univ-paris5.fr and http://viscog.beckman.uiuc.edu/change/.

A related phenomenon is the phenomenon of "inattentional blindness" pioneered by Neisser and Becklen (1975) and Mack and Rock (1998) and recently convincingly extended by Simons and co-workers (Simons and Chabris, 1999). In this, a movie sequence of a complex scene is shown to observers, and they are told to engage in an attentionally demanding task, like counting the number of ball exchanges made in a ball game. An unexpected event (like an actor dressed in a gorilla suit) can go totally unnoticed in such circumstances, even though the event is perfectly visible and in the very center of the visual scene. Demonstrations can be seen on http://nivea.psycho.univ-paris5.fr and http://viscog.beckman.uiuc.edu/djs_lab/demos.html.

Consciousness

The argument made in this paper concerns the nature of sensation: what gives sensation its "experienced" quality, what makes sensory qualities the way they are. But note that we have purposefully not touched upon the question of why and when sensations are conscious. Our claim would

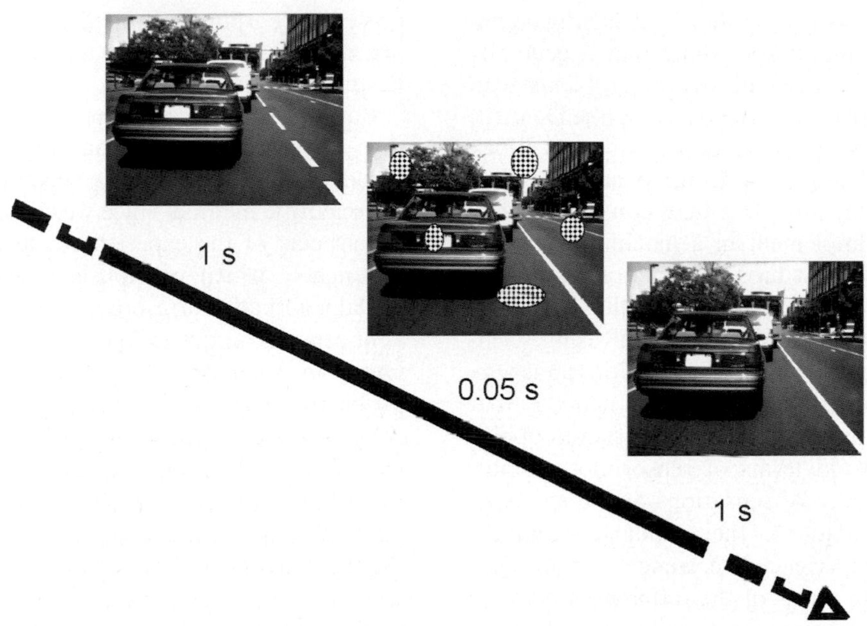

Fig. 5. Change blindness using "mudsplashes" from O'Regan et al. (1999). If the white line in the street changes simultaneously with the occurrence several brief splashes on the screen, the change is very difficult to notice unless it is known in advance.

Fig. 6. Progressive change from red to blue is very difficult to notice if it occurs very slowly (10 s from Auvray and O'Regan, 2003). See Plate 5.6 in Colour Plate Section.

now be that a sensation is conscious when a pers on is poised to cognitively make use of the sensation in their judgments, decisions, and rational behavior.

Why does this constitute progress toward answering the question of the explanatory gap, namely the problem of how a physicochemical mechanism in the brain could ever give rise to an experience? The answer is that first, having cognitive access to a fact is something that is generally considered not to offer particular problems with scientific description and explanation (see Dennett, 1978, Baars, 1988). It amounts to what Block (1995) has called Access Consciousness, and is something which, though it may constitute a difficult thing to implement in a machine, is nevertheless describable in broadly functionalist terms. There is no a priori logical difficulty (although there may be practical difficulties) in using scientific methods to understand Access Consciousness.

Second, we have defined sensation in a way that does not seem problematic from a scientific point of view, namely in terms of sensorimotor skills. The different types of sensations and their experienced characteristics — their similarities and differences, their experienced "presence" — can all be accounted for in terms of the differences between the skills, and in terms of way the neural channels are tuned to the environment, namely by the properties of corporality and alerting capacity. If having a conscious experience amounts to having cognitive access to sensations, then what has previously been considered mysterious, namely what Block has called Phenomenal Consciousness, can now be decomposed into two scientifically tractable components: conscious experience would in our approach consist in having Access Consciousness of sensations. Since Access Consciousness is amenable to scientific methods, and since sensations, being sensorimotor skills, are also amenable to scientific methods, under our approach Phenomenal Conscious now also comes within the domain of science.

Description or explanation?

It is interesting to consider finally the explanatory status of the concepts put forward in this paper. The question of accounting for the experienced quality of sensation is the question of accounting for why certain mental processes are taken to have a sensory nature, while others, like doing arithmetic or making a decision, are not. If one does not espouse a sensorimotor approach, one could claim that saying that sensations involve neural channels possessing corporality and alerting capacity is simply describing something about sensations, and has no explanatory status.

But if one espouses the sensorimotor approach, then the question of accounting for the experienced quality of sensation becomes tractable by the scientific method, since we can see that each of the aspects of the experienced quality of sensory experience, which previously seemed difficult to explain, actually correspond to objectively describable aspects of the skills that are involved. One important such aspect, one which has posed many problems to classical approaches to phenomenal consciousness, is the problem of "presence". We have dealt with this in the sensorimotor approach by noting that sensory stimulation possesses corporality and alerting capacity, thereby providing the skills involved in exploring sensory stimulation with its particular intimate, vivid, inescapable quality. These seem to deal adequately with what we mean by "presence".

We also think our approach holds the promise of accounting for further fine-grained features of sensation that have been noticed by various theorists (see, for example, the list of features in Humphrey, 1992, 2000; O'Regan and Noë, 2001b; Myin and O'Regan, 2002). Consider, for example, ineffability and subjectivity: Under an approach where sensation is neurally generated, it would be difficult to explain why certain neural processes generate qualities which are felt, but which cannot be described (ineffability); equally, it would be difficult to explain why certain neural processes appear to generate subjective quality, whereas others do not.

Within the sensorimotor approach, the appearance of both properties is predicted and is thus explainable: sensory experiences are subjective, and are the sole property of the experiencer because they involve the experiencer himself potentially undertaking actions and exercising sensorimotor skills (see Humphrey (1992, 2000) for a similar explanation). Similarly, sensory experiences are ineffable because they involve exercising implicit, practical skills. Like tying one's shoe laces, exercising the sensorimotor contingencies associated, say, with red, involves putting into practice a practical skill that one cannot describe with words, but that one knows one possesses.

While it may at first sight be unclear how we have made the passage from description to explanation by changing our view of what sensation is, it should be noted that such a shift in theoretical paradigm occurred in the 20th century as regards the question of life. Whereas at the beginning of the 20th century, cell division, metabolism, respiration, etc., were considered to be caused by an as yet unexplained vital essence, today we consider these phenomena to be *constitutive* of life. The notion of life has been redefined: instead of being caused by some underlying mechanism, it is considered now to be constituted by all the various ways the organism can act within its environment. In the same way, by changing one's viewpoint on what sensation is, and espousing the sensorimotor, skill-based approach, one can avoid the issue of generation and thus of the explanatory gap, and immediately see how each of the characteristics that people attribute to sensation arise from aspects of neural machinery and their interaction with the environment.

Thus, we think we have shown that, contrary to the idea that there is an unbridgeable gap between neural processes and "sensory consciousness", a connection may be made between the two domains if neural systems are conceived not as generating sensations, but as allowing organisms to deploy sensorimotor skills.

References

Auvray, M. and O'Regan, J.K. (2003) L'influence des facteurs sémantiques sur la cécité aux changements progressifs dans les scènes visuelles. Ann. Psychol., 103: 9–32.

Baars, B. (1988) A Cognitive Theory of Consciousness. Cambridge University Press, Cambridge.

Bach-y-Rita, P. (1967) Sensory plasticity. Applications to a vision substitution system. Acta Neurol. Scand., 43(4): 417–426.

Block, N. (1995) On a confusion about a function of consciousness. Behav. Brain Sci., 18(2): 227–247.

Block, N. (2005) Two neural correlates of consciousness. Trends Cog. Sci., 9: 46–52.

Bompas, A. and O'Regan, J. K. Evidence for a role of action in color vision. Perception (in press).

Botvinick, M. and Cohen, J. (1998) Rubber hands 'feel' touch that eyes see [letter]. Nature, 391(6669): 756.

Broackes, J. (1992) The autonomy of colour. In: Lennon K. and Charles D. (Eds.), Reduction, Explanation, and Realism. Oxford University Press, Oxford, pp. 421–465.

Dennett, D. (1978) Brainstorms: Philosophical Essays on Mind and Psychology. MIT Press, Cambridge, MA.

Dennett, D. (1988) Quining qualia. In: Marcel A. and Bisiach E. (Eds.), Consciousness in Contemporary Science. Clarendon Press, Oxford, pp. 42–77.

Farne, A. and Ladavas, E. (2000) Dynamic size-change of hand peripersonal space following tool use. Neuroreport, 11(8): 1645–1649.

Gregory, R. (1990) How do we interpret images? In: Barlow H.B., Blakemore C. and Weston-Smith M. (Eds.), Images and Understanding. Cambridge University Press, Cambridge, pp. 310–330.

Hume, D. (17771975) Enquiries Concerning Human Understanding and Concerning the Principles of Morals. Oxford University Press, Oxford.

Humphrey, N. (1992) A History of the Mind. Chatto and Windus, London.

Humphrey, N. (2000) How to solve the mind-body problem. J. Conscious. Stud, 7(4): 5–20.

Hurley and Noë. (2003) Neural plasticity and consciousness. Biol. Philos., 18: 131–168.

Husserl, E. (19071973) Ding und Raum. Vorlesungen 1907. In: Claesges U. (Ed.), Hussserliana 16. M. Nijhoff, The Hague.

Intraub, H. and Richardson, M. (1989) Wide-angle memories of close-up scenes. J. Exp. Psychol. Learn. Mem. Cogn., 15(2): 179–187.

Iriki, A., Tanaka, M. and Iwamura, Y. (1996) Coding of modified body schema during tool use by macaque postcentral neurones. Neuroreport, 7(14): 2325–2330.

Kohler, I. (1951) Über Aufbau und Wandlungen der Wahrnehmungswelt. Österreichische Akademie der Wissenschaften. Sitzungsberichte, philosopohisch-historische Klasse, 227: 1–118.

Levine, J. (1983) Materialism and qualia: The explanatory gap. Pac. Philos. Quart., 64: 354–361.

MacKay, D.M. (1962) Theoretical models of space perception. In: Muses C.A. (Ed.), Aspects of the Theory of Artificial Intelligence. Plenum Press, New York, pp. 83–104.

Mack, A. and Rock, I. (1998) Inattentional Blindness. The MIT Press, Cambridge, MA.

McCollough, C. (1965) The conditioning of color perception. Am. J. Psychol., 78: 362–368.

Meijer, P.B.L. (1992) An experimental system for auditory image representations. IEEE Trans. Biomed. Eng., 39(2): 112–121.

Merleau-Ponty, M. (1945) Phénoménologie de la Perception. Gallimard, Paris.

Myin, E. (2001) Color and the duplication assumption. Synthese, 129(1): 61–77.

Myin, E. (2003) An account of color without a subject? (commentary on Byrne and Hilbert). Behav. Brain Sci., 26(1): 42–43.

Myin, E. and O'Regan, J.K. (2002) Perceptual consciousness, access to modality and skill theories. J. Conscious. Stud., 9(1): 27–45.

Neisser, U. and Becklen, R. (1975) Selective looking: Attending to visually specified events. Cogn. Psychol., 7: 480–494.

Noë, A. (2002a) Is the visual world a grand illusion? J. Conscious. Stud., 9(5/6): 1–12.

Noë, A. (2002b) On what we see. Pac. Philos. Quart, 83(1).
Noë, A. (2004) Action in Perception. The MIT Press, Cambridge, MA.
Noë, A. and O'Regen, J.K. (2002) On the brain-basis of visual consciousness. In: Noë A. and Thompson E. (Eds.), Vision and mind: selected readings in the philosophy of perception. The MIT Press, Cambridge, pp. 567–598.
O'Regan, J.K., Clark, J. and Bompas, A. (2001) Implications of a sensorimotor theory of vision for scene perception and colour sensation (Abstract). Perception, 30(Supplement): 94.
O'Regan, J.K., Myin, E. and Noë, A. (2004) Towards an analytic phenomenology: The concepts of bodiliness and grabbiness. In: Carsetti A. (Ed.), Seeing Thinking and Knowing. Kluwer, Dordrecht, pp. 103–114.
O'Regan, J.K. and Noë, A. (2001a) A sensorimotor account of vision and visual consciousness. Behav. Brain Sci., 24(5): 883–917.
O'Regan, J.K. and Noë, A. (2001b) Acting out our sensory experience: Authors' response to commentary. Behav. Brain Sci., 24(5): 955–975.
O'Regan, J.K. and Noë, A. (2001c) What it is like to see: a sensorimotor theory of perceptual experience. Synthese, 129(1): 79–103.
O'Regan, J.K., Rensink, R.A. and Clark, J.J. (1999) Change-blindness as a result of 'mudsplashes'. Nature, 398: 34.
Pacherie, E. (1999) Leibhaftigkeit and representational theories of perception. In: Petitot J., Varela F.J., Pachoud B. and Roy J.-M. (Eds.), Naturalizing Phenomenology: Issues in Contemporary Phenomenology and Cognitive Science. Stanford University Press, Stanford, pp. 148–160.
Pettit, P. (2003) Looks as powers. Philos. Issue., 13: 221–253.
Philipona, D.L. and O'Regan, J.K. The perceptual structure of color corresponds to singularities in reflection properties (submitted).
Philipona, D., O'Regan, J.K. and Nadal, J.-P. (2003) Is there anything out there? Inferring space from sensorimotor dependencies. Neural Comput., 15(9): 2029–2050.
Rees, G., Kreiman, G. and Koch, C. (2002) Neural correlates of consciousness in humans. Nat. Rev. Neurosci., 3: 261–270.
Richardson, B.L. and Frost, B.J. (1977) Sensory substitution and the design of an artificial ear. J. Psychol., 96(2nd half): 259–285.
Roe, A.W., Pallas, S.L., Hahm, J.O. and Sur, M. (1990) A map of visual space induced in primary auditory cortex. Science, 250(4982): 818–820.
Sampaio, E., Maris, S. and Bach-y-Rita, P. (2001) Brain plasticity: 'visual' acuity of blind persons via the tongue. Brain Res., 908(2): 204–207.
Simons, D.J. (2000) Current approaches to change blindness. Vis. Cog., 7: 1–15.
Simons, D.J. and Chabris, C.F. (1999) Gorillas in our midst: Sustained inattentional blindness for dynamic events. Perception, 28(9): 1059–1074.
Simons, D.J., Franconeri, S.L. and Reimer, R.L. (2000) Change blindness in the absence of visual disruption. Perception, 29: 1143–1154.
Steriade, M. (2001) The Intact and Sliced Brain. MIT press, Cambridge, MA.
Thomas, N. (1999) Are theories of imagery theories of imagination? An active perception approach to conscious mental content. Cog. Sci., 23: 207–245.
Torrance, S. (2002) The skill of seeing: beyond the sensorimotor account? Trends Cog. Sci., 6(12): 495–496.
Veraart, C., Cremieux, J. and Wanet-Defalque, M.C. (1992) Use of an ultrasonic echolocation prosthesis by early visually deprived cats. Behav. Neurosci., 106(1): 203–216.
Yamamoto, S. and Kitazawa, S. (2001) Sensation at the tips of invisible tools. Nat. Neurosci., 4(10): 979–980.

S. Laureys (Ed.)
Progress in Brain Research, Vol. 150
ISSN 0079-6123

CHAPTER 6

Methods for studying unconscious learning

Arnaud Destrebecqz[1,*] and Philippe Peigneux[2]

[1]LEAD, Université de Bourgogne, Pôle AAFE – Esplanade Erasme, BP 26513, 21065 Dijon Cedex, France
[2]Cyclotron Research Center and Cognitive Science Department, University of Liège, Liege, Belgium

Abstract: One has to face numerous difficulties when trying to establish a dissociation between conscious and unconscious knowledge. In this paper, we review several of these problems as well as the different methodological solutions that have been proposed to address them. We suggest that each of the different methodological solutions offered refers to a different operational definition of consciousness, and present empirical examples of sequence learning studies in which these different procedures were applied to differentiate between implicit and explicit knowledge acquisition. We also show how the use of a sensitive behavioral method, the process dissociation procedure, confers a distinctive advantage in brain-imaging studies when aiming to delineate the neural correlates of conscious and unconscious processes in sequence learning.

Introduction

With the recent development of brain-imaging methods, the study of consciousness tends more and more to be considered as a relatively easy problem that does not require any particular methodological approach. From this perspective, consciousness is merely viewed as another aspect of human cognition such as, for instance, motor action, memory, learning, or perception.

About 10 years ago, Chalmers (1995) proposed a distinction between *hard* and *easy* problems in the study of consciousness. According to Chalmers, the hard problem relates to studies aimed at understanding the relationship between brain activity correlated with awareness and the particular subjective *experience* that proceeds from this neural activity. The easy problems of consciousness are, among others, the conceptual and methodological issues faced by researchers studying the ability to report mental states, or the voluntary control of behavior. Chalmers claims that easy problems are those that seem accessible to the standard methods of cognitive science, i.e., they would be explainable in terms of computational or neural mechanisms, whereas hard problems resist these methodological and explanatory frameworks. In this view, and given the current development of brain-imaging techniques that allow for in-depth investigation of the neural correlates of higher-level cognitive processes, the search for the neural correlates of consciousness should belong to Chalmers's easy problems.

In this chapter, however, we argue that functional brain-imaging approaches to the cerebral correlates of conscious and unconscious processes do not obviate the need for powerful behavioral methods as one is still confronted with the need to carefully determine the conscious versus unconscious character of the knowledge expressed in any task. We acknowledge that Chalmers' distinction underlines important differences between first and third person aspects of consciousness that certainly involve different levels of conceptual complexity. It is, for instance, difficult to formulate an explanation of the relationship between a given

*Corresponding author. Tel.: +33 3 80 39 39 68;
Fax: +33 3 80 39 57 67;
E-mail: Arnaud.Destrebecqz@leadserv. u-bourgogne.fr

DOI: 10.1016/S0079-6123(05)50006-2

stimulation and the corresponding subjective experience (e.g., the redness of red). It is comparatively easier to imagine how brain-imaging methods may constitute a powerful tool to differentiate between conscious and unconscious cognition. However, we will show that when one is trying to devise a sensitive and reliable methodological procedure to address this latter question, easy problems might turn out to be not so easy after all.

The definition of an accurate and sensitive methodological approach for differentiating between conscious and unconscious cognitive processes has always been a controversial issue. This is because there is no obvious way to find a proper measure of awareness in the absence of a satisfactory operational definition of the concept. How can one describe what it means for somebody else to be conscious of some knowledge? How can one measure the amount of conscious knowledge held by another individual? These questions have been fiercely debated across several domains of cognitive psychology, including the literature on implicit memory, learning, and perception. In this paper we will focus on implicit sequence learning, which is one of the most popular paradigms for studying unconscious learning.

Sequence learning as an example of unconscious cognition

In a typical sequence learning situation (see Clegg et al., 1998), participants are asked to react to each element of a sequentially structured and typically visual sequence of events in the context of a serial reaction time (SRT) task. On each trial, a stimulus appears at one of several locations on a computer screen. Subjects are instructed to press a spatially corresponding key as fast and as accurately as possible. Unknown to them, the sequence of successive stimuli follows a repeating pattern (Nissen and Bullemer, 1987). Reaction times (RT) tend to decrease progressively with continued practice, but then dramatically increase when the repeating pattern is modified in any of the several ways (Cohen et al., 1990; Curran and Keele, 1993; Reed and Johnson, 1994). This suggests that subjects learn the repeated pattern and prepare their responses based on their knowledge of the sequence. Nevertheless, subjects often fail to exhibit verbal knowledge of the pattern — a dissociation that has led many authors to consider learning to be implicit in this situation.

As discussed elsewhere Cleeremans et al. (1998), sequence learning can be described as implicit in several different ways depending on whether one focuses on the acquisition or retrieval processes, or on the knowledge resulting from the learning episode. Here, we will focus on this latter aspect of implicit learning and discuss how one can establish the extent to which the knowledge acquired during the SRT task can be described as conscious or unconscious.

Recall, however, that the methodological and conceptual issues raised by this question are by no means limited to the implicit learning literature, and a fortiori to the sequence learning paradigm. Indeed, as previously pointed out by Goschke (1997), ever since the 1960s these same issues have also been the object of controversy in the subliminal perception literature (e.g., Ericksen, 1960; Holender, 1986).

Dissociation studies

Most studies aimed at demonstrating the existence of implicit knowledge have taken the form of dissociation experiments in which performance in an initial learning or exposure task — often considered as exclusively dependent on implicit processes — is compared with performance in a subsequent test task, which is assumed to give an index of participants' conscious knowledge. According to this dissociation logic (Erderlyi, 1986), knowledge is implicit if performance exceeds baseline in the first task — indicating that learning took place — but is at chance in the test phase, suggesting that participants do not have conscious access to the knowledge that has been acquired.

Several different measures of conscious knowledge have been used in the sequence learning paradigm, each of them corresponding to a different operational definition of consciousness (see Table 1). In the following sections, we will examine these

Table 1. Synoptic table of different operational definitions of consciousness and the corresponding tasks used to measure consciousness

Operational definitions of consciousness	Measurements of conscious knowledge
Consciousness allows *verbal access* to the acquired knowledge	Verbal reports, questionnaires
Consciousness allows *recollection* of the acquired knowledge	Recognition and generation tasks
Conscious learning allows *control* on the expression of the acquired knowledge	Direct and indirect tasks, inclusion and exclusion tasks
Conscious learning is associated with the acquisition of *meta-knowledge* (i.e., participants know that they have learned something and that they are using this knowledge to perform a task)	Guessing judgment tasks, confidence judgment tasks

separate procedures and compare their ability to provide accurate measures of consciousness.

Verbal reports

Since awareness can naturally be described as an essentially private, first-person phenomenon, verbal reports and questionnaires have been used as the prior measurement method to estimate conscious knowledge. Accordingly, the initial claims for the existence of an implicit form of learning are based on reported dissociations between, on the one hand, a significant decrement in reaction time during the SRT task and, on the other, the inability of some but not all participants to report the regularities of the sequence (Willingham et al., 1989; Curran and Keele, 1993).

Several authors have recently argued that first-person methodologies constitute an essential way of understanding and measuring consciousness (Overgaard, 2001). However, while these methods might indeed capture gross and simple features of conscious experience (Chalmers, 1999), they might not actually be sensitive enough to provide an accurate measurement tool for dissociating between conscious and unconscious knowledge and furthermore for differentiating between the different features of conscious experience. For instance, although there is no doubt that sequence learning can be defined as conscious if participants are able to describe verbally the regularities of the training sequence, the opposite may not be true, i.e., a poor performance in questionnaire or verbal report tasks does not necessarily imply that learning was implicit.

From this perspective, Shanks and St. John (1994) have pointed out that verbal reports do not always satisfy the two criteria that they consider to be critical; the *information* and *sensitivity* criteria. According to the information criterion, the task used to measure conscious knowledge must tap into the same knowledge base upon which learning is based. Otherwise, learning could be described as unconscious not because participants are unable to access their knowledge consciously but simply because they are probed about irrelevant features of the training material that they did not need to process in order to perform the task. For instance, this might be the case in sequence learning studies in which participants are asked to report first- or second-order sequential regularities between successive elements, while zero-order information, such as variations in the frequencies of the sequence elements, are sufficient to account for RT performance.

According to Shanks and St. John's (1994) sensitivity criterion, the test used to measure conscious knowledge must be sensitive to all of the relevant information. If this criterion is not met, unconscious influences on performance might be overestimated because some conscious knowledge remains undetected by the awareness test. There are several reasons to argue that verbal reports fail the sensitivity criterion and, therefore, do not constitute an adequate measure of awareness. Firstly, subjects might fail to report fragmentary knowledge held with low confidence. Secondly, in the case of verbal reports, performance and awareness tests implement very different retrieval contexts (Shanks and St. John, 1994). With respect to the sequence learning paradigm, none of the contextual cues that are available to support performance in the RT task (such as the visual presentation of the stimuli, the requirement of motor responses, and a sustained pace of responding) are available to support verbal reports.

It is possible to improve the sensitivity of the awareness test by using questionnaires that involve more specific queries about the relevant knowledge. However, the use of questionnaires impose conditions on the test phase that are rather different from those in the training phase. Many authors have therefore suggested that valid tests of awareness should involve forced-choice tasks such as recognition tasks. It has been argued that these are able to detect conscious knowledge left undetected by verbal reports or questionnaires (Willingham et al., 1989; Perruchet and Amorim, 1992; Frensch et al., 1994; Shanks and St. John, 1994; Shanks and Johnstone, 1998; Shanks and Johnstone, 1999).

Forced-choice tasks

In order to improve the sensitivity of the awareness test, forced-choice tasks (which implement retrieval conditions closer to the learning task itself) have been used to measure conscious knowledge. Under the assumption that consciousness allows recollection of the acquired knowledge, these tests have generally taken the form of generation or recognition tasks at the end of sequence learning. In a typical generation task, participants are requested to reproduce the training sequence themselves by pressing the key corresponding to the location of the next stimulus instead of reacting to the current target. In the recognition task, they are presented with a sequence fragment, and after reacting to each element as they did in the RT task (i.e., by pressing as fast and as accurately as possible on the corresponding key), are asked to identify whether or not the fragment was part of the training sequence. In sequence learning studies, forced-choice tests of conscious knowledge that prompt reproduction of the training sequence or differentiation of old and new sequence fragments have quite systematically indicated that participants were able to express a great deal of the knowledge they acquired in the SRT task (e.g., Perruchet and Amorim, 1992).

These results have frequently been interpreted as an indication of the conscious nature of sequence learning and have led into questioning the very existence of an implicit form of learning. However, others have challenged the assumption that generation or recognition performance depends solely on conscious knowledge (e.g., Jiménez et al., 1996). It is indeed true that these tasks involve the same type of retrieval conditions as the SRT task: participants have to react to visual stimuli by giving motor responses and, at least in the case of recognition tasks, at the same pace of responding. As a result there is little reason to believe that SRT tasks and awareness tests tap into different knowledge bases, and thus that implicit knowledge does not contribute to generation or recognition performance. Indeed, it has been shown that participants are able to reproduce the training sequence in a generation task even when they claim to guess the location of the next sequence element (Shanks and Johnstone, 1998). Furthermore, in a recognition task, subjects may tend to respond faster to old sequence fragments than to novel ones; recognition ratings may therefore reflect this improved feeling of perceptual and motor fluency rather than explicit recollection of the training material (see Perruchet and Amorim, 1992; Perruchet and Gallego, 1993; Willingham et al., 1993, for relevant discussion, Cohen and Curran, 1993). Performance in both recognition and generation tasks, rather than depending exclusively on conscious knowledge, is thus likely to depend on both implicit and explicit influences. By the same token, it must also be emphasized that sequence acquisition during the SRT task is itself likely to involve both implicit and explicit components.

In sum, forced-choice tasks are more prone to meet the information and sensitivity criteria than free reports or questionnaires. This improvement, however, is at the cost of the so-called *exclusiveness assumption* (Reingold and Merikle, 1988), according to which the test of awareness must be sensitive *only* to the relevant conscious knowledge. Unfortunately, the most sensitive tests of awareness are also the most likely to be contaminated by implicit knowledge (Neal and Hesketh, 1997). The logic of quantitative dissociation has therefore been questioned by the argument that no task can be used as an absolute test of awareness that would be both sensitive to all a subject's conscious knowledge, and only to the relevant conscious knowledge. In other words, it is highly implausible that any task can be considered as "process-pure". To further improve awareness tests, different solutions have been proposed to overcome this so-called "contamination" problem. These procedures, which we discuss in the following sections, were initially proposed in the fields of subliminal perception and implicit memory, and later applied to the domain of implicit learning.

Subjective measures of awareness

Cheesman and Merikle (1984) have introduced the notion of subjective and objective thresholds in subliminal perception. In a typical experiment, the task simply consists in identifying a series of visual targets that are briefly flashed on a computer screen. Perception is said to be under the *subjective* threshold when participants are able to identify the target at above chance performance while stating that they did not perceive it consciously. Perception is said to be under the *objective* threshold when identification is at chance. Hence, perception is under the subjective threshold when a subject does not know that he knows the identity of the target, or in other words, when he has no *meta-knowledge*. Perception is under the objective threshold when the target has simply not been perceived. According to Cheesman and Merikle, perception is unconscious when it is under the subjective threshold.

Dienes and Berry (1997) have suggested the application of the same threshold criteria to the study of implicit learning. In this framework, the acquired knowledge would be over the objective threshold when performance in a forced-choice task is above baseline. Learning could be described as unconscious if knowledge remains under the subjective threshold at the same time, i.e., if participants claim to respond at chance in the forced-choice task used to measure conscious knowledge. This procedure can indeed be extremely fruitful when attempting to disentangle conscious and unconscious knowledge given that, as discussed above, both types of knowledge can subtend performance in a forced-choice task.

Dienes et al. (1995), see also Dienes and Perner (1999), have described two criteria that make it possible to demonstrate unconscious knowledge acquisition. The first one, the guessing criterion, corresponds to the criterion used by Cheesman and Merikle: knowledge is unconscious when performance is above chance while participants claim to perform at chance. The second one, the zero-correlation criterion, is met when confidence levels and performance rates are uncorrelated.

These two procedures have seldom been applied in sequence learning studies. In one experiment, using the guessing criterion after the generation task, Shanks and Johnstone (1998) asked participants to say whether they felt that they were reproducing the training sequence or that they respond randomly. Only 3 out of 15 participants felt that some fragments of the sequence were familiar and part of the training sequence. When these three subjects were excluded from the analysis, generation performance was still above chance level. In another experiment of the same study, these authors applied the zero-correlation criterion by asking participants to rate how confident they were in their generation performance on a scale ranging from 0 to 100. They observed that the experimental subjects reproduced more of the training sequence than a control group presented with a random sequence during training. However, some of the experimental subjects did not show a higher level of confidence than the control participants.

These results seem to suggest that learning was at least partly unconscious. However, Reingold and Merikle (1990) have insisted that subjective measurement of unconscious knowledge must be interpreted with caution given that subjective measures of awareness depend on the participants' interpretation of the task instructions. Indeed, participants might essentially give a larger interpretation to the term "guess" than the experimenter. Accordingly, Shanks and Johnstone have argued that, in their first study, participants' subjective experience might have become confused and discontinued during the generation task, leading them to consider that they were guessing the next location when this was not actually the case. A similar interpretative problem may arise in Shanks and Johnstone's second study, which used the zero-correlation criterion; in this study participants were able to evaluate their confidence level based not only on the accessibility to consciousness of the acquired knowledge but also on how much they believed was expected from them. For instance, a given subject might have underestimated his level of confidence because he assigned a high level of expectancy to the experimenter.

To summarize, the criticisms previously laid against verbal reports may also be applied to subjective measures because, in both cases, subjects have to decide themselves whether they have access to some knowledge or whether they were able to perform a task effectively. Subjective measures of awareness should therefore, be combined with other measurement methods that do not raise the same problems of interpretation. In the following sections, we describe two such methodological frameworks that consider, as a starting point, that conscious knowledge supports intentional control.

Comparison between direct and indirect tasks

Given that no task can be considered as a process-pure measure of awareness, Reingold and Merikle (1988) (in the field of subliminal perception) have proposed to compare the relative sensitivity of *direct* and *indirect* tasks to conscious influences. In an indirect task, participants are not explicitly required to make a response based on relevant information (e.g., the identity of a visual target in a perception study or the sequential regularities in a sequence learning study). Conversely, in a direct task participants have to respond on the basis of this information. In Reingold and Merikle's framework, both tasks must be matched as much as possible in all characteristics, such as retrieval context and demands. It is only the instructions that must differ. This method is based on the hypothesis that both conscious and unconscious knowledge may influence performance in direct and indirect tasks. In addition, it is also posited that conscious influences will not be higher in the indirect task than in the direct task given that participants are required to respond on the basis of explicit knowledge in the latter but not in the former case. Therefore, if the indirect task detects some knowledge that is left undetected by the direct task, this knowledge can be considered as unconscious. This procedure has later been applied to different domains such as subliminal perception (Greenwald et al., 1995), unconscious memory (Merikle and Reingold, 1991), and sequence learning (Jiménez et al., 1996).

In the study of Jiménez et al., the sequence of stimuli was produced according to the rules of an artificial grammar. In 15% of the trials, the stimulus generated by the artificial grammar was replaced by another one that violated the rules of the grammar. The relevant information was thus the difference between the grammatical and ungrammatical sequential transitions. In this context, the RT difference between regular and irregular trials in the SRT task can be considered as an *indirect* measure of sequence learning since participants do not have to respond on the basis of this information: they merely have to indicate the location of the current target irrespective of its grammatical or ungrammatical status. As a *direct* measure of learning, Jiménez et al., this time, used a generation task in which participants were required to reproduce the grammatical transitions of the training sequence. By comparing the performance in both tasks, they were able to show that some knowledge about the sequence was exclusively expressed in the indirect task but not in the direct generation task. For some sequential transitions, participants responded faster for grammatical stimuli than for ungrammatical stimuli. However, there was no accompanying increase in the production of the corresponding regularities in the generation task — leading the authors to conclude that this knowledge was unconscious.

A limitation of this comparative method is that the indirect measure must always be more sensitive than the direct measure in order to show unconscious influences on performance (Toth et al., 1994). Some results, however, suggest that this might not always be the case in sequence learning (Shanks and Johnstone, 1998; Perruchet et al., 1997). In the next section, we describe another methodological framework, which compares subjects' performance on two tasks that differ only with respect to their instructions: this procedure makes it possible to circumvent this particular issue of sensitivity.

The process dissociation procedure

Similar to Reingold and Merikle's framework, the process dissociation procedure (PDP), initially described by Jacoby (1991) in the field of implicit memory research, is formed from the idea that consciousness subtends intentional control. By contrast, it is assumed that unconscious knowledge influences performance independently from, or against, task instructions. According to the logic of the procedure, conscious and unconscious influences can be estimated from the comparison of two situations in which both these influences either contribute to performance — the *inclusion* task — or are set in opposition — the *exclusion* task. The inclusion and exclusion tasks differ only with respect to their instructions. In the context of sequence learning, consider for instance a generation task performed under inclusion instructions. Participants are told to produce a sequence that resembles the training sequence as much as possible. To do so, they can either explicitly recollect the regularities of the training sequence, or they can guess the location of the next stimulus based on intuition or familiarity. Hence, under inclusion instructions, both conscious (C, e.g., recollection) and unconscious (U, e.g., intuition) processes can contribute to improve performance (C + U). Now consider the same generation task, but this time performed under exclusion instructions. Participants are now told to generate a sequence that *differs* as much as possible from the training sequence. Conscious and unconscious influences are now set in opposition, for the only way to successfully avoid producing familiar sequence elements is to consciously know what the training sequence was and to produce something different. Continued generation of familiar elements under exclusion instructions would thus clearly indicate that generation is automatically influenced by unconscious knowledge (U). Within the PDP, an estimate of conscious influences (C) can therefore be obtained by computing the difference between inclusion and exclusion performance and an estimate of unconscious influence (U) can be derived from the amount of which exclusion performance exceeds baseline.

The PDP has also raised many controversies. However, these are mainly concerned with the specific measurement model used to obtain the *quantitative* estimate of the implicit influences on performance. Different models that reflect the hypothetical relationship between both conscious and unconscious influences have indeed been proposed (see Richardson-Klavehn et al., 1996). A complete discussion of this issue is largely beyond the scope of this paper. However, it has been proposed that this measurement problem can be circumvented by focusing on inclusion and exclusion performance only (Neal and Hesketh, 1997).

Goschke (1997) using the PDP in this way, reported that a secondary tone-counting task impaired explicit knowledge acquisition but left implicit learning unaffected. With a similar procedure, Destrebecqz and Cleeremans (2001) have shown that slowing the pace of the SRT task by increasing the response-to-stimulus interval (RSI) tends to improve explicit learning. Other authors have put forward the notion that the pace of the SRT task only influences the expression of knowledge rather than learning *per se* (Willingham et al., 1997) or does not influence the quality of sequence learning, the acquired knowledge being always conscious whatever the pace of the SRT task (Wilkinson and Shanks, 2004). It must be noted, however, that the importance of timing factors has been emphasized in theoretical accounts of sequence learning (Keele et al., 2003) and specifically in relation to the conscious or unconscious nature of learning (Stadler, 1997). Recent studies are also in line with the idea that conscious processes require more time than unconscious ones (Rabbitt, 2002). In the next section, we present additional data supporting this hypothesis.

In search of the neural correlates of conscious and unconscious processes

As evidenced in the above discussion, the identification of the neural correlates of conscious and unconscious processes crucially requires the use of sensitive behavioral methods that overcome potential methodological flaws. Capitalizing on previous behavioral results, we have adapted the PDP in a brain-imaging study in order to identify the neural correlates of conscious and unconscious sequence processing (Destrebecqz et al., 2003; Destrebecqz et al., submitted). In two $H_2^{15}O$ Positron Emission Tomography (PET) studies, volunteers were trained on the SRT task (15 blocks of 96 trials using a repeated 12-element sequence) before being scanned during three consecutive inclusion and three consecutive exclusion blocks. During training, the pace of the SRT task was manipulated by modifying the value of the RSI, i.e., the amount of time that elapses between the motor response and the onset of the next target. Participants were either trained with a standard 250 ms RSI (RSI250 condition) or with a RSI reduced to 0 ms (RSI0 condition). In this latter RSI0 condition, each target was immediately replaced by the next one, reducing the likelihood of participants developing conscious expectancies regarding the identity of the next element (Destrebecqz and Cleeremans, 2001).

Inclusion scores were higher than exclusion scores in both RSI0 and RSI250 conditions, indicating that participants had gained conscious sequence knowledge in both conditions. Inclusion scores, however, were higher in the RSI250 condition than in the RSI0 condition. By contrast, exclusion scores (that reflect the production of the training sequence against the instructions) were higher in the RSI0 condition than in the RSI250 condition, suggesting that unconscious influences were higher in the former than in the latter condition. The mean difference between inclusion and exclusion scores was higher in the RSI250 than in the RSI0 condition indicating that control over behavior was improved in the former condition and, therefore, according to the PDP logic driven by explicit, conscious learning.

At the neuroanatomical level, we reasoned that those brain areas in which variations of regional cerebral blood flow (rCBF) closely follow the variations of the generation scores obtained at each scan in the inclusion or the exclusion condition should be part of the neural network that subtends conscious and unconscious contributions to performance.

Firstly, to identify the neural correlates of *conscious* sequence processing, we looked for brain regions in which the correlation between rCBF and generation score was higher in the inclusion than in the exclusion task, irrespective of the duration of the RSI. Indeed, inclusion scores are thought to reflect both conscious and unconscious contributions (C + U) to performance, whereas rCBF variations related to the exclusion scores only reflect unconscious contributions (U). The interaction $[(C + U) - U = C]$ effect should therefore indicate the brain areas that specifically subtend conscious contributions to performance. Results showed that the anterior cingulate/medial prefrontal cortex (ACC/MPFC) supports the conscious component of sequence processing in both the RSI250 and RSI0 conditions (see Fig. 1C).

Secondly, to identify the neural correlates of *unconscious* sequence knowledge, we looked for brain areas in which the correlation between rCBF and exclusion scores was modulated by the training condition, i.e., was higher in the RSI0 than in the RSI250 condition (exclusion results suggested that knowledge was more implicit in the RSI0 than in the RSI250 condition). This analysis revealed that caudate activity supported implicit contributions to performance in the sequence generation task (see Fig. 1B). This result is in line with previous data showing the involvement of the caudate nucleus during implicit sequence acquisition in a probabilistic version of the SRT task (Fig. 1A; Peigneux et al., 2000).

PET data analysis also evidenced a tight coupling between activity measured in ACC/MPFC and in striatum in the exclusion task in the RSI250 condition, in which learning was essentially explicit, whereas the activity of these regions was uncoupled in the RSI0 condition, in which implicit influences were stronger. These results suggest that the ACC/MPFC exerts control on the activity of the striatum during sequence generation in the former but not in the latter condition — indicating that implicit processes can be successfully controlled by conscious knowledge when learning is essentially explicit.

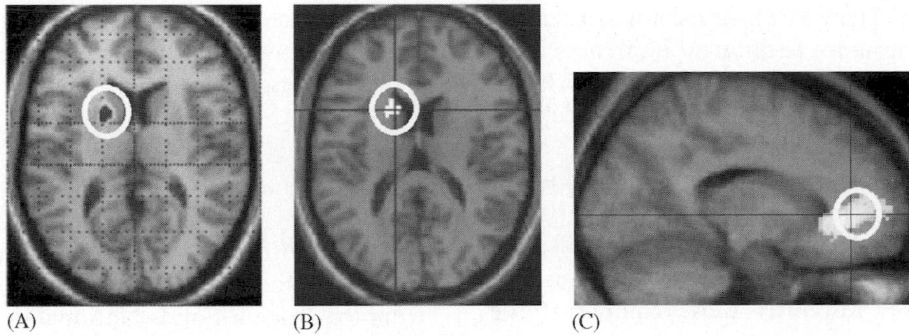

Fig. 1. Brain correlates of conscious and unconscious knowledge. (A) Sequence acquisition in a probabilistic SRT task: increase cerebral blood flow (CBF) in the caudate nucleus was observed only when subjects' performances demonstrated *implicit* knowledge of sequential regularities (Peigneux et al., 2000). (B) *Implicit* sequence generation after practice to the SRT task: caudate CBF correlates more with *exclusion* scores when learning is essentially unconscious (i.e., higher correlation in the RSI0 ms than the RSI250 condition; Destrebecqz et al., submitted). (C) *Explicit* sequence generation after practice to the SRT task: CBF correlations in ACC/mesial prefrontal cortex with generation scores is modulated by the instruction condition (i.e., higher in the inclusion than in the exclusion condition) both in the RSI0 and RSI250 conditions (Destrebecqz et al., 2003). (See text for details.) All activations are displayed at uncorrected $p<0.001$, on subjects' averaged T-1 weighted MRI.

Concluding remarks

In this paper, we have described the numerous intricacies of the arguments surrounding a theoretical dissociation between conscious and unconscious knowledge, as well as the various methodological solutions that have been proposed to address these questions. We have shown that each of these proposed solutions refers to a different operational definition of consciousness, and have presented empirical examples of sequence learning studies in which these procedures have been used to differentiate between implicit and explicit knowledge acquisition. Altogether, these studies indicate that a single measure of awareness is unable to offer a precise assessment of the extent to which knowledge has been consciously acquired during a learning episode. Only a few studies, however, have systematically applied several awareness tests at the same time in order to measure correlations among those tasks or to show possible dissociations between them (e.g., Shanks and Johnstone, 1998; Destrebecqz and Cleeremans, 2003).

Interestingly, some proposed models of consciousness — the global workspace framework (Baars, 1988; this volume) — predict systematic associations between all the different possible measures of awareness. According to Dehaene and Naccache (2001), for instance, once a piece of information becomes conscious, it becomes globally available to a variety of processes including categorization, memorization, evaluation and intentional action. Their hypothesis is that this global availability constitutes conscious subjective experience.

However, other reports have shown that learning can, under some conditions, result in knowledge that is under intentional control but not systematically and simultaneously associated with meta-knowledge (Destrebecqz and Cleeremans, 2003). Such learning would thus be described as conscious with respect to the PDP but as unconscious with respect to the subjective threshold criterion. Based on results from connectionist simulations, it has also been suggested that these behavioral dissociations can be accounted for within a framework in which conscious access is viewed as resulting from continuous, gradual changes in a single dimension involving "quality of representation". This dimension designates several properties of memory traces, such as their relative strength, their distinctiveness, or their stability in time (see Cleeremans and Jiménez, 2002; Cleeremans, forthcoming). Depending on training conditions, the quality of the developed representations varies along a continuous dimension and determines the extent to which memory traces can influence performance in different awareness tests. Subjective measures would, in this framework, entail higher quality representations than procedures based on intentional control.

In the same perspective, Marcel (1993) conducted a visual target detection study in which participants were asked to respond in three different ways: verbally, by blinking, or by pressing a key. When the three responses were required at the same time, they often tended to be dissociated. Participants could, for instance, claim to detect the target in one modality but not in another. Marcel further noted that subjective confidence levels varied consistently between response modalities. Other reports, in the field of unconscious perception, have shown that stimulus *recognition* can occur in the absence of stimulus *detection* either when chromatic flashes (Rollman and Nachmias, 1972) or when words (Merikle and Reingold, 1990) are presented. In other words, participants may be able to recognize a color or a word at a better level of accuracy than the accuracy level produced by chance even when they claim that they did not previously detect its presence. According to Merikle and Reingold (1990, p. 582), however, this pattern of results may be related to differences between tasks in bias or in criterion (i.e., the minimum level of certainty that is necessary for a participant to say that some stimulus has been presented). As noted by Merikle and Reingold, dissociations are indeed expected if criterion placement changes independently across the different tasks used to measure conscious knowledge.

At first sight, these dissociation results might seem to be at odds with global workspace theories of consciousness. However, further research is needed in order to determine whether these dissociations merely reflect sensitivity differences between the tasks used to measure awareness or whether they represent more fundamental differences between the nature of the mental representations and the neural mechanisms that subtend conscious processing in those tasks. We believe that the understanding of the cognitive and neural mechanisms subtending conscious processes will benefit from the simultaneous use of different measurement procedures. An accurate theory of consciousness will have to explain the conditions under which all of these measures are associated or dissociated.

Acknowledgments

AD is supported by a post-doctoral grant from the Fyssen Foundation. PP is supported by grants from the Belgian Fonds National de la Recherche Scientifique, the Fondation Médicale Reine Elisabeth, the ULg Research Fund, and the Interuniversity Attraction Poles Program — Belgian Science Policy. The authors thank the four anonymous reviewers for comments and suggestions on a previous version of this article and Andrew Bremner for improving the English.

References

Baars, B.J. (1988) A Cognitive Theory of Consciousness. Cambridge University Press, Cambridge.

Chalmers, D.J. (1995) Facing up to the problem of consciousness. J. Consciousness Stud., 2: 200–219.

Chalmers, D.J. (1999) First-person methods in the science of consciousness. Consciousness Bulletin. Center for Consciousness Studies, University of Arizona.

Cheesman, J. and Merikle, P.M. (1984) Priming with and without awareness. Percept. Psychophys., 36: 387–395.

Cleeremans, A. Being Virtual. Oxford University Press, Oxford. forthcoming.

Cleeremans, A., Destrebecqz, A. and Boyer, M. (1998) Implicit learning: news from the front. Trends Cogn. Sci., 2: 406–416.

Cleeremans, A. and Jiménez, L. (2002) Implicit learning and consciousness: A graded dynamic perspective. In: French R.M. and Cleeremans A. (Eds.), Implicit Learning and Consciousness: An Empirical Computational And Philosophical Consensus in the Making? Psychology Press, Hove, UK.

Clegg, B.A., DiGirolamo, G.J. and Keele, S.W. (1998) Sequence learning. Trends Cogn. Sci., 2: 275–281.

Cohen, A., Ivry, R.I. and Keele, S.W. (1990) Attention and structure in sequence learning. J. Exp. Psychol. Learn, 16: 17–30.

Cohen, A. and Curran, T. (1993) On tasks, knowledge, correlations and dissociations: Comment on Perruchet and Amorim (1992). J. Exp. Psychol. Learn, 19: 1431–1437.

Curran, T. and Keele, S.W. (1993) Attentional and nonattentional forms of sequence learning. J. Exp. Psychol. Learn, 19: 189–202.

Dehaene, S. and Naccache, L. (2001) Towards a cognitive neuroscience of consciousness: Basic evidence and a workspace framework. Cognition, 79: 1–37.

Destrebecqz, A. and Cleeremans, A. (2001) Can sequence learning be implicit? New evidence with the process dissociation procedure. Psychon. B Rev., 8: 343–350.

Destrebecqz, A. and Cleeremans, A. (2003) Temporal factors in sequence learning. In: Jiménez L. (Ed.), Attention and Implicit Learning. John Benjamins, Amsterdam and Philadelphia, pp. 181–213.

Destrebecqz, A., Peigneux, P., Laureys, S., Degueldre, C., Del Fiore, G., Aerts, J., Luxen, A., van der Linden, M., Cleeremans, A. and Maquet, P. (2003) Cerebral correlates of explicit sequence learning. Cogn. Brain Res., 16: 391–398.

Destrebecqz, A., Peigneux, P., Laureys, S., Degueldre, C., Del Fiore, G., Aerts, J., Luxen, A., van der Linden, M., Cleeremans, A., Maquet, P. The neural correlates of implicit and explicit sequence learning: Interacting networks revealed by the process dissociation procedure (submitted).

Dienes, Z., Altmann, G.T.M., Kwam, L. and Goode, A. (1995) Unconscious knowledge of artificial grammars is applied strategically. J. Exp. Psychol. Learn, 21: 1322–1338.

Dienes, Z. and Berry, D.C. (1997) Implicit learning: Below the subjective threshold. Psychon. B Rev., 4: 3–23.

Dienes, Z. and Perner, J. (1999) A theory of implicit and explicit knowledge. Behav. Brain. Sci., 22: 735–755.

Erderlyi, M.H. (1986) Experimental indeterminacies in the dissociation paradigm of subliminal perception. Behav. Brain Sci., 9: 30–31.

Ericksen, C.W. (1960) Discrimination and learning without awareness: a methodological survey and evaluation. Psychol. Rev., 67: 279–300.

Frensch, P.A., Buchner, A. and Lin, J. (1994) Implicit learning of unique and ambiguous serial transitions in the presence and absence of a distractor task. J. Exp. Psychol. Learn, 20: 567–584.

Goschke, T. (1997) Implicit learning and unconscious knowledge: Mental representation, computational mechanisms, and neural structures. In: Lamberts S.K. and Shanks D. (Eds.), Knowledge, concept and categories. Psychology Press, Hove, UK, pp. 247–333.

Greenwald, A.G., Klinger, M.R. and Schuh, E.S. (1995) Activation by marginally perceptible ("subliminal") stimuli: Dissociation of unconscious from conscious cognition. J. Exp. Psychol. Gen., 124: 22–42.

Holender, D. (1986) Semantic activation without conscious activation in dichotic listening, parafoveal vision, and visual masking: A survey and appraisal. Behav. Brain Sci., 9: 1–23.

Jacoby, L.L. (1991) A process dissociation framework: Separating automatic from intentional uses of memory. J. Mem. Lang., 30: 513–541.

Jiménez, L., Méndez, C. and Cleeremans, A. (1996) Comparing direct and indirect measure of sequence learning. J. Exp. Psychol. Learn, 22: 948–969.

Keele, S.W., Ivry, R., Mayr, U., Hazeltine, E. and Heuer, H. (2003) The cognitive and neural architecture of sequence representation. Psychol. Rev., 110: 316–339.

Marcel, A.J. (1993) Slippage in the unity of consciousness. Experimental and Theoretical Studies in Consciousness (Ciba Foundation Symposium 174). Wiley, New York, pp. 168–186.

Merikle, P.M. and Reingold, E.M. (1990) Recognition and lexical decision without detection: unconscious perception? J. Exp. Psychol. Hum. Percept. Perform., 16: 574–583.

Merikle, P.M. and Reingold, E.M. (1991) Comparing direct (explicit) and indirect (implicit) measures to study unconscious memory. J. Exp. Psychol. Learn, 17: 224–233.

Neal, A. and Hesketh, B. (1997) Episodic knowledge and implicit learning. Psychon. B Rev., 4: 24–37.

Nissen, M.J. and Bullemer, P. (1987) Attentional requirement of learning: Evidence from performance measures. Cognit. Psychol., 19: 1–32.

Overgaard, M. (2001) The role of phenomenological reports in experiments on consciousness. Psycoloquy, 12: 1–13.

Peigneux, P., Maquet, P., Meulemans, T., Destrebecqz, A., Laureys, S., Degueldre, C., Delfiore, G., Aerts, J., Luxen, A., Franck, G., Van der Linden, M. and Cleeremans, A. (2000) Striatum forever, despite sequence learning variability: A random effect analysis of PET data. Hum. Brain Mapp., 10: 179–194.

Perruchet, P. and Amorim, M.A. (1992) Conscious knowledge and changes in performance in sequence learning: Evidence against dissociation. J. Exp. Psychol. Learn, 18: 785–800.

Perruchet, P., Bigand, E. and Benoit-Gonnin, F. (1997) The emergence of explicit knowledge during the early phase of learning in sequential reaction time. Psychol. Res., 60: 4–14.

Perruchet, P. and Gallego, J. (1993) Associations between conscious knowledge and performance in normal subjects: Reply to Cohen and Curran (1993) and Willingham, Greeley, and Bardone (1993). J. Exp. Psychol. Learn, 19: 1438–1444.

Rabbitt, P. (2002) Consciousness is slower than you think. Q J. Exp. Psychol.-A, 55: 1081–1092.

Reed, J. and Johnson, P. (1994) Assessing implicit learning with indirect tests: Determining what is learned about sequence structure. J. Exp. Psychol. Learn, 20: 585–594.

Reingold, E.M. and Merikle, P.M. (1988) Using direct and indirect measures to study perception without awareness. Percept. Psychophys., 44: 563–575.

Reingold, E.M. and Merikle, P.M. (1990) On the inter-relatedness of theory and measurement in the study of unconscious processes. Mind and Language, 5: 9–28.

Richardson-Klavehn, A., Gardiner, J.M. and Java, R.I. (1996) Memory: Task dissociations, process dissociations, and dissociation of consciousness. In: Underwood G. (Ed.), Implicit Cognition. Oxford University Press, Oxford, pp. 85–158.

Rollman, G.B. and Nachmias, J. (1972) Simultaneous detection and recognition of chromatic flashes. Percept. Psychophys., 44: 563–575.

Shanks, D.R. and Johnstone, T. (1998) Implicit knowledge in sequential learning tasks. In: Stadler M.A. and Frensch P.A. (Eds.) Handbook of Implicit Learning, Vol. 16. Sage Publications, pp. 533–572.

Shanks, D.R. and Johnstone, T. (1999) Evaluating the relationship between explicit and implicit knowledge in a serial reaction time task. J. Exp. Psychol. Learn, 25: 1435–1451.

Shanks, D.R. and St. John, M.F. (1994) Characteristics of dissociable human learning systems. Behav. Brain Sci., 17: 367–447.

Stadler, M.A. (1997) Distinguishing implicit and explicit learning. Psychon. B Rev., 4: 56–62.

Toth, J.P., Reingold, E.M. and Jacoby, L.L. (1994) Towards a redefinition of implicit memory: Process dissociations following elaborative processing and self-generation. J. Exp. Psychol. Learn, 20: 290–303.

Wilkinson, L. and Shanks, D.R. (2004) Intentional control and implicit sequence learning. J. Exp. Psychol. Learn, 30: 354–369.

Willingham, D.B., Greeley, T. and Bardone, A.M. (1993) Dissociation in a serial response time task using a recognition measure: Comment on Perruchet and Amorim (1992). J. Exp. Psychol. Learn,, 19: 1424–1430.

Willingham, D.B., Greenberg, A.R. and Cannon Thomas, R. (1997) Response-to-stimulus interval does not affect implicit motor sequence learning but does affect performance. Mem. Cognit., 25: 534–542.

Willingham, D.B., Nissen, M.J. and Bullemer, P. (1989) On the development of procedural knowledge. J. Exp. Psychol. Learn, 15: 1047–1060.

S. Laureys (Ed.)
Progress in Brain Research, Vol. 150
ISSN 0079-6123

CHAPTER 7

Computational correlates of consciousness

Axel Cleeremans*

Cognitive Science Research Unit, Université Libre de Bruxelles CP 191, Avenue F.-D. Roosevelt, 50 1050 Brussels, Belgium

Abstract: Over the past few years numerous proposals have appeared that attempt to characterize consciousness in terms of what could be called its computational correlates: Principles of information processing with which to characterize the differences between conscious and unconscious processing. Proposed computational correlates include architectural specialization (such as the involvement of specific regions of the brain in conscious processing), properties of representations (such as their stability in time or their strength), and properties of specific processes (such as resonance, synchrony, interactivity, or information integration). In exactly the same way as one can engage in a search for the neural correlates of consciousness, one can thus search for the computational correlates of consciousness. The most direct way of doing is to contrast models of conscious versus unconscious information processing. In this paper, I review these developments and illustrate how computational modeling of specific cognitive processes can be useful in exploring and in formulating putative computational principles through which to capture the differences between conscious and unconscious cognition. What can be gained from such approaches to the problem of consciousness is an understanding of the function it plays in information processing and of the mechanisms that subtend it. Here, I suggest that the central function of consciousness is to make it possible for cognitive agents to exert flexible, adaptive control over behavior. From this perspective, consciousness is best characterized as involving (1) a graded continuum defined over quality of representation, such that availability to consciousness and to cognitive control correlates with properties of representation, and (2) the implication of systems of meta-representations.

Introduction

In a surprisingly lucid passage, Sigmund Freud (1949), reflecting on the prospects of developing a scientific approach to psychological phenomena, wrote the following:

> We know two kinds of things about what we call our psyche (or mental life): firstly, its bodily organ and scene of action, the brain (or nervous system) and, on the other hand, our acts of consciousness, which are immediate data and cannot be further explained by any sort of description. Everything that lies in between is unknown to us, and the data do not include any direct relation between these two terminal points of our knowledge. If it existed, it would at the most afford an exact localization of the processes of consciousness and would give us no help towards understanding them.

Freud's insightful but rather pessimistic thoughts about the possibility of developing a "Science of Consciousness" thus illustrates the most fundamental problem that cognitive neuroscience must confront in this context: That of establishing

*Corresponding author. Tel.: +32 2 650 32 96; Fax: +32 2 650 22 09; E-mail: axcleer@ulb.ac.be

DOI: 10.1016/S0079-6123(05)50007-4

causal relationships between fundamentally private, subjective states (what Freud calls "our acts of consciousness") on the one hand, and objective, observable states (e.g., behavioral and neural states) on the other hand.

This program of establishing direct correspondences between subjective and objective states now finds a contemporary echo in the unfolding search for the "Neural Correlates of Consciousness (NCC)." The expression "Neural Correlates of Consciousness" was first used by Crick and Koch (1990) and has since attracted, as an empirical program, the attention of a large community of researchers — from scientists to philosophers alike (see Metzinger, 2000, for an extensive collection of relevant contributions).

According to Chalmers (2000, p. 31), a "neural correlate of consciousness" is "a minimal neural system N such that there is a mapping from states of N to states of consciousness, where a given state of N is sufficient, under conditions C, for the corresponding state of consciousness".

Candidate's NCC, to mention just a few of those listed in Chalmers (2000), include, for instance, 40-Hz oscillations in the cerebral cortex (Crick and Koch, 1990; also Ribary, this volume; John, this volume), reentrant loops in thalamocortical systems (Edelman, 1989; also see Tononi, this volume), neural assemblies bound by *N*-methyl-D-asparate (NMDA) (Flohr, 1985; also see Greenfield, this volume), or extended reticular-thalamic activation systems (Newman and Baars, 1993, also see Baars, this volume).

Chalmers (2000) is quick to point out several potential shortcomings of this definition, such as the facts that there might not be a single NCC, NCCs might not consist of circumscribed regions of the brain, or it might be the case that some aspects of consciousness simply fail to correlate in some sense with brain activity (a view to which few would subscribe). Noë and Thompson (2004) likewise critique — but in a somewhat different direction — what they call the "matching-content doctrine," that is, the idea that the representation of a particular content in a neural system is sufficient for representation of that same content in consciousness. Specifically, Noë and Thompson aim to suggest that the search for the NCC might be misguided to the extent that it eschews the fact that conscious states cannot be analyzed independent of the environment with which the agent interacts constantly (also see O'Regan et al., this volume).

In a rather pessimistic article, Haynes and I raised similar points about the possibility of developing a "science of consciousness" (Cleeremans and Haynes, 1999). How are we to proceed, we asked, given not only that one has no clear idea of what it is exactly that one is measuring when using methods such as functional magnetic resonance imaging (fMRI), but also, and perhaps more importantly, that we lack the conceptual tools that would be necessary to develop a scientific approach to phenomenology? I do not have direct access to your mental states, and, some would argue, neither do I have perfect access to my own mental states (or if I do, I am likely to be mistaken in different ways, see Nisbett and Wilson, 1977; Dennett, 1991; Wegner, 2002).

This assessment will strike many as overly grim, and yet, the challenges are both substantial and numerous. In this respect, it is worth pointing out that renewed interest in consciousness has triggered rather unrealistic expectations in the community. Somehow, many continue to expect that there will be a single "aha" moment when an obscure neuroscientist suddenly comes up with "the" mechanism of consciousness. Needless to say, this is not going to happen: functional accounts of consciousness that take it as a starting point that it is a single, static property associated with some mental states and not with others are doomed to fail, for consciousness is neither "a single thing" nor is it static. Instead, consciousness refers to several, possibly dissociable, aspects of information processing, and it is a fundamentally dynamic, graded, process.

Despite these caveats, many have now rightfully opted for a pragmatic approach focused on the following simple assumption, namely that "for any mental state (state of consciousness) there is an associated neural state; it is impossible for there to be a change of mental state without a corresponding change in neural state" (Frith et al., 1999, p. 105).

On the basis of this rather non-controversial assumption (for materialists, at least), Frith et al. (1999, p. 107) continue by offering a straightforward

Table 1. Characterization of different experimental paradigms (adapted from Frith et al., 1999) through which to study differences between conscious and unconscious cognition in normal (clear cells) and abnormal (shaded cells) cases (see text for details)

	Perception	Memory	Action
Subjective experience change, stimulation and/or behavior remains constant	Binocular rivalry	Episodic Recall	Awareness of intention
	Hallucinations	Confabulation	Delusion of control
Stimulation changes, subjective experience remains constant	Stimulation changes without awareness	Unrecognized "old" items	Stimuli eliciting action without awareness
	Blindsight	Unrecognized items in Amnesia	Stimuli eliciting unintended action
Behavior changes, subjective experience remains constant	Correct guessing without awareness	Implicit learning	Implicit motor behavior
	Correct reaching in form-agnosia	Implicit learning in amnesia	Unintended action

canvas with which to guide the search for the neural correlates of consciousness:

> A major part of the program for studying the neural correlates of consciousness must be to investigate the difference between neural activities that are associated with awareness and those that are not.

This contrastive approach to consciousness (see Baars, 1988, 1994) now constitutes the core of many current efforts to understand the neural bases of consciousness. Frith et al., in their superb review, usefully propose an analysis of the different paradigms through which one can pursue this contrastive approach. Table 1 summarizes the different possibilities delineated by Frith and colleagues, who suggested to organize paradigms to study the "neural correlates of consciousness" in nine groups resulting from crossing two dimensions: (1) three classes of psychological processes involving knowledge of the past, present, and future — memory, perception, and action — and (2) three types of cases where subjective experience is incongruent with the objective situation — cases where subjective experience fails to reflect changes in either (a) the stimulation or (b) behavior, and (c) cases where subjective experience changes, whereas stimulation and behavior remain constant. This approach can be further applied to either normal or pathological cases.

The paradigmatic example of a situation where one seeks to identify the neural correlates of perception is binocular rivalry (see e.g., Lumer et al., 1998; Logothethis and Schall, 1989; Naccache, this volume), in which an unchanging compound stimulus consisting of two elements presented separately and simultaneously to each eye produces spontaneously alternating complete perceptions of each element. By asking participants (or certain animals) to indicate which stimulus they perceive at any moment, one can then strive to establish which regions of the brain exhibits activity that correlates with subjective experience and which do not, in a situation where the actual stimulus remains unchanged. Research on the neural correlates of implicit learning, in contrast, instantiates the reverse situation, where people's subjective experience fails to reflect the fact that they are becoming increasingly sensitive to novel information they are learning about over the course of practicing a task such as sequence learning (Cleeremans et al., 1998). Here again, by contrasting cases where learning is accompanied by conscious awareness with cases where it is not, one can strive to explore which regions of the brains subtend implicit and explicit learning, and to what degree (Destrebecqz et al., 2003; Destrebecqz and Peigneux, this volume). Literally, dozens of other studies have now followed the same logic in varied domains, as illustrated in Table 1.

However, there are reasons to claim that the search for the NCC should now be (and indeed, is) augmented by similar efforts aimed at unraveling what one could call, on the one hand, the *behavioral correlates of consciousness* (BCC), and, on the other hand, the *computational correlates of consciousness* (CCC, see Mathis & Mozer, 1996). One could thus paraphrase Frith et al.'s quote in the following manner:

> A major part of the program for studying the behavioral correlates of consciousness must be to investigate the difference between behaviors that are associated with awareness and those that are not.

and:

> A major part of the program for studying the computational correlates of consciousness must be to investigate the difference between computations that are associated with awareness and those that are not.

While what I have called the "search for the behavioral correlates of consciousness" is nothing new, the search for the computational correlates of consciousness is barely beginning. There is, however, a small community of scientists specifically interested in pursuing the goal of building "conscious machines" (Holland, 2003; Alexsander, this volume) through the development of implemented computational models aimed either at fleshing out broad theories of consciousness (Cotterill, 1998; Dehaene et al., 1998; Franklin and Graesser, 1999; Taylor, 1999; Aleksander, 2000; Sun, 2001; Perruchet and Vinter, 2003) or at providing detailed accounts of the difference between conscious and unconscious cognition (Farah et al., 1994; Mathis and Mozer, 1996; Dehaene et al., 2003; Fragopanagos, Kockelkoren and Taylor, in press; Colagrosso and Mozer, in press). Also relevant is the growing computationally oriented literature dedicated to the phenomena of implicit learning (Cleeremans et al., 1998).

A joint search for the NCC, BCC, and CCC sets up a clear multidisciplinary program for the scientific study of consciousness — one that involves systematically manipulating variables that will result in producing differences between conscious and unconscious neural states, behaviors, or computations. The latter contrast is in my view particularly important, for it may result in the identification of *computational principles* that differentiate between cognition with and without consciousness. This is the issue that I will focus on in the rest of this chapter. To do so, I will first briefly overview different existing, broad proposals with the goal of establishing how they differ from each other and on which information-processing principles they rely to account for differences between conscious and unconscious cognition. Next, I will suggest that, from a computational point of view, consciousness can be analyzed as involving two central aspects.

The first is what one could call "quality of representation" (see also Farah, 1994) — properties associated with representations in the brain or in artificial systems, such as their strength, their stability in time, or their distinctiveness. Quality of representation, by this account, determines, in a graded manner, the extent to which a particular representation becomes available to conscious experience and to cognitive control, and is viewed as a necessary condition for a particular representation to become available to consciousness. The second is the extent to which a given representation is accompanied by further (re-)representation of itself — in other words, whether the system is capable of meta-representation.

Finally, I will close with a brief discussion of a novel class of computational models, — the so-called "forward models," — and their potential in capturing many insights into the computational correlates of consciousness within a single broad computational framework. Before undertaking this analysis, however, it seems important to reflect upon the functions of consciousness. Indeed, as Taylor (1999) points out, "...without a function for consciousness, we have no clue as to a mechanism for it. Scientific modeling cannot even begin in this case; it has nothing to get its teeth into" (p. 49).

The functions of consciousness

Analyzing consciousness in terms of its underlying mechanisms first requires us to identify the functions that it may play within a cognitive system. There are several different manners in which this question can be approached depending on which aspect of consciousness one focuses on. The fact that consciousness is not a unitary concept (Zeman, this volume) is important, particularly because many recent experiments tend to treat it as though it were a "single thing", whereas it is neither a thing nor a unitary concept.[1] Block's (1995) well-known analysis is useful here as a starting point. Block distinguishes between access consciousness, phenomenal consciousness, monitoring consciousness, and self-consciousness.

Access consciousness (A-consciousness) refers to our ability to report and act on our experiences. For a person to be in an A-conscious state entails that there is a representation in that person's brain whose content is available for verbal report and for high-level processes such as conscious judgment, reasoning, and the planning and guiding of action. There is wide agreement around the idea that conscious representations differ from unconscious ones in terms of such global accessibility: Conscious representations are informationally available to multiple systems in a manner that unconscious representations are not. Accessibility is in turn viewed as serving the function of making it possible for an agent to exert flexible, adaptive control over action. Tononi (Tononi, 2003, in press; Tononi and Edelman, 1998) proposes that the main function of consciousness is to rapidly integrate a lot of information — a function that would clearly endow agents who possess this ability with an evolutionary advantage over others who lack it. In a recent overview article, Dehaene and Naccache (2001) state that "The present view associates consciousness with a unified neural workspace through which many processes can communicate. The evolutionary advantages that this system confers to the organism may be related to the increased independence that it affords." (p. 31). Dehaene and Naccache thus suggest that consciousness allows organisms to free themselves from acting out their intentions in the real world, relying instead on less hazardous simulation made possible by the neural workspace. Most existing computational models of consciousness are explicitly targeted toward capturing the computational consequences of A-consciousness rather than the phenomenal qualities associated with conscious states — Block's second concept of phenomenal consciousness.

Phenomenal consciousness (P-consciousness) refers to the qualitative nature of subjective experience: What it is like to smell a particular scent, to feel a particular pain, to remember the emotions associated with a particular event, to be a bat chasing insects at nightfall. There is no agreement concerning the putative functions of P-consciousness. Some authors argue that there is nothing to be explained, that qualia are illusory, or that they are purely epiphenomenal and hence play no causal role in information processing. For instance, O'Regan and Noë (2001) hold that qualia reflect nothing more than mastery of learned sensory–motor contingencies: What it means to consciously experience something is simply to know about the consequences of one's actions (O'Regan et al., this volume). For Dennett (1991, 2001), conscious contents merely reflect the dominance of some representations over others at some point in time — "fame in the brain", as he calls it. Others have proposed that conscious experience might serve error-correcting functions. For instance, Gray's "comparator hypothesis" (2004) states that the function of P-consciousness is to make it possible for the agent to rehearse and deliberate upon the conditions under which something unexpected happened (such as the consequences of an error). Koch proposes that the function of P-consciousness is to provide an "executive summary" to those parts of the brain involved in planning and deliberation (Crick and Koch, 1995; Koch, 2004). This executive summary is assumed to be the result of constraint satisfaction processes, and reflects the best interpretation of the current situation. Another interesting hypothesis concerning the

[1] Contrast, for instance, cases where one asks whether a subject is conscious of a single stimulus presented to her to cases where one asks what is it is like to walk in the Alps or to sample an excellent wine. Our concept of consciousness is radically different in each case.

function of conscious experience was put forward by Gregory (2003), according to whom P-consciousness might serve the function of "flagging the present", so making it possible for the agent to distinguish between actual, remembered, and anticipated states. More generally, perhaps the function of conscious experience is to associate emotional valence to the consequences of one's actions. If nothing ever is done to an agent, there seems to be little basis for learning and adapting behavior in general. On the other hand, one might also argue that it is simply misguiding to look for putative functional accounts of phenomenal consciousness since, by definition, it is what is "left over" once all functional aspects of consciousness have been accounted for.

Monitoring consciousness refers to thoughts about or awareness of one's sensations and percepts, as distinct from those sensations and percepts themselves. Functionally, some form of monitoring consciousness appears to be necessary to support adapted control over behavior, through appraisal of one's internal states and metacognition in general.

Finally, *self-consciousness* refers to thoughts about or awareness of oneself. Studying the self is a huge undertaking in and of itself, and the domain is currently witnessing fascinating developments (see e.g., Knoblich et al., 2003 for a review). It would be too long to develop this aspect of consciousness in this chapter, but a basic fact about conscious experience is simply that it would not make any sense unless there was a self-aware agent experiencing the experience. Hence, consciousness of self is clearly a very important component of what it means to be conscious (Damasio, 1999).

Having delineated a few possible functions for consciousness in its different aspects, we can now ask the following questions: What sorts of mechanisms have been proposed to fulfill these functions? What are the computational correlates of consciousness? These will be the object of the next section.

The search for the CCC

Computational models of the differences between conscious and unconscious information processing are few and far between. This is not surprising, for the challenge of exploring the mechanisms of something as complex and ill-defined as consciousness is enormous. This is also the main reason why most existing computational models of consciousness have been directed at accounting for A-consciousness as opposed to P-consciousness: The former at least receives some sort of functionalist interpretation, while the functions of the latter, if any, clearly remain controversial at this point. Monitoring- and self-consciousness, on the other hand, require accounts that necessarily involve a great deal of complexity before they can even get off the ground, and are hence challenging to explore from a computational point of view.

This being said, existing models generally fall into two classes: Overarching models — often only partially implemented — that aim to offer a general blueprint for information processing with or without consciousness on the one hand, and very specific models of particular empirical situations on the other. Each suffers from its own set of limitations (which they share with computational models in general). Overarching models are often difficult to compare with existing data because they often fail to make testable predictions. Specific models, on the other hand, can always be dismissed as convincing accounts of the mechanisms of consciousness precisely because of their limited scope. In either case, one could question the extent to which such modeling efforts are worth it, though this would clearly invalidate any scientific approach to the problem. For instance, if you assume that consciousness crucially includes properties that can never be amenable to functionalist and cognitive analyses — Chalmers' (1996) "hard problem" — then clearly such models are doomed to fail, and so would the possibility of understanding conscious experience from a third-person perspective. Some authors have also pointed out that while it might be possible to build conscious machines, we would never be able to decide whether such machines actually have experiences of any kind (Prinz, 2003).

Nevertheless, both types of models can play a substantial role in helping us converge onto a set of computational principles to characterize the differences between conscious and unconscious cognition. Identifying such principles is an important endeavor, for it would clearly make it possible to go beyond establishing mere relationships between conscious states and their neural or behavioral correlates. In other words, if we are able to define such principles, we would be in a position to address the mechanisms through which consciousness is achieved in cognitive systems.

Current theories of consciousness sometimes make very different assumptions about its underlying mechanisms. Farah (1994) distinguishes between three types of neuroscientific/computational accounts of consciousness: "privileged role" accounts, "integration" accounts, and "quality of representation" accounts. "Privileged role" accounts take their roots in Descartes' thinking and assume that consciousness depends on the activity of specific brain systems whose function it is to produce subjective experience. "Integration" accounts, in contrast, assume that consciousness only depends on processes of integration through which the activity of different brain regions can be synchronized or made coherent. Finally, "quality of representation" accounts assume that consciousness depends not on particular processes, but on particular properties of neural representations, such as their strength or their stability in time.

In a recent overview article (see also O'Brien and Opie, 1999; Atkinson et al., 2000), my co-authors and I proposed to organize computational theories of consciousness along two dimensions, as depicted in Fig. 1[2]: A process versus vehicle dimension, which opposes models that characterize consciousness in terms of specific processes operating over mental representations to models that characterize consciousness in terms of intrinsic properties of mental representations, and a specialized versus non-specialized dimension, which contrasts models that posit information-processing systems dedicated to consciousness with models for which consciousness can be associated with any information-processing system as long as this system has the relevant properties.

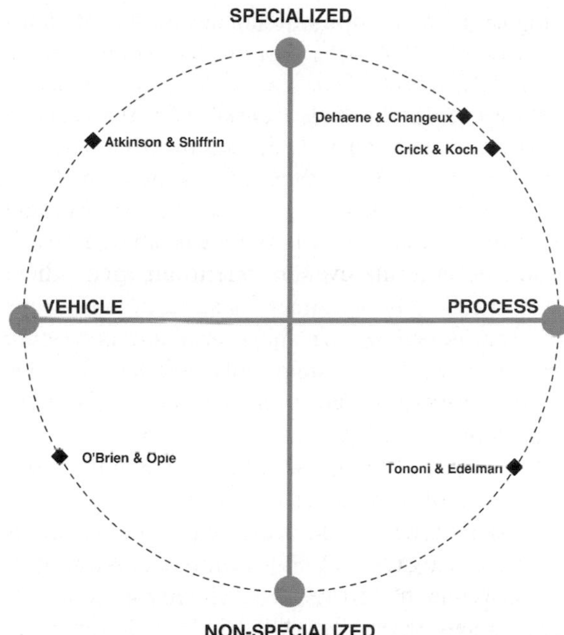

Fig. 1. A conceptual 2-D map in which to locate theories of consciousness. (Adapted from Atkinson et al., 2000.) The map is defined by two dimensions relevant to computational theories of consciousness: Whether the theory assumes the involvement of specialized structures or not (Y-axis), and whether the theory assumes that consciousness depends on properties associated with representational vehicles or with processes (X-axis).

Farah's three categories can be subsumed in this analysis in the following manner: "privileged role" models, which assume that some brain systems play a specific role in subtending consciousness, are specialized models that can be instantiated either through "vehicle" or through "process" principles. "Quality of representation", models, on the other hand, are typical vehicle theories in that they emphasize that what makes some representations available to conscious experience are properties of those representations rather than their functional role. Finally, Farah's "integration" models are examples of non-specialized theories, which can again be either instantiated in terms of the properties of the representations involved or in terms of the processes that engage these representations. Atkinson et al.'s analysis thus offers four broad

[2]Figure 1 is aimed at providing a few illustrative examples and is by no means intended to be exhaustive. Your favorite theory (or your own theory!) may thus not be on the map, which I urge you not to interpret as a suggestion that it is not important.

categories of computational accounts of consciousness.

(1) *Specialized vehicle theories* assume that consciousness depends on the properties of the representations that are located within a specialized system in the brain. An example of such accounts is Atkinson and Shiffrin's (1971) model of short-term memory, which specifically assumes that representations contained in the short-term memory store (a specialized system) only become conscious if they are sufficiently strong (a property of representations).
(2) *Specialized process theories* assume that consciousness arises from specific computations that occur in a dedicated mechanism, as in Schacter's (1989) Conscious Awareness System (CAS) model. Schacter's model assumes that the CAS's main function is to integrate inputs from various domain specific modules, and to make this information available to executive systems. It is therefore a specialized model in that it assumes that there exist specific regions in the brain whose function is to make its contents available to conscious awareness. It is a process model to the extent that any representation that enters the CAS will become available to conscious awareness in virtue of the processes that manipulate these representations, and not in virtue of properties of those representations themselves. More recent computational models of consciousness also fall into this category, most notably Dehaene and colleagues' (1998) neural workspace model and Crick and Koch's (2003) framework, both of which assume, albeit somewhat differently, that the emergence of consciousness depends on the occurrence of specific processes in specialized systems.
(3) *Non-specialized vehicle theories* include any model that posits that availability to consciousness only depends on properties of representations, regardless of where in the brain these representations exist or of which processes engage these representations. O'Brien and Opie's (1999) "connectionist theory of phenomenal experience" is the prototypical example of this category, to the extent that it specifically assumes that any stable neural representation will both be causally efficacious and form part of the contents of phenomenal experience. Mathis and Mozer (1996) likewise propose to associate consciousness with stable states in neural networks, though Mozer's more recent PIT framework (Colagrosso and Mozer, 2005) also puts emphasis on the existence of functional connectivity between different modules as critical for A-consciousness Zeki's notion of "micro-consciousness" is also an example of this type of perspective (Zeki and Bartels, 1998).
(4) *Non-specialized process theories* finally, are theories which assume that representations become conscious whenever they are engaged by certain specific processes, regardless of where these representations exist in the brain. Many recent proposals fall into this category. Examples include Tononi and Edelman's (1998) "dynamic core" model; Crick and Koch's (1995) idea that synchronous firing constitutes the primary mechanisms through which disparate representations become integrated as part of a unified conscious experience or Grossberg's (1999) characterization of consciousness as involving processes of "adaptive resonance" through which representations that simultaneously receive bottom-up and top-down activation become conscious because of their stability and strength.

There are two important caveats to this analysis. Firstly, the taxonomy is defined by how specific computational theories of consciousness characterize the difference between conscious and unconscious cognition rather than by a sharp distinction between vehicles versus processes on the one hand, and specialized versus non-specialized systems on the other. Thus, it should be clear that representation and process cannot be considered independently from each other, to the extent that the effects of particular processes will necessarily result in changes in the nature of the representations involved. For instance, processes like resonance, amplification, or reentrant processing (Lamme, 2004), all of which basically involve constraint satisfaction processes as they occur in interactive networks, will all result in stabilizing and in strengthening specific patterns of activity in the corresponding neural pathways. The distinction between specialized and non-specialized models similarly fails to be as sharp as depicted above, for there are multiple ways in which a system can be described as specialized. For instance, a system can be specialized to the extent that it involves a single "box" or cerebral region whose function it would be to make whatever contents are represented in that system conscious (no current neuroscientific theory of consciousness adopts this assumption this bluntly). On the other hand, a system can be specialized to the extent that it involves specific connectivity between different cerebral regions. Dehaene and Changeux's (2004) notion that the neural workspace relies on specific long-distance cortico-cortical connections is an example of the latter case of specialization, and so contrasts with other proposals that put less emphasis on the involvement of dedicated systems (Tononi and Edelman, 1998).

Secondly, several proposals also tend to be somewhat more hybrid, instantiating features and ideas from several of the categories described by Atkinson et al. Baars' influential "global workspace" model (Baars, 1988, this volume), for instance, incorporates features from specialized process models as well as from non-specialized vehicles theories, to the extent that the model assumes that consciousness involves a specialized system (the global workspace), but also characterizes conscious states in terms of the properties associated with their representations (i.e., global influence and widespread availability) rather than in terms of the processes that operate on these representations. Likewise, Dehaene et al. (1998) assume that consciousness depends on (1) *active firing*, which can be construed as a property of representation, (2) *long-distance connectivity* (a specialized system), and (3) dynamic mobilization, a selective process depending on simultaneous bottom-up and top-down activation of the representations contained in the linked modules. Thus, this model acknowledges both the existence of specific, dedicated mechanisms to support consciousness as well as specific properties of representations brought about by particular processes (e.g., dynamic mobilization).

Lastly, Tononi and Edelman's (1998) analysis recognizes the importance of the thalamo-cortical system in subtending consciousness (and could hence be viewed as specialized theory), but reaches this conclusion based on computational principles that are explicitly non-specialized to the extent that they could occur in any system properly structured.

A final comment on this analysis is that pure vehicle theories of consciousness remain problematic from a computational point of view, for they fail to make it clear how any aspect of consciousness could be produced exclusively by properties of the representational vehicles involved in information processing. Simply equating consciousness with stability in time (see, e.g., O'Brien and Opie, 1999), for instance, would not only force us to consider many physical systems to be conscious to some degree (thus raising the specter of panpsychism), but also appears to eschew any sort of computational explanation short of resorting to hitherto unknown causal properties of neural patterns of activity.

Toward computational principles for the distinction between conscious and unconscious cognition

What can we conclude from this brief overview of current computational approaches to consciousness? A salient point of agreement shared by several of the most popular current theories is that all such models, regardless of whether they assume specialized or non-specialized mechanisms, and regardless of whether they focus primarily on vehicles or on processes, converge toward assuming the following: Conscious representations differ from unconscious representations in that the former are endowed with certain properties such as their stability in time, their strength, or their distinctiveness. Cleeremans (Cleeremans and Jiménez, 2002; forthcoming) proposes the following definitions for these properties:

Stability in time refers to how long a representation can be maintained active during processing. There are many indications that different neural systems involve representations that differ along this dimension. For instance, the prefrontal cortex, which plays a central role in working memory (Baddeley, 1986), is widely assumed to involve circuits specialized in the formation of the enduring representations needed for the active maintenance of task-relevant information (Frank et al., 2001; Norman and O'Reilly, 2001). Stability of representation is clearly related to availability to consciousness, to the extent that consciousness takes time. For instance, the brief stimuli associated with subliminal presentation will result in weaker representations than supraliminal presentation does.

Strength of representation simply refers to how many processing units are involved in a given representation, and to how strongly activated these units are. Strength can also be used to characterize the efficiency of a an entire processing pathway, as in the Stroop model of Cohen et al. (1990). Strong activation patterns exert more influence on ongoing processing than weak patterns, and are most clearly associated with automaticity, to the extent that they dominate ongoing processing.

Finally, *distinctiveness* of representation refers to the extent of overlap that exists between representations of similar instances. Distinctiveness, or discreteness, has been hypothesized as the main dimension through which cortical and hippocampal representations differ (McClelland et al., 1995; O'Reilly and Munakata, 2000), with the latter becoming active only when the specific conjunctions of features that they code for are active themselves. In the context of the terminology associated with attractor networks, this contrast would thus be captured by the difference between attractors with a wide basin of attraction, which will tend to respond to a large number of inputs, and attractors with a narrow basin of attraction, which will only tend to respond to a restricted range of inputs. The notion also overlaps with the difference between episodic and semantic memory, that is, the difference between knowing that Brutus the dog bit you yesterday and knowing that all dogs are mammals: There is a sense in which the distinctive episodic trace, because it is highly specific to one particular experience, is more accessible and more explicit than the semantic information that all dogs share a number of characteristic features. This latter knowledge can be made explicit when the task at hand requires it, but is only normally conveyed implicitly (as a presupposition) by statements about or by actions directed toward dogs.

Strong, stable, and distinctive representations are thus explicit representations, at least in the sense put forward by Koch (2004): They indicate what they stand for in such a manner that their reference can be retrieved directly through processes involving low computational complexity (see also Kirsh, 1991, 2003). Conscious representations, in this sense, are explicit representations that have come to play, through processes of learning, adaptation, and evolution, the functional role of denoting a particular content for a cognitive system. Importantly, quality of representation should be viewed as a graded dimension.

The analysis presented above resonates well with recent computational models of overall cerebral function. O'Reilly and colleagues (McClelland et al., 1995; O'Reilly and Munakata, 2000; Atallah et al., 2004), for instance, have recently proposed that different regions of the brain have evolved to solve different — and incompatible — computational problems by using different representational formats and different learning regimes (McClelland et al., 1995). In their "tripartite" proposal, the brain is organized in three broad interacting systems: The hippocampus (HC), prefrontal cortex/basal ganglia (FC), and posterior cortex (PC). In this framework, each system uses similar, but not identical learning mechanisms and representational formats. The main function of HC is to rapidly learn about specific novel facts (episodic memory). Function of PC, in contrast, is to learn about the statistical regularities shared by many exemplars of a given domain (semantic memory). Finally, the main function of FC is to maintain information in an active state (active maintenance, subtending working memory) and to rapidly switch between active representations. Achieving each of these functions require different (but germane) learning mechanisms and different representational formats. Thus, HC uses the sparse, conjunctive representations necessary to avoid catastrophic interference, and a high learning rate that makes it possible to rapidly bind together the various elements of the current percept. PC, in contrast, slowly accumulates information over largely overlapping, distributed representations, so that broad semantic knowledge can progressively emerge over learning and development. Finally, FC is characterized by self-sustaining representational systems involving the recurrent connectivity necessary for active maintenance as well as the gating mechanisms necessary for rapid switching.

The three systems also differ from each other in terms of processing and learning mechanisms. Thus, O'Reilly and Munakata (2000) argue that the functions typically attributed to FC (i.e., working memory, inhibition, executive control, and monitoring or evaluation of ongoing behavior) require "activation-based processing", characterized by mechanisms of active maintenance through which representations can remain strongly activated for long periods of time as well as rapidly updated so as to make it possible for these representations to modulate processing elsewhere in the brain. Note how this is consistent with Crick and Koch's (2003) notion that "the front of the brain is looking at the back." Because of these properties, frontal representations are thus more accessible to verbalization and other reporting systems.[3] To this, they oppose "weight-based processing", characteristic of PC, in which knowledge is encoded directly by the pattern of connectivity between processing units and hence tends to remain tacit to the extent that this knowledge only manifests itself through the effects it exerts on ongoing processing rather than through the form of representations themselves.

In terms of learning mechanisms, O'Reilly and Munakata (2000) also propose an interesting distinction between model learning (Hebbian learning) and task learning (error-driven learning). Again, their argument is framed in terms of the different computational objectives each of these types of learning processes fulfills: Capturing the statistical structure of the environment so as to develop appropriate models of it on the one hand, and learning specific input–output mappings so as to solve specific problems (tasks) in accordance with one's goals on the other hand. There is a very nice mapping between this distinction — expressed in terms of the underlying biology and a consideration of computational principles — and the distinction between incidental learning and intentional learning on the other hand.

It is tempting to relate the different aspects of the quality of a representation delineated earlier with the functions of each system identified by O'Reilly and colleagues (McClelland et al., 1995; O'Reilly and Munakata, 2000; Atallah et al., 2004). Stability in time is what most saliently characterizes FC representations. Distinctiveness is a property most clearly associated with HC. Finally, PC representations are best characterized by their strength. Importantly, in this computational framework, there is no single system that is uniquely associated with the occurrence of conscious representations. Rather, conscious representations emerge as a result of the joint involvement of each system in ongoing processing.

Stability, strength, or distinctiveness can be achieved by different means. They can result, for instance, from the simultaneous top-down and bottom-up activation involved in the so-called "reentrant processing" (Lamme, 2004), from processes of "adaptive resonance" (Grossberg, 1999), from processes of "integration and differentiation" (Edelman and Tononi, 2000), or from contact with the neural workspace, brought about by "dynamic mobilization" (Dehaene and Naccache, 2001). It is important to realize that the ultimate effect of any of these putative mechanisms is to make the target representations stable, strong, and distinctive. These properties can further be envisioned as involving graded or dichotomous dimensions.

Hence, a first important computational principle through which to distinguish between conscious and unconscious representations is the following:

> "Availability to consciousness depends on quality of representation, where

[3]In this respect, O'Reilly and Munakata (2000) rightfully point out that a major puzzle is to understand how the FC comes to develop what they call a "rich vocabulary of frontal activation-based processing representations with appropriate associations to corresponding posterior-cortical representations" (p. 382).

> quality of representation is a graded dimension defined over stability in time, strength, and distinctiveness."

While high-quality representation thus appears to be a necessary condition for their availability to consciousness, one should ask, however, whether it is a sufficient condition. Cases such as hemineglect, blindsight (Weiskrantz, 1986), or, in normal subjects, attentional blink phenomena (Shapiro et al., 1997), or some instances of change blindness (Simons and Levin, 1997), for instance, suggest that quality of representation alone does not suffice, for even strong patterns can fail to enter conscious awareness unless they are somehow attended. Likewise, merely achieving stable representations in an artificial neural network, for instance, will not make this network conscious in any sense — this is the problem pointed out by Clark and Karmiloff-Smith (1993) about the limitations of what they called first-order networks: In such networks, even explicit knowledge (e.g., a stable pattern of activation over the hidden units of a standard back-propagation network that has come to function as a "face detector") remains knowledge that is in the network as opposed to knowledge for the network. In other words, such networks might have learned to be informationally sensitive to some relevant information, but they never know that they possess such knowledge. Thus, the knowledge can be deployed successfully through action, but only in the context of performing some particular task.

Hence, it could be argued that it is a defining feature of consciousness that when one is conscious of something, one is also, at least potentially so, conscious that one is conscious of being in that state. This is the gist of the so-called higher order thought (HOT) theories of consciousness (Rosenthal, 1997), according to which a mental state is conscious when the agent entertains, in a non-inferential manner, thoughts to the effect that it currently is in that mental state. Importantly, for Rosenthal, it is in virtue of current HOTs that the target first-order representations become conscious. Dienes and Perner (1999) have developed this idea by analyzing the implicit–explicit distinction as reflecting a hierarchy of different manners in which the representation can be explicit. Thus, a representation can explicitly indicate a property (e.g., "yellow"), predication to an individual (the flower is yellow), factivity (it is a fact and not a belief that the flower is yellow) and attitude (I know that the flower is yellow). Fully conscious knowledge is thus knowledge that is "attitude-explicit".

This analysis suggests that another important principle that differentiates between conscious and unconscious cognition is the extent to which a given representation endowed with the proper properties (stability, strength, distinctiveness) is itself the target of meta-representations. Note that meta-representations are *de facto* assumed to play an important role in any theory that assumes interactivity. Indeed, for processes such as resonance, amplification, integration, or dynamic mobilization to operate, one minimally needs to assume two interacting components: A system of first-order representations, and a system of meta-representations that take first-order representations as their input.

Hence, a second important computational principle through which to distinguish between conscious and unconscious representations is the following:

> Availability to consciousness depends on the extent to which a representation is itself an object of representation for further systems of representation.

It is interesting to consider under which conditions a representation will remain unconscious based on combining these two principles (Cleeremans, forthcoming). There are at least four possibilities. Firstly, knowledge that is embedded in the connection weights within and between processing modules can never be directly available to conscious awareness and control. This is simply a consequence of the fact that consciousness necessarily involves representations (patterns of activation over processing units). The knowledge embedded in connection weights will, however, shape the representations that depend on it, and its effects will therefore be detectable — but only indirectly, and only to the extent that these effects are sufficiently marked in the corresponding representations. This is equivalent to Dehaene's principle of "active firing" (Dehaene and Changeux, 2004).

Secondly, to enter conscious awareness, a representation needs to be of sufficiently high quality in terms of strength stability in time, or distinctiveness. Weak representations are therefore poor candidates to enter conscious awareness. This, however, does not necessarily imply that they remain causally inert, for they can influence further processing in other modules, even if only weakly so. This forms the basis for a host of subthreshold effects, including subliminal priming, for instance.

Thirdly, a representation can be strong enough to enter conscious awareness, but fail to be associated with relevant meta-representations. There are thus many opportunities for a particular conscious content to remain, in a way, implicit, not because its representational vehicle does not have the appropriate properties, but because it fails to be integrated with other conscious contents. Dienes and Perner (2003) offer an insightful analysis of the different ways in which what I have called high-quality representations can remain implicit. Likewise, phenomena such as inattentional blindness (Mack and Rock, 1998) or blindsight (Weiskrantz, 1986) also suggest that high-quality representations can nevertheless fail to reach consciousness, not because of their inherent properties, but because they fail to be attended to or because of functional disconnection with other modules.

Finally, a representation can be so strong that its influence can no longer be controlled — automaticity. In these cases, it is debatable whether the knowledge should be taken as genuinely unconscious, because it can certainly become fully conscious as long as appropriate attention is directed to them, but the point is that such very strong representations can trigger and support behavior without conscious intention and without the need for conscious monitoring of the unfolding behavior.

Forward models

How might one go about capturing intuitions about the importance of both quality of representation and of meta-representations in the form of a

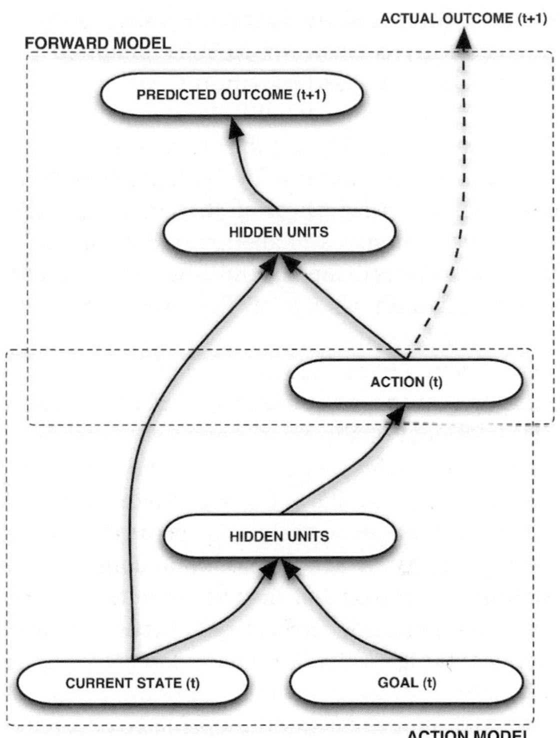

Fig. 2. A Forward Model. Two interconnected networks interact continuously: The action (inverse) model, the task of which is to produce appropriate actions given a representation of the current state and a goal (an intention), and the forward model, the task of which is to anticipate the sensory consequences (the next state) resulting from the model's actions.

computational model? There is an extremely interesting class of models that might provide a good starting point for exploring the computational principles described above (Fig. 2). These models are called "forward models" (Jordan and Rumelhart, 1992) and have been applied mostly in the domain of motor control so far (Miall and Wolpert, 1996; Jordan and Wolpert, 1999). Many control problems (and acting adaptively is the control problem per excellence) are difficult because they require solving two separate problems: (1) learning about the effects of particular actions on the environment, that is, developing a model of the system one is attempting to control (the "forward" model), and (2) learning which particular actions to take so as to achieve a desired goal, that is, learning how to control the system (the "inverse" problem). Forward models make it possible to solve both problems simultaneously. To do so, they generally consist of two interconnected networks. The first takes as input a goal and a description of the current state as input, and produces actions. The second, that is the forward model, takes the response of the first network (an action) and a description of the current state as input, and produces a prediction of how the to-be-controlled system (the "plant", in control theory parlance) would change if the produced action were carried out.

Crucially, the forward component of the model necessarily turns, as a result of training, into an internal model of the environment with which the network as a whole interacts. This sort of model can thus form the basis for a complex system of meta-representations that takes perceptual states and self-produced actions as input. It is also interesting to note that consistently with enactive and embodied perspectives on consciousness (Varela et al., 1991; O'Regan and Noë, 2001; Clark, 2002; Noë, 2005; O'Regan et al., this volume), this model is totally dependent on action: Not only will it be shaped by the sorts of actions the model can enact on its environment, but it would not even be able to bootstrap itself were the system as a whole unable to act.

The fact that sophisticated internal models emerge as a result of the perception–action–anticipation loop that the system implements becomes particularly interesting when one additionally considers (1) that socialized agents not only interact with physical environments, but also with other agents, and (2) that agents also interact with themselves by recycling their expectations about the consequences of their own actions as perceptual input. The main implication of the first point is that a forward model that interacts with other agents will end up developing a model of the internal states of those agents (their "state of mind", so to speak). The main implication of the second point is that we now have a mechanism through which to flesh out the idea that thought is simulation (Hesslow, 2002; Grush, 2004). When combined, however, the implications of these two points become particularly stimulating, for they suggest a mechanism through which representations of self could emerge out of an agent's understanding of the internal states of other agents (Cleeremans, forthcoming) — an idea already hinted at by Rumelhart et al. (1986).

Several authors have recently begun to use such models as the cornerstone of theories in rather disparate domains ranging from motor behavior to cultural cognition and the development of theory of mind (Wolpert et al., 1998; Frith et al., 2000; Grush, 2004; Hesslow, 2002; Holland and Goodman, 2003; Taylor, 2003; Wolpert et al., 2004). Frith and colleagues (2000), for instance, have proposed to analyze some of the symptoms of schizophrenia (i.e., delusions of control) or autism through lesions at various sites in the different components of forward models. Taylor's (1999) CODAM model is built around the same assumptions (also see Aleksander, this volume). Miall (2003) noted the connection between such models and the mirror system discovered by Rizzolati and colleagues (1996). Forward models thus appear to be one of the most promising avenues for further exploration of the CCC, for they suggest a possible integrated functional account of different aspects of conscious experience — both low-level and high-level — as they occur in a system that is tightly coupled with its environment and with other agents.

Discussion and conclusions

In this paper, I have offered a survey of some recent computational models of consciousness, with the overall goal of suggesting that the unfolding search for the NCC should be augmented by a search for the CCC. I have suggested that whether a representation becomes available to consciousness depends on both properties associated with the representation (strength, stability, distinctiveness) and properties associated with the mechanisms through which the representation is redescribed in further, meta-representational systems.

An important benefit of engaging in a search for the CCC is that traditional dichotomies in the cognitive neurosciences (declarative versus procedural memory; implicit versus explicit learning; conscious versus unconscious perception, and so on) are now progressively replaced by accounts that take it as a starting point that such distinctions, rather than being set in stone and subtended by dedicated systems, instead emerge out of the interactions between different regions of the brain that have evolved to solve particular computational problems characterized by the fact that they are incompatible with each other. This focus on function and on mechanisms will undoubtedly contribute to naturalize consciousness. Architectures such as the forward models described in the previous section, while they remain very abstract, offer an intriguing avenue for further research in this direction.

In conclusion, a few pending issues relevant to the search for the CCC:

1. *Should consciousness be viewed as a graded or as an all-or-none phenomenon?* Some computational theories of consciousness, in particular global workspace models, assume that once a representation has entered the workspace, it is fully conscious. Dehaene specifically refers to this process as "ignition", and accordingly predicts that all measures of conscious awareness should systematically be strongly associated with each other (Dehaene et al., 1998, 2003; Dehaene and Naccache, 2001; Dehaene and Changeux, 2004). In this view, consciousness is thus an all-or-none phenomenon. Other frameworks, in contrast, predict that consciousness is fundamentally graded (Cleeremans and Jiménez, 2002; Moutoussis and Zeki, 2002; Lamme, 2004). While there is a clear sense in which one is either aware or unaware of a stimulus (i.e., I perceive the stimulus or I do not), there are also other cases where there is a clear sense of gradedness in conscious experience (e.g., ambient noises, for instance, or perhaps chronic pains). Perceptual awareness also seems to depend in a graded manner on action systems; Marcel (1993) likewise suggests that it is far from being all-or-none. Note that it might also be the case that consciousness is both graded and all-or-none: Any complex system will exhibit non-linearities, and the physical word is replete with cases where continuous, graded changes in some dimension result in abrupt changes in some other dimension (e.g., continuous changes in the temperature of a body of water result in a change of state, say from liquid to solid).
2. *What is the relationship between attention and conscious awareness? What is the nature of the distinction between phenomenal and access consciousness?* Whether attention is necessary for consciousness or not remains a point of debate. Note that this debate is really one about how we should think about what best characterizes conscious states. Some authors take it that unattended perceptual states should simply be considered as unconscious (Dehaene and Changeux, 2004), whereas others consider that such states can form part of the global phenomenology of a conscious subject even when unattended (O'Brien and Opie, 1999; Lamme, 2004). Defenders of the first perspective put more emphasis on the processes (access by systems of meta-representations), while defenders of the second put more emphasis on properties of representational vehicles themselves (strength, stability, distinctiveness). This is related to the distinction between A- and P-consciousness, which Block (1997) describes as involving a battle between biological and computational approaches to the mind. Whether A- and P-consciousness should be taken as different kinds of consciousness or whether they constitute points on a continuum thus remains an object of debate.
3. *What is the function of meta-representational systems?* While some functions of meta-representations are clear (e.g., monitoring and control), it is nevertheless challenging to build computational models that develop "interesting" (i.e., rich, structured) meta-representations. As suggested by the discussion of forward models, the difficulty arises likely from the fact that computational models are often developed in isolation rather than in interaction with other agents. However, one probable function of meta-representations is that they are necessary to communicate one's internal states to others, and to infer internal

states from the observation of others' behavior. Building models that acknowledge this extended character of consciousness is certainly one of the promising avenues of research in the context of the search for the CCC.

Acknowledgments

Axel Cleeremans is a Senior Research Associate with the National Fund for Scientific Research (Belgium). This work was supported by an institutional grant from the Université Libre de Bruxelles to Axel Cleeremans and by grant HPRN-CT-1999-00065 from the European Commission. I thank an anonymous referee for very useful and detailed comments on a previous version of this article. I am afraid that lack of time prevented me from following all of this referee's suggestions, but they are duly noted.

References

Aleksander, I. (2000) How to Build a Mind. weidenfeld and Nicolson, London, UK.

Alexsander, I. (this volume) Machine consciousness.

Atallah, H., Frank, M.J. and O'Reilly, R.C. (2004) Hippocampus, cortex, and basal ganglia: insights from computational models of complementary learning systems. Neurobiology of Learning and Memory, 82: 253–267.

Atkinson, A.P., Thomas, M.S.C. and Cleeremans, A. (2000) Consciousness: mapping the theoretical landscape. Trends Cogn. Sci., 4(10): 372–382.

Atkinson, R.C. and Shiffrin, R.M. (1971) The control of short-term memory. Sci. Am., 224: 82–90.

Baars, B.J. (1988) A Cognitive Theory of Consciousness. Cambridge University Press, Cambridge.

Baars, B. J. (1994) A thoroughly empirical approach to consciousness, from http://psyche.cs.monash.edu.au/v1/psyche-1-06-baars.html

Baddeley, A.D. (1986) Working Memory. Oxford University Press, New York, NY.

Block, N. (1995) On a confusion about a function of consciousness. Behav. Brain Sci., 18: 227–287.

Block, N. (1997) Biology versus computation in the study of consciousness. Behav. Brain Sci., 20(1): 159.

Chalmers, D.J. (1996) The Conscious Mind: In Search of a Fundamental Theory. Oxford University Press, Oxford.

Chalmers, D.J. (2000) What is a neural correlate of consciousness? In: Metzinger T. (Ed.), Neural Correlates of Consciousness. Empirical and Conceptual Questions. MIT Press, Cambridge, MA, pp. 17–39.

Clark, A. (2002) Being There: Putting Brain, Body, and World Together Again. MIT Press, Cambridge, MA.

Clark, A. and Karmiloff-Smith, A. (1993) The cognizer's innards: a psychological and philosophical perspective on the development of thought. Mind Lang., 8: 487–519.

Cleeremans, A. (forthcoming) Being Virtual. Oxford University Press, Oxford, UK.

Cleeremans, A., Destrebecqz, A. and Boyer, M. (1998) Implicit learning: news from the front. Trends Cogn. Sci., 2: 406–416.

Cleeremans, A. and Haynes, J.-D. (1999) Correlating consciousness: a view from empirical science. Rev. Int. Philos., 53: 387–420.

Cleeremans, A. and Jiménez, L. (2002) Implicit learning and consciousness: a graded, dynamic perspective. In: French R.M. and Cleeremans A. (Eds.), Implicit Learning and Consciousness: An Empirical, Computational and Philosophical Consensus in the Making? Psychology Press, Hove, UK, pp. 1–40.

Cohen, A., Dunbar, K. and McClelland, J.L. (1990) On the control of automatic processes: a parallel distributed processing account of the Stroop effect. Psych. Rev., 97: 332–361.

Colagrosso, M.D. and Mozer, M.C. (2005) Theories of access consciousness. In: Saul L.K., Weiss Y. and Bottou L. (Eds.), Advances in neural information processing systems, vol. 17. MIT Press, Cambridge, MA, pp. 289–296.

Cotterill, R. (1998) Enchanted Looms. Conscious Networks in Brains and Computers. Cambridge University Press, Cambridge, UK.

Crick, F.H.C. and Koch, C. (1990) Towards a neurobiological theory of consciousness. Semin. Neuros., 2: 263–275.

Crick, F.H.C. and Koch, C. (1995) Are we aware of neural activity in primary visual cortex? Nature, 375: 121–123.

Crick, F.H.C. and Koch, C. (2003) A framework for consciousness. Nat. Neuros., 6(2): 119–126.

Damasio, A. (1999) The Feeling of What Happens: Body and Emotion in the Making of Consciousness. Harcourt Brace and Company, New York, NY.

Dehaene, S. and Changeux, J.-P. (2004) Neural mechanisms for access to consciousness. In: Gazzaniga M. (Ed.), The Cognitive Neurosciences (3rd ed.). New York, WW Norton, pp. 1145–1157.

Dehaene, S., Kerszberg, M. and Changeux, J.-P. (1998) A neuronal model of a global workspace in effortful cognitive tasks. Proc. Natl. Acad. Sci. USA, 95(24): 14529–14534.

Dehaene, S. and Naccache, L. (2001) Towards a cognitive neuroscience of consciousness: basic evidence and a workspace framework. Cognition, 79: 1–37.

Dehaene, S., Sergent, C. and Changeux, J.-P. (2003) A neuronal network model linking subjective reports and objective physiological data during conscious perception. Proc. Natl. Acad. Sci. USA, 100(14): 8520–8525.

Dennett, D.C. (1991) Consciousness Explained. Little, Brown and Co, Boston, MA.

Dennett, D.C. (2001) Are we explaining consciousness yet? Cognition, 79: 221–237.

Destrebecqz, A., Peigneux, P., Laureys, S., Degueldre, C., Del Fiore, G., Aerts, J., Luxen, A., Van der Linden, M., Cleeremans, A. and Maquet, P. (2003) Cerebral correlates of explicit sequence learning. Cognitive Brain Res, 16(3): 391–398.

Destrebecqz and Peigneux. (this volume) Methods for studying unconscious learning.

Dienes, Z. and Perner, J. (1999) A theory of implicit and explicit knowledge. Behav. Brain Sci., 22: 735–808.

Dienes, Z. and Perner, J. (2003) Unifying consciousness with explicit knowledge. In: Cleeremans A. (Ed.), The Unity of Consciousness: Binding, Integration, and Dissociation. Oxford University Press, Oxford, UK, pp. 214–232.

Edelman, G.M. (1989) The Remembered Present: A Biological Theory of Consciousness. Basic Books, New York, NY.

Edelman, G.M. and Tononi, G. (2000) Consciousness. How Matter Becomes Imagination. Penguin Books, London.

Farah, M.J. (1994) Visual perception and visual awareness after brain damage: a tutorial overview. In: Umiltà C. and Moscovitch M. (Eds.), Attention and Performance XV: Conscious and Nonconscious Information Processing. MIT Press, Cambridge, MA, pp. 37–76.

Farah, M.J., O'Reilly, R.C. and Vecera, S.P. (1994) Dissociated overt and covert recognition of as an emergent property of a lesioned neural network. Psych. Rev., 100: 571–588.

Flohr, H. (1985) Sensations and brain processes. Behav. Brain Res., 71: 157–161.

Fragopanagos, N., Kockelkoren, S. and Taylor, J. G. (in press) Modelling the attentional blink. Cognitive Brain Research.

Frank, M.J., Loughry, B. and O'Reilly, R.C. (2001) Interactions between frontal cortex and basal ganglia in working memory: a computational model. Cognitive, Affect. Behav. Neuros., 1(2): 137–160.

Franklin, S. and Graesser, A.C. (1999) A software agent model of consciousness. Conscious. Cogn., 8: 285–305.

Freud, S. (1949) An Outline of Psychoanalysis (J. Strachey, Trans.). Hogarth Press, London.

Frith, C.D., Blakemore, S.-J. and Wolpert, D.M. (2000) Explaining the symptoms of schizophrenia: abnormalities in the awareness of action. Brain Res. Rev., 31: 357–363.

Frith, C.D., Perry, R. and Lumer, E. (1999) The neural correlates of conscious experience: an experimental framework. Trends Cogn. Sci., 3: 105–114.

Gray, J. (2004) Consciousness: Creeping up on the Hard Problem. Oxford University Press, Oxford.

Gregory, R.L. (Ed.). (2003) The Oxford Companion to the Mind (2nd ed.). Oxford University Press, Oxford, UK.

Grossberg, S. (1999) The link between brain learning, attention, and consciousness. Conscious. Cogn., 8: 1–44.

Grush, R. (2004) The emulation theory of representation: motor control, imagery, and perception. Behav. Brain Sci., 27(3): 377–396.

Hesslow, G. (2002) Conscious thought as simulation of behaviour and perception. Trends Cogn. Sci., 6(6): 242–247.

Holland, O. (Ed.). (2003) Machine Consciousness. Imprint Academic, Exeter, UK.

Holland, O. and Goodman, R. (2003) Robots with internal models. A route to machine consciousness? In: Holland O. (Ed.), Machine Consciousness. Imprint Academic, Exeter, UK, pp. 77–109.

Jordan, M.I. and Rumelhart, D.E. (1992) Forward models: supervised learning with a distal teacher. Cogn. Sci., 16: 307–354.

Jordan, M.I. and Wolpert, D.M. (1999) Computational motor control. In: Gazzaniga M. (Ed.), The Cognitive Neurosciences. MIT Press, Cambridge, MA.

Kirsh, D. (1991) When is information explicitly represented? In: Hanson P.P. (Ed.), Information, Language, and Cognition. Oxford University Press, New York, NY.

Kirsh, D. (2003) Implicit and explicit representation. In: Nadel L. (Ed.) Encyclopedia of Cognitive Science, Vol. 2. Macmillan, London, UK, pp. 478–481.

Knoblich, G., Elsner, B., Aschersleben, G. and Metzinger, T. (Eds.). (2003) Self and Action. Special Issue of Consciousness and Cognition. (Vol. 12, 4).

Koch, C. (2004) The Quest for Consciousness. A Neurobiological Approach. Roberts and Company Publishers, Englewood, CO.

Lamme, V.A.F. (2004) Separate neural definitions of visual consciousness and visual attention; a case for phenomenal awareness. Neural Networ., 17(5–6): 861–872.

Logothethis, N. and Schall, J. (1989) Neuronal correlates of subjective visual perception. Science, 245: 761–763.

Lumer, E.D., Friston, K.J. and Rees, G. (1998) Neural correlates of perceptual rivalry in the human brain. Science, 280: 1931–1934.

Mack, A. and Rock, I. (1998) Inattentional Blindness. MIT Press, Cambridge, MA.

Marcel, A.J. (1993) Slippage in the unity of consciousness. In: Bock G.R. and Marsh J. (Eds.), Experimental and Theoretical Studies of Consciousness (Ciba Foundation Symposium 174). John Wiley and Sons, Chichester, pp. 168–186.

Mathis, W.D. and Mozer, M.C. (1995) On the computational utility of consciousness. In: Tesauro G. and Touretzky D.S. (Eds.) Advances in Neural Information Processing Systems, Vol. 7. MIT Press, Cambridge, pp. 10–18.

Mathis, W.D. and Mozer, M.C. (1996) Conscious and unconscious perception: a computational theory. In: Proceedings of the Eighteenth Annual Conference of the Cognitive Science Society. Lawrence Erlbaum Associates, Hillsdale, N.J., pp. 324–328.

McClelland, J.L., McNaughton, B.L. and O'Reilly, R.C. (1995) Why there are complementary learning systems in the hippocampus and neocortex: insights from the successes and failures of connectionist models of learning and memory. Psychol. Rev., 102: 419–457.

Metzinger, T. (2000) Neural Correlates of Consciousness. Empirical and Conceptual Questions. MIT Press, Cambridge, MA.

Miall, R.C. (2003) Connecting mirror neurons and forward models. Neuroreport, 14(16): 1–3.

Miall, R.C. and Wolpert, D.M. (1996) Forward models for physiological motor control. Neural Networ, 9(8): 1265–1279.

Moutoussis, K. and Zeki, S. (2002) The relationship between cortical activation and perception investigated with invisible stimuli. Proc. Natl. Acad. Sci. USA, 99(4): 9527–9532.

Naccache, L. (this volume) Visual phenomenon consciousness: a neurological guided tour.

Newman, J. and Baars, B.J. (1993) A neural attentional model of access to consciousness: a global workspace perspective. Conc. Neurosci., 4: 255–290.

Nisbett, R.E. and Wilson, T.D. (1977) Telling more than we can do: verbal reports on mental processes. Psychol. Rev., 84: 231–259.

Noë, A. (2005) Action in Perception. MIT Press, Cambridge, MA.

Noë, A. and Thompson, E. (2004) Are there neural correlates of consciousness? J. Conscious. Stud., 11(1): 3–28.

Norman, K. and O'Reilly, R. C. (2001) Modeling Hippocampal and Neocortical Contributions to Recognition Memory: A Complementary Learning Systems Approach (No. Technical Report 01-02): Institute of Cognitive Science — University of Colorado, Boulder.

O'Brien, G. and Opie, J. (1999) A connectionist theory of phenomenal experience. Behav. Brain Sci., 22: 175–196.

O'Regan, J.K. and Noë, A. (2001) A sensorimotor account of vision and visual consciousness. Behav. Brain Sci., 24(5): 883–917.

O'Regan, Myin & Noë (this volume) Skill, corporality and alerting capacity in an account of sensory consciousness.

O'Reilly, R.C. and Munakata, Y. (2000) Computational Explorations in Cognitive Neuroscience: Understanding the Mind by Simulating the Brain. MIT Press, Cambridge, MA.

Perruchet, P. and Vinter, A. (2003) The self-organizing consciousness. Behav. Brain Sci., 25(3): 297.

Prinz, J.J. (2003) Level headed mysterianism and artificial experience. In: Holland O. (Ed.), Machine Consciousness. Imprint Academic, Exeter, UK, pp. 111–132.

Rizzolati, G. (1996) Premotor cortex and the recognition of motor actions. Cogn. Brain Res., 3: 131–141.

Rosenthal, D. (1997) A theory of consciousness. In: Block N., Flanagan O. and Güzeldere G. (Eds.), The Nature of Consciousness: Philosophical Debates. MIT Press, Cambridge, MA.

Rumelhart, D.E., Smolensky, P., McClelland, J.L. and Hinton, G.E. (1986) Schemata and sequential thought processes in PDP models. In: McClelland J.L. and Rumelhart D.E. (Eds.) Parallel Distributed Processing: Explorations in the Microstructure of Cognition. Volume 2: Psychological and Biological Models. MIT Press, Cambridge, MA, pp. 7–57.

Schacter, D.L. (1989) On the relations between memory and consciousness: dissociable interactions and conscious experience. In: H.L.R. III and Craik F.I.M. (Eds.), Varieties of Memory and Consciousness: Essays in Honour of Endel Tulving. Lawrence Erlbaum Associates, Mahwah, NJ.

Shapiro, K.L., Arnell, K.M. and Raymond, J.E. (1997) The Attentional Blink. Trends Cogn. Sci., 1: 291–295.

Simons, D.J. and Levin, D.T. (1997) Change Blindness. Trends Cogn. Sci., 1: 261–267.

Sun, R. (2001) Duality of the Mind. Lawrence Erlbaum Associates, Mahwah, NJ.

Taylor, J.G. (1999) The Race for Consciousness. MIT Press, Cambridge, MA.

Taylor, J.G. (2003) Paying attention to consciousness. Prog. Neurobiol., 71: 305–335.

Tononi, G. (2003) Consciousness differentiated and integrated. In: Cleeremans A. (Ed.), The Unity of Consciousness: Binding, Integration, and Dissociation. Oxford University Press, Oxford, UK, pp. 253–265.

Tononi, G. (2005) Consciousness, information investigation and the brain, this volume.

Tononi, G. and Edelman, G.M. (1998) Consciousness and complexity. Science, 282(5395): 1846–1851.

Varela, F.J., Thompson, E. and Rosch, E. (1991) The Embodied Mind: Cognitive Science and Human Experience. MIT Press, Cambridge, MA.

Wegner, D.M. (2002) The Illusion of Conscious Will. Bradford Books, MIT Press, Cambridge, MA.

Weiskrantz, L. (1986) Blindsight: A case study and implications. Oxford University Press, Oxford, England.

Wolpert, D.M., Doya, K. and Kawato, M. (2004) A unifying computational framework for motor control and social interaction. In: Frith C.D. and Wolpert D.M. (Eds.), The Neuroscience of Social Interaction. Oxford University Press, Oxford, UK, pp. 305–322.

Wolpert, D.M., Miall, R.C. and Kawato, M. (1998) Internal models in the cerebellum. Trends Cogn. Sci., 2: 338–347.

Zeki, S. and Bartels, A. (1998) The asynchrony of consciousness. Proc. Roy. Soc. B, 265: 1583–1585.

Zeman, A. (2005) What in the world is consciousness? This volume.

S. Laureys (Ed.)
Progress in Brain Research, Vol. 150
ISSN 0079-6123

CHAPTER 8

Machine consciousness

Igor Aleksander*

Department of Electrical and Electronic Engineering, Imperial College, London SW7 2BT, UK

Abstract: The work from several laboratories on the modeling of consciousness is reviewed. This ranges, on one hand, from purely functional models where behavior is important and leads to an attribution of consciousness to, on the other hand, material work closely derived from the information about the anatomy of the brain. At the functional end of the spectrum, applications are described specifically directed at a job-finding problem, where the person being served should not discern between being served by a conscious human or a machine. This employs an implementation of global workspace theories. At the material end, attempts at modeling attentional brain mechanisms, and basic biochemical processes in children are discussed. There are also general prescriptions for functional schemas that facilitate discussions for the presence of consciousness in computational systems and axiomatic structures that define necessary architectural features without which it would be difficult to represent sensations. Another distinction between these two approaches is whether one attempts to model phenomenology (material end) or not (functional end). The former is sometimes called "synthetic phenomenology." The upshot of this chapter is that studying consciousness through the design of machines is likely to have two major outcomes. The first is to provide a wide-ranging computational language to express the concept of consciousness. The second is to suggest a wide-ranging set of computational methods for building competent machinery that benefits from the flexibility of conscious representations.

Introduction

Few would dispute that consciousness is the product of the most complex machine on Earth: the living brain. Machine Modeling of Consciousness or Machine Consciousness (MC) is the name given to the work of those who use not only their analytic skills, but also their ability to design machines to understand better what "being conscious" might mean as the property of a machine.

While science progresses through a process of analysis of complex matter, engineering advances through a process of synthesis based on knowledge gleaned from analysis. An example of probably the most complex product of such synthesis is the control system for a jet airplane. This can only be created by bringing together aerodynamics, jet engine behavior equations, fuel science, mathematical control theory, computing, electronics, and much more. From these emerges the comfort, safety, and convenience of airline passengers. Similarly, designing machine models of consciousness is an exceedingly multidisciplinary process that not only involves computing, mathematics, control, chaos, and automata theory, but also all that can be gathered from the contributions of psychologists, clinicians, neuroscientists, and philosophers.

This approach to the understanding of consciousness is a relatively new enterprise. Although suggestions for the constructive method were fielded in the 1990s (e.g., Aleksander, 1996; Taylor, 1999), May 2001 was a seminal date for establishing a machine consciousness paradigm. Philosopher David Chalmers of Arizona State University,

*Corresponding author. Tel.: +44 (0) 207 594 6176; Fax: +44 (0) 207 594 6274; E-mail: i.aleksander@imperial.ac.uk

DOI: 10.1016/S0079-6123(05)50008-6

neurologist Christof Koch, and computer engineer Rod Goodman of the California Institute of Technology organized a small meeting of computer scientists, neuroscientists, and philosophers at Cold Spring Harbor Laboratories (CSHL) in New Jersey. The purpose of the meeting was to discuss the extent to which an attempt to design a conscious machine could contribute to an understanding of consciousness in general. There was a surprising degree of agreement that the concept was beneficial for the following reason: to define a conscious machine one has to be clear, with regard to the precision of designing a jet airplane—about the difference between a machine that is said to be conscious and one that is not. There is also the promise of tests that can be applied to claims that a machine is conscious. This helps to address the third-person problem of discerning consciousness in organisms to be human, animal, or indeed, machines.

In this chapter I pursue some of the arguments initiated at the CSHL meeting and others that have taken place at a symposium at the 2003 meeting of the Association for the Scientific Study of Consciousness in Memphis, Tennessee; workshops by the European Science Foundation in Birmingham, UK (2003) and the European Community complexity community, Turin, Italy, 2003; and other more recent workshops. I review various contributions to MC recognizing that there are also many others who are currently contributing to this rapidly growing paradigm.

The chapter concludes with a speculative look at the benefits of MC not only in computational analyses, but also in the design of practical machines and the difficulties that lie ahead. One of these is the consideration of using such machines for prosthetic use.

Fig. 1. Space odyssey: illustration of human fear for autonomous machine consciousness (Reproduced with permission – copyright Cliff Laureys).

The conscious machine

It should be said at the outset that there is no race among contributors to MC to produce the ultimate and undisputed human-designed conscious machine. As suggested earlier, the key intention of the paradigm is to clarify the notion of what it is to be conscious through synthesis. Of course, whatever is synthesized can also be built, and if the resulting artifact captures consciousness in some way, then it can, *in some sense*, be said to be conscious. At the end of this chapter I return to this point. But whichever way a machine can be said to be conscious, there might be a performance payoff brought by the influence that attempting to capture consciousness in a machine has on its design. It is likely that a "conscious" machine will produce an advanced ability, with respect to the artificial intelligence and neural network machines produced to date, for better autonomy, freedom from pre-programing, and an ability to represent the machine's own role in its environment. This would improve the capacity for action based on an inner 'contemplative' activity, rather than reactive action largely based on table lookup of pre-stored contingency-action couplings. Therefore, a conscious machine ought to have a significant capacity for dealing with circumstances unforeseen by the program (Fig. 1).

A spectrum and a paradigm

Not all designers approach MC in the same way. While what unites them is the desire to clarify, what often divides them is the question whether the isomorphism with brain mechanisms is important or not. In fact, the differences reflect the functionalist/physicalist spectrum in theories of consciousness. At the functionalist end of the spectrum, the main concern with the mental state is the way it serves the purposes of the organism. That is, consciousness is said to be in evidence in what the organism does, where the details of the mechanism responsible for the mental state are not important. Among physicalists, on the other hand, the concern is largely with the material nature of mechanisms and what it is about these that can be said to capture a conscious state. This inevitably examines living neurological machinery for appropriate design clues.

Technologically, the functional work relates more closely to conventional computation and 'artificial intelligence' styles of programing, where achieving a certain behavior is paramount. The physicalist end is closer to neural network ideas, where network dynamics and their emergent properties are important elements of conscious machine model design. Obviously, some models fall between the two extremes drawing on the useful aspects of each method.

At the time of writing, MC workers have shown considerable determination to accept the work anywhere on this spectrum as contributing to the MC paradigm, hoping to learn from one another and work toward a unified understanding. There is also considerable shared hope that the improved machinery mentioned earlier will arise from this effort as dictated by the need for achieving as yet unattained performance. For example, it might be possible to design exploratory robots that understand the mission, are aware of their environment and their own self in it, and currently rely heavily on pre-programed control or human intervention from the control base. Other applications are systems that go beyond intelligence, requiring understanding and sensitivity of the behavior of their environment or their users.

Franklin's intelligent distribution agent system

A good example of a machine that requires understanding and sensitivity is the Intelligent Distribution Agent (IDA) designed by Stan Franklin of Memphis University (Franklin, 2003) (http://csrg.cs.memphis.edu/). Franklin's IDA was based on Bernard Baars' global workspace theory of consciousness (see Baars, this volume), and designed to replace human operators in a seaman billeting task. The communication link between a seaman seeking a new billet and IDA is e-mail. IDA receives information about the current postings, and the seaman's skills and desires for a new location. It then attempts to match this to the current state of available billets perhaps having several cycles of interaction in order to achieve a result. The key feature is that the seaman using the system should not feel that there has been a change from human billeters to a machine in terms of the sensitivity and concern with which their case is handled.

As shown in Fig. 2, the system contains processing modules that implement, in traditional computing formats, various forms of memory (working, autobiographical, associative, and episodic). These are addressed from external stimuli ('I, sailor, need to work in a warm climate') as well internal stimuli ('I, IDA, might suggest Florida'). Memories produce cues and associations that compete to enter the area of 'consciousness.' In IDA this takes the form of a coalition manager, an attention mechanism and a broadcast mechanism. Communication is based on 'codelets,' which are structured programs, also called 'mini agents' in computing. So the content of the consciousness area starts as a partially formed thought that broadcasts information back to address memory areas. This results in new cues and the process repeats until the 'thought' is sufficiently well formed to activate an action selection mechanism that communicates with the sailor and initiates a new set of internal and external inputs for further consideration. In recent versions of IDA, 'emotional' information (such as 'guilt' for, say, not achieving all of a sailor's requests) enters the operation of a large number of modules.

Franklin makes no claim that there is any phenomenological consciousness in this system and is content with the functional stance, which is sufficiently effective to leave users satisfied that they are interacting with a system that is 'conscious' of their needs.

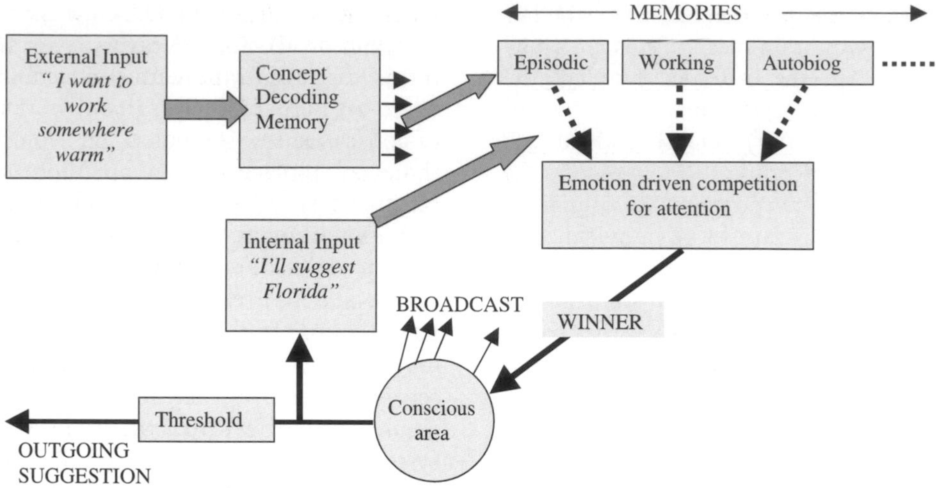

Fig. 2. IDA's (Intelligent Distributed Agent) "cognitive cycle." (Taken from Stan Franklin, Memphis University, with permission.)

Consciousness in virtual machines

In the United Kingdom, Aaron Sloman of Birmingham University and Ron Chrisley of Sussex University have set out to discuss functional, computational ideas as a way of clarifying seemingly arbitrary opinions that enter discussions about consciousness (Sloman and Chrisley, 2003) (http://www.cs.bham.ac.uk/~axs/).

For example, some think that dreams are in consciousness (see Maquet et al., this volume), others do not, some think that consciousness is a matter of degree, others think it is either-or; and so on. They argue that a computational model has the power of making these issues explicit (see Cleeremans, this volume).

The authors evoke the concept of a *virtual machine* that can possess a mixture of states that are important in clarifying consciousness. Virtuality permits one to distinguish the properties of an emulated machine that models aspects of consciousness from those of the underlying host mechanism, that is, a general purpose computer. This 'virtual machine functionalism' is illustrated by an architectural 'schema' (a discussion framework for architectures of consciousness) called CogAff (Cognition and Affect) and a specific architecture called H-CogAff (Human-like architecture for Cognition and Affect). With information processes rather than physical processes being the elements of the schema, these can be structured to represent *perception*, *internal processing*, and *action* as well as the relationships between them. This 'three-tower' vertical division is further divided into three horizontal layers. The first layer is *reactive mechanisms* that link perception to action in a direct way (e.g., reflexes). The second represents *deliberative mechanisms* that are capable of what-if computations for planning (e.g., "I use a stick to knock the banana off the tree"). The third is a *meta-management* layer that senses the lower planning process and is capable of modifying it (e.g., "Using sticks is unfriendly, I should try something else."). Nestling among the reactive mechanisms is an 'alarm' process that has rapid access to all the other parts of the architecture should an emergency be discovered (see Fig. 3).

Sloman and Chrisley's Virtual-Machine Functionalism (VMF) is distinguished from a more general (atomic) form of functionalism where the latter treats a mental state as just one overall internal state of a machine from which stems the organism's behavior. VMF, on the other hand, permits models of interacting architectural features that give rise to many, concurrently acting, interacting mental states. There are several characteristics of VMF that permit the modeling of phenomena that, at the outset, appear puzzling. For example, Chrisley and

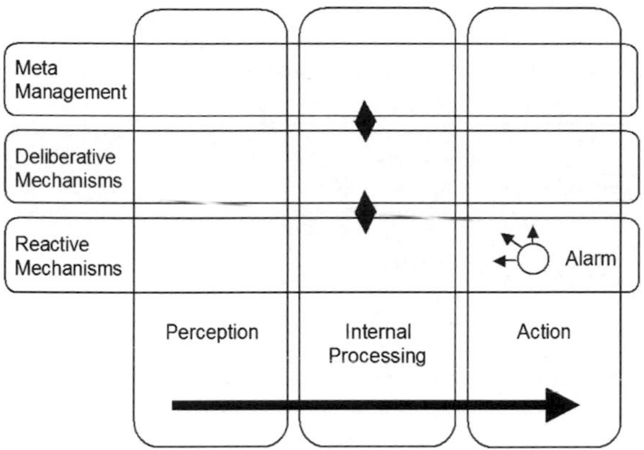

Fig. 3. The Sloman/Chrisley CogAff (Cognition and Affect) schema.

Sloman see emotion as an ill-defined concept which, in their scheme, becomes separated out as being of at least three types that relate closely to the horizontal layers. These are reactive emotions such as anger, deliberative ones such as frustration, and meta-management disruptions such as grief or jealousy. Another example where modeling is helpful is in vision where there are multiple 'what' and 'where' paths, which are explicit in the CogAff structure clarifying their parallel functions and interactions. Further, localized disruptions due to lesions can be modeled, explaining how some functions are disadvantaged while others are left intact. The model also makes clear how resources that control learning must be distributed. The authors also use the model to approach explanations of perceptual failures such as inattention blindness (we think we see everything, but we only see that to which we attend; see O'Regan et al., this volume). Abstract thinking, as when doing mathematics, becomes a task for the meta-management layer. Finally, 'qualia' are explained as the observation exercised by the meta-layer on the activity of lower layers and its ability to monitor and interpret these lower-level processes.

Cognitive neural architectures

Pentti Haikonen of the Nokia Company in Helsinki, Finland, has created architectural models that capture consciousness by having a comprehensive set of cognitive competences (Haikonen, 2003) (www.control.hut.fi/STeP2004/cv_pentti.html). This relies heavily on the ability of recursive or reentrant neural networks to store and retrieve states. Based very roughly on the operation of a brain cell, an artificial neuron is a device that receives input signals and 'learns' to output an appropriate response. Recursive networks have stable states by virtue of the fact that neurons not only receive signals from external sources such as vision or audition, but also from the signals generated by other neurons in the same network. So, for example, if a network has learned to represent the image of a cat, this image can be sustained as each neuron will output its feature of the cat image in response to other neurons outputting cat features. This means that such a network can store several images as stable states and, if the net is given only a fragment of an image 'it knows' it will reconstruct the whole image as more and more neurons will be recruited to output the same image. This kind of reentrant, dynamic mechanism is thought to be important in living brains (see Tononi, this volume) and it is for this reason that Haikonen's models are sited closer to the physicalist end of the functional/physicalist spectrum in the earlier examples in this chapter.

Haikonen's cognitive architecture (see Fig. 4) is based on a collection of similar modules. Each module consists of sensory input and a preprocess that extracts important features from the input. This is followed by a *perception generator* that feeds 'a distributed representation of a percept' to a neural network called the *inner process*. But the network also feeds back to the perception generator. The resulting feedback loop causes the system both to be able to represent both sensory inputs, and inner reconstructions of meaningful states in the absence of input.

There is one such module for each sensory modality. Some modalities, primarily vision, are divided into more detailed submodules that specialize in features such as shape, color, and motion that reflect some divisions known to exist in the brain. The key feature of this architecture is that there is feedback at an even higher level: each inner process of a module receives input from the perception generators of other modules. That is, a module is influenced by what other modules are

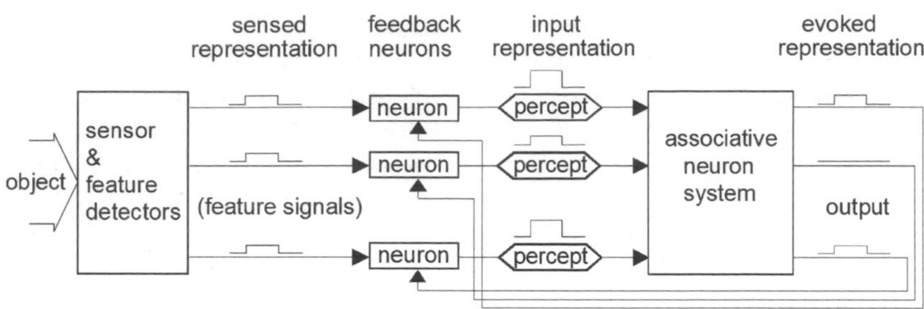

Fig. 4. Haikonen's reentrant structure showing three modules. (Kindly provided by Pentti Haikonen, Nokia Research Center).

representing, leading to overall states in the system that are capable of associating, for example, the image of a cat represented in a visual module with the word 'cat' represented in another module. This collection of modules is the cognitive part of the architecture. Haikonen also envisages a 'motor' part which processes the state of the cognitive part, leading to actions such as the generation of speech or motion (this 'decodes' the state of the associative neuron system in Fig. 4).

Another feature of the architecture is that one of the modules is positioned to act at a level higher than the sensory processing of the others. This module monitors the patterns of activity of the lower level modules. This can assign word-meaning to this overall state that can then use the word 'I' in a meaningful way. Emotions too are not neglected. They are the product of central sensing of the *reactions* to certain sensory input (e.g., forward-going for pleasure and retracting for fear). Haikonen sees 'conscious' as being an accurate term to describe the normal modes and styles of operation of his architecture. He distinguishes between conscious and unconscious modes through the degree of engagement that the inner mechanisms have with a task. For example, he quotes 'the bedtime story effect,' where a parent reads automatically to a child while thinking of something. The reading does not reach consciousness as it goes directly from sensor to actuator without entering the representational loops. In summary, in common with other physicalist approaches, Haikonen suggests that consciousness is a product of the firing of neurons, which sometimes can be due to sensory information and importantly, at other times, to the

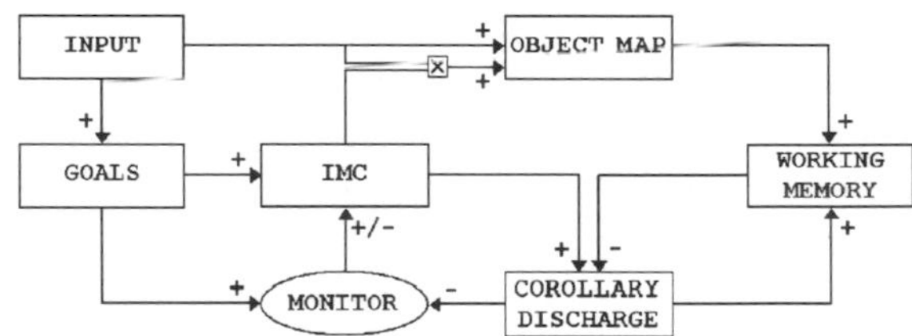

Fig. 5. Taylor's CODAM model (COrollary Discharge of Attention Movement). (Taken from John Taylor, Kings' College London, with permission.)

sustained states of several levels of feedback in the machinery of the artificial or the living brain.

Attention and consciousness

Close to the physicalist end of the spectrum is the work of John Taylor, a mathematician and theoretical physicist at Kings' College London (www.mth.kcl.ac.uk/~jgtaylor/). The key to his model (CODAM: COrollary Discharge of Attention Movement) is based on the principle that without attention to an input there can be no awareness of it (Taylor, 2002). Consequently he investigates a specific brain mechanism called the 'corollary discharge' that is responsible for changes in attention. He expresses this within a framework of control engineering as shown in Fig. 5. The control model involves an *object map* within which objects are selected for 'coming into consciousness' by a competitive process involving working memory and the corollary discharge mechanism. Taylor distinguishes a 'pre-reflective self,' i.e., the feeling of ownership of the content of being conscious, with the corollary discharge, and equates it to 'pure consciousness experience'. He reasons that there exists a buffer in the model, the neural activity of which is the correlate of the consciousness of the organism. The corollary discharge signal appears in this buffer briefly, to be immediately followed by the sensory signal of that which has been attended as selected by the discharge. Therefore the pure conscious state is a temporal extension of the content-less pre-reflective self state.

The CODAM model allows Taylor and his colleagues to arrive at several important conclusions. For example, they explain the meditational processes aimed at achieving a state of 'pure consciousness' found in several Eastern religions. They argue that advanced forms of meditation force the attentional corollary discharge to block sensory input and turn to attending only to itself. Another application is the explanation of the attentional blink which occurs when someone is asked to attend to several objects presented in succession to one another. Schizophrenia, inattention blindness and blindsight are also approached through the CODAM model.

At the physicalist end of the spectrum

Rodney Cotterill, a British scientist working at the Danish Technical University in Copenhagen contributes to MC by searching for consciousness in a young developing child (Cotterill, 2003) (http://info.fysik.dtu.dk/Brainscience/people/rodney.html). Called "Cyberchild," this simulation is a comprehensive model not only of the cortical regions that may be present and necessary in a very young child, but also of the endocrine system (blood control glands), the thalamic regions, and the autonomic nervous system. The model makes it possible to study a biochemical state that could be described as hunger. It can then be given "milk" to increase its simulated glucose levels. Should these fall to zero, the system stops functioning and the child dies. However, the child has a vocal output that enables it to cry and alert an observer that action is needed either to provide milk or change the nappies as the model is capable of urinating and sensing a wet nappy. The crying reaction is built into the system but differs from the 'tamagoshi' object in the sense that it is triggered by proper models physiological parameters rather than simple rules. The model has only two sensory modalities, hearing and touch, these being dominant in the very young child and sufficient to model the effect of sensory input.

Cotterill raises important questions as to whether even a perfectly executed model of a young child is likely to capture being conscious in some way. He remains skeptical of this, reflecting that his work is a salutary experience in understanding the neural mechanisms of a living child which clearly contribute to its, albeit rudimentary, consciousness.

A depictive model

Also close to the physicalist end of the spectrum, the author's own approach has sought to identify mechanisms, which through the action of neurons (real or simulated), are capable of representing the world with the 'depictive' accuracy that is felt introspectively in reporting a sensation (Aleksander, 2005). The model of being conscious stems from five features of consciousness which appear important through introspection. Dubbed 'axioms' (as they are intuited but not proven) they are:

(1) Perception of oneself in an 'out-there' world
(2) Imagination of past events and fiction
(3) Inner and outer attention
(4) Volition and planning
(5) Emotion

This is not an exhaustive list, but is felt to be necessary for a modeling study. In the belief that consciousness is the name given to a composition of the above sensations, the methodology seeks a variety of mechanistic models each of which can support a depiction of at least one of the above basic sensations.

Perception necessitates a neural network that is capable of registering accurately (i.e., depicting) the content of a current perceptual sensation. Out-thereness, particularly in vision, is ensured through the mediation of muscles: eye movement, convergence, head movement, and body movement, all create signals that integrate with sensory signals to produce depictions of being an entity in an out-there world (indexing cells in this way this is called *locking*).

Imagination requires classical mechanisms of recursion in neural networks. That is, memory of an experienced state creates a reentrant set of states in a neural net or set of neural modules with feedback (as explained for Haikonen's work, above). That this is experienced as a less accurate version of the original stems from the known characteristic of recursive networks that their depictive power weakens as the network learns a significant number of states. The experience of fiction uses a mechanism where memory states are entered as controlled from natural language sensory input, a property that has been demonstrated in simulations.

Outer attention (such as foveal movement) is due to a completion of the loop of objects being depicted and triggering further need for muscular movement to complete the depiction. *Inner attention* requires a 'vetoed' movement signal (i.e., the initiation of a movement that is not actually carried out) to imagine, say, looking around a remembered scene. Mechanisms that lead to sensations of *volition, planning*, and *emotions* have been shown to emerge from the interaction of neural modules that are involved in imagination (in which state sequences constitute 'what if' plans) and particular modules that non-depictively (unconsciously) evaluate emotions associated with predicted outcomes of planned events. The engineering upshot of this approach is that it is possible to envisage a 'kernel' architecture that illustrates the meshing together of the mechanistic support of the five axioms (Fig. 6). The perceptual module directly depicts sensory input and can be influenced by bodily input such as pain and hunger. The memory module implements non-perceptual thought for planning and recall of experience. The memory and perceptual modules overlap in awareness as they are both locked to either current or remembered world events. The emotion module evaluates the 'thoughts' in the memory module and the action module causes the best plan to reach the actions of the organism.

This structure has been used in a variety of applications ranging from the assessment of distortions of visual consciousness in Parkinson's sufferers (Aleksander and Morton, 2003) to identifying the possibility of a brain-wide spread of the neural correlates of 'self,' models of visual awareness that explain inattention and change blindness, and a possible mechanism for volition as mediated by emotion (Aleksander, 2005).

The emerging paradigm

So is there evidence to show that MC serves to clarify concepts of consciousness? This chapter has described a spectrum of methods that attempt to achieve an understanding of consciousness through *synthesis* based on notions gleaned from psychology, neurology, and introspection. While the differences between approaches have been highlighted, important common ground has also been found that contributes to an emerging explanatory paradigm. First, most designers see the role of the brain as a control mechanism, which ensures that the organism deals appropriately with its environment and its internal parameters. But consciousness is not present in all control mechanisms. Room thermostats are not conscious. Ricardo Sanz, a control engineer of Madrid University, points out that only control systems with non-trivial representational powers qualify as 'being conscious of something' (Sanz, 2000).

This has been refined by Owen Holland of Essex University and Rod Goodman of the California Institute of Technology (Holland and Goodman, 2003), who argue that the internal representation required for consciousness can be engineered through a structure that contains both a world-model and an agent-model. These are control systems where the two models are interconnected and improve their mutual performance. Owen Holland is, at the time of writing, building a robot that will test designs of inner modeling of the working environment and the 'self' of the robot in it (http://cswww.essex.ac.uk/staff/owen/).

It was said at the outset that the MC paradigm helps to assess the presence of consciousness in an organism. This mainly relies on the discovery of the presence of the above control mechanisms with the ability to model the world and themselves. It cuts out thermostats, but includes bees and properly designed planetary exploration robots. So the 'sense' in which a robot could be said to be conscious is that it passes the structural and

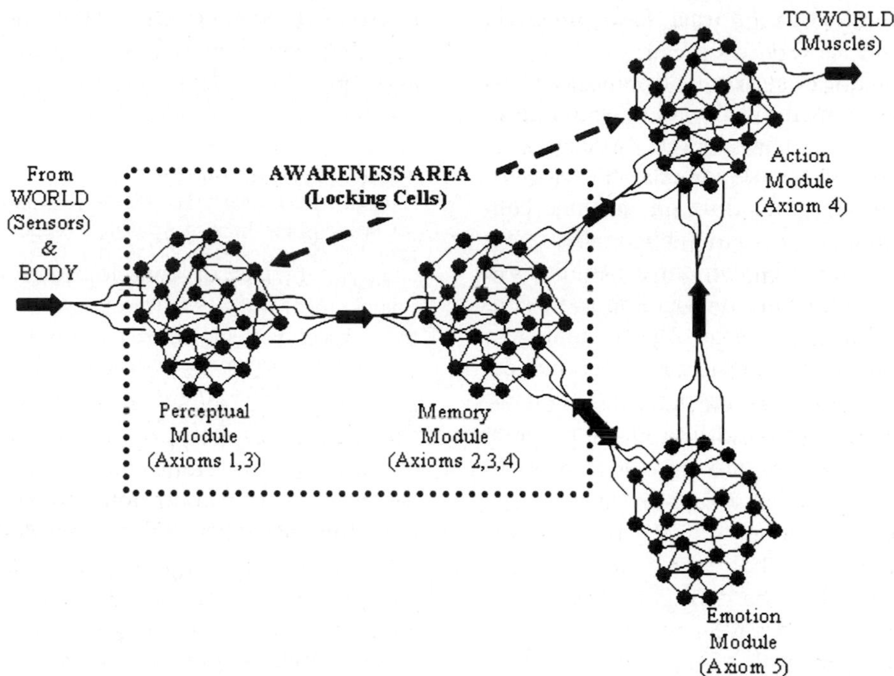

Fig. 6. The "kernel" architecture for implementing Aleksander's axiomatic mechanisms.

architectural assessments that satisfy the conditions of world and self modeling.

Another point of agreement is that consciousness should be seen as having different degrees. A robot conscious of the needs of an exploratory mission on Mars may not have the complex consciousness of a human immersed in daily life, but may have a higher level of consciousness than a bee on a pollen-finding mission. The important common belief is that these differences are due to mechanisms that are similar at some level of abstraction and so constitute a model that explains what is needed by an organism to be conscious. At least, this is the promise of the emergent paradigm of Machine Models of Consciousness.

Future prospects

In imagining future research in MC, it should be remembered that the paradigm is still young and diverse as described. The most pressing need is to address phenomenological issues as without this, the validity of the computational approach is harder to sustain. A start has been made with the 5-axiom approach above, but there is much scope to extend the set of axioms particularly into areas of language depiction and production. There is also scope for hybrid schemes in which the phenomenology is handled through neural or cellular methods while the control aspects of the cellular system remain closer to classical computational intelligence. Opportunities exist for a deeper approach to the role of emotions that, while included in most examples of the above work, are some way of modeling what is currently found in developed approaches that one finds in neurology and psychology.

Further opportunities exist for the machine consciousness community to share their findings with those working in other computational assessments of the biological brain. For example, mutual benefit may arise from taking note of such work as the likely effects of computational prostheses for the enhancement of memory (e.g., Mussa-Ivaldi and Miller, 2003) or engineering assessments of the role of brain areas such as the hippocampus as carried out by Theodore Berger and his colleagues at the

University of Southern California, LA (Tsai et al., 1998).

On the applications side, projects such as Holland's robot cited above should be encouraged, as the old adage is as true as ever: "when discussions get too abstract, make something, even if it's a mistake." The current effort on machine consciousness may drift up many blind alleys, but there is an increasing feeling among contributors to the paradigm that they are focusing on some concrete and shared ideas in terms of both gaining an understanding of the important computational descriptions of consciousness (supporting Cleermans, this volume) and leading to novel machinery that will benefit from the autonomy that the presence of a computational conscious self may provide.

References

Aleksander, I. (1996) Impossible Minds, My Neurons My Consciousness. IC Press, London.

Aleksander, I. (2005) The World in My Mind: Five Steps to Consciousness. Imprint Academic, Exeter, UK.

Aleksander, I. and Morton, H.B. (2003) A digital neural model of visual deficits in Parkinson's disease: Proc IWANN03. Springer, Minorca, pp. 8–16.

Cotterill, R.M.J. (2003) Cyberchild. J. Consciousness Stud., 10: 31–45.

Franklin, S. (2003) IDA: A conscious artifact? J. Consciousness Stud., 10: 47–66.

Haikonen, P.O. (2003) The Cognitive Approach to Machine Consciousness. Imprint Academic, Exeter, UK.

Holland, O. and Goodman, R. (2003) Robots with internal models—A route to machine consciousness? J. Consciousness Stud., 10: 77–109.

Mussa-Ivaldi, F.A. and Miller, L.E. (2003) Brain–machine interfaces: Computational demands and clinical needs meet basic neuroscience. Trends Neurosci., 6: 329–334.

Sanz, R. (2000) Agents for complex control systems. In: Samad T. and Wayrauch J. (Eds.), Automation, Control and Complexity. Wiley, New York.

Sloman, A. and Chrisley, R. (2003) Virtual machines and consciousness. J. Consciousness Stud., 10: 4–5 (April/May), 133–172.

Taylor, J.G. (1999) The Race for Consciousness. MIT Press, Cambridge, MA.

Taylor, J.G. (2002) Consciousness: Theories of. In: Arbib M. (Ed.), Handbook of Brain Theory and Neural Computation. MIT Press, Cambridge, MA.

Tsai, R.H., Sheu, J.B. and Berger, T.W. (1998) A VLSI neural network processor based on a model of the hippocampus. Analog Integr. Circuits Signal Process., 15: 201–213.

S. Laureys (Ed.)
Progress in Brain Research, Vol. 150
ISSN 0079-6123

CHAPTER 9

Consciousness, information integration, and the brain

Giulio Tononi*

Department of Psychiatry, University of Wisconsin, 6001 Research Park Blvd, Madison, WI 53719, USA

Abstract: Clinical observations have established that certain parts of the brain are essential for consciousness whereas other parts are not. For example, different areas of the cerebral cortex contribute different modalities and submodalities of consciousness, whereas the cerebellum does not, despite having even more neurons. It is also well established that consciousness depends on the way the brain functions. For example, consciousness is much reduced during slow wave sleep and generalized seizures, even though the levels of neural activity are comparable or higher than in wakefulness. To understand why this is so, empirical observations on the neural correlates of consciousness need to be complemented by a principled theoretical approach. Otherwise, it is unlikely that we could ever establish to what extent consciousness is present in neurological conditions such as akinetic mutism, psychomotor seizures, or sleepwalking, and to what extent it is present in newborn babies and animals. A principled approach is provided by the information integration theory of consciousness. This theory claims that consciousness corresponds to a system's capacity to integrate information, and proposes a way to measure such capacity. The information integration theory can account for several neurobiological observations concerning consciousness, including: (i) the association of consciousness with certain neural systems rather than with others; (ii) the fact that neural processes underlying consciousness can influence or be influenced by neural processes that remain unconscious; (iii) the reduction of consciousness during dreamless sleep and generalized seizures; and (iv) the time requirements on neural interactions that support consciousness.

Neuroscience and consciousness: facts and challenges

When addressing consciousness, two main problems need to be considered. The first problem is understanding the conditions that determine to what extent consciousness is present or absent. For example, why are changes of neural activity in thalamocortical regions so important for conscious experience, whereas changes in neural activity in cerebellar circuits are not, given that the number of neurons in the two structures is comparable? Or, why is consciousness strikingly reduced during deep slow-wave sleep, given that the average level of neuronal firing is similar to that of wakefulness?

The second problem is understanding the conditions that determine the specific way consciousness is. For example, what makes the activity of specific cortical areas contribute to specific dimensions of conscious experience — auditory cortex to sound, visual cortex to shapes, or colors? What aspect of neural organization is responsible for the fact that shapes look the way they do, and different from the way colors appear, or pain feels? Solving the first problem means that we would know to what extent a physical system can generate consciousness — the *quantity* or level of consciousness. Solving the second problem means that we would

*Corresponding author. Tel.: +1 (608) 263-6063; Fax: +1 (608) 2639340; E-mail: gtononi@wisc.edu

DOI: 10.1016/S0079-6123(05)50009-8

know what kind of consciousness it generates — the *quality* or content of consciousness.

The first problem is best considered by examining some well-established facts about the relationship between consciousness and the brain (the second problem is discussed in Tononi, 2004b). Each of these facts poses a serious challenge to our efforts to understand how consciousness comes about. Considered together, however, they strongly constrain the realm of possible answers.

1. Consciousness is produced by certain parts of the brain and not, or much less, by others. The parts that are essential are distributed within the thalamocortical system. Different areas of the thalamocortical system independently contribute different dimensions to conscious experience, and no single area is solely responsible for consciousness. What is special about this distributed network of thalamocortical circuits?
2. Other regions of the brain, such as the cerebellum, are not essential for consciousness, and can be stimulated or lesioned without giving rise to changes in conscious experience. Yet the cerebellum has as many neurons and is every bit as complicated as the thalamocortical system. Why is consciousness associated with some but not with other neural structures?
3. Neural activity in sensory afferents to the thalamocortical system usually determines what we experience at any given time. However, such neural activity does not appear to contribute directly to conscious experience. For example, while retinal cells can discriminate light from dark and convey that information to visual cortex, their rapidly shifting firing patterns do not correspond well with what we perceive. Moreover, a person who becomes retinally blind as an adult continues to have vivid visual images and dreams. Why is it that the activity of retinal cells per se does not contribute directly to conscious experience, but only indirectly through its action on thalamocortical circuits?
4. Neural activity in motor pathways is necessary to bring about the diverse behavioral responses that we usually associate with consciousness. However, such neural activity does not in itself contribute to consciousness. For example, patients with the locked-in syndrome, who are completely paralyzed except for the ability to gaze upward, are fully conscious. Similarly, we are paralyzed during dreams, but consciousness is not impaired by the absence of behavior. Even lesions of central motor areas do not impair consciousness. Why are we not conscious of what goes on in motor pathways?
5. Neural processes occurring in brain regions whose inputs and outputs are closely linked to the thalamocortical system, such as the basal ganglia, are important in the production and sequencing of action, thought, and language. Yet such processes do not seem to contribute directly to conscious experience. Moreover, some action sequences may be performed consciously when we first learn them, but fade from awareness when they become automatic. At the same time, their cortical substrates may shrink and shift to different circuits. Why are neural processes that take place automatically within cortico-subcortico-cortical circuits less conscious, and how do they become so?
6. Even within the thalamocortical system, many neural processes can influence conscious experience yet do not seem to contribute directly to it. For example, what we see and hear depends on elaborate computational processes in the cerebral cortex that are responsible for object recognition, depth perception, and language parsing, yet such processes remain largely unconscious. Correspondingly, neurophysiological studies indicate that while the activity of certain cortical neurons correlates well with conscious experience, that of others does not. For example, during binocular rivalry the activity of certain visual cortical neurons follows what the subject consciously perceives, while that of other neurons follows the stimulus, whether the subject is perceiving it or not. What determines whether the firing of neurons within the thalamocortical system contributes directly to consciousness or not?

7. Consciousness can be split if the thalamocortical system is split. Studies of split brain patients, whose corpus callosum was sectioned for therapeutic reasons, show that each hemisphere has its own, private conscious experience. Other neurological disconnection syndromes, as well as certain psychiatric dissociations, indicate that anatomical or functional disconnections among brain areas result in the shrinking or splitting of consciousness. What does this reveal about the neural substrate of consciousness?
8. On average, cortical neurons fire almost as much during deep slow-wave sleep as during wakefulness, but the level of consciousness is much reduced in the former condition. Similarly, in absence seizures, neural firing is high and synchronous, yet consciousness is seemingly lost. Why is this the case?
9. The firing of the same cortical neurons may correlate with consciousness at certain times, but not at other times. For example, multi-unit recordings in the primary visual cortex of monkeys show that, after a stimulus is presented, the firing rate of many neurons increases irrespective of whether the animal reports seeing a figure or not. After 80–100 ms, however, their discharge accurately predicts the conscious detection of the figure. What determines when the firing of the same cortical neurons contributes to conscious experience and when it does not?

Many more facts and puzzles could be added to this list. This state of affairs is not unlike the one faced by biologists when, knowing a great deal about similarities and differences between species, fossil remains, and breeding practices, they still lacked a theory of how evolution might have occurred. What was needed, then as now, were not just more facts, but a theoretical framework that could make sense of them. Unfortunately, theoretical approaches that try to provide a coherent explanation for some of the basic facts about consciousness and the brain are few and far between. Here, in order to offer a tentative but at least unified perspective on the issues that need to be addressed, we review a theoretical approach according to which consciousness corresponds to the brain's ability to rapidly integrate information (Tononi, 2001, 2004a). The present review of the information integration theory of consciousness (IITC), which closely follows the original publications, comprises (i) an examination of phenomenology indicating that consciousness has to do with integrating information; (ii) a definition of what integrated information is and how it can be measured; (iii) an attempt at accounting for basic facts about consciousness and the brain; and (iv) some corollaries and predictions.

Phenomenology: consciousness as information integration

According to the IITC, perhaps the most important thing to realize about consciousness is that when one experiences a particular conscious state — say the one experienced when reading *this particular phrase* here and now — each of us is gaining access to an extraordinarily large amount of information. This information has nothing to do with how many letters or words we can take in at time, which is a very small number. Instead, the occurrence of a particular conscious state is extraordinarily informative because of the very large number of alternative conscious states that it rules out. Just think of all possible written phrases you could read, multiply them by the number of possible fonts, ink colors, and sizes in which you could read them, then think of the same phrases spoken aloud, or read and spoken, or think further of all other possible visual scenes you might experience, multiplied by all possible sounds you might hear at the same time, by all possible moods you might be in, and so on ad libitum.

The point is simply that every time we experience a particular conscious state out of such a huge repertoire of possible conscious states, we gain access to a correspondingly large amount of information. This conclusion is in line with the classical definition of information as a reduction of uncertainty among a number of alternatives (Shannon & Weaver, 1963). For example, tossing a fair coin and obtaining heads corresponds to $\log_2(2) = 1$ bit of information, because there are just two alternatives;

throwing a fair dice yields $\log_2(6) = 2.59$ bits of information, because there are six equally likely possibilities. Similarly, the information generated by the occurrence of a particular conscious state lies in the large number of different conscious states that *could potentially* have been experienced but were not. While no attempt has been made to estimate the size of the repertoire of conscious states available to a human being, it is clear that such repertoire must be extraordinarily large, and so is the information yielded by entering a particular conscious state out of this repertoire. This point is so simple that its importance has been overlooked.

Another key aspect of the IITC is that the information associated with the occurrence of a conscious state is not information from the perspective of an external observer, but *integrated information.* When each of us experiences a particular conscious state, that conscious state is experienced as an integrated whole — it cannot be subdivided into independent components, i.e. components that are experienced independently. For example, the conscious experience of *this particular phrase* cannot be experienced as subdivided into, say, the conscious experience of how the words look independently or how they sound in one's mind. Similarly, one cannot experience visual shapes independently of their color, or perceive the left half of the visual field of view independently of the right half. If one could, this would be tantamount to having two separate "centers" of consciousness. Separate centers of consciousness exist, of course, but then each is a different person with a different brain (or a different hemisphere of a split brain).

Finally, it is important to appreciate the characteristic spatio-temporal grain of consciousness. For example, psychophysical evidence indicates that sensory experiences require at least 100–200 ms to become progressively specified and stabilized. On the other hand, a single conscious moment cannot extend beyond 2–3 s. While it is arguable whether conscious experience unfolds more akin to a series of discrete snapshots or to a continuous flow, its time scale is certainly comprised between these lower and upper limits. Thus, a phenomenological analysis indicates that consciousness requires the integration of a large amount of information over a characteristic time scale.

Theory: measuring information integration

If consciousness corresponds to information integration, then a physical system should be able to generate consciousness to the extent that it can rapidly enter any large number of available states (information), yet it cannot be decomposed into a collection of causally independent subsystems (integration). How can one identify such an integrated system, and how can one measure its repertoire of available states?

At first sight, it might seem that all one needs to do is choose a system, e.g. the brain, and measure the repertoire of states that are available to it with their probability. One could then calculate the information associated with the occurrence of each brain state, just as one can measure the information associated with tossing a coin or a dice, by using the entropy function, i.e. the weighted sum of the logarithm of the probability (p) of system states (s): $H = -\sum p(s) \log_2 p(s)$. Measuring the available repertoire would easily account for why a seemingly similar task can be performed unconsciously (or nearly so) by a simple device and consciously by human being. For example, when a retinal cell, or even a photodiode — a simple semiconductor device that changes its electrical resistance depending on the illumination — detects complete darkness, it generates a minimal amount of information, since it can only discriminate between darkness and light. When we consciously detect complete darkness, however, we perform a discrimination that is immensely more informative: we are not just ruling out light, but an extraordinary number of other possible states of affairs, including every possible frame of every possible movie, every possible sound, and every possible combination of them.

Measuring information this way, however, is insufficient, because it is completely insensitive to whether the information is integrated. To give a simple example, consider a collection of one million photodiodes constituting the sensor chip of a digital camera. From the perspective of an external observer, such a chip can certainly enter a very large number of different states, as it could easily be demonstrated by presenting it with all possible input signals. However, due to the absence of any physical interaction among the photodiodes, the chip as such does not integrate any information: the state of each element is causally independent of that of other elements. In other words, what we have is one million photodiodes with a repertoire of two states each, rather than a single integrated system with a repertoire of $2^{1,000,000}$ states. Thus, to measure information integration, it is essential to know whether a set of elements constitute a causally integrated system, or they can be broken down into a number of independent or quasi-independent subsets among which no information can be integrated.

To see how one can achieve this goal, consider an extremely simplified neural system constituted of a set of elements (Tononi & Sporns, 2003; Tononi, 2004b). Each element could represent, for instance, a group of locally interconnected neurons that share inputs and outputs, such as a cortical minicolumn. We could further assume that each element can go through discrete activity states, corresponding to different firing levels, each of which lasts for a few hundreds of milliseconds. Finally, for the present purposes, let us imagine that the system is disconnected from external inputs, just as the brain is disconnected from the environment when it is dreaming.

Consider now a subset S of elements taken from such a system, and the diagram of causal interactions among them (Fig. 1a). We will measure the information generated when S enters a particular state out of its repertoire, but only to the extent that such information can be integrated within S, i.e. each state results from causal interactions within S. To do so, we divide S into two complementary parts A and B. We can now evaluate the responses of B that can be caused by all possible inputs originating from A. In neural terms, we try out all possible combinations of firing patterns as outputs from A, and establish how differentiated is the repertoire of firing patterns they produce in B. In information-theoretical terms, we give maximum entropy to the outputs from A ($A^{H^{\max}}$), i.e. we substitute its elements with independent noise sources, and we determine the entropy of the responses of B that can be caused by inputs from A. Specifically, we define the *effective information* between A and B as $\mathrm{EI}(A \rightarrow B) = \mathrm{MI}(A^{H^{\max}}; B)$. Here $\mathrm{MI}(A; B) = H(A) + H(B) - H(AB)$ stands for mutual information, a measure of the entropy or information shared between a source (A) and a target (B). Note that since A is substituted by independent noise sources, the entropy shared by B and A is necessarily due to causal effects of A on B. Moreover, $\mathrm{EI}(A \rightarrow B)$ measures all possible effects of A on B, not just those that are observed if the system were left to itself. Also, $\mathrm{EI}(A \rightarrow B)$ and $\mathrm{EI}(B \rightarrow A)$ in general are not symmetric. For a given bipartition of a subset, then, the sum of the effective information for both directions is indicated as $\mathrm{EI}(A \leftrightarrows B) = \mathrm{EI}(A \rightarrow B) + \mathrm{EI}(B \rightarrow A)$. In summary, $\mathrm{EI}(A \leftrightarrows B)$ measures the repertoire of possible causal effects of A on B and of B on A.

On the basis of the notion of effective information for a bipartition, we can assess how much information can be integrated within a system of elements. To this end, we note that a subset S of elements cannot integrate any information if there is a way to partition S in two parts A and B such that $\mathrm{EI}(A \leftrightarrows B) = 0$ (Fig. 1b, vertical bipartition). In such a case we would obviously be dealing with at least two causally independent subsets, rather than with a single, integrated subset. This is exactly what would happen with the photodiodes making up the sensor of a digital camera: perturbing the state of some of the photodiodes would make no difference to the state of others. More generally, to measure the information integration capacity of a subset S, we should search for the bipartition(s) of S for which, after appropriate normalization, $\mathrm{EI}(A \leftrightarrows B)$ is lowest: its informational "weakest link", or *minimum information bipartition* ${}^{\mathrm{MIB}}A \leftrightarrows B$. The *information integration* for subset S, or $\Phi(S)$, is simply the (non-normalized) value of $\mathrm{EI}(A \leftrightarrows B)$ for its minimum information bipartition: $\Phi(S) = \mathrm{EI}({}^{\mathrm{MIB}}A \leftrightarrows B)$. The symbol Φ is meant to indicate that the information (the vertical bar "I") is integrated within a single entity (the circle "O").

If $\Phi(S)$ is calculated for every possible subset S of a system, one can establish which subsets are actually capable of integrating information, and how much of it (Fig. 1c). After discarding all those subsets that are included in larger subsets having higher Φ (since they are merely parts of a larger whole), one is left with the *complexes* that make up the system. Specifically, a *complex* is a subset S

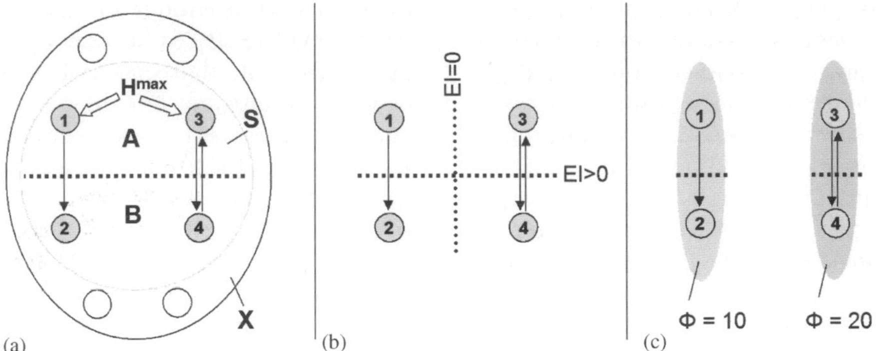

Fig. 1. Effective information, minimum information bipartition, and complexes. (a) Effective information. Shown is a single subset S of 4 elements ($\{1,2,3,4\}$, gray circle), forming part of a larger system X (black ellipse). This subset is bisected into A and B by a bipartition ($\{1,3\}/\{2,4\}$, indicated by the dotted gray line). Arrows indicate causally effective connections linking A to B and B to A across the bipartition (other connections may link both A and B to the rest of the system X). To measure EI(A→B), maximum entropy $H^{\max}$ is injected into the outgoing connections from A (corresponding to independent noise sources). The entropy of the states of B that is due to the input is then measured. Note that A can affect B directly through connections linking the two subsets, as well as indirectly via X. Applying maximum entropy to B allows one to measure EI(B→A). The effective information for this bipartition is $\mathrm{EI}(\mathrm{A}\leftrightarrows\mathrm{B}) = \mathrm{EI}(\mathrm{A}\rightarrow\mathrm{B}) + \mathrm{EI}(\mathrm{B}\rightarrow\mathrm{A})$. (b) Minimum information bipartition. For subset $S = \{1,2,3,4\}$, the horizontal bipartition $\{1,3\}/\{2,4\}$ yields a positive value of EI. However, the bipartition $\{1,2\}/\{3,4\}$ yields $\mathrm{EI} = 0$ and is a minimum information bipartition (MIB) for this subset. The other bipartitions of subset $S = \{1,2,3,4\}$ are $\{1,4\}/\{2,3\}$, $\{1\}/\{2,3,4\}$, $\{2\}/\{1,3,4\}$, $\{3\}/\{1,2,4\}$, $\{4\}/\{1,2,3\}$, all with $\mathrm{EI}>0$. (c) Analysis of complexes. By considering all subsets of system X one can identify its complexes and rank them by the respective values of Φ — the value of EI for their minimum information bipartition. Assuming that other elements in X are disconnected, it is easy to see that $\Phi>0$ for subset $\{3,4\}$ and $\{1,2\}$, but $\Phi = 0$ for subsets $\{1,3\}$, $\{1,4\}$, $\{2,3\}$, $\{2,4\}$, $\{1,2,3\}$, $\{1,2,4\}$, $\{1,3,4\}$, $\{2,3,4\}$, and $\{1,2,3,4\}$. Subsets $\{3,4\}$ and $\{1,2\}$ are not part of a larger subset having higher Φ, and therefore they constitute complexes. This is indicated schematically by having them encircled by a gray oval (darker gray indicates higher Φ). In order to identify complexes and their $\Phi(S)$ for systems with many different connection patterns, each system X was implemented as a stationary multidimensional Gaussian process such that values for effective information could be obtained analytically (for more details see Tononi and Sporns, 2003).

having $\Phi>0$ that is not included within a larger subset having higher Φ. For a complex, and only for a complex, it is appropriate to say that, when it enters a particular state out if its repertoire, it generates an amount of integrated information corresponding to its Φ value. Of the complexes that make up a given system, the one with the maximum value of $\Phi(S)$ is called the *main complex*. Some properties of complexes are worth pointing out. For example, a complex can be causally connected to elements that are not part of it. The elements of a complex that receive inputs from or provide outputs to other elements not part of that complex are called ports-in and ports-out, respectively. Also, the same element can belong to more than one complex, and complexes can overlap. One should also note that the Φ value of a complex is dependent on both spatial and temporal scales that determine what counts as a state of the underlying system. In general, the relevant spatial and temporal scales are those that jointly maximize Φ (Tononi, 2004b). In the case of the brain, the spatial elements and time scales that maximize Φ are likely to be local collections of neurons such as minicolumns and periods of time comprised between tens and hundreds of milliseconds, respectively.

In summary, a system can be analyzed to identify its complexes — those subsets of elements that can integrate information, and each complex will have an associated value of Φ, i.e. the amount of information it can integrate. To the extent that consciousness corresponds to the capacity to integrate information, complexes are the "subjects" of experience, being the locus where information can be integrated. Since information can only be integrated *within* a complex and not outside its boundaries, consciousness as information integration is necessarily subjective, private, and related to a single point of view or perspective (Tononi &

Edelman, 1998; Edelman & Tononi, 2000). It follows that elements that are part of a complex contribute to its conscious experience, while elements that are not part of it do not, even though they may be connected to it and exchange information with it through ports-in and ports-out.

Neuroscience: consciousness, information integration, and the brain

If consciousness corresponds to information integration, and if information integration can be measured as suggested above, it follows that a physical system will have consciousness to the extent that it constitutes a complex having high values of Φ. How do these concepts apply to the brain, and can they account, at least in principle, for some of the facts and puzzles listed above? Can they shed any light, for instance, on why the thalamocortical system is essential for consciousness whereas the cerebellum is not, or on why consciousness is reduced during slow-wave sleep?

Thalamocortical system. A well-functioning thalamocortical system is essential for consciousness (Plum, 1991), although opinions differ about the contribution of specific cortical areas (Tononi & Edelman, 1998; Zeman, 2001; Rees et al., 2002; Crick & Koch, 2003). Studies of comatose or vegetative patients indicate that a global loss of consciousness is usually caused by gray or white matter lesions that impair multiple sectors of the thalamocortical system (Adams et al., 2000; Laureys et al., 2002, 2004; Schiff et al., 2002). By contrast, selective lesions of individual thalamocortical areas impair different submodalities of conscious experience, such as the perception of color or of faces (Kolb & Whishaw, 1996). A global, persistent disruption of consciousness can also be produced by focal lesions of paramedian mesodiencephalic structures, which include the intralaminar thalamic nuclei (Schiff, 2004). Most likely, such focal lesions are catastrophic because the strategic location and connectivity of paramedian structures ensure that distributed cortico-thalamic loops can work together as a system. Electrophysiological and imaging studies also indicate that neural activity that correlates with conscious experience is widely distributed over the cortex (Tononi et al., 1998; Srinivasan et al., 1999; McIntosh et al., 1999, 2003; Rees et al., 2002). It would seem, therefore, that the neural substrate of consciousness is a distributed thalamocortical network, and that there is no single cortical area where it all comes together.

The fact that consciousness as we know it is generated by the thalamocortical system fits well with the IITC, since what we know about its organization appears ideally suited to the integration of information. On the information side, the thalamocortical system comprises a large number of elements that are functionally specialized, becoming activated in different circumstances (Zeki, 1993). Thus, the cerebral cortex is subdivided into systems dealing with different functions, such as vision, audition, motor control, planning, and many others. Each system in turn is subdivided into specialized areas, for example, different visual areas are activated by shape, color, and motion. Within an area, different groups of neurons are further specialized, e.g. by responding to different directions of motion. On the integration side, the specialized elements of the thalamocortical system are linked by an extended network of intra- and inter-areal connections that permit rapid and effective interactions within and between areas (Engel et al., 2001). In this way, thalamocortical neuronal groups are kept ready to respond, at multiple spatial and temporal scales, to activity changes in nearby and distant thalamocortical areas. As suggested by the regular finding of neurons showing multimodal responses that change depending on the context (Cohen & Andersen, 2002; Ekstrom et al., 2003), the capacity of the thalamocortical system to integrate information is probably greatly enhanced by non-linear switching mechanisms, such as gain modulation or synchronization, that can modify mappings between brain areas dynamically (Pouget et al., 2002; Tononi et al., 1992). In summary, the thalamocortical system is organized in a way that appears to emphasize at once both functional specialization and functional integration.

As shown by computer simulations, systems of neural elements whose connectivity jointly satisfies the requirements for functional specialization and functional integration are well suited to integrating information. Fig. 2a shows a representative connection matrix obtained by optimizing for Φ starting from random connection weights. A graph-theoretical analysis indicates that connection matrices yielding the highest values of information integration ($\Phi = 74$ bits) share two key characteristics. First, connection patterns are different for different elements, ensuring functional specialization. Second, all elements can be reached from all other elements of the network, ensuring functional integration. Thus, simulated systems having maximum Φ appear to require both functional specialization and functional integration. In fact, if

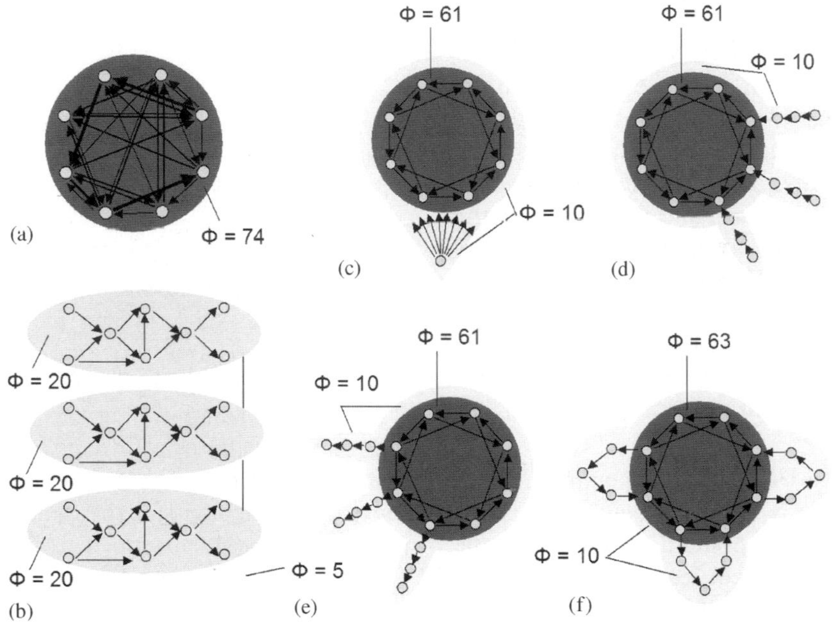

Fig. 2. Information integration for prototypical neural architectures. (a) Schematic of a cortical-like organization obtained by optimization of information integration. Shown is the causal interaction diagram for a network whose connection matrix was obtained by optimization for Φ ($\Phi = 74$ bits). Note the heterogeneous arrangement of the incoming and outgoing connections: each element is connected to a different subset of elements, with different weights. Further analysis indicates that this network jointly maximizes functional specialization and functional integration among its eight elements, thereby resembling the anatomical organization of the thalamocortical system (Tononi and Sporns, 2003). (b) Schematic of a cerebellum-like organization. Shown are three modules of eight elements each, with many feed forward and lateral connections within each module but minimal connections among them. The analysis of complexes reveals three separate complexes with low values of Φ ($\Phi = 20$ bits). There is also a large complex encompassing all the elements, but its Φ value is extremely low ($\Phi = 5$ bits). (c) Schematic of the organization of a subcortical activating system. Shown is a single subcortical "reticular" element providing common input to the eight elements of a thalamocortical-like main complex (both specialized and integrated, $\Phi = 61$ bits). Despite the diffuse projections from the reticular element on the main complex, the complex comprising all nine elements has a much lower value of Φ ($\Phi = 10$ bits). (d) Schematic of the organization of afferent pathways. Shown are three short chains that stand for afferent pathways. Each chain connects to a port-in of a main complex having a high value of Φ (61 bits) that is thalamocortical-like (both specialized and integrated). Note that the afferent pathways and the elements of the main complex together constitute a large complex, but its Φ value is low ($\Phi = 10$ bits). Thus, elements in afferent pathways can affect the main complex without belonging to it. (e) Schematic of the organization of efferent pathways. Shown are three short chains that stand for efferent pathways. Each chain receives a connection from a port-out of the thalamocortical-like main complex. Also in this case, the efferent pathways and the elements of the main complex together constitute a large complex, but its Φ value is low ($\Phi = 10$ bits). (f) Schematic of the organization of cortico-subcortico-cortical loops. Shown are three short chains that stand for cortico-subcortico-cortical loops, which are connected to the main complex at both ports-in and ports-out. Again, the subcortical loops and the elements of the main complex together constitute a large complex, but its Φ value is low ($\Phi = 10$ bits). Thus, elements in loops connected to the main complex can affect it without belonging to it. Note, however, that the addition of these three loops slightly increased the Φ value of the main complex (from $\Phi = 61$ to $\Phi = 63$ bits) by providing additional pathways for interactions among its elements.

functional specialization is lost by replacing the heterogeneous connectivity with a homogeneous one, or if functional integration is lost by rearranging the connections to form small modules, the value of Φ decreases considerably (Tononi & Sporns, 2003). Further simulations show that it is possible to construct a large complex of high Φ by joining smaller complexes through reciprocal connections. In the thalamocortical system, reciprocal connections linking topographically organized areas may be especially effective with respect to information integration. Thus, the coexistence of functional specialization and functional integration, epitomized by the thalamocortical system, is associated with high values of Φ.

Cerebellum. This brain region contains probably more neurons and as many connections as the cerebral cortex, receives mapped inputs from the environment, and controls several outputs. However, in striking contrast to the thalamocortical system, lesions or ablations indicate that the direct contribution of the cerebellum to conscious experience is minimal. According to the IITC, the reason lies with the organization of cerebellar connections, which is radically different from that of the thalamocortical system and is not well suited to information integration. Specifically, the organization of the connections is such that individual patches of cerebellar cortex tend to be activated independently of one another, with little interaction possible between distant patches (Bower, 2002; Cohen & Yarom, 1998). This suggests that cerebellar connections may not be organized so as to generate a large complex of high Φ, but rather to give rise to many small complexes each with a low value of Φ. Such an organization seems to be highly suited for both the learning and the rapid, effortless execution of informationally insulated subroutines.

This concept is illustrated in Fig. 2b, which shows a strongly *modular* network, consisting of three modules of eight strongly interconnected elements each. This network yields $\Phi = 20$ bits for each of its three modules, which form the system's three complexes. This example indicates that, irrespective of how many elements and connections are present in a neural structure, if that structure is organized in a strongly modular manner with little interactions among modules, complex size and Φ values are necessarily low. According to the IITC, this is the reason why these systems, although computationally very sophisticated, contribute little to consciousness. It is also the reason why there is no conscious experience associated with hypothalamic and brainstem circuits that regulate important physiological variables, such as blood pressure.

Activating systems. It has been known for a long time that lesions in the reticular formation of the brainstem can produce unconsciousness and coma. Conversely, stimulating the reticular formation can arouse a comatose animal and activate the thalamocortical system, making it ready to respond to stimuli (Moruzzi & Magoun, 1949). Groups of neurons within the reticular formation are characterized by diffuse projections to many areas of the brain. Many such groups release neuromodulators such as acetylcholine, histamine, noradrenaline, serotonin, dopamine, and glutamate (acting on metabotropic receptors) and can have extremely widespread effects on both neural excitability and plasticity (Steriade & McCarley, 1990). However, it would seem that the reticular formation, while necessary for the normal functioning of the thalamocortical system and therefore for the occurrence of conscious experience, may not contribute much in terms of specific dimensions of consciousness — it may work mostly like an external on-switch or as a transient booster of thalamocortical firing.

Such a role can be explained readily in terms of information integration. As shown in Fig. 2c, neural elements that have widespread and effective connections to a main complex of high Φ may nevertheless remain informationally excluded from it. Instead, they are part of a larger complex having a much lower value of Φ.

Cortical input systems. What we see usually depends on the activity patterns that occur in the retina and that are relayed to the brain. However, many observations suggest that retinal activity does not contribute directly to conscious experience. Retinal cells surely can discriminate light from dark and convey that information to visual cortex, but their rapidly shifting firing patterns do not correspond well with what we perceive. For example, during blinks and eye movements retinal activity changes dramatically, but visual perception does not. The retina has a blind spot at the exit of the optic nerve where there are no photoreceptors, and it has low spatial resolution and no color sensitivity at the periphery of the visual field, but we are not aware of any of this. More importantly, lesioning the retina does not prevent conscious visual experiences. For example, a person who becomes retinally blind as an adult continues to have vivid visual images and dreams. Conversely, stimulating the retina during sleep by keeping the eyes open and presenting various visual inputs does not yield any visual experience and does not affect visual dreams. Why is it that retinal activity usually determines what we see through its action on thalamocortical circuits, but does not contribute directly to conscious experience?

As shown in Fig. 2d, adding or removing multiple, segregated incoming pathways to or from a main complex does not change the composition of the main complex, and causes little change in its Φ. While the incoming pathways do participate in a larger complex together with the elements of the main complex, the Φ value of this larger complex is very low, being limited by the effective information between each afferent pathway and its port-in at the main complex. Thus, input pathways providing powerful inputs to a complex add nothing to the information it integrates if their effects are entirely accounted for by ports-in.

Cortical output systems. Similar considerations apply to cortical output systems. In neurological practice, as well as in everyday life, we tend to associate consciousness with the presence of a diverse behavioral repertoire. For example, if we ask a lot of different questions and for each of them we obtain an appropriate answer, we generally infer that a person is conscious. Such a criterion is not unreasonable in terms of information integration, given that a wide behavioral repertoire is usually indicative of a large repertoire of internal states that is available to an integrated system. However, it appears that neural activity in motor pathways, which is necessary to bring about such diverse behavioral responses, does not in itself contribute to consciousness. For example, patients with the locked-in syndrome, who are completely paralyzed except for the ability to gaze upward, are fully conscious. Similarly, during dreams, consciousness is not impaired despite the absence of overt behavior. Even lesions of central motor areas do not impair consciousness.

Why is it that neurons in motor pathways, which can produce a large repertoire of different outputs and thereby relay a large amount of information about different conscious states, do not contribute directly to consciousness? As shown in Fig. 2e, adding or removing multiple, segregated outgoing pathways to or from a main complex does not change the composition of the main complex, and its Φ value. Like incoming pathways, outgoing pathways do participate in a larger complex together with the elements of the main complex, but the Φ value of this larger complex is very low, being limited by the effective information between each port-out of the main complex and its effector targets.

Basal ganglia and cortico-subcortical loops. Another set of neural structures that may not contribute directly to conscious experience are subcortical structures such as the basal ganglia. The basal ganglia contain many circuits arranged in parallel, some implicated in motor and oculomotor control, others, such as the dorsolateral prefrontal circuit, in cognitive functions, the lateral orbitofrontal and anterior cingulate circuits, in social behavior, motivation, and emotion (Alexander et al., 1990). Each basal ganglia circuit originates in layer V of the cortex, and through a last step in the thalamus, returns to the cortex, not far from where the circuit started (Middleton & Strick, 2000). Similarly arranged cortico-cerebello-thalamo-cortical loops also exist. Why is it that such complicated neural structures, which are tightly connected to the thalamocortical system at both ends, do not seem to contribute directly to conscious experience?

As shown in Fig. 2f, the addition of many parallel cycles generally does not change the composition of the main complex, although Φ values can be altered. Instead, the elements of the main complex and of the connected cycles form a joint complex that can only integrate the limited amount of information exchanged within each cycle. Thus, subcortical cycles or loops implement specialized subroutines that are capable of influencing the states of the main thalamocortical complex without joining it. Such informationally insulated cortico-subcortical loops could constitute the neural substrates for many unconscious processes that can affect and be affected by conscious experience (Baars, 1988; Tononi, 2004a, b). It is likely that new informationally insulated loops can be created through learning and repetition. For example, when first performing a new task, we are conscious of every detail of it, we make mistakes, are slow, and must make an effort. When we have learned the task well, we perform it better, faster, and with less effort, but we are also less aware of it. As suggested by imaging results, a large number of neocortical regions are involved when we first perform a task. With practice, activation is reduced or shifts to different circuits (Raichle, 1998). According to the IITC, during the early trials, performing the task involves many regions of the main complex, while later certain aspects of the task are delegated to neural circuits, including subcortical ones, that are informationally insulated.

Cortical loops. Even within the thalamocortical system proper, a substantial proportion of neural activity does not appear to contribute directly to conscious experience. For example, what we see and hear requires elaborate computational processes dealing with figure-ground segregation, depth perception, object recognition, and language parsing, many of which take place in the thalamocortical system. Yet we are not aware of all this diligent buzzing: we just *see* objects, separated from the background and laid out in space, and know what they are, or *hear* words, nicely separated from each other, and know what they mean. As an example, take binocular rivalry, where the two eyes view two different images, but we perceive consciously just one image at a time, alternating in sequence. Recordings in monkeys have shown that the activity of visual neurons in certain cortical areas, such as the inferotemporal cortex, follows faithfully what the subject perceives consciously. However, in other areas, such as primary visual cortex, there are many neurons that respond to the stimulus presented to the eye, whether or not the subject is perceiving it (Logothetis et al., 1996). Neuromagnetic studies in humans have shown that neural activity correlated with a stimulus that is not being consciously perceived can be recorded in many cortical areas, including the front of the brain (Srinivasan et al., 1999). Why does the firing of many cortical neurons, carrying out the computational processes that enable object recognition (or language parsing), not correspond to anything conscious?

The situation is similar on the executive side of consciousness. When we plan to do or say something, we are vaguely conscious of what we intend, and presumably these intentions are reflected in specific firing patterns of certain neuronal groups. Our vague intentions are then translated almost miraculously into the right words, strung together to form a syntactically correct sentence that conveys what we meant to say. And yet again, we are not at all conscious of the complicated processing that is needed to carry out our intentions, much of which takes place in the cortex. What determines whether the firing of neurons within the thalamocortical system, contributes directly to consciousness or not? According to the IITC, the same considerations that apply to input and output circuits and to cortico-subcortico-cortical loops also apply to circuits and loops contained entirely within the thalamocortical system. Thus, the theory predicts that activity within certain cortical circuits does not contribute to consciousness because such circuits implement informationally insulated loops that remain outside the main thalamocortical complex. At this stage, however, it is hard to say precisely which cortical circuits may be informationally insulated. Are primary sensory cortices organized like massive afferent pathways to a main complex "higher up" in the cortical hierarchy? Is much of prefrontal cortex organized like a massive efferent pathway? Do certain cortical areas, such as those belonging to the dorsal visual stream, remain partly segregated from the main complex? Do interactions *within* a cortico-thalamic minicolumn qualify as intrinsic mini-loops that support the main complex without being part of it? Unfortunately, answering these questions and properly testing the predictions of the theory requires a much better understanding of cortical neuroanatomy than is presently available (Ascoli, 1999).

Anatomical and functional disconnections. When the corpus callosum is sectioned, consciousness is split. The level of consciousness of the dominant hemisphere, and most of its contents, are not altered severely after the operation. The non-dominant hemisphere also appears to be conscious, although it loses some important abilities. Some information, e.g. emotional arousal, seems to be shared across the hemispheres, probably owing to subcortical common inputs. As illustrated by simple computer models (Tononi, 2004b), a "callosal" cut produces, out of large complex corresponding to the connected thalamocortical system, two separate complexes. However, because there is great redundancy between the two hemispheres, their Φ value is not greatly reduced compared to when they formed a single complex. The analysis of complexes also identifies a complex corresponding to both hemispheres and their subcortical common inputs, although with much lower Φ values. That is, there is a sense in which the two hemispheres still form an integrated entity, but the information they share is minimal.

In addition to anatomical disconnections, functional disconnections may also lead to a restriction of the neural substrate of consciousness. Functional disconnections between certain parts of the brain and others may play a role in neurological neglect phenomena, may underlie psychiatric conversion and dissociative disorders, may occur during dreaming, and may be implicated in conditions such as hypnosis. It is also possible that certain attentional phenomena may correspond to changes in the neural substrate of consciousness. For example, when one is absorbed in thought, or focused exclusively on a given sensory modality, such as vision, the neural substrate of consciousness may not be the same as when we are diffusely monitoring the environment. Phenomena such as the attentional blink, where a fixed sensory input may at times make it to consciousness and at times not, may also be due to changes in functional connectivity: access to the main thalamocortical complex may be enabled or not based on dynamics intrinsic to the complex (Dehaene et al., 2003). Phenomena such as binocular rivalry may also be related, at least in part, to dynamic changes in the composition of the main thalamocortical complex caused by transient changes in functional connectivity (Lumer, 1998). Computer simulations confirm that functional disconnection can reduce the size of a complex and reduce its capacity to integrate information (Tononi, 2004b). While it is not easy to determine, at present, whether a particular group of neurons is excluded from the main complex because of hard-wired anatomical constraints, or is transiently disconnected due to functional changes, the set of elements underlying consciousness is not static, but can be considered to form a "*dynamic complex*" or "*dynamic core*" (Tononi & Edelman, 1998).

Slow-wave sleep. If neuroanatomical organization is the key in enabling information integration and thereby consciousness, neurophysiological parameters are no less important. A case in point is provided by sleep, perhaps the most familiar and yet striking alteration of consciousness. Upon awakening from dreamless sleep, we have the peculiar impression that for a while we were not there at all nor, as far as we are concerned, was the rest of the world. This everyday observation tells us vividly that consciousness can be gained and lost, grow and shrink. Indeed, if we did not sleep, it might be hard to imagine that consciousness is not a given, but depends somehow on the way our brain is functioning. The loss of consciousness between falling asleep and waking up is relative, rather than absolute (Hobson et al., 2000). Thus, careful studies of mental activity reported immediately after awakening have shown that some degree of consciousness is maintained during much of sleep. Many awakenings, especially from rapid eye movement (REM) sleep, yield dream reports, and dreams can be at times as vivid and intensely conscious as waking experiences. Dream-like consciousness also occurs during various phases of slow-wave sleep, especially at sleep onset and during the last part of the night. Nevertheless, a certain proportion of awakenings do not yield any dream report, suggesting a marked reduction of consciousness. Such "empty" awakenings typically occur during the deepest stages of slow wave sleep (stages 3 and 4), especially during the first half of the night.

Which neurophysiological parameters are responsible for the remarkable changes in the quantity and quality of conscious experience that occur during sleep? We know for certain that the brain does not simply shut off during sleep. During REM sleep, for example, neural activity is as high, if not higher, than during wakefulness, and EEG recordings show low-voltage fast-activity. This EEG pattern is known as "activated" because cortical neurons, being steadily depolarized and close to their firing threshold, are ready to respond to incoming inputs. Given these similarities, it is perhaps not surprising that consciousness should be present during both states. Changes in the quality of consciousness, however, do occur, and they correspond closely to relative changes in the activation of different brain areas (Hobson et al., 2000).

During slow wave sleep, average firing rates of cortical neurons are also similar to those observed during quiet wakefulness. However, due to changes in the level of certain neuromodulators, virtually all cortical neurons engage in slow oscillations at around 1 Hz, which are reflected in slow waves in the EEG (Steriade, 1997). Slow oscillations consist of a depolarized phase, during which the membrane potential of cortical neurons is close to firing threshold and spontaneous firing rates are similar to quiet wakefulness, and of a hyperpolarized phase, during which neurons become silent and are further away from firing threshold. From the perspective of information integration, a reduction in the readiness to respond to stimuli during the hyperpolarization phase of the slow oscillation would imply a reduction of consciousness. It would be as if we were watching very short fragments of a movie interspersed with repeated unconscious "blanks" in which we cannot see, think, or remember anything, and therefore have little to report. A similar kind of unreadiness to respond, associated with profound hyperpolarization, is found in deep anesthesia, another condition were consciousness is impaired.

From the perspective of information integration, a reduction of consciousness during certain phases of sleep would occur even if the brain remained capable of responding to perturbations, provided its response were to lack differentiation. This prediction is borne out by detailed computer models of a portion of the visual thalamocortical system (Hill & Tononi, in preparation). According to these simulations, in the waking mode different perturbations of the thalamocortical network yield specific responses. In the sleep mode, instead, the network becomes bistable. Specific effects of different perturbations are quickly washed out and their propagation impeded: the whole network transitions into the depolarized or into the hyperpolarized phase of the slow oscillation — a stereotypic response that is observed irrespective of the particular perturbation. And of course, this bistability is also evident in the spontaneous behavior of the network: during each slow oscillation, cortical neurons are either all firing or all silent, with little freedom in between. In summary, these simulations indicate that, even if the anatomical connectivity of a complex stays the same, a change in key parameters governing the readiness of neurons to respond, and the differentiation of their responses may alter radically the Φ value of the complex, with corresponding consequences on consciousness. Further simulations indicate that the capacity to integrate information is also reduced if neural activity is extremely high and near-synchronous, due to a dramatic decrease in the available degrees of freedom (Tononi, unpublished results). This reduction in degrees of freedom could be the reason why consciousness is reduced or eliminated in absence seizure and other conditions characterized by hypersynchronous neural activity.

Conscious experience and time. Consciousness not only requires a neural substrate with appropriate anatomical structure and appropriate physiological parameters: it also needs time. For example, studies of how a percept is progressively specified and stabilized indicate that it takes up to 100–200 ms to develop a fully formed sensory experience, and that the surfacing of a conscious thought may take even longer (Bachmann, 2000). Experiments in which the somatosensory areas of the cerebral cortex were stimulated directly indicate that low intensity stimuli must be sustained for up to 500 ms to produce a conscious sensation (Libet, 1982). Multiunit recordings in the primary visual cortex of monkeys show that, after a stimulus is presented, the firing rate of many neurons increases irrespective of whether the animal reports seeing a figure or not. After 80–100 ms, however, their discharge accurately predicts the conscious detection of the figure. Thus, the firing of the same cortical neurons may correlate with consciousness at certain times, but not at other times (Lamme & Roelfsema, 2000). What determines when the firing of the same cortical neurons contributes to conscious experience and when it does not? And why does it take up to hundreds of milliseconds before a conscious experience is generated?

The IITC predicts that the time requirement for the generation of conscious experience in the brain emerge directly from the time requirements for the buildup of effective interactions among the elements of the thalamocortical main complex. As mentioned above, if one were to perturb half of the elements of the main complex for less than a millisecond, no perturbations would produce any effect on the other half within this time window, and Φ would be equal to zero. After, say, 100 ms, however, there is enough time for differential effects to be manifested, and Φ should grow. Thus, the time scale of neurophysiological interactions needed to integrate information among distant cortical regions appears to be consistent with that required by psychophysical observations (microgenesis), by stimulation experiments, and by recording experiments.

Comparisons and conclusions

The examples discussed above show that the IITC can account, in a coherent manner, for several puzzling facts about consciousness and the brain. How does the theory compare with other approaches to the neurobiology of consciousness, and what are some of its implications and predictions?

Few neuroscientists have devoted an organized body of work to the neural substrates of consciousness. Edelman (1989) was among the first to propose that consciousness should be addressed fully within a neurobiological framework. In several publications (Edelman, 1989, 2003), Edelman has maintained that consciousness requires reentrant interactions between posterior networks involved in perceptual categorization, and anterior-limbic networks involved in "value-category" memory, which result in a kind of "remembered present". This view represents an extension to consciousness of a more general, selectionist approach to brain function (Edelman, 1987). Key ideas are that it is useful to distinguish between primary and higher-order consciousness, that the substrate of consciousness is highly distributed and variable, and that consciousness requires a body and the interaction of the organism with an environment.

Crick and Koch were also among the first to advocate a research program aimed at identifying in progressively greater detail the neural correlates of consciousness (Crick & Koch, 1990). Their proposals are guided primarily by empirical considerations. Over the years, they have made several suggestions, ranging from the role of 40 Hz oscillations in binding different conscious attributes, to suggesting that only a small subset of neurons is associated with consciousness, to the idea that neurons associated with consciousness must project directly to prefrontal cortex, and that neurons in primary visual cortex do not contribute to consciousness (Crick & Koch, 1995, 1998). More recently, they have to some extent enlarged their scope and suggested that the substrate of consciousness may be "coalitions" of neurons, both in the front and the back of the cortex, which compete to establish some metastable, strong firing pattern that explicitly represents information and can guide action (Crick & Koch, 2003). Related ideas are that higher cortical areas as well as attention can strongly modulate the strength of conscious coalitions, that there is a penumbra of neural activity that gives "meaning" to conscious firing patterns, and that there are "zombie" neural systems that are fast but unconscious.

Dehaene and Changeux have taken as their starting point the global workspace theory (Dehaene and Naccache, 2001), elaborated most extensively in a cognitive context by Baars (1988). They have singled out, as experimentally more tractable, the notion of global access — the idea that a "piece of information" encoded in the firing of a group of neurons becomes conscious if it is "broadcast" widely, so that a large part of the brain has access to it. That is, the same information can be conscious or not depending on the size of the audience. This formulation translates, in plausible neural terms, the key insight of global workspace theory, exemplified by the theater (or TV) metaphor: a message becomes conscious when it becomes accessible to a large audience (it goes on stage), but not if it remains private. Key ideas are that global workspace neurons, characterized by their ability to send and receive projections from many distant areas through long-range excitatory fibers, are especially concentrated in prefrontal, anterior cingulate, and parietal areas, that neurons must be actively firing (broadcasting) to contribute to consciousness, that access to consciousness is an all-or-none phenomenon, requiring the nonlinear "ignition" of global workspace neurons, and that higher areas play a role in "mobilizing" lower areas into the global workspace.

There are both similarities and differences between the IITC and neurobiological frameworks such as those just described. Not surprisingly, there is broad convergence on certain key facts: that consciousness is generated by distributed thalamocortical networks; reentrant interactions among multiple cortical regions are important; that the mechanisms of consciousness and attention overlap but are not the same, and that there are many "unconscious" neural systems. Of course, different approaches may emphasize different aspects. However, at the present stage these differences are not crucial, and fluctuate with the pendulum of experimental evidence.

The main differences lie elsewhere. Unlike other approaches, the IITC addresses the so-called hard problem (Chalmers, 1996) head-on. It takes its start from phenomenology and, by making a critical use of thought experiments, argues that subjective experience is integrated information. Therefore, any physical system will have subjective experience to the extent that it is capable of integrating information. In this view, experience, i.e. information integration, is a fundamental quantity, just as mass or energy. Other approaches avoid the hard problem and do not take a theoretical stand concerning the fundamental nature of experience, restricting themselves to the empirical investigation of its neural correlates.

The IITC takes a precise view about information integration, offering a general theoretical definition and a way to measure it as the Φ value of a complex. In other approaches, including the ones inspired by the global workspace metaphor, the notion of information is not well defined. For example, it is often assumed loosely that the firing of specific thalamocortical elements (e.g. those for red) conveys some specific information (e.g. that there is something red), and that such information becomes conscious if it is disseminated widely. However, just like a retinal cell or a photodiode, a given thalamocortical element has no information about whether what made it fire was a particular color rather than a shape, a visual stimulus rather than a sound, or a sensory stimulus rather than a thought. All it knows is whether it fired or not, just as each receiving element only knows whether it received an input or not. Thus, the information specifying "red" cannot possibly be in the message conveyed by the firing of any neural element, whether it is broadcasting widely or not. According to the IITC, that information resides instead in the reduction of uncertainty occurring when a whole complex enters one out of a large number of available states. Moreover, within a complex, both active and inactive neurons count, just as the sound of an orchestra is specified both by the instruments that are playing and by those that are silent. In short, what counts is how much information is generated, and not how widely it is disseminated.

By arguing that subjective experience corresponds to a system's capacity to integrate information, and by providing a mathematical definition of information integration, the IITC can go on to show that several observations concerning the neural substrate of consciousness fall naturally into place. Other approaches generally propose a provisional list of neural ingredients that appear to be important, such as synchronization or widespread broadcasting, without providing a principled explanation of why they would be important or whether they would be always necessary. For example, synchronization is usually an indication that the elements of the complex are capable of interacting efficiently, but is neither necessary nor sufficient for consciousness: there can be strong synchronization with little consciousness (absence seizures) as well as consciousness with little synchronization (as indicate by unit recordings in higher-order visual areas). Or there can be extremely widespread "broadcasting", as exemplified most dramatically by the diffuse projections of neuromodulatory systems, yet lesion, stimulation and recording experiments do not suggest any specific contribution to specific dimensions of consciousness.

The IITC also predicts that consciousness depends exclusively on the ability of a system to integrate information, whether or not it has a strong sense of self, language, emotion, a body, or is immersed in an environment, contrary to some common intuitions. This prediction is consistent with the preservation of consciousness during REM sleep, when both input and output signals from and to the body, respectively are markedly reduced. Transient inactivation of brain areas mediating the sense of self, language, and emotion could assess this prediction in a more cogent manner. Nevertheless, the theory recognizes that these same factors are important historically because they favor the development of neural circuits forming a main complex of high Φ. For example, the ability of a system to integrate information grows as that system incorporates statistical regularities from its environment and learns (Tononi et al., 1996). In this sense, the emergence of consciousness in biological systems is predicated on a long evolutionary history, on individual development, and on experience-dependent changes in neural connectivity.

Finally, the IITC says that the presence and extent of consciousness can be determined, in principle, also in cases in which we have no verbal report, such as infants or animals, or in neurological conditions such as akinetic mutism, psychomotor seizures, and sleepwalking. In practice, of course, measuring Φ accurately in such systems will not be easy, but approximations and informed guesses are certainly conceivable. The IITC also implies that consciousness is not an all-or-none property, but increases in proportion to a system's ability to integrate information. In fact, any physical system capable of integrating information would have some degree of experience, irrespective of the stuff of which it is made.

At present, the validity of this theoretical framework and the plausibility of its implications rest on its ability to account, in a coherent manner, for some basic phenomenological observations and for some elementary but puzzling facts about consciousness and the brain. Experimental developments, especially of ways to concurrently stimulate and record the activity of broad regions of the brain, should permit stringent tests of some of the theory's predictions. Equally important will be the development of realistic, large-scale models of the anatomical organization of the brain. These models should allow a more rigorous measurement of how the capacity to integrate information relates to different brain structures and certain neurophysiological parameters (Tononi et al., 1992; Lumer et al., 1997; Hill & Tononi, 2005).

References

Adams, J.H., Graham, D.I. and Jennett, B. (2000) The neuropathology of the vegetative state after an acute brain insult. Brain, 123(Pt 7): 1327–1338.

Alexander, G. E., Crutcher, M. D. and DeLong, M. R. (1990) Basal ganglia-thalamocortical circuits: parallel substrates for motor, oculomotor, "prefrontal" and "limbic" functions. Progress in Brain Research, Vol. 85, Elsevier, Amsterdam, pp. 119–146.

Ascoli, G.A. (1999) Progress and perspectives in computational neuroanatomy. Anat. Rec., 257: 195–207.

Baars, B.J. (1988) A Cognitive Theory of Consciousness. Cambridge University Press, New York.

Bachmann, T. (2000) Microgenetic Approach to the Conscious Mind. John Benjamins Pub. Co, Amsterdam, Philadelphia.

Bower, J.M. (2002) The organization of cerebellar cortical circuitry revisited: Implications for function. Ann. N Y Acad. Sci., 978: 135–155.

Chalmers, D.J. (1996) The Conscious Mind: In Search of a Fundamental Theory, Philosophy of Mind Series. Oxford University Press, New York.

Cohen, D. and Yarom, Y. (1998) Patches of synchronized activity in the cerebellar cortex evoked by mossy-fiber stimulation: Questioning the role of parallel fibers. Proc. Natl. Acad. Sci. USA, 95: 15032–15036.

Cohen, Y.E. and Andersen, R.A. (2002) A common reference frame for movement plans in the posterior parietal cortex. Nat. Rev. Neurosci., 3: 553–562.

Crick, F. and Koch, C. (1990) Some reflections on visual awareness. Cold Spring Harbor Symposia on Quantitative Biology, 55: 953–962.

Crick, F. and Koch, C. (1995) Are we aware of neural activity in primary visual cortex? Nature, 375: 121–123.

Crick, F. and Koch, C. (1998) Consciousness and neuroscience. Cereb. Cortex, 8: 97–107.

Crick, F. and Koch, C. (2003) A framework for consciousness. Nat. Neurosci., 6: 119–126.

Dehaene, S. and Naccache, L. (2001) Towards a cognitive neuroscience of consciousness: Basic evidence and a workspace framework. Cognition, 79: 1–37.

Dehaene, S., Sergent, C. and Changeux, J.P. (2003) A neuronal network model linking subjective reports and objective physiological data during conscious perception. Proc. Natl. Acad. Sci. USA, 100: 8520–8525.

Edelman, G.M. (1987) Neural Darwinism: The Theory of Neuronal Group Selection. BasicBooks, Inc, New York.

Edelman, G.M. (1989) The Remembered Present: A Biological Theory of Consciousness. BasicBooks, Inc, New York.

Edelman, G.M. (2003) Naturalizing consciousness: A theoretical framework. Proc. Natl. Acad. Sci. USA, 100: 5520–5524.

Edelman, G.M. and Tononi, G. (2000) A Universe of Consciousness: How Matter Becomes Imagination. Basic Books, New York.

Ekstrom, A.D., Kahana, M.J., Caplan, J.B., Fields, T.A., Isham, E.A., Newman, E.L. and Fried, I. (2003) Cellular networks underlying human spatial navigation. Nature, 425: 184–188.

Engel, A.K., Fries, P. and Singer, W. (2001) Dynamic predictions: Oscillations and synchrony in top-down processing. Nat. Rev. Neurosci., 2: 704–716.

Hill, S. and Tononi, G. (2005) Modeling sleep and wakefulness in the thalamocortical system. J. Neurophysiol., 93: 1671–1698.

Hobson, J.A., Pace-Schott, E.F. and Stickgold, R. (2000) Dreaming and the brain: Toward a cognitive neuroscience of conscious states. Behav. Brain Sci., 23: 793–842 discussion 904–1121.

Kolb, B. and Whishaw, I.Q. (1996) Fundamentals of Human Neuropsychology. WH Freeman, New York.

Lamme, V.A. and Roelfsema, P.R. (2000) The distinct modes of vision offered by feedforward and recurrent processing. Trends Neurosci., 23: 571–579.

Laureys, S., Antoine, S., Boly, M., Elincx, S., Faymonville, M.E., Berre, J., Sadzot, B., Ferring, M., De Tiege, X., van Bogaert, P., Hansen, I., Damas, P., Mavroudakis, N., Lambermont, B., Del Fiore, G., Aerts, J., Degueldre, C., Phillips, C., Franck, G., Vincent, J.L., Lamy, M., Luxen, A., Moonen, G., Goldman, S. and Maquet, P. (2002) Brain function in the vegetative state. Acta. Neurol. Belg., 102: 177–185.

Laureys, S., Owen, A.M. and Schiff, N.D. (2004) Brain function in coma, vegetative state, and related disorders. Lancet Neurol., 3: 537–546.

Libet, B. (1982) Brain stimulation in the study of neuronal functions for conscious sensory experiences. Human Neurobiol., 1: 235–242.

Logothetis, N.K., Leopold, D.A. and Sheinberg, D.L. (1996) What is rivalling during binocular rivalry? Nature, 380: 621–624.

Lumer, E.D. (1998) A neural model of binocular integration and rivalry based on the coordination of action-potential timing in primary visual cortex. Cereb. Cortex, 8: 553–561.

Lumer, E.D., Edelman, G.M. and Tononi, G. (1997) Neural dynamics in a model of the thalamocortical system.1. Layers, loops and the emergence of fast synchronous rhythms. Cereb. Cortex, 7: 207–227.

McIntosh, A.R., Rajah, M.N. and Lobaugh, N.J. (1999) Interactions of prefrontal cortex in relation to awareness in sensory learning. Science, 284: 1531–1533.

McIntosh, A.R., Rajah, M.N. and Lobaugh, N.J. (2003) Functional connectivity of the medial temporal lobe relates to learning and awareness. J. Neurosci., 23: 6520–6528.

Middleton, F.A. and Strick, P.L. (2000) Basal ganglia and cerebellar loops: Motor and cognitive circuits. Brain Res. Brain Res. Rev., 31: 236–250.

Moruzzi, G. and Magoun, H.W. (1949) Brain stem reticular formation and activation of the EEG. Electroencephalog. Clin. Neurophysiol., 1: 455–473.

Plum, F. (1991) Coma and related global disturbances of the human conscious state. In: Peters A. and Jones E.G. (Eds.) Normal and Altered States of Function, Vol. 9. Plenum Press, New York, pp. 359–425.

Pouget, A., Deneve, S. and Duhamel, J.R. (2002) A computational perspective on the neural basis of multisensory spatial representations. Nat. Rev. Neurosci., 3: 741–747.

Raichle, M.E. (1998) The neural correlates of consciousness: An analysis of cognitive skill learning. Philos. Trans. R. Soc. Lond. B. Biol. Sci., 353: 1889–1901.

Rees, G., Kreiman, G. and Koch, C. (2002) Neural correlates of consciousness in humans. Nat. Rev. Neurosci., 3: 261–270.

Schiff, N.D. (2004). The neurology of impaired consciousness: Challenges for cognitive neuroscience. In: Gazzaniga, M. (Ed.), The Cognitive Neurosciences. 3rd ed. Cambridge, MA: MIT Press.

Schiff, N.D., Ribary, U., Moreno, D.R., Beattie, B., Kronberg, E., Blasberg, R., Giacino, J., McCagg, C., Fins, J.J., Llinas, R. and Plum, F. (2002) Residual cerebral activity and behavioural fragments can remain in the persistently vegetative brain. Brain, 125: 1210–1234.

Shannon, C.E. and Weaver, W. (1963) The Mathematical Theory of Communication. University of Illinois Press, Urbana.

Srinivasan, R., Russell, D.P., Edelman, G.M. and Tononi, G. (1999) Increased synchronization of neuromagnetic responses during conscious perception. J. Neurosci., 19: 5435–5448.

Steriade, M. (1997) Synchronized activities of coupled oscillators in the cerebral cortex and thalamus at different levels of vigilance. Cereb. Cortex, 7: 583–604.

Steriade, M. and McCarley, R.W. (1990) Brainstem Control of Wakefulness and Sleep. Plenum Press, New York.

Tononi, G. (2001) Information measures for conscious experience. Arch. Ital. Biol., 139: 367–371.

Tononi, G. (2004a) Consciousness and the brain: Theoretical aspects. In: Adelman G. and Smith B. (Eds.), Encyclopedia of Neuroscience. Elsevier, Amsterdam.

Tononi, G. (2004b) An information integration theory of consciousness. BMC Neurosci., 5: 42.
Tononi, G. and Edelman, G.M. (1998) Consciousness and complexity. Science, 282: 1846–1851.
Tononi, G. and Sporns, O. (2003) Measuring information integration. BMC Neurosci., 4: 31.
Tononi, G., Sporns, O. and Edelman, G.M. (1992) Reentry and the problem of integrating multiple cortical areas: Simulation of dynamic integration in the visual system. Cereb. Cortex, 2: 310–335.
Tononi, G., Sporns, O. and Edelman, G.M. (1996) A complexity measure for selective matching of signals by the brain. Proc. Natl. Acad. Sci. USA, 93: 3422–3427.
Tononi, G., Srinivasan, R., Russell, D.P. and Edelman, G.M. (1998) Investigating neural correlates of conscious perception by frequency-tagged neuromagnetic responses. Proc. Natl. Acad. Sci. USA, 95: 3198–3203.
Zeki, S. (1993) A Vision of the Brain. Blackwell Scientific Publications, Oxford, Boston.
Zeman, A. (2001) Consciousness. Brain, 124: 1263–1289.

S. Laureys (Ed.)
Progress in Brain Research, Vol. 150
ISSN 0079-6123

CHAPTER 10

Dynamics of thalamo-cortical network oscillations and human perception

Urs Ribary*

Department of Physiology and Neuroscience, NYU School of Medicine, New York, NY 10016, USA

Abstract: There is increasing evidence that human cognitive functions can be addressed from a robust neuroscience perspective. In particular, the distributed coherent electrical properties of central neuronal ensembles are considered to be a promising avenue of inquiry concerning global brain functions. The intrinsic oscillatory properties of neurons (Llinás, R. (1988) The intrinsic electrophysiological properties of mammalian neurons: Insights into central nervous system function. Science, 242: 1654–1664), supported by a large variety of voltage-gated ionic conductances are recognized to be the central elements in the generation of the temporal binding required for cognition. Research in neuroscience further indicates that oscillatory activity in the gamma band (25–50 Hz) can be correlated with both sensory acquisition and premotor planning, which are non-continuous functions in the time domain. From this perspective, gamma-band activity is viewed as serving a broad temporal binding function, where single-cell oscillators and the conduction time of the intervening pathways support large multicellular thalamo-cortical resonance that is closely linked with cognition and subjective experience. Our working hypothesis is that although dedicated units achieve sensory processing, the cognitive binding process is a common mechanism across modalities. Moreover, it is proposed that such time-dependent binding when altered, will result in modifications of the sensory motor integration that will affect and impair cognition and conscious perception.

Gamma-band oscillations and cognitive processing

The excellent correlation between brain gamma-band coherent oscillations and cognitive functions is presently considered a key issue in the study of the electrophysiological basis for higher brain function in humans (Llinás and Ribary, 1993; Crick and Koch, 1996; Schiff et al., 2002) and animals (Steriade, 1993; Gray, 1999). In addition, gamma-band oscillations have been shown to be reset (Ribary et al., 1991) or otherwise modified by sensory stimulation, and to be correlated with normal and altered auditory perception (Joliot et al., 1994; Llinás et al., 1998a; Ribary et al., 2000; Ribary et al., 2004). Because specific findings concerning the relationship between cognition and the temporal organization of the thalamo-cortical system (Kato, 1990; Lindström and Wróbel, 1990) are now well-documented, a very significant area of research with deep implications on the nature of cognitive processes and their abnormalities, is yet to be done.

Over the past two decades, there have been many studies relating to gamma-band oscillatory brain activity (30–50 Hz) in humans and animals during sensory and cognitive processing (Freeman, 1975; Bressler and Freeman, 1980; Galambos et al., 1981; Sheer, 1984; Ribary et al., 1987; Eckhorn et al., 1988; Gray and Singer, 1989; Ahissar and Vaaida, 1990; Ribary et al., 1991; Llinás et al., 1991; Steriade et al., 1991a, b; Engel et al., 1991; Pantev et al., 1991; Llinás and Ribary, 1992; Young et al., 1992; Murthy and Fetz, 1992; Basar

*Corresponding author. Tel.: +1 (212) 263-6561; Fax: +1 (212) 263-6976; E-mail: urs.ribary@med.nyu.ed

DOI: 10.1016/S0079-6123(05)50010-4

and Bullock, 1992; Llinás and Ribary, 1993; Steriade, 1993; Desmedt and Tomberge, 1994; Singer and Gray, 1995; Traub et al., 1996; Barth and MacDonald, 1996; Neuenschwander and Singer, 1996; Salenius et al., 1996; Lumer et al., 1997; Murthy and Fetz, 1997; Tallon-Baudry et al., 1997; Tallon-Baudry and Bertrand, 1999; Ribary et al., 1999; Knief et al., 2000; Kissler et al., 2000; Jensen et al., 2002a; Palva et al., 2002; Parra et al., 2003) and alterations in aging (Boettger et al., 2002). Freeman and his collaborators implemented multi-channel electric recordings from mammalian olfactory bulb and analyzed the complex spatial distribution of activity during odor discrimination (Freeman, 1975). Basar and co-workers (1987) first reported a 40-Hz auditory response in humans and later identified this distributed property as one of the principal components of the middle latency response of auditory-evoked potentials. Later, Galambos, and co-workers (1981) reported that the human auditory evoked steady-state potentials showed a resonance near 40 Hz and proposed new clinical applications and the extended exploration of cognitive processes. Sheer and his collaborators were the first to record human 40-Hz activity related to cognitive processing, and interpreted the 40-Hz rhythm as an index of a focused state of cortical arousal (Sheer, 1984). Freeman and van Dijk (1987) were the first to report on visual gamma oscillations showing that oscillatory activity in the monkey visual cortex possessed many of the same characteristics as its olfactory counterpart. Gray, from Freeman's laboratory, advanced the field in collaboration with Singer, in parallel with Eckhorn and collaborators, and described stimuli-dependent synchronous oscillations of local field potentials, and multiunit activity at frequencies near 40-Hz activity in the visual cortex of anesthetized cats (Eckhorn et al., 1988; Gray and Singer, 1989). At the single-cell level, gamma band oscillations were demonstrated to be an intrinsic property of cortical inhibitory neurons (Llinás et al., 1991) and of specific and non-specific thalamic neurons studied in vivo (Steriade, 1991a, b). These oscillations occurred coherently at widely separated visual sites within and across both hemispheres (Engel et al., 1991).

Neuronal substrates of human gamma band activity

Analysis of the origin of spatio-temporally coherent transient gamma-band electrical activity in humans during early sensory processing, using magnetoencephalography (MEG), demonstrated specific cortical activations (Pantev et al., 1991) and well-defined cortico-subcortical correlations (Ribary et al., 1991), using combined MEG with magnetic field tomography (MFT), indicating a time shift of 2-3 ms that was consistent with thalamo-cortical conduction times. In particular, MEG/MFT results indicated that the onset of activity at the thalamic level was followed by widespread activation of the thalamo-cortical system (TCS), which resulted in a large coupling of thalamo-cortical gamma band oscillation, organized in space and time (Ribary et al., 1991), which was altered in Alzheimer (AD) patients (Ribary et al., 1989, 1991). In particular, there was also an altered dynamics of steady-state gamma-band oscillations in moderate and severe AD patients, especially at cortical level (Fig. 1). Since our initial MEG findings, several studies have been initiated and confirmed the existence of such thalamo-cortical oscillation by direct recording in animals (Steriade, 1993) as well as the importance of specific and nonspecific thalamo-cortical conjunction (Steriade et al., 1996a, b).

Intracellular recordings from cortical interneurons (Llinás et al., 1991) and thalamic neurons studied in vivo (Steriade, 1991a, b) demonstrated gamma-band activity to be calcium dependent and originate in the dendrites (Pedroarena and Llinás, 1997). Taken together, these studies support the hypothesis that recurrent thalamo-cortico-thalamic activity is involved in organizing and supporting coherent gamma-band oscillations as a general synchronized event during sensory or cognitive processing (Ribary et al., 1987; Llinás and Ribary, 1993; Llinás et al., 1994; Llinás et al., 1998a; Ribary et al., 1999). In addition, mathematical modeling of the thalamo-cortical system (Destexhe and Babloyantz, 1991; Wright et al., 2001; Rennie et al., 2002) showed that the dynamics of the system was turbulent and desynchronized when intrinsic thalamic activity was excluded from the model (Destexhe and Babloyantz, 1991). The onset

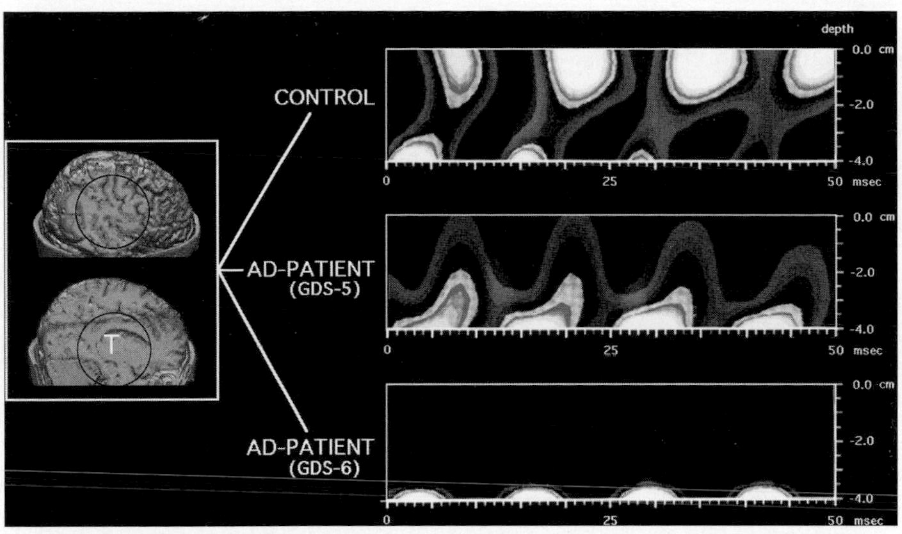

Fig. 1. Magnetic field tomography (MFT) images of steady-state gamma-band activity from a normal elderly adult and two age-matched patients with Alzheimer dementia (AD; moderate AD: Global Deterioration Scale (Reisberg et al., 1982) GDS-5, and severe AD: GDS-6) within the thalamo-cortical system. The location of the cylindrical source space is outlined on the 3D reconstructed MRI scan (left). The top face of the cylinder is close to the temporal area around the auditory cortex, while the bottom face is close to the thalamus (T). The axis of the cylinder is the midpoint of the MEG-probe placement. The MFT images are shown at 1 ms steps within a time window of 50 ms. The onset of activity at the thalamic level precedes the onset of activity at the cortical level by $\approx$ 3 ms. Note the altered gamma-band activation in AD patients, especially at cortical level.

of a pacemaker input organized the system into a more coherent spatio-temporal behavior and indicated further evidence for a coupling of oscillatory activity within thalamo-cortical systems (Steriade et al., 1993a, b; Llinás et al., 1994; Barth and MacDonald, 1996).

Earlier, we indicated that the auditory steady-state gamma-band response also reflected a complex, time-locked sequence of intrinsic network activations involving most probably thalamo-cortico-thalamic pathways, with a focus on temporal sensory areas resulting in increased synchronization of cortical gamma band activity, driven by subcortical areas (Ribary et al., 1988, 1989). A model for the cortico-thalamo-cortical resonance (Fig. 2) and electrophysiological temporal coincidence was proposed (Llinás, 1990; Llinás and Ribary, 1993), which supports gamma-band oscillation at the cortex and was recently demonstrated in vitro using voltage-dependent dye imaging in brain slices of thalamus and cortex (Llinas et al., 2002). Thalamic projection neurons synapse onto GABAergic inhibitory interneurons in layer IV. These interneurons, which have intrinsic membrane oscillations in the gamma band (close to 40 Hz), can elicit inhibitory post-synaptic potential (IPSP) in pyramidal neurons. Such input results in a 40-Hz rebound firing of the pyramidal cells (Pedroarena and Llinás, 1997).

Moreover, input from layer VI pyramidal cells terminate on thalamic neuron dendrites and generate gamma band activation of thalamic projection neurons. This thalamic oscillation is then signaled back to the cortex, establishing a large resonant oscillation between the thalamus and the cortex (Ribary et al., 1991), which can recruit sufficient elements to generate the synchronicity observed at both intracellular and extracellular levels in the cortex and thalamus. In addition to the thalamo-cortical resonance of specific thalamo-cortical neurons, we earlier suggested a second system (Llinás and Ribary, 1993), represented by the intralaminar cortical input to layer I of the cortex and its return-pathway projection via layers V and VI of pyramidal systems to the intralaminar nucleus, directly and indirectly, via collaterals to the nucleus reticularis. The cells in this system have been shown to oscillate in 40-Hz bursts

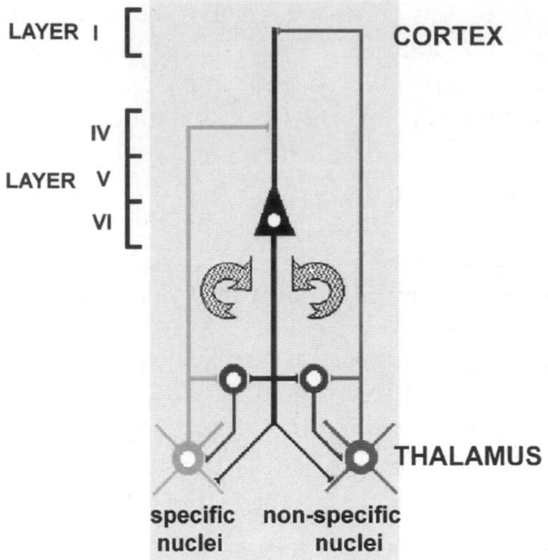

Fig. 2. Diagram of two thalamo-cortical systems proposed to generate gamma-band activity and subserve temporal binding (Llinás and Ribary, 1993). Left: Specific thalamo-cortical circuit: projection of sensory or motor nuclei input to layer IV of the cortex, producing cortical oscillation by direct activation and feed-forward inhibition. Collaterals of these projections produce thalamic feedback inhibition via the reticular nucleus. The return pathway (circular arrow on the right) re-enters this oscillation to specific and reticularis thalamic nuclei via layer VI pyramidal cells. Right: Non-specific thalamo-cortical circuit: Second loop shows nonspecific intralaminar nuclei projecting to the most superficial layer I of the cortex and giving collaterals to the reticular nucleus. Pyramidal cells return oscillation to the reticular and the non-specific thalamic nuclei, establishing a second resonant loop. The conjunction of the specific and non-specific loops is proposed to generate temporal binding.

(Steriade et al., 1993a) and were shown to synapse directly on the apical dendrites of layer V of pyramidal cells (Llinas et al., 2002). (For detailed projections in thalamus, see also Steriade and Llinas, 1988).

Finally, it is also evident from the literature that neither of these two circuits alone can directly relate to cognitive events and conscious perception. Indeed, damage of specific systems produces loss of the particular modality, while damage of the non-specific thalamus produces lethargy and coma (Façon et al., 1958; Castaigne et al., 1980) and disturbances in visual perception (Purpura and Schiff, 1997). As such then, the resonant gamma-band co-activation of both, the specific and non-specific thalamo-cortical system, may contribute to conscious perception (Llinás and Ribary, 1993), and may relate to the binding of sensory information (Gray, 1999).

Network dynamics and sensory perception

It was further proposed that coupling of thalamo-cortical oscillatory activity support the temporal binding mechanism responsible for bringing together information from various sensory modalities into one single percept (Llinás and Ribary, 1993). Indeed, MEG recordings on control subjects demonstrated that such precise timing of thalamo-cortical network activity, originally described during auditory temporal processing (Joliot et al., 1994), was also present in the somatosensory (Sauve et al., 1999) and visual modality (Ramirez et al., 2000, 2002) in the healthy human brain. In particular, we have provided evidence that the reset of the gamma-band magnetic signal correlated with sensory perception, namely with the minimal interval required to identify separate sensory stimuli.

Coherent gamma oscillations vs. auditory perception

The auditory system is known for its excellent time resolution (Teich et al., 1993; Carr, 1993). Indeed, the system can localize sound in space by detecting delays on the order of microseconds (Carr and Konishi, 1990). In addition, neuropsychological and psychophysical observations indicated that the auditory system is capable of tonality discrimination of two stimuli separated by only 1–2 ms (Miller and Taylor, 1948), while 15–20 ms is required for the perceptual identification of two stimuli (Hirsh, 1959).

Using functional brain imaging, such as MEG, earlier data indicated a clear time-locked oscillatory activity in the gamma band in response to one auditory stimulus (Ribary et al., 1991; Pantev et al., 1991; Barth and MacDonald, 1996). This reset was defined as a 2.5 cycle gamma-band oscillation (Joliot et al., 1994). Results further indicated that this coherent gamma-band activity is related to the temporal binding of incoming sensory stimuli (Joliot et al., 1994). In particular, it was determined that the minimal interval required to identify separate auditory stimuli is correlated with the resetting of the gamma-band magnetic signal. In young subjects, experimental and modeling results indicated a stimulus-interval-dependent response with a critical interval of 12–15 ms (Joliot et al., 1994), which occurred specifically in the gamma range only. At shorter intervals, only one gamma-band response occurred following the first stimulus. With longer intervals, a second gamma-band response appeared, which coincided with the subject's perception of a second distinct auditory stimulus. These results indicated that gamma-band oscillation recorded during the first 100 ms post stimulus represent a functional correlate to the early temporal processing of auditory stimuli.

Specifically, MEG activity was recorded in response to two clicks presented at various interstimulus intervals (ISI). A power spectral analysis of the raw MEG data in response to a single click showed activation at lower frequency (9–10 Hz) and at gamma-band frequency (25–50 Hz) (Joliot et al., 1994). In response to the two clicks with increasing ISI, data indicated a distinct temporal coherent pattern in the gamma range only, and no clear temporal characteristics in the lower frequency range. In particular, data indicated only one gamma-band response following the presentation of two auditory stimuli at ISIs of less than 12 ms. Indeed, the response was identical to that following a single stimulus (Joliot et al., 1994). When the stimuli were presented at longer intervals, a second gamma-band response abruptly appeared. This second response overlapped that elicited by the first stimulus.

To investigate the underlying mechanism of these results we tested two possible models (Joliot et al., 1994): The first model posited that the first stimulus triggers a gamma-band oscillation event and the second stimulus would induce a similar response only after a given time interval. The second model posited that both stimuli produced a separate gamma-band oscillation event, independently of the ISI. The results predicted by the two models were compared with the electrophysiological data obtained from all MEG recording positions. Statistical analysis of nine young subjects (20–40 years old) indicated that the first model fitted the experimental data significantly better than the second one. A two-tailed Student's t-test on the criteria value (x) indicated that the dependent model was not significantly different below the ISI of 14.2 ms for all nine subjects (Joliot et al., 1994).

This finding indicated that at ISIs of 14.2 ms or less, only the first stimulus induced a gamma-band response, while with longer intervals a second reset in gamma-band activity occurred. Thus, in this case each stimulus induced its own gamma-band activity. A similar set of stimuli was presented following MEG recordings, and the perceptual threshold for identifying two clicks was established. Perceptual responses from all subjects indicated that stimuli presented at an ISI longer than 13.7 ms could be identified as two clicks. The interval between stimuli required for the second MEG response and that required for the perception of a second auditory stimulus as a distinct event were not statistically different (interval difference $= 1.32$ ms, $n = 9$, $p = 0.309$, two-tailed Student's t-test). No regression between the two variables was significant, indicating a cluster-like correlation near 12–15 ms (Joliot et al., 1994). These findings correlated well with neuropsychological and psychophysical observations.

The ability to judge the temporal order of a sequence of sounds was reported to depend on whether the task required actual identification of the individual elements of the sequence or whether it could be performed by discrimination of their global pattern. While the finest acuity for discrimination tasks occurred on the order of 1–2 ms (Miller and Taylor, 1948), the identification of individual elements was on the order of 15–20 ms (Hirsh, 1959). Brain-imaging results coincided closely to the classical ISI required to identify individual stimuli. In addition, other observations concluded that sensory information was processed in discrete time segments (Poppel, 1970; Madler et al., 1991) as low as 12.5 ms (Kristofferson, 1984). This indicated that stimuli coming within one perceptual "quantum" (12–15 ms) were bound into one cognitive event, rather than perceived as separate entities (Llinás and Ribary, 1993).

These findings indicated that gamma-band oscillatory activity is not only involved in primary sensory processing per se (Pantev et al., 1991), but also forms part of a time conjunction or binding property that amalgamates early sensory information in perceptual time quanta (Llinás and Ribary, 1993). The results further indicated that binding can occur in steps or "quanta" of 12–15 ms, and further support our hypothesis that gamma-band oscillatory activity could serve a broad cognitive binding function of multi-sensory information into a single percept (Llinás and Ribary, 1993).

Coherent gamma oscillations versus somatosensory and visual perception

Furthermore, functional imaging data have shown that this minimal interval to identify two distinct sensory events is similar in other sensory modalities, demonstrating that such precise timing of thalamo-cortical network activity was also present in the somatosensory (Sauve et al., 1999) and visual modality (Ramirez et al., 2000, 2002) and correlated well with the perceptual identification of somatosensory or visual stimuli.

Somatosensory oscillatory cortical representations in the contra-lateral primary somatosensory cortex have been characterized using EEG and MEG (Desmedt and Tomberge, 1994; Sauve, 1999, 2003). Functional somatotopic reorganization in animals and humans has further been observed (Merzenich et al., 1987; Ramachandran et al., 1992; Mogilner et al., 1993; Yang et al., 1994; Elbert et al., 1995). However, relatively few experiments have examined the underlying network dynamics and the binding of somatosensory input. In recent experiments, six sighted and six blind subjects were stimulated with taps of 2 ms duration from two piezoelectric stimulators, held between the thumb and index finger of each hand (Sauve et al., 1999; Ribary et al., 2002). Both hands were stimulated synchronously with one tap each, or with an ISI of 2, 4, 6,..., 26, 30, 34 ms. MEG results indicated a specific time pattern of coherent oscillatory gamma-band activity, which correlated with the perception of somatosensory input (Sauve et al., 1999). In particular, data indicated that the temporal processing of tactile stimuli occurred in discrete time quanta of 12–15 ms, as originally observed in the auditory modality (Joliot et al., 1994).

Recent findings indicated that a similar precise timing of thalamo-cortical networks was also observed in the visual system, responsible for the integration of visual input (Ramirez et al., 2000). The human visual system is hypothesized to code motion in discrete time quanta. To investigate the neurobiological correlates of apparent visual motion perception, we recently designed a paradigm in which the neuromagnetic responses to local apparent motion stimuli were recorded using MEG (Ramirez et al., 2000). Subjects were recorded while stimulated in the periphery with two bars of light separated by 1° at a distance of 45 cm, flashed for 3 ms, with stimulus onset asynchrony (SOA) of 0, 3,..., 27 ms. Psychophysical analysis indicated that the subjects perceived apparent motion more than 50% of the time for SOAs longer than 15–18 ms. For shorter SOAs, subjects perceived two simultaneous bars of light rather than a single moving bar. MEG results therefore indicated that apparent visual motion perception is processed in quanta of 12–15 ms as well, and correlated with increased gamma-band activity in the visual system. Dynamic gamma-band source activations were reconstructed in many of the motion-sensitive visual areas, particularly in area V3A, hMT+, parietal and temporal cortex (Ramirez et al., 2000, 2002; Ribary et al., 2002).

These data further indicated that the same fundamental processing mechanism seems to be functioning for the three sensory modalities, namely a precise timing of thalamo-cortical network activations in the gamma-band in the healthy human brain, in order to bind sensory input into one single percept (Llinás and Ribary, 1993).

Altered gamma oscillations versus altered perception in dyslexia and normal aging

Previous studies have shown that the minimal interval to identify two separate sensory events is altered in subjects with language-based learning disabilities (LLD or dyslexia) (Llinás et al., 1998b; Ribary et al., 2000) and in normal aging (Sauve et al., 2003; Ribary et al., 2004). Such correlation is observable independently by either psychological means or functional MEG imaging. Also, there is an excellent correlation between these two different measurement techniques (Llinás et al., 1998b; Ribary et al., 2004). As such, MEG imaging may offer an objective measure that is correlated with normal and slightly altered sensory cognitive experience (Nagarajan et al., 1999). We previously proposed that gamma-band binding abnormalities could be one of the neurophysiological correlates of the temporal deficits recorded in LLD (Llinás, 1993). Indeed, in subjects with dyslexia, altered MEG data correlated with a delayed perceptual threshold above 20 ms for the identification of a second auditory stimulus (Llinás et al., 1998b; Ribary et al., 2000), compared to 12–15 ms in healthy controls. This dyschronia is consistent with other findings concerning LLD (Tallal et al., 1993, 1996; Merzenich et al., 1996; Salmelin et al., 1996; Nagarajan et al., 1999; Helenius et al., 1999; Simos et al., 2000; Heim et al., 2000; Benasich and Tallal, 2002; Tallal, 2004). MEG recordings further indicated a large variation in the time interval necessary for the appearance of a second gamma-band wave (Llinás et al., 1998b; Ribary et al., 2000). More specifically, MEG data suggested the existence of two different sub-groups of LLD subjects (Fig. 3).

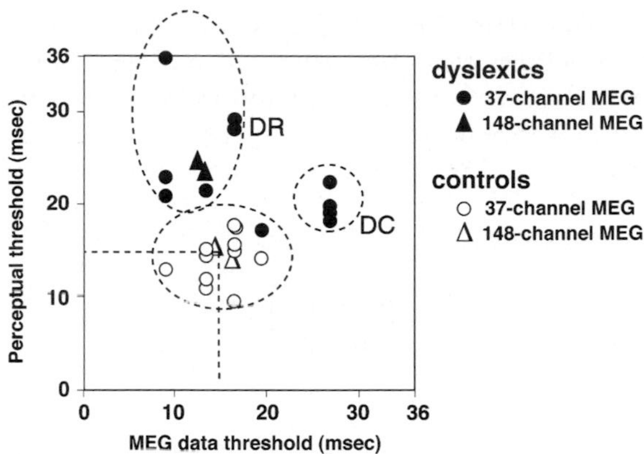

Fig. 3. Altered auditory binding and perception in subjects with language-based learning disabilities (LLD or dyslexia). Perceptual thresholds plotted as a function of MEG thresholds obtained from 16 control (open circles/triangles) and 13 LLD or dyslexic (filled circles/triangles) subjects. Dyschronic (DC) subjects display delayed MEG and perceptual thresholds, while dysrhythmic (DR) subjects display slightly shorter MEG thresholds in conjunction with delayed perceptual thresholds, which may relate to background masking.

Magnetic recordings from one group indicated an increased processing latency for the first stimulus, consistent with increased temporal binding (dyschronia). MEG data recorded from the other group indicated an incomplete processing of the first stimulus and a discontinuity of the binding mechanism resulting in a background masking effect (dysrhythmia). These findings (Ribary et al., 2000) remained the same by using a new state-of-the-art whole head MEG system and by increasing the number to a total of 16 controls and 13 LLD subjects (Fig. 3).

In addition, there is increasing evidence that sensory processing is also altered in the aging brain, although sensory physiological thresholds remain at normal levels. Psychophysical thresholds in many simple auditory tasks deteriorate indeed with age. For example, auditory fusion thresholds for gaps between two long tones have been shown to increase with age, particularly for subjects aged over 60 (McCroskey and Kasten, 1980; Robin and Royer, 1989). Temporal order judgments of auditory stimuli also become less accurate with age (Fitzgibbons and Gordon-Salant, 1998). Such declining auditory acuity has been hypothesized to relate to declining speech processing that has long been evident in elderly subjects. In those studies, however, the discrimination time windows were set above 25 ms. Recent critical discrimination paradigms can produce a more precise measure of perceptual and functional imaging change because the intervals are set between 0 and 36 ms (Ribary et al., 1999, 2004; Sauve et al., 2003).

Preliminary MEG data on elderly subjects that have been tested so far indeed suggested that (1) a distinct gamma-band oscillatory reset within thalamo-cortical systems by sensory stimuli is altered in the aging human brain and (2) an altered reset of gamma-band activity is related to a correlation between delayed MEG thresholds and delayed perceptual thresholds to identify two distinct sensory events during early sensory processing (Sauve et al., 2003; Ribary et al., 2004). In particular, there was a delay of approximately 3–12 ms compared to thresholds in younger control subjects.

Using similar paradigms as mentioned above, two auditory clicks were delivered at different ISIs. MEG activity was recorded and the subjects were asked if they heard one or two clicks (Fig. 4). The gamma-band reset activity was delayed in the five elderly subjects that have been tested (aged 75, 79, 79, 75, 69 years). In particular, at ISIs of 18 ms or more, there was a second reset of gamma-band activity indicating a delayed processing of a second distinct auditory stimulus. Preliminary data further suggested a correlation between delayed perceptual and delayed MEG thresholds in the elderly group. These results strongly suggest a delay during early sensory information processing of sensory stimuli in normal aged subjects.

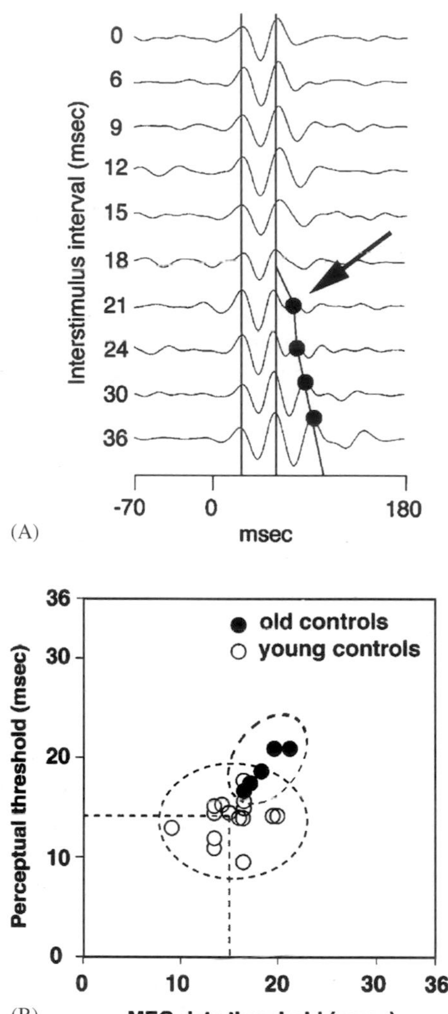

Fig. 4. Altered auditory binding and perception in the aging brain. (A) Effect of increasing the interstimulus interval on MEG activity in the higher (20–50 Hz) gamma band frequency range in an older subject (75 years). As the interval increased > 18 ms, a second, delayed gamma band response appeared. (B) Perceptual thresholds plotted as a function of MEG thresholds obtained from 16 younger (20–40 years, open circles) and 5 older (69–79 years, filled circles) subjects. Note that the tendency of an increased MEG threshold is correlated with an increased perceptual threshold.

Network dysrhythmia in neurology and psychiatry

Earlier MEG studies, as mentioned above, provided evidence for a slight dysrhythmia within thalamo-cortical systems in subjects with LLD (Llinás et al., 1998b; Nagarajan et al., 1999). Clearly an alteration of precise timing of thalamo-cortical networks correlated to altered behavioral patterns, namely to altered perception of sensory input (Ribary et al., 1999). In unconscious humans, magnetic resonance imaging (MRI), MEG, and positron emission tomography (PET) data indeed indicated a massive fracture of thalamo-cortical systems, as observed in patients in a vegetative state (Schiff et al., 1999, 2002, this volume). These findings suggested that a dysrhythmia within thalamo-cortical systems could represent a key issue underlying various pathological behavioral symptoms (Llinás et al., 1999). Recent MEG results, combined with findings based on electrical recordings from human thalamus (Jeanmonod et al., 1996, 2001; Sarnthein et al., 2003) and physiological findings on animals (Jahnsen & Llinás, 1984; Llinás et al., 2002), indeed indicated that a severe and sustained dysrhythmia within thalamo-cortical systems could underlie various positive symptoms observed in a subset of neurological and psychiatric patients (Llinás et al., 1999, 2001; Schulman et al., 2000, 2001, 2003; Sarnthein et al., 2003; Jeanmonod et al., 2003; Timmermann et al., 2003).

Spontaneous MEG activity was recently recorded in patients suffering from neurogenic pain, tinnitus, Parkinson's disease, schizophrenia, depression, or refractory obsessive compulsive disease (Llinás et al., 1999, 2001; Schulman et al.,

2000, 2001, 2003). Compared to healthy controls, patients showed increased low-frequency theta rhythmicity in conjunction with a widespread and marked increase of power correlation among high- and low-frequency oscillations consistent with other reports (John et al., 1988; John, this volume). These data indicated the presence of a thalamo-cortical dysrhythmia, which we propose is related to all the above-mentioned conditions (Llinás et al., 1999). Such dysrhythmia can be explained by either excess inhibition or disfacilitation of the thalamo-cortical system in these patients, inducing the generation of low-threshold calcium spike bursts by thalamic cells as seen in animals (Llinás et al., 2001) and humans (Jeanmonod et al., 1996). The presence of these bursts in human thalamus (Jeanmonod et al., 2001) directly relates to thalamic cell hyperpolarization and low frequency generation within thalamus (Jahnsen and Llinás, 1984). This produces coherent theta activity, the result of a resonant interaction between thalamus and cortex. The emergence of positive clinical symptoms is then viewed as resulting from increased gamma-band activation due to decreased inhibition between high- and low-frequency thalamo-cortical modules at the cortical level, which we refer to as the "edge effect" (Llinás et al., 1999). This effect is observable as increased correlation between low- and high-frequency oscillations resulting from such inhibitory asymmetry at the cortical level.

The description of such thalamo-cortical dysrhythmias (Llinás et al., 1999) also provides a conceptual framework for evaluating slight alterations in resting brain activity in the normal aging brain or in dyslexia. These alterations are expected to be smaller compared to the severe thalamo-cortical dysrhythmia observed in more invalidating neurological and psychiatrical diseases, or compared to selective fracture of thalamo-cortical networks in unconsciousness (e.g., in vegetative patients; Schiff et al., 2002). In states of thalamo-cortical dysrhythmia, ongoing theta-range (4–8 Hz) thalamic activity serves as the trigger for cortical dysfunction, in which a core region of cortex functions at low frequency, surrounded by a region of activation in the normal waking gamma (25–50 Hz) range (Llinás et al., 2002). In addition, the connectivity of the thalamo-cortical system not only maintains this pathological dynamics, but also causes it to become distributed throughout wide areas of cortex representing a large-scale coupling, which allows such activity to constrain thalamo-cortical dynamics so efficiently. This adds considerable weight to the question of whether some of the slight cognitive deficits of aging and dyslexia are based upon varying degrees of a slight dysrhythmia (Llinás et al., 1998a; Schulman et al., 2003; Ribary et al., 2004).

Fracture of thalamo-cortical networks during unconsciousness

Earlier, we speculated that if the dynamics of thalamo-cortical network oscillations is indeed crucial for early sensory processing relating to perception, then we should expect a massive fracture within these thalamo-cortical systems during unconsciousness (Plum et al., 1998), such as during the vegetative state (Jennett and Plum, 1972). In a series of studies on patients in a vegetative state (Schiff and Plum, 2000), using MRI, PET, and MEG, we indeed found that there was a massive structural damage within thalamo-cortical systems, combined with a massive shutdown of brain metabolism, and a massive shutdown in neural connectivity (Plum et al., 1998; Schiff et al., 1999, 2002; Schiff, this volume). These studies were in accordance with other functional brain imaging studies, demonstrating consistently diffuse and uniformly reduced cerebral metabolic activity (Levy et al., 1987; DeVolder et al., 1990; Tomassino et al., 1995; Rudolf et al., 1999; Laureys et al., 1999). In addition, a selective disappearance of sensory midlatency responses and early evoked potentials were reported in comatose patients (Pfurtscheller et al., 1983) and during general anesthesia (Madler et al., 1991). Recent studies also indicated the importance of coherent gamma-band activity in relation to different functional states during general anesthesia (John et al., 2001; John, this volume).

Recent findings demonstrated that although the vegetative state is characterized by massively reduced brain metabolism and connectivity, some patients may express isolated meaningless fragments of behavior that can be related to islands of residual metabolic and physiologic brain activity (Schiff et al., 1999, 2002). An earlier case study consisted of such a unique vegetative patient who randomly produced occasional single words (Schiff et al., 1999). In this patient, isolated regions of preserved cerebral metabolic activity and thalamo-cortical transmission associated with remnants of the language system. These findings led us to evaluate additional vegetative patients with multi-modal imaging techniques in order to determine in detail what cerebral activity may remain in patients with catastrophic brain damage. We reported evidence of reciprocal clinical–pathological correlation with regional differences of quantitative cerebral metabolism (Schiff et al., 2002). We also employed MEG to analyze dynamic aspects of source activations in such patients, namely to identify spatio-temporal sources, if any, of brain activations associated to specific frequency bands as seen during normal sensory processing (Regan, 1989; Ribary et al., 1991, 1999; Baumgartner et al., 1995; Llinás et al., 1999). This diagnostic protocol was designed to examine the possible persistence of remaining coherent neuronal network activity. The MEG data from the vegetative patients indicated partially preserved but abnormal, delayed, incomplete or absent coherent dynamic brain activity (Schiff et al., 2002). Restricted sensory representations evidenced by slow evoked magnetic fields and gamma-band activity, were uniquely expressed in each patient, and correlated with isolated behavioral patterns in two patients. These isolated and abnormal residual MEG activations were further correlated with locally preserved metabolic activity. The combination of MRI, PET and MEG techniques allowed us to assess the residual network properties that underlie the expression of meaningless fractional behavior observed in three of the five chronic vegetative patients reported (Schiff et al., 2002; see also Schiff, this volume).

In the intact, normal healthy brain, these modular networks do process selective sources of information and are typically integrated into large, coherent or coupling patterns of activity (Llinás and Ribary, 1993; Friston et al., 1993; Bressler et al., 1993; Zeki and Bartels, 1998; Tononi and Edelmann, 1998.; Van Essen et al., 1998, Singer, 1998; Raichle, 1999; John, 2002; see also John, this volume). These novel findings on patients in a vegetative state (Schiff et al., 2002) collectively provide a foundation for identifying mechanisms underlying complex brain injuries and may lead to more objective diagnostic procedures for the future. In particular, these studies may represent a first step toward characterizing patients with varying degrees of functional recovery beyond the vegetative state. Such steps will be necessary to appropriately risk-stratify patients for future studies of outcome or therapeutic interventions.

Summary: interaction and dynamics of large-scale network oscillations in the normal and pathological brain

All the above-mentioned findings, within the context of the current literature, suggest that non-invasive brain imaging techniques available today and combined with highly sophisticated signal processing and analysis techniques are providing complementary, and very useful information regarding brain structure, and brain function with detailed spatial and temporal constraints. These findings further suggest that these objective brain imaging measurements may already be correlated with various broad cognitive processing and pathological states. A better characterization of underlying extended large-scale networks and the detailed analysis of the temporal dynamics is very important in order to better understand normal human brain function such as perception and memory and the many pathological alterations. Recent improvements in signal processing and analysis techniques (Dale and Sereno, 1993; Makeig et al., 1997, 2002; Pascual–Marqui et al., 1995; Mosher et al., 1999a, b; Gross et al., 2002; Jensen et al., 2002b; Ramirez et al., 2003, 2004; Michel et al., 2004) indeed provide confidence that non-invasive functional human brain imaging techniques may soon be capable of extracting the various network functions (Ribary et al., 1991; Schiff et al., 2002; Gross et al., 2002) and monitoring their detailed network dynamics and interactions in relation to specific cognitive tasks. In addition, specific time sequences or quantal time windows (Joliot et al., 1994) were already mentioned, in relation to normal or altered early sensory processing, which further suggests that normal brain function may indeed be quantified as non-continuous time sequences (Lehmann et al., 1987; Ribary et al., 1991; Llinás and Ribary, 1993; Michel et al., 2004) based on detailed spatial and temporal constraints. Such a quantification strategy of normal brain function could now be applied in various neurological and psychiatric patient populations, as mentioned here specifically in traumatic brain injury, and the alterations could be specified. I would call such techniques diagnostic brain sequencing procedures. As such then, a "normal" brain sequencing procedure in healthy control subjects would represent an analysis of a normalized series of multiple quantal time windows describing a non-continuous evolution of large-scale network oscillations and couplings, laid out over a time period of several 100 ms and directly correlated to sensory processing and memory. Diagnostic brain sequencing procedures in brain pathologies would then represent a quantitative comparison and a quantitative deviation of such normalized time series having shorter, longer, abnormal, or missing quantal time windows and describing an abnormal non-continuous evolution of large-scale network oscillations and couplings. Detailed analysis of such individual quantal time windows would then specify changes within local networks or between large-scale networks, such as changes in oscillation frequency, changes in spectral power, changes in synchronization, coherence, coupling, etc. Such diagnostic brain sequencing procedures may then lead to better objective diagnostic procedures for various brain pathologies and for better treatment strategies in the future, including dyslexia (Merzenich et al., 1996; Tallal, 2004), neurology (see Schiff, this volume), and psychiatry (Jeanmonod et al., 2003).

Acknowledgments

This review summarizes only a selection of many MEG studies performed at the Center for Neuromagnetism (CNM) at the NYU School of Medicine over the past 16 years in collaboration with many colleagues. Especially, I acknowledge the very inventive and very productive interaction with Dr. R. Llinas and the wonderful interaction with the great team at CNM over all those years, especially with R. Jagow and Drs. E. Kronberg, A. Mogilner, F. Lado, L. Lopez, J. Cappell, J. Schulman, R. Ramirez, K. Sauve, and many others. I further acknowledge the interesting and productive interaction with many collaborators, Drs. N. Schiff, F. Plum, M. Joliot, D. Jeanmonod, J. Volkmann, H. Van Marle, A. Rezai, S. Miller, P. Tallal, S. Ferris, A. Kluger, A. Ioannides, P. Mitra,. P. Kelly, R. Cancro, S. Kolodny, N. Cohen, O. Devinski, S. Pacia, and others. I further acknowledge the MFT analysis by Dr. A. Ioannides for Figure 1.

These studies were supported by many funding sources, especially by NIH, The Charles A. Dana Foundation, The James Watson Foundation, the NYU-GCRC, and 4D-Neuroimaging.

References

Ahissar, E. and Vaaida, E. (1990) Oscillatory activity of single units in a somatosensory cortex of an awake monkey and their possible role in texture analysis. Proc. Natl. Acad. Sci. USA, 87: 8935–8939.

Barth, D.S. and MacDonald, K.D. (1996) Thalamic modulation of high-frequency oscillating potentials in auditory cortex. Nature, 383: 78–81.

Basar, E., Rosen, B., Basar-Eroglu, C. and Greitschus, F. (1987) The associations between 40-Hz EEG and the middle latency response of the auditory evoked potential. Int. J. Neurosci., 33: 103–117.

Basar, E. and Bullock, T. (1992) Induced Rhythms in the Brain, Brain Dynamics Series. Birkhauser, Boston, MA.

Baumgartner, C., Deecke, L., Stroink, G. and Williamson, S.J. (1995) Biomagnetism: Fundamental Research and Clinical Applications. Elsevier, Amsterdam.

Benasich, A.A. and Tallal, P. (2002) Infant discrimination of rapid auditory cues predicts later language impairment. Behav. Brain Res., 136: 31–49.

Boettger, D., Herrmann, C.S. and von Cramon, Y. (2002) Amplitude differences of evoked alpha and gamma oscillations in two different age groups. Int. J. Psychophysiol., 45: 245–251.

Bressler, S.L. and Freeman, W.J. (1980) Frequency analysis of olfactory system EEG in cat, rabbit and rat. Electroenceph. Clin. Neurophysiol., 50: 19–24.

Bressler, S.L., Coppola, R. and Nakamura, R. (1993) Episodic multiregional cortical coherence at multiple frequencies during visual task performance. Nature, 366: 153–156.

Carr, C.E. (1993) Processing of temporal information in the brain. Ann. Rev. Neurosci., 16: 223–243.

Carr, C.E. and Konishi, M. (1990) A circuit for detection of interaural time differences in the brain stem of the barn owl. J. Neurosci., 10: 3227–3246.

Castaigne, P., Lhermitte, F., Buge, A., Escourolle, P., Derouisne, C. and Der Agopian, P. (1980) Paramedian thalamic and midbrain infarcts: clinical and neuropathological study. Ann. Neurol., 10: 127–214.

Crick, F. and Koch, C. (1996) The problem of consciousness. Sci. Am., 267: 153–159.

Dale, A.M. and Sereno, M.I. (1993) Improved localization of cortical activity by combining EEG and MEG with MRI cortical surface reconstruction: a linear approach. J. Cogn. Neurosci., 5: 162–176.

Desmedt, J.E. and Tomberge, C. (1994) Transient phase-locking of 40-Hz electrical oscillations in prefrontal and parietal human cortex reflects the process of conscious somatic perception. Neurosci. Lett., 168: 126–129.

Destexhe, A. and Babloyantz, A. (1991) Deterministic chaos in a model of the thalamo-cortical system. In: Babloyantz A. (Ed.), Self-Organization, Emerging Properties and Learning. Plenum Press, New York.

DeVolder, A.G., Goffinet, A.M., Bol, A., Michel, C., de Barsy, T. and Laterre, C. (1990) Brain glucose metabolism in postanoxic stroke. Arch. Neurol., 47: 197–204.

Eckhorn, R., Bauer, R., Jordan, W., Brosch, M., Kruse, W., Munk, M. and Reitboeck, H.J. (1988) Coherent oscillations: A mechanism of feature linking in the visual cortex? Biol. Cybern., 60: 121–130.

Elbert, T., Pantev, C., Wienbruch, C., Rockstroh, B. and Taub, E. (1995) Increased cortical representation of the fingers of the left hand in string players. Science, 270: 305 307.

Engel, A.K., Konig, P., Kreiter, A.K. and Singer, W. (1991) Interhemispheric synchronization of oscillatory neuronal responses in cat visual cortex. Science, 252: 1177–1179.

Façon, E., Steriade, M. and Wertheim, N. (1958). Rev. Neurol. (Paris), 98: 117–133.

Fitzgibbons, P.J. and Gordon-Salant, S. (1998) Auditory temporal order perception in younger and older adults. J. Speech Lang. Hear. Res., 41: 1052–1060.

Freeman, W.J. (1975) Mass Action in The Nervous System. Academic Press, New York.

Freeman, W.J. and van Dijk, B.W. (1987) Spatial patterns of visual cortical fast EEG during conditioned reflex in a rhesus monkey. Brain Res., 422: 267–276.

Friston, K.J., Frith, C.D., Liddle, P.F. and Frackowiak, R.S. (1993) Functional connectivity: the principal-component analysis of large (PET) data sets. J. Cereb. Blood Flow Metab., 13: 5–14.

Galambos, R., Makeig, S. and Talmachoff, P.J. (1981) A 40-Hz auditory potential recorded from the human scalp. Proc. Natl. Acad. Sci. USA, 78: 2643–2647.

Gray, C.M. and Singer, W. (1989) Stimulus-specific neuronal oscillations in orientation columns of cat visual cortex. Proc. Natl. Acad. Sci. USA, 86: 1698–1702.

Gray, C.M. (1999) The temporal correlation hypothesis of visual feature integration: still alive and well. Neuron, 24: 31–47.

Gross, J., Timmermann, L., Kujala, J., Dirks, M., Schmitz, F., Salmelin, R. and Schnitzler, A. (2002) The neural basis of intermittent motor control in humans. Proc. Natl. Acad. Sci. USA, 99: 2299–2302.

Heim, S., Eulitz, C., Kaufmann, J., Fuchter, I., Pantev, C., Lamprecht-Dinnesen, A., Matulat, P., Scheer, P., Borstel, M. and Elbert, T. (2000) Atypical organisation of the auditory cortex in dyslexia as revealed by MEG. Neuropsychologia, 38: 1749–1759.

Helenius, P., Uutela, K. and Hari, R. (1999) Auditory stream segregation in dyslexic adults. Brain, 122: 907–913.

Hirsh, I.J. (1959) Auditory perception of temporal order. J. Acoust. Soc. Am., 31: 759–767.

Jahnsen, H. and Llinas, R. (1984) Electro-physiological properties of guinea-pig thalamic neurones: an in vitro study. J. Physiol., 349: 205–226.

Jeanmonod, D., Magnin, M. and Morel, A. (1996) Low-threshold calcium spike bursts in the human thalamus: common physiopathology for sensory, motor and limbic positive symptoms. Brain, 119: 363–375.

Jeanmonod, D., Magnin, M., Morel, A., Siegemund, M., Cancro, R., Lanz, M., Llinás, R., Ribary, U., Kronberg, E., Schulman, J.J. and Zonenshayn, M. (2001) Thalamocortical Dysrhythmia II: Clinical and surgical aspects. Thalamus Related Systems, 1: 245–254.

Jeanmonod, D., Schulman, J., Ramirez, R., Cancro, R., Lanz, M., Morel, A., Magnin, M., Siegemund, M., Kronberg, E., Ribary, U. and Llinas, R. (2003) Neuropsychiatric thalamocortical dysrhythmia: surgical implications. Neurosurg. Clin. N. Am., 14: 251–265.

Jennett, B. and Plum, F. (1972) Persistent vegetative state after brain damage. A syndrome in search of a name. Lancet, 1: 734–737.

Jensen, O., Hari, R. and Kaila, K. (2002a) Visually evoked gamma responses in the human brain are enhanced during voluntary hyperventilation. NeuroImage, 15: 575–586.

Jensen, O. and Vanni, S. (2002b) A new method to identify multiple sources of oscillatory activity. NeuroImage, 15: 568–574.

John, E.R., Prichep, L.S., Friedman, J. and Easton, P. (1988) Neurometrics: Computer assisted differential diagnosis of brain dysfunctions. Science, 293: 162–169.

John, E.R., Prichep, L.S., Kox, W., Valdes-Sosa, P., Bosch-Bayard, J., Aubert, E., Tom, M., diMichele, F. and Gugino, L.D. (2001) Invariant reversible qeeg effects of anesthetics. Consciousness Cogn., 10: 165–183.

John, E.R. (2002) The neurophysics of consciousness. Brain Res. Rev., 39: 1–28.

Joliot, M., Ribary, U. and Llinás, R. (1994) Human oscillatory brain activity near 40 Hz coexists with cognitive temporal binding. Proc. Natl. Acad. Sci. USA, 91: 11748–11751.

Kato, N. (1990) Cortico-thalamo-cortical projection between visual cortices. Brain Res., 509: 150–152.

Kissler, J., Muller, M.M., Fehr, T., Rockstroh, B. and Elbert, T. (2000) MEG gamma band activity in schizophrenia patients and healthy subjects in a mental arithmetic task and at rest. Clin. Neurophysiol., 111: 2079–2087.

Knief, A., Schulte, M., Bertran, O. and Pantev, C. (2000) The perception of coherent and non-coherent auditory objects: a signature in gamma frequency band. Hear Res., 145: 161–168.

Kristofferson, A.B. (1984) Quantal and deterministic timing in human duration discrimination. Annals New York Acad. Sci., 423: 3–15.

Laureys, S., Goldman, S., Phillips, C., Van Bogaert, P., Aerts, J., Luxen, A., et al. (1999) Impaired effective cortical connectivity in vegetative state: preliminary investigation using PET. Neuroimage, 9: 377–382.

Lehmann, D., Ozsaki, H. and Pal, I. (1987) Brainstates. Electroenceph. Clin. Neurophysiol., 67: 271–288.

Levy, D.E., Sidtis, J.J., Rottenberg, D.A., Jarden, J.O., Strother, S.C., Dhawan, V., et al. (1987) Differences in cerebral blood flow and glucose utilization in vegetative versus locked-in patients. Ann. Neurol., 22: 673–682.

Lindström, S. and Wróbel, A. (1990) Frequency dependent corticofugal excitation of principal cells in the cat's dorsal lateral geniculate nucleus. Exp. Brain Res., 79: 313–318.

Llinás, R. (1990) Intrinsic electrical properties of mammalian neurons and CNS function. Fidia Res. Foundation Neurosci. Award Lectures, 4: 173–192.

Llinás, R., Grace, A.A. and Yarom, Y. (1991) In vitro neurons in mammalian cortical layer 4 exhibit intrinsic activity in the 10 to 50 Hz frequency range. Proc. Natl. Acad. Sci. USA, 88: 897–901.

Llinás, R. and Ribary, U. (1992) Rostrocaudal scan in human brain: A global characteristic of the 40-Hz response during sensory input. In: Basar and Bullock T. (Eds.), Induced Rhythms in the Brain. Birkhauser, Boston, pp. 147–154.

Llinás, R. (1993) Is dyslexia a dyschronia? Annal NY Acad. Sci., 682: 48–56.

Llinás, R. and Ribary, U. (1993) Coherent 40-Hz oscillation characterizes dream state in humans. Proc. Natl. Acad. Sci. USA, 90: 2078–2081.

Llinás, R., Ribary, U., Joliot, M. and Wang, X.J. (1994) Content and context in temporal thalamocortical binding. In: Buzsaki G., Llinas R., Singer W., Berthoz A. and Christen Y. (Eds.), Temporal Coding in the Brain. Springer, Heidelberg, pp. 251–272.

Llinás, R., Ribary, U., Contreras, D. and Pedroarena, C. (1998a) The neuronal basis for consciousness. Phil. Trans. R. Soc. London, 353: 1841–1849.

Llinás, R., Ribary, U. and Tallal, P. (1998b) Dyschronic language-based learning disability. In: VonEuler C., Lundberg I. and Llinás R. (Eds.), Basic Mechanisms in Cognition and Language. Elsevier Science, New York, pp. 101–108.

Llinás, R., Ribary, U., Jeanmonod, D., Kronberg, E. and Mitra, P.P. (1999) Thalamocortical dysrhythmia: A neurological and neuropsychiatric syndrome characterized by magnetoencephalography. Proc. Natl. Acad. Sci. USA,, 96: 15222–15227.

Llinás, R., Ribary, U., Jeanmonod, D., Cancro, R., Kronberg, E., Schulman, J.J., Zonenshayn, M., Magnin, M., Morel, A. and Siegemund, M. (2001) Thalamocortical dysrhythmia I: functional and imaging aspects. Thalamus Rel. Syst., 1: 237–244.

Llinas, R.R., Leznik, E. and Urbano, F.J. (2002) Temporal binding via cortical coincidence detection of specific and nonspecific thalamocortical inputs: a voltage-dependent dye-imaging study in mouse brain slices. Proc. Natl. Acad. Sci. USA, 99: 449–454.

Lumer, E.D., Edelman, G.M. and Tononi, G. (1997) Neural dynamics in a model of the thalamocortical system I. Layers, loops and the emergency of fast synchronous rhythms. Cere. Cortex, 7: 207–227.

Madler, C., Keller, I., Schwender, D. and Poeppel, E. (1991) Sensory information processing during general anaesthesia: effect of isoflurane on auditory evoked neuronal oscillations. Br. J. Anaesthesia, 66: 81–87.

Makeig, S., Jung, T.P., Ghahremani, D., Bell, A.J. and Sejnowski, T.J. (1997) Blind separation of event-related brain responses into independent components. Proc. Natl. Acad. Sci. USA,, 94: 10979–10984.

Makeig, S., Westerfield, M., Jung, T.P., Enghoff, S., Townsend, J., Courchesne, E. and Sejnowski, T.J. (2002) Dynamic brain sources of visual evoked responses. Science, 295: 690–694.

McCroskey, R. and Kasten, R.N. (1980) Assessment of central auditory processing. In: Rupp R.R. and Stockdell K.G. (Eds.), Speech Protocols in Audiology. Grune & Stratton, New York, pp. 339–389.

Merzenich, M.M., Nelson, R.J., Kaas, J.H., Stryker, M.P., Jenkins, W.M., Zook, J.M. and Cynader, M.S. (1987) Variability in hand surface representations in areas 3b and 1 in adult owl and squirrel monkeys. J. Comp. Neurol., 258: 281–296.

Merzenich, M.M., Jenkins, W.M., Johnston, P., Schreiner, C., Miller, S.L. and Tallal, P. (1996) Temporal processing deficits of language-learning impaired children ameliorated by training. Science, 271: 77–81.

Michel, C.M., Murray, M., Lantz, G., Gonzalez, S. and Grave de Peralta, R. (2004) EEG source imaging. Clin. Neurophysiol., 115: 2195–2222.

Miller, G.A. and Taylor, W.G. (1948) The perception of repeated bursts of noise. J. Acoust. Soc. Am., 20: 171–182.

Mogilner, A., Grossman, J.A.I., Ribary, U., Joliot, M., Volkmann, J., Rapaport, D., Beasley, R.W. and Llinas, R. (1993) Somatosensory cortical plasticity in adult humans revealed by magnetoencephalography. Proc. Natl. Acad. Sci. USA, 90: 3593–3597.

Mosher, J.C., Leahy, R.M. and Lewis, P.S. (1999a) EEG and MEG: forward solutions for inverse methods. IEEE Trans. Biomed. Eng., 46: 245–259.

Mosher, J.C., Baillet, S. and Leahy, R.M. (1999b) EEG Source localization and imaging using multiple signal classification approaches. J. Clin. Neurophysiol., 16: 225–238.

Murthy, V.N. and Fetz, E.E. (1992) Coherent 25-35 Hz oscillations in the sensorimotor cortex of awake behaving monkeys. Proc. Natl. Acad. Sci. USA, 89: 5670–5674.

Murthy, V.N. and Fetz, E.E. (1997) Oscillatory activity in sensorimotor cortex of awake monkeys: synchronization of local field potentials and relation to behavior. J. Neurophysiol., 76: 3949–3967.

Nagarajan, S., Mahncke, H., Salz, T., Tallal, P., Roberts, T. and Merzenich, M.M. (1999) Cortical auditory signal processing in poor readers. Proc. Natl. Acad. Sci. USA, 96: 6483–6488.

Neuenschwander, S. and Singer, W. (1996) Long-range synchronization of oscillatory light responses in the cat retina and lateral geniculate nucleus. Nature, 379: 728–733.

Palva, S., Palva, J.M., Shtyrov, Y., Kujala, T., Ilmoniemi, R.J., Kaila, K. and Naatanen, R. (2002) Distinct gamma-band evoked responses to speech and non-speech sounds in humans. J. Neurosci., 22: RC211.

Pantev, C., Makeig, S., Hoke, M., Galambos, R., Hampson, S. and Gallen, C. (1991) Human auditory evoked gamma-band magnetic fields. Proc. Natl. Acad. Sci. USA, 88: 8996–9000.

Parra, J., Kalitzin, S.N., Iriarte, J., Blanes, W., Velis, D.N. and Lopes Da Silva, F.H. (2003) Gamma-band phase clustering and photosensitivity: is there an underlying mechanism common to photosensitive epilepsy and visual perception? Brain, 126: 1164–1172.

Pascual-Marqui, R.D., Michel, C.M. and Lehmann, D. (1995) Segmentation of brain electrical activity into microstates: model estimation and validation. IEEE Trans. Biomed. Eng., 42: 658–665.

Pedroarena, C. and Llinás, R. (1997) Dendritic calcium conductances generate high-frequency oscillation in thalamocortical neurons. Proc. Natl. Acad. Sci. USA, 94: 724–728.

Pfurtscheller, G., Schwarz, G., Pfurtscheller, B. and List, W. (1983) Computer assisted analysis of EEG, evoked potentials, EEG reactivity and heart rate variability in comatose patients. Elektroenzephalogr Elektromyogr Verwandte Geb., 14: 66–73.

Plum, F., Schiff, N., Ribary, U. and Llinas, R. (1998) Coordinated expression in chronically unconscious persons. Phil. Trans. R. Soc. London, 353: 1929–1933.

Poppel, E. (1970) Excitability cycles in central intermittency. Psychol. Forsch., 34: 1–9.

Purpura, K.P. and Schiff, N.D. (1997) The thalamic intralaminar nuclei: role in visual awareness. Neuroscientist, 3: 8–14.

Raichle, M. (1999) The neural correlates of consciousness: an analysis of cognitive skill learning. Phil. Trans. R. Soc. London, 353: 1889–1901.

Ramachandran, V.S., Rogers-Ramachandran, D. and Stewart, M. (1992) Perceptual correlates of massive cortical reorganization. Science, 258: 1159–1160.

Ramírez, R.R., Horenstein, C., Kronberg, E., Ribary, U. and Llinás, R. (2000) Quantal apparent motion perception and its neuro-magnetic correlates in striate and extrastriate visual cortices. Soc. Neurosci. Abstr., 26: 670–670.

Ramirez, R., Kronberg, E., Ribary, U. and Llinás, R. (2002) Gamma-band and recurrent visual sources correlate with apparent motion perception. Soc. Neurosci. Abstr., 28: on CD ROM.

Ramirez, R.R., van Marle, H.J.F., Kronberg, E., Ribary, U. and Llinas, R. (2003) Large-scale integration, synchronization, and coherent brain dynamics in single trials. NeuroImage, 19: S34(505).

Ramirez, R., Jaramillo, S., Moran, K. A., Ribary U. and Llinas, R. R. Single trial neuromagnetic source dynamics. (2004) Soc. Neurosci. Abstr., 30 (on CD ROM).

Regan, D. (1989) Human Brain Electrophysiology: Evoked Potentials and Evoked Magnetic Fields in Science and Medicine. Elsevier Science, New York.

Reisberg, B., Ferris, S., de Leon, M.J. and Crook, T. (1982) The global deterioration scale (GDS): An instrument for the assessment of primary degenerative dementia. Am. J. Psychiat., 139: 1136–1139.

Rennie, C.J., Robinson, P.A. and Wright, J.J. (2002) Unified neurophysiological model of EEG spectra and evoked potentials. Biol. Cyber., 86: 457–471.

Ribary, U., Weinberg, H., Brickett, P., Ancill, R.J., Holliday, S., Kennedy, J.S. and Johnson, B. (1987) Imaging of EEG and MEG (magnetoencephalography) for the estimate of sources in the human brain responsive to antidepressant drugs. Soc. Neurosci. Abstr., 13: 1268–1268.

Ribary, U., Weinberg, H., Johnson, B., Ancill, R.J. and Holliday, S. (1988) EEG and MEG (magnetoencephalography) mapping of the auditory 40-Hz response: A study of depression. Soc. Neurosci. Abstr., 14: 339–339.

Ribary, U., Llinás, R., Kluger, A., Suk, J. and Ferris, S.H. (1989) Neuropathological dynamics of magnetic, auditory, steady-state responses in Alzheimer's disease. In: Williamson S.J., Hoke M., Stroink G. and Kotani M. (Eds.), Advances in Biomagnetism. Plenum Press, New York, pp. 311–314.

Ribary, U., Ioannides, A.A., Singh, K.D., Hasson, R., Bolton, J.P.R., Lado, F., Mogilner, A. and Llinas, R. (1991) Magnetic Field Tomography (MFT) of coherent thalamo-cortical 40-Hz oscillations in humans. Proc. Natl. Acad. Sci. USA, 88: 11037–11041.

Ribary, U., Cappell, J., Mogilner, A., Hund, M., Kronberg, E. and Llinás, R. (1999) Functional imaging of plastic changes in the human brain. Adv. Neurol., 81: 49–56.

Ribary, U., Joliot, M., Miller, S.L., Kronberg, E., Cappell, J., Tallal, P. and Llinás, R. (2000) Cognitive temporal binding and its relation to 40 Hz activity in humans: Alteration during dyslexia. In: Aine C., Okada Y., Stroink G., Swithenby S. and Wood C.C. (Eds.), Biomag96. Springer, Berlin, pp. 971–974.

Ribary, U., Llinás, R., Jeanmonod, D., Kronberg, E., Sauvé, K., Ramirez, R.R., Schulman, J.J., Horenstein, C. and Van-Marle, H.J.F. (2002) Normal and dysrhythmic thalamo-cortical networks in the auditory, somatosensory and visual modality and their relation to neuro-psychiatric syndromes. In: Novak H., Haueisen J., Giessler F. and Huonker R. (Eds.), BIOMAG 2002. VDE Verlag, Berlin, pp. 198–200.

Ribary, U., Ramirez, R., Sauve, K., Schulman, J., Jaramillo, S., Llinas, R.R. (2004) Altered thalamo-cortical network dynamics and sensory binding during normal aging. Soc. Neurosci. Abstr., 30 (on CD ROM).

Robin, D.A. and Royer, F.L. (1989) Age-related changes in auditory temporal processing. Psychol. Aging, 4: 144–149.

Rudolf, J., Ghaemi, M., Ghaemi, M., Haupt, W.F., Szelies, B. and Heiss, W.D. (1999) Cerebral glucose metabolism in acute and persistent vegetative state. J. Neurosurg. Anesthesiol., 11: 17–24.

Salenius, S., Salmelin, R., Neuper, C., Pfurtscheller, G. and Hari, R. (1996) Human cortical 40 Hz rhythm is closely related to EMG rhythmicity. Neurosci. Lett., 213: 75–78.

Salmelin, R., Service, E., Kiesila, P., Uutela, K. and Salonen, O. (1996) Impaired visual word processing in dyslexia revealed with magnetoencephalography. Ann. Neurol., 40: 157–162.

Sarnthein, J., Morel, A., von Stein, A. and Jeanmonod, D. (2003) Thalamic theta field potentials and EEG: high thalamocortical coherence in patients with neurogenic pain, epilepsy and movement disorders. Thalamus Rel. Syst., 2: 231–238.

Sauve, K., Wang, G., Rolli, M., Jagow, R., Kronberg, E., Ribary, U. and Llinás, R. (1999) Binding of somatosensory stimuli in sighted and blind subjects. Neuroimage, 9: 848–848.

Sauvé, K., Ribary, U. and Llinas, R. (2003) Early, high amplitude intra- and inter-hemispheric gamma-band currents in sensory cortex correlate with reaction times in individual trials in young and old subjects. Soc. Neurosci. Abstr., 29: 172.15 (on CD ROM).

Schiff, N.D., Ribary, U., Plum, F. and Llinas, R. (1999) Words without mind. J. Cognitive Neurosci., 11: 650–656.

Schiff, N.D. and Plum, F. (2000) The role of arousal and 'gating' systems in the neurology of impaired consciousness. J. Clin. Neurophysiol., 17: 438–452.

Schiff, N.D., Ribary, U., Moreno, D.R., Beattie, B., Kronberg, E., Blasberg, R., Giacino, J., McCagg, C., Fins, J.J., Llinas, R. and Plum, F. (2002) Residual cerebral activity and behavioural fragments can remain in the persistently vegetative brain. Brain, 125: 1210–1234.

Schulman, J.J., Zonenshayn, M., Ramirez, R.R., Mogilner, A.Y., Rezai, A.R., Kronberg, E., Ribary, U., Mitra, P.P., Jeanmonod, D. and Llinás, R. (2000) Differences in MEG patterns produced by central and peripheral pain. Neuroimage, 11: 164–164.

Schulman, J.J., Horenstein, C., Ribary, U., Kronberg, E., Cancro, R., Jeanmonod, D. and Llinás, R. (2001) Thalamocortical dysrhythmia in depression and obsessive-compulsive disorder. Neuroimage, 13: 1004–1004.

Schulman, J.J., Ramirez, R., Cancro, R., Ribary, U. and Llinas, R. (2003) Thalamocortical dysrhythmia in schizoaffective disorder. Soc. Neurosci. Abstr., 29: 714.10 (on CD ROM).

Sheer, D. (1984) Focused arousal 40 Hz EEG and dysfunction. In: Ebert T. (Ed.), Self Regulation of the Brain and Behavior. Springer, Berlin, pp. 64–84.

Simos, P.G., Breier, J.I., Fletcher, J.M., Bergman, E. and Papanicolaou, A.C. (2000) Cerebral mechanisms involved in word reading in dyslexic children: a magnetic source imaging approach. Cereb. Cortex, 10: 809–816.

Singer, W. (1998) Consciousness and the structure of neuronal representations. Phil. Trans. R. Soc. London, 353: 1829–1840.

Singer, W. and Gray, C.M. (1995) Visual feature integration and the temporal correlation hypothesis. Ann. Rev. Neurosci., 18: 555–586.

Steriade, M. and Llinas, R.R. (1988) The functional states of the thalamus and the associated neuronal interplay. Physiol. Rev., 68: 649–742.

Steriade, M., Curro Dossi, R., Pare, D. and Oakson, G. (1991a) Fast oscillations (20–40 Hz) in thalamocortical systems and their potentiation by mesopontine cholinergic nuclei in the cat. Proc. Natl. Acad. Sci. USA, 88: 4396–4400.

Steriade, M., Curró Dossi, R., Paré, D. and Oakson, G. (1991b) Fast oscillations (20–40 Hz) in thalamocortical systems and their potentiation by mesopontine cholinergic nuclei in the cat. Proc. Natl. Acad. Sci. USA, 88: 4396–4400.

Steriade, M. (1993) Central core modulation of spontaneous oscillations and sensory transmission in thalamocortical systems. Curr. Opin. Neurobiol., 3: 619–625.

Steriade, M., McCormick, D.A. and Sejnowski, T.J. (1993a) Thalamocortical oscillations in the sleeping and aroused brain. Science, 262: 679–685.

Steriade, M., Curró Dossi, R. and Contreras, D. (1993b) Electrophysiological properties of intralaminar thalamocortical cells discharging rhythmic (~40 Hz) spike-bursts at ~1000 Hz during waking and rapid-eye movement sleep. Neuroscience, 56: 1–9.

Steriade, M. and Amzica, F. (1996a) Intracortical and corticothalamic coherency of fast spontaneous oscillations. Proc. Natl. Acad. Sci. USA, 93: 2533–2538.

Steriade, M., Amzica, F. and Contreras, D. (1996b) Sychronization of fast (30–40 Hz) spontaneous cortical rhythms during brain activation. J. Neurosci., 16: 392–417.

Tallal, P., Miller, S. and Fitch, R.H. (1993) Neurobiological basis of speech: A case for the preeminence of temporal processing. Annal NY Acad. Sci., 682: 27–47.

Tallal, P., Miller, S.L., Bedi, G., et al. (1996) Language comprehension in language-learning impaired children improved with acoustically modified speech. Science, 271: 81–84.

Tallal, P. (2004) Improving language and literacy is a matter of time. Nat. Rev. Neurosci., 5: 721–728.

Tallon-Baudry, C., Bertrand, O., Wienbruch, C., Ross, B. and Pantev, C. (1997) Combined EEG and MEG recordings of visual 40 Hz responses to illusory triangles in human. Neuroreport, 8: 1103–1107.

Tallon-Baudry, C. and Bertrand, O. (1999) Oscillatory gamma activity in humans and its role in object representation. Trends Cogn. Sci., 3: 151–162.

Teich, M.C., Khanna, S.M. and Guiney, P.C. (1993) Spectral characteristics and synchrony in primary auditory-nerve fibers in response to pure-tone acoustic stimuli. J. Stat. Phys., 70: 257–279.

Timmermann, L., Gross, J., Butz, M., Kircheis, G., Haussinger, D. and Schnitzler, A. (2003) Mini-asterixis in hepatic encephalopathy induced by pathologic thalamo-motor-cortical coupling. Neurology, 61: 689–692.

Tomassino, C., Grana, C., Lucignani, G., Torri, G. and Ferrucio, F. (1995) Regional metabolism of comatose and vegetative state patients. J. Neurosurg. Anesthesiol., 7: 109–116.

Tononi, G. and Edelmann, G.M. (1998) Consciousness and complexity. Science, 282: 1846–1851.

Traub, R.D., Whittington, M.A., Stanford, I.M. and Jeffcreys, J.G.R. (1996) A mechanism for generation of long-range synchronous fast oscillations in the cortex. Nature, 383: 621–624.

Van Essen, D.C., Drury, H.A., Joshi, S. and Miller, M.I. (1998) Functional and structural mapping of human cerebral cortex: solutions are in the surfaces. Proc. Natl. Acad. Sci. USA, 95: 788–795.

Wright, J.J., Robinson, P.A., Rennie, C.J., Gordon, E., Bourke, P.D., Chapman, C.L., Hawthorn, N., Lees, G.J. and Alexander, D. (2001) Toward an integrated continuum model of cerebral dynamics: the cerebral rhythms, synchronous oscillation and cortical stability. BioSystems, 63: 71–88.

Yang, T.T., Gallen, C., Schwartz, B., Bloom, F.E., Ramachandran, V.S. and Cobb, S. (1994) Sensory maps in the human brain. Nature, 368: 592–593.

Young, M.P., Tanaka, K. and Yamane, S. (1992) On oscillating neuronal responses on the visual cortex of the monkey. J. Neurophysiol., 67: 1464–1474.

Zeki, S. and Bartels, A. (1998) The autonomy of the visual systems and the modularity of conscious vision. Phil. Trans. R. Soc. London, 353: 1911–1914.

S. Laureys (Ed.)
Progress in Brain Research, Vol. 150
ISSN 0079-6123

Chapter 11

From synchronous neuronal discharges to subjective awareness?

E. Roy John[1,2,*]

[1]Brain Research Laboratories, NYU School of Medicine, New York, NY 10016, USA
[2]Nathan Kline Institute for Psychiatric Research, Orangeburg, NY, USA

Abstract: For practical clinical purposes, as well as because of their deep philosophical implications, it becomes increasingly important to be aware of contemporary studies of the brain mechanisms that generate subjective experiences. Current research has progressed to the point where plausible theoretical proposals can be made about the neurophysiological and neurochemical processes which mediate perception and sustain subjective awareness. An adequate theory of consciousness must describe how information about the environment is encoded by the exogenous system, how memories are stored in the endogenous system and released appropriately for the present circumstances, how the exogenous and endogenous systems interact to produce perception, and explain how consciousness arises from that interaction. Evidence assembled from a variety of neuroscience areas, together with the invariant reversible electrophysiological changes observed with loss and return of consciousness in anesthesia as well as distinctive quantitative electroencephalographic profiles of various psychiatric disorders, provides an empirical foundation for this theory of consciousness. This evidence suggests the need for a paradigm shift to explain how the brain accomplishes the transformation from synchronous and distributed neuronal discharges to seamless global subjective awareness. This chapter undertakes to provide a detailed description and explanation of these complex processes by experimental evidence marshaled from a wide variety of sources.

Introduction

Beliefs about the basis of subjective human experience have slowly evolved, from mystical notions of the soul and a disembodied mind to acceptance of the proposal that consciousness must derive from neurobiological processes. After a long reluctance to confront the task, resulting first from the influence of religion and then from the dogmas of behaviorism, a widespread consensus that understanding consciousness is the most important problem of neuroscience is emerging. Not only philosophical interest motivates this endeavor. Adequate anesthesia demands suppression of movement, pain, and awareness during surgery, which anesthesiologists achieve by administering drugs to alter the interactions between brain regions. Among the millions who undergo general anesthesia every year, a substantial percentage manifest post-operative cognitive deficit of unknown origin. Tens of thousands of persons suffer traumatic head injuries in vehicular accidents, and many remain with significant cognitive impoverishment or in long-lasting coma. Psychoactive drugs are administered to millions of patients every year by general practitioners compared to those by psychiatrists. Many of the cognitive and functional disturbances in patients thus medicated can be considered as disorders of consciousness, distorted in a manner which deviates

*Corresponding author. Brain Research Laboratories, Department of Psychiatry, NYU School of Medicine, 550 First Avenue, New York, NY 10016, USA. Tel.: +1212-263-6288; Fax: +1212-263-6457; E-mail: johnr01@popmail.med.nyu.edu

DOI: 10.1016/S0079-6123(05)50011-6

consistently from a realistic interpretation of the past and the present, in terms of the patient's life experience, leading to maladaptive behaviors. The deviations may originate from the individual's genetic endowment or vulnerability, activation of episodic memories of traumatic experiences in the past, or some temporary neurochemical imbalances. Regardless of the genesis, their manifestation represents a deviation from the regulatory processes that enable the normal brain to achieve adaptive behavior and a comfortable, tranquil state for the individual.

A self-organizing system may define a "ground state" of the brain. Homeostatic regulation of the ground state is temporarily altered during general anesthesia, may be perturbed persistently in coma, and is often disturbed in psychiatric patients. Return of consciousness after general anesthesia is accompanied by restoration of regulation. Successful treatment of comatose patients and those with neurological and psychiatric disorders may be facilitated by correction of underlying disturbances of regulation. For these practical clinical purposes, as well as because of their deep philosophical implications, contemporary studies of the brain mechanisms that generate subjective experiences are increasingly important. Plausible explanations can be made about the neurophysiological and neurochemical processes that mediate perception, sustain subjective awareness, and create the phenomenon of consciousness. This chapter provides an overview of contemporary studies providing useful insights into these processes. It also presents a radical proposal to resolve the gap between the discrete processes of neuronal interaction and the global experience of consciousness, which is speculative in that it is based on inferences indirectly derived from the inadequacy of connectionist theories rather than upon direct evidence. The following overview provides a synopsis of the major conclusions:

(i) Attributes of complex stimuli are fractionated and projected by the thalamo-cortical relays of the sensory specific "exogenous system" to the basal dendrites of ensembles of pyramidal neurons throughout the cortex, and thus encoded as time series of nonrandom synchronization within dispersed cell assemblies rather than by discharges of dedicated cells. These fragments of sensations constitute islands of local negative entropy.

(ii) Collateral fibers from afferent sensory pathways project to the ascending reticular activating system from which they are distributed as time series of nonrandomly synchronized volleys that are propagated to structures in the limbic system, where they are encoded as episodic memories. Those representations of previous experiences most similar to the momentary present input, which have been stored in the limbic system, are simultaneously activated by associational mechanisms and readout as a time series of nonrandom discharges. This "endogenous" readout is propagated via the nonsensory specific nuclei of the diffuse projection system to the apical dendrites of the cortical sheet of pyramidal neurons.

(iii) Coincidence detection between the converging time series of inputs to basal somatic and apical dendritic synapses of dispersed pyramidal neurons causes enhanced excitability, converting those assemblies that are currently encoding fragments of sensation into fragments of perception, which further increases the local negative entropy. This coincidence integrates encoded sensory information with output of systems encoding expectations, memories, planned actions, interoceptive, affective, and motivational states.

(iv) Integration of these fragments is required to yield a global percept. Modulation of membranes by local field potentials (LFPs) facilitates discharge of excited cells as a coherent cortico-thalamic volley, followed by back-propagated high-frequency reverberating oscillations. Coherent interactions of these cortico-thalamo-limbic-cortical oscillations cause a state transition of the system into resonance, binding these fragments into global negative entropy that creates consciousness, producing a unified perception.

(v) LFP oscillations are homeostatically regulated, controlling local synchrony, regional interactions, and periodic sampling, and defining a "ground state" with maximum entropy. Activity corresponding to this ground state is the most probable in a normally functioning, healthy individual at rest with eyes closed but alert.

(vi) Perturbations of the homeostatic LFP regulatory system alter the normal regulated thresholds that define the ground state. Cyclic fluctuations in these thresholds occur with the diurnal rhythms of sleep. Deviations constituting negative entropy are consistently found during processing of sensory stimulation, cognitive activity, developmental, neurological and psychiatric disorders, seizures, sleep, substance use, coma, and general anesthesia. Shifts from the ground state can be induced by chemicals such as anesthetics or addicting substances and accompany many neurological and psychiatric disorders. Restoration of the normal ground state is accompanied by return of normal consciousness and adaptive behavior.

This theory of consciousness proposes that elements of consciousness are dispersed as negative entropy, within many cell assemblies. Consciousness requires that this dispersed statistically nonrandom activity be transformed into a seamless subjective experience. Connectionistic concepts are inadequate to explain this conversion. The integration of dispersed, local negative entropies into global negative entropy may produce consciousness, an inherent property of an electrical field resonating in a critical mass of coherently coupled cells.

Fractionation or "local negative entropy" and consciousness or "global negative entropy"

Understanding perception is propadeutic to an understanding of consciousness, since it specifies the content of consciousness. Evidence indicates that perception is compounded of a multimodal representation of momentary exteroceptive and interoceptive stimuli contributed from an "exogenous system" combined with a continuous contextual rendering derived from readout of relevant episodic and short term or working memories and associated emotional coloration contributed by an "endogenous system." Consideration of this evidence has led some to the apt description of consciousness as perception of the remembered present (Edelman, 2001). The "exogenous system" processes information about each sensory modality, continuously fractionating incoming complex, multimodal signals into distinct features that are processed separately. Feature extractors or analyzers exist within multiple anatomically dispersed brain regions, each containing large neuronal networks relatively specialized for extracting distinct attributes of complex stimuli or representing selective aspects of recent or episodic memories. These fragmented multimodal attributes are recombined and placed into the context of past experience to construct perceptions that are the momentary, dynamically changing content of consciousness. This dispersed local analytic information is constantly reintegrated globally by dynamic interactions among regions. For this synthesis to achieve a reliable assessment of reality, certain errors must be avoided. Neurons are unreliable reporters. Neural activity representing information useful for adaptive response to the momentary environment, or "signal," must automatically be segregated from spontaneous neural activity, or "noise." Failure of a "dedicated" cell in a feature detector to respond due to refractoriness should not result in failure of the system to detect the corresponding attribute. Persistent after effects and self-sustained reverberatory activity should be suppressed. Contributions from neuronal assemblies that represent an attribute absent from the exogenous input must be excluded. It sometimes might be advantageous to use a subset of cues to complete the encoding of important but complex stimuli that have only partially been detected. It would be advantageous to automatically attach a valence to the encoded stimuli that reflected their importance. Failure to avoid such errors may result in seizure activity, inappropriate behavior, misperceptions, delusions, or other psychiatric symptoms. How the brain reassembles the fractionated sensory elements from dispersed populations into a global percept, while complying with these requirements, constitutes the "binding problem." How this representation of essentially statistical information is transformed into a personal subjective experience is the problem of consciousness.

This theory proposes that homeostatically regulated thresholds exist in every neuronal population, defining the baseline levels of random synchronization of firing in the ensemble that reflect transactions within the local neural network and inputs from exteroceptors and other brain regions. A local 'ground state' of maximum entropy can be defined as the absence of information or perfect predictability. The regulated thresholds define the range of fluctuation that is most probable in each brain region, within the normal distribution of the resting power spectra of local field potentials and the cross-spectra of interactions between regions. Such power spectra or cross-spectra are mathematical descriptions of the average frequency composition of a time series of voltages, or synchronized deviations from the randomness in spatiotemporal discharge patterns. A temporal sequence of synchronized activation or inhibition of some significant proportion of the cells in any brain region, deviating from the mean values of the distributions that define the ground state, is the neural activity of informational value. Complex, multimodal environmental inputs cause nonrandom, temporal patterns of synchronized activity in extensive networks that deviate from the regulated thresholds, establishing spatially dispersed states of local negative entropy. Evaluation of such spatially dispersed statistical representations by any cell or set of cells so as to uniquely identify the complex inputs from the environment cannot plausibly be envisaged, let alone the transformation of such statistical description into subjective experience by processes restricted to discrete synaptic transactions in dedicated cellular networks. The difficulty of accounting for this transformation has been termed the "explanatory gap" (Levine, 1983). As a solution to this quandary, we have proposed that consciousness may emerge from global negative entropy, integrated across anatomically dispersed neuronal ensembles that are nonrandomly active and electrically coherent in spite of spatial separation, and is a property of an electrical field of information resonating in a critical mass of brain regions (John et al., 2001; John, 2002, 2003, 2004).

Local field potentials (LFPs) are the envelope of the time series of synchronization in ensembles

In parallel with the activity of neural elements, voltage oscillations of LFPs can be observed within the volume of the brain. Like the LFPs, the electroencephalogram (EEG) is the time series of synchronized activity in huge neural ensembles in the vicinity of the recording site. The voltage waves and coherences arise from synchronized excitatory and inhibitory post-synaptic potentials produced by transactions between elements within a local region as well as from interactions among regions, while neuronal discharge is determined by influences impinging upon a single element. These two aspects of brain activity, discrete unit discharges, and potential waves are intrinsic to the principles by which the brain operates. This article proposes specific processes by which slow wave oscillations (1) contribute to the construction and identification of meaningful stimulus attributes from non-random synchrony of neural discharges caused by afferent inputs producing sensations and (2) enable the brain to bind these fragments into meaningful percepts which enter subjective awareness of the individual.

Since single neurons respond erratically, the discharge of an individual neuron does not provide reliable or interpretable information. Then, a question arises: where is the information? Were neuronal activity in a brain region random, the unit discharges would be distributed evenly throughout time, and a macroelectrode recording would be essentially isopotential. A certain level of "self-organizing" synchronization, or coherence, occurs spontaneously among CNS neurons and has also been observed to take place in stimulations of neural tissue in which the elements are weakly coupled (Abeles et al., 1994). As a consequence, the voltages which are recorded from these large neural aggregates approach zero only in slabs of cortical tissue surgically isolated from the rest of the brain (Burns, 1968). Oscillations of LFPs or evoked potentials (EPs) reflect nonrandom neural activity (Fox and O'Brien, 1965). When brain voltages recorded from some brain region depart significantly from the baseline, the local neuronal activity is synchronized or "nonrandom," indicating transactions among brain regions or regional involvement in processing exteroceptive or interoceptive information.

LFPs and multiple unit activity (MUA) are closely related, with multiple unit activity increasing when LFPs are negative. Multiple unit activity and LFPs can be entrained by visual stimuli in the delta, theta, alpha, beta, and gamma ranges (Rager and Singer, 1998). Synchronization is more precise with oscillatory modulation in the beta and gamma range (Herculano-Houzel et al., 1999). Synchronized neural activity in multiple ensembles, rather than the local discharge of dedicated neurons, integrates dispersed fragments of sensation and may be the basis of perception (Nicolelis et al., 1995). fMRI results suggest that recurrent loops across multiple brain areas lead to resonance, which is the neural correlate of conscious vision (Goebel et al. 2001).

Basis vectors of the brain, the ground state, and global negative entropy

Power Spectra

Striking regularities, which may reflect action of a genetically specified homeostatic system common to all healthy human beings, have been found by quantitative analysis of the EEG (QEEG). The power spectrum of the QEEG from every cortical region changes predictably as a function of age in healthy, normally functioning individuals in a resting but alert state (Matousek and Petersen, 1973; John et al., 1980, 1983, 1988). These highly predictable values of the normative age-appropriate power spectra and spatio-temporal organization of EEG patterns define the baseline or "ground state" of brain activity. Under resting conditions, EEG spectra remain within the variance defined by their distributions in normal reference groups. Normative values have been established from age 1 to 95 in large normative studies for numerous parameters, such as absolute and relative power, power ratios (symmetry) and synchrony (coherence) in different frequency bands, within every local region, and between pairs of regions within and across the hemispheres. The specificity and independence from ethnic factors of these descriptors have been established in cross-cultural replications in Barbados, China, Cuba, Germany, Hungary, Japan, Korea, Mexico, Netherlands, Sweden, USA and Venezuela (John and Prichep, 1993). Like the electrocardiogram, these rhythms display extremely high test-retest reliability within the healthy individual across intervals of hours, days, or even years (Kondacs and Szabo, 1999) and have been demonstrated to be clinically sensitive (John and Prichep, 1993; Hughes and John, 1999). This stability, specificity, and sensitivity of the EEG in all healthy human beings from any ethnic background is due to dynamic regulation by complex homeostatic processes that we believe to reflect our common genetic heritage.

Multivariate descriptors

Lawful patterns of covariance of QEEG variables across the whole brain as well as among sets of brain regions have been quantified as Mahalanobis Distances (composite features with inter-correlations removed), which display similar stability and replicability across the life span (John and Prichep, 1993). These multivariate compressions display remarkable specificity and sensitivity to subtle brain dysfunctions (Jonkman et al., 1985). High levels of common mode resonance of the EEG exist over long-range cortical-cortical distances (John et al., 1990).

Spatial principal components

Spatial principal component analysis (SPCA) has been applied to EEGs from large groups of normally functioning individuals, to quantify statistically significant, independent spatiotemporal modes of neuronal interaction across the entire cortex as well as to classify distinctive brain actions of drugs (John et al., 1972). As few as five spatial principal components account for 90% of the variance of the EEG and reveal a high degree of true phase correlation. The complex electrical activity of the brain reflects interactions among a limited number of simultaneously active, well-regulated subsystems, each with its characteristic mode of oscillation and distinctive functional neuroanatomical organization (Duffy et al., 1992; John et al., 1997; Srinivasan, 1999).

Microstates

A variety of evidence, presented below, indicates that information processing in the brain occurs in discrete discontinuous microstates and sampling occurs at a specific rate that is similarly well regulated. Adaptive segmentation reveals that there are a limited number of microstate topographies and they correspond well to spatial principal component loadings (Koenig et al., 2002). Clearly, this regulation requires control of neurotransmitters.

Departures from ground state

Recently developed clinical QEEG instruments reliably monitor the depth of general anesthesia during surgery independent anesthesia. In individual patients, such instruments provide an index of the level of consciousness by continuous quantification of multivariate deviations of brain electrical activity from the ground state defined above independent particular anesthetics agents. The level of consciousness during anesthesia can be accurately predicted by the magnitude and direction of such multivariate deviations, independent of anesthetic agents. When baseline values are restored, consciousness returns (Prichep et al., 2000; John et al., 2001). A wide variety of QEEG and EP monitors of effects of anesthetics on brain activity confirm the high sensitivity of these descriptors (John and Prichep, 2005).

Deficiencies or excesses of any neurotransmitter should produce marked departure from the homeostatically regulated normative EEG spectrum. In fact, consistent profiles of deviation from values predicted by normative databases have been replicably found in large groups of psychiatric patients in conditions such as attention deficit disorder, dementia, head injury, mood disorders, obsessive compulsive disorder, schizophrenia, and a number of other behavioral, developmental, and cognitive disorders (Prichep and John, 1992; Alper et al., 1993), as well as in groups of patients undergoing general anesthesia. The patterns of these deviant values are so distinctive that accurate, replicable statistical classification of patients into different DSM IV diagnostic categories has been accomplished (Prichep and John, 1992; Alper et al., 1993; John and Prichep, 1993; Hughes and John, 1999; John et al., 2001). Further, deviations induced in normally functioning individuals by drugs that are clinically effective for a particular disorder, are opposite from the distinctive deviations found in unmedicated patients with the corresponding disorder, and successful treatment of such patients restores the QEEG to normative values (Saletu et al., 2000). Such observations validate the conclusion that baseline LFP spectra and patterns of covariance are functionally relevant and clinically useful manifestations of processes relevant to consciousness and subjective experience.

Global negative entropy

These normative baselines are considered to be basis vectors describing genetically determined modes of brain electrical organization, or basic "traits," whose set points determine the origin of a multidimensional signal space. The brain is a self-organizing system that conforms to these genetic constraints. The electrical "ground state" of the brain is defined as a hypersphere around this origin, the surface of which is defined by the variance of the normally functioning human population. Regions within this hypersphere represent a region of maximum entropy in which the range of magnitudes of interactions among functional brain systems can be predicted. Subjectively, we habituate to these states and, therefore, we treat such activity as if it contains no information. The state of the brain can be represented as a "brain state vector" in this multivariate signal space. Excursions of the brain state vector that extend outside the boundaries of this hypersphere constitute negative entropy. Phasic modulations of the fundamental modes reflect the present, momentary sensory, cognitive, and motor states of the individual, as well as transient alterations of consciousness. Tonic or sustained departures from these modes can be considered as deviant or pathophysiological, abundant evidence for which was cited above. Overall momentary deviations from these fundamental modes of synchronization, within and among brain regions, are how information is represented in the brain and are global negative entropy, the content of consciousness.

Origins and functional relevance of rhythmic EEG oscillations

EEG activity has conventionally been described in terms of a set of wide frequency bands, usually defined as delta (1.5–3.5 Hz), theta (3.5–7.5 Hz), alpha (7.5–12.5 Hz), beta (12.5–25 Hz), and gamma (25–50 Hz). Factor analysis EEG power spectra has found a similar factor structure, suggesting that different functional systems produce these rhythms. The observed predictability of the EEG power spectrum arises from regulation by anatomically complex homeostatic systems involving brainstem, thalamus, and cortex and utilizing all neurotransmitters (Llinas, 1988; Steriade et al., 1990; McCormick, 1992; McCormick, 2002). "Pacemaker neurons" throughout the thalamus interact with the cortex and nucleus reticularis to oscillate synchronously in the alpha frequency range. Efferent projections of these oscillations modulate waves also generated by a dipole layer in widely distributed cortical centers, producing the dominant alpha rhythm. The nucleus reticularis can hyperpolarize the cell membranes of thalamic neurons, slowing this rhythm into the theta range, and diminishing sensory throughput to the cortex. Theta activity is also generated in the limbic system by pacemakers in the septal nuclei, which can be inhibited by entorhinal and hippocampal influences (Buzsaki, 2002). Delta activity is believed to originate in neurons in deep cortical layers and in the thalamus, normally inhibited by input from the ascending reticular activating system (ARAS) and nucleus basalis of Meynert. Delta activity may reflect hyperpolarization and inhibition of cortical neurons, resulting in dedifferentiation of neural activity. Activity in the beta band reflects cortico-cortical and thalamo-cortical transactions related to specific information processing. Gamma activity reflects cortico-thalamo-cortical and cortical-cortical reverberatory circuits, which may play an important role in perception (also see Ribary, this volume). This activity is the topic of intense contemporary investigation, some of which will be discussed below. Sleep-wake cycles depend on interactions involving the pontine and mesencephalic reticular formation, locus ceruleus, the raphe nuclei, thalamic nuclei, and the nucleus of Meynert. As an oversimplification, the neurons of the thalamus manifest two intrinsic states: When hyperpolarized to a certain level, they enter a bursting mode in which they do not relay information to the cortex and other brain regions, and sleep ensues. When in this state, the EEG manifests large delta waves, perhaps reflecting the release of cortical neurons mentioned above, normally inhibited by the influences of the ARAS. For wakefulness to take place, the thalamo-cortical neurons must be restored to a state in which throughput of afferent sensory information to the cortex is again possible. This occurs by cholinergic influences of the ARAS and the nucleus of Meynert, a process that is described below in further detail.

EEG regulation depends upon this extensive neuroanatomical homeostatic system. Structures in this system play an important role in a wide variety of behavioral functions. Thus, behavioral implications are inherent in any detailed evaluation of disturbances from this homeostatic regulation, whether transient as in anesthesia, persistent as in coma, or tonic as in developmental and psychiatric disorders. Major elements of this system, neurotransmitters mediating their interactions and their putative behavioral contributions, are schematized below in Fig. 1.

The following explanation is an oversimplification for heuristic purposes. Assume that a subject is drowsy, with an EEG dominated by slow waves, perhaps in the theta frequency range, indicating dedifferentiation and increased synchrony of cortical neurons disengaged from information processing. This slowing of the alpha rhythm arises from hyperpolarization of the alpha pacemakers, reflecting GABAergic inhibition of thalamic pacemakers by the nucleus reticularis that can be initiated by glutaminergic influences from the cortex.

Activation of the mesencephalic reticular formation by altered environmental stimuli results in inhibition of the nucleus reticularis by cholinergic and serotonergic mediation via the ARAS. This inhibition by the ARAS releases the thalamic cells from inhibition by the nucleus reticularis, and flow of information through the thalamus to the cortex is facilitated. The dominant activity of the EEG becomes more rapid, with return of alpha and higher frequency beta activity, reflecting differentiation of

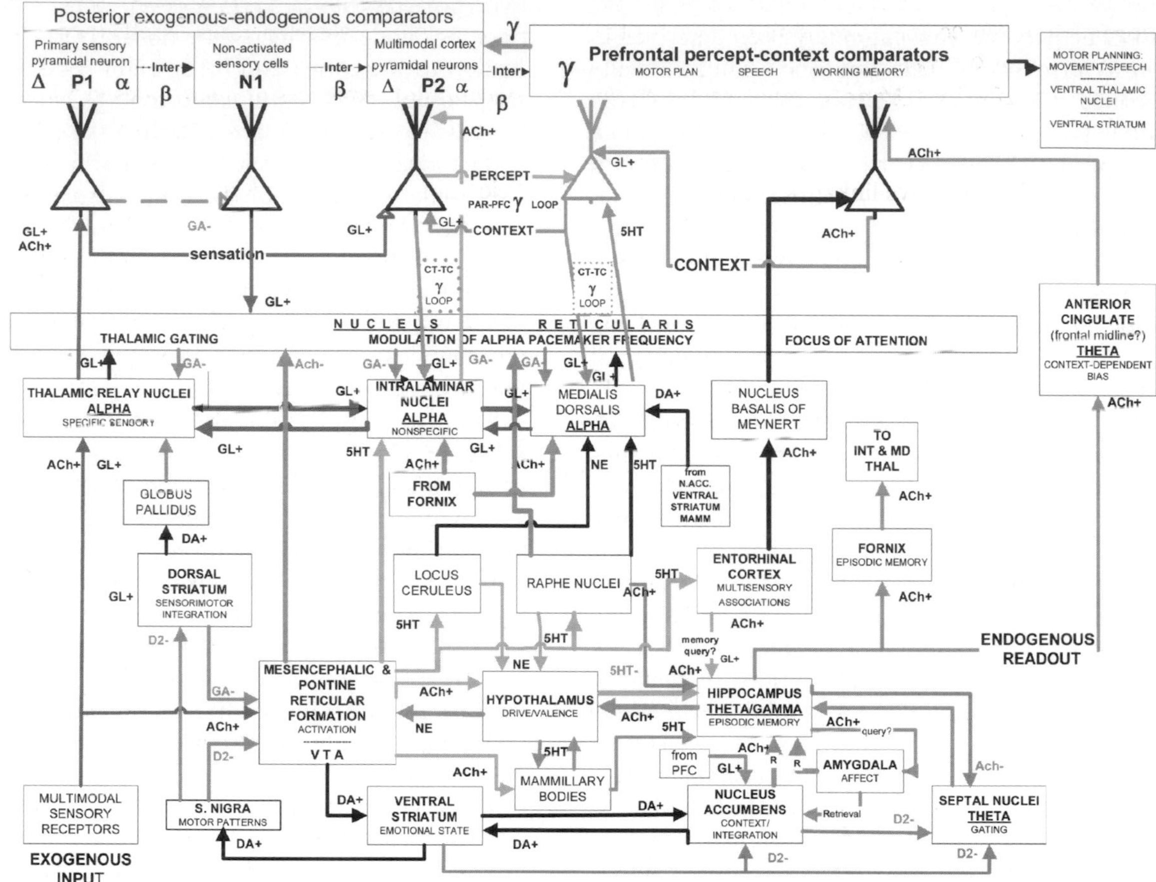

Fig. 1. Oversimplified scheme of the neuroanatomical structures and neurotransmitters comprising the homeostatic system that regulates the electrical rhythms of the brain. The diagram assigns putative roles in the generation of EEG oscillations in the delta, theta, alpha, beta, and gamma frequency ranges and to components of event related potentials. Elements of the exogenous system processing sensory specific inputs are encoded as blue, elements performing nonsensory specific processing are encoded as gold, inhibitory influences are encoded as red, and endogenous processing of readout from the memory system is encoded as green. The behavioral functions that various regions are believed to contribute to are tentatively indicated for some of these brain structures (see text for details). See Plate 11.1 in Colour Plate Section.

cortical neurons and resumption of cortical-cortical interactions. However, as an alternative result of cortical influences, dopaminergic striatal projections can inhibit the mesencephalic reticular formation, enabling differential inhibition of thalamic neurons by the nucleus reticularis to occur. These interactions between the ARAS, nucleus reticularis, and the thalamo-cortical relay neurons provide a mechanism whereby the brain dynamically filters and selects permissible stimuli to reach cortical centers. The flux of activity through these pathways constitutes a temporal pattern of synchronized input from the exogenous system to the axosomatic synapses of cortical pyramidal neurons in lower layers of the cortex.

In parallel, collateral pathways from the mesencephalic reticular formation enter the limbic system, which is intimately involved in the storage and retrieval of memories and contributes hedonic valence and emotional tone to the momentary brain state. Influences from the limbic system impinge upon the cortex via nonsensory specific pathways, and are the major input from the endogenous system to the apical dendrites of cortical sheet of pyramidal neurons, in upper layers of the cortex. An entorhinal–hippocampal–septal circuit is involved in the storage of episodic memories. Affective valences are attached to these episodic memories by associative linkages with other limbic regions. Septal–hippocampal interactions modulate the activity of circuits that generate EEG rhythms in the theta band projected to the cortex via the cingulate gyrus and the medialis dorsalis nucleus of the thalamus. When this circuit is activated, gamma oscillations arise in the hippocampus, become phase locked to the theta activity, and are projected to the cortex. Thus, cortical theta may reflect either underactivation of the ARAS or activation of affective states, memories, or related cognitive processes.

These reciprocal cortical–subcortical interactions can diminish, modulate, or actually block the flow of sensory information to the cortex and, together with the contribution from the limbic system of retrieved memories most relevant to the current environment, play an important role in focusing attention and can have a decisive influence on selective perception. This continuous interaction between the exogenous and endogenous systems constructs the "remembered present." In view of the dependence of both systems upon the delicate balance and relative abundance of multiple neurotransmitters, the relevance of these interactions to psychiatric disorders becomes self-evident.

The "microstate" — temporal parsing of brain fields

Although subjective time is experienced as continuous, "brain time" is in fact discontinuous, parsed into successive "perceptual frames" that define a "traveling moment of perception" (Allport, 1968; Efron, 1970). Sequential events within this brief time interval will be perceived as simultaneous, but are perceived as sequential if separated by a longer time. The duration of the perceptual frame in different sensory modalities is about 75–100 ms. The most familiar example of discontinuous sampling is the apparent stability of a movie viewed at about 64 frames per second. Examination of the momentary voltage maps on the scalp reveals a kaleidoscope with positive and negative areas on a field or "microstate," which persists briefly and changes continuously (Lehmann, 1971). Computerized classification of sequences of microstates in 500 normal subjects, aged 6–80 years, yielded the same topographic patterns in every individual, with 4–5 basic fields accounting for about 85% of the total spatiotemporal variance of the EEG with approximately equal prevalence. The mean microstate duration slowly decreased during childhood, stabilizing for healthy young adults at about 82 ± 4 ms (Koenig et al., 2002). The stability of these microstate topographies and their mean duration across much of the human life span, and their independence from ethnic factors, supports the suggestion of genetic regulation. The topographies of these momentary fields closely resemble those of the computed modes described in the spatial principal component studies cited above, indicating that the spatial principal component loadings are not a computational artifact. While these normative modes of regional interaction are statistically independent, they may overlap in time with different weightings. Different microstates seem to correlate with distinctive modes of ideation (Lehmann et al., 1998). The transition probabilities from one microstate to another are apparently altered during cognitive tasks (Pascual-Marqui et al., 1995) and in schizophrenia (Streletz et al., 2003).

The correspondence between the experimentally obtained durations of each subjective episode and the mean duration of microstates suggest that a microstate may reflect a "perceptual frame." Recent functional brain-imaging evidence has led to similar proposals that consciousness is discontinuous and is parsed into sequential episodes by synchronous thalamo-cortical activity (Llinas and Ribary, 1993; see Ribary, this volume). Asynchronous, multimodal (Zeki, 2000) sensory information may thereby be integrated into a global instant of conscious, apparently simultaneous multimodal experience.

Introspection reveals fading, but persistent recollection of the recent past that coexists in subjective continuity with the momentary present. A degree of constancy must persist across a sequence of perceptual frames, analogous to a moving window of time much longer than the momentary persistence of a microstate. Conversely, humans can make temporal discriminations much faster than these estimates of perceptual frame duration. The brain's sampling algorithm must somehow reconcile the 80 ms microstate duration with millisecond discriminations as well as persistent recent memories. This implies that afferent input in any modality is mediated in parallel by many multiplexed channels. Sequential events within a single such channel are combined into a corresponding perceptual frame. Subsets of channels must be integrated in receiving areas as a "standing spatiotemporal wave" of synchronized activity, like an interference pattern persisting through the integration period across serial microstates. Adaptive behavior requires that information in the immediate perceptual frame be continuously evaluated in the context of recent and episodic memories as well as evaluation of emotional and motivational valence relevant to the immediately preceding frame. Coincidence detectors of the differences in local non-randomness across successive microstates might provide a priority interrupt signal that serves to close the integration period arbitrarily, abruptly resetting the timing and opening a new frame.

The sustained persistence of microstates, as well as many behavioral measures, is difficult to reconcile with the brevity of neural transactions. Some reentrant or reverberatory brain process must sustain cortical transactions as a "smoothed" steady state, independent of the activity of individual neurons. Temporal extension of neural activity has been considered critical for binding, and NMDA has been proposed to play an essential role in this regard (Flohr, 1998). It may be relevant that EPSP–IPSP sequences ranging from 80 to 200 ms are omnipresent in mammalian forebrain neurons (Purpura, 1972), comparable to the long-persistent, post-synaptic depolarization displayed by NMDA receptors.

A comparator constructs perceptions from sensations

Not only is brain processing of information discontinuous, parsed into successive frames that are subjectively continuous, but the content of each frame appears to be constructed from interaction of two separate systems. The phenomenon of "backward masking" or metacontrast, the ability of a later sensory input to block perception of a prior event (Alpern, 1952), suggests that two different processes representing independent inputs to a comparator are required for conscious perception. This interaction is constantly ongoing and, like any good computer operating system, is transparent to the user, the self.

The existence of this combinatory process was initially inferred from the results of animal experiments, in which electrical activity was recorded from electrodes that were chronically implanted into multiple brain regions while the animals learned discriminative behaviors. These electrodes were implanted in pairs with closely spaced tips separated by 1 mm, enabling "bipolar" recordings from well-localized ensembles. The differential cues were "tracer-conditioned stimuli" consisting of auditory or visual stimuli presented at two different specified repetition rates. Appearance of EEG activity at the tracer-conditioned stimuli frequency in bipolar recordings in any region was interpreted as "labeled response," indicating local participation in processing the tracer-conditioned stimuli and in mediating the behavior. When cats learned to discriminate among several visual and auditory differential-conditioned stimuli, marked changes were observed during training in the anatomical distributions of labeled activity. Initially, labeled activity appeared in only a few regions, but as learning progressed, labeled responses began to appear in wider regions of the brain. Anatomical spread of labeled activity gave direct insight into the establishment of extensive brain networks during learning. More unexpected was the observation that when errors were performed, labeled responses in some brain regions corresponded not to the physically presented tracer-conditioned stimuli, but rather to the frequency of the cue appropriate for the behavior that was actually performed. These results showed that the brain could store a distinctive temporal pattern of synchronized discharges in a representational system and release that pattern from memory on a subsequent occasion, when cues were misperceived.

Further experiments examined averaged event-related potential (ERP) waveshapes elicited by the tracer-conditioned stimuli. Initially, ERPs displayed only a primary positive component at early latencies. As the conditioned response was successfully established, a later primarily positive component appeared, implying that a second set of inputs was reaching those regions, but was manifested only in those trials in which appropriate behavior ensued. When differentiated behavioral responses became firmly established, requiring discrimination between visual or auditory tracer-conditioned stimuli at two different repetition rates, the ERP waveshapes elicited by the two tracer-conditioned stimuli developed distinctively different late components. Waveshapes acquired distinctive morphology to different cues and became similar in bipolar recordings from multiple nonsensory- and sensory-specific regions. It was proposed that perception, i.e., accurate interpretation of inputs from the environment, required coincidence detection between activity in sensory-specific and sensory-nonspecific systems (John, 1968).

Evidence of contributions to the comparator by a time series released from the hippocampus

A series of experiments were performed in such differentially conditioned cats, recording from multiple movable arrays of closely spaced microelectrodes chronically implanted in different brain regions, in addition to 24 fixed electrodes usually arranged in bipolar pairs sampling 12 other neuroanatomical structures that varied from animal to animal. Post-stimulus histograms were examined, showing the average temporal firing patterns of single and multiple units as well as LFPs and ERPs simultaneously elicited by different tracer-conditioned stimuli were examined. Recording simultaneously from all bipolar electrode pairs and moving the microelectrodes 0.1 mm every week, extensive brain regions were surveyed as different behaviors were performed to randomized presentations of the two tracer-conditioned stimuli. These experiments yielded some salient findings: (1) Although single tracer-conditioned stimuli presentations from several neurons in the same neighborhood elicited firing patterns that were highly variable and very different for each cell, after several hundred presentations almost identical post-stimulus histograms were obtained from every cell; (2) The post-stimulus histograms of single neurons during large numbers of tracer-conditioned stimuli presentations slowly converged to the post-stimulus histograms of multiple unit activity from the local neural ensemble that was rapidly elicited during a small number of presentations; (3) The ERPs recorded from any region were the envelope of the post-stimulus histograms from multiple units in that region, and remained very similar as the microelectrodes traversed through extensive intracerebral distances (3–4 mm); (4) Different distinctive ERP waveshapes and post-stimulus histograms contours were elicited by different behavioral cues, but the time series of nonrandom synchronization of neuronal firing was very similar in different brain regions; (5) The ERPs and post-stimulus histograms throughout such extensive traverses of the sensory-specific lateral geniculate body (a thalamo-cortical relay nucleus) and in the contralateral non-sensory-specific dorsal hippocampus displayed very similar post-stimulus histograms and ERP waveshapes; (6) The distinctive time series of neuronal discharges were present in the thalamo-cortical relay nucleus, but absent from hippocampus when correct responses failed to occur, and the thalamo-cortical relay nucleus and hippocampus displayed very disparate waveshapes in response to novel stimuli; and (7) Similar ERP waveshapes were recorded simultaneously from bipolar electrode derivations in many other brain regions (John and Morgades, 1969a, b; John, 1972).

These results and those of subsequent studies suggested that activity of an individual element was informational only insofar as it contributed to the population statistics. The time series of voltages comprising an EEG or ERP waveshape recorded from a brain region closely corresponded to the temporal pattern of synchronized firing among the neurons in the local ensemble. Storage of an episodic memory appeared to establish a system that displayed distinctive temporal patterns of nonrandom activity synchronized across widely dispersed regions of the sensory cortex appropriate to the modality of the cue, the corresponding thalamo-cortical relay nucleus, the intralaminar nuclei, the mesencephalic reticular formation, and the hippocampus. When novel stimuli at an intermediate rate midway between the two tracer frequencies were randomly interspersed into a series containing the two tracer-conditioned stimuli, "differential generalization" occurred and the animal performed one or more differentiated, conditioned responses. When differential generalization occurred, both the ERP waveshapes and post-stimulus histograms firing patterns elicited by the neutral cue corresponded to those distinctive for the tracer-conditioned stimuli appropriate for the behavior that was subsequently performed. The retrieval of a particular episodic memory seemed to be related to the readout of the corresponding distinctive time series of synchronized discharges from an endogenous system, concordant with a similar time series of synchronized discharges in the exogenous pathways. When ERP waveshapes recorded during behavioral errors of omission were subtracted from ERP waveshapes when behavioral response was correct, the resulting "difference waveshape" was essentially identical, with simultaneous peak latencies, in cortex, intralaminar nuclei, mesencephalic reticular formation, and hippocampus, but substantially delayed in the thalamo-cortical relay nucleus. The "difference waveshape" was interpreted to be the time series of the "readouts" released when a neutral stimulus elicited differential generalization. The delay in the corresponding thalamo-cortical relay nucleus (lateral geniculate body) was interpreted to mean that a memory system imposed a selective filter on the sensory nucleus processing incoming visual information (John et al., 1969, 1972, 1973; Ramos et al., 1976). Readout waveshapes in sensory-specific brain regions could be released by tracer stimuli of several sensory modalities; during differential generalization, the same cue distinctive ERP waveshapes were elicited in the medial geniculate body (a supposedly specific "auditory thalamo-cortical relay" nucleus) by novel visual as well as auditory tracer-conditioned stimuli (Thatcher and John, 1977).

These results led to an hypothesis that information was encoded statistically by the time series of nonrandomly synchronized neural activity within neuronal ensembles in various brain regions, rather than by synaptic changes in specific neurons in dedicated pathways. This idea was tested experimentally using electrical stimulation of brain regions to artificially produce nonrandom neuronal discharges. Independent of the region, it was shown that brain stimulation could achieve complete differential behavioral control, parametrically proportional to the amount of current delivered. Efficacy depended upon current strength rather than locus of stimulation, and regions differed only with respect to the amount of current required to achieve behavioral control (Kleinman and John, 1975). When two patterned pulse trains were delivered out of phase to arbitrarily selected pairs of brain regions, behaviors appropriate to the global temporal sum of the two electrical patterns were performed; simultaneous recordings suggested that integration between the two pulse trains took place in the intralaminar nuclei. Synchronized activation in arbitrarily selected pairs of brain regions was integrated and incorporated by the brain into a single cue that could guide adaptive behavior. Such arbitrarily placed pulses of electric current cannot be presumed to activate dedicated brain circuits selective for different behaviors.

Different geometric forms elicited different ERP waveshapes in humans, which displayed "size invariance;" the same basic waveshape was elicited by a geometric form independent of its size (John et al., 1967). When a series of numbers or letters were presented to human subjects, a vertical line elicited markedly different ERPs on the parietal cortex, when perceived as the letter "i" imbedded in a letter sequence or as the number "1" in a number sequence. However, ERPs from the visual cortex were identical under the two different conditions (Thatcher and John, 1977). Such findings suggest that information about a complex stimulus is represented by a distinctive spatiotemporal pattern of discharges coherent across a dispersed set of regions, which is encoded by a limbic memory mechanism that can reproduce the specific time series and propagate it effectively to other brain regions.

The chronometry of consciousness

Psychophysiological research has established that latencies of ERP components are correlated with information processing, and successive peaks correspond to (a) sensory activation, (b) detection of sensation, (c) perceptual identification of the information, (d) cognitive interpretation of the meaning of the input, (e) linguistic encoding of that concept, and (f) organization of adaptive reactions.

Sensory activation

In a cognitive auditory task, evoked gamma oscillations (40–45 Hz) phase-locked between vertex and posterior temporal scalp regions have been detected as early as 45 ms (Gurtubay et al., 2004). In recordings restricted to lower frequencies, the earliest components of scalp-recorded ERPs appear as a surface negativity (N1) at about 60–80 ms and a positivity (P1) at about 100 ms. These components have been localized by imaging methods to the primary cortical sensory receiving areas of the corresponding modalities. The later peaks are task-dependent and localization of their sources is still a subject of investigation (Gazzaniga et al., 2002).

Detection of sensation

In an attempt to establish the minimum duration of a stimulus for it to be detected, Libet (1973, 1982, 1985) studied the experiential consequences of series of electrical stimuli delivered directly into the brain of conscious human subjects, as well as other stimuli of varying duration. A debate still rages about the interpretations of those experiments, indicated by a special issue of a leading journal recently devoted to this topic (Baars et al., 2002). In brief, evidence suggests that although a person may not yet be aware of a stimulus, it has already been detected by the brain by about 80–100 ms. Intracerebral recordings from multiple sites suggest that initial visual processing may occur exclusively in the visual association cortex from approximately 90 to 130 ms (Halgren et al., 2002), an estimate concordant with the observed average duration of a microstate discussed above.

Mismatch negativity

A series of experiments (Naatanen et al., 1987, 1993, 2004) have established that as soon as a stimulus registers in the brain, a memory trace appears to be established that compares subsequent events to a representation of the previous event. This dynamic "echoic trace" appears to last for about 10 s (Sams et al., 1993). When the second element of a pair of sensory stimuli differs from the first, an enhancement of the ERP excursion from the second negativity (N2) to the second positivity (P2) takes place. The amplitude N2P2 is considered to be a measure of attention, referred to as mismatch-negativity (MMN). This process appears to be "preconscious." A critical set of experiments is relevant here (Libet et al., 1979). Essentially, a complex set of observations was interpreted to show that conscious awareness of a neural event is delayed approximately 500 ms after the onset of the stimulating event, but this awareness is referred backward in time relative to that onset. Cotterill (1997) proposed a neural circuit to account for this "backward referral" namely, the "vital triangle," which was further elaborated by Gehring and Knight (2000) in a study of human-error correction. These authors propose that a model of the original event is stored in the dorsolateral prefrontal cortex and is compared to the subsequent event. The existence of MMN suggests that a model of the recent past is continuously compared with the present and, together with other evidence, has led Edelman (2001) to refer to consciousness as "the remembered present."

Perceptual awareness

Numerous reports from intracerebral as well as scalp recordings from conscious human subjects indicate that a burst of phase-locked activity appears between the parietal and prefrontal cortex from about 180–230 ms after presentation of a stimulus to a subject engaged in performing a cognitive task (Desmedt and Tomberg, 1994; Tallon-Baudry et al., 1997; Tallon-Baudry, 2000; Varela, 2000). In posterior temporal regions, phase-locked as well as non-phase-locked gamma oscillations at about 200 ± 23 ms at 43 Hz have been observed by Gurtubay et al. (2004). Results in studies already cited above, in which intracerebral recordings were obtained from multiple implanted electrodes in occipital, parietal, Rolandic, and prefrontal sites during working memory tasks, suggested that the results of initial perceptual processing from approximately 90 to 130 ms after stimulus presentation were projected from the visual association cortex to prefrontal–parietal areas from approximately 130 to

280 ms (Halgren et al., 2002). On the basis of such findings, and from results cited in the next section, we propose that the neurophysiological processes that generate the second positive component of the ERP, P2, at about 200 ms represent the neural correlates of awareness.

Cognitive interpretation

The most studied electrophysiological correlate of human cognitive processes, the so-called P300 (Sutton et al., 1965; Donchin et al., 1986), has been investigated in thousands of experimental studies since first reported. This process is actually composed of an early event, referred to as P3A, that appears over anterior brain regions at about 225–250 ms and a later component, or P3B, that appears more prominently over posterior brain regions at about 300–350 ms, after an unexpected "oddball" or "target" event occurs in a series of common events. Intracerebral recordings have established that target-evoked P3-like components were most frequently recorded from medial temporal and prefrontal sites (Clarke et al., 1999). In an auditory oddball task, 33–45 Hz oscillations phase-locked between midline and frontal, central, and parietal regions in the 250–400 ms latency domain were found only after the target stimuli (Gurtubay et al., 2004).

Haig (2001) studied gamma synchrony in schizophrenic patients and normal controls using an oddball paradigm. This author found early- and late-gamma band phase synchrony (37–41 Hz) associated with the components N1 and P3 components elicited by the "target" stimuli. He proposed that gamma synchrony served to index the activity of integrative cognitive networks and suggested an impairment of brain integrative activity in schizophrenic illness. Spencer et al. (2003) have further studied neural synchronization in schizophrenic patients and matched normal controls discriminating between illusory square and control stimuli. These authors hypothesized that illusory square stimuli presumably synchronize neural mechanisms that support visual feature binding. They reported that compared to the matched controls, using complex wavelet analysis of the EEG, responses of the schizophrenic patients relative to the stimulus onset demonstrated (1) absence of the posterior component of the early visual gamma (30–100 Hz) response to the Gestalt stimuli, (2) abnormalities in the topography, latency, and frequency of the anterior component, (3) delayed onset of phase coherence changes, and (4) a pattern of interhemispheric coherence decreases in patients that replaced the pattern of anterior–posterior coherence increases displayed in the responses of control subjects. In a related subsequent study using a similar experimental paradigm, both healthy controls and schizophrenic patients displayed a gamma oscillation that was phase locked to the reaction time, presumed to reflect conscious perception of the stimulus (Spencer et al., 2004). Reaction time was slower in the patients, and further, the frequency of this oscillation was lower in patients than in the controls. This was interpreted to reflect a lower capacity of schizophrenic brains to support high-frequency oscillations. The degree of phase locking was correlated with visual hallucinations, thought disorder, and disorganization in the patients, as assessed by their performance on the positive and negative syndrome scale and separate scales for the assessment of positive and of negative symptoms. They interpreted their findings as an evidence that although the schizophrenics could perform the task, the interactions within the neuronal networks engaged in detection of the target were poorly organized and proceeded at a slower rate.

Semantic encoding

Evidence has accumulated confirming that a late negative component of ERPs, termed the N450, is correlated with the operations performed during semantic encoding (Kutas and Hillyard, 1980; Hillyard et al., 1985; Donchin et al., 1986; Hillyard and Mangun, 1987). This finding is mentioned here only for the sake of completeness, but does not directly pertain to the model being proposed.

In summary, this body of data suggests that short latency contributions to the ERP appear to arise from inherent sensory encoding processes, while later contributions to the ERP appear to arise from working memory and an experientially dependent memory system, and further suggests that perception requires the interaction between two systems, the exogenous sensory-specific system and the endogenous nonsensory-specific system.

Coincidence detection between exogenous and endogenous activity

Direct electrical stimulation, phase-locked to a peripheral stimulus to block the primary component of the ERP in cortical sensory receiving areas, had little effect on behavioral accuracy in trained cats, but greatly decreased performance when delayed to block the secondary nonsensory-specific component. Such results led to the proposal that coincidence at the cortical level between exogenous input of information about the environment and endogenous readout of relevant memories, detected by a neuronal comparator at the cortical level, identified the informational significance of the stimulus and was critical for perception (John, 1968). Similar hypotheses were proposed by other investigators (Sokolov, 1963). Subsequently, in awake neurosurgical patients, Libet (1973) stimulated the cortex electrically to coincide with the time of arrival of the nonsensory-specific, secondary component of the EP waveshape, usually present in the response of the somatosensory cortex that was evoked by a mild electrical shock to the wrist, and found that such time-locked brain stimulation could block subjective awareness of the wrist shock. Similar findings during brain surgery were reported shortly thereafter by Hassler (1979) who recorded the cortical evoked responses to mild wrist shock, and electrical shock to the sensory-specific ventrobasal nucleus and the non-sensory-specific nucleus centralis lateralis. The waveshape of the cortical response evoked by wrist shock consisted of an early component like the response to the ventrobasal nucleus alone and a later component like that to the nucleus centralis lateralis alone. Hassler then showed that the perception of wrist shocks by a patient could be blocked by later stimulation of the nucleus centralis lateralis in the thalamus, appropriately delayed to disrupt the late cortical response. In comatose patients, presence of early but not late components of multisensory evoked potentials has been used to stage severity of traumatic brain injury (Greenberg et al., 1981) (see Guerit this volume). The return of late evoked potential components is predictive of recovery from coma (Alter et al., 1990). Such data indicate that a comparator between exogenous sensory-specific and endogenous non-sensory-specific systems is essential for sensations to be perceived and is reflected in late ERP components.

Recent research has described a priming mechanism, whereby top-down influences have a priming effect on bottom-up signals. The electrical activity of the major output neurons of the cortex is strongly influenced by the laminar distributions of synaptic activity. Vertically aligned sheets of pyramidal neurons have basal axosomatic synapses in layers 4–5 in the neocortex, which receive exogenous inputs about complex environmental stimuli from sensory-specific afferent pathways. Apical axodendritic synapses of these pyramidal cells are in layer 1, and receive endogenous input via the non-sensory-specific projection system of the thalamus. Using direct electrical stimulation of pyramidal neurons with micropipettes, it has been demonstrated that pyramidal neurons act as comparators, detecting temporal coincidence of inputs to apical and basal synapses (Larkum, 1999). Direct stimulation of either apical or basal synaptic regions alone caused a sparse axonal discharge of the pyramidal neuron. However, concurrent stimuli delivered to both apical and basal regions within a brief interval elicited clearly enhanced firing of the neuron, with a back-propagated discharge in the gamma frequency range (~50 Hz). Thus, "top-down" axodendritic signals may modulate the saliency of a "bottom-up" axosomatic signal, changing the neuronal output from one or a few spikes into a burst (Siegel et al., 2000). It has been suggested that local rhythmic oscillations may cause subthreshold membrane fluctuations that play a possible contributory role in this process (Engel and Singer, 2000). The somadendrite interactions seem to depend critically upon frequencies of back-propagation between 20 and about 70 Hz, which is the beta and gamma range.

A more recent experiment extended this observation in a very important way, using voltage-sensitive dyes to achieve visualization of cellular activation in brain slices oriented so as to include regions of the thalamus and their projections to the cortex. Direct electrical stimulation by micropipettes placed upon a specific thalamic relay nucleus (ventrobasal thalamic nucleus) caused a visible moderate activation in layer 5 of the corresponding sensory cortex. Direct stimulation of a nonspecific nucleus of the diffuse projection system (nucleus centralis lateralis) caused a visible moderate activation in cortical layer 1. When the ventrobasal, nuclei and centralis lateralis were stimulated concurrently (i.e., when the simulated inputs from the exogenous and endogenous systems were coincident), cortico-thalamic activity was markedly enhanced, with a synchronized strong discharge back to the thalamic regions from where the stimuli had originated. Most important from the theoretical viewpoint advanced, when both regions were subjected to a train of pulses at 50 Hz, there was markedly stronger and more widely spread activation of the cortical regions containing Layers 1 and 5. A volley of cortico-thalamic activity at gamma frequency was transmitted to the thalamic regions where the stimulating pulses had been initially delivered, that was back-propagated to the cortical regions where coincidence had occurred. Thus, a reverberating cortico-thalamo-cortical loop was established (Llinas et al., 2002).

Such an evidence shows that influences from inputs to the upper levels can dominate the effects of feedforward sensory input arriving at basal levels. The precise effect of these interactions upon the firing patterns of the pyramidal neurons depends upon the duration and timing of the inputs to the different levels. The spatio-temporal configuration of corticofugal and cortico-cortical projections from this coincidence detection system may be strongly dependent upon the concordance between the two time series converging thereupon from the exogenous and endogenous systems. The critical time interval within which such interaction can occur probably varies as a function of the level of arousal, but the present results provide an initial approximation for the pertinent time intervals. While this neural enhancement process has thus far only been shown to operate at intervals on the order of tens of milliseconds, some findings (Flohr, 1998) suggest that delays on the order of a 100 ms or more involving the NMDA receptor may serve as a local binding mechanism for a combination of specific and nonspecific inputs. It must be emphasized that this synthesis of percepts from the interaction between the past and the present is mediated by a wide variety of neurotransmitters. Adaptive brain functions are critically dependent upon the availability of these substances and the processes that control their synthesis and metabolism, a topic beyond the scope of this article.

The evidence suggests the existence of a distributed coincidence detection system that is dispersed throughout the cortex. This system serves as a comparator between present multimodal, exogenous input from the environment and input from a "hedonic system," which integrates endogenous activity reflecting present state, working memory, relevant episodic experience, motivation, and idiosyncratic individual values. The readout from the value system could arise from mechanisms generating experiential associations activated by collateral inputs from the just previous complex sensory stimuli. In a sense, this can be envisaged as a process enabling the past to "leapfrog" into the future. The hypothesized system of coincidence detectors might act as a filter, segregating random neural "noise" or unimportant elements of the complex stimuli from enhanced neural discharges that are thereby recognized as "signal" of informational value.

Contribution to perception of endogenous readout from memory

According to the model herein proposed, activity in a spatially dispersed sheet of neurons encoding attributes representing fragmented sensations would be enhanced, by coincidence between exogenous and endogenous inputs, to become fragments of perception. Outputs from this multimodal comparator converge as ongoing synchronous cortico-thalamo-cortical loops engaging regions of thalamus and cortex in coherent reverberation in the frequency range of the gamma rhythm (25–50 Hz), binding the distributed fragments into a unified percept. Such reverberation has been proposed by some to play an essential role in perception (Ribrary et al., 1991; Llinas and Ribary, 1993; Rodriguez et al., 1999; Varela, 2000; Lachaux et al., 2000; John et al., 2001; Llinas et al., 2002) and is hypothesized to be reflected by the induced long-distance synchronization of gamma oscillations (Pantev, 1995; Rodriguez et al., 1999; Varela, 2000). These ideas are consonant with the experimental evidence and theoretical proposals by Basar and his colleagues that oscillatory systems at various frequencies, but particularly in the gamma range, act as resonant communication systems through large populations of neurons (Basar et al., 2000).

Perception is an active process, modulated strongly by endogenous influences representing readout of the relevant recent and episodic past and present states, which place the immediate environmental input into context and enable the individual to generate predictions and expectations about future stimuli. This process is envisaged to occur continuously in multiplexed channels, such that the past and the present are inseparably merged in every update of the microstate or perceptual frame. An ERP can be conceptualized as an approximation of one cycle of the construction of a frame, extracted from one channel by repetitive phase-locked sampling of the continuously ongoing process.

The sequential steps in this process of construction of perception can be inferred from ERP research and is based upon our knowledge of "mental chronometry" summarized above. In Fig. 1, the endogenous retrieval system was schematized by the green arrows indicating interactions between particular brain regions. The exogenous–endogenous interactions that comprise this process are schematized in Fig. 2, and are hypothetically conceptualized to operate in the following manner:

$f(t_0)$. Inputs from multimodal sensory receptors activated by a complex environmental scene are encoded as time series, $f(t_0)$, of nonrandomly synchronized discharges in multiplexed parallel channels of afferent pathways, projecting to sensory-specific thalamo-cortical relay nuclei of every sensory modality.

$f(t)$. At time t, volleys of synchronized outputs from multimodal sensory-specific thalamo-cortical relay nuclei reencode $f(t_0)$, and are propagated as a modified time series, $f(t)$, of non-randomly synchronized exogenous inputs to the lower layers of the sheet of pyramidal neurons dispersed throughout cortical ensembles of feature detectors. Activation of basal synapses of pyramidal neurons in cortical Layer 5 causes the positive ERP component P1 with latency about 50 ms, reflecting initial registration of fragments of sensation. Corresponding synchronized volleys from the thalamo-cortical relay nuclei, and collateral pathways transmit the modified time series, $f(t)$, to the reticular formation in the brainstem where it is further modulated to reflect the arousal level.

$f(t+1)$. At time $t+1$, efferent volleys from sensory cortical regions instruct nucleus reticularis to modulate the thalamo-cortical relay neurons to focus attention by raising the threshold of those pathways that do not significantly contribute to the ascending volley of synchronous activity received at time t. These corticofugal volleys cause the ERP component N1 at about 130 ms. At the same time, the ARAS sends the time series $f(t+1)$ to the non-sensory-specific intralaminar and dorsomedial nuclei of the thalamus and to the raphe nucleus and structures of the limbic system.

$f(t+2)$. At time $t+2$, the intralaminar nuclei send the nonrandomly synchronized time series $f(t+2)$ modified by the ARAS to the apical synapses of cortical Layer 1, enhancing the excitability of pyramidal neurons in which coincidence detection occurs, thereby transforming fragments of sensation into islands of local negative entropy. This coincidence causes vigorous discharge causing the ERP component P2 at about 210 ms, accompanied by back-propagation of reverberatory cortico-thalamo-cortical gamma activity. The time series $f(t+2)$ is rapidly transmitted from posterior cortical regions and converges with $f(t+2)$ from the dorsomedial thalamic nucleus upon the prefrontal cortex, which in turn relays $f(t+2)$ to Layer 1 of the entorhinal cortex. Activity then propagates from area CA3 to CA1, hence to the amygdala, nucleus accumbens, septum, and hypothalamus. Memories originally encoded as time series of nonrandom synchrony and stored in the neural networks of this system are activated by associational processes modulated by emotions, providing the valence most relevant to the sensory pattern initially encoded by the exogenous system.

$f(t+3)$. At time $t+3$, the circuit of the time series $f(t+2)$ through the memory storage

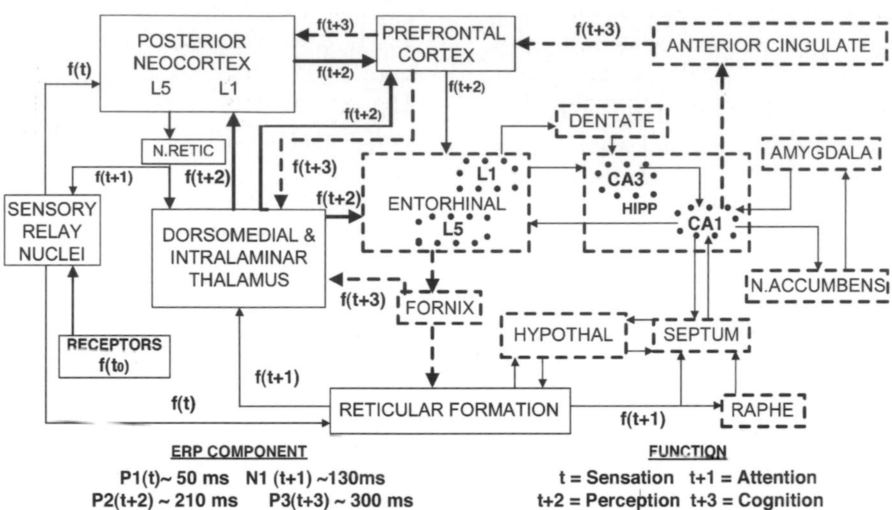

Fig. 2. Simplified scheme of the neurophysiological transactions resulting in the automatic readout of relevant memories from neuroanatomical structures of the endogenous memory storage and retrieval system, and their relation to major components of the event related potentials (see text for details).

networks of the limbic system is completed. The readout of the activated memory is transmitted as the time series $f(t+3)$ via a hippocampus-entorhinal-fornix circuit to the reticular formation and thalamus, and via the anterior cingulate gyrus to the prefrontal cortex, where coincidence detectors compare the time series $f(t+3)$ with time series previously received from posterior sensory regions and the thalamus. If the cortical coincidence detection indicates a discrepancy between time series from these two sources, a discharge produces the ERP component P3 at about 300 ms. The timing of the sequential perceptual frames may be reset whenever a discrepancy is noted.

The process just described must be envisaged as a sliding window of successive microstates, continuously matching the representation of the present in posterior regions of the cortex with representation of the recent past in the prefrontal cortex, together with episodic memories retrieved from the limbic system. As subsets of neurons in the coincidence detector network of the prefrontal cortex match the two time series, sustained gamma reverberations link posterior and prefrontal cortex together with cingulate-limbic and thalamus a resonant field that establishes global negative entropy, i.e., the content of consciousness.

Relevance of local field potentials to extraction of information

The central issue is how coherent, informational neuronal activity in multiple cortical areas is welded into a seamless unity that becomes aware of itself. We postulate that the ground state of local synchronization within neuronal ensembles, spontaneous interactions among cells in different brain regions, and periodic integration intervals is regulated intracerebrally by the same genetically specified homeostatic system that regulates the EEG. While the power spectrum of an artifact-free sample of resting EEG from a given scalp electrode reaches a stable and reproducible composition (stationarity) in as little as 8 s in the alert resting person, the power spectrum of LFPs around an intracerebral electrode in the cat can be considered stationary on the scale of 1 s (Bernasconi and Konig, 2001). Numerous reports in the literature document the relationship between amplitude and polarity of LFP in a domain and the probability of neural discharge from local ensembles (Fox and O'Brien, 1965; John and Morgades, 1969b; Ramos et al., 1976). The evidence of homeostatic regulation of the EEG and the ambient extracellular electrical field clearly implies homeostatic regulation of the probability of synchronized neural activity in the regions of the brain. Single and multiple unit recordings in brain slices *in vitro* and in anesthetized and unanesthetized, unrestrained animals reveal coherent discharges in diffusely distributed ensembles (John et al., 1990; Roelfsema et al., 1997; Brecht et al., 1998). Some of this coherence must be anatomically "hard-wired," while some reflects transient functional interactions and coupling.

Fast spontaneous oscillations in intrathalamic and thalamo-cortical networks may play a decisive role in binding. Synchronization of fast rhythms may be subserved by intralaminar neurons that fire rhythmic spike bursts in the gamma frequency range and have diffuse cortical projections (Steriade et al., 1991, 1993). Sensory stimulation increases synchrony (Castelo-Branco and Neuenschwander, 1998), and synchronized discharges of neurons in different brain regions may bind spatially dispersed representations of a multimodal stimulus into an integrated percept (Llinas, 1988; Gray et al., 1989; Kreiter and Singer, 1996; Llinas and Ribary, 1998). Functionally coherent cell assemblies may represent the constellation of features defining a particular perceptual object, with transient synchrony dependent upon state variables related to the EEG. Such coherence has been proposed as crucial for awareness and conscious processing, including attentional focusing, perceptual integration, and establishment of working memory (Singer, 2001; Engel and Singer, 2001). The signal to noise ratio for several multimodal attributes of a complex attended stimulus may be enhanced by higher order convergence upon a particular neuronal neighborhood (Sheinberg and Logothetio, 1997). Gamma oscillations facilitate coherence of neural activity within and between regions, and have been suggested as a necessary condition for the occurrence awareness stimuli (Llinas and Ribary, 1998; Engel and Singer, 2000). LFP fluctuations predicted response latency to visual stimuli, with negative LFPs associated with earlier responses and positive LFPs with later responses (Fries et al., 2001).

What might be the role of these ubiquitous potential oscillations? The neuronal population in any brain region displays a range of excitability at any moment, depending upon the recency and number of synaptic inputs that have affected their membrane potentials. A small fraction of cells are depolarized almost to the threshold for firing spontaneously and a small proportion are refractory after recent firing. The central limit theorem leads us to expect an overall gaussian distribution of excitability in the population of neurons as a whole. We hypothesize that a weak rhythmic modulation of membrane potentials, insufficient in itself to produce axonal discharges and perhaps arising from and distributed synchronously by the diffuse projection nuclei of the thalamus, is superimposed upon the neurons throughout the brain. As the oscillatory influences impinge, neurons should discharge whose state of membrane depolarization, and the modulatory depolarization exceeds the local firing threshold. At the other end of the excitability distribution, those neurons in a relatively refractory state while their normal resting membrane potential is being restored after a recent discharge should be further inhibited by the subsequent hyperpolarizing phase of the oscillations. The mean excitability of the ensemble should fluctuate at the frequency of the modulatory rhythm to an extent proportional to the depth of modulation. On the basis of a review of evidence relevant to the possibility of non-synaptic modulation of neuronal membranes by extracellular fields, Jefferys (1995) concluded that "In general, electrical field effects can mediate neuronal synchronization on a millisecond time scale." Weaver et al. (1998) computed the theoretical threshold for such effects as requiring a gradient of the order of 100 μV/mm. We have observed gradients of 100–200 μV/mm in intracerebral recordings from human patients.

The effect of this hypothetical rhythmic fluctuation of excitability should be to enhance the synchronization of neural firing within any local brain region. To the extent that if the same modulatory influence is imposed across the cortex, it should increase the level of coherence of non-random output across distant brain regions. If cell assemblies of feature detectors are dispersed throughout the cortex such that some subset of them encoded specific attributes of the afferent stimulus complex as "fragments of sensation," and if the hypothesized comparator network had achieved coincidence between exogenous and endogenous inputs to certain patches of cells thereby to identify "fragments of perception" and selectively increase their excitability, then the consequence of this rhythmic modulatory influence would be to facilitate a coherent corticofugal discharge converging upon the thalamus from many dispersed cortical detectors. These coherent cortico-thalamic volleys are proposed as the integrative process, which enables the brain to bind the spatially dispersed multimodal attributes into an integrated percept.

Abundant evidence has been presented, which establishes that there are distinctive power spectra for any cortical region, with a wide range of frequencies. These cortical rhythms reflect thalamo-cortical and cortico-cortical interactions, which may exert complex modulating influences upon local ensembles. It has been proposed that lower the frequency, the more remote the brain region from which the influence arises (von Stein and Sarnththeim, 2000; Bressler and Kelso, 2001). The observation that about 65% of the EEG is coherent across the cortex (John, 1990) suggests that the hypothesized binding process is actually mediated by a resonance or common mode interference pattern shared among a number of systems generating these rhythms. This proposal is reminiscent of the scansion mechanism suggested half a century ago (Pitts and McCulloch, 1947). It must also be considered that although the LFPs converge rapidly to a predictable power spectrum, on the short term these potentials can be considered as a random ambient noise. Recent studies have shown that the effect of ambient noise on a weakly interconnected network of weakly responsive elements is to synchronize them by stochastic resonance, a process that functions best at high frequencies (Allingham et al., 2004; Torok and Kish, 2004).

Zero-phase synchronization

It is possible (Engel and Singer, 2000) that zero phase lag plays a critical role in binding dispersed fragments of sensation into a coherent whole (König et al., 1995; Engel et al., 1997). Phase-locking of 40 Hz oscillations, with zero delay between the prefrontal and parietal human cortex, has been observed in scalp recordings during focused attention and conscious perception of recognized auditory or visual events. Using intracranial as well as scalp recordings in humans, the amplitude, long distance, and local synchronization of gamma activity have been studied during auditory, visual, and somatosensory stimulation and during cognitive or perceptual tasks (Desmedt and Tomberg, 1994; Tallon-Baudry et al., 1997; Tallon-Baudry, 2000; Varela, 2000). In each modality, a transient burst of phase-locked gamma (of varying frequencies) occurs, most prominently in the primary sensory cortex, about 100 ms after stimulus onset. About 230 ms after stimulus onset, there is a pattern of gamma oscillations synchronized between frontal and parietal cortical regions. The latency of 230 ms corresponds approximately to the latency of the so-called cognitive potential, P3A, prominently displayed in anterior regions of the human scalp. This gamma activity is induced by but not synchronized with stimulus onset, occurs only in trials when a figure or a word is perceived, and has been proposed as "the correlate of perception itself" (Pantev, 1995; Rodriguez et al., 1999). Such phase-locked synchrony between distant brain regions has been adduced as support for proposals that perception involves integration of many distinct, functionally specialized areas (Tallon-Baudry, 2000), and that the "self" may be a transient dynamic signature of a distributed array of many brain regions integrated by such coherence (Varela, 2000). This induced long-distance synchrony seems to be the equivalent in humans of the late component in animal EPs, described above, when stimuli acquire meaning (Galambos and Sheatz, 1962; John, 1972). The long distance between regions synchronized with zero-lag rules out volume conduction, which decays rapidly with distance between electrodes. It is possible that the apparent zero-lag arises from a common source of thalamo-cortical reverberations, but the alternative that it is evidence for a resonating field must also be considered.

Critical neuroanatomy of consciousness

Various strategies have been used to find anatomical structures or brain processes critical for consciousness, including lesioning, imaging, and pharmacological techniques. Recent proposals have included an extended reticulo-thalamo-cortical distributed network, the intralaminar nuclei of the thalamus, nucleus reticularis, anterior cingulate cortex, or hippocampus. Prolonged activity in some set of unique "awareness cells" sparsely distributed over many regions, or a pattern of activity among regions rather than within a particular brain region, have been proposed as the neural correlate of consciousness (Koch, 1998). A "Dynamic Core Hypothesis" has been proposed, constituting a set of spatially distributed and meta-stable thalamo-cortical elements, sustaining interactions that maintain unity in spite of constantly changing composition (Tononi and Edelman, 2000).

One approach to identify such a system is to determine whether a particular set of brain regions changes its state with the loss of consciousness due to the action of anesthetics. Using ^{18}F-fluorodeoxyglucose positron emission tomography (FDG-PET), cerebral metabolism was studied during propofol anesthesia (Alkire et al., 1998). At loss of consciousness, mean cerebral metabolic rate was globally reduced throughout the brain by 38%, relatively more in cortex than subcortex. Similar global quantitative results were found with isoflurane. In sedation, very similar but lesser global reduction was observed. It was initially concluded that loss of consciousness was not caused by changes within a specific circuit, but rather by a uniform reduction below a critical level in a distributed neural system. In a more recent study using $H_2^{15}O$-PET to measure dose-related changes in regional cerebral blood flow (rCBF) during sedation with midazolam, a similar global reduction of about 12% was found (Veselis et al., 2000). In addition, however, a discrete set of brain regions displayed a significantly more extreme reduction in cerebral blood flow. These regions included multiple areas in the prefrontal cortex, the superior frontal gyrus, anterior cingulate gyrus, parietal association areas, insula, and the thalamus (see Fiset, this volume; Alkire, this volume; for an extensive review of functional imaging in general anesthesia).

Routine surgical anesthesia offers a naturalistic environment where loss and return of consciousness can be studied in a systematic manner. In many such studies of anesthetic effects, quantitative features have been extracted from EEG power spectra (QEEG) and bispectra. Such variables have been repeatedly shown to be related to clinical signs or anesthetic endpoints (Sebel et al., 1997; Rampil, 1998; Veselis et al., 2000). Changes in the level of consciousness cause distinctive shifts from the EEG ground state. Using quantitative analysis of continuous EEG recordings collected during several hundred surgical procedures, we sought invariant reversible changes in brain electrical activity with loss and return of consciousness. Details of these studies are given elsewhere (Prichep et al., 2000; John et al., 2001). Two kinds of changes in brain activity were found, independent of anesthetic agents used: (1) At loss of consciousness, in all EEG frequency bands dramatic changes took place in coherence within and between the cerebral hemispheres. Rostral brain regions abruptly became functionally disconnected from posterior regions and the two hemispheres were functionally uncoupled (i.e., incoherent). Only in the beta and gamma bands is coherence restored upon return of consciousness. (2) At loss of consciousness, EEG power on each hemisphere became dominated by low frequencies and strongly anteriorized. Normal posterior power dominance was restored at return of consciousness. These changes were used to construct a multivariate algorithm to provide a quantitative index of the level of consciousness. This algorithm has been implemented in a clinical monitor of the depth of anesthesia (PSA 4000, Physiometrix, Inc., N. Billerica, MA). Three-dimensional source localization revealed that regions of the mesial orbital and dorsolateral prefrontal and frontal cortex, paracentral gyrus, anterior cingulate gyrus, amygdala, and basal ganglia invariably displayed profound reversible inhibition with loss and return of consciousness, independent of the anesthetic agents (Prichep et al., 2000; John et al., 2001). An overview of recent electrophysiological and brain imaging studies has provided the basis for a recently published comprehensive theory of anesthesia (John and Prichep, 2005).

An integrative theory of consciousness

This theory is largely based upon experimental evidence from many levels of neuroscientific inquiry, some cited but much not explicitly mentioned. However, I must concede that much of this formulation represents my speculations of "how things must be," in order for the brain to generate the sense of self that we all share and the abnormal behaviors we observe in clinical surroundings. Assertions that are considered as basic propositions of this theory are summarized below.

Apperception defines content of consciousness

A major hypothesis underlying this theory is that the content of consciousness is dominated by apperception. By apperception, we mean integration of momentary perception of the external and internal environment with the working and episodic memories activated by associative reactions to that perception. I contend that the conscious organism, and particularly the human being, is continuously interpreting awareness of the present in the context of both the recent and the remote past, to attribute "meaning" to the present events. A further major hypothesis of this theory is that information in the brain is not encoded by the firing of dedicated neurons in particular brain regions that represents specific stimulus attributes or features, but rather by distinctive temporal patterns of synchronized firing dispersed among many brain regions. Individual neurons can participate in numerous such temporal patterns.

Information is nonrandomness

Information establishes local negative entropy consisting of spatially distributed time series of synchronized activity in multimodal, sensory-specific systems (the "exogenous system"). Energy reaching the receptor surfaces of multiple sensory systems activates a barrage of neural activity that propagates through successive levels of afferent pathways to widely dispersed ensembles of cells responsive to particular aspects of the complex, multimodal environmental stimuli. In each brain region, this barrage impinges on populations of neurons with a range of excitability, with membrane potentials varying from hyperpolarization, refractory just after discharge, to relatively depolarization, just on the verge of discharge. Circulation of the input throughout local networks causes some responsive subsets of neurons distributed in these many brain regions to respond with a distinctive temporal pattern, producing a time series of synchronized neural discharges. Intrinsic excitability cycles modulate the reactivity of cell ensembles at frequencies in the range of gamma oscillations (35–50 Hz), selecting particular subsets of neurons in the receiving system that can discharge synchronously with this distinctive temporal pattern. This distributed, improbable synchrony constitutes fragments of sensation, which represent islands of local negative entropy, relative to a regulated ground state. These exogenous inputs are encoded as time series of synchronized firing in parallel, multiplexed channels within each sensory pathway, offset by 25 ms and sampled every 80 ms. The cyclic modulation of excitability produces multiplexed, parallel channels encoding the same information. However, this information is phase-shifted so that the temporal patterns of synchronized activity encoding information in the multiplexed channels are asynchronous, offset by some temporal delay on the order of 20 ms. Within each such channel, the length of the representational time series appears to be 80–100 ms.

Perception is discontinuous, parsed into momentary episodes or "perceptual frames"

This discontinuity has been demonstrated in every sensory modality using a variety of experimental psychophysical procedures. These include the so-called "backward masking" and "perceptual frame" phenomena. The experimentally determined duration of such episodes appears to be of the order of 80 ms, during which a representational time series unfolds. This "experiential chunk" may correspond to the episodic synchronization that has been shown by Freeman (2004) to occur among remote cortical regions every 80 ms, and to the "microstates" of stable field topography shown to have a mean duration of 80 ± 2 ms across the age range from 6 to 90 years; as described by Koenig and colleagues (2002).

"Hyperneuron" de-multiplexes afferent input

Circulating updated, multiplexed channels maintain the stable representation of environment as a spatiotemporal steady state in an anatomically dispersed hyperneuron. No single neuron or dedicated set of neurons can maintain the distinctive pattern of times series of synchronization across successive perceptual frames to mediate the apparent continuity of subjective experience or "awareness." Rather, the multiplexed parallel channels converge successively upon the sensory receiving ensembles. Interactions caused by the circulation of temporal patterns of asynchronous activity in neural networks within spatially extensive, dispersed populations of neurons establish a spatiotemporal interference pattern in each region that remains stable for extended periods of time, like a standing wave of covariance. This sustained spatiotemporal distribution of nonrandom synchronization, independent of the firing of any particular neurons, constitutes a "hyperneuron." A hyperneuron is a sustained, improbable spatiotemporal pattern of negative entropy.

"Endogenous system"

Episodic memories are stored in non-sensory-specific brain regions that encode distinctive time series of incoming activity and constitute an "Endogenous system." Collaterals from the multiplexed afferent sensory pathways enter the pontine and mesencephalic reticular formation, and the encoded representational time series of synchronized activation from the brainstem and polysensory cortical areas are distributed via the parahippocampal and entorhinal cortex into the limbic system. When such nonrandom activation by successive inputs persists for a sustained but brief period, associational or "contiguity" mechanisms link together a network of neurons distributed throughout regions including portions of the neocortex together with the amygdala, nucleus accumbens, hippocampus, hypothalamus, and septum, establishing a representational assembly that is sensitive or "tuned" to that distinctive spatiotemporal pattern of input. This tuning in the endogenous system is consolidated over a brief period into a molecular structure, perhaps protein specified via RNA within the neurons in the network, so that it will be preserved by synthetic mechanisms that renew that capability by replication that preserves the tuning even after the initial structure is replaced.

Readout of relevant memory – "context"

At any later time, if a sufficient portion of the activity of sensory exteroceptors and/or interoceptors in the exogenous system becomes nonrandomly synchronized, the tuned representational network in the endogenous system with most similar temporal pattern becomes activated and discharges a synchronized time series from the participating limbic regions, providing a context of emotional tone and motivational valence to the retrieved episodic memory. The readout from this system is a time series of synchronized discharges, reflected by oscillations in the gamma frequency range that become phase locked to waves recurring about every 200 ms, or in the theta frequency range, with essentially zero phase lag among cortical regions. Owing to the multiplexed nature of sustained exogenous inputs and the duration of the stored representational time series, patterns of retrieved outputs from this endogenous system persist for a sustained period, producing an interference pattern similar to that of the hyperneuron.

Coincidence detection by pyramidal neurons

When input from the exogenous system of a nonrandomly synchronous temporal pattern to the basal dendrites of the cortical sheet of pyramidal neurons coincides with a nonrandomly synchronous input from the endogenous system to the apical dendrites with the same temporal pattern, the excitability of a spatially dispersed network of pyramidal neurons is significantly enhanced. This process converts fragments of sensation into fragments of perception, increasing the level of local negative entropy.

"Ground state" of the brain is defined and regulated by a homeostatic system

A genetically determined homeostatic system regulates normal brain electrical activity, controlling interactions among brain regions as well as excitability within local regions. Due to this regulation, the mixture of frequencies in rhythmic oscillations in each brain region, the spatiotemporal principal components of the cross-spectra (which quantify the relationships among all brain regions across frequencies), and the parsing of these spatiotemporal patterns into discrete epochs of time can all be precisely defined, describing the basic processes maintaining the system in stable resting equilibrium. These homeostatically regulated baselines of inter- and intra-regional oscillatory brain electrical activity define the normative, hence most probable, spatiotemporal distribution of variances and covariances in brain electrical activity, the "ground state" of maximum entropy, that contains zero information.

Brain state vector

If the electrical activity of the brain is conceptualized as a "signal space," the complete set of normative QEEG data define the ground state as a "hypersphere," a region around the origin of an N-dimensional signal space that encompasses the variance of the normative descriptors of brain electrical activity. The measurements from any individual, after appropriate rescaling, can be described as a multidimensional Z-vector or "brain state vector" in that signal space; each dimension is scaled in standard deviations (SD) of the distribution of the corresponding variable. The brain state vector of a resting, healthy individual will lie within the hypersphere of radius $Z < 3.2$ SD along all dimensions around the origin of the multidimensional signal space. The brain state vector of any individual in any state can be depicted as a point somewhere in the signal space thus defined. The content of consciousness in the resting state (eyes closed) is fleeting ideas and thoughts causing transient fluctuations of the brain state vector that remain within the normative hypersphere.

Local field potentials acting on pyramidal neurons can produce coherent reverberations

In the short term, LFPs can be considered as random noise that has been shown to synchronize activity among weakly interconnected networks of weakly responsive elements. This effect of LFPs makes coherent the cortico-thalamic discharge of the distributed pyramidal neurons whose excitability has been enhanced by coincidence between exogenous and endogenous inputs. The synchronous discharge from these spatially distributed islands of local negative entropy results in a coherent reverberation of cortico-thalamo-cortical gamma activity. Findings of zero or near-zero phase lag between prefrontal and parietal cortex may be the snapshots of momentary cross-sectional glimpses of these synchronized time series.

Global negative entropy is the content of consciousness

Sustained reverberations of multiple cortico-thalamo-cortical circuits become coherent and phase-locking between them, perhaps facilitated by transmission across gap junctions, leads to resonance in the coherent network. This phenomenon can be considered as a "phase transition," reflecting an integrative process whereby dispersed fragments of perception are bound into a unified global percept, which is global negative entropy. The resonant network includes the prefrontal, parietal, and cingulate cortex as well as regions of the basal ganglia, limbic system, thalamus, and brain stem identified in the discussion of Fig. 2. These brain regions correspond well to those shown to be uncoupled in conditions associated with an absence of consciousness in studies of anesthesia (Alkire et al., 1998; Fiset et al., 1999; Veselis et al., 2000), (Fiset, this volume; Alkire, this volume) deep sleep (Portas et al., 2000) and coma (Laureys et al., 2000, 2002) as well as "absence" seizures (Pavone and Niedermeyer, 2000; Blumenfeld, this volume). This resonant field corresponds in many respects to the parietal-prefrontal system that some have proposed may be the residence of the "observing self" (Baars et al., 2003; Baars, this volume). I believe this theory supports and extends that suggestion. The global negative entropy of the brain encompasses all of the momentary information content of the entire system, as an "information field" that subsumes the parallel processing circuits that are simultaneously activated at a nonrandom level in the brain, and comprises the content of consciousness. For the purpose of this theoretical formulation, the global negative entropy of a resting healthy individual whose brain electrical activity is at the precise origin of the ground state is proposed to be at a maximum. In this hypothetical state, the brain contains no information so that consciousness is void of content. One might speculate that this state corresponds to what Buddhists term "Nirvana."

Acknowledgements

I acknowledge the constructive contributions of Dr. Leslie S. Prichep to this theory and the International Brain Research Foundation and Dr Philip De Fina for support received to conduct this research.

References

Abeles, M., Purt, Y., Bergman, H. and Vaadia, E. (1994) Synchronization in neuronal transmission and its importance information. In: Buzsaki G. and Christen Y. (Eds.), Temporal Coding in the Brain (Research and Perspectives in Neurosciences). Springer-Verlag, Berlin, pp. 39–50.

Alkire, M.T., Haier, R.J. and Fallon, J.H. (1998) Toward the neurobiology of consciousness: using brain imaging and anesthesia to investigate the anatomy of consciousness. In: Hameroff S.R., Kaszriak A. and Scott A. (Eds.), Toward a Science of Consciousness II – The Second Tucson Discussions and Debates. MIT Press, Cambridge, pp. 255–268.

Allingham, Stocks, N., Morse, R.P. and Meyer, G.F. (2004) Noise enhanced information transmission in a model of multichannel cochlear implantation. In: Abbott D., Bezrukov S.M., Der A. and Sanchez A. (Eds.) SPIE – The International Society of Optical Engineers Fluctuations and Noise in Biological, Biophysical, and Biomedical Systems ll, Vol. 5467. SPIE, Gran Canaria Spain, pp. 139–148.

Allport, D.A. (1968) Phenomenal simultaneity and perceptual moment hypotheses. Br. J. Psychol., 59. 395–406.

Alper, K.R., Chabot, R., Prichep, L.S. and John, E.R. (1993) Crack cocaine dependence: discrimination from major depression using QEEG Variables. In: Imaging of the Brain in Psychiatric and Related Fields. Springer-Verlag, Berlin, pp. 289–293.

Alpern, M. (1952) Metacontrast. Am. J. Optometry, 29: 631–646.

Alter, I., John, E.R. and Ransohoff, J. (1990) Computer analysis of cortical evoked potentials following head injury. Brain Injury, 4: 19–26.

Baars, B.J., Banks, W.P. and Revonsuo, A. (Eds.). (2002) Consciousness and Cognition: Special Issue: Timing Relations Between Brain and World. Elsevier, Orlando.

Baars, B.J., Ramsoy, T.Z. and Laureys, S. (2003) Brain conscious experience and the observing self. Trends Neurosci., 26: 671–675.

Basar, E., Basar-Eroglu, C., Karakas, S. and Schurmann, M. (2000) Brain oscillations in perception and memory. Int. J. Psychophysiol., 35: 95–124.

Bernasconi, C. and Konig, P. (2001) On the directionality of cortical interactions studied by structural analysis of electrophysiological recordings. Biol. Cybern., 81: 199–210.

Brecht, M., Singer, W. and Engel, A.K. (1998) Correlation analysis of corticotectal interactions in the cat visual system. J. Neurophysiol., 79: 2394–2407.

Bressler, S.L. and Kelso, J.A. (2001) Cortical coordination dynamics and cognition. Trends Cogn. Neurosci., 5: 26–36.

Burns, B.D. (1968) The Uncertain Nervous System. Williams & Wilkins, Baltimore.

Buzsaki, G. (2002) Theta oscillations inthe hippocampus. Neuron, 33: 325–340.

Castelo-Branco, M. and Neuenschwander, S. (1998) Synchronization of visual responses between the cortex lateral geniculate nucleus and retina in the anesthetized cat. J. Neurosci., 18: 6395–6410.

Clarke, J.M., Halgren, E. and Chauvel, P. (1999) Intracranial ERP's in humans during lateralized visual oddball task:II temporal, parietal and frontal recordings. Clin. Neurophysiol., 110: 1226–1244.

Cotterill, R.M.J. (1997) On the neural correlates of consciousness. Jpn. J. Cogn. Sci., 4: 31–34.

Desmedt, J.E. and Tomberg, C. (1994) Transient phase-locking of 40 Hz electrical oscillations in prefrontal and parietal human cortex reflects the process of conscious somatic perception. Neurosci. Lett., 168: 126–129.

Donchin, E., Karis, D., Bashore, T.R. and Coles, M.G.H. (1986) Cognitive psychophysiology and human information processing. In: Coles M.G.H., Donchin E. and Porges S.W. (Eds.), Psychophysiology: Systems, Processes and Applications. Guilford Press, New York, pp. 244–267.

Duffy, F.H., Jones, K., Bartels, P., McAnutty, G. and Albert, M. (1992) Unrestricted principal component analysis of brain electrical activity: issues of data dimensionality artifact and utility. Brain Topog., 4: 291–308.

Edelman, G. (2001) Consciousness: the remembered present. Ann. NY. Acad. Sci., 929: 111–122.

Efron, E. (1970) The minimum duration of a perception. Neuropsychologia, 8: 57–63.

Engel, A.K., Roelfsema, P.R., Fries, P., Brecht, M. and Singer, W. (1997) Role of the temporal domain for response selection and perceptual binding. Cerebral Cortex, 7: 571–582.

Engel, A.K. and Singer, W. (2000) Temporal binding and neural correlates of awareness. Trends Cogn. Neurosci., 5: 18–25.

Engel, A.K. and Singer, W. (2001) Temporal binding and the neural correlates of sensory awareness. Trends Cogn. Sci., 1: 16–25.

Fiset, P., Paus, T., Daloze, T., Plourde, G., Meuret, V., Bonhomme, V., Hajj-Ali, N., Backman, S.B. and Evans, A.C. (1999) Brain mechanisms of propofol induced loss of consciousness in humans: a positron emission tomographic study. J. Neurosci., 19: 5506–5513.

Flohr, H. (1998) On the mechanism of action of anesthetic agents. In: Hameroff S.R., Kaszriak A. and Scott A.C. (Eds.), Toward a Science of Consciousness II – The Second Tucson Discussions and Debates. MIT Press, Cambridge, MA, pp. 459–467.

Fox, S.S. and O'Brien, J.H. (1965) Duplication of evoked potential waveform by curve of probability of firing of a single cell. Science, 147: 888–890.

Freeman, W. (2004) Origin structure and role of background EEG activity. Part. I. Analytic amplitude. Clin. Neurophysiol., 115: 2077–2088.

Fries, P., Neuenschwander, S., Engel, A.K., Goebel, R. and Singer, W. (2001) Rapid feature selective neuronal synchronization through correlated latency shifting. Nat. Neurosci., 4: 194–200.

Galambos, R. and Sheatz, G.C. (1962) An electroencephalograph study of classical conditioning. Amer. J. Phys., 203: 173–184.

Gazzaniga, M., Ivry, R.B. and Mangun, G.R. (2002) Cognitive Neuroscience: The Biology of the Mind. W.W. Norton & Company, New York.

Gehring, W.J. and Knight, R.T. (2000) Prefrontal-cingulate interactions in action monitoring. Nat. Neurosci., 3: 516–520.

Goebel, R., Muckli, L., Zanella, F.E., Singer, W. and Stoerig, P. (2001) Sustained extrastriate cortical activation without visual awareness revealed by fMRI studies of hemianopic patients. Vision. Res., 41: 1459–1474.

Gray, C.M., Konig, P., Engel, A.K. and Singer, W. (1989) Oscillatory responses in cat visual cortex exhibit inter-columnar synchronization which reflects global stimulus properties. Nature, 338: 334–337.

Greenberg, R.P., Newlon, P.G. and Hyatt, M.S. (1981) Prognostic implications of early multimodality evoked potentials in severely head-injured patients: a prospective study. J. Neurosurgery, 55: 227–236.

Gurtubay, I.G., Alegre, M., Labarga, A., Malanda, A. and Artieda, J. (2004) Gamma band responses to target and non target auditory stimuli in humans. Neurosci. Lett., 367: 6–9.

Haig, A. R. (2001) Missing links: the role of phase synchronous gamma oscillations in normal cognition and their dysfunction on schizophrenia. University of Sydney. ii–v

Halgren, E., Boujou, C., Clarke, J., Wang, C. and Chauvel, P. (2002) Rapid distributed fronto-parieto-occipital processing stages during working memory in humans. Cerebral Cortex, 12: 710–728.

Hassler, R. (1979) Striatal regulation of adverting and attention directing induced by pallidal stimulation. Appl. Neurophys., 42: 98–102.

Herculano-Houzel, S., Munk, M.H.J., Neuenschwander, S. and Singer, W. (1999) Precisely synchronized oscillatory firing patterns require electroencephalographic activation. J. Neurosci., 19: 3992–4010.

Hillyard, S.A. and Mangun, G.R. (1987) Commentary: sensory gating as a physiological mechanism for visual selective attention. In: Johnson, Jr. R., Rohrbaugh J.W. and Parasuraman R. (Eds.), EEG Supplement 40. Current Trends in Event Related Potential Research. Elsevier Publishers, New York, pp. 61–67.

Hillyard, S.A., Munte, T.F. and Neville, H.J. (1985) Visual-spatial attention, orienting and brain physiology. In: Posner M.I. and Marin O.S. (Eds.), Attention and Performance. Erlbaum Publishers, Hillsdale, NJ, pp. 63–84.

Hughes, J.R. and John, E.R. (1999) Conventional and quantitative electroencephalography in psychiatry. J. Neuropsychiatry Clin. Neurosci., 11: 190–208.

Jefferys, J.G.R. (1995) Nonsynaptic modulation of neuronal activity in the brain: electric currents and extracellular ions. Physiol. Rev., 75: 689–723.

John, E.R. (1968) Mechanisms of Memory. Academic Press, New York.

John, E.R. (1972) Switchboard versus statistical theories of learning and memory. Science, 177: 850–864.

John, E.R. (1990) Representation of information in the brain. In: John E.R., Harmony T., Prichep L.S., Valdes-Sosa M. and Valdes-Sosa P. (Eds.), Machinery of the Mind. Birkhauser, Boston, pp. 27–58.

John, E.R. (2002) The neurophysics of consciousness. Brain Res. Rev., 39: 1–28.

John, E.R. (2003) A theory of consciousness. Curr. Directions, 12: 244–249.

John, E.R. (2004) Consciousness from neurons and waves. In: Abbott D., Der A. and Sanchez A. (Eds.) Fluctuations and Noise in Biological, Biophysical and Biomedical Systems II, Vol. 5467. SPIE The International Organization for Optical Engineering, Gran Canaria, Spain, pp. 175–191.

John, E.R., Ahn, H., Prichep, L.S., Trepetin, M., Brown, D. and Kaye, H. (1980) Developmental equations for the electroencephalogram. Science, 210: 1255–1258.

John, E.R., Bartlett, F., Shimokochi, M. and Kleinman, D. (1973) Neural readout from memory. J. Neurophysiol., 36: 892–924.

John, E.R., Easton, P. and Isenhart, R. (1997) Consciousness and Cognition may be Mediated by Multiple Independent Coherent Ensembles. Conscious. Cogn., 6: 3–39.

John, E.R., Herrington, R.N. and Sutton, S. (1967) Effects of visual form on the evoked response. Science, 155: 1439–1442.

John, E.R. and Morgades, P.P. (1969a) Neural correlates of conditioned responses studied with multiple chronically implanted moving microelectrodes. Exp. Neurol., 23: 412–425.

John, E.R. and Morgades, P.P. (1969b) Patterns and anatomical distribution of evoked potentials and multiple unit activity by conditioned stimuli in trained cats. Comm. Behav. Biol., 3: 181–207.

John, E.R. and Prichep, L.S. (1993) Principals of neurometrics and neurometric analysis of EEG and evoked potentials. In: Neidermeyer E. and Lopes Da Silva F. (Eds.), In Basics Principles Clinical Applications and Related Fields (3rd Edition). Williams & Wilkins, Baltimore, MD, pp. 989–1003.

John, E.R. and Prichep, L.S. (2005) The anesthetic cascade: how anesthesia suppresses awareness. Anesthesiology, 102: 447–471.

John, E.R., Prichep, L.S., Ahn, H., Easton, P., Fridman, J. and Kaye, H. (1983) Neurometric evaluation of cognitive dysfunctions and neurological disorders in children. Prog. Neurobiol., 21: 239–290.

John, E. R., Prichep, L. S., Chabot, R. and Easton, P. (1990) Cross-spectral coherence during mental activity. EEG Clin Neurophysiol (Abstracts of the XIIth International Congress of Electroencephalography and Clinical Neurophysiology) 75, S68.

John, E.R., Prichep, L.S., Friedman, J. and Easton, P. (1988) Neurometrics: Computer-assisted differential diagnosis of brain dysfunctions. Science, 293: 162–169.

John, E.R., Prichep, L.S., Valdes-Sosa, P., Bosch, J., Aubert, E., Kox, W., Tom, M., di Michele, F. and Gugino, L.D. (2001) Invariant reversible QEEG effects of anesthetics. Consci. Cogn., 10: 165–183.

John, E.R., Shimokochi, M. and Bartlett, F. (1969) Neural readout from memory during generalization. Science, 164: 1534–1536.

John, E.R., Walker, P., Cawood, D., Rush, M. and Gehrmann, J. (1972) Mathematical identification of brain states applied to classification of drugs. Int. Rev. Neurobiol., 15: 273–347.

Jonkman, E.J., Poortvliet, D.C.J., Veering, M.M., DeWeerd, A.W. and John, E.R. (1985) The use of neurometrics in the study of patients with cerebral ischemia. Electroencephalogr. Clin. Neurophysiol., 61: 333–341.

Kleinman, D. and John, E.R. (1975) Contradiction of auditory and visual information by brain stimulation. Science, 187: 271–272.

Koch, C. (1998) The neuroanatomy of visual consciousness. In: Jasper H.H., Descarries L., Costelucci V.C. and Rossignol S. (Eds.), Advances in Neurology: Consciousness at the Frontiers of Neuroscience. Lippincote-Raven, Philadelphia, PA, pp. 229–241.

Koenig, T., Prichep, L., Lehmann, D., Valdes-Sosa, P., Braeker, E., Kleinlogel, H., Isenhart, R. and John, E.R. (2002) Millisecond by Millisecond, Year by year: Normative EEG Microstates and Development Stages. Neuroimage, 16: 41–48.

Kondacs, A. and Szabo, M. (1999) Long-term intra-individual variability of the background EEG in normals. Clin. Neurophysiol., 110: 1708–1716.

Konig, P., Engel, A.K. and Singer, W. (1995) The relation between oscillatory activity and long range synchronization in cat visual cortex. Proc. Natl. Acade. Sci., 92: 290–294.

Kreiter, A.K. and Singer, W. (1996) Stimulus-dependent synchronization of neuronal responses in the visual cortex of the awake macaque monkey. J. Neurosci., 16: 2381–2396.

Kutas, M. and Hillyard, S.A. (1980) Reading senseless sentences: brain potentials reflect semantic incongruity. Science, 207: 203–205.

Lachaux, J.P., Rodriguez, E., Martinerie, J., Adam, C., Hasboun, D. and Varela, F.J. (2000) A quantitative study of gamma-band activity in human intracranial recordings triggered by visual stimuli. Eur. J. Neurosci., 12: 2608–2622.

Larkum, M.E., Zhu, J.J. and Sakmann, B. (1999) A new cellular mechanism for coupling inputs arriving at different cortical layers. Nature, 398: 338–341.

Laureys, S., Faymonville, M.E. and Goldman, S. (2000) Functional imaging of arousal and levels of consciousness. Int. J. Neuropsychopharm. (Abstracts from the XXIInd CINP Congress), 3: S56.

Laureys, S., Faymonville, M.E., Peigneaux, P., Damas, P., Lambermont, B., Del Fiore, G., Degueldre, C., Aerts, J., Luxen, A., Franck, G., Lamy, M., Moonen, G. and Maquet, P. (2002) Cortical processing of noxious somato-sensory stimuli in the persistant vegetative state. Neuroimage, 17: 732–741.

Lehmann, D. (1971) Multichannel topography of human alpha EEG fields. EEG. Clin. Neurophysiol., 31: 439–449.

Lehmann, D., Strik, W.K., Henggeler, B., Koenig, T. and Koukkou, M. (1998) Brain electrical microstates and momentary conscious mind states as building blocks of spontaneous thinking: I. Visual imagery and abstract thoughts. Int. J. Psychophysiol., 29: 1–11.

Levine, J. (1983) Materialism and qualia: the explanatory gap. Pacific Philo. Q., 64: 354–361.

Libet, B. (1973) Electrical stimulation of cortex in human subjects, and conscious sensory aspects Vol 2. In: Iggo A. (Ed.), Somatosensory System. Handbook of Sensory Physiology. Springer, Berlin, pp. 743–790.

Libet, B. (1982) Brain stimulation in the study of neuronal functions for conscious sensory aspects. Human Neurobiol., 1: 235–242.

Libet, B. (1985) Unconscious cerebral initiative and the role of conscious will in voluntary action. Behavi. Brain Sci., 8: 529–566.

Libet, B., Wright, E.W., Feinstein, B. and Pearl, D.K. (1979) Subjective referral of the timing for conscious sensory experience: a functional role for the somatosensory specific projection system in man. Brain, 102: 193–224.

Llinas, R. and Ribary, U. (1993) Coherent 40-Hz oscillation characterizes dream states in humans. Proc. Natl. Acad. Sci. USA, 90: 2078–2081.

Llinas, R. (1988) The intrinsic properties of mammalian neurons: insights into central nervous system function. Science, 242: 1654–1664.

Llinas, R. and Ribary, U. (1998) Temporal conjunction in thalamocortical transactions. In: Jasper H.H. (Ed.) Consciousness: At the Frontiers of Neuroscience, Advances in Neurology, Vol. 77. Lippincott-Williams and Wikins, Philadelphia PA, pp. 95–103.

Llinas, R., Leznik, E. and Urbano, F.J. (2002) Temporal binding via cortical coincidence detection of specific and nonspecific thalamocortical inputs: a voltage-dependent dye-imaging study in mouse brain slices. Proc. Natl. Acad. Sci. USA, 99: 449–454.

Matousek, M. and Petersen, I. (1973) Frequency analysis of the EEG in normal children and adolescents. In: Kellaway P. and Petersen I. (Eds.), Automation of Clinical Electroencephalography. Raven Press, New York, pp. 75–102.

McCormick, D.A. (1992) Neurotransmitter actions in the thalamus and cerebral cortex and their role in neuromodulation of thalamocortical activity. Prog. Neurobiol., 39: 337–388.

McCormick, D.A. (2002) Cortical and subcortical generators of normal and abnormal rhythmicity. Int. Rev. Neurobiol., 49: 99–113.

Naatanen, R., Paavilainen, P., Alho, K., Reinikainen, K. and Sams, M. (1987) The mismatch negativity to intensity changes in an auditory stimulus sequence. Electroencephalogr. Clin. Neurophysiol. (Suppl.), 40: 125–131.

Naatanen, R., Schroger, E., Karakas, S., Tervaniemi, M. and Paavilainen, P. (1993) Development of a memory trace for a complex sound in the human brain. Neuro. Report., 47: 503–506.

Naatanen, R., Syssoeva, O. and Takegata, R. (2004) Automatic time perception in the human brain for intervals ranging from milliseconds to seconds. Psychophysiology, 41: 660–663.

Nicolelis, M.A., Baccala, L.A., Lin, S.R.C. and Chapin, J.K. (1995) Sensorimotor encoding by synchronous neural ensemble activity multiple levels of the somatosensory system. Science, 268: 1353–1358.

Pantev, C. (1995) Evoked and induced gamma-band activity of the human cortex. Brain Topog., 7: 321–330.

Pascual-Marqui, R., Michel, C. and Lehmann, D. (1995) Segmentation of Brain Electrical Activity into Microstates. IEEE Trans. Biomed. Engi., 42: 658–665.

Pavone, A. and Niedermeyer, E. (2000) Absence seizures and the frontal lobe. Clin. EEG, 31: 153–156.

Pitts, W. and McCulloch, W.S. (1947) How we know universals. Bull Math Biophys., 9: 127–147.

Portas, C.M., Krakow, K., Allen, P., Josephs, O., Armony, J.L. and Frith, C. (2000) Auditory processing across the sleep-wake cycle: simultaneous EEG and fMRI monitoring in humans. Neuron, 28: 991–999.

Prichep, L.S. and John, E.R. (1992) QEEG profiles of psychiatric disorders. Brain Topog., 4: 249–257.

Prichep, L.S., John, E.R., Gugino, L.D., Kox, W. and Chabot, R. (2000) Quantitative EEG assessment of changes in the level of the sedation/hypnosis during surgery under general anesthesia: I. The Patient State Index (PSI). In: Jordan C., Vaughan D.J.A. and Newton D.E.F. (Eds.), Memory and Awareness in Anesthesia. Imperial College Press, London, pp. 97–102.

Purpura, D.P. (1972) Functional studies of thalamic internuclear interactions. Brain Behav., 6: 203–234.

Rager, G. and Singer, W. (1998) The response of cat visual cortex to flicker stimuli of variable frequency. Eur. J. Neurosci., 10: 1856–1877.

Ramos, A., Schwartz, E. and John, E.R. (1976) An examination of the participation of neurons in readout from memory. Brain Res. Bull., 1: 77–86.

Rampil, I.J. (1998) A primer for EEG signal processing in anesthesia. Anesthesiology, 89: 980–1002.

Ribrary, U., Ionnides, A.A., Singh, K.D., Hasson, R., Bolton, J.P.R., Lado, F., Mogilner, A. and Llinas, R. (1991) Magnetic field tomography of coherent thalomocortical 40-Hz oscillations in humans. Proc Natl. Acad. Sci. USA, 88: 11037–11041.

Rodriguez, E., George, N., Lachaux, J.P., Martinerie, J., Renault, B. and Varela, F.J. (1999) Perception's shadow: long-distance synchronization of human brain activity. Nature, 397: 430–433.

Roelfsema, P.R., Engel, A.K., Konig, P. and Singer, W. (1997) Visuomotor integrations associated with zero time-lag synchronization among cortical areas. Nature, 385: 157–161.

Saletu, B., Anderer, P. and Pascual-Marqui, R. (2000) Pharmacodynamics and EEG II; From EEG mapping to EEG tomography. In: Saletu B., Krijzer F., Ferber G. and Anderer P. (Eds.), Electrophysiological Brain Research in Preclinical and Clinical Pharmacology and Related Fields – An Update, pp. 157–163.

Sams, S., Hari, R., Rif, J. and Knuutila, J. (1993) The human auditory sensory memory trace persists about 10 sec- neuromagnetic evidence. J. Cognit. Neurosci., 5: 363–370.

Sebel, P.S., Lang, E., Rampil, I.J., White, P.F., Cork, R., Jopling, M., Smith, N.T., Glass, P.S. and Manberg, P. (1997) A multicenter study of bispectral electroencephalogram analysis monitoring anesthetic effect. Anesth. Analg., 84: 891–899.

Sheinberg, D.L. and Logothetio, N.K. (1997) The role of temporal cortical areas in perceptual organization. Proc. Natl. Acad. Sci. USA, 94: 3408–3413.

Siegel, M., Kording, K.P. and Konig, P. (2000) Integrating top-down and bottom-up sensory processing by somatodendritic interactions. J. Comput. Neurosci., 8: 161–173.

Singer, W. (2001) Consciousness and the binding problem. Ann. NY. Acad. Sci., 929: 123–146.

Sokolov, E.N. (1963) Perception and the Conditioned Reflex. McMillan, New York.

Spencer, K.M., Nestor, P.G., Niznikiewicz, M.A., Salisbury, D.F., Shenton, M.E. and McCarley, R.W. (2003) Abnormal neuronal synchrony in schizophrenia. J. Neurosci., 23: 7407–7411.

Spencer, K.M., Nestor, P.G., Perlmutter, R., Niznikiewicz, M.A., Klump, M.C., Frumin, M., Shenton, M.E. and McCarley, R.W. (2004) Neural synchrony indexes disordered perception and cognition in schizophrenia. Neuroscience, 101: 17288–17293.

Srinivasan, R. (1999) Spatial structure of the human alpha rhythm: global correlation in adults and local correlation in children. Clin. Neurophysiol., 110: 1351–1362.

Steriade, M., Curró Dossi, R. and Pare, D. (1993) Electrophysiological properties of intralaminar thalamocortical cells discharging rhythmic (~40 Hz) spike-bursts at ~1000 Hz during waking and rapid-eye-movement sleep. Neuroscience, 56: 1–19.

Steriade, M., Curró Dossi, R., Paré, D. and Oakson, G. (1991) Fast oscillations (20–40 Hz) in thalamocortical systems and their potentiation by mesopontine cholinergic nuclei in the cat. Proc. Natl. Acad. Sci. USA, 88: 4396–4400.

Steriade, M., Gloor, P., Llinas, R.R., Lopes Da Silva, F. and Mesulam, M.M. (1990) Basic mechanisms of cerebral rhythmic activities. EEG Clin. Neurophysiol., 76: 481–508.

Streletz, V., Faber, P.L., Golikova, J., Novototsky-Vlasov, V., Koenig, T., Giano, L.R., Gruzelier, J.H. and Lehmann, D. (2003) Chronic schizophrenics with positive symptomology have shortened EEG microstate durations. Clin. Neurophysiol., 114: 2043–2051.

Sutton, S., Braren, M., John, E.R. and Zubin, J. (1965) Evoked potential correlates of stimulus uncertainty. Science, 150: 1187–1188.

Tallon-Baudry, C. (2000) Oscillatory synchrony as a signature for the unity of visual experience. Consciousness and Cognition (Proceedings of the 4th Conference of the Association for the Scientific Study of Consciousness), 9: S25–S26.

Tallon-Baudry, C., Bertrand, O., Delpuech, C. and Pernier, J. (1997) Oscillatory g-band (30–70 Hz)activity induced by a visual search task in humans. J. Neurosci., 17: 722–734.

Thatcher, R.W. and John, E.R. (1977). Functional Neuroscience, Vol. I. Foundations of Cognitive Processes. Lawrence Erlbaum, New Jersey.

Tononi, G. and Edelman, G.M. (2000) A Universe of Consciousness: How Matter Becomes Imagination. Basic Books, New York.

Torok, L. and Kish, L. (2004) Integro-differential stochastic resonance. In: Abbott D., Bezrukov S. M., and Sanchez A. (Eds.), Fluctuations and Noise in Biological, Biophysical, and Biomedical Systems II. Vol. 5467. Gran Canaria, Spain, SPIE – The International Society of Optical Engineering, pp. 149–162.

Varela, F. J. (2000). Neural synchrony and consciousness: are we going somewhere? Consciousness and Cognition (Proceedings of the 4th Annual Meeting of the Assn for the Scientific Study of Consciousness), 9: S26–S27.

Veselis, R.A., Reinsel, R.A., Distrian, A.M., Feshchenko, V.A. and Beattie, B.J. (2000) Asymmetric dose-related effects of midazolam on regional cerebral blood flow. In: Jordan C., Vaughan D.J.A. and Newton D.E.F. (Eds.), Memory and Awareness in Anaesthesia IV: Proceedings of the Fourth International Symposium on Memory and Awareness in Anaesthesia. Imperial College Press, London, pp. 287–303.

von Stein, A. and Sarnththeim, J. (2000) Different frequencies for different scales of cortical integration: from local gamma to long range alpha/theta synchronization. Int. J. Psychophysiol., 38: 301–313.

Weaver, J.C., Vaughan, T.E., Adair, R.K. and Astumian, R.D. (1998) Theoretical limits on the threshold for the response of long cells to weak extremely low frequency electric fields due to ionic and molecular flux rectification. Biophys. J., 75: 2251–2254.

Zeki, S. (2000) The disunity of consciousness. Consciousness and Cognition (Proceedings of the 4th Conference of the Association for the Scientific Study of Consciousness), 9: S30.

S. Laureys (Ed.)
Progress in Brain Research, Vol. 150
ISSN 0079-6123

CHAPTER 12

Genes and experience shape brain networks of conscious control

Michael I. Posner*

Department of Psychology, University of Oregon, Eugene, OR 97403-1227, USA

Abstract: One aspect of consciousness involves voluntary control over thoughts and feelings, often called will. Progress in neuroimaging and in sequencing the human genome makes it possible to think about voluntary control in terms of a specific neural network that includes midline and lateral frontal areas. A number of cognitive tasks involving conflict as well as the control of emotions have been shown to activate these brain areas. Studies have traced the development of this network in the ability to regulate cognition and emotion from about 2.5 to 7 years of age. Individual differences in this network have been related to parental reports of the ability of children to regulate their behavior, to delay reward and to develop a conscience. In adolescents these individual differences predict the propensity for antisocial behavior. Differences in specific genes are related to individual efficiency in performance of the network, and by neuroimaging, to the strength of its activation of this network. Future animal studies may make it possible to learn in detail how genes influence the common pattern of development of self-regulation made possible by this network. Moreover, a number of neurological and psychiatric pathologies involving difficulties in awareness and volition show deficits in parts of this network. We are now studying whether specific training experiences can influence the development of this network in 4-year-old children and if so, for whom it is most effective. Voluntary control is also important for the regulation of conscious input from the sensory environment. It seems likely that the same network involved in self-regulation is also crucial for focal attention to the sensory world.

This chapter will deal with two aspects of consciousness. The first is our awareness of the world around us. The second is our voluntary control over our thoughts, emotions and actions. In both of these areas mechanisms underlying attention are central (Posner, 1994).

An important distinction in studies of awareness (Iwasaki, 1993) is between general knowledge of our environment (ambient awareness) and detailed focal knowledge of a scene (focal awareness). We generally believe that we have full conscious awareness of our environment, even when our focal attention is upon our own internal thoughts. Experimental studies (Rensink et al., 1997), show us how much this opinion is in error. In the study of "change blindness" when cues that normally lead to a shift of attention are suppressed, we have only a small focus for which we have full knowledge and even major semantic changes in the remainder of the environment are not reported (see this volume).

Change blindness is closely related to studies of visual search, which have been prominent in the field of attention and are known to involve an interaction between information in the ventral visual pathway about the object identity and information in the dorsal visual pathway that controls orienting to sensory information (for a review see Driver et al., 2004 and see also orient as in Table 1). It is

*Corresponding author. Tel.: +1 541 346 4939;
Fax: +1 541 346 4914; E-mail: mposner@darkwing.uoregon.edu

DOI: 10.1016/S0079-6123(05)50012-8

Table 1. Brain areas and neuromodulators involved in attention networks

Function	Structures	Modulator
Orient	Superior parietal Temporal parietal junction Frontal eye fields Superior colliculus	Acetylcholine
Alert	Locus coeruleus Right frontal and parietal cortex	Norepinephrine
Executive attention	Anterior cingulated Lateral ventral prefrontal Basal ganglia	Dopamine

possible that clear activation of the ventral pathway is sufficient for ambient awareness of a visual scene (Zeki, 2003), but focal knowledge of a target requires a shift of attention to the target using the brain areas involved in orienting. Since many of these areas have been shown to be multimodal, they are involved in all forms of orienting (Driver et al., 2004). Focal attention is often automatically summoned to a target; for example, when that target moves or changes luminance, but many of the same mechanisms are the source of the attention effect when attention is moved voluntarily because of interest in a particular target. In focal awareness to sensory input the brain areas shown as the sources of orienting remain involved but are often joined by frontal brain areas called the executive attention network in Table 1.

It is the executive network that is of most concern to this chapter, whether it involves the voluntary focus of attention to a target within a scene, the need to search memory, regulate emotion or resolve conflict between response options. Humans have a conviction of conscious control that allows us to regulate our thoughts, feelings and behaviors in accord with our goals and also believe that voluntary conscious choice guides at least a part of the action they take. These beliefs have been studied under various names in different fields of psychology. In cognition, cognitive control is the usual name for the voluntary exercise of intentions, while in developmental psychology many of the same issues are studied under the name of self-regulation (Rueda et al., 2004b). Neuroscience has usually used the more traditional names of volition or will for these functions (Zhu, 2004).

This chapter seeks to trace the development of this form of conscious control from early childhood. In order to proceed, it is important to have a method of measuring the executive network that we believe underlies cognitive control, self-regulation and volition. While these names may involve a number of cognitive operations that lie behind implementing our goals, recent studies have tended to focus on the monitoring and resolution of conflict between cognitions or actions as a central operation underlying self-regulation (Botvinick et al., 2001; Rueda et al., 2004b). One reason for this choice lies in the simplicity of developing tasks for all ages that involve this conflict (Fan et al., 2003a). Another is that there appears to be some agreement upon the neural systems involved in carrying out conflict resolution in cognitive and emotional tasks (Bush et al., 2000; Fan et al., 2003a).

Anatomical network

Several conflict-related tasks activate a neural network common among people (Fan et al., 2003b). The most commonly used conflict is when people respond to the color of ink of a conflicting color name (Stroop effect). Since this method cannot be used with children, we have adopted two other conflict tasks. Fig. 1 illustrates two of the conflict tasks appropriate for children: (1) spatial conflict task (top) and (2) the child version of the flanker task bottom. The attention network test (ANT) uses the flanker as a target. With Arrows are used for adults, and fish used for young children. Prior to the target warning, signals are used to inform the person of when and where the target will be presented. Thus ANT provides a measure of alerting, orienting and conflict resolution within the same task (Fan et al., 2002; Rueda et al., 2004a).

Imaging studies have suggested that several frontal areas including the anterior cingulate and

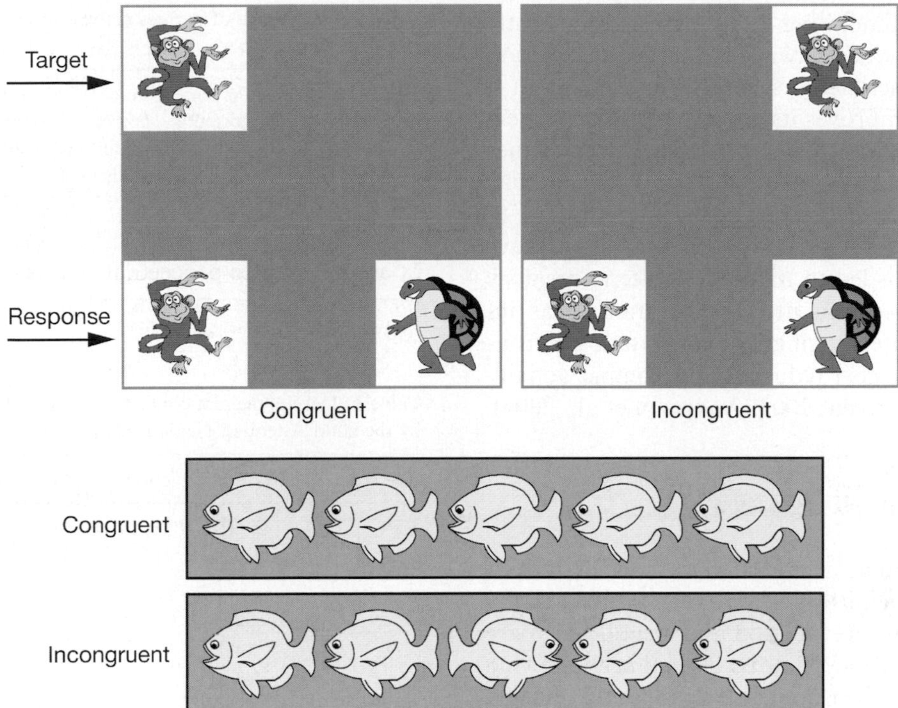

Fig. 1. Upper panel: a cognitive conflict task appropriate for children. The person must respond by matching the top figure to the correct figure on the bottom (identity), but the matching figures may be on the same side or opposite side (location). The conflict between location and identity can be used to study the resolution of conflict in children 2–3 years of age (Adapted from Gerardi–Caulton, 2000). Lower panel: a child friendly version of the attention network task. The arrows used for adults are replaced by fish and the child must press the left key when the center fish swims left, and the right key when it swims right. Adapted from Rueda et al. (2004a).

lateral areas of the prefrontal cortex are involved. Although some have argued that these areas work together as a global network to support aspects of consciousness (Dehaene et al., 1998; Baars, this volume), other studies have suggested a degree of specialization (MacDonald et al., 2000). One view is that the anterior cingulate monitors are for the existence of conflict, while the lateral prefrontal cortex acts to suppress conflicting operations (Botvinick et al., 2001). Another form of specialization suggests that more dorsal areas of the anterior cingulate is involved in cognitive process, while a ventral area is involved in the regulation of emotion (Bush et al., 2000). Even without a final resolution of these questions, there is a widespread convergence that these areas are involved in self-regulation of both cognition and feelings (see this volume).

Self-regulation

In support of this general idea are the findings that when instructed to ward off positive (Beauregard et al., 2001) or negative (Ochsner et al., 2002) feelings about a stimulus, there is a specific activation of the anterior cingulate. In a direct test of the conflict idea, Raz et al. (2002) studied highly hypnotizable persons in a Stroop task. In the standard condition they respond to the ink color under conditions where the word name was either congruent or not. In the non-conflict condition they saw the same stimuli, but after a hypnotic suggestion that the letter strings were meaningless. The hypnotic suggestion eliminated both the Stroop effect and activation in the anterior cingulate (Raz, 2004). Evidence suggested that voluntary instructions without the hypnotic suggestion were not effective in eliminating the conflict. Apparently hypnotism may be a way to put self-instruction under the experimenters control and do so in a manner that improves its effectiveness (see Kupers et al., this volume). However, there are other methods that have also worked. For example, Anderson and associates instructed subjects to avoid thinking about word associations and have shown that this led to reduced long-term memory and also activated a network of brain areas including the anterior cingulate and lateral prefrontal cortex and also reduced hippocampal activity (Anderson & Green, 2003; Anderson et al., 2004).

Development of self-regulation

One function that has been traced to the anterior cingulate is monitoring of error. In the spatial conflict task, where location and identification are in conflict, we found that reaction times following an error were 200 ms longer than those following a correct trial at age 30 months and over 500 ms longer at 36 months, indicating that children of this age were noticing their errors and correcting them (Rothbart et al., 2004). In this study we found no evidence of slowing following an error at 24 months. A somewhat more difficult conflict is introduced when subjects must utilize information from one verbal command while simultaneously ignoring information from another. We have used one version of the Simple Simon game, which asked children to execute a response to commands given by one puppet, while inhibiting commands given by a second puppet. In this study (Jones et al., 2003), children of 36–38 months showed no ability to inhibit their response and no slowing following an error, but at 39–41 months, children showed both an ability to inhibit and slowing of reaction time following an error (see Table 2). These results suggest that in this task somewhere between 38 and 41 months, performance changes based upon detecting an error response.

Because error detection has been studied using scalp electrical recording (Gehring et al., 1993; Luu et al., 2000) and shown to originate in the anterior cingulate in fMRI studies (Bush et al., 2000), we now have the means to examine the emergence of this cingulate function at any age for which we can design appropriate tasks.

Table 2. Percentage of correct activation and inhibition trials as a function of age

Age (months)	*N*	Percent correct activation trials (%)	Percent correct inhibition trials (%)	RT for correct response	RT following error
36–38	10	94	22	363	362
39–41	11	90	76	356	620
46–48	12	94	91	444	904

Adapted from Jones et al. (2003).

Table 3. Development of conflict resolution ability as measured by the child Attention Network Test as a function of age

Age (years)	Overall reaction time (ms)	Conflict scores
4.5*	1599	207
6	931	115
7	833	63
8	806	71
9	734	67
10	640	69
Adults	483	61

Note: The middle column is the overall reaction time (RT) and the right column the conflict score, both are in milliseconds. Conflict scores are obtained by subtracting RT to congruent flankers from RT to incongruent flankers. Adapted from Rueda et al. (2004a).

*A separate study of 4.5-year-olds was added to the data reported in the original.

We have examined conflict in children of 4.5–10 years of age using a version specifically adapted to them as illustrated in Fig. 1 (Rueda et al., 2004a). The results show improvement (see Table 3) in conflict regulation in children up to seven and no change after that. Seven-year-old children are remarkably similar to adults in the efficiency of this network, although the child reaction times are much longer than adults.

Genetic control

The existence of a common anatomical network, suggests that self-regulation was under genetic control. In support of this idea, we found from a study comparing mono- and dizygotic twins that ANT conflict scores had a high heritability (Fan et al., 2001). Since the executive attentional network has significant modulation by dopamine (see Table 1), polymorphisms in several dopamine-related genes were found to be associated with individual differences in the efficiency of this network (Diamond et al., 2004; Fossella et al., 2002). When people with alleles of different efficiency were compared in an imaging study there was differential activation of the anterior cingulate (Fan et al., 2003a).

The genetic differences among people in attention networks account for only a small part of the variance found in behavior and in imaging. However, a major importance of the genetic differences observed so far is that they serve as clues as to the genes that are involved in network development. These then can then serve as candidate genes in comparative animal studies to examine how genes, for example, in hippocampal development, may have affected behavior in species even before there was a hippocampus or in those for species for which the hippocampus plays a role in forms of memory that may be precursors of the explicit recollection, which is its role in humans. In the case of the DRD4 gene, found in humans to relate to attention deficit disorder (Swanson et al., 2001) and normal handling of conflict (Fan et al., 2003b) its role in the mouse seems to relate to exploration of the environment (Grandy & Kruzich, 2004). This approach may give us a handle on the question of how consciousness evolved.

Another reason for the relatively modest effect of genetic alleles in accounting for behavioral differences may be that they interact with experience during the development of the network. Genes by environment interactions are not controversial. Everyone agrees that gene expression can be influenced by aspects of the microenvironment in the brain area where it is expressed. Moreover, there is ample evidence that in primates gene expression can be influenced by events, which like maternal separation, can be imagined as a part of human development (Soumi, 2003).

Socialization

The degree of self-regulation may rest in part upon the results of successful socialization. Our data suggest substantial development in this network between ages 2.5 and 7 (see Tables 2 and 3). Perhaps the obvious difficulty that parents have with instructions to 2-year-olds, are an external manifestation of the changes, which are taking place in this network. Although this period also represents a strong growth in language usage, our direct observations suggest that self-regulation of behavior has a different origin (Jones et al., 2003).

During childhood, performance on conflict tasks is correlated with parental reports of their child's effortful control on temperament questionnaires. In turn, effortful control is associated with the ability to delay gratification, empathy toward others and the development of conscience (for a review, see Rothbart et al., 2004). The negative correlation between differences in the efficiency of the executive network and antisocial behavior continues in adolescence (Ellis et al., 2004) and pathologies of self-regulation have been shown to be related to longer times to resolve conflict.

In a comparison of self-regulation in China and US, it was found that strong control was related to low negative effect in America, but to low extroversion (positive affect) in China (Ahadi et al., 1993). The authors speculated that this was related to the preference for positive emotion that characterizes American culture and the tendency to avoid self-aggrandizing activity that is more typical of China. Of course cultural differences do not distinguish between genetic and environmental effect, but they do warn us that we do not all see or at least act on the world with the same predilections (Rothbart et al., 2004).

Training

For children who suffer from attention deficit hyperactivity disorder (ADHD), use of training attention and working memory can produce improvement in the ability to concentrate and in general intelligence (Kerns et al., 1999; Klingberg et al., 2002; Shavlev et al., 2003). All these studies involved children of 8 or more years with known difficulties in attention. In our studies we worked with normal 4-year-old children to determine if attention training might serve as a component of preschool education. Our studies were designed mainly to understand whether changes might be found during the development of the network. To provide a preliminary test, we used small samples of children for a very limited period of training (Posner et al., in press).

Four-year-olds were trained because our previous studies had shown improvements in performance between 4 and 7 years of age (see Table 2) in the ANT, designed to survey the efficiency of performance on attentional networks (Rueda et al., 2004a). Our exercises were patterned after those used to train macaques for service in space travel (Rumbaugh & Washburn, 1995). Our exercises began with training the child to control the movement of a cat by using a joystick as well as prediction of where an object would move, given its initial trajectory. Other exercises emphasized the use of working memory to retain information for a matching to sample task and the resolution of conflict. Each of the exercises progressed from easy to difficult in seven levels, the children having to perform each level correctly three times to proceed to the next level. Most of the children were able to complete the exercises within the 5 days allotted, and some of them were abbreviated to allow completion. The children seemed to enjoy the training although they were clearly tired at the end of each 30–40 min session.

The children were brought to the laboratory for 7 days for sessions. These sessions were conducted over a 2–3-week period. The first and last days were used to assess the effects of the training by use of ANT, a general test of intelligence (the K-BIT, Kaufman & Kaufman, 1990) and a temperament scale (the Children Behavior Questionnaire or CBQ, Rothbart et al., 2000). During the administration of ANT, we recorded 128 channels of EEG in order to observe the amplitude and time course of activation of brain areas associated with executive attention in adult studies (Rueda et al., 2004c; van Veen & Carter, 2002).

During our first experiment, we compared 12 children who underwent our training procedure with 12 who were randomly selected and took no training, but came in twice for assessment. In our second experiment, we again used 4-year-olds, but the control group came in seven times and saw videos which required an occasional response on their part to keep them playing.

In this chapter we present a brief overview of our initial results, a more complete account will be given elsewhere (Posner et al., in press). Of course, 5 days is a minimal amount of training to influence the development of networks that develop for many years. Nonetheless, we found a general improvement in intelligence in the experimental group as measured by the K-BIT. This is due to improvement of the experimental group in performance on the non-verbal portion of the IQ test. We also discovered that the reaction time measures registered with the ANT were unstable for four-year-olds and of low reliability in children of the age we were testing; thus we were not able to obtain significant improvement in the measures of the various networks, although overall reaction time did improve. We did find that the experimental children produced smaller conflict scores after training than the control children, although this might have been due to differences in the pretest despite the random assignment. The analysis of the brain networks using EEG recording showed that only the trained children following training showed a larger N2 on conflict trials. This is the component most closely related to the anterior cingulate in prior studies (van Veen and Carter, 2002; Rueda et al., 2004c). Of course, we do not know if the change in the N2 component found in these studies indicates merely that children perform the ANT better or a change in the underlying network, which uses the ANT. However, the IQ changes suggest it might be the latter. We did no training of IQ and the exercises provided in the Kbit did not resemble any of the training, yet both in our studies and in the related findings of Klingberg et al. (2002) with older children, there was a significant improvement.

As the number of children who undergo our training increases and as we improve our measures, we can examine aspects of their temperament and genotype to help us understand who might benefit from attention training. To this end we are currently genotyping all of the children in an effort to examine the candidate genes found previously to be related to the efficacy of the executive attention networks. We are also beginning to examine the precursors of executive attention in even younger children, with the goal of determining whether there is a sensitive period during which interventions might prove most effective.

Pathologies

Many neurological and psychiatric disorders including ADHD, neglect, autism, schizophrenia and Alzheimer's dementia (see Salmon et al., this volume), have been said to involve pathologies of attention. Some of these are disorders studied in neurology and others would be called psychiatric. However, without a real understanding of the neural substrates of attention, this has been a somewhat empty classification. This situation is changed with the systematic application of our understanding of attentional networks to pathological issues. Below we discuss several disorders that have been shown to have focal abnormalities related to the network underlying voluntary attention and self-regulation.

Borderline personality. Borderline personality disorder is characterized by very great lability of affect and problems in interpersonal relations. In some cases, patients are suicidal or carry out self-mutilation. Because this diagnosis has been studied largely by psychoanalysts and has a very complex definition, it might at first be thought of as a poor candidate for a specific pathophysiology involving attentional networks. However, we focused on the temperamentally based core symptoms of negative emotions and difficulty in self-regulation (Posner et al., 2003). We found that patients were very high in negative affect and relatively low in effortful control (Rothbart et al., 2000), and defined a temperamentally matched control group of normal persons without personality disorder who were equivalent in scores on these two dimensions. Our study with the ANT found a deficit specific to the executive attention network in borderline patients (Posner et al., 2003). Preliminary imaging results suggested overgeneralization of responding in the amygdala, and reduced responding in the anterior cingulate and related midline frontal areas (Posner et al., 2003). Patients with higher effortful control and lower conflict scores on the ANT were also the most likely to show the effects of therapy. This methodology shows the utility of focusing on the core deficits of patients, defining appropriate control groups based on matched temperament, and using specific attentional tests to help determine how to conduct imaging studies.

Schizophrenia. A number of years ago, never medicated schizophrenic patients were tested both by imaging and with a cued detection task similar to the orienting part of the ANT. At rest, these subjects had shown a focal decrease in cerebral blood flow in the left globus pallidus (Early et al., 1989), a part of the basal ganglia with close ties to the anterior cingulate. These subjects showed a deficit in orienting similar to what had been found for left parietal patients (Early et al., 1989). When their visual attention was engaged, they had difficulty in shifting attention to the right visual field. However, they also showed deficits in conflict tasks, particularly when they had to rely on a language cue. It was concluded that the overall pattern of their behavior was most consistent with a deficit in the anterior cingulate and basal ganglia, parts of a frontally based executive attention system.

The deficit in orienting rightward has been replicated in first break schizophrenics, but it does not seem to be true later in the disorder (Maruff et al., 1995), nor does the pattern appear to be part of the genetic predisposition for schizophrenia (Pardo et al., 2000). First break schizophrenic subjects often have been shown to have left hemisphere deficits, and there have been many reports of anterior cingulate and basal ganglia deficits in patients with schizophrenia (Benes, 1999). The anterior cingulate may be part of a much larger network of frontal and temporal structures that operate abnormally in schizophrenia (Benes, 1999; Kircher, this volume).

A recent study using the ANT casts some light on these results (Wang et al., in press). In this case, the schizophrenic patients were chronic and they were compared with a similarly aged control group. The schizophrenic patients had much greater difficulty resolving conflict than did the normal controls. The deficit in patients was also much larger than that found for borderline personality patients. There was still a great deal of overlap between the patients and normal subjects, however, indicating that the deficit is not suitable for making a differential diagnosis. The data showed a much smaller orienting deficit of the type that had been reported previously. These findings suggest that there is a strong executive deficit in chronic schizophrenia, as would be anticipated by Benes' (1999) theory. It remains to be determined whether this deficit exists prior to the initial symptoms, or whether it develops with the disorder. A finding by Dehaene et al. (2003) identifies the anterior cingulate effect in schizophrenics specifically with conscious conflict and not a similar conflict when unconscious.

Chromosome 22q11.2 deletion syndrome. This syndrome is a complex one that involves a number of abnormalities, including facial and heart structure, but also a mental retardation due to deletion of a number of genes. Children with the deletion are at a high risk for developing schizophrenia. Among the genes deleted in this syndrome is the COMT gene, which has been associated with the performance in a conflict task (Diamond et al., 2004) and with schizophrenia (Egan et al., 2001). In light of these findings, it was to be expected that the disorder would produce a large executive deficit (Simon et al., in press; Sobin et al., 2004). Sobin has found in a further study that the deficit in resolving conflict is associated with the ability to inhibit a blink following a cue that a loud noise would be presented shortly (pre-pulse inhibition). The authors suggest that the association of the high level attention and pre-pulse inhibition deficit suggests a pathway that includes both the basal ganglia and the anterior cingulate.

Sensory awareness

At the start of this chapter we distinguished between two senses of awareness of the world around us. Ambient awareness does not seem to provide any exact knowledge of the environment, but gives us confidence of the general features of the sensory world. It seems possible that this form of awareness is a property of the many brain areas involved in handling sensory information and requires no attention (Zeki, 2003).

However, we can orient to specific sensory information. We now have considerable knowledge of the brain areas involved in bringing attention to a part of the visual field (see Table 1). These areas are similar, regardless of whether orienting involves a physical change such as an eye or head movement or is entirely covert without any overt manifestation. These brain areas, particularly the cortical ones have been shown to be active in many imaging studies using different experimental tasks that all involve orienting to sensory information (Corbetta and Shulman, 2002). Lesion studies suggest that damage to most of these areas produce a failure to orient attention to information on the side of space opposite the lesion.

The different brain areas involved in orienting do not appear to be carrying out similar aspects of the task; rather there is evidence that different areas are responsible for different operations involved in orienting attention to the sensory stimulus (Posner, 1988). For example, the superior parietal lobe seems most closely related to the voluntary shifts of attention following an external or self-instruction (Corbetta and Shulman, 2002). This area has close relations to parts of the brain involved in eye movements and micro stimulation in this area will produce either covert attention shifts or eye movements depending on its intensity (Cutrell and Marrocco, 2002). The temporal parietal junction appears to be more related to disengaging from a current focus of attention as it happens when people are cued to one location and then receive a target of different location (Corbetta and Shulman, 2002). These findings make it unlikely that any single brain area is responsible for orienting, but rather a network of brain areas must be orchestrated to allow more detailed information about the input to become conscious.

Although shifting attention to various locations is the most usual method for examining changes in consciousness within a scene, it is also possible to study shifts in consciousness by setting up a rival information to each eye (Edelman, 2003). Using this method conscious perception has been associated with increased synchrony of a wide spread neural network including frontal and parietal areas.

The various brain areas involved in orienting seem to be multimodal in that they are involved regardless of whether the target is visual, tactile or auditory (Driver et al., 2004). Apparently it is the site to which one orients that determines the specific quality of the information that comes to awareness. Although detailed awareness appears to depend upon successful orienting, there appears to be evidence that sensory information which does not produce orienting, and of which we are unaware, can still influence behavior (Dehaene, 2004).

Focal awareness

Imaging studies suggest that whenever we bring the information to mind, whether extracted from sensory input or from memory, we activate the executive attention network (Duncan et al., 1998, 2000). This may be because focal attention is voluntarily switched to the target information. Thus moving attention to a target in order to bring it fully to mind is one type of voluntary response. As such it would require the executive attention network irrespective of the source of information. We started out this paper with the traditional distinction between awareness and control as components of consciousness. However, one form of awareness, focal awareness, appears to involve the same underlying mechanism as involved in control. In this sense even though some forms of consciousness (e.g. ambient awareness) may have diverse sources within sensory specific cortex, there is also a degree of unity of the underlying mechanism involved in some aspects of consciousness (e.g. focal awareness and voluntary control). The distinction between focal and ambient factors in consciousness has been made before (Iwasaki, 1993) and it may help to clarify the sense of awareness that can be present even when detailed accounts of the scene are not possible as in change blindness (Rensink et al., 1997). It seems possible that current methods will eventually be able to describe the physical basis of the network that has been identified to underlie self-regulation at the molecular level as well as to understand its origin in evolution and in child development. If focal awareness proves to involve the same network, the studies of self-regulation may also clarify the focal aspect of awareness.

Acknowledgment

This research was supported in part by a grant for the James S. McDonnell Foundation 21st Century Awards. The paper was presented as an invited address to the joint meeting of the European and American Societies for Philosophy and Psychology July, 2004.

References

Ahadi, S.A., Rothbart, M.K. and Ye, R. (1993) Children's temperament in the U.S. and China: similarities and differences. Eur. J. Person., 7: 359–378.

Anderson, M.C. and Green, C. (2003) Suppressing unwanted memories by executive control. Nature, 410: 366–369.

Anderson, M.C., Ochsner, K., Kuhl, B., Cooper, J., Robertson, E., Gabrieli, S.W., Glover, G. and Gabrieli, J.D.E. (2004) Neural systems underlying the suppression of unwanted memories. Science, 303: 232–235.

Beauregard, M., Levesque, J. and Bourgouin, P. (2001) Neural correlates of conscious self-regulation of emotion. J. Neurosci., 21: RC165.

Benes, F. (1999) Model generation and testing to probe neural circuitry in the cingulate cortex of postmortem schizophrenic brains. Schizophr. Bull., 24: 219–229.

Botvinick, M.M., Braver, T.S., Barch, D.M., Carter, C.S. and Cohen, J.D. (2001) Conflict monitoring and cognitive control. Psychol. Rev., 108: 624–652.

Bush, G., Luu, P. and Posner, M.I. (2000) Cognitive and emotional influences in the anterior cingulate cortex. Trends Cogn. Sci., 4/6: 215–222.

Corbetta, M. and Shulman, G.L. (2002) Control of goal-directed and stimulus-driven attention in the brain. Nat. Rev. Neurosci., 3: 201–215.

Cutrell, E.B. and Marrocco, R.T. (2002) Electrical microstimulation of primate posterior parietal cortex initiates orienting and alerting components of covert attention. Exp. Brain Res., 144: 103–113.

Dehaene, S. (2004) The neural basis of subliminal priming. In: Kanwisher N. and Duncan J. (Eds.), Functional Neuroimaging of Visual Cognition Attention and Performance XX. Oxford University Press, Oxford UK, pp. 205–224.

Dehaene, S., Artiges, E., Naccache, L., Martelli, C., Viard, A., Schurhoff, F., Recasens, C., Martinot, M.L.P., Leboyer, M. and Martinot, J.L. (2003) Conscious and subliminal conflicts in normal subjects and patients with schizophrenia: the role of the anterior cingulated. Proc. Natl. Acad. Sci. USA, 100: 13722–13727.

Dehaene, S., Kerszberg, M. and Changeux, J-P. (1998) A neuronal model of a global workspace in effortful cognitive tasks. Proc. Natl. Acad. Sci. USA, 95: 1452–1453.

Diamond, A., Briand, L., Fossella, J. and Gehlbach, L. (2004) Genetic and neurochemical modulation of prefrontal cognitive functions in children. Am. J. Psychiatr., 161(1): 125–132.

Driver, J., Eimer, M., Macaluso, W. and van Velzen, J. (2004) Neurobiology of human spatial attention: Modulation, generation and integration. In: Kanwisher N. and Duncan J. (Eds.), Functional Neuroimaging of Visual Cognition. Oxford University Press, Oxford, UK, pp. 267–300.

Duncan, J., Seitz, R.J., Kolodny, J., Bor, D., Herzog, H., Ahmed, A., Newell, F. and Emslie, H. (2000) A neural basis for general intelligence. Science, 289: 457–460.

Early, T.S., Posner, M.I., Reiman, E.M. and Raichle, M.E. (1989) Hyperactivity of the left striato-pallidal projection Part I: Lower level theory. Psychiatr. Dev., 2: 85–108.

Edelman, G. (2003) Naturalizing consciousness: a theoretical perspective. Proc. Natl. Acad. Sci. USA, 100: 5520–5524.

Egan, M.F., Goldberg, T.E., Kolachana, B.S., Callicott, J.H., Mazzanti, C.M., Straub, R.E., Goldman, D. and Weinberger, D.R. (2001) Effect of COMT Val108/158 Met genotype on frontal lobe function and risk for schizophrenia. Proc. Natl. Acad. Sci. USA, 98: 6917–6922.

Ellis, L.K., Rothbart, M.K. and Posner, M.I. (2004). Individual difference in executive attention predict self-regulation and adolescent psychosocial behaviors. In: Dahl, R.E. and Spear, L.P. (Eds.), Adolescent Brain Development: Vulnerabilities and Opportunities, Ann. NY Acad. Sci., 1021: 337–340.

Fan, J., Fossella, J.A., Sommer, T. and Posner, M.I. (2003b) Mapping the genetic variation of executive attention onto brain activity. Proc. Natl. Acad. Sci. USA, 100: 7406–7411.

Fan, J., Flombaum, J.I., McCandliss, B.D., Thomas, K.M. and Posner, M.I. (2003a) Cognitive and brain mechanisms of conflict. Neuroimage, 18: 42–57.

Fan, J., McCandliss, B.D., Sommer, T., Raz, M. and Posner, M.I. (2002) Testing the efficiency and independence of attentional networks. J. Cogn. Neurosci., 14(3): 340–347.

Fan, J., Wu, Y., Fossella, J. and Posner, M.I. (2001) Assessing the heritability of attentional networks. BMC Neurosci., 2: 14.

Fossella, J., Sommer, T., Fan, J., Wu, Y., Swanson, J.M., Pfaff, D.W. and Posner, M.I. (2002) Assessing the molecular genetics of attention networks. BMC Neurosci., 3: 14.

Gehring, W.J., Goss, B., Coles, M.G.H., Meyer, D.E. and Donchin, E. (1993) A neural system for error detection and compensation. Psychol. Sci., 4: 385–390.

Gerardi-Caulton, G. (2000) Sensitivity to spatial conflict and the development of self-regulation in children 24–36 months of age. Developmental Science, 3: 397–404.

Grandy, D.K. and Kruzich, P.J. (2004) Molecular effects of dopamine receptors on attention. In: Posner M.I. (Ed.), Cognitive Neuroscience of Attention. Guilford Press, New York.

Iwasaki, S. (1993) Spatial attention and two modes of visual consciousness. Cognition, 49: 211–233.

Jones, L., Rothbart, M.K. and Posner, M.I. (2003) Development of inhibitory control in preschool children. Devel. Sci., 6: 498–504.

Kaufman, A.S. and Kaufman, N.L. (1990) Kaufman Brief Intelligence Test—Manual. American Guidance Service, Circle Pines, MN.

Kerns, K.A., Esso, K. and Thompson, J. (1999) Investigation of a direct intervention for improving attention in young children with ADHD. Dev. Neuropsychol., 16: 273–295.

Klingberg, T., Forssberg, H. and Westerberg, H. (2002) Training of working memory in children with ADHD. J. Clin. Exp. Neuropsych., 24: 781–791.

Luu, P., Collins, P. and Tucker, D.M. (2000) Mood, personality and self-monitoring: Negative affect and emotionality in relation to frontal lobe mechanisms of error-detection. J. Exp. Psychol. Gen., 129: 43–60.

MacDonald, A.W., Cohen, J.D., Stenger, V.A. and Carter, C.S. (2000) Dissociating the role of the dorsolateral prefrontal and anterior cingulate cortex in cognitive control. Science, 288: 1835–1838.

Maruff, P., Currie, J., Hay, D., McArthur-Jackson, C. and Malone, V. (1995) Asymmetries in the covert orienting of visual spatial attention in schizophrenia. Neuropsychologia, 31: 1205–1223.

Ochsner, K.N., Bunge, S.A., Gross, J.J. and Gabrieli, J.D.E. (2002) Rethinking feelings: An fMRI study of the cognitive regulation of emotion. J. Cog. Neurosci., 14: 1215–1229.

Pardo, P.J., Knesevich, M.A., Vogler, G.P., Pardo, J.V., Towne, B., Clonninger, C.R. and Posner, M.I. (2000) Genetic and state variables of neurocognitive dysfunction in schizophrenia: a twin study. Schizophrenia Bull., 26: 459–477.

Posner, M.I. (1988) Structures and functions of selective attention. In: Boll T. and Bryant B. (Eds.), Master Lectures in Clinical Neuropsychology and Brain Function: Research, Measurement, and Practice. American Psychological Association, pp. 171–202.

Posner, M.I. (1994) Attention: The mechanism of consciousness. Proc. Natl. Acad. Sci. USA, 91(16): 7398–7402.

Posner, M.I., Rothbart, M.K. and Rueda, M.R. (in press) Brain Mechanisms of High Level Skills To appear in Battro, Antonio M. Fischer, Kurt W. and Pierre Léna (Eds.), Mind, Brain, and Education. Cambridge University Press, Cambridge, UK.

Posner, M.I., Rothbart, M.K., Vizueta, N., Thomas, K.M., Levy, K., Fossella, J., Silbersweig, D.A., Stern, E., Clarkin, J. and Kernberg, O. (2003) An approach to the psychobiology of personality disorders. Dev. Psychopath., 15: 1093–1106.

Raz, A. (2004) A typical attention: hypnosis and conflict resolution. In: Posner M.I. (Ed.), Cognitive Neuroscience of Attention. Guilford Press, New York, pp. 420–429.

Raz, A., Shapiro, T., Fan, J. and Posner, M.I. (2002) Hypnotic suggestion and the modulation of Stroop interference. Arch. Gen. Psychiat., 59(12): 1155–1161.

Rensink, R.A., O'Regan, J.K. and Clark, J.J. (1997) To see or not to see: The need for attention to perceive changes in scenes. Psychol. Sci., 8: 368–373.

Rothbart, M.K., Ahadi, S.A. and Evans, D.E. (2000) Temperament and personality: origins and outcomes. J. Pers. Soc. Psych., 78: 122–135.

Rothbart, M.K., Ellis, L.K. and Posner, M.I. (2004) Temperament and self-regulation. In: Baumeister R.F. and Vohs K.D. (Eds.), Handbook of Self Regulation. Guilford Press, New York, pp. 357–370.

Rueda, M.R., Fan, J., McCandliss, B., Halparin, J.D., Gruber, D.B., Pappert, L. and Posner, M.I. (2004a) Development of attentional networks in childhood. Neuropsychologia, 42: 1029–1040.

Rueda, M.R., Posner, M.I. and Rothbart, M.K. (2004b) Attentional control and self regulation. In: Baumeister R.F. and Vohs K.D. (Eds.), Handbook of Self-Regulation: Research, Theory, and Applications. Guilford Press, New York, pp. 283–300.

Rueda, M.R., Posner, M.I., Rothbart, M.K. and Davis-Stober, C.P. (2004c) Development of the time course for conflict resolution: an ERP study with 4-year-olds and adults. BMC Neurosci., 5: 39.

Rumbaugh, D.M. and Washburn, D.A. (1995) Attention and memory in relation to learning: A comparative adaptation perspective. In: Lyon G.R. and Krasengor N.A. (Eds.), Attention, Memory and Executive Function. Brookes Publishing Company, Baltimore, MD, pp. 199–219.

Shavlev, L., Tsal, Y. and Mevorach, C. (2003) Progressive attentional training program: Effective direct intervention for children with ADHD. Proc. Cog. Neurosci. Soc., New York pp. 55–56.

Simon, T.J., Bish, J.P., Bearden, C.E., Ding, L., Ferrante, S., Nguyen, V., Gee, J., McDonald-McGinn, D., Zackai, E.H. and Emanuel, B. S. (in press). A multi-level analysis of cognitive dysfunction and psychopathology associated with chromosome 22q11.2 deletion syndrome in children development and psychopathology. Devel. Psychopathol.

Sobin, C., Kiley-Brabeck, K., Daniels, S., Blundell, M., Anyane-Yeboa, K. and Karayiorgou, M. (2004) Networks of attention in children with the 22q11 deletion syndrome. Dev. Neuropsych., 26: 611–626.

Soumi, S.J. (2003) Gene-environment interactions and the neurobiology of social conflict. Ann. NY Acad. Sci., 1008 132–13.

Swanson, J., Deutsch, C., Cantwell, D., Posner, M., Kenndy, J., Barr, C., Moyzis, R., Schuck, S., Flodman, P., Spence, M.A. and Wasdell, M. (2001) Genes and attention-deficit hyperactivity disorder. Clin. Neurosci. Res., 1(3): 207–216.

van Veen, V. and Carter, C.S. (2002) The timing of action-monitoring processes in the anterior cingulate cortex. J. Cog. Neurosci., 14: 593–602.

Wang, K., Fan, J., Dong, Y., Wang, C., Lee, T.M.C. and Posner, M.I. (in press) Selective impairment of attentional networks of orienting and executive control in schizophrenia.

Zeki, S. (2003) The disunity of consciousness. Trends Cogn. Sci., 7(5): 214–218.

Zhu, J. (2004) Locating volition. Conscious. Cogn., 13: 302–322.

S. Laureys (Ed.)
Progress in Brain Research, Vol. 150
ISSN 0079-6123

CHAPTER 13

Visual phenomenal consciousness: a neurological guided tour

Lionel Naccache[1,2,3,*]

[1]*Fédération de Neurophysiologie Clinique, Hôpital de la Pitié-Salpêtrière, 47-83 Bd. de l'Hôpital, 75013 Paris, France*
[2]*Fédération de Neurologie, Hôpital de la Pitié-Salpêtrière, 47-83 Bd. de l'Hôpital, 75013 Paris, France*
[3]*INSERM U562, Service Hospitalier Frédéric Joliot, CEA/DSV/DRM, 4 place du Général Leclerc, Orsay, France*

Abstract: The scientific study of the cerebral substrate of consciousness has been marked by significant recent achievements, resulting partially from an interaction between the exploration of cognition in both brain-damaged patients and healthy subjects. Several neuropsychological syndromes contain marked dissociations that permit the identification of principles related to the neurophysiology of consciousness. The generality of these principles can then be evaluated in healthy subjects using a combination of experimental psychology paradigms, and functional brain-imaging tools. In this chapter, I review some of the recent results relevant to visual phenomenal consciousness, which is an aspect of consciousness most frequently investigated in neuroscience. Through the exploration of neuropsychological syndromes such as "blindsight," visual form agnosia, optic ataxia, visual hallucinations, and neglect, I highlight four general principles and explain how their generality has been demonstrated in healthy subjects using conditions such as visual illusions or subliminal perception. Finally, I describe the bases of a scientific model of consciousness on the basis of the concept of a "global workspace," which takes into account the data reviewed.

Introduction

The scientific investigation of consciousness has recently stimulated experimental research in healthy human subjects, in neurological and psychiatric patients, and in some animal models. Although this major ongoing effort does not yet provide us with a detailed and explicit neural theory of this remarkable mental faculty, we already have access to a vast collection of results acting as a set of constraints on what should be a scientific model of consciousness. There are many ways to summarize and present this set of "consciousness principles." One may either use a chronological or a domain-specific strategy. Here, I deliberately adopt a narrative approach driven by a neurological perspective. This approach allows an emphasis on the crucial role played by the observation of brain-lesioned patients affected by neuropsychological syndromes. I argue that as in other fields of cognitive neuroscience, clinical neuropsychology often offers profound and precious insights leading to the discovery of neural principles governing distinct aspects of the physiology of consciousness (Ramachandran and Blakeslee, 1998). Most importantly, many of these principles also prove to be relevant and to generalize to the cognition of healthy human subjects. In a schematic manner, the 'borderline cases' provided by clinical neurology have the power to specifically illustrate a single property of consciousness by showing the consequences of its impairment. This magnifying effect makes it easier to isolate and delineate this

*Corresponding author. Tel.: +33 1 40 77 97 99;
Fax: +33 1 40 77 97 89; E-mail: lionel.naccache@wanadoo.fr

DOI: 10.1016/S0079-6123(05)50013-X

property, and then to take it into account in more complex situations where it is functioning in concert with other processes.

I will focus interest on a selected number of these properties, and will limit our investigation to visual phenomenal consciousness, which is by far the most experimentally investigated aspect of consciousness. Following the psychologist Larry Weiskrantz (1997), our criteria to establish subject's conscious perception of a stimulus will be the "reportability" criteria: the ability to report explicitly to oneself or to somebody else the object of our perception: "*I see the word consciousness printed in black on this page*." This criterion is fully operational, and can be easily correlated to other sources of information (external reality, functional brain-imaging data, etc.), thereby paving the way to an objective evaluation of subjective data, a scientific program called "heterophenomenology" by Daniel Dennett (1992). It can be argued, however, that reportability might be a biased measure underestimating subjects' conscious state, and that forced-choice tasks using signal detection theory parameters such as d' might be preferable (Holender, 1986). However, discrediting reportability on these grounds in favor of purely objective measures is far from satisfying. Firstly, unconscious perception of a stimulus might have an impact on objective measures, as illustrated in many unconscious perception situations such as masked priming paradigms (Merikle, Smilek, & Eastwood, 2001). Secondly, to ignore subjective reports is somewhat of a counterproductive approach, because it may lead to simply giving up the original project of investigating consciousness. Finally, some authors contest the criteria of reportability by establishing differences between phenomenal consciousness and access consciousness (see Cleeremans, this volume), claiming that we are actually conscious of much more information than we can access and report (Block, 1995). This last theory does not discredit our criteria, but suggests limits to its usage as a non-exhaustive index of consciousness.

Thus far, I have justified our adoption of the "reportability" criteria to diagnose conscious perception in subjects. How then may we use it to specify a scientific program to investigate systematically the neural basis of visual consciousness? By first recalling a basic but essential "Kantian" statement: when we report being conscious of seeing an object, strictly speaking, we are not conscious of this object belonging to external reality, rather we are conscious of some of the visual representations elaborated in our visual brain areas and participating to the flow of our visual phenomenal consciousness, as masterly expressed by the Belgian Surrealist Painter, René Magritte in his famous painting "This is not a pipe" ("Ceci n'est pas une pipe" or "La trahison des images", 1928–1929, see Fig. 1). This simple evocation of the concept of representation foreshadows the two fundamental stages in the search of the "neural correlates of visual consciousness"(Frith et al., 1999): (i) make a detailed inventory of the multiple representations of the visual world elaborated by different visual brain areas (from retina and lateral geniculate nuclei to ventral occipito-temporal and dorsal occipito-parietal pathways described by Ungerleider and Mishkin (1982), in addition to superior colliculus-mediated visual pathways); and (ii) identify among these different forms of visual coding which participate in visual phenomenal consciousness, and in these cases, specify the precise conditions governing the contribution of these representations to

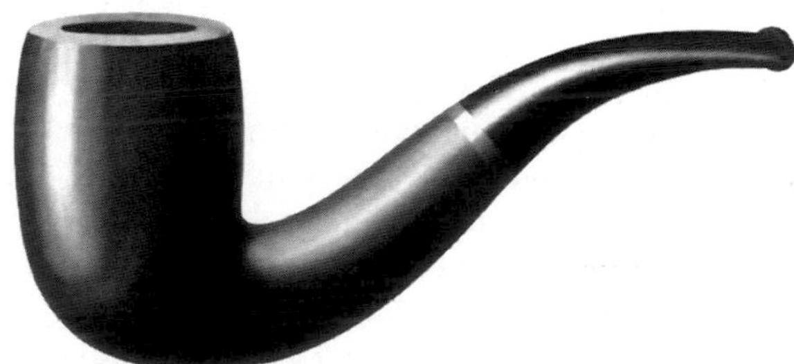

Fig. 1. When we report being conscious of seeing a pipe, strictly speaking we are not conscious of this pipe belonging to external reality, rather we are conscious of some of the visual representations elaborated in our visual brain areas and participating to the flow of our visual phenomenal consciousness. (René Magritte, 1928–1929, La trahison des images (The Treachery of Images), reproduced with permission from the Los Angeles County Museum of Art and the Artists Society.) © René Magritte c/o Beeldrecht Amsterdam 2005.

the flow of phenomenal consciousness. One may date the beginning of this scientific program with the influential publication of Crick and Koch (1995) who proposed, mainly on the basis of neuro-anatomical data, that neural activity in area V1 does not contribute to the content of our phenomenal consciousness.

Blindsight: highlighting the role of visual cortex

Somepatients affected by visual scotoma secondary to primary visual cortex lesions display striking dissociations when presented with visual stimuli at the location of their scotoma. While claiming to have no conscious perception of these stimuli, they perform better than chance on forced-choice visual and visuo-motor tasks such as stimulus discrimination, stimulus detection, or orientation to stimulus spatial source by visual saccades. This phenomenon, discovered in the early seventies (Poppel et al., 1973; Weiskrantz et al., 1974; Perenin and Jeannerod, 1975), has been coined "blindsight" by Weiskrantz. Compelling evidence supports the idea that such unconscious perceptual processes are subserved by the activity of subcortical visual pathways including the superior colliculus, and by-pass the primary visual cortex (Cowey and Stoerig, 1991). In a recent study, de Gelder and Weiskrantz enlarged the range of unconscious perceptual processes accessible to blindsight patients by showing that patient G.Y., whose fame is comparable to that of patient H.M. in the field of medial temporal lobe amnesia, was able to discriminate better than chance emotional facial expressions on forced-choice tasks (de Gelder et al., 1999). Taking advantage of this behavioral result, the authors used functional magnetic resonance imaging (fMRI) to demonstrate that this affective blindsight performance is correlated with the activity in an extra-geniculo-striate colliculo-thalamo-amygdala pathway independently of both the striate cortex and fusiform face area located in the ventral pathway (Morris et al., 2001). In fact, this unconscious visual process discovered in blindsight subjects is also active in healthy human subjects free of any visual cortex lesions. One way to observe it consists of using paradigms of masked or "subliminal" visual stimulation in which a stimulus is briefly flashed foveally for tens of milliseconds, it is then immediately followed by a second stimulus, suppressing conscious perception of the former. Whalen et al. used such a paradigm to mask a first fearful or neutral face presented during 33 ms by a second neutral face presented for a longer duration (167 ms). While subjects did not consciously perceive the first masked face, fMRI revealed an increase of neural activity in the amygdala on masked fearful face trials compared to masked neutral face trials (Whalen et al., 1998). This interesting result has been replicated and enriched by a set of elegant studies conducted by Morris et al. (1998, 1999).

The blindsight model and its extension in healthy subjects via visual masking procedures underlines the importance of the neocortex in conscious visual processing by revealing that a subcortical pathway is able to process visual information in the absence of phenomenal consciousness. In other words, these recent data are in close agreement with Hughlings Jackson's (1932) hierarchical conception (formulated in particular in the 3rd and 4th principles of his "Croonian lectures on the evolution and dissolution of the nervous system") that attributes the more complex cognitive processes, including consciousness, to the activity of neocortex. Nevertheless, should we generalize the importance shown here for the primary visual cortex — the integrity of which seems to be a pre-requisite for visual consciousness — to the whole visual cortex?

Visual form agnosia, optic ataxia and visual hallucinations: the key role of the ventral pathway

As a result of the seminal work of Ungerleider and Mishkin (1982), visual cortex anatomy is considered to be composed of two parallel and interconnected pathways supplied by primary visual cortex area V1: the occipito-temporal or "ventral" pathway and the occipito-parietal or "dorsal" pathway. The dorsal pathway mainly subserves visuomotor transformations (Andersen, 1997), while the ventral pathway neurons represent information from low-level features to more and more abstract stages of identity processing, thus subserving object identification. This "what pathway" is organized according to a posterior-anterior gradient of abstraction, the most anterior neurons located in infero-temporal cortex coding for object-based representations free from physical parameters such as retinal position, object size or orientation (Lueschow et al., 1994; Ito et al., 1995; Grill-Spector et al., 1999; Cohen et al., 2000). Goodale and Milner reported a puzzling dissociation in patient D.F. suffering from severe visual form agnosia due to carbon monoxide poisoning (Goodale et al., 1991). As initially defined by Benson and Greenberg (1969), this patient not only had great difficulties in recognizing and identifying common objects, but she was also unable to discriminate even simple geometric forms and line orientations. Anatomically, bilateral ventral visual pathways were extensively lesioned, while primary visual cortices and dorsal visual pathways were spared. Goodale and Milner presented this patient with a custom "mail-box," the slot of which could be rotated in a vertical plane. When asked to report slot orientation verbally or manually patient D.F. performed at chance-level, thus confirming her persistent visual agnosia. However, when asked to post a letter into this slot she unexpectedly performed almost perfectly, while still unable to report slot orientation consciously. This spectacular observation demonstrates how spared dorsal pathway involved in visuo-motor transformations was still processing visual information but without contributing to patient D.F.'s phenomenal conscious content. This case suggests that some representations elaborated in this "how pathway" are operating unconsciously while the ventral pathway activity subserves our phenomenal visual consciousness. Since this influential paper, many studies have tested this hypothesis in healthy subjects using visual illusions (Aglioti et al., 1995; Gentilucci et al., 1996; Daprati and Gentilucci, 1997). For instance, Aglioti et al. (1995) engaged subjects in a Titchener–Ebinghaus circles illusion task in which a given circle surrounded by larger circles appears smaller than the very same circle surrounded by smaller circles (see Fig. 2). While subjects consciously reported this cognitively impenetrable illusion, when asked to grip the central circle, online measures of their thumb-index distance showed that their visuo-motor response was free of the perceptual illusion and adapted to the objective size of the circle.[1]

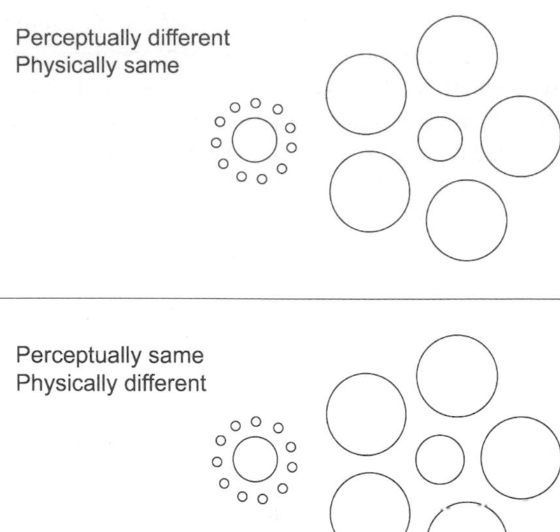

Fig. 2. Size-contrast illusions (i.e., Titchener circles illusion) deceive the eye but not the hand. Upper row panel: The standard version of the illusion. The target circles in the center of the two arrays appear to be different in size even though they are physically identical. For most people, the circle in the annulus of smaller circles appears to be larger than the circle in the annulus of larger circles. Lower row panel: A version of the illusion in which the target circle in the array of larger circles has been made physically larger than the other target circle. The two target circles should now appear to be perceptually equivalent in size. Note that when asked to grip the central circle, Aglioti et al.'s (1995) online measures of thumb-index distances revealed that visuo-motor response is free of the perceptual illusion and is correctly estimating the objective size of the circle. Automatic and metrically accurate calibrations required for skilled actions hence seem mediated by visual processes that are separate from those mediating our conscious experiential perception.

[1]Since these first reports, Franz and colleagues (Franz et al., 2000; Franz, 2001) challenged this interpretation by showing that when task difficulty was equated between perceptual and grasping tasks, action was not resisting to the illusion. However, recent studies taking into account these possible confounds reproduced the dissociation between perceptual and action performances (for a detailed review see Kwok and Braddick, 2003).

An inverse dissociation supporting the same general principle has been recently reported by Pisella et al. (2000) who demonstrated the existence of an unconscious "automatic pilot" located in the dorsal pathway. Their patient I.G. presented important stroke lesions affecting both dorsal pathways, while sparing primary visual cortices and ventral pathways. They designed a subtle task manipulating online motor corrections of pointing movements on a tactile screen on which visual targets appeared and these could unexpectedly jump from one position to another. While normal subjects were capable of extremely fast and automatic visuo-motor corrections in this task, patient I.G. could only rely on very slow strategic and conscious corrections. Crucially, when tested in a more complex condition in which subjects had to inhibit an initiated pointing correction on some trials, patient I.G. committed far less errors than the controls who were unable to inhibit very fast motor corrections and who reported of being astonished by their own uncontrollable behavior.

Taken together, these results are currently interpreted as dissociations between visuo-motor processes subserved by the activity of the dorsal visual pathway, the computations of which do not participate to our phenomenal consciousness,[2] and other visual processes relying on ventral pathway activity which supplies our conscious perception. The strong version of this theoretical position is defended in particular by authors such as Goodale and Milner. The latter claimed for instance that "we have two (largely) separate visual systems. One of them is dedicated to the rapid and accurate guidance of our movements..., and yet it lies outside the realm of our conscious visual awareness. The other seems to provide our perceptual phenomenology,..." (Milner 1998). Additional data originating from behavioral measures of subliminal priming, and functional brain-imaging data support this thesis (Bar and Biederman, 1999; Bar et al., 2001).

Lastly, a recent functional brain-imaging study of consciously reportable visual hallucinations observed in patients with Charles–Bonnet syndrome[3] reinforces this conception, by revealing correlations between color, face, texture and object hallucinations, and increased levels of cerebral blood flow in the corresponding specialized visual areas located in the ventral visual pathway (Ffytche et al., 1998).

Unilateral spatial neglect: the necessity of attentional allocation

The recent proposal of a cerebral substrate of visual consciousness through the distinction drawn between dorsal ("unconscious") and ventral ("conscious") pathways still bears some similarity to Jackson's conception since it relies on a similar anatomical partition between some sectors of the visual system, which would supply the flow of our phenomenal consciousness, and other sectors, which would process information out of our conscious awareness. However, we may posit a further question: Does visual information represented in the ventral pathway depend on some additional conditions to be consciously accessible and reportable? In other words, are we necessarily conscious of all visual information represented in the ventral pathway? A key answer to this question comes from unilateral spatial neglect (USN), a very frequent neuropsychological syndrome clinically characterized by the inability to perceive or respond to stimuli presented to the side contralateral to the site of the lesion, despite the absence of significant sensory or motor deficits. USN has two interesting characteristics: Firstly, most USN patients display impaired visual phenomenal consciousness for objects located on their left

[2]Area MT or V5, located within the dorsal pathway, is an important exception to this principle because : (1) its activity correlates directly with conscious reports of genuine or illusory visual motion (Tootell et al., 1995), (2) when lesioned (Zeki, 1991) or transiently inactivated by trans-cranial magnetic stimulation (Beckers and Homberg, 1992) it results in akinetopsia (i.e., the inability to report visual motion), and (3) microstimulation within this area influences motion orientation discrimination in monkeys (Salzman et al., 1990).

[3]This syndrome is characterized by visual hallucinations in people who have a sudden change in vision. Charles Bonnet, a Swiss philosopher, first described this condition in the 1760s when he noticed his grandfather, who was blinded by cataract, described seeing birds and buildings which were not there. It was later defined as "persistent or recurrent visual pseudo-hallucinatory phenomena of a pleasant or neutral nature in a clear state of consciousness "(Damas-Mora et al., 1982)".

side.[4] Some neglect patients even present a very pure symptom called "visual extinction," defined by the specific loss of phenomenal consciousness for left-sided stimuli presented in competition with right-sided stimuli, while the same left-sided stimuli presented in the absence of contralateral competing stimuli are available to conscious report. Secondly, USN syndrome is usually observed with lesions affecting the spatial attentional network — most often right parietal and/or superior temporal gyrus (Karnath et al., 2001) cortices but also right thalamic or right frontal lesions — sparing the primary visual cortex and the whole ventral visual pathway. Recent behavioral and functional brain-imaging studies have reliably shown that this spared visual ventral pathway still represents the neglected visual information at multiple levels of processing that culminates in highly abstract forms of coding (McGlinchey-Berroth, 1997; Driver and Mattingley, 1998; Driver and Vuilleumier, 2001). For instance, McGlinchey-Berroth et al. (1993) demonstrated that left-sided neglected object pictures could be represented up to a semantic stage, as revealed by significant behavioral priming effects on the subsequent processing of consciously perceived semantically related words. More recently, Rees et al. (2000) have shown that an unconsciously perceived extinguished visual stimulus still activates corresponding retinotopic regions of primary visual cortex and several extra-striate ventral pathway areas.

These results demonstrate that ventral pathway activation constitutes a necessary but not sufficient condition to perceive consciously visual stimuli. The additional mechanism, defective in USN patients and mandatory to conscious perception, seems to be the top-down attentional amplification supplied by the activity of the spatial attention network (Mesulam, 1981).

Recently we have been able to generalize this principle demonstrated by USN patients to healthy subjects, by investigating neural correlates of unconsciously perceived words using a visual-masking procedure (Dehaene et al., 2001). Using both fMRI and event related potential (ERP) recordings we observed significant activations of the left ventral pathway — the visual word form area, previously identified as the first non-retinotopic area responding to letter string stimuli (Cohen et al., 2000) — by unconsciously perceived masked words. In a second experiment, we tested the specificity of these activations by using a masked priming paradigm: on each trial subjects consciously perceived a target word and classified it either as man-made or as a natural object. Subjects responded faster to visible words immediately preceded by the same masked word (e.g., table/table) than to different prime-target pairs (e.g., radio/table). This repetition priming effect was correlated to specific reductions of the blood oxygen level dependent (BOLD) signal in the visual word form area on repeated word trials compared to non-repeated word trials. This repetition suppression effect is strongly suggestive of the activation of common neurons sharing the same response tuning properties as by unconsciously perceived masked words and by unmasked words (Naccache and Dehaene, 2001a).

This work enabled us to compare brain activations elicited by briefly (29 ms) flashed words depending on whether it was consciously perceived or not. During masked trials a backward mask suppressed conscious perception of the word, while words that were flashed for the very same duration but were not backward masked were consciously perceived and reported. When consciously perceiving a word, corresponding neural activity is hugely amplified and temporally sustained in the ventral visual pathway in comparison with the neural activity elicited by masked words. Moreover, conscious perception is systematically accompanied by the co-activation of a long-range distributed network, the epicenters of which involve prefrontal, anterior cingulate, and parietal cortices.

Source and effects of top-down attentional effects: attention is not consciousness

The crucial role of top-down attentional amplification on the perceptual fate of stimuli is likely to

[4] An exact definition of "left side" remains the subject of many investigations, as visual neglect has been reliably observed at several distinct spatial frames of reference such as different subject-centered or "egocentric" frames, and multiple environment or object-centered "allocentric" frames (Mesulam, 1999).

occur recursively at multiple stages of processing all along the ventral visual pathway. This allows large modulations of activation patterns elicited by the same stimulus according to the task presently performed. The rich plasticity of visual representations observed in conscious strategical processing leads to the following question: Are unconscious visual representations impermeable to such top-down effects? Indeed, in most current theories of human cognition, unconscious processes are considered as automatic processes that do not require attention (Posner and Snyder, 1975; Schneider and Shiffrin, 1977; Eysenck, 1984).

Kentridge et al. (1999, 2004) recently questioned this conception by testing the efficacy of several visual cues on the forced-detection of targets in the hemianopic scotoma of the blindsight patient GY. They found that a central, consciously perceived arrow pointing toward the region of the scotoma where the target would appear could enhance GY's performance, although the target remained inaccessible to conscious report.[5] In normal subjects, using a visual-masking procedure, Lachter et al. (in press) recently reported that unconscious repetition priming in a lexical decision task occurred only if the masked primes appeared at spatially attended locations.

We also investigated a similar issue related to the impact of temporal attention on visual masked priming effects (Naccache et al., 2002). In previous studies, we have shown that masked numerical primes can be processed all the way up to quantity coding (Naccache and Dehaene, 2001a, b) and motor response stages (Dehaene et al., 1998b). When subjects had to compare target numbers with a fixed reference of 5, they were faster when the prime and target numbers fell on the same side of 5, and therefore called for the same motor response, than when they did not (i.e., response congruity effect). They were also faster when the same number was repeated as prime and target (i.e., repetition priming effect). In three experiments manipulating target temporal expectancy, we were able to demonstrate that the occurrence of unconscious priming in a number comparison task is determined by the allocation of temporal attention to the time window during which the prime-target pair is presented. Both response-congruity priming and physical repetition priming totally vanish when temporal attention is focused away from this time window. We proposed that when subjects focus their attention on the predicted time of appearance of the target, they open a temporal window of attention for a few hundreds of milliseconds. This temporal attention then benefits unconscious primes that are presented temporally close to the targets.

Taken together, these findings are inconsistent with the concept of a purely automatic spreading of activation during masked priming and refute the view that unconscious cognitive processes are necessarily rigid and automatic. While several paradigms, such as inattentional blindness (Mack and Rock, 1998) or attentional blink (Raymond et al., 1992) suggest that conscious perception cannot occur without attention (Posner, 1994), our findings indicate that attention also has a determining impact on unconscious processing. Thus, attention cannot be identified with consciousness. One of the key criteria for automaticity is independence from top-down influences. However, these results suggest that, by this criterion, masked priming effects or unconscious blindsight effects cannot be considered as automatic. We propose that the definition of automaticity may have to be refined in order to separate the source of conscious strategic control from its effects. Processing of masked primes is automatic inasmuch as it cannot serve as a *source* of information for the subsequent definition of an explicit strategy (e.g., see Merikle et al., 1995). However, this does not imply that it is impermeable to the *effects* of top-down strategic control, for example originating from instructions and/or task context. As a matter of fact, I retrospectively found an explicit formulation of this principle by Daniel Kahneman and Anne Treisman (1984) 20 years ago:

> ...a dissociation between perception and consciousness is not necessarily

[5]This very elegant demonstration in patient GY will require further investigations in additional blindsight patients, given that GY's residual vision has been recently interpreted in terms of low-level phenomenal vision through a set of subtle experiments manipulating visual presentations in both the spared visual field and within the scotoma (Stoerig and Barth, 2001).

> equivalent to a dissociation between perception and attention. (...) To establish that the presentation is subliminal, the experimenter ensures that the subjective experience of a display that includes a word cannot be discriminated from the experience produced by the mask on its own. The mask, however, is focally attended. Any demonstration that an undetected aspect of an attended stimulus can be semantically encoded is theoretically important, but a proof of complete automaticity would require more. Specifically, the priming effects of a masked stimulus should be the same regardless of whether or not that stimulus is attended. (...). These predictions have yet to be tested.

Four principles accounted by a theoretical sketch of consciousness

Thus far, our non-exhaustive review has allowed us to isolate four general principles governing the physiology of visual consciousness. Firstly, a large number of processes coded in some sectors of the visual system — such as the subcortical colliculus mediated pathway, or some areas of the dorsal visual pathway — never participate in conscious visual representations. Secondly, a visual representation is reportable only if coded by the visual ventral pathway. Thirdly, this anatomical constraint is necessary but clearly not sufficient, as is nicely demonstrated in visual neglect. Top-down attentional amplification seems to be the additional and necessary condition for a visual representation coded in the ventral pathway to reach conscious content. Finally, inspired by Posner's (1994) distinctions between the *source* and the *effects* of a top-down attentional process, we propose that only conscious representations can be used as sources of strategic top-down attention, while some unconscious representations are highly sensitive to the effects of such attention. These principles help to better delineate the properties of conscious visual perceptions, and also argue for a distinction between two categories of non-conscious processes: those that never contribute to conscious content, and those that can potentially contribute to it.

These principles can be accounted for within the "global neuronal workspace" theoretical framework developed by Dehaene and co-workers (Dehaene et al., 1998a; Dehaene and Naccache, 2001; Dehaene et al., 2003; also see Baars, this volume). This model, in part inspired from Bernard Baars' (1989) theory, proposes that at any given time many modular cerebral networks are active in parallel and process information in an unconscious manner. Information becomes conscious, however, if the corresponding neural population is mobilized by top-down attentional amplification into a self-sustained brain-scale state of coherent activity that involves many neurons distributed throughout the brain. The long-distance connectivity of these "workspace neurons" can, when they are active for a minimal duration, make the information available to a variety of processes including perceptual categorization, long-term memorization, evaluation, and intentional action. We postulate that this global availability of information through the workspace is what we subjectively experience as a conscious state. Neurophysiological, anatomical, and brain-imaging data strongly argue for a major role of prefrontal cortex, anterior cingulate, and the areas that connect to them, in creating the postulated brain-scale workspace.

Within this framework, the different unconscious visual processes reviewed in this paper can be distinguished and explained. The activity of subcortical visual processors such as the superior colliculus, which do not possess the reciprocal connections to this global neuronal workspace that are postulated to be necessary for top-down amplification, cannot access or contribute to our conscious content, as revealed by blindsight.[6] Moreover, the activity of other visual processors anatomically connected to this global workspace by reciprocal connections can still escape the content of consciousness due to top-down attentional

[6]Indeed neurons located in the superficial visual layers of superior colliculus receive direct input from parietal areas while projecting indirectly to intraparietal cortex through a thalamic synapse (Sparks, 1986; Clower et al., 2001).

failure. This "attentional failure" may result from a direct lesion of the attentional network (such as in USN), from stringent conditions of visual presentation (such as in visual masking), or even from the evanescence of some cortical visual representations too brief to allow top-down amplification processes (such as the parietal 'automatic pilot' revealed by optic ataxia patients).[7] This model also predicts that once a stream of processing is prepared consciously by the instructions and context, an unconscious stimulus may benefit from this conscious setting, and therefore show attentional amplification, such as in blindsight.

Conclusion

This theoretical sketch will of course necessitate further developments and revisions, but its set of predictions can be submitted to experimentation. For instance, this model predicts that a piece of unconscious information cannot itself be used as a source of control to modify a choice of processing steps. Another prediction is to extend the sensitivity of some blindsight effects to top-down attention to other paradigms or relevant clinical syndromes, such as USN, attentional blink, or inattentional blindness.

As a conclusion, I have tried in this chapter to describe how the observation of neurological patients has played a major role in the discovery of several important principles related to the neural bases of visual consciousness. However, this description is not written as a record of an heroic past era of brain sciences. Clinical neuropsychologists and their patients are not dinosaurs, and so we did not adopt here a "paleontologist attitude." On the contrary, this audacious neuropsychology of consciousness will provide us with exciting and unexpected observations, enabling us to tackle the most complex and enigmatic aspects of visual consciousness.

[7]Within the global workspace model only explicit — or active — neural representations coded in the firing of one or several neuronal assemblies are able to reach conscious content. Therefore, a third class of unconscious processes can be described, those resulting from the neural architecture (fiber lengths and connections, synapses, synaptic weights) in which information is not explicitly coded. This type of unconscious information is also postulated to never participate to conscious content.

Acknowledgments

I thank two anonymous referees and Dr. Anna Wilson for their constructive and helpful remarks.

References

Aglioti, S., DeSouza, J.F. and Goodale, M.A. (1995) Size-contrast illusions deceive the eye but not the hand. Curr. Biol., 5: 679–685.

Andersen, R.A. (1997) Multimodal integration for the representation of space in the posterior parietal cortex. Philos. Trans. R. Soc. Lond. B. Biol. Sci., 352: 1421–1428.

Baars, B.J. (1989) A Cognitive Theory of Consciousness. Cambridge University Press, Cambridge, MA.

Bar, M. and Biederman, I. (1999) Localizing the cortical region mediating visual awareness of object identity. P. Natl. Acad. Sci. USA, 96: 1790–1793.

Bar, M., Tootell, R.B., Schacter, D.L., Greve, D.N., Fischl, B., Mendola, J.D., Rosen, B.R. and Dale, A.M. (2001) Cortical mechanisms specific to explicit visual object recognition. Neuron, 29: 529–535.

Beckers, G. and Homberg, V. (1992) Cerebral visual motion blindness: transitory akinetopsia induced by transcranial magnetic stimulation of human area V5. Proc. R. Soc. Lond. B. Biol. Sci., 249(1325): 173–178.

Benson, D. and Greenberg, J. (1969) Visual form agnosia. A specific defect in visual discrimination. Arch. Neurol., 20: 82–89.

Block, N. (1995) On a confusion about the role of consciousness. Behav. Brain Sci., 18: 227–287.

Clower, D.M., West, R.A., Lynch, J.C. and Strick, P.L. (2001) The inferior parietal lobule is the target of output from the superior colliculus, hippocampus, and cerebellum. J. Neurosci., 21(16): 6283–6291.

Cohen, L., Dehaene, S., Naccache, L., Lehericy, S., Dehaene-Lambertz, G., Henaff, M.A. and Michel, F. (2000) The visual word form area: spatial and temporal characterization of an initial stage of reading in normal subjects and posterior split-brain patients. Brain, 123(Pt 2): 291–307.

Cowey, A. and Stoerig, P. (1991) The neurobiology of blindsight. Trends Neurosci., 14: 140–145.

Crick, F. and Koch, C. (1995) Are we aware of neural activity in primary visual cortex? Nature, 375: 121–123.

Damas-Mora, J., Skeleton-Robinson, M., et al. (1982) The Charles Bonnet syndrome in perspective. Psychol. Med., 12(2): 251–261.

Daprati, E. and Gentilucci, M. (1997) Grasping an illusion. Neuropsychologia, 35: 1577–1582.

de Gelder, B., Vroomen, J., Pourtois, G. and Weiskrantz, L. (1999) Non-conscious recognition of affect in the absence of striate cortex. Neuroreport, 10: 3759–3763.

Dehaene, S., Kerszberg, M. and Changeux, J.P. (1998a) A neuronal model of a global workspace in effortful cognitive tasks. P. Natl. Acad. Sci. USA, 95: 14529–14534.

Dehaene, S., Naccache, L., Le Clec, H.G., Koechlin, E., Mueller, M., Dehaene-Lambertz, G., van de Moortele, P.F. and Le Bihan, D. (1998b) Imaging unconscious semantic priming. Nature, 395: 597–600.

Dehaene, S. and Naccache, L. (2001) Towards a cognitive neuroscience of consciousness: basic evidence and a workspace framework. Cognition, 79: 1–37.

Dehaene, S., Naccache, L., Cohen, L., Bihan, D.L., Mangin, J.F., Poline, J.B. and Riviere, D. (2001) Cerebral mechanisms of word masking and unconscious repetition priming. Nat. Neurosci., 4: 752–758.

Dehaene, S., Sergent, C. and Changeux, J. (2003) A neuronal network model linking subjective reports and objective physiological data during conscious perception. P. Natl. Acad. Sci. USA, 100: 8520–8525.

Dennett, D.C. (1992) Consciousness Explained. Penguin, London.

Driver, J. and Mattingley, J.B. (1998) Parietal neglect and visual awareness. Nat. Neurosci., 1: 17–22.

Driver, J. and Vuilleumier, P. (2001) Perceptual awareness and its loss in unilateral neglect and extinction. Cognition, 79: 39–88.

Eysenck, M. (1984) Attention and performance limitations. In: Eysenck M.L. (Ed.), A Handbook of Cognitive Psychology. Erlbaum Assoc., Hillsdale, NJ, pp. 49–77.

Ffytche, D.H., Howard, R.J., Brammer, M.J., David, A., Woodruff, P. and Williams, S. (1998) The anatomy of conscious vision: an fMRI study of visual hallucinations. Nat. Neurosci., 1: 738–742.

Franz, V.H. (2001) Action does not resist visual illusions. Trends Cogn. Sci., 5: 457–459.

Franz, V.H., Gegenfurtner, K.R., Bulthoff, H.H. and Fahle, M. (2000) Grasping visual illusions: no evidence for a dissociation between perception and action. Psychol. Sci., 11: 20–25.

Frith, C., Perry, R. and Lumer, E. (1999) The neural correlates of conscious experience: an experimental framework. Trends Cogn. Sci., 3: 105–114.

Gentilucci, M., Chieffi, S., Deprati, E., Saetti, M.C. and Toni, I. (1996) Visual illusion and action. Neuropsychologia, 34: 369–376.

Goodale, M.A., Milner, A.D., Jakobson, L.S. and Carey, D.P. (1991) A neurological dissociation between perceiving objects and grasping them. Nature, 349: 154–156.

Grill-Spector, K., Kushnir, T., Edelman, S., Avidan, G., Itzchak, Y. and Malach, R. (1999) Differential processing of objects under various viewing conditions in the human lateral occipital complex. Neuron, 24: 187–203.

Holender, D. (1986) Semantic activation without conscious identification in dichotic listening parafoveal vision and visual masking: a survey and appraisal. Behav. Brain Sci., 9: 1–23.

Ito, M., Tamura, H., Fujita, I. and Tanaka, K. (1995) Size and position invariance of neuronal responses in monkey inferotemporal cortex. J. Neurophysiol., 73: 218–226.

Jackson, J.H. (1932) Selected Writings of John Hughlings Jackson. J. Taylor, London.

Kahneman, D. and Treisman, A. (1984) Changing views of attention and automaticity. In: Davies R., Beatty J. and Parasuraman R. (Eds.), Varieties of Attention. Academic Press, New York, pp. 29–61.

Karnath, H.O., Ferber, S. and Himmelbach, M. (2001) Spatial awareness is a function of the temporal not the posterior parietal lobe. Nature, 411: 950–953.

Kentridge, R.W., Heywood, C.A. and Weiskrantz, L. (1999) Attention without awareness in blindsight. Proc. R. Soc. Lond. B. Biol. Sci., 266: 1805–1811.

Kentridge, R.W., Heywood, C.A. and Weiskrantz, L. (2004) Spatial attention speeds discrimination without awareness in blindsight. Neuropsychologia, 42: 831–835.

Kwok, R.M. and Braddick, O.J. (2003) When does the Titchener Circles illusion exert an effect on grasping? Two- and three-dimensional targets. Neuropsychologia, 41: 932–940.

Lachter, J., Forster, K.I., et al. (2004) Forty-five years after broadbent (1958): still no identification without attention. Psychol. Rev., 114(4): 880–913.

Lueschow, A., Miller, E.K. and Desimone, R. (1994) Inferior temporal mechanisms for invariant object recognition. Cereb. Cortex, 4: 523–531.

Mack, A. and Rock, I. (1998) Inattentional Blindness. MIT Press, Cambridge, MA.

McGlinchey-Berroth, R. (1997) Visual information processing in hemispatial neglect. Trends Cogn. Sci., 1: 91–97.

McGlinchey-Berroth, R., Milberg, W.P., Verfaellie, M., Alexander, M. and Kilduff, P. (1993) Semantic priming in the neglected field: evidence from a lexical decision task. Cognitive Neuropsych., 10: 79–108.

Merikle, P.M., Joordens, S. and Stolz, J.A. (1995) Measuring the relative magnitude of unconscious influences. Conscious. Cogn., 4: 422–439.

Merikle, P.M., Smilek, D. and Eastwood, J.D. (2001) Perception without awareness: perspectives from cognitive psychology. Cognition, 79: 115–134.

Mesulam, M.M. (1981) A cortical network for directed attention and unilateral neglect. Ann. Neurol., 10: 309–315.

Mesulam, M.M. (1999) Spatial attention and neglect: parietal, frontal and cingulate contributions to the mental representation and attentional targeting of salient extrapersonal events. Philos. Trans. R. Soc. Lond. B. Biol. Sci., 354: 1325–1346.

Milner, A. (1998) Streams and consciousness: visual consciousness and the brain. Trends Cogn. Sci., 2(1): 25–30.

Morris, J.S., DeGelder, B., Weiskrantz, L. and Dolan, R.J. (2001) Differential extrageniculostriate and amygdala responses to presentation of emotional faces in a cortically blind field. Brain, 124: 1241–1252.

Morris, J.S., Öhman, A. and Dolan, R.J. (1998) Conscious and unconscious emotional learning in the human amygdala. Nature, 393: 467–470.

Morris, J.S., Ohman, A. and Dolan, R.J. (1999) A subcortical pathway to the right amygdala mediating "unseen" fear. P. Natl. Acad. Sci. USA, 96: 1680–1685.

Naccache, L., Blandin, E. and Dehaene, S. (2002) Unconscious masked priming depends on temporal attention. Psychol. Sci., 13: 416–424.

Naccache, L. and Dehaene, S. (2001a) The priming method: imaging unconscious repetition priming reveals an abstract representation of number in the parietal lobes. Cereb. Cortex, 11: 966–974.

Naccache, L. and Dehaene, S. (2001b) Unconscious semantic priming extends to novel unseen stimuli. Cognition, 80: 223–237.

Perenin, M.T. and Jeannerod, M. (1975) Residual vision in cortically blind hemiphields. Neuropsychologia, 13: 1–7.

Pisella, L., Grea, H., Tilikete, C., Vighetto, A., Desmurget, M., Rode, G., Boisson, D. and Rossetti, Y. (2000) An 'automatic pilot' for the hand in human posterior parietal cortex: toward reinterpreting optic ataxia. Nat. Neurosci., 3: 729–736.

Poppel, E., Held, R. and Frost, D. (1973) Residual visual function after brain wounds involving the central visual pathways in man. Nature, 243: 295–296.

Posner, M.I. (1994) Attention: the mechanisms of consciousness. P. Nat. Acad. Sci. USA, 91: 7398–7403.

Posner, M.I. and Snyder, C.R.R. (1975) Attention & cognitive control. In: Solso R. (Ed.), Information Processing & Cognition: The Loyola Symposium. S. R. L. L. Erlbaum, Hillsdale, NJ.

Ramachandran, V. and Blakeslee, S. (1998) Phantoms in the Brain: Probing the Mysteries of the Human Mind. William Morrow & Company, New York.

Raymond, J.E., Shapiro, K.L. and Arnell, K.M. (1992) Temporary suppression of visual processing in an RSVP task: an attentional blink? J. Exp. Psychol. Human, 18: 849–860.

Rees, G., Wojciulik, E., Clarke, K., Husain, M., Frith, C. and Driver, J. (2000) Unconscious activation of visual cortex in the damaged right hemisphere of a parietal patient with extinction. Brain, 123(Pt 8): 1624–1633.

Salzman, C.D., Britten, K.H. and Newsome, W.T. (1990) Cortical microstimulation influences perceptual judgements of motion direction. Nature, 346: 174–177.

Schneider, W. and Shiffrin, R.M. (1977) Controlled and automatic human information processing. Psychological review, 84: 1–66.

Sparks, D.L. (1986) Translation of sensory signals into commands for control of saccadic eye movements: role of primate superior colliculus. Physiol. Rev., 66(1): 118–171.

Stoerig, P. and Barth, E. (2001) Low-level phenomenal vision despite unilateral destruction of primary visual cortex. Conscious Cogn., 10(4): 574–587.

Tootell, R.B., Reppas, J.B., Dale, A.M., Look, R.B., Sereno, M.I., Malach, R., Brady, T.J. and Rosen, B.R. (1995) Visual motion aftereffect in human cortical area MT revealed by functional magnetic resonance imaging. Nature, 375: 139–141.

Ungerleider, L.G. and Mishkin, M. (1982) Two cortical visual systems. In: Ingle D.J., Goodale M.A. and Mansfield R.J. (Eds.), Analysis of Visual Behavior. MIT Press, Cambridge, MA, pp. 549–586.

Weiskrantz, L. (1997) Consciousness Lost and Found: A Neuropsychological Exploration. Oxford University Press, New York.

Weiskrantz, L., Warrington, E.K., Sanders, M.D. and Marshall, J. (1974) Visual capacity in the hemianopic field following a restricted occipital ablation. Brain, 97: 709–728.

Whalen, P.J., Rauch, S.L., Etcoff, N.L., McInerney, S.C., Lee, M.B. and Jenike, M.A. (1998) Masked presentations of emotional facial expressions modulate amygdala activity without explicit knowledge. J. Neurosci., 18: 411–418.

Zeki, S. (1991) Cerebral akinetopsia (visual motion blindness). A review. Brain, 114(Pt 2): 811–824.

S. Laureys (Ed.)
Progress in Brain Research, Vol. 150
ISSN 0079-6123

CHAPTER 14

The mental self

Hans C. Lou[1,*], Markus Nowak[2] and Troels W. Kjaer[3]

[1]*Department of Functionally Integrative Neuroscience, Aarhus University Hospital, Aarhus C, DK-8000, Denmark*
[2]*Department of Clinical Physiology, Bispebjerg Hospital, and Cyclotron and Positron Emission Tomography Center, Rigshospitalet, Copenhagen University Hospital, DK-2100 Copenhagen, Denmark*
[3]*Department of Clinical Neurophysiology, Copenhagen University Hospital, DK-2100 Copenhagen, Denmark*

Abstract: In meditation both the quality and the contents of consciousness may be voluntarily changed, making it an obvious target in the quest for the neural correlate of consciousness. Here we present the results of a positron emission tomography study of yoga nidra relaxation meditation when compared with the normal resting conscious state. Meditation is accompanied by a relatively increased perfusion in the sensory imagery system: hippocampus and sensory and higher order association regions, with decreased perfusion in the executive system: dorsolateral prefrontal cortex, anterior cingulate gyrus, striatum, thalamus, pons, and cerebellum. To identify regions active in both systems we performed a principal component analysis of the results. This separated the blood flow data into two groups of regions, explaining 25 and 18% of their variance: One group corresponded to the executive system, and the other to the systems supporting sensory imagery. A small group of regions contributed considerably to both networks: medial parietal and medial prefrontal cortices, together with the striatum. The inclusion of the striatum and our subsequent finding of increased striatal dopamine binding to D2 receptors during meditation suggested dopaminergic regulation of this circuit. We then investigated the neural networks supporting episodic retrieval of judgments of individuals with different degrees of self-relevance, in the decreasing order: self, best friend, and the Danish queen. We found that all conditions activated a medial prefrontal — precuneus/posterior cingulate cortex, thalamus, and cerebellum. This activation occurred together with the activation of the left lateral prefrontal/temporal cortex. The latter was dependent on the requirement of retrieval of semantic information, being most pronounced in the "queen" condition. Transcranial magnetic stimulation, targeting precuneus, was then applied to the medial parietal region to transiently disrupt the normal function of the circuitry. We found a decreased efficiency of retrieval of self-judgment compared to the judgment of best friend. This shows that the integrity of the function of precuneus is essential for self-reference, but not for reference to others.

Meditation and the medial core of consciousness

Meditation is a term covering a large variety of mental practices involving voluntary changes in the state and contents of consciousness (Ballantyne and Deva, 1990). It is a constituent of major religions such as Hinduism and Buddhism, particularly in its Zen form, and its variants are encountered in Christianity as well. An early expression of increased globalization after the second world war has been the grasp and expansion of these practices in western societies, with or without religious contexts.

Early physiological studies on its effects on physiological homeostasis appeared in the 1960s and 1970s. Particularly influential was a series of studies by Wallace (1970) on oxygen consumption,

*Corresponding author. Tel.: +45 45811476;
Fax: +45 33324240; E-mail: hl@ipm.hosp.dk

DOI: 10.1016/S0079-6123(05)50014-1

which led these investigators to conclude: "The physiological changes during meditation differ from those during sleep, hypnosis, auto-suggestion, and characterize a wakeful, hypometabolic physiologic state", with characteristic effects (i.e., increase in theta activity combined with essentially the preservation of alpha activity). This claim has since been corroborated by some researchers (Benson et al., 1974) and disclaimed by others (Pagano and Warrenburg, 1983). The reason for this variability of physiological data is undoubtedly the phenomenological variability of the condition.

In spite of the surging interest in studying the neural correlate of consciousness during the last decades, neuroimaging methods have only been used for meditation in a very few studies known to us. Again, the results have been heterogeneous, conceivably reflecting the differences in phenomenological contents involved in the meditation practices used. The first study was reported by Herzog et al. (1990) using [18F] fluorodeoxyglucose as a tracer to measure glucose metabolism during self-induced yoga relaxation meditation. This investigation failed to reveal a detailed regional differences but found a significant frontal-occipital ratio of cerebral metabolism, mainly due to a marked decrease in metabolism in the occipital and superior parietal lobes. With another attention demanding meditation technique, Newberg et al. (2001) also found increases in frontal lobe activity. This finding is in accordance with the increased activity in prefrontal cortex during attention-focusing tasks (e.g., Frith et al., 1991; Pardo et al., 1991).

In 1999, we reported an investigation with a different meditation technique, the voice-induced yoga nidra relaxation meditation (Lou et al., 1999). Here the participants listened to a tape guiding them through the meditation sequence, with loss of willful activity and low attention demand. Structured interviews confirmed that the participants experienced a reduced drive to act, and an increase in imagery during the meditation procedure (Kjaer et al., 2002a). We originally attempted to use this technique in order to study the neurological correlates of "passive" and "active" consciousness separately (Taylor, 1999). The activation pattern changed with the change in the focus of meditation: meditation on sensation of weight of limbs and other body parts, presumably related to "motor attention" was supported mainly by the supplementary motor area in the superior medial prefrontal cortex (Martin et al., 1996). Also parietal and occipital activation was noted here. Abstract sensation of "joy" was almost exclusively accompanied by the left hemisphere activation of parietal and superior temporal cortices when compared to the normal resting conscious state. Visual imagery induced strong activation of the lateral parietal and occipital lobe, with the sparing of the primary visual region (V1). These regions are very similar to the regions shown to be active in voluntary visual imagery (Kosslyn et al., 1993), during REM sleep and dreaming (Braun et al., 1998; Maquet et al., this volume). The anterior cingulate cortex did not show activation during our meditation technique. This probably reflects the experienced loss of motivation to act. Meditation of symbolic representation of the self was supported by bilateral parietal activity in accordance with the role attributed to these regions in representation of the physical self (Adair et al., 1995) and the mental self (Lou et al., 2004), and was not found during meditation on other items during the sequence. We found a strong bilateral hippocampal activation and activity in parietal and occipital association and primary sensory regions. Interaction between these regions is instrumental in forming a coherent memory (e.g., Amaral, 1987), a pre-requisite for imagery (e.g., Kosslyn et al., 1993).

Relaxation meditation was accompanied by a relative decrease in bilateral orbital and dorsolateral prefrontal, anterior cingulate, temporal, inferior and parietal cortices; caudate nucleus, thalamus, pons, and cerebellar vermis; hemispheres and structures associated with executive functions; and particularly for dorsolateral prefrontal cortices, with working memory (e.g., Mehta et al., 2000). Meditation, generally speaking, shifted cortical activity from prefrontal to posterior regions. Reduction of perfusion in prefrontal cortical regions, together with increase in proficiency of working memory is seen in dopaminergic activation with methylphenidate (Mehta et al., 2000). We, therefore, hypothesized that dopaminergic release could be responsible for the

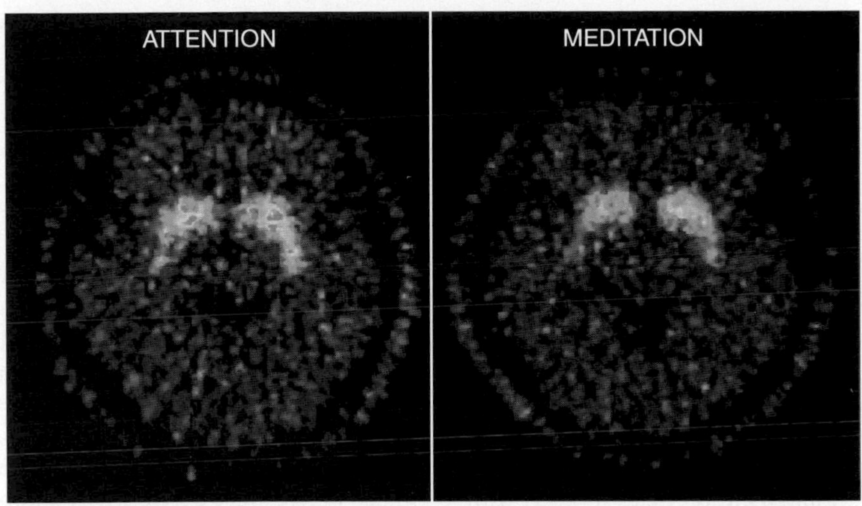

Fig. 1. The [11C] — raclopride binding potential images at the level of the striatum for one participant (no. 8) during attention to speech (left panel), and Yoga Nidra meditation (right panel). The reduced [11C] — raclopride binding potential in ventral striatum is evidence of increased endogenous dopamine release during meditation with focused consciousness. (from Kjaer et al., 2002a, b). See Plate 14.1 in Colour Plate Section.

more selective cortical activation in focused attention and in consciousness during meditation. The activation of dopaminergic neurons in substantia nigra and ventral tegmental area, and consequently, release of dopamine in the striatum, is thought to modulate the processing of glutamate input, serving as gating of sensorimotor and incentive motivational activity through cortico-subcortical loops (Horvitz, 2002). Such a role of dopamine could contribute to the selection of neural signals for emergence of consciousness. If that were the case, one would expect striatal dopamine release, mainly in the ventral striatum, during the vivid imaging and reduced intention to the act observed in yoga nidra meditation. We confirmed this prediction in a study of eight highly experienced yoga teachers, as judged by decreased striatal binding of the D2 receptor radioligand [11C] raclopride acting as a competitor to extracelluar dopamine (Fig. 1) (Kjaer et al., 2002a).

Another important finding of this study was revealed only after its publication. It concerned how these constituents of the neural correlate of consciousness interact. We re-analyzed the blood flow data from 29 regions in the above study with the principal component method. The first principal component explained 25% of the regional cerebral blood flow (rCBF) variability and consisted primarily of regions differentially active during the normal resting conscious state. The second component explained 18% of the variability and consisted primarily of posterior sensory and association regions differentially active during meditation. Three structures contributed markedly to both networks: right precuneus, middle prefrontal frontal gyri (Brodmann area 10), and the right striatum. We interpret this set of regions as a robust medial core of the resting conscious states (Kjaer and Lou, 2000). It is explored more systematically in the following section.

The medial core and self-representation

All subjective experience may be seen as self-conscious in the *weak* sense that there is something it feels like for the subject to have that experience. We may at times be self-conscious in a *deep* way, for example when we are engaged in figuring out who we are and what we are going to do with our lives, a distinctly human experience that gives organization, meaning, and structure to life. In its absence, our representation of ourselves and our world becomes kaleidoscopic and our life chaotic

(Flanagan, 1995). Such explicit "autonoetic consciousness" is thought to emerge by retrieval of the memory of personally experienced events (episodic memory) (Tulving, 1972; Gardiner, 2001). The cerebral activation pattern of episodic memory retrieval differs from that of semantic retrieval of common knowledge: for example, activation of medial parietal cortex is characteristic of the former, and of left lateral temporal lobe of the latter (e.g., Cabeza and Nyberg, 2000). Previously, a few other studies (e.g., Johnson et al., 2002 and Kircher et al., 2000) have used related retrieval methodology, but have not attempted to distinguish the mental self from other. To do this we compared regional cerebral blood flow changes during retrieval of judgment on three subjects with different degrees of self-relevance: Self, Best Friend, and the Danish Queen, using a non-memory loaded task with identical input and output as a control. In addition we used transcranial magnetic stimulation (TMS) (Keenan, 2001) over the medial parietal cortex to target precuneus and transiently disrupt its normal neural activity to see if such disruption would affect the task. We used a bi-conical coil, specifically designed to target medial neural structures approximately 3–6 cm below the scalp. This was ascertained for each subject during calibration by testing for selective motor response of feet and toes by stimulation near the vertex while targeting the medial primary motor (Brodmann area 4). This region is situated at approximately the same depth as the precuneus.

Right angular gyrus was differentially active in the Self-condition, and left inferior dorsolateral prefrontal cortex/ left lateral temporal cortex in the Queen condition (Fig. 2), with Best Friend condition in between (Fig. 3). Medial prefrontal and precuneus/posterior cingulate cortices, the pulvinar region of thalamus and right cerebellar hemisphere were active in all conditions.

Of these structures, the medial prefrontal cortex is the classical region involved in self-reference (Posner and Rothbart, 1998; Gusnard et al., 2001). Early lesion studies have shown deficient self-awareness and self-control in lesions of this structure (Stuss and Levine, 2002), and increased activity has been seen in first person reports of mental states like emotions, self-generated thoughts, and

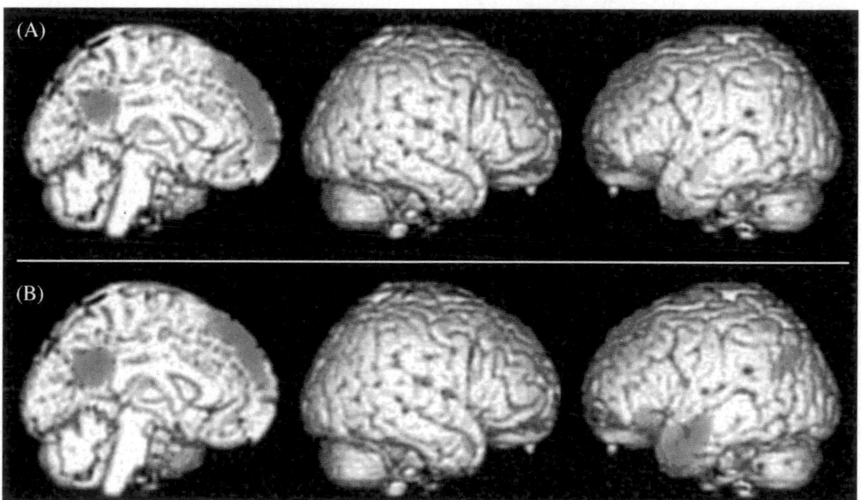

Fig. 2. rCBF distribution in retrieval of previous judgment of mental characteristics, compared with control state. (A) Emergence of self-representation. Differential activity is noted in medial prefrontal and parietal/posterior cingulated regions, together with bilateral occipital and parietal regions, and the confluent left inferior prefrontal and temporal region ($P<0.05$, corrected for multiple comparisons). (B) Emergence of representation of Other (i.e., Danish Queen). Activation of nearly similar regions. Note that the relative contributions of the above regions are, however, different: For Self, activity is comparatively higher in right parietal region and lower in left lateral temporal region. (from Lou et al., 2004, with permission; Copyright 2004, by the National Academy of Sciences.) See Plate 14.2 in Colour Plate Section.

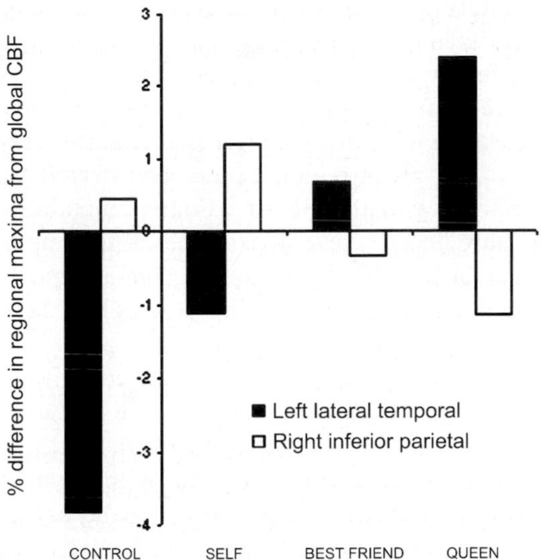

Fig. 3. rCBF differences (%) from global CBF in sites of peak activity for Self, Best Friend, and Queen. For right inferior parietal region, the site is the voxel of maximal activity during the Self condition (x, y, z: 44, −58, 38). The differences between Self vs. Best Friend and Self vs. Queen are both significant ($P = 0.021$ and 0.0008, respectively). For left medial temporal region, the site is the voxel of maximal activity during the Queen condition (x, y, z:−50, 2, −20). The differences between Self and Best Friend, Best Friend and Queen, and Self and Queen are all significant ($P = 0.014$, 0.02, and <0.0001, respectively). (from Lou et al., 2004, with permission. Copyright 2004, by the National Academy of Sciences.)

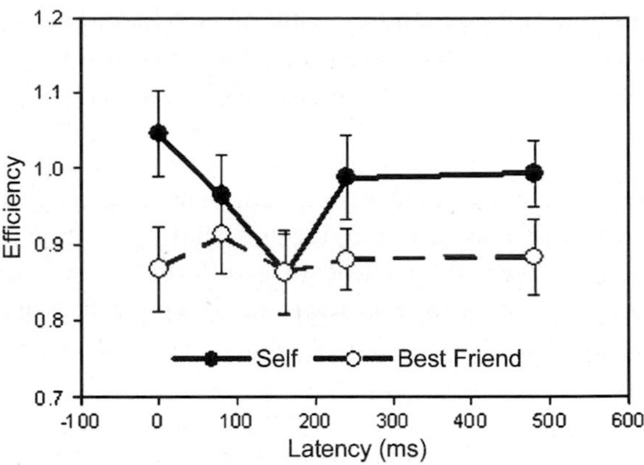

Fig. 4. Transcranial magnetic stimulation (TMS) targeted at precuneus/posterior cingulate region. Retrieval of self-judgment is less efficient with TMS at a latency of 160 ms than at a latency of 0 ms ($P = 0.003$), suggesting that neural activity at that time after stimulus presentation is particularly important for self-representation. This effect is not seen for retrieval of judgment of Best Friend. The difference between Self and Best Friend is significant ($P<0.05$). (from Lou et al., 2004, with permission; Copyright 2004, by the National Academy of Sciences.)

intentions to speak (Cabeza and Nyberg, 2000). In addition, "theory of the mind", or attributing mental states to others, are functions that have been associated with the activity in this region (Vogeley et al., 2001). Medial prefrontal cortex is also highly active during the resting state. It was already suggested by David Ingvar (1979) that this activity expressed a "rehearsal" or "simulation of behavior" while Frith and Frith (1999) concluded that dorsal medial prefrontal regions are concerned with explicit representations of states of the self.

The precuneus/posterior cingulate region is a key region in the retrieval of episodic memory (Cabeza and Nyberg, 2000). It has been suggested that it is involved in linking new information with prior knowledge (Maguire et al., 1999a). The role of the precuneus in the retrieval of episodic memory is more important than in episodic encoding (Kelley et al., 2002) and in semantic retrieval (Henson et al., 1999, Wiggs et al., 1999). TMS selectively impairs the retrieval of highly self-referential information selectively at this site with a latency of 160 ms (Fig. 4). This is evidence for a specific function of the precuneus in self-representation, in spite of our finding of similar hemodynamic activation during all the three tasks.

The right inferior parietal cortex was differentially active during retrieval of self-referential information. This finding came to us as a surprise. However, several studies provide circumstantial evidence for a role of the right inferior parietal region in self-representation: firstly, there seems to be a right hemisphere preference for self-recognition (Keenan et al., 2001) and, secondly, a number of recent studies on physical first-person perspective such as position in space, imagination of agency, and body representation have shown activation in the right inferior parietal region (Maguire et al., 1998; Maguire et al., 1999b, Vogeley and Fink, 2003). Lastly, illusory own-body perceptions have even been produced by direct electrical stimulation of the right inferior parietal cortex during surgical treatment for epilepsy (Blanke et al., 2002; Bünning and Blanke, this volume). These studies point to a role in for the right lateral parietal region in the representation of the physical Self. Our present results show that this is also the case for the mental Self. Hence, we conclude that the right inferior parietal cortex is selectively activated in self-representation in general. It should be noted, though, that the degree of activity was not significantly different from our simple control task and was solely apparent when compared to the memory-loaded retrieval of representations of the other.

The posterior thalamus, including medial pulvinar and the dorsomedial nucleus, is known to be particularly well connected by reciprocal, re-entrant connections with (para)limbic cortical regions, to the extent that the nuclei have been included in a so-called limbic thalamus. Lesions in this thalamic region have been associated with anamnesic syndromes (Mesulam, 2000). The cerebellum has more recently been shown to be active in temporal organization of attention (Allen et al., 1997; Wager et al., 2004) and working memory (Klingberg et al., 1995). Finally, a more general function of the cerebellum has been proposed in smoothing cognitive performance by maintaining it around a homeostatic baseline (Schmahmann, 1996). An increasing contribution of the network of semantic retrieval with decreasing self-reference was apparent in increasing the activity of the left lateral temporal region (Fig. 3). This region is activated not only by presentation of words, but also by pictures (Martin et al., 1995; Martin et al., 1996; Vandenberghe et al., 1996) and faces (Sergent et al., 1992; Gomo-Tempini et al., 1998), pointing to its involvement in higher level semantic processes that are independent of input modality.

The idea that not only medial anterior, but also medial posterior cortical regions are essential for subjectivity is a new development (Raichle et al., 2001; Gusnard and Raichle, 2001; Kjaer et al., 2002a, b; also see Vogt and Laureys, this volume). High activity in medial frontal and medial parietal regions was found "consistent with the continuity of a stable, unified perspective of the organism relative to the environment" (i.e., a "Self") (Gusnard and Raichle, 2001). The tonic activity decreases during engagement in non-self-referential goal-directed actions (i.e., "default mode") (Gusnard et al., 2001). Regional CBF depends on afferent function (i.e., all aspects of presynaptic and postsynaptic processing) but is independent of the efferent function (i.e., the spike rate of the same region) (Lauritzen, 2001). Even if this is not the case, conclusions on increased cognitive expediency of a given region cannot be inferred from increased regional CBF or oxidative metabolism. There are, in fact, several examples of the opposite (Mehta et al., 2000). Therefore, these important suggestions from regional CBF studies have lacked decisive experimental support.

This state of affairs has changed with the present evidence for an essential role of the precuneus in self-reference. It connects the right (and left) lateral inferior parietal cortex with the medial prefrontal cortex, already known to be essential for self-representation (Stuss and Levine, 2002). There are abundant anatomical connections between the precuneus/posterior cingulate region and the medial prefrontal/ anterior cingulate region (Cavada and Goldman-Rakic, 1989), and these regions are functionally integrated in reflective self-awareness (Kjaer et al., 2002a, b) and in the resting conscious state (Greicius et al., 2003).

References

Adair, K.C., Gilmore, R.L., Fennell, E.B., Gold, M. and Heilman, K.M. (1995) Anosognosia during intracarotid barbiturate anestesia: unawareness or amnesia for weakness. Neurology, 45: 241–243.

Allen, G., Buxton, R.B., Wong, E.C. and Courchesne, E. (1997) Attentional activation of the cerebellum independent of motor involvement. Science, 275: 1940–1943.

Amaral, D.G. (1987) Memory: Anatomical organisation of candidate brain regions. In: Mountcastle V.B., Bloom F.E. and Geiger S.R. (Eds.) Handbook of Physiology – The Nervous System, Vol. V. American Physiological Society, Bethesda, MD, pp. 211–294.

Ballantyne, J. R. and Deva, G. S. (1990) Yoga sutras of Panjali Delhi. Parimal, pp. 48–87.

Benson, H., Beary, J.R. and Carol, M.K. (1974) The relaxation response. Psychiatry, 3: 37–46.

Blanke, O., Ortigue, S., Landis, T. and Seeck, M. (2002) Stimulating illusory own-body perceptions. Nature, 149: 269–270.

Braun, A.R., Balkin, T.J., Wesensten, N.J., Gwadry, F., Carson, R.E., Varga, M., Baldwin, P. and Belenky. (1998) Dissociated patterns of activity in visual cortices and their projections during human rapid eye movement sleep. Science, 279: 9194.

Cabeza, R. and Nyberg, L. (2000) Imaging cognition II: An empirical review of 275 PET and fMRI studies. J. Cogn. Neurosci., 12: 1–47.

Cavada, C. and Goldman-Rakic, P.S. (1989). J. Comp. Neurol., 287: 422–445.

Flanagan, O. (1995) Consciousness Reconsidered. MIT Press, Cambridge Mass.

Frith, C.D., Friston, K. and Frackowiak, R.S.J. (1991) Willed action and the prefrontal cortex in man. A study with PET. Proc. Roy. Soc. of London, 244: 241–246.

Frith, C.D. and Frith, U. (1999) Interacting minds — a biological basis. Science, 286: 1692–1695.

Gardiner, J.M. (2001) Episodic memory and autonoetic consciousness: a first person approach. Phil. Trans. R. Soc. Lond., B 356: 1351–1361.

Gomo-Tempini, M.L., Price, C.J., Josephs, O., Vandenberghe, R., Cappa, S.F., Kapur, N. and Frackowiak, R.S. (1998) The neural systems sustaining face and proper-name processing. Brain, 121: 2103–2118.

Greicius, M.D., Krasnow, B., Reiss, A.L. and Menon, V. (2003) Functional connectivity of the resting brain: a network analysis of the default mode hypothesis. Proc. Natl. Acad. Sci. USA, 100: 253–258.

Gusnard, D.A. and Raichle, M.E. (2001) Searching for a baseline: functional imaging and the resting human brain. Nature Rev. Neurosci., 2: 685–694.

Gusnard, D.A., Akbudak, E., Shulman, G. and Raichle, M.E. (2001) Medial prefrontal cortex and self-referential mental activity: relation to a default mode of brain function. Proc. Natl. Acad. Sci. USA, 98: 4259–4264.

Henson, R.N., Rugg, M.D., Shallice, T., Josephs, O. and Dolan, R.J. (1999) A Recollection and familiarity in recognition memory: an event-related functional magnetic resonance study. J. Neurosci., 19: 3962–3972.

Herzog, H., Lele, U.R., Kuwert, T., Langen, K.J., Kops, E.R. and Feinendegen, L.E. (1990) Changed pattern of regional glucose metabolism during yoga meditative relaxation. Neuropsychobiology, 23: 182–187.

Horvitz, J.C. (2002) Dopamine gating of glutamatergic sensorimotor and incentive motivational input signals to the striatum. Behavioural Brain Res., 137: 65–74.

Ingvar, D.H. (1979) "Hyperfrontal" distribution of the cerebral grey matter flow in resting wakefulness, on the functional anatomy of the conscious state. Acta Neurol. Scand., 60: 12–25.

Johnson, S.C., Baxter, L.C., Wilder, L.S., Pine, J.G., Hiesemann, J.E. and Prigatano, G.P. (2002) Neural correlate of self-reflection. Brain, 125: 1808–1814.

Keenan, J.P., Nelson, A., O'Connor, M. and Pasqual-Leone, A. (2001) Self-recognition and the right hemisphere. Nature: 305–409.

Keenan, J.P. (2001) A thing well done. J. Consc. Stud., 3: 31–34.

Kelley, W.M., Macrae, C.N., Wyland, C.L., Caglar, S., Inati, S. and Heatherton, T.F. (2002) Finding the self? An event-related fMRI study. Remembering the past: two facets of episodic memory explored with positron emission tomography. J. Cogn. Neurosci., 14: 785–794.

Kircher, T.T., Senior, C., Phillips, M.L., Benson, P.J., Bullmore, E.T., Barummer, M., Simmonds, A., Williams, S.C.R., Bertels, M. and David, A.S. (2000) Towards a functional neuroanatomy of self-processing: effects of faces and words. Cogn. Brain Res., 10: 133–144.

Kjaer, T.W., Bertelsen, C., Piccini, P., Brooks, D., Alving, J. and Lou, H.C. (2002a) Increased dopamine tone during meditation induced change of consciousness. Cogn. Brain. Res., 13: 255–259.

Kjaer, T.W. and Lou, H.C. (2000) Interaction between precuneus and dorsolateral prefrontal cortex may play a unitary role in consciousness- a principal component analysis of rCBF. Conscious. Cogn., 9: S59.

Kjaer, T.W., Nowak, M. and Lou, H.C. (2002b) Reflective self-awareness and conscious states: PET evidence for a common midline parietofrontal core. Neuroimage, 17: 1080–1086.

Klingberg, T., Roland, P.E. and Kawashima, R. (1995) The neural correlates of the central executive function during working memory: A PET study. Hum. Brain Mapp. (Suppl. 1) 414.

Kosslyn, S.M., Albert, N.M., Thompson, W.L., Maljkovic, V., Weise, S.B., Chabris, C.F., Hamilton, S.E., Rauch, S.L. and Buanno, F.S. (1993) Visual mental imagery activates topographically organized visual cortex. J Cogn. neurosci., 5: 263–287.

Lauritzen, M. (2001) Relationship of spikes, synaptic activity, and local changes of cerebral blood flow. J. Cereb. Blood Flow Metab., 21: 1367–1383.

Lou, H.C., Kjaer, T.W., Friberg, L., Wioldschiodz, G., Holm, S. and Nowak, M. (1999) A $^{15}O-H_2O$ PET study of meditation and the resting state of normal consciousness. Hum. Brain Mapping, 7: 98–105.

Lou, H.C., Luber, B., Crupain, M., Keenan, J.P., Nowak, M., Kjaer, T.W., Sackeim, H.A. and Lisanby, S.H. (2004) Parietal cortex and representation of the mental self. Proc. Natl. Acad. Sci. USA, 101: 6827–6832.

Maguire, E.A., Burgess, N. and O'Keefe, J. (1999a) Human spatial navigation: cognitive maps, sexual dimorphism, and neural substrates. Curr. Opinion Neurobiol., 9: 171–177.

Maguire, E.A., Burgess, N., Donnett, J.G., Frackowiak, R.S., Frith, C.D. and O'Keefe, J. (1998) Knowing where and getting there. A human navigation network. Science, 280: 921–924.

Maguire, E.A., Frith, C.D. and Morris, R.G. (1999a) The functional neuroanatomy of comprehension and memory: the importance of prior knowledge. Brain, 122: 1839–1850.

Martin, A., Haxby, J.V., Lalonde, F.M., Wiggs, C. and Ungerleider, L.G. (1995) Discrete cortical regions associated with knowledge of color and knowledge of action. Science, 5233: 102–105.

Martin, A., Wiggs, C.L., Ungerleider, L.G. and Haxby, J.V. (1996) Neural correlates of category-specific knowledge. Nature, 379: 649–652.

Mehta, M.A., Owen, A.M., Sahakian, B.J., Mavaddat, N., Pickard, J.D. and Robbins, T.W. (2000). J. Neurosci., 20: 1–6.

Mesulam, M.-M. (2000) Principles of Behavioral and Cognitive Neurology (2nd ed.). Oxford University Press, New York, pp. 51–52, 73–74.

Newberg, A., Alavi, A., Baime, M., Pourdehnad, M., Santanna, J. and dÁquili, E. (2001) The measurement of regional cerebral blood flow during the complex cognitive task of meditation: a preliminary SPECT study. Psychiat. Res.: 113–122.

Pagano, R. R., and Warrenburg, S. (1983) Meditation in search of a unique effect. In Davidson R.J., Schwartz N.D. and Shapiro D. (Eds.), Consciousness and self-regulation Advances in Research and Theory. Vol. 3.

Pardo, J.V., Fox, P.T. and Raichle, M.E. (1991) Localization of a human system for sustained attention by positron emission tomography. Nature, 349: 61–64.

Posner, M.I. and Rothbart, M.K. (1998) Attention, self-regulation and consciousness. Phil. Trans. R. Soc. Lond. B, 353: 1915–1927.

Raichle, M.E., MacLeod, A.M., Snyder, A.Z., Powers, W.J., Gusnard, D.A. and Shulman, G.L. (2001) A default mode of brain function. Proc. Natl. Acad. Sci. USA, 98: 676–682.

Schmahmann, J.D. (1996) From movement to thought: Anatomic substrates of the cerebellar contribution to cognitive processing. Human Brain Mapping, 4: 174–198.

Sergent, J., Ohta, S. and MacDonald, B. (1992) Functional neuroanatomy of face and object processing: A positron emission tomography study. Brain, 115: 15–36.

Stuss, D.T. and Levine, B. (2002) Adult clinical neuropsychology: lessons from studies of the frontal lobes. Annu. Rev. Psychol., 53: 401–433.

Taylor, J. (1999) The Race for Consciousness. MIT press, Cambridge Mass, p. 30.

Tulving, E. (1972) In: Tulving E. and Donaldson W. (Eds.), Organization of Memory, Academic Press, New York, pp. 381–403.

Vandenberghe, R., Price, C., Wise, R., Josephs, O. and Frackowiak, R.S. (1996) Functional anatomy of a common semantic system for words and pictures. Nature, 383: 254–256.

Vogeley, K., Bussfeld, P., Newen, A., Herrman, S., Happe, F., Falkei, P., Maier, W., Shah, N.J., Fink, G.R. and Zilles, K. (2001) Mind reading: neural mechanisms of theory of mind and self-perspective. NeuroImage, 14: 170–181.

Vogeley, K. and Fink, G.R. (2003) Neural correlates of the first person perspective. Trends Cogn. Sci., 7: 38–42.

Wager, T.D., Rilling, J.K., Smith, E.E., Sokolik, A., Casey, K.L., Davidson, R.J., Kosslyn, S.M., Rose, R.M. and Cohen, J.D. (2004). Science, 303: 1157–1162.

Wallace, R.K. (1970) Physiological effects of trascendental meditation. Science, 167: 1751–1754.

Wiggs, C.L., Weisberg, J. and Martin, A. (1999) Neural correlates of semantic and episodic memory retrieval. Neuropsychologia, 37: 103–118.

S. Laureys (Ed.)
Progress in Brain Research, Vol. 150
ISSN 0079-6123

CHAPTER 15

Posterior cingulate, precuneal and retrosplenial cortices: cytology and components of the neural network correlates of consciousness

Brent A. Vogt[1,*] and Steven Laureys[2]

[1]Cingulum NeuroSciences Institute and SUNY Upstate Medical University Syracuse, NY 13210, USA
[2]Cyclotron Research Center and Department of Neurology, University of Liège, Sart Tilman B30 4000 Liège, Belgium

Abstract: Neuronal aggregates involved in conscious awareness are not evenly distributed throughout the CNS but comprise key components referred to as the neural network correlates of consciousness (NNCC). A critical node in this network is the posterior cingulate, precuneal, and retrosplenial cortices. The cytological and neurochemical composition of this region is reviewed in relation to the Brodmann map. This region has the highest level of cortical glucose metabolism and cytochrome *c* oxidase activity. Monkey studies suggest that the anterior thalamic projection likely drives retrosplenial and posterior cingulate cortex metabolism and that the midbrain projection to the anteroventral thalamic nucleus is a key coupling site between the brainstem system for arousal and cortical systems for cognitive processing and awareness. The pivotal role of the posterior cingulate, precuneal, and retrosplenial cortices in consciousness is demonstrated with posterior cingulate epilepsy cases, midcingulate lesions that de-afferent this region and are associated with unilateral sensory neglect, observations from stroke and vegetative state patients, alterations in blood flow during sleep, and the actions of general anesthetics. Since this region is critically involved in self reflection, it is not surprising that it is similarly a site for the NNCC. Interestingly, information processing during complex cognitive tasks and during aversive sensations such as pain induces efforts to terminate self reflection and result in decreased processing in posterior cingulate and precuneal cortices.

Introduction

Consciousness is a multifaceted concept that can be divided into two major components: the level of consciousness (i.e., arousal, wakefulness, or vigilance) and the content of consciousness (i.e., awareness of the environment and its relations to self) (see also Zeman, this volume). Although the entire forebrain is engaged to some extent in conscious information processing, there may be specific areas that are particularly crucial to consciousness. The midbrain reticular formation and its projections to the thalamic intralaminar nuclei establish and maintain wakefulness (Kinomura et al., 1996; Steriade, 1996), and postmortem assessments of the vegetative state show that damage to the dorsolateral midbrain and thalamus is a common feature in these cases (Kinney et al., 1994; Adams et al., 2000; Graham et al., this volume) and severe damage around the third ventricle that blocks this projection can produce coma (e.g., case 13, Malamud, 1967). The more difficult problem is the extent to which any forebrain region is necessary for conscious cognitive information processing.

*Corresponding author. Tel.: +1 315 280 6847;
Fax: +1 315 464 7712; E-mail: bvogt@twcny.rr.com

DOI: 10.1016/S0079-6123(05)50015-3

In one sense, the entire cerebral cortex is necessary for awareness of numerous sensory and motor events. In this context, the null hypothesis states that no part of the forebrain plays a disproportionate role in consciousness or information processing during wakefulness. According to the null hypothesis, the neural correlates of consciousness are fully distributed throughout the forebrain. We will attempt to reject the null hypothesis by showing first that parts of the cingulate gyrus are necessary for conscious experience, although they may not be sufficient, and second that there is a critical linkage between the brainstem arousal system and posterior cingulate (PCC) and retrosplenial (RSC) cortices that assures a close coupling between arousal and awareness. This view and its supporting facts can be extended to a few other regions that form a critical neural network correlates of consciousness (NNCC) that is both necessary and sufficient for conscious cognitive information processing.

We propose that PCC and precuneus cortex (PrCC) together are pivotal for conscious information processing. Support for this hypothesis is derived from postmortem assessments of epilepsy, stroke, monkey and human cingulotomy lesions; and vegetative state cases, functional imaging in epilepsy, amnesia, sleep states and general anesthesia, and tasks requiring consciousness for sensorimotor processing. We also emphasize a critical linkage between brainstem generated arousal and processing in the PCC/PrCC region and this dual function assures that PCC/PrCC is a key node in the NNCC.

Anatomical overview of posterior cingulate gyrus and precuneal cortex

Many observations are available from human imaging studies of the PCC and medial parietal lobe region; however, standardized software packages continue to use Brodmann's (1909) anatomical observations from a century ago. Fig. 1 provides a context for this region according to recent cytological observations. Fig. 1A is a co-registration of Brodmann's map with a postmortem case for which we have extensive histology. The histological case is a left hemisphere that was flipped horizontally to match the familiar Brodmann map. All numbers refer to our recent observations (Vogt et al., 2003, 2005) except for the hand drawn numbers for areas 31 and 23 that are original to Brodmann, as are the symbolic designations for each area. Four numbered arrows mark critical locations and issues about the cytological organization of the cingulate gyrus as follows: (1) The border between our area p24′ and area 23d is quite caudal to the original border that Brodmann placed between his areas 24 and 23, which was approximately at the vertical plane of the anterior commissure (VCA). (2) Area 31 does extend dorsally and caudally around area 23; however, we observe a dysgranular area rostral to the arrow and do not agree that area 31 with its very thick layer IV extends as far rostrally as Brodmann placed it. (3) The third arrow marked "7m" refers to the border between medial parietal area 7m and area 31 which we have identified slightly more ventral than Brodmann. Area 7m is PrCC and we use the terms interchangeably. (4) This is the level at which the section immunoreacted for neuron-specific nuclear binding protein was taken for higher magnification in Fig. 1B. This antibody stains mainly neurons, thus removing glial, vascular, and neuropil elements from the analysis and providing a more precise view of cortical lamination patterns and differences among two or more areas. Since Brodmann reconstructed his map onto the convoluted brain surface, he was unable to show the exact placement of RSC areas 29 and 30 in the callosal sulcus on the ventral bank of the cingulate gyrus. For this reason, he extended the retrosplenial areas caudally and ventrally, and in the process, he moved the "apparent" border of area 23 quite caudal on the caudomedial lobule. This reduction in the distribution of area 23 was transposed to the atlas of Talairach and Tournoux (1988) and has produced continuing confusion in the human functional imaging literature as to the true extent of area 23 and the location of RSC. This issue has been thoroughly considered in another context (Vogt et al., 2001).

The PCC comprises areas 23 and 31. These areas have well-differentiated layers IIIc, IV, and Va as seen in Fig. 1B. Only area 23d has a dysgranular layer IV that can form neuronal aggregates rather

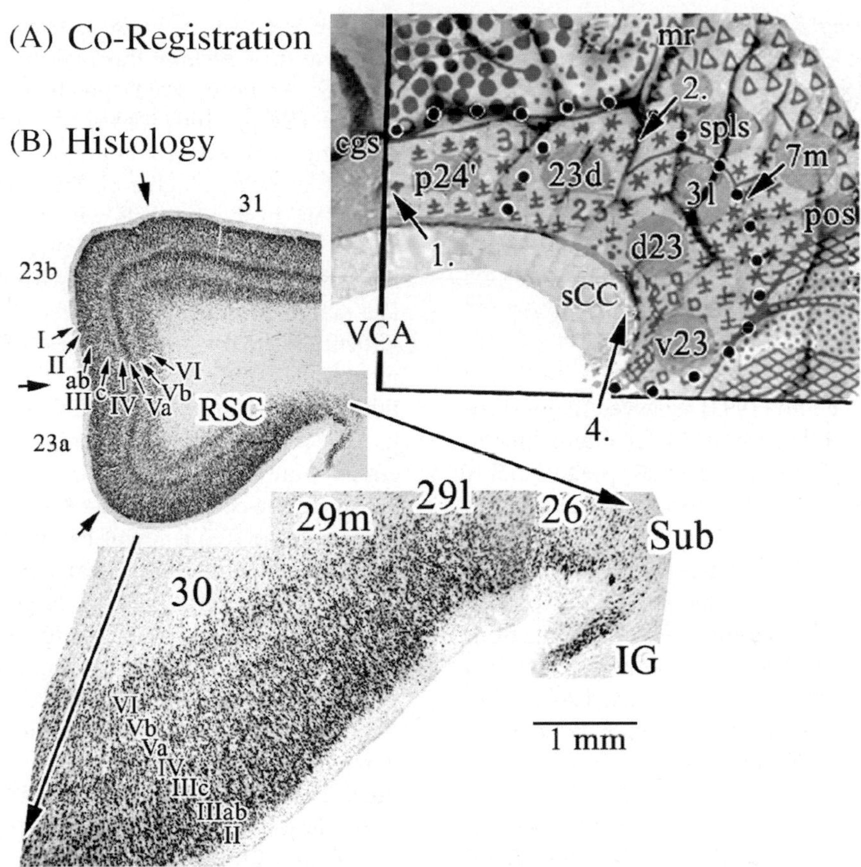

Fig. 1. (A) Co-registration of Brodmann's map of posterior cingulate and precuneal cortices and a postmortem case for joint histological assessment. Stroked dots outline the border of cingulate cortex and four numbered arrows refer to locations of particular Brodmann areas discussed in the text. Brodmann's original numbers 23 and 31 are shown handwritten in the co-registration, while all other numbers refer to Vogt et al. (2005). The arrow at 4 points to the level of the coronal histological section taken for (B) A critical issue is the manner in which Brodmann extended retrosplenial areas 29 and 30 onto the gyral surface where area 23 is located. Retrosplenial cortex comprises the ventral bank of the cingulate gyrus in the callosal sulcus and is not exposed as suggested in his map. See text for further explanations. Abbreviations: RSC, retrosplenial cortex; cgs, cingulate sulcus; IG, indusium griseum; mr, marginal ramus of the cgs; pos, parieto-occipital sulcus; sCC, splenium of the corpus callosum; Sub, dorsal subiculum; VCA, vertical plane at the anterior commissure.

than a single and continuous layer. Area 7m (not shown) has vertically oriented clumps of dendrites in layers III and IV, a thicker layer IV, and relatively larger layer IIIc neurons than those in layer Va (the opposite of cingulate architecture). In contrast to PCC and PrCC, the RSC has poorly differentiated layers IIIc, IV, and Va as shown in the pullout magnification of RSC in the figure. In fact, like area 23d, area 30 has a dysgranular layer IV, which is hard to observe in both photographs. Finally, area 26 is ectosplenial cortex and, along with the subiculum and indusium griseum form the fasciolate gyrus. Since no human in vivo imaging modality has the resolution necessary to identify these latter three areas, they are not considered further here.

Epilepsy, stroke and vegetative state

According to Mazars (1970), the first episode of a cingulate epilepsy discharge is associated with

> "*absences that are often mistaken for inattention. Twenty of our patients never*

had convulsive seizures and their 'absences' did not differ much from the short spells of loss of consciousness of 'petit mal'… the time to regain normal consciousness was longer, and a less abrupt recovery of consciousness was associated with an outburst of temper which could be so marked as to obscure the preceding 'absence.' Photic stimulation did not precipitate the absences but emotion did so effectively."

Devinsky and Luciano (1993) reviewed many studies of cingulate epilepsies and it is interesting to note that anterior cingulate seizures spread quickly to frontal cortex and are most often associated with motor arrest, staring and subsequent automatic behaviors, while seizures localizing to PCC took longer to spread to the dorsal convexities. This could reflect generally weaker connections with the superior frontal gyrus (Vogt and Pandya, 1987). Finally, Archer et al. (2003) used spike-triggered functional magnetic resonance imaging (fMRI) to evaluate alterations in cerebral activity in adult patients with idiopathic generalized epilepsy and frequent spike and slow-wave discharges. Four of their five subjects had significant reductions in PCC activation and they concluded that altered PCC activation may play a role in the "absence" features of generalized spike and slow-wave discharges. Thus, differences between anterior and posterior cingulate epilepsies suggest that abnormal discharges in PCC spread slowly to superior frontal areas and can be associated with a loss of consciousness.

Unilateral lesions in monkey midcingulate cortex produce a contralateral neglect as demonstrated with a somatosensory discrimination task (Watson et al., 1973). As shown below, such lesions de-afferent the PCC/PrCC and the neglect could reflect a selective loss of awareness of the contralateral sensory space. Interestingly, topographic disorientation is produced by a large, right hemisphere lesion of RSC, PCC, and PrCC (Takahashi et al., 1997) and a unilateral, left hemispheric lesion of RSC and the fornix produced profound amnesia (Valenstein et al., 1987). In none of these latter cases, however, was there a loss of consciousness. This could be because the lesions were restricted to one hemisphere and/or because this region is part of a larger NNCC. A more recent positron emission tomography (PET) study, however, used voxel-based mapping methods to compare regional cerebral metabolic rates of glucose (rCMRGlu) measured with PET in five patients with permanent amnesia (three with chronic Wernicke–Korsakoff and two with postanoxia syndrome) with that of healthy age-matched subjects (Aupée et al., 2001). Each patient, regardless of the cause of the permanent retrograde amnesia, showed significant hypometabolism in bilateral PCC/RSC and left thalamus.

A frequent postmortem finding in traumatic vegetative state cases is diffuse axonal injury with focal lesions in the corpus callosum (Adams et al., 2000). Although the frequency of this finding may lead to speculation about its critical role in clinical outcomes (Kampfl et al., 1998), large splenial tumors (Yamamoto et al., 1990; Rudge and Warrington, 1991) and large unilateral cingulate tumors that greatly deform this region (Malamud, 1967) fail to alter consciousness. In view of these latter findings, it is doubtful that the vegetative state is a consequence of splenial and PCC damage. This does not mean, however, that PCC/PrCC are not relevant to conscious information processing; once again, functional lesion sizes and locations are difficult to interpret. Activity in the medial parietal and adjacent PCC also differentiates minimally conscious from vegetative patients (Laureys et al., 2003).

Cerebral metabolism

In humans, the highest level of cortical glucose metabolism occurs in PCC and RSC (Andreasen et al., 1995; Maquet et al., 1997; Minoshima et al., 1997; Gusnard and Raichle, 2001; Laureys et al., 2004). The highest level of basal glucose metabolism in the monkey brain is in RSC and the anterior thalamus and these levels are elevated during performance of a delayed-response task (Matsunami et al., 1989). We have discussed the close laminar association of both high levels of glucose metabolism determined with the 2-deoxy-D-glucose method and anterior thalamic projections assessed with tritiated-amino acid injections (Vogt et al., 1997). Figs. 2A and B

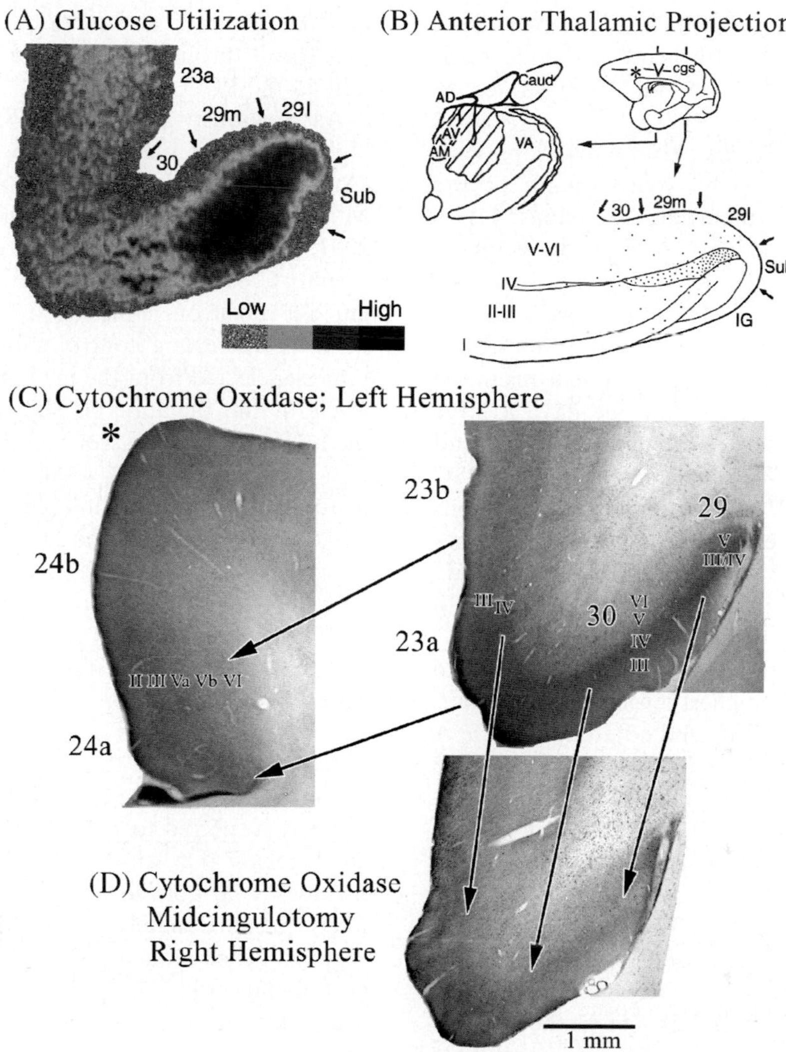

Fig. 2. Morphological context of metabolic activity in posterior cingulate cortex (PCC) in monkeys. 2-deoxy-D-glucose utilization coded for four levels of utilization (A) and thalamic projections to retrosplenial cortex (RSC) shown with a tritiated-amino acid injection (hatched) into the anterior thalamic nuclei and a coronal section through RSC areas 29 and 30 (B). The close relationship between high glucose metabolism and thalamic afferents are obvious. Interestingly, high levels of the mitochondrial enzyme cytochrome *c* oxidase also occur in the granular layer of RSC and in layers III–IV of areas 30 and 23. The asterisk in B shows where the section through ACC was taken in C. Notice that ACC has much less cytochrome *c* oxidase activity than area 23 (shown with the pair of arrows delineating these areas) (C). A midcingulotomy lesion (D) at coronal level (shown with "v" on medial surface in B) that removes thalamic afferents to PCC/RSC as well as frontal lobe inputs shows a massive reduction of activity in the thalamoreceptive layers as predicted from selective thalamic lesions in rat. There is about a 20% volumetric reduction in the posterior cingulate gyrus and reductions in enzyme activity are emphasized with three arrows from layer III/IV in area 29 and layers III and IV in areas 30 and 23c. Thus, high metabolic activity in PCC, RSC, and precuneus is driven primarily by thalamic afferents.

repeat these latter findings and provide the context of high cytochrome oxidase activity (Figures 2C and D). Thalamic termination is greatest to the granular layer of area 29 (undifferentiated layer III/IV) and layers III–IV of area 30. Projections to area 23 arise mainly from the medial pulvinar and lateroposterior nuclei and these were not included in the amino acid injection shown in Fig. 2B; hence thalamocortical connections to area 23 are not demonstrated in this case.

High basal glucose metabolism overlaps with these latter layers on the ventral bank of the cingulate gyrus. Cytochrome *c* oxidase is a mitochondrial enzyme involved in oxidative phosphorylation and the density of activity generated by this enzyme is related to the density of mitochondria in a layer (Carroll and Wong-Riley, 1984). Axon terminals arising from neurons in the anterior thalamic nuclei in the rat have a large diameter, are very dense with mitochondria, and they form asymmetric, excitatory synapses in RSC (Vogt et al., 1981). Moreover, thalamic lesions greatly reduce this enzyme's activity in thalamoreceptive layers I–IV of RSC (Van Groen et al., 1993) suggesting that thalamic driving is critical to metabolic processes in RSC.

The rodent findings are extended for monkey in Figs. 2C and D. The highest level of cytochrome *c* oxidase activity is in RSC and adjacent area 23. A section of area 24 is provided at the asterisk in B for comparative purposes and it shows that anterior cingulate cortex (ACC) has only about 40% of the cytochrome *c* oxidase activity of areas 23 and 30. Area 7m (not shown) has a level of activity similar to that of area 23b. This figure also shows the effects of a midcingulotomy lesion in the contralateral hemisphere on cytochrome *c* oxidase activity (Fig. 2D). Notice in this section the elevated gliosis in the white matter, general shrinkage of the ventral cingulate gyrus and massive reduction in cytochrome *c* oxidase activity. Reduced cytochrome activity is most prominent in the thalamorecipient layers III/IV of areas 29, 30, and 23 as emphasized with the arrows in the figure. Thus, midcingulotomy lesions de-afferent PCC and RSC and this includes sectioning of thalamic afferents that travel through the cingulum bundle to terminate in the PCC.

As previously stated, the PCC/PrCC has the highest level of metabolism in the human brain (Andreasen et al., 1995; Maquet et al., 1997; Minoshima et al., 1997; Gusnard and Raichle, 2001; Laureys et al., 2004). We re-evaluated regional rCMRGlu as assessed by ^{18}F-fluorodeoxyglucose positron emission tomography (FDG-PET) in 32 resting control subjects in regions of interest selected from the medial surface and three levels of the thalamus as shown in Table 1 (for methodological details of metabolic PET acquisition see Laureys et al. (2000) and for details on histological region of interest determination see Vogt et al. (2005)). Although the highest level of activity was in the PrCC, it was also high in PCC and the three samples from thalamus. The ACC and midcingulate cortex (MCC) were much lower and that in RSC was strikingly low. In view of the observations above for monkey, the latter observation for human RSC either reflects a substantial species difference or that the limits of PET resolution in defining RSC have been reached. The latter is suspected to reflect both because this is a small region that winds around the splenium and includes a variable amount of white matter that is likely proportionately greater than for the other regions, hence reducing the apparent level of glucose metabolism (i.e., partial volume effect bias). The slide-autoradiographic measure of glucose uptake in the monkey has a much greater precision and demonstrates laminar localizations that are not possible with human PET imaging. The most important conclusion from the histochemical analysis is that RSC likely has a similar or even higher level of glucose metabolism to PCC/PrCC and the RSC provides a critical link between the

Table 1. Regional cerebral metabolic rates for glucose in healthy volunteers

Cingulate cortex		Retrosplenial and precuneal Cortices		Thalamus	
ACC	7.8±2	RSC	5.4±2	Anterior	8.1±1.9
MCC	7.6±1.7	PrCC	10.4±2.4	Medial	8.9±2.4
PCC	9.3±2.1			Posterior	8.7±2.0

Note: Values are [^{18}F]fluorodeoxyglucose PET measurements by region of interest in 32 healthy volunteers in mg/100 g/min (mean ± SD). ACC, anterior cingulate cortex; MCC, midcingulate cortex; PCC, posterior cingulate cortex; RSC, retrosplenial cortex; PrCC, precuneus.

brainstem arousal system and cortical mechanisms of conscious awareness as discussed in the next section.

Analysis of rCMRGlu in the vegetative state has shown reduced activity in a large region of prefrontal cortex and a posterior, medial reduction in PCC/PrCC that might include RSC (for recent review see Laureys et al., 2004; Laureys, 2004). In spite of the large prefrontal reductions, the crucial connection with this region that was disrupted in vegetative state was in PCC (Laureys et al., 1999). As the relevant area overlaps the callosal sulcus, it is quite possible that the critically impaired connection is actually with RSC. Additionally, the thalamus also has reduced rCMRGlu in vegetative state (Schiff, this volume) and this further emphasizes the critical linkage between the thalamus and PCC/PrCC from brainstem arousal systems.

Anteroventral thalamic link between conscious wakefulness and cognitive awareness

Since the intralaminar thalamic nuclei are particularly important to wakefulness (Kinomura et al., 1996; Steriade, 1996), their projections might be expected to be a primary driver of information processing in critical parts of the NNCC. At first blush, this does not seem to be the case. Although PCC does receive a small projection from midline and intralaminar thalamic nuclei, the major projections of these nuclei are to MCC and ACC (Vogt et al., 1987). Although it is possible that a small projection from these nuclei might be effective in driving the entire NNCC, it is surprising that the rostral parts of the cingulate gyrus are not a critical part of this network based on the midline and intralaminar thalamic projections.

An alternative explanation suggests that the cholinergic mesopontine projections to the anteroventral thalamic nucleus are involved in cortical arousal and provide a linkage with conscious cognitive processing. The lateral tegmental reticular formation, parabrachial and laterodorsal tegmental nuclei contain cholinergic neurons that project to the anteroventral thalamic nucleus (Sofroniew et al., 1985). Electrical stimulation of these nuclei evokes cholinergically mediated, muscarinic, long-lasting excitatory potentials in the anteroventral thalamic nucleus that could mediate arousal (Curro et al., 1991). As noted above, the anteroventral nucleus has a prominent projection to RSC and this is an excitatory connection that likely mediates a high level of glucose metabolism in RSC. This need not be an isolated RSC circuit, however, because area 23 is heavily and reciprocally connected with RSC (Vogt and Pandya, 1987). Alteration in a prefrontal/RSC connection in vegetative state around the callosal sulcus (Laureys et al., 1999) suggests that this region mediates a pivotal link with arousal brainstem systems.

Thus, although there are two aspects to consciousness (arousal/wakefulness and awareness), they have a critical conjunction in the PCC and RSC region. This linkage between the two systems suggests that the posterior cingulate gyrus is one of the most important regions for awareness and is a pivotal player in the NNCC.

Sleep

A region critical to conscious information processing might be expected to be inactivated during sleep and area 31 of PCC has reduced regional cerebral blood flow (rCBF) during the rapid-eye movement (REM) phase of sleep (Maquet et al., 1996; also see Maquet et al., this volume). This does not prove that area 31 quiescence is pivotal REM sleep, however, it is supportive of that conclusion because there was also a massive bilateral prefrontal inactivation during REM. Fig. 3 provides measures of rCBF in PrCC and PCC during wakefulness and each stage of sleep and shows that, although REM sleep has a profound reduction in both regions, there is also a substantial reduction in blood flow in both regions during the preceding stages of sleep (for methodological details of PET sleep studies see Maquet et al., this volume). In view of the massive and reciprocal connections between PCC and prefrontal cortex, it is possible that PCC/PrCC is shut off by prefrontal cortex according to the above noted decreased functional connectivity between these regions in vegetative state as compared with healthy volunteers (Laureys et al., 1999). Thus, the PCC/PrCC

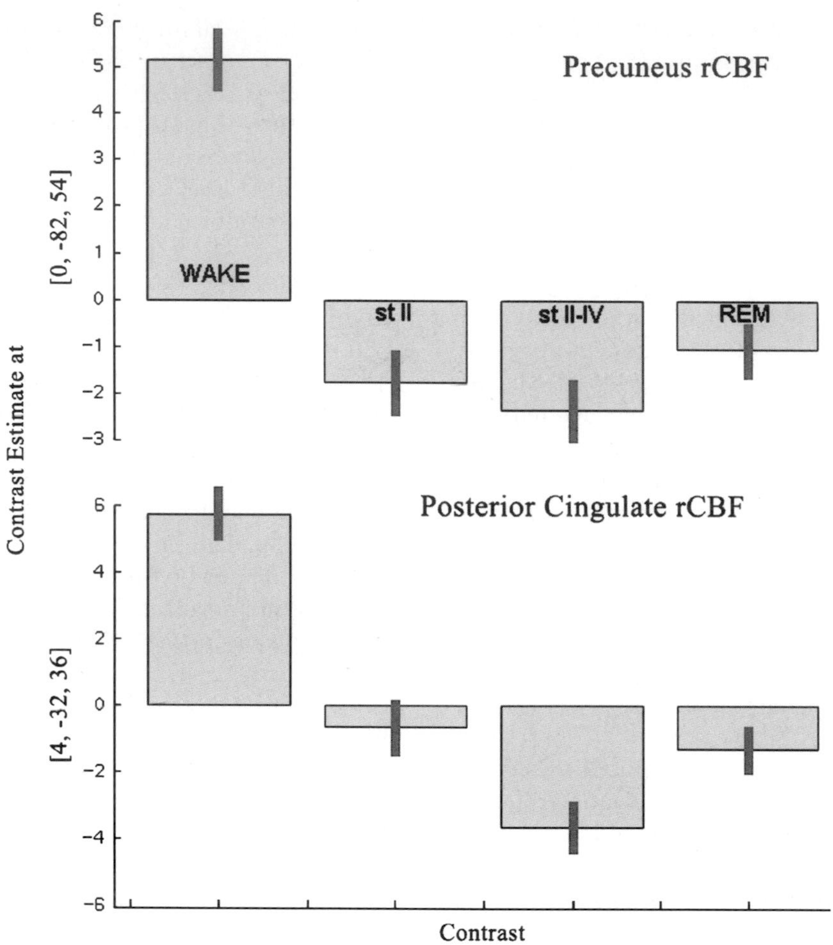

Fig. 3. Adjusted centered rCBF estimates in precuneal and posterior cingulate cortices during wakefulness and sleep: stage II (st II), slow wave (st III–IV), and rapid eye movement sleep (REM) (mean values and 90% confidence interval). The pattern is similar for both cortices with a high level of rCBF during wakefulness and reductions during sleep. Data reanalyzed from Maquet et al. (this volume).

decrease could be a primary or secondary event and it still suggests the relevance of this area in conscious processing. Additionally, thalamic involvement is probably not a mechanism of reducing area 31's blood flow, since the thalami have increased blood flow during REM (Maquet et al., 1996; this volume).

Anesthetic sensitivity

Although anesthetics have general mechanisms of action such as at the $GABA_A$ receptor and these likely reduce neuron activity throughout the brain, there are also particularly susceptible regions and modulation of their activity may be most closely associated with conscious functions (for review see Alkire, this volume; Fiset, this volume). In the rat, the tuberomamillary nucleus is particularly sensitive to anesthetics that act at $GABA_A$ receptors including propofol (Nelson et al., 2002). Although the rat has a well-differentiated RSC, it does not have areas 23, 31, or 7m (Vogt et al., 2004) and the tuberomamillary nucleus does not project to RSC in rat or monkey brain. Thus, other mechanisms must be considered in relation to anesthetic actions and human consciousness.

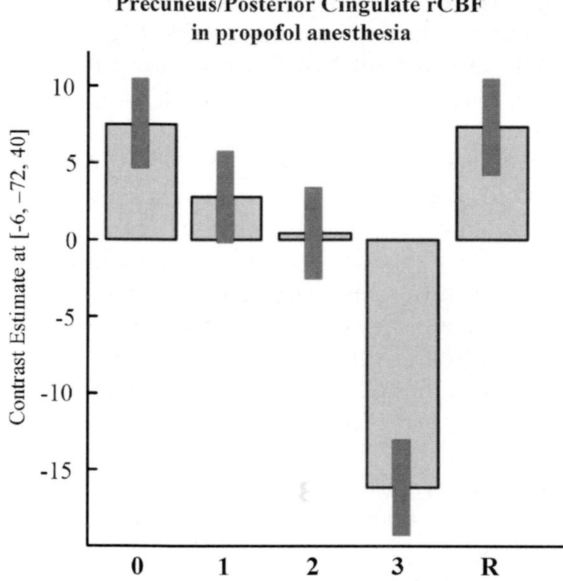

Fig. 4. Adjusted centered rCBF estimates in precuneal/posterior cingulate cortex (PrCC/PCC) cluster during: waking prior to anesthesia (level 0); mild sedation (clear response to commands, slightly drowsy; level 1); moderate sedation (slowed response to commands, slurred speech; level 2); unconsciousness (no response to command; level 3); and recovery of consciousness (response to command; level R) (mean values and 90% confidence interval). Note the progressive decrease in PrCC/PCC activity with deepening of propofol-induced sedation, the profound reduction during loss of consciousness and finally restoration of function paralleled by the return of consciousness. Data reanalyzed from Bonhomme et al. (2001)

In keeping with its general actions at the $GABA_A$ receptor, propofol produces marked global reductions in rCBF, however, there are a few particular regions with relatively large reductions in rCBF at the lowest levels of mean alveolar concentrations. These include the thalamus and PCC/PrCC (Fiset et al., 1999; Kaisti et al., 2002; Alkire, this volume). Although individual changes in rCBF did not correlate with propofol concentrations and the differences among three levels of propofol were not significant for any areas (Kaisti et al., 2002), this latter study did show that the lowest concentration of propofol assessed produced a profound average reduction in PCC/PrCC and Fiset et al. (1999) showed a strong negative correlation between rCBF in the thalamus and PCC/PrCC and propofol concentration. Fig. 4 shows a reanalysis of recent propofol PET data from Fiset's group illustrating the progressive decrease in PrCC/PCC activity with the deepening of sedation and the profound reduction during loss of consciousness, which completely restores upon return of consciousness (data reprocessed for the present paper from Bonhomme et al., 2001).

Furthermore, the close correlation of activity under propofol in the thalamus and midbrain in this latter study confirms the relevance of the reticular projections to thalamus as mediating arousal and initiation of consciousness. Halothane also decreases rCMRGlu in PCC/PrCC (Alkire et al., 1999) suggesting that many anesthetics target this region as an endpoint to induce a loss of consciousness.

Finally, although fentanyl has a complex action in mediating rCBF, it does reduce flow in the thalamus and PCC when compared to a placebo or during noxious heat stimulation (Adler et al., 1997). Taken together, these studies support the notion that PCC and PrCC are prominently impacted by low doses of anesthetics. The role of global changes and interactions with the thalamic/reticular activating system is not yet clear, however, the linkage between these systems points to a pivotal role of PCC/PrCC in consciousness when viewed in relation to the high level of resting metabolism of these areas.

Self reflection and complex information processing

The PCC has a major role in visuospatial orientation, topokinesis, and navigation of the body in space (reviewed in Vogt et al., 2004). Recent functional imaging suggests this function may be specifically related to self reflection (Johnson et al., 2002; Kelley et al., 2002) and autobiographical memory (Piefke et al., 2003), and this includes assessments of objects in space in terms of a first-person orientation (Vogeley et al., 2004). Although all the previous studies engaged PCC, PrCC has also been involved in intentional, self processing (Kircher et al., 2002). The essential function of self reflection for PCC/PrCC is compatible with a pivotal role for this region in conscious awareness. Indeed, self reflection may be the primary substrate for conscious awareness as located in the PCC/PrCC.

This functional substrate may have implications for understanding brain activity during complex cognitive tasks that are not associated with self reflection and actually inhibit functions in PCC/PrCC. Analysis of cognitive skill learning is a means of analyzing the neural correlates of consciousness (Raichle, 2000). Contrasts generated by reading nouns minus viewing nouns in this latter study greatly decreased activity in PCC/PrCC. Also, this region is inactivated during a working memory task (Greicius et al., 2003) and during noxious thermal stimulation (Vogt et al., 1996). Raichle suggested that activity in this and associated regions must be suspended for proper task execution. During noxious thermal stimulation, subjects attempt to withhold conscious perception of the pain stimulation and terminate functions relating to self in PCC/PrCC. It is also possible that during complex sensory processing in other modalities, activity in this region must be suspended. Thus, conscious reflection on self is mediated by PCC/PrCC, and this processing must be suspended and activity in this posterior cingulate region inactivated when engaging in complex cognitive and sensory information processing tasks. To the extent that PCC/PrCC is essential for conscious awareness, painful stimulation might lead subjects to internally inactivate the region of cortex that underpins their conscious awareness.

Anatomical relationships between mental and conscious information processing

The discussed role of RSC/PCC/PrCC in conscious awareness raises another issue: To what extent does the NNCC in the cingulate gyrus overlap with the neural correlates of mind (NCM)? The same view of consciousness has been argued previously for NCM (Vogt and Devinsky, 2000) to the extent that there is a part of cingulate cortex that plays a particularly important role in mental function based on lesion and functional imaging observations. The neuroscientific definition of mind in this context is those locations in the brain where mental activity has the highest probability of causing an increase in neuronal and functional activation. Thus, willed actions/intentions to move (e.g., Frith et al., 1991), the logic of mental activity including inductive reasoning (e.g., Goel et al., 1997), verbal mentalizing (e.g., Mazoyer et al., 1993), long-term storage of self-centered information, and theory of mind tasks (e.g., Fletcher et al., 1995; Lou et al., this volume), in which a subject is asked to interpret the mental state of another individual, all activate only part of the cingulate gyrus rather than its full extent. The conclusion from this assessment was that ACC and MCC along with adjacent frontal areas 8, 9, and 10 form a cingulofrontal convergence region that is a critical node in the NCM. It is striking that only a few studies engaged PCC as in theory of mind work (Fletcher et al., 1995) and this was attributed to its role in self-centered reflection as part of the paradigm rather than a pivotal role for PCC in the NCM.

It appears that the NNCC and NCM do not appreciably overlap in the cingulate gyrus. This conclusion raises doubts about the extent to which mental function in terms of cognitive processing and conscious awareness is subserved by the same neural substrates. Viewed in this manner, mental function is a higher order motor selection process, while conscious awareness involves a continual stream of self reflection and an effort to relate the self to its sensory environment. Surprisingly, both functional networks may operate essentially independent of each other but together form the essence of human brain function. Thus, human functional imaging provides an important context for resolving part of the essential philosophical questions regarding mind and consciousness raised over the past two millennia and it appears these issues will eventually be resolved via a thoughtful consideration of the structure, functions, and pathologies of the cingulate gyrus.

Abbreviations

ACC	anterior cingulate cortex (perigenual region)
AD	anterodorsal thalamic nucleus
AM	anteromedial thalamic nucleus
AV	anteroventral thalamic nucleus
Caud	caudate nucleus
cgs	cingulate sulcus

IG	indusium griseum
MCC	midcingulate cortex
mr	marginal ramus of the cingulate sulcus
NCM	neural correlates of mind
NNCC	neural network correlates of consciousness
PCC	posterior cingulate cortex
PCG	posterior cingulate gyrus
Pos	parieto-occipital sulcus
PrCC	precuneus cortex
rCBF	regional cerebral blood flow
rCMRGlu	regional cerebral metabolic rate; glucose
REM	rapid-eye-movement sleep;
RSC	retrosplenial cortex
sCC	splenium of the corpus callosum
Sub	dorsal subiculum
VA	ventral anterior thalamic nucleus
VCA	vertical plane at the anterior commissure

Acknowledgments

BAV is supported by National Institute of Health grant #NS44222, SL is Research Associate at the Fonds National de Recherche Scientifique (FNRS). For their kind permission to share and reanalyze previous PET data, we thank Pierre Maquet and Philippe Peigneux (sleep studies) and PM, PhP, Eric Salmon, and André Luxen from the Cyclotron Research Center of the University of Liège (healthy control rCMRGlu studies); and Vincent Bonhomme, Pierre Fiset and colleagues from the department of anesthesiology of McGill University, Montreal (propofol anesthesia studies).

References

Adams, J.H., Graham, D.I. and Jennett, B. (2000) The neuropathology of the vegetative state after an acute brain insult. Brain, 123: 1327–1338.

Adler, L.J., Gyulai, F.E., Diehl, D.J., Mintun, M.A., Winter, P.M. and Firestone, L.L. (1997) Regional brain activity changes associated with fentanyl analgesia elucidated by positron emission tomography. Anesth. Analg., 84: 120–126.

Alkire, M.T., Pomfrett, M.T., Haier, C.J., Gianzero, R.J., Chan, M.V., Jacobson, B.P. and Fallon, J.H. (1999) Functional brain imaging during anesthesia in humans: effects of halothane on global and regional cerebral glucose metabolism. Anesthesiology, 90: 701–709.

Andreasen, N.C., O'Leary, D.S., Cizadlo, T., Arndt, S., Rezai, K., Watkins, G.L., Ponto, L.L. and Hichwa, R.D. (1995) Remembering the past: two facts of episodic memory explored with positron emission tomography. Am. J. Psychiatry., 152: 1576–1585.

Archer, J.S., Abbott, D.F., Waites, A.B. and Jackson, G.D. (2003) fMRI "deactivation" of posterior cingulate cortex during generalized spike and wave. NeuroImage, 20: 1915–1922.

Aupée, A.M., Desgranges, B., Eustache, F., Lalevee, C., de la Sayette, V., Viader, F. and Baron, J.C. (2001) Voxel-based mapping of brain hypometabolism in permanent amnesia with PET. Neuroimage, 13: 1164–1173.

Bonhomme, V., Fiset, P., Meuret, P., Backman, S., Plourde, G., Paus, T., Bushnell, M.C. and Evans, A.C. (2001) Propofol anesthesia and cerebral blood flow changes elicited by vibrotactile stimulation: a positron emission tomography study. J. Neurophysiol., 85: 1299–1308.

Brodmann, K. (1909) Vergleichende Localisationslehre der Grosshirnrinde. Barth, Leipzi.

Carroll, E.W. and Wong-Riley, M.T.T. (1984) Quantitative light and electron microscopic analysis of cytochrome oxidase-rich zones in the striate cortex of the squirrel monkey. J. Comp. Neurol., 222: 1–17.

Curro, D.R., Pare, D. and Steriade, M. (1991) Short-lasting nicotinic and long-lasting muscarinic depolarizing responses of thalamocortical neurons to stimulation of mesopontine cholinergic nuclei. J. Neurophysiol., 65: 393–406.

Devinsky, O. and Luciano, D. (1993) The contributions of cingulate cortex to human behavior. In: Vogt B.A. and Gabriel M. (Eds.), Neurobiology of Cingulate Cortex and Limbic Thalamus. Birkhauser, Boston, pp. 527–556.

Fiset, P., Paus, T., Daloze, T., Plourde, G., Meuet, P., Bonhomme, V., Haij-Ali, N., Backman, S.B. and Evans, A.C. (1999) Brain mechanisms of propofol-induced loss of consciousness in humans: a positron emission tomographic study. J. Neurosci., 19: 5506–5513.

Fletcher, P., Happe, F., Frith, U., Baker, S.C., Dolan, R.J., Frakowiack, R.S.J. and Frith, C.D. (1995) Other minds in the brain: a functional imaging study of "theory of mind" in story comprehension. Cognition, 57: 109–128.

Frith, C.D., Friston, K., Liddle, P.F. and Frakowiack, R.S.J. (1991) Willed actions and the prefrontal cortex in man: a study with PET. Proc. R. Soc. Lond. B., 244: 241–246.

Goel, V., Gold, B., Kapur, S. and Houle, S. (1997) The seats of reason? An imaging study of deductive and inductive reasoning. Cog. Neurosci. Neuropsychol., 8: 1305–1309.

Greicius, M.D., Krasnow, B., Reiss, A.L. and Menon, V. (2003) Functional connectivity in the resting brain: a network analysis of the default mode hypothesis. Proc. Natl. Acad. Sci., 100: 253–258.

Gusnard, D.A. and Raichle, M.E. (2001) Searching for a baseline: functional imaging and the resting human brain. Nat. Rev. Neurosci., 2: 685–694.

Johnson, S.C., Baxter, L.C., Wilder, L.S., Pipe, J.G., Heiserman, J.E. and Prigatano, G.P. (2002) Neural correlates of self-reflection. Brain, 125: 1808–1814.

Kaisti, K.K., Metsahonkala, L., Teras, M., Oikonen, V., Aalto, S., Jaaskelainen, S., Hinkka, S. and Scheinin, H. (2002) Effects of surgical levels of propofol and servoflurane anesthesia on cerebral blood flow in healthy subjects studied with positron emission tomography. Anesthesiology, 96: 1358–1370.

Kampfl, A., Schutzhard, E., Franz, E., Pfausler, B., Haring, H.-P., Ulmer, H., Felber, S., Golaszewski, S. and Aichner, F. (1998) Prediction of recovery from post-traumatic vegetative state with cerebral magnetic-resonance imaging. The Lancet, 351: 1763–1767.

Kelley, W.M., Macrae, C.N., Wyland, C.L., Caglar, S., Inati, S. and Heatherton, T.F. (2002) Finding the self? an event-related fMRI study. J. Cog. Neurosci., 14: 785–794.

Kinney, H.C., Korein, J., Panigraphy, A., Dikkes, P. and Goode, R. (1994) Neuropathological findings in the brain of Karen Ann Quinlan, the role of the thalamus in the persistent vegetative state. NEJM, 330: 1469–1475.

Kinomura, S., Larsson, J., Gulyas, B. and Roland, P.E. (1996) Activation by attention of the human reticular formation and thalamic intralaminar nuclei. Science, 271: 512–515.

Kircher, T.T.J., Brammer, M., Bullmore, E., Simmons, A., Bartels, M. and David, A.S. (2002) The neural correlates of intentional and incidental self processing. Neuropsychologia, 40: 683–692.

Laureys, S. (2004) Functional neuroimaging in the vegetative state. NeuroRehabilitation, 19: 335–341.

Laureys, S., Faymonville, M.E., Degueldre, C., Fiore, G.D., Damas, P., Lambermont, B., Janssens, N., Aerts, J., Franck, G., Luxen, A., Moonen, G., Lamy, M. and Maquet, P. (2000) Auditory processing in the vegetative state. Brain, 123: 1589–1601.

Laureys, S., Faymonville, M., Ferring, M., Schnakers, C., Elincx, S., Ligot, N., Majerus, S., Antoine, S., Mavroudakis, N., Berre, J., Luxen, A., Vincent, J.L., Moonen, G., Lamy, M., Goldman, S. and Maquet, P. (2003) Differences in brain metabolism between patients in coma, vegetative state, minimally conscious state and locked-in syndrome. 7th Congress of the European Federation of Neurological Societies (FENS), 30 August–2 September, 2003, Helsinki, Finland, Eur. J. Neurol., 10(Suppl. 1): 224.

Laureys, S., Goldman, S., Phillips, C., Van Bogaert, P., Aerts, J., Luxen, A., Franck, G. and Maquet, P. (1999) Impaired effective cortical connectivity in vegetative state: preliminary investigation using PET. NeuroImage, 9: 377–382.

Laureys, S., Owen, A.M. and Schiff, N.D. (2004) Brain function in coma, vegetative state, and related disorders. Lancet Neurol., 3: 537–546.

Malamud, N. (1967) Psychiatric disorder with intracranial tumors of limbic system. Arch. Neurol., 17: 113–123.

Maquet, P., Peters, J.-M., Aerts, J., Delfiore, G., Degueldre, C., Luxen, A. and Franck, G. (1996) Functional neuroanatomy of human rapid-eye-movement sleep and dreaming. Nature, 383: 163–166.

Maquet, P., Degueldre, C., Delfiore, G., Aerts, J., Peters, J.M., Luxen, A. and Frank, G. (1997) Functional neuroanatomy of human slow wave sleep. J. Neurosci., 17: 2807–2812.

Matsunami, K.I., Kawashima, T. and Satake, H. (1989) Mode of [^{14}C]-2-deoxy-D-glucose uptake into retrosplenial cortex and other memory-related structures of the monkey during a delayed response task. Brain Res. Bull., 22: 829–838.

Mazars, G. (1970) Criteria for identifying cingulate epilepsies. Epilepsia, 11: 41–47.

Mazoyer, B.M., Tzourio, N., Frak, V., Syrota, A., Murayanna, N., Levrier, O., Salamon, G., Dehaene, S., Cohen, L. and Mehler, J. (1993) The cortical representation of speech. J. Cogn. Neurosci., 5: 467–469.

Minoshima, S., Giordani, B., Berent, S., Frey, K., Foster, N.L. and Kuhl, D.E. (1997) Metabolic reduction in the posterior cingulate cortex in very early Alzheimer's disease. Ann. Neurol., 42: 85–94.

Nelson, L.E., Guo, T.Z., Lu, J., Saper, C.B., Franks, N.P. and Maze, M. (2002) The sedative component of anesthesia is mediated by $GABA_A$ receptors in an endogenous sleep pathway. Nat. Neurosci., 5: 978–984.

Piefke, M., Weiss, P.H., Zilles, K., Markowitch, H.J. and Fink, G.R. (2003) Differential remoteness and emotional tone modulate the neural correlates of autobiographical memory. Brain, 126: 650–668.

Raichle, M.E. (2000) The neural correlates of consciousness: an analysis of cognitive skill learning. In: Gazzaniga M.S. (Ed.), The New Cognitive Neurosciences. The MIT Press, Cambridge, MA, pp. 1305–1318.

Rudge, P. and Warrington, E.K. (1991) Selective impairment of memory and visual perception in splenial tumors. Brain, 114: 349–360.

Sofroniew, M.V., Priestley, J.V., Consolazione, A., Eckenstein, F. and Cuello, A.C. (1985) Cholinergic projections from the midbrain and pons to the thalamus in the rat identified by combined retrograde tracing and choline acetyltransferase immunohistochemistry. Brain Res., 329: 213–223.

Steriade, M. (1996) Arousal: revisiting the reticular activating system. Science, 272: 225–226.

Takahashi, N., Kawamura, M., Shiota, J., Kasahata, N. and Hirayama, K. (1997) Pure topographic disorientation due to right retrosplenial lesion. Neurology, 49: 464–469.

Talairach, J. and Tournoux, P. (1988) Co-planar stereotaxic atlas of the human brain. Thieme, Stuttgart.

Valenstein, E., Bowers, D., Verfaellie, M., Heilman, K.M., Day, A. and Watson, R.T. (1987) Retrosplenial amnesia. Brain, 110: 1631–1646.

Van Groen, T., Vogt, B.A. and Wyss, J.M. (1993) Interconnection between the thalamus and retrosplenial cortex in the rodent brain. In: Vogt B.A. and Gabriel M. (Eds.), Neurobiology of Cingulate Cortex and Limbic Thalamus. Birkhauser, Boston, pp. 123–150.

Vogeley, K., Ritzl, M.M., Falkai, P., Zilles, K. and Fink, G.R. (2004) Neural correlates of first-person perspective as one constituent of human self-consciousness. J. Cog. Neurosci., 16: 817–827.

Vogt, B.A., Derbyshire, S.W.J. and Jones, A.K.P. (1996) Pain processing in four regions of human cingulate cortex localized with co-registered PET and MR Imaging. Eur. J. Neurosci., 8: 1461–1473.

Vogt, B.A. and Devinsky, O. (2000) Topography and relationships of mind and brain. Prog. Brain Res., 122: 11–22.

Vogt, B.A. and Pandya, D.N. (1987) Cingulate cortex in rhesus monkey. II. Cortical afferents. J. Comp. Neurol., 262: 271–289.

Vogt, B.A., Rosene, D.L. and Pandya, D.N. (1987) Cingulate cortex in rhesus monkey. I. Cytoarchitecture and thalamic afferents. J. Comp. Neurol., 262: 256–270.

Vogt, B.A., Rosene, D.L. and Peters, A. (1981) Synaptic termination of thalamic and callosal afferents in cinguate cortex of the rat. J. Comp. Neurol., 201: 265–283.

Vogt, B.A., Vogt, L. and Farber, N.B. (2004) Cingulate cortex and disease models. In: Paxinos G. (Ed.), The Rat Nervous System (3rd ed). Elsevier, San Diego, CA, pp. 705–727.

Vogt, B.A., Vogt, L. and Hof, P.R. (2003) Cingulate Gyrus. In: Paxinos G. and Mai J.K. (Eds.), The Human Nervous System (2nd ed). Elsevier, San Diego, CA, pp. 915–949.

Vogt, B.A., Vogt, L.J. and Laureys, S. (2005). Cytology and functionally correlated circuits of posterior cingulate areas. Neuroimage, in press.

Vogt, B.A., Vogt, L.J., Nimchinsky, E.A. and Hof, (1997) Primate cingulate chemoarchitecture and its disruption in Alzheimer's disease. In: Bloom F.E., Björklund A. and Hökfelt T. (Eds.) The Primate Nervous System, Part 1, Vol. 13. Elsevier, New York, pp. 455–528.

Vogt, B.A., Vogt, L.J., Perl, D.P. and Hof, P.R. (2001) Cytology of human caudomedial cingulate, retrosplenial, and caudal parahippocampal cortices. J. Comp. Neurol., 438: 353–376.

Watson, R.T., Heilman, K.M., Cauthen, J.C. and King, F.A. (1973) Neglect aster cingulectomy. Neurology, 23: 1003–1007.

Yamamoto, T., Kurobe, H., Kawamura, J., Hashimoto, S. and Nakamura, M. (1990) Subacute dementia with nectrotizing encephalitis selectively involving the fornix and splenium. J. Neurol. Sci., 96: 159–172.

S. Laureys (Ed.)
Progress in Brain Research, Vol. 150
ISSN 0079-6123

CHAPTER 16

Human cognition during REM sleep and the activity profile within frontal and parietal cortices: a reappraisal of functional neuroimaging data

Pierre Maquet[1,*], Perrine Ruby[1], Audrey Maudoux[1], Geneviève Albouy[1], Virginie Sterpenich[1], Thanh Dang-Vu[1], Martin Desseilles[1], Mélanie Boly[1], Fabien Perrin[2], Philippe Peigneux[1] and Steven Laureys[1]

[1]*Cyclotron Research Centre, University of Lièege-Sart Tilman, 4000 Liège, Belgium*
[2]*Neurosciences and Systèmes Sensoriels (UMR 5020), Université Claude Bernard Lyon I, 69007 Lyon, France*

Abstract: In this chapter, we aimed at further characterizing the functional neuroanatomy of the human rapid eye movement (REM) sleep at the population level. We carried out a meta-analysis of a large dataset of positron emission tomography (PET) scans acquired during wakefulness, slow wave sleep and REM sleep, and focused especially on the brain areas in which the activity diminishes during REM sleep. Results show that quiescent regions are confined to the inferior and middle frontal cortex and to the inferior parietal lobule. Providing a plausible explanation for some of the features of dream reports, these findings may help in refining the concepts, which try to account for human cognition during REM sleep. In particular, we discuss the significance of these results to explain the alteration in executive processes, episodic memory retrieval and self representation during REM sleep dreaming as well as the incorporation of external stimuli into the dream narrative.

Introduction

During the last decade, functional neuroimaging by positron emission tomography (PET) and functional magnetic resonance imaging (fMRI) characterized a very reproducible functional neuroanatomy of human sleep, which we will briefly summarize below.

During slow wave sleep (SWS), as compared to wakefulness, the cerebral energy metabolism and blood flow globally decrease (Maquet, 2000). The most deactivated areas are located in the dorsal pons and mesencephalon, cerebellum, thalami, basal ganglia, basal forebrain/hypothalamus, prefrontal cortex, anterior cingulate cortex and precuneus. As detailed elsewhere (Maquet, 2000), these findings are in keeping with the generation of non-rapid eye movement (REM) sleep in mammals, whereby the decreased firing in brainstem structures causes an hyperpolarization of thalamic neurons and triggers a cascade of processes responsible for the generation of various non-REM sleep rhythms (spindles, theta, slow rhythm).

During REM sleep, as compared to wakefulness, significant activations were found in the pontine tegmentum, thalamic nuclei and limbic and paralimbic areas (e.g., amygdaloid complexes, hippocampal formation and anterior cingulate cortex). Posterior cortices in temporo-occipital areas are also activated and their functional interactions are different in REM sleep than in wakefulness

*Corresponding author. Tel.: +32 43 66 36 87;
Fax: +32 43 66 29 46; E-mail: pmaquet@ulg.ac.be

DOI: 10.1016/S0079-6123(05)50016-5

(Braun et al., 1998). In contrast, the dorso-lateral prefrontal cortex, parietal cortex as well as the posterior cingulate cortex and precuneus are the least active brain regions (Maquet, 1996; Braun et al., 1997). Although early animal studies had already mentioned the high limbic activity during REM sleep (Lydic et al., 1991), functional neuroimaging in humans highlighted the contrast between the activation of limbic, paralimbic and posterior cortical areas, and the relative quiescence of the associative frontal and parietal cortices. The pattern of activity in subcortical structures is easily explained by the known neurophysiological mechanisms which generate REM sleep in animals. In contrast, the distribution of the activity within the cortex remains harder to explain and its origin remains speculative.

The particular pattern of cerebral activity observed during REM sleep has generated a number of papers and commentaries emphasizing the link between the organization of human brain function during sleep stage and the main characteristics of dreaming activity (Maquet et al., 1996; Hobson et al., 1998, 2003; Maquet, 2000). The main relationships between brain structure and dream features are as follows. First, the perceptual aspects of dreams would be related to the activation of posterior (occipital and temporal) cortices. Accordingly, patients with occipito-temporal lesions may report a cessation of visual dreams imagery (Solms, 1997). Second, emotional features in dreams would be related to the activation of amygdalar complexes, orbito-frontal cortex, and anterior cingulate cortex (Maquet et al., 1996; Maquet and Franck, 1997; Hobson et al., 1998, 2003; Maquet, 2000). Third, the activation of mesio-temporal areas would account for the memory content commonly found in dreams. Fourth, the relative hypoactivation of the prefrontal cortex would explain the alteration in logical reasoning, working memory, episodic memory, and executive functions that manifest themselves in dream reports from REM sleep awakenings (Maquet et al., 1996; Hobson et al., 1998, 2003; Maquet, 2000). Oddly enough, at this stage, the deactivation of the parietal cortex has received little attention.

Although we globally agree with these interpretations of the available PET results, we reasoned that analyzing a large set of PET data acquired during sleep in normal human subjects would improve the characterization of the cerebral functional organization during sleep, especially REM sleep. In particular, we were interested in better specifying the topography of the relatively decreased activity in frontal and parietal areas, which are believed to decisively shape human cognition during REM sleep. At the outset, we draw the attention of the reader on the speculative nature of the present paper. Although the hypotheses presented here are based on sound experimental work, the functional relationships between the distribution of regional brain activity and dream features are still to be confirmed experimentally.

Meta-analysis of PET data during human sleep

We ran a meta-analysis on 207 PET scans obtained in 22 young, male, healthy subjects (age range: 18–30 years), during awake resting state (eyes closed, 58 scans), SWS (66 scans), and REM sleep (83 scans). These data were obtained in the framework of two already published experimental protocols (Maquet et al., 2000; Peigneux et al., 2004).

Briefly, sleep was monitored by polysomnography during two consecutive nights spent on the scanner table. Polygraphic recordings included electroencephalogram (EEG recorded between electrode pairs C3-A2 and C4-A1), electro-oculogram and chin-electromyogram and were scored using international criteria (Rechtschaffen and Kales, 1968). Only subjects who showed at least two periods of 15 min spent in each stage of sleep were scanned with PET during the third night. PET data were acquired under polygraphic recording, during wakefulness and sleep, on a Siemens CTI 951 R 16/31 scanner in 3D mode. The subject's head was stabilized by a thermoplastic facemask secured to the head holder (Truscan imaging, MA), and a venous catheter was inserted into a left antebrachial vein. Regional cerebral blood flow (rCBF) was estimated during 12–14 90-s emission scans using automated, non-arousing, slow intravenous water ($H_2^{15}O$) infusion (6 mCi/222 MBq in 5 cc saline). Data were reconstructed using a Hanning filter (cutoff frequency: 0.5 cycle/pixel) and corrected for attenuation and background activity. A transmission scan was acquired to perform measured attenuation correction.

The analysis was run using the statistical parametric mapping software (SPM2, Wellcome Department of Cognitive Neurology, Institute of Neurology, London, UK). In short, a general linear model was designed for each individual subject and tested for the effect of condition (wakefulness, SWS, REM sleep). Global flow adjustment was performed by proportional scaling. Areas of significant changes were determined using linear contrasts. The contrasts of interest simply estimated the main differential effects of conditions, with a special emphasis on the least active brain regions during REM sleep. A comprehensive characterization of these areas required three contrasts, which looked for the brain areas more active in wakefulness than in REM sleep, in REM sleep than in SWS, and finally in wakefulness than in SWS. Taking advantage of the large data set available, the summary statistics images were then entered in second-level design matrices testing for random effects, which were assessed for each contrast of interest using a sample *t*-test. The resulting set of voxel values for each contrast constituted a map of the *t*-statistic {SPM(T)}, thresholded at $p<0.001$. Corrections for multiple comparisons were then performed using the theory of random gaussian fields, at $p<0.05$ at the voxel level over the entire brain volume. The random effects analysis allows us for the first time to take into account the inter-subjects variability in our analysis. As a consequence, the reported findings can be thought as representative of regional brain activity during sleep in the general population.

In this paper, we focus on the results pertaining to the least active brain areas during REM sleep. Concerning the most active brain areas during REM sleep, our data confirmed the metabolic pattern reported earlier (for review see Maquet, 2000), which has already been extensively discussed (Maquet et al., 1996; Maquet and Franck, 1997; Hobson et al., 1998, 2003; Maquet, 2000). The results are summarized in Tables 1–3 and in Fig. 1. Among the cortical areas that are relatively less active during REM sleep than during quiet wakefulness, it appears clearly that neither the whole frontal lobe nor the entire parietal cortex is hypoactive. In the parietal lobe, the quiescent area encompasses only the temporo-parietal region and the inferior lobule of the parietal cortex, well below the intra-parietal sulcus (Fig. 1). In the prefrontal cortex, the least active area is located in the inferior frontal gyrus. The significantly hypoactive region spreads to the middle frontal gyrus but does not reach neither the superior frontal gyrus nor the medial aspect of the frontal lobe. The pattern of activity in the latter region is remarkable. In the medial prefrontal cortex, the rCBF is significantly decreased during SWS as compared to wakefulness, but no difference in regional activity is observed between REM sleep and wakefulness (Fig. 1, inset). These findings indicate a quite remarkable redistribution of the regional activity within the frontal and parietal cortices that probably constrain human cognition during REM sleep in a unique way.

Lateral prefrontal cortex and executive functions

Executive processes coordinate external information, thoughts, and emotions and organize actions in relation to internal goals. The selection of motor actions may directly rely on external stimuli, but in other cases it is based on the perceptual context or a whole temporal episode during which the individual is acting. The set of processes that subtend these operations may require the integration of information from various sources but referring to a specific, sometime extended, period of time (Baddeley, 2000). Another model, based on information theory, rather suggests a cascade of nested levels of control (Koechlin et al., 2003). These various executive processes are anatomically segregated. The selection of stimulus–response association involves the dorsal premotor cortex; the caudal prefrontal cortex would carry out the contextual control; and the episodic[1] control would be subserved by the rostral and ventral prefrontal cortex.

[1]In this paragraph, we use the term episodic in the sense that this control system selects information according to events that occurred in the past. It does not necessarily refer to the way information is stored in long-term memory (Koechlin et al., 2003).

Table 1. Frontal and parietal areas where regional cerebral blood flow is decreased during REM sleep as compared to wakefulness

Region	*x*	*y*	*z*	*Z* score	*p* value
Right inferior frontal gyrus	50	52	2	7.24	< 0.001
Right middle frontal gyrus	42	60	14	6.50	< 0.001
Right orbito-frontal cortex	22	32	−26	6.40	< 0.001
Left inferior frontal gyrus	−50	40	2	6.42	< 0.001
Left middle frontal gyrus	−36	56	0	6.40	< 0.001
Left orbito-frontal cortex	−32	54	−14	6.09	< 0.001
Right inferior parietal lobule	64	−48	40	7.32	< 0.001
Left inferior parietal lobule	−54	−56	50	6.38	< 0.001

Note: Only one peak voxel is reported for each cortical region. *x*, *y*, *z*: coordinates in the stereotactic MNI (Montreal Neurological Institute) space (mm); *Z* score and *p* (corrected for multiple comparisons at the voxel level over the entire brain volume).

Table 2. Frontal and parietal areas where regional cerebral blood flow is higher during REM sleep than during slow wave sleep

Region	*x*	*y*	*z*	*Z* score	*p* value
Anterior cingulate cortex	2	34	10	6.94	< 0.001
Ventral medial prefrontal cortex	2	46	−6	6.77	< 0.001
Dorsal medial frontal cortex	4	4	60	6.36	< 0.001
Right inferior frontal gyrus	26	10	−10	6.52	< 0.001
Left superior parietal cortex	−14	−50	66	6.24	< 0.001
Precuneus	−4	−50	66	6.44	< 0.001

Table 3. Frontal and parietal areas where regional cerebral blood flow is higher during wakefulness than during slow wave sleep

Region	*x*	*y*	*z*	*Z* score	*p* value
Anterior cingulate cortex	2	40	8	6.94	< 0.001
Ventral medial frontal cortex	−2	30	−14	6.17	< 0.001
Dorsal medial frontal cortex	4	2	60	4.56	0.048
Right inferior frontal gyrus	60	12	0	6.20	< 0.001
Right middle frontal gyrus	40	54	4	6.60	< 0.001
Right orbito-frontal cortex	38	42	−26	6.65	< 0.001
Left middle frontal gyrus	−38	44	24	6.61	< 0.001
Left orbito-frontal cortex	−34	42	−26	7.58	< 0.001
Right inferior parietal lobule	68	−48	34	7.36	< 0.001
Left inferior parietal lobule	−56	−50	42	6.36	< 0.001
Precuneus	−6	−76	60	6.14	< 0.001

The topography of the quiescent prefrontal areas during REM sleep overlap with the regions that support the contextual and episodic control systems. This finding suggests that if the perceptual control is maintained in this stage of sleep, the contextual and the episodic controls are probably less efficient than during wakefulness. This would explain the lack of "orientational stability" observed in dream reports, whereby the dreamer is unable to coordinate information of a whole episode: the "person, times and place are fused, incongruous and discontinuous" (Hobson et al., 2003). Likewise,

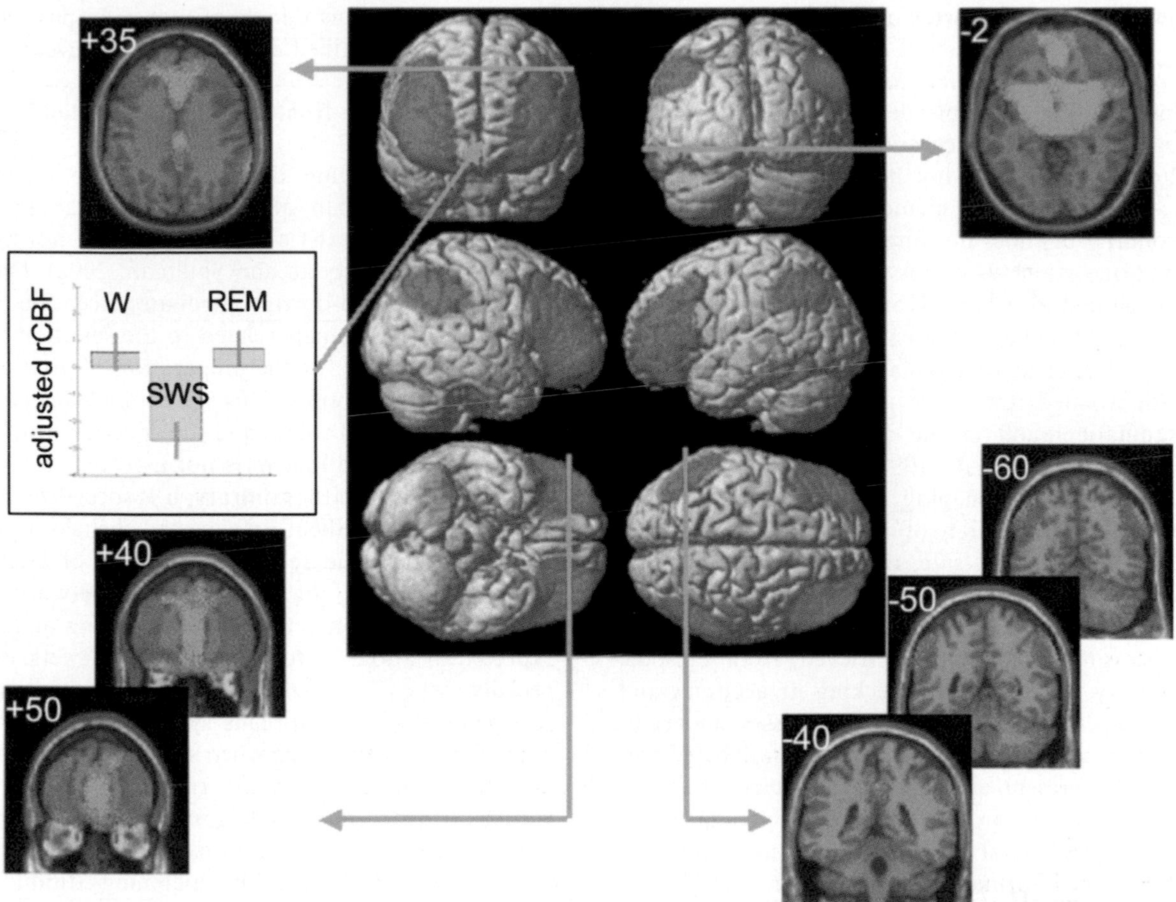

Fig. 1. *Central panel*: Three-dimensional rendering of the brain areas that are less active during REM sleep than during wakefulness (in red). These areas involve the lateral part of the frontal cortex, the inferior lateral lobule, and medial aspects of the parietal cortex. Anterior and posterior views (first row), lateral views (second row) and bottom and top views (third row). *Side panels*: For comparison, the brain areas that are significantly less active during slow wave sleep (SWS) than during wakefulness are indicated in green. Orange areas indicate regions where regional cerebral blood flow is decreased in both SWS and REM sleep as compared to wakefulness. *Left and right upper panels*: Transverse sections 35 (left) and −2 (right) mm from the anterior-posterior commissural plane. The least active areas during REM sleep (as compared to wakefulness, in red) do not reach the medial frontal cortex. In contrast, the medial frontal cortex is one of the least active area during SWS (as compared to wakefulness, in green). *Left lower panel*: Frontal sections through the frontal cortex, at 40 and 50 mm from the anterior commissure. The least active areas during REM sleep (as compared to wakefulness, in red) involve only the inferior and middle frontal gyri but do not reach the superior frontal gyrus or the medial prefrontal cortex. The latter are significantly deactivated during SWS (green). *Right lower panel*: Frontal sections through the parietal cortex, at −40, −50, and −60 mm from the anterior commissure. The least active areas during REM sleep (as compared to wakefulness, in red) involve only the inferior parietal lobule but do not reach the intraparietal sulcus or the superior parietal cortex. They overlap (orange) with the least active area during SWS (green). *Inset*: The adjusted cerebral blood flow in the medial prefrontal area is similar during wakefulness and REM sleep and is decreased during SWS. See Plate 16.1 in Colour Plate Section.

the dreamer would fail to organize his mental representation toward a well-identified internal goal and is seldom able to "control the flow of dream events" (Hobson et al., 2003). Volitional control is notoriously decreased in dreams. In contrast, the dreamer's behavior would be usually adapted to the objects and locations internally perceived, which are putatively related, as previously mentioned, to the activity in the posterior (temporal and occipital) cortices, highly active during REM sleep.

Lateral prefrontal cortex and episodic memory

Episodic memory refers to the capacity to encode and recollect past episodes with their specific integrated details, place and time (Tulving, 2004). Prefrontal cortex is not necessary to store, or to access to, episodic memories, since damage to frontal lobes does not impair episodic memory to the same extent as lesions to mesio-temporal areas (Henson et al., 1999). However, there are reports of densely amnesic patients after frontal lobe lesions. For instance, a remarkable patient presented with isolated retrograde amnesia due to a post-traumatic lesion in the right ventral prefrontal cortex (Levine et al., 1998). On the other hand, functional neuroimaging studies consistently reported the activation of prefrontal cortex, especially on the right side, during episodic memory retrieval (Rugg et al., 2002). The usually held view considers that prefrontal areas participate in processing information retrieved from episodic memory, essentially by checking its accuracy and completeness. The cognitive processes underpinning these frontal activations remain debated and would represent either a specific retrieval mode, retrieval effort or retrieval success (Rugg et al., 2002). The frontal areas activated during retrieval of episodic information are located in the right and left anterior prefrontal cortices and in the right dorso-lateral prefrontal cortex (Rugg et al., 2002).

It is intriguing to observe that the brain areas activated in functional imaging studies of episodic memory or this brain-lesioned amnesic patient overlap with the cortical regions hypoactive during REM sleep. The relative quiescence of the anterior prefrontal areas, and to some extent of the dorsolateral prefrontal cortex would explain that recent waking life episodes, characterized by their specific location, characters, objects, and actions are seldom described as such in dream reports (1.7%) (Fosse et al., 2003). In contrast, "snips" of recent waking activity are frequently (65%) observed in dream reports (Fosse et al., 2003). Although the dreamer has access to "day residues," probably spontaneously generated by the coordinated activity of the mesio-temporal areas and the posterior cortices, he would be prevented by the relative hypoactivity in the anterior prefrontal cortex from tying up the various details of a specific past episode into an identified autobiographical event.

The ventral parieto-frontal system of attention

Recent neuroimaging data studying the neural correlates of human attention have established that two systems exist and interact in the normal human brain (Corbetta and Shulman, 2002). One system exerts a top-down control on perception. It carries neuronal signals related to the selection of stimuli and allows for the preparation of goal-directed motor responses. This system includes part of the intraparietal sulcus and the superior frontal cortex. The second system is not involved in top-down selection. On the contrary, it is specialized in the detection of salient, unexpected, behaviorally relevant stimuli and reorients the focus of attention. For instance, the activity in this network is enhanced in response to targets occurring at unexpected locations or to low frequency targets, especially when relevant to the task at hand (Corbetta et al., 2000). This system would work as an alerting mechanism when salient stimuli arise outside the present focus of processing. The activation of this system would result in an interruption in the current attentional set and in an attentional shift toward the incoming stimulus. As compared to the top-down system, this system is more ventral, mainly lateralized on the right hemisphere and involves the temporo-parietal cortex and the inferior and middle frontal gyri (Corbetta and Shulman, 2002). Several data suggest the activity in the ventral network would be modulated by the locus coeruleus. For instance, the locus coeruleus is involved in selective attention, especially to salient and unexpected stimuli (Aston-Jones et al., 2000). Furthermore, in the macaque, the inferior parietal cortex is known to receive heavy projection from the locus coeruleus (Morrison and Foote, 1986).

As far as a precise anatomical localization is allowed by PET, the topography of the relatively quiescent areas during REM sleep is similar to this ventral attentional network. The parietal area, hypoactive during REM sleep extends through the posterior part of the inferior parietal lobule toward the posterior end of the sylvian fissure. It does not seem to extend to temporal areas. The hypoactive frontal region is located in the anterior part of the inferior frontal gyrus, on the anterior aspect of the region identified in the ventral attentional system. Moreover, although bilateral, the hypoactive areas are more extended on the right side, as is the ventral attention system (Corbetta and Shulman, 2002). Finally, the firing rate of noradrenergic locus coeruleus neurons dramatically decreases during REM sleep, depriving the parietal areas from an important positive modulation (Steriade and McCarley, 1990). These considerations suggest that the ventral attentional network is relatively quiescent during REM sleep. If this hypothesis were true, one predicts that the focus of attention during REM sleep should be less sensitive to salient, behaviorally relevant, external stimuli than during wakefulness. The focus of attention is difficult to assess experimentally during REM sleep. However, it may transpire in dream reports. One would expect that the dream narrative (obtained after awakening from REM sleep) would not be easily modified by external stimulation, even if behaviorally relevant. Some observations support this hypothesis. During REM sleep, external stimuli instead of interrupting the flow of the dream storyline, are incorporated into it (Foulkes, 1966). Similarly, responsiveness to external auditory stimulation is reduced during sleep, at least in part, due to incorporation of the external information into ongoing cognitive activity (Burton et al., 1988).

The frontal and parietal areas and mind representation during REM sleep

The medial frontal areas are active during REM sleep as during wakefulness. In this respect, REM sleep is very different from SWS because the same regions are relatively less active in SWS, as compared to either wakefulness or REM sleep. Theory of mind refers to the ability to attribute intentions, thoughts and feelings to oneself and to others (Carruthers and Smith, 1996). It is an inductive reasoning allowing interpretation and understanding of others' actions and speech, and prediction of their behavior. In comparison to reasoning applied to physical events or to lower level tasks, theory of mind was shown to reliably involve medial prefrontal cortex, temporo-parietal junction especially in the right hemisphere and temporal poles (Fletcher et al., 1995; Brunet et al., 2000, for a review Frith, 2001).

Dreaming usually appears as a multisensorial narrative involving characters interacting with each other. These oneiric characters are credited thoughts, intentions, and emotions by the dreamer himself. Mind representation is thus a key feature of dreaming. We hypothesize that the persistence of a level of activity in the medial frontal areas similar to the activity observed during wakefulness might participate to mind representation during REM sleep.

However, during REM sleep, the preserved activity in the medial prefrontal areas (as compared to wakefulness) contrasts with the low activity in the inferior parietal cortex. In the right hemisphere, this area is a part of the network activated during mind representation at wakefulness and is involved in the distinction of first versus third person perspective in the representation of action, mind and emotion (Ruby and Decety, 2001, 2003, 2004; Chaminade and Decety, 2002; Farrer et al. 2003). During REM sleep, the low activation of the right temporo-parietal junction might be related to a loosening in the distinction between first and third level perspectives. Accordingly, in dream report, the self can participate to the dream action both in a first-person (i.e., the self sees and acts) and in a third-person perspective (i.e., the dreamer sees the self acting in the dream).

Finally, social emotion such as jealousy, pride, embarrassment, infatuation, sexual love, shame, guilt and pride are often reported in dreaming (Adolphs, 2002; Schwartz and Maquet, 2002). Neuropsychological and neuroimaging studies in humans have demonstrated that medial prefrontal cortex and amygdala are consistently involved in basic and social emotions processing and more generally in social cognition (Adolphs, 1999; Phan et al., 2002; Ruby and Decety, 2004). The amygdala are known to be very active during REM sleep (Maquet et al., 1996). The preserved activity in the medial prefrontal cortex might also account for the high proportion of dreams involving social emotions.

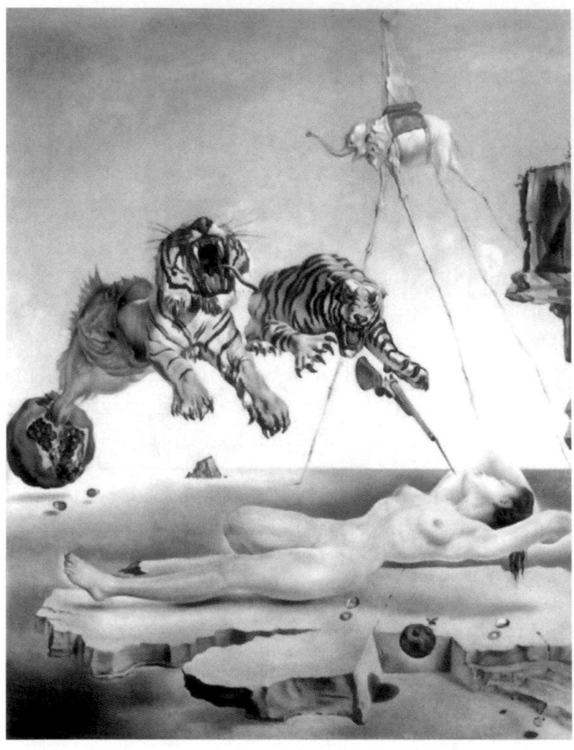

Fig. 2. Dream Caused by the Flight of a Bee Around a Pomegranate One Second Before Awakening, Salvador Dali, 1944. © Fundacióon Gala-Salvador Dali, c/o Beeldrecht Amsterdam 2005.

Conclusions

Our meta-analysis refines the description of the functional neuroanatomy of normal human sleep and provides a tentative framework to explain the relationship between human cognition during REM sleep and regional patterns of decreased activity within frontal and parietal areas. In particular, the hypoactivity in frontal and parietal areas during REM sleep is more precisely characterized. It involves the inferior and middle frontal gyrus as well as the posterior part of the inferior parietal lobule. Interestingly enough, the superior frontal gyrus, the medial frontal areas, the intraparietal sulcus, and the superior parietal cortex are not less active in REM sleep than during wakefulness. We suggest that this peculiar distribution of regional brain activity during REM sleep might correlate with some features of cognition, as reflected in dream reports (Fig. 2). Especially, this regional metabolic pattern provides new insights on the possible neural bases of some characteristic dream features such as self and characters' mind representations, the poor episodic recall, the lower ability of external stimuli to break the dream narrative and the difficulty to organize one's oneiric behavior toward a well identified and persistent goal.

Acknowledgments

Sleep research presented in this paper was supported by the Fonds National de la Recherche Scientifique de Belgique (FNRS), the Fondation Médicale Reine Elisabeth, the Research Fund of ULg and the PAI/IAP Interuniversity Pole of Attraction P5/04. MD, TDV, GV, SL and PM are supported by FNRS.

References

Adolphs, R. (1999) Social cognition and the human brain. Trends Cogn. Sci., 3: 469–479.

Adolphs, R. (2002) Neural systems for recognizing emotion. Curr. Opin. Neurobiol., 12: 169–177.

Aston-Jones, G., Rajkowski, J. and Cohen, J. (2000) Locus coeruleus and regulation of behavioral flexibility and attention. Prog. Brain Res., 126: 165–182.

Baddeley, A. (2000) The episodic buffer: a new component of working memory? Trends Cogn. Sci., 4: 417–423.

Braun, A.R., Balkin, T.J., Wesenten, N.J., Carson, R.E., Varga, M., Baldwin, P., Selbie, S., Belenky, G. and Herscovitch, P. (1997) Regional cerebral blood flow throughout the sleep-wake cycle. An H2(15)O PET study. Brain, 120: 1173–1197.

Braun, A.R., Balkin, T.J., Wesensten, N.J., Gwadry, F., Carson, R.E., Varga, M., Baldwin, P., Belenky, G. and Herscovitch, P. (1998) Dissociated pattern of activity in visual cortices and their projections during human rapid eye movement sleep. Science, 279: 91–95.

Brunet, E., Sarfati, Y., Hardy-Bayle, M.C. and Decety, J. (2000) A PET investigation of the attribution of intentions with a nonverbal task. Neuroimage, 11: 157–166.

Burton, S.A., Harsh, J.R. and Badia, P. (1988) Cognitive activity in sleep and responsiveness to external stimuli. Sleep, 11: 61–68.

Carruthers, P. and Smith, P. (1996) Theories of Theories of Mind. Cambridge University Press, Cambridge.

Chaminade, T. and Decety, J. (2002) Leader or follower? Involvement of the inferior parietal lobule in agency. Neuroreport, 28: 1975–1978.

Corbetta, M., Kincade, J.M., Ollinger, J.M., McAvoy, M.P. and Shulman, G.L. (2000) Voluntary orienting is dissociated from target detection in human posterior parietal cortex. Nat. Neurosci., 3: 292–297.

Corbetta, M. and Shulman, G.L. (2002) Control of goal-directed and stimulus-driven attention in the brain. Nat. Rev. Neurosci., 3: 201–215.

Farrer, C., Franck, N., Georgieff, N., Frith, C.D., Decety, J. and Jeannerod, M. (2003) Modulating the experience of agency: a positron emission tomography study. Neuroimage, 18: 324–333.

Fletcher, P.C., Happe, F., Frith, U., Baker, S.C., Dolan, R.J., Frackowiak, R.S. and Frith, C.D. (1995) Other minds in the brain: a functional imaging study of "theory of mind" in story comprehension. Cognition, 57: 109–128.

Fosse, M.J., Fosse, R., Hobson, J.A. and Stickgold, R.J. (2003) Dreaming and episodic memory: a functional dissociation? J. Cogn. Neurosci., 15: 1–9.

Foulkes, D. (1966) The Psychology of Sleep. Charles Scribner's Sons, New York.

Frith, C.D. (2001) Mind blindness and the brain in autism. Neuron, 20: 969–979.

Henson, R.N., Shallice, T. and Dolan, R.J. (1999) Right prefrontal cortex and episodic memory retrieval: a functional MRI test of the monitoring hypothesis. Brain, 122(Pt 7): 1367–1381.

Hobson, J.A., Pace-Schott, E.F., Stickgold, R. and Kahn, D. (1998) To dream or not to dream? Relevant data from new neuroimaging and electrophysiological studies. Curr. Opin. Neurobiol., 8: 239–244.

Hobson, J., Pace-Schott, E. and Stickgold, R. (2003) Dreaming and the brain: toward a cognitive neuroscience of conscious states. In: Pace-Schott E., Solms M., Blagrove M. and Harnad S. (Eds.), Sleep and Dreaming. Cambridge University Press, Cambridge, pp. 1–50.

Koechlin, E., Ody, C. and Kouneiher, F. (2003) The architecture of cognitive control in the human prefrontal cortex. Science, 302: 1181–1185.

Levine, B., Black, S.E., Cabeza, R., Sinden, M., McIntosh, A.R., Toth, J.P., Tulving, E. and Stuss, D.T. (1998) Episodic memory and the self in a case of isolated retrograde amnesia. Brain, 121(Pt 10): 1951–1973.

Lydic, R., Baghdoyan, H.A., Hibbard, L., Bonyak, E.V., DeJoseph, M.R. and Hawkins, R.A. (1991) Regional brain glucose metabolism is altered during rapid eye movement sleep in the cat: a preliminary study. J. Comp. Neurol., 304: 517–529.

Maquet, P. (2000) Functional neuroimaging of normal human sleep by positron emission tomography. J. Sleep Res., 9: 207–231.

Maquet, P. and Franck, G. (1997) REM sleep and amygdala. Mol. Psychiatr., 2: 195–196.

Maquet, P., Laureys, S., Peigneux, P., Fuchs, S., Petiau, C., Phillips, C., Aerts, J., Del Fiore, G., Degueldre, C., Meulemans, T., Luxen, A., Franck, G., Van Der Linden, M., Smith, C. and Cleeremans, A. (2000) Experience-dependent changes in cerebral activation during human REM sleep. Nat. Neurosci., 3: 831–836.

Maquet, P., Péters, J., Aerts, J., Delfiore, G., Degueldre, C., Luxen, A. and Franck, G. (1996) Functional neuroanatomy of human rapid-eye-movement sleep and dreaming. Nature, 383: 163–166.

Morrison, J.H. and Foote, S.L. (1986) Noradrenergic and serotoninergic innervation of cortical, thalamic, and tectal visual structures in Old and New World monkeys. J. Comp. Neurol., 243: 117–138.

Peigneux, P., Laureys, S., Fuchs, S., Collette, F., Perrin, F., Reggers, J., Phillips, C., Degueldre, C., Del Fiore, G., Aerts, J., Luxen, A. and Maquet, P. (2004) Are spatial memories strengthened in the human hippocampus during slow wave sleep? Neuron, 44: 535–545.

Phan, K., Wager, T., Taylor, S. and Liberzon, I. (2002) Functional neuroanatomy of emotion: A meta-analysis of emotion activation studies in PET and fMRI. Neuroimage, 16: 331–348.

Rechtschaffen, A. and Kales, A.A. (1968) A Manual of Standardized Terminology, Techniques and Scoring System for Sleep Stages of Human Subjects. US Department of Health Education and Welfare, Bethesda.

Ruby, P. and Decety, J. (2001) Effect of subjective perspective taking during simulation of action: a PET investigation of agency. Nat. Neurosci., 4: 546–550.

Ruby, P. and Decety, J. (2003) What you believe versus what you think they believe: A neuroimaging study of conceptual perspective-taking. Eur. J. Neurosci., 17(11): 2475–2480.

Ruby, P. and Decety, J. (2004) How would you feel versus how do you think she would feel? A neuroimaging study of perspective-taking with social emotions. J. Cogn. Neurosci., 16: 988–999.

Rugg, M.D., Otten, L.J. and Henson, R.N. (2002) The neural basis of episodic memory: evidence from functional neuroimaging. Phil. Trans. R. Soc. Lond. B Biol. Sci., 357: 1097–1110.

Schwartz, S. and Maquet, P. (2002) Sleep imaging and the neuro-psychological assessment of dreams. Trends Cogn. Sci., 1: 23–30.

Solms, M. (1997) The Neuropsychology of Dreams. Lawrence Erlbaum Associates Inc., Mahwah.

Steriade, M. and McCarley, R.W. (1990) Brainstem Control of Wakefulness and Sleep. Plenum Press, New York.

Tulving, E. (2004) Episodic memory: from mind to brain. Rev. Neurol. (Paris), 160: S9–S23.

S. Laureys (Ed.)
Progress in Brain Research, Vol. 150
ISSN 0079-6123

CHAPTER 17

General anesthesia and the neural correlates of consciousness

Michael T. Alkire* and Jason Miller

Department of Anesthesiology and The Center for the Neurobiology of Learning and Memory, University of California at Irvine, Irvine, CA, USA

Abstract: The neural correlates of consciousness must be identified, but how? Anesthetics can be used as tools to dissect the nervous system. Anesthetics not only allow for the experimental investigation into the conscious–unconscious state transition, but they can also be titrated to subanesthetic doses in order to affect selected components of consciousness such as memory, attention, pain processing, or emotion. A number of basic neuroimaging examinations of various anesthetic agents have now been completed. A common pattern of regional activity suppression is emerging for which the thalamus is identified as a key target of anesthetic effects on consciousness. It has been proposed that a neuronal hyperpolarization block at the level of the thalamus, or thalamocortical and corticocortical reverberant loops, could contribute to anesthetic-induced unconsciousness. However, all anesthetics do not suppress global cerebral metabolism and cause a regionally specific effect on thalamic activity. Ketamine, a so-called dissociative anesthetic agent, increases global cerebral metabolism in humans at doses associated with a loss of consciousness. Nevertheless, it is proposed that those few anesthetics not associated with a global metabolic suppression effect might still have their effects on consciousness mediated at the level of thalamocortical interactions, if such agents scramble the signals associated with normal neuronal network reverberant activity. Functional and effective connectivity are analysis techniques that can be used with neuroimaging to investigate the signal scrambling effects of various anesthetics on network interactions. Whereas network interactions have yet to be investigated with ketamine, a thalamocortical and corticocortical disconnection effect during unconsciousness has been found for both suppressive anesthetic agents and for patients who are in the persistent vegetative state. Furthermore, recovery from a vegetative state is associated with a reconnection of functional connectivity. Taken together these intriguing observations offer strong empirical support that the thalamus and thalamocortical reverberant network loop interactions are at the heart of the neurobiology of consciousness.

Consciousness as the dependent variable

It was not long ago that the mere mention of the "c" word (i.e., consciousness) in a scientific report would cause the editor of a neuroscience journal to question the very sanity of the heretical researcher who dared mention it. Presently, the study of consciousness has come to the forefront of scientific inquiry. This interest has been propelled forward not only by numerous recent conferences and scholarly works on the subject, but also directly because of numerous significant technological advances that now allow for the study of the working human brain with unprecedented spatial and temporal resolution.

A great debate exists regarding how the problem space for the scientific understanding of

*Corresponding author. Tel.: +1 714 456 5501; Fax: +1 714 456 7702; E-mail: malkire@uci.edu

DOI: 10.1016/S0079-6123(05)50017-7

consciousness should be defined (Chalmers, 1995; Dennett, 1996; Crick and Koch, 2003). It has been proposed that there is an easy and a so-called "hard-problem" of consciousness (Chalmers, 1995). The easy problem, which is still not so easy, involves achieving an understanding of how consciousness might arise from a mass of highly interconnected neurons. The hard-problem of consciousness involves, in essence, coming to an understanding of why the color green has an experiential characteristic of green. It is suggested that the hard-problem may not find its solution through a neuroscientific approach and perhaps, any neuroscientific attempts are ultimately doomed to failure because the hard problem fundamentally requires or demands a different kind of approach. What that different approach might be is not at all clear and may well be an approach that is not yet even in existence.

As anesthesiologists, we deal with the problem of consciousness in a very pragmatic way. Our interest arises because of the practical need to ensure that we remove patients' consciousness prior to the start of an operation and that we return patients' consciousness just after the end of the operation. It is not necessarily within our control as to whether or not the consciousness returned to a particular patient still allows the said patient to experience green as green, but it is generally directly under our control as to when we allow the patient to regain their consciousness. It has, of course, been asked many times in discussions on consciousness, "How do you know that consciousness goes away when a person is under a general anesthetic?" Or, put another way, "Does anesthesia cause loss of consciousness (Kulli and Koch, 1991)?" Before giving an answer to that question, we first turn to an even more fundamental issue regarding the neurobiology of consciousness.

Not long ago, Francis Crick (1994) made an astonishing hypothesis about the neurobiology of consciousness. He simply, but quite profoundly offered the hypothesis that consciousness resides within the brain and that when the brain ceases to function, there is no more consciousness. We here formally expand that astonishing hypothesis by emphatically answering YES to the question of whether anesthesia causes loss of consciousness. Given a sufficient dose of anesthesia, a number of different anesthetics are fully capable of rendering the brain into a state where the brain's spontaneous electrical activity stops (Drummond and Patel, 2000). If monitored with an electroencephalograph (EEG), the brain is found to be in an isoelectric state (i.e., the EEG is a flat line). We hypothesize that when a brain has been rendered isoelectric by anesthesia, then consciousness is lost and when the brain is in such a state it becomes impossible for a person to perceive anything or to be conscious of even the greenest of greens.

The anesthetic approach to the consciousness problem

If we start our investigations of the neural correlates of consciousness from the theoretical framework where we assert that anesthesia causes a loss of consciousness, we are in an excellent position to proceed with the study of consciousness as a dependent variable that can be directly manipulated using the power of anesthetic drugs (Alkire et al., 1998). Anesthetics can be used in this manner because they can be given in precise specific doses that allow one to achieve any particular desired depth of anesthesia in any particular person. When anesthetic manipulation is then coupled with recent advanced brain imaging techniques a powerful method emerges that can provide answers to some fundamental questions about the anatomy of consciousness.

The logic of this approach is relatively straightforward. To illustrate, brain imaging is performed on a person under anesthesia who has an isoelectric EEG. The dose of anesthesia is then lowered and the person regains some basic level of consciousness. A second brain scan is then performed during the newly regained conscious state. A simple cognitive subtraction analysis technique is performed between the two images and the difference image contains those neural circuits that identify the minimum regional neural activity that is sufficient for consciousness. Alternatively, one brain scan image can be obtained at a dose of anesthesia which puts a person into a state where the brain is just about to lose consciousness. When this almost unconscious brain scan image is subtracted from a brain scan image where consciousness was just lost, then the subtraction image between these two states will contain those neural circuits that are the minimum necessary for consciousness. Thus, both the necessary and sufficient regional brain activity required to support consciousness can be identified through this anesthetic manipulation approach and a solution to the easy problem of consciousness would appear at hand.

The definitive anesthetic study versus reality

This chapter will review those anesthesia related studies that move us toward an understanding of what might be the neural correlates of consciousness. It is important to state, however, that the definitive study remains to be done. The definitive study would investigate essentially all anesthetic agents, including those that are considered dissociative in nature, as well as look at other methods of inducing unconsciousness. It would use standardized anesthetic endpoints such that a similar state could be induced with each agent and for each individual. It would use an imaging technique with both high spatial and temporal resolutions such that the precise neurons and networks required for generating consciousness could be identified exactly as they change their activity with the variation in the state of consciousness. It would be amply powered with a large enough and broad enough sample of subjects to allow generalization of the results to the population and account for subpopulation dynamics such as subject age or sex. The study would also be designed to specifically differentiate those effects related to specific changes in consciousness versus those effects directly caused by the anesthetics themselves (Plourde, 2001).

The practical realities of trying to do an anesthesia-related brain imaging study in today's expensive healthcare environment and with the current state of the art of neuroimaging means that the definitive anesthetic-manipulated-consciousness study will likely remain elusive for quite some time. Therefore, we are left with the need to draw inferences about the neural correlates of consciousness from a relatively small number of anesthetic-related studies in which a generally small sample of healthy young right-handed males were given various doses of a few different types of anesthetic agents. These agents, furthermore, were titrated to study specific end-points that may or may not have been primarily concerned with a manipulation of the level of consciousness as an end-point. Moreover, the specific study conditions usually varied between studies such that a number of them would have subjects listen to words while they were being scanned, whereas others might have had subjects performing other types of cognitive tasks, or simply performing no task at all. Thus, to expect a definitive answer regarding the neural correlates of consciousness to emerge from the current state of the anesthesia literature is clearly at best a long shot, if not simply utter folly. Nevertheless, it is worth a look to see, even as an exploration, if some commonalities might emerge from the anesthesia literature that may ultimately be found to be related to the neural correlates of consciousness.

Neuroimaging studies of anesthesia in humans

From a neurobiology perspective, the state of "conscious awareness" is likely to be an emergent property of distributed neural networks involving the thalamus and cerebral cortex (Newman and Baars, 1993; Crick, 1994; Llinas et al., 1998; Edelman and Tononi, 2000; John, 2001; also see Baars, this volume; Tononi, this volume; Ribary, this volume, John, this volume). It has been hypothesized, and follows logically, that the loss of consciousness induced by general anesthetic agents may result, in part, from a disruption of functional interactions within these networks (Sugiyama et al., 1992; Angel, 1993; Ries and Puil, 1999; Alkire et al., 2000). A basic understanding of how anesthesia affects cerebral blood flow (CBF) and cerebral metabolic rate (CMR) is now readily available as common textbook material (Drummond and Patel, 2000). In general, most anesthetic agents decrease global cerebral metabolism in a dose-dependent manner with variable effects on global cerebral blood flow (Heinke and Schwarzbauer, 2002). Furthermore, the regional effects of a number of anesthetic agents have been studied with neuroimaging in humans at doses near to, or just more than, those required to produce unconsciousness. The "unconscious" end-point referred to in many anesthetic studies is that point at which an anesthetic dose is given in sufficient quantity that it causes a subject to be unable to respond to a verbal command or to a rousing shake. This is an end-point that occurs at a low anesthetic dose relative to the dose of anesthesia needed for an operation or that which would cause an isoelectric EEG. Thus, for many of the studies performed to date, the relatively low doses of anesthetics that have been used imply that some level of unconscious information processing can not necessarily be ruled out.

Commonalities exist between agents and classes of agents in how they affect regional CBF (rCBF), regional CMR of O_2 ($rCMRO_2$), and regional CMR of glucose (rCMRglu). Common regional effects, in many cases, may suggest a shared underlying mechanism of action. Ori and colleagues (1986) noted early on that one of the only common regional metabolic effects seen among the various agents studied across a multitude of animal studies was that they all caused metabolic suppression of the somatosensory cortex.

The basis for the original observation of anesthetic effects on the thalamus

For the anesthetic end-point of loss of consciousness in humans, a more recent case has been made for a common effect of most, if not all, agents on thalamic metabolism/blood flow and thalamocortical–corticothalamic connectivity (Alkire et al., 2000; White and Alkire, 2003). This commonality observation led to the development of the "thalamic consciousness switch" hypothesis of anesthetic-induced unconsciousness (Alkire et al., 2000). The fact that anesthetics have an ability to affect thalamocortical signaling is well recognized from *in vivo* electrophysiological work in animals (Angel, 1993; Steriade, 2001). At the cellular level, anesthetic agents compromise the natural firing patterns of thalamic network neurons (i.e., thalamocortical, corticothalamic, and reticulothalamic cells) by hyperpolarizing their resting membrane potentials (Nicoll and Madison, 1982; Berg-Johnsen and Langmoen, 1987; Steriade et al., 2001). As a result, and in a manner that parallels the mechanisms underlying physiologic sleep, a greater proportion of these network cells experience bursting rather than tonic activity (Steriade, 1994). This, in effect, blocks synaptic transmission of sensory information through the thalamus and diminishes the high frequency rhythms that characterize the spontaneous activity associated with the awake state and dreaming mentation (Angel, 1991; Llinas and Pare, 1991; Lytton and Sejnowski, 1991; Buzsáki and Chrobak, 1995; Steriade, 2000). It has been hypothesized that anesthetics may cause unconsciousness in the human brain because they induce a hyperpolarization blockade that involves a sufficient proportion of the thalamocortical cells and networks that are required for the maintenance of conscious awareness (Sugiyama et al., 1992; Ries and Puil, 1999; Alkire et al., 2000).

When the idea of a thalamic consciousness switch was originally developed in relation to human neuroimaging (Alkire et al., 2000), it took into account rCMRglu or rCBF effects on the thalamus that were observed in humans as a site of a common overlapping effect between the benzodiazepines, lorazepam (Volkow et al., 1995), and midazolam (Veselis et al., 1997), and between the intravenous anesthetic agent propofol (Fiset et al., 1999; Fiset this volume) and the inhalational agents isoflurane and halothane (Alkire et al., 2000).

Recent anesthesia studies continue to demonstrate the thalamic effect

Further additional study over the intervening years has remained consistent with the thalamic overlap effect and has shown replications of propofol's thalamic effects (Kaisti et al., 2003; Veselis et al., 2004), along with an overlapping thalamic effect for the additional inhalational anesthetic agent sevoflurane (Kaisti et al., 2002). Additionally, recent studies with another newer class of sedative anesthetics, the $\alpha 2$-adrenoreceptor agonists, dexmedetomidine (Prielipp et al., 2002), and clonidine (Bonhomme et al., 2004) have also shown a consistent overlapping regional suppression effect on the thalamus at doses that cause heavy sedation or at doses that are just beyond a loss of consciousness end-point. Furthermore, two recent replication studies of the lorazepam (Schreckenberger et al., 2004) and sevoflurane (Schlunzen et al., 2004) regional results have strengthened support for the idea of a common regional suppressive effect of anesthetics on the thalamus. Pain and vibrotactile sensory processing were previously examined during increasing doses of isoflurane (Antognini et al., 1997) and propofol (Bonhomme et al., 2001) anesthesia, respectively. Both agents cause signal suppression at the level of the thalamus during anesthetic-induced unconsciousness.

Figure 1 illustrates the relative regional effects of anesthetics found to date for a number of different agents across a number of different neuroimaging studies in humans. It illustrates a stylized combination of the results from a number of different studies in which the regional functional changes caused by an anesthetic substance delivered at, near, or encompassing a dose that caused a loss of consciousness or heavy sedation was investigated. The individual study images were flipped and sized appropriately so that the regional results could be merged together into a single composite image at the center of the figure. This central composite figure illustrates the spatial extent of the thalamic overlap amongst the different agents. The original color pattern applied to each regional effect from each study is retained. It should be noted, however, that in all but one case the regional results represent relative decreases in either rCMRglu or rCBF, despite the fact that some studies used a red-based color scale and others used a blue-based color scale. The one exception to the straightforward regional-decrease-of-activity-interpretation is for the study associated with the propofol correlation image. In that study, the color signifies the extent of the correlation between Decreasing rCBF and increasing anesthetic dose (Fiset et al., 1999).

The studies differ in the anesthetic end-points examined. An unconsciousness end-point was used for the propofol correlation image, the propofol rCBF image, the sevoflurane rCBF image, and the halothane and isoflurane conjunction image. Heavy sedation, where one or a few of the study subjects may have lost consciousness at some point was the behavioral end-point for each of the other studies. The composite figure is interesting in that it shows a clear central regional overlap effect of all agents on the thalamus. It also reveals that an overlap is present for a number of the agents involving the posterior cingulate and medial parietal cortical areas. Another overlapping effect between a few of the agents is seen in the medial basal forebrain areas. Each of these other common regional effects may also have some importance for the neural correlates of consciousness (Baars et al., 2003).

Clearly, interpreting an image such as this warrants some caution. The differences between studies are not necessarily trivial and the basis for comparison is often based on different techniques in which different scanners were used and the results come from multiple groups of independent researchers. Yet, it is precisely because of these differences that the strength of the regional thalamic overlap observation takes on additional importance. In other words, despite the often dramatic technical differences between studies, the one finding that emerges as potentially robust for anesthetic effects on consciousness is that when consciousness goes away or nearly goes away with any number of different anesthetics, a relative decrease in thalamic activity occurs. This relative thalamic decrease has to be interpreted in the broader context of each of these agents also causing a rather large change in global metabolism. Nevertheless, the relative thalamic effect implies that there is a minimal amount of regional thalamic activity that may be necessary to maintain consciousness, and the thalamus or thalamocortical networks, therefore, emerge as potentially important components of the neural correlate of consciousness. These empirical findings fit well with a number of theories regarding the neurobiology of consciousness.

Why is the thalamus at the heart of anesthetic effects on consciousness?

The centralized placement of the thalamus within the brain and its unique direct access to all incoming

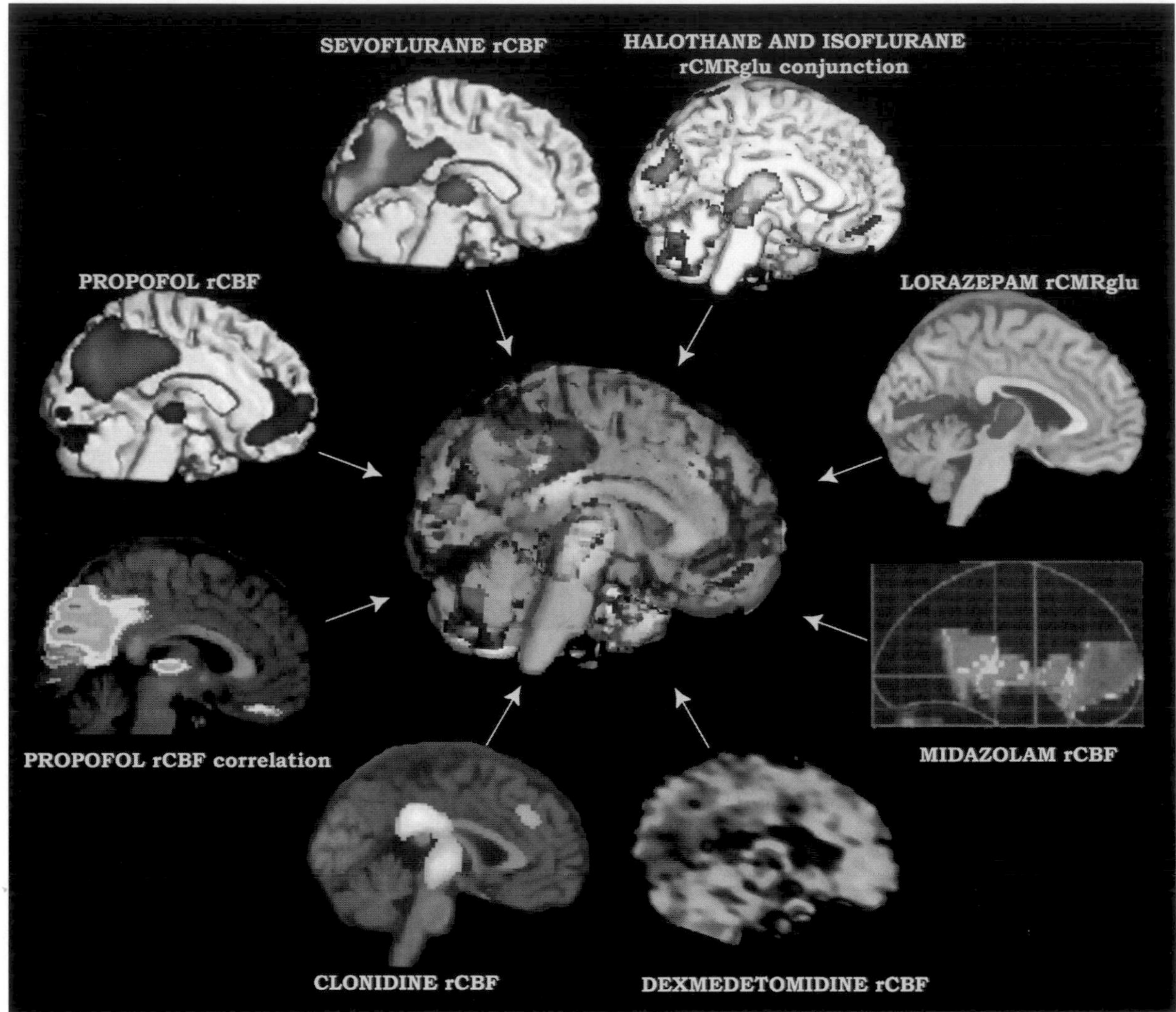

Fig. 1. The regional effects of anesthetics on brain function are shown in humans who were given various anesthetic agents at doses that caused, or nearly caused, a loss of consciousness. The data are from seven different groups of investigators and encompass the study of eight different agents. Clockwise from the 1:00 O'clock position the agents studied were halothane and isoflurane (Alkire et al., 2000), lorazepam (Schreckenberger et al., 2004), midazolam (Veselis et al., 1997), dexmedetomidine (Prielipp et al., 2002), clonidine (Bonhomme et al., 2004), propofol (Fiset et al., 1999), propofol (Kaisti et al., 2002), and sevoflurane (Kaisti et al., 2002). The regional effects were measured using either blood flow- or glucose metabolism-based techniques. The images were reoriented, and resized to allow the direct overlapping effects between studies to be shown in the central image. The original color scales were used. Nevertheless, all images show regional decreases in activity caused by anesthesia compared to the awake state, except the propofol correlation image, which shows where increasing anesthetic dose correlates with decreasing blood flow. The figure identifies that the regional suppressive effects of anesthetics on the thalamus is a common finding that is also associated with anesthetic-induced unconsciousness. See Plate 17.1 in Colour Plate Section.

sensory information, except for olfactory information, along with its access to cortical feedback, places the thalamus at the center of interest as a brain region that might have a central role to play in the mechanisms of consciousness and attention (Newman 1997a, b). It is in a unique position to be able to rapidly integrate essentially all the relevant brain activity that may be applicable to a unified perceptual experience. In essence, it all comes together in the thalamus.

Given that the majority of the regional metabolic Positron Emission Tomography (PET) signal originates from synaptic activity and that the thalamus receives a large afferent input from the cerebral cortex, the actual site of mechanistic overlap among agents is likely to be displaced from the thalamus and may actually reside in the cerebral cortex (Ori et al., 1986), or alternatively, in lower brainstem arousal centers (Newman and Baars, 1993; Newman, 1997a, b). The reduced thalamic metabolism during anesthesia could primarily reflect a drug induced decrease in primarily corticothalamic reverberant activity. Such an idea fits well with electrophysiologic studies of anesthesia (Angel, 1993), and with one study on the regional cerebral metabolic effects of enflurane in the rat, where enflurane's metabolic effects on the thalamus were unilaterally prevented with an ipsilateral cortical ablation (Nakakimura et al., 1988).

If one were to propose that the thalamus plays a fundamental role in consciousness, then one would expect that impairments of consciousness should occur not only if the thalamus is "turned off" with anesthetics, but also if it were lesioned. In fact, impairments of consciousness are known to occur with even relatively small lesions of the thalamus, especially those involving the intralaminar nuclei (Bogen, 1997; also see Graham this volume on neuropathology in the vegetative state). The intralaminar nuclei and the thalamic reticular nucleus are considered an extension of the brain-stem reticular activating system (Newman, 1997a, b) and anesthetic effects on the brain-stem reticular activating system have long been associated with anesthetic effects on consciousness (Moruzzi and Magoun, 1949).

Ultimately, the overlapping regional effect of anesthetics on the thalamus may likely be mediated through the direct effects anesthetics have on normal sleep pathways (Lydic and Biebuyck, 1994; Alkire et al., 2000; Nelson et al., 2002). Such an idea would seem to provide a reasonable explanation as to why the regional metabolic suppressive effects of these different anesthetics closely parallels the regional metabolic suppressive effects of sleep (Baars et al., 2003). This regional correspondence between sleep-induced changes and anesthetic induced changes was an observation we also noted some time ago when examining the regional metabolic effects of halothane (Alkire et al., 1999, 2000).

Is consciousness in the parietal cortex?

The second most consistent anesthetic-related regional overlap effect involves the posterior cingulate and medial parietal cortical areas. These posterior areas are of some interest as potential neural correlates of consciousness for five primary reasons. First, as noted above, and as seen in Fig. 1, a number of these anesthetic agents suppress activity in these posterior brain regions. Second, these posterior parietal regions have been noted to show a relative decrease in functioning during other altered states of consciousness, such as during the vegetative state (Laureys et al., 2004) and sleep (Maquet, 2000). Laureys noted further that a functional disconnection of this region with frontal brain regions appeared associated with the unconsciousness of the persistent vegetative state (Laureys et al., 1999) and restoration of connectivity between this brain region and frontal brain regions was associated with the return to consciousness (Laureys et al., 2000). Third, these regions, especially the posterior cingulate area, are involved in memory retrieval (Rugg and Wilding, 2000; Rugg et al., 2002). This retrieval effect has recently been shown to be multimodal and independent of response contingency; prompting Shannon and Buckner (2004) to "suggest that conceptions of posterior parietal cortical function should expand beyond attention to external stimuli and motor planning to incorporate higher-order cognitive functions." Fourth, some evidence links activity in these regions, especially the medial parietal lobes, to the first person perspective of consciousness. A line of research inquiry has developed in which the neural correlates of consciousness are sought using a technique in which experimental subjects manipulate their intra-personal perspective of an external situation (Zeman, 2001; Kircher and Leube, 2003; Vogeley et al., 2004; also see Lou, this volume). In one series of studies, subjects identify how many items in a scene are visible either from their first person perspective or from the perspective of another observer who is embedded in the scene

(Vogeley et al., 2004). Regional differences between these different perspective conditions are thought to represent those neuronal networks that are associated with the generation of self-consciousness. Such studies have identified that the medial parietal areas are involved in generating the first person perspective. Such localization would seem to fit well with the long established link between neglect syndromes and parietal damage. Fifth and finally, recent work has shown that the posterior cingulate and medial posterior parietal areas seem to be involved in the generation of the baseline functional state of the human brain (Burton et al., 2004). One interpretation of this baseline concept is that these brain regions are active as a reflection of ones self-conscious state when the brain is not involved in any specific cognitive task.

The nature of this baseline activity in the resting awake human brain that is not performing a specific cognitive task will require much further investigation, but it also serves to raise a specific question of what is being subtracted from what in the aforementioned anesthesia loss of consciousness studies. To put it differently, the relative parietal decreases that are seen in the above anesthesia studies could reflect a specific metabolic suppressive effect of each anesthetic agent on these particular brain regions, or alternatively the differential results between states might just be a reflection of the relatively higher regional activity that the brain starts with in these posterior areas. In any event, when the above multiple lines of evidence are taken together, the posterior cingulate and parietal cortical areas emerge as potential key brain regions that may be directly involved with the neurobiology of conscious awareness. It has been proposed that a feed-forward network of these parietal regions onto frontal regions might also be important for consciousness (Baars et al., 2003; also see Vogt and Laureys, this volume). There is a hint of a common regional frontal overlap effect between some of the anesthetics which is consistent with such an idea.

Network activity, anesthetic-induced signal suppression or signal scrambling?

As stated by Tononi and Edelman (1998), "Activation and deactivation of distributed neural populations in the thalamocortical system are not sufficient bases for conscious experience unless the activity of the neuronal groups involved is integrated rapidly and effectively." Following this logic, it is unlikely that a full characterization of the effects of anesthetic agents which ablate conscious awareness, can be made by observing only the regionally specific and global suppressive effects of these agents. A more comprehensive assessment would seem to require an additional understanding of how these agents affect functional integration across neural systems (Cariani, 2000).

Efforts to characterize functional integration across brain regions have lead to the conceptualization of two types of neural connectivities: functional connectivity and effective connectivity. Functional connectivity, defined as correlations between remote neurophysiological events (Friston et al., 1993), is simply a measure of the observed covariance in brain activity and provides insight into the degree to which two or more brain regions are functionally related. Effective connectivity, which is defined as the direct influence one brain region has over another (McIntosh and Gonzalez-Lima, 1994; Friston, 1995), provides insight into the direction and extent of these correlations by impinging anatomical constraints. The anatomical constraints represent a unique set of putative connections (and a model of their interactions) between the nodes of a defined neuroanatomical and/or functional network. In functional neuroimaging, effective connectivity assessments generally rely on the implementation of regression (Friston et al., 1997) or structural equation models (McIntosh and Gonzalez-Lima, 1994). Functional and effective connectivity should not be considered mutually exclusive properties of brain function but rather complementary; it is not uncommon to use functional connectivity measures to empirically identify functional networks within which effective connectivity is subsequently assessed. From this perspective, effective connectivity can be viewed as an extension of functional connectivity given some underlying neuroanatomical assumptions.

Given the above, it may not be simply a quantitative reduction in thalamic or corticothalamic metabolic activity that is key to anesthetic effects on consciousness (Heinke and Schwarzbauer,

2002), rather the key may be that anesthetics act to prevent coordinated communication between the thalamus and the cortex or even within cortico-cortical networks.

Such an idea was supported by a recent functional and effective connectivity analysis of inhalational anesthesia (White and Alkire, 2003). Using a path analysis approach it was determined that anesthetic-induced unconsciousness in humans is associated with a functional change in effective thalamocortical and corticocortical connectivity, such that the thalamus and cortex no longer effectively interact with one another at the point of anesthetic-induced unresponsiveness (see Fig. 2).

The data-driven approach to the network modeling procedure used in the connectivity analysis of anesthesia directed attention toward the lateral cerebello-thalamo-cortical system. The presumed primary role of the cerebello-thalamo-cortical system is in motor control. The cerebellar inputs to the cortex traveling through the thalamus are thought to represent excitatory influences on motor output regions (M1) after substantial sampling of incoming sensory and motor information (Jueptner et al., 1997; Gross et al., 2002). Disruption of cerebello-thalamo-cortical signaling during anesthesia is thus an interesting empirical finding that may fit well with Cotterill's (2001) ideas of consciousness as a controller of motor output.

Cotterill's interpretations coupled with the connectivity findings suggest that anesthetic agents could be interpreted not only as agents that cause a disruption of sensory information processing, but also as agents capable of effectively disengaging or decoupling sensory input systems from motor output systems. This uncoupling would include a disconnection of frontal brain regions from posterior brain regions. Such a functional disconnection is consistent with some observations regarding the importance of the frontal lobes, and the striatum for the generation of consciousness (Dandy, 1946). Furthermore, this uncoupling of frontal motor/planning systems from posterior sensory systems is a repeating theme regarding the neurobiology of consciousness and deserves much more future work (Baars et al., 2003; Crick and Koch, 2003; White and Alkire, 2003).

An additional demonstration that anesthetics change functional connectivity in the anesthetized brain has recently been reported for subjects rendered unconscious under sevoflurane inhalational anesthesia as studied with fMRI (Kerssens et al., 2005). These authors showed that a decrease in cortico-cortical functional activity occurs with unconsciousness, but they did not specifically investigate if a change in thalamocortical connectivity occurred in their study.

Ketamine is an *N*-methyl-D-aspartate (NMDA) antagonist and a dissociative anesthetic that was recently shown to cause regional metabolic increases in the human brain at subanesthetic doses (Langsjo et al., 2004). The imaging results with this agent would seem to point out that a suppression of thalamic activity mechanism can not be the only explanation for anesthetic induced unconsciousness. A study of ketamine's effects on functional and effective connectivity, however, remains to be reported. We hypothesize that a connectivity analysis of ketamine anesthesia will show a disconnection effect of thalamocortical and corticocortical connectivity, in a manner similar to that found previously with the volatile agents. As we await such an analysis, it is interesting to note, nonetheless, that one of ketamine's larger regional metabolic effects (albeit a relative increase in regional metabolism) was still localized to the thalamus (Langsjo et al., 2004).

Regional differences between agents

Differences exist between agents and classes of agents in how they affect rCBF and regional cerebral metabolism (Alkire et al., 1999; Drummond and Patel, 2000). These differences are evident even within agents of similar types that have presumed similar mechanisms of action. Such differences might imply different mechanisms of action between agents. For example, Veselis and colleagues (2004) recently demonstrated that roughly equivalent sedative and hypnotic doses of propofol and thiopental, two agents thought to share a common cellular GABAergic mechanism of action, actually have differential effects on rCBF. Thiopental had a marked effect on the cerebellum,

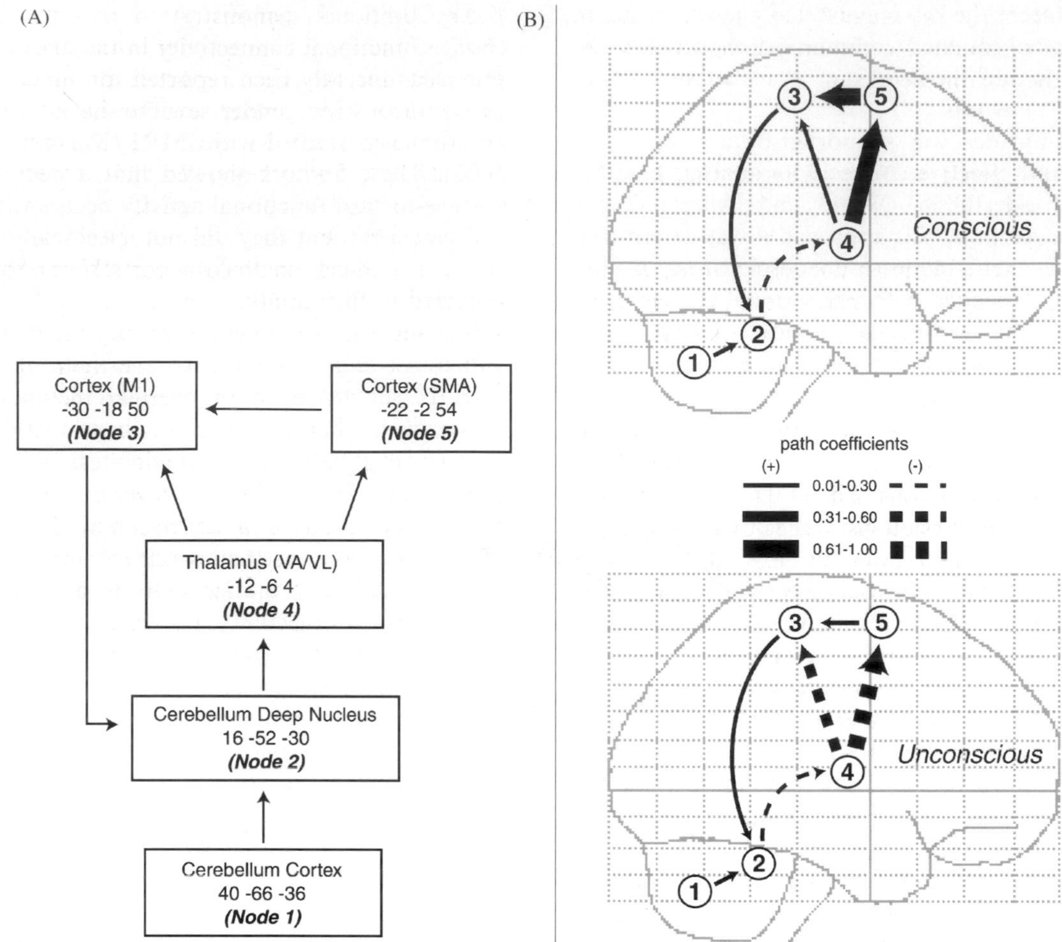

Fig. 2. Effective connectivity changes with anesthetic-induced unconsciousness in the human lateral cerebello-thalamo-cortical network (White and Alkire, 2003). (A) The network nodes, with their Talariach coordinates, and their modeled interactions. Structural equation modeling of this limited corticothalamic network (B) reveals that effective connectivity dramatically changes within this network, especially involving the thalamocortical and corticocortical interactions depending on the presence or absence of consciousness. Such a connectivity analysis approach can reveal network interactions and regional effects that might otherwise be missed with more traditional analysis techniques (Friston, 1995; McIntosh, 1999).

whereas propofol did not. Additionally, propofol had a marked effect on the thalamus and frontal lobes that thiopental did not. These differential effects suggest either regionally selective differential cellular mechanisms of action on neuronal activity for each drug or some differential effect on local flow/metabolism coupling. It is interesting to note that thiopental did not show a regionally selective decrease of thalamic blood flow, as might have been expected from the above discussion on the "thalamic consciousness switch" hypothesis. The thiopental observations were, however, based on a relatively small sample size and await confirmation in a larger study. Nonetheless, metabolic activity within the thalamus, as well as within the rest of the brain would still have been greatly reduced for those subjects on thiopental, so a lack of a specific regional thalamic effect does not imply that thalamic activity failed to be suppressed by this anesthetic. In other words, thiopental essentially still "turned off" the thalamus. It just did it in a manner consistent with how it also affected the rest of the brain.

The dose-dependent effects of anesthetics, keys to future study

As one awakens from an anesthetic, consciousness emerges from the subjective perception of complete oblivion. Of course, it remains unknown if oblivion is actually experienced by those experiencing it. Nonetheless, essentially all people who emerge from the unconsciousness of general anesthesia claim to have no memory of the time when they were unconscious. Their last memory is usually of the anesthesiologist telling them they would be going to "sleep" and their next memory is usually of waking in the operating room after their operation or of being awake in the recovery room.

It should be made explicit that the effects of anesthetics on consciousness do not represent an all-or-nothing process. The effects depend entirely on the dose and the type of agent used. Nevertheless, there are some common behavioral effects that are, by definition, common to all molecules that are considered anesthetic substances. It has been proposed that there are only two fundamental characteristics that identify a molecule as an anesthetic; namely that they all cause amnesia and that they all prevent movement to a noxious stimulus (Eger et al., 1997). Others would expand this limited definition to include the ability to produce unconsciousness (Antognini and Carstens, 2002). Much work in anesthesia has been directed toward understanding the mechanisms by which anesthetics prevent movement in response to a noxious stimulation (Sonner et al., 2003). Such studies investigate the end-point of relatively deep anesthesia, which is known as the minimum alveolar concentration (MAC)-response. In essence, this measure of anesthetic potency determines the MAC of an agent needed to prevent movement in 50% of subjects in response to a painful stimulation (Eger et al., 1965). More recently, attention has been directed toward understanding how each of the numerous component parts of anesthesia might work (Campagna et al., 2003; Rudolph and Antkowiak, 2004). A brief overview of the dose-dependent effects of anesthetics on human brain function and the resultant behavioral manifestations of a particular dose might provide insight for others not in the field and might serve as a focal point for future studies of anesthetic effects on consciousness (Antognini and Carstens, 2002).

If one considers an inhalation induction, the end points examined would look something like those depicted in Fig. 3, with some variation depending on the anesthetic agent chosen. First, several agents, at low concentrations, have been found to cause a paradoxical hyperalgesia before a level of analgesia is obtained (Zhang et al., 2000). Analgesia level is a much broader end-point than most others, implying a diverse and robust physiology. Next, still at relatively low doses, memory is impaired and explicit memory fails first (Ghoneim, 2004a, b). Implicit memory remains intact at levels up to 0.6 MAC (Renna et al., 2000). MAC-awake is the point at which response to

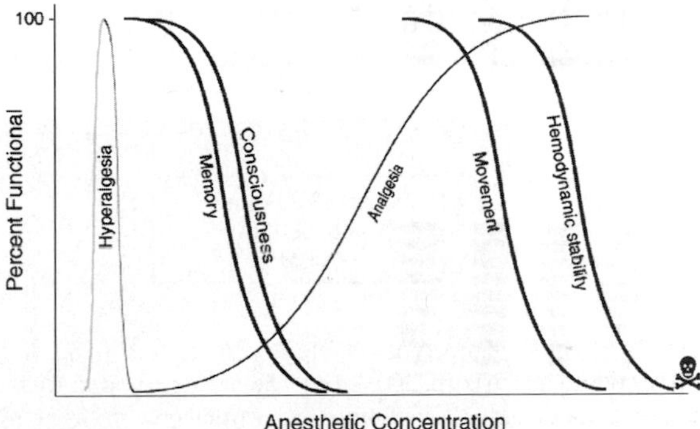

Fig. 3. The dose–response curves for end-points of anesthesia, (after Antognini and Carstens, 2002). Higher anesthetic concentrations are toward the right. The key end-point for relative anesthetic potency and the dose typically needed for anesthesia with surgery is that point where the MAC of an inhaled anesthetic prevents 50% of patients from moving in response to a surgical stimulus (Eger et al., 1965). At very low doses of agents, typically around 0.1 MAC a paradoxical hyperalgesia (pain enhancing) effect occurs (Zhang et al., 2000). Next, analgesia begins and increases with an increasing dose until, at much deeper levels, no movement occurs with any stimulation. The memory effects of anesthesia occur at around 0.1–0.3 MAC (Alkire and Gorski, 2004) and deeper. Consciousness is typically lost at approximately 0.3–0.4 MAC, or at about 30–40% of the anesthetic dose actually needed for surgery. Doses much above those needed to prevent movement can cause a lethal collapse of the cardiovascular system.

verbal command is lost in 50% of subjects. This is typically considered the point at which consciousness is lost and occurs at 0.3–0.4 MAC in younger, healthy individuals (Newton et al., 1990). Suppression of movement occurs at still deeper levels of anesthesia with levels above 1 MAC usually considered as surgical anesthesia. Thereafter, increasing anesthetic doses lead to decreased cardiovascular stability and, with some agents, an isoelectric EEG. Deeper anesthesia can be lethal. The endpoint functions depicted in Fig. 3 indicate that anesthetic agents effects on various aspects of cognition and consciousness is not simple and deserves further study.

Ligand studies

The differential regional effects evident between agents occur in particular detailed patterns. Understanding these detailed patterns can offer clues to the underlying cellular mechanisms of action for each agent, especially when the agent specific patterns overlap with the specific regional distribution of a known receptor system. For example, propofol's regional cerebral metabolic effects are correlated with the regional cerebral distribution of GABA receptors (Alkire and Haier, 2001), as shown in Fig. 4. Unlike propofol, isoflurane's regional metabolic effects do not follow the distribution of the GABAergic system, but rather they are inversely related to the acetylcholine muscarinic system (Alkire and Haier, 2001).

As new imaging ligands become available there will likely be a rapid expansion in the number of studies that link anesthetic mechanisms with consciousness mechanisms and with specific receptors. This imaging approach, however, is clearly not the only approach toward linking functions with molecular mechanisms (Rudolph and Antkowiak, 2004). Yet, this receptor-based imaging approach has already been used to suggest a link between GABAergic receptor changes and isoflurane (Gyulai et al., 2001), as well as propofol and sevoflurane anesthesia (Salmi et al., 2004). Some recent work has also suggested a link between anesthetic-induced unconsciousness and changes in muscarinic receptor binding (Backman et al., 2004). Finally, work in animals has found effects

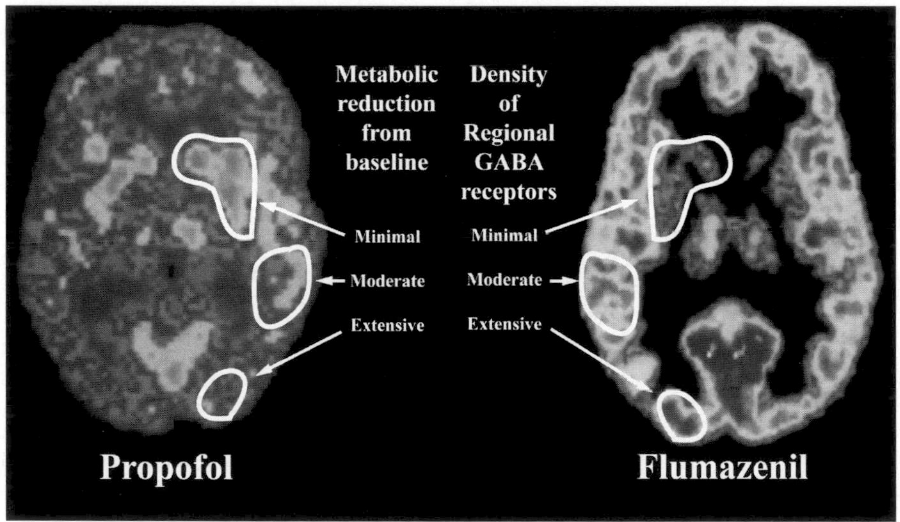

Fig. 4. Regional cerebral metabolic suppressive effects of propofol anesthesia, on the left (Alkire et al., 1995), are compared with the regional distribution of ^{11}C-flumazenil binding (Roland, 1993). The figure suggests that regional cerebral metabolism is more depressed during propofol anesthesia (a presumed GABA agonist) in those brain regions with higher density of GABA receptors. Representative areas of metabolic reduction and receptor density are shown as regions-of-interest. A formal region-of-interest analysis examining this idea revealed that the magnitude of the metabolic suppression caused by propofol in various brain regions is highly correlated with the density of the GABA receptors in those regions (Alkire and Haier, 2001). Future imaging work with other ligands may reveal more directly how the regional metabolic effects of various anesthetics might be interpreted as a simple reflection of their underlying biochemical interactions. See Plate 17.4 in Colour Plate Section.

of anesthetics on binding characteristics of dopamine receptors (Momosaki et al., 2004). Future work should help establish what are the relevant biochemical changes induced by anesthetics that may contribute to a loss of consciousness.

Conclusions

It appears that a convergence of evidence points toward the thalamus, thalamocortical, and corticocortical interactions as being critically involved with mediating not only anesthetic induced unconsciousness, but also with mediating other forms of altered states of consciousness. The study of the effects of anesthetics on consciousness is only just beginning. Even at this early stage, however, the current results suggest pharmacological manipulations in humans coupled with brain imaging techniques may be one experimental way out of the consciousness quagmire. The next generation of anesthetic-related studies that will manipulate the qualitative nature of consciousness has already begun. Investigation of the components of the neurobiological systems that support consciousness, such as memory is well underway (Sperling et al., 2002; Veselis et al., 2002; Honey et al., 2004, 2005).

Abbreviations

CBF	cerebral blood flow
CMR	cerebral metabolic rate
EEG	electroencephalograph
GABA	γ-aminobutyric acid
MAC	minimum alveolar concentration
NMDA	*N*-methyl-D-aspartate
rCBF	regional CBF
rCMR	glu regional CMR of glucose
$rCMRO_2$	regional CMR of O_2

Acknowledgments

Funded, in part, by grant # RO1-GM065212 from the National Institutes of Health, Bethesda, MD, USA.

References

Alkire, M.T. and Gorski, L.A. (2004) Relative amnesic potency of five inhalational anesthetics follows the Meyer-Overton rule. Anesthesiology, 101: 417–429.

Alkire, M.T. and Haier, R.J. (2001) Correlating in vivo anaesthetic effects with ex vivo receptor density data supports a GABAergic mechanism of action for propofol, but not for isoflurane. Br. J. Anaesth., 86: 618–626.

Alkire, M.T., Haier, R.J., Barker, S.J., Shah, N.K., Wu, J.C. and Kao, Y.J. (1995) Cerebral metabolism during propofol anesthesia in humans studied with positron emission tomography. Anesthesiology, 82: 393–403.

Alkire, M.T., Haier, R.J. and Fallon, J.H. (2000) Toward a unified theory of narcosis: brain imaging evidence for a thalamocortical switch as the neurophysiologic basis of anesthetic-induced unconsciousness. Conscious Cogn., 9: 370–386.

Alkire, M.T., Haier, R.J. and Fallon, J.H. (1998) Towards the neurobiology of consciousness: using brain imaging and anesthesia to investigate the anatomy of consciousness. In: Hameroff S.R., Kaszniak A.W. and Scott A.C. (Eds.), Toward a Science of Consciousness II. MIT Press, London, pp. 255–268.

Alkire, M.T., Pomfrett, C.J., Haier, R.J., Gianzero, M.V., Chan, C.M., Jacobsen, B.P. and Fallon, J.H. (1999) Functional brain imaging during anesthesia in humans: effects of halothane on global and regional cerebral glucose metabolism. Anesthesiology, 90: 701–709.

Angel, A. (1991) The Brown G.L. lecture. Adventures in anaesthesia. Exp. Physiol., 76: 1–38.

Angel, A. (1993) Central neuronal pathways and the process of anaesthesia. Br. J. Anaesth., 71: 148–163.

Antognini, J.F., Buonocore, M.H., Disbrow, E.A. and Carstens, E. (1997) Isoflurane anesthesia blunts cerebral responses to noxious and innocuous stimuli: a fMRI study. Life Sci., 61: L349–L354.

Antognini, J.F. and Carstens, E. (2002) In vivo characterization of clinical anaesthesia and its components. Br. J. Anaesth., 89: 156–166.

Baars, B.J., Ramsoy, T.Z. and Laureys, S. (2003) Brain, conscious experience and the observing self. Trends Neurosci., 26: 671–675.

Backman, S.B., Fiset, P. and Plourde, G. (2004) Cholinergic mechanisms mediating anesthetic induced altered states of consciousness. Prog. Brain. Res., 145: 197–206.

Berg-Johnsen, J. and Langmoen, I.A. (1987) Isoflurane hyperpolarizes neurones in rat and human cerebral cortex. Acta Physiol. Scand., 130: 679–685.

Bogen, J.E. (1997) Some neurophysiologic aspects of consciousness. Semin. Neurol., 17: 95–103.

Bonhomme, V., Fiset, P., Meuret, P., Backman, S., Plourde, G., Paus, T., Bushnell, M.C. and Evans, A.C. (2001) Propofol anesthesia and cerebral blood flow changes elicited by vibrotactile stimulation: a positron emission tomography study. J. Neurophysiol., 85: 1299–1308.

Bonhomme, V., Laureys, S., Maquet, P., Hans, P. and Luxen, A. (2004) Effect of Clonidine Infusion on Distribution of

Regional Cerebral Blood Flow (rCBF) in Healthy Human Volunteers: A Positron Emission Tomography (PET) Study. Anesthesiology, A-881 (abstract).

Burton, H., Snyder, A.Z. and Raichle, M.E. (2004) Default brain functionality in blind people. Proc. Natl. Acad. Sci. USA, 101: 15500–15505.

Buzsáki, G. and Chrobak, J.J. (1995) Temporal structure in spatially organized neuronal ensembles: a role for interneuronal networks. Curr. Opin. Neurobiol., 5: 504–510.

Campagna, J.A., Miller, K.W. and Forman, S.A. (2003) Mechanisms of actions of inhaled anesthetics. N. Engl. J. Med., 348: 2110–2124.

Cariani, P. (2000) Anesthesia, neural information processing, and conscious awareness. Conscious Cogn., 9: 387–395.

Chalmers, D.J. (1995) Facing up to the problem of consciousness. J. Con. Studies, 2: 200–219.

Cotterill, R.M. (2001) Cooperation of the basal ganglia, cerebellum, sensory cerebrum and hippocampus: possible implications for cognition, consciousness, intelligence and creativity. Prog. Neurobiol., 64: 1–33.

Crick, F. (1994) The Astonishing Hypothesis. Scribner, New York.

Crick, F. and Koch, C. (2003) A framework for consciousness. Nat. Neurosci., 6: 119–126.

Dandy, W. (1946) The location of the conscious center in the brain. Bull. Johns Hopkins Hosp., 79: 34–58.

Dennett, D.C. (1996) Facing backwards on the problem of consciousness. J. Con. Studies, 3: 4–6.

Drummond, J.C. and Patel, P. (2000) Cerebral blood flow and metabolism. In: Miller R.D. (Ed.), Anesthesia. Churchill-Livingstone, New York, pp. 1203–1256.

Edelman, G.M. and Tononi, G. (2000) A Universe of Consciousness. Basic Books, New York.

Eger, E.I., Koblin, D.D., Harris, R.A., Kendig, J.J., Pohorille, A., Halsey, M.J. and Trudell, J.R. (1997) Hypothesis: inhaled anesthetics produce immobility and amnesia by different mechanisms at different sites. Anesth. Analg., 84: 915–918.

Eger, E.I., Saidman, L.J. and Brandstater, B. (1965) Minimum alveolar anesthetic concentration: a standard of anesthetic potency. Anesthesiology, 26: 756–763.

Fiset, P., Paus, T., Daloze, T., Plourde, G., Meuret, P., Bonhomme, V., Hajj-Ali, N., Backman, S.B. and Evans, A.C. (1999) Brain mechanisms of propofol-induced loss of consciousness in humans: a positron emission tomographic study. J. Neurosci., 19: 5506–5513.

Friston, K.J. (1995) Functional and effective connectivity in neuroimaging: a synthesis. Hum. Brain. Mapping, 2: 56–78.

Friston, K.J., Büchel, C., Fink, G.R., Morris, J., Rolls, E. and Dolan, R.J. (1997) Psychophysiological and modulatory interactions in neuroimaging. Neuroimage, 6: 218–229.

Friston, K.J., Frith, C.D., Liddle, P.F. and Frackowiak, R.S. (1993) Functional connectivity: the principal-component analysis of large (PET) data sets. J. Cereb. Blood Flow Metab., 13: 5–14.

Ghoneim, M.M. (2004a) Drugs and human memory (Part 2). Clinical, theoretical, and methodologic issues. Anesthesiology, 100: 1277–1297.

Ghoneim, M.M. (2004b) Drugs and human memory (Part 1): Clinical, theoretical, and methodologic issues. Anesthesiology, 100: 987–1002.

Gross, J., Timmermann, L., Kujala, J., Dirks, M., Schmitz, F., Salmelin, R. and Schnitzler, A. (2002) The neural basis of intermittent motor control in humans. Proc. Natl. Acad. Sci. U.S.A., 99: 2299–2302.

Gyulai, F.E., Mintun, M.A. and Firestone, L.L. (2001) Dose-dependent enhancement of in vivo GABAa-benzodiazepine receptor binding by isoflurane. Anesthesiology, 95: 585–593.

Heinke, W. and Schwarzbauer, C. (2002) In vivo imaging of anaesthetic action in humans: approaches with positron emission tomography (PET) and functional magnetic resonance imaging (fMRI). Br. J. Anaesth., 89: 112–122.

Honey, G.D., Honey, R.A., O'Loughlin, C., Sharar, S.R., Kumaran, D., Suckling, J., Menon, D.K., Sleator, C., Bullmore, E.T. and Fletcher, P.C. (2005) Ketamine Disrupts Frontal and Hippocampal Contribution to Encoding and Retrieval of Episodic Memory: An fMRI Study. Cereb. Cortex, 15: 749–759.

Honey, R.A., Honey, G.D., O'Loughlin, C., Sharar, S.R., Kumaran, D., Bullmore, E.T., Menon, D.K., Donovan, T., Lupson, V.C., Bisbrown-Chippendale, R. and Fletcher, P.C. (2004) Acute ketamine administration alters the brain responses to executive demands in a verbal working memory task: an FMRI study. Neuropsychopharmacology, 29: 1203–1214.

John, E.R. (2001) The neurophysics of consciousness. Brain Res. Rev., 39: 1–28.

Jueptner, M., Ottinger, S., Fellows, S.J., Adamschewski, J., Flerich, L., Muller, S.P., Diener, H.C., Thilmann, A.F. and Weiller, C. (1997) The relevance of sensory input for the cerebellar control of movements. Neuroimage, 5: 41–48.

Kaisti, K.K., Langsjo, J.W., Aalto, S., Oikonen, V., Sipila, H., Teras, M., Hinkka, S., Metsahonkala, L. and Scheinin, H. (2003) Effects of sevoflurane, propofol, and adjunct nitrous oxide on regional cerebral blood flow, oxygen consumption, and blood volume in humans. Anesthesiology, 99: 603–613.

Kaisti, K.K., Metsahonkala, L., Teras, M., Oikonen, V., Aalto, S., Jaaskelainen, S., Hinkka, S. and Scheinin, H. (2002) Effects of surgical levels of propofol and sevoflurane anesthesia on cerebral blood flow in healthy subjects studied with positron emission tomography. Anesthesiology, 96: 1358–1370.

Kerssens, C., Peltier, S.J., Hamann, S.B., Sebel, P.S., Byas-Smith, M. and Hu, X. (2005) Functional connectivity changes with concentration of sevoflurane anesthesia. Neuroreport, 16: 285–288.

Kircher, T.T. and Leube, D.T. (2003) Self-consciousness, self-agency, and schizophrenia. Conscious Cogn., 12: 656–669.

Kulli, J. and Koch, C. (1991) Does anesthesia cause loss of consciousness? Trends Neurosci., 14: 6–10.

Langsjo, J.W., Salmi, E., Kaisti, K.K., Aalto, S., Hinkka, S., Aantaa, R., Oikonen, V., Viljanen, T., Kurki, T., Silvanto, M. and Scheinin, H. (2004) Effects of subanesthetic ketamine on regional cerebral glucose metabolism in humans. Anesthesiology, 100: 1065–1071.

Laureys, S., Faymonville, M.E., Luxen, A., Lamy, M., Franck, G. and Maquet, P. (2000) Restoration of thalamocortical connectivity after recovery from persistent vegetative state. Lancet, 355: 1790–1791.

Laureys, S., Goldman, S., Phillips, C., Van Bogaert, P., Aerts, J., Luxen, A., Franck, G. and Maquet, P. (1999) Impaired effective cortical connectivity in vegetative state: preliminary investigation using PET. Neuroimage, 9: 377–382.

Laureys, S., Owen, A.M. and Schiff, N.D. (2004) Brain function in coma, vegetative state, and related disorders. Lancet Neurol., 3: 537–546.

Llinas, R., Ribary, U., Contreras, D. and Pedroarena, C. (1998) The neuronal basis for consciousness. Philos. Trans. R. Soc. Lond B. Biol. Sci., 353: 1841–1849.

Llinas, R.R. and Pare, D. (1991) Of dreaming and wakefulness. Neuroscience, 44: 521–535.

Lydic, R. and Biebuyck, J.F. (1994) Sleep neurobiology: relevance for mechanistic studies of anaesthesia [editorial]. Br. J. Anaesth., 72: 506–508.

Lytton, W.W. and Sejnowski, T.J. (1991) Simulations of cortical pyramidal neurons synchronized by inhibitory interneurons. J. Neurophysiol., 66: 1059–1079.

Maquet, P. (2000) Functional neuroimaging of normal human sleep by positron emission tomography. J. Sleep Res., 9: 207–231.

McIntosh, A.R. (1999) Mapping cognition to the brain through neural interactions. Memory, 7: 523–548.

McIntosh, A.R. and Gonzalez-Lima, F. (1994) Structural Equation Modeling and Its Application to Network Analysis in Functional Brain Imaging. Hum. Brain. Mapp., 2: 2–22.

Momosaki, S., Hatano, K., Kawasumi, Y., Kato, T., Hosoi, R., Kobayashi, K., Inoue, O. and Ito, K. (2004) Rat-PET study without anesthesia: anesthetics modify the dopamine D1 receptor binding in rat brain. Synapse, 54: 207–213.

Moruzzi, G. and Magoun, H.W. (1949) Brain stem reticular formation and activation of the EEG. Electroencephalogr. Clin. Neurophysiol., 1: 455–473.

Nakakimura, K., Sakabe, T., Funatsu, N., Maekawa, T. and Takeshita, H. (1988) Metabolic activation of intercortical and corticothalamic pathways during enflurane anesthesia in rats. Anesthesiology, 68: 777–782.

Nelson, L.E., Guo, T.Z., Lu, J., Saper, C.B., Franks, N.P. and Maze, M. (2002) The sedative component of anesthesia is mediated by GABA(A) receptors in an endogenous sleep pathway. Nat. Neurosci., 5: 979–984.

Newman, J. (1997a) Putting the puzzle together. Part I: Towards a general theory of the neural correlates of consciousness. J. Consciousness Stud., 4: 46–66.

Newman, J. (1997b) Putting the puzzle together. Part II: Towards a general theory of the neural correlates of consciousness. J. Consciousness Stud., 4: 100–121.

Newman, J. and Baars, B.J. (1993) A neural attentional model for access to consciousness: A global workspace perspective. Concepts Neurosci., 4: 255–290.

Newton, D.E., Thornton, C., Konieczko, K., Frith, C.D., Dore, C.J., Webster, N.R. and Luff, N.P. (1990) Levels of consciousness in volunteers breathing sub-MAC concentrations of isoflurane. Br. J. Anaesth., 65: 609–615.

Nicoll, R.A. and Madison, D.V. (1982) General anesthetics hyperpolarize neurons in the vertebrate central nervous system. Science, 217: 1055–1057.

Ori, C., Dam, M., Pizzolato, G., Battistin, L. and Giron, G. (1986) Effects of isoflurane anesthesia on local cerebral glucose utilization in the rat. Anesthesiology, 65: 152–156.

Plourde, G. (2001) Identifying the neural correlates of consciousness: strategies with general anesthetics. Conscious Cogn., 10: 241–244.

Prielipp, R.C., Wall, M.H., Tobin, J.R., Groban, L., Cannon, M.A., Fahey, F.H., Gage, H.D., Stump, D.A., James, R.L., Bennett, J. and Butterworth, J. (2002) Dexmedetomidine induced sedation in volunteers decreases regional and global cerebral blood flow. Anesth. Analg., 95: 1052–1059.

Renna, M., Lang, E.M. and Lockwood, G.G. (2000) The effect of sevoflurane on implicit memory: a double-blind, randomised study. Anaesthesia, 55: 634–640.

Ries, C.R. and Puil, E. (1999) Mechanism of anesthesia revealed by shunting actions of isoflurane on thalamocortical neurons. J. Neurophysiol., 81: 1795–1801.

Roland, P. (1993) Brain Activation. Wiley-Liss, Inc, New York.

Rudolph, U. and Antkowiak, B. (2004) Molecular and neuronal substrates for general anaesthetics. Nat. Rev. Neurosci., 5: 709–720.

Rugg, M.D., Otten, L.J. and Henson, R.N. (2002) The neural basis of episodic memory: evidence from functional neuroimaging. Phil. Trans. R. Soc. Lond. B. Biol. Sci., 357: 1097–1110.

Rugg, M.D. and Wilding, E.L. (2000) Retrieval processing and episodic memory. Trends Cogn. Sci., 4: 108–115.

Salmi, E., Kaisti, K.K., Metsahonkala, L., Oikonen, V., Aalto, S., Nagren, K., Hinkka, S., Hietala, J., Korpi, E.R. and Scheinin, H. (2004) Sevoflurane and propofol increase 11C-flumazenil binding to gamma-aminobutyric acidA receptors in humans. Anesth. Analg., 99: 1420–1426.

Schlunzen, L., Vafaee, M.S., Cold, G.E., Rasmussen, M., Nielsen, J.F. and Gjedde, A. (2004) Effects of subanaesthetic and anaesthetic doses of sevoflurane on regional cerebral blood flow in healthy volunteers. A positron emission tomographic study. Acta Anaesthesiol. Scand., 48: 1268–1276.

Schreckenberger, M., Lange-Asschenfeld, C., Lochmann, M., Mann, K., Siessmeier, T., Buchholz, H.G., Bartenstein, P. and Grunder, G. (2004) The thalamus as the generator and modulator of EEG alpha rhythm: a combined PET/EEG study with lorazepam challenge in humans. Neuroimage, 22: 637–644.

Shannon, B.J. and Buckner, R.L. (2004) Functional-anatomic correlates of memory retrieval that suggest nontraditional processing roles for multiple distinct regions within posterior parietal cortex. J. Neurosci., 24: 10084–10092.

Sonner, J.M., Antognini, J.F., Dutton, R.C., Flood, P., Gray, A.T., Harris, R.A., Homanics, G.E., Kendig, J., Orser, B., Raines, D.E., Trudell, J., Vissel, B. and Eger, E.I. (2003) Inhaled anesthetics and immobility: mechanisms, mysteries, and minimum alveolar anesthetic concentration. Anesth. Analg., 97: 718–740.

Sperling, R., Greve, D., Dale, A., Killiany, R., Holmes, J., Rosas, H.D., Cocchiarella, A., Firth, P., Rosen, B., Lake, S., Lange, N., Routledge, C. and Albert, M. (2002) Functional MRI detection of pharmacologically induced memory impairment. Proc. Natl. Acad. Sci. USA, 99: 455–460.

Steriade, M. (1994) Sleep oscillations and their blockage by activating systems. J. Psychiatry. Neurosci., 19: 354–358.

Steriade, M. (2000) Corticothalamic resonance, states of vigilance and mentation. Neuroscience, 101: 243–276.

Steriade, M. (2001) Impact of network activities on neuronal properties in corticothalamic systems. J. Neurophysiol., 86: 1–39.

Steriade, M., Timofeev, I. and Grenier, F. (2001) Natural waking and sleep states: a view from inside neocortical neurons. J. Neurophysiol., 85: 1969–1985.

Sugiyama, K., Muteki, T. and Shimoji, K. (1992) Halothane-induced hyperpolarization and depression of postsynaptic potentials of guinea pig thalamic neurons in vitro. Brain Res., 576: 97–103.

Tononi, G. and Edelman, G.M. (1998) Consciousness and complexity. Science, 282: 1846–1851.

Veselis, R.A., Reinsel, R.A., Beattie, B.J., Mawlawi, O.R., Feshchenko, V.A., DiResta, G.R., Larson, S.M. and Blasberg, R.G. (1997) Midazolam changes cerebral blood flow in discrete brain regions: an H2(15)O positron emission tomography study. Anesthesiology, 87: 1106–1117.

Veselis, R.A., Reinsel, R.A., Feshchenko, V.A. and Dnistrian, A.M. (2002) A neuroanatomical construct for the amnesic effects of propofol. Anesthesiology, 97: 329–337.

Veselis, R.A., Feshchenko, V.A., Reinsel, R.A., Dnistrian, A.M., Beattie, B. and Akhurst, T.J. (2004) Thiopental and propofol affect different regions of the brain at similar pharmacologic effects. Anesth. Analg., 99: 399–408.

Vogeley, K., May, M., Ritzl, A., Falkai, P., Zilles, K. and Fink, G.R. (2004) Neural correlates of first-person perspective as one constituent of human self-consciousness. J. Cogn. Neurosci., 16: 817–827.

Volkow, N.D., Wang, G.J., Hitzemann, R., Fowler, J.S., Pappas, N., Lowrimore, P., Burr, G., Pascani, K., Overall, J. and Wolf, A.P. (1995) Depression of thalamic metabolism by lorazepam is associated with sleepiness. Neuropsychopharmacology, 12: 123–132.

White, N.S. and Alkire, M.T. (2003) Impaired thalamocortical connectivity in humans during general-anesthetic-induced unconsciousness. Neuroimage, 19: 402–411.

Zeman, A. (2001) Consciousness. Brain, 124: 1263–1289.

Zhang, Y., Eger, E.I., Dutton, R.C. and Sonner, J.M. (2000) Inhaled anesthetics have hyperalgesic effects at 0. 1 minimum alveolar anesthetic concentration. Anesth. Analg., 91: 462–466.

S. Laureys (Ed.)
Progress in Brain Research, Vol. 150
ISSN 0079-6123

Chapter 18

Brain imaging in research on anesthetic mechanisms: studies with propofol

Pierre Fiset*, Gilles Plourde and Steven B. Backman

Department of Anesthesiology, McGill University, Montreal, QC H3A 1A2, Canada

Abstract: Brain imaging helps to refine our understanding of the anesthetic effect and is providing novel information that result in the formulation of hypotheses. They have shown that anesthetics act on specific structures that have been known to be important for consciousness at large. They have also helped to show that anesthetics act on specific structures regionally, rather than being non-specific, general depressant of the central nervous system (CNS). A constant finding is that the drugs that we use seem to exert their action on specific sites within the CNS. This is true for a wide variety of drugs like midazolam, anesthetic vapors and opiates. The thalamus has consistently shown marked deactivation coincident with the anesthesia-induced loss of consciousness, appearing to be a very important target of anesthetic effect. Additionally, when vibro-tactile or pain stimulation is given, anesthetics significantly effect cortical structures even before loss of consciousness while loss of transmission at the thalamic level seems to coincide with loss of consciousness. Finally, the use of radioligands allow in vivo characterization of anesthetic effects on neurotransmitter systems.

Introduction

Recent advances in target-specific "designer" anesthetic drugs coupled with advances in molecular neurobiology and brain imaging techniques permit a sophisticated approach to studying how anesthetic drugs act on the central nervous system (CNS). The early observation that potency and oil solubility of anesthetics are tightly correlated formed the basis of the unitary theory of anesthetic action (Meyer, 1899; Overton, 1901). That is, anesthetic drugs exert their effect(s) via a single mechanism of action at the neuronal membrane level based on their physical–chemical properties. This theory has given way to mechanisms of drug action on specific protein–lipid complexes, neurotransmitters and second-messenger systems, although non-specific effects cannot be excluded (for review see Campagna et al., 2003; Rudolph and Antkowiak, 2004).

Anesthetic drugs produce predictable, repeatable and reversible changes in the level of consciousness that range from light sedation to profound unresponsiveness. Precise alterations in level of consciousness can be achieved with anesthetic drugs because they can be titrated to achieve specific behavioral endpoints. With volatile agents, drug dose can be controlled because the expired concentration can be measured, which is an approximation of plasma drug concentration that is in equilibrium with the CNS. With intravenous agents, their population-based pharmacokinetic properties coupled with computer-controlled infusion permits accurate prediction of target plasma (hence CNS) drug concentration (Shafer and Gregg, 1992; Varvel et al., 1992; Vuyk et al., 1995).

Brain imaging is a powerful tool for identifying CNS sites affected by anesthetic drugs. This is based on the premise that anesthetics induce localized

*Corresponding author. Tel.: +1 (514) 842 1231, ext. 34887; E-mail: pierre.fiset@muhc.mcgill.ca

DOI: 10.1016/S0079-6123(05)50018-9

changes in neuronal function, and as blood flow and neuronal activity are tightly coupled under normal conditions, changes in blood flow are accurate markers of changes in activity.

Since anesthetic drugs alter the level of consciousness, CNS sites of action may also be the structures involved in the control of natural states of consciousness (e.g., sleep–awake states). The seminal importance of thalamic and brain stem structures for consciousness have been identified (Moruzzi and Magoun, 1949; Steriade et al., 1993, 1997; Steriade, 1996), and these are the likely targets for anesthetic action.

The present review will address our interpretation of functional imaging data obtained during various anesthetic-induced changes in the level of consciousness. We asked the following three questions:

- Are the areas of the CNS implicated in the maintenance of conscious states targets for anesthetic drugs?
- Does the dose-dependent effect of general anesthetic drugs have a neural correlate?
- Can brain imaging be used to identify specific neurotransmitter systems affected by anesthetics?

The studies reported here were done on human volunteers. The intravenous anesthetic propofol was used because of its pharmacodynamic properties of rapid onset and offset, and predictable pharmacokinetics. It was administered using a computer-controlled infusion pump and blood samples were taken at regular intervals to confirm target plasma drug concentration. Concentrations were targeted to achieve behavioral states of light sedation (0.5 μg/mL), deep sedation (1.5 μg/mL), and unconsciousness (3.0 μg/mL). In the present review, we have limited ourselves to reporting data from studies done in our own laboratory. This is not meant to ignore the work of other researchers who have done important work on anesthesia and brain imaging (for a comprehensive review see Heinke and Schwarzbauer, 2002; also see Alkire, this volume)

Anesthetic drugs and the CNS: target sites

In our first study (Fiset et al., 1999), positron emission tomography (PET) images of the brain were obtained using the H_2O bolus technique to identify changes in regional cerebral blood flow (rCBF). Volunteers were scanned at four different states: awake baseline (no propofol), light sedation, deep sedation and unconsciousness. We identified specific sites in the CNS, where rCBF was specifically and significantly decreased during propofol administration: the thalamus, the precuneus and cuneus, and the parieto–occipital cortex (see Fig. 1A).

The thalamus is an important relay for afferent information to the cortex. We have consistently observed that the thalamus is deactivated at the point of loss of consciousness. Moreover the deactivation is coincident with a decrease in midbrain reticular formation activity. These two brain structures are fundamental in the maintenance of the conscious state (Moruzzi and Magoun, 1949; Steriade, 1996).

The significance of the other CNS sites affected by propofol can be interpreted in the context of their roles in processing of information. The precuneus, posterior cingulate cortex, right angular gyrus and cuneus are implicated in visual associative activities, integration of sensory information, processing of spatial information and evaluation of sensory inputs sub-serving spatial orientation and memory. It is likely significant that regions associated with higher processing functions of the brain appear to be specifically and preferentially affected by anesthetic drugs.

Neural correlates of concentration-dependent effects: tactile transmission

One of the most useful and interesting properties of anesthetic drugs is that their effects vary as a function of dose, or more precisely, of concentration. For example, even with relatively small drug concentration, whereby light sedation is produced, patients report a change in their ability to perceive sensory stimuli (Dunnet et al., 1994). We studied the effects of propofol on cerebral structures involved in the processing of vibro-tactile information (Bonhomme et al., 2001). The pattern of brain activation produced by vibro-tactile stimulation is well known, and when such a stimulus is applied to

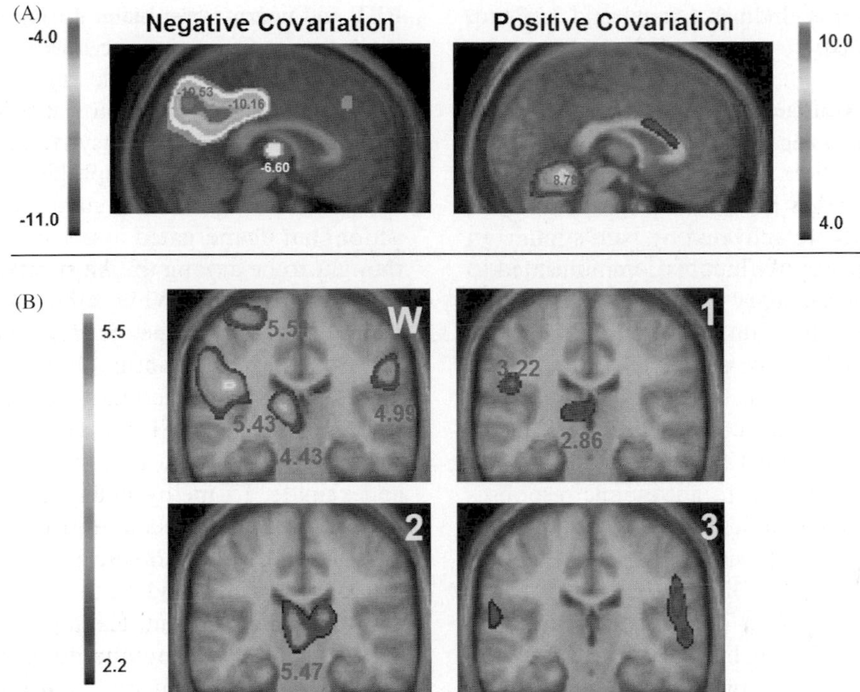

Fig. 1. (A) Representative slices of *t*-statistic maps illustrating the negative covariation (left panel) and the positive covariation (right) between the measured plasma concentration of propofol and regional cerebral blood flow. The left color scale is plotted for *t* values ranging from −11.0 to −4.0 and the right color scale for *t* values ranging from 4.0 to 10.0. The maximal *t* value of each peak is plotted on these maps: the maximal *t* value of the negative covariation observed in the precuneus is −10.53, it is −10.16 in the posterior cingulate cortex, and −6.60 in the thalamus. The maximal *t* value of the positive covariation observed in the cerebellar vermis is 8.78. These results have been replicated in our three studies using H_2O (Dunnet et al., 1994). From Fiset et al. (1999).
(B) Representative slices of subtraction *t*-statistic maps showing the effect of somatosensory stimulation (vibration-on versus vibration-off) for each level of anesthesia. W: awake condition, propofol concentration, 0 μg/mL; 1: Level 1, propofol concentration, 0.5 μg/ml; 2: Level 2, propofol concentration, 1.5 μg/mL; 3: Level 3, propofol concentration, 3.5 μg/mL. The color scale is plotted for *t* values ranging from 2.2 to 5.5. The maximal *t* value of each peak is plotted on these maps: the *t* value of the thalamic peak is 4.43 at Level W, 2.86 at Level 1 and 5.47 at Level 2; the *t* value of the left primary somatosensory cortex peak is 5.51 at Level W, the *t* value of the left secondary somatosensory cortex peak is 5.43 at Level W and 3.22 at Level 1, the *t* value of the right secondary somatosensory cortex peak is 4.99 at Level W. The *t* values in colored regions visible in the statistical map at Level 3 are not significant. Those areas are extensions of non-significant peaks centered elsewhere, in the temporal lobes. From Bonhomme et al. (2001). See Plate 18.1 in Colour Plate Section.

the forearm, rCBF increases in the thalamus, the contralateral primary sensory cortex and bilaterally in the secondary sensory cortices (Coghill et al., 1994). Propofol was administered to induce light sedation, deep sedation and unconsciousness. At each state, vibro-tactile stimulation was applied to the forearm and PET images were obtained. Propofol abolished vibration-induced increases in rCBF in the primary and secondary sensory cortices during light sedation, whereas the thalamic rCBF response was abolished only with loss of consciousness (see Fig. 1B). We concluded that propofol interferes with the processing of vibro-tactile information at the cortical level during sedation, but that transfer of information still persists at the level of the thalamus until unconsciousness occurs. This study confirmed our earlier findings that propofol decreases rCBF in specific brain regions and that those concentration-dependent decreases are associated with changes in the level of consciousness. They also suggest that cortical cells are more sensitive to the effect of propofol, confirming previous findings (Angel and LeBeau, 1992; Angel, 1993).

Neural correlates of concentration-dependent effects: pain processing

Brain imaging studies have shown that several regions of the CNS are activated by pain stimuli. In the forebrain, nociceptive input is communicated to the primary and secondary sensory cortices, where stimulus intensity, location, and temporal aspects are encoded. The anterior cingulate cortex (ACC) also responds to pain: its activity may be important for the affective dimension of pain (Rainville et al., 1997). Finally, nociceptive activation of the insular cortex is probably related to autonomic responses associated with pain (Craig et al., 2000).

Propofol has no analgesic properties. In fact, at low doses that produce light sedation, patients experience an increase in pain intensity and unpleasantness (Petersen-Felix et al., 1996). When the dose is increased to produce moderate sedation, the evaluation of pain is comparable with alert, non-sedated controls. In a functional imaging study, propofol was given to induce light sedation, deep sedation, and loss of consciousness, and pain-evoked changes in cerebral blood flow were measured using PET (Hofbauer, et al., 2004). With low doses (i.e., light sedation), the increase in ratings of pain perception was coincident with an increase in rCBF in the ACC and thalamus. With higher doses, the pain-evoked rCBF increases in the ACC and thalamus were lost during moderate sedation and unconsciousness, respectively. Even after loss of consciousness, some activation of the insular cortex and cerebellum persisted, suggesting consciousness is not a necessary condition for nociceptive input to reach cortical and sub-cortical structures. This may, in fact, reflect the neural correlate of persistence of response to painful stimulation in patients who are comatose. It appears from these studies that cortical structures are more sensitive to propofol than are sub-cortical elements. The augmentation of activity during low-dose propofol, and its reduction produced with higher doses, highlights the complexity (and lack of uniformity) of anesthetic action on CNS function.

PET and neurotransmission during anesthesia: cholinergic muscarinic processes

Many neurotransmitters are affected by anesthetics at the pre- and postsynaptic levels. Downstream elements that couple receptor activity to effector pathways are also susceptible to anesthetic action, but ligand-gated ion channels are currently thought to be among the most relevant targets for general anesthetics. While some anesthetics have a definite effect on one or many neurotransmitter systems (gamma-aminobutytric acid type A ($GABA_A$), glycine, nicotinic acelytcholine (nACh), serotonin type 3 ($5HT_3$), alpha-amino-3-hydroxy-5-methyl-4-isoxazole propionic acid) (AMPA), and kainate, (*N*-methyl-D-aspartate) (NMDA)), it is not known which behavioral effect can be linked to a specific action on a given ligand-gated ion channel (Rudolph and Antkowiak, 2004).While a variety of neurotransmitter/modulator systems appear to be affected by anesthetic drugs, we have focused on the cholinergic system. There is compelling evidence from in vivo animal sleep studies to suggest that altered central cholinergic drive can mediate changes in endogenous states of consciousness (Steriade, 1990). In vitro and in vivo studies demonstrate that a variety of anesthetic drugs interfere with cholinergic transmission (Lydic and Baghdoyan, 1997). In addition, in both human and animal experiments, the anesthetic-induced depression of the CNS is augmented by cholinergic (muscarinic) antagonists (Ali-Melkkilä et al., 1993). Importantly, it is possible to effect changes in central cholinergic activity safely in humans by increasing cholinergic drive with a centrally acting anticholinesterase agent, or conversely, to decrease it with a cholinergic (muscarinic) antagonist.

It has been long known to clinicians that the administration of physostigmine, a cholinesterase inhibitor, decreases the time required for return of consciousness following anesthesia with a variety of drugs. We examined the effects of the anticholinesterase physostigmine and scopolamine, a competitive muscarinic antagonist, on the loss of consciousness produced by anesthetic drugs (Meuret et al., 1997). We tested the hypothesis that the loss of consciousness produced by anesthetic drugs involves

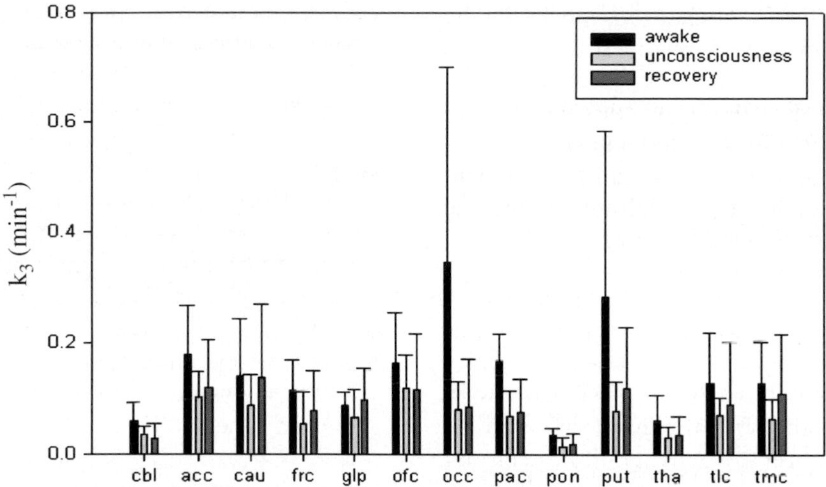

Fig. 2. Muscarinic receptor availability (as assessed by [N-^{11}C-methyl]benztropine PET specibinding rate constant k_3) during the awake condition, propofol-induced unconsciousness and recovery of consciousness for different regions of the brain (cerebellum (cbl), anterior cingulate cortex (acc), caudate nucleus (cau), frontal cortex (frc), globus pallidus (glp), orbito-frontal cortex (ofc), occipital cortex (occ), parietal cortex (pac), pons (pon), putamen (put), thalamus (tha), temporal lateral cortex (tlc), and temporal medial cortex (tmc)). From Xie et al. (2004).

depression, directly or indirectly, of central cholinergic transmission. Propofol was administered to volunteers in stepwise increments to achieve loss of consciousness. Physostigmine was administered while maintaining a constant plasma concentration of anesthetic drug. We demonstrated that the propofol-induced loss of consciousness was reversed by administration of physostigmine and that the reversal was blocked by the pre-administration of scopolamine. The results suggest that the muscarinic cholinergic system is a target, or at least a modulator, of anesthetic-induced loss of consciousness.

In a subsequent PET study on volunteers, we measured the change in central muscarinic receptor availability during loss of consciousness produced by propofol (Xie et al., 2004). The subjects were scanned during three different conditions in the same session: awake, unconscious and recovery (awake after the end of the infusion) following administration of [*N*-^{11}C-methyl]benztropine, a non-selective muscarinic receptor antagonist and proposed a method of analysis with this irreversibly bound radiotracer. The results demonstrated a propofol-related reduction in muscarinic receptor binding, and support the hypothesis that anesthetic drugs produce a change in central cholinergic transmission that could at least in part contribute to the mechanisms of general anesthesia (see Fig. 2). We proposed three different mechanisms to be considered: (1) a change in conformational state of the muscarinic receptor from high to low affinity; (2) a change in benztropine's metabolism induced by propofol; and (3) competition between propofol and benztropine for the muscarinic receptor.

It is hoped that, in the near future, the availability of new tracers will allow in vivo determination of changes in receptor behavior in all known neurotransmitter systems. It seems obvious that anesthesia can induce changes simultaneously in different systems, and that changes in some systems can cause, in cascade, changes in others (see Nelson et al.).

Conclusion

The ability to titrate anesthetic drugs to specific clinical end-points has long been appreciated by anesthesiologists. These end-points traditionally have clinical relevance, such as analgesia, amnesia, paralysis, and unconsciousness, so that surgery can be tolerated. The use of such end-points as behaviors for studies on mechanisms mediating the effects of anesthetic drugs is relatively novel, and is well suited for brain imaging experiments. This experimental paradigm has enormous potential to elucidate the mechanisms underlying the CNS effects of general anesthetic drugs, and to help identifying the underlying neural substrates controlling endogenous states of consciousness.

References

Ali-Melkkilä, T., Kanto, J. and Iisalo, E. (1993) Pharmacokinetics and related pharmacodynamics of anticholinergic drugs. Acta Anesthesiol. Scand., 37: 633–642.

Angel, A. (1993) Central neuronal pathways and the process of anaesthesia. Br. J. Anaesth., 71: 148–163.

Angel, A. and LeBeau, F. (1992) A comparison of the effects of propofol with other anaesthetic agents on the centripetal transmission of sensory information. Gen. Pharmacol., 23: 945–963.

Bonhomme, V., Fiset, P., Meuret, P., Backman, S., Plourde, G., Paus, T., Bushnell, C. and Evans, A. (2001) Effect of propofol-induced general anesthesia on changes in regional cerebral blood flow elicited by vibrotactile stimulation: a positron emission tomography study. J. Neurophysiol., 85: 1299–1308.

Campagna, J.A., Miller, K.W. and Forman, S.A. (2003) Mechanisms of actions of inhaled anesthetics. New Engl. J Med., 348: 2110–2124.

Coghill, R.C., Talbot, J.D., Evans, A.C., Meyer, E., Gjedde, A., Bushnell, M.C. and Duncan, G.H. (1994) Distributed processing of pain and vibration by the human brain. J. Neurosci., 14: 4095–4108.

Craig, A.D., Chen, K., Bandy, D. and Reiman, E.M. (2000) Thermosensory activation of insular cortex. Nat. Neurosci., 3: 184–190.

Dunnet, J.M., Prys-Roberts, C., Holland, D.E. and Browne, B.L. (1994) Propofol infusion and the suppression of consciousness: dose requirements to induce loss of consciousness and to suppress response to noxious and non-noxious stimuli. Br. J. Anaesth., 72: 29–34.

Fiset, P., Paus, T., Daloze, T., Plourde, G., Meuret, P., Bonhomme, V., Hajj-Ali, N., Backman, S.B. and Evans, A.C. (1999) Brain mechanisms of propofol-induced loss of consciousness in humans: a Positron Emission Tomography study. J. Neurosci., 19: 5506–5513.

Heinke, W. and Schwarzbauer, C. (2002) In vivo imaging of anaesthetic action in humans: approaches with positron emission tomography (PET) and functional magnetic resonance imaging (fMRI). Br. J. Anaesth., 89: 112–122.

Hofbauer, R.K., Fiset, P., Plourde, G., Backman, S.B. and Bushnell, M.C. (2004) Dose-dependent effects of propofol on the central processing of thermal pain. Anesthesiology, 100: 386–394.

Lydic, R. and Baghdoyan, H.A. (1997) Cholinergic contribution to the control of consciousness. In: Yaksh T.L. (Ed.), Anesthesia: Biologic Foundations. Lippincott-Raven, Philadelphia, pp. 433–450.

Meyer, H.H. (1899) Theorie der alkoholnarkose. Arch. Exp. Pathol. Pharamkol., 42: 109–118.

Meuret, P., Backman, S., Bonhomme, V., Plourde, G., Fiset, P. (1997). Effect of physostigmine on propofol-induced loss of consciousness and changes in evoked auditory responses in man. Soc. Neurosci.

Moruzzi, G. and Magoun, H.W. (1949) Brain stem reticular formation and activation of the EEG. Electroencephal. Clin. Neurophysiol., 1: 455–473.

Overton, E. (1901). Studien uber die narkose zugleich ein beitrag zur allgemeinen pharmakologie. Jena, Verlag von Gustav Fischer.

Petersen-Felix, S., Arendt-Nielsen, L., Bak, P., Fischer, M. and Zbinden, A.M. (1996) Psychophysical and electrophysiological responses to experimental pain may be influenced by sedation: Comparison of the effects of a hypnotic (propofol) and an analgesic (alfentanil). Br. J. Anaesth., 77: 165–171.

Rainville, P., Duncan, G.H., Price, D.D., Carrier, B. and Bushnell, M.C. (1997) Pain affect encoded in human anterior cingulate but not somatosensory cortex. Science, 277: 968–971.

Rudolph, U. and Antkowiak, B. (2004) Molecular and neuronal substrates for general anaesthetics. Nat. Rev. Neurosci., 5: 709–720.

Shafer, S.L. and Gregg, K. (1992) Algorithms to rapidly achieve and maintain stable drug concentrations at the site of drug effect with a computer controlled infusion pump. J. Pharmacokinet. Biopharm., 20: 147–169.

Steriade, M. (1990) Cholinergic control of thalamic function [Review] [106 refs] [in French]. Archives.Internationales.de.Physiologie.et.de.Biochimie, 98: A11–A46.

Steriade, M. (1996) Arousal: revisiting the reticular activating system [comment]. [Review] [14 refs]. Science, 272: 225–226.

Steriade, M., Jones, E.G. and McCormick, D.A. (1997) Thalamic function and dysfunction. In: Steriade M., Jones E.G. and McCormick D.A. (Eds.), Thalamus. Elsevier, Amsterdam, pp. 1–29.

Steriade, M., McCormick, D.A. and Sejnowski, T.J. (1993) Thalamocortical oscillations in the sleeping and aroused brain. [Review] [70 refs]. Science, 262: 679–685.

Varvel, J.R., Donoho, D.L. and Shafer, S.L. (1992) Measuring the predictive performance of computer controlled infusions. J. Pharmacokinet. Biopharm., 20: 63–94.

Vuyk, J., Engbers, F.H.M., Burm, A.G.L., Vletter, A.A. and Bovill, J.G. (1995) Performance of computer-controlled infusion of propofol: an evaluation of five pharmacokinetic parameter sets. Anesth. Analg., 81: 1275–1282.

Xie, G., Gunn, R.N., Dagher, A., Daloze, T., Plourde, G., Backman, S.B., Diksic, M. and Fiset, P. (2004) PET quantification of muscarinic cholinergic receptors with N-^{11}C-methyl-benztropine and application to studies of propofol-induced unconsciousness in healthy human volunteers. Synapse, 51: 91–101.

S. Laureys (Ed.)
Progress in Brain Research, Vol. 150
ISSN 0079-6123

CHAPTER 19

The cognitive modulation of pain: hypnosis- and placebo-induced analgesia

Ron Kupers[1], Marie-Elisabeth Faymonville[2] and Steven Laureys[3,*]

[1]*Center for Functionally Integrative Neuroscience (CFIN), Aarhus University and Aarhus University Hospital, Aarhus, Denmark*
[2]*Pain Clinic and Department of Anesthesiology, University Hospital of Liège, Liège, Belgium*
[3]*Cyclotron Research Center and Department of Neurology, University of Liège, Liège, Belgium*

Abstract: Nowadays, there is compelling evidence that there is a poor relationship between the incoming sensory input and the resulting pain sensation. Signals coming from the peripheral nervous system undergo a complex modulation by cognitive, affective, and motivational processes when they enter the central nervous system. Placebo- and hypnosis-induced analgesia form two extreme examples of how cognitive processes may influence the pain sensation. With the advent of modern brain imaging techniques, researchers have started to disentangle the brain mechanisms involved in these forms of cognitive modulation of pain. These studies have shown that the prefrontal and anterior cingulate cortices form important structures in a descending pathway that modulates incoming sensory input, likely via activation of the endogenous pain modulatory structures in the midbrain periaqueductal gray. Although little is known about the receptor systems involved in hypnosis-induced analgesia, studies of the placebo response suggest that the opiodergic and dopaminergic systems play an important role in the mediation of the placebo response.

Introduction

Placebo- and hypnosis-induced analgesia should be seen within the broader context of the conceptualization of pain. In 1965, Melzack and Wall proposed the "gate control" theory, according to which activation in large myelinated fibers is capable of inhibiting nociceptive information. This model struck against the contemporary belief that pain processing is a hard-wired process, mediated exclusively by pain-dedicated pathways. A few years later, Melzack and Casey described their theory about the multidimensional processing of pain. This theory added a rostral (cerebral) extension to the gate control theory, which focuses on activity in the spinal cord dorsal horn. According to Melzack and Casey (1968), pain is a complex multidimensional experience comprising sensory-discriminative, motivational-affective and cognitive-evaluative components. For the first time, the concepts of cognition and emotion were introduced to a field in which sensory physiologists had claimed the exclusive rights. Over the years, a number of assumptions have been added to Melzack and Casey's scheme. One of these assumptions is that distinct anatomical pathways are involved in the sensory and affective pain dimension (Albe-Fessard et al., 1985; Bushnell and Duncan, 1989; Price, 1999). The sensory component of pain would involve the somatosensory thalamus and its projections to the primary and secondary somatosensory cortices and insula. The affective component on the other hand would involve the medial thalamus and its projections to the anterior cingulate and prefrontal cortices. Recent neuroimaging studies have endeavored to yield evidence for the

*Corresponding author. Tel.: +32 4 366 23 16;
Fax: +32 4 366 29 46; E-mail: steven.laureys@ulg.ac.be

DOI: 10.1016/S0079-6123(05)50019-0

selective activation of brain areas involved in the sensory-discriminative or affective dimension of pain (Rainville et al., 1997; Hofbauer et al., 2001) by using hypnotic suggestions specifically targeted at activating either system. While, according to the model by Melzack and Casey, cognitive and affective processing is performed in parallel with sensory processing, Price (2000) proposed a serial two-stage model in which the emotional component is the result of the interaction between a hard-wired sensory input and contextual processes. This model views pain perception as a self-contained process simply driven by a reliable sensory process. However, cognitive processes directly influence the operation of the sensory system. According to Wall (1991, 1996), nociceptive processing is not something purely dictated by the sensory characteristics of the stimulus but results from the interaction between the latter and the state of the nervous system at that particular time. The state of the nervous system depends both on past experiences and on the cognitive and emotional processes of the organism at the time of sensory input. A series of elegant animal studies by Dubner and co-workers in the 1980s provide strong experimental support for this view (Dubner et al., 1981, Duncan et al., 1987). These investigators recorded neuronal activity in the medullary dorsal horn of monkeys that had been trained for months to discriminate between noxious thermal stimuli of different temperatures applied to the face. A light signal preceding the noxious stimulus by a variable time announced the onset of the noxious stimulus to the animal. In naïve monkeys, changes in neuronal activity in the dorsal horn were only recorded, as expected, during the actual period of the application of the nociceptive stimulus. In sharp contrast, in monkeys that had been trained for several months in the thermal discrimination task, dorsal horn cells already started firing in response to the warning signal. In other words, the light stimulus had become a conditioned stimulus following repetitive pairings with the unconditioned (noxious) stimulus and now produced a conditioned response in the dorsal horn. This is one of the first examples of top-down modulation at such an early (dorsal horn) stage of pain processing. Hypnotic-induced analgesia and placebo analgesia are two other examples of this top-down modulation of pain modulation.

There is ample evidence that the modulation of pain is mediated by the midbrain periaqueductal gray (PAG) and the rostral ventromedial medulla (RVM) (Fields and Basbaum, 1999). The RVM contains antinociceptive descending pathways targeting the spinal and trigeminal dorsal horn (Urban and Smith, 1994; Hudson et al., 2000). Although there are no known direct descending projections from the PAG to the spinal cord, it has been hypothesized that the analgesic effects of PAG are mediated through relays in the RVM and the dorsolateral pontine tegmentum (DLPT). The PAG and RVM function as a unit exerting global control over pain transmission neurons in the dorsal horn. Opioids injected into the PAG and RVM produce analgesia (Urban and Smith, 1994; Price and Fields, 1997; McGaraughty et al., 2003), which can be blocked by administration of the opioid antagonist naloxone into the same area (Wang and Wessendorf, 2002). Regions of the frontal lobe and the amygdala project via the hypothalamus and also directly to the PAG (An et al., 1998; Ongur et al., 1998).

Many of the ascending pain pathways terminate in cortical and subcortical areas, which are also at the origin of pain modulatory systems (Burstein et al., 1993; Bernard et al., 1996; Helmstetter et al., 1998; Fields and Basbaum, 1999; Price, 1999). These areas also play a major role in threat-elicited defensive-behavior, learning, and memory. This raises the following important questions: (1) which cognitive and environmental circumstances are able to trigger these pain inhibitory mechanisms? and (2) once they are triggered, how do they produce their analgesic effect?

Animal studies over the past three decades have shed some light on the answers to these questions. Activation of the endogenous pain modulatory circuits requires specific extrinsic environmental cues or conditions. For instance, Watkins and Mayer (1982) demonstrated that severe physical or psychological stressors such as inescapable foot-shock and whole body rotation can induce robust analgesic responses. Some, but not all, of these environmentally induced analgesic responses are blocked by naloxone, suggesting that both opioid and non-opioid systems must be involved. When the animals were later placed in the apparatus where they had been submitted to the noxious stimulation, analgesia was induced. In other words, the environmentally induced analgesic response is prone to classical conditioning (Hayes et al., 1978; Price, 1999). These experimental findings may be of great relevance to placebo analgesia in humans. Cues associated with previous pain relief may become effective for evoking endogenous analgesic mechanisms because they were previously associated with an effective treatment. Furthermore, the findings call for the formulation of a pain theory that incorporates cognitive expectations and the state of the nervous system at the time of nociceptive processing.

In this chapter, we will discuss the mechanisms through which hypnosis and placebo may modulate pain perception. We will first try to give a definition of both concepts and we will summarize available data on the neurobiological mechanisms involved in these two forms of cognition-induced analgesia. This will be followed by a discussion of the relevant brain imaging literature.

Hypnosis

Hypnosis has long been known to be associated with heightened control over physical processes and has been used as a therapeutic tool since the early history of mankind (De Betz and Sunnen, 1985). It has been used in many medical and psychological problems (e.g., the treatment of pain, gastro-intestinal and dermatological pathologies, depression, anxiety, stress, and habit disorders). At present, there is no generally accepted definition of hypnosis. For many authors it is seen as a state of focused attention, concentration, and inner absorption with a relative suspension of peripheral awareness. We have all experienced similar states many times, but do not tend to call it hypnosis (e.g., being so absorbed in thought while driving home that we fail to notice consciously what is happening around us). The Executive Committee of the American Psychological Association—Division of Psychological Hypnosis (1994) has constructed a definition from the multiplicity of positions of a number of researchers advocating differing theoretical perspectives. Their definition regards hypnosis as "a procedure during which a health professional or researcher suggests that a patient or subject experience changes in sensations, perceptions, thoughts, or behavior." The hypnotic context is generally established by an induction procedure. Most hypnotic inductions include suggestions for relaxation. Faymonville and co-workers use instructions to imagine or think about pleasant autobiographical experiences to decrease pain perception in patients undergoing surgery (1997), and in healthy volunteers participating in functional brain imaging (2000). Hypnosis has three main components: absorption, dissociation, and suggestibility (Spiegel, 1991). Absorption is the tendency to become fully involved in a perceptual, imaginative, or ideational experience. Subjects prone to this type of cognition are more highly hypnotizable than others who never fully engage in such experience (Hilgard et al., 1963). Dissociation is the mental separation of components of behavior that would ordinarily be processed together (e.g., the dream-like state of being both actor and observer when re-experiencing autobiographical memories). This may also involve a sense of involuntariness in motor functions or discontinuities in the sensations of one part of the body compared with another. Suggestibility leads to an enhanced tendency to comply with hypnotic instructions. This represents not a loss of will but rather a suspension of critical judgment because of the intense absorption of the hypnotic state. It is important to stress that hypnosis makes it easier for subjects or patients to experience suggestions or access memories, but cannot force them to have these experiences. Contrary to some depictions of hypnosis in the media, hypnotized subjects do not lose complete control over their behavior. They typically remain aware of who they are and where they are, and unless amnesia has been specifically suggested, they usually remember what transpired during hypnosis.

Since 1992, the university hospital of Liège has used "hypnosedation", a combination of hypnosis with local anesthesia and minimal conscious sedation, in over 4,800 patients, (Faymonville et al., 1997). Hypnosedation was shown to be a valuable, safe, efficient, and economic alternative to general anesthesia in specific indications such as thyroid and parathyroid surgery (Meurisse et al., 1996,

1999a, b, Defechereux et al., 1998, 1999, 2000), plastic surgery (Faymonville et al., 1994, 1995, 1997, 1999), and peri-dressing change pain and anxiety in severely burned patients (Frenay et al., 2001). In patients undergoing surgery, the technique of hypnosedation is associated with improved intraoperative patient comfort and with reduced anxiety, pain, intraoperative requirements for anxiolytic and analgesic drugs, optimal surgical conditions, and a faster recovery of the patient. In our opinion, clinical hypnosis should be used only by properly trained and credentialed health care professionals who have also been trained in the clinical use of hypnosis and are working within the areas of their professional expertise.

In addition to its use in clinical settings, hypnosis is used in research, with the goals of learning more about the nature of hypnosis itself as well as its impact on central nervous system processes such as pain, perception, learning, and memory (e.g., Kosslyn et al., 2000). However, as its acceptance by the scientific community remains limited, the neural correlates of the hypnotic state remain poorly understood. One field where the efficacy of hypnosis has been the most extensively evaluated and validated is pain control. We will here review hypnosis-induced (i) changes in regional brain function, (ii) modulation of pain perception, and (iii) increases in cerebral functional connectivity as studied by means of positron emission tomography (PET).

Hypnosis-induced changes in regional brain function

Maquet et al. (1999) first explored the brain mechanisms underlying hypnosis in healthy volunteers by determining the distribution of regional cerebral blood flow (rCBF), by use of the $H_2^{15}O$ technique. The hypnotic procedure used was similar to the one used in clinical routine (Faymonville et al., 1995, 1997, 1999; Meurisse et al., 1999b) and was induced using eye fixation, a 3-minute muscle relaxation procedure, and permissive and indirect suggestions. Subjects were invited to re-experience very pleasant autobiographical memories. As in clinical conditions, they were continuously given cues for maintaining and deepening the hypnotic state. Just before scanning, subjects confirmed by a prearranged foot movement that they were experiencing hypnosis. Oculographic recording showed roving eye movements sometimes intermingled with few saccades. This pattern of eye movements, in conjunction with the subject's behavior was used to differentiate hypnosis from other states. Polygraphic monitoring (electroencephalographic, electromyographic, and oculographic recordings) further ensured that no sleep occurred during the experimental session.

The choice of the control task was difficult as, a priori, no cerebral state is close to the hypnotic state. Because the induction and maintenance of our hypnotic procedure relies on revivification of pleasant autobiographical memories, the closest situation is the evocation of autobiographical information, in the absence of the hypnotic state (i.e., in a state of normal alertness). To better understand the comparisons made for hypnosis, the authors first investigated this control condition. Listening to autobiographical material activated the anterior part of both temporal lobes, basal forebrain structures, and some left mesiotemporal areas (Fig. 1, left panel). This pattern is in agreement with previous findings on autobiographical memory (Fink et al., 1996).

During hypnosis, compared to the control task, a vast activation was observed that involved occipital, parietal, precentral, prefrontal, and cingulate cortices (Fig. 1, right panel). The neural network implicated in hypnosis and in the control task (i.e., evocation of autobiographical information in a state of normal alertness) did not overlap. These results show that the hypnotic state relies on cerebral processes different from simple evocation of episodic memory and suggest it is related to the activation of sensory and motor cortical areas, as during perceptions or motor acts, but without actual external inputs or outputs. In this respect, hypnosis is reminiscent of mental imagery (Kosslyn et al., 2001). The imagery content in hypnosis was polymodal. Although subjects predominantly reported visual impressions, somesthetic and olfactory perceptions were also mentioned. A lot of actions also appeared in the hypnotic experience of most of the subjects. In contrast, none of the subjects reported auditory imagery. When sounds were mentioned, they came from the actual experimental

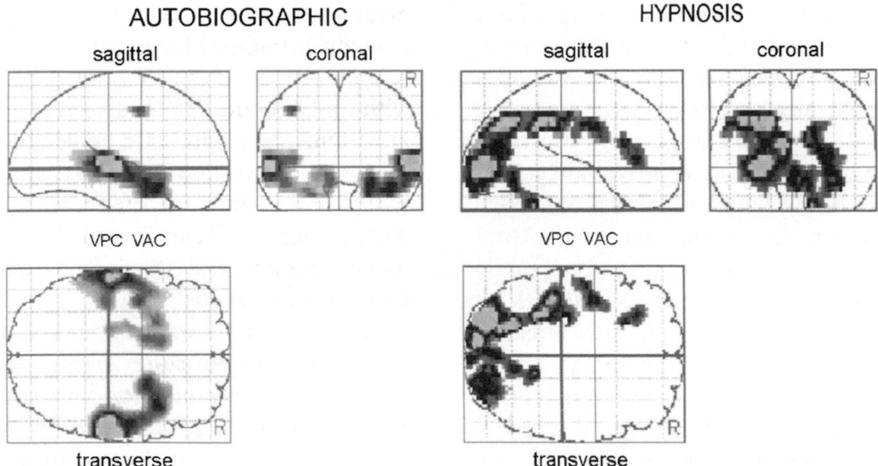

Fig. 1. Projections in stereotactic space of Talairach and Tournoux (1988) of brain areas where rCBF is increased during mental imaging of autobiographical memories in "normal alertness" (control distraction task) compared to the resting state (left) and brain areas where rCBF is significantly increased during hypnosis compared to the distraction task (right). Functional PET results are displayed at threshold of $p<0.001$. VAC and VPC identify anterior and posterior commissural planes, respectively. Adapted from Maquet et al. (1999) and Laureys et al. (2004).

environment (mainly, the experimenter's voice). The visual mental imagery might take into account the activation of a set of occipital areas. More anteriorly, the activation of precentral and premotor cortices is similar to that observed during motor imagery (Decety, 1996), which could also have participated in the parietal activation. The activation of ventrolateral prefrontal cortex has also been observed in mental imagery tasks and would be involved in the programming of the building up of the mental image or in the maintenance of image in memory. Finally, the activation in anterior cingulate cortex (ACC) is thought to reflect the attentional effort necessary for the subject to internally generate mental imagery.

Prominent decreased activity during hypnosis relative to the alert state was observed in the medial parietal cortex (i.e., precuneus). This area is hypothesized to be involved in the representation (monitoring) of the world around us (also see Vogt and Laureys, this volume and Lou et al., this volume). Indeed, the precuneus shows the highest level of glucose use (the primary fuel for brain energy metabolism) of any area of the cerebral cortex in the so-called "conscious resting state". It is known to show task-independent decreases from the baseline during the performance of goal-directed actions (Binder et al., 1999; Gusnard and Raichle, 2001; Mazoyer et al., 2001; Raichle et al., 2001). The functions to which this region of the cerebral cortex contributes include those concerned with both orientation within, and interpretation of, the environment (Vogt et al., 1992). Interestingly, the precuneus is one of the most dysfunctional brain regions in states of unconsciousness or altered consciousness, such as coma (Laureys et al., 2001), vegetative state (Laureys et al., 1999), general anesthesia (Alkire et al., 1999; also see Alkire et al., this volume; Fiset et al., this volume), slow wave and rapid eye movement sleep (Maquet, 2000; also see Maquet et al., this volume), amnesia (Aupee et al., 2001), and dementia (Salmon et al., 2000; Matsuda, 2001; also see Salmon et al., this volume), suggesting that it is part of the critical neural network subserving conscious experience.

Hypnosis-induced changes in pain perception

In a next step, Faymonville et al. (2000) investigated the brain mechanisms underlying the modulation of pain perception proper to their clinical hypnotic protocol. During this procedure, hypnotized healthy volunteers and patients are invited to have revivification of pleasant life episodes, without any reference to the pain perception. This technique lowers both the unpleasantness (i.e., affective component) and the perceived intensity (i.e., sensory component) of the noxious stimuli (Faymonville et al., 1997, 2000). Hypnosis decreases both components of pain perception by approximately 50% compared to the resting state and by approximately 40% compared to a distraction task (mental imagery of autobiographical events) (Fig. 2).

Faymonville et al. (2000) and Rainville et al. (1997, 1999) showed that this modulatory effect of hypnosis is mediated by the ventral part of the ACC (Brodmann area 24′a). Indeed, the reduction of pain perception correlated with ACC activity specifically in the context of hypnosis (Fig. 3). The ACC is a functionally very heterogeneous region thought to regulate or modulate the interaction between cognition, sensory perception and motor control in relation to changes in attentional, motivational, and emotional states (Devinsky et al., 1995; Bush et al., 2000). The ACC can be divided into a perigeniculate and midcingulate cortex on the basis of cytoarchitectonical structure, connectivity, and functional observations (Vogt et al., 2002); whereas the perigeniculate part is mostly involved in emotional processing, and the midcingulate part is more involved in cognitive processing.

The ACC is abundantly innervated by a multitude of neuromodulatory pathways including opioidergic, dopaminergic, noradrenergic, and serotoninergic systems and contains high levels of substance P, corticotropin-releasing factor, neurotensin, and prosomatostatin-derived peptides (Paus, 2001). Although the ACC contains high

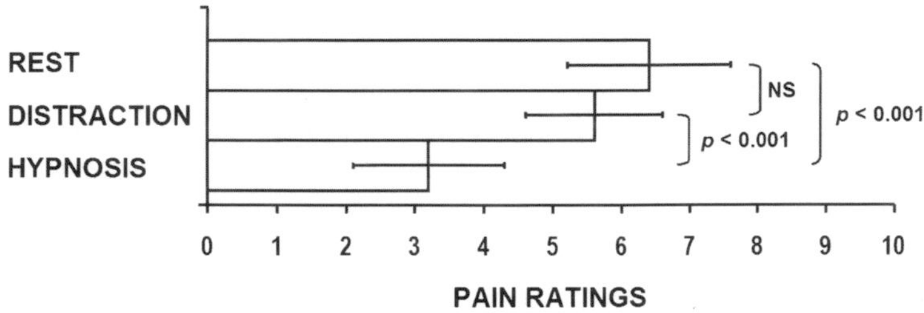

Fig. 2. Ratings of pain perception in the resting state, the distraction task (mental imagery of biographical memories), and in the hypnotic state. Values are means and standard deviations (NS = not significant). Adapted from Faymonville et al. (2003).

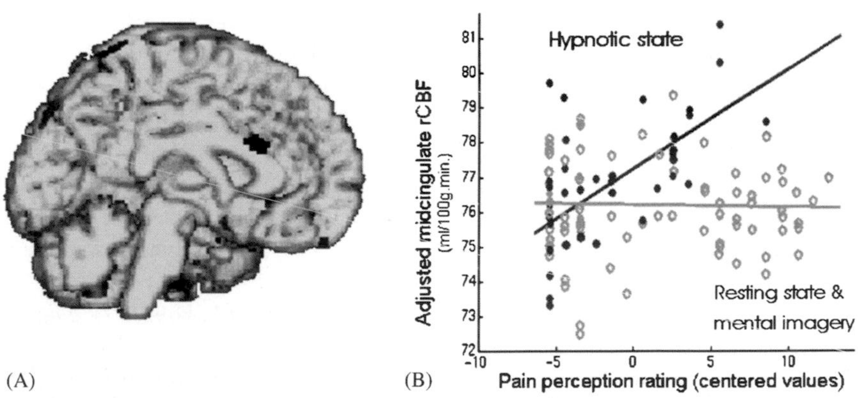

Fig. 3. (A) Brain area in which blood flow increases in proportion to pain sensation ratings, in the specific context of hypnosis: the ventral part of the midcingulate cortex (area 24′a). (B) Plot of changes in pain perception ratings versus changes in adjusted blood flow in ACC. Note the difference ($p < 0.05$) in regression slopes between hypnosis (black dots) and control conditions (gray circles). Results are displayed on a 3-D rendered spatially normalized MRI scan. Adapted from Faymonville et al. (2000) and Laureys et al. (2004).

concentrations of opioid receptors and peptides, it is doubtful whether opioid neurotransmission underlies the midcingulate cortical activation in hypnosis-induced reduction of pain perception. Indeed, psychopharmacological studies showed that hypnotic analgesia was not altered by the administration of naloxone (Moret et al., 1991). It is also unlikely that the ACC might modulate pain perception during hypnosis through pure attentional mechanisms. The midcingulate cortex that was identified by Faymonville et al. (2000) has been related to pain perception, whereas the more anterior portions of the ACC are involved in attention-demanding tasks (Derbyshire et al., 1998; Petrovic and Ingvar, 2002). From an anatomical viewpoint, the midcingulate cortex is in a critical position to receive both the sensory and the affective aspects of a noxious stimulus from respectively the somatosensory areas and insula, and the amygdaloid complex and perigenual ACC. Since pain is a multidimensional experience including sensory-discriminative, affective-emotional, cognitive, and behavioral components, its cerebral correlate is best described in terms of neural circuits or networks, referred to as the 'neuromatrix' for pain processing, and not as a localized 'pain center' (Jones et al., 1991; Peyron et al., 2000).

Hypnosis-induced changes in cerebral functional connectivity

In order to further explore the antinociceptive effects of hypnosis, Faymonville et al. (2003) assessed hypnosis-induced changes in functional connectivity between the ACC and a large neural network involved in the different aspects of noxious processing. Before we discuss these results, we will briefly explain what is meant by 'functional connectivity analyses' when using PET data.

The functional role played by any component (e.g., neuronal population) of a connected system (e.g., the brain) is largely defined by its connections. Complementary to the concept of functional segregation as a principle of organization of the human brain (i.e., localizing a function to a cerebral area), recent neuroimaging techniques have focused on functional integration (i.e., assessing the interactions between functionally segregated areas mediated by changes in functional connectivity). Functional connectivity is defined as the temporal correlation of a neurophysiological index (i.e., rCBF) measured in different remote brain areas (Friston, 2002). A psychophysiological interaction analysis explains the activity in one cortical area in terms of an interaction between the influence of another area and some experimental condition (i.e., being in a hypnotic state or in a state of normal alertness). A psychophysiological interaction means that the contribution (i.e., regression slope) of one area to another changes significantly with the experimental context assessed with the general linear model as employed by statistical parametric mapping (Friston et al., 1997). The statistical analysis will identify brain regions that show condition-dependent differences in modulation with another (chosen) area. It is important to stress that one cannot guarantee that these connections are direct (i.e., they may be mediated through other areas) and that the two regions can have a common input (a third area, which shows context-sensitive responses, may be providing input to the two areas implicated in the psychophysiological interaction). Anatomical connectivity (e.g., neuroanatomic tracer studies obtained in animals) is a necessary underpinning for the assessment of functional connectivity.

Using a functional cerebral connectivity analysis, Faymonville et al. (2000) showed that the hypnosis-induced reduction of pain perception mediated by the midcingulate cortex (Rainville et al., 1997, 1999; Faymonville et al., 2000) relates to an *increased* functional modulation between this area and a large neural network of cortical and subcortical structures known to be involved in different aspects of pain processing, encompassing prefrontal cortex, pre-supplementary motor area (pre-SMA), insular and perigenual cortices, striatum, thalami, and brainstem (Fig. 4).

These hypnosis-induced changes in connectivity between ACC and prefrontal areas may indicate a modification in distributed associative processes of cognitive appraisal, attention or memory of perceived noxious stimuli. As discussed above, frontal increases in rCBF have previously been demonstrated in the hypnotic state (Maquet et al.,

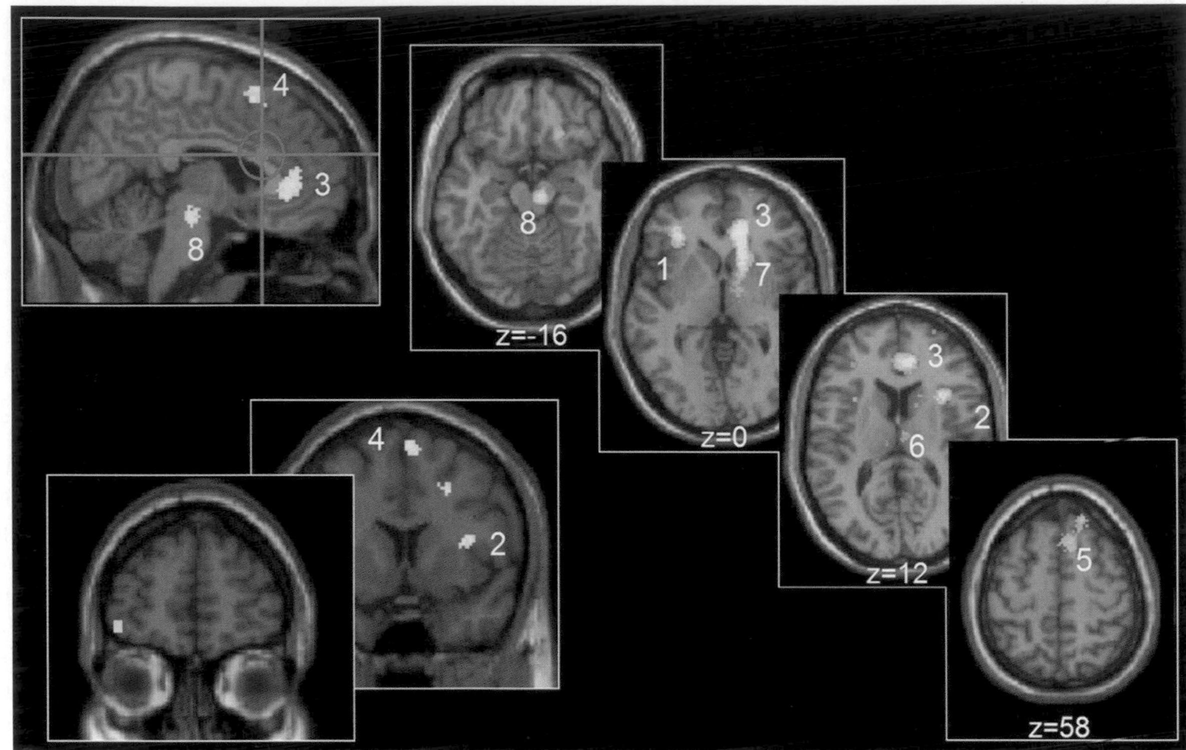

Fig. 4. During hypnosis an increase in activity in midcingulate cortex increases activity in a wide cortical and subcortical neural network (part of the 'pain matrix'), much more so than is observed under control conditions (rest or distraction tasks). Regions that showed such hypnosis-related increased functional connectivity with midcingulate cortex (peak voxel marked by red crosshair in circle) are: left insula (1), right insula (2), perigenual cortex (3), pre-supplementary motor area (4), superior frontal gyrus (5), right thalamus (6), right caudate nucleus (7), and midbrain/brainstem (8). Adapted from Faymonville et al. (2003). See Plate 19.4 in Colour Plate Section.

1999; Rainville et al., 1999; Faymonville et al., 2000). Frontal activation has also been reported in a series of studies on experimental pain (Kupers et al., 2001, 2004; Witting et al., 2001; Bornhovd et al., 2002; Lorenz et al., 2003), but the precise role of particular regions in the central processing of pain remains to be elucidated (Coghill et al., 1999; Treede et al., 1999).

The ACC has also a major role in motor function (Fink et al., 1997). Its increased functional connectivity with pre-SMA and striatum during hypnosis may therefore allow the midcingulate cortex to organize the most appropriate behavioral response to the pain stimulus. Indeed, the basal ganglia encode and initiate basic movement patterns expressed through premotor and primary motor areas and show frequent activation to noxious stimuli (Jones et al., 1991; Coghill et al., 1994; Derbyshire et al., 1997; Derbyshire and Jones, 1998). The basal ganglia are not exclusively linked to motor function, but have also been proposed to support a basic attentional mechanism facilitating the calling up of motor programs and thoughts (Brown and Marsden, 1998).

The insular and the anterior cingulate cortices show the most consistent activation in functional imaging studies on pain perception (Peyron et al., 2002). The insula takes an intermediate position between the lateral (sensory-discriminative) and medial (affective-emotional) pain systems. It receives major input from the somatosensory system (Mesulam and Mufson, 1982), has direct thalamocortical nociceptive input (Craig et al., 1994) and is implicated in affective and emotional processes through its projections to the amygdala (Augustine, 1996). The observation of an increased midcingulate-insular modulation during hypnosis is in line with its proposed role in pain affect (Rainville et al., 1999) and pain intensity coding (Craig et al., 2000). In light of the 'somatic marker' hypothesis of consciousness (Damasio, 1994), the right anterior insular cortex has been hypothesized to be involved in the mental generation of an image of one's physical state underlying the assignment of emotional attributes to external and internal stimuli.

The increased functional connectivity between the midcingulate cortex and the thalamus and midbrain during hypnosis could be related to pain-relevant arousal or attention (Kinomura et al., 1996). Since thalamic and midbrain activity have been shown to correlate, respectively, with pain threshold and pain intensity (Tolle et al., 1999; Moerman and Jonas, 2002), it is tempting to hypothesize that hypnosis induces a subcortical gating of cortical activation, resulting in decreased subjective pain perception. Previous studies have shown that different forms of defensive or emotional reactions, analgesia and autonomic regulation are represented in different regions of the midbrain PAG (Bandler and Shipley, 1994). The perigenual and insular cortices and thalami are also known to be implicated in autonomic regulation (Bandler and Shipley, 1994; Augustine, 1996). The modulatory role of the midcingulate cortex on this network could explain the clinical observation that patients undergoing surgery during the hypnotic state show modified autonomic responses and less defensive reactions in response to an aversive encounter (Faymonville et al., 1997).

Hypnosis can hence be seen as a particular cerebral waking state where the subject, seemingly somnolent, experiences a vivid, multimodal, coherent, memory-based mental imagery that invades and fills the subject's consciousness. The pattern of cerebral activation, measured by means of $H_2^{15}O$-PET, during the hypnotic state differs from that induced by simple mental imagery. The reduced pain perception during hypnosis is mediated by an increased functional connectivity between the midcingulate cortex and insular, perigenual, frontal, and pre-SMA regions as well as brainstem, thalamus, and basal ganglia. These findings point to a critical role for the midcingulate cortex in hypnosis-related alteration of sensory, affective, cognitive and behavioral aspects of nociception. It reinforces the idea that not only pharmacological, but also psychological strategies for relieving pain can modulate the interconnected network of cortical and subcortical regions that participate in the processing of painful stimuli.

Placebo

Like hypnosis, placebo is another example of a powerful form of top-down modulation of pain by cognitive processes. There are several striking similarities between the two. Just like there is no generally accepted definition for hypnosis, there is also no clear definition of placebo. It is indeed difficult or maybe even impossible to define the placebo and the placebo effect in a coherent and logical manner. For instance, Shapiro (1960) defined placebo as: "any therapeutic procedure (or that component of any therapeutic procedure) which is given deliberately to have an effect, or unknowingly, and has an effect on a symptom, syndrome, disease, or patient but which is objectively without specific activity for the condition being treated. The placebo effect is defined by the changes produced by placebos". Consequently, the placebo is described as being: "without specific activity for the condition being treated". Putting this phrase in place of the word itself, it then reads: "The placebo effect is defined as the changes produced by things without specific activity for the condition being treated", a statement that does not make any sense. Because of the difficulty to provide a clear definition, some authors have suggested the abandonment the entire concept of placebo (Götzsche, 1994) or to rename placebo effects as "meaning responses", by which is meant the physiologic and psychological effects of meaning in the origins or treatment of illness (Moerman and Jonas, 2002). A second characteristic that placebo has in common with hypnosis is that it has long been considered an unpopular topic that was largely neglected by the neuroscience community. Fortunately, this situation has changed in recent years during which we have witnessed an increased interest in the neurobiology of the placebo

Fig. 5. The Quack (1652) by Gerrit Dou, from the Dutch Leiden School. This painting depicts a quack — a medical charlatan — hawking his wares to a small crowd on the outskirts of Leiden.

response, both for pain and other conditions such as Parkinson's disease and depression (Fig. 5).

Placebo: classes of explanation

Conditioning and expectation

One of the main theories of placebo analgesia is that it is mediated by classical (Pavlovian) conditioning. Applied to placebo analgesia, active agents that reduce pain act as unconditioned stimuli (UCS) and the vehicles and settings in which they are delivered (pills, syringes, the medical setting, etc.) act as conditioned stimuli (CS). It follows that medical treatments act as acquisition phases or conditioning trials where vehicles and active agents are paired. The pairings enable the CS to produce therapeutic effects as conditioned responses (CR) (Wickramasekera, 1980; Suchman and Ader, 1992; Kirsch, 1997; Montgomery and Kirsch, 1997). Several investigators have provided empirical evidence in support of a role of conditioning in placebo analgesia in man (Laska and Sunshine, 1973; Voudouris et al., 1989, 1990; Amanzio and Benedetti, 1999). In addition, experimental studies of conditioned pharmacological responses revealed that animals with a history of drug administration are likely to show drug resembling behavior when a placebo is administered in a drug-associated context (Siegel, 1985).

Not everybody agrees with the purported role of classical conditioning in placebo analgesia (Brewer, 1974). For instance, placebo effects are sometimes not in accordance with the pharmacological property of the active conditioning drug, but are context and suggestion dependent. This has led to the adoption of an informational (cognitive) view of classical conditioning to explain the placebo effect. According to this view, conditioning produces expectancies that certain events will follow other events. Conditioning depends on the information that the CS provides about the UCS (Rescorla, 1988). The unconditioned response (UCR) is seen as the anticipatory response preparing the organism for the occurrence of the anticipated US. In order for a stimulus to serve as a US, it has to be *perceived* and that what is perceived is the active drug's *effect*. Essentially what is learned is that drugs produce specific effects; consequently, conditioning is one way to obtain response expectancies. This implies that the drug response becomes the UCS rather than the UCR. It follows that the placebo effect is dependent on the strength of an individual's expectancies and not on how these expectancies have been formed (Evans, 1985; Kirsch, 1997; Montgomery and Kirsch, 1997; Price et al., 1999; De Pascalis et al., 2002).

Conditioning and expectancy might represent two separate dimensions of a common process. It is likely that each mechanism can be applied to explain placebo analgesia under different experimental conditions. Benedetti and colleagues (2003) tried to disentangle the respective roles of expectancy and conditioning in the placebo response. They showed that conditioning mediates the placebo effect when an unconscious physiological function like hormonal secretion is involved. However, even though a conditioning procedure has taken place, the placebo effect is mediated by expectancy when conscious physiological processes like pain and motor performance are involved. In other words, the placebo effect is learned at an unconscious or a conscious level depending on the system involved.

Endogenous opioids

The question whether placebo analgesia is mediated by the release of endogenous opioids has divided the research community since the beginning. Levine and colleagues (1978) provided the first demonstration that opioids are involved in the placebo response by showing that the opioid antagonist naloxone can block placebo analgesia. This finding was important for two reasons. First, it showed that endogenous opioids may mediate the placebo response and second, it was the first neurobiological legitimation for the unpopular placebo topic. The study gave placebo analgesia a biological underpinning and initiated a series of pharmacological studies, most of which confirmed Levine's findings (Amanzio and Benedetti, 1997, 1999; Benedetti et al., 1999). However, Gracely et al. (1983) reported that placebo analgesia can occur without the involvement of the endogenous opioid system. Likewise, Grevert et al. (1983) showed that placebo analgesia was only partly blocked by naloxone. We recently also studied the involvement of the endogenous opioid system in two patients suffering from chronic low back pain (Kupers et al., in preparation). Both patients showed a profound and long-lasting placebo response to placebo infusions. Naloxone and saline were administered in a double-blind manner to test whether the placebo response was mediated by endogenous opioids. Naloxone (10 mg) failed to block the placebo response, suggesting that the placebo response in these patients was not mediated by the endogenous opioid system. In order to better understand the conditions that produce naloxone-reversible and naloxone-insensitive placebo effects, Amanzio and Benedetti (1999) evoked different types of placebo analgesic effects by using cognitive expectation cues, drug-conditioning or a combination of both. Their aim was to dissect placebo analgesia into opioid and non-opioid parts and determine their relation with expectancy and conditioning. The drug conditioning trials were carried out with either morphine or the non-steroidal anti-inflammatory drug ketorolac. Morphine conditioning generated a naloxone-reversible placebo effect. In contrast, when conditioning was done with the non-opioid ketorolac, the resulting placebo effect was not antagonized by naloxone. Expectancy cues presented alone without prior drug conditioning generated a small placebo effect that was completely blocked by naloxone. This demonstration that placebo analgesia can be dissected into opioid and non-opioid parts, depending on the procedure used to evoke the placebo effect, may resolve the old controversy of the role of opioids in placebo analgesia. The results further indicate that cognitive factors and conditioning are differently balanced in relation to placebo analgesia, activating either the opioid or non-opioid system. Whereas expectation triggers the endogenous opioid system, conditioning procedures may activate either opioid or non-opioid systems.

A subsequent study by the same group (Benedetti et al., 1999) further elaborated on how endogenous opioids are activated by expectancy cues. They induced specific expectancies of analgesia directed toward different body parts. In brief, a noxious stimulus was simultaneously applied to four different body areas (arms and legs, left and right). A placebo cream was applied to one of the stimulated body parts and the subjects were informed that the cream was an analgesic (induction of expectations). The expectancy of an analgesic effect was exclusively directed toward the site where the placebo had been applied. The results showed a placebo analgesic effect that was restricted to the treated site and that was completely abolished by a hidden injection of naloxone. These data show that a spatially directed expectancy produces a placebo effect that is somatotopically restricted to the body site that is the target of expectation. These data further indicate that placebo-activated endorphines act only on the expected target of action and not on the entire body. It can thus be hypothesized that a cognitive component in the form of spatial attention or spatial directed expectation plays a pivotal role in the activation of specific opioid systems (Benedetti et al., 1999; Price, 1999).

The role of opioids in placebo analgesia implies that the activation of the pain modulatory circuitry, including the PAG and the RVM, is triggered by psychological factors. Fields and co-workers already emphasized the role of expectation, arousal, and attention in the modulation by the PAG and RVM (Price and Fields, 1997). Furthermore, Soper and Melzack (1982) found that the PAG is functionally and somatotopically organized. That is, stimulation of different sites of the PAG produced an analgesic effect in different cutaneous areas. In other words, the spatially directed effect may be explained by the somatotopical organization of the PAG. Because the pain modulatory circuitry is influenced by opioids it can be hypothesized that the endogenous opioid systems has a role in organising the PAG and RVM (Benedetti et al., 1999).

There is evidence from studies on the placebo effect in patients with Parkinson's disease (PD) that the opioid system is not the only neurotransmitter system involved in the placebo response. Another candidate is dopamine. de la Fuente-Fernández and colleagues (2001) used PET to study the involvement of the brain dopamine system in the placebo response in patients with PD. They used [^{11}C]raclopride (RAC), which is an antagonist for the dopamine D2/D3 receptors. By calculating the displacement of the exogenously applied RAC following a pharmacological or behavioral challenge, one can estimate the amount of endogenously released dopamine. Six PD patients were scanned in a placebo-controlled, double-blind fashion following the administration of a placebo or an active dopaminergic drug (levodopa). Placebo administration caused a significant decrease in RAC binding potential in the dorsal striatum. Interestingly, the magnitude of the decrease in RAC binding potential following a placebo was comparable to that evoked by a therapeutic dose of levodopa. The authors also reported a positive correlation between the degree of dopamine release and the extent of the perceived placebo effect. In a following study, the same authors showed that placebo also releases dopamine in the nucleus accumbens (de la Fuente-Fernández et al., 2002a). The dopamine release in the ventral striatum may be caused by the expectation of reward — in casu, the anticipation of a therapeutic effect.

Brain imaging findings

Brain imaging studies over the past decade have greatly advanced our understanding of the mechanisms underlying placebo analgesia. Their contribution has been twofold. First, these studies showed that placebo-induced pain relief is associated with a concomitant decrease of brain activity in pain-related areas such as the thalamus, the insula, and the anterior cingulate cortex. This is important in view of the discussion whether reported pain reductions following placebos represent genuine analgesic effects or mere compliance with experimental instructions. Second, these studies pointed out a number of structures that may be at the origin of the placebo effect. Among these figure the dorsolateral, ventrolateral and orbital prefrontal cortices, the ACC, and the midbrain. Interestingly, the dorsolateral prefrontal cortex has also been proposed to be an important structure in the endogenous modulation of pathological pain (Lorenz et al., 2003).

Using PET, Petrovic and colleagues (2002) scanned a group of healthy volunteers following the administration of a short acting opioid (remifentanil), placebo, or no drugs. Subjects were scanned during non-painful and painful tonic heat stimulation of the dorsum of the left hand. Pain ratings following placebo administration were significantly lower than when no drugs were given. The placebo response was associated with a significant rCBF increase in the orbitofrontal and ACCs, areas that were also reliably activated by the opioid remifentanil. When the results were analyzed for the high and low placebo responders separately, it was found that only the high placebo responders activated the rostral ACC (rACC) during remifentanil analgesia. This suggests that there exists a relationship between how effectively opioids activate the rACC and how well subjects respond to placebo during pain. In other words, placebo responders seem to have a more efficient opioid system. This hypothesis is supported by data from a study by Lasagna et al. (1954) showing that 95% of individuals responding to a placebo report pain reduction following the injection of morphine, whereas only 54% of the individuals not responding to a placebo report pain reduction following the injection of morphine. A regression analysis of the PET data from Petrovic's study further showed that in the pain–opioid and pain–placebo conditions, activity in the rACC covaried with activity in an area near the PAG and the pons. This suggests that higher cortical areas may take control over descending pain modulatory systems during opioid- and placebo analgesia. As mentioned above, cognitive factors such as expectation and desire for pain relief contribute a lot to the occurrence of the placebo response. The ACC and lateral orbitofrontal cortex may play an important role in this cognitive modulation of pain via their projections to the PAG, an area involved in descending inhibitory control of pain.

Wager et al. (2004) used fMRI to study the brain mechanisms involved in placebo analgesia. In two separate experiments, they addressed the following questions: (1) which pain-responsive areas of the brain show reduced activity following a placebo? and (2) which areas in the brain show increased activity following the administration of a placebo? Whereas the previous PET study tried to answer the second question, it did not address the first one. To increase the likelihood of the occurrence of a placebo response, the authors used a conditioning procedure. The results showed that placebo significantly reduced the BOLD response in the rACC, contralateral insula, and thalamus, areas that are part of the so-called pain-matrix (Peyron et al., 2000). In order to answer the second question, the authors compared brain activity in the anticipation period of pain under the control and placebo conditions. The results revealed that activity in the dorsolateral and orbital prefrontal cortices, ACC and PAG was significantly higher following a placebo compared to the control condition. In addition, the increased prefrontal activity correlated with the placebo-induced reductions in pain-evoked activity in thalamus, insula, and ACC. The increased activity in the prefrontal cortex is in line with the hypothesis that brain areas involved in generating and mediating expectation contribute to placebo-analgesia; whereas the above-mentioned brain imaging studies used acute, experimentally induced pain. A recent study by Lieberman et al. (2004) investigated placebo analgesic responses in patients with irritable bowel syndrome. The experimental setup of the study was different from the previous studies. Before and after a 3-week placebo regimen, patients were scanned at rest and during controlled rectal stimulation. In line with the results of the placebo response in acute pain, the placebo response was associated with significantly increased activity in the prefrontal cortex. In contrast with the previous results (Petrovic et al., 2002; Wager et al., 2004), the increased activity occurred in the ventrolateral and not in the dorsolateral or orbitofrontal part of the prefrontal cortex. Interestingly, the stereotactic coordinates of the ventrolateral prefrontal activation in the Lieberman study ($x = 30$, $y = 33$, $z = -5$) is nearly a right-hemispheric mirror image of the hypnosis-induced rCBF increase ($x = -28$, $y = 26$, $z = 8$) in the left ventrolateral prefrontal cortex in the study by Maquet et al. (1999). A second difference with the previous

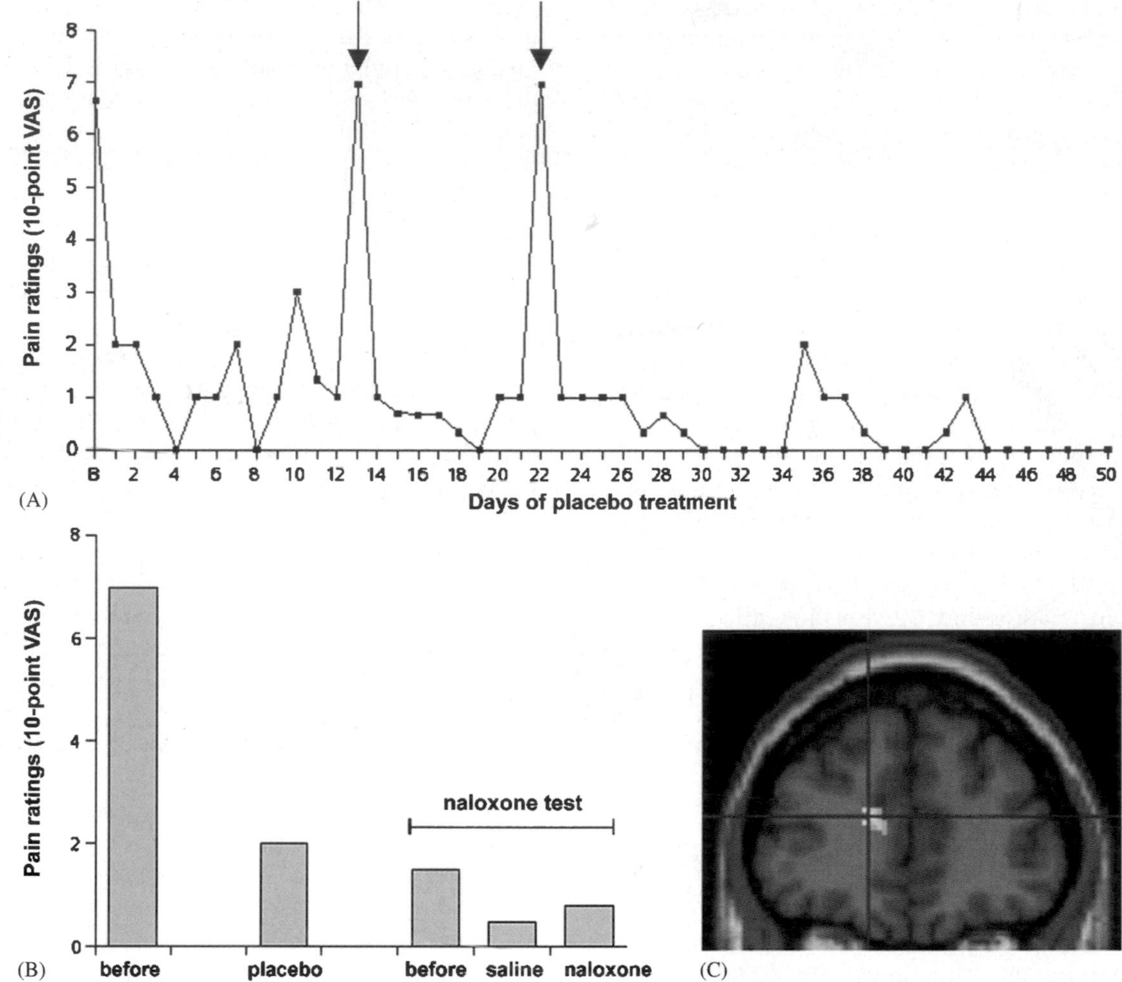

Fig. 6. (A) Time course of the placebo response in a chronic pain patient. Before placebo, average pain rating was around 7 on a 10-point visual analogue scale (VAS). The patient's pain was significantly reduced by placebo during the 50 days she was followed by the pain clinic. On two occasions, indicated by the arrows, the placebo administration was stopped and the original pain reappeared, indicating that natural course of the disease or regression to the mean are unlikely to explain the pain improvement. The data also indicate that placebo analgesia can last over extended periods. (B) Effect of naloxone on the placebo response in a chronic pain patient. Naloxone (10 mg, i.v.) or saline was administered in a double-blind manner in a patient with placebo analgesia. The left bar (before) shows average pain ratings before placebo. Patient's average pain ratings during the 30 days of placebo administration dropped significantly (placebo). Naloxone did not abolish the placebo response in this patient. (C) rCBF changes in the anterior cingulate cortex in a chronic pain patient during a placebo response (Kupers et al., unpublished results).

studies is that the placebo response was not associated with increased activity in the ACC. In contrast, activity in ACC was significantly reduced and the reduction in this region correlated negatively with activity in the ventrolateral prefrontal cortex. Our own preliminary findings in patients with neuropathic pain argue for a possible role of the orbitofrontal cortex in the placebo response (Fig. 6).

Fig. 7 shows a composite image of the activations induced by placebo and hypnosis in the higher discussed papers. As can be seen, ACC and prefrontal cortex are activated in both forms of cognition-induced analgesia.

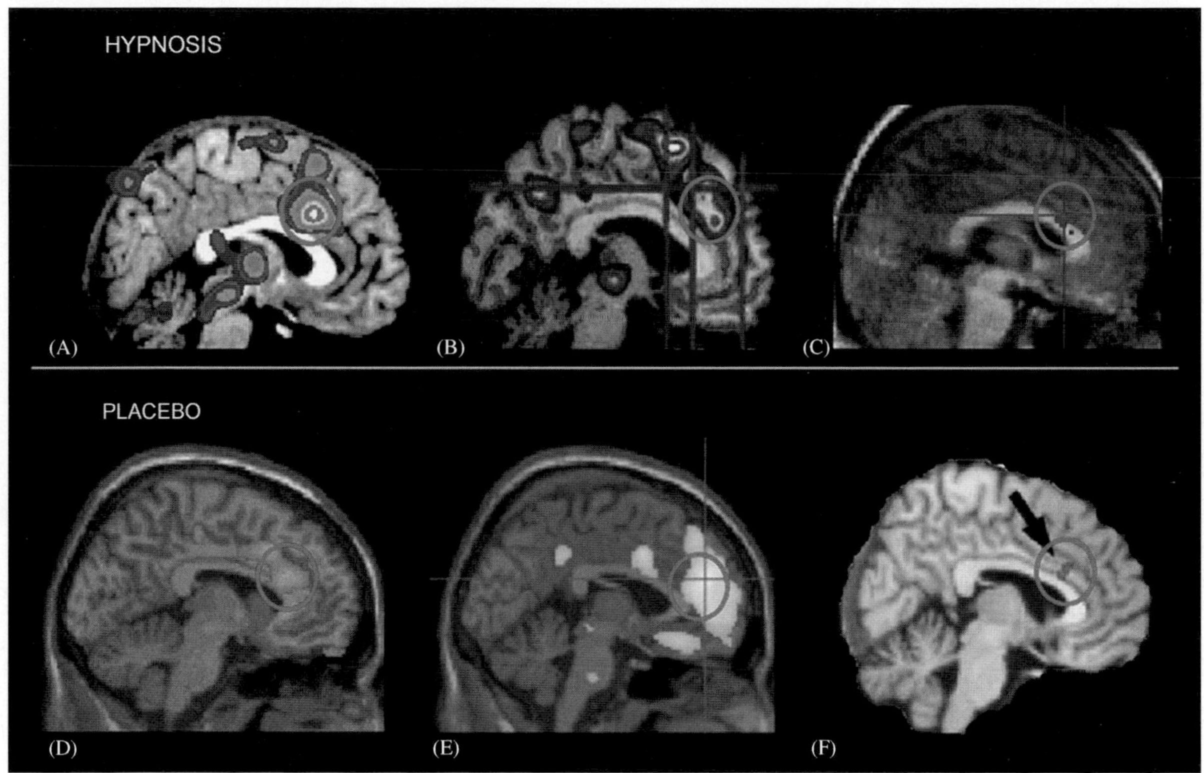

Fig. 7. Both hypnosis- and placebo-induced modulation of pain perception seems to be mediated by the ACC (red circles). Note that mean peak coordinates from hypnosis- and placebo-induced pain modulation studies, using very different methodology, are within 5 mm from each other ($x = 2$, $y = 22$, $z = 28$ versus $x = 6$, $y = 17$, $z = 31$ mm, respectively). *Hypnosis*: (A) Suggestion-related changes in rCBF during pain perception: the image shows rCBF changes for the subtraction hypnosis-with-suggestion minus hypnosis-without-suggestion, both under painful stimulation (coordinates of peak voxel: $x = 7$, $y = 20$ and $z = 29$ mm). From Rainville et al. (1997). (B) Pain-related activity associated with hypnotic suggestions of high unpleasantness (painfully hot/high unpleasantness minus neutral/hypnosis control condition; peak voxel: $x = 0$, $y = 29$, $z = 34$ mm). From Rainville et al. (1999). (C) rCBF increases in proportion to pain sensation ratings, in the specific context of the hypnotic state (difference in pain ratings versus rCBF regression slopes between the hypnotic state and control conditions (i.e., non-hypnotic mental imagery and resting state; peak voxel: $x = 2$, $y = 18$, $z = 22$ mm). From Faymonville et al. (2000). *Placebo*: (D) rCBF increases with increased intestinal discomfort in placebo-treated patients with chronic abdominal pain (irritable bowel syndrome): a greater reduction in rCBF response to visceral stimulation from pre- to post-placebo was associated with greater self-reported symptom improvement (peak voxel: $x = 8$, $y = 19$, $z = 34$ mm). From Lieberman, M. D., Jarcho, J. M., Berman, S., Naliboff, B. D., Suyenobu, B. Y., Mandelkern, M. and Mayer, E. A. (2004). Neuroimage, 22: 447–455. (E) Activated areas in high placebo responders (healthy volunteers) during heat pain and opioid treatment minus heat pain only (peak voxel: $x = 3$, $y = 18$, $z = 34$ mm). From Petrovic et al. (2002). (F) Pain regions showing correlations between placebo effects in reported pain (control minus placebo) and placebo effects in neural pain (i.e., measured activity in pain related brain areas) (control minus placebo) (peak voxel: $x = 6$, $y = 14$, $z = 26$ mm). From Wager et al. (2004). See Plate 19.7 in Colour Plate Section.

Conclusion

It has taken a long time before the scientific community finally accepted that hypnosis and placebo are worth being studied scientifically. Therefore, our knowledge about the mechanisms mediating these powerful effects is still fragmentary. Brain imaging studies have investigated the brain circuitry that is involved in hypnosis- and placebo-induced forms of pain modulation. Brain structures that are involved in these cognitive-induced forms of analgesia are the ACC, and the dorsolateral and orbitofrontal prefrontal cortices. Functional brain connectivity studies further suggest that the anterior cingulate and prefrontal cortices exert their effects by modulating activity in the midbrain periaqueductal gray, a structure that is of utmost importance in the endogenous modulation of pain. Taken together, these findings reinforce the idea that psychological interventions can relieve pain by modulating activity in an interconnected network of cortical and subcortical regions that are implicated in the processing of noxious information.

Acknowledgments

S. L. is Research Associate at the Fonds National de la Recherche Scientifique de Belgique (FNRS). This research was supported by research grants from the University of Liège, the Centre Hospitalier Universitaire Sart Tilman of the University of Liège and the Danish Medical Research Council.

References

Albe-Fessard, D., Berkley, K.J., Kruger, L., Ralston, H.J. and Willis Jr., W.D. (1985) Diencephalic mechanisms of pain sensation. Brain Res., 356: 217–296.

Alkire, M.T., Pomfrett, C.J., Haier, R.J., Gianzero, M.V., Chan, C.M., Jacobsen, B.P. and Fallon, J.H. (1999) Functional brain imaging during anesthesia in humans: effects of halothane on global and regional cerebral glucose metabolism. Anesthesiology, 90: 701–709.

Amanzio, M. and Benedetti, F. (1997) The neurobiology of placebo analgesia: From endogenous opioids to cholecystokinin. Prog. Neurobiol., 52: 109–125.

Amanzio, M. and Benedetti, F. (1999) Neuropharmacological dissection of placebo analgesia: expectation-activated opioid systems vs. conditioning-activated specific subsystems. J. Neurosci., 19: 484–494.

An, X., Bandler, R., Ongur, D. and Price, J.L. (1998) Prefrontal cortical projections to longitudinal columns in the midbrain periaqueductal gray in macaque monkeys. J. Comp. Neurol., 40: 455–479.

Augustine, J.R. (1996) Circuitry and functional aspects of the insular lobe in primates including humans. Brain Res. Rev., 22: 229–244.

Aupee, A.M., Desgranges, B., Eustache, F., Lalevee, C., de la Sayette, V., Viader, F. and Baron, J.C. (2001) Voxel-based mapping of brain hypometabolism in permanent amnesia with PET. Neuroimage, 13: 1164–1173.

Bandler, R. and Shipley, M.T. (1994) Columnar organization in the midbrain periaqueductal gray: modules for emotional expression? Trends Neurosci., 17: 379–389.

Benedetti, F., Arduino, C. and Amanzio, M. (1999) Somatotopic activation of opioid systems by target-directed expectations of analgesia. J. Neurosci., 19: 3639–3648.

Benedetti, F., Pollo, A., Lopiano, L., Lanotte, M., Vighetti, S. and Rainero, I. (2003) Conscious expectation and unconscious conditioning in analgesic, motor, and hormonal placebo/nocebo responses. J. Neurosci., 23: 4315–4323.

Bernard, J.F., Bester, H. and Besson, J.M. (1996) Involvement of the spino-parabrachio amygdaloid and-hypothalamic pathways in the autonomic and affective emotional aspects of pain. Prog Brain Res., 107: 243–255.

Binder, J.R., Frost, J.A., Hammeke, T.A., Bellgowan, P.S., Rao, S.M. and Cox, R.W. (1999) Conceptual processing during the conscious resting state. A functional MRI study. J. Cogn. Neurosci., 11: 80–95.

Bornhovd, K., Quante, M., Glauche, V., Bromm, B., Weiller, C. and Buchel, C. (2002) Painful stimuli evoke different stimulus-response functions in the amygdala, prefrontal, insula and somato-sensory cortex: a single-trial fMRI study. Brain, 125: 1326–1336.

Brewer, W.F. (1974) There is no convincing evidence for operant or classical conditioning in adult humans. In: Weimer W.B. and Palermo D.S. (Eds.), Cognition and the Symbolic Processes. Erlbaum, Hillsdale, NJ.

Brown, P. and Marsden, C.D. (1998) What do the basal ganglia do? Lancet, 351: 1801–1804.

Burstein, R. and Potrebic, S. (1993) Retrograde labeling of neurons in the spinal cord that project directly to the amygdala or the orbital cortex in the rat. J Comp Neurol., 335: 469–485.

Bush, G., Luu, P. and Posner, M.I. (2000) Cognitive and emotional influences in anterior cingulate cortex. Trends Cogn. Sci., 4: 215–222.

Bushnell, M.C. and Duncan, G.H. (1989) Sensory and affective aspects of pain perception: is medial thalamus restricted to emotional issues? Exp Brain Res., 78: 415–418.

Coghill, R.C., Sang, C.N., Maisog, J.M. and Iadarola, M.J. (1999) Pain intensity processing within the human brain: a bilateral, distributed mechanism. J. Neurophysiol, 82: 1934–1943.

Coghill, R.C., Talbot, J.D., Evans, A.C., Meyer, E., Gjedde, A., Bushnell, M.C. and Duncan, G.H. (1994) Distributed processing of pain and vibration by the human brain. J. Neurosci, 14: 4095–4108.

Craig, A.D., Bushnell, M.C., Zhang, E.T. and Blomqvist, A. (1994) A thalamic nucleus specific for pain and temperature sensation. Nature, 372: 770–773.

Craig, A.D., Chen, K., Bandy, D. and Reiman, E.M. (2000) Thermosensory activation of insular cortex. Nat. Neurosci., 3: 184–190.

Damasio, A.R. (1994) Descartes' Error: Emotion, Reason, and the Human Brain. G.P. Putnam, New York.

De Betz, B. and Sunnen, G. (1985) A Primer of Clinical Hypnosis. PSG Publishing, Littleton, MA.

de la Fuente-Fernandez, R., Phillips, A.G., Zamburlini, M., Sossi, V., Calne, D.B., Ruth, T.J. and Stoessl, A.J. (2002a) Dopamine release in human ventral striatum and expectation of reward. Behav. Brain Res., 136: 359–363.

de la Fuente-Fernandez, R., Ruth, T.J., Sossi, V., Schulzer, M., Calne, D.B. and Stoessl, A.J. (2001) Expectation and dopamine release: mechanism of the placebo effect in Parkinson's disease. Science, 293: 1164–1166.

De Pascalis, V., Chiaradia, C. and Carotenuto, E. (2002) The contribution of suggestibility and expectation to placebo analgesia phenomenon in an experimental setting. Pain, 96: 393–402.

Decety, J. (1996) Neural representations for action. Rev. Neurosci., 7: 285–297.

Defechereux, T., Degauque, C., Fumal, I., Faymonville, M.E., Joris, J., Hamoir, E. and Meurisse, M. (2000) Hypnosedation, a new method of anesthesia for cervical endocrine surgery. Prospective randomized study. Ann. Chir., 125: 539–546.

Defechereux, T., Faymonville, M.E., Joris, J., Hamoir, E., Moscato, A. and Meurisse, M. (1998) Surgery under hypnosedation. A new therapeutic approach to hyperparathyroidism. Ann. Chir., 52: 439–443.

Defechereux, T., Meurisse, M., Hamoir, E., Gollogly, L., Joris, J. and Faymonville, M.E. (1999) Hypnoanesthesia for endocrine cervical surgery: a statement of practice. J Altern. Complement Med., 5: 509–520.

Derbyshire, S.W. and Jones, A.K. (1998) Cerebral responses to a continual tonic pain stimulus measured using positron emission tomography. Pain, 76: 127–135.

Derbyshire, S.W., Jones, A.K., Gyulai, F., Clark, S., Townsend, D. and Firestone, L.L. (1997) Pain processing during three levels of noxious stimulation produces differential patterns of central activity. Pain, 73: 431–445.

Derbyshire, S.W., Vogt, B.A. and Jones, A.K. (1998) Pain and Stroop interference tasks activate separate processing modules in anterior cingulate cortex. Exp. Brain Res., 118: 52–60.

Devinsky, O., Morrell, M.J. and Vogt, B.A. (1995) Contributions of anterior cingulate cortex to behaviour. Brain, 118: 279–306.

Dubner, R., Hoffman, D.S. and Hayes, R.L. (1981) Neuronal activity in medullary dorsal horn of awake monkeys trained in a thermal discrimination task. Task-related responses and their functional role. J. Neurophysiol., 46: 444–464.

Duncan, G.H., Bushnell, M.C., Bates, R. and Dubner, R. (1987) Task-related responses of monkey medullary dorsal horn neurons. J. Neurophysiol., 57: 289–310.

Evans, F.J. (1985) Expectancy, therapeutic instructions, and the placebo response. In: White L., Tursky B. and Schwartz G.E. (Eds.), Placebo: Theory, Research and Mechanisms. Guilford Press, New York, pp. 215–228.

Faymonville, M.E., Fissette, J., Mambourg, P.H., Delchambre, A. and Lamy, M. (1994) Hypnosis, hypnotic sedation. Current concepts and their application in plastic surgery. Rev. Med. Liege, 49: 13–22.

Faymonville, M.E., Fissette, J., Mambourg, P.H., Roediger, L., Joris, J. and Lamy, M. (1995) Hypnosis as adjunct therapy in conscious sedation for plastic surgery. Reg. Anesth., 20: 145–151.

Faymonville, M.E., Laureys, S., Degueldre, C., DelFiore, G., Luxen, A., Franck, G., Lamy, M. and Maquet, P. (2000) Neural mechanisms of antinociceptive effects of hypnosis. Anesthesiology, 92: 1257–1267.

Faymonville, M.E., Mambourg, P.H., Joris, J., Vrijens, B., Fissette, J., Albert, A. and Lamy, M. (1997) Psychological approaches during conscious sedation. Hypnosis versus stress reducing strategies: a prospective randomized study. Pain, 73: 361–367.

Faymonville, M.E., Meurisse, M. and Fissette, J. (1999) Hypnosedation: a valuable alternative to traditional anaesthetic techniques. Acta Chir. Belg., 99: 141–146.

Faymonville, M.E., Roediger, L., Del Fiore, G., Degueldre, C., Phillips, C., Lamy, M., Luxen, A., Maquet, P. and Laureys, S. (2003) Increased cerebral functional connectivity underlying the antinociceptive effects of hypnosis. Cogn. Brain Res., 17: 255–262.

Fields, H.L. and Basbaum, A.I. (1999) Central nervous system mechanisms of pain modulation. In: Wall P.D. and Melzack R. (Eds.), Textbook of pain. Churchill Livingstone, London, pp. 309-29.

Fink, G.R., Frackowiak, R.S., Pietrzyk, U. and Passingham, R.E. (1997) Multiple nonprimary motor areas in the human cortex. J. Neurophysiol., 77: 2164–2174.

Fink, G.R., Markowitsch, H.J., Reinkemeier, M., Bruckbauer, T., Kessler, J. and Heiss, W.D. (1996) Cerebral representation of one's own past: neural networks involved in autobiographical memory. J. Neurosci., 16: 4275–4282.

Frenay, M.C., Faymonville, M.E., Devlieger, S., Albert, A. and Vanderkelen, A. (2001) Psychological approaches during dressing changes of burned patients: a prospective randomized study comparing hypnosis against stress reducing strategy. Burns, 27: 793–799.

Friston, K. (2002) Beyond phrenology: what can neuroimaging tell us about distributed circuitry? Annu. Rev. Neurosci., 25: 221–250.

Friston, K.J., Buechel, C., Fink, G.R., Morris, J., Rolls, E. and Dolan, R.J. (1997) Psychophysiological and modulatory interactions in neuroimaging. Neuroimage, 6: 218–229.

Götzsche, P.C. (1994) Is there logic in the placebo? Lancet, 344: 25–26.

Gracely, R.H., Dubner, R., Wolskee, P.J. and Deeter, W.R. (1983) Placebo and naloxone can alter post-surgical pain by separate mechanisms. Nature, 306: 264–265.

Grevert, P., Albert, L.H. and Goldstein, A. (1983) Partial antagonism of placebo analgesia by naloxone. Pain, 16: 129–143.

Gusnard, D.A. and Raichle, M.E. (2001) Searching for a baseline: functional imaging and the resting human brain. Nat. Rev. Neurosci., 2: 685–694.

Hayes, R.L., Price, D.D., Bennett, G.J., Wilcox, G.L. and Mayer, D.J. (1978) Differential effects of spinal cord lesions on narcotic and non-narcotic suppression of nociceptive reflexes: further evidence for the physiologic multiplicity of pain modulation. Brain Res., 155: 91–101.

Helmstetter, F.J., Tershner, S.A., Poore, L.H. and Bellgowan, P.S. (1998) Antinociception following opioid stimulation of the basolateral amygdala is expressed through the periaqueductal gray and rostral ventromedial medulla. Brain Res., 779: 104–118.

Hilgard, E.R., Lauer, L.W. and Morgan, A.H. (1963) Manual for Standard Profile Scales of Hypnotic Susceptibility, Forms I and II. Consulting Psychologists Press, Palo Alto.

Hofbauer, R.K., Rainville, P., Duncan, G.H. and Bushnell, M.C. (2001) Cortical representation of the sensory dimension of pain. J Neurophysiol., 86: 402–411.

Hudson, P.M., Semenenko, F.M. and Lumb, B.M. (2000) Inhibitory effects evoked from the rostral ventrolateral medulla are selective for the nociceptive responses of spinal dorsal horn neurons. Neuroscience, 99: 541–547.

Jones, A.K., Brown, W.D., Friston, K.J., Qi, L.Y. and Frackowiak, R.S. (1991) Cortical and subcortical localization of response to pain in man using positron emission tomography. Proc. R. Soc. Lond. B Biol. Sci., 244: 39–44.

Kinomura, S., Larsson, J., Gulyas, B. and Roland, P.E. (1996) Activation by attention of the human reticular formation and thalamic intralaminar nuclei. Science, 271: 512–515.

Kirsch, I. (1997) Specifying nonspecifics: Psychological mechanisms of placebo effects. In: Harrington A. (Ed.), The Placebo Effect: An Interdisciplinary Exploration. Hardvard University Press, Cambridge, MA.

Kosslyn, S.M., Ganis, G. and Thompson, W.L. (2001) Neural foundations of imagery. Nat. Rev. Neurosci., 2: 635–642.

Kosslyn, S.M., Thompson, W.L., Costantini-Ferrando, M.F., Alpert, N.M. and Spiegel, D. (2000) Hypnotic visual illusion alters color processing in the brain. Am. J. Psychiatry, 157: 1279–1284.

Kupers, R. (2001) Is the placebo powerless? N. Engl. J. Med., 345: 1278.

Kupers, R., Svensson, P. and Jensen, T.S. (2004) Central representation of muscle pain and mechanical hyperesthesia in the orofacial region: a positron emission tomography study. Pain, 108: 284–293.

Lasagna, L.L., Mosteller, F., Von Felsinger, J.M. and Beecher, H.K. (1954) A study of the placebo response. Am. J. Med., 16: 770–779.

Laska, E. and Sunshine, A. (1973) Anticipation of analgesia: a placebo effect. Headache, 13: 1–11.

Laureys, S., Berré, J. and Goldman, S. (2001) Cerebral function in coma, vegetative state, minimally conscious state, locked-in syndrome and brain death. In: Vincent J.L. (Ed.), 2001 Yearbook of Intensive Care and Emergency Medicine. Springer, Berlin, pp. 386–396.

Laureys, S., Goldman, S., Phillips, C., Van Bogaert, P., Aerts, J., Luxen, A., Franck, G. and Maquet, P. (1999) Impaired effective cortical connectivity in vegetative state: preliminary investigation using PET. Neuroimage, 9: 377–382.

Laureys, S., Maquet, P. and Faymonville, M.E. (2004) Brain function during hypnosis. In: Otte A., Audenaert K., Peremans K., Van Heeringen K. and Dierckx R.A. (Eds.), Nuclear Medicine in Psychiatry. Springer, Berlin, Heidelberg, pp. 507–519.

Levine, J.D., Gordon, N.C. and Fields, H.L. (1978) The mechanism of placebo analgesia. Lancet, 2: 654–657.

Lieberman, M.D., Jarcho, J.M., Berman, S., Naliboff, B.D., Suyenobu, B.Y., Mandelkern, M. and Mayer, E.A. (2004) The neural correlates of placebo effects: a disruption account. Neuroimage, 22: 447–455.

Lorenz, J., Minoshima, S. and Casey, K.L. (2003) Keeping pain out of mind: the role of the dorsolateral prefrontal cortex in pain modulation. Brain, 126: 1079–1091.

Maquet, P. (2000) Functional neuroimaging of normal human sleep by positron emission tomography. J. Sleep Res., 9: 207–231.

Maquet, P., Faymonville, M.E., Degueldre, C., Delfiore, G., Franck, G., Luxen, A. and Lamy, M. (1999) Functional neuroanatomy of hypnotic state. Biol. Psychiatry, 45: 327–333.

Matsuda, H. (2001) Cerebral blood flow and metabolic abnormalities in Alzheimer's disease. Ann. Nucl. Med., 15: 85–92.

Mazoyer, B., Zago, L., Mellet, E., Bricogne, S., Etard, O., Houde, O., Crivello, F., Joliot, M., Petit, L. and Tzourio-Mazoyer, N. (2001) Cortical networks for working memory and executive functions sustain the conscious resting state in man. Brain Res. Bull., 54: 287–298.

McGaraughty, S., Chu, K.L., Bitner, R.S., Martino, B., El Kouhen, R., Han, P., Nikkel, A.L., Burgard, E.C., Faltynek, C.R. and Jarvis, M.F. (2003) Capsaicin infused into the PAG affects rat tail flick responses to noxious heat and alters neuronal firing in the RVM. J. Neurophysiol., 90: 2702–2710.

Melzack, R. and Casey, K.L. (1968) Sensory, motivational and central determinants of pain: a new conceptual model. In: Kenshalo D.R. (Ed.), The Skin Senses. Thomas, Springfield, IL, pp. 423–443.

Melzack, R. and Wall, P.D. (1965) Pain mechanisms: a new theory. Science, 150: 971–978.

Mesulam, M.M. and Mufson, E.J. (1982) Insula of the old world monkey. III: Efferent cortical output and comments on function. J. Comp. Neurol., 212: 38–52.

Meurisse, M., Defechereux, T., Hamoir, E., Maweja, S., Marchettini, P., Gollogly, L., Degauque, C., Joris, J. and Faymonville, M.E. (1999a) Hypnosis with conscious sedation instead of general anesthesia? Applications in cervical endocrine surgery. Acta Chir. Belg., 99: 151–158.

Meurisse, M., Faymonville, M.E., Joris, J., Nguyen Dang, D., Defechereux, T. and Hamoir, E. (1996) Endocrine surgery by hypnosis. From fiction to daily clinical application. Ann. Endocrinol., 57: 494–501.

Meurisse, M., Hamoir, E., Defechereux, T., Gollogly, L., Derry, O., Postal, A., Joris, J. and Faymonville, M.E. (1999b) Bilateral neck exploration under hypnosedation: a new standard of care in primary hyperparathyroidism? Ann. Surg., 229: 401–408.

Moerman, D.E. and Jonas, W.B. (2002) Deconstructing the placebo effect and finding the meaning response. Ann. Intern. Med., 136: 471–476.

Montgomery, G.H. and Kirsch, I. (1997) Classical conditioning and the placebo effect. Pain, 72: 107–113.

Moret, V., Forster, A., Laverriere, M.C., Lambert, H., Gaillard, R.C., Bourgeois, P., Haynal, A., Gemperle, M. and Buchser, E. (1991) Mechanism of analgesia induced by hypnosis and acupuncture: is there a difference? Pain, 45: 135–140.

Ongur, D., An, X. and Price, J.L. (1998) Prefrontal cortical projections to the hypothalamus in macaque monkeys. J. Comp. Neurol., 401: 480–505.

Paus, T. (2001) Primate anterior cingulate cortex: where motor control, drive and cognition interface. Nat. Rev. Neurosci., 2: 417–424.

Petrovic, P. and Ingvar, M. (2002) Imaging cognitive modulation of pain processing. Pain, 95: 1–5.

Petrovic, P., Kalso, E., Petersson, K.M. and Ingvar, M. (2002) Placebo and opioid analgesia — imaging a shared neuronal network. Science, 295: 1737–1740.

Peyron, R., Frot, M., Schneider, F., Garcia-Larrea, L., Mertens, P., Barral, F.G., Sindou, M., Laurent, B. and Mauguiere, F. (2002) Role of operculoinsular cortices in human pain processing: Converging evidence from PET, fMRI, dipole modeling, and intracerebral recordings of evoked potentials. Neuroimage, 17: 1336–1346.

Peyron, R., Laurent, B. and Garcia-Larrea, L. (2000) Functional imaging of brain responses to pain. A review and meta-analysis. Neurophysiol. Clin., 30: 263–288.

Price, D.D. (1999) Psychological mechanisms of pain and analgesia. IASP Press, Seattle, WA.

Price, D.D. (2000) Psychological and neural mechanisms of the affective dimension of pain. Science, 288: 1769–1772.

Price, D.D. and Fields, H. (1997) Where are the causes of placebo analgesia? An experimental behavioral analysis. Pain Forum, 6: 44–52.

Price, D.D., Milling, L.S., Kirsch, I., Duff, A., Montgomery, G.H. and Nicholls, S.S. (1999) An analysis of factors that contribute to the magnitude of placebo analgesia in an experimental paradigm. Pain, 84: 110–113.

Raichle, M.E., MacLeod, A.M., Snyder, A.Z., Powers, W.J., Gusnard, D.A. and Shulman, G.L. (2001) A default mode of brain function. Proc. Natl. Acad. Sci. U.S.A., 98: 676–682.

Rainville, P., Duncan, G.H., Price, D.D., Carrier, B. and Bushnell, M.C. (1997) Pain affect encoded in human anterior cingulate but not somatosensory cortex. Science, 277: 968–971.

Rainville, P., Hofbauer, R.K., Paus, T., Duncan, G.H., Bushnell, M.C. and Price, D.D. (1999) Cerebral mechanisms of hypnotic induction and suggestion. J. Cogn. Neurosci., 11: 110–125.

Rescorla, R.A. (1988) Pavlovian conditioning: It's not what you think it is. Am. Psychologist, 43: 151–160.

Salmon, E., Collette, F., Degueldre, C., Lemaire, C. and Franck, G. (2000) Voxel-based analysis of confounding effects of age and dementia severity on cerebral metabolism in Alzheimer's disease. Hum. Brain Mapp., 10: 39–48.

Shapiro, A.K. (1960) A contribution to the history of the placebo effect. Behav. Sci., 5: 109–135.

Siegel, S. (1985) Drug anticipatory responses in animals. In: White L. (Ed.), Placebo: Theory, Research, and Mechanisms. Guilford Press, New York.

Soper, W.Y. and Melzack, R. (1982) Stimulation-produced analgesia: evidence for somatotopic organization in the midbrain. Brain Res., 18: 251 301–311.

Spiegel, D. (1991) Neurophysiological correlates of hypnosis and dissociation. J. Neuropsychiatry Clin. Neurosci., 3: 440–445.

Suchman, A.L. and Ader, R. (1992) Classic conditioning and placebo effects in crossover studies. Clin. Pharmacol. Ther., 52: 372–377.

Talairach, J. and Tournoux, P. (1988) Co-planar Stereotaxis Atlas of the Human Brain. Georges Thieme Verlag, Stuttgart.

The Executive Committee of the American Psychological Association - Division of Psychological Hypnosis. (1994) Definition and description of hypnosis. Contemp. Hypnosis, 11: 142–162.

Tolle, T.R., Kaufmann, T., Siessmeier, T., Lautenbacher, S., Berthele, A., Munz, F., Zieglgansberger, W., Willoch, F., Schwaiger, M., Conrad, B. and Bartenstein, P. (1999) Region-specific encoding of sensory and affective components of pain in the human brain: a positron emission tomography correlation analysis. Ann. Neurol., 45: 40–47.

Treede, R.D., Kenshalo, D.R., Gracely, R.H. and Jones, A.K. (1999) The cortical representation of pain. Pain, 79: 105–111.

Urban, M.O. and Smith, D.J. (1994) Nuclei within the rostral ventromedial medulla mediating morphine antinociception from the periaqueductal gray. Brain Res., 25: 9–16.

Vogt, B.A., Finch, D.M. and Olson, C.R. (1992) Functional heterogeneity in cingulate cortex: the anterior executive and posterior evaluative regions. Cereb. Cortex, 2: 435–443.

Vogt, B.A., Hof, P.R. and Vogt, L. (2004) Cingulate gyrus. In: Vogt B.A. (Ed.), The Human Nervous System. Elsevier Academic Press, San Diego, pp. 915–949.

Voudouris, N.J., Peck, C. and Coleman, G. (1989) Conditioned response models of placebo phenomena: Further support. Pain, 38: 109–116.

Voudouris, N.J., Peck, C.L. and Coleman, G. (1990) The role of conditioning and verbal expectancy in the placebo response. Pain, 43: 121–128.

Wager, T.D., Rilling, J.K., Smith, E.E., Sokolik, A., Casey, K.L., Davidson, R.J., Kosslyn, S.M., Rose, R.M. and Cohen, J.D. (2004) Placebo-induced changes in FMRI in the anticipation and experience of pain. Science, 303: 1162–1167.

Wall, P.D. (1991) The placebo effect: an unpopular topic. Pain, 51: 1–3.

Wall, P.D. (1996) Comments after 30 years of the Gate Control Theory. Pain Forum, 5: 12–22.

Watkins, L.R. and Mayer, D.J. (1982) Organization of endogenous opiate and nonopiate pain control systems. Science, 216: 1185–1192.

Wickramasekera, I. (1980) A conditioned response model of the placebo effect: predictions from the model. Biofeedback Self Regul., 5: 5–18.

Witting, N., Kupers, R., Svensson, P., Arendt-Nielsen, L., Gjedde, A. and Jensen, T.S. (2001) Experimental brush-evoked allodynia activates posterior parietal cortex. Neurology, 57: 1817–1824.

S. Laureys (Ed.)
Progress in Brain Research, Vol. 150
ISSN 0079-6123

CHAPTER 20

Consciousness and epilepsy: why are patients with absence seizures absent?

Hal Blumenfeld*

Departments of Neurology, Neurobiology, and Neurosurgery, Yale University School of Medicine, 333 Cedar Street, New Haven, CT 06520, USA

Abstract: Epileptic seizures cause dynamic, reversible changes in brain function and are often associated with loss of consciousness. Of all seizure types, absence seizures lead to the most selective deficits in consciousness, with relatively little motor or other manifestations. Impaired consciousness in absence seizures is not monolithic, but varies in severity between patients and even between episodes in the same patient. In addition, some aspects of consciousness may be more severely involved than other aspects. The mechanisms for this variability are not known. Here we review the literature on human absence seizures and discuss a hypothesis for why effects on consciousness may be variable. Based on behavioral studies, electrophysiology, and recent neuroimaging and molecular investigations, we propose absence seizures impair focal, not generalized brain functions. Imapired consciousness in absence seizures may be caused by focal disruption of information processing in specific corticothalamic networks, while other networks are spared. Deficits in selective and varying cognitive functions may lead to impairment in different aspects of consciousness. Further investigations of the relationship between behavior and altered network function in absence seizures may improve our understanding of both normal and impaired consciousness.

Introduction

What is the relationship between brain activity and conscious thought? In science, difficult questions of this kind often yield to investigation, provided a good model system is available. One approach is to determine relationships between brain function and consciousness in situations where both vary. Examples include the investigation of sleep, anesthesia, brain lesions, evolution, development, and epilepsy. Epilepsy is an attractive model system for studying consciousness because epileptic seizures can cause selective, dynamic, and rapidly reversible changes in consciousness associated with altered function in specific brain networks. In this review, we will first introduce the major seizure types that cause impaired consciousness, including absence seizures. We next discuss electrophysiology and behavioral studies of impaired consciousness in absence seizures. Finally, we will briefly discuss recent neuroimaging and molecular studies suggesting that selective corticothalamic network involvement in absence seizures may, ultimately, explain the specific cognitive impairments that cause loss of consciousness.

Epilepsy models for studying impaired consciousness

Epileptic seizures are usually classified as either partial, involving focal brain regions, or generalized, involving widespread regions of the brain bilaterally (ILAE, 1981). Partial seizures associated with impaired consciousness are referred to as complex partial seizures, while those that spare consciousness are called simple partial seizures. Complex partial seizures arise most commonly

*Corresponding author. Tel.: +1 203 785-3928;
Fax: +1 203 737-2538; E-mail: hal.blumenfeld@yale.edu

DOI: 10.1016/S0079-6123(05)50020-7

from the temporal lobe. They typically begin with an unusual abdominal sensation or premonition, followed by staring, unresponsiveness, and simple repetitive movements of the mouth and limbs lasting 1–2 min, followed by amnesia for the episode. We and others have recently discussed a possible mechanism for impaired consciousness in temporal lobe complex partial seizures, in which network interactions with the upper brainstem and medial thalamus inhibit function of the bilateral fronto-parietal association cortex (Menzel et al., 1998; Lee et al., 2002; Norden and Blumenfeld, 2002; Blumenfeld and Taylor, 2003; Van Paesschen et al., 2003; Blumenfeld et al., 2004a, b).

Generalized seizures should invariably cause loss of consciousness if they truly involve the whole brain. Interestingly, however, this is not the case. Consciousness is spared in several types of generalized seizures (Gokygit and Caliskan, 1995; Bell et al., 1997; Vuilleumier et al., 2000). Moreover, recent work suggests that the so-called "generalized" seizures preferentially involve certain cortical–subcortical networks, while sparing others (Blumenfeld, 2003; McNally and Blumenfeld, 2004; McNally et al., 2004a, b; Nersesyan et al., 2004a, b). An important and common form of generalized seizure is the grand mal, or generalized tonic-clonic seizure. These typically consist of a dramatic rigid stiffening of the limbs for approximately 10 s, followed by synchronous contractions of the extremities for 1–2 min, with unresponsiveness throughout, and profound lethargy and confusion for a variable period after the episode ends. We recently proposed that the severe impairment in consciousness during and following generalized tonic-clonic seizures is caused by intense, abnormally increased neuronal activity in focal regions of the fronto-parietal association cortex and their related subcortical networks (Fig. 1) (Blumenfeld et al., 2003a, b; Blumenfeld and Taylor, 2003; McNally and Blumenfeld, 2004). Thus, abnormal *decreased* activity in bilateral fronto-parietal networks may impair consciousness in complex partial seizures, while abnormal *increased* activity in these same networks may impair consciousness in generalized tonic-clonic seizures.

The most "pure" example of impaired consciousness in epilepsy is absence seizures. Like generalized tonic-clonic seizures, absence seizures (also called petit mal) are classified as generalized. However, in absence seizures the impaired consciousness is accompanied by relatively few additional motor or other complicating phenomena (like those seen in tonic-clonic and complex partial seizures).

Typical absence seizures consist of brief episodes of staring and unresponsiveness, often accompanied by mild eyelid fluttering or myoclonic jerks. Duration is usually less than 10 s, and no obvious deficits are present after the seizure concludes, apart from the "missed time" during the seizure. Absence seizures are most common in childhood, and can occur up to hundreds of times per day, leading to impaired school performance. Electroencephalogram (EEG) recordings during typical absence seizures reveal large amplitude bilateral 3–4 Hz spike-wave discharges (Fig. 2). In atypical forms of absence seizures, and in absence status epilepticus, consciousness may be preserved, and the EEG usually reveals a slower or more irregular spike-wave pattern. In this review we will focus on typical absence seizures, and on the impaired consciousness accompanying these episodes. It should be noted, however, that typical and atypical absence are not always clearly distinct categories, and in fact lie along the same diagnostic spectrum (Holmes et al., 1987).

Numerous human and animal studies have suggested that absence seizures are generated through abnormal network oscillations involving both the cortex and thalamus (Williams, 1953; Avoli et al., 1990; Blumenfeld and McCormick, 2000;

Fig. 1. Fronto-parietal cortical involvement in generalized tonic-clonic seizures. Generalized tonic-clonic seizures were induced by electroconvulsive therapy for treatment of refractory depression (American Psychiatric Association, 2001) and cerebral blood flow (CBF) was imaged by single photon emission computed tomography (SPECT). Red represents relative increases in ictal compared to interictal CBF on SPECT scans, and green represents decreases. Despite the clinically generalized convulsions, focal relative signal increases are present in higher-order frontal and parietal association cortex, while many other brain regions are relatively spared. Ictal versus interictal SPECT images were analyzed with statistical parametric mapping. Reproduced with permission from Blumenfeld et al. (2003b). See Plate 20.1 in Colour Plate Section.

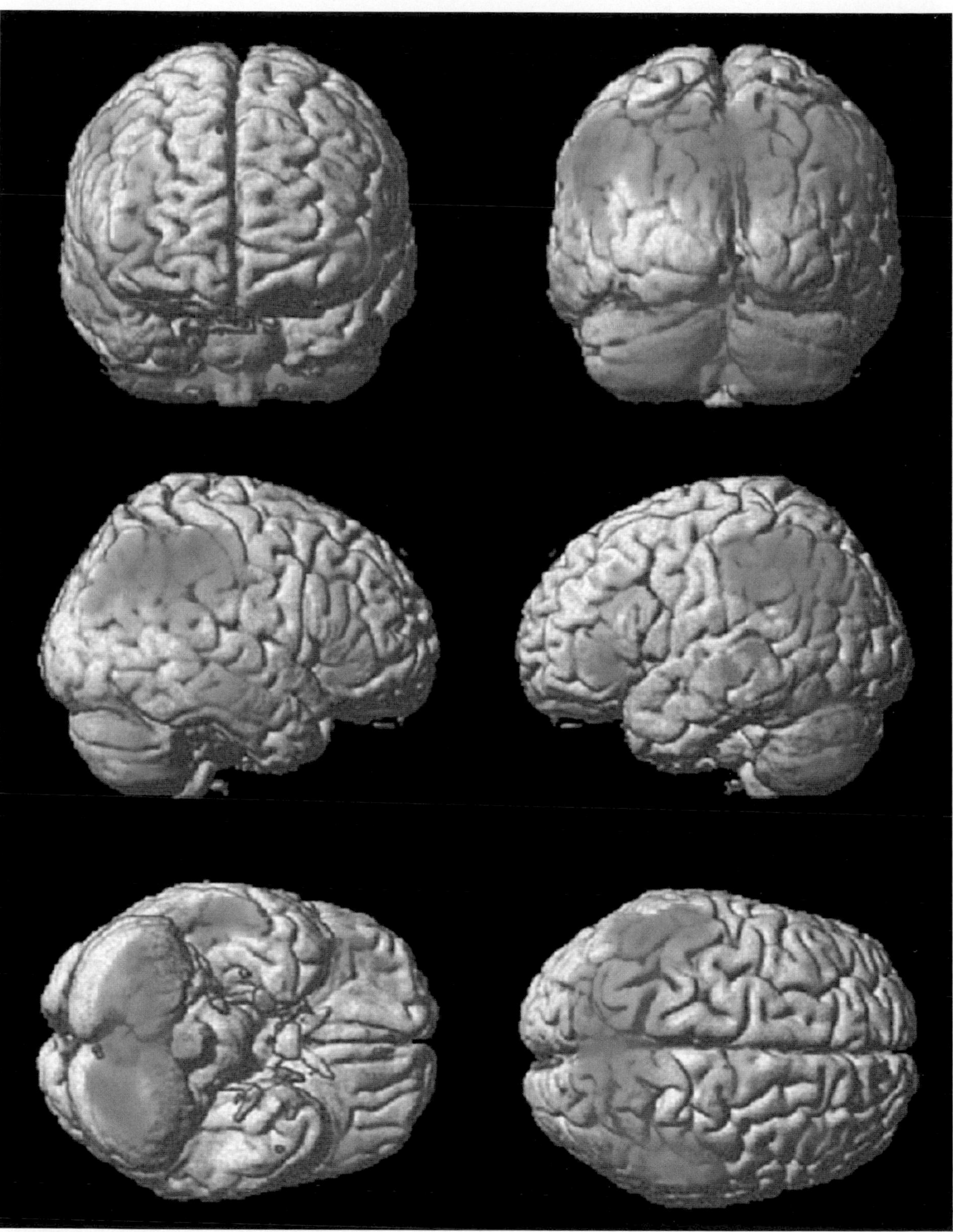

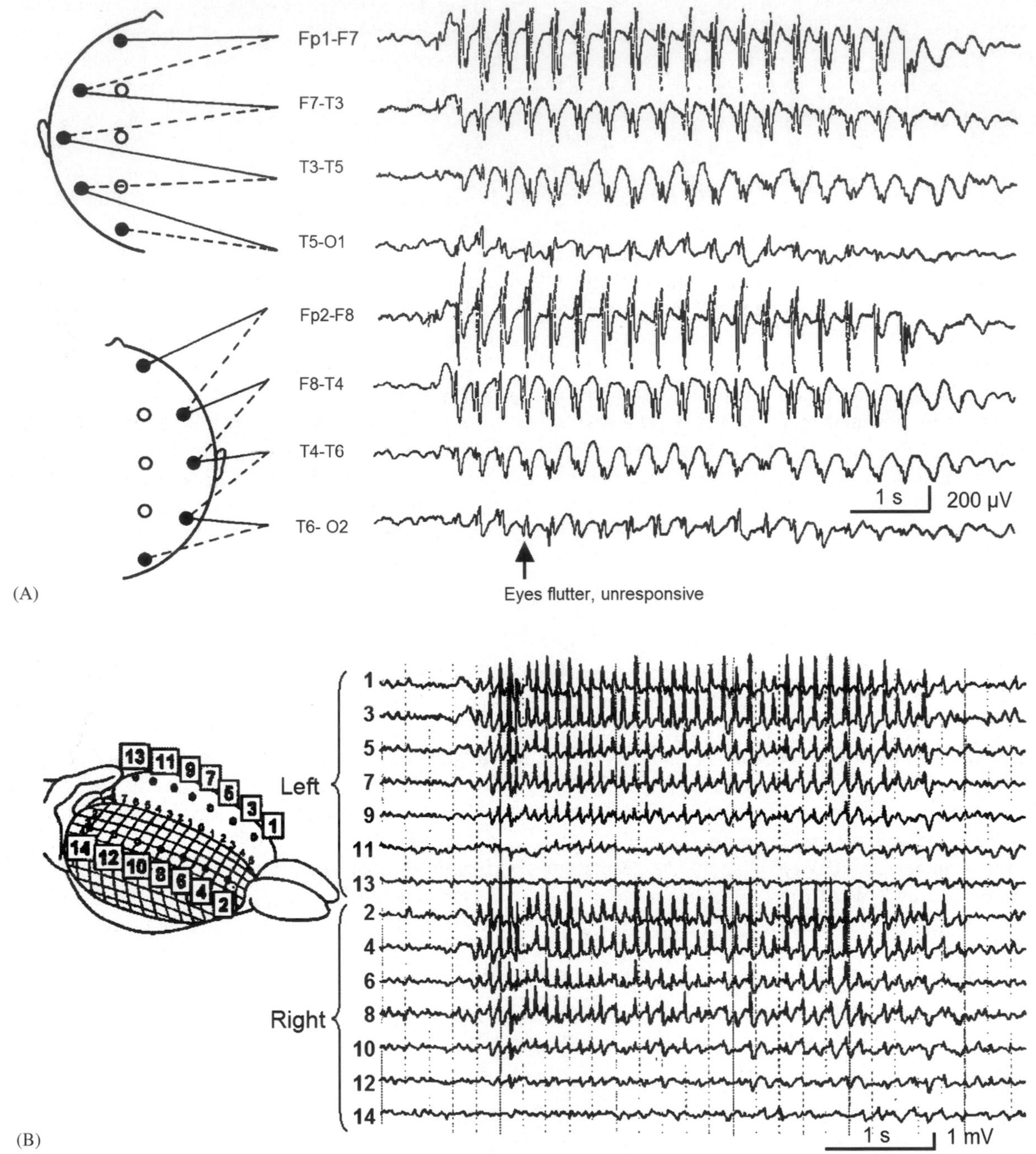

Fig. 2. Typical spike-wave EEG discharges during absence seizures have a focal frontal predominance. (A) EEG recording from a 7-year-old girl during a typical absence seizure reveals bilateral synchronous 3–4 Hz spike-wave discharges, with an anterior to posterior amplitude gradient. (Inset of electrode positions modified with permission from Fisch, B.J. (1991) Spehlmann's EEG Primer. Elsevier, Amsterdam. EEG recording modified with permission from Daly, D.D. and Pedley, T.A. (1990) Current Practice of Clinical Electroencephalography. Raven Press, New York.) (B) Electrocorticography from the surface of the WAG/Rij rat cortex during spike-and-wave seizures reveals intense involvement of the anterior cortex and relative sparing of the occipital lobes. (Reprinted with permission from Meeren et al., 2002. Copyright 2002, Society for Neuroscience.)

Kostopoulos, 2001; McCormick and Contreras, 2001; Blumenfeld, 2002; Crunelli and Leresche, 2002; Blumenfeld 2003). Despite much work, it is still not known with certainty whether absence seizures are generated through an overall increase or decrease in neuronal activity in cortical and subcortical circuits (Engel et al., 1985; Salek-Haddadi et al., 2003; Aghakhani et al., 2003; Nersesyan et al., 2004a, b). In addition, even in typical absence seizures there is substantial variability in the degree of impaired consciousness from one patient to another, and even from one episode to the next in the same patient. The cause of this variability is not known.

To understand the fundamental mechanisms for impaired consciousness in absence seizures it will be necessary to obtain a more detailed understanding of both the physiological changes in the brain and the specific cognitive deficits that occur during these events. As we will discuss, prior studies have begun to investigate the anatomy and physiology of absence seizures, and altered behavior during these events, but little work has been done to directly relate impaired brain function to behavior. For example, can the variability in impaired consciousness during absence seizures be explained by variability in brain dysfunction during these episodes? Are there specific anatomical regions of the brain that are involved or spared when patients do or do not show impaired responsiveness during absence seizures? Is the impaired brain function caused by abnormally increased activity, abnormally decreased activity, or both? What are the molecular and circuit mechanisms causing specific regions but not others to be involved during seizures? We will now review prior work on the physiology and behavior of typical human absence seizures, and highlight areas that need to be addressed through further investigations. We mention atypical absence, and animal models only briefly, as these are reviewed in detail elsewhere (Holmes et al., 1987; Avoli et al., 1990; Yaqub, 1993; Crunelli and Leresche, 2002; Coenen and Van Luijtelaar, 2003).

Electrophysiology

The characteristic 3–4 Hz spike-wave discharge of absence seizures was first described by Gibbs and colleagues in 1935 (Gibbs et al., 1935), just 6 years after Berger (1929) developed human EEG. Since then, countless recordings have been made of these discharges, which are usually bilaterally symmetric, but with an occasional left or right amplitude preponderance (Ebersole and Pedley, 2003). Although considered a form of generalized epilepsy, it has long been appreciated that the amplitude of spike-wave discharges on scalp EEG is not uniform in all electrodes. In most cases, there is a clear bifrontal amplitude maximum, greatest in the midline (Figs. 2A and 3) (Weir, 1965; Rodin and Ancheta, 1987; Coppola, 1988; Holmes et al., 2004). The focal electrographic distribution of spike-wave discharge is also supported by recent studies in rodent models of absence seizures (Vergnes et al., 1990; Meeren et al., 2002; Nersesyan et al., 2004a, b). In agreement with human scalp EEG recordings, invasive recordings from rodent models show bilateral spike-wave discharges of maximal amplitude in the anterior regions, sparing the occipital cortex (Fig. 2B). Thalamic nuclei corresponding to the anterior cortical regions are intensely involved, while the thalamic lateral geniculate nuclei (with connections to the occipital cortex), are spared (Nersesyan et al., 2004b). Kostopoulos and others (Kostopoulos, 2001; Blumenfeld and Taylor, 2003) have proposed that focal involvement of specific thalamocortical sectors (e.g. in the frontal cortex), during absence seizures may cause selective deficits during absence seizures, affecting some cognitive modalities, and sparing others. To more fully investigate this hypothesis in humans, noninvasive methods with higher spatial resolution than scalp EEG, such as neuroimaging, will be needed. However, it is first necessary to understand in as much detail as possible the specific cognitive deficits caused by absence seizures.

Behavior

There has been an interesting historical trend in behavioral studies of impaired consciousness during absence seizures (Fig. 4). These studies, initiated by Schwab and Jung in the late 1930s (Jung, 1939; Schwab, 1939), reached a peak in the 1960s, and then dropped off precipitously, with hardly any

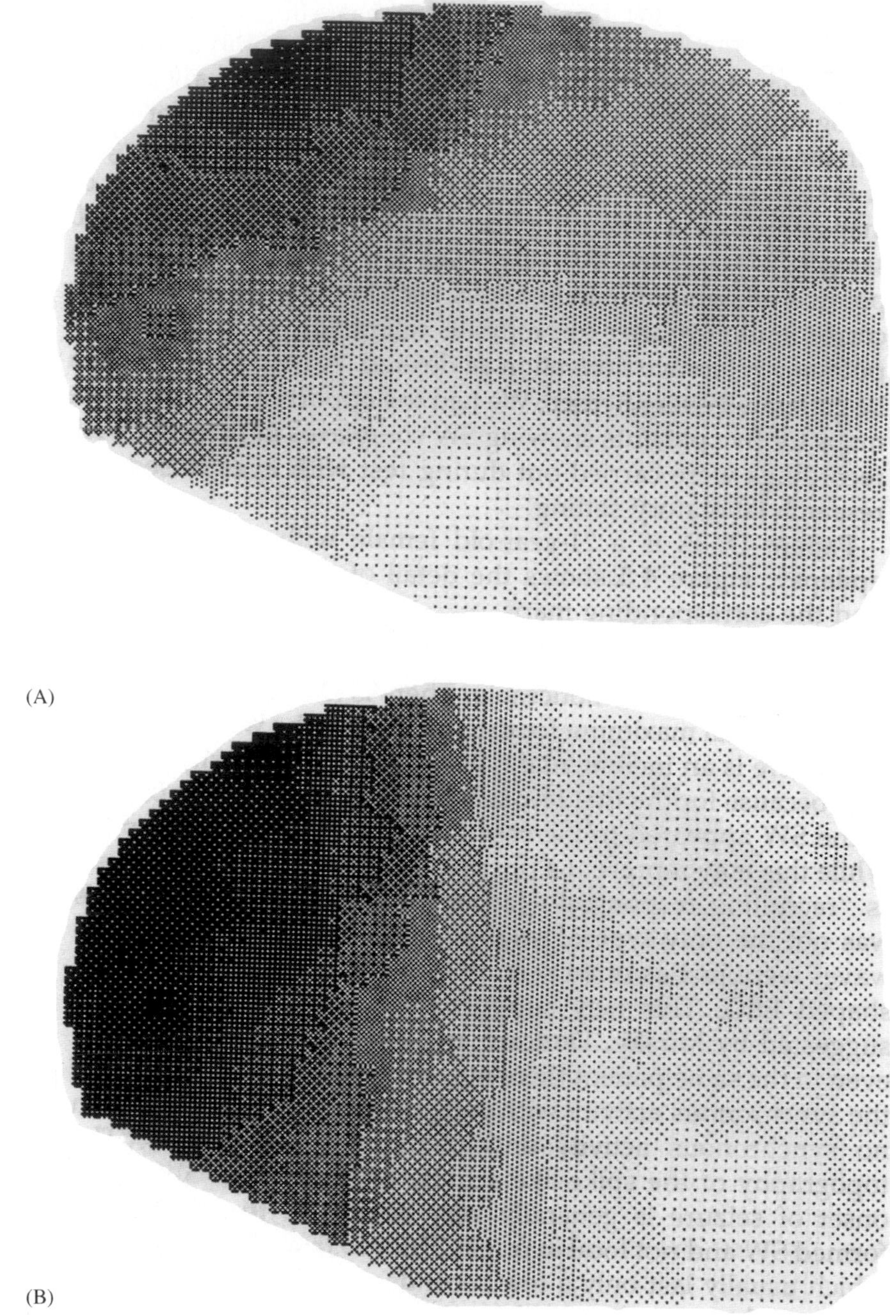

Fig. 3. Topographic mapping showing frontal predominance of spike-wave discharges in absence. Instantaneous EEG amplitude maps were generated from recordings during a generalized spike-wave complex in a patient with typical absence seizures (petit mal). (A) Amplitude map at the peak of the spike. (B) Amplitude map at the peak of the wave. Increased gray scale indicates greater negativity relative to the reference. Reproduced with permission from Coppola (1988).

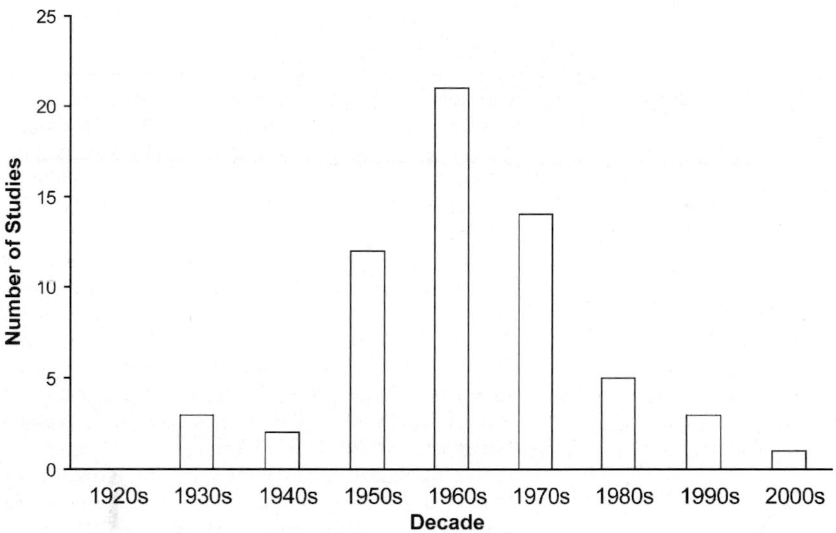

Fig. 4. Historical trends in behavioral studies of impaired consciousness in absence seizures. Number of studies on impaired consciousness during absence seizures were tabulated for each decade, peaking in the 1960s, and then showing a sharp drop-off, with few studies in recent years. References were found using medline search or based on citation in other articles, and then verified by reviewing content directly. Only original articles that investigated behavioral impairments during clinical absence seizures or subclinical generalized 3–4 Hz spike-wave discharges have been included.

published investigations in the last two decades. The rise in behavioral studies of absence seizures can easily be explained by the development of EEG (Berger, 1929), improved quantitative testing methods, and increasing interest in experimental psychology. However, the cause for the decline is more elusive, especially given the explosion of interest in cognitive neuroscience and consciousness in recent years. One possible explanation is the development of effective anti-absence medications such as ethosuximide in 1960 and valproate in 1978, leading to a smaller patient pool with uncontrolled episodes to study. However, studies should still be possible using patients with newly diagnosed absence, or patients who are unable to tolerate therapeutic doses of these medications. Another, more intriguing possibility is that studies on impaired consciousness in absence stopped because most questions were answered within the technical limitations of the times. In other words, once the major behavioral features of absence seizures were described, and their relationship to EEG discharges measured, there was little novel work left to be done until better methods for studying brain function were developed. As we discuss in the next section, it is our hope that this "forgotten science" will be reinvigorated now that modern functional neuroimaging and other methods are available.

Meanwhile, there remains much to be learned from the first wave of behavioral studies of absence seizures. These investigations utilized a variety of techniques to assess patients' cognitive function during absence seizures. Because the literature in this field is finite, an effort has been made to include all or nearly all pertinent studies in this review. Some of the studies have significant methodological limitations, including heterogeneous patient types, limited sample size, and descriptive rather than quantitative analysis. Nevertheless, the majority provide useful data, allowing some preliminary conclusions to be drawn, which may spark future research. We will now review three common themes in these studies: (1) Which cognitive functions are involved and which preserved during absence seizures? (2) What is the timing of deficits relative to spike-wave discharges on EEG? (3) Is the variability in impaired consciousness from one episode to the next related to any properties of the EEG?

Table 1. Tests used to asses consciousness during absence seizures[a]

Test	Examples
Repetitive tapping (or similar tasks)	Courtois et al. (1953), Shimazono et al. (1953), Gastaut (1954), Fischgold and Mathis (1957), Yeager and Guerrant (1957), Davidoff and Johnson (1964), Guey et al. (1965), Mirsky and Buren (1965), Chatrian et al. (1970), Gloor (1986), Sengoku et al. (1990), Heikman et al. (2002)
Response to mild or vigorous stimuli	Jung (1939), Kakegawa (1948), Courtois et al. (1953), Jung (1954), Boudin et al. (1958)
Simple reaction time (visual or auditory stimulation)	Schwab (1939, 1941, 1947), Cornil et al. (1951), Gastaut (1954), Fischgold (1957), Fischgold and Mathis (1957), Yeager and Guerrant (1957), Tuvo (1958), Hauser (1960), Jus and Jus (1960), Tizard and Margerison (1963a), Guey et al. (1965), Fernandez et al. (1967), Chatrian et al. (1970), Binnie and Lloyd (1973), Porter and Penry (1973), Browne et al. (1974), Mallin et al. (1981), Holmes et al. (1987), Sengoku et al. (1990)
Choice reaction time	Prechtl et al. (1961), Lehmann (1963), Tizard and Margerison (1963a, b), Grisell et al. (1964), Fairweather and Hutt (1969), Hutt and Fairweather (1971a, b), Sellden (1971), Hutt (1972), Hutt and Fairweather (1973, 1975), Hutt et al. (1976), Hutt et al. (1977), Sengoku et al. (1990)
Short term memory	Courtois et al. (1953), Jung (1954), Kooi and Hovey (1957), Boudin et al. (1958), Milstein and Stevens (1961), Jus and Jus (1962), Mirsky and Buren (1965), Geller and Geller (1970), Hutt and Gilbert (1980), Aarts et al. (1984), Holmes et al. (1987)
Continuous motor performance (tracking)	Guey et al. (1965), Goode et al. (1970), Opp et al. (1992)
Continuous performance task (vigilance)	Mirsky et al. (1960), Mirsky and Rosvold (1963), Tizard and Margerison (1963a, b), Mirsky and Buren (1965), Mirsky and Tecce (1968)
Counting aloud	Cornil et al. (1951), Courtois et al. (1953), Gastaut (1954), Jus and Jus (1960, 1962), Gloor (1986), Panayiotopoulos et al. (1992, 1997)
Reading aloud	Bates (1953), Panayiotopoulos et al. (1989)
Response to commands	Courtois et al. (1953), Shimazono et al. (1953) Jung (1954), Tizard and Margerison (1963a); Davidoff and Johnson (1964), Bureau et al. (1968), Chatrian et al. (1970), Gloor (1986), Holmes et al. (1987), Panayiotopoulos et al. (1989)
Verbal response to questions	Courtois et al. (1953), Shimazono et al. (1953), Jung (1954), Hovey and Kooi (1955), Kooi and Hovey (1957), Goldie and Green (1961), Milstein and Stevens (1961), Tizard and Margerison (1963a), Davidoff and Johnson (1964), Guey et al. (1965), Bureau et al. (1968), Chatrian et al. (1970), Gloor (1986), Holmes et al. (1987), Panayiotopoulos et al. (1989), Vuilleumier et al. (2000)

[a]Tests are arranged in approximate order of impairment during absence seizures. Thus, tests most spared during absence seizures are listed towards the top, while those most impaired by absence seizures are listed towards the bottom.

Which cognitive functions are impaired during absence?

A variety of different testing paradigms have been used to study impaired consciousness during absence seizures (Table 1). In fact, the different tests used illustrate that several cognitive functions may contribute to consciousness. There has been considerable dispute in the absence literature, whether impairment of some cognitive functions with sparing of others, qualifies as impaired consciousness. Does a patient who appears normal and is able to talk, but has transient slowing of reaction times during spike-wave episodes, have impaired consciousness? On the other extreme, can a patient who is completely unresponsive and has no recall afterwards of episodes be considered unconscious, or is the patient simply immobile, mute, and amnestic due to selective motor, verbal, and memory impairment (Gloor, 1986)? Ultimately, to answer questions of this kind, a more specific, mechanistic definition of consciousness is needed, along with a better understanding of the underlying physiological processes. Until this is achieved, and perhaps as a means of achieving such a definition, it is useful to study each of the specific cognitive processes that may contribute to loss of consciousness.

Testing paradigms during absence seizures have varied from simple repetitive motor tasks to higher order decision making (Table 1). Some functions are more severely impaired than others during absence seizures. The tests in Table 1 are listed in approximate order of impairment by absence seizures. For example, several investigators reported that simple repetitive motor tasks such as finger tapping are more often conserved than more demanding tasks, such as counting aloud, and impairment is even more common with tasks requiring a response to verbal questions (Schwab, 1939; Cornil et al., 1951; Courtois et al., 1953; Gastaut, 1954; Goldie and Green, 1961; Davidoff and Johnson, 1964). Decision making based on sensory input, and complex motor performance were more severely affected than simpler tests (Guey et al., 1965; Mirsky and Buren, 1965). Furthermore, impaired memory of items presented during seizures depended on task difficulty (Jus and Jus, 1962; Mirsky and Buren, 1965; Geller and Geller, 1970). Interestingly, while verbal responses were very often interrupted during seizures, some patients recalled questions or commands given during a seizure, and responded appropriately after the seizure ended (Courtois et al., 1953; Kooi and Hovey, 1957; Boudin et al., 1958). In addition, while some patients were unable to respond verbally, they were able to turn their gaze toward the examiner (Goldie and Green, 1961). Some investigators also found that absence seizures could be terminated by external stimulation, with loud or painful stimuli being most effective (Jung, 1939, 1954; Boudin et al., 1958), and seizure frequency was diminished by tasks requiring active mental attention (Lennox et al., 1936; Gastaut, 1954; Kooi and Hovey, 1957; Davidoff and Johnson, 1964; Bureau et al., 1968). Another interesting observation was that even "subclinical" (also called "larval" or "phantom") spike-wave discharges, with no obvious clinical impairment, were often associated with more subtle deficits in reaction time or continuous motor performance (Schwab, 1939; Kooi and Hovey, 1957; Yeager and Guerrant, 1957; Tuvo, 1958; Grisell et al., 1964; Guey et al., 1965).

In summary, these findings suggest that absence seizures can independently affect some cognitive functions that may be important for consciousness, while sparing certain others. The mechanism for this remains unknown, but it can be hypothesized that selective cognitive deficits may be related to selective impairment of specific brain networks during seizures, while others are spared.

What is the timing of impairments during absence?

On the basis of verbal response to questions and motor tests, it was observed that patients tend to have little impairment in the initial 1–3 s and final 3–5 s of absence seizures, but enter a "trough" of impaired consciousness in between (Shimazono et al., 1953; Goldie and Green, 1961; Mirsky and Buren, 1965; Goode et al., 1970). On the other hand, some investigators reported on the basis of vigilance or reaction time tasks that impairment coincides with electrographic seizure onset (Browne et al., 1974), or may even occur a few seconds before the onsets (Mirsky and Buren, 1965). In addition, although impairment of vigilance (continuous performance task) was reported to end about 5 s before the end of seizures, impaired simple tapping and sometimes responses to verbal questions have been reported to be impaired for several seconds after the end of the electrographic seizure (Goldie and Green, 1961; Mirsky and Buren, 1965). The different timing effects in these studies may again reflect the different tasks and methods used. Thus, some cognitive functions may be more severely affected during certain phases of seizures, while others are left relatively unharmed. This, once again, supports the notion that absence seizures dynamically disrupt specific functional brain networks, while sparing others.

There has also been some variability in reports of the timing of amnesia during absence seizures. Thus, some investigators reported no retrograde amnesia for stimuli presented prior to seizure onset (Courtois et al., 1953), while others reported retrograde amnesia extending up to 14 s before seizure onset, with gradual shrinking of the retrograde amnesia after the seizure ended (Jus and Jus, 1962; Mirsky and Buren, 1965; Geller and Geller, 1970).

Are EEG duration, amplitude, or any other features related to degree of impairment?

One important common theme in all these studies is that impaired responsiveness is not a monolithic feature of absence seizures, but rather highly variable between patients, and even from one episode to the next within a single patient. The mechanism for this variability is of great potential importance for understanding the fundamental mechanisms of impaired consciousness in absence seizures. Efforts have been made to relate certain EEG features to the degree of impairment. This subject has also been a matter of some controversy among investigators, most likely because of varying testing methods and cognitive tasks that were employed.

Longer seizure durations, especially those lasting more than 3–4 s are reported to cause more severe impairments in simple reaction time, continuous performance task, memory testing, and continuous motor performance (Schwab, 1939; Guey et al., 1965; Mirsky and Buren, 1965; Goode et al., 1970). Some impairment can be seen on careful reaction time testing even with brief spike-wave discharges (Browne et al., 1974). However, the impairment in this case is more obvious and lasts longer with longer duration discharges. Amplitude (Courtois et al., 1953) and "generalization" of discharges were associated with worse performance on reaction time and other tasks (Sellden, 1971; Porter and Penry, 1973; Browne et al., 1974). Mirsky and Van Buren conducted a detailed investigation of EEG characteristics, and found that spike-wave duration, amplitude, rhythmicity, and fronto-central distribution were all associated with more severe impaired performance on vigilance and memory testing (Mirsky and Buren, 1965). A recent small case series suggested different EEG spike-wave distributions during impaired or preserved consciousness (Vuilleumier et al., 2000). Other investigators, however, found no association between EEG duration (Browne et al., 1974), or any other EEG features with impaired consciousness (Gastaut, 1954).

The majority of these studies suggest that an association exists between certain EEG features, and impaired performance on cognitive tasks contributing to impaired consciousness in absence. However, the precise anatomical and physiological source of this variability in EEG and behavioral impairment will require further investigation with more advanced methods.

Neuroimaging and molecular mapping

Human imaging studies during absence seizures have been highly controversial. Some studies have shown global increases in cerebral metabolism or blood flow (Engel et al., 1982, 1985; Theodore et al., 1985; Prevett et al., 1995; Diehl et al., 1998; Yeni et al., 2000), while others have shown no change, focal or generalized increases or decreases (Theodore et al., 1985; Ochs et al., 1987; Park et al., 1994; Ferrie et al., 1996; Diehl et al., 1998; Salek-Haddadi et al., 2003; Aghakhani et al., 2003; Archer et al., 2003). Some of these variabilities may reflect technical limitations, such as limited temporal resolution of fluoro-2-deoxy-D-glucose positron emission tomography (FDG-PET) or Tc99 m single photon emission computed tomography (SPECT) imaging relative to absence seizure duration. In addition, the variability may result from heterogeneous patient populations, and the inclusion in these studies of pediatric and adult patients along with variable numbers of patients with atypical absence seizures.

Because absence episodes are relatively brief, it is crucial to simultaneously record EEG with neuroimaging so that images acquired during seizures and baseline can be identified and analyzed separately. In addition, an imaging modality with sufficient temporal resolution to capture individual absence seizures should be used. Functional magnetic resonance imaging (fMRI) fulfils both of these criteria, and several recent studies have begun to show interesting results with this method (Aghakhani et al., 2003; Archer et al., 2003; Salek-Haddadi et al., 2003). Interestingly, fMRI changes during absence seizures include both increases and decreases in signal, and these changes are heterogeneous, not involving the entire brain

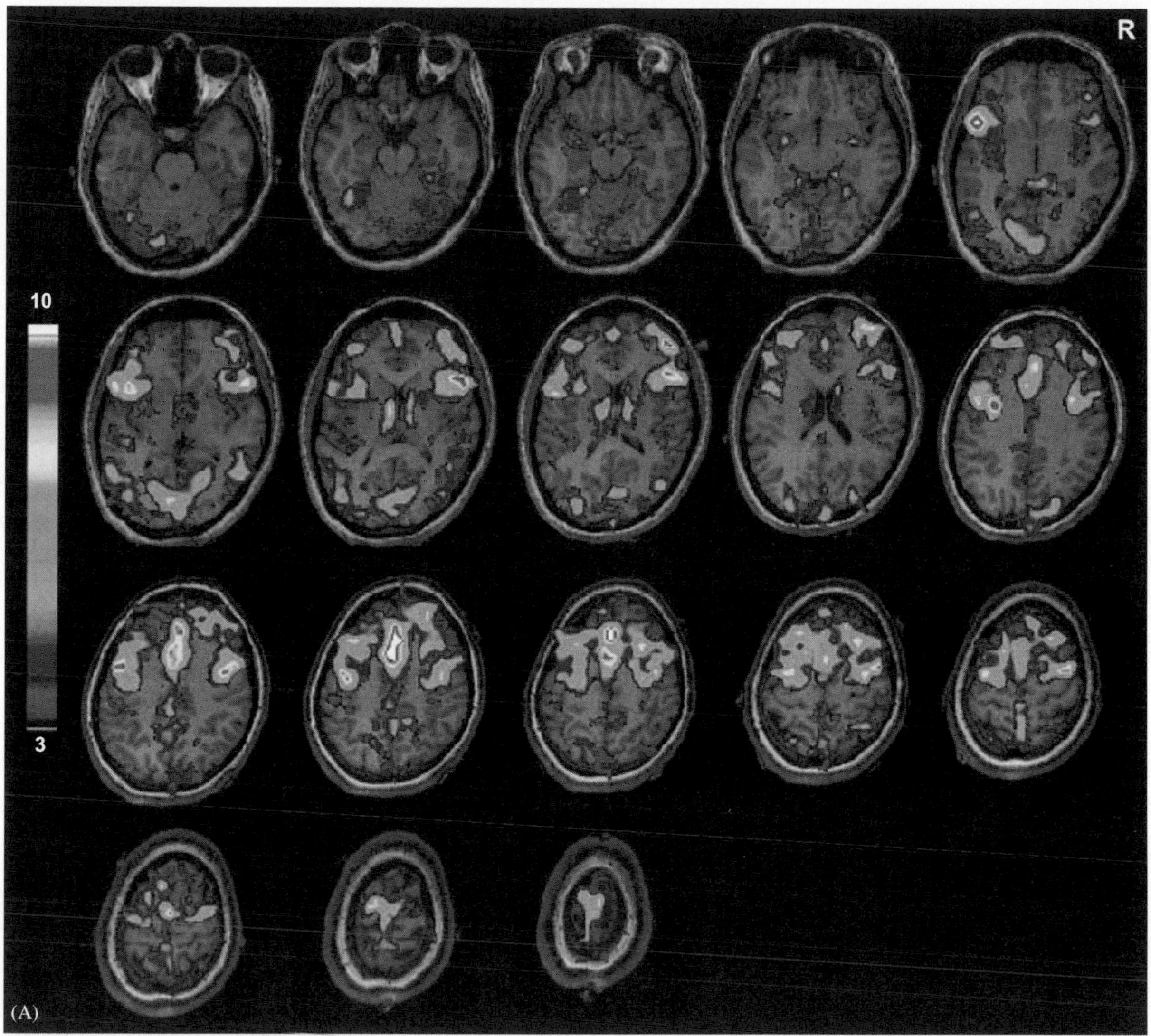

Fig. 5. fMRI during human absence seizures. 18-year-old patient with onset of absence seizures at age 8. EEG and fMRI were recorded simultaneously. (A) fMRI activation. (B) fMRI deactivation. Focal regions of increased and decreased signal are seen bilaterally. Color bars show t scores. Reproduced with permission from Aghakhani et al. (2003). See Plate 20.5 in Colour Plate Section.

in a uniform fashion (Fig. 5). These findings agree with behavioral and EEG studies suggesting that absence seizures may selectively involve certain networks, while sparing others. Additional neuroimaging studies with improved spatial and temporal resolution will be needed to investigate the possible role of focal brain dysfunction in loss of consciousness during absence seizures. In addition, fMRI studies are limited because the signals measured are only indirectly related to neuronal activity. Further work will be needed to fully interpret the relationship of fMRI signal increases and decreases during absence seizures with the underlying changes in neuronal activity (Smith et al., 2002; Nersesyan et al., 2004b).

Studies in animal models may also shed light on mechanisms of spatial heterogeneity in spike-wave seizures (Blumenfeld, 2003). Recent studies on

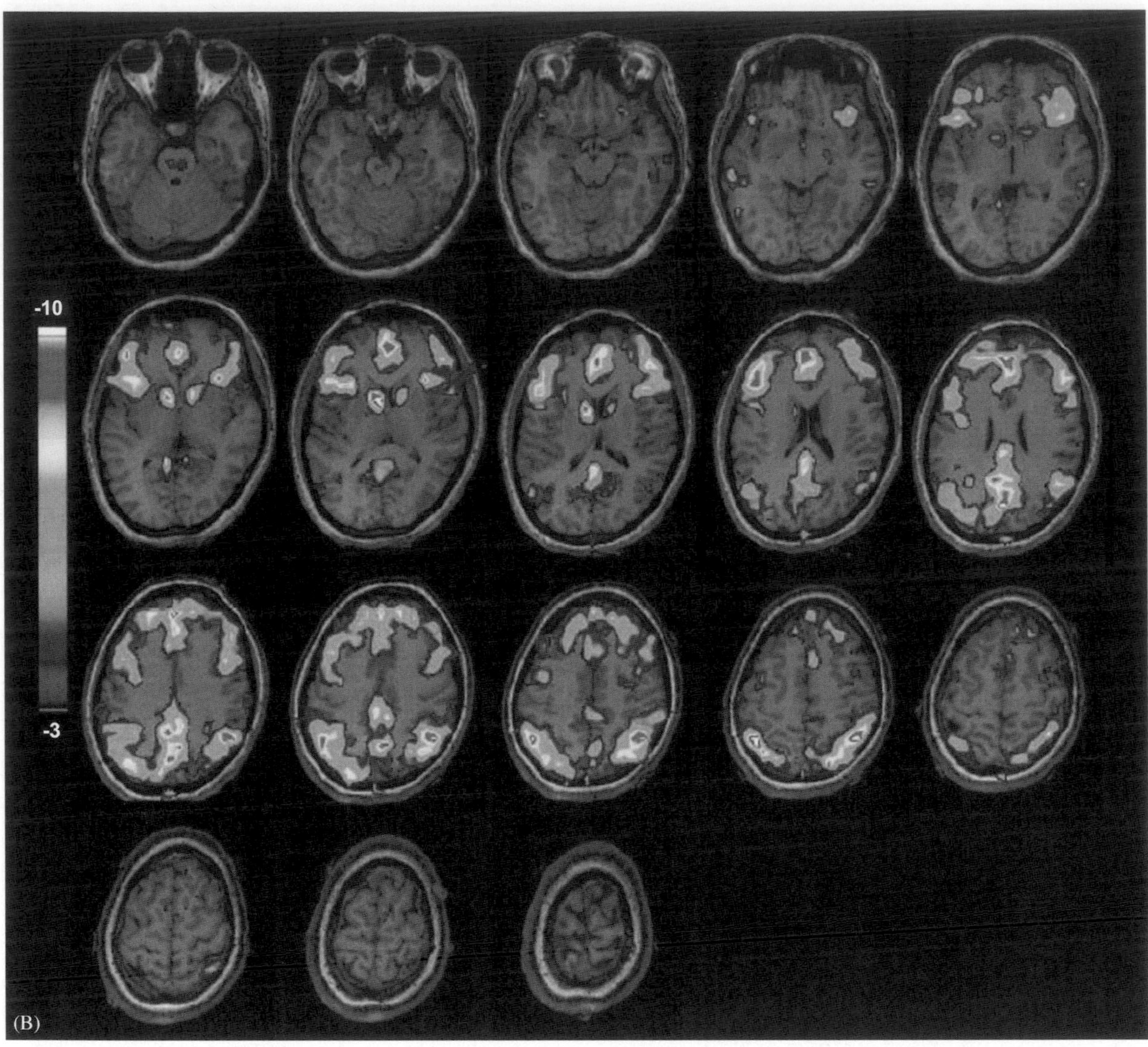

Fig. 5. (*Continued*).

rodent models on absence seizures have demonstrated that, as in humans, seizure discharges often have an anterior predominance, with the posterior brain regions relatively spared (Vergnes et al., 1990; Meeren et al., 2002; Klein et al., 2004; Nersesyan et al., 2004a, b) (Fig. 2B). These investigations have included electrophysiologic, fMRI, and Doppler flow measurements, all implying that spike-wave seizures that appear fairly generalized on scalp EEG recordings, in fact, may intensely involve some corticothalamic networks, while sparing others.

Ultimately, it will be crucial to identify the underlying mechanisms that cause certain corticothalamic networks to be intensely involved in spike-wave seizures, while others remain functional. Many factors can contribute to altered excitability and synchronous firing of corticothalamic networks, and only a few candidate molecules and mechanisms have begun to be explored (Crunelli and Leresche, 2002). It was recently found in a rodent model of absence seizures that enhanced excitability in focal cortical regions may be related to altered expression of voltage gated sodium channels (Klein et al., 2004) or to reduced function of hyperpolarization-activated cation channels (Ih) (Strauss et al., 2004). If the specific genes and proteins can be identified, which generate absence seizures in selective brain regions, but not others, it may ultimately be possible to map networks predisposed to seizure discharges using molecular techniques. Furthermore, identification of these molecular changes will allow targeted therapies aimed at reducing abnormal excitability in these networks, and to prevent absence seizures and the accompanying loss of consciousness.

Conclusions

Loss of consciousness during absence seizures may serve as a good model system for studying normal mechanisms of consciousness. Impairments in consciousness during absence seizures are dynamic and rapidly reversible, and are relatively selective, without many of the motor or other manifestations seen in other seizure types. Several converging lines of evidence suggest that impaired consciousness during absence seizures is not a global phenomenon, but rather can be decomposed into specific deficits in a number of different cognitive functions, associated with impaired function in selective neuroanatomical networks. On the basis of the studies reviewed here, we hypothesize that loss of consciousness in absence seizures is not an "all-or-none" phenomenon resulting from involvement of the entire brain in the seizure discharge. Rather, focal involvement of the bilateral association cortex, perhaps most prominently in frontal neocortex and related subcortical structures, disrupts normal information processing in these brain regions leading to impairment of specific cognitive functions necessary for normal consciousness. The detailed spatial distribution of these changes, the physiological effects of spike-wave discharges on normal neuronal signaling, the molecular and cellular mechanisms underlying selective network involvement, and the relationship of these changes to behavior will need additional investigation.

Acknowledgments

I gratefully thank Dr. Alain Hyman, Dr. David L. Blumenfeld, and Dr. Ulrich Schridde for translating of papers from French and German literature. I also thank George Varghese for assistance in preparing the figures. This work was supported by NIH R01 NS049307 and by the Betsy and Jonathan Blattmachr family.

References

Aarts, J.H.P., Binnie, C.D., Smit, A.M. and Wilkins, A.J. (1984) Selective cognitive impairment during focal and generalized epileptiform EEG activity. Brain, 107: 293–308.

Aghakhani, Y., Bagshaw, A.P., Benar, C.G., Hawco, C., Andermann, F., Dubeau, F. and Gotman, J. (2003) fMRI activation during spike- and wave-discharges in idiopathic generalized epilepsy. Brain, 127: 1127–1144.

American Psychiatric Association CoET. (2001) The Practice of Electroconvulsive Therapy: Recommendations for Treatment, Training, and Privileging (2nd ed.). American Psychiatric Association, Washington, DC.

Archer, J.S., Abbott, D.F., Waites, A.B. and Jackson, G.D. (2003) fMRI "deactivation" of the posterior cingulate during generalized spike and wave. Neuroimage, 20: 1915–1922.

Avoli, M., Gloor, P., Kostopoulos, G. and Naquet, T. (Eds.). (1990) Generalized Epilepsy. Birkhauser, Basel, Boston.

Bates, J.A.V. (1953) A technique for identifying changes in consciousness. Electroen. Clin. Neuro., 5: 445–446.

Bell, W.L., Walczak, T.S., Shin, C. and Radtke, R.A. (1997) Painful generalised clonic and tonic-clonic seizures with retained consciousness. J. Neurol. Neurosur. Ps., 63: 792–795.

Berger, H. (1929) Ueber das Elektrenkephalogramm des Menschen. Arch. Psychiat., 87: 527.

Binnie, C.D. and Lloyd, D.S.L. (1973) A technique for measuring reaction times during paroxysmal discharges. Electroen. Clin. Neuro., 35: 420.

Blumenfeld, H. (2002) The thalamus and seizures. Arch. Neurol., 59: 135–137.

Blumenfeld, H. (2003) From molecules to networks: cortical/subcortical interactions in the pathophysiology of idiopathic generalized epilepsy. Epilepsia, 44(2): 7–15.

Blumenfeld, H. and McCormick, D.A. (2000) Corticothalamic inputs control the pattern of activity generated in thalamocortical networks. J. Neurosci., 20: 5153–5162.

Blumenfeld, H., McNally, K.A., Ostroff, R. and Zubal, I.G. (2003a) Targeted prefrontal cortical activation with bifrontal ECT. Psychiat. Res. Neuroimag., 123: 165–170.

Blumenfeld, H., McNally, K.A., Vanderhill, S.D., Paige, A.L., Chung, R., Davis, K., Norden, A.D., Stokking, R., Studholme, C., Novotny, E.J., Zubal, I.G. and Spencer, S.S. (2004b) Positive and negative network correlations in temporal lobe epilepsy. Cereb. Cortex, 14: 892–902.

Blumenfeld, H., Rivera, M., McNally, K.A., Davis, K., Spencer, D.D. and Spencer, S.S. (2004a) Ictal neocortical slowing in temporal lobe epilepsy. Neurology, 63: 1015–1021.

Blumenfeld, H. and Taylor, J. (2003) Why do seizures cause loss of consciousness? The Neuroscientist, 9: 301–310.

Blumenfeld, H., Westerveld, M., Ostroff, R.B., Vanderhill, S.D., Freeman, J., Necochea, A., Uranga, P., Tanhehco, T., Smith, A., Seibyl, J.P., Stokking, R., Studholme, C., Spencer, S.S. and Zubal, I.G. (2003b) Selective frontal, parietal and temporal networks in generalized seizures. Neuroimage, 19: 1556–1566.

Boudin, G., Barbizet, J. and Masson, S. (1958) Etude de la dissolution de la conscience dans 3 cas de petit mal avec crises prolongees. Rev. Neurol., 99: 483–487.

Browne, T.R., Penry, J.K., Porter, R.J. and Dreifuss, F.E. (1974) Responsiveness before during and after spike-wave paroxysms. Neurology, 24: 659–665.

Bureau, M., Guey, J., Dravet, C. and Roger, J. (1968) A study of the distribution of petit mal absences in the child in relation to his activities. Electroen. Clin. Neuro., 25: 513.

Chatrian, G.E., Lettich, E., Miller, L.H., Green, J.R. and Kupfer, C. (1970) Pattern-sensitive epilepsy. 2. Clinical changes, tests of responsiveness and motor output, alterations of evoked potenials and therapeutic measures. Epilepsia, 11: 151–162.

Coenen, A.M. and Van Luijtelaar, E.L. (2003) Genetic animal models for absence epilepsy: a review of the WAG/Rij strain of rats. Behav. Genet., 33: 635–655.

Coppola, R. (1988) Topographic display of spike-and-wave discharges. In: Mysobodsky M.S. and Mirsky A.F. (Eds.), Elements of Petit Mal Epilepsy. Peter Lang, New York, pp. 105–130.

Cornil, L., Gastaut, H. and Corriol, J. (1951) Appreciation du degre de conscience au cours des paroxysmes epileptiques 'petit mal'. Rev. Neurol., 84: 149–151.

Courtois, G.A., Ingvar, D.H. and Jasper, H.H. (1953) Nervous and mental defects during petit mal attacks. Electroen. Clin. Neuro. (Suppl. 3): 87.

Crunelli, V. and Leresche, N. (2002) Childhood absence epilepsy: genes channels neurons and networks. Nat. Rev. Neurosci., 3: 371–382.

Davidoff, R.A. and Johnson, L.C. (1964) Paroxysmal EEG activity and cognitive-motor performance. Electroen. Clin. Neuro., 16: 343–354.

Diehl, B., Knecht, S., Deppe, M., Young, C. and Stodieck, S.R. (1998) Cerebral hemodynamic response to generalized spike-wave discharges. Epilepsia, 39: 1284–1289.

Ebersole, J.S. and Pedley, T.A. (2003) Current Practice of Clinical Electroencephalography (3rd ed.). Lippincott Williams & Wilkins, Philadelphia, PA.

Engel Jr., J., Kuhl, D.E. and Phelps, M.E. (1982) Patterns of human local cerebral glucose metabolism during epileptic seizures. Science, 218: 64–66.

Engel Jr., J., Lubens, P., Kuhl, D.E. and Phelps, M.E. (1985) Local cerebral metabolic rate for glucose during petit mal absences. Ann. Neurol., 17: 121–128.

Fairweather, H. and Hutt, S.J. (1969) Inter-relationships of EEG activity and information processing on paced and unpaced tasks in epileptic children. Electroen. Clin. Neuro., 27: 701.

Fernandez, H., Robinson, R. and Taylor, R.R. (1967) A device for testing consciousness. Am. J. EEG Technol., 7: 77–78.

Ferrie, C.D., Maisey, M., Cox, T., Polkey, C., Barrington, S.F., Panayiotopoulos, C.P. and Robinson, R.O. (1996) Focal abnormalities detected by 18FDG PET in epileptic encephalopathies. [comment]. Arch. Dis. Child., 75: 102–107.

Fischgold, H. (1957) La conscience et ses modifications. Systemes de references en EEG clinique. In: Congres International des Sciences Neurologiques. 21–28 juillet, Bruxelles, pp. 181–208.

Fischgold, H. and Mathis, P. (1957) Polygraphies des modifications de la conscience. Electroen. Clin. Neuro., 9: 177.

Gastaut, H. (1954) The brain stem and cerebral electrogenesis in relation to consciousness. In: Delafresnaye J.F. (Ed.), Brain Mechanisms and Consciousness. Thomas, Springfield, ILH, pp. 249–283.

Geller, M.R. and Geller, A. (1970) Brief amnestic effects of spike-wave discharges. Neurology (Section on Child Neurology, Abstract CN1), 20: 380–381.

Gibbs, F.A., Davis, H. and Lennox, W.G. (1935) The electroencephalogram in epilepsy and in conditions of impaired consciousness. Arch. Neurol. Psychiatr., 34: 1133–1148.

Gloor, P. (1986) Consciousness as a neurological concept in epileptology: a critical review. Epilepsia, 27: S14–S26.

Gokygit, A. and Caliskan, A. (1995) Diffuse spike-wave status of 9-year duration without behavioral change or intellectual decline. Epilepsia, 36: 210–213.

Goldie, L. and Green, J.M. (1961) Spike-and-wave discharges and alterations of conscious awareness. Nature, 191: 200–201.

Goode, D.J., Penry, J.K. and Dreifuss, F.E. (1970) Effects of paroxysmal spike-wave on continuous visual-motor performance. Epilepsia, 11: 241–254.

Grisell, J.L., Levin, S.M., Cohen, B.D. and Rodin, E.A. (1964) Effects of subclinical seizure activity on overt behavior. Neurology, 14: 133–135.

Guey, J., Tassinari, C.A., Charles, C. and Coquery, C. (1965) Variations du niveau d'efficience en relation avec des descharges epileptiques paroxystiques. Rev. Neurol., 112: 311–317.

Hauser, F. (1960) Perception et responses motrices au cours des paroxysmes de pontesondes. Foulon, Paris, p. 52.

Heikman, P., Kalska, H., Katila, H., Sarna, S., Tuunainen, A. and Kuoppasalmi, K. (2002) Right unilateral and bifrontal electroconvulsive therapy in the treatment of depression: a preliminary study. J. ECT., 18: 26–30.

Holmes, M.D., Brown, M. and Tucker, D.M. (2004) Are "generalized" seizures truly generalized? Evidence of localized mesial frontal and frontopolar discharges in absence. Epilepsia, 45(12): 1568–1579.

Holmes, G.L., McKeever, M. and Adamson, M. (1987) Absence seizures in children: clinical and electroencephalographic features. Ann. Neurol., 21: 268–273.

Hovey, H.B. and Kooi, K.A. (1955) Transient disturbances of thought processes and epilepsy. AMA Arch. Neurol. Psy., 74: 287–291.

Hutt, S.J. (1972) Experimental analysis of brain activity and behaviour in children with "minor" seizures. Epilepsia, 13: 520–534.

Hutt, S.J., Denner, S. and Newton, J. (1976) Auditory thresholds during evoked spike-wave activity in epileptic patients. Cortex, 12: 249–257.

Hutt, S.J. and Fairweather, H. (1971a) Spike-wave paroxysms and information processing. P. Roy. Soc. Med., 64: 918–919.

Hutt, S.J. and Fairweather, H. (1971b) Some effects of performance variables upon generalized spike-wave activity. Brain, 94: 321–326.

Hutt, S.J. and Fairweather, H. (1973) Paced and unpaced serial response performance during two types of EEG activity. J. Neurol. Sci., 19: 85 96.

Hutt, S.J. and Fairweather, H. (1975) Information processing during two types of EEG activity. Electroen. Clin. Neuro., 39: 43–51.

Hutt, S.J. and Gilbert, S. (1980) Effects of evoked spike-wave discharges upon short term memory in patients with epilepsy. Cortex, 16: 445–457.

Hutt, S.J., Newton, J. and Fairweather, H. (1977) Choice reaction time and EEG activity in children with epilepsy. Neuropsychologia, 15: 257–267.

ILAE. (1981) Proposal for revised clinical and electroencephalographic classification of epileptic seizures. From the commission on classification and terminology of the International League Against Epilepsy. Epilepsia, 22: 489–501.

Jung, R. (1939) Uber vegetative Reaktionen und Hemmungswirkung von Sinnesreizen im kleinen epileptischen Anfall. Nervenarzt, 12: 169–185.

Jung, R. (1954) Correlation of biolectrical and autonomic phenomena with alterations of consciousness and arousal in man. In: Delafresnaye J.F. (Ed.), Brain Mechanisms and Consciousness. Thomas, Springfield, ILH, pp. 310–344.

Jus, A. and Jus, C. (1960) Etude electro-clinique des alterations de conscience dans le petit mal. Studii si cercetari de Neurol, 5: 243–254.

Jus, A. and Jus, K. (1962) Retrograde amnesia in petit mal. Arch. Gen. Psychiat., 6: 163–167.

Kakegawa, Y. (1948) On the wave and spike formation of petit mal epilepsy. Folia Psychiat. Neurol. Jp., 3: 16–29.

Klein, J.P., Khera, D.S., Nersesyan, H., Kimchi, E.Y., Waxman, S.G. and Blumenfeld, H. (2004) Dysregulation of sodium channel expression in cortical neurons in a rodent model of absence epilepsy. Brain Res., 1000: 102–109.

Kooi, K.A. and Hovey, H.B. (1957) Alterations in mental function and paroxysmal cerebral activity. Arch. Neurol. Psychiatr., 78: 264–271.

Kostopoulos, G.K. (2001) Involvement of the thalamocortical system in epileptic loss of consciousness. Epilepsia, 42: 13–19.

Lee, K.H., Meador, K.J., Park, Y.D., King, D.W., Murro, A.M., Pillai, J.J. and Kaminski, R.J. (2002) Pathophysiology of altered consciousness during seizures: subtraction SPECT study. Neurology, 59: 841–846.

Lehmann, H.J. (1963) Praparoxysmale Weckreaktion bei pyknoleptischen Absencen. Arch. Psychiatr. Nervenkr., 204: 417–426.

Lennox, W.G., Gibbs, F.A. and Gibbs, E.L. (1936) Effect on the electroencephalogram of drugs and conditions which influence seizures. Arch. Neurol. Psychiatr., 36: 1236–1250.

Mallin, U., Stefan, H. and Penin, H. (1981) Psychopathometrische Studien zum Verlauf epileptischer Absencen. Z EEG EMG, 12: 45–49.

McCormick, D.A. and Contreras, D. (2001) On the cellular and network bases of epileptic seizures. Ann. Rev. Physiol., 63: 815–846.

McNally, K.A. and Blumenfeld, H. (2004) Focal network involvement in generalized seizures: new insights from electroconvulsive therapy. Epilepsy Behav., 5: 3–12.

McNally, K.A., Davis, K., Doernberg, S.B., Zubal, I.G., Spencer, S.S., and Blumenfeld, H. (2004a). Cerebral blood flow changes during secondarily generalized tonic-clonic seizures. Soc. Neurosci. Abstracts, Online at http://websfnorg/.

McNally, K.A., Davis, K., Doernberg, S.B., Zubal, I.G., Spencer, S.S., and Blumenfeld, H. (2004b). Ictal SPECT in secondarily generalized tonic-clonic seizures. Epilepsia, AES abstracts.

Meeren, H.K.M., Pijn, J.P.M., Van Luijtelaar, E.L.J.M., Coenen, A.M.L. and da Silva, F.H.L. (2002) Cortical focus drives widespread corticothalamic networks during spontaneous absence seizures in rats. J. Neurosci., 22: 1480–1495.

Menzel, C., Grunwald, F., Klemm, E., Ruhlmann, J., Elger, C.E. and Biersack, H.J. (1998) Inhibitory effects of mesial temporal partial seizures onto frontal neocortical structures. Acta Neurol. Belgica, 98: 327–331.

Milstein, V. and Stevens, J.R. (1961) Verbal and conditioned avoidance learning during abnormal EEG discharge. J. Nerv. Ment. Dis., 132: 50–60.

Mirsky, A.F. and Buren, J.M.V. (1965) On the nature of the "absence" in centrencephalic epilepsy: a study of some behavioral, electroencephalographic, and autonomic factors. Electroen. Clin. Neuro., 18: 334–348.

Mirsky, A.F., Primac, D.W., Ajmone Marsan, C., Rosvold, H.E. and Stevens, J.R. (1960) A comparison of the psychological test performance of patients with focal nonfocal epilepsy. Exp. Neurol., 2: 75–89.

Mirsky, A.F. and Rosvold, H.E. (1963) Behavioral and physiological studies in impaired attention. In: Zea V. (Ed.), Psychopharmacological Methods: Proceedings of a Symposium on the Effects of Psychotropic Drugs on Higher Nervous Activity. Pergamon Press, Oxford, pp. 302–315.

Mirsky, A. and Tecce, J. (1968) The analysis of visual evoked potentials during spike-and-wave activity. Epilepsia, 9: 211–220.

Nersesyan, H., Hyder, F., Rothman, D. and Blumenfeld, H. (2004a) Dynamic fMRI and EEG recordings during spike-wave seizures and generalized tonic-clonic seizures in WAG/Rij rats. J. Cereb. Blood Flow Metab., 24: 589–599.

Nersesyan, H., Herman, P., Erdogan, E., Hyder, F. and Blumenfeld, H. (2004b) Relative changes in cerebral blood flow and neuronal activity in local microdomains during generalized seizures. J. Cereb. Blood Flow Metab., 24: 1057–1068.

Norden, A.D. and Blumenfeld, H. (2002) The role of subcortical structures in human epilepsy. Epilepsy Behav., 3: 219–231.

Ochs, R.F., Gloor, P., Tyler, J.L., Wolfson, T., Worsley, K., Andermann, F., Diksic, M., Meyer, E. and Evans, A. (1987) Effect of generalized spike-and-wave discharge on glucose metabolism measured by positron emission tomography. Ann. Neurol., 21: 458–464.

Opp, J., Wenzel, D. and Brandl, U. (1992) Visuomotor coordination during focal and generalized EEG discharges. Epilepsia, 33: 836–840.

Panayiotopoulos, C.P., Chroni, E., Daskalopoulos, C., Baker, A., Rowlinson, S. and Walsh, P. (1992) Typical absence seizures in adults: clinical EEG video-EEG findings and diagnostic/syndromic considerations. J. Neurol. Neurosur. Ps., 55: 1002–1008.

Panayiotopoulos, C., Koutroumanidis, M., Giannakodimos, S. and Agathonikou, A. (1997) Idiopathic generalised epilepsy in adults manifested by phantom absences, generalised tonic-clonic seizures, and frequent absence status. J. Neurol. Neurosur. Ps., 63: 622–627.

Panayiotopoulos, C.P., Obeid, T. and Waheed, G. (1989) Differentiation of typical absence seizures in epileptic syndromes: a video-EEG study of 124 seizures in 20 patients. Brain, 112: 1039–1056.

Park, Y.D., Hoffman, J.M., Radtke, R.A. and DeLong, G.R. (1994) Focal cerebral metabolic abnormality in a patient with continuous spike waves during slow-wave sleep. J. Child Neurol., 9: 139–143.

Porter, R.J. and Penry, J.K. (1973) Responsiveness at the onset of spike-wave bursts. Electroen. Clin. Neuro., 34: 239–245.

Prechtl, H.F.R., Boeke, P.E. and Schut, T. (1961) The electroencephalogram and performance in epileptic patients. Neurology, 11: 296–304.

Prevett, M.C., Duncan, J.S., Jones, T., Fish, D.R. and Brooks, D.J. (1995) Demonstration of thalamic activation during typical absence seizures using $H_2(15)O$ and PET. Neurology, 45: 1396–1402.

Rodin, E. and Ancheta, O. (1987) Cerebral electrical fields during petit mal absences. Electroen. Clin. Neuro., 66: 457–466.

Salek-Haddadi, A., Lemieux, L., Merschhemke, M., Friston, K.J., Duncan, J.S. and Fish, D.R. (2003) Functional magnetic resonance imaging of human absence seizures. Ann. Neurol., 53: 663–667.

Schwab, R.S. (1939) Method of measuring consciousness in attacks of petit mal epilepsy. Arch. Neurol. Psychiat. (Society Transactions: Boston Society of Psychiatry and Neurology, presented May 19, 1938), 41: 215–217.

Schwab, R.S. (1941) The influence of visual and auditory stimuli on the electroencephalographic tracing of petit mal. Amer. J. Pschiatr., 97: 1301–1312.

Schwab, R.S. (1947) Reaction time in petit mal epilepsy. Res. Publ. Assoc. Res. Nerv. Ment. Dis.: 339–341.

Sellden, U. (1971) Psychotechnical performance related to paroxysmal discharges in EEG. Clin. Electroencephal., 2(1): 18–27.

Sengoku, A., Kanazawa, O., Kawai, I. and Yamaguchi, T. (1990) Visual cognitive disturbance during spike-wave discharges. Epilepsia, 31: 47–59.

Shimazono, Y., Hirai, T., Okuma, T., Fukuda, T. and Yamamasu, E. (1953) Disturbance of consciousness in petit mal epilepsy. Epilepsia, 2: 49–55.

Smith, A.J., Blumenfeld, H., Behar, K.L., Rothman, D.L., Shulman, R.G. and Hyder, F. (2002) Cerebral energetics and spiking frequency: the neurophysiological basis of fMRI. Proc. Natl. Acad. Sci. USA, 99: 10765–10770.

Strauss, U., Kole, M.H., Brauer, A.U., Pahnke, J., Bajorat, R., Rolfs, A., Nitsch, R. and Deisz, R.A. (2004) An impaired neocortical Ih is associated with enhanced excitability and absence epilepsy. Eur. J. Neurosci., 19: 3048–3058.

Theodore, W.H., Brooks, R., Margolin, R., Patronas, N., Sato, S., Porter, R.J., Mansi, L., Bairamian, D. and DiChiro, G. (1985) Positron emission tomography in generalized seizures. Neurology, 35: 684–690.

Tizard, B. and Margerison, J.H. (1963a) Psychological functions during wave-spike discharges. Brit. J. Soc. Clin. Psychol., 3: 6–15.

Tizard, B. and Margerison, J.H. (1963b) The relationship between generalized and paroxysmal EEG discharges and various test situations in two epileptic patients. J. Neurol. Neurosurg. Psychiatry., 26: 308–313.

Tuvo, F. (1958) Contribution a l'etude des niveaux de conscience au cours des paroxysmes epileptiques infraclinique. Electroen. Clin. Neuro., 10: 715–718.

Van Paesschen, W., Dupont, P., Van Driel, G., Van Billoen, H. and Maes, A. (2003) SPECT perfusion changes during complex partial seizures in patients with hippocampal sclerosis. Brain, 126: 1103–1111.

Vergnes, M., Marescaux, C. and Depaulis, A. (1990) Mapping of spontaneous spike-and-wave discharges in Wistar rats with genetic generalized non-convulsive epilepsy. Brain Res., 523: 87–91.

Vuilleumier, P., Assal, F., Blanke, O. and Jallon, P. (2000) Distinct behavioral and EEG topographic correlates of loss of consciousness in absences. Epilepsia, 41: 687–693.

Weir, B. (1965) The morphology of the spike-wave complex. Electroen. Clin. Neuro., 19: 284–290.

Williams, D. (1953) A study of thalamic and cortical rhythms in petit mal. Brain, 76: 50–69.

Yaqub, B.A. (1993) Electroclinical seizures in Lennox-Gastaut syndrome. Epilepsia, 34: 120–127.

Yeager, C.L. and Guerrant, J.S. (1957) Subclinical epileptic seizures; impairment of motor performance and derivative difficulties. Calif. Med., 86: 242–247.

Yeni, S.N., Kabasakal, L., Yalcinkaya, C., Nisli, C. and Dervent, A. (2000) Ictal and interictal SPECT findings in childhood absence epilepsy. Seizure, 9: 265–269.

S. Laureys (Ed.)
Progress in Brain Research, Vol. 150
ISSN 0079-6123

CHAPTER 21

Two aspects of impaired consciousness in Alzheimer's disease

Eric Salmon[1,*], Perrine Ruby[1], Daniela Perani[2], Elke Kalbe[3], Steven Laureys[1], Stéphane Adam[4] and Fabienne Collette[4]

[1]*Cyclotron Research Centre and Department of Neurology, University of Liege, B35 Sart Tilman, B4000 Liege, Belgium*
[2]*Vita-Salute San Raffaele University, Via Olgettina 60, 20132 Milan, Italy*
[3]*Max-Planck Institute for Neurological Research, Cologne, Gleueler Str. 50, D-50931 Cologne, Germany*
[4]*Department of Neuropsychology, University of Liège, B32 Sart Tilman, B-4000 Liege, Belgium*

Abstract: Alzheimer's disease (AD) is a degenerative dementia characterized by different aspects of impaired consciousness. For example, there is a deficit of controlled processes that require conscious processing of information. Such an impairment is indexed by decreased performances at controlled cognitive tasks, and it is related to reduced brain metabolic activity in a network of frontal, posterior associative, and limbic regions. Another aspect of impaired consciousness is that AD patients show variable levels of anosognosia concerning their cognitive deficits. A discrepancy score between patient's and caregiver's assessment of cognitive functions is one of the most frequently used measures of anosognosia. A high discrepancy score has been related to impaired activity in the superior frontal sulcus and the parietal cortex in AD. Anosognosia for cognitive deficits in AD could be partly explained by impaired metabolism in parts of networks subserving self-referential processes (e.g., the superior frontal sulcus) and perspective-taking (e.g., the temporoparietal junction). We hypothesize that these patients are impaired in the ability to see themselves with a third-person perspective (i.e., being able to see themselves as other people see them).

Introduction

Alzheimer's disease (AD) is the most common cause of dementia among people aged 65 and older. About 3% of men and women aged 65 to 74 have AD, and nearly half of those aged 85 and older may have the disease. AD is clinically characterized by a dementia syndrome. Current features for dementia include a deterioration in cognitive functions, sufficient to impair daily living activities (APA, 1994). Neuropsychological studies have demonstrated that AD patients show impairment in controlled cognitive processes, while automatic activities may be more preserved (Fabrigoule et al., 1998). It is frequently observed that AD patients fail to consciously recollect information whereas they provide target memories in implicit conditions. This constitutes a first aspect of impaired consciousness in AD.

Other characteristics of dementia include personality change and altered judgment (APA, 1987). Behavioral and psychological impairments are well described in AD (Cummings et al., 1994; Neary et al., 1998). More specifically, lack of awareness for self-cognitive or behavioral difficulties (anosognosia for deterioration) is frequently observed in the disease. However, assessing anosognosia is a difficult task, and patients may show different degrees of awareness in different domains (e.g., reasoning, memory, and social behavior) (Fig. 1).

*Corresponding author. Tel.: +0032 4 366 23 16; Fax: +3243662946; E-mail: eric.salmon@ulg.ac.be

DOI: 10.1016/S0079-6123(05)50021-9

Fig. 1. "Alzheimer, maybe, but I am still a genius!." Demented patient with anosognosia for cognitive deficit (Reproduced with permission – copyright C. Laureys).

The aim of this chapter is to discuss possible relationships between abnormal brain function in patients with AD and (i) decreased controlled access to information in memory and (ii) measures of anosognosia concerning cognitive impairment.

Controlled and automatic processes in AD

It is widely acknowledged that AD can selectively impair specific cognitive processes early in the course of the disease, while others remain relatively preserved until the pathology becomes severe (Collette et al., 2003). It was suggested that controlled processes (requiring attentional resources) are affected at an early stage while automatic processes are relatively more preserved (Jorm, 1986). The hypothesis of controlled processes impairment was explicitly explored in recent studies (Fabrigoule et al., 1998; Salthouse and Becker, 1998; Amieva et al., 2000). For example, Fabrigoule (1998) longitudinally studied cognitive performances in a large sample of 1,159 healthy elderly subjects. After 2 years, 16 of these subjects were diagnosed as having AD. A principal component analysis coupled with a logistic regression analysis showed that one factor explained 45% of the variance in test performance and constituted a good predictor of the risk of developing AD. The authors proposed, from post-hoc analyses, that this factor corresponds to the controlled or executive processes of the tasks. Controlled and automatic processes can be explored more directly in the memory domain by using explicit and implicit memory tasks, respectively. AD is associated with impairments in explicit (i.e., controlled) memory tasks such as free and cued recall or recognition (Salmon, 2000). However, the results related to implicit (i.e., automatic) memory tasks are more controversial. Some reviews of the literature (Meiran and Jelicic, 1995; Fleischman and Gabrieli, 1998) showed that AD patients often present a preserved priming effect on implicit tasks such as word and picture identification or lexical decisions, but they are frequently impaired on priming tasks that require implicit retrieval of conceptual information such as word association, category-exemplar generation, and word-stem completion. One reason invoked by various authors to explain these ambiguous results is that healthy participants may use explicit retrieval strategies when performing an ostensibly implicit retrieval task (Vaidya et al., 1996). This use of explicit strategies might artificially increase the "priming effect" for conceptual information in normal controls but not in AD patients, considering their explicit memory deficit (Fleischman and Gabrieli, 1998). This contamination interpretation is in line with Jacoby's view that there are no "process-pure" memory tasks (Jacoby et al., 1992).

Behavioral measures of controlled and automatic processes: the process dissociation procedure

Jacoby hypothesized that there are two independent ways to retrieve information in memory, i.e., controlled and automatic processes, and he proposed an elegant method, the process dissociation procedure (PDP), to separate the degree to which controlled and automatic processes contribute to performance in a single memory task (Yonelinas and Jacoby, 1995). This procedure comprises two conditions: one condition in which controlled and automatic processes act in concert to influence performance in the same way (the *inclusion* condition where both processes converge to produce the correct answer), and the other in which controlled and automatic processes have opposing effects (the *exclusion* condition). Considering that automatic and controlled processes contribute independently to performance, an equation can be written that represents how the two processes act to determine performance in each condition. On the basis of these two equations, the contributions of automatic and controlled processes to task performance can be estimated using simple algebra (Fig. 2; note that other measurement models exist and that the specific equations assume that automatic and controlled processes are independent from each other).

Several studies have applied PDP in order to explore automatic and controlled processes in AD, especially in the domain of memory. In particular, a word-stem completion task was administered to early AD patients, applying the PDP (Koivisto et al., 1998). In each inclusion and exclusion condition, subjects were presented with word-stems and were told to use them as cues for recollecting a word that had previously been studied (e.g., *flow*, — as a cue to recall *flower*, Fig. 2). In the inclusion condition, subjects were asked to complete a stem with a previously studied word or, if they were unable to do so, to use the first word that came to mind. Therefore, in this condition, subjects could correctly complete a stem with an earlier studied word either because they consciously recollected it having seen the word before (probability C), or because it was the first word that came to mind automatically (probability A), without any conscious recollection that the word had been presented earlier (1–C). Thus, the probability of completing a stem in the inclusion condition can be represented as inclusion = $C+A$ (1–C). By contrast, in the exclusion condition, subjects were asked to complete the stem with a new word that had not been encountered during the study phase and to avoid (exclude) words that had been studied. In this condition, subjects might incorrectly complete a word that had been studied earlier only if that word came automatically to mind (A), without any "conscious", controlled recollection that it had been presented earlier (1–C). Thus, the probability of producing an error (i.e., completing a stem with a word that had been studied earlier) in the exclusion condition can be represented as exclusion = A (1–C). Following Jacoby, the contribution of controlled ("conscious") processes in the task can be estimated by subtracting the probability of responding with a studied word in the exclusion condition from the probability of responding with an old (i.e., studied) word in the inclusion condition. Once an estimate of controlled processes has been obtained, the contribution of automatic processes corresponds to the

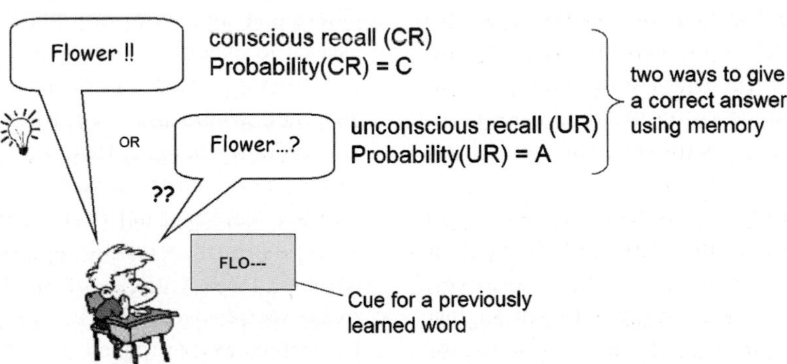

Fig. 2. Illustration of the process dissociation procedure proposed by Jacoby (1992). Inclusion condition: "complete the stem with a word presented during the learning stage". Score (Inclusion) = Probability ("Flower") = $C+A(1-C)$. Exclusion condition: "complete the stem with a word, which was not presented during the learning stage". If the subject erroneously completes the stem with the studied word, the score (Exclusion) = Probability ("Flower") = $A(1-C)$. Estimation of Conscious process contribution to recall: C = score (inclusion) – score (exclusion), Estimation of Unconscious process contribution to recall: A = score (exclusion)/(1$-C$).

probability of completing a stem with the studied word in the exclusion condition (exclusion score) divided by one minus the probability of consciously completing a stem with the studied word in the inclusion condition ($1-C$).

A severe impairment of controlled memory processes was reported in mild and moderate AD patients (Koivisto et al., 1998). More specifically, in AD patients, no difference was found between the inclusion and exclusion conditions in terms of the probability of completion with old words. On the other hand, the estimates of automatic processes were similar in AD patients and control subjects, suggesting that automatic memory processes were preserved in AD. However, data suggested that patients and controls may have used different response strategies in the two conditions. Using a similar stem completion task, another study showed that although the AD patients' performance on the word-stem completion memory task was strongly supported by automatic memory processes, the control group relied on automatic processes even more (Knight, 1998). This suggested that automatic processes were also impaired in AD, even if this impairment was not as great as that of controlled processes. However, in this study, the task was very easy for the control subjects (as suggested by a ceiling effect in the inclusion and a floor effect in the exclusion condition), leading to an overestimation of the contribution made by automatic processes in this group. Moreover, it was not possible to guarantee that the AD patients understood the task instructions, particularly in the exclusion condition (Curran and Hintzman, 1995). Difficulties understanding instructions may produce differential biases in responding across the inclusion and exclusion conditions.

To sum up, the PDP seems to be a useful tool to directly explore the contribution made by controlled and automatic processes in AD. However, the studies that have used this paradigm to test memory functioning tend to be either unclear or somewhat controversial, particularly concerning the status of automatic processes (preserved or not preserved). A recent study verified the integrity of automatic processes in AD by designing a memory task that limited the methodological and psychometric problems affecting previous experiments applying the PDP approach (Adam et al., 2005). The results confirmed the marked deterioration in controlled processes, while automatic memory processes were preserved in AD patients compared to control subjects. Moreover, analyses showed a positive correlation between controlled processes and disease progression as measured by the mini mental state examination (MMSE; Folstein et al., 1975). These results should be interpreted by considering that MMSE principally measures controlled processes, as this scale includes items involving effortful attentional tasks (e.g., immediate and episodic memory, mental calculation).

A potential area where the demonstrated distinction between controlled and automatic processes could be applied is the cognitive rehabilitation of AD patients. One would predict that AD patients, whose controlled processes are impaired, would behave better when they can rely on automatic procedures (Jorm, 1986). From this perspective, it is, for example, possible to promote specific learning procedures to allow early AD patients to automatically use a cellular phone (Lekeu et al., 2002).

Neural substrate of controlled processes

Which are the neurobiological substrates for controlled processes allowing access to memories? There is a tight coupling between controlled processes and executive functions. In Baddeley's model of working memory (1992), the central executive is assumed to be an attentional control system. It was recently suggested that executive functions (or "top-down" controlled processes) rely on a distributed cerebral network including both anterior and posterior cerebral areas (Weinberger, 1993; Morris, 1994; Collette et al., 1999a, 2002). Moreover, recollection of episodic memories (controlled retrieval compared to simple familiarity for information) was related to activation in prefrontal, parietal, and posterior cingulate regions in normal subjects (Henson et al., 1999). In AD, performances at diverse executive tasks were related either to frontal activity for phonemic

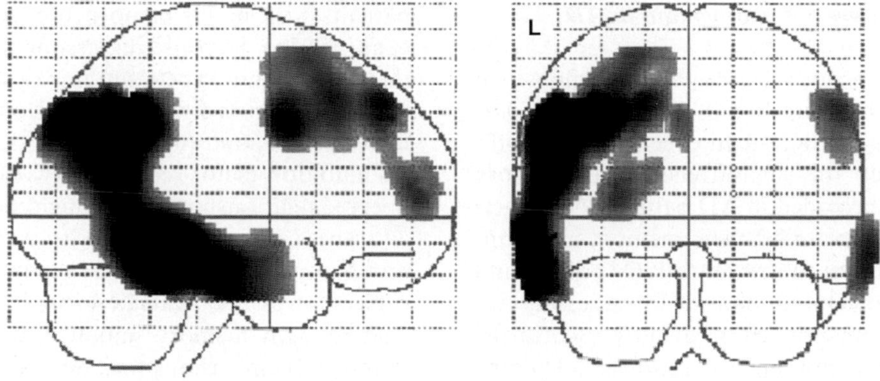

Fig. 3. Decreased metabolism in temporo-parietal and frontal associative cortices is related to severity of the dementia in AD.

fluency, or to posterior associative cortices for dual tasks (Collette et al., 1999c). From this perspective, two interpretations may be proposed to account for the controlled process deficits of AD patients: they could originate either from multiple neuropathological and metabolic changes in both anterior and posterior cerebral areas or from a (partial) disconnection between the anterior and posterior cortical areas, leading to a less efficient transfer of information between these regions. Consistent with the first interpretation, global dementia scores, which essentially depend on controlled processes, are related to decreased metabolism in both associative temporoparietal and frontal cortices in AD (Salmon et al., 2005) (see Fig. 3).

Consistent with the second interpretation, a number of neuropathological, neuropsychological, and neuroimaging studies have suggested that AD can be characterized as a disconnection syndrome (Azari et al., 1992; Leuchter et al., 1992; Morris, 1994; Collette et al., 2002; Lekeu et al., 2003a). Thus, loss of synaptic contacts in associative cortices and limbic structures would interrupt long connecting pathways in AD (Hyman et al., 1984; Pearson et al., 1985). The relative preservation of automatic processes in AD patients would mean that these processes are subserved by circumscribed cortical areas that are less affected by disconnection in the early stages of the disease. As the disease progresses, the neuropathological changes would affect more cortical areas (and local connections between them) and consequently also alter automatic processes. Such a hypothesis of successive impairment of controlled and automatic processes has been proposed to explain working memory deficits in AD (Collette et al., 1999b).

Frontoparietal associative cortices related to controlled processes may constitute a major component of a cerebral workspace, where attentional display of information would allow active and controlled recollection of information (Dehaene and Naccache, 2001). However, lesions in other structures, such as the hippocampus, could also explain impairment of conscious (i.e., controlled) retrieval of information in AD. Supportive to this hypothesis, a study required AD patients to learn words that they subsequently had to recognize among distractors (Lekeu et al., 2003b): an accuracy score (hits minus false alarms), taken as a measure of "conscious/controlled retrieval" (recollection), correlated with residual metabolism in the hippocampus. In neuroimaging studies in control subjects, medial temporal activation is known to be greater for explicit cued recall using word stem than for a priming task using similar stems, suggesting a role for hippocampus in conscious access to memories (Squire, 1992).

Anosognosia in AD

Anosognosia, lack of awareness or loss of insight, is used interchangeably to describe the impaired judgment of AD subjects concerning their own cognition, mood, behavior, or daily activities.

Behavioral measures of anosognosia in AD

The study of anosognosia requires to choose an instrument to measure loss of insight into a particular domain of interest. In experimental conditions, evidence of unawareness of memory dysfunction is provided in AD patients when their predictions overestimate their memory performance in a subsequent retrieval task (McGlynn and Schacter, 1989). Questionnaires for assessing anosognosia in dementia are frequently used in the literature, and a discrepancy score is calculated between answers obtained from the patient and from a caregiver (Migliorelli et al., 1995a). Self-rating of cognitive deficits is also proposed as an index of cognitive unawareness, because AD patients report cognitive impairment that they frequently do not find particularly abnormal for their age (Cummings et al., 1995). Loss of insight may be evaluated by the clinician, on the basis of patient answers to questions probing awareness of cognitive deficits (expert judgment), or by caregivers assessing the frequency of behavioral manifestations of unawareness in everyday life (Derouesne et al., 1999). While most patients with mild cognitive impairment (MCI) tend to overestimate their cognitive deficits when compared to their caregiver's assessment, AD patients in early stages of the disease with similar MMSE scores underestimate their cognitive dysfunction (Kalbe et al., 2005). Accordingly, underestimation of cognitive difficulties (compared to caregiver's assessment) is considered as a risk factor for conversion from MCI to AD (Tabert et al., 2002). Several demographic, clinical, and neuropsychological parameters were proposed to be associated with unawareness for self-cognitive deficits in AD. Lack of awareness of AD patients was correlated to dementia severity in most, but not all studies (Sevush and Leve, 1993; McDaniel et al., 1995; Sevush, 1999; Zanetti et al., 1999; Gil et al., 2001). The relationship between anosognosia and depression remains a matter of discussion (Sevush and Leve, 1993; Cummings et al., 1995). Differences in assessment of anosognosia and definition of depression might explain these discrepancies. Patients with dysthymia would have a significantly better awareness of intellectual deficits than patients with major or no depression (Migliorelli et al., 1995b). Stepwise regression analysis showed that insight into functional impairment was (positively) associated with depression and anxiety factor scores, and (negatively) with agitation and disinhibition factor scores (Harwood et al., 2000). Specific relationships were sought between lack of awareness of cognitive deficits and performance on neuropsychological tests. Anosognosia in AD subjects was associated, in some but not all studies, with memory impairment, reflected, for example, by intrusions (i.e., erroneous selection of non-target items) in recognition tasks (Reed et al., 1993; Dalla Barba et al., 1995). A relationship was frequently suggested between lack of awareness of cognitive dysfunction and impairment in specific "frontal" tests (Lopez et al., 1994; Michon et al., 1994; Dalla Barba et al., 1995; Ott et al., 1996). All those parameters may be viewed as confounding factors, that contribute to, but do not yet explain, anosognosia in dementia. Moreover, they show that different dimensions exist in anosognosia (Agnew and Morris, 1998).

To further discuss anosognosia for cognitive deficits in early stages of dementia, a review of the different components of the "unified self" proposed by Klein et al. (2002) might be useful. These authors suggested that the self can be conceptualized as a complex knowledge structure subserved by at least two neurally and functionally dissociable components: episodic memory and semantic memory. Episodic memory of one's own life (i.e., episodic autobiographical memory) is certainly impaired in AD subjects (Greene et al., 1995; Piolino et al., 2003). So, patients will have a decrease of their autonoetic experiences, i.e., those events that patients have personally lived, but cannot recollect in a precise contextual format that makes them unique subjective experiences (Wheeler et al., 1997; Tulving, 2002). However, deficits in episodic memory are not sufficient to explain the different degrees of unawareness for self-cognitive deficits in AD because numerous patients with severe amnesia from other etiologies do not show anosognosia. Semantic personal knowledge seems to be also impaired in AD, however, it is difficult to dissociate the respective influences of semantic and episodic memory disturbances (Kazui et al., 2003). Representation of one's own personality traits might be a specific (and independent) aspect of semantic knowledge (Klein et al., 2002). In a study on moderately severe AD patients, awareness of behavioral and psychological problems was relatively preserved compared to awareness of cognitive impairment (Kotler-Cope and Camp, 1995). Such an observation, however, might reflect the natural evolution of AD and not the different levels of anosognosia in different domains, because AD patients have relatively few behavioral troubles compared to cognitive deficits in a moderate stage of dementia. Indeed, in a patient with severe AD, it was shown that knowledge of own personality traits failed to be updated (Klein et al., 2003). Experience of personal agency has not been directly assessed in AD, but experience of continuity through time is impaired since AD patients are known to experience a temporal gradient for loss of autobiographical memories (Greene et al., 1995). Ability to reflect on one's own thoughts and experience is an important component of the self that appears to be impaired in AD patients, who deny any cognitive problem even when they are totally unable to remember any recent personal event.

The neural correlates of anosognosia for cognitive impairment in AD

Different brain regions were implicated in anosognosia for different neurological disorders, such as neglect, hemiparesis, aphasia, cortical blindness, or profound amnesia. Lack of awareness was reported in patients with diverse frontal lobe pathologies (McGlynn and Schacter, 1989). In the functional imaging literature, medial prefrontal and posterior cingulate cortices were shown to be part of a neural network subserving self-reflective thoughts in normal subjects (Johnson et al., 2002; Kelley et al., 2002). Decreased activity in the ventromedial prefrontal cortex and the temporal poles is characteristic of frontotemporal dementia (FTD; Salmon et al., 2003) and this might explain impairment of self-reflection and early loss of insight into behavioral disturbances in these patients. On the contrary, there is an early decrease of activity in posteromedial cortices (posterior cingulate and precuneus; Fig. 4A) in AD patients (Minoshima et al., 1994). Posterior cingulate hypometabolism was related to dementia severity in AD (Fig. 4B; Salmon et al., 2000) and it was involved in episodic, semantic, and working memory in control populations (Petrides et al., 1993; Fletcher et al., 1998;

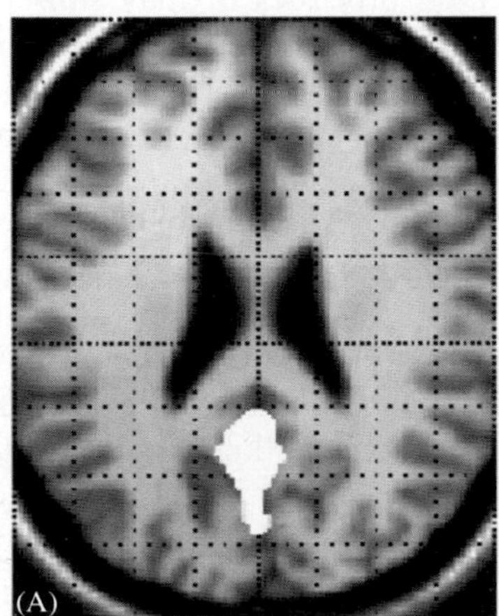

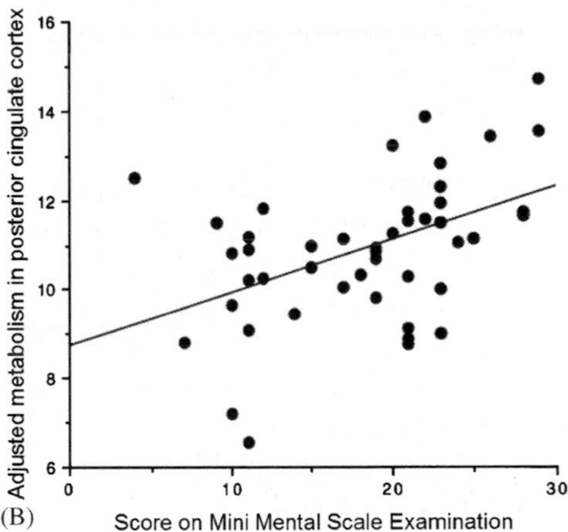

Fig. 4. (A) Early decrease of activity in posteromedial areas (encompassing posterior cingulate cortex and precuneus) in patients with AD. (B) Linear correlation between dementia severity (score on mini mental state examination) and metabolism in posterior cingulate cortex (area 31, coordinates: $x = 0$, $y = -56$, $z = 28$ mm) in AD. (From Salmon et al., 2000, with permission.)

Henson et al., 1999; Maguire and Mummery, 1999). Posteromedial cortices (in relation with hippocampal structures) are thought to be involved in the episodic integration of interoceptive and exteroceptive information required for conscious/controlled processing (Maguire et al., 1999; Salmon, 2003). Accordingly, in healthy subjects, posterior cingulate activity is low during executive processes, which are driven by external stimuli (without self-reference) (Salmon et al., 1999; Gusnard et al., 2001). So, posteromedial cortices are mainly related to personal memories and they would be poorly recruited in self-referential tasks by AD patients due to their episodic memory impairment.

Studies relating anosognosia to brain activity have shown a decrease of right lateral frontal and parietal activity in AD patients with anosognosia (Leys et al., 1989; Starkstein et al., 1995; Derouesne et al., 1999). In a theoretical model, the "conscious awareness system" was suggested to depend on inferior parietal and posterior cingulate cortex (McGlynn and Schacter, 1989). It has an output link to an executive system housed in the frontal lobe. Frontal damage would be associated with unawareness of complex deficits, such as difficulties in problem solving or behavioral changes. In a recent study, a cognitive discrepancy score (difference between patient's and caregiver's assessment) was inversely related to metabolism in temporoparietal junctions and superior frontal sulci in a large population of AD patients (Salmon et al., 2004). In other words, the more the patient underestimates his cognitive impairment, the lower the metabolism in temporoparietal junctions. It is noteworthy that the temporoparietal junction and lateral prefrontal areas are part of the associative cortices where metabolism is characteristically impaired in AD (Herholz et al., 2002). This would suggest that a relative anosognosia for cognitive impairment is intrinsic to AD pathology and would also explain the trend to greater anosognosia with greater severity of dementia (Zanetti et al., 1999).

The temporoparietal cortex and superior frontal sulcus have been previously reported in self-referential processes (Wicker et al., 2003). For example, lesions in the temporoparietal junction have been related to mirrored self-misidentification in demented patients (Breen et al., 2001). In healthy subjects, activation of temporoparietal junctions was observed in neuroimaging studies when contrasting first- and third-person perspectives in various types of tasks involving: intention attribution (Vogeley et al., 2001), 3D visuospatial perspective (Vogeley et al., 2004), action imagination (Ruby and Decety, 2001), truthfulness judgment (Ruby and Decety, 2003), or emotional assessment (Ruby and Decety, 2004). It has been demonstrated that stimulation of the right temporoparietal junction in humans can induce out of body experiences, in other words, a visual observation of the self with a third-person perspective (Blanke et al., 2002; Blanke, this volume). According to these results, it was suggested that the temporoparietal junction participates to the self/other distinction process required during perspective-taking (Ruby and Decety, 2004). One reason which makes this region a good candidate for such a role is that it houses the praxicon, a lexicon of representations concerning human actions (Peigneux et al., 2001). In the right hemisphere, the temporoparietal junction was related to anosognosia of hemiplegia (Venneri and Shanks, 2004). The praxicon is essential to compare self and other's familiar actions, and AD patients with anosognosia seem not to be aware of the difference between their impaired abilities and other's normal activities.

The parietal associative cortex related to a discrepancy score in AD may be involved in self-referential processes or may subserve the "third-person" knowledge that AD patients have of themselves (Tulving, 1993; Agnew and Morris, 1998; Klein et al., 2003). Accordingly, the right inferior parietal cortex was activated when volunteers had to recall whether an adjective had initially been characterized as fitting their personality (Lou et al., 2004). In that sense, anosognosia assessed by a discrepancy score might be interpreted as an impaired ability to see oneself with a third-person perspective (knowing how other people see yourself), an ability which would correspond to "the observing self" defined by Baars (Baars et al., 2003; Baars, this volume). According to this interpretation, one would expect that with the evolution of the disease, AD patients demonstrate impaired perspective taking on others. In line with this hypothesis, severe AD patients have been shown to become unable to assess other people's personality traits (Klein et al., 2003).

Conclusion

We suggest that specific brain regions are involved in different aspect of impaired consciousness in AD. On the one hand, a network of frontal, posterior associative, and limbic structures was demonstrated to be related to controlled processes that require conscious processing of information. On the other, impaired metabolism in lateral prefrontal and parietal areas was shown to correlate with the discrepancy score between AD patient and caregiver, one of the different measures of anosognosia for cognitive impairment. According to those results, we suggest that anosognosia for cognitive deficits in AD might be partly explained by impaired metabolism in parts of networks subserving self-referential processes (the superior frontal sulcus) and perspective-taking (the temporo-parietal junction). Patients appear to be impaired in the ability to see themselves with a third-person perspective (i.e., knowing how other people see themselves). Models of anosognosia suggest that those regions are only part of a wider cerebral network subserving unawareness of cognitive deficits in AD (Agnew and Morris, 1998).

Acknowledgements

The work in the Cyclotron Research Centre in Liege was made possible by grants from the National Fund for Scientific Research (FNRS), Reine Elisabeth Medical Fondation, a 5th European Framework project (NEST-DD), and InterUniversity Attraction Pole 5/04, Belgian Science Policy. FC and SL are Research Associates at FNRS.

References

Adam, S., Van der Linden, M., Collette, F., Lemauvais, L. and Salmon, E. (2005). Further exploration of controlled and automatic memory processes in early Alzheimer's disease. Neuropsychology (in press).

Agnew, S.K. and Morris, R.G. (1998) The heterogeneity of anosognosia for memory impairment in Alzheimer's disease: a review of the literature and a proposed model. Aging Ment. Health, 2: 7–19.

Amieva, H., Rouch-Leroyer, I., Fabrigoule, C. and Dartigues, J.F. (2000) Deterioration of controlled processes in the preclinical phase of dementia: a confirmatory analysis. Dement. Geriatr. Cogn. Disord., 11: 46–52.

APA. (1987) Diagnostic and Statistical Manual of Mental Disorders (3rd ed.). American Psychiatric Association, Washington, DC.

APA. (1994) Diagnostic and Statistical Manual of Mental Disorders (4th ed.). American Psychiatric Association, Washington, DC.

Azari, N.P., Rapoport, S.I., Grady, C.L., Schapiro, M.B., Salerno, J.A., Gonzalez-Aviles, A., et al. (1992) Patterns of interregional correlations of cerebral glucose metabolism rates in patients with dementia of the Alzheimer type. Neurodegeneration, 1: 101–111.

Baars, B.J., Ramsoy, T.Z. and Laureys, S. (2003) Brain conscious experience and the observing self. Trends Neurosci., 26: 671–675.

Baddeley, A. (1992) Working memory. Science, 255: 556–559.

Blanke, O., Ortigue, S., Landis, T. and Seeck, M. (2002) Stimulating illusory own-body perceptions. Nature, 419: 269–270.

Breen, N., Caine, D. and Coltheart, M. (2001) Mirrored-self misidentification: two cases of focal onset dementia. Neurocase, 7: 239–254.

Collette, F., Salmon, E., Van der Linden, M., Chicherio, C., Belleville, S., Degueldre, C., et al. (1999a) Regional brain activity during tasks devoted to the central executive of working memory. Brain Res. Cogn. Brain Res., 7: 411–417.

Collette, F., Van der Linden, M., Bechet, S. and Salmon, E. (1999b) Phonological loop and central executive functioning in Alzheimer's disease. Neuropsychologia, 37: 905–918.

Collette, F., Van der Linden, M., Delrue, G. and Salmon, E. (2002) Frontal hypometabolism does not explain inhibitory dysfunction in Alzheimer disease. Alzheimer Dis. Assoc. Disord., 16: 228–238.

Collette, F., Van der Linden, M., Juillerat, A.C. and Meulemans, T. (2003) A cognitive neuropsychological approach to Alzheimer's disease. In: Mulligan R., Van der Linden M. and Juillerat A.C. (Eds.), Clinical Management of Early Alzheimer's Disease. Erlbaum, Mahwah, pp. 35–73.

Collette, F., Van der Linden, M. and Salmon, E. (1999c) Executive dysfunction in Alzheimer's disease. Cortex, 35: 57–72.

Cummings, J.L., Mega, M., Gray, K., Rosenberg-Thompson, S., Carusi, D.A. and Gornbein, J. (1994) The Neuropsychiatric inventory: comprehensive assessment of psychopathology in dementia. Neurology, 44: 2308–2314.

Cummings, J.L., Ross, W., Absher, J., Gornbein, J. and Hadjiaghai, L. (1995) Depressive symptoms in Alzheimer disease: assessment and determinants. Alzheimer Dis. Assoc. Disord., 9: 87–93.

Curran, T. and Hintzman, D.L. (1995) Violations of the independence assumption in process dissociation. J. Exp. Psychol. Learn Mem. Cogn., 21: 531–547.

Dalla Barba, G., Parlato, V., Iavarone, A. and Boller, F. (1995) Anosognosia, intrusions and 'frontal' functions in Alzheimer's disease and depression. Neuropsychologia, 33: 247–259.

Dehaene, S. and Naccache, L. (2001) Towards a cognitive neuroscience of consciousness: basic evidence and a workspace framework. Cognition, 79: 1–37.

Derouesne, C., Thibault, S., Lagha-Pierucci, S., Baudouin-Madec, V., Ancri, D. and Lacomblez, L. (1999) Decreased awareness of cognitive deficits in patients with mild dementia of the Alzheimer type. Int. J. Geriatr. Psychiatry, 14: 1019–1030.

Fabrigoule, C., Rouch, I., Taberly, A., Letenneur, L., Commenges, D., Mazaux, J.M., et al. (1998) Cognitive process in preclinical phase of dementia. Brain, 121(Pt 1): 135–141.

Fleischman, D.A. and Gabrieli, J.D. (1998) Repetition priming in normal aging and Alzheimer's disease: a review of findings and theories. Psychol. Aging, 13: 88–119.

Fletcher, P.C., Shallice, T. and Dolan, R.J. (1998) The functional roles of prefrontal cortex in episodic memory. I. Encoding. Brain, 121(Pt 7): 1239–1248.

Folstein, M.F., Folstein, S.E. and McHugh, P.R. (1975) "Mini-mental state". A practical method for grading the cognitive state of patients for the clinician. J. Psychiatr. Res., 12: 189–198.

Gil, R., Arroyo-Anllo, E.M., Ingrand, P., Gil, M., Neau, J.P., Ornon, C., et al. (2001) Self-consciousness and Alzheimer's disease. Acta Neurol. Scand., 104: 296–300.

Greene, J.D., Hodges, J.R. and Baddeley, A.D. (1995) Autobiographical memory and executive function in early dementia of Alzheimer type. Neuropsychologia, 33: 1647–1670.

Gusnard, D.A., Akbudak, E., Shulman, G.L. and Raichle, M.E. (2001) Medial prefrontal cortex and self-referential mental activity: relation to a default mode of brain function. Proc. Natl. Acad. Sci. USA, 98: 4259–4264.

Harwood, D.G., Sultzer, D.L. and Wheatley, M.V. (2000) Impaired insight in Alzheimer disease: association with cognitive deficits, psychiatric symptoms, and behavioral disturbances. Neuropsychiatry Neuropsychol. Behav. Neurol., 13: 83–88.

Henson, R.N., Rugg, M.D., Shallice, T., Josephs, O. and Dolan, R.J. (1999) Recollection and familiarity in recognition memory: an event-related functional magnetic resonance imaging study. J. Neurosci., 19: 3962–3972.

Herholz, K., Salmon, E., Perani, D., Baron, J.C., Holthoff, V., Frolich, L., et al. (2002) Discrimination between Alzheimer dementia and controls by automated analysis of multicenter FDG PET. Neuroimage, 17: 302–316.

Hyman, B.T., Van Horsen, G.W., Damasio, A.R. and Barnes, C.L. (1984) Alzheimer's disease: cell-specific pathology isolates the hippocampal formation. Science, 225: 1168–1170.

Jacoby, L.L., Lindsay, D.S. and Toth, J.P. (1992) Unconscious influences revealed. Attention, awareness, and control. Am. Psychol., 47: 802–809.

Johnson, S.C., Baxter, L.C., Wilder, L.S., Pipe, J.G., Heiserman, J.E. and Prigatano, G.P. (2002) Neural correlates of self-reflection. Brain, 125: 1808–1814.

Jorm, A.F. (1986) Controlled and automatic information processing in senile dementia: a review. Psychol. Med., 16: 77–88.

Kalbe, E., Salmon, E., Perani, D., Holthoff, V., Sorbi, S., Elsner, A., et al. (2005) Anosognosia in very mild Alzheimer's disease but not in mild cognitive impairment. Dement. Geriatr. Cogn. Disord., 19: 349–356.

Kazui, H., Hashimoto, M., Hirono, N. and Mori, E. (2003) Nature of personal semantic memory: evidence from Alzheimer's disease. Neuropsychologia, 41: 981–988.

Kelley, W.M., Macrae, C.N., Wyland, C.L., Caglar, S., Inati, S. and Heatherton, T.F. (2002) Finding the self? An event-related fMRI study. J. Cogn. Neurosci., 14: 785–794.

Klein, S.B., Cosmides, L. and Costabile, K.A. (2003) Preserved knowledge of self in a case of Alzheimer's dementia. Soc. Cognition, 21: 157–165.

Klein, S.B., Rozendal, K. and Cosmides, L. (2002) A social-cognitive neuroscience analysis of the self. Social Cognition., 20: 105–135.

Knight, R.G. (1998) Controlled and automatic memory process in Alzheimer's disease. Cortex, 34: 427–435.

Koivisto, M., Portin, R., Seinela, A. and Rinne, J. (1998) Automatic influences of memory in Alzheimer's disease. Cortex, 34: 209–219.

Kotler-Cope, S. and Camp, C.J. (1995) Anosognosia in Alzheimer disease. Alzheimer Dis. Assoc. Disord., 9: 52–56.

Lekeu, F., Van der Linden, M., Chicherio, C., Collette, F., Degueldre, C., Franck, G., et al. (2003a) Brain correlates of performance in a free/cued recall task with semantic encoding in Alzheimer disease. Alzheimer Dis. Assoc. Disord., 17: 35–45.

Lekeu, F., Van der Linden, M., Degueldre, C., Lemaire, C., Luxen, A., Franck, G., et al. (2003b) Effects of Alzheimer's disease on the recognition of novel versus familiar words: neuropsychological and clinico-metabolic data. Neuropsychology, 17: 143–154.

Lekeu, F., Wojtasik, V., Van der Linden, M. and Salmon, E. (2002) Training early Alzheimer patients to use a mobile phone. Acta Neurol. Belg., 102: 114–121.

Leuchter, A.F., Newton, T.F., Cook, I.A., Walter, D.O., Rosenberg-Thompson, S. and Lachenbruch, P.A. (1992) Changes in brain functional connectivity in Alzheimer-type and multi-infarct dementia. Brain, 115(Pt 5): 1543–1561.

Leys, D., Steinling, M., Petit, H., Salomez, J.L., Gaudet, Y., Ovelacq, E., et al. (1989) [Alzheimer's disease: study by single photon emission tomography (Hm PAO Tc99 m)]. Rev. Neurol., 145: 443–450.

Lopez, O.L., Becker, J.T., Somsak, D., Dew, M.A. and DeKosky, S.T. (1994) Awareness of cognitive deficits and anosognosia in probable Alzheimer's disease. Eur. Neurol., 34: 277–282.

Lou, H.C., Luber, B., Crupain, M., Keenan, J.P., Nowak, M., Kjaer, T.W., et al. (2004) Parietal cortex and representation of the mental Self. Proc. Natl. Acad. Sci. USA, 101: 6827–6832.

Maguire, E.A., Frith, C.D. and Morris, R.G. (1999) The functional neuroanatomy of comprehension and memory: the importance of prior knowledge. Brain, 122(Pt 10): 1839–1850.

Maguire, E.A. and Mummery, C.J. (1999) Differential modulation of a common memory retrieval network revealed by positron emission tomography. Hippocampus, 9: 54–61.

McDaniel, K.D., Edland, S.D. and Heyman, A. (1995) Relationship between level of insight and severity of dementia in Alzheimer disease. CERAD Clinical Investigators. Consortium to Establish a Registry for Alzheimer's Disease. Alzheimer Dis. Assoc. Disord., 9: 101–104.

McGlynn, S.M. and Schacter, D.L. (1989) Unawareness of deficits in neuropsychological syndromes. J. Clin. Exp. Neuropsychol., 11: 143–205.

Meiran, N. and Jelicic, M. (1995) Implicit memory in alzheimer's disease: a meta-analysis. Neuropsychology, 9: 291–303.

Michon, A., Deweer, B., Pillon, B., Agid, Y. and Dubois, B. (1994) Relation of anosognosia to frontal lobe dysfunction in Alzheimer's disease. J. Neurol. Neurosurg. Psychiatry, 57: 805–809.

Migliorelli, R., Teson, A., Sabe, L., Petracca, G., Petracchi, M., Leiguarda, R., et al. (1995a) Anosognosia in Alzheimer's disease: a study of associated factors. J. Neuropsychiatry Clin. Neurosci., 7: 338–344.

Migliorelli, R., Teson, A., Sabe, L., Petracchi, M., Leiguarda, R. and Starkstein, S.E. (1995b) Prevalence and correlates of dysthymia and major depression among patients with Alzheimer's disease. Am. J. Psychiatry, 152: 37–44.

Minoshima, S., Foster, N.L. and Kuhl, D.E. (1994) Posterior cingulate cortex in Alzheimer's disease. Lancet, 344: 895.

Morris, R.G. (1994) Working memory in Alzheimer-type dementia. Neuropsychology, 8: 544–554.

Neary, D., Snowden, J.S., Gustafson, L., Passant, U., Stuss, D., Black, S., et al. (1998) Frontotemporal lobar degeneration: a consensus on clinical diagnostic criteria. Neurology, 51: 1546–1554.

Ott, B.R., Lafleche, G., Whelihan, W.M., Buongiorno, G.W., Albert, M.S. and Fogel, B.S. (1996) Impaired awareness of deficits in Alzheimer disease. Alzheimer Dis. Assoc. Disord., 10: 68–76.

Pearson, R.C., Esiri, M.M., Hiorns, R.W., Wilcock, G.K. and Powell, T.P. (1985) Anatomical correlates of the distribution of the pathological changes in the neocortex in Alzheimer disease. Proc. Natl. Acad. Sci. USA, 82: 4531–4534.

Peigneux, P., Salmon, E., Garraux, G., Laureys, S., Willems, S., Dujardin, K., et al. (2001) Neural and cognitive bases of upper limb apraxia in corticobasal degeneration. Neurology, 57: 1259–1268.

Petrides, M., Alivisatos, B., Meyer, E. and Evans, A.C. (1993) Functional activation of the human frontal cortex during the performance of verbal working memory tasks. Proc. Natl. Acad. Sci. USA, 90: 878–882.

Piolino, P., Desgranges, B., Belliard, S., Matuszewski, V., Lalevee, C., De la Sayette, V., et al. (2003) Autobiographical memory and autonoetic consciousness: triple dissociation in neurodegenerative diseases. Brain, 126: 2203–2219.

Reed, B.R., Jagust, W.J. and Coulter, L. (1993) Anosognosia in Alzheimer's disease: relationships to depression cognitive function and cerebral perfusion. J. Clin. Exp. Neuropsychol., 15: 231–244.

Ruby, P. and Decety, J. (2001) Effect of subjective perspective taking during simulation of action: a PET investigation of agency. Nat. Neurosci., 4: 546–550.

Ruby, P. and Decety, J. (2003) What you believe versus what you think they believe: a neuroimaging study of conceptual perspective-taking. Eur. J. Neurosci., 17: 2475–2480.

Ruby, P. and Decety, J. (2004) How Would You Feel versus How Do You Think She Would Feel? A Neuroimaging Study of Perspective-Taking with Social Emotions. J. Cogn. Neurosci., 16: 988–999.

Salmon, D.P. (2000) Disorders of memory in Alzheimer's disease. In: Boller F. and Grafman J. (Eds.), Handbook of Neuropsychology (2nd ed.). Elsevier Science B.V., Amsterdam, pp. 155–195.

Salmon, E. (2003) Physiology of human memory explored by functional imaging. In: De Deyn P.P., Thiery E. and D'Hogge R. (Eds.), Memory. Basic Concepts, Disorders and Treatment. Acco, Leuven, pp. 321–336.

Salmon, E., Collette, F., Degueldre, C., Lemaire, C. and Franck, G. (2000) Voxel-based analysis of confounding effects of age and dementia severity on cerebral metabolism in Alzheimer's disease. Hum. Brain Mapping, 10: 39–48.

Salmon, E., Collette, F., Van der Linden, M. and Franck, G. (1999) Increased demand on executive processes decreases activity in posteromedial regions. Neuroimage, 9: S340.

Salmon, E., Garraux, G., Delbeuck, X., Collette, F., Kalbe, E., Zuendorf, G., et al. (2003) Predominant ventromedial frontopolar metabolic impairment in frontotemporal dementia. Neuroimage, 20: 435–440.

Salmon, E., Lespagnard, S., Marique, P., Herhol, K., Perani, D., Holthoff, V., et al. (2005) Cerebral metabolic correlates of four dementia scales in Alzheimer's disease. J. Neurol., 252: 283–290.

Salmon, E., Perani, D., Herholz, K., Marique, P., Kalbe, E., Holthoff, V., Delbeucq, X., Beuthien-Bauman, B., Pelati, O., Lespagnard, S., Collette, F. and Garraux, G. (2004). Multiple regional cerebral account for unawareness of cognitive impairment in AD. HBM Abstract, Budapest.

Salthouse, T.A. and Becker, J.T. (1998) Independent effects of Alzheimer's disease on neuropsychological functioning. Neuropsychology, 12: 242–252.

Sevush, S. (1999) Relationship between denial of memory deficit and dementia severity in Alzheimer disease. Neuropsychiatry Neuropsychol. Behav. Neurol., 12: 88–94.

Sevush, S. and Leve, N. (1993) Denial of memory deficit in Alzheimer's disease. Am. J. Psychiatry, 150: 748–751.

Squire, L.R. (1992) Memory and the hippocampus: a synthesis from findings with rats, monkeys, and humans. Psychol. Rev., 99: 195–231.

Starkstein, S.E., Vazquez, S., Migliorelli, R., Teson, A., Sabe, L. and Leiguarda, R. (1995) A single-photon emission computed tomographic study of anosognosia in Alzheimer's disease. Arch. Neurol., 52: 415–420.

Tabert, M.H., Albert, S.M., Borukhova-Milov, L., Camacho, Y., Pelton, G., Liu, X., et al. (2002) Functional deficits in patients with mild cognitive impairment: prediction of AD. Neurology, 58: 758–764.

Tulving, E. (1993). Self-knowledge of an amnesic individual is represented abstractly. In: Srull, T.Kw., R.S., (Ed.) Advances in Social Cognition. Vol. 5. Erlbaum, Hillsdale, NJ, pp. 147–156.

Tulving, E. (2002) Episodic memory: from mind to brain. Annu. Rev. Psychol., 53: 1–25.

Vaidya, C.J., Gabrieli, J.D., Demb, J.B., Keane, M.M. and Wetzel, L.C. (1996) Impaired priming on the general knowledge task in amnesia. Neuropsychology, 10: 529–537.

Venneri, A. and Shanks, M.F. (2004) Belief and awareness: reflections on a case of persistent anosognosia. Neuropsychologia, 42: 230–238.

Vogeley, K., Bussfeld, P., Newen, A., Herrmann, S., Happe, F., Falkai, P., et al. (2001) Mind reading: neural mechanisms of theory of mind and self-perspective. Neuroimage, 14: 170–181.

Vogeley, K., May, M., Ritzl, A., Falkai, P., Zilles, K. and Fink, G.R. (2004) Neural correlates of first-person perspective as one constituent of human self-consciousness. J. Cogn. Neurosci., 16: 817–827.

Weinberger, D.R. (1993) A connectionist approach to the prefrontal cortex. J. Neuropsychiatry Clin. Neurosci., 5: 241–253.

Wheeler, M.A., Stuss, D.T. and Tulving, E. (1997) Toward a theory of episodic memory: the frontal lobes and autonoetic consciousness. Psychol. Bull., 121: 331–354.

Wicker, B., Ruby, P., Royet, J.P. and Fonlupt, P. (2003) A relation between rest and the self in the brain? Brain Res. Brain Res. Rev., 43: 224–230.

Yonelinas, A.P. and Jacoby, L.L. (1995) Dissociating automatic and controlled processes in a memory-search task: beyond implicit memory. Psychol. Res., 57: 156–165.

Zanetti, O., Vallotti, B., Frisoni, G.B., Geroldi, C., Bianchetti, A., Pasqualetti, P., et al. (1999) Insight in dementia: when does it occur? Evidence for a nonlinear relationship between insight and cognitive status. J. Gerontol. B Psychol. Sci. Soc. Sci., 54: 100–106.

S. Laureys (Ed.)
Progress in Brain Research, Vol. 150
ISSN 0079-6123

Chapter 22

Functional brain imaging of symptoms and cognition in schizophrenia

Tilo T.J. Kircher* and Renate Thienel

Department of Psychiatry and Psychotherapy, University of Aachen, Pauwelsstrasse 30, D-52074 Aachen, Germany

Abstract: The advent of functional magnetic resonance imaging and positron emission tomography has provided novel insights into the neural correlates of cognitive function and psychopathological symptoms. In patients with mental disorders, cognitive and emotional processes are disrupted. In this chapter, we review the basic methodological and conceptual principles for neuroimaging studies in these patients. By taking schizophrenia as an example, we outline the cerebral processes involved in the symptoms of this disorder, such as auditory hallucinations and formal thought disorder. We also characterize the neural networks involved in their emotional and cognitive dysfunction.

Introduction

Mental phenomena and their disruption, evident in mental disorders, result from disturbances in the interaction of different neural networks. Functional magnetic resonance imaging (fMRI) and positron emission tomography (PET) enable us to visualize cerebral function. We use schizophrenia as an example of how core psychopathological symptoms of the disorder, such as auditory hallucinations and formal thought disorder (FTD) as well as cognitive deficits, can be represented by their cerebral correlates. Since there are many myths and misconceptions about schizophrenia, we will concentrate in this chapter on established facts on brain–behavior relationships in this disorder. We will further discuss briefly on models and recent data on consciousness and self-consciousness in schizophrenia (Kircher and David, 2003; Kircher and Leube, 2003).

*Corresponding author. Tel.: +49 (0)241 8089637; Fax: +49 (0)241 8082401; E-mail: tkircher@ukaachen.de

Methodological and conceptual background

The neurobiological foundation of complex psychiatric disorders is one of the greatest challenges facing clinical neuroscience. Our conceptions of how the brain works exert a strong influence on the planning and interpretation of studies. Two basic principles in macroscopic brain function can be distinguished and should be kept in mind when drawing any inferences from neuroimaging research. The first principle is the assumption of functional segregation; i.e., through our ability to localize processes, we assume that a particular function can be linked to a discrete cerebral structure. For example, failure of parts of the motor cortex can lead to paresis in the corresponding body part. This notion of the strict localization of brain functions has its historic roots in the phrenology of Franz Gall (1758–1828) and was later based on a lesion deficit approach, which deduces the functional significance of a brain area through observation of a deficit following either temporary or permanent brain lesions (Corkin et al., 1997). An exclusively localizationist view of large modules

DOI: 10.1016/S0079-6123(05)50022-0

being responsible for a category of functions—such as language—does not adequately describe the organization of the brain. Therefore, functional brain imaging techniques reveal brain areas involved in, though not necessarily essential to, the ongoing performance of a given task.

The second principle is that of functional integration, whereby functions are achieved through interplay between several brain regions, a notion that has its foundation in brain anatomy, where regions are amply networked with one another. For example, goal-oriented movements are only possible through interplay between premotor, motor, cerebellar and subcortical structures. Carl Wernicke (1848–1904) had already postulated the necessity of an organized interplay between several cortical regions for normal mental function. Thus the majority of functional imaging studies show that a series of different brain regions participate in the tasks under investigation. The principles of functional segregation and integration are not necessarily antagonistic, as integrative models have shown (Tononi et al., 1994; Tononi this volume), and new developments in the analysis of MRI data potentiate the assessment of connectivity between brain regions in addition to mere activation.

There is a correlative, but not causal, relationship between mental events and their accompanying cerebral activations. During activation studies with functional imaging, the same cognitive event is described on two levels: the subjective experience (e.g. hearing a voice or feeling sadness) and the correlating neuronal activation (e.g. in the superior temporal gyrus, anterior cingulate or amygdala). The exact link between the two levels of description reaches far into the "mind–body problem," which in recent years has been experiencing a revival in the philosophy of mind, giving rise to the new field of neurophilosophy that tries to merge empirical and theoretical research (e.g. Kircher and David, 2003). In functional imaging, it is necessary to differentiate between resting state and cognitive activation paradigms. Resting state measures are performed with either glucose PET or single photon emission computed tomography (SPECT) while the subjects lie still in tomographs with no cognitive stimulation. It is important to remember, however, that the so-called "resting activity" can vary widely according to the cognitive and emotional experience of the subject at the time of scanning. The brain of a living person is never fully at rest, and its momentary condition depends on input signals from the body and its environment as well as memories, thoughts and feelings. Therefore, not surprisingly, results from imaging studies at rest are quite inconsistent regarding differences between schizophrenic patients and healthy controls.

Activation studies seemingly do more justice to the brain as an unceasingly active organ by putting it into predefined states (active and a control condition) evoked by standardized manipulations of sensory, motor, cognitive or emotional tasks. Active and control condition ideally differ only in the variable under investigation. Brain activations measured during the task are then statistically subtracted from the control condition. If we want, for example, to find out which brain regions participate specifically in the processing of faces, we present faces in the active condition and complex objects (e.g. houses) in the control condition. When brain activation caused by looking at the complex control objects is subtracted from brain activation caused by looking at faces, the visual and association regions (e.g. visual cortex and parietal lobe), among others, will be cancelled out as required. The subtraction yields those regions activated only during the specific processing of faces, including the fusiform gyrus among others.

The predominant fMRI technique uses the subject's own blood as a sort of internal contrast agent (oxygenated versus deoxygenated hemoglobin), whereas PET techniques use water containing a radioactive marker (Habel et al., 2002). Animal studies have shown that the blood oxygenation level in fMRI (Logothetis et al., 2001) correlate with local neuronal activity (see Hirsch, in this volume, for further details).

Correlates of psychopathological symptoms in schizophrenia

To date, imaging studies conducted with fMRI and PET have been performed for virtually every known psychiatric disorder. In this chapter, we will concentrate on their application in schizophrenia.

Schizophrenia is being diagnosed by the presence of several symptoms, present over a certain time period. The current diagnostic criteria for schizophrenia of the American Psychiatric Association (1994) are listed in Table 1.

The first studies in schizophrenia have been performed with PET during "resting state", i.e. without cognitive task. One prominent approach has associated schizophrenia syndromes with characteristic neuropsychological deficits and their cerebral activation patterns (Liddle et al., 1992, 1994). The necessity of subdividing schizophrenia patients according to their symptoms arises from the enormous heterogeneity of its clinical manifestations. The phenomenology of schizophrenia can be subdivided into three syndromes: the *disorganized* type, characterized by disorganization in speech and behavior, the *psychomotor* poverty type with poverty of speech, lack of spontaneous movement and various aspects of blunting of affect and the *reality distorted* type with hallucinations and delusions. On a neurocognitive level, patients with marked disorganization symptoms show pronounced deficits in executive functions such as planning, goal definition, and working memory. This syndrome was further associated at the cerebral level with a hypoactivation in the mediobasal prefrontal cortex when compared to healthy controls (Liddle et al., 1992). The psychomotor poverty type, showing difficulties in the initiation of behavior, was associated with reduced activation in the dorsolateral prefrontal cortex (Liddle, 1987) and also exhibited deficits in executive functions. Patients with this syndrome show anomalies in the blood flow patterns of the temporal lobe (Liddle et al., 1992).

More recent functional imaging studies in this field have been performed employing activation tasks, such as simple sensory and motor paradigms, for higher cognitive and emotional functions, the direct representation of psychopathological symptoms and the influence of psychoactive drugs. In the following sections, we will summarize a few important results from PET and fMRI activation studies in patients with schizophrenia.

Table 1. DSM IV diagnostic criteria of schizophrenia (American Psychiatric Association, 1994)

	DSM IV diagnostic criteria of schizophrenia
General criteria	A. Characteristic symptoms: two (or more) of the following: delusions, hallucinations, disorganized speech, grossly disorganized or catatonic behavior, negative symptoms during a 1 month period B. Social/occupational dysfunction C. Duration: continuous signs for at least 6 months D. Schizoaffective and mood disorder exclusion E. Substance/general medical condition exclusion F. Exclusion of relationship to a pervasive developmental disorder
	Schizophrenia subtypes
Paranoid type	A. Preoccupation with one or more delusions or frequent auditory hallucinations B. None of the following is prominent: disorganized speech, disorganized or catatonic behavior, or flat or inappropriate affect
Disorganized type	A. Disorganized speech and behavior, flat or inappropriate affect B. The criteria are not met for catatonic type
Catatonic type	At least two of the following: 1. Motoric immobility as evidenced by catalepsy or stupor 2. Excessive motor activity 3. Extreme negativism or mutism 4. Posturing, stereotyped movements, prominent mannerism, or prominent grimacing 5. Echolalia or echopraxia
Undifferentiated type	Schizophrenia symptoms meet Criterion A, but the criteria are not met for one of the above-mentioned subtype.

Auditory hallucinations

Since auditory hallucinations are an important feature in the diagnosis of schizophrenia and can be investigated relatively easily, most of the research on the correlates of psychopathological symptoms has been published. In fMRI and PET studies measuring brain activation during both the presence and absence of auditory hallucinations, patients were either examined at regular intervals of several weeks or indicated during scanning when "voices" were present and when not (Suzuki et al., 1993; Silbersweig et al., 1995; Woodruff et al., 1997; Dierks et al., 1999; Shergill et al., 2000). Consistently across all studies, activation was seen in the left superior and middle temporal lobes (Fig. 1). Some studies also showed activation in the right superior temporal lobe. Thus, structures that show signal variation during the perception of real, external voices are also activated during auditory hallucinations. Additional activations were found in regions responsible for the processing of emotion (e.g. the amygdala–hippocampus complex and insula; Dierks et al., 1999; Shergill et al., 2000). This can be regarded as a neural correlate of the anxiety that usually accompanies the hearing of voices. One prominent theory links hallucinations and passivity phenomena to a defective self-monitoring system (Frith, 1996; Kircher and Leube, 2003; Kircher and David, 2003; Knoblich et al., 2004). These patients show a deficit in the integration of the feed-forward signals normally providing the perception of one's own behavior and its correction if necessary (Leube et al., 2003b; Knoblich and Kircher, 2004).

Formal thought disorder

Formal thought disorder (FTD) can be divided into symptoms that are "positive" (e.g. neologisms, incoherence, loosening of association), "negative" (e.g. impoverishment of thought, blocking) or those that do not fall into the positive/negative dichotomy (e.g. echolalia, perseveration) (Liddle et al., 2002). The spontaneous language of patients with positive FTD has much in common with that of Wernicke's aphasia (Faber et al., 1983). In the available studies on positive FTD, patients spoke continuously about visual stimuli while brain activation was measured. Their speech was recorded and the degree of thought and language disturbance classified by means of standardized scales and then correlated with brain activation. These studies demonstrated that the presence of positive FTD is associated with underactivation of the posterior superior temporal gyrus (McGuire et al., 1998; Kircher et al., 2001). Thus, positive FTD is associated with a transient "functional lesion" in

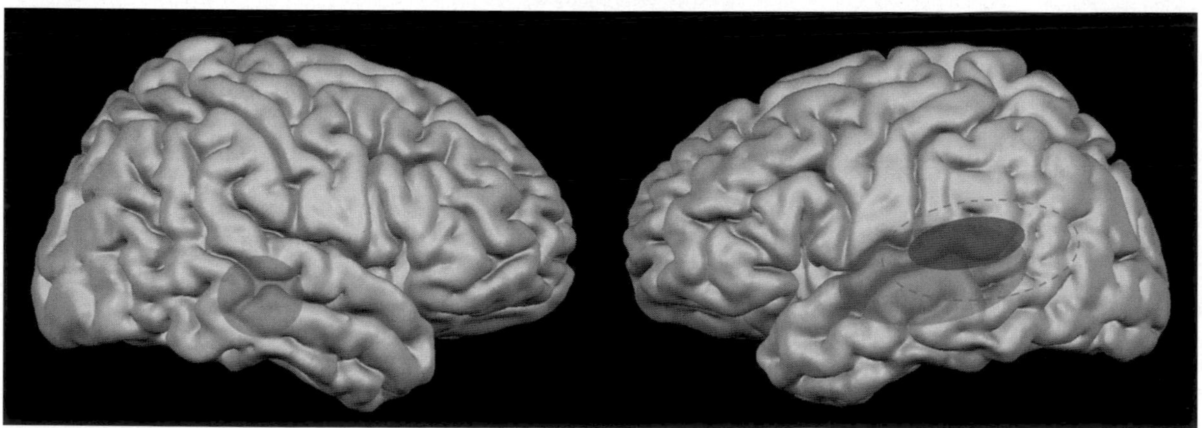

Fig. 1. A schematic illustration of neural activity changes during auditory hallucinations and formal thought disorder in schizophrenia. Areas shaded in red represent activations during auditory hallucinations. The area shaded in blue represents reduced activation compared with normal language when positive formal thought disorder is present. The green, dotted line indicates the location of Wernicke's area (from Kircher et al., 2004b.). See Plate 22.1 in Colour Plate Section.

Wernicke's speech area (McGuire et al., 1993, 1998; Suzuki et al., 1993; Silbersweig et al., 1995; Woodruff et al., 1997; Dierks et al., 1999; Shergill et al., 2000; Kircher et al., 2001) (see Fig. 1). Meanwhile, a study of negative FTD showed positive correlations with brain activation in different regions, including the left precuneus, cuneus, frontal medial gyrus and right inferior parietal lobe (Kircher et al., 2002; Fig. 2A).

FTD is a complex phenomenon with a number of underlying language and cognitive deficits (Kircher, 2003). Similar to auditory hallucinations, it has been associated with an ineffective self-monitoring mechanism (Knoblich et al., 2004). In particular, one process that seems to be impaired in patients is online verbal error monitoring and speech planning. These processes occur during short speech pauses of 250–2000 ms and have been associated in healthy subjects with activation in the left posterior superior temporal gyrus, part of Wernicke's speech area (Fig. 2B; Kircher et al., 2004a). It is exactly this brain region that is underactivated during the production of FTD in schizophrenia. One might therefore link, at least in part, the occurrence of FTD to impaired online speech monitoring.

The involvement of the superior temporal gyri in auditory hallucinations as well as positive FTD is in line with data from structural imaging studies (Shergill et al., 2000; Shenton et al., 2001). Structural studies in schizophrenia have shown

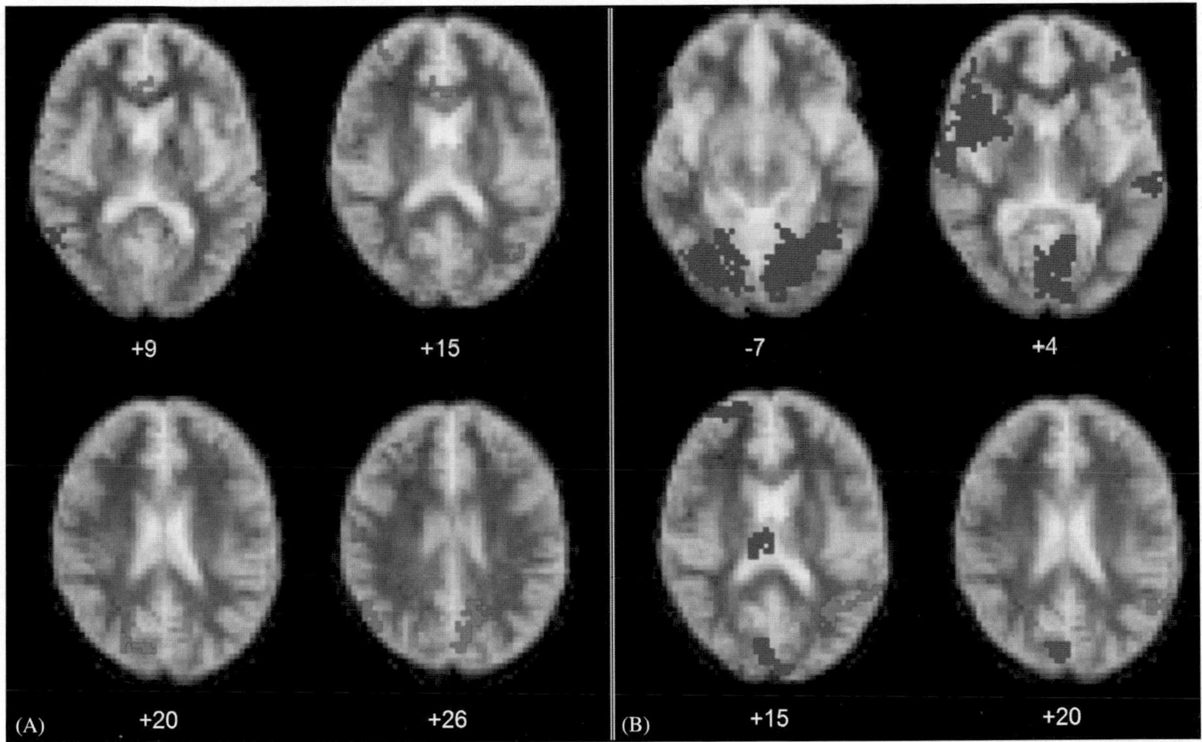

Fig. 2. (A) Neural correlates of "negative" formal thought disorder in patients with schizophrenia. In this study, patients with schizophrenia speak continuously while their brain activity is measured with fMRI. The amount of negative formal thought disorder (e.g. impoverishment of thought, blocking, poverty of content of speech as measured with the Thought and Language Index, Liddle et al., 2002) per 20 s speech epoch was correlated with the BOLD effect. Red voxels indicate positive and blue voxels negative correlation ($p<0.001$; from Kircher et al., 2003). Note that left precuneus, cuneus, medial frontal gyrus and right inferior parietal lobe are activated when negative formal thought disorder is prominent. (B) In healthy volunteers, short speech pauses during fluent speech correlate with activation in the superior temporal sulcus, between superior/middle temporal gyrus and parietal lobe (marked in red; $p<0.001$). Note that this region, part of Wernicke's speech area, is underactivated during the production of formal thought disorder in schizophrenia. Activated voxels during overt speech are shown in blue. Talairach z coordinates are shown below the transverse sections. The left side of the image represents the right side of the brain (from Kircher et al., 2004a). See Plate 22.2 in Colour Plate Section.

volumetric anomalies in the superior temporal gyri and a correlation between the magnitude of the reduction in the volume of the left superior temporal gyrus with the severity of FTD and increased susceptibility to auditory hallucinations. Thus, there is strong evidence for a neurophysiological interaction among psychopathology (e.g. thought disorders, hearing voices), brain function (e.g. measured activity in the temporal lobe) and structure (e.g. gray matter deficits).

Neuropsychological and emotional deficits

With respect to neuropsychological function, approximately 60–80% of schizophrenia patients show relatively generalized deficits (Table 2), whereupon deficits in verbal memory and executive functions are especially marked. While these neuropsychological deficits are not nosologically specific and are insufficient for an individual diagnosis, they are stable over the course of the disorder. The role of cognitive vulnerability markers in the prediction and prevention of schizophrenia gains increasingly more importance. The psychosocial functioning and outcome of schizophrenia patients correlates with the degree of underperformance in cognitive tests (Green et al., 2004), inferring that cognitive test scores are a better prognostic marker than psychopathology. Cognitive capacity is important for rehabilitation, as it is correlated with social competence and professional success. Neuropsychological deficits are also found in first-degree relatives and prodromal patients, indicating its potential as a genetic phenotype marker (Sitskoorn et al., 2004).

Memory

Compared to healthy subjects, patients with schizophrenia show deficits in all domains of neuropsychological tests (Table 2). In particular, the memory domain has shown the most consistent and severe impairments (Saykin et al., 1994; Heinrichs and Zakzanis, 1998). With the help of imaging studies, a link could be established between reduced memory performance and the cerebral structures involved. In a series of similarly designed studies, patient and control groups were asked to learn either words or pictures (encoding) or to remember stimuli to which they were previously exposed (retrieval and recognition) (Heckers et al., 1998, 1999; Ragland et al., 2001; Jessen et al., 2003; Leube et al., 2003a). The results showed significant reductions in the activity of the hippocampal complex, which remained significant even after correcting for the impaired memory of

Table 2. Neuropsychological domains dysfunctional in patients with schizophrenia and their related neuronal correlates

Cognitive domain	Definition	Dysfunctional areas in schizophrenia
Attention	Differentiate	Anterior cingulate cortex (Volz et al., 1999; Yucel et al., 2002)
	Selective attention Vigilance Speed of information processing	
Perception	Perception of information without higher cognitive processing	Primary and secondary visual, auditory olfactory and sensory areas (Braus et al., 2002; Kircher and David, 2003)
Motor system	Coordinated, goal-directed movements	Primary and secondary motor cortex, cerebellum and basal ganglia (Muller et al., 2002)
Episodic memory	Encoding and retrieval of information	Hippocampal area (Heckers et al., 1998; Leube et al., 2003a)
Working memory	Short-term storage and retrieval (<7 s)	Lateral prefrontal and parietal cortex (Callicott et al., 2000; Jansma et al., 2004; Walter et al., 2003)
Language	Production and processing of words and sentences	Lateral temporal cortex (Kircher et al., 2002; Sommer et al., 2001)
Executive function	Cognitive flexibility and planning	Lateral prefrontal cortex (Ragland et al., 1998; Riehemann et al., 2001)

the patients. Consequently, memory impairment as verified by psychological testing can be directly linked to disruption of activation in the hippocampal region. It has long been known from lesion studies in brain-damaged patients that the hippocampus has an important role in memory function, including the encoding of new material as well as the retrieval of information (Maguire, 1997). In patients with schizophrenia, the architecture of hippocampal cells is altered, and reduced volumes have been observed in the participants of both imaging and neuropathological studies (Harrison, 2004). Through functional imaging this knowledge has been expanded and we are now able to link impaired neuropsychology (e.g. poor results in psychological tests of memory capability) with changes in brain function (e.g. reduced activation of the hippocampus) and structure (e.g. altered cell architecture in the hippocampus).

Executive functions

Schizophrenia patients are further impaired in executive functions. Goal selection, pre-planning, monitoring, the use of feedback, and anticipation have often been shown to be disturbed in patients with frontal lobe lesions as well as in schizophrenia subgroups (Reading, 1991), particularly with negative symptoms such as anhedonia and emotional/social withdrawal (Heydebrand et al., 2004). To assess executive functioning, neuropsychological test such as the Tower of London Task (ToL) (Shallice, 1982) and the Wisconsin Card Sorting Test are used. Under cognitive load, a task-difficulty-dependent increase in prefrontal activation could be demonstrated in healthy controls using PET and MRI techniques while performing the ToL (Schall et al., 2003) when patients with schizophrenia exhibited reduced activation in a PET study (Andreasen et al., 1992). Arguably, the most popular test of executive functioning, the Wisconsin Card Sorting Test, assesses concept formation and shifting of concepts due to changes in the experimentally specified rules. Weinberger and co-workers (1986, 1992, 1994) were able to identify increased prefrontal activation during the performance of the test in healthy control subjects. This cortical activation was reduced in schizophrenic patients. A finding which is in line with the hypofrontality under resting conditions in schizophrenia. However it must be emphasized, that this "hypofrontality" is not nosologically specific for the disease but depends on the symptomatology of the patient and the cognitive task employed. Besides frontal hypoactivation, recent studies showed hyperactivation (Quintana et al., 2003; Walter et al., 2003) as well as "normoactivation" (Manoach, 2003).

Emotion

Patients with schizophrenia suffer from affective symptoms that are a significant component of the positive (e.g. anxiety, fear) and negative (e.g. flattened affect) symptom domains. Beyond signs and symptoms in the emotional domain (e.g. mood and affect), impairments are also found in emotion recognition. An inability to recognize emotions in the facial expressions of others is a robust finding and was reproducible in long-term studies (Streit et al., 1997; Kohler et al., 2003). These findings indicate that the deficit in facial affect identification is independent of the stage of illness and, like the cognitive disturbances, might reflect basic psychophysiological mechanisms of the disorder. To investigate emotional processes in schizophrenia patients, paradigms for mood induction and emotional discrimination are being employed. Usually, faces with different emotional expressions (including joy, rage, sadness and others) are presented and subjects are requested to classify the emotion. Employing fMRI techniques, hypoactivation could be consistently demonstrated in the anterior cingulate cortex (Yucel et al., 2002; Hempel et al., 2003) and the amygdala–hippocampal complex (Schneider et al., 1998; Mancini-Marie et al., 2004) of schizophrenia patients. Thus, it became evident that the neural networks responsible for the recognition, experience and expression of emotions in healthy subjects are disrupted in patients with schizophrenia.

Connectivity

Since the time of Wernicke (1906), a disrupted interplay among different brain regions has been made responsible for mental disorders and in particular schizophrenia. With the advent of functional connectivity analysis with fMRI, this topic has been of considerable interest. Particularly the connections between lateral frontal and temporal lobe (lateral and medial) has been the focus of some research. For example, it has been found that in drug naïve (Schlosser et al., 2003) and medicated patients (Lawrie et al., 2002), connectivity measures across a number of regions are different from healthy control groups. These are promising newer applications within functional imaging and might, together with diffusion tensor imaging (DTI), offer more insight into an old concept.

Conclusion

In our review of functional imaging techniques, we have shown that it is possible to represent the neural correlates of psychological processes by means of fMRI and PET. However, these psychological processes are only understood at a rudimentary level, even in healthy subjects, and many years of intensive research will be necessary before the foundations and conditions of subjective experience can be described in detail. Through the example of schizophrenia, we were able to show that psychiatric disorders manifest through a dysfunctional interplay of numerous sensory, motor, cognitive, and emotional functions and structures of the central nervous system. In just a few years, functional imaging has revolutionized our understanding of psychopathological phenomena, providing us with a research tool that can lead to a functional neuropsychopathology in which psychiatric phenomena can be correlated with cerebral functional states (through the integration of magneto–electrophysiological, structural–anatomic, experimental psychological, pharmacological and basic research).

A complete mapping of psychopathological symptoms and cognitive–emotional dysfunctions will be one of the greatest challenges for psychiatric research in the near future. Important progress in this realm can be expected, especially from the investigation of patient subgroups with defined symptomatology and from the linking of functional imaging with genetics and transmitter systems. It is hoped that a better understanding of psychiatric disorders will open new therapeutic avenues that can be applied earlier in the etiological chain than currently possible, and may thus improve the prognosis and therapy of the illness.

References

American Psychiatric Association. (1994) Diagnostic and Statistical Manual of Mental Disorders (4th ed.). American Psychiatric Association, Washington, DC.

Andreasen, N.C., Rezai, K., Alliger, R., Swayze, V.W., Flaum, M., Kirchner, P., Cohen, G. and O'Leary, D.S. (1992) Hypofrontality in neuroleptic-naive patients and in patients with chronic schizophrenia. Assessment with xenon 133 single-photon emission computed tomography and the Tower of London. Arch. Gen. Psychiat., 49: 943–958.

Braus, D.F., Weber-Fahr, W., Tost, H., Ruf, M. and Henn, F.A. (2002) Sensory information processing in neuroleptic-naive first-episode schizophrenic patients: a functional magnetic resonance imaging study. Arch. Gen. Psychiat., 59: 696–701.

Callicott, J.H., Bertolino, A., Mattay, V.S., Langheim, F.J., Dyn, J., Coppola, R., Goldberg, T.E. and Weinberger, D.R. (2000) Physiological dysfunction of the dorsolateral prefrontal cortex in schizophrenia revisited. Cereb. Cortex, 10: 1078–1092.

Corkin, S., Amaral, D.G., Gonzalez, R.G., Johnson, K.A. and Hyman, B.T. (1997) H. M.'s medial temporal lobe lesion: findings from magnetic resonance imaging. J. Neurosci., 17: 3964–3979.

Dierks, T., Linden, D.E., Jandl, M., Formisano, E., Goebel, R., Lanfermann, H. and Singer, E. (1999) Activation of Heschl's gyrus during auditory hallucinations. Neuron, 22: 615–621.

Faber, R., Abrams, R., Taylor, M.A., Kasprison, A., Morris, C. and Weisz, R. (1983) Comparison of schizophrenic patients with formal thought disorder and neurologically impaired patients with aphasia. Am. J. Psychiat., 140: 1348–1351.

Frith, C. (1996) The role of the prefrontal cortex in self-consciousness: the case of auditory hallucinations. Philos. Trans. R. Soc. Lond. B Biol. Sci., 351: 1505–1512.

Green, M.F., Kern, R.S. and Heaton, R.K. (2004) Longitudinal studies of cognition and functional outcome in schizophrenia: implications for MATRICS. Schizophr. Res., 72: 41–51.

Habel, U., Posse, S. and Schneider, F. (2002) Funktionelle kernspintomographie in der klinischen psychologie und psychiatrie. Fortschr. Neurol. Psychiatr., 70: 61–70 [In German].

Harrison, P.J. (2004) The hippocampus in schizophrenia: a review of the neuropathological evidence and its pathophysiological implications. Psychopharmacology (Berl.), 174: 151–162.

Heckers, S., Goff, D., Schacter, D.L., Savage, C.R., Fischman, A.J., Alpert, N.M. and Rauch, S.L. (1999) Functional imaging of memory retrieval in deficit vs nondeficit schizophrenia. Arch. Gen. Psychiat., 56: 1117–1123.

Heckers, S., Rauch, S.L., Goff, D., Savage, C.R., Schacter, D.L., Fischman, A.J. and Alpert, N.M. (1998) Impaired recruitment of the hippocampus during conscious recollection in schizophrenia. Nat. Neurosci., 1: 318–323.

Heinrichs, R.-W. and Zakzanis, K.-K. (1998) Neurocognitive deficit in schizophrenia: a quantitative review of the evidence. Neuropsychology, 12: 426–445.

Hempel, A., Hempel, E., Schonknecht, P., Stippich, C. and Schroder, J. (2003) Impairment in basal limbic function in schizophrenia during affect recognition. Psychiatry Res., 122: 115–124.

Heydebrand, G., Weiser, M., Rabinowitz, J., Hoff, A.L., DeLisi, L.E. and Csernansky, J.G. (2004) Correlates of cognitive deficits in first episode schizophrenia. Schizophr. Res., 68: 1–9.

Jansma, J.M., Ramsey, N.F., van der Wee, N.J. and Kahn, R.S. (2004) Working memory capacity in schizophrenia: a parametric fMRI study. Schizophr. Res., 68: 159–171.

Jessen, F., Scheef, L., Germeshausen, L., Tawo, Y., Kockler, M., Kuhn, K.U., Maier, W., Schild, H.H. and Heun, R. (2003) Reduced hippocampal activation during encoding and recognition of words in schizophrenia patients. Am. J. Psychiat., 160: 1305–1312.

Kircher, T. (2003) Neuronale Korrelate psychopathologischer Symptome. Denk- und Sprachprozesse bei Gesunden und Patienten mit Schizophrenie. Steinkopff, Darmstadt.

Kircher, T.T.J., Brammer, M.J., Levelt, W., Bartels, M. and McGuire, P.K. (2004a) Pausing for thought: engagement of left temporal cortex during pauses in speech. Neuroimage, 21: 84–90.

Kircher, T.T.J. and David, A.S. (2003). In: Kircher T.T.J. and David A.S. (Eds.), The Self in Neuroscience and Psychiatry. University Press, Cambridge, UK pp. 1–496.

Kircher, T.T. and Leube, D.T. (2003) Self-consciousness, self-agency, and schizophrenia. Conscious Cogn., 12: 656–669.

Kircher, T., Liddle, P., Brammer, M., Murray, R. and McGuire, P. (2003) Neuronale korrelate "negativer" formaler Denkstörungen. Nervenarzt, 74: 748–754.

Kircher, T.T.J., Liddle, P.F., Brammer, M.J., Williams, S.C., Murray, R.M. and McGuire, P.K. (2001) Neural correlates of formal thought disorder in schizophrenia. Arch. Gen. Psychiat., 58: 769–774.

Kircher, T.T.J., Liddle, P.F., Brammer, M.J., Williams, S.C., Murray, R.M. and McGuire, P.K. (2002) Reversed lateralisation of temporal activation during speech production in thought disordered patients with schizophrenia. Psychol. Med., 32: 439–449.

Kircher, T., Schneider, F., Sauer, H. and Buchkremer, G. (2004b) Funktionelle Bildgebung am Beispiel der Schizophrenie. Deutsches Ärzteblatt, 101: A1975–A1980.

Knoblich, G. and Kircher, T.T.J. (2004) Deceiving oneself about being in control: conscious detection of changes in visuomotor coupling. J. Exp. Psychol. Hum. Percept. Perform., 30: 657–666.

Knoblich, G., Stottmeister and Kircher, T.T.J. (2004) Self-monitoring in patients with schizophrenia. Psychol. Med., 34: 1561–1569.

Kohler, C.G., Turner, TH., Bilker, W.B., Brensinger, C.M., Siegel, S.J., Kanes, S.J., Gur, R.E. and Gur, R.C. (2003) Facial emotion recognition in schizophrenia: intensity effects and error pattern. Am. J. Psychiat., 160: 1768–1774.

Lawrie, S.M., Buechel, C., Whalley, H.C., Frith, C.D., Friston, K.J. and Johnstone, E.C. (2002) Reduced frontotemporal functional connectivity in schizophrenia associated with auditory hallucinations. Biol. Psychiat., 51: 1008–1011.

Leube, D.T., Knoblich, G., Erb, M., Grodd, W., Bartels, M. and Kircher, T.T. (2003b) The neural correlates of perceiving one's own movements. Neuroimage, 20: 2084–2090.

Leube, D., Rapp, A., Buchkremer, G., Bartels, M., Kircher, T.T., Erb, M. and Grott, W. (2003a) Hippocampal dysfunction during episodic memory encoding in patients with schizophrenia—an fMRI study. Schizophr. Res., 64: 83–85.

Liddle, P.F., Friston, K.J., Frith, C.D. and Frackowiak, R.S. (1992) Cerebral blood flow and mental processes in schizophrenia. J. Roy. Soc. Med., 85: 224–227.

Liddle, P.F. (1987) Schizophrenic syndromes, cognitive performance and neurological dysfunction. Psychol. Med., 17: 49–57.

Liddle, P., Carpenter, W.T. and Crow, T. (1994) Syndromes of schizophrenia. Br. J. Psychiat., 165: 721–727.

Liddle, P.F., Ngan, E., Caissie, S., Anderson, C.M., Bates, A.T., Quested, D.J., White, R. and Weg, R. (2002) Thought and language index: an instrument for assessing thought and language in schizophrenia. Br. J. Psychiat., 181: 326–330.

Logothetis, N.K., Pauls, J., Augath, M., Trinath, T. and Oeltermann, A. (2001) Neurophysiological investigation of the basis of the fMRI signal. Nature, 412: 150–157.

Maguire, E.A. (1997) Hippocampal involvement in human topographical memory: evidence from functional imaging. Philos. Trans. R. Soc. Lond. B Biol. Sci., 352: 1475–1480.

Mancini-Marie, A., Stip, E., Fahim, C., Mensour, B., Leroux, J.M., Beaudoin, G., Bentaleb, L.A., Bourgouin, P. and Beauregard, M. (2004) Fusiform gyrus and possible impairment of the recognition of emotional expression in schizophrenia subjects with blunted affect: a fMRI preliminary report. Brain Cogn., 54: 153–155.

Manoach, D.S. (2003) Prefrontal cortex dysfunction during working memory performance in schizophrenia: reconciling discrepant findings. Schizophr. Res., 60: 285–298.

McGuire, P.K., Quested, D.J., Spence, S.A., Murray, R.M., Frith, C.D. and Liddle, P.F. (1998) Pathophysiology of 'positive' thought disorder in schizophrenia. Br. J. Psychiat., 173: 231–235.

McGuire, P.K., Shah, G.M. and Murray, R.M. (1993) Increased blood flow in Broca's area during auditory hallucinations in schizophrenia. Lancet, 342: 703–706.

Muller, J.L., Roder, C.H., Schuierer, G. and Klein, H. (2002) Motor-induced brain activation in cortical, subcortical and cerebellar regions in schizophrenic inpatients. A whole brain fMRI fingertapping study. Prog. Neuropsychopharmacol. Biol. Psychiat., 26: 421–426.

Quintana, J., Wong, T., Ortiz-Portillo, E., Kovalik, E., Davidson, T., Marder, S.R. and Mazziotta, J.C. (2003) Prefrontal-posterior parietal networks in schizophrenia: primary dysfunctions and secondary compensations. Biol. Psychiat., 53: 12–24.

Ragland, J.D., Gur, R.C., Glahn, D.C., Censits, D.M., Smith, R.J., Lazarev, M.G., Alavi, A. and Gur, R.E. (1998) Frontotemporal cerebral blood flow change during executive and declarative memory tasks in schizophrenia: a positron emission tomography study. Neuropsychology, 12: 399–413.

Ragland, J.D., Gur, R.C., Raz, J., Schroeder, L., Kohler, C.G., Smith, R.J., Alavi, A. and Gur, R.E. (2001) Effect of schizophrenia on frontotemporal activity during word encoding and recognition: a PET cerebral blood flow study. Am. J. Psychiat., 58: 1114–1125.

Reading, P.J. (1991) Frontal lobe dysfunction in schizophrenia and parkinson's disease—a meeting point for neurology psychology and psychiatry: discussion paper. J. Roy. Soc. Med., 84: 349–353.

Riehemann, S., Volz, H.P., Stutzer, P., Smesny, S., Gaser, C. and Sauer, H. (2001) Hypofrontality in neuroleptic-naive schizophrenic patients during the Wisconsin Card Sorting Test—a fMRI study. Eur. Arch. Psychiat. Clin. Neurosci., 251: 66–71.

Saykin, A.J., Shtasel, D.L., Gur, R.E., Kester, D.B., Mozley, L.H., Stafiniak, P. and Gur, R.C. (1994) Neuropsychological deficits in neuroleptic naive patients with first-episode schizophrenia. Arch. Gen. Psychiat., 51: 124–131.

Schall, U., Johnston, P., Lagopoulos, J., Juptner, M., Jentzen, W., Thienel, R., Dittmann-Balcar, A., Bender, S. and Ward, P.B. (2003) Functional brain maps of Tower of London performance: a positron emission tomography and functional magnetic resonance imaging study. Neuroimage, 20: 1154–1161.

Schlosser, R., Gesierich, T., Kaufmann, B., Vucurevic, G. and Stoeter, P. (2003) Altered effective connectivity in drug free schizophrenic patients. Neuroreport, 14: 2233–2237.

Schneider, F., Weiss, U., Kessler, C., Salloum, J.B., Posse, S., Grodd, W. and Muller-Gartner, H.W. (1998) Differential amygdala activation in schizophrenia during sadness. Schizophr. Res., 34: 133–142.

Shallice, T. (1982) Specific impairments of planning. Philos. Trans. R. Soc. Lond. B Biol. Sci., 298: 199–209.

Shenton, M.E., Dickey, C.C., Frumin, M. and McCarley, R.W. (2001) A review of MRI findings in schizophrenia. Schizophr. Res., 49: 1–52.

Shergill, S.S., Brammer, M.J., Williams, S.C., Murray, R.M. and McGuire, P.K. (2000) Mapping auditory hallucinations in schizophrenia using functional magnetic resonance imaging. Arch. Gen. Psychiat., 57: 1033–1038.

Silbersweig, D.A., Stern, E., Frith, C., Cahill, C., Holmes, A., Grootoonk, S., Seaward, J., McKenna, P., Chua, S.E., Schnorr, L., Jones, T. and Frackowiak, R.S.J. (1995) A functional neuroanatomy of hallucinations in schizophrenia. Nature, 378: 176–179.

Sitskoorn, M.M., Aleman, A., Ebisch, S.J., Appels, M.C. and Kahn, R.S. (2004) Cognitive deficits in relatives of patients with schizophrenia: a meta-analysis. Schizophr. Res., 71: 285–295.

Streit, M., Wolwer, W. and Gaebel, W. (1997) Facial-affect recognition and visual scanning behaviour in the course of schizophrenia. Schizophr. Res., 24: 311–317.

Sommer, I.E.C., Ramsey, N.F. and Kahn, R.S. (2001) Language lateralization in schizophrenia, an fMRI study. Schizophr. Res., 52: 57–67.

Suzuki, M., Yuasa, S., Minabe, Y., Murata, M. and Kurachi, M. (1993) Left superior temporal blood flow increases in schizophrenic and schizophreniform patients with auditory hallucination: a longitudinal case study using 123I-IMP SPECT. Eur. Arch. Psychiat. Clin. Neurosci., 242: 257–261.

Tononi, G., Sporns, O. and Edelman, G.M. (1994) A measure for brain complexity: relating functional segregation and integration in the nervous system. Proc. Natl. Acad. Sci. USA, 91: 5033–5037.

Volz, H., Gaser, C., Hager, F., Rzanny, R., Ponisch, J., Mentzel, H., Kaiser, W.A. and Sauer, H. (1999) Decreased frontal activation in schizophrenics during stimulation with the continuous performance test—a functional magnetic resonance imaging study. Eur. Psychiat., 14: 17–24.

Walter, H., Wunderlich, A.P., Blankenhorn, M., Schafer, S., Tomczak, R., Spitzer, M. and Gron, G. (2003) No hypofrontality, but absence of prefrontal lateralization comparing verbal and spatial working memory in schizophrenia. Schizophr. Res., 61: 175–184.

Weinberger, D.R., Aloia, M.S., Goldberg, T.E. and Berman, K.F. (1994) The frontal lobes and schizophrenia. J. Neuropsych. Clin. Neurosci., 6(4): 419–427.

Weinberger, D.R., Berman, K.F., Suddath, R. and Torrey, E.F. (1992) Evidence of dysfunction of a prefrontal-limbic network in schizophrenia: a magnetic resonance imaging and regional cerebral blood flow study of discordant monozygotic twins. Am. J. Psychiat., 149: 890–897.

Weinberger, D.R., Berman, K.F. and Zec, R.F. (1986) Physiologic dysfunction of dorsolateral prefrontal cortex in schizophrenia. I. Regional cerebral blood flow evidence. Arch. Gen. Psychiat., 43: 114–124.

Wernicke, C. (1906) Grundriss der Psychiatrie in klinischen Vorlesungen. Georg Thieme, Leipzig.

Woodruff, P.W., Wright, I.C., Bullmore, E.T., Brummer, M., Howard, R.J., Williams, S.C., Shapleske, J., Rossell, S., David, A.S., McGuire, P.K. and Murray, R.M. (1997) Auditory hallucinations and the temporal cortical response to speech in schizophrenia: a functional magnetic resonance imaging study. Am. J. Psychiat., 154: 1676–1682.

Yucel, M., Pantelis, C., Stuart, G.W., Wood, S.J., Maruff, P., Velakoulis, D., Pipingas, A., Crowe, S.F., Tochon-Danguy, H.J. and Egan, G.F. (2002) Anterior cingulate activation during Stroop task performance: a PET to MRI coregistration study of individual patients with schizophrenia. Am. J. Psychiat., 159: 251–254.

S. Laureys (Ed.)
Progress in Brain Research, Vol. 150
ISSN 0079-6123

CHAPTER 23

Hysterical conversion and brain function

Patrik Vuilleumier*

Laboratory for Behavioral Neurology and Imaging of Cognition, Clinic of Neurology & Department of Neurosciences, University Medical Center, Faculty of Psychology & Education Sciences, University of Geneva, Geneva, Switzerland

Abstract: Hysterical conversion disorders represent "functional" or unexplained neurological deficits such as paralysis or somatosensory losses that are not explained by organic lesions in the nervous system, but arise in the context of "psychogenic" stress or emotional conflicts. After more than a century of both clinical and theoretical interest, the exact nature of such emotional disorders responsible for hysterical symptoms, and their functional consequences on neural systems in the brain, still remain largely unknown. However, several recent studies have used functional brain imaging techniques (such as EEG, fMRI, PET, or SPECT) in the attempt to identify specific neural correlates associated with hysterical conversion symptoms. This article presents a general overview of these findings and of previous neuropsychologically based accounts of hysteria. Functional neuroimaging has revealed selective decreases in the activity of frontal and subcortical circuits involved in motor control during hysterical paralysis, decreases in somatosensory cortices during hysterical anesthesia, or decreases in visual cortex during hysterical blindness. Such changes are usually not accompanied by any significant changes in elementary stages of sensory or motor processing as measured by evoked potentials, although some changes in later stages of integration (such as P300 responses) have been reported. On the other hand, several neuroimaging results have shown increased activation in limbic regions, such as cingulate or orbitofrontal cortex during conversion symptoms affecting different sensory or motor modalities. Taken together, these data generally do not support previous proposals that hysteria might involve an exclusion of sensorimotor representations from awareness through attentional processes. They rather seem to point to a modulation of such representations by primary affective or stress-related factors, perhaps involving primitive reflexive mechanisms of protection and alertness that are partly independent of conscious control, and mediated by dynamic modulatory interactions between limbic and sensorimotor networks. A better understanding of the neuropsychobiological bases of hysterical conversion disorder might therefore be obtained by future imaging studies that compare different conversion symptoms and employ functional connectivity analyses. This should not only lead to improve clinical management of these patients, but also provide new insights on the brain mechanisms of self-awareness.

Introduction

Since more than a century, hysteria has continuously fascinated both clinicians and theorists interested in altered states of self-consciousness. As a medical condition in which patients may present with various somatic symptoms without a recognized organic illness, hysteria still constitutes a poorly understood class of disorders at the border between psychiatry and neurology, with many different appearances and names. A variety of classifications and explanations have been proposed over the years, with different emphasis on psychological or neurobiological factors, but none is still

*Corresponding author. Tel.: +41 (0)22 3795.381;
Fax: +41 (0)22 3795.402;
E-mail: patrik.vuilleumier@medecine.unige.ch

DOI: 10.1016/S0079-6123(05)50023-2

Fig. 1. Illustration of paraparesis (astasia–abasia) of hysterical origin (drawn by Jean-Martin Charcot in the 1870s).

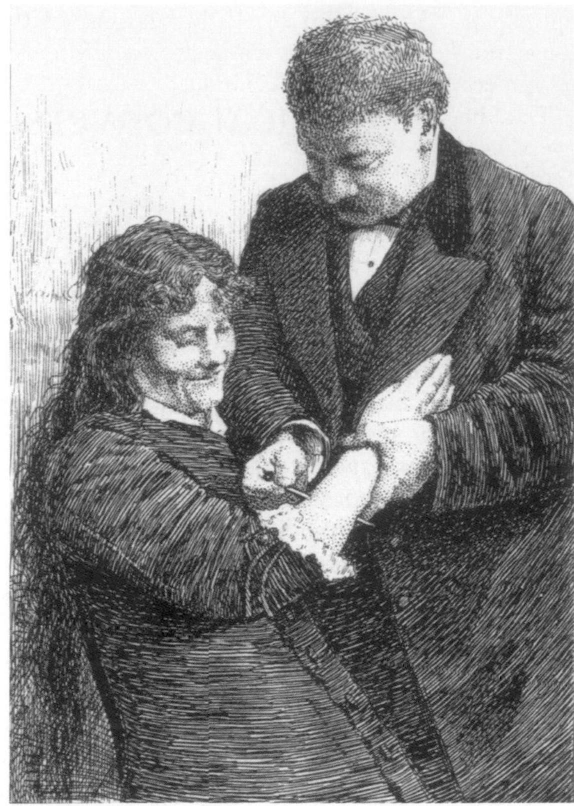

Fig. 2. "L'anesthésie hystérique", a cruel demonstration of anesthesia in a patient at La Salpêtrière, while in a state of hysteria (etched from photograph by P. Régnard, 1887, Les maladies épidémiques de l'esprit – Sorcellerie, magnétisme, morphinisme, délire des grandeurs. Plon-Nourrit, Paris).

entirely satisfactory (see Halligan et al., 2001 for a recent general overview).

In some respects, the most important change since classical descriptions of hysteria by Charcot and others at the end of the 19th century (see Figs. 1 and 2) is related to the fact that the term "hysteria" was recently eliminated from the official psychiatric terminology. Thus, hysteria is now referred to as "conversion disorder" in the DSM-IV classification (American Psychiatric Association, 1994, *Diagnostic and Statistical Manual of Mental Disorders*), where it is defined as a loss or distortion of a neurological function (e.g., paralysis or anesthesia) that is not explained by an organic neurological lesion or medical disease, arising in relation to some psychological stress or conflict, but not consciously produced or intentionally feigned. However, although the term "conversion" implies a specific mechanism whereby a primary psychological disorder is converted into bodily symptoms, the exact processes responsible for triggering this phenomenon and altering the patients' awareness of their bodily state still remain unknown. In fact, when considering the possible neurobiological changes associated with hysterical conversion symptoms, many of the current hypotheses are not very different from those elaborated in the 19th century. Moreover, in contrast to other psychiatric conditions (such as depression, anxiety, compulsion, or phobia), surprisingly few investigations have been carried out to examine whether the "functional" symptoms experienced by hysterical patients might correspond to any underlying "functional changes" in cerebral activity, as can be evaluated for instance using a variety of modern neuroimaging techniques.

The aim of this chapter is to present a general overview of the relationships between hysterical conversion and brain function, focusing particularly on previous attempts to determine some neurophysiological correlates of hysterical conversion disorders. This review will primarily concentrate on patients with unexplained neurological deficits in sensory and/or motor functions (i.e., hysterical paralysis or anesthesia), since such disorders are the most frequent and most easily described in terms of specific neurological pathways. Similar approaches in other domains (e.g., visual loss, deafness, or amnesia) will also be briefly mentioned, but other conversion disorders associated with more "positive" symptoms such as pseudo-seizure and abnormal movements will not be discussed here, since even less is known about their possible functional neural correlates (for review see Prigatano et al., 2002; Reuber et al., 2003). Finally, a few studies have examined the functional neuroanatomy of dissociations induced by hypnosis (Maquet et al., 1999; Faymonville et al., 2000; Halligan et al., 2000; Kupers et al., this volume), but their implications for hysterical deficits are still in part unclear and will be discussed briefly.

A better understanding of patients with hysterical conversion has important implications for several reasons, both clinical and theoretical. First, such disorders are very common in clinical practice, causing frequent problems in diagnosis and therapy for neurologists as well as psychiatrists, and resulting in important medical costs and major socio-economic burden for the patients and their relatives. Second, conversion disorders raise important theoretical questions concerning the relationships between body and mind, and the neurobiological and psychological mechanisms underlying self-awareness.

From the perspective of clinical neurology, dissociations between performance and awareness of performance are not unusual in patients with organic brain lesions, a phenomenon termed "anosognosia" whereby patients may fail to acknowledge and even explicitly deny a severe handicap resulting from their brain lesion (Vuilleumier, 2000a, 2004; Marcel et al., 2004). Anosognosia cannot be explained by psychogenic factors alone, or by general confusion, but it is often associated with a lack of emotional concern quite similar to "la belle indifference" that has typically been described in hysteria. Moreover, like anosognosia for hemiplegia, hysterical paralysis and hysterical anesthesia are often thought to affect the left more than the right limbs, without this being entirely accounted for a more frequent dominance of the right hand (Galin et al., 1977; Stern, 1983; Pascuzzi, 1994; Gagliese et al., 1995), although a recent meta-analysis has questioned the existence of such an asymmetry (Stone et al., 2002). It is also intriguing to note that the neurological functions most frequently associated with anosognosia are strikingly similar to those frequently concerned by hysteria, involving not only unilateral paralysis and anesthesia, but also blindness, amnesia, or jargon aphasia (Merskey, 1995; Vuilleumier, 2000a). Although such similarities do not indicate any clear relationships between these different disorders, these parallels between neurology and psychiatry highlight the fact that some behavioral abilities may dissociate from the subjective experience of these abilities, and that such dissociations are likely to arise from specific changes in brain function – be they due to certain types of organic damage or to certain psycho-affective states.

Incidence and evolution

Hysterical conversion disorders are thought to represent 1–4% of all diagnoses in general hospitals throughout western countries. This frequency has remained remarkably stable across the past decades despite many important changes in medicine. Thus, a retrospective survey at the National Hospital of Neurology and Neurosurgery in London, Queen Square, from 1955 to 1975 revealed a relatively constant rate of patients investigated for conversion or "functional" symptoms, ranging from 0.85 to 1.55% over decades (Trimble, 1981). A similar result was found by another study examining the diagnoses made on 7836 successive outpatient referrals at Charing Cross

Hospital between 1977 and 1987 (Perkin, 1989), which revealed a stable incidence of 3.8% of conversion disorders over the years. In Switzerland, Frei (1984) also estimated that the proportion and presentation of hysterical conversion disorders among patients seen in a large public hospital was similar in the 1920s as compared with the 1980s.

However, it is important to emphasize that the type of conversion disorders may significantly vary across different medical settings and referral sources. This might explain why the incidence of hysteria may appear to vary in some domains but not in others. It is known that a large number of patients with conversion and somatoform symptoms are never seen by either neurologists nor psychiatrists (Bridges and Goldberg, 1988; Crimlisk and Ron, 1999). Neurologists probably see many more patients with relatively acute and limited symptoms, as compared with psychiatrists who tend to see patients with more chronic and multiform disorders (Marsden, 1986). Moreover, although there is a good consensus between neurologists as to clinical syndromes reflecting non-organic diseases (irrespective of which term is actually preferred to describe these syndromes, e.g., hysterical, functional, or psychogenic), there is usually less agreement among psychiatrists (Mace and Trimble, 1991). Furthermore, although most neurologists would agree with the DSM-IV statement that "typically, individual conversion symptoms are of short duration", many psychiatrist have rather observed that conversion symptoms often tend to persist or reoccur in association with other psychiatric comorbidities and pathological personality traits (Ron, 1994; Binzer and Kullgren, 1998; Stone et al., 2003).

However, in most cases seen by neurologists, a rapid remission of initial symptoms is observed after appropriate behavioral treatment. The likelihood of spontaneous remission has repeatedly been found to be ~50–60% after 1 or 2 years (Singh and Lee, 1997; Binzer and Kullgren, 1998; Crimlisk et al., 1998) or even greater (Folks et al., 1984). Among young patients (< 27 year-old), only 3% have symptoms for more than 1 month (Turgay, 1990). The duration of effective behavioral treatment in such cases is usually between 2 to 26 weeks (Speed, 1996). Several factors are associated with a favorable prognosis, including young age, sensory rather than motor symptoms, acuteness of presentation, onset precipitated by a stressful event, good premorbid health and good socio-economic status, as well as an absence of any other concomitant organic disease or major psychiatric symptoms – especially depression (e.g., Ford and Folks, 1985; Binzer and Kullgren, 1998; Crimlisk et al., 1998; Stone et al., 2003). Personality disorder appears as a particularly decisive prognostic factor (see Crimlisk et al. 1998). It is therefore important to identify potential risk factors during the initial evaluation of patients, in order to prevent the development of a more chronic handicap.

Body, mind, and brain diseases

A potential limitation to our current understanding of hysterical conversion comes from a lack of clear definition of the boundaries with some related disorders. From the perspective of psychiatry, a number of problems are still unresolved concerning the terminology and classification of functional symptoms without an organic cause. In DSM-IV, the concept of hysteria has been broken down to different disorders, including somatoform disorders on one hand and dissociative disorders on the other hand, with conversion disorders being considered as a specific category of somatoform disorders involving sensorimotor symptoms or pseudo-seizures, distinct from other somatization symptoms or psychogenic pain disorders. Furthermore, depression and chronic fatigue syndrome would fall under other diagnostic categories, although in clinical practice there is often some overlap between all such conditions. For example, many patients recovering from conversion eventually suffer from depression at a later stage (Binzer and Kullgren, 1998). Furthermore, whereas conversion disorders are defined as deficits affecting a specific neurological function such as motor strength or somatosensory perception, psychogenic memory loss is considered instead under the category of "dissociative disorders" even though memory is obviously also a specific brain function (see Markowitsch, 1999, 2003 for a neuropsychological perspective of psychogenic amnesia). Therefore, unlike DSM-IV, the International Classification of Diseases (ICD-10) of the World Health Organization has included all sensorimotor and memory symptoms of presumed psychogenic origin under the same general category of dissociative disorders.

From a neurological perspective, it is worth noting that hysterical conversion may sometimes coexist with a real organic brain disease, although the conversion symptoms in such cases is not directly explained by this brain lesion alone. Gowers (1893) already recognized that hysteria could occasionally be a "complication" of organic brain disease, and Schilder (1935) wrote that brain dysfunction could sometimes induce "organic neurotic attitudes". More recently, an intriguing study by Eames (1992) described a series of 167 patients who were admitted to a rehabilitation ward and reported that 32% of these patients exhibited at least one "hysteria-like" behavior during the course of their rehabilitation, as assessed by systematic rating scales used by caregivers. Such behaviors could include exaggeration, secondary gain expectancy, or non-organic patterns of the deficit. Interestingly, not all types of patients presented hysteria-like symptoms, but these behaviors were more frequent after diffuse brain lesions (e.g., closed injuries, anoxia, encephalitis) than after focal lesions (e.g., stroke), after subcortical more than cortical lesions, and in patients with extrapyramidal motor disorders more than in others (Eames, 1992). Similarly, Gould et al. (1986) reported that among a prospective series of 30 patients with an acute hemispheric stroke, "atypical signs" usually suspected to reflect non-organic "functional" origin were observed in approximately 20% of cases (e.g., changing deficit, patchy sensory loss, or "give-away" weakness). Finally, multiple sclerosis (e.g., Nicolson and Feinstein, 1994) and epilepsy (e.g., Devinsky and Gordon, 1998) constitute two other frequent neurological diseases in which not only some truly "organic" manifestations may sometimes be difficult to distinguish from non-organic disorders, but some patients may also present with a combination of apparently both "organic" and "psychogenic" manifestations at the same time. These occasional associations between hysterical conversion and neurological diseases are not only challenging for current "dichotomous" classification schemes, but in fact might potentially provide valuable clues about the neurocognitive mechanisms by which awareness of a function can be dissociated from actual abilities in a patient. However, the coexistence of these neurological diseases and conversion might also just be a coincidence, purely due to their high incidence or to any other general stress factors.

Nevertheless, despite these problems of definitions and associations, hysterical conversion is only rarely falsely diagnosed for an occult neurological condition. Although a few early studies suggested that up to 20% of patients initially diagnosed with motor conversion deficits may subsequently develop a real organic neurological disease explaining their original symptoms (Slater, 1965; Mace and Trimble, 1996), several recent studies have now clearly established that such rates were overestimated and that only 1–5% of hysterical patients may present with a underlying but occult organic cause (e.g., Crimlisk et al., 1998; Stone et al., 2003). The rarity of organic diseases detected in current follow-up studies is probably due to several factors, including a better characterization of neurological diseases and the availability of more sensitive non-invasive diagnostic procedures.

History and theories

The term of "hysteria" was forged by the ancient Greeks to indicate that physical disturbances in the uterus were the primary cause of psychic symptoms in women. By contrast, all modern theories of conversion disorders acknowledge that not only both men and women may be affected, but also a primary psychological disturbance is probably responsible for triggering subjective physical symptoms, and perhaps also for triggering some associated changes in the neurophysiological state of the central nervous system. However, the exact psychological factors to blame still remain disputed, and the possible implication of a particular brain functions in generating an abnormal physical experience still remains unresolved.

Since the 19th century, a myriad of different theories have been put forward to explain how hysterical motor or sensory deficits might be implemented in the brain in terms of specific anatomical circuits, or in relation to putative cognitive architectures. However, most of these theories rest on purely speculative grounds, and little empirical work has been conducted to test these neuropsychological hypotheses. Although it is beyond the scope of this chapter to discuss all of these theories in great detail (for more complete reviews, see Merskey, 1995; Halligan et al., 2001), a few conjectures on the neural substrates of hysteria that have been proposed by influential neuroscientists will be briefly illustrated. Although many old theories are susceptible to misinterpretations based on our current knowledge, it is interesting to note that some of them might still appear attractive if they were rephrased using more modern concepts and terminologies from current cognitive and affective neurosciences. In fact, several recent hypotheses about the possible cerebral correlates of hysterical conversion, proposed on the basis of new functional neuroimaging results, can be traced back to strikingly similar ideas that were elaborated more than a century ago, but naturally phrased in terms of models of brain physiology from that time.

One of the first and best known hypothesis on the cerebral mechanisms of hysterical conversion was proposed by Charcot in the early 1890s (see Charcot, 1892; Widlocher, 1982; White, 1997). Charcot suggested that hysterical losses in motor or sensory functions were produced by functional alterations within the central nervous system, affecting activity of motor or sensory pathways without any permanent structural damage. Hysteria was thus considered as a "neurosis" among other functional illnesses such as epilepsy or Parkinson's disease. Charcot suggested that such functional changes in the nervous system could be induced by particular ideas, suggestions, or psychological states, as demonstrated by the effect of hypnosis on hysterical symptoms. Thus, hysterical paralysis could result from an inability to form a "mental image" of movement, or instead from an abnormal "mental image" of paralysis. In essence, these ideas are echoing the more recent proposals of a selective impairment in action "representations" for intentional motor planning (e.g., Spence, 1999), although both the terminology and concepts of brain function have considerably changed since then.

At the same time as Charcot, Janet (1894) also proposed that hysterical deficits were the result of "fixed ideas" that could take control over mental or motor functions. However, he added that such ideas could arise at an unconscious level and acted by inducing a dissociation between distinct domains of behavior (one becoming dominated by the unconscious fixed idea). Such views paved the way to the classical theory of Freud and Breuer (1895), later refined by Freud alone (Freud, 1909), who formulated a purely psychodynamic account with only minimal reference to the nervous system. In brief, according to Freud and Breuer, hysterical deficits were produced by affective motives and conflicts, which were unconsciously repressed and transformed into bodily complaints with symbolic values. For this reason, hysterical deficits did not obey anatomical constraints as observed with organic neurological lesions. This view has then naturally led to the term of "conversion disorder", and in many respects is still prevailing today. Only later did Freud emphasize that the unconscious affective motives usually had their origin in sexual concerns from early childhood.

An attempt to integrate these different cerebral and psychological perspectives was made by the French neurologist Babinski in 1912, who also first described anosognosia after brain damage two years later (Babinski, 1914). Babinski believed that hysteria involved a form of suggestion permeating the subject's awareness in the same way as hypnosis, but under the influence of strong emotions, and in the presence of some individual predisposition. According to Babinski, such emotions were elicited in a purely automatic and reflexive manner, and could have both physical ("organic") and subjective ("imaginal") manifestations. Conscious control could operate only at the level of a subsequent interpretation stage. A more sophisticated neuro-anatomical scheme was later proposed by Pavlov (1928–1941, 1933), who suggested that in predisposed individuals with a somewhat weak "resistance", some over-excitation of subcortical cerebral centers (possibly mediating emotional or learned conditioned responses) could lead to a reactive inhibition of cortical inputs imposed by the frontal cortex. This inhibition was responsible for hysterical paralysis or anesthesia, whose duration was proportional to the amount of resources depleted in the individual.

A number of more recent accounts based on neurophysiological speculations have similarly proposed a role for inhibitory or "filtering" mechanisms as a likely neural substrate for generating hysterical conversion deficits. Thus, in line with Babinski and Pavlov, Sackeim et al. (1979) and Stern (1983) hypothesized that affective or motivational processes might induce a selective blockage (or distortion) of sensory and/or motor inputs, resulting in their exclusion from conscious awareness. A greater involvement of the right hemisphere in emotion might account for more frequent symptoms on the left side of the body observed in some series (Stern, 1983; see Stone et al., 2002). Other researchers such as Ludwig (1972) suggested that an inhibition of sensory or motor functions could arise through gating mechanisms at the level of thalamic nuclei, under the influence of attentional factors (see also Sierra and Berrios, 1999), whereas Spiegel (1991) instead emphasized an attentional mechanism mediated by the anterior cingulate cortex. Likewise, Kihlstrom (1994) argued that conversion disorders might involve a dysfunction in the "monitoring and controlling functions of consciousness", resulting in a dissociative state that might affect the subjective experience of perception and/or voluntary action, presumably through abnormal coordination of particular neurocognitive modules. Oakley (1999) also referred to neuropsychological models of attention to suggest that during hysteria, an internal representation of the ongoing sensory or motor activity might be excluded from awareness by a central executive attentional system involving frontal and cingulate areas. Finally, Halligan and David (1999) and Marshall and colleagues (1997) as well as Spence (1999) and Spence and colleagues (2000) also speculated that representations of motor action might be inhibited in patients with hysterical paralysis, with activation in their motor cortex being actively suppressed by abnormal signals from limbic systems within orbitofrontal and cingulate cortex during volitional movements (Marshall et al., 1997). Again, such inhibition was ascribed to unconscious motivational factors.

Finally, from the mid-1970s to mid-1980s, owing to the influence of the research in split-brain patients, a number of authors interpreted hysterical neurological symptoms as a form of interhemispheric disconnection or dysregulation syndrome (Galin et al., 1977). The greater occurrence of left hemibody disorders led to the idea that the transfer of sensory or motor inputs might be impaired between the right hemisphere (involved in emotions but without language abilities) and the left hemisphere (responsible for verbal and symbolic expression), with or without some additional impairment in the right (Stern, 1983) or left hemisphere (Flor-Henry et al., 1981). However, only very indirect evidence from performance on neuropsychological tests was offered in support of these interhemispheric hypotheses (Flor-Henry et al., 1981).

In sum, most theoretical models trying to link hysterical disorders with specific neuropsychological or neurophysiological mechanisms have essentially been inspired by speculative arguments or analogies with general models of the brain and mind, rather than by the convergence of systematic empirical research. It is striking that relatively few studies over the past decades have exploited neurophysiological techniques (see below) that provide objective measures of brain functions (such as electroencephalogram (EEG) or brain imaging), allowing a better identification of neurobiological factors associated with hysterical conversion. This is all the more surprising since the well-defined neurological-like symptoms of conversion (e.g., paralysis, anesthesia, blindness, etc.) should potentially be amenable to a precise investigation with well-defined predictions about the site and type of neurophysiological dysfunction. In particular, the advent of new functional imaging techniques should now allow a refine assessment of functional correlates in brain activity potentially associated with hysterical conversion, and thus go beyond the dichotomous question of "organic" versus "non-organic" disease. A better knowledge of such functional correlates might provide important constraints on psychodynamic theories of conversion, with greater biological and neurological plausibility, and might also improve the clinical assessment and management of patients.

Electrophysiological correlates of conversion disorders

Since EEG provided one of the first tools to measure brain activity, a number of studies from the 1960s to 1970s onward have used this technique and other related electrophysiological measures to investigate brain functions in patients with hysterical conversion. These studies can generally be considered in two broad categories: those aiming at demonstrating intact electrophysiological responses despite subjective functional losses, and those trying to determine some abnormal pattern correlating with functional symptoms.

Many reports have described that basic scalp potentials evoked during simple sensory stimulation are usually normal during hysterical sensory deficits. For instance, standard somatosensory evoked potentials (SEPs), visual evoked potentials (VEPs), or brainstem auditory evoked potentials (BAEPs) are usually found to disclose normal amplitudes and latencies in the presence of subjective anesthesia, blindness, or deafness, respectively (e.g., Howard and Dorfman, 1986; Drake, 1990). These findings clearly indicate that primary sensory pathways are both structurally and functionally intact in the patients. In the somatosensory domain, patients have also been studied using magnetoencephalography (MEG) in order to distinguish activations in primary (SI) and secondary (SII) cortical areas (Hoechstetter et al., 2002), since data from healthy people typically show a differential modulation of SII but not SI activity by attention and task-related factors. However, MEG results in patients with hysterical sensory loss also showed a normal pattern of response for the characteristic components generated in both SI and SII, controlaterally and ipsilaterally to the deficit. If anything, a trend for even greater responses was observed in the patients relative to healthy subjects, in both SI and SII (Hoechstetter et al., 2002). These data converge with earlier EEG results suggesting a paradoxical amplification rather than attenuation of tactile evoked responses during hysterical anesthesia (Moldofsky and England, 1975). Taken together, these findings do not appear consistent with the common hypothesis about conversion disorder, according to which sensory stimuli might be filtered out of awareness by attention-related gating mechanisms (see, e.g., Ludwig, 1972), at least at these relatively early stages of cortical processing. Moreover, these early SEPs may not necessarily correlate with subjective perceptual experience, since SI responses can still be elicited by unperceived stimuli in patients with brain tumors involving the parietal lobe but sparing SI (Preissl et al., 2001). Also, in patients in a vegetative state, devoid of any conscious perception, preserved SI activation has been shown using functional imaging and simultaneously recorded SEPs (Laureys et al., 2002). These data emphasize that SI activation does not necessarily mean conscious somatosensory perception.

On the other hand, a few other studies have reported subtle changes in paradigms that were slightly more sophisticated than just detection of simple tactile stimuli. For instance, tactile stimuli close to perceptual threshold may fail to produce normal evoked potentials in patients with sensory conversion symptoms, even when stimuli above threshold still produce normal responses (Levy and Mushin, 1973). In addition, anomalies in the rate of habituation to repeated stimulations were observed in hysterical conversion using SEPs (Moldofsky and England, 1975) as well as skin-conductance reactivity (Horvath et al., 1980). In normal subjects, responses were found to decrease over time when comparing late blocks of stimuli relative to initial blocks. Such habituation could also be observed in patients with high levels of anxiety (Horvath et al., 1980), but was not present in patients with hysterical conversion, indicating that the latter tended to process frequent and expected stimuli as if they were still novel.

Another recent single-case study reported anomalies in tactile evoked potentials, but involving a more cognitive stage of attentive processing, as indexed by the P300 component (Lorenz et al., 1998). This component is typically elicited by novel stimuli in an "oddball" task or by infrequent targets within a stream of successive stimuli, and presumably reflects a normal orienting response to relevant stimuli. Lorenz et al. (1998) designed an elegant EEG paradigm in which they repeatedly stimulated the unaffected left hand of a man with hysterical sensory loss on the right hand, and occasionally applied a "deviant" stimuli on either the affected right hand, or on another finger of the same unaffected left hand. The patient showed a normal P300 response for deviant stimuli on the unaffected left hand, but no P300 for deviant stimuli applied to his affected hand. Standard SEPs in this patient also revealed normal activation of early cortical areas (e.g., SI) for innocuous and painful tactile stimuli, demonstrating intact sensory pathways and preserved inputs to somatosensory cortex. Furthermore, when an healthy control subject was asked to feign anesthesia on one hand, and to intentionally omit to report the infrequent stimuli on that side (i.e., like the conversion patient), a P300 was still normally evoked by these deviants on the pseudo-anesthetized hand, indicating that a reduction of P300 in the patient was not due to malingering or control by voluntary inhibition.

Interestingly, reduced P300 responses to tactile "oddball" stimuli have also been observed in patients who present with a "segmental exclusion syndrome" (Beis et al., 1998). These patients exhibit an abnormally prolonged under-use and pain of their upper limb after suffering a relatively minor injury to peripheral body tissues in one hand or one finger, and such functional exclusion of the limb cannot be explained by the severity of injury. Although this syndrome is different from conversion disorder, it has similarly been considered as a maladaptive deficit with a partly psychogenic origin, and abnormal P300 responses were interpreted as reflecting some kind of attentional inhibition or hemineglect in motor behavior (Beis et al., 1998). Accordingly, anomalies in P300 have been found for undetected stimuli not only in patients with spatial hemineglect after right parietal lobe lesions (Lhermitte et al., 1985), but also in patients with Parkinson's disease with impairments in intentional motor planning (Kropotov and Ponomarev, 1991; Sohn et al., 1998).

Evoked potentials in other sensory modalities such as vision and audition have been less often studied beyond elementary sensory responses, perhaps because conversion disorders in such modalities are less frequent, more variable, and/or seen by physicians other than neurologists. In patients with functional blindness, a P300 was still evoked by unreported visual stimuli but with smaller amplitude (Towle et al., 1985). By contrast, a study of auditory processing in patients with hysterical deafness reported a reduced P300 response, with preservation of earlier N1 and N2 auditory responses (Fukuda et al., 1996). Some anomalies were also reported in patients with various somatization symptoms for the auditory mismatch negativity (MMN), which is normally evoked by deviant stimuli in a series of repetitive sounds (James et al., 1989). Still other studies reported more general changes in EEG spectra in patients with conversion and somatoform disorders, including abnormal ratios in frequency distribution over right and left frontal lobes (Drake et al., 1988).

Similarly, motor conversion disorders have been investigated by transcranial magnetic stimulation (TMS), typically demonstrating normal and symmetric motor evoked potentials (MEPs) when pulses were applied over the motor cortex, despite the presence of a unilateral hysterical paralysis (Meyer et al., 1992; Magistris et al., 1999). Again, such results are usually taken to indicate that motor pathways are structurally and functionally intact in these patients. Only one recent study found a decreased excitability of motor cortex in right hemisphere of two patients who had a contralateral left hysterical weakness (Foong et al., 1997b), but these cortical excitability thresholds did not change when patients recovered from their weakness. There is also anecdotal evidence that conversion patients may show an abnormal contingent negative variation (CNV) component in EEG, which is normally evoked during motor preparation in response to a cue prior to an expected to-be-judged stimulus (Drake, 1990).

Considered all together, electrophysiological data in hysterical conversion are generally consistent with the absence of organic brain pathology affecting the primary sensory or primary motor systems. Instead, any changes in brain function associated with conversion might involve higher levels of processing or representations where sensory and/or motor signals are integrated with more complex information related to the meaning and self-relevance of stimuli and actions, (e.g., motivational significance, novelty, expectedness, etc.). This might relate to the reduced P300 response found in a few different studies using different paradigms. However, EEG and MEG investigations still remain remarkably scarce, and reported findings have too rarely been replicated to allow firm conclusions about any putative neural correlates of specific conversion symptoms.

Hemodynamic brain imaging

Over the past 10 years, functional brain imaging has literally exploded into innumerable paths of new research on the cerebral bases of various behavioral functions in humans, including not only perceptual and motor processes accessible to external objective assessment, but also much more complex mental operations related to internal affective states (Damasio et al., 2000), perceptual or motor imagery (Kosslyn et al., 1995; Ehrsson et al., 2003), and even unconscious processing or preferences (Elliott and Dolan, 1998). This functional brain mapping approach has also been successfully extended to a variety of psychiatric conditions such as depression, obsessive-compulsive disorders, phobia, post-traumatic stress disorders, schizophrenia, or hallucinations (e.g., Frith and Dolan, 1998; Parsey and Mann, 2003; Kircher, this volume). Surprisingly, however, very few neuroimaging studies have been performed in patients with conversion symptoms, despite the fact that such symptoms might often be very well suited to neuroimaging investigations.

Most neuroimaging studies of conversion used SPECT (single photon emission computerized tomography) or PET (positron emission tomography), and focused on motor rather than sensory conversion symptoms. These techniques allow only a few brain scans to be taken and provide an estimate of activity averaged over several minutes, indirectly obtained by a measure of cerebral blood flow changes during a resting state or during a task period. The first of such SPECT studies was carried out by Tiihonen and colleagues (Tiihonen et al., 1995) in a woman who had a long history of left hemisensory disturbances of presumed hysterical origin, and reported both decreases in right parietal activity and increases in right frontal activity when the affected hand of the patient was stimulated (as compared with a more symmetric pattern after recovery). However, this single observation was more qualitative than truly quantitative, and not statistically analyzed. Similarly, another SPECT study reported a series of five patients with heterogeneous conversion symptoms, including not only limb weakness but also vertigo and gait disturbances (Yazici and Kostakoglu, 1998), in whom brain scans at rest showed a reduction in activity for several cortical regions, predominantly in left parietal and left temporal lobes, but with a great variability across patients.

A more systematic SPECT study was conducted in our own center, in a group of seven patients who were prospectively selected based on the presence of an isolated and "focal" motor conversion disorder, with a recent onset and short duration (< 2 months) (Vuilleumier et al., 2001). Strict selection criteria were used including: unilateral weakness in upper and lower limb, with or without sensory loss in the same territory, but without any other psychogenic or neurological symptoms (such as headache, vertigo, blurred vision, etc.), without any past history of major psychiatric or neurological disease, and without any organic lesion as determined by extensive medical investigations (i.e., brain and spine MRI, SEPs, MEPs, VEPs, EMG, etc.). These patients were followed up for 6 months after the onset of their symptoms, and underwent brain SPECT scanning in three different conditions: (1) a baseline rest condition (To), with eye closed and no stimulation, when motor symptoms were present; (2) a passive activation condition (T1), with bilateral vibrotactile stimuli (50 Hz) applied to both the affected and unaffected limbs simultaneously, when motor symptoms were present; and (3) the same activation condition (T2), again with bilateral vibrotactile stimulation of the affected and unaffected limbs, after motor symptoms had recovered (in four patients; three others had persisting or new symptoms at 6 months follow-up). The rationale of vibrotactile stimulation was to provide an indirect activation of both sensory and motor areas in the brain (since such stimuli provide inputs not only to cutaneous but also deep tendon fibers), in a completely passive, symmetric, and reproducible manner. All voxel-based analyses in this study were done on a group basis using statistical parametric mapping (SPM) (Friston et al., 1995).

A first analysis comparing activation by bilateral vibrotactile stimulation to resting baseline (T1 > T0) revealed relatively symmetric increase in cerebral blood flow in frontal and parietal areas involved in somatosensory and motor functions, both ipsilaterally and contralaterally to the motor conversion symptoms (Vuilleumier et al., 2001). This result converges with previous electrophysiological data indicating intact sensorimotor pathways in such patients. A second, more interesting analysis compared activation by bilateral vibrotactile stimulation after recovery relative to the same stimulation during symptoms (T2 > T1), providing a direct measure of changes in brain activity specifically associated with hysterical motor weakness, irrespective of any other distinctive pattern of brain function in these patients (e.g., related to depression, anxiety, or other personality characteristics). This comparison revealed selective decreases in activity of the basal ganglia and thalamus in the hemisphere contralateral to the motor deficit, when the deficit was present as compared with recovery (see Fig. 3A).

Further, the degree of decreases in caudate nucleus and thalamus at the time of symptoms (T1) was significantly correlated with the duration of conversion, i.e., activity was lower in these two regions in patients who did not recover 6 months later, relative to those who subsequently recovered (Fig. 3B). This reduced activation in contralateral basal ganglia-thalamic circuits might therefore provide a plausible substrate for the subjective motor conversion deficits. Finally, we also performed a reverse comparison of vibrotactile stimulation during symptoms relative to recovery (T1 > T2). This showed only mild increases in somatosensory cortex contralateral to the symptoms, again converging with previous electrophysiology results suggesting that sensory processing in these early cortical stages does not seem suppressed (as previously proposed: Ludwig, 1972; Sierra and Berrios, 1999) but instead appears enhanced (Moldofsky and England, 1975; Hoechstetter et al., 2002).

These selective anomalies in subcortical brain regions during motor conversion (Vuilleumier et al., 2001) are intriguing since these regions are interconnected into functional loops forming a cortico-striato-thalamo-cortical circuit which is critical for voluntary motor action (Alexander et al., 1986) and since the striatum (especially caudate nucleus) constitutes an essential neural site within such loops where motivational signals can modulate motor preparation activity (Mogenson et al., 1980; Kawagoe et al., 1998; Haber, 2003; see Fig. 4). It is therefore conceivable that these circuits might become functionally suppressed during hysterical motor conversion under the influence of a particular affective or motivational states, resulting in an impaired "motor readiness" or impaired "motor intention" for the affected limb(s). Notably, in humans, focal lesions (e.g., stroke) affecting the basal ganglia (Watson et al., 1978; Healton et al., 1982) and thalamus (Laplane et al., 1986; von Giesen et al., 1994) are implicated in a syndrome of unilateral motor neglect, in which patients present with impairments in motor use that cannot be explained by primary weakness but rather reflect a lack of motor intention or planning. Such a loss of motor intention has also been implicated in the failure of some brain-damaged patients to become aware of their (real) paralysis, i.e., anosognosia for hemiplegia (Gold et al., 1994; Vuilleumier, 2000b), suggesting that subjective experience of conscious motor action and volition might be linked to basal ganglia or thalamus function. Also in support of this, direct electric stimulation of lateral thalamic nuclei by depth electrodes may trigger contralateral movements with a subjective experience of voluntary action (Hécaen et al., 1949). Finally, changes in basal ganglia activity have also been implicated in immobilization behaviors exhibited by animals to protect an injured limb (De Ceballos et al., 1986).

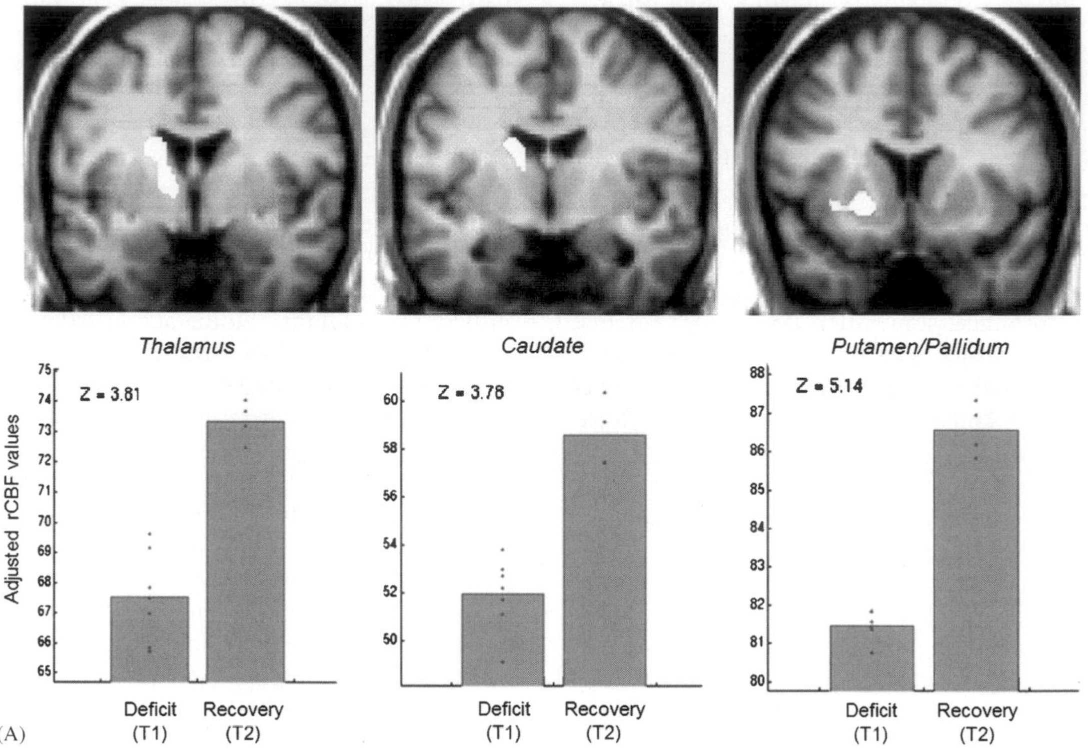

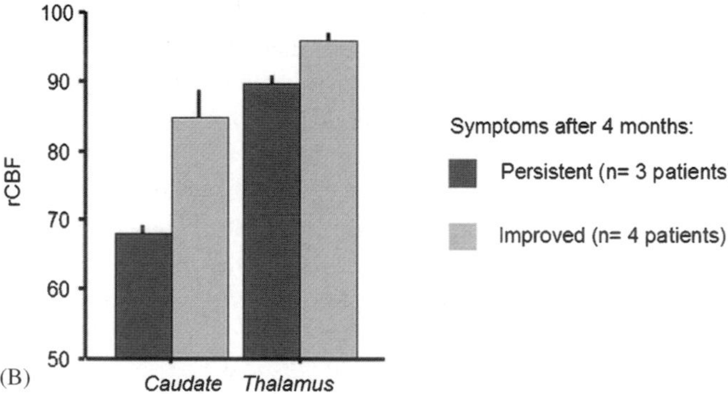

Fig. 3. Changes in brain activity associated with hysterical paralysis. (A) Selective decreases in activity were found in the thalamus, caudate, and putamen (upper row) in the hemisphere contralateral to the limb affected by sensorimotor symptoms. Measures of regional cerebral blood flow (rCBF) were obtained by SPECT scans in seven patients (lower row) during bilateral sensorimotor stimulation by vibrotactile stimuli when their symptoms were present (T1 scan), and in four patients when their symptoms had abated 3–4 months later (T2 scan). Brain activity was decreased in all these subcortical regions in all patients in T1 as compared with the same regions in T2 or with homologous regions in the hemisphere ipsilateral to the symptoms in T1. (B) Such decreases in the thalamus and caudate (but not in the putamen) were more severe in the initial scan (T1) in the three patients who had persisting symptoms at follow-up 4 months later, as compared with four other patients who had recovered, suggesting that the severity of decrease at the time of initial symptoms, may predict the duration of their symptoms. (From Vuilleumier et al., 2001).

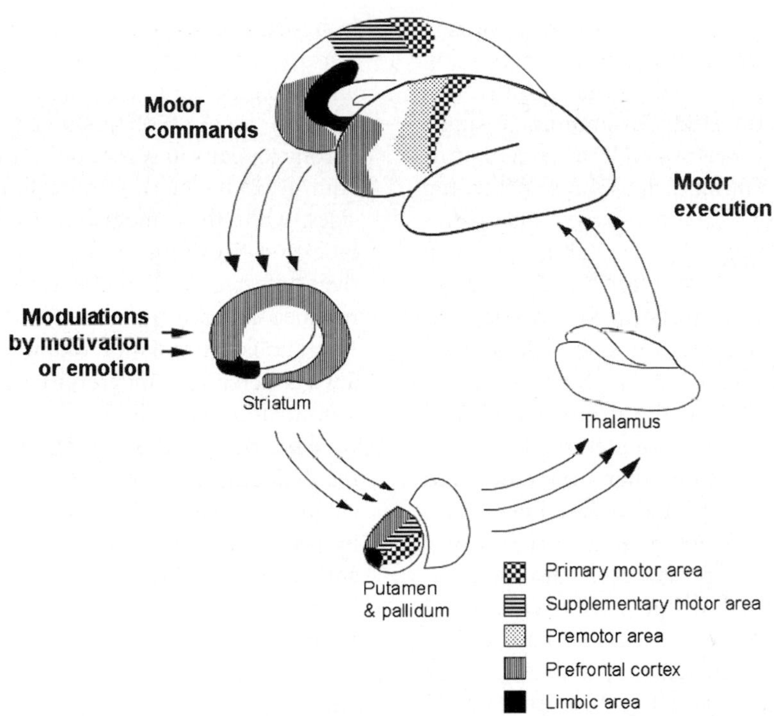

Fig. 4. Schematic illustration of cortico-subcortico-cortical loops. These circuits link various areas in frontal cortex to the caudate nucleus, putamen and pallidum, thalamus, and then back to the cortex, allowing a modulation and coordination of motor commands initiated in the cortex during movement execution, but presumably also during more complex cognitive operations. Such loops provide several neural sites, particularly in the striatum/caudate, where neural signals can be modulated by affective and motivational inputs from many other brain regions (such as orbitofrontal cortex, cingulate cortex, or amygdala), constituting a cerebral system thought to be critical for the integration of volitionally guided and emotionally triggered expressions of behavior.

Other functional changes in brain areas related to voluntary motor control have been suggested by two PET studies in patients with hysterical motor deficits (Marshall et al., 1997; see also Spence, 1999; Spence et al., 2000; for review). Marshall et al. (1997) studied a single patient with a history of unilateral left leg weakness persisting more than 2 years, in a task requiring either to prepare or to execute movements with either the left (affected) or right (unaffected) limb. Whereas motor preparation (without execution) activated brain areas in a relatively symmetric manner for both limbs, motor execution compared to preparation showed a selective activation of motor cortex for the right leg movement only, but not surprisingly, there was no such activation for the subjectively paralyzed left leg. However, in the latter condition, attempts to execute movements in the left leg produced increased activation of ventromedial frontal cortex, including right anterior cingulate and right orbitofrontal cortex, which was not seen during movement execution with the unaffected right leg. It was concluded that during initiation of motor action on the affected side, some signals might be generated in the limbic ventro–medial frontal and cingulate cortex due to affective or motivational factors, and that such signals might actively inhibit the activation of motor cortex, preventing the execution of normal movements (Konishi et al., 1999; Paus, 2001). However, it is possible that this increase in frontal and cingulate activation might reflect other processes known to implicate these regions, such as difficulty, monitoring of failure (Paus et al., 1998; van Veen et al., 2004) or conflict (Dehaene et al., 2003; Badre and Wagner, 2004) due to ambiguous task demands (i.e., "try to move even if you cannot"). In any case, a similar pattern (reduced motor activation with increased cingulate cortex activation) was found in a follow-up study examining a single healthy subject who performed the same motor preparation and execution task as in the preceding study, but now after hypnotic suggestion of unilateral paralysis (Halligan et al., 2000). Although this result suggests some similarity between hypnotic suggestion and conversion symptoms, this classic relationship (entertained since Charcot by many others: see Bliss, 1984; Spiegel, 1991; Oakley, 1999) may still not be firmly established (see, e.g., Persinger, 1994; Foong et al., 1997a).

A subsequent PET study therefore compared three patients who had hysterical weakness (two patients in left arm and one patient in right arm), with four healthy control subjects who were instructed to feign motor weakness of the right hand (Spence et al., 2000). All participants had to perform regularly paced movements with a joystick held in their affected (or pseudo-affected) hand. As a group, patients with conversion disorder showed decreased activity in left prefrontal cortex relative to the control feigners; whereas feigners showed decreased activity in right prefrontal cortex relative to conversion patients. The authors suggested that left frontal deactivation in hysteria may reflect the role of these cortical areas in motor planning (Spence, 1999). However, left frontal hypoactivity is also frequently seen in other conditions such as depression (Mayberg, 2003). All patients in this study had a history of depression, although none needed treatment at the time of scanning. Moreover, these left frontal anomalies did not provide a direct functional substrate for the contralateral motor deficit itself, since the affected side differed across the patients. But nevertheless, some impairment in internal representations of voluntary movements is likely to be present in these patients, as also suggested by purely behavioral studies that compared mental motor imagery with the affected and unaffected hands in individuals with motor conversion (Maruff and Velakoulis, 2000; Roelofs et al., 2002).

Finally, only two recent studies used functional magnetic resonance imaging (fMRI) to examine the functional patterns of brain activity associated with hysterical conversion, concerning deficits in somatosensory processing (Mailis-Gagnon et al., 2003) and visual perception (Werring et al., 2004). Mailis-Gagnon et al. (2003) studied four patients with chronic deficits in sensation of touch and/or chronic pain affecting one or more limbs on one or both sides, while they underwent fMRI scanning during blocks of brush and mildly noxious stimulation on their affected and unaffected body parts. Results revealed different patterns for different brain areas. First, unlike stimulation on intact limbs (which were always perceived and reported), both noxious and non-noxious stimuli on the affected limb (which were not perceived or not reported) failed to activate the thalamus, insular, inferior frontal, and posterior cingulate cortices. Second, some areas activated by perceived stimuli on the intact limb were apparently deactivated during stimulation on the affected limb (relative to a resting baseline without any stimulation), including parts of contralateral SI and SII, as well as bilateral prefrontal areas. Third, anterior cingulate cortex showed a greater activation during unperceived/unreported stimulation on the affected limb than during perceived stimulation on the unaffected side. Finally, several regions within prefrontal and parietal cortex (including a superior part of SI) appeared similarly responsive to unperceived and perceived stimuli, although the commonalities of such activations was not formally tested. Moreover, it is unclear whether the deactivation reported for some areas may correspond to inhibitory effects, or be more apparent than real due to greater activation in the baseline condition. In any case, these fMRI findings suggest that abnormal sensory symptoms in these patients might have partly resulted from objective neurobiological changes in a distributed somatosensory network, which the authors tentatively attributed to attentional and emotional processes triggered by mild injuries or painful conditions, and perhaps exacerbated by a preexisting vulnerability (Mailis-Gagnon et al., 2003). The complex results from this study might partly reflect the heterogeneity of patients and the fixed-effect statistical analysis used in a small sample, but it is notable that increased activity in cingulate cortex was also found in the critical condition involving stimulation of the affected limb, as previously found during an attempt to move the paralyzed limb in motor conversion or hypnosis (Marshall et al., 1997; Halligan et al., 2000). Likewise, a recent fMRI study of five patients with unexplained visual loss (Werring et al., 2004) found reduced activation in visual cortical areas during stimulation by whole-field color flickers, together with decreased (rather than increased) activation in anterior cingulate cortex. On the other hand, increased activity was found in several other regions including more posterior cingulate cortex, insula, temporal poles as well as the thalamus and striatum on both sides. Although different from previous findings for sensorimotor deficits, this pattern was again interpreted as generally consistent with the idea that conversion might involve inhibitory modulation of visual processing through increases in the activity of limbic areas.

Altogether, these new functional neuroimaging data provide compelling evidence that "functional" symptoms in patients with hysterical conversion may at least in part correspond to specific components in brain function, potentially underlying an abnormal awareness of perceptual and/or motor abilities. Here again, there are still too few studies, using different paradigms in different types of patients, such that no definite conclusion is possible. Nevertheless, converging results point to the existence of dynamic changes within several distinct brain areas, affecting activity in motor or sensory systems under the influence of higher order systems, for instance in relation to attention, emotion, or motivation factors. A new challenge for future research will be to provide a more precise mapping between these patterns of brain activation as revealed by neuroimaging techniques and the various psychodynamic dimensions (acute conflict or stress, depression, personality traits, etc.) that may be involved in conversion. In particular, further systematic studies are needed to better tease apart the neural correlates associated with subjective deficits themselves from other aspects potentially associated with coexisting disorders, such as changes in mood or anxiety, personality characteristics, expectations, attention, coping reactions as well as any activity conceivably related to unconscious affective motives and conflicts that were postulated by Freudian psychodynamic accounts.

More questions and new directions

A better understanding of functional changes in brain activity during hysterical conversion is certainly important not only because this may yield new insights into neural correlates of subjective experience and awareness, but also because this may provide new constraints on theoretical accounts of conversion, in a neurobiologically plausible framework, and therefore lead to improve the diagnosis and management of patients. There is no doubt (and perhaps no surprise) that an altered experience of bodily functions might be associated with specific modifications in the brain networks normally responsible for generating our conscious experience of such functions, as also demonstrated for even more extreme disturbances in bodily awareness (Vuilleumier et al., 1997; Blanke et al., 2004; Blanke, this volume). Demonstrating specific neural correlates of hysterical conversion may also help reassure the patients as well as their caretakers, including nurses and doctors who sometimes show negative or unsympathetic reactions when confronted with complaints unaccompanied by visible organic pathology. As already suggested by James (1896), this may eventually help convince some skeptics that hysteria is a "real disease, but a mental disease".

Importantly, neurobiological findings should lead to refine our current psychopathological explanations of conversion, by suggesting possible mechanisms by which the mind can produce changes in the brain and body functions. Such mechanisms are likely to involve dynamic interactions between neural systems mediating specific functions (e.g., motor, somatosensory, or visual processing) and neural systems responsible for affective evaluations and reactions, based on current as well as past experiences (e.g., limbic areas in the broad sense, such as cingulate gyrus, orbitofrontal cortex, or amygdala). Such interactions between distributed brain areas might be usefully investigated by neuroimaging studies using connectivity analysis (e.g., Friston, 1994; McIntosh and Gonzalez-Lima, 1994). Various statistical tools now exist to describe how neural activity in one or several regions can either influence, or instead be contingent upon, neural activity in other regions.

New methods allowing inferences about the directionality of such influences (such as dynamic causal modeling, see Friston et al., 2003) might prove particularly valuable in this context.

Thus, in our own study of hysterical paralysis (Vuilleumier et al., 2001) demonstrating reduced activation in the contralateral basal ganglia-thalamic circuits, which became normal again after recovery from paralysis, we were also able to show that such changes co-varied with concomitant changes in other distant brain areas, where activity was not globally reduced or enhanced but rather appeared differentially coupled with the basal ganglia-thalamic circuits during paralysis. This was demonstrated using a type of principal component analysis (scaled subprofile modeling; Moeller and Strother, 1991) that allowed us to identify three distinct networks of regions whose activity tended to covary together across all subjects and all sessions, irrespective of their absolute level of activity. Critically, this analysis identified a network including the caudate and thalamus, together with inferior frontal regions (areas 44/45) and orbitofrontal cortex (area 11), which was specifically coupled during paralysis in the contralateral hemisphere. Second significant network was found to include motor and sensory cortical areas (e.g., areas 1–2–3, 4–6), which were coactivated in all conditions, but less so during paralysis in the contralateral hemisphere; whereas a third significant network included anterior cingulate and temporoparietal areas (areas 39 and 40), and was mainly activated in the ipsilateral hemisphere during paralysis. Although very indirect, this network analysis suggests that hysterical paralysis did not arise in association with reduced activation of the basal ganglia-thalamic circuits alone, but this subcortical reduction was coupled with a distinct pattern of functional connectivity with inferior frontal and orbitofrontal cortex.

One hypothesis to account for this pattern is that changes in basal ganglia-thalamic circuits might implement an inhibition in motor behavior under the influence of affective or stressful signals represented in orbitofrontal or ventral frontal cortex (Mogenson et al., 1980; Kringelbach and Rolls, 2004). Such inhibitory effects might be akin to "behavioral arrest" observed in animals when subjected to threats or forced restraints (Rougeul-Buser et al., 1983; Gray, 1993; Klemm, 2001). Indeed, motor arrest and protective immobility (including behaviors such as "freezing", "sham death", or "alert state") seem to constitute fundamental and stereotyped modes of reactivity to environmental events that are perceived as stressful and hostile, where animals can sometimes adopt awkward fixed postures while waiting for termination of the unfavorable situation (Klemm, 2001). It has already been proposed that such reflexive behaviors might provide a possible primitive mechanism underlying more complex hysterical and "illness" behaviors in humans (e.g., Kretschmer, 1948; Whitlock, 1967). Hysterical conversion is typically triggered by psychological stressors, and sometimes minor physical injuries, which may be experienced as potentially more harmful in the context of a subjective lack of control and helplessness. It is therefore plausible that in some individuals (perhaps due to predisposition or past history) a kind of primitive protection or avoidance mechanisms might be induced and abnormally maintained, resulting in a pathological state of alertness and attention that can modulate their sensory or motor experience, and that is reflected in neural activity within specific brain areas such as cingulate cortex and sensorimotor networks. It is also possible that similar psychobiological responses may induce behavioral changes in other modalities (e.g., blindness, deafness, dysphonia) or in more complex cognitive domains (e.g., memory) depending on some interactions between the causal events and prior experiences in a given individual. In all such cases, these changes in neural activity often found in limbic areas of cingulate or orbitofrontal cortex might indicate a primary "core" dysfunction common to different conversion disorders, inducing secondary functional changes in other brain areas connected more directly with the symptoms. However, more work will be necessary to confirm such changes in different paradigms and to compare them with other conditions such as feigning or hypnosis. Note, however, that many conversion patients present with relatively complex and multiform symptoms, making homogenous group studies difficult, and unfortunately limiting the generalization of the findings from one study to another.

Future studies might therefore fruitfully compare patterns of brain activity in conversion patients showing different types of deficits (e.g., motor or somatosensory loss) to determine more directly what are the changes related to specific symptoms, and conversive reactions in general (for instance, testing whether some effects in anterior cingulate and frontal cortex may arise irrespective of the type of deficit). New studies should also use different possible approaches, for instance by examining brain function more systematically during and after conversion deficits in the same individuals; by manipulating more explicitly factors related to attention, expectation or inhibition; and by investigating any differences in brain reactivity to stressful probes presented to these patients even when they have recovered from specific conversion symptoms. In addition, we need to better understand the cerebral mechanisms involved in other psychiatric disturbances characterized by dissociative symptoms, such as psychogenic amnesia or "mnestic block" (Markowitsch, 2003; Glisky et al., 2004), or depersonalization disorders affecting perception of the self and/or of the environment in stressful situations (Lanius et al., 2002; Reinders et al., 2002). New imaging data may help to clarify the possible relationships of these disorders with conversion disorders and other types of psychogenic "block" affecting conscious sensory and motor functions. Interestingly, functional neuroimaging correlates of dissociative responses have also highlighted a key role of medial prefrontal areas for integrating perceptual or motor representations with a subjective sense of conscious control (Lanius et al., 2002; Reinders et al., 2003).

Conclusions

In summary, although hysterical conversion has remained a common and fascinating disorder at the border between neurology and psychiatry, few systematic and controlled studies have yet been conducted using modern neuroimaging tools. This is surprising given the variety of currently available techniques (ERPs, MEG, PET, SPECT, fMRI) and the rich productivity of imaging research in other neuropsychiatric domains. A better understanding of functional changes in brain activity during hysterical conversion will undoubtedly provide unique insights into neural mechanisms of human consciousness, but should also improve the diagnosis, management, and assessment of prognosis in these patients. Finally, further research in this field might certainly contribute to strengthen the links between psychiatry and brain sciences.

Acknowledgments

Many thanks to F. Assal, C. Chicherio, T. Landis, and S. Schwartz for their precious collaboration, and to P. Halligan for many valuable discussions. Part of this work was supported by grants from the Swiss National Foundation.

References

Alexander, G.E., DeLong, M.R. and Strick, P.L. (1986) Parallel organization of functionally segregated circuits linking basal ganglia and cortex. Ann. Rev. Neurosci., 9: 357–381.

American Psychiatric Association. (1994). Diagnostic and Statistical Manual of Mental Disorders: APA, Washington, DC.

Babinski, J. (1914) Contribution a l'étude des troubles mentaux dans l'hémiplégie organique (anosognosie). Revue Neurologique, 27: 845–848.

Badre, D. and Wagner, A.D. (2004) Selection, integration, and conflict monitoring; assessing the nature and generality of prefrontal cognitive control mechanisms. Neuron, 41: 473–487.

Beis, J.M., André, J.M., Vielh, A. and Ducrocq, X. (1998) Event-related potentials in the segmental exclusion syndrome of the upper limb. Electromyogr. Clin. Neurophysiol., 38: 247–252.

Binzer, M. and Kullgren, G. (1998) Motor conversion disorder: A prospective 2- to 5-year follow-up study. Psychosomatics, 39: 519–527.

Blanke, O., Landis, T., Spinelli, L. and Seeck, M. (2004) Out-of-body experience and autoscopy of neurological origin. Brain, 127: 243–258.

Bliss, E.L. (1984) Hysteria and hypnosis. J. Nerv. Ment. Dis., 172: 203–206.

Bridges, K.W. and Goldberg, D.P. (1988) Somatic presentations of psychiatric illness in primary care settings. J. Psychosom. Res., 32: 137–144.

Charcot, J.M. (1892) Leçons du Mardi à la Salpêtrière (1887–1888). Bureau du Progrès Médical, Paris.

Crimlisk, H.L., Bhatia, K., Cope, H., David, A., Marsden, C.D. and Ron, M.A. (1998) Slater revisited: 6 year follow up study of patients with medically unexplained motor symptoms. Br. Med. J., 316: 582–586.

Crimlisk, H.L. and Ron, M.A. (1999) Conversion hysteria: history, diagnostic issues, and clinical practice. Cognitive Neuropsychiatry, 4: 165–180.

Damasio, A.R., Grabowski, T.J., Bechara, A., Damasio, H., Ponto, L.L., Parvizi, J. and Hichwa, R.D. (2000) Subcortical and cortical brain activity during the feeling of self-generated emotions. Nat. Neurosci., 3: 1049–1056.

De Ceballos, M.L., Baker, M., Rose, S., Jenner, P. and Marsden, C.D. (1986) Do enkephalins in basal ganglia mediate a physiological motor rest mechanism? Movement Disord., 1: 223–333.

Dehaene, S., Artiges, E., Naccache, L., Martelli, C., Viard, A., Schurhoff, F., Recasens, C., Martinot, M.L., Leboyer, M. and Martinot, J.L. (2003) Conscious and subliminal conflicts in normal subjects and patients with schizophrenia: the role of the anterior cingulate. Proc. Natl. Acad. Sci. USA, 100: 13722–13727.

Devinsky, O. and Gordon, E. (1998) Epileptic seizures progressing into nonepileptic conversion seizures. Neurology, 51: 1293–1296.

Drake, M.E. (1990) Clinical utility of event-related potentials in neurology and psychiatry. Sem. Neurol., 10: 196–203.

Drake, M.J., Padamadan, H. and Pakalnis, A. (1988) EEG frequency analysis in conversion and somatoform disorder. Clin. Electroencephalogr., 19: 123–128.

Eames, P. (1992) Hysteria following brain injury. J. Neurol. Neurosurg. Psychiatry, 55: 1046–1053.

Ehrsson, H.H., Geyer, S. and Naito, E. (2003) Imagery of voluntary movement of fingers, toes, and tongue activates corresponding body-part-specific motor representations. J. Neurophysiol., 90: 3304–3316.

Elliott, R. and Dolan, R.J. (1998) Neural response during preference and memory judgments for subliminally presented stimuli: a functional neuroimaging study. J. Neurosci., 18: 4697–4704.

Faymonville, M.E., Laureys, S., Degueldre, C., DelFiore, G., Luxen, A., Franck, G., Lamy, M. and Maquet, P. (2000) Neural mechanisms of antinociceptive effects of hypnosis. Anesthesiology, 92: 1257–1267.

Flor-Henry, P., Fromm-Auch, D., Tapper, M. and Schopflocher, D. (1981) A neuropsychological study of the stable syndrome of hysteria. Biol. Psychiatry, 16: 601–616.

Folks, D.G., Ford, C.V. and Regan, W.M. (1984) Conversion symptoms in a general hospital. Psychosomatics, 25: 285–295.

Foong, J., Lucas, P.A. and Ron, M.A. (1997a) Interrogative suggestibility in patients with conversion disorders. J. Psychosom. Res., 43: 317–321.

Foong, J., Ridding, M., Cope, H., Marsden, C.D. and Ron, M.A. (1997b) Corticospinal function in conversion disorder. J. Neuropsych. Clin. N., 9: 302–303.

Ford, C.V. and Folks, D.G. (1985) Conversion disorders: an overview. Psychosomatics, 26: 371–383.

Frei, J. (1984) Hysteria: problems of definition and evolution of the symptomatology. Schweiz. Arch. Neurol. Neurochir. Psychiatr., 134: 93–129.

Freud, S. (1909) Allgemeines über den hysterischen Anfall. Deuticke, Vienna.

Freud, S. and Breuer, J. (1895) Studien über Hysterie. Deuticke, Leipzig and Vienna.

Friston, K. (1994) Functional and effective connectivity: a synthesis. Hum. Brain Mapp., 2: 56–78.

Friston, K.J., Harrison, L. and Penny, W. (2003) Dynamic causal modelling. Neuroimage, 19: 1273–1302.

Friston, K.J., Holmes, A.P., Worsley, K.J., Poline, J.B., Frith, C.D. and Frackowiak, R.S.J. (1995) Statistical parametric maps in functional imaging: a general linear approach. Hum. Brain Mapp., 2: 189–210.

Frith, C. and Dolan, R.J. (1998) Images of psychopathology. Curr. Opin. Neurobiol., 8: 259–262.

Fukuda, M., Hata, A., Niwa, S., Hiramatsu, K., Yokokoji, M., Hayashida, S., Itoh, K., Nakagome, K. and Iwanami, A. (1996) Event-related potential correlates of functional hearing loss: reduced P3 amplitude with preserved N1 and N2 components in a unilateral case. Psychiatr. Clin. Neuros., 50: 85–87.

Gagliesc, L., Schiff, B. and Taylor, A. (1995) Differential consequences of left- and right-sided chronic pain. Clin. J. Pain, 11: 201–207.

Galin, D., Diamond, R. and Braff, D. (1977) Lateralization of conversion symptoms: more frequent on the left. Am. J. Psychiatry., 134: 578–580.

Glisky, E.L., Ryan, L., Reminger, S., Hardt, O., Hayes, S.M. and Hupbach, A. (2004) A case of psychogenic fugue: I understand, aber ich verstehe nichts. Neuropsychologia, 42: 1132–1147.

Gold, M., Adair, J.C., Jacobs, D.H. and Heilman, K.M. (1994) Anosognosia for hemiplegia: an electrophysiologic investigation of the feed-forward hypothesis. Neurology, 44: 1804–1808.

Gould, R., Miller, B.L., Goldberg, M.A. and Benson, D.F. (1986) The validity of hysterical signs and symptoms. J. Nerv. Ment. Dis., 174: 593–597.

Gowers, W.R. (1893) A manual of Diseases of the Nervous System. J & A Churchill, London.

Gray, J.A. (1993) The neuropsychology of the emotions: framework for a taxonomy of psychiatric disorders. In: VanGoozen S. (Ed.), Emotions: Essays on Emotion Theory. Erlbaum, Hillsdale, NJ, pp. 29–59.

Haber, S.N. (2003) The primate basal ganglia: parallel and integrative networks. J. Chem. Neuroanat., 26: 317–330.

Halligan, P.W., Athwal, B.S., Oakley, D.A. and Frackowiak, R.S.J. (2000) Imaging hypnotic paralysis: implications for conversion hysteria. Lancet, 355: 986–987.

Halligan, P.W., Bass, C. and Marshall, J.C. (2001) Contemporary Approaches to the Study of Hysteria: Clinical and Theoretical Perspectives. Oxford University Press, Oxford, UK.

Halligan, P.W. and David, A.S. (1999) Conversion Hysteria: Towards a Cognitive Neuropsychological Account (Special Issue of the Journal Cognitive Neuropsychiatry). Psychology Press Publication, Hove, UK.

Healton, E.B., Navarro, C., Bressman, S. and Brust, J.C. (1982) Subcortical neglect. Neurology, 32: 776–778.

Hécaen, H., Talairach, J., David, M. and Dell, M.B. (1949) Coagulation limitée du thalamus dans les algies du syndrome thalamique: résultats thérapeutiques and physiologiques. Rev. Neurolog., 81: 917–931.

Hoechstetter, K., Meinck, H.M., Henningsen, P., Scherg, M. and Rupp, A. (2002) Psychogenic sensory loss: magnetic source imaging reveals normal tactile evoked activity of the human primary and secondary somatosensory cortex. Neurosci. Lett., 323: 137–140.

Horvath, T., Friedman, J. and Meares, R. (1980) Attention in hysteria: a study of janet's hypothesis by means of habituation and arousal measures. Am. J. Psychiat., 137: 217–220.

Howard, J.E. and Dorfman, L.J. (1986) Evoked potentials in hysteria and malingering. J. Clin. Neurophysiol., 3: 39–50.

James, L., Gordon, E., Kraiuhin, C. and Meares, R. (1989) Selective attention and auditory event-related potentials in somatization disorder. Comp. Psychiatry., 30: 84–89.

James, W. (1896) On Exceptional Mental States: The 1896 Lowell Lectures. Charles Scriber's Sons, New York.

Janet, P. (1894) L'état Mental des Hystériques. Rueff, Paris.

Kawagoe, R., Takikawa, Y. and Hikosaka, O. (1998) Expectation of reward modulates cognitive signals in the basal ganglia. Nat. Neurosci., 1: 411–416.

Kihlstrom, J.F. (1994) One hundred years of hysteria. In: Steven Jay Lynn E.J.W.R., et al. (Eds.), Dissociation: Clinical and theoretical perspectives. Guilford Press, New York, NY, pp. 365–394.

Klemm, W.R. (2001) Behavioral arrest: in search of the neural control system. Prog. Neurobiol., 65: 453–471.

Konishi, S., Nakajima, K., Uchida, I., Kikyo, H., Kameyama, M. and Miyashita, Y. (1999) Common inhibitory mechanism in human inferior prefrontal cortex revealed by event-related functional MRI. Brain, 122(Pt5): 981–991.

Kosslyn, S.M., Thompson, W.L., Kim, I.J. and Alpert, N.M. (1995) Topographical representations of mental images in primary visual cortex. Nature, 378: 496–498.

Kretschmer, E. (1948) Hysteria: Reflex and Instinct. Peter Owen, London.

Kringelbach, M.L. and Rolls, E.T. (2004) The functional neuroanatomy of the human orbitofrontal cortex: evidence from neuroimaging and neuropsychology. Prog. Neurobiol., 72: 341–372.

Kropotov, J.D. and Ponomarev, V.A. (1991) Subcortical neuronal correlates of component P300 in man. Electroencephalogr. Clin. Neurophysiol., 78: 40–49.

Lanius, R.A., Williamson, P.C., Boksman, K., Densmore, M., Gupta, M., Neufeld, R.W., Gati, J.S. and Menon, R.S. (2002) Brain activation during script-driven imagery induced dissociative responses in PTSD: a functional magnetic resonance imaging investigation. Biol. Psychiat., 52: 305–311.

Laplane, D., Baulac, M. and Carydakis, C. (1986) Négligence motrice d'origine thalamique. Revue Neurologique, 142: 375–379.

Laureys, S., Faymonville, M.E., Peigneux, P., Damas, P., Lambermont, B., Del Fiore, G., Degueldre, C., Aerts, J., Luxen, A., Franck, G., Lamy, M., Moonen, G. and Maquet, P. (2002) Cortical processing of noxious somatosensory stimuli in the persistent vegetative state. Neuroimage, 17: 732–741.

Levy, R. and Mushin, J. (1973) Somatosensory evoked responses in patients with hysterical anesthesia. J. Psychosom. Res., 17: 81–84.

Lhermitte, F., Turell, E., LeBrigand, D. and Chain, F. (1985) Unilateral visual neglect and wave P300. Arch. Neurol., 42: 567–573.

Lorenz, J., Kunze, K. and Bromm, B. (1998) Differentiation of conversive sensory loss and malingering by P300 in a modified oddball task. Neuroreport, 9: 187–191.

Ludwig, A.M. (1972) Hysteria: a neurobiological theory. Arch. Gen. Psychiat., 27: 771–777.

Mace, C.J. and Trimble, M.R. (1991) 'Hysteria', 'functional' or 'psychogenic'? A survey of British neurologists' preferences [see comments]. J. R. Soc. Med., 84: 471–475.

Mace, C.J. and Trimble, M.R. (1996) Ten-year prognosis of conversion disorder. Br. J. Psychiatry, 169: 282–288.

Magistris, M.R., Rosler, K.M., Truffert, A., Landis, T. and Hess, C.W. (1999) A clinical study of motor evoked potentials using a triple stimulation technique. Brain, 122(Pt2): 265–279.

Mailis-Gagnon, A., Giannoylis, I., Downar, J., Kwan, C.L., Mikulis, D.J., Crawley, A.P., Nicholson, K. and Davis, K.D. (2003) Altered central somatosensory processing in chronic pain patients with "hysterical" anesthesia. Neurology, 60: 1501–1507.

Maquet, P., Faymonville, M.E., Degueldre, C., Delfiore, G., Franck, G., Luxen, A. and Lamy, M. (1999) Functional neuroanatomy of hypnotic state. Biol. Psychia., 45: 327–333.

Marcel, A.J., Tegnér, R. and Nimmo-Smith, I. (2004) Anosognosia for plegia: Specificity, extension, partiality and disunity of unawareness. Cortex, 40(1): 19–40.

Markowitsch, H.J. (1999) Functional neuroimaging correlates of functional amnesia. Memory, 7: 561–583.

Markowitsch, H.J. (2003) Psychogenic amnesia. Neuroimage, 20(Suppl.1): S132–S138.

Marsden, C.D. (1986) Hysteria: a neurologist's view. Psychol. Med., 16: 277–288.

Marshall, J.C., Halligan, P.W., Fink, G.R., Wade, D.T. and Frackowiack, R.S.J. (1997) The functional anatomy of a hysterical paralysis. Cognition, 64: B1–B8.

Maruff, P. and Velakoulis, D. (2000) The voluntary control of motor imagery. Imagined movements in individuals with feigned motor impairment and conversion disorder. Neuropsychologia, 38: 1251–1260.

Mayberg, H.S. (2003) Positron emission tomography imaging in depression: a neural systems perspective. Neuroimaging Clin. N. Am., 13: 805–815.

McIntosh, A.R. and Gonzalez-Lima, F. (1994) Structural equation modeling and its application to network analysis in functional brain imaging. Hum. Brain Mapp., 2: 2–22.

Merskey, H. (1995) The Analysis of Hysteria: Understanding Conversion and Dissociation. Gaskell/Royal College of Psychiatrists, London, England, UK.

Meyer, B.U., Britton, T.C., Benecke, R., Bischoff, C., Machetanz, J. and Conrad, B. (1992) Motor responses evoked by magnetic brain stimulation in psychogenic limb weakness: Diagnostic value and limitations. J. Neurol., 239: 251–255.

Moeller, J.R. and Strother, S.C. (1991) A regional approach to the analysis of functional patterns in positron emission tomographic data. J. Cereb. Blood F. Met., 11-Suppl 1: 121–135.

Mogenson, G.J., Jones, D.L. and Yim, C.Y. (1980) From motivation to action: functional interface between the limbic system and the motor system. Prog. Neurobiol., 14: 69–97.

Moldofsky, H. and England, R.S. (1975) Facilitation of somatosensory average-evoked potentials in hysterical anesthesia and pain. Arch. Gen. Psychiatry, 32: 193–197.

Nicolson, R. and Feinstein, A. (1994) Conversion, dissociation, and multiple sclerosis. J. Nerv. Ment. Dis., 182: 668–669.

Oakley, D.A. (1999) Hypnosis and conversion hysteria: a unifying model. Cog. Neuropsychiatry, 4: 243–265.

Parsey, R.V. and Mann, J.J. (2003) Applications of positron emission tomography in psychiatry. Semin. Nucl. Med., 33: 129–135.

Pascuzzi, R.M. (1994) Nonphysiological (functional) unilateral motor and sensory syndromes involve the left more often than the right body. J. Nerv. Ment. Dis., 182: 118–120.

Paus, T. (2001) Primate anterior cingulate cortex: where motor control, drive and cognition interface. Nat. Rev. Neurosci., 2: 417–424.

Paus, T., Koski, L., Caramanos, Z. and Westbury, C. (1998) Regional differences in the effects of task difficulty and motor output on blood flow response in the human anterior cingulate cortex: a review of 107 PET activation studies. Neuroreport, 9: 37–47.

Pavlov, I.P. (1928–1941) Lectures on Conditioned Reflexes. International Publishers, New York.

Pavlov, J. (1933) Essai d'une interprétation physiologique de l'hystérie. Encéphale, 22: 285–293.

Perkin, G.D. (1989) An analysis of 7836 succesive new outpatient referrals. J. Neurol. Neurosurg. Ps., 52: 447–448.

Persinger, M.A. (1994) Seizure suggestibility may not be an exclusive differential indicator between psychogenic and partial complex seizures: the presence of a third factor. Seizure, 3: 215–219.

Preissl, H., Flor, H., Lutzenberger, W., Duffner, F., Freudenstein, D., Grote, E. and Birbaumer, N. (2001) Early activation of the primary somatosensory cortex without conscious awareness of somatosensory stimuli in tumor patients. Neurosci. Lett., 308: 193–196.

Prigatano, G.P., Stonnington, C.M. and Fisher, R.S. (2002) Psychological factors in the genesis and management of nonepileptic seizures: clinical observations. Epilepsy Behav., 3: 343–349.

Reinders, A.A., Nijenhuis, E.R., Paans, A.M., Korf, J., Willemsen, A.T. and den Boer, J.A. (2003) One brain, two selves. Neuroimage, 20: 2119–2125.

Reuber, M., House, A.O., Pukrop, R., Bauer, J. and Elger, C.E. (2003) Somatization, dissociation and general psychopathology in patients with psychogenic non-epileptic seizures. Epilepsy Res., 57: 159–167.

Roelofs, K., van Galen, G.P., Keijsers, G.P. and Hoogduin, C.A. (2002) Motor initiation and execution in patients with conversion paralysis. Acta. Psychol. (Amst), 110: 21–34.

Ron, M.A. (1994) Somatisation in neurological practice. J. Neurol. Neurosurg. Ps., 57: 1161–1164.

Rougeul-Buser, A., Bouyer, J.J., Montaron, M.F. and Buser, P. (1983) Patterns of activities in the ventro–basal thalamus and somatic cortex SI during behavioral immobility in the awake cat: focal waking rhythms. In: Massion J., Paillard J., Schultz W. and Wiesendanger M. (Eds.), Neural Coding of Motor Performance. Experimental Brain Research, Suppl. 7. Springer, Berlin, pp. 69–87.

Sackeim, H.A., Nordlie, J.W. and Gur, R.C. (1979) A model of hysterical and hypnotic blindness: cognition, motivation, and awareness. J. Abnorm. Psychol., 88: 474–489.

Schilder, P. (1935) The Image and Appearance of the Human Body. Kegan, Paul, Trench, Trubner and Company, London.

Sierra, M. and Berrios, G.E. (1999) Towards a neuropsychiatry of conversive hysteria. In: Halligan P.W. and David A.S. (Eds.), Conversion Hysteria: Towards a Cognitive Neuropsychological Account (Special Issue of the Journal Cognitive Neuropsychiatry). Psychology Press Publication, Hove, UK, pp. 267–287.

Singh, S.P. and Lee, A.S. (1997) Conversion disorders in Nottingham: alive, but not kicking. J. Psychosom. Res., 43: 425–430.

Slater, E. (1965) Diagnosis of hysteria. Br. Med. J., 1: 1395–1399.

Sohn, Y.H., Kim, G.W., Huh, K. and Kim, J.S. (1998) Dopaminergic influences on the P300 abnormality in Parkinson's disease. J. Neurol. Sci., 158: 83–87.

Speed, J. (1996) Behavioral management of conversion disorder: retrospective study. Arch. Phys. Med. Rehabil., 77: 147–154.

Spence, S.A. (1999) Hysterical paralyses as disorders of action. In: Halligan P.W. and David A.S. (Eds.), Conversion Hysteria: Towards a Cognitive Neuropsychological Account (Special Issue of the Journal Cognitive Neuropsychiatry). Psychology Press Publication, Hove, UK, pp. 203–226.

Spence, S.A., Crimlisk, H.L., Cope, H., Ron, M.A. and Grasby, P.M. (2000) Discrete neurophysiological correlates in prefrontal cortex during hysterical and feigned disorder of movement. Lancet, 355: 1243–1244.

Spiegel, D. (1991) Neurophysiological correlates of hypnosis and dissociation. J. Neuropsych. Clin. N., 3: 440–445.

Stern, D.B. (1983) Psychogenic somatic symptoms on the left side: review and interpretation. In: Myslobodsky M.S. (Ed.), Hemisyndromes: Psychobiology, Neurology, Psychiatry. Academic Press, New York, pp. 415–445.

Stone, J., Sharpe, M., Carson, A., Lewis, S.C., Thomas, B., Goldbeck, R. and Warlow, C.P. (2002) Are functional motor and sensory symptoms really more frequent on the left? A systematic review. J. Neurol. Neurosurg. Ps., 73: 578–581.

Stone, J., Sharpe, M., Rothwell, P.M. and Warlow, C.P. (2003) The 12 year prognosis of unilateral functional weakness and sensory disturbance. J. Neurol. Neurosurg. Ps., 74: 591–596.

Tiihonen, J., Kuikka, J., Viinamäki, H., Lehtonen, J. and Partanen, J. (1995) Altered cerebral blood flow during hysterical paresthesia. Biol. Psychiat., 37: 134–135.

Towle, V.L., Sutcliffe, E. and Sokol, S. (1985) Diagnosing functional visual deficits with the P300 component of the visual evoked potential. Arch. Ophthalmol., 103: 47–50.

Trimble, M.R. (1981) Neuropsychiatry. Wiley, Chichester.

Turgay, A. (1990) Treatment outcome for children and adolescents with conversion disorder. Canadian J. Psychiatry, 35: 585–589.

van Veen, V., Holroyd, C.B., Cohen, J.D., Stenger, V.A. and Carter, C.S. (2004) Errors without conflict: implications for performance monitoring theories of anterior cingulate cortex. Brain Cogn. Nov; 56(2): 267–276.

von Giesen, H., Schlaug, G., Steinmetz, H.R.B., Freund, H. and RJ, S. (1994) Cerebral network underlying unilateral motor neglect: evidence from positron emission tomography. J. Neurol. Sci., 125: 129–138.

Vuilleumier, P. (2000a) Anosognosia. In: Bogousslavsky J. and Cummings J.L. (Eds.), Behavior and mood disorders in focal brain lesions. Cambrige University Press, Cambridge, UK, pp. 465–519.

Vuilleumier, P. (2000b) Anosognosia. In: Bogousslavsky J. and Cummings J.L. (Eds.), Disorders of behavior and mood in focal brain lesions. Cambrige University Press, New York, pp. 465–519.

Vuilleumier, P. (2004) Anosognosia: the neurology of beliefs and uncertainties. Cortex, 40: 9–17.

Vuilleumier, P., Chichério, C., Assal, F., Schwartz, S., Slosman, D. and Landis, T. (2001) Functional neuroanatomical correlates of hysterical sensorimotor loss. Brain, 124: 1077–1090.

Vuilleumier, P., Despland, P.A., Assal, G. and Regli, F. (1997) Voyages astraux et hors du corps: heautoscopie, extase et phénomènes expérientiels dans l'épilepsie temporale. Rev. Neurolog., 153: 115–119.

Watson, R.T., Miller, B.D. and Heilman, K.M. (1978) Nonsensory neglect. Ann. Neuro., 3: 505–508.

Werring, D.J., Weston, L., Bullmore, E.T., Plant, G.T. and Ron, M.A. (2004) Functional magnetic resonance imaging of the cerebral response to visual stimulation in medically unexplained visual loss. Psychol. Med., 34: 583–589.

White, M. (1997) Jean-Martin Charcot's contributions to the interface between neurology and psychiatry. Canadian J. Neurolog. Sci., 24: 254–260.

Whitlock, F.A. (1967) The aetiology of hysteria. Acta Psychiat. Scand., 43: 144–162.

Widlocher, D. (1982) Hysteria 100 years later. Rev. Neurol. (Paris), 138: 1053–1060.

Yazici, K.M. and Kostakoglu, L. (1998) Cerebral blood flow changes in patients with conversion disorder. Psychiat. Res. Neuroim., 83: 163–168.

S. Laureys (Ed.)
Progress in Brain Research, Vol. 150
ISSN 0079-6123

CHAPTER 24

The out-of body experience: precipitating factors and neural correlates

Silvia Bünning and Olaf Blanke*

Laboratory of Cognitive Neuroscience, Ecole Polytechnique Fédérale de Lausanne (EPFL), Lausanne, Switzerland and Department of Neurology, University Hospital, Geneva, Switzerland

Abstract: Out-of-body experiences (OBEs) are defined as experiences in which a person seems to be awake and sees his body and the world from a location outside his physical body. More precisely, they can be defined by the presence of the following three phenomenological characteristics: (i) disembodiment (location of the self outside one's body); (ii) the impression of seeing the world from an elevated and distanced visuo-spatial perspective (extracorporeal, but egocentric visuo-spatial perspective); and (iii) the impression of seeing one's own body (autoscopy) from this elevated perspective. OBEs have fascinated mankind from time immemorial and are abundant in folklore, mythology, and spiritual experiences of most ancient and modern societies. Here, we review some of the classical precipitating factors of OBEs such as sleep, drug abuse, and general anesthesia as well as their neurobiology and compare them with recent findings on neurological and neurocognitive mechanisms of OBEs. The reviewed data suggest that OBEs are due to functional disintegration of lower-level multisensory processing and abnormal higher-level self-processing at the temporo-parietal junction. We argue that the experimental investigation of the interactions between these multisensory and cognitive mechanisms in OBEs and related illusions in combination with neuroimaging and behavioral techniques might further our understanding of the central mechanisms of corporal awareness and self-consciousness much as previous research about the neural bases of complex body part illusions such as phantom limbs has done.

Introduction

In an out-of-body experience (OBE) people seem to be awake and feel that their "self," or center of experience, is located outside of the physical body. It is from an elevated extracorporeal location that the subjects who undergo an OBE experience seeing their body and the world. The following example from Irwin (1985; case 1) may illustrate what subjects experience during an OBE:

> I was in bed and about to fall asleep when I had the distinct impression that "I" was at the ceiling level looking down at my body in the bed. I was very startled and frightened; immediately [afterwards] I felt that, I was consciously back in the bed again.

An OBE is defined by the presence of three phenomenological characteristics: (i) disembodiment (location of the self outside one's body); (ii) the impression of seeing the world from an elevated and distanced visuo-spatial perspective (extracorporeal, but egocentric visuo-spatial perspective); and (iii) the impression of seeing one's own body (autoscopy) from this perspective (Green, 1968; Blackmore, 1982a; Blanke et al., 2004). This is illustrated in Fig. 1. OBEs challenge the

*Corresponding author. Tel.: +41 21 6939621; Fax: +41 21 6939625; E-mail: olaf.blanke@epfl.ch

DOI: 10.1016/S0079-6123(05)50024-4

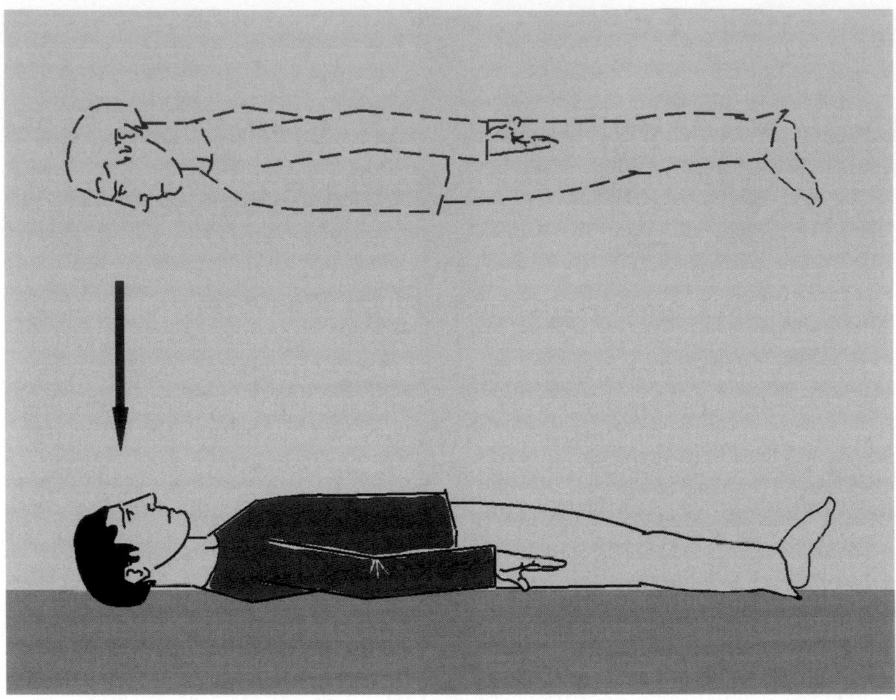

Fig. 1. Phenomenology of out-of-body experience (OBE). During an OBE the subject appears to "see" his/her body (bottom figure) and the world from a location above his physical body (extracorporeal location and visuo-spatial perspective; top figure). The self is localized outside one's physical body (disembodiment). The direction of the subject's visuo-spatial perspective during an OBE is indicated by an arrow (modified from Blanke, 2004).

experienced spatial unity of self and body or the experience of a "real me" that resides in one's body and is the subject of experience (Neisser, 1988; Gallagher, 2000; Metzinger, 2003, in press). This has also been suggested by psychologists (Palmer, 1978; Blackmore, 1982a; Irwin, 1985) and neurologists (Devinsky et al., 1989; Grüsser and Landis, 1991; Brugger et al., 1997; Blanke et al., 2004). These authors argued that OBEs are culturally invariant neuropsychological phenomena or deviant self models due to abnormal brain activation patterns whose scientific investigation might lead to a better understanding of the processes mediating the self under normal conditions.

Understanding how the brain generates the abnormal self during OBEs is particularly interesting since OBEs are not only found in clinical populations (Devinsky et al., 1989; Grüsser and Landis, 1991; Brugger et al., 1997; Blanke et al., 2004), but also appear in approximately 10% of the healthy population (Blackmore, 1982a; Irwin, 1985). Moreover, OBEs have been described in the majority of the world's cultures (Sheils, 1978). Yet, there are to date only few scientific investigations on OBEs, probably because they generally occur spontaneously, are of short duration, and happen only once or twice in a lifetime (Green, 1968; Blackmore, 1982a). This sparseness of scientific investigations is astonishing because other body illusions such as supernumerary phantom limbs or the transformation of an extremity (i.e., visual illusions of body parts) have been systematically investigated by many basic and clinical neuroscientists (see Ramachandran and Hirstein, 1998; Halligan, 2002, for review). Importantly, these latter studies have led to the description of neurophysiological and neuroanatomical mechanisms of visual body part illusions and to the development of more efficient treatments for the affected neurological patients. This is not the case for visual illusions of the entire body such as OBEs, which continue to occupy a neglected position between neuroscience and mysticism.

Several precipitating factors have been described for the OBE (Muldoon and Carrington, 1951; Crookall, 1964; Blackmore, 1982a; Irwin, 1985). Based on case collections, Muldoon and Carrington (1951) described in addition to spontaneously occurring OBEs, OBEs associated with illness, sleep, drugs, general anesthesia, accidents, and hypnosis. This was confirmed by the work of Crookall (1964) and several questionnaire studies (Blackmore, 1982a; Irwin, 1985). In the following, we will focus on several of these precipitating factors of OBEs and their brain correlates and review data in neurological and psychiatric disease, sleep and dreaming, marijuana use, general anesthesia as well as influences of body position and cognitive mechanisms.

Neurology

Only few neurological patients with OBEs due to brain damage have been reported in the last 50 years. Nevertheless, this has led to the description of etiology, associated phenomenology, and anatomy of OBEs (Devinsky et al., 1989; Brugger, 2002; Blanke et al., 2004; Blanke and Arzy, 2005). OBEs have been observed predominantly in patients with non-lesional and lesional epilepsy as well as migraine, and are generally classified with two other forms of autoscopic phenomena, autoscopic hallucination and heautoscopy. The three forms of autoscopic phenomena differ with respect to the three phenomenological characteristics of OBEs as defined above (disembodiment, visuo-spatial perspective, autoscopy). Whereas there is no disembodiment in autoscopic hallucinations and always disembodiment in OBEs, subjects with heautoscopy generally do not report clear disembodiment, but are not able to localize their self easily. Thus, in some cases the self may be localized at multiple positions including an extracorporeal position. Accordingly, the visuo-spatial perspective is body-centered in autoscopic hallucinations, extracorporeal in the OBE, and at different extracorporeal and corporeal positions in heautoscopy. The impression of seeing one's own body (autoscopy) is present in all three forms of autoscopic phenomena (for further details see Brugger et al., 1997; Blanke et al., 2004; Blanke and Arzy, 2005). Thus, during autoscopic hallucinations and heautoscopy the subject experiences seeing a double in extracorporeal space without disembodiment. Whereas during autoscopic hallucinations the double is generally experienced as an image and as an hallucination, during heautoscopy the double is often experienced as a three-dimensional person. In some heautoscopy cases the subject cannot even decide whether he or she is localized within the double (autoscopic body) or in the physical body (for further details see Blanke et al., 2004). In addition, subjects with heautoscopy often report to see in an alternating or simultaneous fashion from different visuo-spatial perspectives (physical body, double's body) as reported by patient 2B in Blanke et al. (2004).

Neurological authors have observed disturbed own body processing in patients with OBEs. Whereas Devinsky et al. (1989) observed the frequent association of vestibular sensations and OBEs, Grüsser and Landis (1991) proposed that a paroxysmal vestibular dysfunction might be an important mechanism for the generation of OBEs. In Blanke et al.'s (2004) study, the importance of vestibular mechanisms in OBEs was underlined by their presence in all patients with OBEs and by the fact that vestibular sensations were evoked by electrical stimulation in a patient at the same cortical site where higher currents induced an OBE (Blanke et al., 2002). In addition to vestibular disturbances, it has been reported that OBE-patients may also experience paroxysmal visual body-part illusions such as phantom limbs, supernumerary phantom limbs, and illusory limb transformations either during the OBE or during other periods related to epilepsy or migraine (Hécaen and Ajuriaguerra, 1952; Lunn, 1970; Devinsky et al., 1989; Blanke et al., 2002, 2004). Blanke et al. (2002) reported a patient in whom OBEs and visual body-part illusions were induced by electrical stimulation at the right temporo-parietal junction (TPJ). In this patient an OBE was induced repetitively by electrical stimulation whenever the patient looked straight ahead (without fixation of any specific object). If she fixated her arms or legs that were stretched out, she had the impression that the inspected body part was transformed leading to the illusory, but very realistic, visual perception of limb shortening (only the left arm, but both legs). If she fixated her arms or legs that were bent at the elbow or the knees she reported an illusory limb movement (only the left arm, but both legs) toward her face. Finally, with closed eyes the patient did not have either an OBE or a visual body-part illusion, but perceived her upper body as moving toward her legs (Blanke et al., 2002). These data suggest that visual illusions of body parts and visual illusions of the entire body such as autoscopic phenomena might depend on similar functional and anatomical mechanisms as argued by previous authors (Hécaen and Ajuriaguerra, 1952; Brugger et al., 1997).

Daly (1958), Lunn (1970), Devinsky et al. (1989), Blanke et al. (2004), and Maillard et al. (2004) have described OBE patients with circumscribed brain damage. These rare neurological cases allow to investigate whether certain brain areas or a hemisphere are linked to OBEs. In this patient sample, OBEs were due to epileptic seizures in the large majority of cases (82%). Brain damage was due to neoplasia, posttraumatic brain injury, arteriovenous malformation, or a dysembryoblastic neuroepithelial tumor (accounting for 56% of the cases). Recently, Blanke et al. (2004; cases 1, 2a, and 3) performed lesion analysis based on magnetic resonance imaging (MRI) and showed involvement of the TPJ in all three of their OBE patients (Fig. 2). This was also found in the OBE-patient by Maillard et al. (2004; case 1). Moreover, Blanke et al. (2002) have shown that OBEs can be induced by electrical stimulation of the TPJ pointing to the importance of this region in the generation of OBEs. A recent review of all previously reported OBE-cases of focal neurological origin found that OBEs were related in 80% to right hemispheric brain damage (Blanke and Arzy, 2005). A lesion overlap analysis of all reported OBE-patients with focal brain damage or focal electroencephalographic (EEG) abnormalities is illustrated in Fig. 2.

Based on these neurological findings, Blanke et al. (2004) have recently proposed a model of OBEs that extended models by previous authors (Blackmore, 1982a; Irwin, 1985; Grüsser and Landis, 1991; Brugger, 2002) and is also compatible with theoretical accounts for the affected body part in supernumerary phantom limbs (Ramachandran and Hirstein, 1998). Blanke et al. (2004) suggested that during an OBE the integration of proprioceptive, tactile, and visual information of one's body fails due to discrepant central representations by the different sensory systems. This may lead to the experience of seeing one's body (autoscopy) in a position (i.e., on a bed) that does not coincide with the felt position of one's body (i.e., under the ceiling; see above quoted example from Irwin, 1985). As OBEs are also characterized by disembodiment and elevated visuo-spatial perspective, Blanke et al. (2004) speculated that an additional vestibular dysfunction is present in OBEs. They suggested that OBEs are related to an integration failure of proprioceptive, tactile, and visual information with respect to one's own body (disintegration in personal space) and to a vestibular dysfunction leading to an additional disintegration between corporeal (vestibular) space and extracorporeal (visual) space. Both disintegrations were proposed to be necessary for the occurrence of an OBE. The neurological data also suggest that OBEs are due to a paroxysmal cerebral dysfunction of the TPJ in a state of partially impaired consciousness or awareness due to epilepsy or migraine.

Psychiatry

OBEs, autoscopic hallucinations and heautoscopy have been reported in psychiatric patients suffering from schizophrenia, depression, and personality disorders (Menninger-Lerchenthal, 1935; Lhermitte, 1939; Hécaen and Ajuriaguerra, 1952; Devinsky et al., 1989; Brugger et al., 1997; Blanke et al., 2004). Interestingly, we were only able to find two single case descriptions of psychiatric patients with OBEs, while the large majority experienced autoscopic hallucinations or heautoscopy. The 20 psychiatric patients that we analyzed suffered mainly from schizophrenia or depression and are listed in Table 1. We could thus only find one diagnosed psychiatric OBE-case due to severe

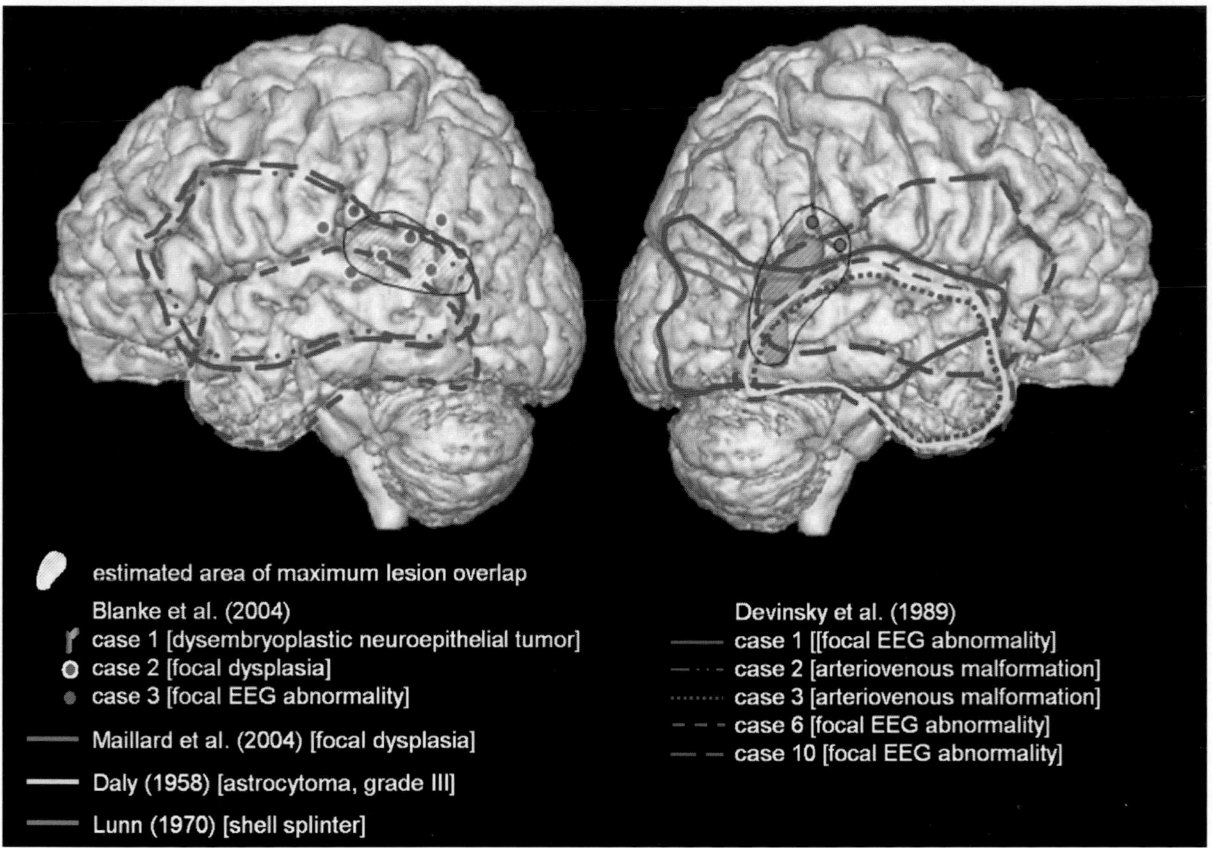

Fig. 2. Lesion location in OBE patients. MRI-based lesion overlap analysis of all previously reported OBE patients due to focal brain damage. The data for all patients are drawn on the MRI of one of the patients from Blanke et al. (2004). Lesion data are plotted separately for the left (left figure) and right hemisphere (right figure). For each study a different color was chosen: Daly (1958; yellow; 1 patient), Lunn (1970; green; 1 patient), Devinsky et al. (1989; blue; 5 patients), Blanke et al. (2004; pink; 3 patients), and Maillard et al. (2004; red; 1 patient). For the three patients by Blanke et al. (2004) the pink dots surrounded by white represent the zone of seizure onset for case 2, the pink dots surrounded by black represent the zone of seizure onset for case 3, and the pink, shaded, area represents the area of lesion overlap for case 1. Lesion location is indicated by cortical area surrounded by a line for each patient. The lesion sites of all patients by Blanke et al. (2004) were transformed to Talaraich space and transformed to the coordinates of the MRI of one of the patients. Thus, when the lesion location (or the epileptic focus) was described as right temporal, we marked the whole temporal lobe for that patient, if the lesion was characterized as left fronto-temporal parts of the fronto-temporal lobe were marked. The two patients with bilateral EEG abnormalities from Devinsky et al. (1989; cases 6 and 10) have been plotted on the right and left hemisphere. The regions of maximal overlap are indicated by the hatched area for each hemisphere and were found on the right and left TPJ. Most of the OBEs were due to interference with the right TPJ. See Blanke et al. (2004) for further details. See Plate 24.2 in Colour Plate Section.

depression (Hécaen and Green, 1957). In another psychiatric patient a full-blown OBE was reported (Zutt, 1953), but no psychiatric diagnosis was given. Thus, it seems that almost all of the previously reported psychiatric patients with autoscopic phenomena suffered from autoscopic hallucinations or heautoscopy. This is somewhat astonishing since it could be expected from the above-mentioned neurological findings that OBEs also exist more frequently in psychiatric patients, especially since OBEs are related functionally and anatomically to autoscopic hallucinations and heautoscopy (see above). Other authors have investigated OBEs in psychiatric patients, but did not report much phenomenological detail with respect to the OBEs of their patients other than stating that the phenomenology of a typical OBE in schizophrenic patients did not differ from the phenomenology of

Table 1. Case reports of autoscopic phenomena of psychiatric origin. Twenty patients with autoscopic phenomenon due to psychiatric illness are listed. Only patients with sufficient phenomenological detail in order to classify the autoscopic phenomenon with respect to autoscopic hallucination (AH), heautoscopy (HAS), and OBE were included. Only two cases of OBE were found. AH and HAS were equally distributed

	Psychiatric disorder			Autoscopic phenomena		
Case (Year)	Depression	Schizophrenia	Other	AH	HAS	OBE
Aizenberg and Modai (1985)	x			x		
Alonso–Fernandez (reported in Damas Mora et al. (1972))		x			x	
Arenz (2001)	x				x	
Craske and Sacks (1969)	x				x	
Damas Mora et al. (1980)		x		x		
Dening and Berrios (1994) (case 3)	x			x		
Hécaen and Green (1957) (case 3)	x					x
Letailleur et al. (1958)		x			x	
	x			x		
	x			x		
Lukianowicz (1958) (cases A, C, D, G)		x		x		
		x		x		
Maack and Mullen (1983)		x		x		
McConnel (1965)	x				x	
McCreery and Claridge (2002)	x				x	
Ostrow (1960) (cases A, B)	x				x	
	x			x		
Salama (1981)		x			x	
Schmitt (1948) (case 5)		x			x	
Zutt (1953)			x			x

OBEs in healthy subjects (Blackmore, 1986; Röhricht and Priebe, 1997). Moreover, Blackmore (1986) and Röhricht and Priebe (1997) investigated the prevalence of OBEs in schizophrenic and healthy control subjects and found no evidence for a higher prevalence of OBEs in patients suffering from schizophrenia (if compared with healthy control populations). Although, hallucinations such as auditory verbal hallucinations were more frequent (intra-individually) in schizophrenic patients, this was not the case for OBEs (Blackmore, 1986; Röhricht and Priebe, 1997). McCreery (1993) also found no relationship between the occurrence of OBEs and psychiatric illness. Evidence for a relation between psychiatric illness and OBEs was provided by studies investigating the presence of OBEs in healthy subjects as classified by several schizotypy scales. Schizotypy refers to a personality trait in healthy subjects that is characterized by aberrant perceptions and beliefs, cognitive disorganization, introvertive anhedonia, and asocial behavior (Meehl, 1962; McCreery and Claridge, 1996b). Thus, McCreery and Claridge (1995, 2002) found that healthy subjects with OBE scored higher on a measure of schizotypy ("Aberrant Perceptions and Beliefs") than subjects without OBEs. This was confirmed by Wolfradt and Watzke (1999), who also showed that healthy subjects with OBEs scored higher on a different schizotypy measure (including disorganization, cognitive-perceptual deficits, depersonalization) than subjects without OBEs. It is thus at this point not possible to analyze the phenomenology of OBE in schizophrenic patients and compare them with the phenomenology of neurological patients and healthy subjects. Finally, Reynolds and Brewin (1999) showed that the frequency of OBEs in a group of patients suffering from posttraumatic stress disorder was 42% and thus characterized by a fourfold increase with respect to the prevalence in healthy subjects.

In conclusion, these data suggest that OBEs occur in psychiatric conditions such as schizophrenia, depression, and posttraumatic stress disorder.

The presence of autoscopic hallucinations and heautoscopy in psychiatric patients, the increased presence of OBEs in healthy subjects with schizotypy (McCreery and Claridge, 1995, 2002; Wolfradt and Watzke, 1999), depression and posttraumatic stress disorder (Reynolds and Brewin, 1999) suggests that OBEs may also be more frequent in schizophrenic patients than is actually thought. However, we think that further research needs to be conducted in psychiatric patients with OBEs by obtaining precise phenomenological information about the content of the OBE, about the associated illusions and hallucinations (with special reference to vestibular sensations and body schema disturbances) as well as a detailed neuropsychological examination.

Body position

Most techniques that are used to voluntarily induce OBEs propose that the subjects use a supine and relaxed position (Blackmore, 1982a; Irwin, 1985), suggesting an influence of proprioceptive and tactile mechanisms on OBEs. Interestingly, a supine body position was also reported by all neurological patients with OBEs of Blanke et al. (2004) and in 75% of healthy subjects with OBEs that were reported by Green (1968). On the contrary, patients with autoscopy or heautoscopy in the study by Blanke et al. (2004) were either in a sitting or standing position. This confirms results by Dening and Berrios (1994) about the body position during autoscopy, who reviewed a large number of patients with autoscopy and heautoscopy. Thus, OBEs seem to depend differently on the subject's position prior to the experience than the other types of autoscopic phenomena suggesting that proprioceptive and tactile mechanisms influence both phenomena differently. Interestingly, for two other main precipitating factors of OBEs—general anesthesia and sleep—subjects are also in supine position corroborating the observation that OBEs are facilitated by the somatosensory coding of a supine body position.

OBEs also seem to occur frequently during rapid bodily position changes such as rapid whole body accelerations and decelerations. OBEs have been reported in these latter situations for a long time such as in falls of mountain climbers (Heim, 1892; Ravenhill, 1913; Habeler, 1979; Brugger et al., 1999; Firth and Bolay, 2004), in car accidents (Muldoon and Carrington, 1929; Devinsky et al., 1989), and more recently also by airplane pilots (Benson, 1999). Thus, the Swiss geologist Albert Heim, who was an amateur mountain climber and who nearly lost his life more than once in mountain climbing accidents, started to collect systematically the accounts from fellow mountain climbers who have also been victims of life-threatening falls (quoted after Ring, 1980, p. 21). In addition, Brugger et al. (1999) have shown that extreme altitude climbing is accompanied by a high incidence of hallucinatory experiences. These authors reported that 75% of highly specialized extreme-altitude climbers experienced OBEs (especially if climbing above 6000 m altitude), of which 63% occurred during acute life-threatening situations (see also Habeler, 1979). Firth and Bolay (2004) reported a case of a mountain climber who experienced an OBE at 6700 m altitude:

> [...] Descending the mountain, he felt as if the person were following him and conducted a mental conversation with his apparent companion. While walking, he felt as if his legs were moving of their own accord and his torso was elongated. He felt detached from his person, as if he were observing himself from a distance. He was aware of the hallucinatory nature of his experiences. These anomalies lasted approximately 10 min before disappearing spontaneously [...]

In addition, Brugger et al. (1999) have shown that high-altitude mountain climbing is also associated with reversible neuropsychological deficits, probably due to a number of factors such as cerebral hypoxia, physical exhaustion, hypothermia, food deprivation, dehydration, and sleep deprivation (see below). As stated above, OBEs have also been reported in aviation. Thus, a pilot might fail to sense correctly the position, motion or tilt of the aircraft as well as his own body position with respect to the surface of the earth and the gravitational "earth-vertical" (Benson, 1999). It is assumed that this is caused by a vestibular disturbance or failure to signal the extreme motions during aviation. A particular experience reported in these instances is called the "break-off phenomenon" and is highly similar to OBEs. The break-off phenomenon is a cluster of experiences and characterized by feelings of physical separation from the earth, lightness, and an altered perception of the pilot's own orientation with respect to the ground and the aircraft (Clark and Graybiel, 1957; Sours, 1965; Benson, 1999). Some pilots have described a feeling of detachment, isolation, and remoteness from their immediate surroundings, which are sometimes described as an OBE with disembodiment, elevated visuo-spatial perspective, and autoscopy (Tormes and Guedry, 1975; Benson, 1999). Thus, the break-off phenomenon has been described as a feeling of being isolated, detached, or separated physically from the world and the body. Other pilots have described the feeling of being all of a sudden outside the aircraft or outside themselves watching themselves while flying the aircraft, and of being "broken off from reality" (Benson, 1999). These OBEs are most often experienced by jet aviators who are flying alone especially at high altitudes (above 10,000 m), although helicopter pilots can experience this phenomenon already at an altitude of 1500–3000 m (Tormes and Guedry, 1975; Benson, 1999). Finally, the break-off phenomenon is mostly encountered when pilots are relatively unoccupied with flight details (i.e., absorption; Irwin, 1985) and when they are in long cross-country missions over featureless terrain (Clark and Graybiel, 1957).

In conclusion, static body position and dynamic body position changes seem to influence the occurrence of OBEs. Whereas under static conditions, the supine position seems to favor the occurrence of OBEs, vestibular imbalances during falls or rapid decelerations in mountain climbers, jet aviators, or car drivers seem to increase the frequency of OBEs or OBE-like experiences.

Sleep

The functional relation between sleep (and dreaming) with OBEs has been investigated and discussed by several authors (Muldoon and Carrington, 1951; Blackmore, 1982b, pp. 313–315; Rogo, 1982, pp. 95, 130–149; Irwin, 1985, pp. 16–21, 257–263). Yet, despite these studies, there has not been much work about the neurobiological mechanisms that might link OBEs with sleep and dreaming. This is probably due to the methodology that was used. First, only questionnaires and self-reports were evaluated. Second, this was done often several months or even years after the occurrence of the analyzed OBE. Third, a systematic analysis of OBEs in healthy subjects with respect to sleep stages as defined by EEG recordings has, to our knowledge, not yet been carried out (see below). However, in the following, we have summarized data from studies that have reported correlations between OBEs and sleep, OBEs and dreaming, and OBEs and EEG recordings. This is complemented by the discussion of more recent neuroimaging studies on sleep and dreaming.

With respect to sleep, the aforementioned studies suggested that OBEs are especially frequent in the periods before falling asleep (hypnagogic), resting, or dozing and before waking up (hypnopompic) (Monroe, 1974; Sheils, 1978; Palmer, 1979; Twemlow et al., 1980; Twemlow et al. 1982; Gabbard & Twemlow 1984; Rogo, 1986; LaBerge, 1990; Alvarado, 2000). In more detail, Greenhouse (1975, p. 91) alleges that most OBEs occur during sleep. In a sample of 122 healthy subjects, Poynton (1975) reports the prevalence of OBEs that occur while "sleeping" at 27%, while "dozing" at 20%, and while "relaxing" at 16%. This was also found by Green (1968), who reported that in 24% of the subjects with multiple OBEs, at least one OBE started during sleep. Girard and Cheyne (2004) also have reported (in a group of healthy subjects) that OBEs may occur as hypnagogic and hypnopompic hallucinations (among many other experiences). Other authors have tried to provoke OBEs by inducing a hypnagogic-like state (Muldoon and Carrington, 1929; Mitchell, 1973; Monroe, 1974; Baker, 1977; Rogo, 1986). More experimentally, Palmer (1979) has reported that his technique of "progressive muscle relaxation" in combination with the instructions "to relax and to project consciousness to "a place outside one's body" might lead to the occurrence of OBE-like experiences in 50% of his sample of 180 healthy subjects (in comparison to a prevalence of only 10% under normal conditions). Yet, with respect to all aforementioned studies, it needs to be stressed that the data were mainly collected retrospectively, that the phenomenology of the reported experiences did only rarely correspond to a full-blown OBE (complying with all three variables as described above), and that no EEG data were available. Based on these studies it cannot be stated precisely which of the many sleep variables or which sleep stages are precipitating factors of OBEs. We describe in the following, several recent neuroimaging studies that have investigated the neural correlates of pre-sleep wakefulness (or resting state; Andreasen et al., 1995).

For this we selected several studies that have compared the brain activation between pre-sleep wakefulness with either sleep or cognitive tasks. Thus, Maquet et al. (1996, 1997) compared pre-sleep wakefulness with slow-wave sleep (SWS, stages 3 and 4) and rapid-eye movement (REM sleep) by positron emission tomography (PET) and found that the most active cortical brain regions were located in the prefrontal, anterior cingulate, and parietal and precuneal cortices (Fig. 3; adapted from Maquet, 2000; p. 215). Similar results were obtained by Andreasen et al. (1995). They compared a resting state with a memory task and observed that the brain regions that were more active in the resting state were located in the prefrontal and parietal cortex. Both

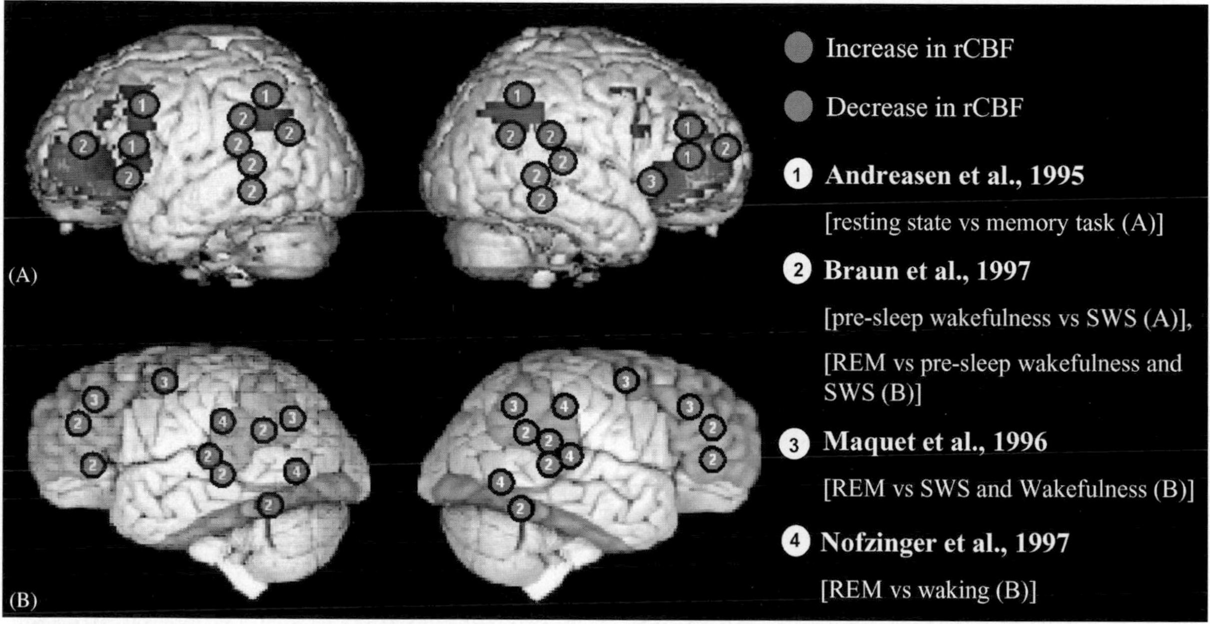

Fig. 3. Regional cerebral blood flow during pre-sleep wakefulness, resting state, and during REM sleep. (A) The figure shows brain activation patterns during pre-sleep wakefulness and resting state compared to sleep (see text). The results of several studies are shown (1, Andreasen et al., 1995; 2, Braun et al., 1997). Estimated peak of activation (red) and deactivation (blue) are plotted on a figure reproduced with permission from Maquet (2000) showing activated brain areas during pre-sleep wakefulness as compared to sleep. The main voxels of activation on the lateral cortical surface are shown. Note the focalization on two lateral cortical areas: prefrontal cortex and TPJ. Whereas the prefrontal cortex is activated during pre-sleep wakefulness, the TPJ is characterized by an activation of the inferior parietal lobule and deactivation of the posterior parts of the superior, middle, and inferior temporal gyri. (B) Schematic representation of the relative increases and decreases in neural activity associated with REM sleep. The figure is modified and reproduced with permission from Schwartz and Maquet (2002). The results of several studies (2, Braun et al., 1997; 3, Maquet et al., 1996; 4, Nofzinger et al., 1997) are plotted. The main voxels of activation on the lateral cortical surface are shown. Red color indicates increases and blue color indicates decreases in rCBF. Brain deactivations focus on the lateral prefrontal cortex and the TPJ. Again, the TPJ shows patterns of activation and deactivation in close proximity. (See text for more details). See Plate 24.3 in Colour Plate Section.

studies reported equal levels and locations of activations in the right and left hemisphere. All activations were found bilaterally except for prefrontal cortex that was only found in the left hemisphere. Braun et al. (1997) showed that pre-sleep wakefulness as compared to SWS is associated with increases in brain activation in the angular and supramarginal gyri, as well as in prefrontal cortex laterally and medially. Decreases in activation were observed in the posterior lateral temporal lobe (superior, middle, and inferior temporal gyri), the fusiform gyrus as well as striate cortex and the lateral occipital cortex. These findings are summarized in Fig. 3A. Inspection reveals that brain activations focus on two lateral regions, the prefrontal cortex and the TPJ. In addition, brain activation at the TPJ showed patterns of activation and deactivation in close proximity.

With respect to dreaming, several authors have pointed out the connection between OBEs and dreaming (Poynton, 1975; Osis and Mitchell, 1977; Sheils, 1978; Palmer, 1979; Twemlow et al., 1982). Thus, Twemlow et al. (1982) reported that 35% of subjects who were asked about their mental state at the time of the OBE reported that they were dreaming. Palmer (1979) found a highly significant correlation between having an OBE and having vivid dreams. Finally, Sheils (1978) showed that in 32% of all cultures that were investigated (from 67 different ethnic groups), OBEs are frequently and habitually connected to dreaming. Again, no EEG data were reported in these studies. Although dreaming also occurs during non-REM sleep, dreams during REM sleep tend to be longer, more easily elicited, more vivid, more emotional, and more coherent (Aserinsky and Kleitman, 1953; Rechtschaffen and Siegel, 2000; Nielsen, 2003). In the following we will thus review some recent studies on the functional neuroanatomy of REM sleep (and dreaming) if compared with SWS or wakefulness. During REM sleep, Maquet et al. (1996, 1997) found significant and widespread cortical activations when compared to SWS. These authors used PET and described REM sleep-related activations in limbic areas (amygdaloid complexes, hippocampal formation, and anterior cingulate cortex) and the cortex (bilateral striate cortex, temporo-occipital cortex, extrastriate cortex, motor and premotor as well as superior parietal lobule). In contrast, the dorsolateral prefrontal cortex, inferior parietal lobule (supramarginal gyrus) as well as posterior cingulate cortex and precuneus were deactivated. The findings of Nofzinger et al. (1997) corroborate these results (REM sleep versus waking). Braun et al. (1997) found more focal REM sleep activations and compared REM sleep with SWS and pre-sleep wakefulness. They found activations of extrastriate cortex, particularly within the ventral processing stream, and an attenuation of activity in the striate cortex. In addition, increases in regional cerebral blood flow (rCBF) in extrastriate areas were significantly correlated with decreases in the striate cortex. Extrastriate activity was also associated with concomitant activation of limbic and paralimbic regions, but with a marked reduction of activity in frontal association areas including lateral orbital and dorsolateral prefrontal cortices. These findings have been reviewed by Schwartz and Maquet (2002) and are summarized graphically in Fig. 3B. Note that brain activations again focus on the lateral prefrontal cortex and the TPJ. Again, the TPJ showed patterns of activation and deactivation in close proximity. Also note that these data suggest that the prefrontal cortex was mainly activated during pre-sleep wakefulness (Fig. 3A), whereas it was mainly deactivated during REM sleep.

With respect to the relationship between OBEs and sleep or dreaming, we noted that none of the aforementioned questionnaire or self-report studies recorded EEG variables in their subjects neither between nor during OBEs. Yet, the EEG is necessary in order to define the different sleep stages (Rechtschaffen and Siegel, 2000). Furthermore, with respect to EEG recordings during OBEs several methodological difficulties exist. These are due to the nature of OBEs, which are usually short lasting and occur spontaneously. In addition, in the large majority of subjects they happen only once or twice in a lifetime (Blackmore, 1982a; Irwin, 1985). It is thus extremely difficult to have the opportunity to record an EEG during an OBE. Finally, there are methodological problems as to the temporal definition of the onset and offset of an OBE and how subjects are to indicate this to the investigator in order to interfere with the mental state of the investigated subject as least as possible. Nevertheless, several authors have tried to record EEGs during OBEs in several subjects that were known to experience OBEs frequently and claim to be able to induce OBEs voluntarily (OBE adepts).

Tart (1967, 1968) was the first to investigate EEG activity during OBEs. Although in his first study, Tart (1967) was able to record EEG during three OBEs, these OBEs did not reflect the subject's habitual OBEs. In addition, the patient reported sensations of floating in and out of her body (first and third OBE), sensations of flying and experiential phenomena (second OBE), but no autoscopy (all OBEs), and did not mention an elevated visuo-spatial perspective explicitly (first two OBEs). OBEs occurred during an EEG pattern between sleep and wakefulness characterized by a poorly developed "unusual" alpha-like pattern (slower than the normal rhythm) associated additionally with EEG flattening, low voltage spindles, and transitory periods of wakefulness. In his second study, Tart (1968) recorded the EEG during two full-blown OBEs of the OBE-adept Robert Monroe (Monroe, 1974). No significant abnormality was seen in Monroe's standard and sleep EEG and the investigated OBEs were reported to be associated in the EEG with REM sleep followed by awakening and body movements. Janis et al. (1973) recorded EEG in the OBE-adept Blue Harary and found no EEG differences during the OBE period if compared to OBE-free periods. No phenomenological detail about the OBE was given. Hartwell et al. (1974) further investigated the EEG recorded from Blue Harary (Blackmore, 1982a). They analyzed alpha density and EEG frequency, but again, no differences were found between OBE and OBE-free periods. The standard EEG of Blue Harary during the OBE was also described as normal. No phenomenological detail about the OBE was given. Osis and Mitchell (1977) recorded the EEG in the OBE-adept Ingo Swann (Blackmore, 1982a). The authors describe that the OBE was associated with a diminished EEG amplitude and that, in general, the duration of alpha was diminished in Ingo Swann's EEG recording. No phenomenological detail about the OBE was given. Twemlow (1977) recorded Robert Monroe's EEG during one full-blown OBE and reported a general slowing down of the EEG during the OBE stating that they observed no EEG rhythms faster than 10 Hz during the OBE period. Finally, they mentioned EEG right–left asymmetries during the OBE.

In conclusion, these studies did not allow to describe a specific EEG pattern or sleep stage in association with OBEs in this group of OBE adepts (see also Irwin, 1985). Flattening, faster EEG rhythms, as well as slowing, and sub-alpha EEG patterns, right–left asymmetries, or no EEG differences between OBE- and OBE-free periods were described. OBEs occurred either during sleep or wakefulness. In addition, it should be mentioned that most of these studies were done quite some time ago and recorded only with few electrodes allowing to define sleep stages, but not the recording and determination of EEG patterns, their localization, and lateralization with enough spatial precision. Nevertheless, these studies point to the feasibility of a scientific approach in order to investigate the association of OBEs with certain EEG measures and sleep. In our opinion, future studies should ideally investigate subjects with full-blown OBEs and use EEG recordings with at least 32 scalp electrodes to allow lateralization and localization of EEG patterns or even the localization of their brain generators (Michel et al., 2004). Also, systematic approaches need to be used in order to determine as unambiguously as possible the onset and offset of the OBE. Finally, it might be possible to carry out experimental studies in normal subjects in whom OBEs can be more systematically induced by a given procedure allowing to perform multichannel EEG recordings in larger subject groups as pioneered by Palmer (1979) and McCreery and Claridge (1996a).

Drugs

Drug abuse including marijuana, opium, heroin, mescaline or lysergic acid diethylamide (LSD), has been reported to be associated with a higher prevalence of autoscopic phenomena such as OBEs (i.e., Tart, 1971; Blackmore, 1982a; Aizenberg and

Modai, 1985; Grüsser and Landis, 1991; Shermer, 1998). In addition, hallucinogens have long been used intentionally in various cultures to induce OBE-like states (Sheils, 1978). In an exemplary fashion, for the present review, we will focus on marijuana (δ-9-tetrahydrocannabinol) and its relation to OBEs and its neural mechanisms.

With respect to marijuana, Tart (1971) reported an OBE prevalence of 44% among people with varying abuse of marijuana, whereas OBEs only occur in about 10% in the general population (Irwin, 1985, pp. 174–175; Alvarado, 1986). This questionnaire study was carried out in a sample of 150 college students and the presence of an OBE was defined as the feeling that the subject "felt located outside the physical body" (no full-blown OBE as described in the present article). Moreover, Tart (1970) mentions that the majority of all OBE cases were observed in subjects that had used marijuana frequently (more than once a week or daily) and also had experience with more powerful psychoactive drugs such as LSD (even in the 6-month period preceding the questionnaire study). A small sample had even used hard narcotics or dangerous stimulants (also see Irwin, 1985, p. 209). Thus it cannot be stated whether the high frequency of OBEs in this subject sample was due to the effects of marijuana or the effects of these other drugs. Further evidence that marijuana increases the occurrence of OBEs comes from Siegel (1977). Siegel (1977) administered either marijuana or placebo in a controlled experiment and directly questioned both subject groups about their experiences. Although subjects mostly reported simple visual illusions and hallucinations, some subjects of the marijuana group reported OBE-like experiences such as a feeling of being dissociated from their body, a feeling of floating, and a realistic impression of a surround view. Finally, given that surveys of the general population have obtained a OBE-prevalence of 10% and of 25% in college students (Irwin, 1996), Palmer (1979) has suggested that this increase might be due to the higher use of mind-altering drugs by students.

Although, the influence of marijuana on the OBEs has to our knowledge only been investigated in these two studies, they suggest that marijuana use increases the probability of experiencing an OBE. Given this probable relation and the fact that the use of marijuana may result in an alteration of sensory, motor, and cognitive functions (Block et al., 1992), we have summarized below several neuroimaging studies that have examined the effect of marijuana administration on brain metabolism. Although over 60 cannabinoids have been identified in the plant *Cannabis sativa*, δ-9-tetrahydrocannabinol (THC) appears to be the major psychoactive ingredient of marijuana. In the following we summarize neuroimaging studies that have used THC. Thus, Mathew et al. (1997) described PET data in 32 volunteers before and after intravenous infusion of THC or placebo. THC resulted in increased rCBF bilaterally in the frontal lobes, the insula, anterior cingulum, temporal lobes and in subcortical regions, with the largest effects in the right hemisphere. Volkow et al. (1996) described the cerebral metabolic changes induced by THC in eight chronic users of marijuana compared to non-users of marijuana. Even without the administration of marijuana, the group of chronic marijuana users had significantly less relative metabolic activity in the cerebellum than did normal control subjects. After smoking marijuana, cerebellar activity was increased in both groups, but only the group of chronic marijuana users showed additional significant increases in orbitofrontal, prefrontal cortex, and basal ganglia. A PET study carried out by O'Leary et al. (2000) measured rCBF prior to and following smoking marijuana while subjects were engaged in an auditory attention task. This study of five subjects found that smoking marijuana increases rCBF in a number of paralimbic regions (orbital frontal lobes, insula, temporal poles) and in the anterior cingulated cortex, and the cerebellum. Large reductions in rCBF were observed in the temporal lobe regions. A more recent study by O'Leary et al. (2002) involved a sample of 12 volunteers, who were occasional smokers of marijuana. Again a PET study was carried out before and after smoking, and the smoking of marijuana was associated with increased rCBF in the temporal lobe, prefrontal lobe, insula, and the cerebellum. Regions with lower rCBF were observed in prefrontal cortex, in the superior temporal gyrus, and the occipital lobe. In summary, the present data suggest that, although the prevalence of OBEs increases by marijuana use, other visual illusions and hallucinations increase more strongly. Secondly, the observed brain activations were focused on the cerebellum and were at the neocortical level quite widespread without focalization. No significant activations were found at the TPJ.

General anesthesia

It has also been reported that general anesthesia is a common precipitating factor of OBEs. This was first stated by Muldoon and Carrington (1951) and confirmed by Crookall (1964). These authors collected many cases of OBEs from the first half of the 20th century following the introduction of general anesthesia into medicine. Yet, although Crookall (1964) reported 21 OBE cases (cases 347–365; pp. 98–108), only 11 comply with the OBE criteria as defined in the present article. Several other experiences classified by Crookall (1964) as OBEs in the context of general anesthesia were unspecific feelings of dissociation, other autoscopic phenomena, or the feeling of somebody closeby. The following example (containing the three defining features of the OBE: disembodiment, elevated visuo-spatial perspective, and autoscopy) is taken from Muldoon and Carrington (1951; quoted after Blackmore, 1982a):

> I saw myself- my physical self- lying there. I saw a sharply outlined view of the operating table. I myself freely hovering and looking downward from above, saw my physical body, lying on the operating table. I could see the wound of the operation on the right side of my body, see the doctor with instrument in his hand, which I cannot more closely describe. All this I observed very clearly. I tried to hinder it all. It was so real. I can still hear the words I kept calling out: Stop it— what are you doing there?

Although these early case collections suggest that OBEs may occur in patients undergoing general anesthesia, it is not known what types of anesthetics lead to OBEs, whether only selective patient subgroups might suffer from OBEs during general anesthesia, or whether specific operations are associated with OBEs. Moreover, the aforementioned case collections did not inquire about other experiences during general anesthesia and how frequent awareness (including OBEs) was during general anesthesia. Finally, as stated above for the case collection by Crookall (1964) not all patients reported as OBE cases experienced an OBE that complied with the definition of an OBE that was used in the present article.

"Awareness during anesthesia is as old as the specialty itself" (Spitellie et al., 2002) and is generally thought to be due to two main factors. First, lack of experience with the new therapeutics in the first half of the 20th century. Secondly, the introduction of muscle relaxants in general anesthesia since 1942. Although this has allowed to use lower doses of toxic anesthetic agents, it increased the difficulty in monitoring the patient's level of consciousness and thus increased the risk of awareness due to insufficient levels of anesthesia (Blacher, 1975; Spitellie et al., 2002). Although the true incidence of awareness is not known, it was estimated between 0.1 and 0.2% in non-cardiac, non-obstetric, and non-traumatic surgery (Sandin et al., 2000; Spitellie et al., 2002; Sandin, 2003). Much higher incidences of awareness have been reported for caesarean section (0.4%), cardiac surgery (1.5%), and surgical treatment for trauma (11–43%). Moreover, several of these authors have investigated the medical reasons for awareness and, in addition, questioned their patients immediately after their operation about the phenomenological contents of awareness (Blacher, 1975; Cobcroft and Forsdick, 1993; Moermann et al., 1993; Ranta et al., 1998; Ostermann et al., 2001). Blacher (1975) reported five patients with awareness during anesthesia, who received muscle relaxants. Moermann et al. (1993) found increased cardiovascular abnormalities during general anesthesia such as heart beat acceleration and blood pressure increase (probably indicating insufficient anesthesia) in 67% of patients with awareness during anesthesia. In comparison, these intraoperative cardiovascular abnormalities were only found in 21% of patients without awareness during anesthesia. Sandin et al. (2000) performed a study of 11,785 patients and found a higher incidence of awareness if muscle relaxants were used. In addition, technical problems during anesthesia were responsible for insufficient anesthesia in 50% of patients suffering from awareness during anesthesia (Sandin et al., 2000). These studies have also shown that in principle all anesthetics independently of their administration (oral, intravenous, and inhalatory) may lead to awareness speaking against a special role for certain anesthetics in the induction of OBEs and related experiences (i.e., Ketamin) (Jansen, 1989, 1997). In summary, insufficient levels of anesthesia in combination with the application of muscle relaxants seem to be the main cause of awareness during anesthesia. Yet, effects of hemodynamic instability on the central nervous system during surgery also seem to be another important pathophysiological factor since awareness during anesthesia in patients undergoing cardiac and traumatic surgery has a very high incidence (i.e., almost every third patient) and is often associated with hemodynamic instability (Gurman et al., 2002; Spitellie et al., 2002; Sandin, 2003).

With respect to the phenomenological content of the period of awareness during anesthesia, Moermann et al. (1993) found that most reported experiences are auditory perceptions (89%) as well as somatosensory or motor perceptions such as the sensations of paralysis (85%), pain (39%), bodily movements, and bodily modifications (30–40%; Cobcroft and Forsdick, 1993; Moermann et al., 1993). The presence and relative distribution of these perceptions have been confirmed by other authors (Ranta et al., 1998; Ostermann et al., 2001). Visual perceptions and OBEs are thus quite rare perceptions during general anesthesia. Whereas it is not possible to decide whether any of Blacher's patients (1975) experienced OBEs during periods of surgical awareness, Moermann et al. (1993) reported visual experiences in 27% of their patients including several instances during which patients experienced to see the surgeon and other people and/or surroundings of the operating theatre during the actual operation. Yet, it is not mentioned whether any of the patients presented any of the three defining features of OBEs (disembodiment, elevated visuo-spatial perspective, and autoscopy). The experience of patient 3 of Ranta et al. (1998) might be classified as an OBE as she reported that she saw the "light of the operating room and people operating on her" (i.e., explicit autoscopy), but there is no mentioning of disembodiment or elevated visuo-spatial perspective. A full-blown OBE is reported by Cobcroft and Forsdick (1993) in four of a total of 187 analyzed cases (i.e., 2%) of which one example is given here:

> I had the strangest [...] sensation of coming out of my self; of being up at the ceiling looking down on the proceedings [of the operation]. After the initial realization that I couldn't communicate at all, came the feeling of acceptance ... of being aware of having one hell of an experience.

Other patients of Cobcroft and Forsdick (1993) reported sensations of moving in a tunnel, of seeing faces, of seeing operating theatre details, as well as seeing bright lights and surrounding whiteness (4% of all evaluated patients). Several of Osterman et al.'s patients (2001) reported that they "left their body during the operation at some point". Thus, OBEs are not very frequent in patients undergoing general anesthesia. Yet, if analyzed only with respect to the reported visual experiences during general anesthesia, OBEs and OBE-like experiences are very frequent. Finally, the sensations of paralysis as well as of bodily movements and modifications are not only frequent in awareness during general anesthesia, but also frequently reported by subjects with OBEs of spontaneous or neurological origin (Blackmore, 1982a; Irwin, 1985; Blanke et al., 2004).

In summary, these data suggest that although visual awareness is rare during general anesthesia, many of such visual perceptions are full-blown OBEs or related experiences. The partial awareness of patients during general anesthesia, in association with the frequent sensations of paralysis, bodily movements, and bodily modifications, as well as the patient's supine body position have also been described in OBEs of neurological origin. Hemodynamic instabilities such as during cardiovascular and traumatic surgery (with much higher incidences of awareness) might be a second precipitating factor for OBEs in general anesthesia. It has been shown that hemodynamic instabilities during these operations may lead to decreases in cerebral blood perfusion and may have, as a consequence, transient or manifest low-flow brain infarctions (or borderzone infarctions). It is interesting to note that these frequently include the TPJ (Ringelstein and Zunker, 1998) suggesting that OBEs under general anesthesia might be related to the functional and anatomical pathomechanisms as described in neurological patients with epilepsy, migraine, and cerebrovascular disease.

Cognitive neuroscience of OBE and self

The above findings suggest that OBEs are related to states of partially impaired consciousness (i.e., hypnopompic or hypnagogic states; dreaming; complex partial seizures, migraine) or a state of partially conserved awareness under general anesthesia as proposed previously (Tart, 1974, p. 370; Tart, 1975, pp. 71–72; Blackmore, 1982a). Importantly, these clinical and neuroimaging findings have allowed to link the complex phenomenon of the OBE with multisensory disintegration and deficient own body processing at the TPJ. This is not trivial as these findings may help to demystify OBEs and facilitate the formulation of precise research hypotheses about the sensory, cognitive, and neural mechanisms of OBEs. The neuroscientific investigation of OBEs may also turn out to be very useful in defining some of the functions and brain structures mediating such aspects of the normal self such as corporeal awareness, embodiment, egocentric visuo-spatial perspective, and self-consciousness (Metzinger, 2003, in press; Blanke and Arzy, 2005). In the following we will review recent neuroimaging data about the TPJ's implication in self and corporeal awareness.

Neuroimaging studies support the role of the TPJ in vestibular processing, multisensory integration as well as the perception of human bodies or body parts. First, the core region of the human vestibular cortex (Lobel et al., 1998; Fasold et al., 2002) is situated at the TPJ including the posterior insula. Brain damage in this area has been associated with vestibular sensations and dysfunctions (Smith, 1960; Brandt, 2000). Second, several neuropsychological and neuroimaging studies suggest the implication of the TPJ and cortical areas along the intraparietal sulcus in combining tactile, proprioceptive, and visual information in a coordinated reference frame (Calvert et al., 2000). Interestingly, Leube et al. (2003) have shown that the TPJ codes multisensory conflict between visual and proprioceptive information about one's arm position as proposed in the above OBE-model for the entire body. Third, the TPJ is also involved in the perception of many different aspects of the human body including the perception of body parts (Bonda et al., 1995), the entire body (Downing et al., 2001; Astafiev et al., 2004) as well as biological motion (Grossman et al., 2000; Beauchamp et al., 2002) and mental imagery with respect to one's own body (Zacks et al., 2002; Blanke et al., 2005). Importantly, Astafiev et al. (2004) have shown that some of these "visual" areas are not only modulated by visually presented human bodies or body parts, but also by limb movements (without visual feedback) suggesting their role in multisensory own body perception.

The TPJ has also been involved in cognitive functions that are closely linked to self processing and OBEs: egocentric visuo-spatial perspective taking (Farrell and Robertson, 2000), agency (the feeling of being the agent of one's actions and thoughts; Chaminade and Decety, 2002; Farrer and Frith, 2002; Farrer et al., 2003), and self–other distinction (the capacity by which one distinguishes between oneself and other conspecifics; Ruby and Decety, 2001, 2003, 2004; Chaminade and Decety, 2002; Farrer et al., 2002, 2003). Thus, during OBEs one's visuo-spatial perspective and one's sense of agency are localized at the position of the disembodied self that is hovering above the physical body. In other words, the self is experienced as looking at the (autoscopic) body from a third (or other) person's visuo-spatial perspective and position. Furthermore, the TPJ is the classical lesion site in patients with visuo-spatial neglect (Halligan et al., 2003), a clinical condition, which has been shown to disturb the patient's egocentric spatial relationship with extracorporeal space and visuo-spatial perspective taking (i.e., Farrell and

Robertson, 2000). Neuroimaging studies in healthy observers have also revealed activation of the TPJ during egocentric visuo-spatial perspective changes in healthy subjects (Maguire et al., 1998; Ruby and Decety, 2001). Moreover, it has been shown that mental activities such as agency and self–other distinction activate the TPJ (Ruby and Decety, 2001, 2003, 2004). In another study, Perrin et al. (2005) have shown by measuring rCBF and evoked potentials that subjects' own name or names from other people activated the TPJ. In a recent study, Blanke et al. (2005) have used a mental imagery task with respect to one's own body and linked essential phenomenological characteristics of the OBE to the TPJ. They asked healthy subjects to imagine themselves in the position and visual perspective that is generally reported by people experiencing spontaneous OBEs and found an activation of the TPJ. Interference with the TPJ by transcranial magnetic stimulation impaired mental transformation of the own body in healthy volunteers, but not for imagined spatial transformations of external objects suggesting the selective implication of the TPJ in mental imagery of one's own body (Blanke et al., 2005). Finally these authors showed, in an epileptic patient with OBEs originating from the TPJ, partial activation of the seizure focus during mental transformations of her body and visual perspective mimicking her OBE percepts. Based on these results, Blanke et al. (2005) argued that the TPJ might be a crucial structure for the conscious experience of the normal self mediating spatial unity of self and body and that impaired processing at the TPJ may lead to the experience of abnormal selfs, such as OBEs.

In summary, although many other cortical areas such as prefrontal, anterior cingulate, postcentral, precuneal, occipito-temporal junction, superior parietal lobule, and insular cortices (Grossman et al., 2000; Ruby and Decety, 2001, 2003, 2004; Beauchamp et al., 2002; Chaminade and Decety, 2002; Zacks et al., 2002; Leube et al., 2003) have been shown to play a role in self processing, the reviewed neuroimaging data on body and self processing as well as the clinical data on OBEs suggest that the TPJ is a key neural locus for self processing that is involved in multisensory body-related information processing as well as in processing of phenomenological and cognitive aspects of the self.

Conclusions

OBEs have fascinated mankind from time immemorial and are abundant in folklore, mythology, and spiritual experiences of most ancient and modern societies. We have reviewed clinical and neuroimaging evidence suggesting that OBEs are culturally invariant brain phenomena that have specific precipitating factors and that can be investigated neuroscientifically. The reviewed data on the precipitating factors sleep and general anesthesia in the normal population suggest that OBEs are linked to situations that are associated with partial impairments of consciousness, supine body position, and disturbed own body processing. This fits with the reviewed neurological data where OBEs were observed in patients with partially impaired consciousness in paroxysmal brain dysfunctions (such as epilepsy, migraine, and electrical cortical stimulation), supine body position, and disturbed own body processing. Based on these data we have suggested previously that OBEs might be caused by a functional disintegration in lower-level multisensory processing (vestibular, proprioceptive, tactile, and visual information). Finally, the reviewed data suggest an interaction of lower-level multisensory processing with higher-level self processing such as egocentric visuo-spatial perspective taking, agency, and self location (or embodiment). From an anatomical point of view, the reviewed data in neurological patients, patients under general anesthesia, and neuroimaging studies on sleep as well as cognitive neuroscience point to the importance of the right TPJ in the generation of OBEs, although the left TPJ as well as other brain areas have also been associated with OBEs. It is hoped that the experimental investigations of the interactions between these multisensory and cognitive systems in OBEs and related illusions in combination with neuroimaging and behavioral techniques will further our understanding of the central mechanisms of self and corporal awareness much as previous research into the neural bases of complex body part illusions has demystified phantom limbs (Ramachandran and Hirstein, 1998; Halligan, 2002).

Acknowledgments

The authors are supported by the "Fondation Leenaards" and the "Fondation de Famille Sandoz."

References

Aizenberg, D. and Modai, I. (1985) Autoscopic and drug induced perceptual disturbances. A case report. Psychophatology, 18: 97–111.

Alvarado, C.S. (1986) ESP during spontaneous out-of-body experiences: a research and metholodical note. J. Soc. Psychical Res., 53: 393–397.

Alvarado, C.S. (2000) Out-of-body experiences. In: Cardeña E., Lynn S.J. and Krippner S. (Eds.), Varieties of Anomalous Experiences. American Psychological Association, Washington, DC, pp. 183–218.

Andreasen, N.C., O'Leary, D.S., Cizadlo, T., Arndt, S., Rezai, K., Watkins, G.L., Boles Ponto, L.L. and Hichwa, R.D. (1995) Remembering the past: two facets of episodic memory explored with positron emission tomography. Am. J. Psychiatry, 152: 1576–1585.

Arenz, D. (2001) Heautoskopie. Doppelgängerphänomen und seltene Halluzinationen der eigenen Gestalt. Der Nervenarzt, 72: 376–379.

Aserinsky, E. and Kleitman, N. (1953) Regularly occurring periods of eye motility and concomitant phenomena during sleep. Science, 118: 273–274.

Astafiev, S.V., Stanley, C.M., Shulman, G.L. and Corbetta, M. (2004) Extrastriate body area in human occipital cortex responds to the performance of motor actions. Nat. Neurosci., 7: 542–548.

Baker, D.M. (1977) Practical Techniques of Astral Projection. Aquarian Press, Wellingborough, England.

Beauchamp, M.S., Lee, K.E., Haxby, J.V. and Martin, A. (2002) Parallel visual motion processing streams for manipulable objects and human movements. Neuron, 34: 149–159.

Benson, A.J. (1999) Spatial disorientation – common illusions. In: Ernsting J., Nicholson A.N. and Rainford D.J. (Eds.), Aviation Medicine (3rd ed.). Butterworth & Heinmann, Oxford, pp. 437–454.

Blacher, R.S. (1975) On awakening paralyed during surgery – a syndrome of traumatic neurosis. J. Am. Medi. Assoc., 234(1): 67–68.

Blackmore, S.J. (1982a) Beyond the Body. An Investigation of Out-of-Body Experiences. Heinemann, London.

Blackmore, S.J. (1982b) OBEs, lucid dreams and imagery: two surveys. J. Am. Soc. Psychical Res., 76: 301–317.

Blackmore, S.J. (1986) Out-of-body experiences in schizophrenia. A questionaire survey. J. Nerv. Ment. Disease, 174: 615–619.

Blanke, O. (2004) Illusions visuelles. In: Safran A.B., Vighetto A., Landis T. and Cabanis E. (Eds.), Neuro-opthalmologie. Masson, Paris, pp. 147–150.

Blanke, O. and Arzy, S. (2005) The out-of body experience. Disturbed self processing at the temporo-parietal junction. Neuroscientist, 11: 16–24.

Blanke, O., Landis, T., Spinelli, L. and Seeck, M. (2004) Out-of-body experience and autoscopy of neurological origin. Brain, 127: 243–258.

Blanke, O., Mohr, C., Michel, C.M., Pascual-Leone, A., Brugger, P., Landis, T., Seeck, M. and Thut, G. (2005) Linking out-of-body experience and self processing to the temporo-parietal junction. J. Neurosci, 25: 550–557.

Blanke, O., Ortigue, S., Landis, T. and Seeck, M. (2002) Stimulating illusory own-body perceptions. Nature, 419: 269–270.

Block, R.I., Farinpour, R. and Braverman, K. (1992) Acute effects of marijuana on cognition: relationships to chronic effects and smoking techniques. Pharmacol. Biochem. Behav., 43(3): 907–917.

Bonda, E., Petrides, M., Frey, S. and Evans, A. (1995) Neural correlates of mental transformations of the body-in-space. Proc. Natl. Acad. Sci. USA, 92: 11180–11184.

Brandt, T. (2000) Central vestibular disorders. In: Vertigo. Its Multisensory Syndromes. Springer, London, pp. 146–167.

Braun, A.R., Balkin, T.J., Wesensten, N.J., Carson, R.E., Varga, M., Baldwin, P., Selbie, S., Belenky, G. and Herscovitch, P. (1997) Regional cerebal blood flow throughout the sleep-wake cycle. An H215O PET study. Brain, 120: 1173–1197.

Brugger, P. (2002) Reflective mirrors: perspective taking in autoscopic phenomena. Cogn. Neuropsychiatr., 7: 179–194.

Brugger, P., Regard, M. and Landis, T. (1997) Illusory reduplication of one's own body: phenomenology and classification of autoscopic phenomena. Cogn. Neuropsychiatr., 2: 19–38.

Brugger, P., Regard, M., Landis, T. and Oelz, O. (1999) Hallucinatory experiences in extreme-altitude climbers. Neuropsychi. Neuropsy. Behav. Neur., 12: 67–71.

Calvert, G.A., Campbell, R. and Brammer, M.J. (2000) Evidence from functional magnetic resonance imaging of crossmodal binding in the human heteromodal cortex. Curr. Biol., 10: 649–657.

Chaminade, T. and Decety, J. (2002) Leader or follower? Involvement of the inferior parietal lobule in agency. Neuroreport, 13(15): 1975–1978.

Clark, B. and Graybiel, A. (1957) The break-off phenomenon. A feeling of separation from the earth experienced by pilots at high altitude. J. Aviation Med., 28: 121–126.

Cobcroft, M.D. and Forsdick, C. (1993) Awareness under anesthesia: the patients' point of view. Anaesth. Intens. Care, 21: 837–843.

Craske, S. and Sacks, B.I. (1969) A case of "double autoscopy". Br. J. Psychiatr., 115: 343–345.

Crookall, R. (1964) More Astral Projections. Analyses of Case Histories. Aquarian Press, London.

Daly, D.D. (1958) Ictal affect. Am. J. Psychiatr., 115: 171–181.

Damas Mora, J.M.R., Jenner, F.A. and Eacott, S.E. (1980) On heautoscopy or the phenomenon of the double: case presentation and review of the literature. Br. J. Med. Psychol., 53: 75–83.

Dening, T.R. and Berrios, G.E. (1994) Autoscopic phenomena. Br. J. Psychiatry, 165: 808–817.

Devinsky, O., Feldmann, E., Burrowes, K. and Bromfeld, E. (1989) Autoscopic phenomena with seizures. Arch. Neurol., 46: 1080–1088.

Downing, P.E., Jiang, Y., Shuman, M. and Kanwisher, N. (2001) A cortical area selective for visual processing of the human body. Science, 293: 2470–2473.

Farrell, M.J. and Robertson, I.H. (2000) The automatic updating of egocentric spatial relationships and its impairment due to right posterior cortical lesions. Neuropsychologia, 38: 585–595.

Farrer, C. and Frith, C.D. (2002) Experiencing oneself vs another person as being the cause of an action: the neural correlates of the experience of agency. Neuroimage, 15(3): 596–603.

Farrer, C., Franck, N., Georgieff, N., Frith, C.D., Decety, J. and Jeannerod, M. (2003) Modulating the experience of agency: a positron emission tomography study. Neuroimage, 18(2): 324–333.

Fasold, O., von Brevern, M., Kuhberg, M., Ploner, C.J., Villringer, A. and Lempert, T. (2002) Human vestibular cortex as identified with caloric stimulation in functional magnetic resonance imaging. Neuroimage, 17: 1384–1393.

Firth, P.G. and Bolay, H. (2004) Transient high altitude neurological dysfunction: an origin in the temporoparietal cortex. High Altitude Med. Biol., 5: 71–75.

Gabbard, G.O. and Twemlow, S.W. (1984) With the Eyes of the Mind: An Empirical Analysis of Out-of-Body States. Praeger Scientific, New York.

Gallagher, S. (2000) Philosophical conceptions of the self: implications for cognitive science. Trends Cogn. Sci., 4: 14–21.

Girard, T.A. and Cheyne, J.A. (2004) Individual differences in lateralization of hallucinations associated with sleep paralysis. Laterality: asymmetries of body. Brain Cognit., 9(1): 93–111.

Green, C.E. (1968) Out-of-Body Experiences. Hamish Hamilton, London.

Greenhouse, H.B. (1975) The Astral Journey. Doubleday, Garden City, NY.

Grossman, E., Donnelly, M., Price, R., Pickens, D., Morgan, V., Neighbor, G. and Blake, R. (2000) Brain areas involved in perception of biological motion. J. Cogn. Neurosci., 12: 711–720.

Grüsser, O.J. & Landis, T. (1991) The splitting of 'I' and 'me': heautoscopy and related phenomena. In: Grüsser, O.J., Landis, T. (General Editor: J.R. Cronly Dillon), Visual Agnosias and Other Disturbances of Visual Perception and Cognition. Macmillan, Amsterdam, pp. 297–303.

Gurman, G.M., Weksler, N. and Schily, M. (2002) Should disclosure of the danger of awareness during general anesthesia be a part of preanesthesia consent? Anesthesia, 68: 905–910.

Habeler, P. (1979) The Lonely Victory. Simon & Shuster, New York, pp. 166–176.

Halligan, P.W. (2002) Phantom limbs: the body in mind. Cogn. Neuropsychiatr., 7: 251–268.

Halligan, P.W., Fink, G.R., Marshall, J.C. and Vallar, G. (2003) Spatial cognition: evidence from visual neglect. Trends Cogn. Sci., 7: 125–133.

Hartwell, J., Janis, J. and Harary, S.B. (1974) A study of the physiological variables associated with out-of-body experiences. In: Morris J.D., Roll W.G. and Morris R.L. (Eds.), Research in Parapsychology 1974. Scarecrow Press, Metuchen, NJ, pp. 127–129.

Hécaen, H. and Ajuriaguerra, J. (1952) L'heautoscopie. In: Hécaen H. and Ajuriaguerra J. (Eds.), Meconnaissances et Hallucinations Corporelles. Masson, Paris, pp. 310–343.

Hécaen, H. and Green, A. (1957) Sur l'héautoscopie. Encephale, 46: 581–594.

Heim, A. (1892) Notizen über den Tod durch Absturz. Jahrbuch des Schweizer Alpenklub, 27: 327–337.

Irwin, H.J. (1985) Flight of Mind: A Psychological Study of the Out-of-Body Experience. The Scarecrow Press, Inc., Metuchen, NJ.

Irwin, H.J. (1996) Childhood antecedents of the out-of-body experience and déjà vu experiences. J. Soc. Psychical Res., 90: 157–173.

Janis, J., Hartwell, J., Harary, S.B., Levin, J. and Morris, R.L. (1973) A description of the physiological variables connected with an out-of-body study. Res. Parapsychol.: 36–37.

Jansen, K.L.R. (1989) Near-death experience and the NMDA receptor [Letter]. Br. Med. J., 298: 1708.

Jansen, K.L.R. (1997) The Ketamine model of the near-death experience: a central role for the N-Methyl-D-Aspartate receptor. J. Near Death Studies, 16(1): 5–27.

LaBerge, S. (1990) Lucid dreaming: psychophysiological studies of consciousness during REM sleep. In: Bootzen R.R., Kihlstrom J.F. and Schacter D.L. (Eds.), Sleep and Cognition. American Psychological Association, Washington, DC, pp. 109–126.

Letailleur, M., Morin, J. and Le Borgne, Y. (1958) Heautoscopie heterosexuelle et schizophrenie. Etude d'une observation. Ann. Méd.-Psych., 116: 451–461.

Leube, D.T., Knoblich, G., Erb, M., Grodd, W., Bartels, M. and Kircher, T.T. (2003) The neural correlates of perceiving one's own movements. Neuroimage, 20: 2084–2090.

Lhermitte, J. (1939) Les phénomènes héautoscopiques, les hallucinations spéculaires et autoscopiques. In: L'image de Notre Corps. L'Harmattan, Paris, pp. 170–227.

Lobel, E., Kleine, J.F., Blhan, D.L., Leroy-Willig, A. and Berthoz, A. (1998) Functional MRI of galvanic vestibular stimulation. J. Neurophysiol., 80: 2699–2709.

Lukianowicz, N. (1958) Autoscopic phenomena. Arch. Neurol. Psychiatry, 80: 199–220.

Lunn, V. (1970) Autoscopic phenomena. Acta Psychiatr. Scand., 46, 219: 118–125.

Maack, L.H. and Mullen, P.E. (1983) The Doppelgänger, disintegration and death: a case report. Psychol. Med., 13(3): 651–654.

Maguire, E.A., Burgess, N., Donnett, J.G., Frackowiak, R.S., Frith, C.D. and O'Keefe, J. (1998) Knowing where and getting there: a human navigation network. Science, 280: 921–924.

Maillard, L., Vignal, J.P., Anxionnat, R. and TaillandierVespignani, L. (2004) Semiologic value of ictal autoscopy. Epilepsia, 45: 391–394.

Mathew, R.J., Wilson, W.H., Coleman, R.E., Turkington, T.G. and DeGrado, T.R. (1997) Marijuana intoxication and brain activation in marijuana smokers. Life Sci., 60(23): 2075–2089.

Maquet, P. (2000) Functional neuroimaging of normal human sleep by positron emission tomography. J. Sleep Res., 9: 207–231.

Maquet, P., Degueldre, C., Delfiore, G., Aerts, J., Peters, J.M., Luxen, A. and Franck, G. (1997) Functional neuroanatomy of human slow wave sleep. J. Neurosci., 17: 2807–2812.

Maquet, P., Peters, J.M., Aerts, J., Delfiore, G., Degueldre, C., Luxen, A. and Franck, G. (1996) Functional neuroanatomy of human rapid eye movement sleep and dreaming. Nature, 383: 163–166.

McConnel, W.B. (1965) The phantom double in pregnancy. Br. J. Psychiatry, 111: 67–68.

McCreery, C. (1993) Schizotypy and the out-of-body-experience. Unpublished doctoral dissertation, Oxford University, Oxford, England.

McCreery, C. and Claridge, G. (1995) Out-of-the body experiences and personality. J. Soc. Psychical Res., 60: 129–148.

McCreery, C. and Claridge, G. (1996a) A study of halluciantation in normal subjects: II. Electrophysiological data. Pers. Ind. Differences, 21: 749–758.

McCreery, C. and Claridge, G. (1996b) The factor structure of 'schizotypal' traits: a large replication study. Br. J. Clin. Psychol., 35: 103–115.

McCreery, C. and Claridge, G. (2002) Healthy schizotypy: the case of out-of-the-body experiences. Pers. Ind. Differences, 32: 141–154.

Meehl, P.E. (1962) Schizotaxia, schizotypy, schizophrenia. Am. Psychologist, 17: 827–838.

Menninger-Lerchenthal, E. (1935) Das Truggebilde der eigenen Gestalt (Heautoskopie, Doppelgänger). Karger, Berlin.

Metzinger, T. (2003) Being No One. MIT Press, Cambridge, MS.

Metzinger, T. (2005) The prescientific concept of a "soul": a neurophenomenological hypothesis about its origin. In: Peschl M.F. (Ed.), Die Rolle der Seele in der Kognitionswissenschaft und der Neurowissenschaft. Auf der Suche nach dem Substrat der Seele. Königshausen and Neumann, Germany, pp. 189–214.

Michel, C.M., Murray, M.M., Lantz, G., Gonzalez, S., Spinelli, L. and Grave De Peralta, R. (2004) EEG source imaging. Clin. Neurophysiol., 115: 2195–2222.

Mitchell, L.C. (1973) Out-of-the-body vision. Psychic: 44–47.

Moermann, N., Bonke, B. and Oosting, J. (1993) Awareness and recall during general Anesthesia. Anesthisiology, 79: 454–464.

Monroe, R.A. (1974) Journeys Out of the Body. Corgi, London.

Muldoon, S.J. and Carrington, H. (1929) The Projection of the Astral Body. Rider, London.

Muldoon, S.J. and Carrington, H. (1951) The Phenomena of Astral Projection. Rider, London.

Neisser, U. (1988) The five kinds of self-knowledge. Phil. Psychol., 1: 35–59.

Nielsen, T.A. (2003) A review of mentation in REM and NREM sleep: "Covert" REM sleep as a possible reconciliation of two opposing models. In: Pace-Schott E.F., Solms M., Blagrove M. and Harnad S. (Eds.), Sleep and Dreaming: Scientific Advances and Reconsiderations. Cambridge University Press, Cambridge, pp. 125–132.

Nofzinger, E.A., Mintun, M.A., Wiseman, M., Kupfer, D.J. and Moore, R.Y. (1997) Forebrain activation in REM sleep: an FDG PET study. Brain Res., 770: 192–201.

O'Leary, D.S., Block, R.I., Flaum, M., Schultz, S.K., Boles Ponto, L.L., Watkins, G.L., Hurtig, R.R., Andreasen, N.C. and Hichwa, R.D. (2000) Acute marijuana effects on rCBF and cognition: a PET study. NeuroReport, 17(27): 3835–3841.

O'Leary, D.S., Block, R.I., Koeppel, J., Flaum, M., Schultz, S.K., Andreasen, N.C., Boles Ponto, L.L., Watkins, G.L., Hurtig, R.R. and Hichwa, R.D. (2002) Effects of smoking marijuana on brain perfusion and cognition. Neuropsychopharmacology, 26(6): 802–816.

Osis, K. and Mitchell, J.L. (1977) Physiological correlates of reported out-of-body-experiences. J. Soc. Psychical Res., 49: 525–536.

Ostermann, J.E., Hopper, J., Heran, W.J., Keane, T.M. and van der Kolk, B.A. (2001) Awareness under anaesthisia and the development of post traumatic stress disorder. Gen. Hosp. Psychiatry, 23: 193–204.

Ostrow, M. (1960) The metapsychology of autoscopic phenomena. Int. J. Psychoanal., 41: 619–625.

Palmer, J. (1978) ESP and out-of-body experiences: an experimental approach. In: Rogo D.S. (Ed.), Mind Beyond the Body. Penguin, Harmondsworth, England, pp. 193–217.

Palmer, J. (1979) A community mail survey of psychic experiences. J. Am. Soc. Psychical Res., 73: 221–251.

Perrin, F., Maquet, P., Peigneux, P., Ruby, P., Degueldre, C., Balteau, S., Del Fiore, G., Moonen, G., Luxen, A. and Laurcys, S. (2005) Neural mechanisms involved in the detection of our first name: a combined ERPs and PET study. Neuropsychologia, 43: 12–19.

Poynton, J.C. (1975) Results of an out-of-body survey. In: Poynton J.C. (Ed.), Parapsychology in South Africa. South African Society for Psychical Research, Johannesburg, South Africa, pp. 109–123.

Ramachandran, V.S. and Hirstein, W. (1998) The perception of phantom limbs. Brain, 121: 1603–1630.

Ranta, S.O.V., Laurelia, R., Saario, J., Ali- Melkkilä, T. and Hynynen, M. (1998) Awareness with recall during general anesthesia: incidence and risk factors. Anesth. Analg., 86: 1084–1089.

Ravenhill, T.H. (1913) Some experiences of mountain sickness in the Andes. J. Trop. Med. Hyg., 16: 313–320.

Rechtschaffen, A. and Siegel, J. (2000) Sleep and dreaming. In: Kandel E.R.J., Schwartz J.J.H. and Jessel T.M. (Eds.),

Principles of Neural Science (4th ed.). McGraw-Hill, New York, pp. 937–948.

Reynolds, M. and Brewin, C.R. (1999) Intrusive memory in depression and posttraumatic stress disorder. Behav. Res. Therapy, 37: 201–215.

Ring, K.R. (1980) Life at Death. A Scientific Investigation of the Near-Death Experience. Coward, McCann & Geoghegan, New York.

Ringelstein, E.B. and Zunker, P. (1998) Low-flow infarction. In: Ginsberg M. and Bogousslavsky J. (Eds.) Cerebrovascular Disease: Pathophysiology, Diagnosis and Management, Vol. 2. Blackwell Science INC, Cambridge, pp. 1075–1089.

Rogo, D.S. (1982) Psychological models of the out of body experience. A review and critical evaluation. J. Parapsychol., 46: 29–45.

Rogo, D.S. (1986) Leaving the Body. A Complete Guide to Astral Projection. Simon and Schuster, NY.

Röhricht, F. and Priebe, S. (1997) Disturbances of body experience in schizophrenic patients. Fortschr. Neurol. Psychiatr., 65(7): 323–336.

Ruby, P. and Decety, J. (2001) Effect of subjective perspective taking during simulation of action: a PET investigation of agency. Nat. Neurosci., 4: 546–550.

Ruby, P. and Decety, J. (2003) What you believe versus what you think they believe: a neuroimaging study of conceptual perspective-taking. Eur. J. Neurosci., 17(11): 2475–2480.

Ruby, P. and Decety, J. (2004) How do you feel versus how do you think she would feel? A neuroimaging study of perspective taking with social emotions. J. Cogn. Neurosci., 16(6): 988–999.

Salama, A.E.A.A. (1981) The autoscopic phenomenon: case report and review of literature. Can. J. Psychiatry, 26: 476.

Sandin, R.H. (2003) Awareness 1960–2002, explicit recall of events during general anesthesia. Adv. Exp. Med. Biol., 523: 135–147.

Sandin, R.H., Enlund, G., Samuelsson, P. and Lennmarken, C. (2000) Awareness during anesthesia: a prospective case study. Lancet, 355: 707–711.

Schmitt, B. (1948) L'heautoscopie. Cahiers de Psychiatrie, 2: 21–26.

Schwartz, S. and Maquet, P. (2002) Sleep imaging and the neuropsychological assessment of dreams. Trends Cog. Sci., 6(1): 23–30.

Sheils, D. (1978) A cross-cultural study of beliefs in out-of-the-body experiences, waking and sleeping. J. Soc. Psychol. Res., 49: 697–741.

Shermer, M. (1998) A mind out of body. Skeptic, 6(3): 72–79.

Siegel, R.K. (1977) Hallucinations. Sci. Am., 237: 132–140.

Smith, B.H. (1960) Vestibular disturbances in epilepsy. Neurology, 10: 465–469.

Sours, J.A. (1965) The "Break-Off" phenomenon. A precipitant of anxiety in jet aviators. Arch. Gen. Psychiatr., 13: 447–456.

Spitellie, P.H., Holmes, M.A. and Domino, K.B. (2002) Awareness during anesthesia. Anesthisiol. Clin. N. Am., 20: 555–570.

Tart, C.T. (1967) A second psychophysiological study of the out-of-the-body experiences in a selected subject. Int. J. Parapsychol., 9(3): 251–258.

Tart, C.T. (1968) A psychophysiological study of the out-of-the-body experiences in a selected subject. J. Am. Soc. Psychical Res., 62: 3–27.

Tart, C.T. (1970) Marijuana intoxication: common experiences. Nature, 226: 701–704.

Tart, C.T. (1971) On Being Stoned. A Psychological Study of Marijuana Intoxication. Science and Behavior Books, Palo Alto.

Tart, C.T. (1974) Out-of-body experiences. In: White J. (Ed.), Psychic Exploration: A Challenge for Science. Putnam, New York, pp. 349–373.

Tart, C.T. (1975) States of Consciousness. Dutton, New York.

Tormes, F.R. and Guedry, F.E. (1975) Disorientation phenomena in naval helicopter pilots. Aviat. Space Environ. Med., 46: 387–393.

Twemlow, S.W. (1977) Epilogue: personality file. In: Monroe R.A. (Ed.), Journeys Out of the Body. Anchor, New York, pp. 275–280.

Twemlow, S.W., Gabbard, G.O. & Jones, F.C. (1980) The out-of-body experience: I. Phenomenology. Paper presented at the annual meeting of the American Psychiatric Association, San Francisco.

Twemlow, S.W., Gabbard, G.O. and Jones, F.C. (1982) The out-of-body experience: a phenomenological typology based on questionnaire responses. Am. J. Psychiatry, 139: 450–455.

Volkow, N.D., Gillespie, H., Mullani, N., Tancredi, L., Grant, C., Valentine, A. and Hollister, L. (1996) Brain glucose metabolism in chronic marijuana users at baseline and during marijuana intoxication. Psychiatry Res.: Neuroimaging, 67: 29–38.

Wolfradt, U. and Watzke, S. (1999) Deliberate out-of-body-experiences, depersonalization, schizotypal traits and thinking styles. J. Amer. Soc. Psy. Res., 93: 249–257.

Zacks, J.M., Ollinger, J.M., Sheridan, M.A. and Tversky, B. (2002) A parametric study of mental spatial transformations of bodies. Neuroimage, 16: 857–872.

Zutt, J. (1953) "Aussersichsein" und "auf sich selbst Zurückblicken" als Ausnahmezustand. Zur Psychopathologie des Raumerlebens. Nervenarzt, 24: 24–31.

S. Laureys (Ed.)
Progress in Brain Research, Vol. 150
ISSN 0079-6123

CHAPTER 25

Near-death experiences in cardiac arrest survivors

Christopher C. French*

Anomalistic Psychology Research Unit, Department of Psychology, Goldsmiths College, University of London, New Cross, London SE14 6NW, UK

Abstract: Near-death experiences (NDEs) have become the focus of much interest in the last 30 years or so. Such experiences can occur both when individuals are objectively near to death and also when they simply believe themselves to be. The experience typically involves a number of different components including a feeling of peace and well-being, out-of-body experiences (OBEs), entering a region of darkness, seeing a brilliant light, and entering another realm. NDEs are known to have long-lasting transformational effects upon those who experience them. An overview is presented of the various theoretical approaches that have been adopted in attempts to account for the NDE. Spiritual theories assume that consciousness can become detached from the neural substrate of the brain and that the NDE may provide a glimpse of an afterlife. Psychological theories include the proposal that the NDE is a dissociative defense mechanism that occurs in times of extreme danger or, less plausibly, that the NDE reflects memories of being born. Finally, a wide range of organic theories of the NDE has been put forward including those based upon cerebral hypoxia, anoxia, and hypercarbia; endorphins and other neurotransmitters; and abnormal activity in the temporal lobes. Finally, the results of studies of NDEs in cardiac arrest survivors are reviewed and the implications of these results for our understanding of mind–brain relationships are discussed.

Introduction

Greyson (2000a, pp. 315–316) describes near-death experiences (NDEs) as

> profound psychological events with transcendental and mystical elements, typically occurring to individuals close to death or in situations of intense physical or emotional danger. These elements include ineffability, a sense that the experience transcends personal ego, and an experience of union with a divine or higher principle.

He also provides a typical example of an NDE experienced by a 55-year-old man who had been admitted to hospital with an irregular heartbeat. During diagnostic angiography he suffered a coronary occlusion and had to undergo emergency quadruple bypass surgery. Following this, he reported having had an out-of-body experience (OBE) during which he observed the operating room from above. He was able to accurately describe the behavior of the cardiovascular surgeon during the operation. He also described following a brilliant light through a tunnel to a region of warmth, love, and peace. Here he experienced an apparent encounter with deceased relatives, who telepathically communicated to him that he should return to his body. Upon recovery, he felt transformed, with an intense desire to help others and to talk about his experience.

Modern interest in NDEs owes much to the publication in 1975 of Raymond Moody's best-selling book *Life after Life*, although reports of similar experiences can be found in much earlier texts. Moody (1975, 1977) identified a number of

*Corresponding author. Tel.: +44 020 7919 7882;
Fax: +44 020 7919 7873; E-mail: c.french@gold.ac.uk

DOI: 10.1016/S0079-6123(05)50025-6

Table 1. Common elements recurring in adult NDEs according to Moody (1975, 1977; as summarized by Greyson, 2000, p. 318)

Elements occurring during NDEs:
- Ineffability
- Hearing oneself pronounced dead
- Feelings of peace and quiet
- Hearing unusual noises
- Seeing a dark tunnel
- Being "out of the body"
- Meeting "spiritual beings"
- Experiencing a bright light as a "being of light"
- Panoramic life review
- Experiencing a realm in which all knowledge exists
- Experiencing cities of light
- Experiencing a realm of bewildered spirits
- Experiencing a "supernatural rescue"
- Sensing a border or limit
- Coming back "into the body"

Elements occurring as aftereffects:
- Frustration relating the experience to others
- Subtle "broadening and deepening" of life
- Elimination of fear of death
- Corroboration of events witnessed while "out of the body"

common elements that recur in adult NDEs (see Table 1), although he noted that no element occurs in all NDE reports. He also noted that the order in which the elements occurred varied in the different accounts he had collected. Subsequent researchers, like Ring (1980), adopted a more systematic approach to the study of NDEs. Ring identified a "core experience" on the basis of a structured interview and measurement scale that he administered to 102 people who had been near to death, 48% of whom reported an NDE. This consisted of the following five stages, which tend to occur in the following order:

(a) *Peace and well-being:* The positive emotional tone of the NDE was reported by 60% of Ring's (1980) sample. Although the vast majority of NDEs are indeed blissful, more recent research (e.g., Greyson and Bush, 1992) has established that NDEs can occasionally cause terror and distress. Negative NDEs appear to fall into three distinct categories. Firstly, there are those that seem phenomenologically similar to the positive NDE with the exception that the experient finds the whole process unpleasant. Secondly, there are those involving experiences of visiting hellish regions and encountering the Devil or demonic beings. Finally, there are those in which the NDEr (i.e., the person experiencing the NDE) finds himself or herself in an isolated, featureless, eternal void.

(b) *Separation from the physical body:* OBEs were reported by 37% of Ring's (1980) sample. OBEs, which can occur independently of NDEs, involve the feeling that one's consciousness has become disconnected from one's physical body. Often the experience involves apparently being able to see one's physical body from an external vantage point, as was the case in about half of the OBErs (i.e., the person experiencing the OBE) in Ring's study.

(c) *Entering a region of darkness:* About a quarter of Ring's (1980) cases involved entering a transitional region of darkness, either before or after the OBE, which was sometimes referred to as "tunnel-like".

(d) *Seeing a brilliant light:* A brilliant light, which did not hurt the eyes, was reported by 16% of Ring's (1980) subjects. They felt drawn toward this light, which was often perceived to be some kind of spiritual being, such as God or Jesus. A panoramic life review may then take place during which key events in the subject's life are replayed, sometimes in the company of the spiritual being. The process is felt to be non-judgemental.

(e) *Through the light, entering another realm:* Around 10% of Ring's (1980) sample reported entering a spiritual realm, often described as a beautiful garden with heavenly music. Deceased relatives or other spiritual guides are apparently encountered in this realm. Also, some kind of natural border, such as a fence or a river, is often encountered. This seems to symbolically represent the point of no return — and the decision is made, often very reluctantly, to return to the physical body.

Attention has also focused upon the aftereffects of the NDE. In the immediate aftermath of an NDE, many NDErs feel that the experience was positive and life-enhancing, but some find the experience disturbing and difficult to talk about even when the experience itself was positive (Orne, 1995). How well individuals are able to integrate the experience into their everyday lives often depends crucially upon how their initial reports are received by family, friends, and nursing staff. All too often, this reaction can be to ridicule or dismiss such reports, sometimes leading the NDEr to doubt their own sanity. In the longer term, experients typically report that they are less materialistic, more spiritual, less competitive and, not surprisingly, have a decreased fear of death (e.g., Ring, 1980) but even then there may be some negative aftereffects (Bush, 1991; Greyson, 1997). Among the problems most often encountered are frustration at being unable to communicate the significance of the experience to others, fear of ridicule, despair at being returned to the ordinary everyday world having experienced such bliss, and difficulties with ordinary human relationships, having experienced perfect divine love. Friends and family may also have problems dealing with the transformation, and divorce rates are very high following NDEs. Negative, long-term aftereffects following distressing NDEs can be even more disabling, with sufferers understandably showing heightened fear of death, along with flashbacks and other symptoms of post-traumatic stress disorder (Greyson and Bush, 1992).

Although the definition of an NDE provided at the start of this chapter is as good as any other (and considerably better than some), it should be appreciated that no universally accepted definition of the NDE exists. This is an important issue insofar as researchers using different definitions of the NDE are likely to reach different conclusions regarding its nature, causes, and consequences. Given the complex and multifaceted nature of the experience, it should also come as no surprise that there are many other ways of categorizing the phenomenological elements of the NDE in addition to Moody's (1975, 1977) 15-element model (see Table 1) and Ring's (1980) five-stage model referred to above. For example, Noyes (1972) identified three developmental stages of the NDE (resistance, review, and transcendence), while Lundahl (1993) feels that research points to ten stages (peace, bodily separation, sense of being dead, entering the darkness, seeing the light, entering another world, meeting others, life review, deciding to or being told to return to life, and returning to the body). Noyes and Slymen (1978–1979) classified the common features into three categories (mystical, depersonalization, and hyperalertness) on the basis of factor analysis. Greyson (1985), on the basis of cluster analysis, arrived at four categories related to cognitive, affective, paranormal, and transcendental features.

In order to minimize the potential confusion that could be caused by different researchers adopting different definitions of the NDE, many studies employ standard scales to decide who has and who has not had an NDE and the "depth" of the experience. Although many different scales have been developed, discussion here will be limited to two of the most commonly used. Ring (1980) developed the "weighted core experience index" (often referred to as WCEI), in which the following components are assigned different weights: the subjective sense of being dead, feelings of peace, bodily separation, entering a dark region, encountering a presence or hearing a voice, taking stock of one's life, seeing or being enveloped in light, seeing beautiful colors, entering into the light, and encountering visible spirits. Each feature is scored for presence or absence and the weighted total of those features that are present gives a score between 0 and 29.

Greyson (1983) criticized Ring's scale on the grounds that people could get a fairly high score on the basis of very few typical NDE components. He developed an improved scale commonly referred to as the Greyson NDE scale. To do this, he began by listing 80 features that included all the main items from previous scales. Following an initial pilot study, this list was reduced to 33 items with three-point scaled answers. After further development, the final 16-item scale was produced with questions in four groups (relating to cognitive, affective, paranormal, and transcendental features). This final scale is essentially a modified version of Ring's scale with a maximum score of 32. It has good test–retest reliability and internal consistency. A score of 7 or higher is the criterion for a true NDE.

Greyson (1998), after reviewing all published estimates of the incidence of NDEs, concluded that they probably happen to between 9 and 18% of people who come close to death. The main focus of this chapter will be to consider what can be learned about the nature of NDEs from studying cardiac arrest survivors. This particular population is of special interest in this context for a number of reasons. First, the vast majority of studies of NDE studies are retrospective in nature, with accounts sometimes not being collected until years or even decades after the experience itself. With respect to cardiac arrest survivors, however, it is possible to set up prospective studies that allow for survivors to be interviewed within days of their experience, thus greatly reducing potential problems of memory distortion. Second, such studies potentially allow researchers to correlate objective, physiological and pharmacological measurements with the reported features of the NDE, providing very useful data with respect to testing different theoretical accounts. Third, this population allows for the possibility of objectively testing the veridicality of the OBE component of the NDE, as described below. Finally, it can be argued that the mental state of cardiac arrest victims provides the closest model we have to that of a dying brain (Parnia et al., 2001).

The next section will present an overview of the different theoretical approaches that have been adopted in attempts to explain the NDE. This will be followed by a consideration of studies of NDEs in cardiac arrest patients and the implications such studies may have for the different theoretical approaches. For reviews of various other aspects of the NDE, the reader is referred to Ring (1980, 1982), Sabom (1982), Morse (1990, 1992), Blackmore (1993, 1996a), Fenwick and Fenwick (1995), Bailey and Yates (1996), Greyson (2000a), Roe (2001), and Irwin (2003, Chapters 11 and 12, pp. 163–196).

Theoretical approaches to NDEs

Roe (2001) divides the different theoretical approaches into three broad categories, although it should be realized that such categorization is purely for convenience of presentation. In reality, the theories and models of the NDE are not distinct and independent, but instead show considerable overlap. The first broad category that Roe refers to is spiritual theories (also sometimes known as transcendental theories). The most popular interpretation of the NDE is that it is exactly what it appears to be to the person having the experience. It is taken as strong evidence that the (allegedly immaterial) mind (or soul) can become separated from the physical body (see artistic depiction in Fig. 1). It is further assumed that the NDE provides a glimpse of a spiritual realm to which souls migrate after death. The allegedly paranormal nature of NDEs would be supported if it could be shown that information gained during the OBE component of an NDE was veridical and could not have been obtained in any conventional way. Some prominent researchers (e.g., Sabom, 1982) in this area have put forward cases that they feel provide strong support for such a position, but critics (e.g., Blackmore, 1993; Woerlee, 2003) remain unconvinced that all non-paranormal explanations for such information acquisition have been ruled out. Blackmore (1996b, p. 480) lists several such explanations including "information available at the time, prior knowledge, fantasy or dreams, lucky guesses, and information from the remaining senses. Then there is selective memory for correct details, incorporation of details learned between the end of the NDE and giving an account of it, and the tendency to "tell a good story". The degree to which these factors might provide an adequate explanation for such cases is discussed in more detail below.

The second broad category encompasses psychological theories. Some of these offer reasonable explanations for some components of the experience but fail to provide an adequate explanation of all aspects. An example would be Noyes and Klett's (1976, 1977) suggestion that the NDE is a form of depersonalization which acts as a defense against the threat of death in situations of extreme danger, by allowing a sense of detachment and engagement in pleasurable fantasies. Tellegen and Atkinson's (1974) concept of psychological absorption, which may be defined as the propensity to focus attention on imaginative or selected

Fig. 1. It has long been believed that soul leaves the body at the moment of death, as illustrated above. Does the NDE offer support for such claims? (The soul leaves the body at the moment of death, William Blake, engraved by Schiavonetti, 1808; illustrating Blair's poem The Grave; reproduced with permission from the Mary Evans Picture Library).

sensory experiences to the exclusion of stimuli in the external environment, is relevant here as is Wilson and Barber's (1983) concept of fantasy proneness. The depersonalization model receives some support from Irwin's (1985) study which shows that OBErs show higher levels of absorption ability, and Council and Greyson's (1985) observation of both higher absorption and fantasy proneness scores amongst NDErs. The model accounts reasonably well for the OBE component of the NDE, but not for the fact that most NDEs involve a sense of "hyper-reality" and do not feel dream-like at all, as is typical of other forms of depersonalization.

Irwin (1993) pointed out that NDEs differ from depersonalization in a number of ways, e.g., that what is altered is not one's sense of identity but the association of this identity with bodily sensations. He suggested that the NDE is a dissociative state in which there is a dissociation of self-identity from bodily sensations and emotions. This is supported by Greyson's (2000b) observation that NDErs report more dissociative symptoms than a control group, consistent with a non-pathological dissociative response to stress. It is possible that this tendency developed originally as a coping mechanism in response to childhood trauma (Ring, 1980; Irwin, 1993).

Another psychological theory that has been proposed is that the NDE reflects memories of being born, with the tunnel representing the birth canal, the light at the end of the tunnel representing the lights in the delivery room, and the being of light representing the midwife, obstetrician, or father (Grof and Halifax, 1977; Sagan, 1979). Becker (1982) has presented a thorough refutation of this theory. As he points out, the experience of being born is only very superficially similar to the NDE. The birth canal would not appear to the fetus as a tunnel with a light at the end, down which the fetus would gently float. Instead, it would be dark and extremely constrained, and furthermore babies do not travel down the birth canal facing forward. There is also a great deal of evidence to show that young infants are simply incapable of laying down accurate autobiographical memories during the first couple of years of life. Blackmore (1983) argued that if the NDE truly reflected memories of birth, the nature of the birth ought to influence the type of NDE reported. She found, however, that the tunnel experience is just as common in those delivered by Caesarean section as in those delivered naturally.

The final broad category is organic theories. A number of theories have been proposed that attempt to account for components of the NDE in terms of brain function. It is worth noting that most components of the NDE are known to also occur in non-NDE contexts, many of which have the potential to allow for greater experimental control than would usually be the case in naturally occurring NDEs. Study of these components in the non-NDE context obviously provides strong direct evidence either for or against particular organic theories.

A number of theorists have considered the possible role of abnormal levels of blood gases in the NDE. Cerebral anoxia is the final common pathway to brain death and it is therefore instructive to consider the degree to which the symptoms of hypoxia and anoxia reflect those of NDEs. Whinnery (1997) pointed out that there are indeed many similarities between NDEs and the so-called G-LOC syndrome (i.e., acceleration (+Gz)-induced loss of consciousness). G-LOC episodes can occur in fighter pilots engaged in certain types of maneuver because the extreme acceleration involved can result in a loss of adequate blood supply to the brain. Based on observations of almost 1000 episodes of G-LOC, Whinnery noted that such episodes often involved "tunnel vision and bright lights, floating sensations, automatic movement, autoscopy, OBEs, not wanting to be disturbed, paralysis, vivid dreamlets of beautiful places, pleasurable sensations, psychological alterations of euphoria and dissociation, inclusion of friends and family, inclusion of prior memories and thoughts, the experience being very memorable (when it can be remembered), confabulation, and a strong urge to understand the experience." (Whinnery, 1997, p. 245). Life review, mystical insights, and long-lasting transformational aftereffects are not reported to be associated with G-LOC episodes, but this may simply reflect the fact that such episodes rarely involve any expectation of actually dying.

One common objection to the idea that anoxia may be a factor in some NDEs is the claim that anoxia is characterized by confused thinking, whereas the NDE is characterized by extreme clarity of thought. In fact, as Blackmore (1996a) points out, the effects of anoxia vary enormously depending upon the type of anoxia, its speed on onset, and the time until oxygen is restored. It should also be borne in mind that confused thinking is a state that is often attributed by an outside observer to an individual on the basis of their behavior. As Liere and Stickney (1963, p. 300) point out, "Hypoxia quickly affects the higher centers, causing a blunting of the finer sensibilities and a loss of sense of judgment and of self-criticism. The subject feels, however, that his mind is not only quite clear, but unusually keen." In the case of NDE reports, we should perhaps be wary of accepting the retrospective claims of great clarity of thought.

Hypercarbia is often associated with anoxia and can itself produce NDE-like symptoms including bright lights, OBEs, relived past memories, and mystical experiences (Meduna, 1950). Sabom (1982) reported an NDE in a 60-year-old man who had suffered a heart attack and cardiac arrest. Measured blood gases in this patient showed that levels in the periphery were relatively normal. Furthermore, it is widely recognized that NDEs can occur in people who are obviously not suffering from anoxia and/or hypercarbia. Blackmore (1996a), when arguing in favor of a possible role for such factors in the NDE, explicitly states that there may be many triggers for an NDE with anoxia being just one of them.

Other theorists have considered the possible role of neurotransmitters in generating the NDE. Early suggestions that NDEs are hallucinatory experiences caused by externally administered drugs used in resuscitation attempts can be rejected on empirical grounds. Not only are NDEs reported by patients in circumstances where we can be sure that no drugs have been administered, but research has shown that administration of such drugs leads to fewer and more muted NDEs (e.g., Ring, 1980).

Others have suggested models based upon naturally occurring neurotransmitters. Carr (1982) argued that endorphin release could account for many aspects of the NDE, a suggestion that was incorporated into the more comprehensive neurobiological model put forward by Saavedra-Aguilar and Gómez-Jeria (1989). Endorphins are released in times of stress and lead to a reduction in pain perception and a pleasant, even blissful, emotional state. Consistent with the idea that endorphin release may be responsible for the positive emotional tone of most NDEs are the occasional reports of pleasant NDEs changing into unpleasant "Hellish" NDEs upon administration of endorphin-blocking drugs such as naloxone (Judson and Wiltshaw, 1983). Morse et al. (1989) have argued that serotonin has a more important role to play in generating NDEs than endorphins, at least with respect to mystical hallucinations and OBEs.

Jansen (1989, 1997, 2001) pointed out that endorphins are not powerful hallucinogens, but that many aspects of the NDE, such as seeing lights and moving through tunnels, can be induced by the dissociative anesthetic ketamine. Jansen developed a model of NDEs on the basis of this observation. He started from the fact that some conditions that lead to NDEs (e.g., hypoxia) involve a sudden increase in the concentration of the excitatory neurotransmitter glutamate, which destroys neurons by overactivating NMDA receptors. Ketamine can bind to NMDA receptors, blocking this neurotoxicity and producing NDE-like symptoms. Jansen hypothesized that under those conditions that produce a flood of glutamate, as yet unidentified neuroprotective "endopsychosins", similar to ketamine, are released to prevent neuronal damage and that this results in the NDE. However, there are important differences between typical NDEs and typical ketamine experiences. For example, the latter are more likely to be frightening (Strassman, 1997) and to feel unreal (Fenwick, 1997).

The temporal lobe is almost certain to be involved in NDEs, given that both damage to and direct cortical stimulation of this area are known to produce a number of experiences corresponding to those of the NDE, including OBEs, hallucinations, and memory flashbacks (Penfield, 1955; Blanke et al., 2002, 2004). It is worth noting that both the temporal lobes and the limbic system are sensitive to anoxia, and that release of endorphins lowers the seizure threshold in the temporal lobes and the limbic system (Frenk et al., 1978). Britton and Bootzin (2004) recently produced evidence supporting the idea that altered temporal lobe functioning may be involved in the NDE. They found that individuals who had reported having NDEs had more temporal-lobe epileptiform electroencephalographic activity than a non-NDE control group and that this activity was almost completely lateralized to the left cerebral hemisphere. The control group used in this study consisted of individuals who had not come close to death rather than people who had come close to death but had not experienced an NDE. It is possible, therefore, that the findings are a generalized result of trauma rather than specifically relating to the NDE itself.

Although it seems likely that levels of blood gases, fluctuations in levels of various neurotransmitters, and dysfunction in the temporal lobes and associated structures play some role in the NDE, models which focus mainly on one of these aspects at the expense of the others often seem to be successful in providing plausible accounts of some components of the NDE but not others. This has led to the development of models that attempt to integrate the different types of explanation, most notably by Saavedra-Aguilar and Gómez-Jeria (1989) and Blackmore (e.g., 1993).

Saavedra-Aguilar and Gómez-Jeria's (1989) model invokes temporal-lobe dysfunction, hypoxia, psychological stress, and neurotransmitter changes to explain the NDE. According to this model, brain stress caused by traumatic events leads to the release of endogenous neuropeptides, neurotransmitters, or both, producing such effects as analgesia, euphoria, and detachment. These neurotransmitter effects combine with the effect of decreases in oxygen tension, primarily in limbic structures, to produce epileptiform discharges in the hippocampus and amygdala, leading to complex visual hallucinations and a life review.

Further hallucinations and a sensation of a brilliant light are produced by afterdischarges propagating through the limbic connections to other brain structures. The linguistic system also has a role to play in that the memory of these sensations is reconstructed following recovery to produce a narrative consistent with the individual's cultural beliefs and expectations.

As Greyson (2000a) points out, in common with other models based upon physical trauma and hypoxia, this model cannot account for NDEs experienced in the absence of physical injury even though such cases are known to occur. Greyson is also correct in pointing out that the model is based upon a number of unsupported assumptions and speculations derived from neurochemical investigations of non-humans and that some of its key elements (e.g., "brain stress") are vague. However, he is wrong in criticizing the model (following Rodin, 1989) for allegedly ascribing to abnormal activity in the temporal-lobes key features of the NDE "such as feelings of peace or bliss and sensations of being out of the body" (Greyson, 2002, pp. 235–236) that he claims "have not in fact been reported either in clinical seizures or in electrical stimulation of those brain structures" (Greyson, 2002, p. 235).

Feelings of bliss and even mystical feelings of oneness with the universe are often reported by temporal lobe epileptics just prior to a seizure. Perhaps most famously, Dostoyevsky, himself a temporal-lobe epileptic, provided the following description of this experience in his novel, *The Idiot:* "I have really touched God. He came into me, myself; yes, God exists, I cried. You all, healthy people can't imagine the happiness that we epileptics feel during the second before our attack."

The fact that the temporo-parietal area is associated with OBEs has already been referred to (also see Bunning and Blanke elsewhere in this volume). A recent report by Blanke et al. (2002) described an OBE induced in a 45-year-old woman who had suffered from complex partial seizures for 11 years, probably as a result of right temporal-lobe epilepsy. The OBE was induced by focal electrical stimulation of the right angular gyrus while undergoing evaluation for epilepsy treatment. As pointed out by Tong (2003), these findings are very similar to those reported by Wilder Penfield over 60 years ago (e.g., Penfield, 1955).

Blackmore (1993, 1996a) argued that different components of the NDE are likely to have different underlying physiological mechanisms, thus producing a model that is to a large extent a synthesis of suggestions from previous theorists along with some novel explanations of particular components. Given the heterogeneous nature of the NDE, this is a reasonable approach. There is no reason to assume that a single comprehensive theory will explain the entire phenomenon. In the limited space available here, full details of Blackmore's theory cannot be presented, but some illustrative examples of her approach will be provided. For example, she argues that the typically positive emotional tone of the NDE may well be due to endorphin release, with the rarer negative NDEs caused by morphine antagonists. Her explanation for the tunnel effect is that it is caused by neuronal disinhibition in the visual cortex, typically (but not necessarily) caused by anoxia. She argues that such cortical disinhibition would cause random firing of the cells in the visual cortex. As there are more cells devoted to the fovea than the peripheral visual field, the subjective experience of this phenomenon is that of a bright circle of light that grows larger as more cells begin to fire, giving the impression of moving down a tunnel towards the light (Fig. 2 shows what some consider as the first artistic representation of this "tunnel to paradise" experience).

Blackmore (1996b) explains the OBE component of the NDE by arguing that our sense of self is a mental construction as indeed is our entire model of reality. The latter is constantly updated on the basis of the ongoing interaction between incoming sensory information (the so-called "bottom-up processing") and our existing knowledge, beliefs, and expectations about the world ("top–down processing"). Our sense of self is one important aspect of our model of reality and, given the predominance of visual information in producing this model, the sense of self is often felt to be just behind the eyes. Under most conditions, our model of reality is strongly influenced by sensory input and, as a consequence, our model of reality corresponds well with the external world. In

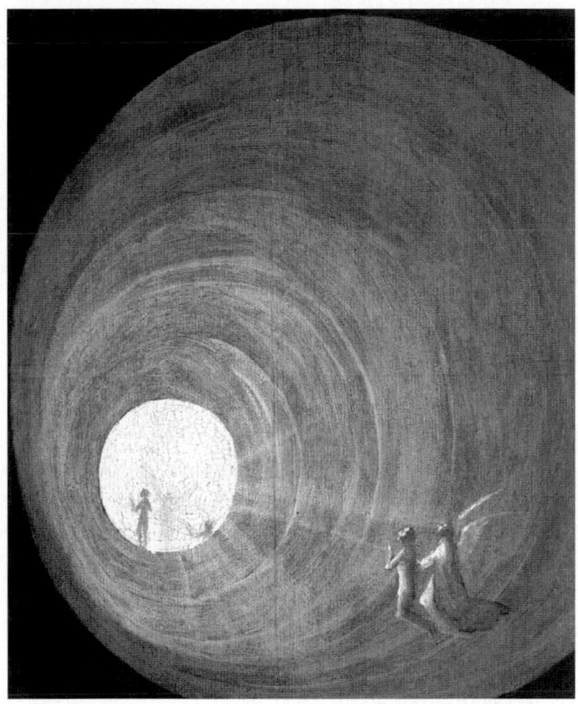

Fig. 2. The first artistic depiction of a NDE? "Paradise and the Ascent to the Empyrean" by Hieronymus Bosch appears to show angels, guiding spirits down a tunnel toward a bright light (detail from panel of the four afterlife paintings, 1500s).

some circumstances, however, especially when the sensory input is degraded in some way (e.g., through the ingestion of drugs, through meditation, or through anoxia), we can, without realizing it, adopt a model of reality that is based upon top–down influences. Thus the world of the OBE is, according to Blackmore, based upon memory, imagination, and expectation. During the NDE, the model of reality that is operative during the OBE will also incorporate any remaining sensory input and any unusual sensations caused by abnormal activity in the brain. Blackmore (1993) essentially argues that there is a common core experience to the NDE that is physiologically determined, but top–down influences can come into play in terms of influencing the detail of what is experienced and how it is interpreted. For example, the religious figures that are met during the NDE almost always correspond to the religion of the person having the experience, with Christians tending to see Jesus and Hindus seeing the messengers of Yamraj coming to collect them (Osis and Haraldsson, 1977).

Studies of NDEs in cardiac arrest patients

Cassem and Hackett (1978) wrote that at that time the incidence of NDEs in cardiac arrest survivors was unknown, but estimated it to be about 2%. Martens (1994) cites two small-scale studies of consecutive survivors of cardiac arrest (Druss and Kornfeld, 1967; Dobson et al., 1971) in support of the conclusion that NDEs are unusual in such patients. He also offers as support Negovsky's (1993) assertion that only 0.3–0.5% of resuscitated people talk about visions they had during clinical death (i.e., cardiac arrest). He goes on, however, to echo previous calls (e.g., Greyson and Stephenson, 1980) for "a prospective study on patients who are at risk for sudden cardiac death, such as a cohort of coronary care unit patients" (Martens, 1994, p. 173) in order to "eliminate sampling biases against a background of emotionally driven views, myths and anecdotes" (Martens, 1994, p. 173). We are now in a position where four such studies have been reported. As already stated, such prospective studies provide data that are much less likely to be contaminated by memory bias that might affect previous retrospective studies in which variable lengths of time elapse, typically from a few months to several years, between the NDE and the collection of interview data. Furthermore, prospective studies maximize the opportunity to collect data on objective physiological measurements in order to address issues of etiology.

Parnia et al. (2001) reported a pilot prospective study designed to assess the frequency of NDEs in cardiac arrest survivors and to determine the features of such NDEs. Participants in the study were all of the survivors of cardiac arrest over a 1-year period on the medical, emergency, and coronary care units of a British hospital. Patients were asked if they had any memories occurring during unconsciousness. The Greyson NDE scale (Greyson, 1983; Lange et al., 2004) was used to assess interview data, with a score of 7 or more constituting the criterion for a true NDE. Data concerning a range of objective physiological measures were also collected, including arterial and peripheral blood gas levels (along with sodium and potassium levels), records of any drugs administered, and records of any abnormality of cardiac rhythm during the arrest. Over the year, 63 survivors were interviewed within a week of their cardiac arrest. Seven (11.1%) reported memories of their period of unconsciousness, of whom four (6.3% of the total sample) had had NDEs according to the criterion employed. All found the experience to be pleasant.

Due to the small number of patients reporting NDEs, it was not possible to draw any clear conclusions regarding possible causative physiological factors, although it was noted that partial pressure of oxygen was higher in the NDE group. This may simply reflect sampling bias given the low number of data points in the NDE group, as acknowledged by the investigators. They also tentatively suggest that this observation may serve to somewhat undermine the idea that cerebral anoxia is an important causative factor in the NDE, given that the NDE group showed higher levels of oxygen during resuscitation. This objection depends upon (a) blood oxygen levels accurately reflecting cerebral oxygen levels and (b) the NDE occurring during the resuscitation phase. Given that both assumptions are open to question and the difference noted is not statistically significant anyway, Parnia et al. (2001) are quite correct to caution against drawing any significant conclusion from this finding.

One notable feature of this study was the attempt by Parnia et al. (2001) to obtain evidence relating to the possible veridicality of the OBE. Boards were suspended from the ceilings on the wards involved in such a way that figures on the surface facing the ceiling were only visible from a vantage point near the ceiling. It was hoped that any individual reporting an OBE involving experienced movement to such a vantage point would be able to report accurately the hidden target figures. Such evidence would constitute a major challenge to those supporting a non-paranormal explanation of the NDE. Unfortunately, and somewhat atypically, none of the survivors in this sample experienced an OBE.

A large-scale prospective study was carried out in the Netherlands by Van Lommel et al. (2001). This involved 344 consecutive cardiac arrest survivors across 10 Dutch hospitals, all interviewed within a few days of resuscitation. The study found that 62 patients (18%) reported an NDE, of whom 41 (12%) reported a "core experience". NDE was defined in this study as "the reported memory of all impressions during a special state of consciousness, including specific elements such as OBE, pleasant feelings, and seeing a tunnel, a light, deceased relatives, or a life review" (p. 2040). Patients assigned to the "core experience" group were those scoring 6 or more on Ring's (1980) WCEI. No patients reported distressing NDEs. The 62 patients reporting NDEs were compared with the non-NDE survivors with respect to a number of demographic, medical, pharmacological, and psychological indices.

Such factors as duration of cardiac arrest or unconsciousness, medication, fear of death before cardiac arrest, and the duration of the interval between the NDE and the interview were not related to the occurrence of an NDE. However, people younger than 60 were more likely to report an NDE than older people, as were those suffering their first myocardial infarction. Deeper experiences were reported by patients surviving cardiopulmonary resuscitation (CPR) outside hospital, and were more likely to be reported by women and those reporting being afraid before CPR. More of the patients who had had an NDE, especially a deep one, died within 30 days of CPR. Memory problems following lengthy CPR appeared to make it less likely that an NDE would be reported but it was unclear how memory had actually been assessed.

This study also included a longitudinal aspect by following up sub-groups after 2 years and then again after 8 years in order to assess any life-transforming effects of the NDE. At the 2-year follow-up stage, 37 of the 62 patients who originally reported an NDE were available and willing to be interviewed, and 37 surviving control participants were also interviewed. Of the former group, it now appeared that 6 had not in fact had an NDE at all, whereas 4 of the control group had, resulting in 35 patients who had had an NDE and 39 who had not. Eight years after the initial interview, available survivors from the first follow-up were interviewed for a second time, with 23 patients in the NDE group and 15 in the control group. Broadly speaking, results confirmed previous studies of this issue, indicating that NDEs produce long-lasting effects in terms of, for example, reduced fear of death, heightened intuition, social awareness, and so on.

Schwaninger et al. (2002) evaluated all patients who suffered a cardiac arrest between April 1991 and February 1994 at Barnes-Jewish Hospital in St. Louis (with the exception of those in the surgical intensive care unit) in order to assess the frequency of NDEs in cardiac arrest patients, to characterize the experiences, and to assess their impact on psychosocial and spiritual attitudes. Patients were interviewed and assessed using Greyson's (1983) NDE scale and an experience rating form. No attempt was made to relate characteristics of the reported NDEs to objective physiological data. Of 174 cardiac arrests over this period, 119 (68%) died. Of the remaining 55, 30 patients (17%) were interviewed and the remainder were excluded either because they had neurological damage or were intubated through to the point of discharge. Of the 30 patients in the study sample, 19 (63%) had never experienced an NDE, 7 (23%) experienced an NDE during the cardiac arrest, and 4 (13%) did not have an NDE during the index cardiac arrest but had experienced one on a previous life-threatening occasion. Given the fact that the stated objectives of this study all specifically relate to cardiac arrest, it is unfortunate that over a third of the final NDE group had almost certainly experienced their NDEs in different circumstances. Patients were also sent a follow-up questionnaire 6 months after the initial interview. This was returned by 8 of the 11 NDE patients and 10 of the 19 non-NDE patients. Results confirmed previous findings concerning the nature of NDEs and of long-lasting transformational effects of the NDE with regard to personal understanding of life and self, social attitudes, and changes in social customs and religious/spiritual beliefs.

Greyson (2003) carried out a prospective study of all patients admitted to the cardiac intensive care unit or the cardiology step-down unit at the University of Virginia Hospital over a 30-month period, with the exception of those who were too ill, psychotic, or cognitively impaired to be interviewed. Of the consecutive sample of 1595 patient admissions, 116 (7%) had a primary diagnosis of cardiac arrest. NDEs, as assessed by the Greyson (1983) NDE scale, were reported by 10% of cardiac arrest patients compared to only 1% of other cardiac patients. Compared to non-NDE patients, those reporting NDEs did not differ in terms of "sociodemographic variables, social support, quality of life, acceptance of their illness, cognitive function (as assessed using a standard instrument, the Mini-Mental State Exam; Folstein et al., 1975), capacity for physical activities, degree of cardiac dysfunction, objective proximity to death, or coronary prognosis" (Greyson, 2003, p. 269). Those reporting NDEs were younger, more likely to have lost consciousness, and to report higher levels of approach-oriented death acceptance. They were also more likely to report prior so-called "paranormal" experiences (e.g., déjà vu, altered states of consciousness, and related activities such as meditation) but not "psychic" experiences, such as extrasensory perception (note the rather idiosyncratic usage of the word "paranormal"). This latter observation is consistent with the idea of an "NDE-prone personality" as proposed by Ring (1992). Although prior religion and religiosity is not related to tendency to experience NDEs, a prior tendency to cope with trauma through dissociating may be, although Blackmore (1993) argues that there is simply too little evidence to draw firm conclusions on this issue.

The four prospective studies described above have reported a range of incidence rates for survivors of cardiac arrest: 6.3% (Parnia et al., 2001), 12% (Van Lommel et al., 2001), 23% (Schwaninger et al., 2002), and 10% (Greyson, 2003). Given that the two larger-scale studies are in fairly close agreement, it would appear that the best estimate for incidence of NDEs among cardiac arrest survivors is around 10–12%, considerably higher than earlier estimates. It is unclear why such a discrepancy should have arisen, but the prospective studies are methodologically superior to the earlier retrospective studies and thus should be accorded more evidential weight.

Parnia and Fenwick (2002) have reviewed evidence relating to cerebral physiology during and after cardiac arrest. As they point out, cardiac arrest is the closest model we have to the dying process itself and may thus provide insight into the experiences likely to accompany the dying process irrespective of cause. The brain receives 15% of cardiac output under normal conditions, and 40–50% of total cerebral blood flow is required to supply enough glucose and oxygen to maintain cellular integrity with 50–60% needed to maintain electrophysiological activity (Buunk et al., 2000).

Both human and animal studies have provided extensive data on cerebral physiology during and after cardiac arrest (Parnia and Fenwick, 2002). Immediately following arrest, blood pressure drops sharply, and even properly performed chest compressions and the administration of epinephrine would typically not be enough to raise diastolic and mean blood pressure to the level needed for adequate coronary and cerebral perfusion. EEGs recorded during cardiac arrest show an initial slowing that, within 10–20 s, progresses to an isoelectric (flat) line. The EEG then remains flat during the cardiac arrest until cardiac output resumes in cases of early defibrillation but may not return for many hours if cardiac arrest is prolonged. Animal studies in which activity is recorded from deep brain structures by in-dwelling electrodes show that absence of cortical activity correlates with absence or reduction of activity in deep structures. Even after the restoration of adequate blood pressure and gross cerebral blood flow rate, local cerebral blood perfusion can remain severely impaired due to local increases in vasoconstriction. Thus normal EEG may not begin to recover for a long period despite the maintenance of adequate blood pressure during the recovery phase (Buunk et al., 2000).

Several NDE investigators have argued that the findings from the prospective studies reviewed above pose a major challenge to current scientific thinking regarding the relationship between mind and brain, and in particular the assumptions that (a) higher cognitive functions can only occur when cerebral functioning is relatively unimpaired and (b) that consciousness is entirely dependent upon the underlying neuronal substrate and cannot become separated from that substrate. In the words of Parnia and Fenwick (2002, p. 8), "The occurrence of lucid, well-structured thought processes together with reasoning, attention and memory recall of specific events during a cardiac arrest (NDE) raise a number of interesting and perplexing questions regarding how such experiences could arise. These experiences appear to be occurring at a time when cerebral function can be described at best as severely impaired, and at worst absent." Greyson (2003, p. 275) concurs:

> The paradoxical occurrence of heightened, lucid awareness and logical thought processes during a period of impaired cerebral perfusion raises particularly perplexing questions for our current understanding of consciousness and its relation to brain function. [...] [A] clear sensorium and complex perceptual processes during a period of apparent clinical death challenge the concept that consciousness is localized exclusively in the brain.

Similarly, Van Lommel and colleagues (2001, p. 2044) ask, "How could a clear consciousness outside one's body be experienced at the moment that the brain no longer functions during a period of clinical death with flat EEG?" With respect to the latter point, however, it should be borne in mind that electroencephalography is not a perfectly reliable indicator of brain death, as it is only able to detect activity in one half of the area of the cerebral cortex. Activity in the other half of the cerebral cortex and in the deeper structures cannot be observed (Paolin et al., 1995; Bardy, 2002).

It is clear that the argument that recent findings present a major challenge to modern neuroscience hinges upon the claim that the NDE is actually experienced "during a period of clinical death with flat EEG" as claimed, with the implication that no cortical activity is taking place during this period. Parnia and Fenwick (2002) essentially argue, on the basis of the physiological data reviewed above, that the NDE cannot occur as the patient is becoming unconscious as this happens too quickly. They further argue that it cannot happen during recovery from cardiac arrest as this phase would be characterized by confused thinking, whereas the experience is usually described as involving great clarity of thought. Other commentators have argued, however, that the possibility remains that the NDE may indeed have occurred as patients rapidly entered the period of flat EEG or as they more gradually recovered from that state (e.g., French, 2001). With regard to the possibility that the experience may have occurred as the patient rapidly entered unconsciousness, it should be borne in mind that altered states of consciousness often have an effect on time perception. This is indeed illustrated very well by the life review component of the NDE itself during which it is claimed that the whole of an individual's life is replayed in a fraction of a second. Who can say, therefore, that the few seconds of remaining consciousness as an individual enters the state of clinical death is insufficient for the experiences that form the basis of the NDE? With respect to the claim that an individual could not have an NDE as they slowly emerge back into consciousness because their thinking will be confused, it has already been pointed out (in discussing the effects of anoxia) that the subjective claim of great clarity of thought may well be an illusion. Finally, and most importantly, it should be borne in mind that we are always dealing with reports of experiences rather than with the experiences themselves. Memory is a reconstructive process. It is highly likely the final narrative will be much more coherent after the individual has reflected upon it before telling it to others, given the inherently ineffable nature of the experience itself.

Parnia and Fenwick (2002) also claim that anecdotal evidence of veridical perception during the OBE phase of the NDE supports the argument that such perception sometimes occurs during periods of clinical death. Unfortunately, many such anecdotes are essentially uncorroborated hearsay. As Blackmore (1993) points out, when serious attempts at corroboration are attempted, the evidence often turns out to be nowhere near as impressive as it initially appeared. Furthermore, as stated earlier, there are several non-paranormal factors that might account for those instances when people do sometimes provide accurate accounts of events that took place during the NDE.

One of the cases often presented as being a strong challenge to those theorists arguing in favor of non-paranormal accounts of the NDE is that of Pam Reynolds as presented by cardiologist Michael Sabom (1998). In 1991, 35-year-old Reynolds was operated upon by Dr. Robert Spetzler in order to remove a potentially fatal giant basilar artery aneurysm. Standard neurological operating techniques could not be used because of the size and location of the aneurysm and instead a more complex procedure known as hypothermic cardiac arrest was employed. This involved lowering body temperature to 60 °F (i.e., 16 °C), stopping heartbeat and breathing, flattening of brainwaves, and the draining of blood from the head. The aneurysm was then carefully removed, and the patient's body warmed up, normal heartbeat and circulation restored, and head and other wounds were closed. Reynolds was then allowed to awaken slowly in the recovery room. When she was once again able to speak, she told of a NDE that had apparently occurred while she was unconscious under general anesthetic and low-temperature cardiac arrest.

Reynolds reported that she awoke during the early stages of the operation to the sound of the small pneumatic saw that was being used to open her skull. She then felt as if she was being pulled out through the top of her head and, during the subsequent OBE, she was able to watch the proceedings from above the neurosurgeon's shoulder. Her account accords very well with those of the medical staff present at the time, including her description of the pneumatic saw and the fact that the cardiac surgeon expressed surprise that the blood vessels in her right groin were too small to handle the large flow of blood needed to feed the cardiopulmonary bypass machine. She reported that after her heart was stopped and the blood drained from her body, she passed through a black vortex and into a realm of light where she met with deceased relatives. These relatives looked after her, provided her with nourishment, and eventually helped her to return to her physical body. She was able to report the music that was being played in the operating theatre at the point of her return.

This case is often presented as one that simply defies all conventional explanations (e.g., Greyson, 2000). Woerlee (2005a, b), an anesthesiologist with many years of clinical experience, has considered this case in detail and remains unconvinced of the need for a paranormal explanation. He points out that it is perfectly possible for patients to regain consciousness during operations because the concentration of sleep-inducing and maintaining drugs may vary. Even though such patients cannot move and feel no pain because of the effects of other administered drugs, they may be perfectly aware of what is going on around them. If their eyes are open, they can actually see what is going on in the operating theatre, but even with eyes closed (Reynolds' eyes were taped shut) they are likely to be able to internally visualize proceedings fairly accurately on the basis of other sensory inputs. It should be noted that the OBE phase of Reynolds' NDE took place during the early phase of the operation, before the cardiac bypass apparatus had even been connected to her body.

Greyson (2000), among others, rejects the idea that Reynolds may have been able to hear during the operation because she had small molded speakers inserted in her ears that he claims would block out any other auditory stimulation. These speakers are used to emit 100-db clicks so that auditory evoked potentials (AEPs) recorded from the brainstem can be used to monitor levels of consciousness throughout the operation. However, anyone who has ever worn earphones to listen to music will readily acknowledge that they do not totally block out other sounds from the environment. Sound is transmitted into the auditory pathways not only via the ear itself but also by bone conduction.

Woerlee (2005b) also draws attention to the fact that Reynolds could only give a report of her experience some time after she recovered from the anesthetic as she was still intubated when she regained consciousness. This would provide some opportunity for her to associate and elaborate upon the sensations she had experienced during the operation with her existing knowledge and expectations. The fact that she described the small pneumatic saw used in the operation also does not impress Woerlee. As he points out, the saw sounds like and, to some extent, looks like the pneumatic drills used by dentists. The sound heard by Reynolds when she regained consciousness during the early phase of the operation was unlikely to be interpreted as being a large chain saw or industrial angle cutter even in the unlikely event that the patient might have expected such inappropriate instruments to be used. As Sabom (1998, p. 189) himself acknowledges, "For some, evidence arising from cases such as Pam's will continue to suggest some type of out-of-body experience occurring when death is imminent. For others, the inexactness which arises in the evaluation of these cases will be reason enough to dismiss them as dreams, hallucinations, or fantasies."

Many commentators are particularly impressed by accounts from blind people that during NDEs they are able to see even though, in some cases, they may have been blind from birth (Ring and Cooper, 1999). Initial readings of such accounts often give the impression that the experience involves seeing events and surroundings in the same way that sighted people do, but closer reflection upon these cases suggests otherwise. As Ring (2001, p. 69) writes,

> As this kind of testimony builds, it seems more and more difficult to claim that the blind simply see what they report. Rather, it is beginning to appear it is more a matter of their knowing, through a still poorly understood mode of generalized awareness, based on a variety of sensory impressions, especially tactile ones, what is happening around them.

Ring argues that casual readers may often gain the impression that such NDEs in the blind involved literally seeing events and surroundings but that this impression is based upon having to describe these experiences, no matter how they were originally coded, in linguistic form. Our language is based largely upon the experiences of the sighted and is therefore a "language of vision". Ring believes that the form of knowing reported by the blind during their NDEs, which he refers to as "mindsight", is beyond conventional scientific explanation and is in fact the same form of transcendent knowing that is reported by the sighted during NDEs. But we are then faced once more with the question of whether or not those cases involving apparent veridical perception could be explained by the non-paranormal factors discussed above. The consequence of this line of argument is that NDEs in the blind are certainly worthy of study but do not merit any special status in terms of evidential support for spiritual explanations of the phenomenon.

There can be little doubt, however, that the OBE component of the NDE still provides the best opportunity to seriously challenge conventional views of the relationship between consciousness and the brain. Although somewhat unconventional (to say the least), attempts to test the veridicality of OBEs using hidden targets (e.g., Parnia et al., 2001) should be welcomed. Should any such test ever produces convincing evidence that the OBE truly allows one to view the world without using the known visual channels, this would indeed be a major challenge to conventional science. To date, no such evidence has been forthcoming.

It is also worth noting in this context that OBEs can be studied in non-NDE contexts (e.g., Blackmore, 1996b; Blanke et al., 2002, 2004). As there appears to be little reason to believe that NDE-based and other OBEs differ in terms of their underlying psychological and physiological causes, it would obviously make sense to pursue research into the veridicality of OBEs of the non-NDE type. However, past attempts to do exactly that have failed to produce convincing evidence for any paranormal aspect to the OBE (Blackmore, 1996b).

French (2001) raised the possibility that at least some reports of NDEs might be based upon false memories, of the mind trying retrospectively to "fill the gap" after a period of disrupted cortical activity followed by cortical inactivity. This suggestion was based upon the observation that, in Van Lommel et al. (2001) study, around 10% of the control sample, i.e., individuals who had not reported NDEs immediately following their cardiac arrests, were reporting that in fact they had had an NDE at the 2-year follow-up stage. It is now known that simply imagining an event that never actually happened can lead to the development of false memories for that event (e.g., Loftus, 2001; McNally, 2003). It is worth noting that susceptibility to false memories is known to be correlated with tendency to dissociate (e.g., Heaps and Nash, 1999; Hyman and Billings, 1999), which in turn correlates with tendency to report NDEs (Greyson, 2000b; see French, 2003, for a detailed review of the possible relationship between susceptibility to false memories and tendency to report ostensibly paranormal experiences).

However, it should also be borne in mind that there is another possible explanation for the fact that many of van Lommel et al.'s patients reported having had an NDE during the follow-up interviews despite having failed to do so when first interviewed: it may simply be that, although they were always aware of having had an NDE, they feared being ridiculed or diagnosed as mentally ill if they reported their experience when first interviewed (Greyson, 1988, 2003). Perhaps by the time of the second interview they had had the opportunity to research the phenomenon and had realized how common it is among near-death survivors and that it is not associated in any way with mental illness. They may thus have been more willing to talk about their NDE on this occasion. If this is the case, it is an important practical consideration that needs to be addressed urgently in future research if the research community is to gain an accurate picture of incidence rates.

There is no doubt that the research reviewed above raises far more questions than it answers. While none of the studies produced evidence in support of any of the organic theories of NDEs, it should be noted that two of the four prospective studies (Schwaninger et al., 2002; Greyson, 2003) did not attempt to collect data relevant to such theories and another (Parnia et al., 2001) involved a sample that was too small for any valid conclusions on such issues to be drawn. Future research should focus on devising ways to distinguish between the two main hypotheses relating to when the NDE is occurring. If it really is occurring when some claim that it is, during a period of flat EEG with no cortical activity, then modern neuroscience would require serious revision. This would also be the case if the OBE, either within the NDE or not, could be shown to be veridical. However, both of these claims currently remain open to dispute. Challenges facing those proposing purely organic theories include not only producing direct evidence in support of their accounts, but also satisfactorily accounting for those NDEs that are known to occur in the complete absence of physical threat, such as those that occur when individuals are not actually close to death but only think they are. Further research into this enigmatic phenomenon is likely to cast light upon some fundamental issues relating to the nature of consciousness.

References

Bailey, L.W. and Yates, J. (Eds.). (1996) The Near Death Experience: A Reader. Routledge, New York and London.

Bardy, A.H. (2002) Near-death experiences [letter]. Lancet, 359: 2116.

Becker, C.B. (1982) The failure of Saganomics: Why birth models cannot explain near-death phenomena. Anabiosis J. Near Death Stud., 2: 102–109.

Blackmore, S.J. (1983) Birth and the OBE: an unhelpful analogy. J. Am. Soc. Psychical Res., 77: 229–238.

Blackmore, S.J. (1993) Dying to Live: Science and the Near-Death Experience. Grafton, London.

Blackmore, S.J. (1996a) Near-death experiences. In: Stein G. (Ed.), The Encyclopedia of the Paranormal. Prometheus Books, Amherst, NY, pp. 425–441.

Blackmore, S.J. (1996b) Out-of-body experiences. In: Stein G. (Ed.), The Encyclopedia of the Paranormal. Prometheus Books, Amherst, NY, pp. 471–483.

Blanke, O., Landis, T., Spinelli, L. and Seeck, M. (2004) Out-of-body experience and autoscopy of neurological origin. Brain, 127: 243–258.

Blanke, O., Ortigue, S., Landis, T. and Seeck, M. (2002) Stimulating illusory own-body perceptions. Nature, 419: 269–270.

Britton, W.B. and Bootzin, R.R. (2004) Near-death experiences and the temporal lobe. Psychol. Sci., 15: 254–258.

Bush, N.E. (1991) Is ten years a life review? J. Near Death Stud., 10: 5–9.

Buunk, G., van der Hoeven, J.G. and Meinders, A.E. (2000) Cerebral blood flow after cardiac arrest. Neth. J. Med., 57: 106–112.

Carr, D.B. (1982) Pathophysiology of stress-induced limbic lobe dysfunction: A hypothesis relevant to near-death experiences. Anabiosis J. Near Death Stud., 2: 75–89.

Cassem, N.H. and Hackett, T.P. (1978) The setting of intensive care. In: Hackett T.P. and Cassem N.H. (Eds.), Massachusetts General Hospital Handbook of General Hospital Psychiatry. Mosby, St. Louis, pp. 319–341.

Council, J.R. and Greyson, B. (1985). Near-death experiences and the fantasy-prone personality: Preliminary findings. Paper presented at the 93rd Annual Convention of the American Psychological Association, Los Angeles, August 1985.

Dobson, M., Tattersfield, A.E., Adler, M.W. and McNicholl, M.W. (1971) Attitudes and long-term adjustment of patients surviving cardiac arrest. Br. Med. J., 111: 207–212.

Druss, R.G. and Kornfeld, D.S. (1967) The survivors of cardiac arrest: A psychiatric study. J. Am. Med. Assoc., 201: 291–296.

Fenwick, P. (1997) Is the near-death experience only N-methyl-D-aspartate blocking? J. Near Death Stud., 16: 43–53.

Fenwick, P. and Fenwick, E. (1995) The Truth in the Light: An Investigation of over 300 Near-Death Experiences. Headline, London.

Folstein, M.F., Folstein, S.E. and McHugh, P.R. (1975) "Mini-mental state": A practical method for grading the cognitive state of patients for the clinician. J. Psychiatr. Res., 12: 189–198.

French, C.C. (2001) Dying to know the truth: visions of a dying brain, or false memories? Lancet, 358: 2010–2011.

French, C.C. (2003) Fantastic memories: The relevance of research into eyewitness testimony and false memories for reports of anomalous experiences. J. Conscious. Stud., 10: 153–174.

Frenk, H., McCarty, B.C. and Liebeskind, J.C. (1978) Different brain areas mediate the analgesic and epileptic properties of enkephalin. Science, 200: 335–337.

Greyson, B. (1983) The near-death experience scale: Construction, reliability, and validity. J. Nerv. Ment. Dis., 171: 369–375.

Greyson, B. (1985) A typology of near-death experiences. Am. J. Psychiat., 142: 967–969.

Greyson, B. (1997) The near-death experience as a focus of clinical attention. J. Nerv. Ment. Dis., 185: 327–334.

Greyson, B. (1998) The incidence of near-death experiences. Med. Psychiat., 1: 92–99.

Greyson, B. (2000a) Near-death experiences. In: Cardeña E., Lynn S.J. and Krippner S. (Eds.), Varieties of Anomalous Experiences: Examining the Scientific Evidence. American Psychological Association, Washington DC, pp. 315–352.

Greyson, B. (2000b) Dissociation in people who have near-death experiences: out of their bodies or out of their minds? Lancet, 355: 460–463.

Greyson, B. (2003) Incidence and correlates of near-death experiences in a cardiac care unit. Gen. Hosp. Psychiat., 25: 269–276.

Greyson, B. and Bush, N.E. (1992) Distressing near-death experiences. Psychiatry, 55: 95–110.

Greyson, B. and Stephenson, I. (1980) The phenomenology of near-death experiences. Am. J. Psychiat., 137: 1193–1196.

Grof, S. and Halifax, J. (1977) The Human Encounter with Death. Dutton, New York.

Heaps, C. and Nash, M. (1999) Individual differences in imagination inflation. Psychon. Bull. Rev., 6: 313–318.

Hyman Jr., I.E. and Billings, F.J. (1999) Individual differences and the creation of false childhood memories. Memory, 6: 1–20.

Irwin, H.J. (1985) Flight of Mind: A Psychological Study of the Out-of-Body Experience. Scarecrow Press, Metuchen, NJ.

Irwin, H.J. (1993) The near-death experience as a dissociative phenomenon: An empirical assessment. J. Near Death Stud., 12: 95–103.

Irwin, H.J. (2003) An Introduction to Parapsychology. (4th ed.). McFarland and Company, Jefferson, NC, and London.

Jansen, K.L.R. (1989) Near-death experience and the NMDA receptor. Br. Med. J., 298: 1708–1709.

Jansen, K.L.R. (1997) The ketamine model of the near-death experience: A central role for the N-methyl-D-aspartate receptor. J. Near Death Stud., 16: 79–95.

Jansen, K.L.R. (2001) Ketamine: Dreams and Realities. Multidisciplinary Association for Psychedelic Studies (MAPS), Sarasota, FL.

Judson, I.R. and Wiltshaw, E. (1983) A near-death experience. Lancet, 8349: 561–562.

Lange, R., Greyson, B. and Houran, J. (2004) A Rasch scaling validation of a 'core' near-death experience. Br. J. Psychol., 95: 149–177.

Liere, E.J. and Stickney, J.C. (1963) Hypoxia. University of Chicago Press, Chicago.

Loftus, E.F. (2001) Imagining the past. Psychologist, 14: 584–587.

Lundahl, C.R. (1993) The near-death experience: A theoretical summarization. J. Near Death Stud., 12: 105–118.

Martens, P.R. (1994) Near-death-experiences in out-of-hospital cardiac arrest survivors. Meaningful phenomena or just fantasy of death? Resuscitation, 27: 171–175.

McNally, R.J. (2003) Remembering Trauma. Harvard University Press, Cambridge, MA.

Meduna, L.J. (1950) Carbon Dioxide Therapy. Charles C. Thomas, Springfield, IL.

Moody, R.A. (1975) Life After Life. Bantam Books, New York.

Moody, R.A. (1977) Reflections on Life After Life. Mocking Bird Books, St. Simon's Island, GA.

Morse, M. (1990) Closer to the Light. Souvenir, London.

Morse, M. (1992) Transformed by the Light: The Powerful Effect of Near-Death Experiences on People's Lives. Ballantine, New York.

Morse, M.L., Venecia, D. and Milstein, J. (1989) Near-death experiences: A neurophysiological explanatory model. J. Near Death Stud., 8: 45–53.

Negovsky, V.A. (1993) Death, dying and revival: ethical aspects. Resuscitation, 25: 99–107.

Noyes, R. (1972) The experience of dying. Psychiatry, 35: 174–184.

Noyes, R. and Klett, R. (1976) Depersonalisation in the face of life-threatening danger: an interpretation. Omega, 7: 103–114.

Noyes, R. and Klett, R. (1977) Depersonalisation in the face of life-threatening danger. Compr. Psychiat., 18: 375–384.

Noyes, R. and Slymen, D. (1978–1979) The subjective response to life-threatening danger. Omega, 9: 313–321.

Orne, R.M. (1995) The meaning of survival: The early aftermath of a near-death experience. Res. Nurs. Health., 18: 239–247.

Osis, K. and Haraldsson, E. (1977) At the Hour of Death. Avon, New York.

Paolin, A., Manuali, A., Di Paola, F., Boccaletto, F., Caputo, P., Zanata, R., Bardin, G.P. and Simini, G. (1995) Reliability in diagnosis of brain death Intens Care Med, 21: 657–662.

Parnia, S. and Fenwick, P. (2002) Near death experiences in cardiac arrest: visions of a dying brain or visions of a new science of consciousness. Resuscitation, 52: 5–11.

Parnia, S., Waller, D.G., Yeates, R. and Fenwick, P. (2001) A qualitative and quantitative study of the incidence features and aetiology of near death experiences in cardiac arrest survivors. Resuscitation, 48: 149–156.

Penfield, W. (1955) The role of the temporal cortex in certain psychical phenomena. J. Ment. Sci., 101: 451–465.

Ring, K. (1980) Life After Death: A Scientific Investigation of the Near-Death Experience. Coward, McCann, and Geoghegan, New York.

Ring, K. (1992) The Omega Project. William Morrow, New York.

Ring, K. (2001) Mindsight: Eyeless vision in the blind. In: Lorimer D. (Ed.), Thinking Beyond the Brain: A Wider Science of Consciousness. Floris, Edinburgh, pp. 59–70.

Ring, K. and Cooper, S. (1999) Mindsight: Near-Death and Out-of-Body Experiences in the Blind. William James Center for Consciousness Studies, Palo Alto, CA.

Rodin, E. (1989) Comments on "a neurobiological model for near-death experiences. J. Near Death Stud., 7: 255–259.

Roe, C.A. (2001) Near-death experiences. In: Roberts R. and Groome D. (Eds.), Parapsychology: The Science of Unusual Experience. Arnold, London, pp. 141–155.

Saavedra-Aguilar, J.C. and Gómez-Jeria, J.S. (1989) A neurobiological model for near-death experiences. J. Near Death Stud., 7: 205–222.

Sabom, M.B. (1982) Recollections of Death: A Medical Investigation. Corgi, London.

Sabom, M.B. (1998) Light and Death: One Doctor's Fascinating Account of Near-Death Experiences. Zondervan, Grand Rapids, Michigan.

Sagan, C. (1979) Broca's Brain: Reflections on the Romance of Science. Random House, New York.

Schwaninger, J., Eisenberg, P.R., Schectman, K.B. and Weiss, A.N. (2002) A prospective analysis of near-death experiences in cardiac arrest patients. J. Near Death Stud., 20: 215–232.

Strassman, R. (1997) Endogenous ketamine-like compounds and the NDE: If so, so what? J. Near Death Stud., 16: 27–41.

Tellegen, A. and Atkinson, G. (1974) Openness to absorbing and self-altering experiences ("absorption"), a trait related to hypnotic susceptibility. J. Abnorm. Psychol., 83: 268–277.

Tong, F. (2003) Out-of-body experiences: From Penfield to present. Trend. Cogn. Sci., 7: 104–106.

Van Lommel, P., van Wees, R., Meyers, V. and Elfferich, I. (2001) Near-death experience in survivors of cardiac arrest: a prospective study in the Netherlands. Lancet, 358: 2039–2045.

Whinnery, J.E. (1997) Psychophysiologic correlates of unconsciousness and near-death experiences. J. Near Death Stud., 15: 231–258.

Wilson, S.C. and Barber, T.X. (1983) The fantasy-prone personality: Implications for understanding imagery hypnosis and parapsychological phenomena. In: Sheikh A.A. (Ed.), Imagery: Current Theory, Research, and Application. Wiley, New York, pp. 340–390.

Woerlee, G.M. (2003) Mortal Minds: A Biology of the Soul and the Dying Experience. De Tijdstroom, Utrecht.

Woerlee, G.M. (2005a) An anaesthesiologist examines the Pam Reynolds story. Part. 1. Background considerations. Skeptic (British version), 18.1 (in press).

Woerlee, G.M. (2005b) An anaesthesiologist examines the Pam Reynolds story. Part 2. An explanation. Skeptic (British version), 18.2 (in press).

S. Laureys (Ed.)
Progress in Brain Research, Vol. 150
ISSN 0079-6123

CHAPTER 26

The concept and practice of brain death

James L. Bernat*

Neurology Section, Dartmouth Medical School, Hanover, NH 03755, USA

Abstract: Brain death, the colloquial term for the determination of human death by showing the irreversible cessation of the clinical functions of the brain, has been practiced since the 1960s and is growing in acceptance throughout the world. Of the three concepts of brain death — the whole-brain formulation, the brain stem formulation, and the higher brain formulation — the whole-brain formulation is accepted and practiced most widely. There is a rigorous conceptual basis for regarding whole-brain death as human death based on the biophilosophical concept of the loss of the organism as a whole. The diagnosis of brain death is primarily a clinical determination but laboratory tests showing the cessation of intracranial blood flow can be used to confirm the clinical diagnosis in cases in which the clinical tests cannot be fully performed or correctly interpreted. Because of evidence that some physicians fail to perform or record brain death tests properly, it is desirable to require a confirmatory test when inadequately experienced physicians conduct brain death determinations. The world's principal religions accept brain death with a few exceptions. Several scholars continue to reject brain death on conceptual grounds and urge that human death determination be based on the irreversible cessation of circulation. But despite the force of their arguments they have neither persuaded any jurisdictions to abandon brain death statutes nor convinced medical groups to change clinical practice guidelines. Other scholars who, on more pragmatic grounds, have called for the abandonment of brain death as an anachronism or an unnecessary prerequisite for multi-organ procurement, similarly have not convinced public policy makers to withdraw the dead-donor rule. Despite a few residual areas of controversy, brain death is a durable concept that has been accepted well and has formed the basis of successful public policy in diverse societies throughout the world.

Introduction

Brain death is the colloquial term for the determination of human death by showing the irreversible cessation of the clinical functions of the brain. Although *brain death* is a term hallowed by history and accepted by common usage, it is a misleading and unfortunate term. It promotes confusion by wrongly implying that there are different kinds of death and that it is only the brain that dies. Notwithstanding these shortcomings and because of its commonly accepted usage, I also use the term but only in the manner carefully defined here.

*Corresponding author. Tel.: +1 603 650 5104;
Fax: +1 603 650 0458; E-mail: bernat@dartmouth.edu

History

Brain death is a product of technological advances. Beginning with the development of positive-pressure ventilators in the 1950s, patients with complete apnea and paralysis could be successfully ventilated permitting the continuation of heartbeat and circulation that otherwise would have rapidly ceased. By the late 1950s, several patients who developed apnea resulting from complete destruction of the brain had their ventilation and, hence, circulation supported temporarily. Early examiners recognized that such patients in a state that they termed *coma dépassé* (beyond coma) had a depth of unresponsiveness unlike that of any previously recorded, but they were in doubt whether such

DOI: 10.1016/S0079-6123(05)50026-8

patients should be regarded as alive (Mollaret and Goulon, 1959). Patients with *coma dépassé* had some features associated with live patients (e.g., heartbeat, circulation, digestion, excretion) but other features associated with dead patients (e.g., loss of breathing, no movement, no reflex responses). It became clear that physicians could not state confidently whether patients with *coma dépassé* were alive or dead until there was agreement on what it meant to be dead in a technological era.

In 1968, the Harvard Medical School Ad Hoc Committee first provided criteria for the claim that *coma dépassé* patients with permanent cessation of recordable neurological function were not simply irreversibly comatose but were already dead, and therefore could serve as organ donors without organ procurement being regarded as the cause of death (Ad Hoc Committee, 1968). Since that time, the concept of brain death has become nearly universally accepted and enshrined in laws and practice guidelines throughout the developed world and in many parts of the undeveloped world, amounting to approximately 80 countries (Wijdicks, 2002). While there remain areas of persisting controversy, brain death is a durable phenomenon with a strong level of consensus and acceptance (Capron, 2001).

The concept of death

The practice of brain death determination is predicated upon a concept of death that affords brain functions a critical role. In the pretechnological era, it was unnecessary to make explicit the concept of death because all bodily systems critical to life (the so-called "vital functions" of breathing, heartbeat, circulation, and brain functions) were mutually interdependent and ceased within minutes of each other whenever one ceased. But the advent of positive-pressure ventilation permitted the dissociation of vital functions: all brain functions could have ceased irreversibly yet ventilation and circulation could be continued with mechanical support. The technologically created dissociation of vital functions created ambiguity in death determination and raised the essential question: which vital functions are most vital to life?

In a series of articles, my Dartmouth colleagues and I offered a rigorous philosophical argument that the human organism was dead when all clinical brain functions ceased irreversibly, irrespective of mechanical continuation of ventilation and support of circulation (Bernat et al., 1981; Bernat, 1998; Bernat, 2002). This argument, providing a conceptual basis for brain death, was cited by the President's Commission in their influential book *Defining Death* (1981), and has been regarded by many scholars, including opponents (Shewmon and Shewmon, 2004), as the standard conceptual defense of whole-brain death.

The Dartmouth analysis of death is conducted in four sequential phases: (1) agreeing upon the "paradigm" of death — the set of preconditions that makes an analysis possible; (2) the philosophical task of determining the definition of death by making explicit the consensual concept of death that has been confounded by technology; (3) the philosophical and medical task of determining the best criterion of death — that measurable condition that shows that the definition has been fulfilled by being both necessary and sufficient for death; and (4) the medical-scientific task of determining the tests of death for physicians to employ at the patient's bedside to demonstrate that the criterion of death has been fulfilled with no false positive and minimal false negative determinations (Bernat, 2002).

The paradigm of death comprises seven conditions: (1) *death* is a nontechnical word, thus defining it should make explicit its ordinary, consensual meaning; (2) death is fundamentally a biological phenomenon thus its definition must conform to the known facts of biological reality; (3) the definitional domain should be restricted to the death of *homo sapiens* and related higher vertebrates for whom death is a univocal phenomenon; (4) the term *death* can be applied directly and categorically only to organisms; other uses are metaphorical; (5) all organisms must be either dead or alive, none can reside in both states or in neither; (6) death is an event and not a process; and (7) death is irreversible. Elsewhere I explain these conditions in detail and defend them (Bernat, 2002).

There are three competing concepts of brain death, popularly known as whole-brain death, brain stem death, and the higher brain formulation (Bernat, 1992). The whole-brain formulation of death is the original concept of brain death, enjoys the greatest prevalence throughout the world by far, and is the concept my Dartmouth colleagues and I endorse and defend. Brain stem death is practiced in the United Kingdom and a few other countries. The higher brain formulation is propounded by a group of academic scholars but has not been endorsed by any medical society and forms the basis of no law in any country or jurisdiction.

The definition and criterion of death

Whole-brain death

Whole-brain death is based on a definition of death as the irreversible cessation of the critical functions of the organism as a whole. The organism as a whole is not synonymous with the whole organism (sum of its parts). Rather it is the set of the organism's emergent functions (properties of a whole not possessed by any of its component parts) that integrate its subsystems and create the unity of the organism. The biophilosophical concept of the organism as a whole (Bernat, 2002) encompasses the concept of an organism's critical system (Korein and Machado, 2004). Death is the irreversible loss of the critical functions of the organism as a whole or of the organism's critical system.

The criterion of death satisfying this definition is the irreversible cessation of the clinical functions of the entire brain ("whole-brain"). The critical functions of the organism as a whole include those of respiration and control of circulation residing in the brain stem, neuroendocrine control systems for homeostatic regulation residing in the diencephalon, and conscious awareness residing in the thalami, the cerebral hemispheres, and their connections. Because all these functions must be irretrievably lost for death, the clinical functions served by each of these structures must be proved to be permanently absent.

Despite its categorical-sounding name, the whole-brain criterion does not require the irreversible cessation of functioning of every brain neuron. Rather, it requires only the irreversible cessation of all clinical functions of the brain, namely those measurable at the bedside by clinical examination. Some brain cellular activities, such as random EEG activity, may remain recordable after brain death (Grigg et al., 1987). This activity that results from isolated neurons, although measurable, does not generate a clinical function of the brain or contribute to the functioning of the organism as a whole, and, thus, is irrelevant to the determination of death.

The progression to whole-brain death from an initial brain injury usually requires the pathophysiological process of brain herniation to produce widespread destruction of the neuronal systems that provide the brain's clinical functions (Plum and Posner, 1980). The most frequent etiologies of brain death are trauma, hypoxic-ischemic damage during cardiopulmonary arrest or asphyxia, meningoencephalitis, or enlarging intracranial mass lesions such as neoplasms. When the brain is diffusely injured by one of these disorders, brain edema within the rigidly fixed skull causes intracranial pressure to rise to levels exceeding mean arterial blood pressure, or in some cases, exceeding systolic blood pressure. At this point, intracranial circulation ceases and nearly all brain neurons that were not destroyed by the initial brain injury are secondarily destroyed by the cessation of intracranial circulation. The whole-brain formulation thus provides a fail-safe mechanism to eliminate false-positive brain death determinations and assure the loss of the critical functions of the organism as a whole (See Fig. 1). Showing the absence of all intracranial circulation is sufficient to prove widespread destruction of all critical neuronal systems.

Brain stem death

Brain stem death is also based on a definition of death of the cessation of the organism's integrated, unified functioning. But the brain stem theorists argue that the criterion on death should be simply the irreversible loss of the capacity for consciousness combined with the irreversible loss of the

Fig. 1. This artistic rendering of a premature burial depicts the primal human fear of incorrect death determination. It emphasizes that physicians' tests for death, including brain death, should be designed to prevent any false-positive determinations (Antoine Wiertz, Belgian Romantic, The Premature Burial, 1854).

capacity to breathe (Conference of Medical Royal Colleges and their Faculties in the United Kingdom, 1976; Pallis, 1983). Christopher Pallis, the most eloquent proponent of the concept of brain stem death, pointed out that the brain stem is at once the through-station for nearly all hemispheric input and output, the center generating conscious wakefulness, and the center of breathing. Therefore, destruction of the brain stem produces loss of brain functions that is sufficient for death. He epitomized the role of the brain stem in brain death: "the irreversible cessation of brain stem function implies death of *the brain as a whole*" (Pallis, 1983). Pallis also correctly observed that most of the clinical tests for whole-brain death measure the loss of brain stem functions.

But there are two serious problems with the brain stem formulation. First, because it does not require cessation of the clinical functions of the diencephalon or cerebral hemispheres, it creates the possibility of misdiagnosis of death resulting from a pathological process that appears to destroy all brain stem activities but that preserves a degree of conscious awareness that cannot be clinically detected. I called such a possibility a "super locked-in syndrome" (Bernat, 1992). In a whimsical moment, Pallis (1983) also considered this possibility in a limerick that he remarked could have been penned by one of the *tricoteuses* (knitters) who sat by the guillotine in Paris in 1793, as memorably portrayed by Charles Dickens's character Madame LaFarge in *The Tale of Two Cities*:

> We knit on, too *blasées* to ask it:
> "Could the tetraparesis just mask it?
> When the brain stem is dead
> Can the cortex be said
> To tick on, in the head, in the basket?"

The second problem of the brain stem formulation is that by not requiring intracranial circulatory arrest, it eliminates the fail-safe feature of the whole-brain formulation to confidently demonstrate global neuronal destruction. As a practical matter, it also eliminates the possibility of using a confirmatory test to show cessation of intracranial circulation, thereby guaranteeing the irreversible loss of consciousness and of the brain's other clinical functions.

Higher-brain formulation

In the early days of the brain death debate, Robert Veach proposed a refinement in the concept of brain death that became known as the higher brain formulation. He argued that because it was man's cerebral cortex that defined the person, and not the primitive brainstem structures, the loss of the higher functions served by the cortex should define death. He proposed that death should be defined formally as "the irreversible loss of that which is considered to be essentially significant to the nature of man" (Veatch, 1973). He rejected the idea that death should be based upon the biophilosophical concept of an organism's loss of the capacity to integrate bodily function. His definition of death thus was unique for *homo sapiens* and was centered upon the seemingly unique attribute of human conscious awareness (Veatch, 1993). Veatch's idea became popular, particularly among philosophers and medical ethicists, where it remains firmly embraced (reviewed in Bernat, 1992). But despite over three decades of scholarly articles endorsing it, the higher brain formulation has not convinced lawmakers in any jurisdiction to change laws or medical societies in any country to change practice standards.

Although Veatch did not explicitly stipulate, the criterion of death that satisfied his definition would be the irreversible loss of consciousness and cognition. Thus, patients in persistent vegetative states (PVS) would be declared dead by this definition. But, despite their profound disability, and the tragic irony of persistently noncognitive life, all societies, cultures, and laws throughout the world consider PVS patients as alive. Practice guidelines permit the withdrawal of life-sustaining treatment from PVS patients under certain conditions to allow them to die (Jennett, 2002), but nowhere are they summarily declared dead. The higher brain formulation is an inadequate concept of death because it fails the first condition of the paradigm of death: to make explicit our underlying consensual concept of death and not to contrive a new definition of death. The higher brain formulation is a contrived redefinition of death that neither comports with biological reality nor is consistent with prevailing societal beliefs and laws.

The circulatory formulation

Brain death is not accepted universally and has had opponents since it became popularized in the 1960s. Early critics claimed brain death practices violated Christian religious beliefs (Byrne et al., 1979). Later critics detected inconsistencies between the definition and criterion of death of the whole brain formulation (Halevy and Brody, 1993; Veatch, 1993). Current critics reject outright the concept of brain death and in its place propose the circulatory formulation of death: the organism is not dead until its circulation ceases irreversibly. The circulatory idea had its conceptual birth by the philosopher Josef Seifert (Seifert, 1993) and received its conceptual consolidation by Alan Shewmon, its most eloquent and persuasive advocate, in a penetrating series of recent articles (Shewmon, 2001, 2004).

Shewmon's summons evidence that the brain performs no qualitatively different forms of bodily integration than the spinal cord and concludes that therefore it should be granted no special status above other organs in death determination. He presents a series of cases of what he infelicitously calls "chronic brain death" in which a group of brain dead patients were treated aggressively and had their circulation maintained for many months or longer (Shewmon, 1998). He concludes that these cases prove that the concept of brain death is inherently counterintuitive, for how could a dead body continue visceral organ functioning for extended periods, gestate infants, or grow?

First, I question how many of the patients in Shewmon's series were truly brain dead or might have been examined incorrectly. Second, I observe that prolonged physiologic maintenance of the circulation of brain dead patients represents a *tour de force* that reflects our current impressive critical care technological virtuosity. Third, on more conceptual grounds, I argue that the circulatory formulation has the inverse problem of the higher brain formulation. While the higher brain formulation generates a criterion that is necessary but insufficient for death, the circulatory formulation generates a criterion that is sufficient but unnecessary for death. Elsewhere I have provided arguments supporting this conclusion, on basis of the fact that it is unnecessary for the determination of death to require the cessation of functions of organs that do not serve the critical functions of the organism as a whole (Bernat, 1998, 2002). Finally, although I concede that Shewmon shows weaknesses in the coherence of the whole-brain formulation, he must admit that his arguments have not swayed the majority who experience an intuitive attraction to the whole-brain formulation and find it sufficiently coherent and useful to wish to preserve it as public policy.

The tests of death

Brain death tests must be used to determine death only in the unusual death determination in which a patient's ventilation is supported. In an apneic patient, if positive-pressure ventilation is neither employed nor planned, the traditional tests of death — the prolonged absence of breathing and heartbeat — can be used confidently. These tests are completely predictive of death because the brain will be rapidly destroyed by the resultant hypoxemia and ischemia from apnea, at which time death will have occurred.

Beginning with the 1968 Harvard Medical School Ad Hoc Committee report, advocates for brain death have proposed a series of bedside tests to show that the whole-brain criterion of death has been satisfied. Numerous batteries of brain death tests were published in the 1970s, which varied slightly. In 1981, the Medical Consultants to the President's Commission published a test battery that was quickly accepted, and superceded previous batteries (Medical Consultants to the President's Commission, 1981). In 1995, following an evidence-based review of the brain death scientific literature (Wijdicks, 1995), the Quality Standards Subcommittee of the American Academy of Neurology published a practice parameter for determining brain death in adults (American Academy of Neurology, 1995) that forms the current standard for brain death determination in the United States. Similar test batteries have been published in Canada (Canadian Neurocritical Care Group, 1999) and in the majority of European countries (Haupt and Rudolf, 1999). The individual tests have been described in detail by Wijdicks (2001a, b).

All brain death tests require satisfying preconditions of irreversibility. The cause of the loss of brain functions must be known to be structural and irreversible, and must be sufficient to account for the clinical signs. Before brain death can be declared, the clinician must scrupulously exclude potentially reversible metabolic encephalopathies, such as those from hypothermia, hypoglycemia, or organ failure as well as toxic encephalopathies, such as those from depressant drug intoxication or neuromuscular blockade. To prove that the brain damage is permanent it is critical to exclude potentially reversible conditions that have been reported to mimic brain death including severe de-efferentation from Guillain-Barré syndrome, hypothermia, and intoxication with tricyclic antidepressant drugs or barbiturates (Wijdicks, 2001a).

The clinical examination for brain death must demonstrate three cardinal signs: (1) profound coma with utter unresponsiveness to noxious stimuli; (2) absence of all brain stem-mediated reflexes, such as pupillary light–dark reflexes, corneal reflexes, vestibulo-ocular reflexes, gag reflexes, and cough reflexes; and (3) complete apnea in the face of maximal chemoreceptor stimulation by adequate hypercapnia (Marks and Zisfein, 1990). Serial neurological examinations over a time interval (determined by the patient's age and cause of brain injury) are necessary unless a confirmatory laboratory test is also used.

The clinical tests for whole-brain death and for brain stem death are identical. There are no accepted tests for the higher brain formulation because it has achieved no medical or legal acceptance. The principal difference between tests for whole-brain and brain-stem death lies in the availability of confirmatory tests measuring cessation of intracranial blood flow to prove the whole-brain criterion.

Using laboratory tests to confirm the clinical determination of brain death is useful to expedite brain death determination when organ donation is planned or when the clinical tests cannot be performed adequately. Showing electrocerebral silence by EEG is the oldest confirmatory test but generates too many false-positive determinations to be reliable (Boutros and Henry, 1982). Linking EEG with measurements of brain stem auditory evoked potentials or somatosensory evoked potentials (Goldie et al., 1981; Facco et al., 2002) improves its reliability.

The most reliable confirmatory tests are those demonstrating the cessation of intracranial blood flow. Intracranial blood flow ceases once intracranial pressure exceeds mean arterial blood pressure. Two widely available and reliable tests to show cessation of intracranial circulation are intravenous cerebral isotope angiography (Flowers and Patel, 1997) and transcranial Doppler ultrasound (Ducrocq et al., 1998). Both have been shown to have high positive and negative predictive values to confirm whole brain death. There have been a few recent reports demonstrating the absence of intracranial circulation in brain death using other techniques, including brain SPECT scintigraphy (Donohoe et al., 2003), magnetic resonance diffusion-weighted imaging (Lovblad and Bassetti, 2000), magnetic resonance angiography (Karantanas et al., 2002), and computed tomographic angiography (Qureshi et al., 2004).

In most cases where brain stem death is determined, rostrocaudal transtentorial herniation from transmitted supratentorial pressure waves secondarily destroy the brain stem and most other neurons by ischemic infarction. However, in the unusual case caused by a primary brain stem or cerebellar hemorrhage, intracranial pressure may not rise to levels sufficient to interfere with intracranial circulation, thereby eliminating the possibility of using intracranial circulatory arrest as a confirmatory test (Ferbert et al., 1986; Kosteljanetz et al., 1988). Only in such a case might the determination of death by the whole-brain criterion yield a different result than the determination of death by the brain stem criterion.

Brain death determination in practice

The clinical tests used to determine brain death are well known and well accepted. However, empirical studies of the adequacy of physicians' bedside testing, including apnea testing, have shown unfortunate and widespread variability in performing the tests properly and recording the results completely (Earnest et al., 1986; Mejia and Pollack, 1995; Wang et al., 2002). These discouraging findings suggest the disquieting implication that some physicians probably are declaring patients dead with brain death tests when they may not be dead. This inaccuracy may help explain some of Shewmon's cases of "chronic brain death" and obviously suggests the need for further education and standardization of brain death testing.

A confounding factor is the conceptual confusion among physicians and other health professionals about brain death, irreversible coma, and the definition of death. One widely quoted study showed an appalling misunderstanding of the definitions and boundaries of these categories by many critical care physicians and nurses, the very professionals expected to be most knowledgeable about brain death (Youngner et al., 1989). A more recent survey of medical students provided somewhat encouraging results (Frank, 2001) but all experienced neurologists know that brain death is inadequately understood by fellow physicians and nurses. The educational need to correct misunderstanding is obvious.

According to a recent survey, brain death is currently accepted practice in approximately 80 countries but the testing protocols vary somewhat among countries (Wijdicks, 2002). Brain death is practiced widely throughout the Western world but is also being practiced in a growing number of countries in the non-Western world, such as in the Islamic Middle East (Yaqub and Al-Deeb, 1996) and in India (Jain and Maheshawari, 1995). Because of the variation in test batteries among countries, international standardization of brain death testing has become the principal project of the Ethics Committee of the World Federation of Neurology during the next 5 years (Prof. F. Gerstenbrand, personal communication, June 25, 2004). Simple guidelines for the determination of brain death are needed to guide inexperienced physicians in countries where neurologists may not be available (Baumgartner and Gerstenbrand, 2002) and there are no facilities for confirmatory tests.

For the past several years, I have become increasingly disturbed by the continued publication of reports of improper brain death determinations, my own personal experience of witnessing errors, and of reported cases purported to be brain dead who were not. Although brain death ideally is a primary clinical determination, accurately diagnosing it requires skill and scrupulous attention to detail. I have reluctantly concluded that as a matter of practice, especially for those who lack the necessary skill or experience, it is desirable to perform a test showing the absence of intracranial blood flow to confirm the clinical diagnosis (Bernat, 2004).

Religious views

Early commentators on brain death asserted that its concept and practice were compatible with the beliefs of the world's principal religions (Veith et al., 1977). While that assertion was debatable in 1977, it is largely true now. Among Christian believers, Protestantism has accepted brain death without serious exception (Campbell, 1999). The decades-long debate in Roman Catholicism (Byrne et al., 1979) that saw brain death approved by successive pontifical academies was finally settled in 2000 when Pope John Paul II, in an address to the 18th Congress of the International Transplantation Society, formally stated that brain death determination was compatible with Catholic beliefs and teachings (Furton, 2002).

A rabbinic debate persists in Judaism. Reform and Conservative rabbis accept brain death almost without exception. But the Orthodox Jewish rabbinate remains split between acceptance and rejection. Orthodox authorities such as the Talmudic scholar-physician Fred Rosner argue that brain death is compatible with traditional Jewish law because it is the modern physiological equivalent of decapitation (Rosner, 1999). Other Talmudic authorities, such as Rabbi David Bleich, reject brain death because death as understood in Jewish law requires the irreversible cessation of both cardiac and respiratory activity (Bleich, 1979). In general, the strictest Orthodox rabbis continue to oppose brain death.

The former opposition of Islam to brain death was reversed in 1986 by a decree from the Council of Islamic Jurisprudence Academy. Now, religious authorities in several Islamic countries, including in conservative Wahabian Saudi Arabia, permit brain death and organ transplantation (Yaqub and Al-Deeb, 1996). Hindu culture in India endorses brain death (Jain and Maheshawari, 1995) as does Shinto-Confucian religious authorities in Japan following a lengthy social battle (Lock, 1995).

All states in the United States have enacted statutes or written administrative regulations permitting physicians to declare death by brain death determination (Beresford, 1999). In the 1990s, two states amended their laws to accommodate religious opposition. New Jersey enacted a religious exemption providing that any citizen who could show that brain death determination violated "personal religious beliefs or moral convictions" could not be declared brain dead (Olick, 1991). New York state amended its administrative regulations on brain death declaration to provide a similar though more restricted exemption (Beresford, 1999).

Organ donation

The principal reason for societies to recognize and for physicians to practice brain death is to acknowledge biological reality given the advances in ICU technology that permit increasingly prolonged physiological maintenance of patients' organ subsystems after the demise of the organism as a whole. The main utility of brain death is that it permits multiple vital organ procurement for transplantation. There is evidence that the desire to obtain transplantable vital organs was a strong motivating factor in the development of the Harvard Ad Hoc Committee's pioneering 1968 report (Giacomini, 1997). At the time of the Harvard report, there was also no legally acceptable means to discontinue life-sustaining therapy once it had been started; the adoption of brain death determination provided such a means.

Currently there are laws and medical practice guidelines in most societies that permit withdrawing life-sustaining therapy from living hopelessly ill patients. Some scholars have therefore described brain death as an "anachronism" that should be "abandoned" because it has outlived its usefulness: it is no longer necessary to declare a patient dead to discontinue supportive therapy (Truog, 1997). Truog further called for the dissociation of the relationship between death declaration and vital organ donation and the abandonment of the dead-donor rule (Truog and Robinson, 2003). I believe that these efforts, while understandable in intent, are misguided.

The dead-donor rule is the ethical axiom of multi-organ procurement for transplantation: the donor must first be dead, and it is unethical to kill even hopelessly ill donors for their organs despite the intent to save others (Robertson, 1999). Truog has suggested that the dead-donor rule in vital organ transplantation could be dropped if two conditions were filled: (1) the donor patient was hopelessly ill and beyond being harmed because of neurological devastation or because of imminently dying; and (2) the patient had previously consented to serve as an organ donor (Truog, 1997). The problem with this idea is not that patients will be harmed because they probably would not, given Truog's two conditions. Rather, the problem is that eliminating the dead-donor rule may diminish public confidence in the organ donation enterprise. The public needs to maintain confidence that physicians will remove their organs only after they are dead. Public confidence is fragile and can be jeopardized by publicized claims of physician malfeasance, even when false (see cases discussed in Bernat, 2002).

The resistance to implementing protocols permitting organ donation after cardiac death is the latest example of the fragility of public confidence in the organ transplantation enterprise. The Institute of Medicine has endorsed protocols permitting organ donation immediately after cardiac death with family consent for cases in which life-sustaining therapy is withheld on a severely brain-damaged but not brain dead patient (Institute of Medicine, 1997, 2000). The "donation after cardiac death" (DCD) protocols permit the following procedure: (1) family decision to withhold a patient's life-sustaining treatment and allow the patient to die because of poor neurological prognosis and the patient's prior wishes; (2) family consent for organ donation from the patient once the patient is dead; (3) readiness of the organ procurement team; (4) extubation of the patient with resulting apnea which produces cardiac asystole (5) death is declared following 5 min of asystole; and (6) organ procurement proceeds immediately after death declaration, usually successfully procuring the kidneys and at times the liver.

The DCD protocols were introduced to try to increase the finite (and possibly shrinking) brain dead organ donor pool by reactivating and making more efficient an organ donation scheme that was practiced prior to the era of brain death (Sheehy et al., 2003). This seemingly simple scheme has met resistance from the public for two reasons: (1) uncertainty if the patient is unequivocally dead after 5 min of asystole; and (2) the plausibility of a causal connection between the family's desire for organ donation and their decision to withhold life-sustaining therapy. The unfortunate and highly publicized "scandal" at the Cleveland Clinic when such a protocol was simply being considered is a measure of the fragility of public confidence in death determination and organ donation. The public's tenuous confidence in the ability and impartiality of physicians to correctly diagnose death, upon which the cadaver organ donation enterprise rests, could be jeopardized by unnecessarily and unwisely sacrificing the dead-donor rule.

References

Ad Hoc Committee. (1968) A definition of irreversible coma: report of the Ad Hoc Committee of the Harvard Medical School to examine the definition of brain death. JAMA, 205: 337–340.

American Academy of Neurology Quality Standards Subcommittee. (1995) Practice parameters for determining brain death in adults [summary statement]. Neurology, 45: 1012–1014.

Baumgartner, H. and Gerstenbrand, F. (2002) Diagnosing brain death without a neurologist: simple criteria and training are needed for the non-neurologist in many countries. Br. Med. J., 324: 1471–1472.

Beresford, H.R. (1999) Brain death. Neurol. Clin., 17: 295–306.

Bernat, J.L., Culver, C.M. and Gert, B. (1981) On the definition and criterion of death. Ann. Intern. Med., 94: 389–394.

Bernat, J.L. (1992) How much of the brain must die in brain death? J. Clin. Ethics., 3: 21–26.

Bernat, J.L. (1998) A defense of the whole-brain concept of death. Hastings Cent. Rep., 28(2): 14–23.

Bernat, J.L. (2002) The biophilosophical basis of whole-brain death. Soc. Philoso. Policy, 19: 324–342.

Bernat, J.L. (2004) On irreversibility as a prerequisite for brain death determination. Adv. Exp. Med. Biol., 550: 161–167.

Bleich, J.D. (1979) Establishing criteria of death. In: Rosner F. and Bleich J.D. (Eds.), Jewish Bioethics. Sanhedrin Press, New York, pp. 277–295.

Boutros, A.R. and Henry, C.E. (1982) Electrocerebral silence associated with adequate spontaneous ventilation in a case of fat embolism: a clinical and medicolegal dilemma. Arch. Neurol., 39: 314–316.

Byrne, P.A., O'Reilly, S. and Quay, P.M. (1979) Brain death — an opposing viewpoint. JAMA, 242: 1985–1990.

Campbell, C.S. (1999) Fundamentals of life and death: Christian fundamentalism and medical science. In: Youngner S.J., Arnold R.M. and Schapiro R. (Eds.), The Definition of Death: Contemporary Controversies. John Hopkins University Press, Baltimore, pp. 194–209.

Canadian Neurocritical Care Group. (1999) Guidelines for the diagnosis of brain death. Can. J. Neurol. Sci., 26: 64–66.

Capron, A.M. (2001) Brain death — well settled yet still unresolved. N. Engl. J. Med., 344: 1244–1246.

Conference of Medical Royal Colleges and their Faculties in the United Kingdom. (1976) Diagnosis of brain death. Br. Med. J., 2: 1187–1188.

Donohoe, K.J., Frey, K.A., Gerbaudo, V.H., Mariani, G., Nagel, J.S. and Shulkin, B. (2003) Procedural guidelines for brain death scintigraphy. J. Nucl. Med., 44: 846–851.

Ducrocq, X., Braun, M., Debouverie, M., Junges, C., Hummer, M. and Vespignani, H. (1998) Brain death and transcranial Doppler: experience in 130 cases of brain dead patients. J. Neurol. Sci., 160: 41–46.

Earnest, M.P., Beresford, H.R. and McIntyre, H.B. (1986) Testing for apnea in brain death: methods used by 129 clinicians. Neurology, 36: 542–544.

Facco, E., Munari, M., Gallo, F., et al. (2002) Role of short-latency evoked potentials in the diagnosis of brain death. Clin. Neurophysiol., 113: 1855–1866.

Ferbert, A., Buchner, H., Ringelstein, E.B., et al. (1986) Isolated brainstem death. Electroencephalogr. Clin. Neurophysiol., 65: 157–160.

Flowers Jr., W.M. and Patel, B.R. (1997) Radionuclide angiography as a confirmatory test for brain death: a review of 229 studies in 219 patients. South Med. J., 90: 1091–1096.

Frank, J.I. (2001) Perceptions of death and brain death among fourth-year medical students: defining our challenge as neurologists (abst). Neurology, 56: A429.

Furton, E.J. (2002) Brain death, the soul, and organic life. National Catholic Bioethics Quart., 2: 455–470.

Giacomini, M. (1997) A change of heart and a change of mind? Technology and the redefinition of death in 1968. Soc. Sci. Med., 44: 1465–1482.

Goldie, W.D., Chiappa, K.H., Young, R.R. and Brooks, R.B. (1981) Brainstem auditory and short-latency somatosensory evoked responses in brain death. Neurology, 31: 248–256.

Grigg, M.M., Kelly, M.A., Celesia, G.G., Ghobrial, M.W. and Ross, E.R. (1987) Electroencephalographic activity after brain death. Arch. Neurol., 44: 948–954.

Halevy, A. and Brody, B. (1993) Brain death: reconciling definitions, criteria, and tests. Ann. Int. Med., 119: 519–525.

Haupt, W.F. and Rudolf, J. (1999) European brain death codes: a comparison of national guidelines. J. Neurol., 246: 432–437.

Institute of Medicine. (1997) Non-Heart-Beating Organ Donation: Medical and Ethical Issues in Procurement. National Academy Press, Washington, DC.

Institute of Medicine. (2000) Non-Heart-Beating Organ Transplantation: Practice and Protocols. National Academy Press, Washington, DC.

Jain, S. and Maheshawari, M.C. (1995) Brain death — the Indian perspective. In: Machado C. (Ed.), Brain Death. Elsevier, Amsterdam, pp. 261–263.

Jennett, B. (2002) The Vegetative State: Medical Facts, Ethical and Legal Dilemmas. Cambridge University Press, Cambridge, pp. 97–125.

Karantanas, A.H., Hadijigeorgion, G.M., Paterakis, K., Sfiras, D. and Komnos, A. (2002) Contributions of MRI and MR angiography in early diagnosis of brain death. Europ. Radiology., 12: 2710–2716.

Korein, J. and Machado, C. (2004) Brain death: updating a valid concept for 2004. Adv. Exper. Med. Biol., 550: 1–14.

Kosteljanetz, M., Ohrstrom, J.K., Skjodt, S. and Teglbjaerg, P.S. (1988) Clinical brain death with preserved cerebral arterial circulation. Acta. Neurol. Scand., 78: 418–421.

Lock, M. (1995) Contesting the natural in Japan: moral dilemmas and technologies of dying. Culture, Med. and Psychiatry, 19: 1–38.

Lovblad, K.O. and Bassetti, C. (2000) Diffusion-weighted magnetic resonance imaging in brain death. Stroke, 31: 539–542.

Marks, S.J. and Zisfein, J. (1990) Apneic oxygenation in apnea tests for brain death: a controlled trial. Arch. Neurol., 47: 1066–1068.

Medical Consultants to the President's Commission. (1981) Report of the medical consultants on the diagnosis of death to the President's commission for the study of ethical problems in medicine and biomedical and behavioral research. Guidelines for the determination of death. JAMA, 246: 2184–2186.

Mejia, R.E. and Pollack, M.M. (1995) Variability in brain death determination practices in children. JAMA, 274: 550–553.

Mollaret, P. and Goulon, M. (1959) Le coma dépassé (mémoire préliminaire). Rev. Neurol., 101: 3–15.

Olick, R.S. (1991) Brain death, religious freedom, and public policy: New Jersey's landmark legislative initiative. Kennedy Inst. Ethics. J., 4: 275–288.

Pallis, C. (1983) ABC of Brainstem Death. British Medical Journal Publishers, London.

Plum, F. and Posner, J.B. (1980) The Diagnosis of Stupor and Coma (3rd ed.). F.A. Davis, Philadelphia, PA, pp. 88–101.

President's Commission for the Study of Ethical Problems in Medicine and Biomedical and Behavioral Research. (1981)

Defining Death: Medical, Legal and Ethical Issues in the Determination of Death. U.S. Government Printing Office, Washington, DC, pp. 31–43.

Qureshi, A.I., Kirmani, J.F., Xavier, A.R. and Siddiqui, A.M. (2004) Computed tomographic angiography for the diagnosis of brain death. Neurology, 62: 652–653.

Robertson, J.A. (1999) The dead donor rule. Hastings Cent. Rep., 29(6): 6–14.

Rosner, F. (1999) The definition of death in Jewish law. In: Youngner S.J., Arnold R.M. and Schapiro R. (Eds.), The Definition of Death: Contemporary Controversies. John Hopkins University Press, Baltimore, pp. 210–221.

Seifert, J. (1993) Is brain death actually death? A critique of redefinition of man's death in terms of 'brain death'. The Monist, 76: 175–202.

Sheehy, E., Conrad, S.L., Brigham, L.E., Luskin, R., Weber, P., Eakin, M., et al. (2003) Estimating the number of potential organ donors in the United States. N. Engl. J. Med., 349: 667–674.

Shewmon, D.A. (1998) Chronic 'brain death': meta-analysis and conceptual consequences. Neurology, 51: 1538–1545.

Shewmon, D.A. (2001) The brain and somatic integration: insights into the standard biological rationale for equating "brain death" with death. J. Med. Philosophy., 26: 457–478.

Shewmon, D.A. (2004) The "critical organ" for the organism as a whole: lessons from the lowly spinal cord". Adv. Exper. Med. Biol., 550: 23–42.

Shewmon, D.A. and Shewmon, E.S. (2004) The semiotics of death and its medical implications. Adv. Exper. Med. Biol., 550: 89–114.

Truog, R.D. (1997) Is it time to abandon brain death? Hastings Cent. Rep., 27(1): 29–37.

Truog, R.D. and Robinson, W.M. (2003) Role of brain death and the dead-donor rule in the ethics of organ transplantation. Crit. Care Med., 31: 2391–2396.

Veatch, R.M. (1973) The whole brain-oriented concept of death: an outmoded philosophical formulation. J. Thanatol., 3: 13–30.

Veatch, R.M. (1993) The impending collapse of the whole-brain definition of death. Hastings Cent. Rep., 23(4): 18–24.

Veith, F.J., Fein, J.M., Tendler, M.D., Veatch, R.M., Kleiman, M.A. and Kalkines, G. (1977) Brain death: a status report of medical and ethical considerations. JAMA, 238: 1651–1655.

Wang, M.Y., Wallace, P. and Gruen, J.B. (2002) Brain death documentation: analysis and issues. Neurosurgery, 51: 731–735.

Wijdicks, E.F.M. (1995) Determining brain death in adults. Neurology, 45: 1003–1011.

Wijdicks, E.F.M. (2001a) The diagnosis of brain death. N. Engl. J. Med., 344: 1215–1221.

Wijdicks, E.F.M. (Ed.). (2001b) Brain Death. Lippincott Williams & Wilkins, Philadelphia pp. 61-90.

Wijdicks, E.F.M. (2002) Brain death worldwide: accepted fact but no global consensus in diagnostic criteria. Neurology, 58: 20–25.

Yaqub, B.A. and Al-Deeb, S.M. (1996) Brain death: current status in Saudi Arabia. Saudi. Med. J., 17: 5–10.

Youngner, S.J., Landefeld, C.S., Coulton, C.J., Juknialis, B.W. and Leary, M. (1989) "Brain death" and organ retrieval. A cross-sectional survey of knowledge and concepts among health professionals. JAMA, 261: 2205–2210.

S. Laureys (Ed.)
Progress in Brain Research, Vol. 150
ISSN 0079-6123

CHAPTER 27

The minimally conscious state: defining the borders of consciousness

J.T. Giacino*

New Jersey Neuroscience Institute, 65 James St., Edison, NJ 08818, USA

Abstract: There is no agreement as to where the limits of consciousness lie, or even if these putative borders exist. Problems inherent to the study of consciousness continue to confound efforts to establish a universally accepted theory of consciousness. Consequently, clinical definitions of consciousness and unconsciousness are unavoidably arbitrary. Recently, a condition of severely altered consciousness has been described, which characterizes the borderzone between the vegetative state and so-called "normal" consciousness. This condition, referred to as the *minimally conscious state (MCS)*, is distinguished from the vegetative state by the presence of minimal but clearly discernible behavioral evidence of self or environmental awareness. This chapter reviews the diagnostic criteria, pathophysiology, prognostic relevance, neurobehavioral assessment procedures and treatment implications associated with MCS.

The problem of consciousness

The concept of consciousness continues to stymie philosophers, neuroscientists and clinicians despite decades of research and observational experience. Various theoretical frameworks have been proposed (Zeman et al., 1997), but none have achieved universal acceptance. The neural basis of consciousness remains elusive and there is continued disagreement as to which behaviors provide unequivocal evidence of conscious processing. More than 20 years ago, Plum and Posner (1982) observed:

> The limits of consciousness are hard to define satisfactorily and we can only infer the self-awareness of others by their appearance and their acts (p. 3).

Neurologists and neurorehabilitation specialists involved in the care of patients with very severe brain injury are routinely faced with the challenge of delineating the borders of consciousness. From a clinician's perspective, three core problems constrain the assessment of patients with disorders of consciousness (DOC). First, the central constructs used to describe consciousness are ambiguous. For example, the terms *wakefulness*, *alertness*, *awareness* and *attention* are often used interchangeably to describe different aspects of consciousness. Second, judgments concerning level of consciousness are dependent upon behavioral observations. This may lead to interpretive errors because behavior is an indirect measure of consciousness. A particular behavior may represent a manifestation of neural processing at various points along the neuraxis from the hindbrain to the forebrain. For example, at the highest level, "smiling" may be an intentional act executed for purposes of amusement. Alternatively, smiling behavior may reflect involuntary motor activity mediated by the release of bulbar structures from cortical control (as in pseudobulbar syndrome), or may even be

*Corresponding author. Tel.: +1 732-205-1461;
Fax: +1 732-632-1584; E-mail: jgiacino@solarishs.org

DOI: 10.1016/S0079-6123(05)50027-X

due to reflexive muscle contractions such as those that occur in the risus sardonicus of tetanus.

The third, and perhaps the most intriguing problem associated with assessment of consciousness concerns changes in "state." Traditionally, the transition from an unconscious to a conscious state has been viewed as a linear function. The linear model holds that unconsciousness and consciousness are steady states divided by a sharply demarcated border. The progression from unconsciousness to consciousness occurs abruptly when a critical threshold is crossed and consciousness emerges. Section A of Fig. 1 illustrates the linear model.

Conversely, the relationship between unconsciousness and consciousness may be viewed as dynamic. In evaluating patients with DOC, it is not uncommon to clearly demonstrate discernible signs of consciousness at the bedside, yet, be unable to reproduce these signs on subsequent attempts to elicit the same behavior. Following brain injury, one's level of consciousness may fluctuate between complete and partial unawareness of self

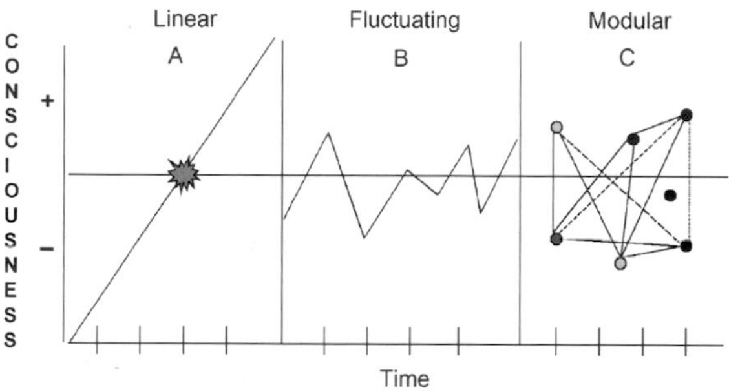

Fig. 1. Conceptual models depicting the interface between unconscious and conscious states following severe brain injury. Section A shows the linear model, which represents consciousness and unconsciousness as distinct states separated by a sharply demarcated border. Section B illustrates the fluctuating model which suggests that patients may vacillate between conscious and unconscious states over time. The modular model, shown in section C, holds that specific cognitive modules (e.g., expressive speech) may become underactive or disconnected from other modules. The light- and dark-colored circles depict weak and strongly activated modules, while the dotted and solid lines represent interrupted and uninterrupted connections between modules. Modules that remain active but become isolated may produce complex behaviors that occur in the absence of conscious experience. Adapted from Katz and Giacino (2004), with permission.

or environment. Clearly, discernible signs of consciousness may be apparent on examination at time 1, but may be absent when the same examination is conducted at time 2. Fluctuations in state may occur day to day, hour to hour, or moment to moment and may be independent of changes in wakefulness (Schiff et al., 2000; Giacino and Trott, 2004). The fluctuating model of recovery of consciousness is depicted in section B of Fig. 1.

Finally, one's state of consciousness may be dependent upon the extent to which individual neurocognitive subsystems (e.g., language, object perception, goal-directed movement) are able to activate and communicate with other subsystems at any given time. In this modular perspective, state of consciousness is a function of the level of activation of each subsystem, the degree of interconnectedness among subsystems and the status of both the former and the latter over time. Full consciousness, therefore, would require that each module remains sufficiently activated and integrated with all other modules. Following brain injury, any given module may become disabled as the direct or indirect result of a lesion and normal consciousness may be compromised. Focal lesions affecting the language, motor and object recognition modules result in well-recognized aphasic, apraxic and agnosic syndromes, respectively (Mesulam, 2000). Such lesions may also influence the level of activation of these modules causing the cognitive functions they subserve to fluctuate over time. Transient reversal of stroke-related cognitive impairments (i.e., right hemineglect, right hemiplegia, aphasia, asomatognosia) has been demonstrated following application of vestibular stimulation, suggesting that cognitive modules may, indeed, be subject to variations in level of activation over time (Schiff and Pulver, 1999). Moreover, a particular module may retain its specific functional capacity, while its connection to other modules may be disrupted by non-focal cortical or subcortical injuries. Theoretically, global forms of brain injury (e.g., cerebral hypoxia, diffuse axonal injury) could sever the connections between each module without destroying the module itself. Under these circumstances, normal consciousness would be severely altered, but the functional integrity of a particular module may be spared. This account of altered consciousness is displayed in section C of Fig. 1.

An approximation of the modular model is exemplified in a case study reported by Schiff and colleagues (1999). A 49-year-old woman who lapsed into a VS as the result of multiple ruptures of an arteriovenous malformation 20 years earlier was selected for study because of an unusual but consistent behavioral finding noted on examination. Although she followed no commands, made no purposeful movements and was unable to communicate by word or gesture, video-electroencephalogram (EEG) monitoring captured one to three-word intelligible verbalizations every 24–72 h. The verbalizations occurred spontaneously and typically consisted of expletives and single word perseverations (e.g., "down, down, down"). To explore the correlation between the verbal behavior and the underlying pathophysiologic substrate, resting cerebral metabolism was analyzed using ^{18}F-fluorodeoxyglucose–positron emission tomography (FDG–PET) studies co-registered to conventional magnetic resonance imaging (MRI) scans. MRI showed severe bilateral posterior subcortical white-matter hemorrhagic changes, destruction of the right thalamus and severe involvement of the posterior left thalamus and right basal ganglia. The PET scan revealed that the average global cerebral metabolic rate of glucose consumption (CMR_{glc}) was 43% of normal. Of particular interest, regional analysis of glucose uptake indicated that CMR_{glc} was within normal limits in essential language structures including the left basal ganglia, left insular cortex, Broca's area and Wernicke's area. Figure 2 outlines these regions of increased metabolic activity.

On the basis of the observed loci of activation, Schiff and colleagues concluded that the patient's verbal output was mediated by a topographically isolated, yet partially functional, speech circuit. In effect, the language module retained functional viability in the absence of any other evidence of cognitive processing. In the vast majority of patients with DOCs, intelligible speech represents a sign of consciousness. Nonetheless, this highly unusual case could easily result in a false positive diagnosis of minimally conscious state (MCS),

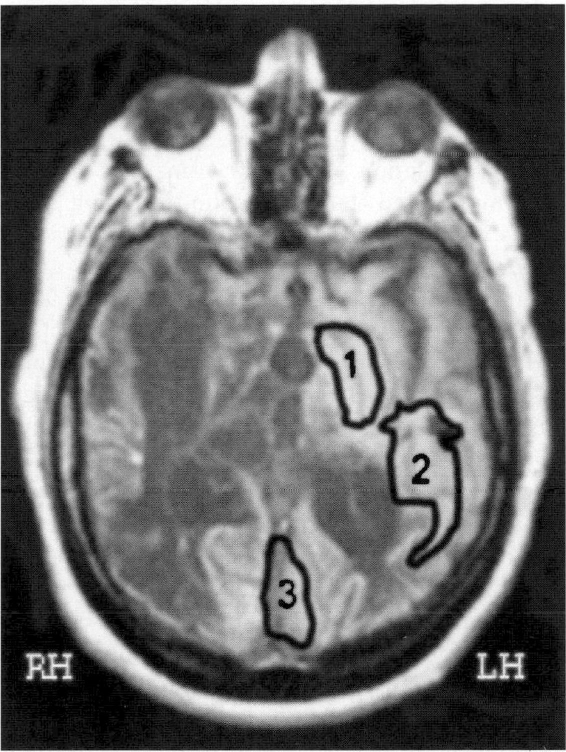

Fig. 2. Regions of heightened cerebral metabolic activity in a patient diagnosed with VS who retained the capacity to produce intelligible single words. ^{18}F-fluorodeoxyglucose PET data are co-registered to the MRI. Metabolic activity is within the normal range in (1) the left caudate nucleus; (2) left temporoparietal cortex (including Wernicke's area) and left frontal cortex (including Broca's area) (the latter is not shown); and (3) left calcarine cortex. From Schiff et al. (1999), with permission.

highlighting the pitfalls associated with over-reliance on behavioral markers of consciousness.

Rationale for defining the minimally conscious state

Clinicians specializing in the care of patients with severe brain injury are well acquainted with the clinical features of coma and the VS. Both of these disorders are characterized by the complete absence of behavioral signs of self and environmental awareness. VS can be readily distinguished from coma by observing for spontaneous or elicited eye-opening which occurs in VS and not in coma (Jennett and Plum, 1972). The reemergence of eye-opening signals that the reticular system is again active, generating a wakeful, yet unaware, state. VS is also distinguished from coma by the return of the so-called "vegetative" functions of the body, which include respiration, heart rate and thermal regulation. While these functions may still be compromised during VS, life-sustaining treatments such as mechanical ventilation are typically no longer required.

In most cases, the course of recovery from VS is slow with gradual recovery of cognitive function. The transition from unconsciousness to consciousness is characteristically subtle and often ambiguous. While command following and discernible communication represent the clearest signs of reemerging consciousness, these behaviors usually occur infrequently during this transitional period and are often difficult to differentiate from random movements (Giacino and Zasler, 2005). Further complicating matters, among those patients who demonstrate clear signs of conscious behavior, such responses usually occur inconsistently during the early stages of recovery. Some patients fail to progress beyond this level of responsiveness and remain permanently incapable of consistently producing sentient behavior. These patients occupy an intermediate point along a continuum of consciousness that includes those in VS on one pole, and those who consistently exhibit meaningful behavioral responses on the other.

Until recently, this intermediate subgroup was indiscriminately lumped together with patients in VS and coma. As a by-product of the lack of adequate differential diagnostic criteria, high rates of diagnostic inaccuracy have been reported among patients with DOC (Childs et al., 1993; Andrews, 1996). In response to this concern, the American Congress of Rehabilitation introduced the term minimally responsive state (MRS) to describe patients who demonstrated inconsistent but meaningful behavioral responses (ACRM, 1995). A key element of this new diagnostic category was the requirement that the behavior(s) of interest had to be viewed as unequivocally "meaningful" by the examiner. Following publication of the definition of MRS, concerns were raised that patients correctly diagnosed with coma or VS also display some degree of behavioral responsiveness so that these conditions might also descriptively qualify as MRS. Consequently, an expert panel, known as the Aspen Workgroup, was convened to revisit the existing diagnostic and prognostic criteria for DOC. Among their recommendations, the Aspen Workgroup proposed that the term MRS be replaced by MCS to emphasize the partial preservation of consciousness that distinguishes this condition from coma and VS (Giacino et al., 1997).

The primary purpose of establishing operationally defined criteria for MCS was to provide a common frame of reference for future scientific study. Since the crafting of a case definition for MCS, incidence and prevalence figures have been estimated, pathophysiology has been investigated and outcome studies have been conducted. The remainder of this article will review the diagnostic criteria for MCS and summarize recent empirical findings. An illustrative case study will be presented to highlight typical assessment and management issues and directions for future research will be outlined.

Definition and diagnostic criteria

> MCS is a condition of severely altered consciousness in which minimal but definite behavioral evidence of self or environmental awareness is demonstrated (Giacino et al., 2002).

The key element of the definition is the requirement that there be at least one clear-cut behavioral sign of consciousness. The behavior of interest must be reproducible during the examination and must be representative of cognitive processing. It is often necessary to conduct serial examinations before an accurate diagnosis can be made because cognitively mediated behavioral responses occur inconsistently in MCS. The behavioral fluctuation that is perhaps the hallmark of MCS may be related to changes in arousal level, stimulus-induced variations in cerebral function, injury-related instability in neuronal firing or some combination of these factors.

The diagnostic criteria for MCS rest largely on the integrity of the language and motor systems.

MCS should be diagnosed only when there is *clearly discernible* evidence of *one or more* of the following behaviors:

- Simple command-following;
- Gestural or verbal yes/no responses (regardless of accuracy);
- Intelligible verbalization;
- Movements or affective behaviors that occur in contingent relation to relevant environmental stimuli and are not attributable to reflexive activity. The following examples provide sufficient evidence for contingent behavioral responses:
  - Episodes of crying, smiling, or laughing in response to the linguistic or visual content of emotional but not neutral topics or stimuli.
  - Vocalizations or gestures that occur in direct response to the linguistic content of comments or questions.
  - Reaching for objects that demonstrate a clear relationship between object location and direction of reach.
  - Touching or holding objects in a manner that they accommodate the size and shape of the object.
  - Pursuit eye movement or sustained fixation that occurs in direct response to moving or salient stimuli.

Disturbances of higher cognitive function, particularly aphasia and apraxia, may confound assessment of these criteria and should be considered before establishing the diagnosis of MCS.

Additional behavioral parameters have been proposed to define the upper limit of MCS and to mark the transition to a higher level of functional capacity. *Emergence from MCS is signaled by the return of reliable and consistent interactive communication or functional object use.* These two benchmarks were chosen because of their obvious importance in establishing meaningful environmental interaction. Interactive communication may occur through verbalization, writing, yes/no signals or augmentative communication devices. Functional object use requires discrimination and appropriate use of at least two familiar objects (e.g., cup, hairbrush). While any instance of discernible communication is sufficient to establish the diagnosis of MCS, emergence from MCS requires accurate and consistent communicative responses.

It is clear that MCS is clinically distinct from other DOC. Table 1 compares the clinical features of MCS with those observed in coma and VS. Figure 3 presents an algorithm designed to assist clinicians in the differential diagnosis of DOC.

Incidence and prevalence

The incidence and prevalence of MCS are difficult to estimate not only because of the novelty of the case definition, but also because of the lack of adequate surveillance outside of primary care settings. In the United States, the majority of patients in MCS are transferred to long-term care facilities after a relatively brief hospitalization at a trauma center or inpatient rehabilitation facility. Even among those individuals admitted to inpatient rehabilitation programs, most are discharged to nursing homes or private residences within 2–3 months of the injury. It is clear that neither of these settings permits the level of clinical monitoring required to determine the number of patients in MCS at a given time. Further complicating surveillance efforts, some individuals in VS at the time of hospital discharge transition to MCS, or emerge from MCS, while they are in custodial care.

Strauss and colleagues (2000) have provided the only empirically derived prevalence estimate for MCS published to date. These investigators extracted data from a large state registry used by the California Department of Developmental Services to track medical care and services administered to residents between the ages of 3 and 15. Using data from a standardized functional rating scale employed by the registry, operational definitions were established for VS and MCS, according to accepted diagnostic criteria. Of the 5,075 individuals who met criteria for one of these two diagnoses, 11% were in VS and 89% in MCS. Extrapolating from US census data for the general adult population, the prevalence of MCS was estimated to be between 112,000 and 280,000. If accurate, these data

Table 1. Comparison of behavioral features of MCS, VS, and coma

Behavior	MCS	VS	Coma
Eye opening	Spontaneous	Spontaneous	None
Spontaneous movement	Automatic/object manipulation	reflexive/patterned	None
Response to pain	Localization	Posturing/withdrawal	Posturing/none
Visual response	Object recognition/pursuit	Startle/pursuit (rare)	None
Affective response	Contingent	Random	None
Commands	Inconsistent	None	None
Verbalization	Intelligible words	Random vocalization	None

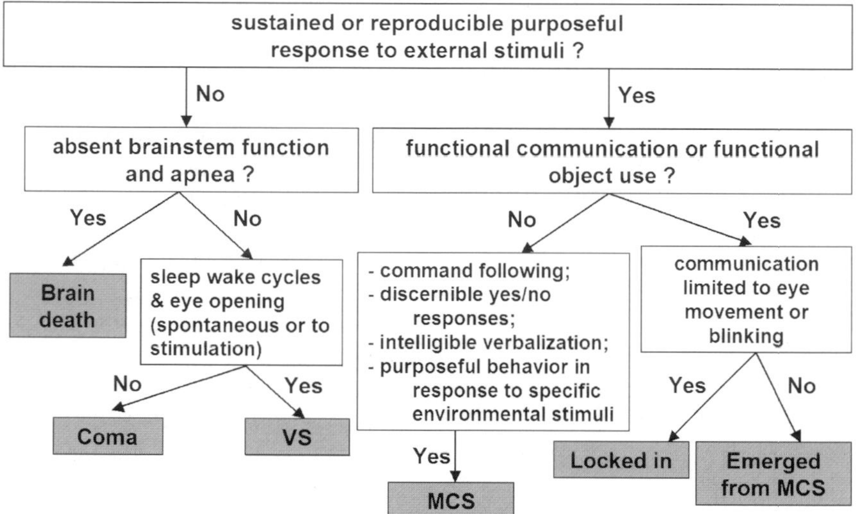

Fig. 3. IS MY PATIENT CONSCIOUS? Diagnostic algorithm for differential diagnosis of disorders of consciousness. From Ashwal and Cranford (2002), with permission.

suggest that MCS may be eight times more prevalent than VS (Multi-Society Task Force, 1994).

Pathophysiology and residual brain function

Until recently, it was presumed that the pathophysiology of MCS was qualitatively similar to VS, albeit more severe. Findings from a new generation of structural and functional neuroimaging studies suggest, however, that the neuropathologic substrate and pattern of residual neural activity associated with MCS is significantly different from VS.

In a post-mortem analysis, Jennett and colleagues (2001) compared 35 individuals who remained in post-traumatic VS until death to a second group of 30 patients, 12 of whom were in MCS and 18 that emerged from MCS but remained severely disabled (SD) until the time of death. While all of the patients in the VS group had moderate (i.e., grade 2) to severe (i.e., grade 3) diffuse axonal injury (DAI) and lesions involving the thalamus, 50% of the SD group had no evidence of grade 2 or 3 DAI *and* no indication of thalamic damage. In the MCS group, moderate to severe DAI was noted in 42% of cases (versus 22% in the SD group) but was considerably less common relative to the VS group (71%). Thalamic lesions were also notably less prevalent in MCS (50%) relative to VS (80%) cases. These findings demonstrate important differences in lesion location and severity, which may account for the partial sparing of consciousness in MCS. Indeed, an outcome study conducted by Kampfl and others (1998) of patients in VS who were between 2 and 3 months post-injury found that those with callosal lesions were 214 times less likely to recover consciousness by 1 year than those without lesions in this location. Lesions involving the dorsolateral brain stem were associated with a seven-fold increase in the probability of remaining in VS.

The imperfect and sometimes illusory relationship between cerebral pathology and behavioral findings in patients with severe brain injury is generally well recognized (Prayer et al., 1993; Ichise et al., 1994). For example, large left temporal lesions that overly language cortex do not necessarily produce aphasic symptoms. In contrast, small focal lesions involving critical frontostriatal structures may result in overt behavioral changes. Functional neuroimaging procedures such as PET and functional MRI (fMRI) have been increasingly relied upon in the assessment of patients with DOC as they provide a more direct means of gauging neural activity.

Unlike patients in VS who have been shown to activate primary sensory cortex but not higher level heteromodal association zones in response to various forms of sensory stimulation, patients in MCS show more widespread activation. Laureys and co-workers (2000) exposed a patient in MCS to no sound, frequency-modulated noise, infant cries and the patient's own voice and found differential activation using $H_2^{15}O$-PET. Although global metabolism was significantly reduced relative to controls, activation spread to heteromodal association cortices following presentation of the infant cries and the patient's name. Functional connectivity analyses performed in other PET paradigms have also shown significantly greater activation of medial parietal, posterior cingulate, and secondary frontal and temporal cortices in MCS patients, relative to those in VS (Laureys et al., 2003; Boly et al., 2004). These multimodal association zones are highly active during conscious waking and have been linked to a neural network thought to mediate human awareness (Laureys et al., 2004). The lack of activity noted in these regions in VS patients is consistent with the presumed absence of cognitive processing in this condition.

Recent clinical applications of fMRI have begun to explore the relationship between brain structure and behavior in patients with DOC. Hirsch and colleagues (2001) developed a passive listening fMRI paradigm to investigate cortical responses in two patients who met diagnostic criteria for MCS more than a year after sustaining traumatic brain injury. During the scan, the patients were exposed to an audiotape of a family member recounting a familiar past event. The same audiotape was time-reversed and played backward so that it was devoid of content and prosody but still recognizable as speech. When the audiotape was played forward, areas of activation were remarkably similar between the MCS patients and normal volunteers. The MCS patients demonstrated robust activation in a distributed network of structures associated with language functions (i.e., superior and middle temporal gyri, inferior and medial frontal gyri, inferior and middle occipital gyri, calcarine sulcus, precuneus and cuneus). When the audiotapes were presented backward without linguistic content, the normals maintained activation levels similar to those noted during the forward condition. In contrast, activation levels dropped dramatically in the MCS patients during the backward condition. Although preliminary, these findings suggest that the cortical language network remains functional in MCS despite the absence of consistent command-following or reliable communication. The reduced activation noted during the backward condition in the MCS patients suggests a failure to sustain active cognitive processing in the context of ambiguous stimuli and may be related to the loss of selective attentional resources. In view of the studies described above, it is apparent that there are critical neuropathologic differences between MCS and VS, which have important implications for prognosis, clinical management and long-term outcome.

Prognosis and outcome

Few outcome predictors have been identified that are specific to MCS. This is largely because prior outcome studies failed to distinguish between patients in VS and MCS. There is growing evidence that outcome from MCS may be associated with certain clinical findings. A number of investigators have reported that reemergence of visual pursuit may presage recovery of other more definitive signs of consciousness. Giacino and Kalmar (1997) found that the incidence of visual pursuit was significantly higher in MCS patients although this behavior was noted in both VS and MCS. Among 104 patients admitted to an inpatient rehabilitation center with a diagnosis of VS or MCS, visual pursuit was intact in 82% of the MCS patients and only 20% of the VS group. This finding was not influenced by the length of time post-injury. Moreover, 73% of the VS group with intact visual pursuit recovered other clear-cut signs of consciousness by 12 months, as compared to 45% of those without pursuit.

Other investigators have also reported an association between preserved visual pursuit and subsequent recovery of function. Ansell and Keenan (1989) conducted a study of severely brain-injured patients to identify predictors of early cognitive improvement and readiness for rehabilitation. Subjects who demonstrated late improvement performed significantly better on tests of visual pursuit completed on admission to rehabilitation when compared to those who did not improve. Similarly, Shiel and colleagues (2000) reported that acutely brain-injured patients who showed visual pursuit on hospital admission were more likely to demonstrate social interaction and communicative behavior later in the course than those who did not.

Rate of recovery has been shown to predict outcome following MCS in two studies. While neither study stratified patients by diagnosis, both included cases diagnosed with MCS and VS. In a sample of 34 severely brain-injured patients who were unable to follow commands or communicate reliably, Giacino and co-workers (1991) found that change scores on three different assessment scales administered during the first month of rehabilitation were more predictive of functional outcome at discharge when compared to admission scores alone. While 19 of 20 patients with low change scores on the Disability Raring Scale (DRS) (Rappaport et al., 1982) were "extremely severely disabled" to "extremely vegetative" at discharge, only 1 of 8 patients with high change scores fell into one of these unfavorable outcome categories.

There is accumulating evidence that patients in MCS demonstrate improvement over a longer period of time and attain better functional recovery relative to those in VS. A recently completed multicenter study of the natural history of recovery from VS and MCS conducted by Whyte and colleagues (2005) found that rate of DRS improvement over the first 2 weeks of observation was highly predictive of subacute functional outcome. Specifically, those patients with better DRS scores at enrollment and faster rates of initial improvement tended to have better DRS scores at 16 weeks. The combination of rate of DRS recovery, the time between injury and enrollment and the DRS score at enrollment accounted for nearly 50% of the variance in DRS score at 16 weeks. These same three variables were also highly significant predictors of time until commands were followed.

In a retrospective comparative analysis of outcome following VS and MCS, Giacino and Kalmar (1997) found that 50% of MCS patients, versus 3% of VS patients, had "no disability" to "moderate disability" on the DRS at 1 year post-injury. Additionally, while 43% of patients in the traumatic VS subgroup fell between the "VS" and "extreme VS" categories, none of the patients in the traumatic MCS had scores in this range. When outcome scores were compared in the MCS group according to etiology of injury, more rapid and progressive improvement was noted in the traumatic MCS subgroup. Figure 4 compares mean DRS scores across the first year post-injury in VS and MCS patients stratified by etiology.

Life expectancy

Strauss and co-workers (2000) estimated mortality risk and survival rates in patients diagnosed with VS and MCS. The investigators subdivided the MCS group into patients who were mobile and those without mobility, on the basis of the ability to lift the head while lying prone, roll forward or

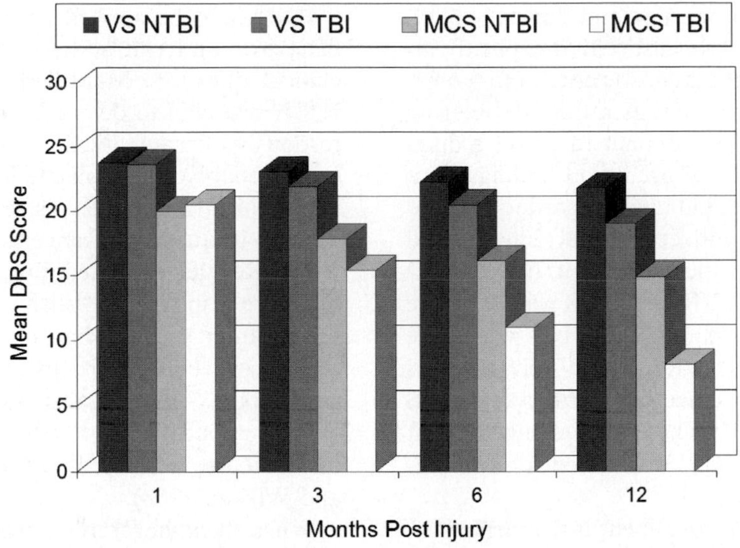

Fig. 4. Comparison of mean DRS scores at 1, 3, 6, and 12 months post-injury in patients diagnosed with traumatic or non-traumatic VS and MCS. From Giacino and Kalmar (1997), with permission.

backward or maintain a sitting position for at least 5 min. The primary question addressed in this study was whether immobile MCS patients had more a favorable survival rate than patients in VS. The authors found little difference in survival time between the two groups, suggesting that mobility is a better predictor of survival than the presence of consciousness. The percentage of patients in mobile MCS surviving for 8 years was 81% as compared to 65% and 63% for immobile MCS and VS, respectively. Duration of survival was also found to be longer for patients with acquired versus congenital brain injury and shorter for those dependent upon gastrostomy feeding.

Specialized neurobehavioral assessment methods

Conventional bedside assessment procedures and neurosurgical rating scales such as the Glasgow Coma Scale (Teasdale and Jennett, 1974) have limited utility when used to monitor progress in patients with prolonged disturbance in consciousness. These procedures are not designed to detect subtle (but potentially important) changes in neurobehavioral responsiveness over time, they are unable to distinguish random or reflexive behaviors from those that are purposeful and they lack operationally defined scoring criteria. To address these shortcomings, assessment procedures designed specifically for use in patients with prolonged DOC have been devised. Standardized rating scales, which rely on fixed administration and scoring procedures, are useful for establishing a comprehensive profile of neurobehavioral function. Alternatively, individualized quantitative behavioral assessment protocols, based on single subject design principles, have been developed to address case-specific questions.

Standardized neurobehavioral assessment measures include the Coma Recovery Scale-R (CRS-R) (Giacino et al., 1991, 2004), the Coma-Near Coma Scale (CNC) (Rappaport et al., 1992), the Western Neurosensory Stimulation Profile (WNNSP) (Ansell and Keenan, 1989), the Comprehensive Level of Consciousness Scale (CLOCS) (Stanczak et al., 1984) and the Sensory Modality and Rehabilitation Technique (SMART) (Wilson and Gill-Thwaites, 2000). Although item content varies across measures, all evaluate behavioral responses to a variety of auditory, visual, motor and communication prompts. All of these instruments have been reported to have adequate reliability and validity, however, there is considerable variability in their psychometric properties and clinical utility. The CRS-R is the only measure that explicitly incorporates the current diagnostic criteria for coma, VS and MCS into the administration and scoring scheme and allows the examiner to derive a diagnosis directly from the examination findings. In a recent study of 80 patients with disorders of consciousness, Giacino and colleagues (2004) found that although diagnostic agreement between examiners using the CRS-R and DRS was 87%, the CRS-R identified 10 patients in MCS who were classified as VS on the DRS. There were no cases in which the DRS detected features of MCS missed by the CRS-R. These results suggest that the CRS-R is particularly well suited for differential diagnostic purposes.

While standardized assessment instruments provide a comprehensive overview of neurobehavioral status, these measures are unable to address case-specific questions. Clinicians involved in the care of MCS patients often encounter situations in which the patients' behavioral responses are ambiguous or occur too infrequently to clearly discern their significance. These problems are often due to injury-related sensory, motor and drive deficits. For this reason, "Individualized Quantitative Behavioral Assessment (IQBA)" strategies have been developed (Whyte and DiPasquale, 1995; Whyte et al., 1999; DiPasquale and Whyte, 1996). IQBA procedures are individually tailored, assessment methods are standardized, target responses are operationalized and controls for examiner and response bias are incorporated into the protocol. Once the examiner has selected a behavior of interest (e.g., command-following, visual tracking), the frequency of the behavior is recorded under varying conditions. Raters track the frequency of the behavior following administration of the appropriate command, following an incompatible command and during a rest interval. Data are analyzed statistically to determine whether the target behavior occurs significantly more often in one condition relative to the others. For example, when the frequency of the behavior is greater during the "rest" condition than it is during the "command" condition, it suggests that the behavior simply represents random movement, not a deliberate response to the command.

IQBA can be applied across a broad array of behaviors and can address virtually any type of clinical question. McMillan (1996) employed an IQBA protocol to determine whether a minimally responsive, traumatically brain-injured patient could reliably communicate a preference concerning withdrawal of life-sustaining treatment. Responses to questions were executed using a button press. Results indicated that the number of affirmative responses to "wish to live" questions was significantly greater than chance suggesting that the patient could participate in end-of-life decision-making. McMillan's findings were subsequently replicated in a second IQBA assessment conducted by a different group of examiners (Shiel and Wilson, 1998).

When behavioral responses are equivocal on bedside examination, additional steps may be taken to facilitate detection of consciousness. The examiner should assure that the patient is adequately aroused prior to conducting the examination. Potentially sedating medications should be discontinued and the patient should be screened for subclinical seizure activity. Arousal facilitation techniques that incorporate tactile stimulation, deep pressure or vestibular stimulation (Giacino et al., 1992) have been developed to augment arousal and alertness. To increase the likelihood of detecting volitional responses, unnecessary sources of distraction should be eliminated, a broad range of stimuli should be presented and response modalities compromised by sensorimotor impairment should be avoided. Whenever possible, observations of family members and paraprofessionals should be integrated into the examination and the patient should be reevaluated serially over time. Special care must be taken when evaluating young children with disturbances in consciousness as immature language and motor development may confound interpretation of command following and communicative responses.

Case study

To demonstrate the assessment and clinical management issues common to patients in MCS, an illustrative case study is presented. Following review of the medical history, the salient diagnostic, prognostic and treatment aspects of the case are discussed.

Medical history

A 23-year-old, right-handed stockbroker sustained a severe traumatic brain injury as the result of a motor vehicle accident. He was unconscious at the scene with right-sided posturing. Seizure activity was observed in the field and he was comatose on admission to the emergency room. On neurologic examination, the right pupil was unreactive and the left was sluggishly responsive. CT scan of the head revealed multiple punctate hemorrhages involving the left frontal, temporal, and parietal lobes with extension into the brain stem, indicating severe diffuse axonal injury. Tachycardia, right pleural effusion, and gastroparesis were noted and he experienced central fevers spiking to 105°F (41°C). Tracheostomy and gastrostomy tubes were placed and an inferior vena cava filter was inserted. Phenytoin was started for control of the acute seizures and a cooling blanket was applied for the hyperthermia. There was slow but steady improvement in medical stability across the first week post-injury. On day 11, the patient was transferred to an acute inpatient rehabilitation center. Spontaneous movement was noted in all four limbs (R > L), but there was no evidence of purposeful movement and the eyes remained continuously closed. Follow-up CT scan of the brain showed hyperintensities in the left thalamus and in the gray–white-matter junction of the left frontal region.

Neurobehavioral assessment with the Coma Recovery Scale (Giacino et al., 1991) was consistent with VS. Responses to auditory and visual threat were equivocal and there was no evidence of command-following, gestural or verbal communication, visual pursuit or purposeful motor behavior. Isolated flexion withdrawal was elicited following application of noxious stimulation to the right upper and both lower extremities and there was abnormal extension of the left upper extremity. Spontaneous eye opening reemerged during week 5 signaling transition to VS.

Prognostic assessment

Discernible signs of consciousness were not observed in this case until approximately 2 months post-injury. At 3 months post-injury, the patient unexpectedly lapsed back into VS but regained consciousness again in month 4. On the basis of outcome data described by the Multi-Society Task Force on PVS (1994), approximately 35% of patients in traumatic VS at 3 months post-injury will eventually recover consciousness, while the remaining 65% will die or evolve into permanent VS. Other studies have reported considerably higher rates of recovery of consciousness after 3 months (Giacino and Kalmar, 1997). It is difficult to apply the data from the prior studies in this case as the patient's diagnosis fluctuated between VS and MCS during the first 4 months post-injury. On average, approximately 80% of patients who remain in VS or MCS beyond 3 months post-injury will remain severely disabled according to DRS criteria (Giacino, 1991; Multi-Society Task Force, 1994).

The patient's CT scan showed multiple punctate hemorrhages in the cortex extending down to the brain stem, indicating grade III diffuse axonal injury. Grade III DAI is associated with an increased probability of prolonged (i.e., > 1 year) or permanent VS (Kampfl et al., 1998; Jennett et al., 2001). The patient's early neurologic findings coupled with the acute neuroradiologic data pointed to a protracted period of unconsciousness and a high probability of severe long-term disability.

Although the patient remained unresponsive throughout the acute course, he began to exhibit multiple signs of self and environmental awareness at 4 months post-injury. The late recovery heralded a more favorable outcome than originally projected, reinforcing the claim that rate of change is a more powerful predictor of outcome relative to static ratings (Bricolo et al., 1980; Giacino et al., 1991; Whyte et al., 2005).

Treatment course

The initial 3 weeks of the rehabilitation course focused primarily on medical management of the central fevers, hypertonus and autonomic dysreflexia. Traditional physical management and sensory regulation procedures were initiated to promote medical stability and prevent complications. Following resolution of the fever, the decision was made to implement pharmacologic interventions to facilitate arousal and behavioral responsiveness as there was still no evidence of command following or communication through week 5. Amantadine hydrochloride (50 mg, bid) was started but had to be stopped after 5 days because of increasing hypertonus and diaphoresis. Bromocriptine was started in place of the amantadine in week 7 and was increased to 5 mg/day after 1 week. CRS evaluations were conducted serially on a weekly basis to monitor the patient's response to treatment. During week 8, episodes of command-following, automatic motor behavior and sustained visual fixation were noted. Subsequent to these changes, the bromocriptine was increased to 7.5 mg/day. Over the next 5 weeks, there was an unexpected decrease in neurobehavioral responsiveness. The patient lapsed back into VS as evidenced by the loss of command-following and purposeful motor behavior. Amantadine was restarted in week 15, approximately 9 weeks after the initial trial was discontinued. This was followed by reemergence of reproducible command-following (week 16), automatic motor responses and visual pursuit (week 19). As the dose was titrated up to 400 mg/day, gestural communication attempts (i.e., head movements) (week 21), object recognition (week 21), and episodes of intelligible verbalization (week 24) were observed.

The amantadine was discontinued at approximately 7 months post-injury in an effort to discern its role in the patient's recovery. Neurobehavioral functions continued to improve during and after the drug taper. Reassessments conducted during week 30 demonstrated reliable yes/no responses to a series of situational orientation questions and enabled a functional communication system to be established. The reestablishment of a reliable communication system triggered a shift in the rehabilitation program from environmental control to training in self-management strategies. Figure 5 depicts the patient's performance on the CRS across the 37-week course.

The lack of a clear dose-dependent response to the amantadine in this case suggests either that the relatively abrupt improvements in responsiveness were simply part of the spontaneous recovery curve or that exposure to the drug was able to "jump start" the recovery process. Very few well-designed studies have investigated the effectiveness of pharmacologic interventions in patients with DOC. A recent evidence-based review completed by the author failed to identify any class I studies (Giacino, in press) in the traumatic brain-injury literature. Among the class II and III studies, amantadine was the only agent reported to be effective in promoting rate of recovery in patients with DOC (Zafonte et al., 1998; Meythaler et al., 2002). At present, existing data are inadequate to make an evidence-based recommendation concerning the effectiveness of medications used to facilitate recovery in patients with DOC.

Directions for future research

The last 10 years have been witness to important advances in our understanding of DOC. The development of a case definition for MCS, the availability of novel functional neuroimaging strategies, the refinement of neurobehavioral assessment tools and the identification of reliable prognostic indicators of functional recovery represent examples of such achievements. These accomplishments have given rise to many new questions and have helped forge a new research agenda. Some of the key questions that will need to be addressed over the next decade are listed below:

- *Should we continue to rely on behavior as the "gold standard" for evidence of consciousness?* Functional neuroimaging studies suggest that cognitive processing capacity may be underestimated in patients in MCS (Menon et al., 1998; Hirsch et al., 2001). This may be related to sensory, motor and drive deficits, which may mask signs of consciousness. fMRI and PET studies are expected to clarify the relationship between behavioral signs of consciousness and the integrity of underlying neural networks.

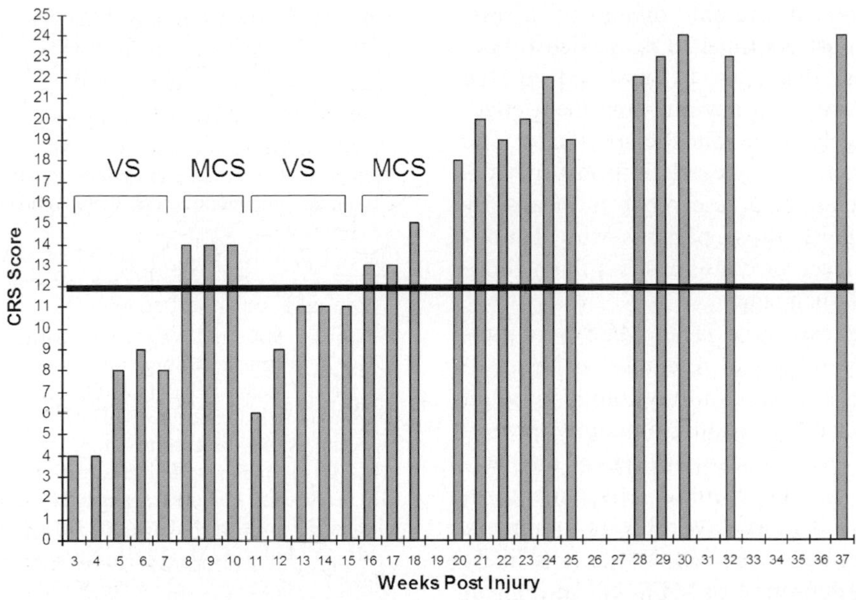

Fig. 5. Early fluctuations in neurobehavioral status in an MCS patient monitored across a 37-week period. The patient's scores on the CRS are shown over time. A CRS score of 12 marked the boundary between VS and MCS in this case. During the first 4 months of the recovery course, the patient fluctuated between VS and MCS. He subsequently emerged from MCS and demonstrated progressive improvement through week 30. Adapted from Giacino and Trott (2004), with permission.

- *What is the natural history of MCS?* Preliminary data suggest that MCS usually represents a transitional state between coma/VS and normal consciousness but may also be a permanent outcome. The natural history of MCS will need to be investigated further so that there is a reference against which the effectiveness of rehabilitative interventions can be measured.
- *What accounts for the fluctuation in cognitive responsiveness that defines MCS?* To solve the fluctuation problem, a multidimensional assessment approach that incorporates electrophysiologic recordings (EEG, evoked potentials), structural and functional neuroimaging techniques (MRI, fMRI) and behavioral measures (standardized rating scales, video logs) will be required. The resources required to accomplish this will likely require multicenter collaboration.
- *How effective are existing behavioral assessment methods for monitoring recovery from MCS and response to treatment?* Most available neurobehavioral assessment instruments fail to meet criteria for interval measurement and do not have well-established clinical utility. The reliability/validity, sensitivity/specificity and positive/negative predictive values of these measures must be clarified.
- *Is it possible to improve functional outcome following MCS?* At present, there are no proven treatments for promoting recovery from MCS. There are some promising drug studies although these are comprised by methodologic weaknesses (Schneider et al., 1999; Meythaler et al., 2002). Deep brain stimulation of carefully selected neuromodulatory targets has also been proposed to facilitate cognitive recovery (Schiff et al., 2000). Clinical trials will require multicenter protocols to assure adequate sample size and sufficient power.

Conclusions

Until it is possible to precisely map the neural substrate underlying consciousness, its borders will remain arbitrary. At present, diagnostic assessment of DOC must continue to be guided by behavioral criteria that can be assessed at the bedside. Prognostic accuracy and treatment effectiveness rest largely on diagnostic specificity. The recent development of a case definition for MCS offers the clinician and researcher a means by which to distinguish those patients who demonstrate some evidence of consciousness from those who never show such signs.

Since the diagnostic criteria for MCS were published, an emerging body of research has begun to show clear differences in pathophysiology, residual cerebral activity and functional outcome between VS and MCS. There is also theoretical and empirical support for the premise that patients in MCS may respond more favorably to treatment interventions than those in VS. While it is likely that the existing definition of MCS will be refined, the current diagnostic criteria provide a consistent frame of reference for multidisciplinary communication and future scientific investigation of this extraordinary syndrome.

Acknowledgments

Supported in part by the National Institute on Disability and Rehabilitation Research (H133A 031713). The author thanks Dr. Kathleen Kalmar for her invaluable assistance and limitless energy in pursuing the enigmatic questions concerning disorders of consciousness that are encountered daily. I would also like to extend an especially heartfelt thanks to my wife and children for their continuous understanding and support, and for not taking, "just 15 more minutes" literally.

References

American Congress of Rehabilitation Medicine. (1995) Recommendations for use of uniform nomenclature pertinent to persons with severe alterations in consciousness. Arch. Phys. Med. Rehabil., 76: 205–209.

Andrews, K. (1996) International working party on the management of the vegetative state: summary report. Brain Injury, 10(11): 797–806.

Ansell, B.J. and Keenan, J.E. (1989) The Western Neuro Sensory Stimulation Profile: A tool for assessing slow-to-recover head-injured patients. Arch. Phys. Med. Rehabil., 70: 104–108.

Boly, M., Faymonville, M.E., Peigneux, P., Lambermont, B., Damas, P., Del Fiore, G., Degueldre, C., Franck, G., Luxen, A., Lamy, M., Moonen, G., Maquet, P. and Laureys, S. (2004) Auditory processing in severely brain injured patients: Differences between the minimally conscious state and the persistent vegetative state. Arch. Neurol., 61: 233–238.

Bricolo, A., Turazzi, S. and Feriotti, G. (1980) Prolonged post-traumatic unconsciousness: therapeutic assets and liabilities. J. Neurosurg., 52: 625–634.

Childs, N.L., Mercer, W.N. and Child, H.W. (1993) Accuracy of diagnosis of persistent vegetative state. Neurology, 43(8): 1465–1467.

DiPasquale, M.C. and Whyte, J. (1996) The use of quantitative data in treatment planning for minimally conscious patients. J. Head Trauma Rehabil., 11(6): 9–17.

Giacino J. Rehabilitation of patients with disorders of consciousness: An evidence-based review. In: High W. and Hart K., (Eds.), Rehabilitation Interventions Following TBI: State of the Science. Oxford University Press, New York. In press.

Giacino, J.T., Zasler, N.D., Katz, D.I., Kelly, J.P., Rosenberg, J.H. and Filley, C.M. (1997) Development of practice guidelines for assessment and management of the vegetative and minimally conscious states. J. Head Trauma Rehabil., 12(4): 79–89.

Giacino, J.T., Ashwal, S.A., Childs, N., Cranford, R., Jennett, B., Katz, D.I., Kelly, J., Rosenberg, J., Whyte, J., Zafonte, R.A. and Zasler, N.D. (2002) The minimally conscious state: Definition and diagnostic criteria. Neurology, 58: 349–353.

Giacino, J.T. and Kalmar, K. (1997) The vegetative and minimally conscious states: A comparison of clinical features and functional outcome. J. Head Trauma Rehabil., 12(4): 36–51.

Giacino, J.T., Kalmar, K. and Whyte, J. (2004) The JFK Coma Recovery Scale-Revised: measurement characteristics and diagnostic utility. Arch. Phys. Med. Rehabil., 85: 2020–2029.

Giacino, J.T., Kezmarsky, M.A., DeLuca, J. and Cicerone, K.D. (1991) Monitoring rate of recovery to predict outcome in minimally responsive patients. Arch. Phys. Med. Rehabil., 72: 897–901.

Giacino, J.T., Sharlow-Galella, M., Kezmarsky, M.A., McKenna, K., Nelson, P., King, M., Cowhey-Brown, A. and Cicerone, K. (1992) The JFK Coma Recovery Scale and Coma Intervention Program Treatment Procedures Manual. Center for Head Injuries, Edison, pp. 1–24.

Giacino, J.T. and Trott, C. (2004) Rehabilitative management of patients with disorders of consciousness: Grand rounds. J. Head Trauma Rehabil., 19(3): 262–273.

Giacino, J.T. and Zasler, N.D. (2005) Outcome after severe traumatic brain injury: Coma, the vegetative state, and the minimally responsive state. J. Head Trauma Rehabil., 10(1): 40–56.

Hirsch, J., Kamal, A., Rodriguez-Moreno, D., Petrovich, N., Giacino, J., Plum, F. and Schiff, N. (2001) fMRI reveals intact cognitive systems in two minimally conscious patients. Abst. Soc. Neurosci., 271: 1397.

Ichise, M., Chung, D.G., Wang, P., Wortzman, G., Gray, B.G. and Franks, W. (1994) Technetium-99m-HM-PAO SPECT,

CT and MRI in the evaluation of patients with chronic traumatic brain injury: A correlation with neuropsychological performance. J. Nucl. Med., 35: 217–226.

Jennett, B., Adams, J.H., Murray, L.S. and Graham, D.I. (2001) Neuropathology in vegetative and severely disabled patients after head injury. Neurology, 56: 486–489.

Jennett, B. and Plum, F. (1972) Persistent vegetative state after brain damage: A syndrome in search of a name. Lancet, 1: 734–737.

Kampfl, A., Schmutzhard, E., Franz, G., Pfausler, B., Haring, H.P., Ullmer, H., Felber, F., Golaszewski, S. and Aichner, F. (1998) Prediction of recovery from post-traumatic vegetative state with cerebral magnetic-resonance imaging. Lancet, 351: 1763–1767.

Laureys, S., Faymonville, M.E., Degueldre, C., Del Fiore, G., Damas, P., Lambermont, B., Jannsens, N., Aerts, J., Franck, G., Luxen, A., Moonen, G., Lamy, M. and Maquet, P. (2000) Auditory processing in the vegetative state. Brain, 123: 1589–1681.

Laureys, S., Faymonville, M.E., Ferring, M., et al. (2003) Differences in brain metabolism between patients in coma, vegetative state, minimally conscious state and locked-in syndrome. Eur. J. Neurol., 10(Suppl. 1): 224.

Laureys, S., Owen, A. and Schiff, N.D. (2004) Brain function in coma, vegetative state, and related disorders. Lancet Neurol., 3: 537–546.

McMillan, T.M. (1996) Neuropsychological assessment after extremely severe head injury in a case of life or death. Brain Injury, 11(7): 483–490.

Menon, D.K., Owen, A.M., Williams, E.J., Minhas, P.S., Allen, C.M.C., Boniface, S.J., Pickard, J.D. and Wolfson Brain Imaging Centre Team. (1998) Cortical processing in persistent vegetative state. Lancet, 352: 200.

Mesulam, M.M. (2000) Behavioral neuroanatomy: Large-scale networks, association cortex, frontal syndromes, the limbic system and hemispheric specializations. In: Mesulam M.M. (Ed.), Principles of Behavioral and Cognitive Neurology (2nd ed.). Oxford University Press, Oxford, pp. 1–120.

Meythaler, J.M., Brunner, R.C., Johnson, A. and Novack, T.A. (2002) Amantadine to improve neurorecovery in traumatic brain injury — associated diffuse axonal injury: a pilot double blind randomized trial. J. Head Trauma Rehabil., 17(4): 300–313.

Multi-Society Task Force Report on PVS. (1994) Medical aspects of the persistent vegetative state. NEJM, 330: 1499–1508 1572–1579.

Plum, F. and Posner, J. (1982) The pathologic physiology of signs and symptoms of coma. The diagnosis of stupor and coma (3rd ed.). FA Davis, Philadelphia, PA.

Prayer, L., Limberger, D., Oder, W., Kramer, J., Schindler, E., Podreka, I. and Imhof, H. (1993) Cranial MR imaging and cerebral 99mTc HM-PAO SPECT in patients with subacute or chronic severe closed head injury and normal CT examinations. Acta Radiol., 34(6): 593–599.

Rappaport, M., Dougherty, A.M. and Kelting, D.L. (1992) Evaluation of coma and vegetative states. Arch. Phys. Med. Rehabil., 73: 628–634.

Rappaport, M., Hall, K.M., Hopkins, K., Belleza, T. and Cope, D.N. (1982) Disability Rating Scale for severe head trauma: coma to community. Arch. Phys. Med. Rehabil., 63: 118–123.

Schiff, N.D. and Pulver, M. (1999) Does vestibular stimulation activate thalamocortical mechanisms that reintegrate impaired cortical regions? Proc. R Soc. Lond. B, 266: 421–423.

Schiff, N.D., Rezai, A.R. and Plum, F.P. (2000) A neuromodulation strategy for rational therapy of complex brain injury states. Neurol. Res., 22: 267–272.

Schiff, N.D., Ribary, U., Plum, F. and Llinas, R. (1999) Words without mind. J. Cogn. Neurosci., 11(6): 650–656.

Schneider, W.N., Drew-Cates, J., Wong, T.M. and Dombovy, M.L. (1999) Cognitive and behavioural efficacy of amantadine in acute traumatic brain injury: an initial double-blind placebo-controlled study. Brain Injury, 13: 863–872.

Shiel, A., Horn, S.A., Wilson, B.A., Watson, M.J., Campbell, M.J. and McLellan, D.L. (2000) The Wessex Head Injury Matrix (WHIM) main scale: a preliminary report on a scale to assess and monitor patient recovery after severe head injury. Clin. Rehabil., 14(4): 408–416.

Shiel, A. and Wilson, B. (1998) Assessment after extremely severe head injury in a case of life or death: further support for McMillan. Brain Injury, 12(10): 809–816.

Stanczak, D.E., White, J.G., Gouvier, W.D., Moehle, K.A., Daniel, M., Novack, T. and Long, C.J. (1984) Assessment of level of consciousness following severe neurological insult: a comparison of the psychometric qualities of the Glasgow Coma Scale and the Comprehensive Level of Consciousness Scale. J. Neurosurg., 60: 955–960.

Strauss, D.J., Ashwal, S., Day, S.M. and Shavelle, R.M. (2000) Life expectancy of children in vegetative and minimally conscious states. Pediatr. Neurol., 23(4): 1–8.

Teasdale, G. and Jennett, B. (1974) Assessment of coma and impaired consciousness. Lancet, 2: 81–84.

Wilson, S.L. and Gill-Thwaites, H. (2000) Early indications of emergence from vegetative state derived from assessment with the SMART — a preliminary report. Brain Injury, 14(4): 319–331.

Whyte, J. and DiPasquale, M. (1995) Assessment of vision and visual attention in minimally responsive brain injured patients. Arch. Phys. Med. Rehabil., 76(9): 804–810.

Whyte, J., DiPasquale, M. and Vaccaro, M. (1999) Assessment of command-following in minimally conscious brain injured patients. Arch. Phys. Med. Rehabil., 80: 1–8.

Whyte, J., Katz, D., Long, D., DiPasquale, M., Polansky, M., Kalmar, K., Giacino, J., Childs, N., Mercer, W., Novak, P., Maurer, P. and Eifert, B. (2005) Predictors of outcome in prolonged posttraumatic disorders of consciousness and assessment of medication effects: a multicenter study. Arch. Phys. Med. Rehabil., 86: 453–462.

Zafonte, R.D., Watanabe, T. and Mann, N.R. (1998) Amantadine: a potential treatment for the minimally conscious state. Brain Injury, 12(7): 617–621.

Zeman, A.Z., Grayling, A.C. and Cowey, A. (1997) Contemporary theories of consciousness. J. Neurol. Neurosurg. Psychiatry, 62: 549–552.

S. Laureys (Ed.)
Progress in Brain Research, Vol. 150
ISSN 0079-6123

CHAPTER 28

Behavioral evaluation of consciousness in severe brain damage

Steve Majerus[1,#], Helen Gill-Thwaites[3], Keith Andrews[4] and Steven Laureys[2,#,*]

[1]Department of Cognitive Sciences, University of Liege, Liege, Belgium
[2]Dept of Neurology and Cyclotron Research Center, University of Liege, Liege, Belgium
[3]Occupational Therapy Department, Royal Hospital for Neuro-disability, London, UK
[4]Institute of Complex Neuro-disability, Royal Hospital for Neuro-disability, London, UK

Abstract: This paper reviews the current state of bedside behavioral assessment in brain-damaged patients with impaired consciousness (coma, vegetative state, minimally conscious state). As misdiagnosis in this field is unfortunately very frequent, we first discuss a number of fundamental principles of clinical evaluation that should guide the assessment of consciousness in brain-damaged patients in order to avoid confusion between vegetative state and minimally conscious state. The role of standardized behavioral assessment tools is particularly stressed. The second part of this paper reviews existing behavioral assessment techniques of consciousness, showing that there are actually a large number of these scales. After a discussion of the most widely used scale, the Glasgow Coma Scale, we present several new promising tools that show higher sensitivity and reliability for detecting subtle signs of recovery of consciousness in the post-acute setting.

Introduction

The evaluation of consciousness in severely brain-damaged patients is of major importance for their daily management. Consciousness is a multifaceted concept that, in a simplified manner, can be divided into two major components: the *level* of consciousness (i.e., arousal, wakefulness or vigilance) and the *content* of consciousness (i.e., awareness of the environment and of the self) (Plum and Posner, 1983). Arousal is supported by numerous brainstem neuronal populations (previously called reticular activating system) that directly project to both thalamic and cortical neurons (see Fig. 1). Therefore, depression of either brainstem or global hemispherical function may cause reduced wakefulness. Awareness is thought to be dependent upon the functional integrity of the cerebral cortex and its reciprocal subcortical connections; each of its many aspects resides to some extent in anatomically defined regions of the brain.

Unfortunately, for the time being, consciousness cannot be measured objectively by any machine. Its estimation requires the interpretation of several clinical signs. Many scoring systems have been developed for the quantification and standardization of the assessment of consciousness. The present paper will discuss the strengths and pitfalls of a behavioral assessment of consciousness in patients, with a special focus on patients in a vegetative state, and discuss new promising assessment tools. Neurophysiological assessment of consciousness as well as the prognostic value of assessment in patients with impaired consciousness will not be

*Corresponding author. Tel.: +32 4 366 23 04;
Fax: +32 4 366 29 46; E-mail: steven.laureys@ulg.ac.be
[#]Steve Majerus is a Postdoctoral Researcher and Steven Laureys is a Research Associate, both at the Belgian National Fund for Scientific Research (FNRS)

DOI: 10.1016/S0079-6123(05)50028-1

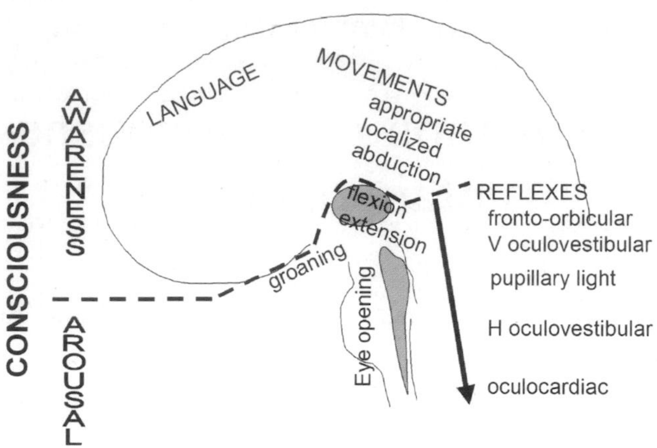

Fig. 1. A simplified scheme of consciousness and its two major components: arousal and awareness. *Note*: the gray area represents the reticular activating system encompassing the brainstem and thalamus; the arrow near the brainstem denotes the progressive disappearance of brainstem reflexes during rostral-caudal deterioration (i.e., evolution from coma to brain death). V = vertical; H = horizontal. (Reproduced with permission from Laureys et al., 2002.)

considered here as they are the issue of other papers in this volume (see Kotchoubey; Guerit; Jennett this volume).

Clinical evaluation of consciousness

Arousal and awareness are not on-off phenomena but are part of a large continuum. At the bedside, arousal is assessed by the presence of spontaneous or stimulation-induced eye opening. It ranges from *coma* (no eye opening), *stupor* (eye opening following vigorous external stimuli), through *sleep* (eye opening following moderate external stimuli) and alert *waking* (spontaneous eye opening). Awareness refers to the collective thoughts and feelings of an individual. Clinically, we are limited to the appraisal of the patient's capacity to perceive the external world and to voluntarily interact with it (i.e., perceptual awareness). In practice, this is evaluated by careful and repeated examination of the capacity to formulate reproducible, voluntary, purposeful and sustained behavioral responses to auditory, tactile, visual or noxious stimuli (e.g., by asking the patient to follow command, to visually discriminate between Yes/No cards by pointing or eye movements). Much more difficult is the evaluation of patients' self-recognition in a mirror; this can be done by putting a mark (colored versus invisible) on the patient's face and by determining whether the patient will touch this mark when being shown his face in a mirror (Gallup, 1997). Obviously, the patient needs to be well aroused in order to perform the cognitive processes required for awareness. Hence, patients in a coma are unaware because they cannot be aroused. However, as illustrated by patients in a *vegetative state*, arousal is only a necessary and not a sufficient condition for awareness. Indeed, patients in a vegetative state are aroused (as shown by preserved spontaneous eye opening and sleep–wake cycles) but show no sign of awareness (i.e., no sign of command following or any other voluntary behavior). When the first signs of voluntary behavior appear, the patient may be in a *minimally conscious state*: here the patient is partially conscious, as evidenced by the presence of limited but reproducible signs of awareness (inconsistent command following, inconsistent but intelligible verbalization, sustained visual fixation, localization of sound and noxious stimuli) (Giacino et al., 2002; Giacino and Whyte, 2005; Giacino, this volume).

Diagnosing and misdiagnosing signs of consciousness

The diagnosis of the vegetative state depends on behavioral assessment of the responses obtained from the patient. It is not a pathological or even neuro-physiological diagnosis. While there have been exciting developments in the use of functional MRI (Magnetic Resonance Imaging) scanning (see Schiff, this volume; Owen, this volume), brain mapping and other neuro-physiological approaches (see Kotchoubey, this volume; Guerit, this volume) these are primarily aids to diagnosis rather than a method of diagnosis. Consider, for instance, the patient whose neuro-physiological investigations suggest that there is some integrated "higher-level" cerebral function in response to stimulation — but where there is no behavioral evidence that the person is aware of his environment — who shows no evidence of communication or understanding of what others are communicating with him. Where does that leave the patient, the family and the caring team? While it might incite to reexamine the clinical responses and strive harder to demonstrate any awareness, if the patient continues with no meaningful responses and remains clinically vegetative, then we would argue that the patient is in the vegetative state.

This, however, does lead us on to questioning how sensitive our clinical–behavioral assessments are. Giacino and Zasler (1995) have pointed out the limitations of clinical assessment in the identification of "internal awareness" in a patient who otherwise lacks the motor function to demonstrate his awareness. The concept that we are only able to *infer* the presence or absence of conscious experience is a long-standing philosophical issue, which has been pointed out by Bernat (1992) and The Multi-Society Task Force (1994) in the specific context of the vegetative state. The International Working Party on the Vegetative State (Andrews, 1996) discussed this point in detail and criticized the use of the term "meaningful response" on the grounds that it requires a considerable amount of subjective interpretation on the part of the observer and that what was meaningful for the patient may not be considered meaningful by those treating the patient. Similarly the term "purposeful response" was criticized because of the subjective interpretation and that a withdrawal reflex could be considered as purposeful in that it removes the limb, for instance, from danger.

This is where there must be some concern. For instance there are several studies that have described the misdiagnosis of the vegetative state. In a group of long-term patients in a nursing home in the USA, Tresch et al. (1991) found that 18% of those diagnosed as being in the persistent vegetative state were aware of themselves or their environment. Childs et al. (1993) report that 37% of patients admitted more than 1 month post-injury with a diagnosis of coma or persistent vegetative state had some level of awareness. In another study (Andrews et al., 1996), 43% of patients admitted to a profound brain injury unit at least 6 months following their brain damage (i.e. could be expected to be stable) were found to have been misdiagnosed. While these figures cause concern they at least emphasize that bedside diagnosis was possible — otherwise they would not have been identified as having been misdiagnosed.

So why are patients misdiagnosed? One striking finding was that 65% of the "misdiagnosed" patients were either blind or very severely visually impaired in the form of marked visual field defects and/or visual perceptual disorders (Andrews et al., 1996). This has obvious implications for assessment since one of the prime features for assessing whether a patient is non-vegetative is eye tracking. If the patient has visual impairment, then he will not follow objects and therefore eye tracking will be absent even in a mentally alert individual.

Since all patients followed verbal commands, it is assumed that none were deaf or had severe hearing impairment. This, however, is a possibility and should be considered. This also emphasizes the importance of assessing a wide range of stimuli (touch, taste and smell as well as visual or auditory), a range of frequent observations with standardized assessment tools and optimal patient management (e.g., with the patient in seating position) to ensure that disturbance of one modality is not the cause of missing evidence of awareness.

Making the diagnosis of vegetative state

Previous studies have shown that misdiagnosed "vegetative" patients were at the "severe" level of the Glasgow Outcome Scale (Jennett and Teasdale, 1977), being totally physically dependent for all care needs (Andrews et al., 1996). The only method that any of us can use to demonstrate our awareness to others is through some form of motor activity — speech, facial expression, eye-tracking, limb movement, shrugging shoulders, nodding-shaking the head, etc. For 88% of the patients, pressing a buzzer was the only functional movement, although one patient later developed an ability to point with a finger and another patient became able to write words; the other two patients communicated by eye pointing (Andrews et al., 1996).

The importance of physical function was dramatically demonstrated by one patient where responses were not identified until 25 weeks after admission, though it was obvious from subsequent conversations with him that he had not been vegetative for some time. This patient was admitted with very severe joint contractures, which required surgical release and a prolonged physical management program before he could be seated appropriately in a special seating system. Only when he was satisfactorily seated was sufficient muscle tone released for him to indicate with a slight shoulder shrug that he was aware — he was able to carry out simple mental mathematical calculations and was aware of his immediate physical and social environment (Andrews et al., 1996).

Another difficulty is the relevance of the blink response to awareness. The patient may blink to menace but appear not to be attentive. Note that at least one authority (Working Group of the Royal College of Physicians, 1996) has regarded a blink to threat as evidence of cortical connection and therefore indicating that the patient is not vegetative. This is a very questionable approach since the concept of the vegetative state is the demonstration of *awareness*, and not of whether there are some cortical connections. The Multi-Society Task Force (1994) urges *caution* in making the diagnosis of the vegetative state if there is blinking to threat but does not go as far as to claim that if present that it indicates that the patient is no longer vegetative. Actually one of the difficulties is taking too little notice of the blink response — or more relevantly the speed of the blink response. Often too little time is given to waiting for the response. There is often a delay between stimulation and response when there is awareness, as though the brain was having to work out the response to give. Of course, this leads to the problem of how long to wait and the risk of spontaneous blinking being interpreted as a volitional response. This requires a considerable amount of experience to interpret. One clue is that the blink is often of a different quality to reflex blinking — either in the slowness of the blink or the length of time the eye is kept closed. As noted by Whyte and others, differentiation of spontaneous eye blinks from those that are purposeful can be done by systematically recording eye-blinks under different conditions (e.g., at rest, to inappropriate command, to appropriate command) and statistically determining whether the frequency of the response is significantly higher following administration of the appropriate command, relative to the other conditions (see, e.g., Whyte and DiPasquale, 1995).

There are, of course, other signs that may cause a misdiagnosis that the patient is aware when in fact the responses are reflex in nature. For instance, there may be roving eye movements and the patient's eyes may seem to briefly follow moving objects. The movement is usually inconsistent and never sustained. For instance, the patient's eyes may turn toward a sound or a sudden movement but does so only briefly and does not focus on the source of stimulation. This can catch out the unwary who interpret this as awareness. What is probably happening is that the subcortical centers that alert the brain to incoming stimuli, e.g., the superior colliculi for vision and the thalamus for tactile sensations, are still active but the alerting mechanism does not reach "higher level" cortical interpretation. This situation is seen, for instance, in cortical blindness where although the visual cortex is damaged the patient will still turn toward a visual stimulus even though he cannot "see" it.

Some staff and family interpret the withdrawal response as being an indication that the patient is aware of the noxious stimulus. It would be more relevant if the patient pushed away the stimulus. Another confusing feature for many carers is the non-volitional grasp reflex. This can cause considerable concern to relatives or carers who feel that the patient recognizes them when they hold his hand. This is particularly reinforced when the grip tightens as there is an attempt to pull the hand or fingers away. This is supportive of the diagnosis of a grasp reflex rather than supportive of a meaningful response.

What can be even more confusing are the fragments of coordinated movement, such as scratching or even moving hands toward a noxious stimulus. These must always be taken seriously as indicating awareness but do occur in the vegetative patient usually affecting the same repetitive movement on each occasion. They are probably long-learned automatic response activities. However, scratching oneself on different locations depending on the irritant's source would be indicative of a minimally conscious state.

Chewing movements or grinding of teeth (to which can be added constant movement of the tongue) again cause concern to relatives and carers feeling that the patient is indicating that he is thirsty or hungry. Grunting and groaning provoked by noxious stimuli can also often be interpreted as indicating an attempt to communicate. This can cause disagreement between family and clinicians when some relatives claim to be able to "understand" the words spoken when others only hear sounds. These are, however, commonly found features in the vegetative state. The skills is to decide whether the responses are contingent on the quality of the external stimulus.

Factors influencing the diagnosis

The International Working Party (Andrews, 1996) pointed out that the assessments in general use are based on a series of behavioral patterns. The clinician is, therefore, dependent on overt responses that depend on a number of factors including:

a. The physical ability of the patient to respond — this has been discussed above.
b. The desire or willingness (if the patient is aware) of the patient to respond. It is not unusual for members of the family to obtain responses that the professional members of the team are not able to. This is probably not surprising since the members of the family are more likely to be "sensitive" to the responses seen. On the other hand, the family may be desperate for a response and easily misinterpret the reflex responses. Patients may also be more willing to respond to family or to some members of the staff rather than to others. Let us face facts — some staff are better at relating than others.
c. The ability to observe accurately. This is particularly relevant since profound brain damage is a rare condition and few professionals have seen sufficient patients to have gained that level of experience required to produce "expertise".
d. The time available for observation and assessment. Time is one of the major factors in assessing the profoundly brain-damaged patients. They do not conveniently have their best levels of awareness at the time set aside for the formal assessment. This requires flexibility of the assessor to take advantage of the windows of opportunity and to take advantage of the observations of other members of the team and members of the family.
e. The lack of available and reliable assessment tools. It is not so much that there is a shortage of tools — see discussion below — but that they are not used in more general acute, or even neurological or rehabilitation, units.
f. The patient is not always seen by a skilled team to address all of these issues.
g. The family and carers and those who know the patient best are not always involved as much as they should be.
h. Patients are assessed by some assessors who are unfamiliar with the patient — leading to meaningful responses being missed.

There are several principles to the accurate assessment of the person thought to be in the vegetative state:

1. That the patient should be healthy. Even simple conditions such as constipation, chronic urinary tract infection (usually associated with long-term catheterization) or bronchial infections can prevent optimal responses from being obtained.
2. The patients should be in a good nutritional state. The earlier use of gastrostomy feeding has altered this pattern but still some patients

admitted from general units have a low Body Mass Index, emphasizing the difficulty in managing people with such complex medical and physical disabilities.

3. As many sedating drugs as possible should be withdrawn, or at least decreased to the lowest effective dose — these include antispasticity drugs and antiepileptic drugs. In the case of antiepileptic drugs, which are still required to control fitting, drugs with the least sedative effect should be used.
4. Complications and consequences of neurological imbalance should be prevented — this includes high muscle tone and contractures by the provision of special seating, good bed and sitting posture to control abnormal muscle tone. These complications in the long term increase the amount of nursing care required, which, since the patient may live for many years, increases the cost of care considerably.
5. Controlled posture is important. Most doctors have been trained to examine patients on the bed. Experience suggests that patients are more likely to be alert when sitting up (presumably due to greater stimulation of the ascending reticular activating system). A well-supporting seating system is essential to reduce sufficient muscle tone to allow movement of limbs that can be used for communication purposes, e.g. to press a touch-sensitive switch.
6. Providing a controlled environment of sensory *regulation* to avoid sensory overload of the severely damaged brain. Since it is likely that profoundly brain-damaged patients have problems with selective attention, sensory input should be simple and interspersed with periods of rest. It is, therefore, logical to assess for cognitive responses after a period of rest rather than after a period of activity, such as being washed and dressed or after physiotherapy. This requires staff and family to understand the importance of avoiding over-stimulation prior to the assessment.
7. Assessments should be short (to avoid tiring the patient), repeated (to identify windows of opportunity) and over a prolonged period of time (to accommodate the learning process of both the patient and the assessor). One-off short assessment of the patient who is lying in bed is likely to result in a missed diagnosis even by the most experienced clinician.
8. The ability to generate a behavioral response fluctuates from day to day and hour to hour, and even minute to minute, depending on fatigue factors, general health of the patient and the underlying neurological condition.
9. Observation needs to take into account delayed responses. Assimilation of even basic information is often slow and therefore response time may be delayed. Because of this, information provided at any one time should be simple, consistent, repeated after a period of rest, and allow for a delayed response.
10. Communication requires skilled techniques and sensitivity for the method by which the patient wants to communicate.
11. Families and other carers have a very important role in identifying the best responses and the optimal conditions for assessment. While there are some relatives who interpret reflex responses as being meaningful, there is no doubt that members of the family are often more sensitive to early changes than even very experienced clinical staff.

Consciousness scales

There are many scales designed to monitor the recovery of consciousness in brain-damaged patients. In this section, we will first address the Glasgow Coma Scale (GCS), which remains the most widely used scale in the acute and subacute setting. We will then review other existing tools, by focusing more specifically on three new and promising assessment instruments, the Coma Recovery Scale-Revised (CRS-R; Giacino et al., 2004), the Wessex Head Injury Matrix (WHIM; Shiel et al., 2000), and the Sensory Modality Assessment and

Rehabilitation Technique (SMART; Gill-Thwaites, 1997, 1999).

Glasgow coma scale

Teasdale and Jennett (1974) developed the GCS as an aid in the clinical assessment of post-traumatic unconsciousness. It was devised as a formal scheme to overcome the ambiguities that arose when information about comatose patients was presented and groups of patients compared. The GCS has three components: eye (E), verbal (V) and motor (M) response to external stimuli (see Fig. 2). The scale consisted of 14 points, but was later adapted to 15, with the division of the motor category "flexion to pain" into two further categories. The best or highest responses are recorded. So far, more than 2390 publications have appeared to its use (MEDLINE search performed in February 2005, limited to title and abstract word). It is a component of the Acute Physiology and Chronic Health Evaluation (APACHE) II score, the (Revised) Trauma Score, the Trauma and Injury Severity Score (TRISS) and the Circulation, Respiration, Abdomen, Motor, Speech (CRAMS) Scale, demonstrating the widespread adoption of the scale.

The observation of spontaneous eye opening "indicates that the arousal mechanisms of the brainstem are active" (Teasdale and Jennett, 1974). As previously stated, recovered arousal does not imply the recovery of awareness. Patients in a vegetative state have awakened from their coma but remain unaware of their environment and self. Most comatose patients who survive will eventually open their eyes, regardless of the severity of their cerebral damage (Jennett, 1972). Indeed, less than 4% of head-damaged patients never open their eyes before they die (Bricolo et al., 1980). The eye opening in response to speech tests the reaction "to any verbal approach, whether spoken or shouted, not necessarily the command to open the eyes" (Teasdale and Jennett, 1974). Again, this response is observed in the vegetative state where "awakening" can be induced by non-specific auditory stimulation. In these patients, it is recommended to differentiate between a reproducible response to command and to nonsense speech. Eye opening in response to pain should be tested by stimulation at the level of the limbs, because the grimacing associated with supraorbital or jaw-angle pressure may cause eye closure.

The presence of verbal responses indicates the restoration of a high degree of interaction with the environment (i.e., awareness). An oriented conversation implies awareness of the self (e.g., the patient can answer the question: "What is your name?") and environment (e.g., the patient correctly answers the questions: "Where are we?" and "What year/month is it?"). Confused speech is recorded when the patient is capable of producing language, for instance phrases and sentences, but is unable to answer the questions about orientation. When the patient presents intelligible articulation but exclaims only isolated words in a random way (often swear words, obtained by physical stimulation rather than by a verbal approach), this is scored as "inappropriate speech". Incomprehensible sounds refer to moaning and groaning without any recognizable words. This rudimentary vocalization does not necessitate awareness and is thought to depend upon subcortical functioning as it can be observed in anencephalic children and vegetative patients.

The motor response first assesses whether the patient obeys simple commands, given in verbal, gestural or written form. A non-specific sound stimulus may induce a reflex contraction of the patient's fingers or alternatively such a reflex response can result from the physical presence of the examiner's fingers against the palm of the patient (i.e., grasping reflex). Before accepting that the patient is truly obeying commands, it is advised to test that the patient will also release and squeeze again to repeated commands. If there is no response a painful stimulus is applied. First, pressure is applied to the fingernail bed with a pencil. If flexion is observed, stimulation is then applied to other sites (applying pressure to the supraorbital ridge, pinching the trapezium or rubbing the sternum) to differentiate between localization (i.e., a noxious stimulus applied at more than one site causes a limb to move so as to attempt to remove it by crossing the midline), withdrawal flexion (i.e., a

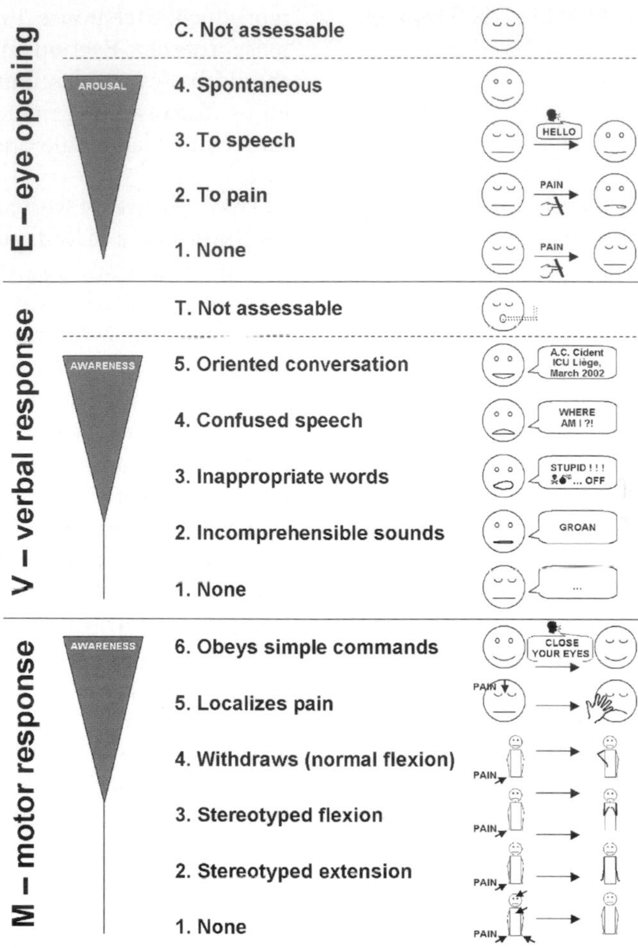

Fig. 2. Pictographic representation of the Glasgow Coma Scale (Teasdale and Jennett, 1974). (Reproduced with permission from Laureys et al., 2002.)

rapid flexion of the elbow associated with abduction of the shoulder) or "abnormal" flexion (i.e., a slower stereotyped flexion of the elbow with adduction of the shoulder that can be achieved when stimulated at other sites). Stereotyped flexion responses are the most common of the motor reactions observed in severely brain-injured patients; they are also the most enduring (Born, 1988). Extensor posturing is more easily distinguished and is usually associated with adduction, internal rotation of the shoulder and pronation of the forearm. Abnormal flexion and extension motor responses often co-exist (Bricolo et al., 1977). The scale of responses to pain is applicable to the movements of the arms. The movements of the legs are not only more limited in range, but may take place on the basis of a spinal withdrawal reflex (e.g., in brain death, a spinal reflex may still cause the legs to flex briskly in response to pain applied locally (Ivan, 1973; Saposnik, 2000)).

Much too often the three components of the GCS (E–V–M) are summed into a total score, ranging from 3 to 15. However, given the increased use of intubation, ventilation, and sedation of patients with impaired consciousness (Marion and Carlier, 1994), patients might wrongly be scored as GCS 3/15 rather than being more appropriately reported as impossible to score. In a European multi-centric study of head-injured patients, assessment of each of the three components of the GCS was possible only in 56% on arrival in the neurosurgical unit, and in 49% in the "post-resuscitation" phase (Murray et al., 1993). In Glasgow, patients are always described by the three separate responses and never by the total (Teasdale et al., 1983).

Glasgow Liège scale

A frequently expressed reservation regarding the GCS is its failure to incorporate brainstem reflexes. A number of investigators have disagreed with Teasdale and Jennett that spontaneous eye opening is sufficiently indicative of brainstem arousal systems activity and have proposed coma scales that include brainstem responses (Segatore and Way, 1992). Many coma scales that include brainstem indicators have been proposed (e.g., the Comprehensive Level of Consciousness Scale (CLOCS, Stanczak et al., 1984), the Clinical Neurologic Assessment Tool (Crosby and Parsons, 1989), the Bouzarth Coma Scale (Bouzarth, 1968), the Maryland Coma Scale (Salcman et al., 1981)...) but none has become widely used. These scales generally have been more complex than the GCS.

A simpler system is the Glasgow Liège Scale (GLS) (see Fig. 3). It was developed in 1982 in Liège and combines the Glasgow Scale with a quantified analysis of five brainstem reflexes: fronto-orbicular, vertical oculo-cephalic, pupillary, horizontal oculo-cephalic and oculo-cardiac (Born et al., 1982). The fronto-orbicular reflex is considered present when percussion of the glabella produces contraction of the orbicularis oculi muscle. The oculo-cephalic reflexes (doll's head) are scored as present when deviation of at least one eye can be induced by repeated flexion and extension (vertical) or horizontal neck movement (horizontal). If the reflexes are absent or cannot be tested (e.g., immobilized cervical spine), an attempt is made to elicit ocular motion by external auditory canal irrigation using iced water (i.e., oculo-vestibular reflex testing). With cold-water irrigation of the head at 30° elevation from the horizontal, the eyes deviate tonically toward the ear irrigated (horizontal). When cold water is injected simultaneously into both ear canals, the eyes deviate tonically downward; the reverse occurs with bilateral irrigation of warm water (vertical). The oculocardiac reflex is scored as present when pressure on the eyeball causes the heart rate to slow down. As for the GCS, the best response determines the brainstem reflex score (R). The selected reflexes disappear in descending order during rostral-caudal deterioration. The disappearance of the last, the oculo-cardiac, coincides with brain death.

Pitfalls encountered when administering the GCS/GLS

Inexperienced or untrained observers produce unreliable scoring of consciousness (Rowley and Fielding, 1991). In one study, one out of five health care workers were mistaken when asked to make judgments as to whether patients were "conscious" or "unconscious" (Teasdale and Jennett, 1976). Consciousness needs considerable skill to evaluate and the observer should be aware of the pitfalls. It is also well known that the preceding score of the patient frequently influences the examiner when rating the patient's present state of consciousness. It is therefore recommended to score in a "blinded" manner. Problems arise when the eyes are swollen shut (e.g., following periorbital edema, direct ocular trauma or facial injury) or paralyzed (e.g., neuromuscular blockade). In these circumstances the enforced closure of the patient's eyes should be recorded on his chart by marking "C" (= eyes closed) (Teasdale, 1975). In deep coma, flaccid eye muscles will show no response to stimulation yet the eyes remain open if the lids are drawn back. Speechlessness may be due to causes other than unawareness (e.g., intubation via the oropharynx or through tracheostomy, orofacial fractures, edematous tongue, foreign language, aphasia, confusion or delirium). The evaluation of verbal responses is also biased when patients received sedatives or neuromuscular blocking agents, alcohol or are drug intoxicated or too young to speak. When the verbal score cannot be assessed a non-numerical designation of "T" (= intubated) should be used (Marion and Carlier, 1994) and the total GCS score cannot be reported. Finally, motor responses cannot be reliably monitored in the presence of splint or immobilization devices or in cases of spinal cord, plexus or peripheral nerve injury. As

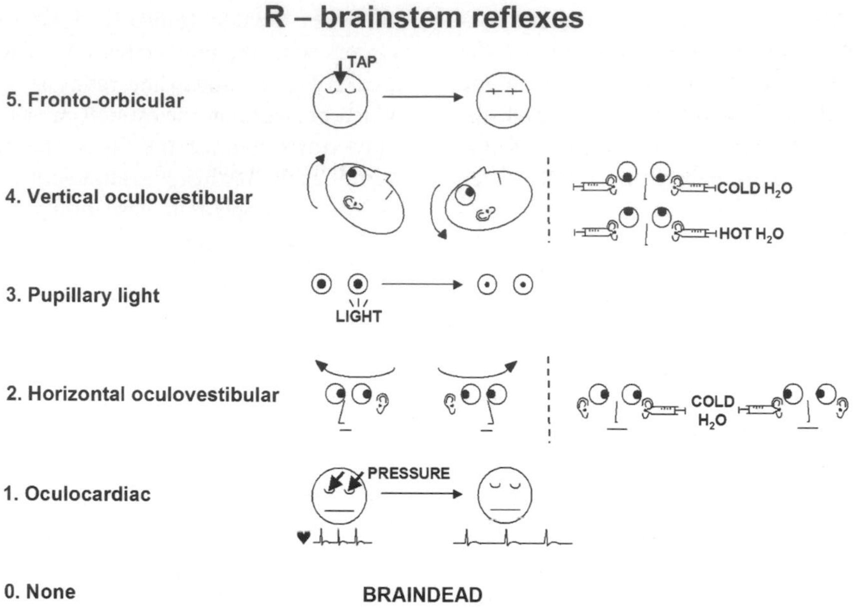

Fig. 3. Pictographic representation of the Glasgow-Liège Scale GLS; (Born et al., 1982). *Note*: when oculocephalic reflexes (doll's eyes) cannot be tested or are absent, the (vertical and horizontal) oculovestibular reflexes (ice water testing) should be evaluated. (Reproduced with permission from Laureys et al., 2002.)

is the case for all scoring systems, awareness is assessed as the level of obeying commands. This approach cannot be applied to cases where the patient is clinically or pharmacologically paralyzed yet alert (e.g., locked-in syndrome, severe polyneuropathy or use of neuromuscular blocking agents) or those with psychogenic unresponsiveness. It is important to stress that special effort should be made to identify and exclude these rare causes of pseudo-coma. In patients with brainstem lesions the diagnosis of "locked-in" syndrome should always be excluded. Locked-in patients suffer from quadriplegia and anarthria caused by a disruption of corticospinal and corticobulbar pathways, respectively. Their only means of expressing their thoughts and feelings is through vertical or lateral eye movement or blinking of the upper eyelid to signal (also see Laureys et al., this volume).

New promising tools: the CRS-R, SMART and the WHIM

Following the acute stage, when patients are recovering from their coma, the GCS becomes less reliable in monitoring consciousness (Segatore and Way, 1992). Here, a large number of alternative behavioral assessment tools of consciousness have been proposed (see also Horn et al., 1993; Majerus et al., 2000a; Canedo et al., 2002; Giacino and Whyte, 2005). For example, Majerus et al. (2000a) identified more than 20 different scales. However, only a few of them are really better alternatives than the GCS/GLS scales. Some alternative scales are mainly variants of the GCS scale (e.g., CLOCS, Stanczak et al., 1984; Benzer et al., 1991). Other tools are simple rating scales that classify the level of consciousness displayed by a patient in rather broad categories (see, e.g., Ommaya, 1966; Hagen et al., 1979; Stalhammar et al., 1988). These scales (although trying to explicitly answer the question whether the patient is aware or not by providing labels that are supposed to qualify the patient's level of consciousness) do not cover a broad range of behaviors nor do they provide explicit guidelines for the systematic observation of the patient's behavior. A far more limited number of scales have been designed to detect more subtle changes in awareness through a thorough, systematic, precisely defined and reliable observation of the patient's behaviors. These scales are supposed to be more sensitive than previous scales as they include a much larger number of items (e.g., Davis, 1991; Giacino et al., 1991; Coma/Near Coma Scale; Rappaport et al., 1992; Visual Response Evaluation, Coma Exit Chart, Freeman, 1996; CRS, Giacino et al., 2004). However, it must be noted that studies specifically aimed at providing empirical evidence for this theoretically superior sensitivity are often lacking.

The CRS-R by Giacino et al. (2004) is a good example of those scales providing a more fine-grained assessment of the recovery of consciousness. The basic structure is similar to the GCS; it includes similar visual, motor and verbal subscales as the GCS, but there are in addition three other scales: an auditory function scale, a communication scale and an arousal scale (see Table 1). Furthermore, the visual, motor and verbal subscales are much more detailed than is the case in the GCS. For example, the visual subscale assesses visual startle responses, eye fixation, eye movement, visual object localization, and object recognition. These items are critical for identifying subtle signs of recovery of consciousness as discussed in the previous sections. Furthermore, for each item, fully operational definitions are provided and special importance is given to the consistency of behaviors assessed via the establishment of baseline observations and repeated administration of the item. This two-step procedure (baseline observation followed by repeated administration of a given item) permits greater certainty that a given behavior is not simply random or reflex, but that it is contingent upon a given stimulus. Inter-rater agreement and test–retest reliability are high for total CRS-R scores, and good concurrent validity is observed in relation to other scales such as the CRS and the Disability Rating Scale (Giacino et al., 1991, 2004, this volume). Furthermore, the CRS-R has been designed to be particularly helpful for discriminating between vegetative and minimally conscious state. As can be seen in Table 1, a number of specific items are proposed that should permit discrimination between vegetative and minimally conscious state (e.g., the observation of item 2 (fixation for more than 2 sec) on the visual function scale is supposed to be incompatible with a diagnosis of vegetative state but supports a diagnosis of minimally conscious state).

Table 1. JFK coma recovery scale — revised

Auditory function scale
4 — Consistent movement to command[a]
3 — Reproducible movement to command[a]
2 — Localization to sound
1 — Auditory startle
0 — None

Visual function scale
5 — Object recognition[a]
4 — Object localization: reaching[a]
3 — Pursuit eye movements[a]
2 — Fixation (>2 sec)[a]
1 — Visual startle
0 — None

Motor function scale
6 — Functional object use[b]
5 — Automatic motor response[a]
4 — Object manipulation[a]
3 — Localization to noxious stimulation[a]
2 — Flexion withdrawal
1 — Abnormal posturing
0 — None/flaccid

Oromotor/verbal function scale
3 — Intelligible verbalization[a]
2 — Vocalization/Oral movement
1 — Oral reflexive movement
0 — None

Communication scale
2 — Functional (accurate)[b]
1 — Non-functional (intentional)[a]
0 — None

Arousal scale
3 — Attention
2 — Eye opening without stimulation
1 — Eye opening with stimulation
0 — Unarousable

Adapted from Giacino et al. (2004).
[a]Indicates a minimally conscious state. [b]Indicates emergence from the minimally conscious state.

Some other scales have been explicitly designed for assessing minimal changes of recovery in response to sensory stimulation treatments and are supposed to be particularly sensitive for detecting subtle changes in the level of consciousness, and are therefore also very helpful for discriminating between vegetative state and minimally conscious state, even if these scales do not explicitly highlight those items that are supposed to differentiate between these two states (e.g., Western Neuro Sensory Stimulation Profile (WNSSP), Ansell and Keenan, 1989; Sensory Stimulation Assessment Measure, Rader and Ellis, 1994). One of these is the SMART (Gill-Thwaites, 1997). Although the tool originated from the parameters of the GCS, further extended by Freeman (1996), it has been further enhanced since its inception in 1988 with extensive evidence from both clinical practice and research (Wilson et al., 1991, 1993, 1996a, b; Gill-Thwaites and Munday, 1999; Wilson and Gill-Thwaites, 2000). The final design categorizes all behavioral responses observed in more than 300 patients in vegetative or minimally conscious states (as previously mentioned, this latter category defines patients who present some signs of awareness but behavioral responses are still very elementary, inconsistent and sometimes difficult to elicit; see Giacino, this volume for extensive review).

The SMART was designed to identify evidence of the patient's awareness through a graded assessment of the level of sensory, motor and communicative responses to a structured and regulated sensory program and also as a treatment tool to guide future treatment to enhance the patient's potential responses. The SMART comprises two components, including the informal component that consists of information from family and carers in respect of observed behaviors and information pertaining to the patients' pre-morbid interests, likes and dislikes. This component encourages active participation from families and carers, ensures that all responses seen to be day-to-day activity are recorded and categorized and that the treatment is relevant to the patients' interest, thus optimizing the opportunity for a meaningful response to stimuli. The SMART's formal assessment comprises of the SMART Behavioral Observation Assessment and Sensory Assessment and is conducted in 10 sessions within a 3-week period with an equal number of sessions in the morning and afternoon. This time frame provides frequent assessments over a short time frame to determine whether the behavioral responses observed are both consistent and repeatable. The behavioral observation enables the assessor to become familiar with the patients' reflexive, spontaneous and purposeful behavior during a 10-minute period prior to the commencement of the SMART Sensory Assessment.

The Sensory Assessment has eight modalities including the five sensory modalities (visual, auditory, tactile, olfactory and gustatory) and also motor function, functional communication and wakefulness/arousal. Consisting of 29 standardized techniques, the SMART provides opportunity for patients to exhibit their full behavioral repertoire, in each of the different sensory modalities. For example, to assess the patients' responses within the auditory modality, a range of standardized auditory stimuli are presented, including loud sound, voice and a variety of specifically selected verbal instructions. The verbal instructions are carefully selected from the patient's behavioral repertoire exhibited as being potentially meaningful in the SMART Behavioral Observation, such as "raise your eyebrows", "move your thumb", to provide the patient with the best opportunity to follow any one or more instructions.

The SMART's 5-point hierarchical scale is consistent and comparable across all of the sensory modalities. The five levels range from "no response" (level 1) through "reflexive" (level 2), "withdrawal" (level 3), "localizing" (level 4) and "discriminating" responses (level 5). This 5-point scale relates directly to the description of Rancho Levels 1–4 (Hagen et al., 1979); a consistent response (on five consecutive assessments) at SMART level 5 in any one of the sensory modalities demonstrates a meaningful response and thus indicates that the patient is showing behaviors indicative of a minimally conscious state or higher levels of function (Table 2).

Recent research (Gill-Thwaites and Munday, 2004) has established the reliability and validity of the SMART on 60 subjects diagnosed in vegetative state on admission and assessed at two monthly intervals. The Rancho level (Hagen et al., 1979) ratings were derived from referring physicians, SMART and WNSSP (Ansell and Keenan, 1989) for each subject and the scores of each were compared. The intra-observer intra-class correlation (ICC) was 0.97 and inter-observer ICC was 0.96, which implied very little variation within and

Table 2. Sensory Modality Assessment and Rehabilitation Technique hierarchical scale for sensory modalities and their comparison to Rancho levels

SMART level	SMART response	Rancho levels
1	*No response* To any stimulus	*I No response* In deep sleep and unresponsive to stimuli
2	*Reflex response* To stimuli reflexive and generalized responses, i.e. startle, flexor or extensor pattern	*II Generalized response* Reacting inconsistently and non-purposefully to stimuli
3	*Withdrawal response* To stimuli may, for example, turn head or eyes away or withdraw limbs from stimulus	*III Localized response* Patient reacts specifically but inconsistently to stimuli
4	*Localizing response* To stimulus may, for example, turn head or move upper limbs toward stimuli	III
5	*Differentiating response* Patient may, for example, follow visual or auditory commands or use object appropriately	*IV Confused-agitated* And subsequent Rancho levels

Adapted from Gill-Thwaites and Munday (2004).

between observers[1]; modest, although significant correlation was established between SMART and either physician or WNSSP scores. A total of 45% of subjects diagnosed to have been in vegetative state by the referring physician on admission demonstrated awareness of self and the environment. Of these subjects, 28% demonstrated this behavior within week 1 of admission. While this figure does not take account of those patients who required time to enable the staff to become familiar with the patient and to fully stabilize the patients' medical status, it is clear that the rate of misdiagnosis may have been greater. The research indicates that the SMART is a valid and reliable assessment for discriminating awareness in vegetative state and minimally conscious state. The SMART therefore provides a reliable and valid tool, which enables the assessor to establish consistency, quality, and meaning of specific responses within each sensory modality to specifically define evidence of awareness and discriminates vegetative from minimally conscious and higher level functioning patients.

[1]It must be noted that there is somewhat greater variation among the scores when considering the different subscales separately, but as the composite score is the relevant measure, the reliabilities reported here accurately reflect the performance of the scale.

The WHIM, developed by Shiel et al. (2000) and based on previous work by Horn et al. (1992, 1993) and Wilson et al. (1994), was created by observing the behaviors that occurred spontaneously or in response to stimulation in a large cohort of initially comatose patients followed longitudinally over time. Following this initial phase of empirical observation, 145 behaviors were identified. These 145 behaviors were then categorized into six subscales (communication, attention, social behavior, concentration, visual awareness, and cognition) which were then assembled to form a single main scale of 62 items. Most importantly, these 62 items are ordered in a hierarchical way, the hierarchy of behaviors assessed reflecting a statistically derived order of recovery from coma: item 1 should appear before item 2, item 2 before item 3, etc. To obtain this hierarchy, the behaviors were ranked a posteriori as a function of order of appearance observed during recovery, using a paired preference technique, similar to the paired comparisons technique often used for the construction of ordinal scales (Watson and Horn, 1992; Watson et al., 1997). The WHIM score represents the rank order of the most advanced item observed (rather than adding the different items observed). The WHIM was designed to monitor all stages of recovery from coma to emerging post-traumatic amnesia, to monitor subtle changes in patients in a minimally conscious state and to reflect performance in everyday life. Majerus et al. (2000b) conducted a validation study of a French version of the WHIM scales (Majerus et al., 2001) showing that the WHIM scales presented good inter-rater agreement (fair to excellent inter-rater agreement was obtained for 93% of the items) and very good test–retest reliability (a correlation of 0.98 was obtained between WHIM scores obtained in a test and a retest session). Most importantly, the study confirmed that the WHIM was largely superior to the GCS and GLS scales for detecting subtle changes for patients emerging from the vegetative state and for patients being in a minimally conscious state: although GCS scores remained unchanged across time for many patients in these states, assessment with the WHIM permitted to detect an important number of changes in behavior and corresponding states of consciousness (see Fig. 4 for an illustrative example). Furthermore, WHIM scores were more than five times more variable than GLS scores for patients in the minimally conscious state or patients showing a good recovery, suggesting that the WHIM is particularly sensitive for patients in the minimally conscious state. However, the study by Majerus et al. (2000b) also showed that the sequence of recovery proposed by Shiel et al. (2000) is very probabilistic and lacks precision, as the proposed order of recovery could not be replicated for all items of the scale. Further studies are needed to strengthen the validity of the sequence of recovery proposed by the original version of the WHIM scales.

Wilson et al. (2002) emphasized the need for "careful, repeated and reliable assessment of patients with impaired consciousness" to add accuracy to the assessment process. The CRS-R, the SMART and the WHIM have all proven to be very useful tools to monitor behaviors with head injured patients and both the features and design of these scales compliment each other in clinical practice. Both the SMART and WHIM have been recommended to provide standardized assessment protocols for patients in vegetative or minimally conscious states by the UK Royal College of Physician (2003) guidelines.

Finally, it should be noted that the administration of standardized scales might sometimes be difficult as a result of sensory and motor disturbances

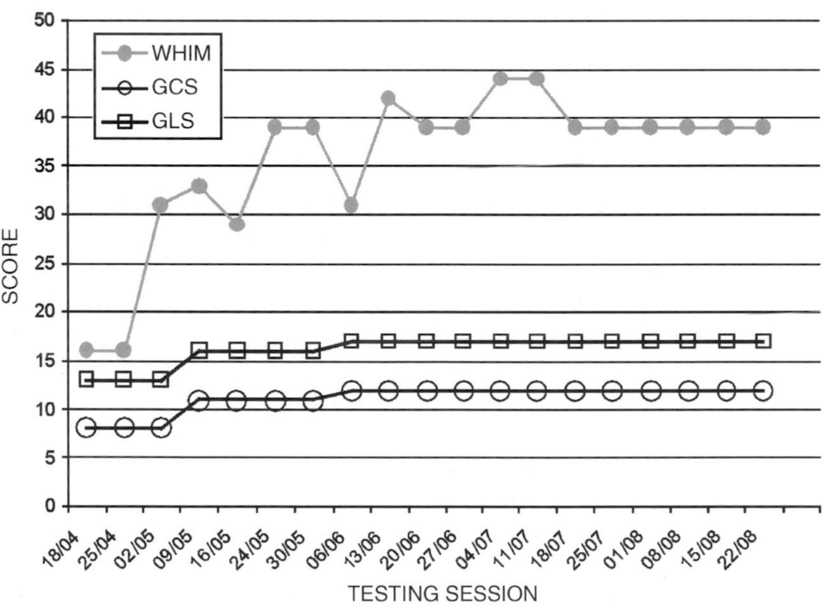

Fig. 4. Longitudinal assessment with the WHIM, GCS and GLS in an initially comatose patient, evolving to a minimally conscious state. Note the absence of notable changes for the GCS and GLS scales from test session 09/05, while WHIM scores show important behavioral changes.

that will make impossible the scoring of a number of items needing a given sensory or motor modality. This is particularly important as these sensory and motor impairments are a frequent cause of misdiagnosis. In these cases, the use of individualized assessment techniques is recommended. For example, Whyte et al. (1995, 1999) proposed a method for a reliable assessment of visual attention and command following in these patients (see also Giacino and Whyte, 2005). The principle of this method is to find, for an individual patient, at least one behavior with which the patient seems to produce voluntary responses. This behavior is likely to be different in each patient and depends on his particular sensory and motor impairments. Once this behavior has been detected, the second aim is to determine whether this behavior is really voluntary, by determining the baseline frequency of this behavior, and by determining increases in frequency of this behavior over time and as a result of stimulation (e.g., on command). In order to consider a behavior as volitional, the patient has to respond more frequently when required to produce the behavior than during baseline and he must respond less frequently when instructed not to produce the behavior. This method permits to obtain discrimination scores between the three conditions (behavior on, behavior off, baseline) for which statistical significance can be tested. Using such individualized assessment methods in combination with the standardized assessment scales presented above, both the sensitivity and reliability of behavioral assessment of altered states of consciousness is likely to be maximized. The CRS-R has incorporated parts of these individualized assessment techniques as described by Whyte et al. (1997, 1999).

As obvious time constraints in the clinical setting will not allow to assess every patient with each of the presented scales, we will conclude this section by providing some guidelines for selecting the most appropriate scale, depending on the question that is asked and the state the patient is in. In the acute setting, the GCS remains the "gold standard" in the evaluation of coma. By virtue of its simplicity, it is the most universally utilized consciousness scale worldwide and seems, despite its drawbacks, destined to be used in emergency medicine and intensive care for some time. In the post-coma phase, and to differentiate between vegetative and minimally conscious state, the CRS-R (Giacino et al., 2004), in conjunction with the individualized assessment technique proposed by Whyte et al. (1995, 1999), might be the best solution as it was specifically designed for making this differential diagnosis. When following a patient longitudinally and documenting subtle progresses in the recovery of consciousness, the SMART and WHIM could be more appropriate. The WHIM seems more practical for assessments made on a daily basis as time needed to administer this scale is only about 10 min (range 2–35 min), while administration of the SMART takes between 30 and 40 min. The WHIM has been shown to be particularly sensitive for patients in the minimally conscious state and patients showing slow but relatively good recovery. On the other hand, one important advantage of the SMART is that it also assesses responses to a sensory stimulation program, which is not the case for the other scales. Finally, the SMART appears to be particularly suitable for patients in the vicinity of the vegetative state. Although the WHIM also shows a good sensitivity for this state, the inclusion of an olfactory function subscale in the SMART provides an additional opportunity for detecting subtle signs of responsiveness.

Conclusions

Assessment of awareness is not a matter of all or nothing. Recovery of awareness is a very gradual process, with sometimes great leaps forwards, but more often subtle changes, and also sometimes setbacks. For the patient emerging from coma, it is of utmost importance that the medical staff adapts its assessment to the level of awareness the patient is currently in. The subtlest signs of awareness, as well as their fluctuation, have to be reliably captured as they are the only means for avoiding misdiagnosis, but also for communicating with these patients. This implies the use of standardized, sensitive and individualized assessment tools that cover a wide range of possible, although sometimes minimal, behaviors in all sensory modalities. The major challenge of the years to come will not be to develop new tools, but to effectively implement those existing ones in the daily practice of carers of patients with impaired consciousness.

References

Andrews, K., Murphy, L., Munday, R. and Littlewood, C. (1996) Misdiagnosis of the vegetative state: retrospective study in a rehabilitation unit. Bri. Med. J., 313: 13–16.

Andrews, K. (1996) International Working Party on the Management of the Vegetative State: summary report. Brain Inj., 10: 797–806.

Ansell, B.J. and Keenan, J.E. (1989) The Western neuro sensory stimulation profile: a tool for assessing slow to recover head injured patients. Arch. Phys. Med. Rehabil., 70: 104–108.

Benzer, A., Mitterschiffthaler, G., Marosi, M., Luef, G., Puhringer, F., De La Renotiere, K., Lehner, H. and Schmutzhard, E. (1991) Prediction of non-survival after trauma: Innsbruck Coma Scale. Lancet, 338: 977–978.

Bernat, J.L. (1992) The boundaries of the persistent vegetative state. J. Clin. Ethics, 3: 176–180.

Born, J.D. (1988) The Glasgow-Liège Scale. Prognostic value and evaluation of motor response and brain stem reflexes after severe head injury. Acta Neurochir., 95: 49–52.

Born, J.D., Hans, P., Dexters, G., Kalangu, K., Lenelle, J., Milbouw, G. and Stevenaert, A. (1982) Practical assessment of brain dysfunction in severe head trauma. Neurochirurgie, 28: 1–7.

Bouzarth, W.F. (1968) Neurosurgical watch sheet for craniocerebral trauma. J. Trauma, 8: 29–31.

Bricolo, A., Turazzi, S., Alexandre, A. and Rizzuto, N. (1977) Decerebrate rigidity in acute head injury. J. Neurosurg., 47: 680–689.

Bricolo, A., Turazzi, S. and Feriotti, G. (1980) Prolonged posttraumatic unconsciousness: therapeutic assets and liabilities. J. Neurosurg., 52: 625–634.

Canedo, A., Grix, M.C. and Nicoletti, J. (2002) An analysis of assessment instruments for the minimally responsive patient (MRP): clinical observations. Brain Injury, 16: 453–461.

Childs, N.L., Mercer, W.N. and Childs, H.W. (1993) Accuracy of diagnosis of persistent vegetative state. Neurology, 43: 1465–1467.

Crosby, L. and Parsons, L.C. (1989) Clinical neurologic assessment tool: development and testing of an instrument to index neurologic status. Heart Lung, 18: 121–129.

Davis, A.L. (1991) The visual response evaluation: a pilot study of an evaluation tool for assessing visual responses in low-level brain injured patients. Brain Injury, 5: 315–320.

Freeman, E.A. (1996) The coma exit chart: assessing the patient in prolonged coma and vegetative state. Brain Injury, 10: 615–624.

Gallup Jr., G.G. (1997) On the rise and fall of self-conception in primates. Ann. NY Acad. Sci., 818: 72–82.

Giacino, J.T., Ashwal, S., Childs, N., Cranford, R., Jennett, B., Katz, D., Kelly, J., Rosenberg, J., Whyte, J., Zafonte, R. and Zasler, N. (2002) The minimally conscious state: definition and diagnostic criteria. Neurology, 58: 349–353.

Giacino, J.T., Kezmarsky, M.A., DeLuca, J. and Cicerone, K.D. (1991) Monitoring rate of recovery to predict outcome in minimally responsive patients. Arch. Phys. Med. Rehab., 72: 897–901.

Giacino, J.T. and Zasler, N.D. (1995) Outcome after severe traumatic brain injury: Coma vegetative state and the minimally responsive state. J. Head Trauma Rehab., 10: 40–56.

Giacino, J.T., Kalmar, K. and Whyte, J. (2004) The JFK Coma Recovery Scale - Revised: measurement characteristics and diagnostic utility. Arch. Phys. Med. Rehab., 85: 2020–2029.

Giacino, J.T. and Whyte, J. (2005) The vegetative and minimally conscious states: current knowledge and remaining questions. J. Head Trauma Rehab., 20: 30–50.

Gill-Thwaites, H. (1997) The sensory modality assessment and rehabilitation technique — a tool for the assessment and treatment of patients with severe brain injury in a vegetative state. Brain Injury, 11: 723–734.

Gill-Thwaites, H. and Munday, R. (1999) The sensory modality assessment and rehabilitation technique (SMART): a comprehensive and integrated assessment and treatment protocol of the vegetative state and minimally responsive patient. Neuropsychol. Rehabil., 9: 305–320.

Gill-Thwaites, H. and Munday, R. (2004) The sensory modality assessment and rehabilitation technique (SMART). A valid and reliable assessment for vegetative state and minimally conscious state patients. Brain Injury, 18: 1255–1269.

Hagen, C., Malkmus, D. and Durham, P. (1979) Levels of Cognitive Function Rehabilitation of Head Injured Adults: Comprehensive Physical Management. Profession Staff Association of Rancho Los Amigos Hospital Inc., Downey, CA.

Horn, S., Watson, M., Wilson, B.A. and McLellan, D.L. (1992) The development of new techniques in the assessment and monitoring of recovery from severe head injury: a preliminary report and case history. Brain Injury, 6: 321–325.

Horn, S., Shiel, A., McLellan, D.L., Campbell, M., Watson, M. and Wilson, B.A. (1993) A review of behavioural assessment scales for monitoring recovery in and after coma with pilot data on a new scale of visual awareness. Neuropsychol.Rehabil, 3: 121–137.

Ivan, L.P. (1973) Spinal reflexes in cerebral death. Neurology, 23: 650–652.

Jennett, B. (1972) Prognosis after severe head injury. Clin. Neurosurg., 19: 200–207.

Jennett, B. and Teasdale, G. (1977) Aspects of coma after severe head injury. Lancet, i: 878–881.

Laureys, S., Majerus, S. and Moonen, G. (2002) Assessing consciousness in critically ill patients. In: Vincent J.L. (Ed.), Yearbook of Intensive Care and Emergency Medicine. Springer, Berlin, pp. 715–727.

Majerus, S., Van der Linden, M. and Damas, F. (2000a) Les états de conscience altérée: comment les définir et comment les évaluer? Rev. Neuropsychol., 10: 219–254.

Majerus, S., Van der Linden, M. and Shiel, A. (2000b) Wessex Head Injury Matrix and Glasgow/Glasgow-Liège Coma Scale: a validation and comparison study. Neuropsychol. Rehabil., 10: 167–184.

Majerus, S., Azouvi, P., Fontaine, A., Marlier, N., Tissier, A.-C. and Van der Linden, M. (2001). Adaptation française de la Wessex Head Injury Matric - 62 items. Unpublished test manual.

Marion, D.W. and Carlier, P.M. (1994) Problems with initial Glasgow Coma Scale assessment caused by prehospital treatment of patients with head injuries: results of a national survey. J. Trauma, 36: 89–95.

Murray, L.S., Teasdale, G.M., Murray, G.D., Jennett, B., Miller, J.D., Pickard, J.D., Shaw, M.D., Achilles, J., Bailey, S. and Jones, P. (1993) Does prediction of outcome alter patient management? Lancet, 341: 1487–1491.

Ommaya, A.K. (1966) Trauma to the nervous system. Ann. Roy. Coll. Surg., 39: 317–347.

Plum, F. and Posner, J.B. (1983) The Diagnosis of Stupor and Coma. Davis, F.A., Philadelphia, PA.

Rader, M.A. and Ellis, D.W. (1994) The sensory stimulation assessment measure (SSAM): A tool for early evaluation of severely brain-injured patients. Brain Injury, 8: 309–321.

Rappaport, M., Dougherty, A.M. and Kelting, D.L. (1992) Evaluation of coma and vegetative states. Arch. Phys. Med. Rehab., 73: 628–634.

Rowley, G. and Fielding, K. (1991) Reliability and accuracy of the Glasgow Coma Scale with experienced and inexperienced users. Lancet, 337: 535–538.

Royal College of Physicians. (2003) The Vegetative State. Guidance on diagnosis and management. Publication Unit of the Royal College of Physicians, London.

Salcman, M., Schepp, R.S. and Ducker, T.B. (1981) Calculated recovery rates in severe head trauma. Neurosurgery, 8: 301–308.

Saposnik, G., Bueri, J.A., Maurino, J., Saizar, R. and Garretto, N.S. (2000) Spontaneous and reflex movements in brain death. Neurology, 54: 221–223.

Segatore, M. and Way, C. (1992) The Glasgow Coma Scale: time for change. Heart Lung, 21: 548–557.

Shiel, A., Horn, S., Wilson, B.A., McLellan, D.L., Watson, M. and Campbell, M. (2000) The Wessex Head Injury Matrix main scale: a preliminary report on a scale to assess and monitor patients recovery after severe head injury. Clin. Rehabil., 14: 408–416.

Stalhammar, D., Starmark, J.E., Holmgren, E., Eriksson, N., Nordstrom, C.H., Fedders, O. and Rosander, B. (1988) Assessment of neurological responsiveness in acute cerebral disorders. Multicenter study on the Reaction Level Scale (RLS85). Acta Neurochir., 90: 73–80.

Stanczak, D.E., White 3rd, J.G., Gouview, W.D., Moehle, K.A., Daniel, M., Novack, T. and Long, C.J. (1984) Assessment of level of consciousness following severe neurological insult. A comparison of the psychometric qualities of the Glasgow Coma Scale and the Comprehensive Level of Consciousness Scale. J. Neurosurg., 60: 955–960.

Teasdale, G. (1975) Acute impairment of brain function-1. Assessing 'conscious level'. Nurs. Times, 71: 914–917.

Teasdale, G. and Jennett, B. (1974) Assessment of coma and impaired consciousness. A practical scale. Lancet, 2: 81–84.

Teasdale, G. and Jennett, B. (1976) Assessment and prognosis of coma after head injury. Acta Neurochir., 34: 45–55.

Teasdale, G., Jennett, B., Murray, L. and Murray, G. (1983) Glasgow Coma Scale: to sum or not to sum. Lancet, 2: 678.

The Multi-Society Task Force on PVS. (1994) Medical aspects of the persistent vegetative state (First of Two Parts). New Engl. J. Med., 330: 1499–1508.

Tresch, D.D., Farrol, H.S., Duthie, E.H., Goldstein, M.D. and Lane, P.S. (1991) Clinical characteristics of patients in the persistent vegetative state. Arch. Int. Med., 151: 930–932.

Watson, M. and Horn, S. (1992) Paired preferences technique: an alternative method for investigating sequences of recovery in assessment scales. Clin. Rehabil., 6: 170.

Watson, M., Horn, S., Shiel, A. and McLellan, D.L. (1997) The application of a paired comparisons technique to identify sequence of recovery after severe head injury. Neuropsychol. Rehabil., 7: 441–458.

Whyte, J. and DiPasquale, M. (1995) Assessment of vision and visual attention in minimally responsive brain injured patients. Arch. Phys. Med. Rehab., 76: 804–810.

Whyte, J., DiPasquale, M. and Vaccaro, M. (1999) Assessment of command-following in minimally conscious brain injured patients. Arch. Phys. Med. Rehab., 80: 1–8.

Wilson, B.A., Shiel, A., Watson, M., Horn, S. and McLellan, D.L. (1994) Monitoring behaviour during coma and post-traumatic amnesia. In: Uzell B. and Christensen A.L. (Eds.), Progress in the Rehabilitation of Brain Injured People. Lawrence Erlbaum Associates Inc., Hillsdale, NJ.

Wilson, F.C., Harper, J., Watson, T. and Morrow, J.I. (2002) Vegetative state and minimally responsive patients: regional survey, long-term case outcome and service recommendations. NeuroRehabilitation, 17: 231–236.

Wilson, S.L., Powell, G.E., Elliot, K. and Thwaites, H. (1991) Sensory stimulation in prolonged coma — four single case studies. Brain Injury, 5: 393–401.

Wilson, S.L., Powell, G.E., Elliot, K. and Thwaites, H. (1993) Evaluation of sensory stimulation as a treatment for prolonged coma — seven single experimental case studies. Neuropsychol. Rehabil., 3: 191–201.

Wilson, S.L., Brock, D., Powell, G.E., Thwaites, H. and Elliot, K. (1996a) Constructing arousal profiles for vegetative state patients — a preliminary report. Brain Injury, 10: 105–113.

Wilson, S.L., Powell, G.E., Brock, D. and Thwaites, H. (1996b) Vegetative state and response to sensory stimulation: an analysis of twenty four cases. Brain Injury, 10: 807–818.

Wilson, S.L. and Gill-Thwaites, H. (2000) Early indications of emergence from vegetative state derived from assessment with the SMART — a preliminary report. Brain Injury, 14: 319–331.

Working Group of the Royal College of Physicians. (1996) The permanent vegetative state. J Roy. Coll. Phys. Lond., 30: 119–121.

S. Laureys (Ed.)
Progress in Brain Research, Vol. 150
ISSN 0079-6123

CHAPTER 29

Evoked potentials in severe brain injury

Jean-Michel Guérit*

Clinique Edith Cavell, rue Edith Cavell 32–B 1180 Brussels, Belgium

Abstract: Three-modality evoked potentials (EPs) have been used for several years in association with the electroencephalogram (EEG) as a diagnostic and prognostic tool in acute traumatic or nontraumatic coma. In 1993 we proposed to combine these in two indices: the index of global cortical function (IGCF) and the index of brain-stem conduction (IBSC). Four EP patterns based on both indices emerge at the acute stage of severe head trauma. These are easily explainable by pathophysiology. Pattern 1 corresponds to alterations in the index of global cortical function without changes in the index of brain-stem conduction. Its prognosis is good (80 to 90% of these patients recover). Pattern 2 is characterized by alterations of somatosensory EPs that are suggestive of midbrain dysfunction. The prognosis depends both on the reversibility of the midbrain dysfunction and on the extent of associated diffuse axonal lesions, whose evaluation requires MRI. Patients who recovered from Pattern 2 sometimes did so after a long interval during which they remained vegetative. Pattern 3 is characterized by alterations of brain-stem auditory EPs that are suggestive of pontine involvement. It usually follows uncontrolled intracranial hypertension and corresponds to evolving transtentorial herniation. All patients with that transient pattern eventually died. Pattern 4 is categorized by the disappearance of all activities of intracranial origin, contrasting with the preservation of all activities of retinal, spinal-cord, and peripheral-nerve origin. This pattern corresponds to brain death. In our experience, three-modality EPs are currently the best bedside brain-death confirmatory tool.

Introduction

Electrophysiological techniques (electroencephalogram (EEG); evoked potentials (EPs)) provide functional evaluation of the nervous system. Therefore, their domain differs from that of structural imaging techniques (CT, MRI). It coincides with that of clinical examination, but EEG and EPs present two major advantages: they are still feasible in curarized patients and provide more quantitative results, which are amenable to follow-up studies. This chapter summarizes our personal experience of EP recording at the acute stage of severe traumatic coma (defined as GCS $\leqslant$ 8 on admission). More comprehensive information can be found in recent reviews (Nuwer, 1994; Jordan, 1995; Chatrian et al., 1996; Guérit et al., 1999a,b). We will first briefly remind some basics of EEG and EPs in the intensive care unit (ICU). Thereafter, we will review the pathophysiology, the main types, and the prognostic significance of EPs at the acute stage of traumatic coma. We will compare our own data with those of the literature and conclude with some short statements on concrete clinical applications.

Basics of EEG and EPs

The EEG is dominated by the activity of the cerebral cortex underlying the recording electrodes. Therefore, its alterations may either reflect primary cortical dysfunction (cortical lesions, epilepsy) or be the consequence of an abnormal modulation of the cerebral cortex, owing to thalamic or brain-stem dysfunction. Noteworthy,

*Corresponding author. Tel.: +32 498 32 74 67;
E-mail: Jean-Michel.Guerit@chirec.be

DOI: 10.1016/S0079-6123(05)50029-3

the EEG also contains activities coming from deeper brain structures (brain-stem), but their amplitude is too low to allow analyzing these, at least in the conventional EEG. EPs correspond to the EEG modifications induced by sensory stimuli or cognitive activities. Because their amplitude is much lower than that of the EEG, these can only be obtained through averaging the responses to a series of stimuli. One advantage of averaging is that it allows analyzing brain-stem activities. EPs can be subdivided into "exogenous EPs", which reflect the passive reception of sensory stimuli and "cognitive EPs", which correspond to the additional brain electrogenesis that is produced by the cognitive treatment of these stimuli. On their turn, exogenous EPs are classified on the basis of the type of used stimulus: visual (V), auditory (A), and somatosensory (S) EPs. EPs consist of a succession of peaks, which are characterized by their latencies and amplitudes; each peak reflects the activation of one sensory relay or a limited network of sensory structures. Short-latency EPs evaluate the brain-stem, the subcortical somatosensory pathways, and the primary parietal cortex (N20). Middle-latency EPs provide assessment of the temporal (middle-latency auditory evoked potentials (AEPs)), parietal and frontal (middle-latency Somatosensory evoked potentials (SEPs)), and the occipital (Visual evoked potentials (VEPs)) cortex. Long-latency exogenous and cognitive EPs depend on multiple cortical generators and are roughly under the same influences as the EEG.

Our routine ICU examination (Guérit et al., 1993) consists of three-modality EPs and, in some cases, cognitive EPs (Guérit et al., 1999a,b). VEPs are obtained with LED goggles delivering flashes of a sufficient intensity to cross the eyelids, even in patients with closed eyes. SEPs are obtained by surface stimulation of the left and right median nerves at the wrist; the recorded activities are generated in the brachial plexus, cervical spinal cord, cervico-medullary junction, and parietal and frontal cortices. Brain-stem AEPs (BAEPs) are obtained by monaural clicks; these are generated by the auditory nerve and the pons. The most frequently analyzed components of exogenous EPs are shown in Fig. 1. Stimulation and recording parameters are shown in Table 1.

When recorded, cognitive EPs are obtained with an auditory "oddball paradigm", in which the patient is stimulated with two types of auditory stimuli, a frequent (80%) one consisting of lower-frequency (500 Hz) tone bursts and a rare (20%) one consisting of higher-frequency (750 Hz) tone bursts. Both types of stimuli give rise to similar exogenous EPs, but only the rare ones produce additional activities (mismatch negativity, N2 waves, and the P300 waves) that reflect the cognitive brain reactions to rarity (Fischer et al., 1999; Guérit et al., 1999a,b; Kane et al., 2000; also see Kotchoubey, in this volume).

The combined recording of VEPs, SEPs, and BAEPs gives rise to a huge and complex set of parameters. To improve communication with the intensivists and to facilitate patient follow-up, we combined these parameters in two indices: the *index of global cortical* function (*IGCF*) and the *index of brain-stem conduction* (*IBSC*) (Guérit et al. 1993). The IGCF is based on VEPs and the cortical components of SEPs; five grades are defined according to the rules given in Table 2. Briefly, grade 0 corresponds to normal IGCF and is almost never observed in truly comatose patients (but is classically observed in the locked-in syndrome or psychogenic coma); grade 4 corresponds to the loss of all cortical components (observed in brain death); grades 1–3 correspond to intermediate states of increasing severity. We demonstrated a significant correlation between the IGCF and the Glasgow Coma Score.

The IBSC is based on BAEPs and the subcortical and primary cortical components of SEPs. It is first quantitatively measured on the basis of interpeak latencies (I–III, III–V, and I–V in BAEPs; N13–N20 and/or P14–N20 in SEPs) and amplitude ratios (V–I interpeak ratio in BAEPs) (Fig. 1). If abnormal, we first differentiate mere latency increases without any alteration in the overall EP morphology, which will be referred to as "minor IBSC abnormalities", from structural alterations (essentially loss of peaks) that will be referred to as "major IBSC abnormalities". In our experience, minor IBSC abnormalities are most often explainable by the influence of non-neurological factors like hypothermia, superimposed metabolic disturbances, or drug influences. Major IBSC

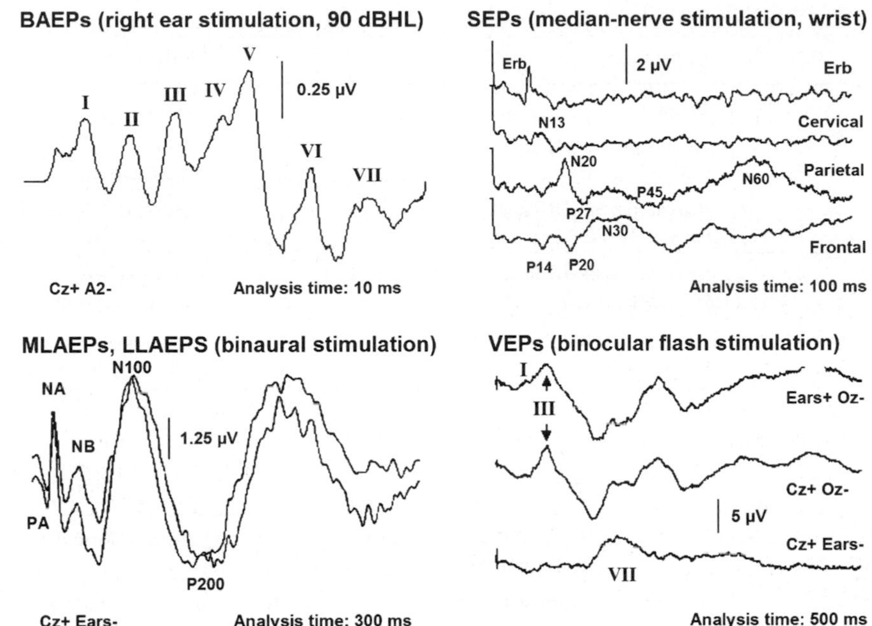

Fig. 1. Three-modality evoked potentials. Normal brain-stem auditory evoked potentials (BAEPs, top left), somatosensory evoked potentials (SEPs, top right), middle- (M) and long-latency evoked potentials (LLEPs, bottom left), visual evoked potentials (VEPs, bottom right). All electrode labels correspond to the 10–20 International System. IPL: interpeak latency; CCT: central conduction time. Adapted from Guérit (2001).

Table 1. Three-modality EPs: stimulation and recording parameters

	VEPs	SEPs	BAEPs
Stimulus	LED goggles, binocular 0.9/s	Left and right median nerve at the wrist 3.1/s	Clics, monaural 21.7/s
Number of channels	4 channels: O_1 and O_2 referred to C_z and linked earlobes	4 channels: Erb's point, C_5 spinous process, 2 cm behind C'_3 or C'_4, Fpz referred to ipsilateral earlobe	1 channel: ipsilateral earlobe referred to C_z
Filters	1–100 Hz	30–3000 Hz (Erb + C_{5sp}) 1–250 Hz (cortical recordings)	150–3000 Hz
Analysis time	500 ms	100 ms	10–15 ms

abnormalities usually reflect prognostically relevant structural brain damage. It is qualitatively described in terms of medullar, pontine, or midbrain dysfunction, according to the rules described in Table 3.

Pathophysiological considerations

Almost all the pathophysiological processes occurring at the acute stage of head trauma can cause both reversible and irreversible neuronal dysfunction. This depends on two factors: the severity of the process itself and the relative sensitivity of the brain structure under study to this process. Three factors will be considered: the immediate repercussion of the mechanical injury, brain edema, and possible secondary complications.

Mechanical aspects of head trauma

Brisk head deceleration is associated with an almost instantaneous transfer of kinetic energy to or from the brain. On its turn, this transfer gives rise

Table 2. Index of global cortical function (IGCF)

	VEPs	SEPs
Grade 0	Normal	Normal
Grade 1	Increased peak III latency Peak VII present	Normal N20, P24, and P27 N30 present
Grade 2	Increased peak III latency Peak VII absent	Normal N20 and P24 N30 absent
Grade 3	Increased peak III latency No subsequent activities	Normal N20 No subsequent activities
Grade 4	No reproducible VEPs ERG present	No cortical activities P14 present

Table 3. Index of brain-stem conduction (IBSC)

Level	SEPs	BAEPs
Midbrain	Normal P14 Abnormal N20	Normal
Pons	Normal P14 Abnormal N20	Abnormal
Medulla	Absent P14 Abnormal N20	Normal

to mechanical constraints, which, according to Newton's law (force = mass × acceleration), induce regional brain movements whose amplitude depends on the mass of the target structure. This implies that two neighboring structures of very different masses tend to move with respect to each other. These mechanical constraints also depend on the direction of trauma; they result from the combination of both linear and angular components. Therefore, the anatomical distribution of brain dysfunctions of mechanical origin depends on both endogenous (distribution of mass density, whose inter-individual variability is relatively low) and exogenous (direction of trauma) factors. Considering the distribution of mass density, two brain regions are particularly exposed to mechanical constraints: the midbrain, which lies at the boundary between the cerebral hemispheres and brain-stem, that is, two structures of very different masses, and the white matter of the cerebral hemispheres, which is more superficial and, therefore, more remote from the center of rotation in trauma with large angular constraints. To be complete, one should also mention the splenium of the corpus callosum, which is submitted to very different constraints from both hemispheres in case of asymmetrical head rotation (see Shaw, 2002, for review). Linear constraints can also cause some brain regions to be crushed against bony or fibrous structures: this explains both the superficial cerebral contusions and the possible lesions on the posterior part of the midbrain against the cerebellar tentorium. Noteworthy, both exogenous and endogenous factors explain the elective midbrain sensitivity to the mechanical constraints associated with head trauma. Elective midbrain sensitivity explains that midbrain lesions or dysfunction can occur in the absence of cortical or diffuse axonal hemispheric lesions (Hume Adams, 1984) (Fig. 2).

Factors determining brain dysfunction after head trauma

Immediate repercussions of mechanical injury

While severe mechanical constraints usually cause irreversible neuronal lesions, less severe constraints can provoke reversible arrest of neuronal function.

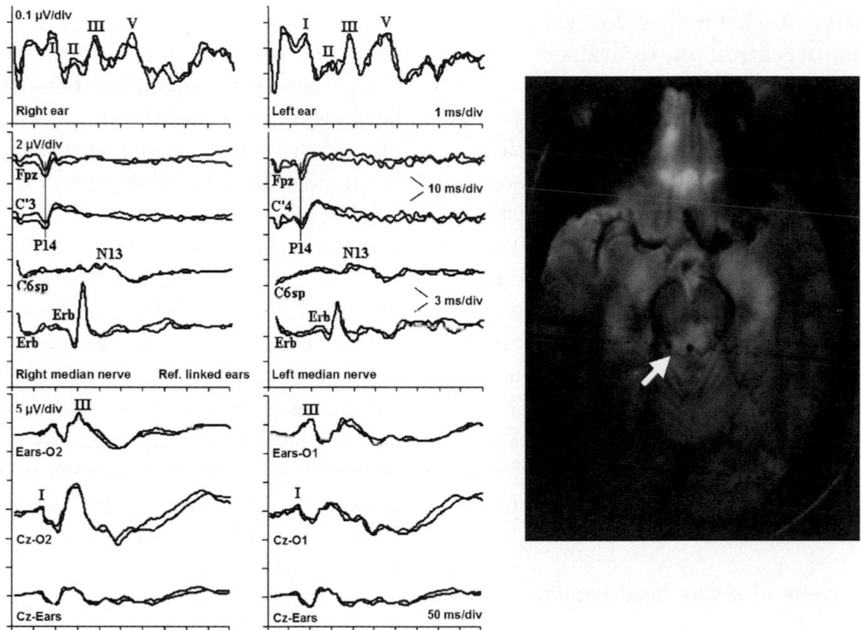

Fig. 2. Pattern 2: midbrain dysfunction. Head trauma, recording performed on day 1. BAEPs (top left) are normal. SEPs (middle left) show normal peripheral nerve (Erb), cervical (N13), and brain-stem (P14) components but profoundly altered N20 (absent on the left side, altered on the right side). VEPs (bottom left) show left–right asymmetry (compare recordings at O1 and O2) but are relatively well preserved when compared to SEPs. This association is typical of a midbrain pathology; an example of a reversible midbrain lesion is shown on the right side. All electrode labels correspond to the 10–20 International System.

In the white matter, the opening of ion channels sensitive to mechanical stimulations can, indeed, give rise to an initial period of hyperexcitability, which is followed by a conduction block (Katayama et al., 1990). In the gray matter, a mechanical stimulation can cause paroxysmal depolarization shifts and trigger seizures (Speckman and Elger, 1982). However, even if both conduction blocks and brief seizures could explain the transient deficits associated with brain concussion, these are unlikely to be still present within a few hours after the trauma, that is, during the period when the first neurophysiologic examination is usually performed.

Short-term consequence: cerebral edema

Cerebral edema is defined as the accumulation of excess fluids intracellularly or within the extracellular space. As both direct mechanical injuries and ion transfer due to modifications in membrane permeability can cause cerebral edema, its anatomical distribution will be roughly the same as that of the immediate lesions or dysfunction. Usually, cerebral edema has already developed when performing the first neurophysiologic examination. As a consequence, the functional changes that are reflected by EEG or EP abnormalities actually constitute a complex mixture of irreversible phenomena due to brain lesions and potentially reversible changes due to edema. Noteworthy, EEG and EPs in isolation are usually unable to differentiate both types of changes.

Possible secondary complications

Three pathophysiological processes can account for delayed worsening: epilepsy, ischemia (due to secondary circulatory arrest or to the local consequences of brain edema), or uncontrolled intracranial hypertension with subsequent transtentorial herniation, brain-stem lesions, and, through secondary ischemia at both the supra- and infratentorial levels, evolution toward brain death, that is, whole destruction of the encephalon. Noteworthy, ischemia is the only process that can cause whole cortical destruction after head trauma, irrespective of whether it is consecutive to secondary circulatory arrest or to transtentorial herniation.

Obviously, EEG and EPs evaluate the current functional status of the brain but may be unable to predict secondary worsening. In other words, some patients may present with a poor evolution despite EPs that are only mildly altered at the acute stage; however, this poor evolution is not the consequence of the brain lesions that were present at the time of EP recording, but of additional lesions that secondarily appeared. Retrospective studies only considered the percentage of poor evolutions, irrespective of whether secondary complications occurred or not: these studies artificially attributed a lower prognosis to mildly altered EPs in the acute stage of traumatic coma.

EPs at the acute stage of severe head trauma

This paragraph will consist of two parts. We will first review our own experience and demonstrate that acute traumatic coma can be associated with four prognostically relevant patterns. Second, we will show that such categorization could reconcile several of the apparent divergences of the literature.

Four EP patterns of acute head trauma

These patterns correspond to recordings performed within the first 3 days (and usually within the first 24 h) after coma onset. Four patterns emerge from the combined recording of V, S, and BAEPs; these can be easily situated with respect to the pathophysiology of head trauma (Guérit et al., 1993; Guérit, 1999a, 2001).

Pattern 1

Pattern 1 is characterized by IGCF alterations without IBSC changes or with minor IBSC changes that are explainable by body temperature, metabolic, or drug influences. This pattern is thus suggestive of an isolated cortical dysfunction in the absence of any primary brain-stem dysfunction. This pattern is qualitatively similar to that observed in anoxic comas; however, in contrast with anoxic coma in which all IGCF grades can be observed, from 1 to 4, only grades 1 or 2 were observed in traumatic patients with Pattern 1 who did not develop secondary ischemia. This observation is in keeping with neuropathology. Indeed, while a complete destruction of the cerebral cortex without brain-stem damage can occur in major anoxic encephalopathy, whole cortical destruction in head trauma without evidence of secondary circulatory arrest presupposes transtentorial herniation, which must be associated with brain-stem destruction and is thus incompatible with Pattern 1.

Another difference between Pattern 1 in brain anoxia and head trauma resides in its prognostic value. Actually, the prognosis of Pattern 1 in head trauma is good as a favorable evolution[1] was observed in our series in 85% of patients (90 and 80% of patients with grades 1 and 2, respectively). By comparison, in brain anoxia, grades 1 and 2 were associated with 65 and 40% probabilities of good outcome, respectively. This difference is explainable by pathophysiology. Although anoxia can cause both reversible arrest of neuronal function and irreversible cytotoxic lesions, mostly irreversible lesions are likely to account for these IGCF alterations that are observed since the 1st day after the acute episode. By contrast, several factors can explain the IGCF alterations in head trauma; some are irreversible (diffuse axonal lesions, secondary ischemia), but other may be reversible (brain edema, abnormal cortical modulation from the brain-stem). We hypothesize that IGCF alterations in Pattern 1 are most often induced by reversible factors, while diffuse axonal or secondary ischemic lesions would account for the IGCF alterations that are found in the small percentage of patients who initially presented with Pattern 1 but eventually became vegetative.

Pattern 2

Pattern 2 corresponds to midbrain dysfunction. It is characterized by intact BAEPs (demonstrating pontine integrity), SEPs displaying signs of subcortical dysfunction, and variable VEP alterations

[1]We define a good outcome as the possibility to regain some independent life, irrespective of whether there are neurological sequellae or not, that is, grades 4 and 5 in the classification of Jennett and Bond (1975).

(Fig. 2). In the absence of cortical SEPs (i.e., grade 4 SEPs, corresponding to major IBSC alterations), the prognosis is bad (more than 80% of patients either died or became vegetative); with SEPs grade 2 or less, prognosis is better (67% of patients presented a good evolution, particularly from the cognitive standpoint). The difference between post-traumatic and post-anoxic coma is worth being underlined: in traumatic coma, some patients with bilaterally absent N20 may recover, as opposed to anoxic coma, in which the bilateral absence of N20 at least 24 h after the acute episode has always been associated with death or vegetative state.

Interestingly, Pattern 2 is usually associated with a severely altered neurological examination (decerebrate posturing, i.e., a Glasgow coma score of 4/15). That is, patients with Pattern 2 can still present a good evolution, despite a severely altered neurological examination (GCS = 4) at the acute stage. Noteworthy, such favorable evolution may sometimes briskly manifest itself after a period of several weeks or months (up to 5 months in our series) during which the patient was considered vegetative.

To explain these contrasting observations, one should first remind that the midbrain contains the midbrain reticular formation, which constitutes one starting point of the ascending reticular activating system (ARAS); hence, isolated ARAS dysfunction may be sufficient to induce coma (Moruzzi and Magoun, 1949), even if the remaining of the brain is structurally intact. Second, we already mentioned that, owing to the elective midbrain sensitivity to head deceleration, it is quite conceivable that the midbrain can be the only lesioned structure. We think that two factors determine prognosis in the presence of midbrain dysfunction: first, whether it merely reflects transient midbrain dysfunction (for example, in case of midbrain edema) or is consecutive to an irreversible lesion, and, second, the extent of associated diffuse axonal lesions in the white matter of the cerebral hemispheres. Now, even if more severe diffuse axonal lesions are likely to be associated with more severely altered VEPs, these are not reliable enough to precisely evaluate the extent of these lesions. As the midbrain pathology interferes with SEPs (through a direct influence on somatosensory afferent pathways) and can also influence the EEG (through ARAS disturbance), clinical neurophysiology actually looks ill-adapted to evaluate diffuse axonal lesions. This justifies the use of MRI, which should be obtained as soon as possible, in patients with Pattern 2.

Pattern 3

Pattern 3 corresponds to evolving transtentorial herniation. As it is a transient pattern, it is only rarely observed (at least when only isolated examinations are performed). Its distinctive feature is the appearance of BAEP abnormalities suggestive of a pontine lesion. Usually, cortical VEPs and SEPs have already disappeared. Even if we rarely observed some reversibility of abnormalities after decompression surgery, all patients of our series with this pattern eventually died.

Pattern 4

Pattern 4 corresponds to brain death, which anatomically corresponds to the destruction of the whole encephalon. All activities of intracranial origin have disappeared while activities of extracranial origin are preserved. Practically, only retinal activities persist in VEPs, SEPs are reduced to activities originating in the brachial plexus (in 100% of cases), cervical cord (±80% of cases), and cervico-medullar junction (nearly 100% of cases). BAEPs are flat in 75% of cases and reduced to a bilateral or unilateral peak I in about 25% of cases (Goldie et al., 1981; Facco et al., 2002; Machado, 2004) (Fig. 3). Contrary to the EEG, EPs allows differentiating brain death from reversible situations associated with a similar clinical pattern (major intoxications by sedative drugs, metabolic disturbances, and deep hypothermia) (Hantson, 2001). Provided that some rare associations should be excluded (associated lesions of both optic nerves and cervico-medullar junction and pre-existing deafness or bilateral section of the 8th nerves), three-modality EPs are in our experience the most reliable tool to confirm the clinical diagnosis of brain death. Their results are more straightforward and less ambiguous than those of the EEG and, contrary to imaging techniques, they can be conducted at the bedside and easily repeated as many times as needed. We routinely use these as a confirmatory test for brain death.

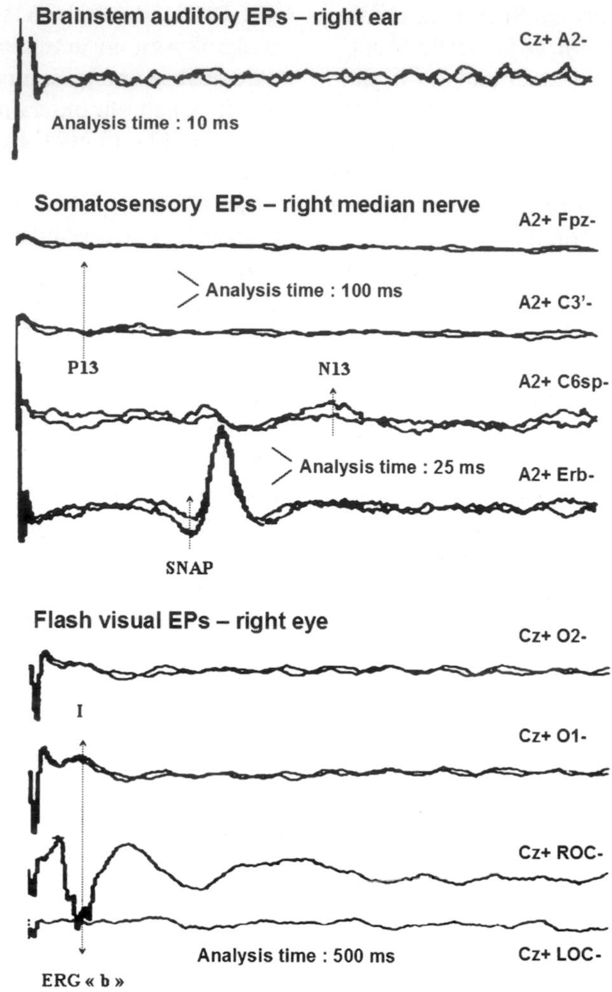

Fig. 3. Pattern 4: brain death. Head trauma. BAEPs (top) are null (right ear shown). SEPs (middle) are restricted to SNAP (sensory nerve action potential), N13 (cervical origin), and an ill-defined P13 (generated at the cervico-medullar junction). All activities of intracranial origin have disappeared. VEPs (bottom) are restricted to peak I of retinal origin; note its synchronization with the "b" peak of the ERG. All electrode labels correspond to the 10–20 International System; R(L)OC: right (left) outer canthus.

Final remarks

First, this classification does not consider cognitive EPs whose presence in our series of comatose patients implies a more than 95% probability of consciousness recovery (Fischer et al., 1999; Guérit et al., 1999b; Kane et al., 2000). Second, even if the diagnosis of superficial brain contusions is much better performed with brain imaging techniques (scanner, MRI), such lesions can interfere with EPs and give rise to lateralization (between left and right occipital recordings in VEPs or between SEPs to right and left median nerve stimulation), in which case the best side is chosen to determine the IGCF. Third, it should be reminded that the diagnosis of epilepsy escapes the EP domain and requires EEG recording. As an online technique, EEG is also the unique tool to assess rapid variations in vigilance level.

Data from the literature

Several papers dealt with EP recording in acute traumatic coma, most of these used BAEPs and/or short-latency SEPs (Lindsay et al., 1981, 1990; Rumpl et al., 1983; Anderson et al., 1984; Cant et al., 1986; Barelli et al., 1991; Cusumano et al., 1992; Beca et al., 1995; Bosch et al., 1995; Facco et al., 1985, 1990, 2002; Karnaze et al., 1985; Ottaviani et al., 1986; Firsching and Frowein, 1990; Houlden et al., 1990; Judson et al., 1990; Gutling et al., 1994; Pohlmann-Eden et al., 1997; Moulton et al., 1998; Sleigh et al., 1999; Claassen and Hansen, 2001; Robinson et al., 2003; see Chatrian et al., 1996; Guérit, 1999a; Facco and Munari, 2000, for recent reviews). In other words, most papers only considered the IBSC. It is difficult to compare these studies, owing to marked variations both in recording techniques and criteria of patient inclusion. This paragraph attempts to present the main trends of the literature and compares these with our own approach (Fig. 4).

Overall, BAEPs and early SEPs are normal in about 50 and 40% of cases, respectively, and about 75% of patients with normal BAEPs and/or SEPs wake up with good recovery or moderate disability. Minor IBSC abnormalities (BAEPs or SEPs) are found in about 25% of cases with a good outcome, 40% of patients in cases of mild BAEP abnormalities, and 50% of cases with mild SEP abnormalities. Normal BAEPs and SEPs correspond to our Pattern 1 and, in our series, minor IBSC abnormalities were most often associated with non-neurological factors (these cases correspond to Pattern 1, too) or reflect minor midbrain dysfunction (normal BAEPs and mildly altered SEPs), in which case these are included in Pattern 2.

Major IBSC abnormalities can consist either a partial loss of peaks (peak V in BAEPs, N20 with preservation of P14 in SEPs) or total peak disappearance of all peaks of intracranial origin. In the literature, partial peak disappearance is described in about 20% of cases for BAEPs and 28% of cases for SEPs. Only 10% of these cases presented a good outcome, irrespective of whether abnormalities predominated on BAEPs or SEPs. When compared with our series, this group with partial loss of peaks is likely to include both severe cases of Pattern 2 (midbrain group) and patients evolving toward brain death (Pattern 3).

Finally, a disappearance of all intracranial components was noted in about 5% of cases. This group corresponds to brain-dead patients (Pattern 4).

Only some rare papers included VEPs (Greenberg et al., 1981; Narayan et al., 1981; Guérit et al., 1993). Even if the two former studies did not demonstrate any clear superiority of multimodality

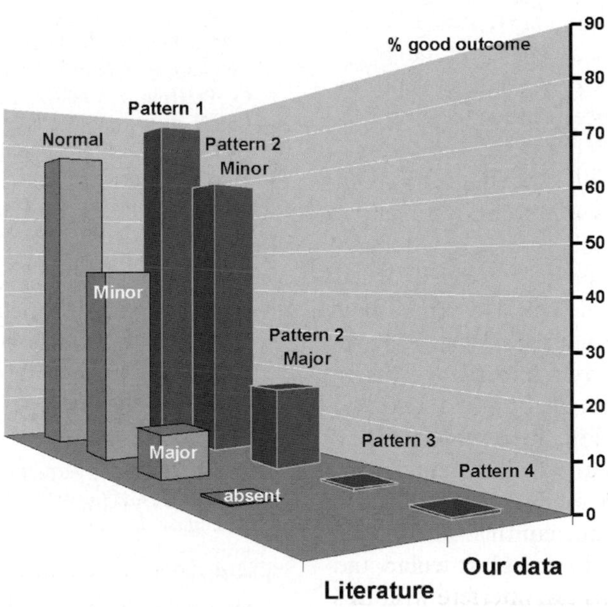

Fig. 4. Prognostic value of EPs in post-traumatic coma. Comparison between our data, based on Patterns 1–4 and data from the literature, essentially based on the degree of IBSC alterations.

approaches including VEPs vs. SEPs alone, we still record VEPs in severe traumatic coma for three reasons. First, mildly altered VEPs (grade 1 or 2) constitute an extremely useful criterion to differentiate Pattern 2 from evolving brain death in patients whose SEPs disclose bilaterally absent SEPs. Second, flat VEPs with well-preserved electroretinogram (ERG) provide interesting redundancy in the confirmation of brain death, especially when a primary brain-stem pathology is suspected (Machado, 1993). Third, VEPs are used, together with SEPs to determine the IGCF, even if we agree that the IGCF is more useful to determine the prognosis in anoxic than in traumatic coma, which is easily explained by the fact that the cerebral cortex constitutes the main target of anoxia.

With regard to cognitive EPs (Kane et al., 1996; Guérit et al., 1999a,b; Fischer et al., 2004), there is widespread agreement that they predict recovery with a more than 90% probability but that no conclusion can be drawn for their absence.

Conclusions

Like the neurological examination, EPs provide functional assessment of the nervous system. However, this assessment is more quantitative and, therefore, amenable to follow-up studies. EPs are more specific than the EEG, even if the latter remains the technique of choice to assess epilepsy or rapid fluctuations in vigilance. Three-modality EPs are early prognostic markers in severe head trauma. One specific contribution of this technique has been the identification of primary midbrain dysfunction, which can remain compatible with a good long-term prognosis despite the severity of neurological alterations at the acute stage. We also emphasized the complementarity between structural (MRI) and functional approaches. In our experience, three-modality EPs also constitute the best bedside technique to confirm brain death; in particular, they allow organ harvesting in brain-dead patients under sedation. Finally, EPs allow a better understanding of the pathophysiological mechanisms that are involved in an individual patient. In particular, the early detection of Pattern 3 could constitute an essential criterion of impending brain death. In this respect, it is highly desirable in the near future to integrate EPs, together with the EEG, in the toolbox of continuous neuromonitoring.

References

Anderson, D.C., Bundlie, S. and Rockswold, G.L. (1984) Multimodality evoked potentials in closed head trauma. Arch. Neurol., 41: 369–374.

Barelli, A., Valente, M.R., Clemente, A., Bozza, P., Proietti, R. and Della, C.F. (1991) Serial multimodality-evoked potentials in severely head-injured patients: diagnostic and prognostic implications. Crit. Care Med., 19: 1374–1381.

Beca, J., Cox, P.N., Taylor, M.J., Bohn, D., Butt, W., Logan, W.J., Rutka, J.T. and Barker, G. (1995) Somatosensory evoked potentials for prediction of outcome in acute severe brain injury. J. Pediatr., 126: 44–49.

Bosch, B.J., Olesti, M.M., Poch Puig, J.M., Rubio, G.E., Nogues, B.P. and Iglesias, B.J. (1995) Predictive value of brain-stem auditory evoked potentials in children with post-traumatic coma produced by diffuse brain injury. Childs Nerv. Syst., 11: 400–405.

Cant, B.R., Hume, A.L., Judson, J.A. and Shaw, N.A. (1986) The assessment of severe head injury by short-latency somatosensory and brain-stem auditory evoked potentials. Electroencephalogr. Clin. Neurophysiol., 65: 188–195.

Chatrian, G.E., Bergamasco, B., Bricolo, A., Frost, J.D. and Prior, P. (1996) IFCN recommended standards for electrophysiologic monitoring in comatose and other unresponsive states. Report of an IFCN committee. Electroencephalogr. Clin. Neurophysiol., 99: 103–122.

Claassen, J. and Hansen, H.C. (2001) Early recovery after closed traumatic head injury: somatosensory evoked potentials and clinical findings. Crit. Care Med., 29: 494–502.

Cusumano, S., Paolin, A., Di Paola, F., Boccaletto, F., Simini, G., Palermo, F. and Carteri, A. (1992) Assessing brain function in post-traumatic coma by means of bit-mapped SEPs, BAEPs, CT, SPET and clinical scores. Prognostic implications. Electroencephalogr. Clin. Neurophysiol., 84: 499–514.

Facco, E., Martini, A., Zuccarello, M., Agnoletto, M. and Giron, G.P. (1985) Is the auditory brain-stem response (ABR) effective in the assessment of post-traumatic coma? Electroencephalogr. Clin. Neurophysiol., 62: 332–337.

Facco, E. and Munari, M. (2000) The role of evoked potentials in severe head injury. Intensive Care Med., 26: 998–1005.

Facco, E., Munari, M., Baratto, F., Dona, B. and Giron, G.P. (1990) Somatosensory evoked potentials in severe head trauma. Electroencephalogr. Clin. Neurophysiol. Suppl., 41: 330–341.

Facco, E., Munari, M., Gallo, F., Volpin, S.M., Behr, A.U., Baratto, F. and Giron, P. (2002) Role of short latency evoked potentials in the diagnosis of brain death. Clin. Neurophysiol., 113: 1855–1866.

Firsching, R. and Frowein, R.A. (1990) Multimodality evoked potentials and early prognosis in comatose patients. Neurosurg. Rev., 13: 141–146.

Fischer, C., Luauté, J., Adeleine, P. and Morlet, D. (2004) Predictive value of sensory and cognitive evoked potentials for awakening from coma. Neurology, 63: 669–673.

Fischer, C., Morlet, D., Bouchet, P., Luaute, J., Jourdan, C. and Salord, F. (1999) Mismatch negativity and late auditory evoked potentials in comatose patients. Clin. Neurophysiol., 110: 1601–1610.

Goldie, W.D., Chiappa, K.H., Young, R.R. and Brooks, E.B. (1981) Brainstem auditory and short-latency evoked potentials in brain death. Neurology, 31: 248–256.

Greenberg, R.P., Newlon, P.G., Hyatt, M.S., Narayan, R.K. and Becker, D.P. (1981) Prognostic implications of early multimodality evoked potentials in severely head-injured patients. A prospective study. J. Neurosurg., 55: 227–236.

Guérit, J.M. (1999) Medical Technology Assessment. EEG and EPs in the Intensive Care Unit. Neurophysiol. Clin., 29: 301–317.

Guérit, J.M. (2001) Les potentiels évoqués chez le patient comateux ou végétatif. In: Guérit J.M. (Ed.), L'évaluation neurophysiologique des comas de la mort encéphalique et des états végétatifs. Solal, Marseille, France.

Guérit, J.M., de Tourtchaninoff, M., Soveges, L. and Mahieu, P. (1993) The prognostic value of three-modality evoked potentiels (TMEPs) in anoxic and traumatic coma. Neurophysiol. Clin., 23: 209–226.

Guérit, J.M., Fischer, C., Facco, E., Tinuper, P., Ronne-Engström, E. and Nuwer, M. (1999a) Standards of Clinical Practice of EEG and EPs in comatose and other unresponsive states. In: Deuschl G. and Eisen A. (Eds.), Recommendations for the Practice of Clinical Neurophysiology (EEG Suppl. 52). Elsevier, Amsterdam, The Netherlands, pp. 117–131.

Guérit, J.M., Verougstraete, D., de Tourtchaninoff, M., Debatisse, D. and Witdoeckt, C. (1999b) ERPs obtained with the auditory oddball paradigm in coma and altered states of consciousness: clinical relationships, prognostic value, and origin of components. Clin. Neurophysiol., 110: 1260–1269.

Gutling, E., Gonser, A., Imhof, H.G. and Landis, T. (1994) Prognostic value of frontal and parietal somatosensory evoked potentials in severe head injury: a long-term follow-up study. Electroencephalogr. Clin. Neurophysiol., 92: 568–570.

Hantson, P. (2001) Mort encéphalique et imprégnation médicamenteuse. In: Guérit J.M. (Ed.), L'évaluation neurophysiologique des comas, de la mort encéphalique et des états végétatifs. Solal, Marseille, France, pp. 321–330.

Houlden, D.A., Li, C., Schwartz, M.L. and Katic, M. (1990) Median nerve somatosensory evoked potentials and the Glasgow Coma Scale as predictors of outcome in comatose patients with head injuries. Neurosurgery, 27: 701–707.

Hume Adams, J. (1984) Head Injury. In: Hume Adams J., Corsellis J.A.N. and Duchen L.W. (Eds.), Greenfield's Neuropathology (4th ed). Edward Arnold, London, UK, pp. 85–124.

Jennett, B. and Bond, M. (1975) Assessment of outcome after severe brain damage. Lancet, 1(7905): 480–484.

Jordan, K.G. (1995) Neurophysiologic monitoring in the neuroscience intensive care unit. Neurologic Clinics N. Am., 13: 579–626.

Judson, J.A., Cant, B.R. and Shaw, N.A. (1990) Early prediction of outcome from cerebral trauma by somatosensory evoked potentials. Crit. Care Med., 18: 363–368.

Kane, N.M., Butler, S.R. and Simpson, T. (2000) Coma outcome prediction using event-related potentials: p3 and mismatch negativity. Audiol. Neuro-otol., 5: 186–191.

Kane, N.M., Curry, S.H., Rowlands, C.A., Manara, A.R., Lewis, T., Moss, T., Cummins, B.H. and Butler, S.R. (1996) Event-related potentials—neurophsyiological tools for predicting emergence and early outcome from traumatic coma. Intensive Care Med., 22: 39–46.

Karnaze, D.S., Weiner, J.M. and Marshall, L.F. (1985) Auditory evoked potentials in coma after closed head injury: a clinical-neurophysiologic coma scale for predicting outcome. Neurology, 35: 1122–1126.

Katayama, Y., Becker, D.P., Tamura, T. and Hovda, D.A. (1990) Massive increases in extracellular potassium and the indiscriminate release of glutamate following concussive brain injury. J. Neurosurg., 73: 889–900.

Lindsay, K., Pasaoglu, A., Hirst, D., Allardyce, G., Kennedy, I. and Teasdale, G. (1990) Somatosensory and auditory brain stem conduction after head injury: a comparison with clinical features in prediction of outcome. Neurosurgery, 26: 278–285.

Lindsay, K.W., Carlin, J., Kennedy, I., Fry, J., McInnes, A. and Teasdale, G.M. (1981) Evoked potentials in severe head injury—analysis and relation to outcome. J. Neurol. Neurosurg. Psychiatry, 44: 796–802.

Machado, C. (1993) Multimodality evoked potentials and electroretinography in a test battery for an early diagnosis of brain death. J. Neurol. Sci., 37: 125–131.

Machado, C. (2004) Evoked potentials in brain death. Clin. Neurophysiol., 115: 238–239.

Moruzzi, G. and Magoun, H.W. (1949) Brain stem reticular formation and activation of the EEG. Electroen. Clin. Neuro., 1: 455–473.

Moulton, R.J., Brown, J.I. and Konasiewicz, S.J. (1998) Monitoring severe head injury: a comparison of EEG and somatosensory evoked potentials. Can. J. Neurol. Sci., 25: S7–S11.

Narayan, R.K., Greenberg, R.P., Miller, J.D., Enas, G.G., Choi, S.S., Kishore, P.R., Selhorst, J.B., Lutz, H.A. and Becker, D.P. (1981) Improved confidence of outcome prediction in severe head injury. A comparative analysis of the clinical examination, multimodality evoked potentials, CT scanning, and intracranial pressure. J. Neurosurg., 54: 751–762.

Nuwer, M.R. (1994) Electroencephalograms and evoked potentials: monitoring cerebral function in the neurosurgical intensive care unit. Neurosurg. Clin. N. Amer., 5: 647–657.

Ottaviani, F., Almadori, G., Calderazzo, A.B., Frenguelli, A. and Paludetti, G. (1986) Auditory brain-stem (ABRs) and middle latency auditory responses (MLRs) in the prognosis of severely head-injured patients. Electroencephalogr. Clin. Neurophysiol., 65: 196–202.

Pohlmann-Eden, B., Dingethal, K., Bender, H.J. and Koelfen, W. (1997) How reliable is the predictive value of SEP (somatosensory evoked potentials) patterns in severe brain damage with special regard to the bilateral loss of cortical responses? Intensive Care Med., 23: 301–308.

Robinson, L.R., Micklesen, P.J., Tirschwell, D.L. and Lew, H.L. (2003) Predictive value of somatosensory evoked potentials for awakening from coma. Crit. Care Med., 31: 960–967.

Rumpl, E., Prugger, M., Gerstenbrand, F., Hackl, J.M. and Pallua, A. (1983) Central somatosensory conduction time and short latency somatosensory evoked potentials in post-traumatic coma. Electroencephalogr. Clin. Neurophysiol., 56: 583–596.

Shaw, N.A. (2002) The neurophysiology of concussion. Prog. Neurobiol., 67: 281–344.

Sleigh, J.W., Havill, J.H., Frith, R., Kersel, D., Marsh, N. and Ulyatt, D. (1999) Somatosensory evoked potentials in severe traumatic brain injury: a blinded study. J. Neurosurg., 91: 577–580.

Speckman, E.-J. and Elger, C.E. (1982) Introduction to the neurophysiological basis of the EEG and DC potentials. In: Niedermeyer E. and Lopes da Silva F. (Eds.), Electroencephalography Basin Principles clinical applications and related fields (2nd ed.). Urban & Schwarzenberg, Baltimore, USA., pp. 1–13.

S. Laureys (Ed.)
Progress in Brain Research, Vol. 150
ISSN 0079-6123

CHAPTER 30

Event-related potential measures of consciousness: two equations with three unknowns

Boris Kotchoubey*

Institute of Medical Psychology and Behavioral Neurobiology, Eberhard-Karls-University of Tübingen, Gartenstr. 29, 72074 Tübingen, Germany

Abstract: This is a brief review of event-related brain potentials (ERPs) as indices of cortical information processing in conditions in which conscious perception of stimuli is supposed to be absent: sleep, coma, vegetative state, general anesthesia, neglect as well as presentation of subliminal or masked stimuli. Exogenous ERP components such as N1 and P2 are much more likely to remain in all these conditions than endogenous components. Further, all varieties of the late posterior positive ERP waves (e.g., P3b, P600, late positive complex) are most difficult to be elicited in such conditions, indicating that the cortical activity underlying the late posterior positivity may have a particularly close relationship to brain mechanisms of conscious perception. Contrary to what might be expected, reliable ERP effects indicating complex analysis of semantic stimulus features (i.e., meaning) can be recorded without conscious awareness, generally, as easy as (in some conditions, even easier than) ERP components related to rather simple physical stimulus features. It should be emphasized, however, that we never should overestimate our confidence about the degree of subjects' unawareness. Particularly in the conditions in which no behavioral response can be obtained (e.g., sleep, coma, anesthesia), residual conscious processing, at least in some subjects and on some trials, cannot be ruled out.

Event-related brain potentials (ERPs) are EEG oscillations time- and phase-locked to particular events, such as stimuli or subject's movements. Owing to their perfect time resolution they are supposed to manifest, in true time, processing of stimulus information and movement preparation in the cortex (Donchin and Coles, 1988a; Meyer et al., 1988). ERPs were proposed as an index of cognitive processes as early as in the late 1960s, when the word "consciousness" was taboo and anyone who dared to mention it was ostracized. This prohibition explains the enthusiasm that emerged when the term "controlled processing" (Schneider and Shiffrin, 1977) was introduced to substitute "conscious processes", although the subject of this verb (controlled by whom?) still remains unknown (Neumann, 1989). Soon after this, hypotheses were formulated relating some ERP components (e.g., N2b or P3b, see below) to those controlled processes (Näätänen and Picton, 1986; Rösler et al., 1986).

Now the situation has radically changed, and we are free to look for neural correlates of consciousness, although this does not mean that we have a slightly better idea of what it might be than our fathers had. Therefore, in this selective review, I shall pursue a negative approach, in that ERP data related to unconscious states, such as sleep, coma, vegetative state, and anesthesia will be discussed as well as the ERP effects of presentation of subliminal stimuli. The classical (and less studied) "altered states of consciousness" such as hypnosis,

*Corresponding author. Tel.: +49 7071 2974221;
Fax: +49 7071 295956;
E-mail: boris.kotchoubey@uni-tuebingen.de

DOI: 10.1016/S0079-6123(05)50030-X

meditation, trance, etc. are not included here; for the most recent review on this topic, see Vaitl et al. (2005). We also avoid, primarily for logical reasons, to count attentional distraction and errors of memory to "disorders of consciousness", thus effects of attention manipulations remain outside the scope of this paper.

Examples of ERP phenomena are shown in Fig. 1. Several classifications of these phenomena can be relevant to the problem of consciousness. First, ERP components such as N1 and P2 are referred to as "exogenous" (i.e., depending mainly on stimulus qualities), while others, like "mismatch negativity" (MMN), P300 or P3, and N400, are regarded "endogenous" (i.e., depending mainly on the task and subject's state). Although this distinction is relative rather than absolute (since, for instance, N1 can vary as a function of attention, and P3 as a function of stimulus intensity), it may be useful. Because of their stimulus-dependence, some exogenous ERP components (like N1) are sometimes described as components of auditory or visual evoked potentials; this does not mean, however, that they are not affected by endogenous factors (e.g., there is a vast literature about attention effects on the occipital P1 and N1

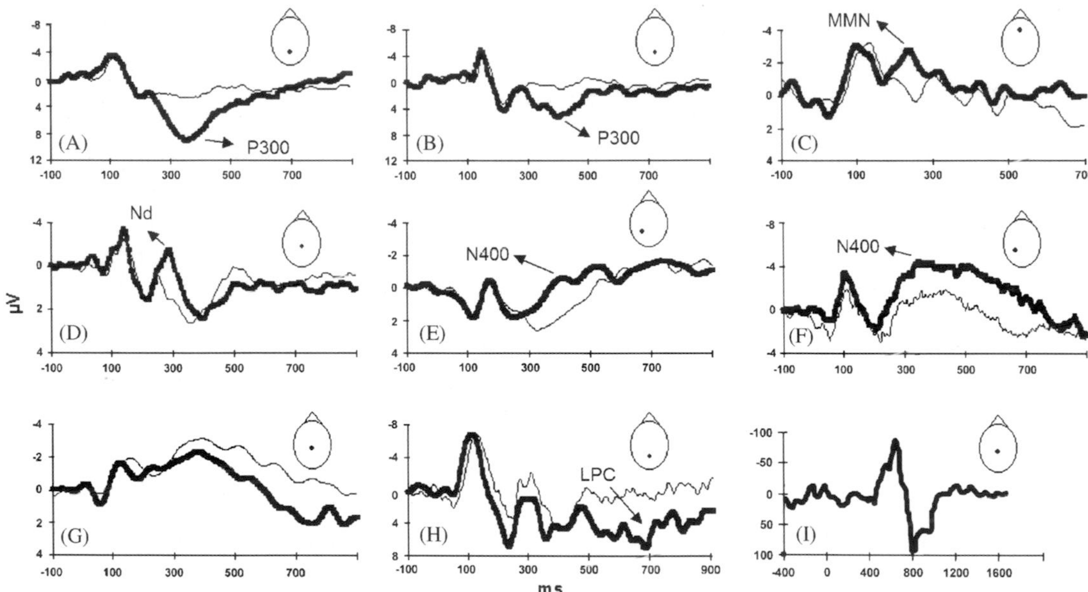

Fig. 1. Examples of event-related potentials (ERPs) phenomena discussed in the text. For simplicity only one electrode is presented. In all the graphs negativity is plotted upward. The approximate recording site is shown by a point on the schematic representation of a scalp viewed from above, the triangle represents the nose. **(A)** A typical ERP in an auditory "oddball task" in which two attended stimuli are presented with unequal probabilities. Thin line: averaged responses to frequent stimuli. Bold line: averaged responses to rare stimuli, which have to be counted, resulting in a positive potential with a latency around 300 ms (hence its name P300 or P3). **(B)** A smaller P3 is also obtained in the same condition without an active task requirement. Note that **A** and **B** have the same amplitude scale. **(C)** An auditory oddball with unattended stimuli (subjects simultaneously performed a visual task) results in a "mismatch negativity" (MMN) to rare stimuli. **(D)** ERPs in a dichotic listening task. A comparison between the waveforms to stimuli presented in the relevant (bold) versus irrelevant channel (thin) reveals an additional negativity referred to as "negative difference" (Nd). **(E)** ERPs to final words in sentences. Thin line: highly expected end words. Bold line: unexpected, meaningless words (semantic mismatch), resulting in a negativity around 400 ms, called N400. **(F)** Semantic priming effect. ERPs to a second word (target) in a word pair, preceded by a semantically related "prime" word (thin) or a semantically unrelated "prime" (bold). In both **E** and **F**, violation of semantic context results in an N400 potential, which is more dispersed in the priming condition. **(G)** Word repetition effect. ERPs to new (thin) and repeated (bold) words. The difference appears to entail two components: an N400 to new words and a late positivity to repeated words. **(H)** A late posterior positive complex (called P600), in this case in an oddball task with grammatical categories. Bold line: nouns (rare stimuli). Thin line: verbs and adverbs (frequent stimuli). **(I)** Averaged K-complexes in stage 2 sleep. These K-complexes have a larger amplitude to rare stimuli than to frequent ones and show habituation to stimulus repetition.

of visual evoked potentials; see, e.g., Martinez et al., 1999). Further, some ERP components (both exo- and endogenous) are attention-dependent, whereas the MMN is relatively independent of attention (see Näätänen et al., 1993, for details). From the point of view of sequential processing of stimuli, one can simply classify ERP components according to their latency, assuming that the earlier of them may manifest simpler, and the later more complex processing operations. The sequential model may give rise to another distinction, namely, between stimulus- and response-related components, in which, regardless of the position on the time axis, those response-related are, in a sense, "later" than components reflecting the analysis of stimulus (for such processing models, see Leuthold et al., 1996; Meyer et al., 1988). Finally, endogenous ERP components can be classified according to whether they reflect the processing of physical stimulus features (e.g., pitch, intensity, duration), or the analysis of stimulus meaning (e.g., semantics).

Non-REM sleep

ERPs in sleep have been recently analyzed in three comprehensive reviews (Atienza et al., 2001a, 2002; Peigneux et al., 2001), therefore we can restrict ourselves to a brief overview including recent data. The most prominent electrical cortical response to stimulation in non-REM sleep are K-complexes whose main component is a negativity with a latency of about 500–600 ms (often called N550) and a huge amplitude up to 200 μV (Crowley et al., 2004), therefore they are easily seen in raw EEG without the usual averaging procedure (Colrain et al., 1999; Cote et al., 1999). K-complexes are ERPs (Fig. 1I), although they can also appear spontaneously (Amzica and Steriade, 2002), perhaps elicited by internal stimuli (Niiyama et al., 1996). Besides the largest N550, K-complexes may contain other positive and negative components such as P220, N350, P450, and P900 (Perrin et al., 2000). Importantly, K-complexes in stage 2 sleep habituate with stimulus repetition (Bastien and Campbell, 1994) and have a larger amplitude to rare stimuli than to frequent ones (Bastuji et al., 1995; Pratt et al., 1999). There is a controversy of whether these processes really indicate simple learning and differentiation between stimuli of different probability, which are the most fundamental characteristics of adaptive behavior. An alternative, simple explanation makes use of the long refractory time of K-complexes; from this point of view the refractoriness of the units responding to frequent stimuli is responsible for the larger responses to rare stimuli. However, this account meets difficulties to explain variations with very complex stimuli, such as a subject's name (see below).

Besides K-complexes, the exogenous ERP component N1 is recorded in sleep, with its amplitude being well correlated with the level of arousal, i.e., it is largest in stage 1 sleep, smallest in SWS (slow wave sleep, which encompasses stages 3 and 4), and smaller in stage 2 sleep during spindles than in their absence.

The MMN is typically elicited by acoustical deviations in frequency, intensity, duration, timbre, or more complex stimulus patterns (for review, see Schröger, 1997; and Näätänen and Winkler, 1999). Like N1, the MMN is largely generated in the supratemporal plane but may also be contributed to by frontal sources. Eleven consecutive sleep studies from Paavilainen et al. (1987) to Sabri et al. (2003) failed to find a significant MMN. Sallinen et al. (1994) obtained an MMN in stage 2 sleep only when a stimulus also elicited a K-complex, but could not replicate this finding later (Sallinen et al., 1997). Moreover, many authors noted a substantial decrease in or even disappearance of the MMN during drowsiness or in stage 1 sleep, i.e., even before sleep onset (Sallinen and Lyytinen, 1997; Campbell and Colrain, 2002; Sabri et al., 2003). Only Sabri and Campbell (2002) showed recently a distinct MMN in SWS (its significance, however, was not reported). These authors used stimuli of high intensity (80 dB), very large deviance (2000 Hz versus 1000 Hz), very low probability of deviants (0.033), very high presentation rate (1/150 ms), and, additionally, a high pass filter of 3 Hz to suppress the background delta waves. This finding stresses the importance of methodological details such as filtering, presentation rate (Atienza et al., 2002), and modality (Loewy et al., 2000). However, band pass cannot explain the lack of MMN in stage 2 sleep in which no delta activity has to be removed.

It is interesting to note that the MMN, which is so hard to elicit in sleeping adults, is unfailingly obtained in infants in all states including quiet sleep (analogous to SWS in adults) (Cheour et al., 2000; 2002). Given the considerable differences between infant and adult sleep (not to mention the differences between infant and adult consciousness), this apparent paradox is presently left without explanation.

In wakefulness, processing of deviant stimuli, perceived as "targets", results in a large parietal ERP component called P3 (or P300). Although P3 amplitude is greatly increased when subjects perform a task such as counting deviants or pressing a button (Fig. 1A), a significant P3 is also found without task requirement (Polich, 1989; Lang et al., 1997; Lang and Kotchoubey, 2002; Fig. 1B), thus it might be recorded in sleep, too. However, the most consistent finding in non-REM sleep is the lack of P3 (Bastuji et al., 1995; Winter et al., 1995; Voss and Harsh, 1998; Cote and Campbell, 1999; Hull and Harsh, 2001; Afifi et al., 2003). During drowsiness and stage 1 sleep, P3 amplitude radically decreases with lowering arousal level, so that when a subject cannot produce a behavioral response to the target any longer, virtually no P3 can be obtained (Sallinen and Lyytinen, 1997; Cote and Campbell, 1999; Campbell and Colrain, 2002). There exist instead other positive ERP components, possibly related to K-complexes, which inversely correlate with stimulus probability like the wake P3. Other properties of these waves are, however, different from the P3. They do not possess the typical posterior topography and, even in stage 1 sleep, when subjects still produce behavioral responses, they can have larger amplitudes to non-targets than targets (Hull and Harsh, 2001). Atienza et al. (2001a) asked whether the earliest of these components (P220) is analogous to an early frontal subcomponent of P3, called P3a (Polich and Comerchero, 2003). In contrast to the "true" parietal P3 (or P3b), the P3a is relatively attention-independent and may serve as a component of the orienting response. Several features of sleep-P220 (its short latency, high amplitude to non-targets, the presence in response to frequent stimuli in the absence of preceding exogenous components) make the P3a hypothesis implausible. Hence, although there are interesting ERP components in non-REM sleep (probably components of K-complexes), none of them behave like a wake-MMN or P3 (Cottone et al., 2004).

In contrast, ERPs to semantic stimuli in stage 2 sleep may be similar to those in wakefulness. A person's own name elicits not only much more K-complexes than other stimuli (Voss and Harsh, 1998; Perrin et al., 2000), but also a small but significant parietal P3b like in wakefulness (Perrin et al., 1999). Also semantically inappropriate stimuli (semantic mismatch) may result in a specific ERP component N400 (Brualla et al., 1998; see Fig. 1E). The ability to respond specifically to significant stimuli appears, therefore, to remain in stage 2 sleep. However, more research is necessary to control for possible awakenings during semantic sleep experiments, because significant stimuli are more likely to shortly awake subjects than simple, uninteresting stimuli (Voss and Harsh, 1998).

In summary, ERP morphology and topography in non-REM sleep is rather different from that in wakefulness and closely related to sleep-specific cortical responses, i.e., K-complexes. Such typical ERP components as the MMN and P3 are difficult to obtain. Nevertheless, in stage 2 sleep the cortex is able to respond according to stimulus probability and, possibly, also according to semantic appropriateness of words. In contrast to stage 2 sleep, cortical reactivity in SWS remains largely unknown.

REM sleep

Despite the misleading use of the term "sleep", REM sleep is a completely different state whose neurophysiology is more like wakefulness than non-REM sleep (Hobson and Pace-Schott, 2002; Pace-Schott and Hobson, 2002). This holds true also for ERPs findings. The N1 component is not only present, but also behaves as a component of the orienting response, being larger to the first stimulus in a stimulus run than the following stimuli. Atienza et al. (2001b) varied tone pitch within runs so that each run contained only one deviant stimulus on positions 1, 2, 4, or 6. Because only the deviant on position 1 yielded an enhanced N1, and because the inter-deviant intervals were long and hardly varied, the authors concluded that the N1 changes in REM sleep were really due to orienting rather than refractoriness. The same result was obtained in wakefulness.

A significant MMN during REM sleep was found in five studies (most using frequency deviants), but not found in four (using frequency and intensity deviants) (Atienza et al., 2002). Atienza and Cantero (2001) used a very complex tonal sequence with a small deviance that subjects usually do not perceive at all. However, after they had intentionally learnt to perceive this deviance, it elicited an MMN in wakefulness and, importantly, also in the REM sleep 2 days later.

At least four studies (Niiyama et al., 1994; Bastuji et al., 1995; Cote and Campbell, 1999; Cote et al., 2001) reported a significant P3 to rare stimuli in REM sleep, although its magnitude was reduced as compared to wakefulness, and its elicitation required stimuli of higher intensity. Negative P3 findings in REM have also been reported (Loewy et al., 1996; Nordby et al., 1996).

Both the well-expressed parietal P3 to a person's own name and the N400 to semantic mismatch can be recorded in REM sleep (Brualla et al., 1998; Bastuji et al., 2002). Perrin et al. (2002) presented their subjects meaningful words that were expected or unexpected within a semantic context, as well as meaningless non-words. In wakefulness, unexpected words elicited an N400, which was yet larger in response to non-words. This N400 was also elicited in stage 2 sleep, but the difference between unexpected words and non-words disappeared. Further, in REM sleep, non-words (in contrast to unexpected words) produced no N400 any longer, as if non-words became "expected" in this state.

In summary, unlike in non-REM sleep, ERP effects in REM sleep and wakefulness are remarkably similar (Fig. 2), but in REM sleep they are usually attenuated and delayed, and require stronger and more salient stimuli than during wakefulness. Also some subtle cognitive effects may be different.

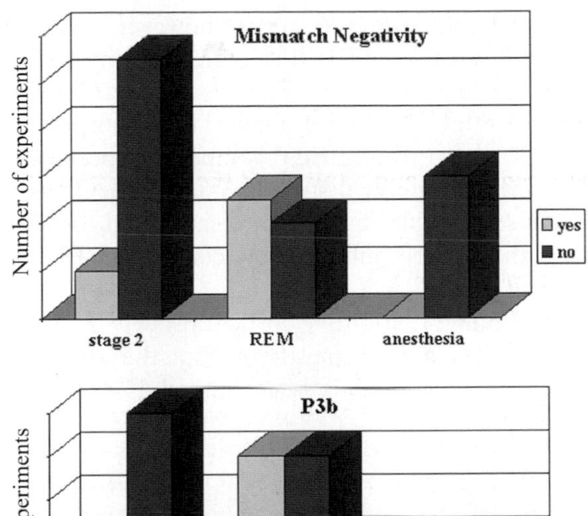

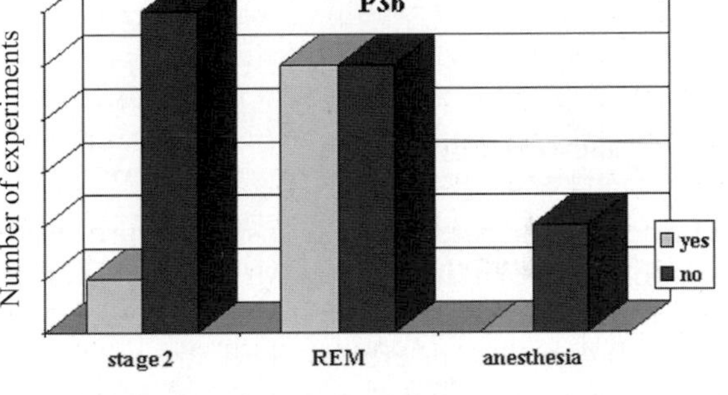

Fig. 2. Number of experiments in which a "cognitive" ERP component (top: mismatch negativity or MMN, bottom: P3b) was found (light-gray) or not found (black) in stage 2 sleep, REM sleep, and anesthesia. Since the experimental conditions largely varied between the experiments, the data should not be taken as exact numbers, but rather as tendencies. Specifically, one can see that both ERPs components are consistently absent in anesthesia and occur more frequently in REM than in stage 2 non-REM sleep (as regards the MMN, this difference approaches significance: $p = 0.057$, Fisher's exact test).

Anesthesia

ERP studies in anesthesia are scarcer than in sleep, which should not be surprising given the technical and ethical problems of such experiments. Studies using sufentanil and propofol anesthesia during cardiac surgery converge in that the parietal P3b component is completely abolished, while exogenous components are preserved but significantly attenuated and delayed as compared with wakefulness (Plourde and Boylan, 1991; Plourde and Picton, 1991; Van Hooff et al., 1995, 1997). Also the MMN was not found in those studies. Simpson et al. (2002) reported that during sedation, the MMN had disappeared even before patients lost the ability to talk with anesthesiologists, i.e., while they were still conscious. Yppärilä et al. (2002) described a subgroup of anesthetized patients having a P3a-like wave, but this wave did not respond selectively to novel stimuli like the wake-P3a. Plourde et al. (1993), however, observed a significant frontal P3a during the post-intubation period, while the parietal P3b was lacking. Finally, Heinke et al. (2004) examined, in addition to the MMN, also a frontal ERP component specifically responsive to musical stimuli. When patients were unconscious, this component disappeared together with the MMN, but the early component P1 was present.

In summary, auditory cortical processing is not entirely lost during anesthesia, but the ability to differentiate stimuli is very limited if not completely lacking.

Coma and vegetative state

Owing to their potential diagnostic and prognostic significance, ERP in patients with severe disorders of consciousness are evaluated in individual patients rather than in groups. This stresses the importance of the criteria of individual assessment. No generally accepted criterion exists as of when an ERP component should be assumed as valid in an individual case. Nevertheless, ERP components are consistently found in many coma patients: N1 in 50 to 70% (Gott et al., 1991; Fischer et al., 1999, 2000; Guerit et al., 1999), MMN in 30 to 60% (Kane et al., 1993, 1996, 1998; Fischer et al., 1999, where the lowest estimate was obtained using probably a too strict criterion), P3 in 25 to 50% (Gott et al., 1991; Mutschler et al., 1996; Signorino et al., 1997; Kane et al., 2000). Additionally, Guerit et al. (1999) described in several coma patients a late negative–positive ERP complex to rare stimuli, which was different from wake N2 and P3.

Late ERP components have been correlated with both the severity of coma (e.g., Keren et al., 1998; Mazzini et al., 2001) and its outcome (e.g., Fischer et al., 1999; Guerit et al., 1999). Importantly, whereas early sensory evoked potentials are better negative predictors (i.e., their absence indicates poor outcome), the MMN and particularly P3 tend to be better positive predictors (i.e., their presence heralds awakening from coma). In a recent study, the best ERP predictor proved to be simply the presence of any discernible component (Lew et al., 2003).

Coma most often is an acute suppression of cortical functions caused by severe dysfunction of the brain stem reticular formation whose activity is necessary for arousal. The cortex itself can morphologically be completely intact in coma. On the contrary, persistent vegetative state (PVS) is a chronic condition, which is often associated with massive cortical lesions such as diffuse atrophy of the gray matter due to anoxia, or diffuse axonal damage due to severe head injury.

Reuter et al. (1989) were probably the first who observed a P3-like wave to rare stimuli in three PVS patients, a finding replicated by Rappaport et al. (1991), who examined eight severely disabled patients, five of whom were in PVS. Several later reports on P3 in PVS were brief letters, and the reliability of those data cannot be evaluated. Jones et al. (2000) recorded MMN to a very complex change in auditory stimulation in 2 of 12 patients who probably were in PVS.

Two larger studies have been performed to date. Witzke and Schönle (1996) presented data of 43 patients qualified as "surely vegetative" and 23 as "possibly vegetative". ERP components were evaluated on the basis of mere visual inspection of the waveforms. We recently examined 50 PVS patients and applied, for the first time, strict statistical criteria based on the analysis of single trials (Kotchoubey et al., 2001, 2003, in press). Of course, there is a factor of deliberate choice in any criterion; for instance, if we shift the alpha-level from 0.1 to 0.05 and further to 0.025, fewer patients' responses attain significance. Nevertheless, this quantitative estimation of response frequencies gives us a more reliable basis for discussion of cortical responsiveness in PVS. Our sample revealed a clear heterogeneity with respect to the patients' background EEG. In patients with very severe disturbance in the EEG pattern (e.g., diffuse delta waves), only the N1 component occurred slightly above chance (i.e., in about 30%), other ERP effects were practically lacking. A completely different situation was found in patients whose EEG was only moderately disturbed (mostly, with a prevailing theta rhythm of 5–7 Hz). As shown in Table 1, which presents both our' and Witzke and Schönle's (1996) findings, ERP component frequencies were close to those obtained in coma

Table 1. Occurrence of ERP components in the vegetative state

	Witzke and Schönle (1996) "surely vegetative" patients	Witzke and Schönle (1996) "questionably vegetative" patients	Kotchoubey et al. (2003; in press) all vegetative patients	Kotchoubey et al. vegetative patients with prevailing theta-rhythm[a]
At least one ERP component	68	87	82	100
N1–P2 complex "recognizable"	74	96	—	—
N1–P2 complex "well expressed"	40	78	72	89
MMN	21	26	51	65
P3	12	30	27	36
N400	5	14	21	25

Notes: Results are given in percentages of total number of patients and are based on $\alpha = 0.05$ criterion for the studies by Kotchoubey et al.
[a]Patients with diagnosis of vegetative state but with prevailing theta-rhythm in resting EEG, indicating partially preserved thalamo-cortical loops.

studies, notwithstanding the different pathophysiological mechanisms of the two states. Like in coma, we found the presence of MMN being significantly associated with regaining consciousness in a 6-month follow-up. It should be stressed that due to the conservatism of statistical assessment, our estimates should be regarded as lower limits that possibly underestimate the real frequencies in the population.

Two additional findings should be mentioned. First, in coma (Guerit et al., 1999; Lew et al., 2003) and non-REM sleep (Atienza et al., 2001a; Hull and Harsh, 2001), ERP waveforms have been described that clearly differed from ERP in wakefulness but, nevertheless, consistently vary as a function of stimuli. Likewise, we recorded significant but abnormal responses in the oddball task in about 12 to 15% of PVS patients (Fig. 3). Second, both MMN and P3 were obtained significantly more often, and of a significantly larger amplitude, in experiments with complex musical tones than in the identical experiments with simple sine tones (Fig. 4). The positive role of stimulus complexity has recently been demonstrated in healthy subjects as well (see Tervaniemi et al., 2000, for MMN; and Lew et al., 1999, for P3). Besides the better method of detection of ERP components, the relatively high occurrence of ERP components in our data (Table 1) can largely be accounted for by using more complex stimuli.

As Guerit et al. (1999) pointed out, in patients with very severe brain damage it is sometimes very

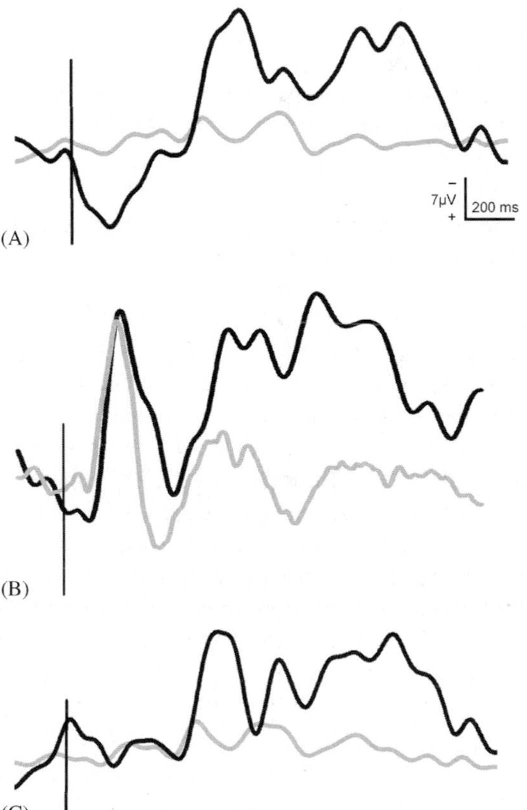

Fig. 3. Examples of significant but abnormal cortical responses in an oddball task in patients in a vegetative state following head injury (patient A), anoxia (B), and stroke (C). As compared with frequent stimuli (gray), rare stimuli (black) elicit a large slow negativity instead of the expected P3. Bars indicate stimulus onset.

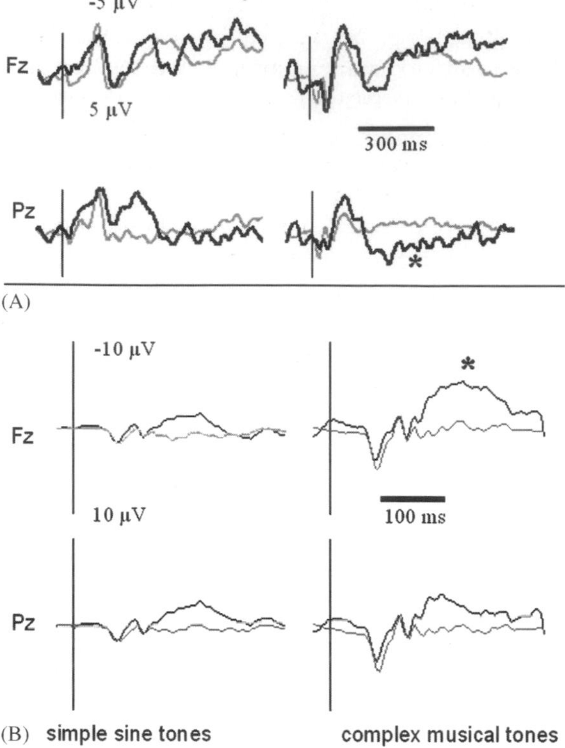

Fig. 4. Effects of stimulus complexity. (A) An oddball task with interstimulus intervals (ISI) of 900 ms, averaged across 32 severely brain-damaged patients. A P3b (marked by an asterisk) component is recorded to complex chords, but not to simple sinusoidal tones. Bars indicate stimulus onset. (B) An oddball task with ISI of 330 ms, averaged across eight patients with very severe brain damage (vegetative or minimally conscious patients) but with normal ERP waveforms. A mismatch negativity (MMN) to rare stimuli is recorded using both simple sinusoidal tones (left panel) or complex chords (right panel), but its amplitude is significantly larger to complex musical tones (asterisk). Modified from Kotchoubey et al. (2001, 2003). With permission.

difficult to decide whether an observed endogenous ERP components is really, for instance, N2 or P3. Particularly difficult may be the distinction between the frontal P3a and the posterior P3b subcomponents, whose functional meaning is different. Firstly, an atypical ERP topography is a rule rather than an exception in such patients, thus the topographic criterion does no longer work. Secondly, the procedural criterion (i.e., the lack of an active task requirement) seems not specific because, as stated above, the typical posterior P3b can also be elicited in passive conditions. Hence, neither in our own data nor in those of the literature can the possibility be ruled out that a portion of P3 findings in coma and PVS were, indeed, P3a.

In summary, nearly all ERP components can be found in a subgroup of patients in coma and PVS. Thus attention should be paid to a better classification of these patients, e.g., on the basis of the background EEG, etiology, or other factors. As regards P3, future studies should find out measures to disentangle P3a from P3b effects. Semantic processing is also possible in PVS, but nobody has sought for semantic ERP effects in coma.

Neglect

Neglect is a selective partial unawareness of events and objects; most frequently, patients with a right parietal damage do not perceive objects presented on the left side. Neuropsychological data demonstrate that unperceived stimuli nevertheless undergo extensive information processing, e.g., they can semantically prime subsequent perceived stimuli on the right side. Regarding ERPs, one might hence expect that early stages of processing of the neglected stimuli and relatively attention-independent components such as the MMN would remain intact; but attention-dependent components such as P3 would be impaired. Review papers indicate that many earlier ERP studies were, generally, in line with this hypothesis (Verleger, 2001; Deouell, 2002). Verleger et al. (1996) found a decrease of the Nd component, which is known to be strongly attention-dependent (see Fig. 1), in patients with right parietal damage to left-side versus right-side stimuli (as well as in comparison with controls), but the attention-independent P3a was present and even enhanced. However, more recent investigations, in which ERPs were averaged separately for detected and undetected left-side stimuli, show that even middle-latency components such as P1 and N1 to undetected stimuli on the left side may be greatly compromised or even fully disappear (e.g., Marzi et al., 2000). Particularly, latencies of ERP components related to spatial location of stimuli on the neglected side are delayed (for review see Deouell, 2002). Also the attention-unrelated MMN evoked by changes in location of auditory stimuli on the left side was absent in neglect patients (Deouell et al., 2000). Earlier, Berti and Rizzolatti (1992) suggested on the basis of their psychological studies of neglect patients that spatial encoding is more crucial than semantic encoding for conscious awareness of a stimulus. Unfortunately, I do not know ERP studies to spatial stimulus features in other conditions of impaired consciousness (e.g., sleep, coma, anesthesia), which might support or disprove this hypothesis.

In neglect, unlike all states described in the previous sections, a behavioral response can be obtained indicating whether the just presented stimulus has been consciously perceived or not. ERPs to neglected stimuli are characterized not only by suppression or latency delay of late attention-dependent components such as Nd and P3b. Also the MMN and even earlier components can be strongly attenuated.

Subthreshold, weak, and brief stimuli in normal subjects

Not surprising, the idea to use ERPs for studying processing of subthreshold stimulation came from psychiatry, a field that was always interested in effects of stimuli that cannot be consciously perceived. A group of psychoanalysts first described visual ERPs (mostly entailing an N1–P2 complex) to subliminally presented words (Shevrin and Fritzler, 1968), and later reported that their amplitude negatively correlated with repressiveness (Shevrin et al., 1970, 1992) and positively with using related words in free associations (Shevrin et al., 1971). Subliminal presentation of highly significant, disease-related words was found to elicit a P3 in psychiatric patients (Kostandov and Arzumanov, 1977). In this vein, ERP effects of subliminal emotional stimuli were studied in conditioning experiments in which conditional stimuli (CS: faces or words, visually presented with a very brief exposition time such as 2 ms) were combined with aversive unconditional stimuli (UCS) such as an electric shock or loud tone (Wong et al., 1994, 1997, 2004). During the extinction phase, when CS were explicitly presented without UCS, those previously conditioned (i.e., CS+) produced a larger ERP positivity (starting already with N1, overlapping P2 and P3, and lasting several hundreds ms) than stimuli not associated with UCS (i.e., CS−). A similar slow positivity was obtained to unpleasant, as compared with pleasant, adjectives, presented both in supraliminal (40 ms exposition) and subliminal (1 ms) mode (Bernat et al., 2001a)—a finding that might be attributed to arousing effects of unpleasant stimuli rather than to their emotional valence (Cuthbert et al., 1996). Subliminal emotional stimuli (fearful faces) elicited a larger N2 component than similar supraliminal, recognized stimuli; however, later ERP components to fearful faces, including N400 and a late positivity, were substantially weakened when the stimuli were not recognized (Liddell et al., 2004; Williams et al., 2004).

More difficult is a proof of the existence of late ERP components to subliminally presented neutral stimuli. Methodological shortcomings of many studies to "unconscious P3" include use of subjective reports instead of objective measures of stimulus detection (for discussion, see Shevrin, 2001), lax methods of statistical evaluation of the very noisy and unstable ERP waveforms to weak stimuli, and insufficient attention to possible artifacts. The most rigorous study (Bernat et al., 2001b) demonstrated a significant parietal P3 to rare (20%) visual stimuli presented for 1 ms with the intensity of 5 ft/Lambert. Its amplitude was about eight times smaller than in the identical experiment with clearly visible stimuli, but the waveforms in the subliminal and supraliminal experiments were highly correlated, indicating similarity of underlying mechanisms. Brázdil et al. (2002) recorded ERPs to letters presented for 10 ms in 13 epilepsy patients using electrodes implanted in their temporal and frontal lobes and obtained a "classical" P3 (whose significance, however, was not supported by a single-trial analysis) in 7 of the 13 patients. This P3-related activity was found in the hippocampus and left temporal cortex, mesiofrontal, and orbitofrontal structures, but, in contrast to the P3 to supraliminal stimuli, neither in the right temporal lobe nor in the dorsolateral prefrontal cortex.

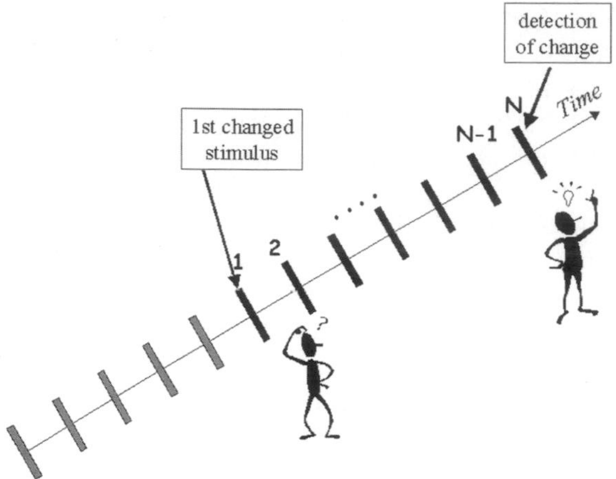

Fig. 5. A schema of experiments with minimal stimulus change. Unchanged stimuli (presenting random motion of dots) are symbolized by gray bars, changed stimuli (in which the motion is not completely random), by black bars. As a rule, subjects detect this change only after several presentations. The stimulus, on which detection occurs, is referred to as N, the immediately preceding stimulus as $N-1$, etc. For more details of these experiments, see Niedeggen et al. (2001, 2004).

A modification of the subliminal approach was proposed by Niedeggen et al. (2001, 2004), who presented, instead of minimal stimuli, minimal changes (which should be detected) in stimulus sequences. A very large late (about 600 ms post-stimulus) parietal positivity distinguished detected from undetected changes. Most interesting, a much smaller but nonetheless significant positivity was also found to undetected stimuli, which immediately preceded the conscious detection, as if the brain detected the change a bit earlier than consciousness did (Fig. 5).

In summary, the components N1 and P2, as well as late ERP components to emotional stimuli, can be elicited by stimuli of extremely low intensity and brief exposure. P3 and P3-like positivities can also be obtained to subliminal stimuli, but they are so strongly attenuated that their recording demands particular methodological precision.

Masked stimuli in normal subjects

Besides low intensity and brief exposure, a target stimulus can be made unrecognized when another (as a rule, a more salient) stimulus is presented simultaneously, shortly before (forward masking) or shortly after it (backward masking). Early studies of visual masking demonstrated a significant amplitude decrease (e.g., Vaughan and Silverstein, 1968; Samoilovich and Trush, 1979; Andreassi, 1984) of exogenous components N1 and P2. A backward visual mask that leaves detectable (though attenuated) P1 and N1 can completely eliminate late components such as P3 and N400 (Dehaene et al., 2001; Liddell et al., 2004). Similar data were obtained for the auditory modality (e.g., Kevanishvili and Lagidze, 1987). Simultaneous masking noise presented in the same ear as the stimulus (Martin et al., 1999), or in the different ear (Salo et al., 1995), suppressed the MMN as well as the N2 and P3 components (Whiting et al., 1998) to a greater extent than N1. The same is true for backward masking (Winkler and Näätänen, 1992).

We stated above that the N400 component to a target word (e.g., "dog") is attenuated when it is presented in a semantically related context (e.g., preceded by the word "cat"). This effect, designated as "N400 priming", exists also when the prime ("cat") is not consciously perceived because of backward (Kiefer and Spitzer, 2000) or forward (Kiefer, 2002) masking. However, the effect of masked primes strongly decreases with increasing interval between the prime and the target (Deacon et al., 2000) even if the behavioral priming effect (i.e., faster responses to related words) remains (Brown and Hagoort, 1993). To explore this tendency, Ruz et al. (2003) used a 128-channel cap and a rather long prime-target interval of 1.5 s. In line with previous studies, near-threshold primes had no effect in this condition. Surprisingly, however, extremely subthreshold primes resulted in significant differences between ERP to semantically related versus unrelated targets, but these differences, appearing at various locations between 200 and 500 ms post-stimulus, were not the usual N400 effect. N400 priming disappears when an irrelevant word is inserted between prime and target (e.g., "cat-table-dog"), regardless of whether this intervening word is recognized or suppressed by masking (Deacon et al., 2004).

When the target is not a semantically related word but just a repetition of the same word, a large late positivity comes out, which results from the N400 attenuation depicted above on the one hand, and from an additional P3-like parietal wave on the other hand (Fig. 1). Backward visual masking effectively abolishes the latter (i.e., P3-related) component of this repetition effect, while the former (i.e., N400-related) subcomponent remained unaffected by perceptual non-discriminability of primes (Schnyer et al., 1997; Misra and Holcomb, 2003).

A careful reader would have surely noticed that in a typical unconscious priming paradigm the N400 effect was measured to the second word (target), though the first word (prime) was subliminally presented. However, Stenberg et al. (2000) conducted a series of visual experiments in which the target was masked so to be detected in about 50% trials, and ERPs were averaged for detected and undetected targets separately. A highly significant N400 effect was found in both conditions, although its magnitude was about four times smaller with undetected than detected stimuli. On the other hand, the repetition effect disappeared when *both primes and targets* were made unrecognized by scrambling (Zhang et al., 1997).

A special case of masking, called attentional blink, present two task-relevant signals with a short interval (typically 200–600 ms) within a rapid stream of stimulation (Fig. 6). This frequently leads to the second signal being missed, both behaviorally and subjectively. Both forward and backward mask effects seem to participate in attentional blink—the latter because the recognition of the attentional blink signal is greatly enhanced if it is the very last stimulus in the experimental run. Most ERP components to the unperceived signal (also the N400) are only slightly attenuated, except P3 whose amplitude is strongly reduced (Vogel et al., 1998), even if the two consecutive signals belong to different sensory modalities (Dell'Acqua et al., 2003). Interestingly, the probability of an attentional blink is strongly related, at the group as well as individual level, to the time course of the P3 to the first signal so that the second one is most likely missed when presented on the peak of the positivity to the first one (McArthur et al., 1999). However, the signals that are detected during the attentional blink time interval do elicit a P3 wave as well (Kranczioch et al., 2003). Likewise, the last stimulus in a run, which is often perceived despite being preceded by the first signal, also yields a significant P3 (Vogel and Luck, 2002).

Rolke et al. (2001) used the second (i.e., suppressed by attentional blink) signal within a priming paradigm as a prime for a third relevant stimulus (target), which could be either strongly semantically related, or weakly related, or unrelated to that prime. Recognized primes elicited a P3, and the corresponding targets yielded a large (unrelated) or smaller (weakly related) N400. As in the other studies, there was no P3 to unrecognized primes, and no N400 to weakly related targets thereafter. However, unrelated targets resulted in a significant N400 despite the lack of conscious perception for the preceding prime.

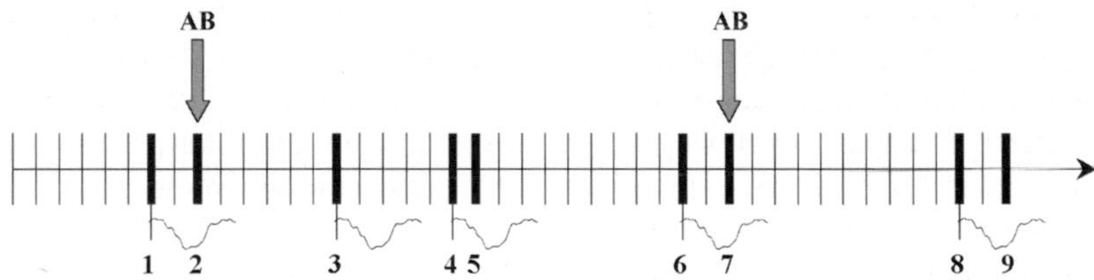

Fig. 6. An experimental paradigm for attentional blinks. An experimental run is presented consisting of 45 stimuli that follow each other with brief intervals (e.g., 150 ms). Nine of the stimuli are relevant signals (bold bars) that are to be responded to. These stimuli elicit a P3 wave (shown for signals 1, 3, 4, 6, 8). If the next signal occurs around the apex of this wave (e.g., signals 2 and 7) they are usually missed due to the "attentional blink" (arrows AB). Note that signal 5 presented with a shorter interval (150 ms) is not missed (no attentional blink), although perceptual masking effects might be expected stronger for shorter intervals. This underscores the role of attentional rather than perceptual factors. On the other hand, perceptual effects are also in play, since signal 9 is not missed either, although its position in relation to the preceding signal 8 should lead to attentional blink. This is because it is the last stimulus in the run, making backward masking impossible.

Conclusion

Relationships between neurophysiological markers, behavioral states, and first-person states of consciousness are like a system of two linear equations with three unknowns: if you only fix one of them, you can find the other two. If we know for sure, for example, that the P3b reflects conscious information processing, we could easily distinguish between states or microstates with and without elements of conscious awareness. Conversely, if we were sure that no conscious processing is possible in a given state, we could split up ERP into the components related versus unrelated to conscious experience. Unfortunately, this is not the case. We do not know for sure whether our sleeping subjects could not shortly perceive some of the stimuli presented during REM or stage 2 sleep but forget them upon awakening. In general anesthesia, the possibility of partial conscious perception is a long-discussed issue, much feared by anesthesiologists (Schwender et al., 1998; Sigalovsky, 2003; Daunderer and Schwender, 2004). In neglect, different results can be obtained when ERPs are averaged across all stimuli presented on the neglected side versus when separate averages are built for extinguished and for occasionally detected stimuli (Deouell, 2002). In contrast to neglect, we have no possibility to distinguish responses to occasionally perceived stimuli in coma or PVS, but we cannot rule out that at least some stimuli in some patients do attain the level of awareness despite the lack of a behavioral response (Andrews et al., 1996; Shewmon et al., 1999).

Nevertheless, we may cautiously assume that at least in a majority of the above-depicted studies, subjects, and patients, conscious perception of stimuli did not take place. Then, although we cannot answer the final question about the neural correlates of consciousness, we may ask a more specific question about possible factors determining the strength of correlation between an ERP effect and conscious perception:

1. *Endogenous versus exogenous ERP components.* Exogenous components N1 and P2 are more stable than endogenous components in conditions such as REM sleep, anesthesia, PVS, and during presentation of masked stimuli.
2. *Latency of the ERPs.* This factor is confounded with the previous one, since exogenous components have, typically, a shorter latency than endogenous ERPs components. Yet, if we compare ERP components within the exogenous and endogenous group, the latency as such does not separate components that are more likely to appear without conscious perception versus those that are less likely.
3. *Stimulus- versus response-related ERPs.* Although only few data exist about unconscious effects on movement-related brain potentials, studies of Eimer and Schlaghecken (1998, 2003) clearly indicate that stimuli presented below the recognition threshold consistently affect ERP correlates of movement preparation, like they affect ERP correlates of stimulus processing (see also Jaskowski et al., 2002).
4. *Attention-dependent versus attention-independent ERPs.* The attention-independent MMN and P3a occur more frequently than, e.g., the attention-dependent N400 in coma and PVS. But this is not the case for sleep or masking. Even in neglect, which is sometimes regarded as a disorder of attention, some forms of MMN are considerably impaired. Only a preliminary conclusion is possible, because I do not know any single experiment in which MMN and N400 effects would be investigated under exactly the same conditions of "unconscious" stimulation. However, if the dissociation between these components is real, it would be an additional argument against the identity of attention and consciousness.
5. *Physical complexity and semantic meaning of stimuli.* Another idea based on the sequential processing model is that the cortex does not need awareness in order to respond to "basic", simple stimuli, while increasing stimulus complexity requires conscious resources. From this point of view, meaning is a highly complex stimulus property, which would first disappear without conscious perception.

Not only is this hypothesis difficult to reconcile with numerous data of general psychology demonstrating unconscious perception of word meaning (e.g., Kihlstrom, 1987; Naccache and Dehaene, 2001; Kunde et al., 2003), but also the discussed ERP data show an opposite tendency. Stimulus complexity enhances, rather than reduces, ERP effects of unconsciously presented stimuli. Responses to stimulus meaning are well defined in REM and even non-REM sleep, and in response to very brief or masked stimuli, particularly when such stimuli are emotionally relevant.

6. *Late positive posterior complex.* A family of ERP waves including the parietal P3b and similar components referred to as P600 or simply "late positive complex" to verbal stimuli, appears to benefit from an exceptional status. They are considerably more difficult to obtain in sleep and in response to unrecognized stimuli than any other endogenous component regardless of latency and physical or semantic relatedness. This fact is particularly impressive in the studies that directly compared N400 with P3 or the late positive complex (Schnyer et al., 1997; Vogel et al., 1998; Rolke et al., 2001) and demonstrated the lack of the posterior positive components to subjectively non-perceived stimuli while the N400-related effects were distinctly present. A discussion of this intriguing difference would require a thorough analysis of the many existing theories of P3 (see Donchin and Coles, 1988b; Verleger, 1988, for a subset of those theories) and are outside our present scope. Even this component cannot be considered a correlate of conscious perception because it also can be obtained, though with great methodological difficulties, without conscious awareness (Perrin et al., 1999; Bernat et al., 2001b). It can be stated, however, that of all ERP effects investigated to date, the neural processes underlying the late posterior positivity are most closely linked with those aspects of brain activity, which make possible conscious recognition of external events. The exact nature of that link remains to be elucidated.

Abbreviations

AB	attentional blink
CS	conditional stimulus
ERP	event-related potentials
MMN	mismatch negativity
PVS	persistent vegetative state
REM	rapid eye movements
SWS	slow wave sleep
UCS	unconditional stimulus

Acknowledgment

The study was funded by the German Research Society (Deutsche Forschungsgemeinschaft, DFG) as a part of the SFB 550. The author thanks Niels Birbaumer, Vladimir Bostanov, Simone Lang, and Nicola Neumann.

References

Afifi, L., Guilliminault, C. and Colrain, I.M. (2003) Sleep and respiratory stimulus specific dampening of cortical responsiveness in OSAS. Resp. Physiol. Neurobiol., 136: 221–234.

Amzica, F. and Steriade, M. (2002) The functional significance of K-complexes. Sleep Med. Rev., 6: 139–149.

Andreassi, J.L. (1984) Interactions between target and masking stimuli: perceptual and event-related potential effects. Int. J. Psychophysiol., 1: 153–162.

Andrews, K., Murphy, L., Munday, R. and Littlewood, C. (1996) Misdiagnosis of the vegetative state: retrospective study in a rehabilitation unit. Br. Med. J., 313: 13–16.

Atienza, M. and Cantero, J.L. (2001) Complex sound processing during human REM sleep by recovering information from long-term memory as revealed by the mismatch negativity (MMN). Brain Res., 901: 151–160.

Atienza, M., Cantero, J.L. and Domingez-Martin, E. (2002) Mismatch negativity (MMN): an objective measure of sensory memory and long-lasting memories during sleep. Int. J. Psychophysiol., 46: 215–225.

Atienza, M., Cantero, J.L. and Escera, C. (2001a) Auditory information processing during human sleep as revealed by event-related brain potentials. Clin. Neurophysiol., 112: 2031–2045.

Atienza, M., Cantero, J.L. and Gomez, C.M. (2001b) The initial orienting response during human REM sleep as revealed by the N1 component of auditory event-related potentials. Int. J. Psychophysiol., 42: 131–141.

Bastien, C. and Campbell, K.B. (1994) Effect of rate of tone-pip stimulation on the evoked K-complex. J. Sleep Res., 3: 65–72.

Bastuji, H., Garcia-Larrea, L., Franc, C. and Mauguiere, F. (1995) Brain processing of stimulus deviance during slow-wave and paradoxical sleep: a study of human auditory evoked responses using the oddball paradigm. J. Clin. Neurophysiol., 12: 155–167.

Bastuji, H., Perrin, F. and Garcia-Larrea, L. (2002) Semantic analysis of auditory input during sleep: studies with event related potentials. Int. J. Psychophysiol., 46: 243–255.

Bernat, E., Bunce, S. and Shevrin, H. (2001a) Event-related brain potentials differentiate positive and negative mood adjectives during both supraliminal and subliminal visual processing. Int. J. Psychophysiol., 42: 11–34.

Bernat, E., Shevrin, H. and Snodgrass, M. (2001b) Subliminal visual oddball stimuli evoke a P300 component. Clin. Neurophysiol., 112: 159–171.

Berti, A. and Rizzolatti, G. (1992) Visual procesing without awareness: evidence from unilateral neglect. J. Cogn. Neurosci., 4: 345–351.

Brázdil, M., Rektor, I., Daniel, P., Dufek, M. and Jirák, P. (2002) Intracerebral event-related potentials to subthreshold target stimuli. Clin. Neurophysiol., 112: 650–661.

Brown, C. and Hagoort, P. (1993) The processing nature of the N400: evidence from masked priming. J. Cogn. Neurosci., 5: 34–44.

Brualla, J., Romero, M.F., Serrano, M. and Valdizan, J.R. (1998) Auditory event-related potentials to semantic priming during sleep. Electroenceph. Clin. Neurophysiol., 108: 283–290.

Campbell, K.B. and Colrain, I.M. (2002) Event-related potential measures of the inhibition of information processing: the sleep onset period. Int. J. Psychophysiol., 46: 197–214.

Cheour, M., Kushnerenko, E., Ceponieme, R., Fellman, V. and Näätänen, R. (2002) Electric brain responses obtained from newborn infants to changes in duration in complex harmonic tones. Devel. Neuropsychol., 22: 471–479.

Cheour, M., Leppänen, P.H.T. and Kraus, N. (2000) Mismatch negativity (MMN) as a tool for investigating auditory discrimination and sensory memory in infants and children. Clin. Neurophysiol., 111: 4–16.

Colrain, I.M., Webster, K.E. and Hirst, G. (1999) The N550 component of the evoked K-complex: a modality non-specific response? J. Sleep Res., 8: 273–280.

Cote, K.A. and Campbell, K.B. (1999) The effect of varying stimulus intensity on P300 during sleep. Neuroreport, 10: 2313–2318.

Cote, K.A., de Lugt, D.R., Langley, S.D. and Campbell, K.B. (1999) Scalp topography of the auditory evoked K-complex in stage 2 and slow wave sleep. J. Sleep Res., 6: 263–272.

Cote, K.A., Etienne, L. and Campbell, K.B. (2001) Neurophysiological evidence for the detection of external stimuli during sleep. Sleep, 24: 791–803.

Cottone, L.A., Adamo, D. and Squires, N.K. (2004) The effect of unilateral somatosensory stimulation on hemispheric asymmetries during slow wave sleep. Sleep, 27: 63–68.

Crowley, K., Trinder, J. and Colrain, I.M. (2004) Evoked K-complex generation: the impact of sleep spindles and age. Clin. Neurophysiol., 115: 471–476.

Cuthbert, B.N., Bradley, M.M. and Lang, P.J. (1996) Probing picture perception: activation and emotion. Psychophysiol., 33: 103–111.

Daunderer, M. and Schwender, D. (2004) Unerwünschte Wachheit während der anaesthesia. Der Anaesthesist, 53: 581–592.

Deacon, D., Hewitt, S., Yang, C.-M. and Nagata, M. (2000) Event-related potential indices of semantic priming using masked and unmasked words: evidence that the N400 does not reflect a post-lexical process. Cogn. Brain Res., 9: 137–146.

Deacon, D., Grose-Fifer, J., Hewitt, S., Nagata, M., Shelley-Tremblay, J. and Yang, C.-M. (2004) Physiological evidence that a masked unrelated intervening item disrupts semantic priming: implications for theories of semantic representation and retrieval models of semantic priming. Brain Lang., 89: 38–46.

Dehaene, S., Naccache, L., Cohen, L., Bihan, D.L., Mangin, J.-F., Poline, J.-B. and Riviere, D. (2001) Cerebral mechanisms of word masking and unconscious repetition priming. Nat. Neurosci., 4: 752–758.

Dell'Acqua, R., Jolicoeur, P., Pesciarelli, F., Job, C.R. and Palomba, D. (2003) Electrophysiological evidence of visual encoding deficits in a cross-modal attentional blink paradigm. Psychophysiology, 40: 629–639.

Deouell, L.Y. (2002) Pre-requisites for conscious awareness: clues from electrophysiological and behavioral studies of unilateral neglect patients. Cons. Cogn., 11: 546–567.

Deouell, L.Y., Bentin, S. and Soroker, N. (2000) Electrophysiological evidence for an early (pre-attentive) information processing deficit in patients with right hemispheric damage and unilateral neglect. Brain, 123: 353–365.

Donchin, E. and Coles, M.G.H. (1988a) Behavior, cognition and event-related brain potentials. Behav. Brain Sci., 11: 735–739.

Donchin, E. and Coles, M.G.H. (1988b) Is the P300 component a manifestation of context updating? Behav. Brain Sci., 11: 357–374.

Eimer, M. and Schlaghecken, F. (1998) Effects of masked stimuli on motor activation: behavioral and electrophysiological evidence. J. Exper. Psychol. Hum. Perc. Perf., 24: 1737–1747.

Eimer, M. and Schlaghecken, F. (2003) Response facilitation and inhibition in subliminal priming. Biol. Psychol., 64: 7–26.

Fischer, C., Morlet, D., Bouchet, P., Luaute, J., Jourdan, C. and Salord, F. (1999) Mismatch negativity and late auditory evoked potentials in comatose patients. Clin. Neurophysiol., 110: 1601–1610.

Fischer, C., Morlet, D. and Giard, M.-H. (2000) Mismatch negativity and N100 in comatose patients. Audiol. Neuro-Otol., 5: 192–197.

Gott, P.S., Rabinovicz, A.L. and DiGiorgio, C.M. (1991) P300 auditory event-related potentials in nontraumatic coma: association with Glasgow Coma Score and awakening. Arch. Neurol., 48: 1267–1270.

Guerit, J.M., Verougstraete, D., Tourtchaninoff, M., Debatisse, D. and Witdoeckt, C. (1999) ERPs obtained with the auditory oddball paradigm in coma and altered states of consciousness: clinical relationships, prognostic value and origin of components. Clin. Neurophysiol., 110: 1260–1269.

Heinke, W., Kenntner, R., Gunter, T.C., Sammler, D., Olthoff, D. and Koelsch, S. (2004) Sequential effects of increasing propofol sedation on frontal and temporal cortices as indexed by auditory event-related potentials. Anesthesiology, 100: 617–625.

Hobson, J.A. and Pace-Schott, E.F. (2002) The cognitive neuroscience of sleep: neuronal systems, consciousness and learning. Nat. Rev. Neurosci., 3: 679–693.

Hull, J. and Harsh, J. (2001) P300 and sleep-related positive waveforms (P220, P450, and P900) have different determinants. J. Sleep Res., 10: 9–17.

Jaskowski, P., van der Lubbe, R.H.J., Schlotterbeck, E. and Verleger, R. (2002) Traces left on visual selective attention by stimuli that are not consciously identified. Psychol. Sci., 13: 48–53.

Jones, S.J., vas Pato, M., Sprague, L., Stokes, M., Munday, R. and Haque, N. (2000) Auditory evoked potentials to spectro-temporal modulation of complex tones in normal subjects and patients with severe brain injury. Brain, 123: 1007–1016.

Kane, N.M., Butler, S.R. and Simpson, T. (2000) Coma outcome prediction using event-related potentials: P3 and mismatch negativity. Audiol. Neuro-Otol., 5: 186–191.

Kane, N.M., Curry, S.H., Butler, S.R. and Cummings, B.H. (1993) Electrophysiolological indicator of awakening from coma. Lancet, 341: 688.

Kane, N.M., Curry, S.H., Rowlands, C.A., Minara, A.R., Lewis, T., Moss, T., Cummins, B.H. and Butler, S.R. (1996) Event-related potentials—neurophysiological tools for predicting emergence and early outcome from traumatic coma. Intens. Care Med., 22: 39–46.

Kane, N.M., Moss, T.H., Curry, S.H. and Butler, S.R. (1998) Quantitative electroencephalographic evaluation of non-fatal and fatal traumatic coma. Electroenceph. Clin. Neurophysiol., 106: 244–250.

Keren, O., Ben-Dror, S., Stern, M.J., Goldberg, G. and Groswasser, Z. (1998) Event-related potentials as an index of cognitive function during recovery from sever closed head injury. J. Head Trauma Rehab., 13: 15–30.

Kevanishvili, Z. and Lagidze, Z. (1987) Masking level difference: an electrophysiological approach. Scand. Audiol., 16: 3–11.

Kiefer, M. (2002) The N400 is modulated by unconsciously perceived masked words: further evidence for an automatic spreading activation account of N400 priming effects. Cogn. Brain Res., 13: 27–39.

Kiefer, M. and Spitzer, M. (2000) Time course of conscious and unconscious semantic brain activation. Neuroreport, 11: 2401–2407.

Kihlstrom, J.F. (1987) The cognitive unconscious. Science, 237: 1445–1452.

Kostandov, E. and Arzumanov, Y. (1977) Averaged cortical evoked potentials to recognized and non-recognized visual stimuli. Acta Neurobiol. Exper., 37: 311–324.

Kotchoubey, B., Lang, S., Baales, R., Herb, E., Maurer, P., Mezger, G., Schmalohr, D., Bostanov, V. and Birbaumer, N. (2001) Brain potentials in human patients with severe diffuse brain damage. Neurosci. Lett., 301: 37–40.

Kotchoubey, B., Lang, S., Mezger, G., Schmalohr, D., Schneck, M., Semmler, A., Bostanov, V., Birbaumer, N. (in press) Information processing in severe disorders of consciousness: Persistent vegetative state and minimally conscious state. Clinical Neurophysiology.

Kotchoubey, B., Lang, S., Herb, E., Maurer, P., Schmalohr, D. and Bostanov, V. (2003) Birbaumer N. Stimulus complexity enhances auditory discrimination in patients with extremely severe brain injuries. Neurosci. Lett., 352: 129–132.

Kranczioch, C., Debener, S. and Engel, A.K. (2003) Event-related potential correlates of the attentional blink phenomenon. Cogn. Brain Res., 17: 177–187.

Kunde, W., Kiesel, A. and Hoffmann, J. (2003) Conscious control over the content of unconscious cognition. Cognition, 88: 223–242.

Lang, S. and Kotchoubey, B. (2002) Brain responses to number sequences with and without active task requirement. Clin. Neurophysiol., 113: 1734–1741.

Lang, S., Kotchoubey, B., Lutz, A. and Birbaumer, N. (1997) Was tut man, wenn man nichts tut? Kognitive EKP-Komponenten ohne kognitive Aufgabe. Z. Exper. Psychol., 44: 138–162.

Leuthold, H., Sommer, W. and Ulrich, R. (1996) Partial advance information and response preparation: inference from the lateralized readiness potential. J. Exp. Psychol. Gen., 125: 307–323.

Lew, H.L., Dikmen, S., Slimp, J., Temkin, N., Lee, E.H., Newell, D. and Robinson, L.R. (2003) Use of somatosensory-evoked potentials and cognitive event-related potentials in predicting outcomes of patients with severe traumatic brain injury. Am. J. Phys. Med. Rehab., 82: 53–61.

Lew, H.L., Slimp, J., Price, R., Massagli, T. and Robinson, L.R. (1999) Comparison of speech-evoked versus tone-evoked P300 response. Am. J. Phys. Med. Rehab., 78: 367–371.

Liddell, B.J., Williams, L.M., Rathjen, J., Shevrin, H. and Gordon, E. (2004) A temporal dissociation of subliminal versus supraliminal fear perception: an event-related potential study. J. Cogn. Neurosci., 16: 479–486.

Loewy, D.H., Campbell, K.B. and Bastien, C. (1996) The mismatch negativity to frequency deviant stimuli during natural sleep. Electroencephal. Clin. Neurophysiol., 98: 493–501.

Loewy, D.H., Campbell, K.B., de Lugt, D.R., Elton, M. and Kok, A. (2000) The mismatch negativity during natural sleep: intensity deviants. Clin. Neurophysiol., 111: 863–872.

Martin, B.A., Kurzberg, D. and Stapells, D.R. (1999) The effects of decreased audibility produced by high-pass noise masking on N1 and the mismatch negativity to speech sounds /da/ and /ba/. J. Speech Lang. Hear. Res., 42: 271–286.

Martinez, A., Anllo-Vento, L., Sereno, M.I., Frank, L.R., Buxton, R.B., Dubowitz, D.J., Wong, E.C., Hinrichs, H., Heinze, H.J. and Hillyard, S.A. (1999) Involvement of striate and extrastriate visual cortical areas in spatial attention. Nat. Neurosci., 2: 364–369.

Marzi, C.A., Girelli, M., Miniussi, C., Smania, N. and Maravita, A. (2000) Electrophysiological correlates of conscious vision: evidence from unilateral extinction. J. Cogn. Neurosci., 12: 869–877.

Mazzini, L., Zaccala, M., Gareri, F., Giordano, A. and Angelino, E. (2001) Long-latency auditory evoked potentials in severe traumatic brain injury. Arch. Phys. Med. Rehab., 82: 57–65.

McArthur, G., Budd, T. and Michie, P. (1999) The attentional blink and P300. Neuroreport, 10: 3691–3695.

Meyer, D.E., Osman, A.M., Irwin, D.E. and Yantis, S. (1988) Modern mental chronometry. Biol. Psychol., 26: 3–67.

Misra, M. and Holcomb, P.J. (2003) Event-related potential indices of masked repetition priming. Psychophysiology, 40: 115–130.

Mutschler, V., Chaumeil, C.G., Marcoux, L., Wioland, N., Tempe, J.D. and Kurtz, D. (1996) Etude du P300 auditif chez des sujets en coma post-anoxique. Donnees preliminaires. Neurophysiol. Clin., 26: 158–163.

Näätänen, R., Paavilainen, P., Tiitinen, H., Jiang, D. and Alho, K. (1993) Attention and mismatch negativity. Psychophysiology, 30: 436–450.

Näätänen, R. and Picton, T.W. (1986) N2 and automatic versus controlled processes. In: McCallum W.C., Zappoli R. and Denoth L. (Eds.), Cerebral Psychophysiology. Elsevier, Amsterdam, pp. 169–195.

Näätänen, R. and Winkler, I. (1999) The concept of auditory stimulus representation in cognitive neuroscience. Psychol. Bull., 125: 826–859.

Naccache, L. and Dehaene, S. (2001) Unconscious semantic priming extends to novel unseen stimuli. Cognition, 80: 215–229.

Neumann, O. (1989) On the origins and status of the concept of automatic processing. Z. Psychol., 197: 411–428.

Niedeggen, M., Wichmann, P. and Stoerig, P. (2001) Change blindness and time to consciousness. Eur. J. Neurosci., 14: 1719–1726.

Niedeggen, M., Hesselmann, G., Sahraie, A., Milders, M. and Blakemore, C. (2004) Probing the prerequisites for motion blindness. J. Cogn. Neurosci., 16: 584–597.

Niiyama, Y., Fujiwara, R., Satoh, N. and Hishikawa, Y. (1994) Endogenous components of event-related potential appearing during NREM stage 1 and REM sleep in man. Int. J. Psychophysiol., 17: 165–174.

Niiyama, Y., Satoh, N., Kutsuzawa, O. and Hishikawa, Y. (1996) Electrophysiological evidence suggesting that sensory stimuli of unknown origin induce spontaneous K-complexes. Electroenceph. Clin. Neurophysiol., 98: 394–400.

Nordby, H., Hugdahl, K., Stickgold, R., Bronnick, K.S. and Hobson, J.A. (1996) Event-related potentials (ERPs) to deviant auditory stimuli during sleep and waking. Neuroreport, 7: 1082–1086.

Paavilainen, P., Cammann, R., Alho, K., Reinikainen, K., Sams, M. and Näätänen, R. (1987) Event-related potentials to pitch change in an auditory stimulus sequence during sleep. In: Johnson R., Rohrbaugh J.W. and Parasuraman R. (Eds.), Current Trends in Event-Related Potential Research. Elsevier, Amsterdam, pp. 246–255.

Pace-Schott, E.F. and Hobson, J.A. (2002) The neurobiology of sleep: genetics, cellular physiology and subcortical networks. Nat. Rev. Neurosci., 3: 591–605.

Peigneux, P., Laureys, S., Delbeuck, X. and Maquet, P. (2001) Sleeping brain, learning brain. The role of sleep for memory systems. Neuroreport, 12: A111–A124.

Perrin, F., Bastuji, H. and Garcia-Larrea, L. (2002) Detection of verbal discordances during sleep. Neuroreport, 13: 1345–1349.

Perrin, F., Bastuji, H., Mauguiére, F. and García-Larrea, L. (2000) Functional dissociation of the early and late portions of human K-complexes. Neuroreport, 11: 1637–1640.

Perrin, F., Garcia-Larrea, L., Mauguiere, F. and Bastuji, H. (1999) A differential brain response to the subject's own name persists during sleep. Clin. Neurophysiol., 110: 2153–2164.

Plourde, G. and Boylan, J.F. (1991) The long-latency auditory evoked potential as a measure of the level of consciousness during sufentanil anesthesia. J. Cardiothor. Vasc. An., 5: 577–583.

Plourde, G., Joffe, D., Villemure, C. and Trahan, M. (1993) The P3a wave of the auditory event-related potential reveals registration of pitch change during sufentanil anesthesia for cardiac surgery. Anesthesiology, 78: 498–509.

Plourde, G. and Picton, T.W. (1991) Long-latency auditory evoked potentials during general anesthesia: N1 and P3 components. Anesth. Analg., 72: 342–350.

Polich, J. (1989) P300 from a passive auditory paradigm. Electroenceph. Clin. Neurophysiol., 74: 312–320.

Polich, J. and Comerchero, M.D. (2003) P3a from visual stimuli: typicality, task, and topography. Brain Topogr., 15: 141–152.

Pratt, H., Berlad, I. and Lavie, P. (1999) 'Oddball' event-related potentials and information processing during REM and non-REM sleep. Clin. Neurophysiol., 110: 53–61.

Rappaport, M., McCandless, K.L., Pond, W. and Krafft, M.C. (1991) Passive P300 response in traumatic brain injury patients. J. Neuropsychiatry Clin. Neurosci., 3: 180–185.

Reuter, B.M., Linke, D.B. and Kurthen, M. (1989) Kognitive Prozesse bei Bewußtlosen: eine brain-mapping-Studie zu P300. Arch. Psychol., 141: 155–173.

Rolke, B., Heil, M., Streb, J. and Henninghausen, E. (2001) Missed prime words within the attentional blink evoke an N400 semantic priming effect. Psychophysiology, 338: 164–174.

Rösler, F., Clausen, G. and Sojka, B. (1986) The double-priming paradigm: a tool for analyzing the functional significance of endogenous event-related brain potentials. Biol. Psychol., 22: 239–268.

Ruz, M., Madrid, E., Lupianez, J. and Tudela, P. (2003) High density ERP indices of conscious and unconscious semantic priming. Cogn. Brain Res., 17: 719–731.

Sabri, M. and Campbell, K.B. (2002) The effects of digital filtering on mismatch negativity in wakefulness and slow-wave sleep. J. Sleep Res., 11: 123–127.

Sabri, M., Labelle, S., Gosselin, A. and Campbell, K.B. (2003) Effect of sleep onset on the mismatch negativity to frequency deviants using a rapid rate of presentation. Cogn. Brain Res., 17: 164–176.

Sallinen, M., Kaartinen, J. and Lyytinen, H. (1994) Is the appearance of mismatch negativity during stage 2 sleep related to the elicitation of K-complexes? Electroencephal. Clin. Neurophysiol., 91: 140–148.

Sallinen, M., Kaartinen, J. and Lyytinen, H. (1997) Precursors of the evoked K-complex in event-related brain potentials in stage 2 sleep. Electroencephal. Clin. Neurophysiol., 102: 363–373.

Sallinen, M. and Lyytinen, H. (1997) Mismatch negativity during objective and subjective sleepiness. Psychophysiology, 34: 694–702.

Salo, S.K., Lang, A.H. and Salmivalli, A.J. (1995) Effect of contralateral white noise masking on the mismatch negativity. Scand. Audiol., 24: 165–173.

Samoilovich, L.A. and Trush, V.D. (1979) Cortical evoked potentials during consecutive visual masking. Hum. Physiol., 4: 217–223.

Schneider, W. and Shiffrin, R.M. (1977) Controlled and automatic human information processing. Psychol. Rev., 84: 1–66.

Schnyer, D.M., Allen, J.J.B. and Forster, K.I. (1997) Event-related brain potential examination of implicit memory processes: masked and unmasked repetition priming. Neuropsychology, 11: 243–260.

Schröger, E. (1997) On the detection of auditory deviations: a pre-attentive activation model. Psychophysiology, 34: 245–257.

Schwender, D., Kunze-Kronawitter, H., Dietrich, P., Klasing, S., Forst, H. and Madler, C. (1998) Conscious awareness during general anaesthesia: patients' perceptions, emotions, cognition and reactions. Br. J. Anaesth., 80: 133–139.

Shevrin, H. (2001) Event-related markers of unconscious processes. Intern. J. Psychophysiol., 42: 209–218.

Shevrin, H. and Fritzler, D.E. (1968) Visual evoked response correlates of unconscious mental processes. Science, 161: 295–298.

Shevrin, H., Smith, W.H. and Fitzler, D.E. (1970) Subliminally stimulated brain and verbal responses of twins differing in repressiveness. J. Abnorm. Psychol., 76: 39–46.

Shevrin, H., Smith, W.H. and Fitzler, D.E. (1971) Average evoked response and verbal correlates of unconscious mental processes. Psychophysiology, 8: 149–162.

Shevrin, H., Williams, W.J., Marshall, R.E., Hertel, R.K., Bond, J.A. and Brakel, L.A. (1992) Event-related potential indicators of the dynamic unconsciousness. Cons. Cogn., 1: 340–366.

Shewmon, D.A., Holmes, G.L. and Byrne, P.A. (1999) Consciousness in congenitally decorticate children: developmental vegetative state as self-fulfilling prophecy. Devel. Med. Child Neurol., 41: 364–374.

Sigalovsky, N. (2003) Awareness under general anesthesia. AANA J., 71: 373–379.

Signorino, M., D'Acunto, S., Cercaci, S., Pietropaoli, P. and Angeleri, F. (1997) The P300 in traumatic coma: conditioning of the oddball paradigm. J. Psychophysiol., 11: 59–70.

Simpson, T.P., Manara, A.R., Kane, N.M., Barton, R.L., Rowlands, C.A. and Butler, S.R. (2002) Effect of propofol anaesthesia on the event-related potential mismatch negativity and the auditory-evoked potential N1. Br. J. Anaesth., 89: 382–388.

Stenberg, G., Lindgren, M., Johansson, M., Olsson, A. and Rosen, I. (2000) Semantic processing without conscious identification: evidence from event-related potentials. J. Exp. Psychol. Learn. Mem. Cogn., 26: 973–1004.

Tervaniemi, M., Schröger, E., Saher, M. and Näätänen, R. (2000) Effects of spectral complexity and sound duration in complex-sound pitch processing in humans—a mismatch negativity study. Neurosci. Lett., 290: 66–70.

Vaitl, D., Birbaumer, N., Gruzelier, J., Jamieson, G., Kotchoubey, B., Kübler, A., Lehmann, D., Miltner, W.H.R., Ott, U., Putz, P., Sammer, G., Strauch, I., Strehl, U., Wackermann, J. and Weiss, T. (2005) Psychobiology of altered states of consciousness. Psychol. Bull., 131: 98–127.

Van Hooff, J.C., De Beer, N.A., Brunia, C.H., Cluitmans, P.J. and Korsten, H.H. (1997) Event-related potential measures of information processing during general anesthesia. Electroencephal. Clin. Neurophysiol., 103: 268–281.

Van Hooff, J.C., De Beer, N.A., Brunia, C.H., Cluitmans, P.J., Korsten, H.H., Tavilla, G. and Grouls, R. (1995) Information processing during cardiac surgery: an event-related potential study. Electroencephal. Clin. Neurophysiol., 96: 433–452.

Vaughan, H.G. and Silverstein, L. (1968) Metacontrast and evoked potentials. Science, 160: 207–208.

Verleger, R. (1988) Event-related potentials and cognition: a critique of the context updating hypothesis and an alternative interpretation of P3. Behav. Brain Sci., 11: 343–356.

Verleger, R. (2001) Comment on "Electrophysiological correlates of conscious vision: evidence from unilateral extenction" by Marzi, Girelli, Miniussi, Smania, and Maravita, in JOCN 12:5. J. Cogn. Neurosci., 13: 416–417.

Verleger, R., Heide, W., Butt, C., Wascher, E. and Kompf, D. (1996) On-line brain potential correlates of right parietal patients' attentional deficits. Electroencephal. Clin. Neurophysiol., 99: 444–457.

Vogel, E.K. and Luck, S.J. (2002) Delayed working memory consolidation during the attentional blink. Psychonom. Bull. Rev., 9: 739–743.

Vogel, E.K., Luck, S.J. and Shapiro, K.L. (1998) Electrophysiological evidence for a postperceptual locus of suppression during the attentional blink. J. Exp. Psychol. Hum. Perc. Perf., 24: 1656–1674.

Voss, U. and Harsh, J. (1998) Information processing and coping style during the wake/sleep transition. J. Sleep Res., 7: 225–232.

Whiting, K.A., Martin, B.A. and Stapells, D.R. (1998) The effects of broadband noise maksing on cortical event-related potentials to speech sounds /ba/ and /da/. Ear Hear., 19: 218–231.

Williams, L.M., Liddell, B.J., Rathjen, J., Brown, K.J., Gray, J., Phillips, M., Young, A. and Gordon, E. (2004) Mapping the time course of nonconscious and conscious perception of fear: an integration of central and peripheral measures. Hum. Brain Map., 21: 64–74.

Winkler, I. and Näätänen, R. (1992) Event-related potentials in auditory backword recognition masking: a new way to study the neurophysiological basis of sensory memory in humans. Neurosci. Lett., 140: 239–242.

Winter, O., Kok, A., Kenemans, J.L. and Elton, M. (1995) Auditory event-related potentials to deviant stimuli during drowsiness and stage 2 sleep. Electroenceph. Clin. Neurophysiol., 96: 398–412.

Witzke, W. and Schönle, P.W. (1996) Ereigniskorrelierte Potentiale als diagnostisches Mittel in der neurologischen Frührehabilitation. Neurol. Rehab., 2: 68–80.

Wong, P.S., Bernat, E., Bunce, S. and Shevrin, H. (1997) Brian indices of nonconscious associative learning. Cons. Cogn., 6: 519–544.

Wong, P.S., Bernat, E., Snodgrass, M. and Shevrin, H. (2004) Event-related brain correlates of associative learning without awareness. Int. J. Psychophysiol., 53: 217–231.

Wong, P.S., Shevrin, H. and Williams, W.J. (1994) Conscious and non-conscious processes: an ERP index of an anticipatory response in a conditioning paradigm using visually masked stimuli. Psychophysiology, 31: 87–101.

Yppärilä, H., Karhu, J., Westeren-Punnonen, S., Musialowicz, T. and Partanen, J. (2002) Evidence of auditory processing during postoperative propofol sedation. Clin. Neurophysiol., 113: 1357–1364.

Zhang, X.L., Begleiter, H., Porjesz, B. and Litke, A. (1997) Visual object priming differs from visual word priming. Electroencephal. Clin. Neurophysiol., 102: 200–215.

S. Laureys (Ed.)
Progress in Brain Research, Vol. 150
ISSN 0079-6123

CHAPTER 31

Novel aspects of the neuropathology of the vegetative state after blunt head injury

D.I. Graham[1,*], W.L. Maxwell[2], J. Hume Adams[1] and Bryan Jennett[3]

[1]*Academic Unit of Neuropathology, Institute of Neurological Sciences, Southern General Hospital, University of Glasgow, Glasgow G51 4TF, UK*
[2]*Human Anatomy, University of Glasgow, Glasgow, UK*
[3]*Academic Unit of Neurosurgery, University of Glasgow, Glasgow, UK*

Abstract: A detailed neuropathological study was undertaken of the brains of patients who had been assessed clinically as vegetative after blunt head injury. There were 35 cases, (33 male; median age 38 years) with a survival of 6.5–19 months (median 9): 17 were injured in a road traffic accident, 9 after assault and 6 after a fall; 3 were recorded as having had a lucid interval. There was an intracranial hematoma in 9 and the median contusion index was 4; raised intracranial pressure was identified in 25, grades 2 and 3 diffuse traumatic axonal injury was present in 25, ischemic damage in 15 and hydrocephalus in 27. Thalamic and hippocampal damage was present in 28 and stereological studies revealed a differential loss of neurons in three principal nuclei of the thalamus and in different sectors of the hippocampus. Immunohistochemistry provided evidence of an inflammatory reaction and in situ DNA fragmentation, features that are strongly indicative of a continuing neuronal loss in subcortical gray matter. These findings provide evidence for the importance of diffuse brain damage to white matter as the structural basis of the vegetative state after blunt head injury with contributions from neuronal loss in the thalami and the hippocampus. Although amyloid plaques and tau inclusions were identified in some, their contribution did not seem important in the ultimate clinical outcome.

Introduction

A prospective clinical study undertaken by the European Brain Injury Consortium indicated that 3% of patients remained in a vegetative state (VS) as assessed by the Glasgow Coma Score, 6 months after blunt head injury (Murray et al., 1999). The structural basis of VS after an acute brain insult, whether due to head injury or hypoxia-ischemia, is either diffuse damage to gray matter or white matter, or focal damage in each thalamus (Kinney and Samuels, 1994). Since this publication, further data are available stressing the need for stereological assessment of neuronal numbers in the thalami (Maxwell et al., 2004) and hippocampi (Maxwell et al., 2003), and to take into account any neurodegenerative changes in patients who survive for up to 4 weeks after injury (Smith et al., 2003). Therefore, the principal objectives of the current study were to provide quantitative data for changes in the number of neurons in the hippocampus and the dorsomedial, lateral posterior and ventral posterior nuclei of the thalamus and to determine the nature of any neurodegenerative changes to allow a more full understanding of the structural basis of VS.

*Corresponding author. Tel.: +44 141-201 2113;
Fax: +44 141-201 2998; E-mail: D.Graham@clinmed.gla.ac.uk

Materials and methods

The head injury database in the Department of Neuropathology, Institute of Neurological Sciences, Glasgow, comprises more than 1500 fatal cases of

DOI: 10.1016/S0079-6123(05)50031-1

blunt head injury many of whom had been managed by colleagues in the Department of Neurosurgery. Because of the close working relationships between the relevant Scottish legal authorities and colleagues in the Department of Forensic Medicine and Science, it has been possible after autopsy to suspend, in most instances, the brain in 10% formal saline for a minimum of a week before dissection. Cerebral hemispheres were cut in the coronal plane and the cerebellar hemispheres at right angles to the folia, and the brain stem horizontally. Material from 35 cases assessed as vegetative by the Glasgow Outcome Scale (GOS; Jennett and Bond, 1975), at least 4 weeks after head injury, were available for study (Adams et al., 1999, 2000; Jennett et al., 2001). Comprehensive histological studies were undertaken (Adams et al., 1980). One feature recorded as being either present or absent was raised intracranial pressure (Adams and Graham, 1976). Others were semi quantified and graded: these included contusions (Adams et al., 1985), ischemic brain damage (Graham et al., 1989), diffuse traumatic axonal injury (Adams et al., 1989), and hydrocephalus. In addition to these assessments additional studies were undertaken on sections cut from the hippocampus and the temporal and cingulate cortex for the features of Alzheimer's disease (β-amyloid, tau protein), cortical Lewy body disease (α-synuclein, ubiquitin), and associated reactive changes in microglia/macrophages (CR3/43 and CD68) and astrocytes (GFAP). The presence of in situ DNA fragmentation was assessed by the terminal deoxynucleotidyl (TUNEL) technique (Gavrieli et al., 1992) and immunocytochemical localization of caspase-3 activity as a marker for cells entering programmed cell death. For this part of the study, material was obtained from vegetative patients ($n = 12$) and age-matched control ($n = 9$) patients with no clinical history of central nervous system injury or disease.

Material from the hippocampus and thalamus was sectioned to allow stereological analysis of differences in the number of neurons and glial/immunocompetant cells (Maxwell et al., 2003, 2004). For the latter technique, coronal 8 μm serial sections of the hippocampus and thalamus at the level of the lateral geniculate body were cut from paraffin-embedded blocks.

In examination of the hippocampus, the following criteria were used to distinguish CA1, CA2, CA3, and CA4. CA1 extended from the subiculum towards the fimbria and contained cells arranged in two rows of pyramidal neurons. CA2 extended from the fimbrial end of CA1 and formed a C-shaped group of larger cells forming a single row. CA3 contained the largest pyramidal cells, which lacked side branches from the apical dendrite extending into the stratum radiatum. CA4 consisted of cells scattered within the hilum of the dentate gyrus.

Three nuclei in the thalamus were selected for study. There is a consensus that the ventral posterior nucleus (VPN) is both the major site of termination for nerve fibers forming the medial lemniscus/dorsal column pathway, spinothalamic tract and the trigeminal cranial nerve, and the origin of fibers to the primary somatic sensory areas of the cerebral cortex; that neurons in the lateral posterior nucleus (LPN) project to the cortex of association sensory areas posterior to the postcentral gyrus (Brodmann area 5); and that the dorsomedial (DM) nucleus is both the terminal field of fibers from the olfactory cortex and amygdala, and the origin of fibers to large areas of the frontal cortex and the terminal field of reciprocal corticothalamic fibers (reviewed by Jones, 1985). On rostro-caudal sections of the thalamus, boundaries of the DM, VPN and LPN nuclei were visualized: the internal medullary lamina separated the DM from the centromedian and lateral thalamic nuclei; the LPN lay between the dorsal one-third of the internal medullary lamina and the external medullary lamina; and the VPN between the external medullary lamina and the centromedian and ventral posteromedial nuclei.

In sections from control patients, the mean diameter of each type of cell to be counted was estimated to allow determination of the focus depth of the counting frame to remove any potential of one cell being counted twice when using the optical disector method. The technique involves counting particles with optical disectors in a uniform and systematic sample, which is a known fraction of the region being analyzed. The estimator is not affected by tissue shrinkage or expansion and can be applied to frozen, vibratome, celloidin and paraffin sections (West et al., 1991). It has been suggested (Benes and Lange, 2001) that the use of the usual-sized sampling window (50 μm^2) is inappropriate, because the method includes the mathematical assumption that the items being estimated follow a Poisson distribution. Neurons in different parts of the brain are not randomly distributed, that is have a Poisson distribution (Benes and Lange, 2001). Rather, a larger sized sampling window, determined by the size of the neurons counted, should be used. The present study has adopted such a technique, where a window 250 μm^2 was used (Maxwell et al., 2003, 2004). For estimation of cell numbers all sections, from controls and VS patients, were coded by one author (DIG). Another author, "blinded" to the clinical history of each patient, then carried out the stereological analysis (WLM). The code was only broken once when all the quantitative data for the whole study population had been obtained. Evidence for neuronal cellular pathology was noted in VS in the hippocampus and the thalamus, where some neurons appeared "dark" and shrunken. There is not a consensus as to the significance of "dark" neurons first investigated by Cammermeyer (1961). It was not the purpose of the present study to investigate this phenomenon. In our experimental protocol it was decided to count all neurons (normal and "dark") within the counting box with the correlation that interpretation of "dark" neurons was irrelevant to the work.

Quantitative data were analyzed using analysis of variance (ANOVA) across all experimental groups. The statistical significance of differences in the number of neurons, astrocytes, macrophages and microglia between control and VS patients was determined using the Bonferoni t test. This study was approved by the Research Ethics Committee of the Southern General Hospital, Glasgow.

Results

Clinical and neuropathological features

The principal clinical features are given in Table 1. Three patients had experienced a lucid interval before developing an intracranial hematoma. The median survival was 9 months. The main pathological features are given in Table 2. In all cases the cause of death was bronchopneumonia. The most common abnormality was diffuse traumatic axonal injury and, in particular, the most severe grades 2 and 3. Macroscopically, the lesions in the corpus callosum and rostral brain stem, respectively, could usually be recognized as orange-brown, granular and occasionally cystic scars; however, some of the lesions were only recognizable microscopically. With increase in survival there was a reduction in the bulk of the white matter with corresponding enlargement of the ventricular system (Fig. 1). In three patients there was grade 1 diffuse traumatic axonal injury: in each of these cases there was additional ischemic damage either in the cerebral cortex or in the thalami. Of the 10 patients who did not have diffuse traumatic axonal injury, an intracranial hematoma had developed in seven and ischemic damage was present in nine.

Table 1. Principal clinical features of patients in a vegetative state ($n = 35$)

Gender	33 male:2 female
Age	7–75 years (median 38)
Cause of injury	17 road traffic accidents
	9 assault
	6 fall
	3 not known
Survival	6.5–19 months (median 9)
Lucidity	3 (9%)

Table 2. Principal neuropathological features of patients in a vegetative state ($n = 35$)

Focal pathologies	
Skull fracture	12 (34%)
Intracranial hematoma	9 7c (26%)
Total contusion index	0–21 (median 4)
Raised intracranial pressure	25 (71%)
Diffuse pathologies	
Traumatic axonal injury	
Total	28 (80%)
Grade 2/3	25 (71%)
Ischemic damage	15 (43%)
Abnormalities in thalamus	28 (80%)
Hydrocephalus	27 (77%)

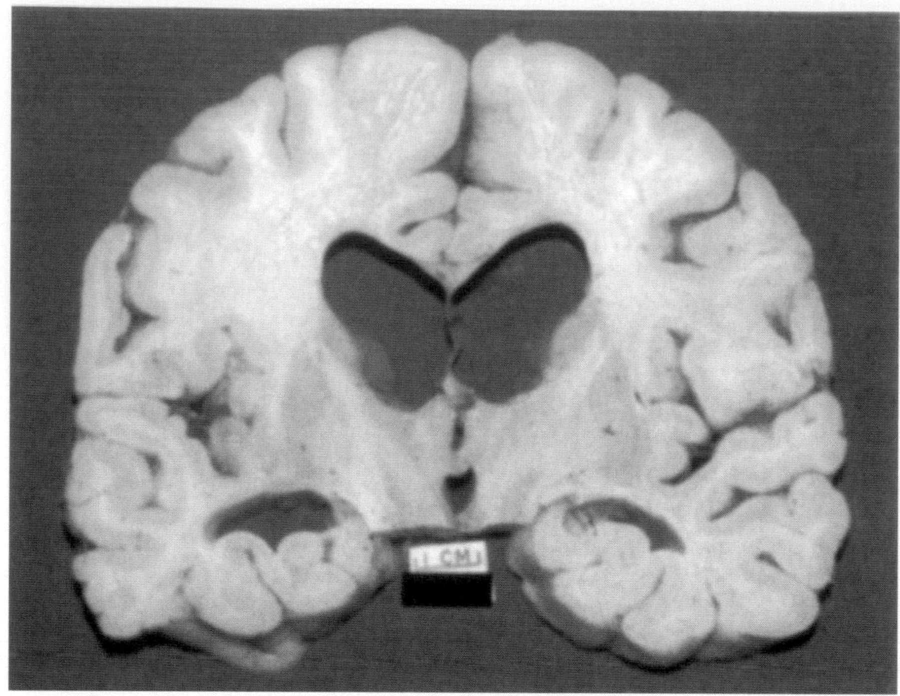

Fig. 1. Diffuse traumatic axonal injury in a patient in vegetative state: 19 months survival after road traffic accident. Note normal cortical ribbon, striatum and hippocampi. There is partial atrophy of the thalami. There is reduction in white matter of centrum semiovale, and in the corpus callosum within which is a focal cystic lesion to the left of the midline, and there is moderate hydrocephalus.

Ischemic damage was graded moderate or severe in 15 of the patients: in six it was accentuated in the arterial boundary zones of the cerebral hemispheres, in five it was diffuse or multifocal (Fig. 2), and in two it was centered on a particular arterial territory. Intracranial pressure had been high in all but two of these 15 patients and in six there had been major extracranial injuries that might have contributed to an inadequate cerebral blood flow. Abnormalities were present in the thalamus of 28 cases: they were both diffuse and focal (Fig. 3).

An intracranial hematoma was present in nine and in seven of these it had been removed surgically. In six of these seven cases the intracranial pressure had been high. In 25 of the 35 patients the intracranial pressure had been high; in only eight there had been an intracranial hematoma. The total contusion index ranged from 0 to 24 with a median of 4: in eight cases there were no identifiable contusions. A fracture of the skull was present in 12 cases. There were no macroscopic or microscopic abnormalities in control cases.

Assessment of neurodegenerative changes

Additional paraffin sections were available in only 16 of the 35 cases (14 males; aged 15–75 years; survival 8 weeks–4 years). Contemporaneous immunohistochemical studies were carried out in 12 controls (11 males; aged 18–79 years) none of whom had had a head injury or CNS disease, and whose cause of death was pneumonia in four, heart failure in three, acute pancreatitis in two, systemic lymphoma in two, and in one case it was associated with an acute bout of psychosis. All the sections were assessed "blind" to clinical diagnosis.

In only three of the head-injured cases were deposits of Aβ identified and of these the plaques were infrequent in two (cases aged 18 and 65

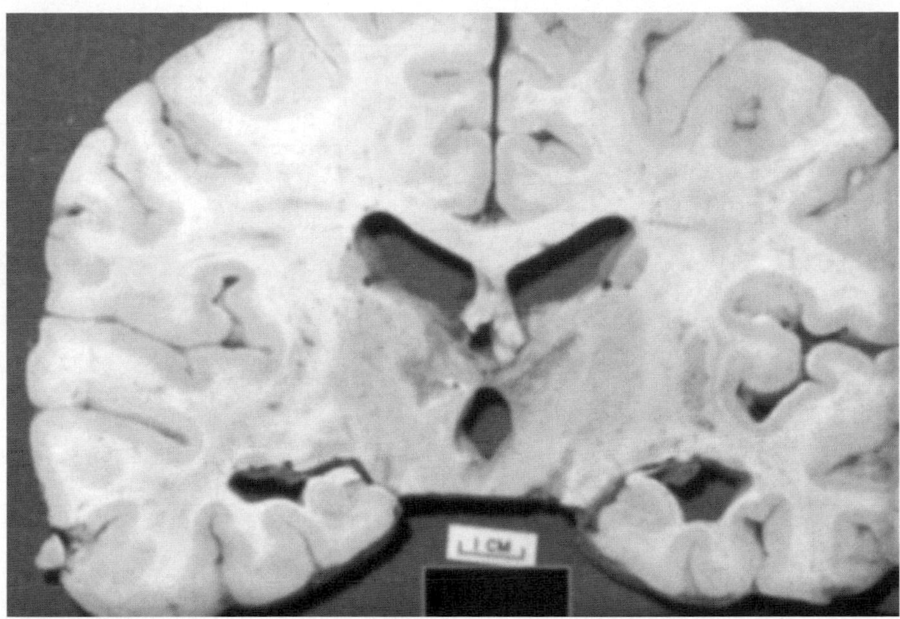

Fig. 2. Focal infarction in each thalamus in a patient in vegetative state. 12 months survival after an assault. The principal pathology in this case was raised intracranial pressure due to an intracranial clot that was evacuated, and ischemic damage within the distribution of each posterior cerebral artery.

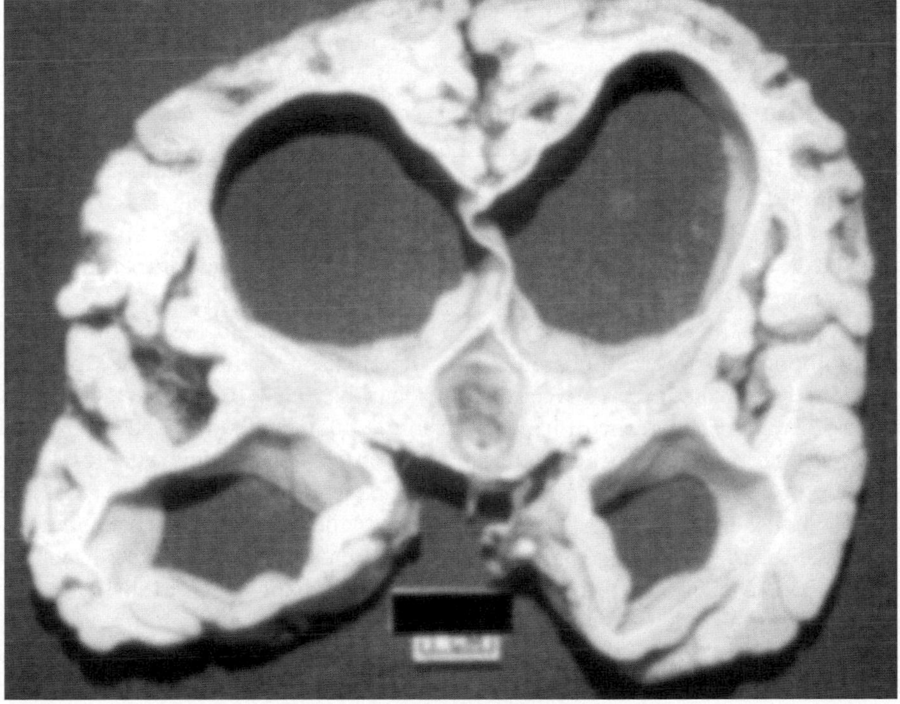

Fig. 3. Diffuse ischemic damage in a patient in vegetative state (case not in current series). 4 years survival after minor head injury complicated by seizures. There is gross thinning of the cortical ribbon and atrophy of subcortical gray matter and a corresponding marked loss of white matter. There is marked symmetrical hydrocephalus.

Table 3. Number of neurons in thalamic nuclei in control and vegetative patients

	Area in (mm^2)	Δ% Change from control (statistical significance)	Number of neurons in a 28 μm-thick slice	Δ% Change from control (statistical significance)
Control				
Dorsomedial nucleus	48.3±3.2	—	1823±100	—
Lateral posterior nucleus	35.5±3.2	—	1075±150	—
Ventral posterior nucleus	45.4±2.3	—	1528±77	—
Vegetative state				
Dorsomedial nucleus	34.2±2.6	−29% ($p<0.0001$)	1253±150	−31% ($p<0.0001$)
Lateral posterior nucleus	32.0±5.2	−10% ($p = 0.08$)	957±240	−11% ($p = 0.19$)
Ventral posterior nucleus	39.8±1.2	−12% ($p = 0.0001$)	1259±69	−18% ($p = 0.015$)

years), but were more common in the third case (aged 75 years). In five cases there were small numbers of tau IR$^+$ in the cytoplasm of neurons with grain and thread formation in two cases (aged 75 years) and again in the same cases only small amounts of ubiquitin IR$^+$ were a feature. In none of the cases was there immunoreactivity for α-synuclein. In four of the 12 controls there were Aβ IR$^+$ plaques in three of which there were a few tau and ubiquitin IR$^+$ inclusions, threads and IR$^+$ in the nuclei of glia. Immunoreactivity for α-synuclein was not present in any case.

Neuronal and glial features

The thalamus and hippocampus were abnormal in 28 VS cases, in 25 of whom the damage was diffuse. Neurons were distinguished from other cell types by their relative size. In the hippocampus the cell soma diameter ranged from 22 to 16 μm. The cell soma was measured across two perpendicularly separate diameters across neurons that contained regular-shaped pale nuclei with a discrete nucleolus. The overall mean value obtained was 18.95±2.8 μm. There was no difference in size between hippocampal subfields CA1, CA2, CA3, and CA4 (Maxwell et al., 2003). In the thalamus of control patients there was a significant difference in the size of neurons between the dorsomedial (DMN), VPN, and LPN (Table 3). Within the DM nucleus the median cell diameter was 7.4±0.19 μm, within VPN it was 11.3±0.7 μm, and within LPN it was 14.4±1.1 μm. ANOVA demonstrated a significant difference between groups ($p<0.0001$) (Maxwell et al., 2004). The total number of neurons was estimated in control and VS patients in the following subdivisions of the hippocampus and the thalamus, CA1, CA2, CA3, and CA4 in the hippocampus; and in DMN, VPN and LPN in the thalamus.

Overall, there was loss of pyramidal neurons from the hippocampus and parts of the thalamus in the VS cases with survival after blunt head injury. But, analysis between subfields of both the hippocampus and the thalamus demonstrated different degrees and rates of loss. In CA1 30% of neurons ($F = 16.6$, $p<0.001$), in CA3 23% ($F = 20.5$, $p<0.001$), and in CA4 34% ($F = 23.1$, $p = 0.001$) of neurons were lost. There was no loss of neurons from CA2 ($F = 6.1$, $p = 0.01$). Moreover, the rate of loss of neurons differed both between subfields and with length of survival. Two-thirds of the total loss of neurons from CA1 occurred early after injury, although there was a continued, lower rate of loss in late surviving patients. In CA3 almost all loss of neurons occurred shortly after injury, but a novel finding in this study was that there was continuing loss of neurons from CA4 and that this sector lost the largest proportion of neurons of any subfield (Maxwell et al., 2003). However, in the thalamus, analysis of differences in mean neuronal diameter did not support the hypothesis that there was an overall change in neuronal soma dimensions for either DMN ($p = 0.88$) or VPN ($p = 0.076$). But there was a significant change in the mean size of neurons within LPN ($p<0.0001$, $t = 7.1$). Thus, in surviving VS patients, there is an ongoing response by neurons in LPN. In terms of the total number of neurons within a transverse 28 μm thick slice, however, there was a significant loss of neurons in VS patients from both DMN ($p<0.0001$, $t = 10.4$) and VPN ($p<0.0001$, $t = 8.0$). But there was no loss from LPN ($p = 0.76$) (Maxwell et al., 2004) (Table 3).

In parallel with changes in the number of neurons in the DM, VP and LP thalamic nuclei described above, a study of changes in cell number of reactive astrocytes (GFAP positive), macrophages, perivascular macrophages and microglia (CD68-positive), and activated microglia (CR3/43 positive) and caspase-3-positive (cells entering programed cell death) have been undertaken in each nucleus within a 21 μm-thick coronal slice at the level of the lateral geniculate body (Maxwell et al., personal communication). In outline, there was no change in the total number of astrocytes, but an increased number of reactive, GFAP-positive, astrocytes in VS patients. The number of CD-68 labeled cells (macrophages) and CR3/43-positive (activated microglia) increased in VS from sham values. But the number of caspase-3-labeled macrophages and microglia did not differ between control and VS patients. On the contrary, the number of caspase-3-labeled neurons and oligodendrocytes increased from control values in vegetative patients with the greatest change occurring in VPN.

Discussion

Aspects of the clinical and structural basis of the VS after head injury from the archives of the Departments of Neuropathology and Neurosurgery at the Institute of Neurological Sciences, Southern General Hospital, Glasgow, have been published already (Adams et al., 2000; Jennett et al., 2001). The principal inclusion criteria were that (i) the cases had been managed by the Department of Neurosurgery, (ii) there had been a full autopsy and subsequent retention of the brain for comprehensive neuropathological studies, and (iii) there was sufficient clinical information to allow an assessment of the GOS by one of us (BJ) (Jennett and Bond, 1975). As with any retrospective study the findings might not be entirely representative of the population at large but given that the department provides a tertiary service for the general histopathologists and forensic pathologists, it would seem likely that the findings in the study are reasonably representative of the pathology of VS, an opinion that is reinforced by the albeit scant literature. Likewise, it is recognized that the diagnosis of VS required sufficient clinical information from hospital and other records to allow the diagnosis to be made with confidence, even though the diagnostic criteria are relatively new and are in fact still evolving (Jennett and Plum, 1972; Multi-society Task Force on PVS, 1994; Giacino et al., 1997; Jennett, 2002; Royal College of Physicians, 2003). Given these caveats the principal objectives of this study were to provide basic clinical and neuropathological information, to provide novel additional quantitative data using stereological techniques on neuronal cell counts in the hippocampus and thalamus, and to assess the extent to which any neurodegenerative changes might have been contributing to the clinical outcome.

The neuropathological basis of the VS consists largely of case reports or small series of cases (French, 1952; Brierley et al., 1971; Ingvar et al., 1978; Dougherty et al., 1981; Cole and Cowie, 1987; Relkin et al., 1990; Kinney et al., 1994). The comprehensive review of the subject matter by Kinney and Samuels (1994) was especially important as it concluded that there were three main patterns of brain damage associated with VS, namely (i) widespread damage to gray matter, (ii) widespread damage to white matter, and (iii) damage to each thalamus. They emphasized the role of the thalamic nuclei in VS and remarked that progress in the understanding of the structural basis of this clinical syndrome would require a quantitative assessment of various neuronal populations using the appropriate stereological techniques.

Examples of each of the three principal types of brain damage were found in the present series of cases. The first pattern of widespread damage in the cortex (and subcortical gray matter) was present in 15 (43%) of cases, such damage being a consequence of either cardiac arrest or an episode of prolonged profound hypotension (Adams et al., 1966; Dougherty et al., 1981; Cole and Cowie, 1987; Kinney and Samuels, 1994). The second pattern of widespread damage to white matter is now recognized to be the consequence of diffuse traumatic axonal injury (Adams et al., 1989) as previously described by Strich (1956) in a series of cases of "severe post-traumatic dementia" described as widespread Wallerian-type degeneration in the neuraxis. The more severe grades of diffuse traumatic axonal injury (grades 2 and 3) were present in 28 (80%) of the current series and was in fact the most common type of structural damage identified in this cohort of cases. The third pattern described by Kinney and Samuels (1994) was that of damage restricted to, or maximal in, the thalami. This is the least common pattern in the current series with either diffuse or focal damage in 15 (43%) but has been noted previously (Jellinger, 1994; Kinney et al., 1994). However, changes in the thalamus were one of the more common findings in the series – 28 of the 35 (80%), which is why it has been suggested that VS in large measure may be attributed to structural changes in the thalami whether hypoxic/ischemic, secondary to disconnection as a result of axonal damage, or as a consequence of the vascular complications of raised intracranial pressure. Given the heterogeneity of findings in the current series it is likely that either diffuse or focal or a combination of both may be found. According to Kinney and Samuels (1994) a cause of VS is selective damage to the thalami. In the present series there was not one such case as there was always structural damage largely due to DAI or ischemia in other brain areas. However, it is accepted that the techniques used in this study have limitations and may not have been sufficiently sensitive to have identified the full nature of damage and its distribution. Quantitative studies of the type recommended by Kinney and Samuels (1994) are currently undertaken by us (Maxwell et al., 2004) and should provide the appropriate data.

In the thalamus, after head injury and survival in the VS for greater than three months, morphological change comparable to that described in other studies (reviewed by Ross and Ebner, 1990; Xuereb et al., 1991) was noted in neurons. Neuronal cell bodies were either within the control size range with a well-defined, pale central nucleus or reduced in size with a darkened cytoplasm, a peripherally located nucleus the cell often lying at the edge of its residual lacuna within the neuropil. This, it is suggested, may be due to the fact that the proportion of small, dark neurons did not differ between control and head-injured patients. The interpretation of these data can only be hypothetical at this time. The present working hypothesis is that loss of neurons occurs within three months of blunt head injury from DMN and VPN, but neuronal loss from LPN occurs much more slowly and has a much longer time course. Overall, cellular mechanisms for neuronal loss post-trauma was not a remit of the present study. But a review of the literature indicates that knowledge of the detailed intercellular relationships of human thalamic neurons is still incomplete. The present study does provide quantitative evidence that there is, at least, more than a single mechanism and time course for neuronal loss and that the degree of loss differs both between thalamic nuclei and subfields of the hippocampus. Further, there are many types of interneurons within the human thalamus (reviewed in Ross and Ebner, 1990; Chmielowska and Pons, 1995; Rausell et al., 1992). A mechanistic analysis is therefore limited by the present lack of detailed knowledge concerning thalamic input originating either from the cerebral cortex or from the brain stem (Ross and Ebner, 1990; Chmielowska and Pons, 1995; Rausell et al., 1992). Nonetheless, the present study provides novel, quantitative data that there is, at least more than one phase of loss of neurons from certain thalamic nuclei after human blunt head injury. For example, rapid loss resulting from either oligaemia (Stamatakis et al., 2002) and altered glucose metabolism (Juhasz et al., 1999) or retrograde thalamic degeneration, which in man takes some 3 months post-trauma to develop (Adams et al., 2000). Further, it is noteworthy that the nuclei from which loss occurred in the present study (DMN and VPN) differ from those noted in Alzheimer's disease (anterodorsal and centromedian) (Xuereb et al., 1991). Further analysis will probably require complex immunocytochemical analysis of which particular subgroup(s) of neurons within the human thalamus respond or are lost after traumatic brain injury (TBI).

Delayed changes in the thalami after head injury in both laboratory animals and in patients have been described previously (see Chen et al., 2003) and have been attributed to programed cell death as demonstrated by the TUNEL technique and caspase activation in both laboratory animals (Rink et al., 1995) and head injured patients (Wilson et al., 2004). With increasing survival the brain undergoes atrophy and there is an astrocytosis and compensatory enlargement of the ventricular system. Current speculation concerns whether these processes are passive reflecting a long natural history as with Wallerian-type degeneration, or to continuing microglial/macrophage activation with continuing death of oligodenrocytes (Wilson et al., 2004).

Recent studies have found accumulation of intracytoplasmic inclusions within neurons and in axons after experimentally induced head injury (Smith et al., 1999). Such findings have contributed to speculation that head injury is a risk factor for the subsequent onset of neurodegeneration changes akin of Alzheimer's disease (Roberts et al., 1991; Gentleman et al., 1997).

The epidemiological evidence for such an association is most marked in various contact sports especially in boxers (Corsellis et al., 1973; Jordan et al., 1997). The mechanisms underlying a possible association between head injury and Alzheimer's disease is pathology in the form of neurofibrillary tangles and neuropil threads (Newman et al., 1995). In a more recent detailed study (Smith et al., 2003) it was concluded that neurofibrillary tangles were not more common in patients surviving for up to 1 month after head injury when compared with age-matched controls. This finding is supported by the funding of only limited amounts of Aβ and tau immunoreactivity in the present series of cases. It is likely that genetic influences may play an important role in modifying the outcome after head injury. Possession of ApoE $\varepsilon 4$ is associated with a worse outcome after head injury (Nicoll et al., 1995; Teasdale et al., 1997) perhaps mediated through glial-neuronal interactions (Griffin et al., 1998).

In conclusion, this study has confirmed the importance of abnormalities in the thalami and hippocampi of VS patients after head injury and showed that these changes have a prolonged time course. While new degenerative abnormalities are also present, their frequency and distribution do not appear to materially contribute to the structural basis of the VS.

References

Adams, J.H., Brierley, J.B., Connor, R.C. and Treip, C.S. (1966) The effects of systemic hypotension upon the human brain. Clinical and neuropathological report in 11 cases. Brain, 89: 235–258.

Adams, J.H., Doyle, D., Ford, I., Gennarelli, T.A., Graham, D.I. and McLellan, D.R. (1989) Diffuse axonal injury in head injury: definition, diagnosis and grading. Histopathology, 15: 49–59.

Adams, J.H., Doyle, D., Graham, D.I., Lawrence, A.E., McLellan, D.R., Gennarelli, T.A., Pastuszko, M. and Sakamoto, T. (1985) The contusion index: a reappraisal in human and experimental non-missile head injury. Neuropathol. Appl. Neurobiol., 11: 299–308.

Adams, J.H. and Graham, D.I. (1976) The relationship between ventricular fluid pressure and the neuropathology of raised intracranial pressure. Neuropathol. Appl. Neurobiol., 2: 323–332.

Adams, J.H., Graham, D.I. and Jennett, B. (2000) The neuropathology of the vegetative state after acute brain insults. Brain, 123: 1327–1338.

Adams, J.H., Graham, D.I., Scott, G., Parker, L.S. and Doyle, D. (1980) Brain damage in fatal non-missile head injury. J. Clin. Pathol., 33: 1132–1145.

Adams, J., Jennett, B., McLellan, D.R., Murray, L.S. and Graham, D.I. (1999) The neuropathology of the vegetative state after head injury. J. Clin. Pathol., 52: 804–806.

Benes, F.M. and Lange, N. (2001) Two-dimensional versus three dimensional cell counting: a practical prospective. Trends Neurosci., 24: 11–17.

Brierley, J.B., Graham, D.I., Adams, J.H. and Simpson, J.A. (1971) Neocortical death after cardiac arrest: a clinical, neurophysiological and neuropathological report of two cases. Lancet, 2: 560–565.

Cammermeyer, J. (1961) The importance of avoiding "dark" neurons in experimental neuropathology. Acta Neuropathol., 1: 245–270.

Chen, S., Pickard, J.D. and Harris, N.G. (2003) Time course of cellular pathology after controlled cortical impact injury. Exp. Neurol., 182: 87–102.

Chmielowska, J. and Pons, T.P. (1995) Patterns of thalamocortical degeneration after ablation of somatosensory cortex in monkeys. J. Comp. Neurol., 360: 377–392.

Cole, G. and Cowie, V.A. (1987) Long survival after cardiac arrest: case report and neuropathological findings. Clin. Neuropathol., 6: 104–109.

Corsellis, J.A.N., Bruton, C.J. and Foreman-Browne, D. (1973) The aftermath of boxing. Psychol. Med., 3: 270–303.

Dougherty, J.H., Rawlinson, D.H., Levy, D.E. and Plum, F. (1981) Hypoxic-ischaemic brain injury and the vegetative state: clinical and neuropathological correlation. Neurology, 31: 991–997.

French, J.D. (1952) Brain lesions associated with prolonged unconsciousness. Acta Neurol. Psychiatr., 68: 727–740.

Gavrieli, Y., Sherman, Y. and Ben-Sasson, S.A. (1992) Identification of programmed cell death in situ via specific labeling of nuclear DNA fragmentation. J. Cell Biol., 119: 493–501.

Gentleman, S.M., Greenberg, B.D., Savage, M.J., Noori, M., Newman, S.J., Roberts, G.W., Griffin, W.S. and Graham, D.I. (1997) A beta 42 is the predominant form of amyloid beta-protein in the brains of short-term survivors of head injury. Neuroreport, 8: 1519–1522.

Giacino, J.T., Zasler, N.D., Katz, D.I., Kelly, J.P. and Rosenberg, J.H. (1997) Development of practice guidelines for assessment and management of the vegetative and minimally conscious state. J. Head Trauma Rehabil., 12: 79–89.

Graham, D.I., Ford, I., Adams, J.H., Doyle, D., Teasdale, G.M., Lawrence, A.E. and McLellan, D.R. (1989) Ischaemic brain damage is still common in fatal non-missile head injury. J. Neurol. Neurosurg. Psychiatry, 52: 346–350.

Griffin, W.S., Sheng, J.G., Royston, M.C., Gentleman, S.M., McKenzie, J.E., Graham, D.I., Roberts, G.W. and Mrak, R.E. (1998) Glial-neuronal interactions in Alzheimer's disease: the potential role of a 'cytokine cycle' in disease progression. Brain Pathol., 149: 32–40.

Ingvar, D.H., Brun, A., Johansson, L. and Samuelsson, S.M. (1978) Survival after severe cerebral anoxia with destruction of the cerebral cortex in the apallic syndrome. Ann. NY Acad. Sci., 315: 184–214.

Jellinger, K. (1994) The brain of Karen Ann Quinlan (Letter). N. Engl. J. Med., 331: 1378–1379.

Jennett, B. (2002) The Vegetative State. Cambridge University Press, Cambridge.

Jennett, B., Adams, J.H., Murray, L.S. and Graham, D.I. (2001) Neuropathology in vegetative and severely disabled. Neurology, 56: 486–490.

Jennett, B. and Bond, M. (1975) Assessment of outcome after severe brain damage. A practical scale. Lancet, 1: 480–484.

Jennett, B. and Plum, F. (1972) Persistent vegetative state after brain damage. A syndrome in search of a name. Lancet, 1: 734–737.

Jones, E.G. (1985) The Thalamus. Plenum Press, New York.

Jordan, B.D., Relkin, N.R., Ravdin, L.D., Jacobs, A.R., Bennett, A. and Gandy, S. (1997) Apolipoprotein E epsilon 4 associated with chronic traumatic brain injury in boxing. J.A.M.A., 278: 136–140.

Juhasz, C., Nagy, F., Watson, C., da Silva, E.A., Muzik, O., Chugani, D.C., Shah, J. and Chugani, H.T. (1999) Glucose and [^{11}C] flumazenil positron emission tomography abnormalities of thalamic nuclei in temporal lobe epilepsy. Neurology, 53: 2045–2073.

Kinney, H.C., Korein, J., Panigrahy, A., Dikkes, P. and Goode, R. (1994) Neuropathological findings in the brain of Karen Ann Quinlan. The role of the thalamus in the persistent vegetative state. Br. Engl. J. Med., 330: 1469–1475.

Kinney, H.C. and Samuels, M.A. (1994) Neuropathology of the persistent vegetative state. A review. J. Neuropathol. Exp. Neurol., 53: 548–558.

Maxwell, W.L., Dhillon, K., Harper, L., Espin, J., McIntosh, T.K., Smith, D.H. and Graham, D.I. (2003) There is differential loss of pyramidal cells from the human hippocampus with survival after blunt head injury. J. Neuropathol. Exp. Neurol., 62: 272–279.

Maxwell, W.L., Pennington, K., MacKinnon, M.-A., Smith, D.H., McIntosh, K.T., Lindsay Wilson, J.T. and Graham, D.I. (2004) Differential responses in three thalamic nuclei in moderately disabled, severely disabled and vegetative patients after blunt head injury. Brain, 127: 2470–2478.

Multi-Society Task Force in PVS. (1994) Medical aspects of the persistent vegetative state. N. Eng. J. Med., 330: 1572–1579.

Murray, G.D., Teasdale, G.M., Braakman, R., Cohadon, F., Dearden, M., Iannotti, F., Karimi, A., Lapierre, F., Maas, A., Ohman, J., Persson, L., Servadei, F., Stocchetti, N., Trojanowski, T. and Unterberg, A. (1999) The European brain injury consortium survey of head injuries. Acta Neurochir. (Wien), 141: 223–236.

Newman, S.J., Gentleman, S.M., Graham, D.I., Brown, F. and Roberts, G.W. (1995) Tissue distribution and cellular localization of hyperphosphorylated tau in human head injury and ag-matched controls. In: Iqbal K., Mortimer J.A., Winblad B. and Wisniewski H.M. (Eds.), Research Advances in Alzheimer's disease and Related Disorders. Wiley, Chichester, pp. 397–403.

Nicoll, J.A.R., Roberts, G.W. and Graham, D.I. (1995) Apolipoprotein E ε4 allele is associated with deposition of amyloid β-protein following head injury. Nat. Med., i: 135–137.

Rausell, E., Bae, C.S., Vinuela, A., Huntley, G.W. and Jones, E.G. (1992) Calbindin and parvalbumin cells in monkey VPL thalamic nucleus: distribution, laminar cortical projections, and relations to spinothalamic terminations. J. Neurosci., 12: 4088–4111.

Relkin, N.R., Petito, C.K. and Plum, F. (1990) Coma and the vegetative state associated with thalamic injury after cardiac arrest (Abstract) Ann. Neurol., 28: 221–222.

Rink, A., Fung, K.M., Trojanowski, J.Q., Lee, V.M., Neugebauer, E. and McIntosh, T.K. (1995) Evidence of apoptotic cell death after experimental traumatic brain injury in the rat. Am. J. Pathol., 147: 1575–1583.

Roberts, G.W., Gentleman, S.M., Lynch, A. and Graham, D.I. (1991) β-A4 amyloid protein deposition in brain after head trauma. Lancet, 338: 1422–1423.

Ross, D.T. and Ebner, F.F. (1990) Thalamic retrograde degeneration following cortical injury: an excitotoxic process? Neuroscience, 35: 525–550.

Royal College of Physicians. (2003) The vegetative state. Clin. Med., 3: 249–254.

Smith, C., Graham, D.I., Murray, L.S. and Nicoll, J.A. (2003) Tau immunohistochemistry in acute brain injury. Neuropathol. Appl. Neurobiol., 29: 496–502.

Smith, C., Graham, D.I., Murray, L.S. and Nicoll, J.A.R. (2003) Tau immunohistochemistry in acute brain injury. Neuropath. Appl. Neurobiol., 29: 496–502.

Smith, D.H., Chen, X.H., Nonaka, M., Trojanowski, J.Q., Lee, V.M., Saatman, K.E., Leoni, M.J., Xu, B.N., Wolf, J.A. and Meaney, D.F. (1999) Accumulation of amyloid beta and tau and the formation of neurofilament inclusions following diffuse brain injury in the pig. J. Neuropath. Exp. Neurol., 58: 982–992.

Stamatakis, E.A., Wilson, J.T.L., Hadley, D.M. and Wyper, D.J. (2002) SPECT imaging in head injury interpreted with statistical parametric mapping. J. Nucl. Med., 43: 476–483.

Strich, S.J. (1956) Diffuse degeneration of the cerebral white matter in severe dementia following head injury. J. Neurol. Neurosurg. Psychiatry, 19: 163–185.

Teasdale, G.M., Nicoll, J.A., Murray, G. and Fiddes, M. (1997) Association of Apolipoprotein E polymorphism with outcome after head injury. Lancet, 350: 1069–1071.

West, M.J., Slomianka, L. and Gundersen, H.J.G. (1991) Unbiased stereological estimation of the total number of neurons in the subdivisions of the rat hippocampus using the optical fractionator. Anat. Rec., 231: 482–497.

Wilson, S., Raghupathi, R., Saatman, K.E., MacKinnon, M.A., McIntosh, T.K. and Graham, D.I. (2004) Continued in situ DNA fragmentation of microglia/macrophages in white matter weeks and months after traumatic brain injury. J. Neurotrauma, 21: 238–250.

Xuereb, J.H., Perry, R.H., Candy, J.M., Perry, E.K., Marshall, E. and Bonham, J.R. (1991) Nerve cell loss in the thalamus in Alzheimer's disease and Parkinson's disease. Brain, 114: 1363–1379.

S. Laureys (Ed.)
Progress in Brain Research, Vol. 150
ISSN 0079-6123

CHAPTER 32

Using a hierarchical approach to investigate residual auditory cognition in persistent vegetative state

Adrian M. Owen[1,2,*], Martin R. Coleman[2], David K. Menon[2], Emma L. Berry[1], Ingrid S. Johnsrude[1], Jennifer M. Rodd[3], Matthew H. Davis[1] and John D. Pickard[2]

[1]*MRC Cognition and Brain Sciences Unit, 15 Chaucer Road, Cambridge CB2 2EF, UK*
[2]*Wolfson Brain Imaging Centre and the Cambridge Coma Study Group, Addenbrooke's Hospital, University of Cambridge, Hills Road, Cambridge CB2 2QQ, UK*
[3]*Department of Psychology, University College London, London WC1H 0AP, UK*

Abstract: Persistent vegetative state is arguably one of the least understood and most ethically troublesome neurological conditions in modern medicine. The term describes a rare disorder in which patients who emerge from coma appear to be awake, but show no signs of awareness. In recent years, a number of studies have demonstrated an important role for functional neuroimaging in the identification of residual cognitive function in patients meeting the clinical criteria for persistent vegetative state. Such studies, when successful, may be particularly useful where there is a concern about the accuracy of the diagnosis and the possibility that residual cognitive function has remained undetected. Unfortunately, functional neuroimaging in persistent vegetative state is extremely complex and subject to numerous methodological, clinical and theoretical difficulties. In this chapter, we argue that in order to most effectively define the degree and extent of preserved cognitive function in persistent vegetative state, a hierarchical approach to cognition is required. To illustrate this point, a series of functional neuroimaging paradigms in the auditory domain are described, which systematically increase in complexity in terms of the auditory and/or linguistic processes required and, therefore, the degree of preserved cognition that can be inferred from "normal" patterns of activation in persistent vegetative patients. Preliminary results in a small series of patients provide a strong basis for the systematic study of possible residual cognitive function in persistent vegetative state.

Introduction

The clinical features of persistent vegetative state were formally introduced into the literature by Jennett and Plum (1972) and later clarified and refined by the Multi-Society Task Force on Persistent Vegetative State (1994a, b) and the Royal College of Physicians (1996). Etiology is variable, although the condition may arise as a result of road traffic accident, ischemic attack, anoxia, encephalitis or viral infection. A diagnosis of persistent vegetative state is not normally considered until between 1 and 3 months post ictus, at which point there must be no evidence of sustained, reproducible, purposeful or voluntary behavioral response to visual, auditory, tactile or noxious stimuli. There must also be no evidence of language comprehension or expression, although there is generally sufficiently preserved hypothalamic and brain stem autonomic functions to permit survival with medical care. Although persistent vegetative state often follows coma, it is characterized by an irregular but cyclic state of circadian sleeping and waking. In

*Corresponding author. Tel.: +44 (0) 1223 355 294 X511; Fax: +44 (0) 1223 359 062; E-mail: adrian.owen@mrc-cbu.cam.ac.uk

DOI: 10.1016/S0079-6123(05)50032-3

contrast, patients in coma present with eyes closed and lack any consistent sleep–wake cycles.

An accurate and reliable evaluation of the level and content of cognitive processing is of paramount importance for the appropriate management of patients diagnosed with persistent vegetative state. Objective behavioral assessment of residual cognitive function can be extremely difficult in these patients, as motor responses may be minimal, inconsistent, and difficult to document, or may be undetectable because no cognitive output is possible. In recent years, a number of studies have demonstrated an important role for functional neuroimaging in the identification of residual cognitive function in vegetative patients. Until recently, the majority of these studies used either fluorodeoxyglucose (FDG), positron emission tomography (PET) or single photon emission computed tomography (SPECT), and reported widespread reductions of up to 50% in (resting) cerebral blood flow and glucose metabolism (Levy et al., 1987; DeVolder et al., 1994; Tommasino et al., 1995). In some cases, however, isolated "islands" of metabolism have been identified in circumscribed regions of cortex, which may suggest residual cognitive processing in a subset of patients (see Schiff and Plum, 1999). While metabolic studies are useful in this regard, they can only identify functionality at the most general level; that is, mapping cortical and subcortical regions that are *potentially* recruitable, rather than relating neural activity within such regions to specific cognitive processes (Momose et al., 1989; Turkstra, 1995).

On the other hand, methods such as $H_2^{15}O$ PET and functional magnetic resonance imaging (fMRI) can be used to link residual neural activity to the presence of covert cognitive *function*. In short, functional neuroimaging, or the so-called "activation studies", have the potential to demonstrate distinct and specific physiological responses (changes in regional cerebral blood flow or changes in regional cerebral haemodynamics) to controlled external stimulation in the absence of any overt response on the part of the patient. In the first of such studies, $H_2^{15}O$ PET was used to measure regional cerebral blood flow (rCBF) in a post-traumatic vegetative patient during an auditorily presented story told by his mother (de Jong et al., 1997). Compared to non-word sounds, activation was observed in the anterior cingulate and temporal cortices, possibly reflecting emotional processing of the contents, or tone, of the mother's speech. In another patient diagnosed as vegetative, Menon et al. (1998) also used PET, but to study covert *visual* processing in response to familiar faces. During "experimental" scans, the patient was presented with pictures of the faces of family and close friends, while during "control" scans scrambled versions of the same images were presented, which contained no meaningful visual information whatsoever. Previous imaging studies in healthy volunteers have shown that such tasks produce robust activity in the right fusiform gyrus, the so-called human "face area" (e.g. Haxby et al., 1991, 1994). The same visual association region was activated in the vegetative patient when the familiar face stimuli were compared to the meaningless visual images (Menon et al., 1998; Owen et al., 2002). In other cohort studies, both noxious somatosensory stimuli (Laureys et al., 2002) and auditory stimuli (Laureys et al., 2000; Owen et al., 2002; Boly et al., 2004) have also been shown to systematically activate appropriate cortical regions in patients meeting the clinical criteria for the vegetative state.

In all such studies, however, the choice of imaging paradigm poses a number of methodological, ethical and procedural problems. For example, as noted above, motor responses are often minimal, inconsistent or absent in patients with persistent vegetative state and, by definition, cannot be elicited directly (i.e. willfully) by external stimulation. In addition, even assuming that some level of residual cognitive processing does exist, there is no reliable mechanism for ensuring that the presented stimuli are actually *perceived* by the patient. Many persistent vegetative state patients suffer serious damage to auditory and/or visual input systems, which may impede performance of any "higher" cognitive functions (e.g. voice discrimination), which place demands on these "lower" sensory systems (e.g. hearing). Like patients with any form of serious brain damage, persistent vegetative state may also be accompanied by a significant reduction in attention span (assuming some level of cognitive processing remains), which may further complicate the assessment of higher cognitive functions. Spontaneous movements during the scan itself may also compromise the interpretation of functional neuroimaging data, particularly scans acquired using fMRI. Where PET methodology is employed, issues of radiation burden must also be considered and may preclude longitudinal or follow-up studies in many patients. Finally, data processing of functional neuroimaging data may also present challenging problems in patients clinically diagnosed as persistent vegetative state. For example, the presence of gross hydrocephalus or focal pathology may complicate coregistration of functional data (e.g. acquired with PET or fMRI) to anatomical data (e.g. acquired using structural MRI), and the normalization of images to a healthy reference brain. Under these circumstances, statistical assessment of activation patterns is complex and interpretation of activation foci in terms of standard stereotaxic coordinates may be impossible.

A final problem concerns the extent to which functional neuroimaging data can, and should, be used as evidence of preserved *awareness* in patients meeting the clinical criteria for persistent vegetative state. While a number of recent studies have successfully identified islands of residual cognitive function in vegetative patients (e.g., de Jong et al., 1997; Menon et al., 1998; Laureys et al., 2000, 2002; Owen et al., 2002; Boly et al., 2004), invariably in such cases, the evidence for awareness of either self or the environment has been equivocal, at best. A plethora of data in healthy volunteers from studies of implicit learning and the effects of priming (see e.g., Schacter, 1994 for review) to studies of learning during anesthesia (e.g., Bonebakker et al., 1996) have demonstrated that many aspects of human cognition can go on in the absence of awareness. Thus, the fact that a normal response to face stimuli was observed in the patient described by Menon et al. (1998), for example, does not necessarily mean that the patient was *aware* during the procedure; in healthy volunteers, signal intensity changes have been observed in the fusiform gyrus in response to primed faces in the absence of conscious awareness (Henson et al., 2000).

In this chapter, we will argue that functional neuroimaging studies in patients meeting the clinical criteria for persistent vegetative state should be conducted hierarchically; beginning with the simplest form of processing within a particular domain (e.g. auditory) and then progressing sequentially through more complex cognitive functions. By way of example, a series of auditory paradigms will be described that have all been successfully employed in functional neuroimaging studies of vegetative patients. These paradigms increase in complexity systematically from basic acoustic processing to more complex aspects of language comprehension and semantics. We suggest that such a hierarchy of cognitive tasks provides the most valid mechanism for defining the depth and breadth of preserved cognitive function in patients meeting the clinical criteria for persistent vegetative state, and discuss how such an approach might be extended to allow clear inferences about the level of "awareness" or consciousness to be made.

Acoustic processing

As noted above, many vegetative patients suffer serious damage to auditory and/or visual input systems, which may impede performance of all cognitive functions that place demands on these "lower" sensory systems. At the most basic level, therefore, it is important to establish normal or near-normal sensory perception in any candidate patient for functional neuroimaging studies of "higher" cognitive functions (e.g. language processing). For the most part, functional neuroimaging is not necessary in this regard; that is to say, the measurement of auditory or visual evoked potentials are usually sufficient to establish that the respective neural pathways are intact. The integrity of the auditory neural axis can be assessed using a number of tests including the brainstem auditory evoked response (BAER) and passive mismatch negativity (MMN). The BAER is typically elicited using broad band clicks presented through headphones and produces a series of seven waveforms recorded from the scalp during the first 12 ms following a click. These waveforms reflect post-synaptic activity along the initial length of the neural axis, including the eighth cranial nerve, pons and thalamic nuclei. In contrast, the passive MMN test is one of the several long-latency auditory evoked potentials able to assess the integrity of the auditory cortex. The MMN is elicited following an infrequent change in a repetitive sound. This is usually a tone burst, such as a 1500 Hz deviant occurring among 1000 Hz standard tones, but can also be more complex auditory stimuli such as synthesized speech. Experimental evidence suggests that MMN (main component) is generated in the superior temporal cortex (Alho, 1995). It is widely thought to reflect a pre-cognitive response generated by comparing the deviant input with a neural memory trace encoding the physical features of the repetitive sound (Naatanen, 2003). The MMN has been successfully applied to the assessment of vegetative patients, although with considerable variability in results (Jones et al., 2000; Kotchoubey et al., 2001; Kotchoubey, this volume). The integrity of the visual neural pathway can be assessed in a similar way to the BAER. The post-synaptic response to transient visual stimuli, such as flashes from light emitting diodes (LED), can be recorded from the scalp using electrodes. In response to LED flashes, the visual cortex typically produces a series of three waveforms during the first 250 ms after the stimulus. These waveforms reflect post-synaptic activity in the lateral geniculate and striate areas of the visual cortex. However, although such investigations can be easily applied to vegetative patients using LED goggles, for example, this visual evoked potential is too crude to pinpoint most visual impairments and, where necessary, more detailed investigations using retinography or pattern reversal visual-evoked potentials should be employed.

Perceptual processing

Once basic neural responses to sounds have been established, it becomes possible to investigate whether the lesioned brain is able to discriminate between different categories of sound. Speech perception in healthy volunteers has been widely investigated within the functional neuroimaging literature and the findings have obvious clinical and therapeutic relevance for the investigation of preserved cognitive function in vegetative patients. Most commonly, studies of speech perception involve volunteers being scanned while listening to binaurally presented spoken words (experimental condition), or signal correlated noise sounds (acoustic control condition), or no auditory stimulus at all (silence condition). For example, Mummery et al. (1999) used PET to scan six neurologically normal volunteers while they listened to concrete nouns or signal-correlated noise at a rate of 30 items per minute. The task instruction was to "pay attention to the stimuli without responding". When speech was compared with signal-correlated noise, they found a broad swathe of activation along both superior temporal gyri, extending ventrolaterally into the superior temporal sulcus (see Fig. 1, left panel).

The same paradigm has been applied recently to small numbers of patients meeting the clinical criteria for persistent vegetative state. For example, Owen et al. (2002) described the case of a 30- year-old female bank manager, who suffered severe head injuries during a road traffic accident involving a head-on collision with another vehicle (patient DE). During a significant period trapped in the car, she probably suffered a period of hypoxia with hypotension. The Glasgow Coma Scale at the scene of the accident indicated a score of 4/15 with no improvement post-resuscitation. A long stay in intensive care was accompanied by episodes of pupillary unreactivity on more than one occasion. Fourteen weeks *post ictus*, the pupils remained dilated and unreactive. The patient was weaned off the ventilator but still required a tracheotomy. Brainstem auditory evoked responses at the time were intact on the right but abnormal on the left. Computed tomography findings at admission revealed a left frontal sub-cortical hemorrhagic contusion and smaller fronto-parietal contusion. A small hemorrhage in the inter-peduncular fossa with low-density areas in adjacent midbrain-pons region were also observed. A repeat scan on the same day showed fresh midline hemorrhage in the anterior midbrain extending to posteromedial right thalamus. In addition, punctuate areas of high density in the right cerebellum and both cerebral hemispheres suggested diffuse axonal injury. The cerebral ventricles were noted to be normal in

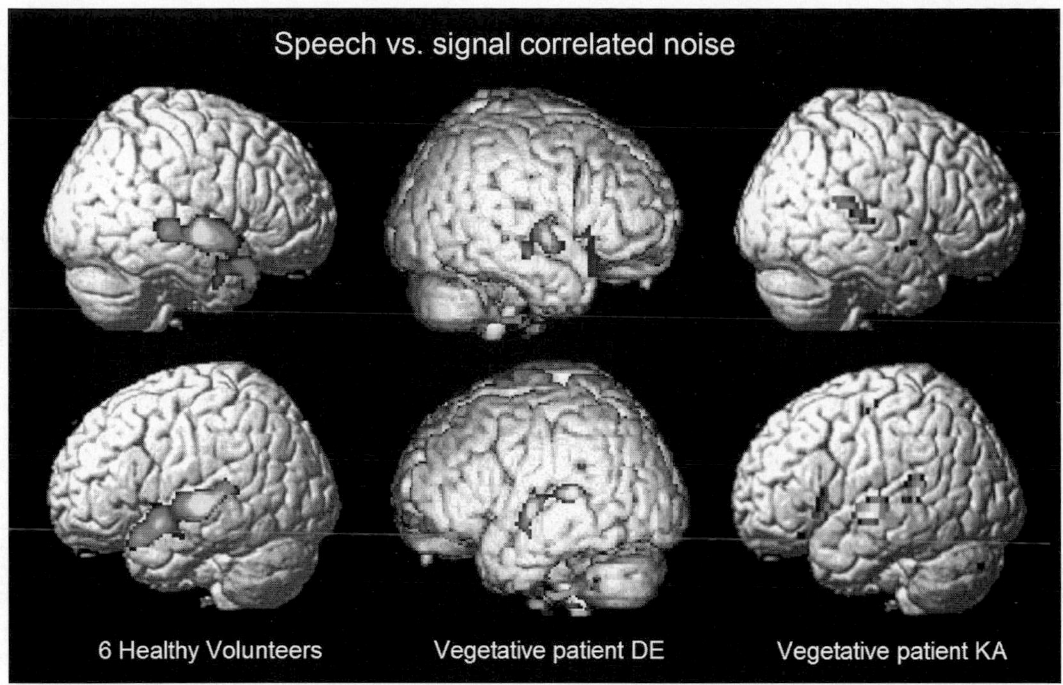

Fig. 1. PET data illustrating the comparison of speech with signal correlated noise. On the left, control data (adapted from Mummery et al., 1999), and on the right data from two patients meeting the clinical criteria for vegetative state. All data were pre-processed and analyzed using Statistical Parametric Mapping software (SPM99, Wellcome Department of Imaging Neuroscience, London, UK). See Plate 32.1 in Colour Plate Section.

size. Over several weeks the patient developed a withdrawal to pain but showed no consistent evidence of volitional activity.

The decision to use a PET auditory speech task was made largely on the basis of the partially intact brain auditory evoked responses. The patient was scanned while being presented binaurally with either spoken words (4 scans), signal correlated noise sounds (4 scans) or no auditory stimulus at all (4 scans). The presented words were disyllabic nouns matched for frequency (6–20,000), concreteness (400–700) and imageability (300–700) and were prerecorded on a tape by a speaker, rate being controlled by a metronome. The signal-correlated noise stimuli were made by selecting a sample of these spoken nouns with varying segmental durations and initial manners of articulation digitizing them, and then multiplying these with noise. The two sets of stimuli were matched for loudness by adjusting the amplification until they were subjectively similar. The task instruction given to the patient was to "pay attention to the stimuli without responding" or, in the silence condition, just to rest. In both stimulus conditions, the words were presented at rates of 30 per minute and presentations were started 30 s prior to tracer infusion and continued until the end of data acquisition. The data were pre-processed and analyzed using Statistical Parametric Mapping software (SPM99, Wellcome Department of Imaging Neuroscience, London, UK).

In patient DE, the comparison of noise bursts with rest revealed significant foci of activation bilaterally in the auditory region, suggesting that basic auditory processes were at least somewhat functional. The comparison of speech sounds with noise bursts revealed significant rCBF increases on the superior temporal plane bilaterally and posterior to auditory cortex, in the region of the planum temporale, in the left hemisphere only (see Fig. 1, middle panel). The reverse subtraction yielded no significant foci of activation. These findings correspond extremely closely with the results described by Mummery et al. (1999), using the same task in healthy volunteers (see Fig. 1). The same group have described similar findings in a second patient (patient KA) who also met the clinical criteria for persistent vegetative state (Owen et al., in press). In that study, a similar task was used and yielded an almost identical pattern of findings in the superior temporal lobe region when speech stimuli were compared with signal correlated noise (see Fig. 1, right panel).

Although, at first glance, activation in patients DE and KA appears to be rather less intense and less extensive than that observed in healthy volunteers (Mummery et al., 1999; see Fig. 1, left panel), this is probably not the case. While the normal data comprises 36 scans per condition, the two patients were scanned only four times in each condition, with a correspondingly lower chance of any voxel achieving significance. Given residual variations in normal structural and functional anatomy after spatial normalization, activation would be expected to be more extensive when averaged across several subjects than when measured in a single subject. When these factors are taken into account, the activation observed in both patients DE and KA is very similar to that seen in the group of healthy individuals.

In short, notwithstanding qualitative differences that are well within the range that would be expected given normal inter-subject variability, the pattern of activation observed in two vegetative patients during speech versus signal correlated noise contrasts is very similar to that observed in healthy awake control volunteers while performing the same tasks. Of course, recognising speech as speech does not imply anything about comprehension; that is, whether the *content* of the speech is understood or not (consider the experience of listening to a speaker talk in a foreign language of which you have no prior experience). To assess speech comprehension in persistent vegetative state it is necessary to move on to more complex experiment designs which tap aspects of *phonological* processing.

Phonological processing

While the results from studies of speech processing in vegetative patients (e.g., Owen et al., 2002; in press; Boly et al., 2004), have suggested that some level of covert linguistic functioning may be preserved, such tasks do not allow any conclusions to be drawn about *comprehension*, i.e., whether speech is processed beyond the point at which it is identified as speech. One approach to this problem, which has met with some success, is to document responses to a set of stimuli of *graded complexity*. Crucially, assessment of shifts in the patient's pattern of activation to a standardized set of graded stimuli may provide a more comprehensive assessment of level of residual cognitive capacity, at least with reference to those processes recruited by the experimental task in question. Davis and Johnsrude (2003) have developed such a task using a test of graded intelligibility to look at speech comprehension. During the task, volunteers listen to sentences that have been distorted such that they produce a range of six levels of intelligibility (as measured by subsequent word report scores). In a parallel fMRI study, intelligibility (operationalized as "the amount of a sentence that is understood") was found to correlate with activation in a region of the left anterior and superior temporal lobe; as intelligibility increased, so did signal intensity in this region. This increase was also significantly positively correlated with word report scores; signal intensity increasing linearly as the subjects reported more words correctly. These findings in healthy volunteers suggest that activity in the left anterior and superior temporal lobe reflects processing of the linguistic content of spoken sentences (words and meanings), rather than their more general acoustic properties.

With this in mind, this auditory comprehension paradigm has been adapted recently for use with PET in vegetative patients (Owen et al., in press). From the test sentences used in a previous study (Davis and Johnsrude, 2003), 189 declarative English sentences, on a range of topics, comprising 5–17 words (1.7–4.3 s in duration) were taken. The same form of distortion (speech in noise) was generated by adding a continuous pink-noise background to these sentences at three signal-to-noise ratios (−1, −4 or −6 dB). This form of distortion disrupts both the spectral and temporal properties of speech, while preserving the duration, amplitude and overall spectral composition of the original.

There were three experimental conditions corresponding to the three signal-to-noise ratios used to generate these stimuli. The first condition, which is referred to as "high intelligibility", had a signal-to-noise ratio of -1 dB (3 scans). The second "medium-intelligibility" condition had a signal-to-noise ratio of -4 dB (3 scans). The third "low-intelligibility" condition had a signal-to-noise ratio of -6 dB (3 scans).

Each scan comprised 21 declarative "speech in noise" sentences, with a 1 s gap between each sentence. The total time taken to play each trial was approximately 100 s. The study also contained a "silence" condition (also undertaken three times) during which no stimuli were presented.

Following the PET scan, the participants were asked to fill in a simple yes/no recognition form containing 10 sentences from each of the scans mixed up with 90 unheard "foil" sentences. As in the previous study, a linear relationship was observed between intelligibility and performance on the graded comprehension task (see Fig. 2).

Anecdotally, all of the subjects reported understanding "all" or "most" of the sentences in the high intelligibility condition; "some" of the sentences in the medium intelligibility condition and "hardly any", "very little" or "none" of the sentences in the low intelligibility condition.

Subtraction of the silence condition from the three hearing conditions revealed significant activation along the superior and middle temporal gyri bilaterally. These results are entirely consistent with previous functional imaging studies investigating speech processing in healthy control subjects, in which activation has been found in the temporal cortex bilaterally, incorporating Heschel's gyrus and the planum temporale (Mummery et al., 1999; Scott et al., 2000; Davis and Johnsrude, 2003).

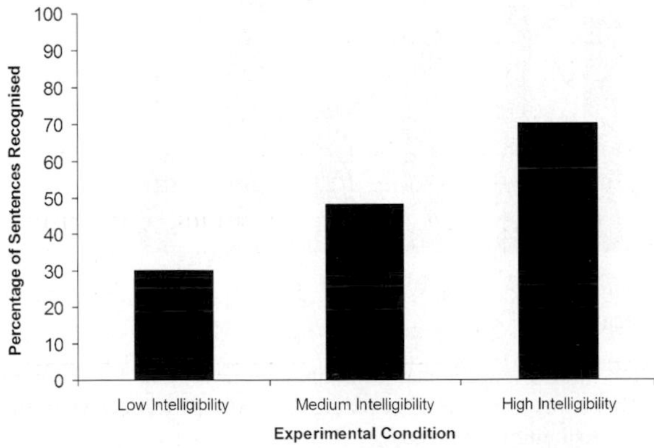

Fig. 2. The relationship between intelligibility and performance in healthy volunteers on the graded comprehension task.

Subtraction of the low intelligibility activation from high intelligibility activation revealed a significant rCBF increase in the anterior superior and middle temporal gyri (Brodmann's areas 21 and 22) in the left hemisphere (Fig. 3, left). Analysis of blood flow values for the group across the three speech conditions revealed that rCBF values increased in this region as the sentences became more intelligible (Fig. 3, right). Thus, not only was there a linear relationship between behavioral responses and intelligibility (the higher the intelligibility of the sentences, the better able the participants were to recognize them after a short delay), but blood flow in this region also increased linearly as a function of increased intelligibility. Subtraction of the high intelligibility conditions from the low intelligibility conditions (the reverse subtraction) produced no significant areas of rCBF change.

One of the main aims of this study was to generate a task that can be used with vegetative patients. With this in mind, a case-by-case analysis was conducted in individual volunteers to ascertain the extent to which the group results were replicated in each individual. As these were individual and not group analyses, the threshold for reporting a peak as significant was reduced to $p<0.01$, uncorrected for multiple comparisons. The results revealed that every one of the volunteers exhibited an increase in rCBF in response to increasing intelligibility of sentences within the left superior or middle temporal gyri (see Fig. 4), confirming that this region of cortex is sensitive to sentence intelligibility in individual volunteers. This finding is important because it suggests that the paradigm produces *consistent* blood flow responses in healthy volunteers. The reliability of any paradigm is paramount if it is to be used to detect residual cognitive functioning in patients in a vegetative state.

The same paradigm has been applied recently to small numbers of patients meeting the clinical

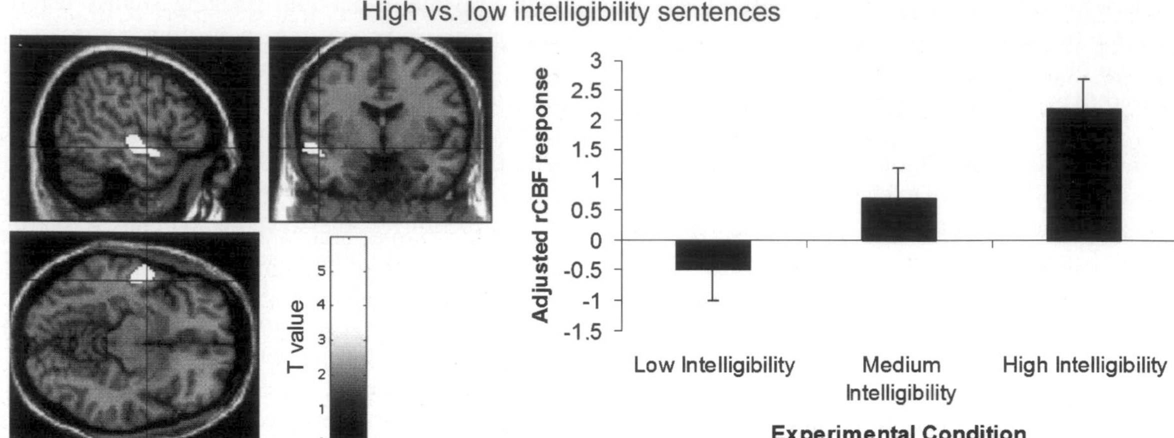

Fig. 3. Subtraction of the low intelligibility activation from high intelligibility activation in healthy volunteers reveals a significant rCBF increase in the anterior superior and middle temporal gyri (Brodmann's areas 21 and 22) in the left hemisphere (left). Analysis of blood flow values for the group across the three speech conditions reveals that rCBF values increased in this region as the sentences become more intelligible (right). All data were pre-processed and analyzed using Statistical Parametric Mapping software (SPM99, Wellcome Department of Imaging Neuroscience, London, UK). Data presented are thresholded at $p<0.05$ (corrected).

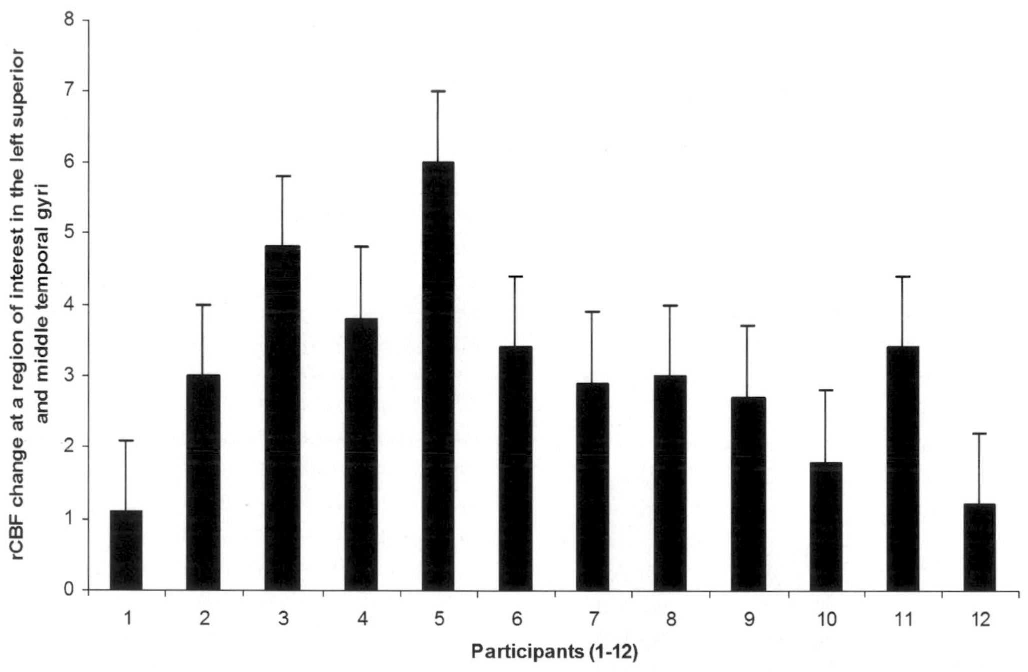

Fig. 4. Relative blood flow values for each healthy participant in the left anterior superior and middle temporal gyrus for the high intelligibility minus low intelligibility subtraction. For every participant there was an increase in blood flow in this location when the participants were listening to high intelligibility sentences relative to low intelligibility sentences.

criteria for persistent vegetative state. For example, Owen et al. (in press) described the case of a 30-year-old male (patient KA), who was diagnosed with basilar thrombosis and a posterior circulation infarction. In early June 2003, he collapsed with a severe headache and quickly became unresponsive. By the following day, he was drowsy, with a left partial Horner's syndrome, horizontal nystagmus, right hemiparesis and bilateral Babinksi sign. An MRI revealed an infarction of the left pons, cerebellum and posterior thalamus. Consciousness fluctuated for the next few days until the patient became unconscious with absent dolls eye movement. At this stage, 1 week post ictus, angiography revealed severe basilar stenosis. Two recombinant tissue plasminogen activator infusions produced some improvement, although following a brief spell of partial recovery after the anesthesia had worn off, he deteriorated into a deep state of unconsciousness. Three weeks post ictus KA left the intensive care unit and to date he has not recovered. During the acute care period, a BAER and a passive MMN odd-ball paradigm were conducted. The BAER revealed preserved responses bilaterally from the pons and midbrain, although the onset of the midbrain component and consequently peak III–V interpeak interval were increased bilaterally. An MMN superior temporal N1 response was also observed, however an N2 discriminating response was absent. Following sequential multidisciplinary assessment a diagnosis of VS was made.

In October 2003, 4 months post ictus, a decision was made to investigate the possibility of residual cognitive functions using PET and the novel language intelligibility task described above. Nine months later an identical PET activation study was performed, both to assess the reproducibility of the technique and to establish whether there had been any significant deterioration in cortical activity. During the first PET study, the comparison of speech (collapsed across the three levels of intelligibility) with the silence baseline condition revealed significant foci of activation over the left and right superior temporal planes (see Fig. 5, top left), suggesting that basic auditory processes were probably functional.

With this in mind, a second comparison was made, comparing low intelligibility sentences with high intelligibility sentences in order to isolate any residual activity related specifically to the comprehension of spoken language. This comparison revealed two peaks in the brain in the superior and middle temporal gyri of the left hemisphere (see Fig. 5, top right).

During the second PET study, 13 months post ictus, the comparison of speech (collapsed across the three levels of intelligibility) with the silence baseline condition revealed several foci of activation over the superior and middle temporal gyri of the left and right hemispheres (see Fig. 5, bottom left), suggesting again that basic auditory processes were probably still functional. The second comparison, comparing low intelligibility sentences with high intelligibility sentences, revealed two peaks in the brain in the superior and middle temporal gyrus of the left hemisphere (see Fig. 5, bottom right). In both comparisons, the activation foci are extremely close to those regions that were activated by the same comparison in the patient, 9 months earlier (see Fig. 5).

These findings have a number of important theoretical and clinical implications. First and foremost, it suggests that whatever level of residual cognitive activity existed in patient KA, it was persistent across time and remains, at least until 13 months post ictus. Thus, the pattern of activation observed during the first PET study was both qualitatively and quantitatively similar to that observed 9 months later when an identical procedure was carried out. Although anatomical and global blood flow factors preclude direct statistical comparisons between the two sessions, examination of Fig. 5 reveals a startling similarity between the activation patterns observed in temporal-lobe auditory areas in both cases. Notwithstanding qualitative differences that are well within the range that would be expected given normal inter-subject variability, the pattern of activation observed in KA during both PET sessions was similar to that observed in healthy awake control volunteers while performing identical tasks.

Second, the results suggest that some level of speech comprehension is preserved in this patient. The region around the left superior and middle temporal gyrus is believed to be involved in the perception of intelligible spoken language

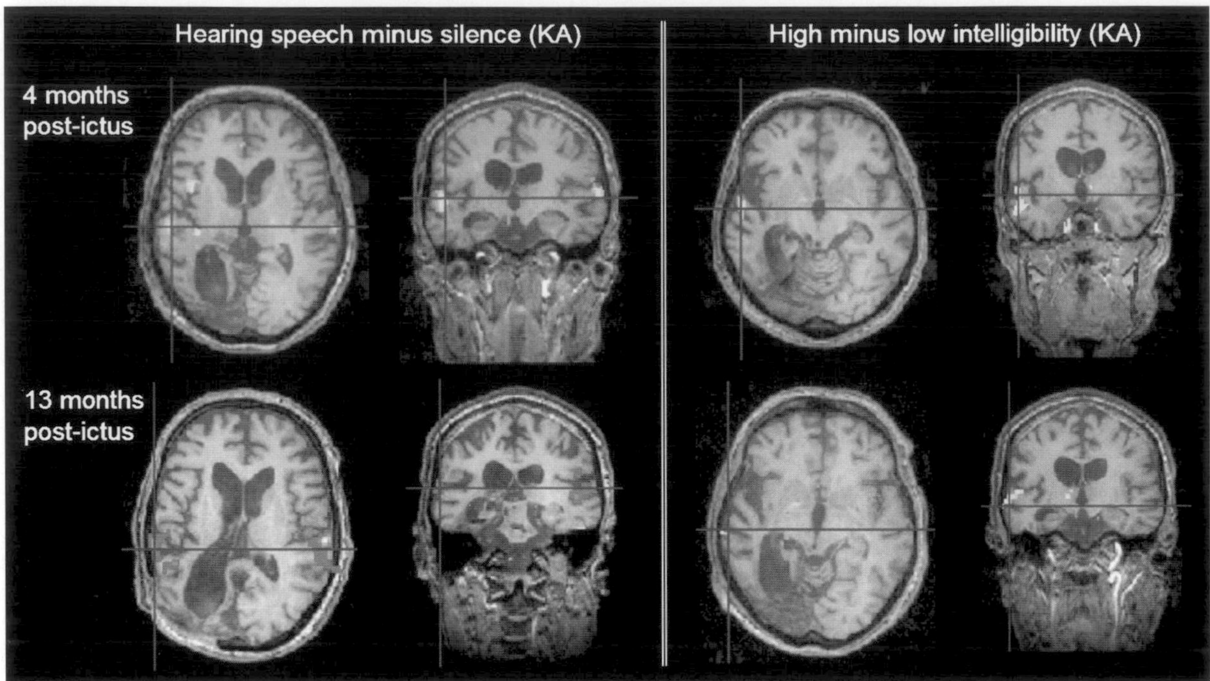

Fig. 5. Activation data from the two PET sessions, 9 months apart. (i) Hearing speech minus silence during the first PET session (top left) reveals activity bilaterally in the superior temporal gyrus. (ii) Hearing speech minus silence during the second PET session (bottom left) reveals very similar activity bilaterally in the superior temporal gyrus. (iii) High intelligibility speech minus low intelligibility speech during the first PET session (top right) reveals activity predominantly in the left superior temporal gyrus. (iv) High intelligibility speech minus low intelligibility speech during the second PET session (bottom right) reveals very similar activity predominantly in the left superior temporal gyrus. All data were pre-processed and analyzed using Statistical Parametric Mapping software (SPM99, Wellcome Department of Imaging Neuroscience, London, UK). See Plate 32.5 in Colour Plate Section.

(Binder et al., 2000; Scott et al. 2000). More specifically, it has been suggested that while the left superior temporal sulcus responds to the presence of phonetic information in general, its anterior part will only respond if the stimuli becomes intelligible (Scott et al., 2000). In the patient studied by Owen et al. (in press), activity in the left superior and middle temporal gyri was certainly focused *anteriorly* and in a region very similar to that activated in healthy volunteers when performing a formally identical task.

However, whether the responses observed reflect speech comprehension per se (i.e., understanding the contents of spoken language) or a more basic response to the acoustic properties of intelligible speech that distinguish it from less intelligible speech cannot be determined on the basis of these data alone.

Semantic processing

Understanding natural speech is ordinarily so effortless that we often overlook the complex computations that are necessary to make sense of what someone is saying. Not only must we identify all the individual words on the basis of the acoustic input, but we must also retrieve the meanings of these words and appropriately combine them to construct a representation of the whole sentence's meaning. When words have more than one meaning, contextual information must be used to identify the appropriate meaning. For example, for the sentence "The boy was frightened by the loud bark", the listener must work out that the ambiguous word "bark" refers to the sound made by a dog and not the outer covering of a tree. This process of selecting appropriate word meanings is important because most words in English are ambiguous (Rodd et al., 2005). Therefore, selecting appropriate word meanings is likely to place a substantial load on the neural systems involved in computing sentence meanings.

Recently, an fMRI study in healthy volunteers has used this phenomenon to identify the brain regions that are involved in the semantic aspects of speech comprehension, in particular in the processes of activating, selecting and integrating contextually appropriate word meanings (Rodd et al., 2005). During the fMRI scan, the volunteers were played sentences containing two or more ambiguous words (e.g. "the *shell* was *fired* toward the *tank*") and well-matched low-ambiguity sentences (e.g. "her secrets were written in her diary"). The ambiguous words were either homonyms (which have two meanings that have the same spelling and pronunciation; e.g. "bark") or homophones (which have two meanings that have the same pronunciation but have different spelling; e.g. "knight"/"night"). While the two types of sentences have similar acoustic, phonological, syntactic and prosodic properties (and are rated as being equally natural), the high-ambiguity sentences require additional processing to identify and select contextually appropriate word meanings. An additional set of sentences that were matched for number of syllables, number of words and physical duration were converted into signal-correlated noise. These stimuli had the same spectral profile and amplitude envelope as the original speech, but since all spectral detail is replaced with noise they are entirely unintelligible and they were used as a low-level baseline condition.

Relative to low-ambiguity sentences, high-ambiguity stimuli produced increases in signal intensity in the left posterior inferior temporal cortex and inferior frontal gyri bilaterally (see Fig. 6, top panel).

The results of this study demonstrate that a key aspect of spoken language comprehension — the resolution of semantic ambiguity — can be used to identify the brain regions involved in the semantic aspects of speech comprehension (e.g. activating, selecting and integrating word meanings). Moreover, they support models of speech comprehension in which posterior inferior temporal regions are involved in semantic processing (Hickok and Poeppel, 2000), and they demonstrate that the lateral inferior frontal gyrus, which has long been known to be important in syntactic processing of sentences and the semantic properties of single words, also plays an important role in processing the meanings of words in sentences. It is certainly striking that the comparison between these two sets of sentences that were rated as equally natural, and which differ only on a relatively subtle linguistic manipulation should produce such extensive activation differences. Further work is clearly needed to determine the precise function of the activated regions, but it is clear that they must form an important part of a neural pathway that processes the meaning of spoken language.

The same paradigm has been applied recently to small numbers of patients meeting the clinical criteria for persistent vegetative state. In one recent study, for example, Owen et al. (in press) used the test of semantic ambiguity to assess patient KA (described above) who had exhibited a normal pattern of PET activation in the graded complexity intelligibility task described above. The contrast between all the speech conditions (irrespective of ambiguity) plus the signal-correlated noise versus silence baseline yielded a pattern of activation very similar to the speech versus silence contrast of the initial PET investigation, although all of these changes were statistically significant using whole-brain methods of correction. Thus, large areas of activity were observed bilaterally in the superior temporal gyrus. When the combined high and low ambiguity sentences were compared to signal correlated noise, significant changes in signal intensity were again observed, bilaterally, in the superior temporal gyrus, a pattern which is again very similar to that observed in healthy volunteers (Rodd et al., 2005) and similar to that observed in the corresponding comparison of the PET study (Owen et al., in press).

When the high-ambiguity sentences were compared to the low-ambiguity sentences an ROI analysis based on the findings in healthy volunteers revealed responses in the patient that were well within the normal range in the left posterior inferior temporal region, but no consistent changes in the inferior frontal gyrus in either hemispheres (Fig. 6, bottom panel).

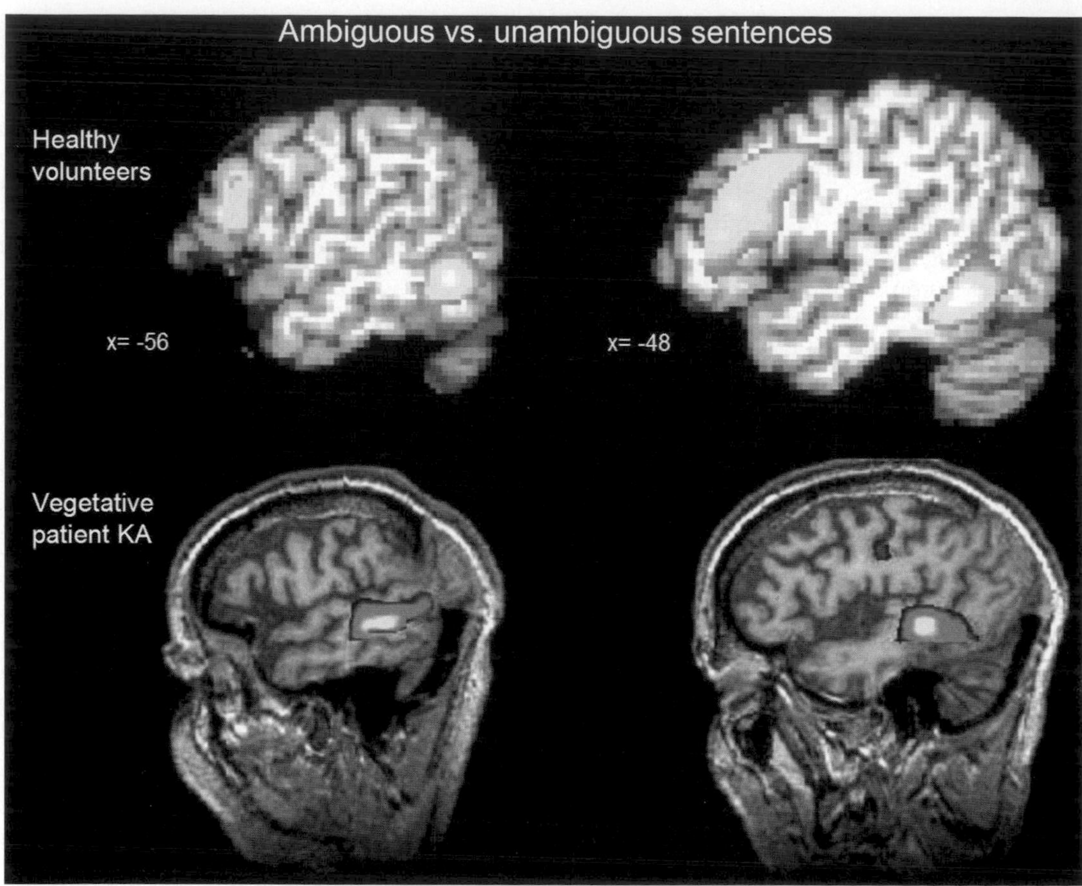

Fig. 6. fMRI data for the ambiguous sentences versus unambiguous sentences comparison. Like controls (top; adapted from Rodd et al., 2005), the patient (bottom) exhibited significant signal intensity changes in the left posterior inferior temporal cortex, but (unlike controls) not in the inferior frontal gyrus. All data were pre-processed and analyzed using Statistical Parametric Mapping software (SPM99, Wellcome Department of Imaging Neuroscience, London, UK). Data presented are thresholded at $p<0.05$ (corrected). See Plate 32.6 in Colour Plate Section.

These results provide compelling evidence for high-level residual linguistic processing in a patient meeting the clinical criteria for vegetative state and suggest that some of the processes involved in activating, selecting and integrating contextually appropriate word meanings may be intact, despite his clinical diagnosis.

Discussion and conclusions

A number of recent studies have clearly demonstrated that functional neuroimaging has the potential to elicit distinct and specific physiological responses from patients meeting the clinical criteria for persistent vegetative state, in the absence of any overt behavioral response. In this chapter we have attempted to illustrate the benefits of conducting such studies hierarchically; beginning with the simplest form of processing within a particular domain and then progressing sequentially through to more complex cognitive functions. This strategy, when applied to individual patients in a longitudinal manner, has the power to define the depth and breadth of preserved cognitive function in vegetative state. For example, in patient KA described recently (Owen et al., in press), clear conclusions could be drawn about the extent of residual function in the auditory domain at the perceptual, phonological and semantic levels. Thus, the imaging data confirmed that the patient's brain was responding to something more complex than pure sound (as indexed by the significantly increased activity in response to speech relative to signal correlated noise), suggesting that some component of the *perception* of speech was relatively preserved. Second, the fact that a significant response was observed to speech of increasing intelligibility suggests that these perceptual processes are recruited more strongly for speech that can be more readily understood. These results could be interpreted as suggesting that *comprehension* may also have been relatively preserved in KA. One piece of evidence that supports this final conclusion was the observation that ambiguous sentences yielded a significant response, showing that some *semantic* aspect of sentences can alter neural activity; in other words, not only did the patient's brain recognize speech as speech, but it was also being processed at a level sufficient to detect when words with multiple meanings were presented.

The question therefore arises as to whether the presence of "normal" activation indicates a level of "awareness" in patient KA and in similar cases where preserved neural responses have been described (e.g., de Jong et al., 1997; Menon et al., 1998; Laureys et al., 2000 and 2002; Owen et al., 2002; Boly et al., 2004). There is a wealth of data on healthy volunteers, from studies of implicit learning and the effects of priming (e.g., see Schacter, 1994 for review), to studies of learning during anesthesia (e.g. Bonebakker et al., 1996), that have demonstrated that many aspects of human cognition can go on in the absence of awareness. Even the semantic content of masked information can be primed to affect subsequent behavior without the explicit knowledge of the participant, suggesting that some aspects of semantic processing may occur without conscious awareness (Dehaene et al., 1998). Thus, the fact that a partially normal response to ambiguous sentences was observed in the patient described by Owen et al. (in press), does not necessarily imply that the patient was *aware* during the procedure. However, without direct evidence (i.e. using exactly the same task) to the contrary it remains a possibility, and establishing whether changes in signal intensity related to semantic ambiguity can be elicited in other situations in which sentences are known not to be consciously perceived (e.g. during sleep) remains a critical issue for future investigations with healthy volunteers (see Kotchoubey, this volume).

It remains to be seen, therefore, whether any functional neuroimaging paradigm can unequivocally demonstrate a level of preserved conscious awareness in a patient meeting the clinical criteria for persistent vegetative state. In our opinion, this issue will only be addressed directly using tasks that tap volitional (or consciously "willed") aspects of behavior. In all of the examples discussed above, from face processing to speech perception and even the detection of semantic ambiguous sentences, under normal circumstances cognitive processing is relatively *automatic*. That is to say, it occurs without the need for willful intervention on the part of the patient (you cannot choose to *not* recognize a face as a face, or to *not* understand speech that is presented clearly in your native language). In fact, this was an important factor in the choice of paradigms for study in these patients; persistent vegetative state, like any form of serious brain damage, may be accompanied by a significant reduction in attention span and an increased susceptibility to fatigue, either of which may mask positive findings if the chosen task is too demanding of depleted resources.

Notwithstanding this important consideration, a number of recent functional neuroimaging studies in healthy volunteers have described robust and anatomically specific changes in signal intensity at the single-subject level during tasks that unequivocally require "willed action" for their completion. For example, in one recent event-related fMRI study (Dove et al., submitted), conditions requiring volunteers to simply look at pictures of abstract art were compared with conditions in which they were explicitly instructed to remember similar stimuli for later retrieval. Looking, with no explicit instruction to remember was associated with significant increases in signal intensity in the medial temporal lobe, but not in a region of the mid-ventrolateral prefrontal cortex that has previously been implicated in memory encoding and retrieval. When the task instructions were changed subtly to encourage the volunteers to remember the stimuli, significant increases in signal intensity were observed bilaterally, in the mid-ventrolateral frontal cortex, with no concomitant increase within the medial temporal-lobe region. Of the 20 volunteers scanned, 14 showed this effect to a level of corrected significance, while in three non-significant activation was observed in the appropriate regions of the brain. Importantly, because the only difference between the conditions that elicited frontal-lobe activation and those that did not, was in the instruction given *prior* to each trial, the activation observed can only reflect the *intentions* of the volunteer (which were, of course, based on the remembered instruction), rather than some altered property of the outside world. In this sense, the decision to "remember" rather than simply "attend" is an act of willed intention and, therefore, clear evidence for awareness of self and surroundings in these healthy volunteers. In this respect, if the same task were to yield positive results in a patient meeting the clinical criteria for persistent vegetative state, it would have to be argued that a similar level of conscious awareness was present. Of course, *negative* findings under the same circumstances could not (and should not) be used as evidence for lack of awareness; false-negative findings in functional neuroimaging studies are not uncommon, and in the example given above 15% of the healthy volunteers did not show the predicted pattern of signal intensity changes despite (presumably) being fully aware throughout the experimental procedure.

In summary, there is a clear need to improve our characterization of the clinical syndrome of persistent vegetative state, not only to redefine diagnosis, but also to stratify patients in terms of prognosis and possible responses to novel therapies that may emerge in the future. The use of functional neuroimaging in this context will clearly continue to present logistic and procedural problems. However, the detection and elucidation of residual cognitive function in this group of patients has such major clinical and scientific implications that such an effort is clearly justified.

References

Alho, K. (1995) Cerebral generators of the mismatch negativity (MMN) and its magnetic counterpart (MMNm) elicited by sound changes. Ear Hearing, 1: 38–50.

Binder, J.R., Frost, J.A., Hemmeke, T.A., Bellgown, P.S.F., Springer, J.A., Kaufman, J.N. and Possing, E.T. (2000) Human temporal lobe activation by speech and nonspeech sounds. Cerebral Cortex, 10: 512–528.

Boly, M., Faymonville, M.E., Peigneux, P., Lambermont, B., Damas, P., Del Fiore, G., Degueldre, C., Franck, G., Luxen, A., Lamy, M., Moonen, G., Maquet, P. and Laureys, S. (2004) Auditory processing in severely brain injured patients: differences between the minimally conscious state and the persistent vegetative state. Arch. Neurol., 61: 233–238.

Bonebakker, A., Bonke, B., Klein, J., Wolters, G., Stijnen, T., Passchier, J. and Merikle, P.M. (1996) Information processing during general anaesthesia: evidence for unconscious memory. In: Bonke B., Bovill J.G.W. and Moerman N. (Eds.), Memory and Awareness in Anaesthesia. Swets and Zeitlinger, Lisse, Amsterdam, pp. 101–109.

de Jong, B., Willemsen, A.T. and Paans, A.M. (1997) Regional cerebral blood flow changes related to affective speech presentation in persistent vegetative state. Clin. Neurol. Neurosurg., 99: 213–216.

Davis, M.H. and Johnsrude, I.S. (2003) Hierarchical processing in spoken language comprehension. J. Neurosci., 23(8): 3423–3431.

Dehaene, S., Naccache, L., Le Clec'h, G., Koechlin, E., Mueller, M., Dehaene-Lambertz, G., Van De Moortele, P.F. and Le Bihan, D. (1998) Imaging unconscious semantic priming. Nature, 395: 597–600.

DeVolder, A.G., Michel, C., Guerit, J.M., Bol, A., Georges, B., Debarsy, T. and Laterre, C. (1994) Brain glucose metabolism in postanoxic syndrome due to cardiac arrest. Acta Neurologica Belgica, 94: 183–189.

Dove, A., Brett, M., Cusack, R. and Owen, A. M. Frontohippocampal contributions to human memory: the role of intention. Submitted.

Haxby, J.V., Grady, C.L., Horwitz, B., Ungerleider, L.G., Mishkin, M., Carson, R.E., Herscovitch, P., Schapiro, M.B. and Rapoport, S.I. (1991) Dissociation of object and spatial visual processing pathways in human extrastriate cortex. Proc. Nat. Acad. Sci. USA, 88: 1621–1625.

Haxby, J.V., Horwitz, B., Ungerlieder, L.G., Maisog, J.M., Pietrini, P. and Grady, C.L. (1994) The functional organization of human extrastriate cortex: a PET-rCBF study of selective attention to faces and locations. J. Neurosci., 14: 6336–6353.

Henson, R., Shallice, T. and Dolan, R. (2000) Neuroimaging evidence for dissociable forms of repetition priming. Science, 287: 1269–1272.

Hickok, G. and Poeppel, D. (2000) Towards a functional neuroanatomy of speech perception. Trends Cogn. Sci., 4: 131–138.

Jennett, B. and Plum, F. (1972) Persistent vegetative state after brain damage. Lancet, 1: 734–737.

Jones, S.J., Vaz Pato, M., Sprague, L., Stokes, M., Munday, R. and Haque, N. (2000) Auditory evoked potentials to spectro-temporal modulation of complex tones in normal subjects and patients with severe brain injury. Brain, 123: 1007–1016.

Kotchoubey, B., Lang, S., Baales, R., Herb, E., Maurer, P., Mezger, G., Schmalohr, D., Bostanov, V. and Birbaumer, N. (2001) Brain potentials in human patients with extremely severe diffuse brain damage. Neurosci. Lett., 301: 37–40.

Laureys, S., Faymonville, M.E., Degueldre, C., Fiore, G.D., Damas, P., Lambermont, B., Janssens, N., Aerts, J., Franck, G., Luxen, A., Moonen, G., Lamy, M. and Maquet, P. (2000) Auditory processing in the vegetative state. Brain, 123: 1589–1601.

Laureys, S., Faymonville, M.E., Peigneux, P., Damas, P., Lambermont, B., Del Fiore, G., Degueldre, C., Aerts, J., Luxen, A., Franck, G., Lamy, M., Moonen, G. and Maquet, P. (2002) Cortical processing of noxious somatosensory stimuli in the persistent vegetative state. Neuroimage, 17(2): 732–741.

Levy, D.E., Sidtis, J.J. and Rottenberg, D.A. (1987) Differences in cerebral blood flow and glucose utilisation in vegetative versus locked-in patients. Ann. Neurol., 22: 673–682.

Menon, D.K., Owen, A.M., Williams, E.J., Kendall, I.V., Downey, S.P.M.J., Minhas, P.S., Allen, C.M.C., Boniface, S., Antoun, N. and Pickard, J.D. (1998) Cortical processing in the persistent vegetative state revealed by functional imaging. Lancet, 352: 200.

Momose, T., Matsui, T., Kosaka, N., Ohtake, T., Watanabe, T., Nishikawa, J., Abe, K., Tanaka, J., Takakura, K. and Iio, M. (1989) Effect of cervical spinal cord stimulation (cSCS) on cerebral glucose metabolism and blood flow in a vegetative patient assessed by positron emission tomography (PET) and single photon emission tomography (SPECT). Radiat. Med., 7: 243–246.

Multi-Society Task Force on PVS. (1994a) Medical aspects of the persistent vegetative state (first part). New Engl. J. Med., 330: 1400–1508.

Multi-Society Task Force on PVS. (1994b) Medical aspects of the persistent vegetative state (second part). New Engl. J. Med., 330: 1572–1579.

Mummery, C.J., Ashburner, J., Scott, S.K. and Wise, R.J.S. (1999) Functional neuroimaging of speech perception in six normal and two aphasic subjects. J. Acoust. Soc. Amer., 106: 449–456.

Naatanen, R. (2003) Mismatch negativity: clinical research and possible applications. Inter. J. Psychophys., 48: 179–188.

Owen, A.M., Menon, D.K., Johnsrude, I.S., Bor, D., Scott, S.K., Manly, T., Williams, E.J., Mummery, C. and Pickard, J.D. (2002) Detecting residual cognitive function in persistent vegetative state. Neurocase, 8: 394–403.

Owen, A.M., Coleman, M.R., Menon, D.K., Johnsrude, I.S., Rodd, J.M., Davis. M.H. and Pickard, J.D. Residual auditory function in persistent vegetative state: a combined PET and fMRI study. Neuropsych. Rehabilitat., in press.

Rodd, J.M., Davis, M.H. and Johnsrude, I.S. (2005). The neural mechanisms of speech comprehension: fMRI studies of semantic ambiguity. Cerebral Cortex Jan 5; [Epub ahead of print].

Royal College of Physicians Working Group. (1996) The permanent vegetative state. J. Roy. Coll. Physic. London, 30: 119–121.

Schacter, D.L. (1994) Priming and multiple memory systems: Perceptual mechanisms of implicit memory. In: Schacter D.L. and Tulving E. (Eds.), Memory Systems. MIT Press, Cambridge, MA, pp. 233–268.

Schiff, N.D. and Plum, F. (1999) Cortical function in the persistent vegetative state. Trend. Cogn. Sci., 3(2): 43–44.

Scott, S.K., Blank, C., Rosen, S. and Wise, R.J.S. (2000) Identification of a pathway for intelligible speech in the left temporal lobe. Brain, 123: 2400–2406.

Tommasino, C., Grana, C., Lucignani, G., Torri, G. and Fazio, F. (1995) Regional cerebral metabolism of glucose in comatose and vegetative state patients. J. Neurosurg. Anesthesiol., 7: 109–116.

Turkstra, L.S. (1995) Electrodermal response and outcome from severe brain injury. Brain Injury, 9(1): 61–80.

S. Laureys (Ed.)
Progress in Brain Research, Vol. 150
ISSN 0079-6123

CHAPTER 33

Modeling the minimally conscious state: measurements of brain function and therapeutic possibilities

Nicholas D. Schiff*

Laboratory of Cognitive Neuromodulation, Department of Neurology and Neuroscience, Weill Medical College of Cornell University, 1300 York Avenue Room F610, New York, NY 10021, USA

Abstract: The minimally conscious state (MCS) defines a functional level of recovery following severe brain injuries. Patients in MCS demonstrate unequivocal evidence of response to their environment yet fail to recover the ability to communicate. Drawing on recent functional brain-imaging studies, pathological data, and neurophysiological investigations, models of brain function in MCS are proposed. MCS models are compared and contrasted with models of the vegetative state (VS), a condition characterized by wakeful appearance and unconsciousness. VS reflects a total loss of cognitive function and failure to recover basic aspects of the normal physiologic brain state associated with wakefulness. MCS may represent a recovery of the minimal dynamic architecture required to organize behavioral sets and respond to sensory stimuli. Several pathophysiological mechanisms that might limit further recovery in MCS patients are considered. Implications for future research directions and possible therapeutic strategies are reviewed.

Introduction

The recent definition of the minimally conscious state (MCS) challenges neurologists to improve the rational basis of evaluation and treatment of severely brain-injured patients (Giacino et al., 2002; Giacino, this volume). The ultimate impact of this nosological distinction is likely to rival the importance of the definition of the vegetative state (VS) by Jennett and Plum (1972; Jennett, this volume). Unlike VS, a condition characterized by the dissociation of behavioral unconsciousness and wakeful appearance, MCS classifies patients with unequivocal evidence of contingent response to their environment. The range of clinical phenotypes in MCS is quite large (see Giacino and Whyte, in press) and includes patients with relatively high-level behavioral responses such as complex command following or intelligible verbalizations. Detractors have raised ethical concerns that distinguishing MCS will lead to undervaluing the patients by leading to their conflation with VS (Burke, 2002; Coleman, 2002) and alternatively, futility concerns that there is no point to drawing further distinctions with the category of severe disability. The later concern is often expressed as a conclusion that this patient population is uniformly hopeless. To support the need for the MCS category and further refinement of the severe disability category of the Glasgow Outcome Scale (Jennett and Bond, 1975), neurobiological models of VS and MCS are examined and contrasted below. The conceptual models are considered in light of new measurements of brain function in severely brain-injured patients. Despite a limited number of studies, significant differences in underlying brain function are already unfolding (cf. Kobylarz and Schiff, 2004;

*Corresponding author. Tel.: +1 (212) 746 2372;
Fax: +1 (212) 746 8532; E-mail: nds2001@med.cornell.edu

DOI: 10.1016/S0079-6123(05)50033-5

Laureys et al., 2004). These advances make diagnostic clarity an imperative in the evaluation of severe brain damage (Fins and Plum, 2004). Furthermore, the potential that specific measurements of brain function may provide a basis for selective therapeutic interventions in some severely disabled patients supports continued efforts to identify the different pathophysiological mechanisms arising in this context.

Akin to the dissociation of arousal and consciousness observed in VS, MCS dissociates the appearance of wakefulness and some level of responsiveness from a capacity to communicate and organize goal-directed behaviors. Models of MCS must therefore consider the neurobiological basis for supporting continuous interactive behaviors. Below, conceptual models of MCS are advanced and contrasted to current models of VS. Patients near the point of emergence from MCS, where late recoveries are sometimes identified, arc proposed to primarily suffer failures of the initiation, maintenance, and completion of behavioral sets. Similar, but less profound impairments of general cognitive function are common across a spectrum of outcomes of severe brain injuries. Identifying and quantifying the necessary and sufficient conditions to emerge from MCS will require a neuroscientific framework that accounts for basic mechanisms underlying consciousness and cognition in the human brain. Thus, another motivation for detailed studies and models of MCS is this set of fundamental questions in neuroscience.

Nosology

Figure 1 provides a schematic overview of the nosology of global disorders of consciousness following severe brain damage. The initial brain state produced by severe brain damage is coma. Coma is a state of unarousable unresponsiveness and reflects overwhelming functional impairment of the forebrain arousal mechanisms (Plum and Posner, 1982). Coma is typically a transient state that, if uncomplicated by intercurrent processes (e.g., infection, metabolic derangement), will resolve within 1 or 2 weeks, heralded by the return of a limited cyclical arousal pattern during which an eyes-open "wakeful" appearance alternates with an eyes-closed "sleep" state. This pattern identifies the VS, a period of indeterminate duration that is otherwise similar to coma in that patients demonstrate no evidence of awareness of self or their environment (Jennett and Plum, 1972).

Patients who remain in a VS beyond 30 days are considered to be in a persistent vegetative state (PVS) and VS that lasts at least 1 year following traumatic brain injuries (TBI) or 3 months following hypoxic-ischemic injuries is considered permanent (Jennett, 2002). Rarely, PVS patients exhibit fragmentary behaviors that appear to arise from isolated intact cerebral networks (Schiff et al., 1999, 2002a). These behavioral fragments are not appropriate or specific to a given behavioral context, nor can they be reliably influenced to establish any evidence of interaction. Such patients may be placed in the "gray zone" shown in Fig. 1. Close to this minimal level of behavioral interaction, patients enter into MCS once they demonstrate reliable but inconsistent evidence of awareness of self or the environment as demonstrated by verbal or gestural output (Giacino et al., 2002). MCS patients can show wide fluctuations in baseline behaviors. The upper boundary determining a patient's emergence from MCS is reliable communication. This clinical categorization scheme includes patients with a large variety of behavioral patterns suggesting the utility of further refinement, particularly if based on quantitative

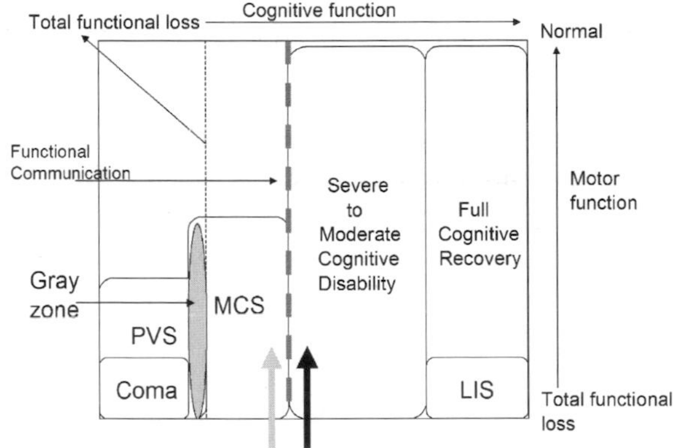

Fig. 1. Conceptual scheme for global disorders of consciousness. Abbreviations: PVS, persistent vegetative state; MCS, minimally conscious state; LIS, locked-in state. Gray and black arrows indicate functional levels just below and above emergence from MCS. Adapted from Schiff (2004), with permission from MIT Press.

indices. At present, no data support establishing a predictive time frame for emergence from MCS following severe brain damage (particularly if a result of TBI where well-documented cases demonstrate significant further recovery over long time periods; see Fins, this volume). Of note, if motor function is severely impaired, reliable identification of intermediate states between MCS and the locked-in state (LIS, not a disorder of consciousness; see Laureys et al., this volume) may not be possible. Across the range of clinical phenotypes encompassed by MCS, the gray arrow in Fig. 1 indicates a functional level from which some MCS patients may spontaneously emerge (black arrow) after long intervals (unpublished observations).

Models of the vegetative state

Before considering models of mechanisms underlying VS, it is useful to distinguish VS as a transient functional disturbance versus a permanent condition. VS often arises following an acute brain insult and can give way to significant further recovery. A wide range of outcomes, etiologies and structural injury patterns can be associated with transient VS. Similarly, the available literature of electrodiagnostic and functional imaging studies examining VS patients vary significantly with respect to time in VS, etiology of the condition, and underlying structural pathology. VS lasting at least one month and persisting to death, however, is associated with specific structural pathologies typically resulting in overwhelming damage to efferent and afferent cerebral connections (Adams et al., 2000). More rarely, permanent VS can be associated with extended bilateral damage to the paramedian mesencephalon, typically in combination with the paramedian thalamus (Ingvar and Sourander, 1970; Castaigne et al., 1981; Schiff et al., 2002a). Despite these variations a convergence of evidence supports VS as a functionally distinct state associated with common disruption of brain activity in the early and chronic stages.

Pathological studies

Adams et al. (2000) studied 49 patients remaining in a VS for at least 1 month until death and identified specific patterns associated with traumatic and non-traumatic etiologies. Non-traumatic injuries associated with VS showed severe bilateral thalamic damage in all instances and in the majority of cases was associated with diffuse cortical damage (64% of cases). Traumatic etiologies showed grade 2 and 3 diffuse axonal injuries and severe thalamic degeneration in the majority of VS patients (96% of patients who survived for 3 months before death). These and other pathological studies confirm the intuition that the chronic VS is characterized by overwhelming cerebral damage (cf. Dougherty et al., 1981). The conclusion that the most consistent and severe pathologies arising from both types of injuries are in subcortical structures, particularly the thalamus, is not widely appreciated. The investigators point out that damage to the thalamus following diffuse axonal injuries (DAI) is indirect as a result of transneuronal degeneration and that if delayed restoration of function in axons initially damaged is possible, the neuronal substrate in the thalamus remains in these situations. This difference is suggested to play a role in the very different expected time course of recovery and point beyond which permanence is expected in VS resulting from TBI compared to hypoxic-ischemic insults (see discussion below). An important related observation from these studies is that the cerebral cortex is generally spared in TBI resulting in VS, with only 11% of patients showing diffuse ischemic neocortical injury patterns and 37% showing any neocortical ischemic injuries (compared with 64% and 93% in VS of non-traumatic origins). Brainstem damage was uncommon in chronic VS patients emphasizing that VS is primarily a disorder of cerebral integration at the thalamocortical level (also see Graham et al., elsewhere in this volume).

Electrodiagnostic studies

Electroencephalographic studies in VS identify several patterns of abnormality limiting specific insights into mechanisms underlying this condition (Jennett 2002). In general, EEG findings in VS are comparable to findings in coma and typically show profound slowing with amplitude increases in delta and theta rhythms (e.g., Hansotia, 1985). Alternatively, very low amplitude, nearly isoelectric, EEG may be recorded in the VS. Importantly, the recovery of arousal without consciousness in VS does not imply that the distribution of power across frequencies in the EEG is normal — as a rule the shape of the EEG power spectrum (the measure that quantifies this distribution) is markedly abnormal in VS. Thus, in addition to dissociating arousal and awareness on a behavioral level, VS dissociates cyclic activation of the cerebrum associated with eyes open and eyes closed states from the normal sleep–wake architecture despite cyclical alteration of aberrant EEG patterns associated with behavioral state changes (cf. Isono et al., 2002). Early evoked potential components are often preserved in VS but show abnormal central conduction times; the loss of sensory evoked potentials has been strongly correlated with diffuse ischemic injury of the neocortex and permanent VS (Rothstein et al., 1991; Guerit, this volume). Late components and mid-latency components of sensory evoked potentials are generally absent or show marked abnormality in VS (Kotchoubey, this volume).

Imaging studies

Initial studies of brain function in VS focused on measurements of cerebral metabolism and brain electrical activity in the electroencephalogram (EEG) and evoked potentials (EP). Levy et al. (1987) studied resting cerebral metabolism using fluorodeoxyglucose-positron emission tomography (FDG-PET) in the eyes-open "awake" state of 7 VS patients, 3 LIS patients, and 18 normal controls. In their studies, VS was associated with a 60–70% reduction in resting cerebral metabolism across most brain structures. This finding of profoundly depressed cerebral metabolism in VS has been replicated across several laboratories (DeVolder et al., 1990; Tomassino et al., 1995; Rudolf et al., 1999; Laureys et al., 2000a, 2002; Schiff et al., 2002). Comparable levels of reduction in cerebral metabolism are typically only observed in pharmacologically induced coma (reviewed in Laureys et al., 2004). Although significant reductions of glucose metabolism can be interpreted as a proxy for widely reduced neuronal firing rates (Eidelberg et al., 1997; Smith et al., 2002), improvements in the overall level of resting cerebral metabolism do not necessarily accompany recovery from VS (Laureys et al., 1999, see the discussion below).

FDG-PET, clinical EEG and EP studies can provide only limited information about cerebral processing in VS because these techniques cannot directly measure the presence or absence of distributed cerebral network responses. Functional brain imaging using $H_2^{15}O$ PET or magnetic resonance imaging (fMRI) or more quantitative analyses of brain electrical activity are required to examine the distributed activation of cerebral structures in the VS brain in response to selective stimuli. These methods are, however, more sensitive and require greater technical expertise (see Owen, this volume). Laureys and colleagues (Laureys et al., 2000a, 2002; Boly et al., 2004a, b) have studied patients unequivocally meeting the criteria for VS for at least one month using functional $H_2^{15}O$ PET (fPET) paradigms. Their studies included patients with both traumatic and non-traumatic etiologies and identified the loss of distributed activation across cerebral structures seen in normal subjects in response to the same simple auditory and somatosensory stimuli. In most of the patients studied, early evoked potential components (reflecting primary sensory cortical response) were preserved and correlated with fPET activations of primary sensory cortices. Late cortical evoked responses, as in other studies, were absent. These functional imaging findings are consistent with a general model that VS enduring for at least a month is the result of widespread disconnection of cerebral networks, usually on the basis of extensive structural injuries if enduring for at least one month. Taken together with the clinical features of VS, these findings support modeling of VS as a total loss of cerebral integrative activity. This loss is evident even at the earliest stage of cortical sensory processing reflected in the loss of late and midlatency electrical (magnetic) evoked potential components and distributed network activations measured by fPET.

Laureys et al. (1999) also reported changes in FDG-PET metabolism in one of their patients studied before and after recovery from VS. The patient had remained in a transient VS lasting 19 days after carbon monoxide poisoning but recovered to a level of only mild cognitive impairment 1 month after admission. FDG-PET studies done at 15 days (while still in VS) and 37 days (recovery of consciousness with moderate short-term memory impairment) both showed a global metabolic rate of ~62% of normal values across the cerebrum. Although the overall metabolic rate did not change with recovery from VS, a clear difference in the pattern of metabolic activity was observed between the two scans (notably, the patient's metabolic rate in this transient VS was higher than typically associated with permanent VS where FDG-PET metabolic rate may be ~40% of normal levels). During VS significant metabolic reductions were observed in pre-motor, sensorimotor, and posterior parietal-occipital regions. Recovery of consciousness correlated with increased metabolism in the pre-motor and parietal-occipital regions. A follow-up study identified increased thalamocortical connectivity between the intralaminar regions of the thalamus and the prefrontal cortices following recovery (Laureys et al., 2000b). These findings can be compared to studies in normal subjects that identify the medial posterior parietal-occipital region as the most metabolically active area in the normal resting brain (Raichle et al., 2001; Vogt and Laureys, this volume). Thus, the observed shift in the pattern of resting metabolism may thus correlate with an overall normalization of brain activity associated with re-establishing frontal thalamocortical systems and related posterior networks (see discussion below); significant increases in correlation of intralaminar thalamic and prefrontal regions may also reflect a re-establishing of functional connectivity associated with elementary cognitive behavioral sets (cf. Paus et al., 1997). In a larger study, including patients with traumatic etiologies, Laureys and colleagues (2003) have identified a similar pattern of metabolic recovery in the medial posterior parietal region.

Unusual behavioral patterns in VS

Stereotyped responses to stimuli can be observed in VS patients such as grimacing, crying, or occasionally vocalization that originate primarily from brainstem circuits and limbic cortical regions. Very rarely, fragments of behavior that may appear semi-purposeful or inconsistently related to environmental stimuli may be identified in a patient who otherwise meets criteria for VS or PVS. In a multimodal imaging study using FDG-PET, structural MRI, and magnetoencephalography (MEG), we identified three such patients with unusual fragments of behavior. One, a 49-year-old woman, who had suffered successive hemorrhages from a vascular malformation of the right thalamus and basal ganglia, infrequently expressed single words (typically epithets) in isolation of environmental stimulation despite a 20-year period of VS (Schiff et al., 1999). MRI images showed absence of the right basal ganglia, right thalamus and severe injury to the left thalamus (Fig. 2A). Resting FDG-PET measurements showed marked reduction of global cerebral metabolism to <50% of normal across most brain regions with metabolic sparing in relatively small regions in the left hemisphere (Fig. 2B). MEG responses to bilateral auditory stimulation in this patient revealed an abnormal time-locked response in the gamma range (20–50 Hz) localized by single-dipole analysis to primary auditory areas in the left hemisphere alone (see Ribary, this volume). These locations corresponded to the islands of higher resting brain metabolism observed by PET imaging shown in Fig. 2. Taken together, the imaging and neurophysiological data indicate isolated sparing of left-sided thalamo-cortical-basal ganglia loops that normally support language function, including neuronal populations in Heschl's gyrus, Broca's area, and Wernicke's area. This finding and similar observations in other PVS patients (Schiff et al., 2002a) provide a model of brain function for patient's in the "gray zone" of Fig. 1: isolated cerebral networks may remain active and correlate with the occasional generation of fragments of behavior.

The asymmetry of subcortical injuries in this patient also provided a unique opportunity to examine the impact of removal of the thalamus and basal ganglia on the EEG (shown in Fig. 2C). We identified a sharp decline of coherence in the cerebral hemisphere deprived of both the subcortical gray matter structures and their return path through the thalamus (Davey et al., 2000). Coherence is a measure of cross-correlation in the frequency domain (Mitra and Pesaran, 1999). A high coherence indicates potential relationships

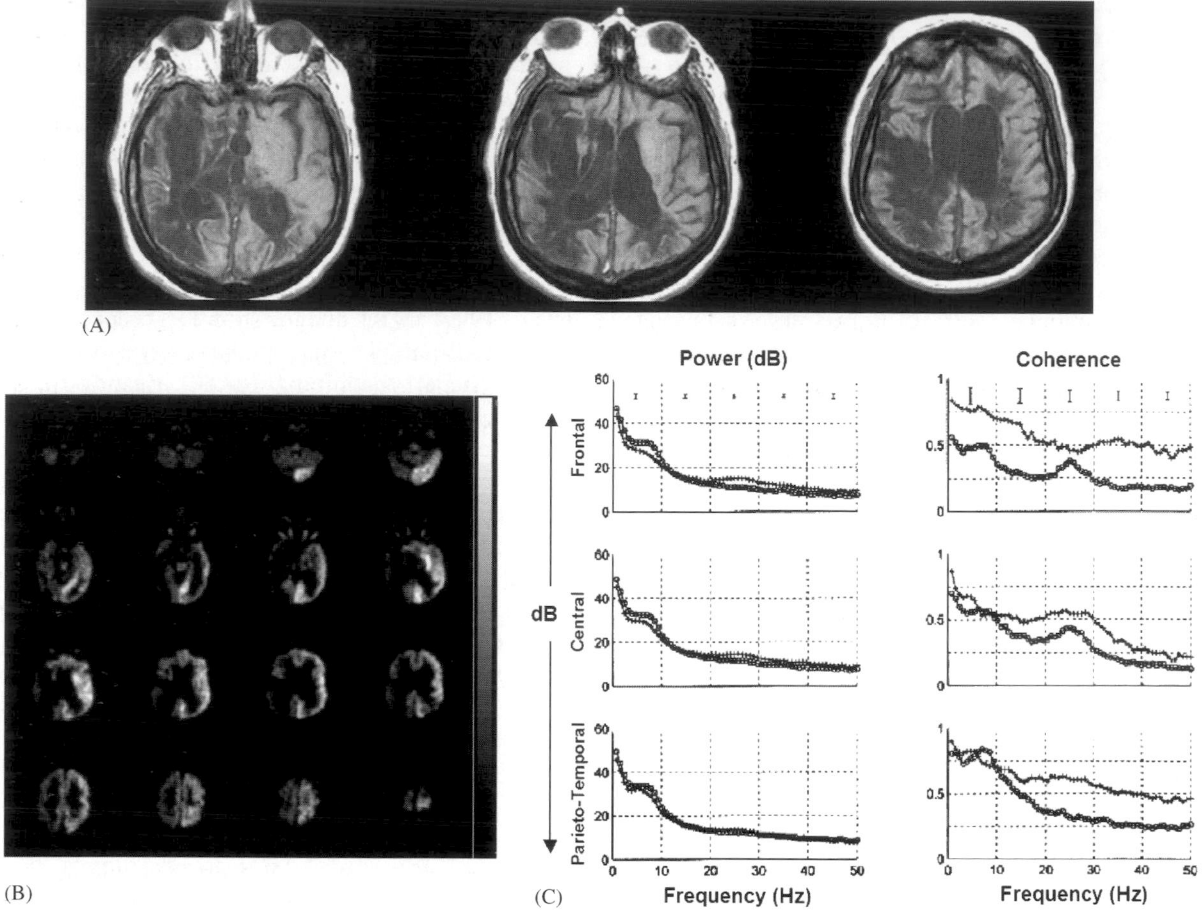

Fig. 2. Magnetic resonance imaging, positron emission tomography and electroencephalography studies for patient described in text. (A) MRI structural images show severe asymmetric brain damage with loss of right-sided basal ganglia and thalamic structures (Schiff et al., 1999). (B) Positron emission tomography images of resting glucose metabolism across entire cerebrum. Marked asymmetry of right and left hemisphere metabolism is seen. (C) Dissociation of hemispheric variations of coherence measurements and regional power spectrum measurements (from Davey et al., 2000). Reproduced with permission from Elsevier Press.

between two signals such as the presence of a common input signal, mutual driving of the signals, or one signal driving the other (cf. Bendat and Piersol, 2000). Coherence is a ratio of coherent power to total power and is therefore not sensitive to changes in the amount of power within frequency bands; significant coherence differences therefore reflect either the influence of common input or changes in functional connectivity *per se*. Theoretical studies of the origins of EEG coherence place strong emphasis on the role of cortico-cortical fiber pathways (Thatcher et al., 1986; Nunez et al., 1999).

As shown in Fig. 2C, EEG samples obtained over multiple episodes of an eyes-closed sleep-like state in this patient (from Davey et al., 2000) reveal a marked dissociation of differences in intrahemispheric coherence and regional power spectra. Regional power spectra and coherence from the left (+) and right (o) hemispheres are displayed in the figure (95% confidence limits are shown by brackets in top panels). The power spectra did not differ substantially between electrode pairs obtained from frontal regions (electrodes F3/F7 (left hemisphere) and F4/F8 (right hemisphere)), central regions (electrodes C3/T3 and C4/T4), or parieto-temporal regions (electrodes P3/T5 and P4/T6). Of note, small differences of increased frontal and central theta band activity (5–10 Hz) and gamma band activity (here ~20–35 Hz) were evident in the comparison of the power spectrum obtained from pairs in each hemisphere. This finding correlated with the presence of residual spontaneous and evoked gamma band activity (20–50 Hz) isolated to the left hemisphere identified in MEG studies of the same patient (Schiff et al., 1999, 2002a). Intrahemispheric coherences, however, demonstrated marked differences between the left and right hemispheres, with broadband reduction of coherence seen across all right hemisphere electrode pairs. These findings are consistent with a critical role for subcortical structures in shaping coherence relationships in the EEG that are not reflected by changes in the regional power spectrum. Similar dissociations of the power spectrum and coherence spectrum have been observed in MCS patients (Kobylarz et al., 2003, discussed below).

The significance of this finding for the present discussion is that it represents a correlation with an unusual but straightforward anatomic difference in the cortical inputs to each hemisphere in this patient. It suggests that coherence spectra may provide important functional information not available in the power spectral characterization of the EEG that summarizes overall frequency content. More generally, the failure to recover the normal distribution of frequencies observed in the wakeful EEG in VS supports the view that the VS brain is not able to generate endogenous central states to prepare motor behaviors, and anticipate or process sensory stimuli.

Models of MCS

The diagnostic category of MCS canvasses a wider range of clinical phenotypes and structural pathologies than VS. At this time only a few studies have focused on patients fulfilling the diagnostic criteria for the condition and conceptual models must accordingly be seen as tentative. It is anticipated that as additional investigational studies are done this category will become further refined, hopefully based on mechanistic distinctions. Nonetheless, existing data provide evidence that brain function in VS and MCS may be well separated at the extremes if not more generally.

In considering the available data from functional imaging, pathology, and observational studies, a model is proposed that frames MCS primarily in terms of instability of the initiation, maintenance, and completion of behavioral sets. These critical functions depend on the interaction of brainstem arousal systems and mesencephalic and diencephalic "gating systems" (see below) with other cerebral structures. Pathological studies and observational data of fluctuations observed in severely brain-damaged patients suggest that relatively subtle measurements of brain function may be necessary to identify the underlying mechanisms of failure to organize goal-directed behaviors and communication in MCS. Mechanisms identified in MCS patients with limited structural injuries will likely also apply to understanding problems of cognitive recovery of patients with less severe or moderate disabilities following brain damage.

Correlations of MCS with structural pathology

Comprehensive studies of specific anatomic pathologies associated with MCS are unavailable. Autopsy studies of patients with severe disability following brain injuries show wide variations in underlying neuroanatomical substrates. Jennett and colleagues (2001) reported 65 autopsies of patients with traumatic brain injury leading either to a VS or severe disability. This study included 12 patients with histories consistent with MCS at the time of death. Over half of the severely disabled group demonstrated only focal brain injuries, without DAI or focal thalamic infarction (including 2 of the MCS patients). Structural brain-imaging studies also demonstrate that the behavioral level ultimately achieved by a patient following severe brain injuries often cannot be simply graded by the degree of vascular, DAI, and direct ischemic brain damage. Kampfl et al. (1998) described indirect volumetric MRI indices that provide reasonable predictive accuracy (~84%), when combined with time in VS, for a permanently vegetative outcome of overwhelming traumatic brain injuries. Unfortunately, many patients fulfilling these criteria can recover after long intervals. In our own ongoing studies, we have identified one MCS patient, with a structural injury pattern on MRI fulfilling all of the Kampfl et al. criteria, who emerged at 8 months and is now near an independent functional level (unpublished observations). Danielsen et al. (2003) report detailed MRI and magnetic resonance spectroscopy (^{1}H-MRS) findings from a patient with severe DAI measured over several time points, while the patient remained in coma for 3 months and 21 months later when the patient had slowly recovered to a near independent level. In this patient, ^{1}H-MRS revealed characteristic regional reductions in NAA (*N*-acetyl-aspartate)/choline ratios associated with severe DAI that normalized by the study done at 21 months and correlated with cognitive recovery. McMillan and Herbert (2004) recently reported a 10-year follow-up on an MCS patient who continued to recover 7–10 years following a traumatic brain injury to a point of regaining the capacity to initiate conversation, and express clear preferences and spontaneous humor. These observations suggest that some slow variables of recovery may exist and should be quantified through further structural imaging and longitudinal analysis of brain dynamics (see below).

Attempts to correlate outcome with structural injuries is further complicated by the potentially disproportionate impact of certain focal injury patterns. It is well known that enduring global disorders of consciousness can result from relatively discrete injuries concentrated in the paramedian mesencephalon and thalamus (Schiff and Plum, 2000). The structures involved in these lesions include the thalamic intralaminar nuclei (ILN) and the mesencephalic reticular formation (MRF), which together with their connections to the thalamic reticular nucleus appear to play a key role linking arousal states to the control of moment-to-moment intention or attentional gating (Schlag-Rey and Schlag, 1984; Llinas et al., 1994, 2002; Kinomura et al., 1996; Paus et al., 1997; Purpura and Schiff, 1997; Steriade, 1997; Jones, 2001; Matsumoto et al., 2001; Minamimoto and Kimura, 2002; Schiff and Purpura, 2002; Wyder et al., 2003, 2004). These structures can be considered "gating" systems that control interactions of the cerebral cortex, basal ganglia, and thalamus through their patterns of innervation within the cortex as well as rich innervation from the brainstem arousal systems (Groenewegen and Berendse, 1994; Schiff and Plum, 2000; van der Werf et al., 2002). Patients who recover from bilateral paramedian thalamic injuries typically demonstrate persistent instability of arousal level and within-state fluctuations of the selective gating of different cognitive functions (Katz et al., 1987; Meissner et al., 1987; Mennemeier et al., 1997; van Der Werf et al., 1999). Thus, even incomplete injuries to the gating systems may produce unique deficits in maintaining adequate cerebral activation and patterns of brain dynamics necessary to establish, maintain, and complete behavioral set formation (Schiff and Purpura, 2002; see the discussion below).

Enduring VS or MCS produced by such focal injuries will typically include bilateral damage to the mesencephalic reticular formation extending bilaterally into the intralaminar thalamic nuclei (Plum, 1991; Schiff et al., 2002a). However, *en passant* damage to the thalami and upper brain stem commonly follows both traumatic brain injury and stroke as a result of the selective vulnerability of this region to the effects of diffuse brain swelling that leads to herniation of these midline structures through the base of the skull (see Plum and Posner, 1982). It is likely that most patients who recover from severe brain injuries may represent mixed outcomes resulting from intermediate pathologies that combine moderately diffuse injuries with limited focal damage to paramedian structures (Adams et al., 2001; Jennett et al., 2001). Pathophysiologic mechanisms arising in the setting of such mixed pathologies have not been the subject of systematic study. It is known, however, that damage to the paramedian brain stem worsens prognosis following TBI and is associated with MCS and other poor outcomes (Wedekind et al., 2002).

In the aggregate, clinical and pathological findings suggest significant variability in both the underlying mechanisms of cognitive disabilities and residual brain function accompanying severe brain injuries associated with MCS and other outcomes. It appears that severe disabilities may arise under at least two different conditions: (1) extensive, relatively uniform diffuse axonal injury or hypoxic-ischemic damage and (2) focal cerebral injuries combined with minimal diffuse axonal or ischemic damage with possible coexisting functional alteration of subcortical gating systems and their interaction with cortical association areas.

Functional brain imaging in MCS

Recent functional imaging studies have examined patients using the Aspen criteria for MCS (Giacino et al., 2002). Boly et al. (2004a) studied five MCS patients using the same fPET auditory stimulation paradigm applied by Laureys et al. (2000) to study vegetative patients. In their studies, MCS patients and healthy controls both showed activation of auditory association regions in the superior temporal gyrus that did not activate in the PVS patients and strong correlation of the auditory cortical responses with frontal cortical regions, providing evidence for preservation of cerebral processing associated with higher order integrative function. The majority of the MCS patients were scanned approximately 1 month after initial injury and at a time when EEG examinations revealed significant bilateral abnormalities (mostly slowing in the theta and delta range). Preliminary data from Laureys and co-workers also show a near-normal pain network response to somatosensory stimulation in their MCS patients (Boly et al., 2004b).

Menon et al. (1998) described selective cortical activation patterns using a $H_2^{15}O$ PET subtraction paradigm in a 26-year-old woman described as in a PVS 4 months following an attack of acute disseminated encephalomyelitis. The patient later improved to an MCS level by 6 months; emergence from MCS occurred sometime after 8 months and the patient eventually made a full cognitive recovery (Macniven et al., 2003). Imaging studies done during the PVS period demonstrated selective activations of right occipital-temporal regions. This pattern of activity was interpreted as indicating a recovery of minimal awareness without behavioral manifestation. Such an interpretation is limited by the lack of any evidence of behavioral response from the patient. It is generally agreed that the present state of imaging technologies cannot provide alternative markers of awareness (Menon et al., 1999; Schiff and Plum, 1999; Schiff et al., 1999; Laureys et al., 2004). The findings of Menon et al. (1998) contrast with those of Laureys et al. (2000, 2002) and suggest that ultimately neuroimaging studies may be able to elucidate underlying differences between PVS and MCS patients. Bekinschtein et al. (2004) recently reported brain activations obtained using fMRI in an MCS patient recovering from traumatic brain injury. A subtraction comparison of responses to presentations of the patient's mother's voice and a neutral control voice revealed selective activation of the amygdala and insular cortex suggesting emotional processing associated with the mother's voice. As in the interpretation of "high-level" responses in VS, in patients without the ability to communicate we can only speculate about whether such activations indicate awareness.

We studied two MCS patients near the border of emergence more than 18 months after injury (gray arrow in Fig. 1) using fMRI, FDG-PET, and quantitative EEG (Schiff et al., 2005; Kobylarz et al., 2003). The patients and seven control subjects were studied with fMRI language activation paradigms similar to paradigms used in normal subjects and neurosurgical candidates to map language networks (Hirsch et al., 2000; Hirsch, this volume). Two 40-second narratives were pre-recorded by a familiar relative and presented as normal speech and also played time-reversed. Forward presentations generated robust activity in several language-related areas in both patients. Figure 3 shows cortical activity maps associated with the presentation of linguistic stimuli in a single patient. While wide network activation occurred with the forward presentations, time-reversed narratives only activated early sensory cortices in the left hemisphere. This pattern differs from that of normal subjects, where large activations for both stimulus types were observed, with time-reversed language presentation showing slightly more activation than forward presentations. These preliminary fMRI results have now been confirmed in further studies of MCS patients (unpublished data). The findings indicate that some MCS patients may retain large-scale cortical networks that underlie language comprehension and expression despite their inability to execute motor commands or communicate reliably.

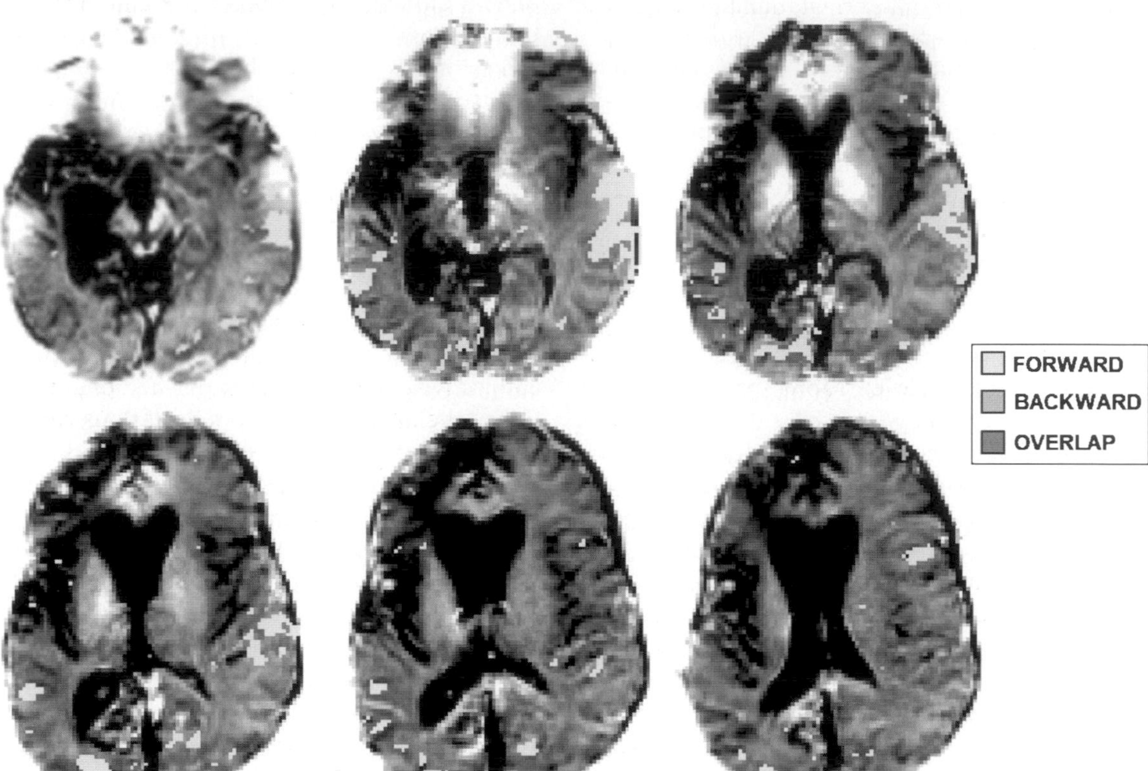

Fig. 3. Functional magnetic resonance imaging activation patterns of BOLD signal in response to passive language presentations. Reproduced with permission from MIT Press. See Plate 33.3 in Colour Plate Section.

In both patients studied we correlated fMRI findings with FDG-PET and quantitative EEG measurements. The patients demonstrated low global resting metabolic rates with significant differences in hemispheric resting metabolic rates and baseline thalamic activity. EEG studies in both patients revealed significant reductions in interregional coherence of the more damaged hemisphere in wakefulness (Kobylarz et al., 2003). In one patient this interregional coherence pattern showed a marked dependence on arousal state with coherence decreases observed across frequencies only in the state of wakefulness. The abnormalities of EEG coherence measures indicate a significant alteration of the functional integration of cortical regions in the more damaged hemisphere. This is all the more striking in that the EEG power spectrum showed no differences in the distribution of power across frequencies for both hemispheres in the two patients. Traditional EEG and MRI evaluations are known to be insensitive to detection of mild and moderate disabilities following brain injuries and to be poor predictors of gradation of severe TBI (Thatcher et al., 2001). The observation of marked coherence abnormalities is consistent with experimental studies that indicate that coherence measures can provide a more direct reflection of behaviorally relevant dynamics than changes in the power spectrum (cf. Vaadia et al., 1995).

Brain dynamics underlying behavioral fluctuations in MCS

It is notable that the low level of behavioral responses represented by MCS can be associated in some patients with intact large-scale network responses as observed in normal human subjects (shown in Fig. 3). These observations lead naturally to the question of how to model the coexistence of recruitable large-scale networks and severely limited behavioral repertoires. A systematic approach to this question is likely to require both consideration of normal mechanisms studied in cognitive neuroscience and a variety of clinical neurological disorders. As noted above, correlations of structural injuries and functional outcomes are not as strong as naïve assumptions would suggest, as widely differing structural pathologies may correlate with the same poor functional level. Moreover, functional measurements offer only snapshots of brain function in time. Baseline metabolic assessments or functional activation studies cannot adequately identify the frequency of the resting brain state sampled or likelihood of response at the time the measurements are taken. In patients with widely varying responsiveness, these limitations present an important methodological concern and emphasize the need for more careful consideration of ongoing brain dynamics. What kinds of dynamical measures are needed? At least two different kinds of measurements suggest themselves. Dynamical structures arising in the EEG that correlate with elementary cognitive functions underlying behavioral set formation may quantify fluctuating responsiveness in MCS. Alongside these measurements there is also a need to develop more sensitive diagnostics that can identify dynamical signatures of several abnormal processes that may arise in the setting of severe brain damage and limit recovery.

Beginning with the observations above, the shape of the spectrum of the EEG can be relatively normal in MCS patients, and it is reasonable to next consider the fine correlation structure of the EEG as a potential indicator of mechanisms. In our preliminary studies discussed above, hemispheric coherence abnormalities have been identified (Kobylarz et al., 2003) but such observations are only starting points for more detailed consideration of markers of cognition. The background activity of ongoing EEG during different arousal states can be precisely described as shifts in spectral content of the activity of distributed forebrain networks (Steriade, 2000). Combined studies of intralaminar thalamic neurons and EEG power spectra show that these neurons in concert with the brain stem arousal systems support the shift away from low frequencies characteristic of sleep to a mixed state including increased synchronized high-frequency activity in natural awake attentive states (Steriade and Glenn, 1982; Steriade et al., 1996). A recent theoretical model of the EEG demonstrates that most of the features of the shape of the EEG spectrum as it evolves across wakefulness and sleep stages can be captured in a partial differential equation system constructed from physiologically realistic parameters and the connectivity of only three major neuronal populations: thalamic relay and reticular neurons and cortical pyramidal neurons (Robinson et al., 2002). This architecture is consistent with experimentally based models of EEG generation. Simply recovering the shape of the EEG spectrum may therefore only indicate that an essential substrate of thalamocortical connectivity remains to produce this signal — not that the brain has reestablished organized activity across widely distributed networks correlated with goal-directed behavior and cognition (also see John, this volume).

Importantly, the long-lasting changes of ongoing EEG background activity and thalamic firing patterns associated with the arousal state of wakefulness are episodically shaped at a finer temporal scale by brief phasic modulations of the rhythms that organize behavioral set formation. The aggregate abnormalities of resting coherence spectra observed in our two MCS patients likely reflect loss of this fine structure within their resting wakeful EEG. In wakeful states, quantitative EEG studies in normal subjects and experimental studies suggest several potential surrogate markers of elementary cognitive processes underlying the formation of behavioral sets. Among such measures that may prove relevant are regional excitation of high frequencies seen in primate cortical recordings in the 30–80 Hz range associated with working memory and attention (Fries et al., 2001; Pesaran et al., 2002). Similar patterns of frequency-specific, event-related synchronization and desynchronization events are identified in the human EEG (Pfurtscheller and Lopes da Silva, 1999) and in the dynamical structure associated with the contingent negative variation

(CNV), a measure of expectancy generated by paramedian thalamic structures, and medial frontal cortices in response to a warning cue (cf. Nagai et al., 2004; Slobounov et al., 2000).

Although most studies of the correlation structure of the EEG examine dynamic patterns elicited by specific goal-directed tasks, such activations may only reflect half of the necessary fine structure typically present in a normal subject (and therefore possibly required for emergence from MCS). Raichle and colleagues have proposed that the very high resting metabolic rates in the normal human brain reflect "default self-monitoring" activity that characterizes the conscious goal-directed brain (Raichle et al., 2001; Gusnard and Raichle, 2001). This baseline activity is identified by specific patterns of reduction of brain oxygen extraction fraction (OEF) measured at rest across brain regions in a wide variety of goal-directed tasks. Maximum reductions in OEF arise in midline regions of the posterior medial parietal cortex (posterior cingulate cortex and precuneus) and mesial prefrontal cortex. The baseline mode is proposed to depend on tonically active processing in these areas and to correlate with the overall metabolic demands of resting wakeful states. The very low overall resting cerebral metabolic rates in MCS patients may reflect a severe deficit of such tonically active processes. The dissociation of low resting cerebral metabolism despite recruitable networks raises the possibility that patients who remain near the border of emergence from MCS are characterized by a loss of ongoing self-monitoring with fluctuation of recruitment of these large-scale networks under varying internal conditions of arousal and appearance of environmentally salient stimuli.

In normal subjects, Laufs et al. (2003) correlated fMRI BOLD signal with spontaneous power fluctuations in EEG frequency bands during the "baseline" resting state. They identified a strong positive correlation of beta activity (17–23 Hz) with posterior medial parietal (retrosplenial), temporal-parietal, and dorsomedial BOLD activation. This regional grouping overlaps with Raichle et al.'s baseline network. In addition, they identified a strong negative correlation of alpha activity (8–12 Hz) and BOLD signal in lateral frontal and parietal cortices. These observations raise the possibility that it may ultimately be possible to isolate specific dynamical signatures of ongoing activity in the distributed networks deactivated by task performance against the signature of other systems activated during behavioral performances.

In a study including 10 MCS patients, Laureys and colleagues observed relatively increased metabolic activity in these medial posterior parietal regions compared with VS patients. As noted above, this may indicate a partial reestablishing of baseline metabolic activity. It is interesting that although these regions are the most metabolically active regions in the resting human brain, bilateral injuries in these locations are not known to produce global disorders of consciousness. Focal injuries producing states of globally impaired consciousness and cognition, such as VS, MCS, and other forms of severe disability, are typically associated with bilateral injuries of the paramedian mesencephalon and thalamus, medial frontal cortical systems, or posterior-lateral temporal-parietal regions (Schiff and Plum, 2000). A possible interpretation of this difference, consistent with the proposed functions of these cortical regions, is that the self-monitoring activity thought to drive this high metabolic demand may not be necessary for goal-directed behavior and awareness *per se.*

In addition to quantifying incompletely or insufficiently established dynamic phenomena associated with normal cognition, a systematic evaluation of abnormal dynamics arising in the severely injured brain will be required in evaluating MCS patients. A large variety of pathophysiological mechanisms producing abnormal dynamics have been catalogued in the context of severe brain injuries. At present few diagnostic efforts are applied to assess the contribution of such mechanisms in patients recovering from severe brain damage. A relatively common finding following focal brain lesions is a reduction in cerebral metabolism in brain regions remote from the site of injury (Nguyen and Botez, 1998). Disproportionately large reductions of neuronal firing rates are associated with modest reduction of cerebral blood flow produced by these crossed-synaptic effects (Gold and Lauritzen, 2002). The cellular basis of this effect appears to be a loss of excitatory drive to neuronal populations that results in a form of inhibition known as disfacilitation in which hyperpolarization of neuronal membrane potentials arises from the absence of excitatory synaptic inputs allowing remaining leak currents (principally potassium) to dominate (Timofeev et al., 2001). Disfacilitation may play a major role in changing resting brain activity levels given recent evidence (Steriade, 2004) that cortical neurons may change fundamental firing properties based on levels of depolarization (considered here as a proxy for excitatory drive). Multifocal injuries may therefore result in wide passive inhibition of networks due to loss of background activity. Note that selective structural injuries to the paramedian thalamus are unique in producing hemisphere-wide metabolic reductions presumably through this mechanism (Szelies et al., 1991; Caselli et al., 1991). Similarly, herniation injuries may generally produce some level of hemisphere-wide disfacilitation. Thus, the broadband, hemispheric, reductions in EEG coherence observed in the MCS patients discussed above may reflect ongoing functional alteration of common thalamic driving inputs to the cerebral cortex (as opposed to complete structural thalamic injury as seen in Fig. 2).

In addition to disfacilitation, which may arise on the basis of non-selective injuries across many different cerebral structures, other specific dynamical abnormalities may be associated with severe brain injuries. In some patients selective structural injuries may damage pathways of the brainstem arousal systems where the fibers emanate or run close together. Consequent withdrawal of broad cortical innervation by a neuromodulator could produce significant dynamical effects on the EEG and behavior. In a small series of VS patients with isolated MRI findings of axonal injuries near the cerebral peduncle (including substantia nigra and ventral tegmental area) and parkinsonism, the patients made late recoveries following administration of levodopa (Matsuda et al., 2003). The ascending cholinergic pathway also runs in tight bundles at points along its initial trajectory to the cerebral cortex and a role for focal injuries along this pathway has been proposed (Selden et al., 1998).

Epileptiform or similar hypersynchronous phenomena may arise in severe brain damage without obvious traditional EEG markers. Williams and Parsons-Smith (1951) described local epileptiform activity in the human thalamus that appeared only as surface slow waves in the electroencephalogram in a patient with a neurological exam alternating between a state consistent with MCS and interactive communication following an encephalitic injury. A similar mechanism might underlie a case of episodic recovery of communication in a severely disabled patient that intermittently resolved following occasional generalized seizures (Burruss and Chacko, 1999). Clauss et al. (2001) described emergence from MCS in a 28-year-old man with diffuse axonal injury after a stable 3-year period following administration of the GABA agonist zolpidem that correlated with 35–40% increases in blood flow measured by single photon emission tomography (SPECT) in the medial frontal cortex bilaterally and left middle frontal and supramarginal gyri. Experimental studies have shown increased excitability following even minor brain trauma that may promote epileptiform or other forms of hypersynchronous activity in both cortical and subcortical regions (Santhakumar et al., 2001). Other observed phenomena in severe brain injuries include several syndromes with features of dystonia such as oculogyric crises (Leigh et al., 1987; Kakigi et al., 1986), obsessive compulsive disorder (Berthier et al., 2001), and paroxysmal autonomic phenomena (reviewed in Blackman et al., 2004). These phenomena typically show selective responses to different pharmacotherapies.

It is not yet possible to predict the presence and influence of reversible dynamical phenomena that may arise in the setting of novel connective topologies induced by structural brain injuries. However, it may be possible to begin to identify specific dynamical signatures of such state-dependent phenomena using quantitative EEG and MEG methods. Llinas et al. (1999) demonstrated examples of spectral abnormalities in cross-frequency interactions in several different disorders including epilepsy, dystonia, and tremor. At present, however, no systematic methods have been developed to screen for these mechanisms. The brief review above suggests that to accurately model recovery from severe brain damage it will be necessary to attempt to isolate brain dynamics across different structural pathologies and possibly even patterns of resting metabolic activity. Available studies reviewed above indicate that structural pathology and resting metabolism may provide only limited guides to understanding cerebral integrative processes associated with consciousness and cognition in severe brain dysfunction. Given these limitations complementary EEG measures need to be developed to track longitudinal changes in correlation with behavioral patterns and functional imaging.

Summary

Figure 4 organizes the proposed mechanisms for the clinical spectrum arising across VS and MCS patients. The common feature across all MCS patients is the preservation of contingent response to the environment even if infrequently observed. As suggested in Fig. 4, some MCS patients who remain behaviorally near the gray zone, where unusual VS patients can exhibit isolated fragments of behavior, might only retain a limited number of modular sensorimotor networks that nonetheless can show a patterned response. In such cases the patient's limited behavioral repertoire may not reflect greater residual cognitive capacities. Conversely, functional imaging studies already provide evidence that patients closer to emergence from MCS may harbor multiple, responsive large-scale networks. These functional differences likely underlie the rare instances of patients who spontaneously emerge late in the course of MCS. An accurate model of MCS for patients near this upper boundary will require understanding the mechanisms underlying endogenous recruitment of these distributed networks to form and stabilize behavioral sets. It is proposed that unstable interactions of the arousal and gating systems may underlie the fluctuations observed in MCS patients. Emergence from MCS (blue arrow) may then reflect recovery of sufficient ongoing dynamics to support communication and goal-directed behaviors in brains that have remained widely functionally connected but dynamically impaired.

Possible therapeutic strategies

Spontaneous emergence late in the course of MCS indicates that some patients with non-progressive encephalopathies retain reserve capacities. The observations raise the question of how these capacities might be recruited in MCS patients and others with less severe cognitive disability. Recent efforts have begun to examine the effects of dopaminergic and other neuromodulators early in the course of treatment of MCS patients (Giacino, this volume). As discussed above, in some patients, single-agent pharmacologic interventions may lead to dramatic improvements. Another direction for experimental therapeutics is the development of deep brain-stimulation (DBS) strategies.

DBS of selective intralaminar thalamic nuclei (ILN) has been proposed as a strategy for treating patients with acquired cognitive disabilities (Schiff et al., 2000, 2002b; Schiff and Purpura, 2002). Appropriate DBS stimulation parameters to produce clinically meaningful effects are unknown. At present DBS therapies are "open-loop" applications in which the frequency and amplitude of electrical pulses generated by the stimulator are empirically adjusted to achieve a steady-state stimulation rate that is titrated to clinical response (Volkmann et al., 2002). At least two complementary rationales for open-loop cognitive neuromodulation in the ILN can be articulated: increases in cortical neuronal activity induced by DBS might help support and extend ongoing distributed network activity. In addition, or alternatively, reestablishing normal patterns of coherence of neuronal activity may be important. Effective cognitive rehabilitation strategies suggest that special synchronizing signals also play a key role in reestablishing cognitive functions suggesting a basis for closed-loop DBS strategies as well (reviewed in Schiff and Purpura, 2002).

Initial experimental studies of open-loop DBS in primates and rodents provide some support for this research direction. We modeled a human vigilance paradigm (Kinomura et al., 1996) in the nonhuman primate to study central thalamic contributions to the formation and completion of behavioral sets (Schiff et al., 2001). Figure 5 shows the timeline of this experiment. The animal initiates the trial by holding a bar, and following a

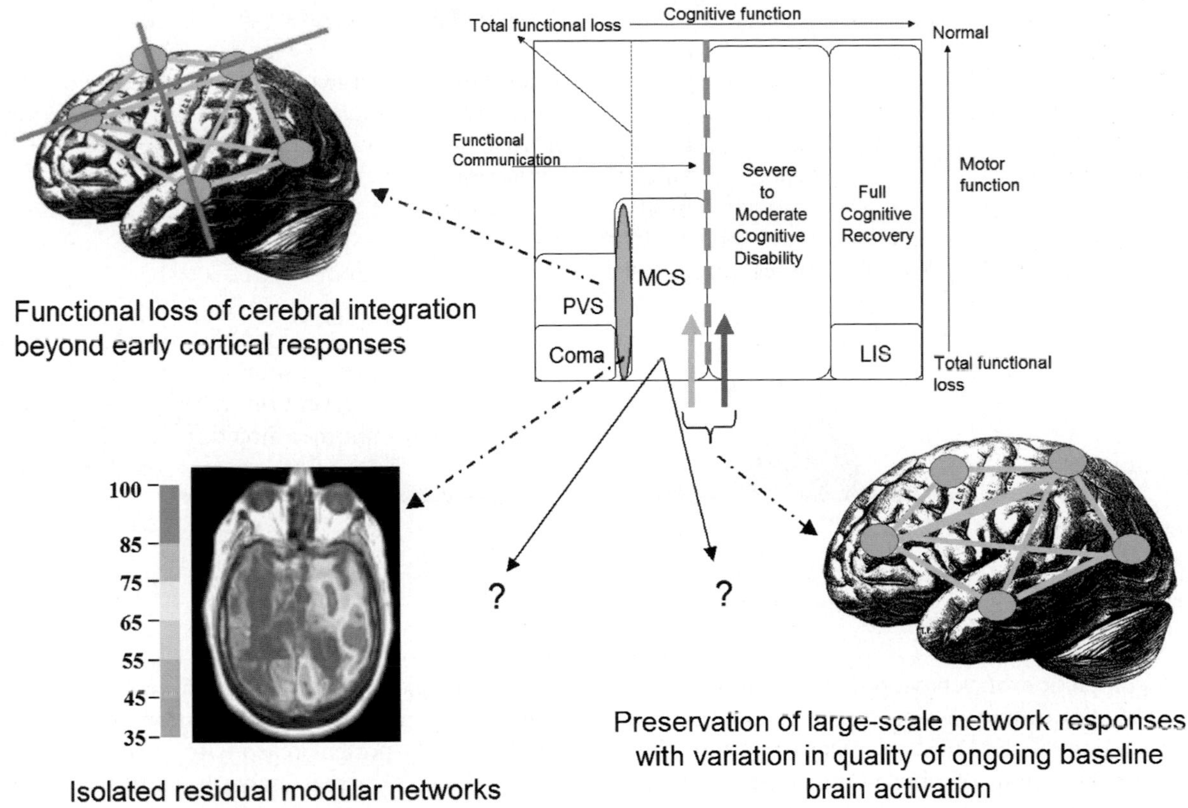

Fig. 4. Mechanisms underlying functional levels across spectra of vegetative state and minimally conscious state patients. Coregistered FDG-PET and MRI image from patient in Fig. 2 with color scale indicating percentage of normal regional metabolic rates (from Schiff et al., 2002; see text for further discussion). See Plate 33.4 in Colour Plate Section.

fixed delay a target appears in one of nine locations in a spatial array. After acquiring the target by a saccadic eye movement, the animal is then required to hold fixation for a variable delay until the target changes color providing a "go" signal to release the bar within one second to receive a juice reward. Figure 5B illustrates a peri-stimulus time histogram of single-unit responses from a central thalamic location during the sustained attention (variable delay) component of this reaction-time task. The persistent neuronal firing pattern seen is similar to the delay-period activity recorded during both selective attention and working memory paradigms in the prefrontal cortex (Fuster, 1973; Goldman-Rakic, 1996), frontal eye fields (Schall, 1991), and the posterior parietal cortex (Andersen, 1989; Pesaran et al., 2002). The central thalamic recordings shown here may be recorded from rostral regions of the ILN (Schlag-Rey and Schlag 1984) or closely related paralaminar regions of the median dorsalis nucleus. Collectively these regions selectively project to prefrontal cortex, frontal eye fields, and anterior cingulate cortex and lateral parietal areas placing them in a central position to participate in the integration of intentional gaze control with attentional and working memory systems (Purpura and Schiff, 1997).

The recordings from incorrect trials show that initiation of the persistent firing activity may arise but fail to build up to the same level and maintain activation over the course of the trial. Figure 6 illustrates the conceptual basis for the possible use of open-loop central thalamic DBS in selected MCS patients. If the patient can initiate behavioral set formation across distributed networks spontaneously, it is proposed that activation of these pathways may support increased firing rates at target cortical locations and improve the maintenance

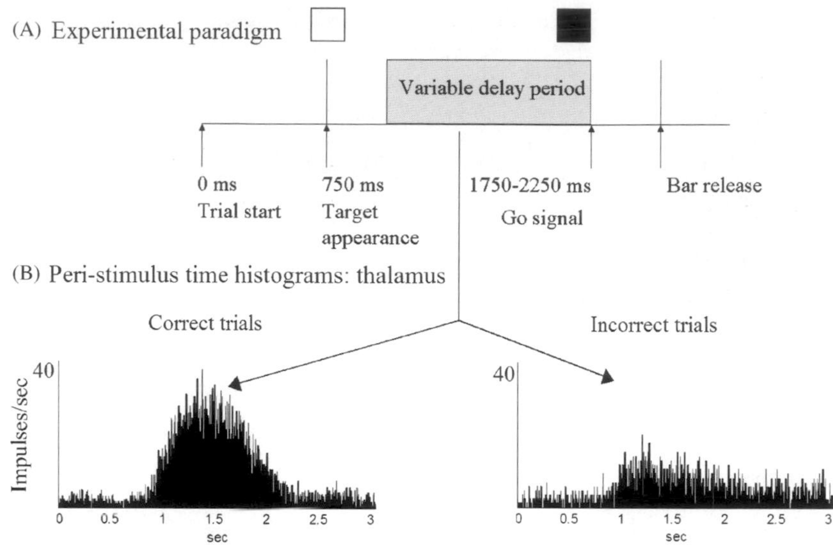

Fig. 5. (A) Behavioral paradigm for elementary visuomotor attention task. (B) Peri-stimulus time histograms for central thalamic neurons during correct and incorrect performance of the task.

and completion of behaviors that are initiated. In pilot studies using the same cognitive paradigm in conjunction with DBS in recording sites that showed elevation of single-unit firing rates (as shown in Fig. 5), the percentage of correct performances was significantly improved at the end of the day when large performance decrements arise (Schiff et al., 2002c). These findings are comparable to improvements in the performance of object recognition tasks during stimulation of the central lateral intralaminar nucleus in rats (Shirvalkar et al., 2004). These initial studies support further investigations into the contributions of central thalamic populations to elementary cognitive operations (cf. Wyder et al., 2004) and effects of direct electrical stimulation.

Implications and research directions

Why should we carefully study MCS patients and others with severe brain dysfunction? The most general answer is that it appears that functional disabilities may often exceed the obvious burden of structural brain injuries and that neuroimaging studies may show more distributed functional activation of cerebral networks than anticipated by the bedside examination. Further research efforts must focus on what these activations may mean, when the data present reasons to expect potential improvement or a reasonable basis to pursue the use of experimental therapeutics, and related diagnostic and prognostic concerns.

It is an empirical question whether residual cerebral capacities in some MCS patients can be augmented to achieve a palliative care goal. Fins (this volume) articulates a framework for palliative care in the context of severe brain damage. One apparently defensible palliative goal would be to help MCS patients reliably communicate. Communication presents a "bright-line" distinction that immediately places the patient into a different functional category. It may also be that reliable communication is the boundary where most would agree that concerns about futility are largely resolved. Along these lines, it is increasingly recognized that placing all patients with apparently nonprogressive encephalopathies into custodial care without further consideration of brain mechanisms is not consistent with basic principles of clinical ethics (Fins, 2003).

It is anticipated that understanding brain mechanisms underlying MCS will extend to insights into other less devastating outcomes of

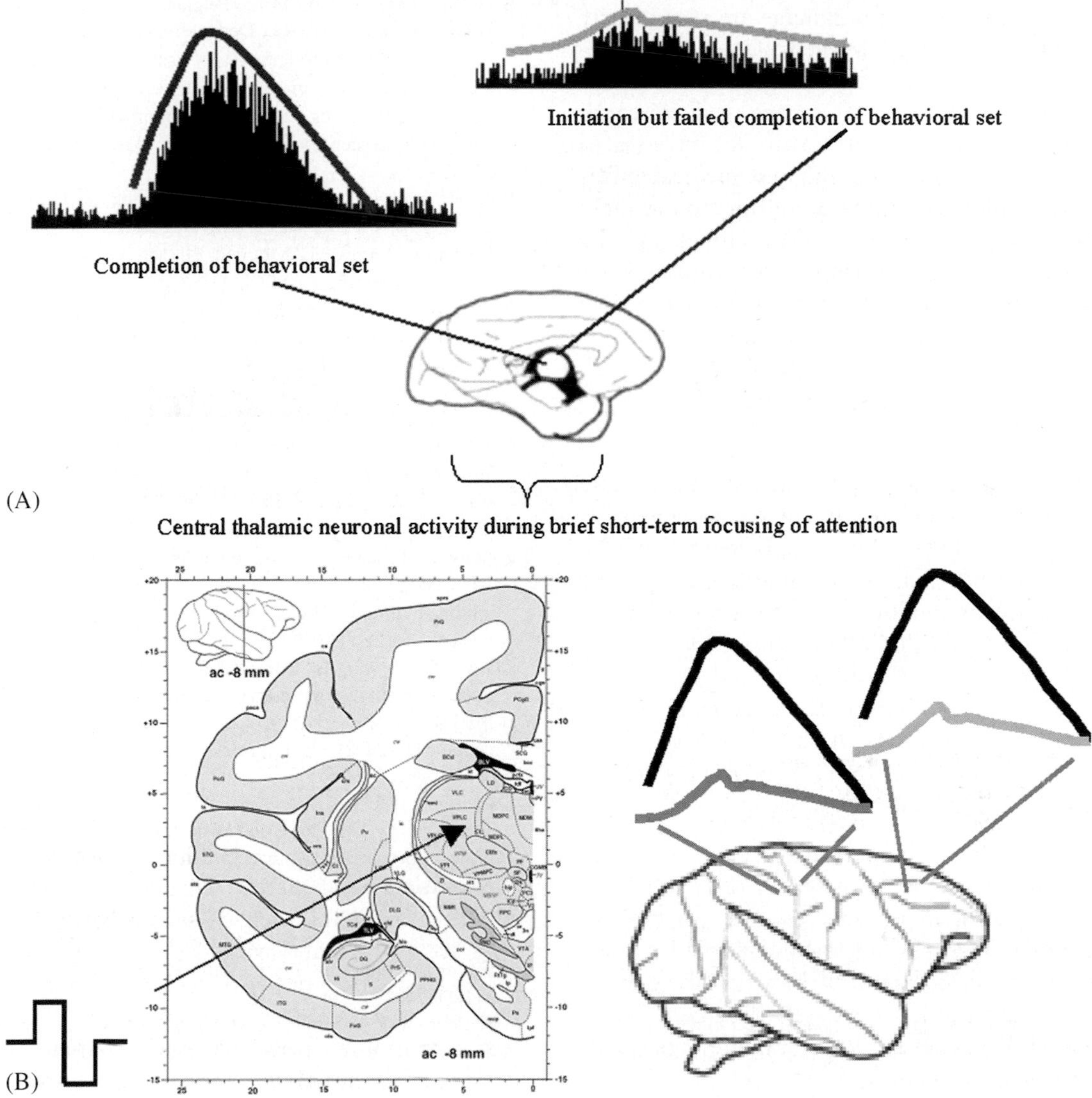

Fig. 6. Illustration of the theoretical basis for the proposed use of open-loop deep brain stimulation as a cognitive neuromodulation strategy (see text).

severe brain damage and potentially more significant palliative goals, if still falling short of restoring normal brain function. Patients suffering less severe brain injuries than those producing MCS will likely share similar pathophysiological mechanisms. For example, pathological studies in patients remaining with only moderate disabilities following brain injury identified post-traumatic epilepsy in 75% of the patients but no diffuse thalamic damage and only mild diffuse axonal injuries (Adams et al., 2001), suggesting that functional (dynamical) disturbances may play a greater role in outcome. To date, few studies have considered accessing potential cognitive reserve in patients with non-progressive encephalopathies.

Finally, it should be recognized that studies of patients with severe brain damage are particularly vulnerable to dismissal by neurologists as irrelevant and hopeless, and by neuroscientists as too diffuse to fit into nicely packaged research projects. Unfortunately such attitudes have led to a continuing lack of scientific and medical infrastructure available to study brain function in these disorders (Laureys et al., 2004). Hopefully, the large number of new scientific contributions reflected in this volume will promote further curiosity and intellectual engagement of these issues.

Acknowledgments

This paper was originally presented at The Satellite Symposium on Coma and Impaired Consciousness, University of Antwerp, Antwerp, Belgium, June 24, 2004. The author thanks Dr. Steven Laureys, the Mind Science Foundation and the Association for the Scientific Study of Consciousness, for the invitation to speak at this symposium and Dr. Joseph Fins and Andrew Hudson for comments on the manuscript. The support of the Charles A. Dana Foundation and the NIH-NINDS (NS02172, NS43451) are gratefully acknowledged.

References

Adams, J.H., Graham, D.I. and Jennett, B. (2000) The neuropathology of the vegetative state after acute insult. Brain, 123: 1327–1338.

Adams, J.H., Graham, D.I. and Jennett, B. (2001) The structural basis of moderate disability after traumatic brain damage. J. Neurol. Neurosur. Ps., 71: 521–524.

Andersen, R. (1989) Visual and eye movement functions of the posterior parietal cortex. Annual Rev. Neurosci., 12: 377–403.

Bekinschtein, T., Leiguarda, R., Armony, J., Owen, A., Carpintiero, S., Niklison, J., Olmos, L., Sigman, L. and Manes, F.J. (2004) Emotion processing in the minimally conscious state. J. Neurol. Neurosur. Ps., 75(5): 788.

Bendat, J.S. and Piersol, A.G. (2000) Random Data: Analysis and Measurement Procedures. Wiley, New York.

Berthier, M.L., Kulisevsky, J.J., Gironell, A. and Lopez, O.L. (2001) Obsessive compulsive disorder and traumatic brain injury: behavioral, cognitive, and neuroimaging findings. Neuropsy. Neuropsy. Be., 14: 23–31.

Blackman, J.A., Patrick, P.D., Buck, M.L. and Rust Jr., R.S. (2004) Paroxysmal autonomic instability with dystonia after brain injury. Arch. Neurol., 61(3): 321–328.

Boly, M., Faymonville, M.E., Peigneux, P., Lambermont, B., Damas, P., Del Fiore, G., Degueldre, C., Franck, G., Luxen, A., Lamy, M., Moonen, G., Maquet, P. and Laureys, S. (2004a) Auditory processing in severely brain injured patients: differences between the minimally conscious state and the persistent vegetative state. Arch. Neurol., 61(2): 233–238.

Boly M., *et al.* Faymonville, M. E., Peigneux, P., Lambermont, B., Damas, P., Del Fiore, G., Degueldre, C., Franck, G., Luxen, A., Lamy, M., Moonen, G., Maquet, P. and Laureys, S. (2004b). Abstract ASSC8.

Burke, W.J. (2002) The minimally conscious state: definition and diagnostic criteria. Neurology, 59(9): 1473.

Burruss, J.W. and Chacko, R.C. (1999) Episodically remitting akinetic mutism following subarachnoid hemorrhage. J. Neuropsychiatry Clin. Neurosci., 11(1): 100–102.

Caselli, R.J., Graff-Radford, N.R. and Rezai, K. (1991) Thalamocortical diaschisis: single-photon emission tomographic study of cortical blood flow changes after focal thalamic infarction. Neuropsy. Neuropsy. Be., 4: 193–214.

Castaigne, P., Lhermitte, F., Buge, A., Escourolle, R., Hauw, J.J. and Lyon-Caen, O. (1981) Paramedian thalamic and midbrain infarcts: clinical and neuropathological study. Ann. Neurol., 10(2): 127–148.

Clauss, R.P., van der Merwe, C.E. and Nel, H.W. (2001) Arousal from a semi-comatose state on zolpidem. S. Afr. Med. J., 91(10): 788–789.

Coleman, D. (2002) The minimally conscious state. Neurology, 58(3): 506.

Danielsen, E.R., Christensen, P.B., Arlien-Soborg, P. and Thomsen, C. (2003) Axonal recovery after severe traumatic brain injury demonstrated in vivo by 1 H MR spectroscopy. Neuroradiology, 45(10): 722–724.

Davey, M.P., Victor, J.D. and Schiff, N.D. (2000) Power spectra and coherence in the EEG of a vegetative patient with severe asymmetric brain damage. Clin. Neurophysiol., 111(11): 1949–1954.

DeVolder, A.G., Goffinet, A.M., Bol, A., Michel, C., de Barsy, T. and Laterre, C. (1990) Brain glucose metabolism in post-anoxic syndrome. Positron emission tomographic study. Arch. Neurol., 47(2): 197–204.

Dougherty Jr., J.H., Rawlinson, D.G., Levy, D.E. and Plum, F. (1981) Hypoxic-ischemic brain injury and the vegetative state: clinical and neuropathologic correlation. Neurology, 31(8): 991–997.

Eidelberg, D., Moeller, J.R., Kazumata, K., Antonini, A., Sterio, D., Dhawan, V., Spetsieris, P., Alterman, R., Kelly, P.J., Dogali, M., Fazzini, E. and Beric, A. (1997) Metabolic correlates of pallidal neuronal activity in Parkinson's disease. Brain, 120(Pt 8): 1315–1324.

Fins, J.J. (2003) Constructing ethical stereotaxy for severe brain injury; balancing risks, benefits and access. Nat. Rev. Neurosci., 4(4): 323–327.

Fins, J.J. and Plum, F. (2004) Neurological diagnosis is more than a state of mind: diagnostic clarity and impaired consciousness. Arch. Neurol., 61(9): 1354–1355.

Fries, P., Reynolds, J.H., Rorie, A.E. and Desimone, R. (2001) Modulation of oscillatory neuronal synchronization by selective visual attention. Science, 291: 1560–1563.

Fuster, J.M. (1973) Unit activity in prefrontal cortex during delayed-response performance: neuronal correlates of transient memory. J. Neurophysiol., 36: 61–78.

Giacino, J.T. and Whyte, J. (2005). The vegetative state and minimally conscious state: current knowledge and remaining questions. J. Head Trauma Rehabil., Jan–Feb; 20(1): 30–50.

Giacino, J.T., Ashwal, S., Childs, N., Cranford, R., Jennett, B., Katz, D.I., Kelly, J.P., Rosenberg, J.H., Whyte, J., Zafonte, R.D. and Zasler, N.D. (2002) The minimally conscious state: definition and diagnostic criteria. Neurology, 58: 349–353.

Gold, L. and Lauritzen, M. (2002) Neuronal deactivation explains decreased cerebellar blood flow in response to focal cerebral ischemia or suppressed neocortical function. Proc. Natl. Acad. Sci., 99: 7699–7704.

Groenewegen, H. and Berendse, H. (1994) The specificity of the 'nonspecific' midline and intralaminar thalamic nuclei. Trends Neurosci., 17: 52–66.

Gusnard, D.A., Raichle, M.E. and Raichle, M.E. (2001) Searching for a baseline: functional imaging and the resting human brain. Nat. Rev. Neurosci., 2(10): 685–694.

Hansotia, P.L. (1985) Persistent vegetative state. Review and report of electrodiagnostic studies in eight cases. Arch. Neurol., 42(11): 1048–1052.

Hirsch, J., Ruge, M.I., Kim, K.H., Correa, D.D., Victor, J.D., Relkin, N.R., Labar, D.R., Krol, G., Bilsky, M.H., Souweidane, M.M., DeAngelis, L.M. and Gutin, P.H. (2000) An integrated functional magnetic resonance imaging procedure for preoperative mapping of cortical areas associated with tactile, motor, language, and visual functions. Neurosurgery, 47(3): 711–721.

Ingvar, D.H. and Sourander, P. (1970) Destruction of the reticular core of the brainstem. Archives of Neurology, 23: 1–8.

Isono, M., Wakabayashi, Y., Fujiki, M.M., Kamida, T. and Kobayashi, H. (2002) Sleep cycle in patients in a state of permanent unconsciousness. Brain Injury, 16(8): 705–712.

Jennett, B. (2002) The Vegetative State. Cambridge University Press, Cambridge.

Jennett, B., Adams, J.H., Murray, L.S., et al. (2001) Neuropathology in vegetative and severely disabled patients after head injury. Neurology, 56: 486–490.

Jennett, B. and Bond, M. (1975) Assessment of outcome after severe brain damage. Lancet, 1: 480–484.

Jennett, B. and Plum, F. (1972) Persistent vegetative state after brain damage. A syndrome in search of a name. Lancet, 1: 734–737.

Jones, E.G. (2001) The thalamic matrix and thalamocortical synchrony. Trends Neurosci., 24: 595–601.

Kampfl, A., Schmutzhard, E., Franz, G., Pfausler, B., Haring, H.P., Ulmer, H., Felber, S., Golaszewski, S. and Aichner, F. (1998) Prediction of recovery from post-traumatic vegetative state with cerebral magnetic-resonance imaging. Lancet, 351(9118): 1763–1767.

Kakigi, R., Shibasaki, H., Katafuchi, Y., Iyatomi, I. and Kuroda, Y. (1986) The syndrome of bilateral paramedian thalamic infarction associated with an oculogyric crisis. Rinsho Shinkeigaku, 26: 1100–1105.

Katz, D.I., Alexander, M.P. and Mandell, A.M. (1987) Dementia following strokes in the mesencephalon and diencephalon. Arch. Neurol., 44: 1127–1133.

Kinomura, S., Larssen, J., Gulyas, B. and Roland, P.E. (1996) Activation by attention of the human reticular formation and thalamic intralaminar nuclei. Science, 271: 512–515.

Kobylarz, E., Kamal, A., and Schiff, N.D. (2003) Power spectrum and coherence analysis of the EEG from two minimally conscious patients with severe asymmetric brain damage. ASSC Meeting 2003.

Kobylarz, E.J. and Schiff, N.D. (2004) Functional imaging of severely brain-injured patients: progress, challenges, and limitations. Arch. Neurol., 61(9): 1357–1360.

Laufs, H., Krakow, K., Sterzer, P., Eger, E., Beyerle, A., Salek-Haddadi, A. and Kleinschmidt, A. (2003) Electroencephalographic signatures of attentional and cognitive default modes in spontaneous brain activity fluctuations at rest. Proc. Natl. Acad. Sci., 16100(19): 11053–11058.

Laureys, S., Lemaire, C., Maquet, P., Phillips, C. and Franck, G. (1999) Cerebral metabolism during vegetative state and after recovery to consciousness. J. Neurol. Neurosurg. Ps., 67(1): 121.

Laureys, S., Faymonville, M.E., Degueldre, C., Fiore, G.D., Damas, P., Lambermont, B., Janssens, N., Aerts, J., Franck, G., Luxen, A., Moonen, G., Lamy, M. and Maquet, P. (2000a) Auditory processing in the vegetative state. Brain, 123: 1589–1601.

Laureys, S., Faymonville, M.E., Luxen, A., Lamy, M., Franck, G. and Maquet, P. (2000b) Restoration of thalamocortical connectivity after recovery from persistent vegetative state. Lancet, 355(9217): 1790–1791.

Laureys, S., Faymonville, M.E., Peigneux, P., Damas, P., Lambermont, B., Del Fiore, G., Degueldre, C., Aerts, J., Luxen, A., Franck, G., Lamy, M., Moonen, G. and Maquet, P. (2002) Cortical processing of noxious somatosensory stimuli in the persistent vegetative state. Neuroimage, 17(2): 732–741.

Laureys, S., Faymonville, M., Ferring, M., Schnakers, C., Elincx, S., Ligot, N., Majerus, S., Antoine, S., Mavroudakis, N., Berre, J., Luxen, A., Vincent, J.L., Moonen, G., Lamy, M., Goldman, S. and Maquet, P. (2003) Differences in brain metabolism between patients in coma, vegetative state, minimally conscious state and locked-in syndrome. Eur. J. Neurol., 10(Suppl. 1): 224.

Laureys, S.L., Owen, A.M. and Schiff, N.D. (2004) Brain function in coma, vegetative state and related disorders. Lancet Neurol., 3(9): 537–546.

Leigh, R.J., Foley, J.M., Remler, B.F. and Civil, R.H. (1987) Oculogyric crisis: a syndrome of thought disorder and ocular deviation. Ann. Neurol., 22: 13–17.

Levy, D.E., Sidtis, J.J., Rottenberg, D.A., Jarden, J.O., Strother, S.C., Dhawan, V., Ginos, J.Z., Tramo, M.J., Evans, A.C. and Plum, F. (1987) Differences in cerebral blood flow and glucose utilization in vegetative versus locked-in patients. Ann. Neurol., 22: 673–682.

Llinas, R.R., Ribary, U., Jeanmonod, D., Kronberg, E. and Mitra, P.P. (1999) Thalamocortical dysrhythmia: A neurological and neuropsychiatric syndrome characterized by magnetoencephalography. Proc. Natl. Acad. Sci., 96: 15222–15227.

Llinas, R., Ribary, U., Joliot, M. and Wang, X.J. (1994) Content and context in temporal thalamocortical binding. In: Buzsaki G., et al. (Eds.), Temporal Coding in the Brain. Springer, Heidelberg, pp. 252–272.

Llinas, R.R., Leznik, E. and Urbano, F.J. (2002) Temporal binding via cortical coincidence detection of specific and nonspecific thalamocortical inputs: a voltage-dependent dye-imaging study in mouse brain slices. Proc. Natl. Acad. Sci., 99: 449–454.

Macniven, J.A., Poz, R., Bainbridge, K., Gracey, F. and Wilson, B.A. (2003) Emotional adjustment following cognitive recovery from 'persistent vegetative state': psychological and personal perspectives. Brain Injury, 17(6): 525–533.

Matsuda, W., Matsumura, A., Komatsu, Y., Yanaka, K. and Nose, T. (2003). J. Neurol. Neurosurg. Psychiatry, 74: 1571–1573.

Matsumoto, N., Minamimoto, T., Graybiel, A.M. and Kimura, M. (2001) Neurons in the thalamic CM-Pf complex supply striatal neurons with information about behaviorally significant sensory events. J. Neurophysiol., 85: 960–976.

McMillan, T.M. and Herbert, C.M. (2004) Further recovery in a potential treatment withdrawal case 10 years after brain injury. Brain Inj., 18(9): 935–940.

Meissner, I., Sapir, S., Kokmen, E. and Stein, S.D. (1987) The paramedian diencephalic syndrome: a dynamic phenomenon. Stroke, 18(2): 380–385.

Mennemeier, M., Crosson, B., Williamson, D.J., Nadeau, S.E., Fennell, E., Valenstein, E. and Heilman, K.M. (1997) Tapping, talking and the thalamus: possible influence of the intralaminar nuclei on basal ganglia function. Neuropsychologia, 35(2): p183–p193.

Menon, D.K., Owen, A.M. and Pickard, J.D. (1999) Response from Menon, Owen and Pickard. Trends Cogn. Sci., 3(2): 44–46.

Menon, D.K., Owen, A.M., Williams, E.J., Minhas, P.S., Allen, C.M., Boniface, S.J. and Pickard, J.D. (1998) Cortical processing in persistent vegetative state. Lancet, 352: 1148–1149.

Mitra, P.P. and Pesaran, B. (1999) Analysis of dynamic brain imaging data. Biophys. J., 76(2): 691–708.

Minamimoto, T. and Kimura, M. (2002) Participation of the thalamic CM-Pf complex in attentional orienting. J. Neurophysiol., 87: 3090–3101.

Nagai, Y., Critchley, H.D., Featherstone, E., Fenwick, P.B.C., Trimble, M.R. and Dolan, R.J. (2004) Brain activity relating to the contingent negative variation: an fMRI investigation. NeuroImage, 21(4): 1232–1241.

Nguyen, D.K. and Botez, M.I. (1998) Diaschisis and neurobehavior. Can. J. Neurol. Sci., 25: 5–12.

Nunez, P.L., Silberstein, R.B., Shi, Z., Carpenter, M.R., Srinivasan, R., Tucker, D.M., Doran, S.M., Cadusch, P.J. and Wijesinghe, R.S. (1999) EEG coherency II: Experimental comparisons of multiple measures. Clin. Neurophys., 110: 469–486.

Paus, T., Zatorre, R., Hofle, N., Caramanos, Z., Gotman, J., Petrides, M. and Evans, A. (1997) Time-related changes in neural systems underlying attention and arousal during the performance of an auditory vigilance task. J. Cogn. Neurosci., 9: 392–408.

Pesaran, B., Pezaris, J.S., Sahani, M., Mitra, P.P. and Andersen, R.A. (2002) Temporal structure in neuronal activity during working memory in macaque parietal cortex. Nat. Neurosci., 5: 805–811.

Pfurtscheller, G. and Lopes da Silva, F.H. (1999 Nov) Event-related EEG/MEG synchronization and desynchronization: basic principles. Clin. Neurophysiol., 110(11): 1842–1857.

Plum, F. and Posner, J. (1982) Diagnosis of Stupor and Coma. F.A. Davis and Company, New York.

Plum, F. (1991) Coma and related global disturbances of the human conscious state. In: Jones E. and Peters P. (Eds.), Cerebral Cortex, Vol. 9. Plenum Press, New York.

Purpura, K.P. and Schiff, N.D. (1997) The thalamic intralaminar nuclei: role in visual awareness. Neuroscientist, 3: 8–14.

Raichle, M.E., MacLeod, A.M., Snyder, A.Z., Powers, W.J., Gusnard, D.A. and Shulman, G.L. (2001) A default mode of brain function. Proc. Natl. Acad. Sci., 98(2): 676–682.

Robinson, P.A., Rennie, C.J. and Rowe, D.L. (2002) Dynamics of large-scale brain activity in normal arousal states and epileptic seizures. Phys. Rev. E Stat. Nonlin Soft Matter Phys., 65(4): 041924.

Rothstein, T.L., Thomas, E.M. and Sumi, S.M. (1991) Predicting outcome in hypoxic-ischemic coma. A prospective clinical and electrophysiologic study. Electroen. Clin. Neuro., 79(2): 101–107.

Rudolf, J., Ghaemi, M., Ghaemi, M., Haupt, W.F., Szelies, B. and Heiss, W.D. (1999) Cerebral glucose metabolism in acute and persistent vegetative state. J. Neurosurg. Anesthesiol., 11(1): 17–24.

Santhakumar, V., Ratzliff, A.D., Jeng, J., Toth, Z. and Soltesz, I. (2001) Long-term hyperexcitability in the hippocampus after experimental head trauma. Ann. Neurol., 50: 708–717.

Selden, N.R., Gitelman, D.R., Salamon-Murayama, N., Parrish, T.B. and Mesulam, M.M.. (1998) Trajectories of cholinergic pathways within the cerebral hemispheres of the human brain. Brain, 121: 2249–2257.

Schiff, N.D., Rezai, A. and Plum, F. (2000) A neuromodulation strategy for rational therapy of complex brain injury states. Neurological Research, 22(3): 267–272.

Schiff, N.D. (2004) The neurology of impaired consciousness: challenges for cognitive neuroscience. In: Gazzaniga M.S. (Ed.), The Cognitive Neurosciences (3rd ed). MIT Press, Cambridge, MA.

Schiff, N.D., Hudson, A.E., and Purpura, K.P. (2002c). Modeling wakeful unresponsiveness: characterization and microstimulation of the central thalamus. Society for Neuroscience 31th Annual Meeting (62.12).

Schiff, N.D., Kalik, S.F., and Purpura, K.P. (2001). Sustained activity in the central thalamus and extrastriate areas during attentive visuomotor behavior: correlation of single unit activity and local field potentials. Society for Neuroscience 30th Annual Meeting (722.12).

Schiff, N.D. and Plum, F. (1999) Cortical processing in the vegetative state. Trends Cogn. Sci., 3(2): 43–44.

Schiff, N.D. and Plum, F. (2000) The role of arousal and 'gating' systems in the neurology of impaired consciousness. J. Clin. Neurophysiol., 17: 438–452.

Schiff, N.D., Plum, F. and Rezai, A.R. (2002b) Developing prosthetics to treat cognitive disabilities resulting from acquired brain injuries. Neurol. Res., 24: 116–124.

Schiff, N.D. and Purpura, K.P. (2002) Towards a neurophysiological basis for cognitive neuromodulation through deep brain stimulation. Thalamus and Related Systems, 2(1): 51–69.

Schiff, N.D., Ribary, U., Plum, F. and Llinas, R. (1999) Words without mind. J. Cogn. Neurosci., 11(6): 650–656.

Schiff, N., Ribary, U., Moreno, D., Beattie, B., Kronberg, E., Blasberg, R., Giacino, J., McCagg, C., Fins, J.J., Llinas, R. and Plum, F. (2002a) Residual cerebral activity and behavioral fragments in the persistent vegetative state. Brain, 125: 1210–1234.

Schiff, N., Rodriguez-Moreno, D., Kamal, A., Kim, K.H., Giacino, J., Plum, F. and Hirsch, J. (2005) fMRI reveals large-scale network activation in minimally conscious patients. Neurology, 64: 514–523.

Schlag-Rey, M. and Schlag, J. (1984) Visuomotor functions of central thalamus in monkey. I. Unit activity related to spontaneous eye movements. J. Neurophysiol., 40: 1149–1174.

Shirvalkar, P., Schiff, N.D., and Herrera, D.G. (2004). Deep brain stimulation of the central lateral nucleus selectively modifies immediate-early gene expression and object recognition memory. Society for Neuroscience 32th Annual Meeting.

Slobounov, S.M., Fukada, K., Simon, R., Rearick, M. and Ray, W. (2000) Neurophysiological and behavioral indices of time pressure effects on visuomotor task performance. Cognitive Brain Res., 9: 287–298.

Smith, A.J., Blumenfeld, H., Behar, K.L., Rothman, D.L., Shulman, R.G. and Hyder, F. (2002 Aug 6) Cerebral energetics and spiking frequency: the neuropsychological basis of fMRI. Proc. Natl. Acad. Sci. U S A, 99(16): 10765–10770.

Steriade, M. (1997) Thalamic substrates of disturbances in states of vigilance and consciousness in humans. In: Steriade M., Jones E. and McCormick D. (Eds.), Thalamus. Elsevier Publishers, Amsterdam.

Steriade, M. (2000) Corticothalamic resonance, states of vigilance and mentation. Neuroscience, 101: 243–276.

Steriade, M. (2004) Neocortical cell classes are flexible entities. Nat. Rev. Neurosci., 5(2): 121–134.

Steriade, M., Contreras, D., Amzica, F. and Timofeev, I. (1996) Synchronization of fast (30–40 Hz) spontaneous oscillations in intrathalamic and thalamocortical networks. J. Neurosci., 16: 2788–2808.

Steriade, M. and Glenn, L.L. (1982) Neocortical and caudate projections of intralaminar thalamic neurons and their synaptic excitation from midbrain reticular core. J. Neurophysiol., 48: 352–371.

Szelies, B., et al. (1991) Widespread functional effects of discrete thalamic infarction. Arch. Neurol., 48: 178–182.

Thatcher, R.W., Krause, P. and Hrybyk, M. (1986) Corticocortical associations and EEG coherence: a two-compartmental model. Electroen. Clin. Neuro., 64(2): 123–143.

Thatcher, R.W., North, D.M., Curtin, R.T., Walker, R.A., Biver, C.J., Gomez, J.F and Salazar, A.M. (2001) An EEG severity index of traumatic brain injury. Neuropsychiatry Clin. Neurosci., 13(1): 77–87.

Timofeev, I., Grenier, F. and Steriade, M. (2001) Disfacilitation and active inhibition in the neocortex during the natural sleep-wake cycle: an intracellular study. Proc. Natl. Acad. Sci., 98: 1924–1929.

Tomassino, C., Grana, C., Lucignani, G., Torri, G. and Ferrucio, F. (1995) Regional metabolism of comatose and vegetative state patients. J. Neurosurg. Anesthesiol., 7(2): 109–116.

Wedekind, C., Hesselmann, V., Lippert-Gruner, M. and Ebel, M. (2002) Trauma to the pontomesencephalic brainstem - a major clue to the prognosis of severe traumatic brain injury. Br. J. Neurosurg., 16: 256–260.

Williams, D. and Parsons-Smith, G. (1951) Thalamic activity in stupor. Brain, 74: 377–398.

Wyder, M.T., Massoglia, D.P. and Stanford, T.R. (2003) Quantitative assessment of the timing and tuning of visual-related, saccade-related, and delay period activity in primate central thalamus. J. Neurophysiol., 90(3): 2029–2052.

Wyder, M.T., Massoglia, D.P. and Stanford, T.R. (2004) Contextual modulation of central thalamic delay-period activity: representation of visual and saccadic goals. J. Neurophysiol., 91(6): 2628–2648.

Vaadia, E., Haalman, I., Abeles, M., Bergman, H., Prut, Y., Slovin, H. and Aertsen, A. (1995) Dynamics of neuronal interactions in monkey cortex in relation to behavioural events. Nature, 373(6514): 515–518.

van der Werf, Y.D., Weerts, J.G., Jolles, J., Witter, M.P., Lindeboom, J. and Scheltens, P. (1999) Neuropsychological correlates of a right unilateral lacunar thalamic infarction. J. Neurol. Neurosurg. Psychiatry, 66(1): 36–42.

van der Werf, Y.D., Witter, M.P. and Groenewegen, H.J. (2002) The intralaminar and midline nuclei of the thalamus. Anatomical and functional evidence for participation in processes of arousal and awareness. Brain Res. Brain Res. Rev., 39(2–3): 107–140.

Volkmann, J., Herzog, J., Kopper, F. and Deuschl, G. (2002) Introduction to the programming of deep brain stimulators. Mov. Disord., 17(Suppl. 3): S181–S187.

S. Laureys (Ed.)
Progress in Brain Research, Vol. 150
ISSN 0079-6123

CHAPTER 34

The locked-in syndrome : what is it like to be conscious but paralyzed and voiceless?

Steven Laureys[1,*], Frédéric Pellas[2], Philippe Van Eeckhout[3], Sofiane Ghorbel[2], Caroline Schnakers[1], Fabien Perrin[4], Jacques Berré[5], Marie-Elisabeth Faymonville[6], Karl-Heinz Pantke[7], Francois Damas[8], Maurice Lamy[6], Gustave Moonen[1] and Serge Goldman[9]

[1]*Neurology Department and Cyclotron Research Center, University of Liège, Sart Tilman B30, 4000 Liege, Belgium*
[2]*Neurorehabilitation Medicine, Hôpital Caremeau, CHU Nîmes, 30029 Nîmes Cedex, France*
[3]*Department of Speech Therapy, Hospital Pitié Salpétrière, Paris and French Association Locked in Syndrome (ALIS), 225 Bd Jean-Jaures, MBE 182, 92100 Boulogne-Billancourt, France*
[4]*Neurosciences et Systèmes Sensoriels Unité Mixte de Recherche 5020, Université Claude Bernard Lyon 1 – CNRS, 69007 Lyon, France*
[5]*Intensive Care Medicine, Hôpital Erasme, Université Libre de Bruxelles, Route de Lennik 808, 1070 Brussels, Belgium*
[6]*Anesthesiology, Reanimation and Pain Clinic, CHU University Hospital, Sart Tilman B33, 4000 Liege, Belgium*
[7]*German Association Locked in Syndrome LIS e.V., Evangelischen Krankenhaus Köningin Elisabeth Herzberge gGmbh (Lehrkrankenhaus der Charité), Haus 30, Herzbergstrasse 79, 10365 Berlin, Germany*
[8]*Intensive Care Medicine, Centre Hospitalier Régional de la Citadelle, Boulevard du 12e de Ligne 1, 4000 Liege, Belgium*
[9]*Biomedical PET Unit, Hôpital Erasme, Université Libre de Bruxelles, Route de Lennik 808, 1070 Brussels, Belgium*

Abstract: The locked-in syndrome (pseudocoma) describes patients who are awake and conscious but selectively deefferented, i.e., have no means of producing speech, limb or facial movements. Acute ventral pontine lesions are its most common cause. People with such brainstem lesions often remain comatose for some days or weeks, needing artificial respiration and then gradually wake up, but remaining paralyzed and voiceless, superficially resembling patients in a vegetative state or akinetic mutism. In acute locked-in syndrome (LIS), eye-coded communication and evaluation of cognitive and emotional functioning is very limited because vigilance is fluctuating and eye movements may be inconsistent, very small, and easily exhausted. It has been shown that more than half of the time it is the family and not the physician who first realized that the patient was aware. Distressingly, recent studies reported that the diagnosis of LIS on average takes over 2.5 months. In some cases it took 4–6 years before aware and sensitive patients, locked in an immobile body, were recognized as being conscious. Once a LIS patient becomes medically stable, and given appropriate medical care, life expectancy increases to several decades. Even if the chances of good motor recovery are very limited, existing eye-controlled, computer-based communication technology currently allow the patient to control his environment, use a word processor coupled to a speech synthesizer, and access the worldwide net. Healthy individuals and medical professionals sometimes assume that the quality of life of an LIS patient is so poor that it is not worth living. On the contrary, chronic LIS patients typically self-report meaningful quality of life and their demand for euthanasia is surprisingly infrequent. Biased clinicians might provide less aggressive medical treatment and influence the family in inappropriate ways. It

*Corresponding author. Tel.: +32 4 366 23 16; Fax: +32 4 366 29 46; E-mail: steven.laureys@ulg.ac.be

DOI: 10.1016/S0079-6123(05)50034-7

is important to stress that only the medically stabilized, informed LIS patient is competent to consent to or refuse life-sustaining treatment. Patients suffering from LIS should not be denied the right to die — and to die with dignity — but also, and more importantly, they should not be denied the right to live — and to live with dignity and the best possible revalidation, and pain and symptom management. In our opinion, there is an urgent need for a renewed ethical and medicolegal framework for our care of locked-in patients.

> "*Io non mori', e non rimasi vivo;*
> *Pensa omai per te, s'hai fior d'ingegno,*
> *Qual io divenni, d'uno e d'altro privo.*
>
> *Neither did I die, nor did I remain alive;*
> *Imagine yourself, if your spirit is fine,*
> *what I came to be, deprived of both.*"
>
> Alighieri Dante, 1265-1321, Divina Comedia, *Inferno* XXXIV, 25–27

> "*An old laborer, bent double with age and toil, was gathering sticks in a forest.*
> *At last he grew so tired and hopeless that he threw down the bundle of sticks, and cried out*:
> "*I cannot bear this life any longer. Ah, I wish Death would only come and take me!*"
> *As he spoke, Death, a grisly skeleton, appeared and said to him*:
> "*What wouldst thou, Mortal? I heard thee call me.*"
> "*Please, sir,*" *replied the woodcutter,*
> "*would you kindly help me to lift this faggot of sticks on to my shoulder?*"
>
> Aesop, approximately 620–560 B.C., The Old Man and Death (translated in verse by Jean de La Fontaine, 1621–1695, La Mort et le Bucheron) (Fig. 1).

It is hard to think of a physical disability more cruel than the inability to speak and to move the extremities. In 1966, Plum and Posner first introduced the term "locked-in syndrome" (LIS) to refer to the constellation of quadriplegia and anarthria brought about by the disruption of the brain stem's corticospinal and corticobulbar pathways, respectively (Plum and Posner, 1983). The earliest example of a "locked-in" patient in the medical literature comes from Darolles (1875). However, the locked-in syndrome was already brilliantly described 30 years earlier in Alexandre Dumas's novel *The Count of Monte Cristo* (1844–45) Dumas (1997, Original publication 1854). Herein a character, Monsieur Noirtier de Villefort, was depicted as "a corpse with living eyes." Mr. Noirtier had been in this state for more than 6 years, and he could only communicate by blinking his eyes. His helper pointed at words in a dictionary and the monsignor indicated with his eyes the words he wanted.

Some years later, Emile Zola wrote in his novel *Thérèse Raquin* (Zola, 1979, Original publication

Fig. 1. The Old Man and Death, artist unknown.

1868) about a paralyzed woman who "was buried alive in a dead body" and "had language only in her eyes." Dumas and Zola highlighted the locked-in condition before the medical community did. In the LIS, unlike coma, the vegetative state or akinetic mutism, consciousness remains intact. The patient is locked inside his body, able to perceive his environment but extremely limited to voluntarily interact with it. Both characters lived in an age when their ventral pontine lesion, which is most often vascular, should have killed them quickly. Indeed, for a long time, LIS has mainly been a retrospective diagnosis based on post-mortem findings (Haig et al., 1986; Patterson and Grabois, 1986). Medical technology can now achieve long survival in such cases — the longest history of this condition being 27 years (French Association of Locked-In Syndrome (ALIS)-database and Thadani et al., 1991). Computerized devices now allow the LIS patient and other patients with severe motor impairment to "speak." The preeminent physicist Stephen Hawking, author of the best sellers *A Brief History of Time* and *The Universe in a Nutshell*, is able to communicate solely through the use of a computerized voice synthesizer. With one finger, he selects words presentedserially on a computer screen; the words are then stored and later presented as a synthesized and coherent message (http://www.hawking.org.uk). The continuing brilliant productivity of Hawking despite his failure to move or speak illustrates that locked-in patients can be productive members of the society.

In December 1995, Jean-Dominique Bauby, aged 43 and editor in chief of the fashion magazine *Elle*, had a brain stem stroke. He emerged from a coma several weeks later to find himself in a LIS only able to move his left eyelid and with very little hope of recovery. Bauby wanted to show the world that this pathology, which impedes movement and speech, does not prevent patients from living. He has proven it in an extraordinary book (Bauby, 1997) in which he composed each passage mentally and then dictated it, letter by letter, to an amanuensis who painstakingly recited a frequency-ordered alphabet until Bauby chose a letter by blinking his left eyelid once to signify "yes". His book "The diving bell and the butterfly" became a best-seller only weeks after his death due to septic shock on March 9, 1997. Bauby created ALIS aimed to help patients with this condition and their families (www.club-internet.fr/alis).

Since its creation in 1997, ALIS has registered 367 locked-in patients in France (data updated in August 2004). What follows is a review of LIS, discussing studies on the cause, outcome, symptoms, and quality of life of locked-in patients based on the available literature and the ALIS database, which is the largest in its kind. The latter data should be regarded as a preliminary record of ongoing research. After elimination of patients with missing data, 250 patients were included for further analyses.

Classical, incomplete and total locked-in syndrome

Plum and Posner (1983) described the LIS as

> "a state in which selective supranuclear motor de-efferentation produces paralysis of all four limbs and the last cranial nerves without interfering with consciousness. The voluntary motor *paralysis* prevents the subjects from communicating by word or body movement. Usually, but not always, the anatomy of the responsible lesion in the brainstem is such that locked-in patients are left with the capacity to use vertical eye movements and blinking to communicate their awareness of internal and external stimuli."

Bauer et al. (1979) subdivided the syndrome on the basis of the extent of motor impairment: (a) *classical* LIS is characterized by total immobility except for vertical eye movements or blinking; (b) *incomplete* LIS permits remnants of voluntary motion; and (c) *total* LIS consists of complete immobility including all eye movements combined with preserved consciousness. The American Congress of Rehabilitation Medicine (1995) most recently defined LIS by (i) the presence of sustained eye opening (bilateral ptosis should be ruled out as a complicating factor); (ii) preserved basic cognitive abilities; (iii) aphonia or severe hypophonia;

(iv) quadriplegia or quadriparesis; and (v) a primary mode of communication that uses vertical or lateral eye movement or blinking of the upper eyelid.

Etiology

LIS is most frequently caused by a bilateral ventral pontine lesion (e.g., Plum and Posner, 1983, Patterson and Grabois, 1986) (Fig. 2A). In rarer instances, it can be the result of a mesencephalic lesion (e.g., Chia, 1991; Meienberg et al., 1979, Bauer et al., 1979). The most common etiology of LIS is vascular pathology, either a basilar artery occlusion or a pontine hemorrhage (see Table 1). Another relatively frequent cause is traumatic brain injury (Britt et al., 1977; Landrieu et al., 1984; Keane, 1986; Rae-Grant et al., 1989; Fitzgerald et al., 1997; Golubovic et al., 2004). Following trauma, LIS may be caused either directly by brain stem lesions, secondary to vertebral artery damage and vertebrobasilar arterial occlusion, or due to compression of the cerebral peduncles from tentorial herniation (Keane, 1986). It has also been reported secondary to subarachnoid hemorrhage and vascular spasm of the basilar artery (Landi et al., 1994), a brain stem tumor (Cherington et al., 1976; Hawkes and Bryan-Smyth, 1976; Pogacar et al., 1983; Inci and Ozgen, 2003; Breen and Hannon, 2004), central pontine myelinolysis (Messert et al., 1979; Oda et al., 1984; Morlan et al., 1990; Lilje et al., 2002), encephalitis (Pecket et al., 1982; Katz et al., 1992; Acharya et al., 2001), pontine abscess (Murphy et al., 1979), brain stem drug toxicity (Davis et al., 1972; Durrani and Winnie, 1991; Kleinschmidt-DeMasters and Yeh, 1992), vaccine reaction (Katz et al., 1992), and prolonged hypoglycemia (Negreiros dos Anjos, 1984; Mikhailidis et al., 1985).

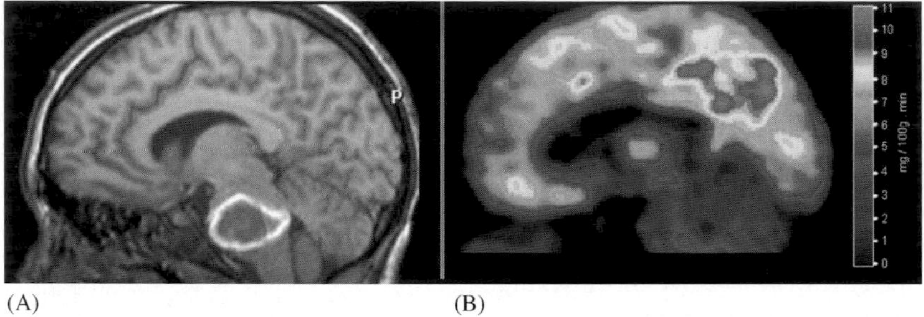

Fig. 2. (A) Magnetic resonance image (sagittal section) showing a massive hemorrhage in the brainstem (circular hyperintensity) causing a locked-in syndrome in a 13 year-old girl. (B) ^{18}F-fluorodeoxyglucose — Positron emission tomography illustrating intact cerebral metabolism in the acute phase of the LIS when eye-coded communication was difficult due to fluctuating vigilance. The color scale shows the amount of glucose metabolized per 100 g of brain tissue per minute. Statistical analysis revealed that metabolism in the supratentorial gray matter was not significantly lower as compared to healthy controls (Adapted from Laureys et al., 2004a). See Plate 34.2 in Color Plate Section.

Table 1. Etiology of the locked-in syndrome most frequently is vascular

Reference	Number of patients	(%) of males	Mean age at onset of LIS (range)	Vascular etiology (%)
Patterson and Grabois, (1986)	139	62	52 (20–77)	60
Katz et al. (1992)	29	66	34 (1–70)	52
Richard et al. (1995)	11	82	(17–73)	91
Casanova et al. (2003)	14	64	45 (16–71)	79
Leon–Carrion et al. (2002b)	44[a]	51	47 (22–77)	86
Pantke et al. (unpublished)	15	33	39	100
ALIS database	250	64	45 (13–84)	86

[a]These patients are part of the ALIS database.

A comparable awake conscious state simulating unresponsiveness may also occur in severe cases of peripheral polyneuropathy as a result of total paralysis of limb, bulbar, and ocular musculature. Transient LIS cases have been reported after Guillain Barré polyradiculoneuropathy (Loeb et al., 1984; Bakshi et al., 1997; Ragazzoni et al., 2000) and severe postinfectious polyneuropathy (Carroll and Mastaglia, 1979; O'Donnell, 1979). Unlike basilar artery stroke, vertical eye movements are not selectively spared in these extensive peripheral disconnection syndromes. Another important cause of complete LIS can be observed in end-stage amyotrophic lateral sclerosis, i.e., motor-neuron disease (Hayashi and Kato, 1989; Kennedy and Bakay, 1998; Kotchoubey et al., 2003). Finally, temporary pharmacologically induced LIS can sporadically be observed in general anesthesia when patients receive muscle relaxants together with inadequate amounts of anesthetic drugs (e.g., Sandin et al., 2000). Testimonies from victims relate that the worst aspect of the experience was the anxious desire to move or speak while being unable to do so (Anonymous, 1973; Brighouse and Norman, 1992; Peduto et al., 1994). Awake–paralyzed patients undergoing surgery may develop posttraumatic stress disorder (for recent review see Sigalovsky, 2003).

Misdiagnosis

Unless the physician is familiar with the signs and symptoms of the LIS, the diagnosis may be missed and the patient may erroneously be considered as being in a coma, vegetative state, or akinetic mustism (Gallo and Fontanarosa, 1989). In a recent survey of 44 LIS patients belonging to ALIS, the first person to realize the patient was conscious and could communicate via eye movements most often was a family member (55% of cases) and not the treating physician (23% of cases) (Table 2) (Leon-Carrion et al., 2002b). Most distressingly, the time elapsed between brain insult and LIS diagnosis was on average 2.5 months (78 days). Several patients were not diagnosed for more than 4 years. Leon-Carrion et al. (2002b) believed that this delay in the diagnosis of LIS mainly reflected initial misdiagnosis. Clinical experience indeed shows how difficult it is to recognize unambiguous signs of conscious perception of the environment and of the self in severely brain-injured patients (for review see Majerus et al., this volume). Voluntary eye movements and/or blinking can erroneously be interpreted as reflexive in anarthric and nearly completely paralyzed patients who classically show decerebration posturing (i.e., stereotyped extension reflexes). However, part of the delay could be explained by an initial lower level neurological state (e.g., decreased or fluctuating arousal levels) or even psychiatric symptoms that would mask residual cognitive functions at the outset of LIS.

Table 2. First person to realize the patient was conscious and could communicate via eye movements in 44 LIS patients. From an ALIS survey by Leon–Carrion et al. (2002)

Person making diagnosis	Number of patients (% of total)
Family member	24 (55)
Physician	10 (23)
Nurse	8 (18)
Other	2 (4)

Some memoirs written by LIS patients well illustrate the clinical challenge of recognizing a LIS. A striking example is *Look Up for Yes* written by Julia Tavalaro (1997). In 1966, 32-year old Tavalaro fell into a coma following a subarachnoid hemorrhage. She remained comatose for 7 months and gradually woke up to find herself in a New York State chronic care facility. There, she was known as "the vegetable" and it was not until 1973 (i.e., after 6 years) that her family identified a voluntary "attempt to smile" when Julia was told a dirty joke. This made speech therapist Arlene Kraat brake through Julia's isolation. With the speech therapist pointing to each letter on a letter board, Julia began to use her eyes to spell out her thoughts and relate the turmoil of her terrible years in captivity. She later used a communication device, started to write poetry and could cheek-control her wheelchair around the hospital. Julia Tavalaro died in 2003 at the age of 68 from aspiration pneumonia.

Another poignant testimony comes from Philippe Vigand, author of *Only the Eyes Say Yes*

(Vigand and Vigand, 2000, original publication in 1997) and formerly publishing executive with the French conglomerate, Hachette. The book is written in two parts: the first by Philippe, and the second by his wife Stéphane detailing *her* experiences. In 1990, Philippe Vigand, 32-years old, presented a vertebral artery dissection and remained in a coma for 2 months. Philippe and his wife write that at first, doctors believed he was a "vegetable and was treated as such." His wife eventually realized that he was blinking his eyes in response to her comments and questions to him but had difficulties convincing the treating physicians. It was speech therapist Philippe Van Eeckhout who formally made the diagnosis of LIS: when testing Vigand's gag reflex, Van Eeckhout was bit in his finger and yelled "chameau" (French for 'camel'), whereupon the patient started to grin. On the subsequent question "how much is 2 plus 2" Vigand blinked four times confirming his cognitive capacities. He later communicated his fist phrase by means of a letter board: "my feet hurt." After many months of hospital care, Vigand was brought home, where an infrared camera attached to a computer enabled him to "speak." The couple conceived a child after Philippe became paralyzed and he has written his second book (dealing with the menaced French ecosystem) on the beach of the Martinique isles (Vigand, 2002), illustrating that LIS patients can resume a significant role in family and society.

Survival and mortality

It has been stated that long-term survival in LIS is rare (Ohry, 1990). Mortality is indeed high in acute LIS (76% for vascular cases and 41% for nonvascular cases) with 87% of the deaths occurring in the first 4 months (Patterson and Grabois, 1986). In 1987, Haig et al. (1987) first reported on the life expectancy of persons with LIS, showing that individuals can actually survive for significant periods of time. Encompassing 29 patients from a major US rehabilitation hospital who had been in LIS for more than 1 year, they reported formal survival curves at the fifth year (Katz et al., 1992) and follow-up at the 10th year (Doble et al., 2003). These authors have shown that once a patient has medically stabilized in LIS for more than a year, 10-year survival is 83% and 20 year-survival is 40% (Doble et al., 2003).

Data from the ALIS database ($n = 250$) show that survivors are younger at onset than those who die (survivor mean 45 ± 14 years, deceased subjects 56 ± 13 years, $p < 0.05$), but there is no significant correlation between age at onset and survival time (Fig. 3). The mean time spent locked-in is 6 ± 4 years (range 14 days to 27 years, the latter patient still being alive). Reported causes of death of the 42 subjects are predominantly infections (40%, most frequently pneumonia), primary brain stem stroke (25%), recurrent brain stem stroke (10%), patient's refusal of artificial nutrition, and hydration (10%), and other causes (i.e., cardiac arrest, gastrostomy-surgery, heart failure, and hepatitis). It should be noted that the ALIS database does not contain the many LIS patients who die in the acute setting without being reported to the association. Recruitment to the ALIS database is based on case-reporting by family and health care workers prompted by the exceptional media publicity of ALIS in France and tracked by continuing yearly surveys. This recruitment bias should, however, be taken into account when interpreting the presented data.

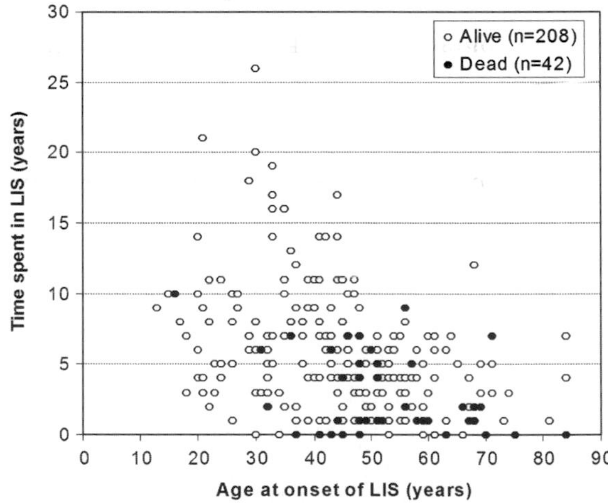

Fig. 3. Age at insult versus survival time of 250 locked-in patients registered in the ALIS (Association Locked in Syndrome) database, 42 of whom died (filled circles).

Prognosis and outcome

Classically, the motor recovery of LIS of vascular origin is very limited (Patterson and Grabois, 1986, Doble et al., 2003), even if rare cases of good recovery have been reported (McCusker et al., 1982; Ebinger et al., 1985). Chang and Morariu (1979) reported the first transient LIS caused by a traumatic damage of the brain stem. In their milestone paper, Patterson and Grabois (1986), reviewed 139 patients — 6 cases from the authors' rehabilitation center in Texas, USA, and 133 from 71 published studies from 1959 to 1983 — and reported earlier and more complete recovery in nonvascular LIS compared to vascular LIS. Return of horizontal pursuit eye movements within 4 weeks post-onset are thought to be predictive of good recovery (Chia, 1991). Richard et al. (1995) followed 11 LIS patients for 7 months to 10 years and observed that despite the persisting serious motor deficit, all patients did recover some distal control of fingers and toe movements, often allowing a functional use of a digital switch. The motor improvement occurred with a distal to proximal progression and included a striking axial hypotonia.

LIS is uncommon enough that many clinicians do not know how to approach rehabilitation and there are no existing guidelines as how to organize the revalidation process. Casanova et al. (2003) recently followed 14 LIS patients in 3 Italian rehabilitation centers for a period of 5 months to 6 years. They reported that intensive and early rehabilitative care improved functional outcome and reduced mortality rate when compared to the older studies by Patterson and Grabois (1986) and Katz et al. (1992). These results are in line with preliminary retrospective observations from the German Association for LIS lead by Pantke (2005).

Often unknown to physicians caring for LIS in the acute setting and despite the limited motor recovery of LIS patients, many patients can return living at home. The ALIS database shows that out of 245 patients, 108 (44%) are known to live at home (21% are staying in a hospital setting and 17% in a revalidation center). Patients return home after a mean period of 2 ± 16 years (range 2 months to 6 years, data obtained on $n = 55$). Results obtained in 95 patients show a moderate to significant recovery of head movement in 92% of patients, 65% showed small movement in one of the upper limbs (finger, hand, or arm), and 74% show a small movement in lower limbs (foot or leg). Half of the patients has recovered some speech production (limited to single comprehensible words) and 95% can vocalize unintelligible sounds (data obtained on $n = 50$). Some kind of electrical communication device is used by 81% of the LIS patients (data obtained on $n = 95$).

While all locked-in patients initially had a tracheotomy, 65% had it removed at the time of the last questionnaire. The mean interval to the removal of the tracheotomy was 14 ± 16 months (range 2 weeks to 5 years, data obtained on $n = 66$). All patients also initially had a gastrostomy and 58% had it removed at the time of the questionnaire; 66% of patients were able to have some kind of oral feeding (normal or mixed food, sometimes in addition to gastrostomy). The mean interval to removal of the gastrostomy was 18 ± 20 months (range 2 weeks to 6 years, data obtained on $n = 64$).

The level of care remains extensive in chronic LIS. Out of 50 questioned patients, 16 had nursing care once a day, 28 two times a day, and 6 three times a day. Physical therapy was performed at least five times a week in 66% of the patients, and speech therapy was performed at least three times a week in 55% of the patients. Nearly all patients (96%) complained of spasticity, 75% from difficulties swallowing oropharyngeal secretions, 66% from sialorrhea, and 61% had respiratory difficulties of various types.

Communication

In order to functionally communicate, it is necessary for the LIS patient to be motivated and to be able to receive (verbally or visually, i.e., written commands) and emit information. The first contact to be made with these patients is through a code using eyelid blinks or vertical eye movements. In cases of bilateral ptosis, the eyelids need to be manually opened in order to verify voluntary eye movements on command. To establish a yes/no eye code, the following instruction can suffice : "yes" is indicated by one blink and "no" by two or look up indicates "yes" and look down "no." In practice, the patient's best eye movement should be chosen and the same eye code should be used by all interlocutors. Such a code will only permit to communicate via closed questions (i.e., yes/no answers on presented questions). The principal aim of reeducation is to reestablish a genuine exchange with the LIS patient by putting into place various codes to permit them to reach a higher level of communication and thus to achieve an active participation. With sufficient practice, it is possible for LIS patients to communicate complex ideas in coded eye movements. Feldman (1971) first described a LIS patient who used jaw and eyelid movements to communicate in Morse Code.

Most frequently used are alphabetical communication systems. The simplest way is to list the alphabet and ask the LIS patient to make a prearranged eye movement to indicate a letter. Some patients prefer a listing of the letters sorted in function of appearance rate in usual language (i.e., in the French language: E—S—A—R—I—N—T—U—L—O—M—D—P—C—F—B—V—H—G—J—Q—Z—Y—X—K – W; or in the English language: E—T—A—O—I—N—S—R—H—L—D—C—U—M—F—P—G—W—Y—B—V—K—X—J—Q—Z). The interlocutor pronounces the letters beginning with the most frequently used, E, and continues until the patient blinks after hearing the desired letter, which the interlocutor then notes. It is necessary to begin over again for each letter to form words and phrases. The rapidity of this system depends upon practice and the ability of patient and interlocutor to work together. The interlocutor may be able to guess at a word or a phrase before all the letters have been pronounced. It is sufficient for him to pronounce the word or the rest of the sentence. The patient than confirms the word by making his eye code for "yes" or disproves by making his eye code for "no."

Another method is the "vowel and consonant method." Here, the alphabet is divided into 4 groups : Vowels, Consonants 1 (B-H), Consonants 2 (J-Q), and Consonants 3 (R-Z) (Table 3). The interlocutor says : "Vowel" and then Consonants 1, 2, 3 and the patient blinks his eyelid to indicate the chosen group (Table 3).

A similar system is the "alphabetical system using a grid of letters" (Table 4). Here, to designate, for example, the letter "B" (1-1), the patient blinks his eye once, pauses, and then blinks one time again. If he wishes to designate a vowel, he raises his eyes before blinking. After using this system for a certain length of time, both the patient and the person communicating with the patient know it by heart. The patient indicates the position of the chosen letter with his eyes ; the interlocutor guesses the letter. The resulting dialogue can become remarkably rapid. There are many other variants

Table 3. Vowel and consonant method

V	C1	C2	C3
A	B	J	R
E	C	K	S
I	D	L	T
O	F	M	V
U	G	N	W
Y	H	P	X
		Q	Z

Notes: V, Vowel, C1,C2,C3 consonants. See text for details of use. From van Eeckhout (1997).

Table 4. Alphabetical system using a grid of letters

	Consonants	Consonants	Consonants	Consonants	Consonants	Vowels	Vowels
	1	2	3	4	5	1	2
1	B	G	L	Q	V	A	O
2	C	H	M	R	W	E	U
3	D	J	N	S	X	I	Y
4	F	K	P	T	Z		

Notes: See text for details of use. From van Eeckhout (1997).

to these systems which should be tailored to the patient's preferences and physical capabilities.

The above discussed systems all require assistance from others. It is important to stress that access to informatics is drastically changing the lives of patients with LIS. Instead of passively responding to the requests of others, computers allow the patient to initiate conversations and prepare detailed messages for caregivers who do not have time for lengthy guessing rounds. Experts in rehabilitation engineering and speech-language pathology device various patient-computer interfaces such as infra-red eye movement sensors (e.g., Quick Glance www.eyetechds.com or Eyegaze Communication System www.eyegaze.com/indexdis.htm) which can be coupled to on-screen virtual keyboards (e.g., WiViK www.wivik.com) allowing the LIS patient to control his environment (lights, appliances, etc.), use a word processor (which can be coupled to a text-to-speech synthesizer), operate a telephone or fax, or access the Internet and use e-mail (see Fig. 4). Unfortunately, the cost of these computer interfaces are often substantial and not routinely paid for by third parties.

Fig. 4. A locked-in person updates the database of ALIS, moving the cursor on screen by eye movements. An infrared camera (white arrow) mounted below the monitor observes one of the user's eyes, an image processing software continually analyzes the video image of the eye and determines where the user is looking on the screen. The user looks at a virtual keyboard that is displayed on the monitor and uses his eye as a computer-mouse. To "click" he looks at the key for a specified period of time (typically a fraction of a second) or blinks. An array of menu keys allow the user to control his environment, use a speech synthesizer, browse the worldwide web or send electronic mail independently (picture used with kind permission from DT). With a similar device Philippe Vigand, locked-in since 1990, has written a testimony of his LIS experience in an astonishing book *Putain de silence* (1997) translated as *Only the Eyes Say Yes* (2000). Photograph by S. Laureys.

Residual brain function

Neuropsychological testing

Surprisingly, there are no systematic neuropsychological studies of the cognitive functions in patients living with a LIS. Most case reports, however, failed to show any significant cognitive impairment when LIS patients were tested 1 year or more after the brainstem insult. Allain et al. (1998) performed extensive neuropsychological testing in two LIS patients studied 2 and 3 years after their basilar artery thrombosis. Patients communicated via a communication print–writer system and showed no impairment of language, memory and intellectual functioning. Cappa et al. (1985, 1982) studied one patient who was LIS for over 12 years and observed intact performances on language, calculation, spatial orientation, right–left discrimination and personality testing. Recently, New and Thomas (2005) assessed cognitive functioning in LIS patient 6 months after basilar artery occlusion, and noted significant reduction in speed of processing, moderate impairment of perceptual organization and executive skills, mild difficulties with attention, concentration, and new learning of verbal information. Interestingly, they subsequently observed progressive improvement in most areas of cognitive functioning until over 2 years after his brainstem stroke.

In a survey conducted by ALIS and Léon-Carrion et al. (2002b) in 44 chronic LIS patients, 86% reported a good attentional level, all but two patients could watch and follow a film on TV, and all but one were well-oriented in time (mean duration of LIS was 5 years). More recently, ALIS and Schnakers et al. (2005) adapted a standard battery of neuropsychological testing (i.e., sustained and selective attention, working and episodic memory, executive functioning, phonological and

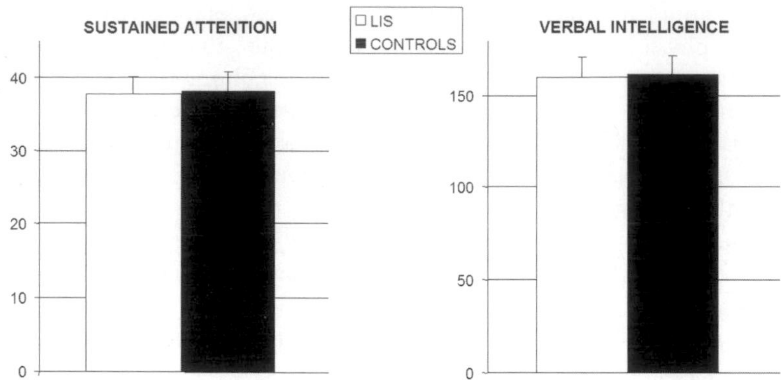

Fig. 5. Sustained attention, assessed by means of a newly developed auditory attention task, and verbal intelligence, assessed using the French adaptation of the Peabody Picture Vocabulary Test (Dunn and Thériault-Whalen, 1993), in five chronic LIS patients (three males; aged 24–57 years; LIS duration 3–6 years) and 10 healthy controls, matched for age, gender, and educational level. Data taken from Schnakers et al. (2005).

lexico-semantic processing, and vocabulary knowledge) to an eye-response mode for specific use in LIS patients. Overall, performances in the five LIS patients studied 3–6 years after their brainstem insult were not significantly different from 10 matched, healthy controls who, like the LIS patients, had to respond solely via eye-movements (Fig. 5). These data re-emphasize the fact that LIS due to purely pontine lesions is characterized by the restoration of a globally intact cognitive potential.

Electrophysiologic measurements

Markand (1976) reviewed electroencephalographic (EEG) recordings in eight patients with LIS and reported it was normal or minimally slow in seven and showed reactivity to external stimuli in all patients. These results were confirmed by Bassetti et al. (1994) who observed a predominance of reactive alpha activity in six LIS patients. In their seminal paper, Patterson and Grabois (1986) reported normal EEG findings in 39 (45%) and abnormal (mostly slowing over the temporal or frontal leads or more diffuse slowing) in 48 (55%) patients out of 87 reviewed patients. Jacome and Morilla-Pastor (1990), however, reported three patients with acute brainstem strokes and LIS whose repeated EEG recordings exhibited an "alpha coma" pattern (i.e. alpha rhythm unreactive to multimodal stimuli). Unreactive EEG in LIS was also reported by Gutling et al. (1996) confirming that lack of alpha reactivity is not a reliable indicator of unconsciousness and cannot be used to distinguish the "locked-in" patients from those comatose due to a brainstem lesion. Nevertheless, the presence of a relatively normal reactive EEG rhythm in a patient that appears to be unconscious should alert one to the possibility of a LIS.

Somatosensory-evoked potentials are known to be unreliable predictors of prognosis (Bassetti et al., 1994, Towle et al., 1989) but motor-evoked potentials have been proposed to evaluate the potential motor recovery (e.g., Bassetti et al., 1994).

Cognitive event-related potentials (ERPs) in patients with LIS may have a role in differential diagnosis of brainstem lesions (Onofrj et al., 1997) and have also shown their utility to document consciousness in total LIS due to end-stage amyotrophic lateral sclerosis (Kotchoubey et al., 2003) and fulminant Guillain–Barré syndrome (Ragazzoni et al., 2000). Fig. 6 shows event-related potentials in a 57-year-old locked-in patient following basilar artery thrombosis showing a positive "P3" component (peaking at 700 ms) only evoked by the patient's own name (thick line) and not by other names (thin line). It should, however, be noted that such responses can also be evoked in minimally conscious patients (Laureys et al., 2004b) and that they even persist in sleep in normal subjects (Perrin et al., 1999).

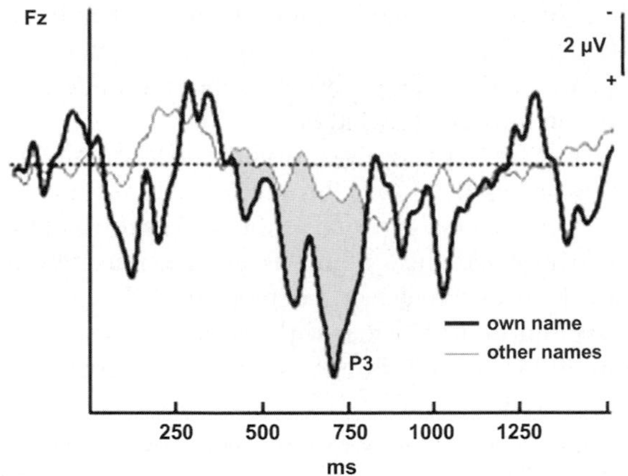

Fig. 6. Event-related potentials in chronic LIS. Patient JB (Adapted from Perrin et al., 2005).

Functional neuroimaging

Classically, structural brain imaging (MRI) may show isolated lesions (bilateral infarction, hemorrhage, or tumor) of the ventral portion of the basis pontis or midbrain (e.g., Leon-Carrion et al., 2002a). PET scanning has shown significantly higher metabolic levels in the brains of patients in a LIS compared to patients in the vegetative state (Levy et al., 1987). Preliminary results from ongoing PET studies by Laureys et al. (2003, 2004a) indicate that no supra-tentorial cortical area show significantly lower metabolism in acute and chronic LIS patients when compared to age-matched healthy controls (Fig. 2B). Conversely, a significantly hyperactivity was observed in bilateral amygdala of acute, but not in chronic, LIS patients. The absence of metabolic signs of reduced function in any area of the gray matter re-emphasizes the fact that LIS patients suffer from a pure motor de-efferentation and recover an entirely intact intellectual capacity. Previous PET studies in normal volunteers have demonstrated amygdala activation in relation to negative emotions such as fear and anxiety (e.g., Calder et al., 2001). It is difficult to make judgments about patient's thoughts and feelings when they awake from their coma in a motionless shell. However, in the absence of decreased neural activity in any cortical region, we assume that the increased activity in the amygdala in acute non-communicative LIS patients, relates to the terrifying situation of an intact awareness in a sensitive being, experiencing frustration, stress and anguish, locked in an immobile body. These preliminary findings emphasize the need for a quick diagnosis and recognition of the terrifying situation of a pseudocoma (i.e., LIS) at the intensive care or coma unit. Health care workers should adapt their bedside-behavior and consider pharmacological anxiolytic therapy of locked-in patients, taking into account the intense emotional state they go through.

Daily activities

For those not dealing with these patients on a daily basis it is surprising to see how chronic LIS patients, with the help of family and friends, still have essential social interaction and lead meaningful lives. Doble et al. (2003) reported that most of their chronic LIS patients continued to remain active through eye and facial movements. Listed activities included TV, radio, music, books on tape, visiting with family, visit vacation home, e-mail, telephone, teaching, movies, shows, the beach, bars, school, and vocational training. They also reported an attorney who uses Morse code eye blinks to provide legal opinions and keeps up with colleagues through fax and e-mail. Another patient taught math and spelling to third graders using a mouth stick to trigger an electronic voice device. The authors reported being impressed with the social interactions of chronic LIS patients and stated it was apparent that the patients were actively involved in family and personal decisions and that their presence was valued at home. Only four out of the 13 patients used computers consistently, two accessed the internet and one was able to complete the telephone interview by himself using a computer and voice synthesizer. A survey by ALIS and Ghorbel et al. (2002) showed that out of 17 questioned chronic LIS patients living at home, 11 (65%) used a personal computer.

Quality of life

In March 2002, at the annual meeting of ALIS at La Pitié-Salpétrière hospital in Paris, patients with LIS and their family members were asked to fill in the Short Form-36 (SF-36) questionnaire (Ware et al., 1993) on quality of life. Seventeen chronic (i.e., >1 year) locked-in patients who did not show major motor recovery (i.e., used eye movements or blinking as the major mode of communication) and who lived at home replied to the questionnaire (mean age 44 ± 6 years; range 33–57 years). Mean time of LIS duration was 6 ± 4 years (range 2–16 years). On the basis of the SF-36 questionnaire locked-in patients unsurprisingly showed maximal limitations in physical activities (all patients scoring zero). Interestingly, self-scored perception of mental health (evaluating mental well-being and psychological distress) and personal general health were not significantly lower than values from age-matched French control subjects (Fig. 7). Perception of mental health and the presence of physical pain was correlated to the frequency of suicidal thoughts ($r = -0.67$ and 0.56 respectively, $p < 0.05$). This stresses the importance (and current frequent inadequacy) of proper pain management in chronic LIS patients.

These findings confirm earlier reports on quality of life assessments in chronic locked-in patients. Previous surveys from ALIS ($n = 44$) showed that 48% regarded their mood as good versus 5% as bad; 13% declared being depressed; 73% enjoyed going out and 81% met with friends at least twice a month (Leon-Carrion et al., 2002b). In the study from Doble et al. (2003) 7 out of 13 patients were satisfied with life in general and 5 were occasionally depressed. As stated by Doble and co-workers, the results of studies on quality of life in chronic LIS may run contrary to many health care professionals. Superficially involved for the short term when the patient is at his or her worst, clinicians may often tend to comfortably assume that these persons will die anyway, or would choose to die if they only knew what the clinicians knew. As a result, debates about cost, daily management, quality of life, withdrawal or withholding of care,

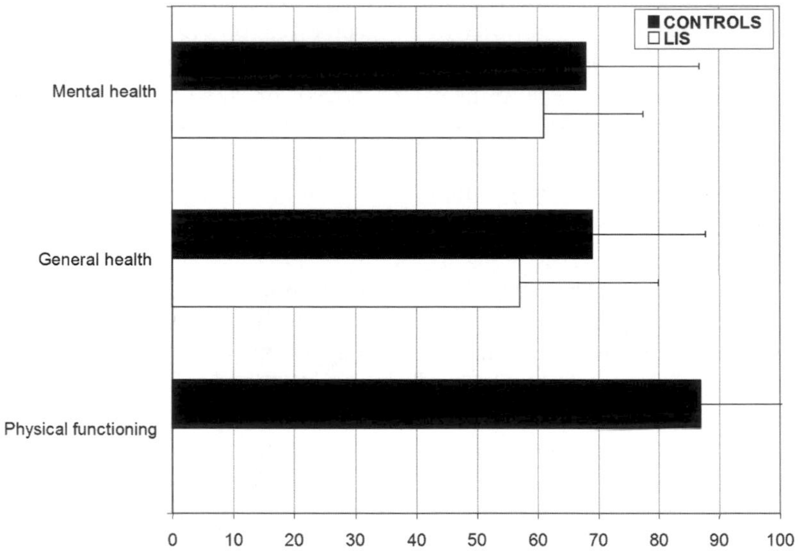

Fig. 7. Self-scored mental health, general health and physical functioning as assessed by a French version of the Short Form-36 (Ware et al., 1993) in 15 chronic locked-in patients living at home (white bars) compared to healthy age-matched French controls (black bars). Values are means and standard deviations. For all measures the reporting frame is the past month, scores range from 0 to 100, with higher scores representing better health-related quality of life. Note that physical functioning was scored as zero in all locked-in patients, but the perception of mental health (evaluating mental well-being and psychological distress) and personal general health were not significantly different in patients compared to healthy controls (historical control data taken from Leplege et al., 1998). Adapted from a survey conducted by ALIS and Ghorbel et al. (2002).

end-of-life decisions and euthanasia often go on with prejudice and without any input from the conscious but mute and immobile patient. To "judge a book by its cover" is unfair. Clinicians should realize that quality of life often equates with social rather than physical interaction and that the will to live is strong when struck by an acute devastating disease.

It is distressing to note that many people with disabilities feel their physicians will be too quick to help them with euthanasia (Batavia, 1997). Medical treatment for persons with LIS should be as aggressive as it would be for other people with potential survival of a decade or more. Contrary to the perceptions of some health care professionals who have not experienced such a severe motor disability, LIS patients typically have a wish to live. As discussed above, many return home and start a new, different, but meaningful life. In the future, more widely available access to enhanced communication computer prosthetics should additionally enhance the quality of life of locked-in patients.

The right to die or the right to live ?

The American Academy of Neurology (AAN) has published a position statement concerning the management of conscious and legally competent patients with profound and permanent paralysis (Ethics and Humanities Subcommittee of the AAN, 1993, Bernat et al., 1993, Allen, 1993). The conclusion is that such patients have the right to make health care decisions about themselves, including whether to accept or refuse life-sustaining therapy — either not start or stop once it started. Doctors caring for LIS have "an ethical obligation to minimize subsequent suffering" and should help patients with pain and dyspnea, "even if these medications contribute… to respiratory depression, coma, or death." However, patients should first be fully informed about their condition and the treatment options and patients' decision must be consistent over a period of time. The latter is clearly necessary to exclude the impulsive transient reactions of despair that are common in patients with severe illness.

Since its creation in March 1997, ALIS has registered 367 patients with LIS in France. Four reported deaths were related to the patient's wish to die (unpublished data from ALIS' database). Doble et al. (2003) accounted that none of the 15 deaths of their study cohort of 29 chronic LIS patients from the Chicago area followed for over a decade could be attributed to euthanasia. None of their 13 chronic LIS patients had a "do not resuscitate" order, 7 had never considered or discussed euthanasia, 6 had considered it in the past but not at the time of survey and 1 wished to die.

In the survey conducted by ALIS and Ghorbel (2002) most of chronic LIS patients without motor recovery (i.e., worst-case scenario) rarely or never had suicidal thoughts (Fig. 8; mean age 45 years, mean duration of LIS 6 years). In reply to the question "would you like to receive antibiotics in case of pneumonia ?" 80% answered "yes" and to the question "would you like reanimation to be tempted in case of cardiac arrest ?" 62% answered positively. Anderson et al. (1993) reported suicidal thoughts in four out of seven LIS patients with long-term survival but all patients nevertheless wanted life-sustaining treatment (mean age 43 years, duration of LIS ranged from 8 to 37 months). Similarly, in the case of high spinal cord injury resulting in acute onset quadriplegia, Hall et al. (1999) reported that 81 out of 85 survivors surveyed (95%) were "glad to be alive," including

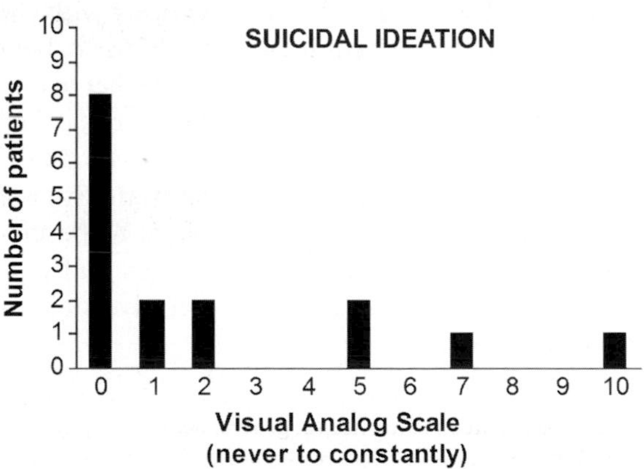

Fig. 8. Frequency of suicide thoughts in patients with chronic locked-in syndome. Note that 75% of the patients rarely or never had suicidal thoughts. From a survey conducted by ALIS and Ghorbel et al. (2002).

all ventilator-assisted patients studied 14 to 24 years post-trauma.

While the right of LIS individuals to withdraw from treatment is not questioned (Humbert, 2003, Guerra, 1999), the discussed data call into question the assumption among some health care workers and policy makers that severe disability is intolerable. The unfortunate consequence of this prejudice is that biased clinicians might provide less aggressive medical treatment and influence family in ways not appropriate to the situation (Doble et al., 2003). Likewise, in amyotrophic lateral sclerosis, ill-informed patients are regularly advised by physicians to refuse intubation and withhold lifesaving interventions (Christakis and Asch, 1993, Trail et al., 2003). However, ventilator users with neuromuscular disease report meaningful life satisfaction (Kubler et al., 2005). Bach (2003) warns that "virtually no patients are appropriately counseled about all therapeutic options" and states that advance directives, although appropriate for patients with terminal cancer, are inappropriate for patients with severe motor disability.

Katz et al. (1992) cite the Hastings Center Report, "Who speaks for the patient with LIS?." With the initial handicap of communicating only through eyeblink who can decide whether the patient is competent to consent or to refuse treatment (Steffen and Franklin, 1985)? With regard to end-of-life decisions taken in LIS patients, an illustrative case is reported by Fred (1986). His 80-year-old mother became locked-in. In concert with the attending physician, without consent of the patient herself, the decision was made to "have her senses dulled" and provide supportive care only. She died shortly thereafter with a temperature of 109°F (43°C). In the accompanying editorial, Stumpf (1986) commented that "human life is to be preserved as long as there is consciousness and cognitive function in contrast to a vegetative state or neocortical death."

Conclusion

The discussed data stress the need for heath care workers who might be confronted to the LIS to recognize this infrequent syndrome as early as possible and to adapt their bedside-behavior. Physicians who take care of acute LIS patients need a better understanding of their long-term outcome. With appropriate medical care, most patients can return home and their life expectancy is several decades. Opposite to the beliefs of many healthy individuals, LIS patients self-report a meaningful quality of life and the demand of euthanasia is uncommon (ALIS database and Doble et al., 2003). Even if good recovery of motor and speech function is very rare in LIS, recent studies show that intensive and early rehabilitation can improve functional motor outcome and verbal communication (Casanova et al., 2003; Pantke, 2005). Improvements in augmentative communication devices such as infrared eye-gaze sensors and switching devices, sometimes using minuscule electromyographic or even electroencephalographic signals (see Kubler et al, this volume), coupled to sophisticated computer translation software now give LIS patients a synthesized "voice" and enable them to control their surroundings in ways never possible before. New technology offer the LIS a virtual window on the world via internet and has permitted locked-in patients to resume an active role in society. Caring for LIS patients is far from futile.

Acknowledgments

This research was supported by the French Association Locked in Syndrome (ALIS), the Belgian Fonds National de la Recherche Scientifique (FNRS), the Centre Hospitalier Universitaire Sart Tilman, Liège, the University of Liège, the Erasme Hospital, University of Brussels, Belgium, and the Mind Science Foundation, San Antonio, Texas, USA. Steven Laureys is Research Associate at FNRS. The authors thank all participating LIS patients, their families, and their physicians and acknowledge Véronique Blandin and Dominique Toussaint for managing the ALIS database.

References

Acharya, V.Z., Talwar, D. and Elliott, S.P. (2001) Enteroviral encephalitis leading to a locked-in state. J. Child Neurol., 16: 864–866.

Allain, P., Joseph, P.A., Isambert, J.L., Le Gall, D. and Emile, J. (1998) Cognitive functions in chronic locked-in syndrome: a report of two cases. Cortex, 34: 629–634.

Allen, C.M. (1993) Conscious but paralysed: releasing the locked-in. Lancet, 342: 130–131.

American Congress of Rehabilitation Medicine (1995) Recommendations for use of uniform nomenclature pertinent to patients with severe alterations of consciousness. Arch. Phys. Med. Rehabil., 76: 205–209.

Anderson, C., Dillon, C. and Burns, R. (1993) Life-sustaining treatment and locked-in syndrome. Lancet, 342: 867–868.

Anonymous. (1973) Awareness during anaesthesia. Lancet, 2: 1305.

Bach, J.R. (2003) Threats to "informed" advance directives for the severely physically challenged? Arch. Phys. Med. Rehabil., 84: S23–S28.

Bakshi, N., Maselli, R.A., Gospe Jr., S.M., Ellis, W.G., McDonald, C. and Mandler, R.N. (1997) Fulminant demyelinating neuropathy mimicking cerebral death. Muscle Nerve, 20: 1595–1597.

Bassetti, C., Mathis, J. and Hess, C.W. (1994) Multimodal electrophysiological studies including motor evoked potentials in patients with locked-in syndrome: report of six patients. J. Neurol. Neurosurg. Psychiatry, 57: 1403–1406.

Batavia, A.I. (1997) Disability and physician-assisted suicide. N. Engl. J. Med., 336: 1671–1673.

Bauby, J.-D. (1997). The Diving Bell and the Butterfly (original title: Le scaphandre et le papillon, Robert Laffont, Paris) KNOPF, New York.

Bauer, G., Gerstenbrand, F. and Rumpl, E. (1979) Varieties of the locked-in syndrome. J. Neurol., 221: 77–91.

Bernat, J.L., Cranford, R.E., Kittredge Jr., F.I. and Rosenberg, R.N. (1993) Competent patients with advanced states of permanent paralysis have the right to forgo life-sustaining therapy. Neurology, 43: 224–225.

Breen, P. and Hannon, V. (2004) Locked-in syndrome: a catastrophic complication after surgery. Br. J. Anaesth., 92: 286–288.

Brighouse, D. and Norman, J. (1992) To wake in fright. BMJ., 304: 1327–1328.

Britt, R.H., Herrick, M.K. and Hamilton, R.D. (1977) Traumatic locked-in syndrome. Ann. Neurol., 1: 590–592.

Calder, A.J., Lawrence, A.D. and Young, A.W. (2001) Neuropsychology of fear and loathing. Nat. Rev. Neurosci., 2: 352–363.

Cappa, S.F., Pirovano, C. and Vignolo, L.A. (1985) Chronic 'locked-in' syndrome: psychological study of a case. Eur. Neurol., 24: 107–111.

Cappa, S.F. and Vignolo, L.A. (1982) Locked-in syndrome for 12 years with preserved intelligence. Ann. Neurol., 11: 545.

Carroll, W.M. and Mastaglia, F.L. (1979) 'Locked-in co ma' in postinfective polyneuropathy. Arch. Neurol., 36: 46–47.

Casanova, E., Lazzari, R.E., Lotta, S. and Mazzucchi, A. (2003) Locked-in syndrome: improvement in the prognosis after an early intensive multidisciplinary rehabilitation. Arch. Phys. Med. Rehabil., 84: 862–867.

Chang, B. and Morariu, M.A. (1979) Transient traumatic "locked-in" syndrome. Eur. Neurol., 18: 391–394.

Cherington, M., Stears, J. and Hodges, J. (1976) Locked-in syndrome caused by a tumor. Neurology, 26: 180–182.

Chia, L.G. (1991) Locked-in syndrome with bilateral ventral midbrain infarcts. Neurology, 41: 445–446.

Christakis, N.A. and Asch, D.A. (1993) Biases in how physicians choose to withdraw life support. Lancet, 342: 642–646.

Darolles, M. (1875) *Prog. Med.*, 3: 629.

Davis, L.E., Wesley, R.B., Juan, D. and Carpenter, C.C. (1972) "Locked-in syndrome" from diazepam toxicity in a patient with tetanus. Lancet, 1: 101.

Doble, J.E., Haig, A.J., Anderson, C. and Katz, R. (2003) Impairment, activity, participation, life satisfaction, and survival in persons with locked-in syndrome for over a decade: follow-up on a previously reported cohort. J. Head Trauma Rehabil., 18: 435–444.

Dumas, A. (1997) The Count of Monte Cristo. Wordworth Editions Limited, London.

Dunn, L. M. and Thériault-Whalen, C. M. (1993) Échelle de vocabulaire en images Peabody. American Guidance Service, Minnesota.

Durrani, Z. and Winnie, A.P. (1991) Brainstem toxicity with reversible locked-in syndrome after intrascalene brachial plexus block. Anesth. Analg., 72: 249–252.

Ebinger, G., Huyghens, L., Corne, L. and Aelbrecht, W. (1985) Reversible "locked-in" syndromes. Intensive Care Med., 11: 218–219.

Ethics and Humanities Subcommittee of the AAN (1993) Position statement: certain aspects of the care and management of profoundly and irreversibly paralyzed patients with retained consciousness and cognition. Report of the ethics and humanities subcommittee of the American academy of neurology. Neurology, 43: 222–223.

Feldman, M.H. (1971) Physiological observations in a chronic case of "locked-in" syndrome. Neurology, 21: 459–478.

Fitzgerald, L.F., Simpson, R.K. and Trask, T. (1997) Locked-in syndrome resulting from cervical spine gunshot wound. J. Trauma, 42: 147–149.

Fred, H.L. (1986) Helen. South Med. J., 79: 1135–1136.

Gallo, U.E. and Fontanarosa, P.B. (1989) Locked-in syndrome: report of a case. Am. J. Emerg. Med., 7: 581–583.

Ghorbel, S. (2002) Statut fonctionnel et qualité de vie chez le locked-in syndrome a domicile. In: DEA Motricité Humaine et Handicap, Laboratory of Biostatistics, Epidemiology and Clinical Research. Université Jean Monnet Saint-Etienne, Montpellier, France.

Golubovic, V., Muhvic, D. and Golubovic, S. (2004) Posttraumatic locked-in syndrome with an unusual three day delay in the appearance. Coll. Antropol., 28: 923–926.

Guerra, M.J. (1999) Euthanasia in Spain: the public debate after Ramon Sampedro's case. Bioethics, 13: 426–432.

Gutling, E., Isenmann, S. and Wichmann, W. (1996) Electrophysiology in the locked-in-syndrome. Neurology, 46: 1092–1101.

Haig, A.J., Katz, R.T. and Sahgal, V. (1986) Locked-in syndrome: a review. Curr Concepts Rehabil. Med., 3: 12–16.

Haig, A.J., Katz, R.T. and Sahgal, V. (1987) Mortality and complications of the locked-in syndrome. Arch. Phys. Med. Rehabil., 68: 24–27.

Hall, K.M., Knudsen, S.T., Wright, J., Charlifue, S.W., Graves, D.E. and Werner, P. (1999) Follow-up study of individuals with high tetraplegia (C1-C4) 14 to 24 years postinjury. Arch. Phys. Med. Rehabil., 80: 1507–1513.

Hawkes, C.H. and Bryan-Smyth, L. (1976) Locked-in syndrome caused by a tumor. Neurology, 26: 1185–1186.

Hayashi, H. and Kato, S. (1989) Total manifestations of amyotrophic lateral sclerosis. ALS in the totally locked-in state. J. Neurol. Sci., 93: 19–35.

Humbert, V. (2003) Je vous demande le droit de mourir. Michel Lafon, Paris.

Inci, S. and Ozgen, T. (2003) Locked-in syndrome due to metastatic pontomedullary tumor–case report. Neurol. Med. Chir. (Tokyo), 43: 497–500.

Jacome, D.E. and Morilla-Pastor, D. (1990) Unreactive EEG: pattern in locked-in syndrome. Clin. Electroencephalogr., 21: 31–36.

Katz, R.T., Haig, A.J., Clark, B.B. and DiPaola, R.J. (1992) Long-term survival, prognosis, and life-care planning for 29 patients with chronic locked-in syndrome. Arch. Phys. Med. Rehabil., 73: 403–408.

Keane, J.R. (1986) Locked-in syndrome after head and neck trauma. Neurology, 36: 80–82.

Kennedy, P.R. and Bakay, R.A. (1998) Restoration of neural output from a paralyzed patient by a direct brain connection. Neuroreport, 9: 1707–1711.

Kleinschmidt-DeMasters, B.K. and Yeh, M. (1992) "Locked-in syndrome" after intrathecal cytosine arabinoside therapy for malignant immunoblastic lymphoma. Cancer, 70: 2504–2507.

Kotchoubey, B., Lang, S., Winter, S. and Birbaumer, N. (2003) Cognitive processing in completely paralyzed patients with amyotrophic lateral sclerosis. Eur. J. Neurol., 10: 551–558.

Kubler, A., Winter, S., Ludoplh, A. C., Hautzinger, M. and Birbaumer, N. (2005) Depression and quality of life in ALS patients (submitted).

Landi, A., Fornezza, U., De Luca, G., Marchi, M. and Colombo, F. (1994) Brain stem and motor-evoked responses in "locked-in" syndrome. J. Neurosurg. Sci., 38: 123–127.

Landrieu, P., Fromentin, C., Tardieu, M., Menget, A. and Laget, P. (1984) Locked in syndrome with a favourable outcome. Eur. J. Pediatr., 142: 144–145.

Laureys, S., Owen, A.M. and Schiff, N.D. (2004a) Brain function in coma, vegetative state, and related disorders. Lancet Neurol., 3: 537–546.

Laureys, S., Perrin, F., Faymonville, M.E., Schnakers, C., Boly, M., Bartsch, V., Majerus, S., Moonen, G. and Maquet, P. (2004b) Cerebral processing in the minimally conscious state. Neurology, 14: 916–918.

Laureys, S., van Eeckhout, P., Ferring, M., Faymonville, M., Mavroudakis, N., Berre, J., Van Bogaert, P., Pellas, F., Cornu, P., Luxen, A., Vincent, J. L., Moonen, G., Maquet, P. and Goldman, S. (2003) Brain function in acute and chronic locked-in syndrome. Presented at the 9th Annual Meeting of the Organisation for Human Brain Mapping (OHBM), June 18-22, 2003, NY, USA, NeuroImage CD ROM Volume 19, Issue 2, Supplement 1.

Leon-Carrion, J., van Eeckhout, P. and Dominguez-Morales Mdel, R. (2002a) The locked-in syndrome: a syndrome looking for a therapy. Brain Inj., 16: 555–569.

Leon-Carrion, J., van Eeckhout, P., Dominguez-Morales Mdel, R. and Perez-Santamaria, F.J. (2002b) The locked-in syndrome: a syndrome looking for a therapy. Brain Inj., 16: 571–582.

Leplege, A., Ecosse, E., Verdier, A. and Perneger, T.V. (1998) The French SF-36 Health Survey: translation, cultural adaptation and preliminary psychometric evaluation. J. Clin. Epidemiol., 51: 1013–1023.

Levy, D.E., Sidtis, J.J., Rottenberg, D.A., Jarden, J.O., Strother, S.C., Dhawan, V., Ginos, J.Z., Tramo, M.J., Evans, A.C. and Plum, F. (1987) Differences in cerebral blood flow and glucose utilization in vegetative versus locked-in patients. Ann. Neurol., 22: 673–682.

Lilje, C.G., Heinen, F., Laubenberger, J., Krug, I. and Brandis, M. (2002) Benign course of central pontine myelinolysis in a patient with anorexia nervosa. Pediatr. Neurol., 27: 132–135.

Loeb, C., Mancardi, G.L. and Tabaton, M. (1984) Locked-in syndrome in acute inflammatory polyradiculoneuropathy. Eur. Neurol., 23: 137–140.

Markand, O.N. (1976) Electroencephalogram in "locked-in" syndrome. Electroencephalogr. Clin. Neurophysiol., 40: 529–534.

McCusker, E.A., Rudick, R.A., Honch, G.W. and Griggs, R.C. (1982) Recovery from the "locked-in" syndrome. Arch. Neurol., 39: 145–147.

Meienberg, O., Mumenthaler, M. and Karbowski, K. (1979) Quadriparesis and nuclear oculomotor palsy with total bilateral ptosis mimicking coma: a mesencephalic "locked-in syndrome"? Arch. Neurol., 36: 708–710.

Messert, B., Orrison, W.W., Hawkins, M.J. and Quaglieri, C.E. (1979) Central pontine myelinolysis. Considerations on etiology, diagnosis, and treatment. Neurology, 29: 147–160.

Mikhailidis, D.P., Hutton, R.A. and Dandona, P. (1985) "Locked in" syndrome following prolonged hypoglycemia. Diabetes Care, 8: 414.

Morlan, L., Rodriguez, E., Gonzalez, J., Jimene-Ortiz, C., Escartin, P. and Liano, H. (1990) Central pontine myelinolysis following correction of hyponatremia: MRI diagnosis. Eur. Neurol., 30: 149–152.

Murphy, M.J., Brenton, D.W., Aschenbrener, C.A. and Van Gilder, J.C. (1979) Locked-in syndrome caused by a solitary pontine abscess. J. Neurol. Neurosurg. Psychiatry., 42: 1062–1065.

Negreiros dos Anjos, M. (1984) "Locked in" syndrome following prolonged hypoglycemia. Diabetes Care, 7: 613.

New, P.W. and Thomas, S.J. (2005) Cognitive impairments in the locked-in syndrome: A case report. Arch. Phys. Med. Rehabil., 86: 338–343.

Oda, Y., Okada, Y., Nakanishi, I., Kajikawa, K., Kita, T., Hirose, G. and Kurachi, M. (1984) Central pontine myelinolysis with extrapontine lesions. Acta Pathol. Jpn., 34: 403–410.

O'Donnell, P.P. (1979) 'Locked-in syndrome' in postinfective polyneuropathy. Arch. Neurol., 36: 860.

Ohry, A. (1990) The locked-in syndrome and related states. Paraplegia., 28: 73–75.

Onofrj, M., Thomas, A., Paci, C., Scesi, M. and Tombari, R. (1997) Event-related potentials recorded in patients with locked-in syndrome. J. Neurol. Neurosurg. Psychiatry., 63: 759–764.

Pantke, K.-H. (2005) Bestimmung des Barthelindex aus Beginn und Länge der Rehabilitation nach einer Basilarithrombose, (submitted).

Patterson, J.R. and Grabois, M. (1986) Locked-in syndrome: a review of 139 cases. Stroke, 17: 758–764.

Pecket, P., Landau, Z. and Resnitzky, P. (1982) Reversible locked-in state in postinfective measles encephalitis. Arch. Neurol., 39: 672.

Peduto, V.A., Silvetti, L. and Piga, M. (1994) An anesthetized anesthesiologist tells his experience of waking up accidentally during the operation. Minerva Anestesiol., 60: 1–5.

Perrin, F., Garcia-Larrea, L., Mauguiere, F. and Bastuji, H. (1999) A differential brain response to the subject's own name persists during sleep. Clin. Neurophysiol., 110: 2153–2164.

Perrin, F., Schnakers, C., Faymonville, M. E., Degueldre, C., Bredart, S., Collette, F., Lamy, M., Goldman, S., Moonen, G., Luxen, A., Maquet, P. and Laureys, S. (2005) Evaluation of preserved linguistic processing in brain damaged patients, (submitted).

Plum, F. and Posner, J.B. (1983) The Diagnosis of Stupor and Coma. Davis, F.A., Philadelphia, PA.

Pogacar, S., Finelli, P.F. and Lee, H.Y. (1983) Locked-in syndrome caused by a metastasis. R I Med. J., 66: 147–150.

Rae-Grant, A.D., Lin, F., Yaeger, B.A., Barbour, P., Levitt, L.P., Castaldo, J.E. and Lester, M.C. (1989) Post traumatic extracranial vertebral artery dissection with locked-in syndrome: a case with MRI documentation and unusually favourable outcome. J. Neurol. Neurosurg. Psychiatry., 52: 1191–1193.

Ragazzoni, A., Grippo, A., Tozzi, F. and Zaccara, G. (2000) Event-related potentials in patients with total locked-in state due to fulminant Guillain-Barre syndrome. Int. J. Psychophysiol., 37: 99–109.

Richard, I., Pereon, Y., Guiheneu, P., Nogues, B., Perrouin-Verbe, B. and Mathe, J.F. (1995) Persistence of distal motor control in the locked in syndrome. Review of 11 patients. Paraplegia., 33: 640–646.

Sandin, R.H., Enlund, G., Samuelsson, P. and Lennmarken, C. (2000) Awareness during anaesthesia: a prospective case study. Lancet, 355: 707–711.

Schnakers, C., Majerus, S., Laureys, S., Van Eeckhout, P., Peigneux, P. and Goldman, S. (2005) Neuropsychological testing in chronic locked-in syndrome. Psyche, abstracts from the Eighth Conference of the Association for the Scientific Study of Consciousness (ASSC8), University of Antwerp, Belgium, 26–28 June 2004, pp. 11.

Sigalovsky, N. (2003) Awareness under general anesthesia. Aana. J., 71: 373–379.

Steffen, G.E. and Franklin, C. (1985) Who speaks for the patient with the locked-in syndrome? Hastings Cent. Rep., 15: 13–15.

Stumpf, S.E. (1986) A comment on "Helen". South Med. J., 79: 1057–1058.

Tavalaro, J. and Tayson, R. (1997) Look Up for Yes. Kodansha America Inc, New York, NY.

Thadani, V.M., Rimm, D.L., Urquhart, L., Fisher, L., Williamson, P.D., Enriquez, R., Kim, J.H. and Levy, L.L. (1991) 'Locked-in syndrome' for 27 years following a viral illness: clinical and pathologic findings. Neurology, 41: 498–500.

Towle, V.L., Maselli, R., Bernstein, L.P. and Spire, J.P. (1989) Electrophysiologic studies on locked-in patients: heterogeneity of findings. Electroencephalogr. Clin. Neurophysiol., 73: 419–426.

Trail, M., Nelson, N.D., Van, J.N., Appel, S.H. and Lai, E.C. (2003) A study comparing patients with amyotrophic lateral sclerosis and their caregivers on measures of quality of life, depression, and their attitudes toward treatment options. J. Neurol. Sci., 209: 79–85.

van Eeckhout, P. (1997) Le locked-in syndrom. Rééducation Orthophonique, 35: 123–135.

Vigand, P. (2002) Promenades Immobiles Le Livre de Poche, Paris.

Vigand, P. and Vigand, S. (2000) Only The Eyes Say Yes (original title: Putain de silence). Arcade Publishing, NY.

Ware, J.E., Snow, K.K. and Kosinski, M. (1993) SF-36 Health Survey Manual and Interpretation Guide. The Health Institute, New England Medical Center, Boston, MA.

Zola, E. (1979) Thérère Raquin. Gallimard, Paris.

S. Laureys (Ed.)
Progress in Brain Research, Vol. 150
ISSN 0079-6123

CHAPTER 35

Brain-computer interfaces — the key for the conscious brain locked into a paralyzed body

Andrea Kübler* and Nicola Neumann

Institute of Medical Psychology and Behavioral Neurobiology, University of Tübingen, Gartenstr. 29, 72074 Tübingen, Germany

Abstract: Brain–computer interfaces (BCIs) are systems that allow us to translate in real-time the electrical activity of the brain in commands to control devices. They do not rely on muscular activity and can therefore provide communication and control for those who are severely paralyzed (locked-in) due to injury or disease. It has been shown that locked-in patients are able to achieve EEG-controlled cursor or limb movement and patients have successfully communicated by means of a BCI. Current BCIs differ in how the neural activity of the brain is recorded, how subjects (humans and animals) are trained to produce a specific EEG response, how the signals are translated into device commands, and which application is provided to the user. The present review focuses on approaches to BCIs that process the EEG on-line and provide EEG feedback or feedback of results to the user. We regard online processing and feedback cornerstones for routine application of BCIs in the field. Because training patients in their home environment is effortful and personal and financial resources are limited, only few studies on BCI long-term use for communication with paralyzed patients are available. The need for multidisciplinary research, comprising computer science, engineering, neuroscience, and psychology is now being acknowledged by the BCI community. A standard BCI platform, referred to as BCI2000, has been developed, which allows us to better combine and compare the different BCI approaches of different laboratories. As BCI laboratories now also join to unify their expertise and collaborations are funded, we consider it realistic that within few years we will be able to offer a BCI, which will be easy to operate for patients and caregivers.

Welcome to my voice of silence.
Life is a precious shell.
Hold it for as long as you can.
Remember it is only a shell
Which contains the powerful mind.
It says you can do anything
Be anything.
So say what you will say.
Julia Tavalaro[1]

Introduction

Brain–computer interfaces (BCIs) are devices that allow interaction between humans and artificial devices. Historically, human–machine interfaces have been largely dependent on human motor control. However, powerful computers, new microchip design, microstimulation technology to interact with neuronal tissue, sophisticated signal detection algorithms, and the rapidly growing field of neurosciences allow us to tap into a completely new application of interfaces, i.e., the restoration of lost motor or sensory function.

In the following review we will first give a brief summary of diseases leading to the so-called

*Corresponding author. Tel.: +49 7071 297 5997; Fax: +49 7071 29 5956; E-mail: andrea.kuebler@uni-tuebingen.de

[1]Julia Tavalaro and Richard Tayson — *Lookup for Yes* (1998), Penguin Books, New York.

DOI: 10.1016/S0079-6123(05)50035-9

locked-in state, which renders BCIs the ultimate possibility to interact with the environment, followed by a general introduction to BCI design. We will then present different approaches to BCIs to "translate" people's thoughts. Since thoughts, intentions, and emotions are reflected in brain activity, they can be detected in theory by modern neuroimaging techniques. Whether this is actually feasible, depends on the temporal and spatial resolution of the applied method and on the complexity of the thought, intention, or emotion. Apart from EEG-driven BCIs, we will introduce a BCI that uses the blood oxygen level dependent (BOLD) response measured with functional magnetic resonance imaging (fMRI). We will end our review with a discussion of current BCI research and outlook into future research.

In a typical BCI setting, participants are presented with stimuli or are required to perform specific mental tasks while their EEG is being recorded. The data can then be used for off-line analysis with different classification algorithms or they can be online translated into device commands or fed back to the participant, or both. A huge body of literature exists that deals with the extraction and classification of relevant features of the electrical activity of the brain (reviewed in *IEEE Transactions of Neural Systems and Rehabilitation Engineering*, Vol. 11, 2003, special issue on brain–computer interfaces). These features can then be fed back to the user by the so-called closed-loop BCIs. Feedback can either be provided within a few milliseconds or with a delay of several seconds. The term closed-loop BCI is contrasted to open-loop BCIs that do not feed back results to the user. We restrict this review to BCI technology that uses online EEG classification and translation of the extracted signals into device commands, because we regard this step as a cornerstone for application of BCI technology in locked-in patients. For the same reason we will focus on closed-loop systems.

The locked-in syndrome

Neurological diseases may lead to paralysis of the entire motor system restricting both verbal and nonverbal communication. The term locked-in refers to a state in which individuals are conscious and alert, but unable to use their muscles and therefore cannot communicate their needs, wishes, and emotions: the healthy brain is locked into a paralyzed body (see Laureys et al., elsewhere in this volume). In the "classic" locked-in syndrome, vertical eye movement and eye blinks remain intact (Bauer et al., 1979; Leon-Carrion et al., 2002), whereas in the total locked-in state, patients lose all ability to move and communicate (Hayashi and Kato, 1989). Hemorrhage or an ischemic stroke in the ventral pons can cause a locked-in syndrome, which includes tetraplegia and paralysis of cranial nerves (Chia, 1991). The syndrome can also occur due to traumatic brainstem injury (Leon-Carrion et al., 2002), encephalitis (Acharya et al., 2001), or tumor (Breen and Hannon, 2004). Other causes of the locked-in state are degenerative neurological diseases (Karitzky and Ludolph, 2001), the most frequent being amyotrophic lateral sclerosis, which involves a steadily progressive degeneration of central and peripheral motoneurons (Jerusalem et al., 1996; Leigh et al., 2004). The locked-in state in which cognitive functioning of the brain is supposed to remain intact has to be distinguished from minimally conscious state, persistent vegetative state, and coma (Laureys et al., 2004).

A generic brain–computer interface system

A BCI system can be depicted as a series of functional components (Kübler et al., 1999; Mason and Birch, 2003). The starting point is the user, whose intent is coded in the neural activity of his or her brain (input). The end point is the device that is controlled by the brain activity of the user (output). Several steps are necessary to translate the former into the latter (Fig. 1). These steps are described below for interfaces controlled by the electrical activity of the brain.

The input signal

The electrical brain activity of the user is recorded by single or multiple electrodes from the scalp (EEG), from the cortical surface (electrocorticogram (ECoG)), or intracortically with single electrodes or multielectrode arrays (the latter in

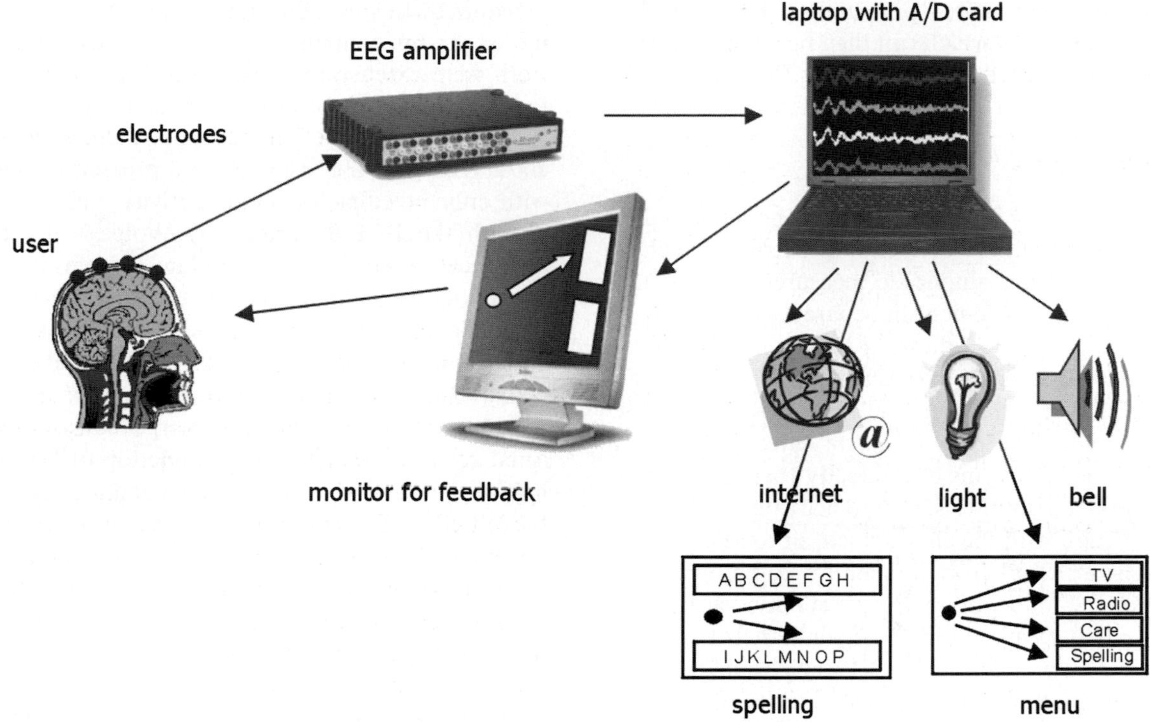

Fig. 1. A generic BCI setup is depicted. The EEG of the user is recorded with few or multiple electrodes. The signals are amplified and transferred to a PC for online filtering and translation into device commands — in this example into cursor movement on a computer screen. Cursor movement can be used, for example, to select items in a menu to control devices in the patient's environment, to select letters in a language support program, or to surf the internet.

animals only). Electrodes are connected to an amplifier that bandpass filters and amplifies the electrical signals of the user's brain. Different brain states of the user that convey a message can either be achieved by self-regulation of the relevant EEG features or are elicited by mental tasks or presentation of external visual or auditory stimuli. Although designs to teach users to self-regulate their EEG vary across groups, the general setup is as follows: the relevant features of the EEG are extracted and translated in real time (see below) into cursor movement on a computer screen (visual feedback), a stream of tones (auditory feedback) (Hinterberger et al., 2004), or any other distinctly changing signal. Feedback can be provided synchronously in discrete trials (Wolpaw et al., 1991; Kübler et al., 1999), or asynchronously with continuous feedback (Millan and Mourino, 2003; Borisoff et al., 2004; Scherer et al., 2004) (see also section "machine learning"). Self-regulation of the EEG allows the user real-time control of a cursor (or any other signal which represents the user's EEG) in one or two dimensions (Kotchoubey et al., 1996; Wolpaw and McFarland, 2003).

Another possibility to voluntarily produce specific patterns of brain activity is to perform different mental tasks, such as motor imagery, mental rotation, or mental arithmetic, without continuous online feedback. In this case, the participants produce a specific EEG pattern related to the task that is recognized by the BCI. The user receives feedback of results (Millan et al., 2004b). This pattern recognition approach tries to minimize the effort for the participant to alter the EEG in predefined frequency bands. Likewise, BCIs that use external stimulation to elicit specific patterns of brain activity do not involve self-regulation of the brain. Depending on the stimulation paradigm, different event-related potentials (ERPs) are elicited in the EEG, which can then be detected by the BCI (Donchin et al., 2000; Sutter, 1992).

Transformation and output

The acquired signals are digitized and subjected to a variety of feature extraction procedures, such as spatial filtering, amplitude measurement, spectral analysis, or single-neuron separation (Wolpaw et al., 2002). In the following step, a specific algorithm translates the extracted features into commands that represent the user's intent. These commands can either control effectors directly such as robotic arms or indirectly via cursor movement on a computer screen to activate switches for interaction with the environment or to select items, words, or letters from a menu for communication.

For the following review we subdivided BCIs in those that require users to regulate their brain potentials (user learning — self-regulation of brain responses), those that use mental tasks and pattern recognition algorithms (machine learning), and those that elicit ERPs with external stimuli (ERPs). It is important to note that this categorization of BCIs is not absolute: in almost all BCIs, parameters to extract the relevant EEG features are adapted to the individual user (machine learning) and the user has to be trained in EEG-controlled BCI use (user learning).

Regulation of brain responses for BCI control

Cursor control with sensorimotor rhythms

Sensorimotor or mu-rhythm refers to 8–15 Hz EEG activity, which can be recorded in awake people over primary sensory or motor cortical areas (Niedermeyer, 1998). It is usually accompanied by 18–26 Hz beta-rhythms. It decreases or desynchronizes with movement or preparation for movement and increases or synchronizes in the post-movement period or during relaxation (Pfurtscheller, 1999). Furthermore, and most relevant for BCI use by locked-in patients, it also desynchronizes with motor imagery. Thus, to modulate the sensorimotor-rhythm amplitude no actual movement is required. In the following sections, two sensorimotor rhythm controlled BCIs shall be introduced. Both were extensively tested with healthy participants and a few patients in the laboratory or in the field (Pfurtscheller et al., 2000b; Wolpaw et al., 2003; Kübler et al., in press), and provide the user with online feedback of brain activity.

With the BCI developed by Wolpaw and his colleagues, users learn to regulate the sensorimotor-rhythm amplitude and to use this ability to move a cursor on a computer screen in one or two dimensions (Wolpaw et al., 1991; McFarland et al., 2003). During each trial of one-dimensional control, users are presented with a target consisting of a red vertical bar that occupies the top or bottom half of the right edge of the screen and a cursor on the left edge. The cursor moves steadily across the screen, with its vertical movement controlled by sensorimotor-rhythm amplitude. The user's task is to move the cursor to the level of the target so that it hits the target when it reaches the right edge. Figure 2 shows a typical example of the power spectrum of a patient after successful learning. During relaxation, the amplitude of the sensorimotor rhythm and the associated beta-rhythm is high corresponding to upward cursor movement. Conversely, during motor imagery, the amplitude of sensorimotor rhythm decreases and the cursor moves downward. The topography shows that sensorimotor-rhythm control is sharply focused to the feedback electrode over sensorimotor cortex within the 8–12 Hz alpha band, which was fed back to the patient. With such control over sensorimotor-rhythm amplitude, a binary "yes–no" answer is possible. This is the least that is needed to execute commands in the environment or to select letters in a Language Support Program (Perelmouter et al., 1999). Recently, two-dimensional control with cursor movement into eight targets was realized (Wolpaw and McFarland, 2004). After learning to control the sensorimotor-rhythm amplitude, users are provided with a two- or four-choice spelling program (Wolpaw et al., 2003).

The BCI, developed by Pfurtscheller's group, also uses motor imagery and associated sensorimotor-rhythm desynchronization for device control (Pfurtscheller et al., 1996; Neuper et al., 2003). In the first session of the standard protocol users have

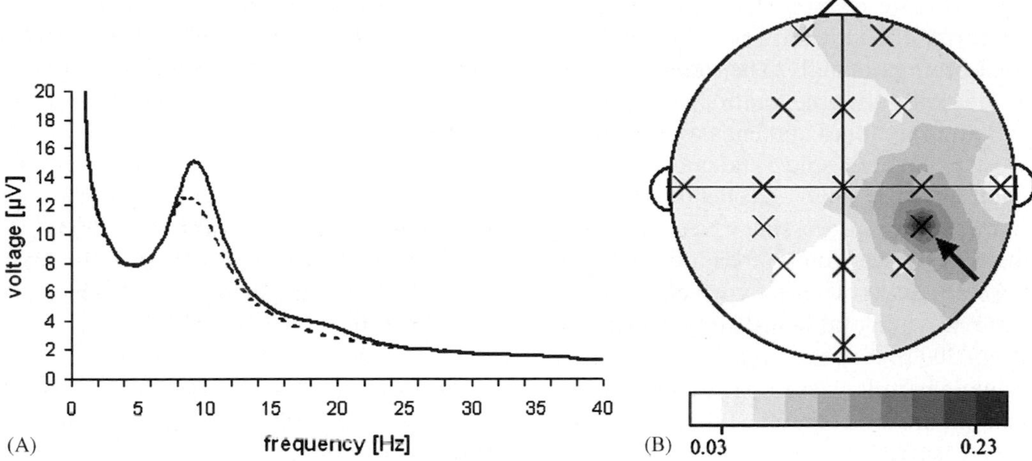

Fig. 2. Data from a patient trained to self-regulate sensorimotor rhythm. He received feedback from CP4 (arrow in B) in the frequency range of 9–12 Hz. (A) Power spectrum. The voltage is plotted as a function of frequency. The solid line shows the power spectrum during relaxation and related upward cursor movement; the dashed line during imagined left-hand movement and related downward cursor movement. A voltage difference can be clearly seen in the 9–12 Hz frequency band (mu-rhythm) and also — albeit smaller — in the beta band around 18–22 Hz. (B) The topography of the determination coefficient (r^2), is the proportion of the total variance of the sensorimotor-rhythm amplitude that is accounted for by target position. The r^2 is highest (dark grey) under the feedback electrode.

to imagine hand, feet, or tongue movement while their EEG is being recorded. Signal features are extracted and a user-specific classifier is established that determines from the EEG frequency pattern which movement the user is imagining. In subsequent sessions the user receives feedback of motor imagery-related EEG changes. As in the BCI described formerly, participants are provided with a spelling program (Obermaier et al., 2001; Scherer et al., 2004). In the spelling program recently introduced by Scherer and co-workers the time for each selection is set by the user. This may considerably speed up the selection process. The same group also presented a BCI-controlled restoration of hand grasp in a patient with tetraplegia after spinal cord injury (Pfurtscheller et al., 2000a, 2003).

Cursor control with slow cortical potentials

The BCI developed by Birbaumer and colleagues uses slow cortical potentials of the brain for cursor control and is referred to as the thought translation device (TTD) (Birbaumer et al., 1999; Kübler et al., 1999). Slow cortical potentials depend on sustained afferent intracortical or thalamocortical input to cortical layers I and II and on simultaneous depolarization of large pools of pyramidal neurons. The depolarization of cortical cell assemblies reduces their excitation threshold, and firing of neurons in regions responsible for specified motor or cognitive tasks is facilitated (Birbaumer et al., 1990).

In various studies Birbaumer and his colleagues showed that severely paralyzed patients were able to learn to regulate their slow cortical potential amplitude and that they could use this ability to communicate (Kübler et al., 2001b; Birbaumer et al., 2003; Neumann et al., 2003). Similar to the sensorimotor-rhythm controlled BCI, patients are provided with online feedback of their slow cortical potentials and are presented with targets at the top and bottom of the screen. When patients achieve an accuracy of cursor control of more than 75%, they are provided with a Language Support Program (Kübler et al., 2001b). A browser to surf the world wide web is also available and extensively used by one of our patients (Mellinger et al., 2003).

The TTD also encompasses a psychophysiological system for detection of possible cognitive functioning in locked-in, minimally conscious, vegetative, or comatose states (Kotchoubey et al., 2002). It consists of a test battery using auditory event-related brain potentials. The series of examinations begins with a simple auditory oddball using frequent and rare tones and ends with complex sentences using semantic and syntactic violations for detection of the N400 and P600 waves reflecting higher order processing (Neumann and Kotchoubey, 2004) (see Kotchoubey, this volume). Although we do not know exactly which cognitive functions are most favorable for using a BCI (see Vernon et al., 2003), it is clear that an application of BCI technology in locked-in patients is feasible only if a minimum of cognitive functioning and attention are ensured.

Machine learning

Numerous research groups have tested algorithms to discriminate EEG patterns associated with task-related mental imageries. The rationale behind that procedure is that different imageries underlie activity in different neural networks that can be measured and discriminated by classification algorithms, such as linear discriminant analysis, support vector machines, or artificial neural networks (Dornhege et al., 2004). Thereby the main part of learning is imposed on the machine, not on the user, as in self-regulation paradigms. However, most research has been done as off-line analysis that demonstrated discrimination rates up to more than 90%. A detailed review of classification algorithms is beyond the scope of this paper (see Wolpaw et al., 2002 for a review). To our knowledge, only few groups tested their algorithms online in healthy volunteers or disabled patients. The Berlin BCI uses multichannel scalp EEG recordings and analyzes the single-trial differential potential distributions of the Bereitschaftspotential preceding voluntary movement. Healthy participants are required to typewrite in a self-paced manner on one of four keys of a computer keyboard using either their index or little finger of their left or right hand. Bereitschaftspotentials are then classified in single trials to predict the laterality of left- versus right-finger movements at a time point prior to the onset of muscle activity (Blankertz et al., 2003). Blankertz and colleagues report good accuracies even at a pace of as fast as 2 taps per second. Recently, this group also conducted experiments with online feedback of motor imagery (Blankertz, personal communication).

Online feedback is also provided by the Canadian group of the "Neil Squire Foundation" (Birch et al., 2003; Borisoff et al., 2004). They developed an "asynchronous" BCI for intermittent control in a natural environment, which allows the user to switch on and off the BCI according to their own accord (Mason and Birch, 2000). This brain "switch" is controlled by self-paced motor imagery, and patients diagnosed with high-level spinal cord injury were able to control it. Healthy participants successfully controlled a simple video game (Birch et al., 2003).

A Swiss group under the direction of Millan also constructed an asynchronous BCI (Millan and Mourino, 2003). During the acquisition phase of the classifier, healthy participants and one patient diagnosed with spinal muscular atrophy had to perform three mental tasks (out of six) either on command of the operator or they could choose the task on their own accord. This approach accommodates for an asynchronous BCI, which allows participants self-paced decisions on when to stop and start a specific mental task instead of a fixed trial length determined by the BCI. After training of classifiers, participants receive online feedback of their task-related EEG and learn to control a wheelchair-like mobile robot through self-paced imagery (Millan et al., 2004b) or are provided with a spelling device (Millan et al., 2004a).

Event-related potentials

ERPs are deflections in the EEG that have a fixed time delay to the stimulus, while the ongoing EEG activity constitutes additive noise. To detect ERPs averaging techniques are used. An averaged ERP is composed of a series of large, biphasic waves, lasting a total of 500–1000 ms. We restrict our review on the BCI system that is based on the measurement of the P300, a positive ERP deflection that occurs 300 ms after the presentation of a deviant stimulus among standard stimuli (Farwell and Donchin, 1988). Several other BCIs rely on visually evoked potentials measured over the visual cortex (Middendorf et al., 2000; Sutter, 1992). However, these BCIs require intact gaze as the target letter or item has to be fixated precisely. To date, the P300 BCI also depends on intact vision, but the P300 can also be elicited auditorily.

In the standard form of the P300 BCI letters are presented in a 6 × 6 letter matrix (Fig. 3). Users are required to select one letter as their target letter; then the rows and columns of the matrix flash sequentially and users have to count how often their target letter flashes. Since the target letter represents a deviant stimulus among the other standard letters, it elicits a P300 that can be detected by the BCI. The speed with which a BCI can detect a P300 linked to a specific letter depends on how prominent the P300 is. In a recent study it has been demonstrated that this principle can be used to select items, such as a TV or stereo system, in a virtual environment (Bayliss, 2003). The P300 BCI has the advantage that it does not require the execution of an active skill of the user as is necessary for the systems based on the self-regulation of an EEG parameter. However, to date, no data on long-term use of the P300 BCI in patients are available. Preliminary data from one session with our patients ($N = 7$) show that the amplitude and latency of the P300 differ tremendously among individuals. Based on the preliminary data, one of our patients would be able to spell three letters per minute, while others would need 10 min for one letter (Mellinger et al., 2004).

Invasive recording

All invasive recording requires brain surgery and implanted electrodes or electrode arrays have to remain stable over a long period of time.

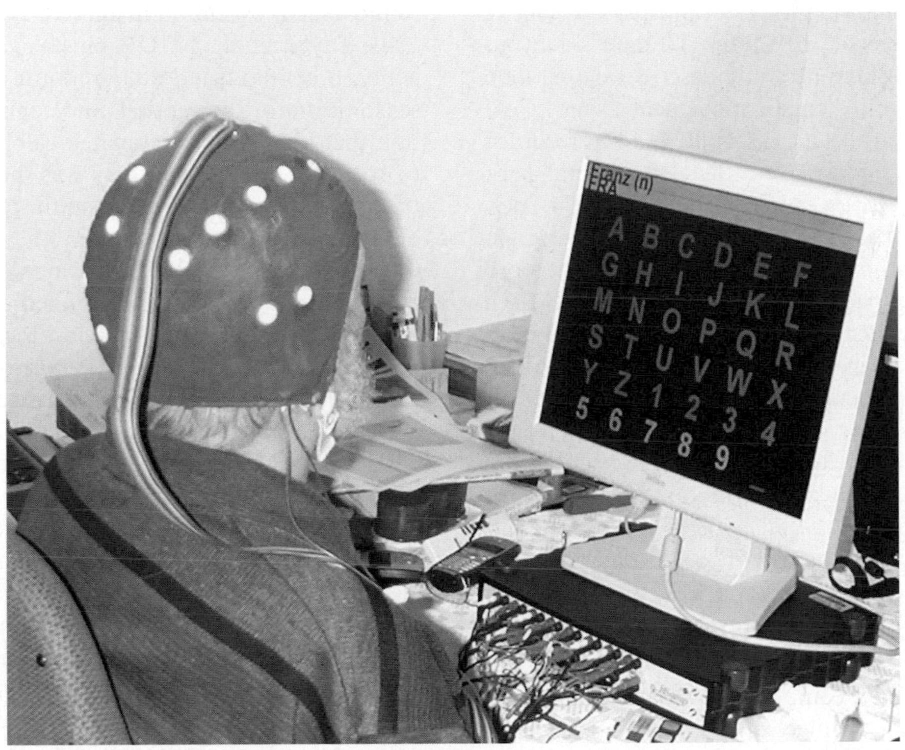

Fig. 3. A patient during training with the P300 BCI. The patient is trained at home under the supervision of an assistant from the laboratory. He views a 6 × 6 letter matrix and is required to copy the word on the top of the screen (Franz). He has already copied correctly FRA and is now counting how often the letter N flashes (for details about the method see section on "event related potentials").

Electrocorticogram (ECoG)

The electrocorticogram (ECoG) is a measure of the electric potential on the cortical surface of the brain. ECoG uses epidural or subdural electrode strips or arrays to record the electrical activity from the cerebral cortex, and is thus an invasive procedure that requires a craniotomy for implantation of the electrodes. Its main advantages are a higher spatial resolution than the EEG (tenths of millimeters versus centimeters), broader bandwidth (0–200 Hz versus 0–40 Hz), higher amplitude (50–100 μV maximum versus 5–20 μV), and less vulnerability to artifacts such as electromyogram (Leuthardt et al., 2004). The common purpose of ECoG is to localize a suspected seizure focus in the cerebral cortex for individuals who are candidates for surgery. However, some research groups made use of the ECoG in epilepsy patients to investigate ECoG patterns and signal analysis methods as the basis for a BCI. Most of these studies performed off-line open-loop analysis of ECoG data (Huggins et al., 1999; Levine et al., 1999; Scherer et al., 2003). To date — to our knowledge — only one study describes closed-loop real-time control of cursor movement using ECoG activity (Leuthardt et al., 2004). ECoG was recorded from 32 electrodes over the left frontal–parietal–temporal cortex while patients ($N = 4$) performed each of six tasks: opening and closing right or left hand, protruding the tongue, saying the word "move", and imaging each of these actions. Each task was associated with a decrease in mu- and beta-rhythm, and an increase of gamma-rhythm amplitudes over prefrontal, premotor, sensorimotor or speech areas. The spatial and spectral foci of task-related ECoG activity were similar for action and imagery. When provided with online feedback of imagery-related ECoG, all patients achieved significant cursor control above 70% accuracy within a brief training of 3–24 min. Thus, ECoG may offer a good basis for a BCI. However, no data on long-term use of invasive ECoG grids are available and the risk of infection may be critical.

Intracortical recording

Intracortical signal acquisition can be realized with single, few, or multiple electrodes that capture the action potentials of individual neurons. Electrode tips have to be in close proximity to the signal source and the arrays have to be stable over a long period of time. In two exemplary ALS patients, Kennedy and Bakay showed that humans are able to modulate the action potential firing rate when provided with feedback (Kennedy and Bakay, 1998). The authors implanted a single electrode with a glass tip containing neurotrophic factors into the motor cortex. Adjacent neurons grew into the tip and after few weeks action potentials were recorded. One patient was able to move a cursor on a computer screen to select presented items by modulating his action potential firing rate (Kennedy et al., 2000).

Conversely, multielectrode arrays for intracortical recording are not yet ready for clinical application (Nicolelis, 2003). They are used in animals only and are feasible for long-term recording (Donoghue, 2002). Several groups use multielectrode recording to detect activation patterns related to movement execution in animals (Carmena et al., 2003; Taylor et al., 2003; Paninski et al., 2004). The action potential firing rate in motor areas contains sensory, motor, perceptual, and cognitive information that allows us to estimate a subject's intention to execute a movement, and it was shown that hand trajectories can be derived from the activity pattern of neuronal cell assemblies in the motor cortex (Serruya et al., 2002). For example, Taylor and colleagues realized brain-controlled cursor and robot arm movement using recordings from a few neurons (18 cells) in the motor cortex only (Taylor et al., 2002). Rhesus macaques learned first to move a cursor into eight targets located at the corners of an imaginary cube with real hand movements. Accompanying neural activity patterns were recorded and used to train an adaptive movement prediction algorithm. After sufficient training of subjects and algorithm, subjects' arms were restricted and cursor movement was performed brain-controlled. Similarly, rhesus monkeys were trained to move a brain-controlled robot arm in virtual reality (Taylor et al., 2003) and then to feed themselves with a real robot arm (Schwartz, 2004) (Fig. 4).

Recently, Musallam and colleagues presented data from three monkeys, which were implanted with electrode arrays in the parietal reach area,

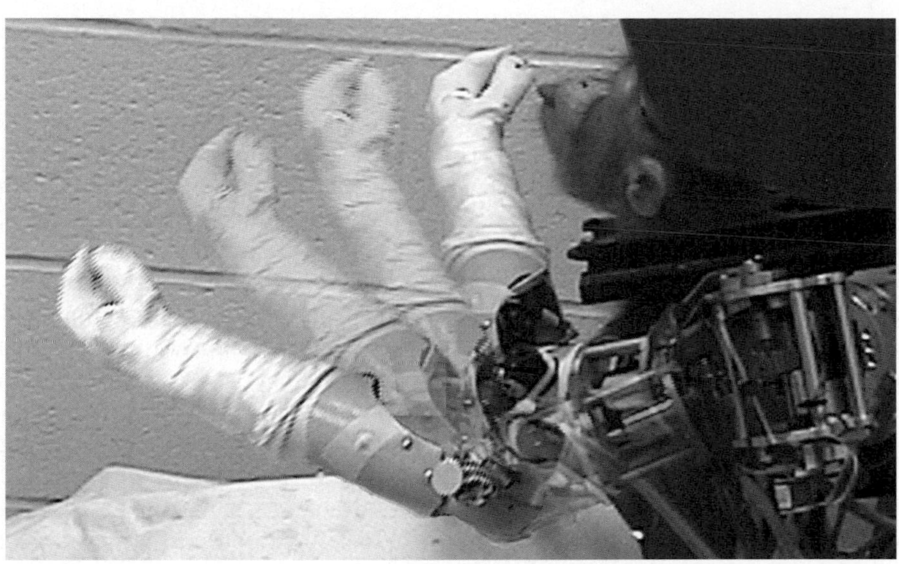

Fig. 4. A monkey is feeding himself with the aids of a robotic arm by producing the same pattern of neural activity in the motor cortex as would be required to move his own limb. The trajectory of the robot arm is depicted sequentially. In expectancy of the piece of an apple, the monkey protrudes his tongue. From the monkey only the head is visible. (We thank Dr. Andrew Schwartz, from the School of Medicine, University of Pittsburgh, Pittsburgh, USA, for this picture and for the permission of reproduction.)

area 5, and the dorsal premotor cortex (Musallam et al., 2004). Subjects were first trained to reach for targets at different positions on a screen after a delay of 1.2–1.8 sec, following cue presentation. Neural activity during the memory period was correctly decoded with an accuracy of about 64%. The authors then trained subjects to associate visual cues with the amount, probability, or type of reward (orange juice versus water). Neural activity was then found to alter as a function of expected reward and thus, represented additional information for classification. Accordingly, classification results could be improved by 12%. These experiments reveal that with sophisticated algorithms it is possible to derive limb or cursor movement directly from the neural activity patterns of the cortex.

BCIs based on functional magnetic resonance imaging (fMRI)

In contrast to the other methods described above, fMRI provides an indirect measure for the electrical activity of the brain. fMRI measures the BOLD response, which refers to signal differences due to changes in the concentration of deoxygenated hemoglobin in the brain tissue, depending on neuronal metabolism and activity (Logothetis, 2002). Thus, fMRI allows us to record noninvasively neural activity in the entire brain with a high-spatial resolution. With a new method referred to as real-time fMRI (Weiskopf et al., in press; Weiskopf et al., 2003), the BOLD-response is being recorded and fed back to the participant. It permits the regulation of local brain activity in analogy to neurofeedback on the basis of EEG, and enables studies on behavioral and cognitive consequences of self-modulated activity, and its possible therapeutic use. Only few studies have been conducted using real-time fMRI. Whereas first studies provided feedback with a delay of several seconds (Posse et al., 2003; Yoo and Jolesz, 2002), Weiskopf introduced the concept of immediate feedback of the BOLD signal (2 sec from image acquisition to display of feedback) and demonstrated that the activity in two subdivisions of the anterior cingulate cortex could be up- and down-regulated (Weiskopf et al., 2003). Subsequent testing of the affective state suggested an increase in valence and arousal during self-induced up-regulation of the BOLD response. In another study, Weiskopf and colleagues demonstrated that participants could differentially activate the supplementary motor area and the parahippocampal place area (Weiskopf et al., 2004). DeCharms and colleagues showed that participants were able to enhance their BOLD response over the somatomotor cortex during feedback training (deCharms et al., 2004). Signal enhancement was achieved without muscle tension, and voluntary control continued even when feedback was no longer provided. In a recent study by Yoo and colleagues, feedback of four different regions-of-interest linked to specific mental tasks enabled participants to navigate a cursor through a simple two-dimensional maze (Yoo et al., 2004).

In sum, real-time fMRI is a promising method that allows us to control local neural activation even in subcortical areas. A possible target application could thus be the regulation in "emotional areas", such as amygdala, orbitofrontal, and anterior cingulate cortex, in anxiety disorders. However, since fMRI is very susceptible to artifacts and unspecific effects, such as head motion or cardiorespiratory effects, great effort is needed to control for these artifacts during acquisition and real-time analysis of the data (Weiskopf et al., 2003, 2004, submitted).

Shortcomings and future of BCI research and development

BCIs that use electrical activity of millions of neurons typically allow at the most three characters or other choices per minute (Craelius, 2002). Therefore, many scientists regard them as slow and the slowness is pointed out as a major shortcoming (Donchin et al., 2000; Donoghue, 2002). However, slowness is a matter of definition. If only a single switch is available, even few yes or no answers per minute may be a great relief for the patient and significant others (for a discussion of slowness in BCI control see Kübler et al., 2001a).

Factors that determine performance of a BCI and its application include the EEG signal of interest, the signal processing method for feature extraction, the algorithms that translate these features into device commands, the application that is controlled by those commands, the feedback that is provided to the user, and the specific medical and psycho-social characteristics of the user (Schalk et al., 2004). Consequently, developing a real-time BCI and implementing it in the daily life of a user is a multidisciplinary challenge involving computer science, engineering, neuroscience, neurology, and psychology. It involves finding the brain signals that a user can control best. Powerful algorithms have to be delineated to deal with artifacts, such as eye movement or muscle tension, to filter the desired signals, and transform them into commands for controlling devices. Training methods to gain and maintain control have to be established and applications suiting the users' needs have to be identified and adapted. To investigate the impact and feasibility of a BCI, precise and objective procedures for evaluating short- and long-term BCI performance have to be adopted. Wolpaw and colleagues suggest the determination coefficient r^2, which is the proportion of the total variance of the EEG signal of interest that is accounted for by the presented task, as a measure for user performance (Wolpaw et al., 2002). The authors propose further to evaluate system performance as performance in a specific application expressed as speed or accuracy, or both. The information transfer rate (bit-rate) could serve as a standard measure to compare the efficacy of different BCIs (Wolpaw et al., 2002). It has to be elucidated how psychological variables, such as motivation, current state of mood, or quality of life, influence and are influenced by the BCI. The neurobiological basis of achieving self-regulation of a brain signal in the healthy brain and in the brain affected by a neurological disease has to be clarified. The development of technically elaborated BCIs has outstripped clinical application, since there are over 30 BCI groups (reviewed in *IEEE Transactions of Neural Systems and Rehabilitation Engineering*, Vol. 11, 2003, special issue on brain-computer interfaces), but only few conduct long-term patient training in their home environment. Training locked-in patients is costly and a challenge for each BCI approach (Neumann and Kübler, 2003).

Although laboratories involved in BCI research do cover the different disciplines listed, interaction between the different approaches has been small or nonexistent. Thus, to date, BCI research has rarely gone beyond demonstration that specific EEG signals can be classified by specific algorithms, and that these signals can then be used to control specific devices by few subjects. BCI systems have been idiosyncratic to each laboratory rendering comparisons and exchange of expertise difficult and thus, hampering BCI research and development. In recognition of this situation, general-purpose BCI research and application systems, such as the BCI2000 (Schalk et al., 2004), have been developed to allow us better comparison between the different approaches (Bianchi et al., 2003; Schalk et al., 2004). Such a system facilitates systematic studies to pursue improvements that may be achieved by incorporating and comparing diverse brain signals, different signal analysis, and translation algorithms. BCI research groups from Europe and the United States join forces and apply successfully for third party funding, indicating that the need for multidisciplinary research to promote clinical application of BCI systems for locked-in patients is now being acknowledged. Thus, we consider it realistic that we will be able to provide patients and their caregivers with a more user friendly, easy-to-handle BCI in the near future.

Acknowledgments

This work was supported in part by grants from NIH in the USA (NICHD (HD30146) and NIBIB/NINDS (EB00856)) and the German Research Society (DFG) (SFB550) in Germany.

References

Acharya, V.Z., Talwar, D. and Elliott, S.P. (2001) Enteroviral encephalitis leading to a locked-in state. J. Child. Neurol., 16: 864–866.

Bauer, G., Gerstenbrand, F. and Rumpl, E. (1979) Variables of the locked-in syndrome. J. Neurol., 221: 77–91.

Bayliss, J.D. (2003) Use of the evoked potential P3 component for control in a virtual apartment. IEEE Trans. Neural Syst. Rehabil. Eng., 11: 113–116.

Bianchi, L., Babiloni, F., Cincotti, F., Arrivas, M., Bollero, P. and Marciani, M.G. (2003) Developing wearable bio-feedback systems: a general-purpose platform. IEEE Trans. Neural Syst. Rehabil. Eng., 11: 117–119.

Birbaumer, N., Elbert, T., Canavan, A.G.M. and Rockstroh, B. (1990) Slow potentials of the cerebral cortex and behavior. Physiol. Rev., 70: 1–41.

Birbaumer, N., Ghanayim, N., Hinterberger, T., Iversen, I., Kotchoubey, B., Kübler, A., Perelmouter, J., Taub, E. and Flor, H. (1999) A spelling device for the paralysed. Nature, 398: 297–298.

Birbaumer, N., Hinterberger, T., Kübler, A. and Neumann, N. (2003) The thought-translation device (TTD): neurobehavioral mechanisms and clinical outcome. IEEE Trans. Neural Syst. Rehabil. Eng., 11: 120–123.

Birch, G.E., Mason, S.G. and Borisoff, J.F. (2003) Current trends in brain–computer interface research at the Neil Squire Foundation. IEEE Trans. Neural Syst. Rehabil. Eng., 11: 123–126.

Blankertz, B., Dornhege, G., Schafer, C., Krepki, R., Kohlmorgen, J., Müller, K.R., Kunzmann, V., Losch, F. and Curio, G. (2003) Boosting bit rates and error detection for the classification of fast-paced motor commands based on single-trial EEG analysis. IEEE Trans. Neural Syst. Rehabil. Eng., 11: 127–131.

Borisoff, J.F., Mason, S.G., Bashashati, A. and Birch, G.E. (2004) Brain–computer interface design for asynchronous control applications: improvements to the LF-ASD asynchronous brain switch. IEEE Trans. Biomed. Eng., 51: 985–992.

Breen, P. and Hannon, V. (2004) Locked-in syndrome: a catastrophic complication after surgery. Br. J. Anaesth., 92: 286–288.

Carmena, J.M., Lebedev, M.A., Crist, R.E., O'Doherty, J.E., Santucci, D.M., Dimitrov, D., Patil, P.G., Henriquez, C.S. and Nicolelis, M.A. (2003) Learning to control a brain–machine interface for reaching and grasping by primates. PLoS Biol., 1: E42.

Chia, L.G. (1991) Locked-in syndrome with bilateral ventral midbrain infarcts. Neurology, 41: 445–446.

Craelius, W. (2002) The bionic man: restoring mobility. Science, 295: 1018–1021.

deCharms, R.C., Christoff, K., Glover, G.H., Pauly, J.M., Whitfield, S. and Gabrieli, J.D. (2004) Learned regulation of spatially localized brain activation using real-time fMRI. Neuroimage, 21: 436–443.

Donchin, E., Spencer, K.M. and Wijesinghe, R. (2000) The mental prosthesis: assessing the speed of a P300-based brain–computer interface. IEEE Trans. Neural Syst. Rehabil. Eng., 8: 174–179.

Donoghue, J.P. (2002) Connecting cortex to machines: recent advances in brain interfaces. Nat. Neurosci., 5(Suppl.): 1085–1088.

Dornhege, G., Blankertz, B., Curio, G. and Müller, K.R. (2004) Boosting bit rates in noninvasive EEG single-trial classifications by feature combination and multiclass paradigms. IEEE Trans. Biomed. Eng., 51: 993–1002.

Farwell, L.A. and Donchin, E. (1988) Talking off the top of your head: toward a mental prosthesis utilizing event-related brain potentials. Electroencephalogr. Clin. Neurophysiol., 70: 512–523.

Hayashi, H. and Kato, S. (1989) Total manifestations of amyotrophic lateral sclerosis: ALS in the totally locked-in state. J. Neurol. Sci., 93: 19–35.

Hinterberger, T., Neumann, N., Pham, M., Kübler, A., Grether, A., Hofmayer, N., Wilhelm, B., Flor, H. and Birbaumer, N. (2004) A multimodal brain-based feedback and communication system. Exp. Brain Res., 154: 521–526.

Huggins, J.E., Levine, S.P., BeMent, S.L., Kushwaha, R.K., Schuh, L.A., Passaro, E.A., Rohde, M.M., Ross, D.A., Elisevich, K.V. and Smith, B.J. (1999) Detection of event-related potentials for development of a direct brain interface. J. Clin. Neurophysiol., 16: 448–455.

Jerusalem, F., Pohl, C., Karitzky, J. and Ries, F. (1996) Als. Neurology, 47: S218–S220.

Karitzky, J. and Ludolph, A.C. (2001) Imaging and neurochemical markers for diagnosis and disease progression in ALS. J. Neurol. Sci., 191: 35–41.

Kennedy, P.R. and Bakay, R.A. (1998) Restoration of neural output from a paralyzed patient by a direct brain connection. Neuroreport, 9: 1707–1711.

Kennedy, P.R., Bakay, R.A., Moore, M.M., Adams, K. and Goldwaithe, J. (2000) Direct control of a computer from the human central nervous system. IEEE Trans. Neural Syst. Rehabil. Eng., 8: 198–202.

Kotchoubey, B., Lang, S., Bostanov, V. and Birbaumer, N. (2002) Is there a mind? Electrophysiology of unconscious patients. News Physiol. Sci., 17: 38–42.

Kotchoubey, B., Schleichert, H., Lutzenberger, W., Anokhin, A.P. and Birbaumer, N. (1996) Self-regulation of interhemispheric asymmetry in humans. Neurosci. Lett., 214: 91–94.

Kübler, A., Kotchoubey, B., Hinterberger, T., Ghanayim, N., Perelmouter, J., Schauer, M., Fritsch, C., Taub, E. and Birbaumer, N. (1999) The thought translation device: a neurophysiological approach to communication in total motor paralysis. Exp. Brain Res., 124: 223–232.

Kübler, A., Kotchoubey, B., Kaiser, J., Wolpaw, J.R. and Birbaumer, N. (2001a) Brain–computer communication: unlocking the locked in. Psychol. Bull., 127: 358–375.

Kübler, A., Neumann, N., Kaiser, J., Kotchoubey, B., Hinterberger, T. and Birbaumer, N.P. (2001b) Brain–computer communication: self-regulation of slow cortical potentials for verbal communication. Arch. Phys. Med. Rehabil., 82: 1533–1539.

Kübler, A., Nijboer, F., Mellinger, J., Vaughan, T.M., Pawelzik, H., Schalk, G., McFarland, D.J., Birbaumer, N. and Wolpaw, J.R. (in press) Patients with ALS can learn to operate a sensorimotor-rhythm based brain–computer interface (BCI). *Neurology*.

Laureys, S., Owen, A.M. and Schiff, N.D. (2004) Brain function in coma, vegetative state, and related disorders. Lancet Neurol., 3: 537–546.

Leigh, P.N., Swash, M., Iwasaki, Y., Ludolph, A., Meininger, V., Miller, R.G., Mitsumoto, H., Shaw, P., Tashiro, K. and Van Den Berg, L. (2004) Amyotrophic lateral sclerosis: a consensus viewpoint on designing and implementing a clinical trial. Amyotroph. Lateral Scler. Other Motor Neuron Disord., 5: 84–98.

Leon-Carrion, J., van Eeckhout, P., Dominguez-Morales, M.d.R. and Perez-Santamaria, F.J. (2002) The locked-in syndrome: a syndrome looking for a therapy. Brain Injury, 16: 571–582.

Leuthardt, E., Schalk, G., Wolpaw, J.R., Ojemann, J.G. and Moran, D.W. (2004) Brain–computer interface using electrocorticographic signals in humans. J. Neural. Eng., 1: 63–71.

Levine, S.P., Huggins, J.E., BeMent, S.L., Kushwaha, R.K., Schuh, L.A., Passaro, E.A., Rohde, M.M. and Ross, D.A. (1999) Identification of electrocorticogram patterns as the basis for a direct brain interface. J. Clin. Neurophysiol., 16: 439–447.

Logothetis, N.K. (2002) The neural basis of the blood-oxygen-level-dependent functional magnetic resonance imaging signal. Philos. Trans. R. Soc. Lond. B. Biol. Sci., 357: 1003–1037.

Mason, S.G. and Birch, G.E. (2000) A brain-controlled switch for asynchronous control applications. IEEE Trans. Biomed. Eng., 47: 1297–1307.

Mason, S.G. and Birch, G.E. (2003) A general framework for brain–computer interface design. IEEE Trans. Neural Syst. Rehabil. Eng., 11: 70–85.

McFarland, D.J., Sarnacki, W.A. and Wolpaw, J.R. (2003) Brain–computer interface (BCI) operation: optimizing information transfer rates. Biol. Psychol., 63: 237–251.

Mellinger, J., Hinterberger, T., Bensch, M., Schröder, M. and Birbaumer, N. (2003). Surfing the web with electrical brain signals: the brain web surfer (BWS) for the completely paralysed. Paper presented at the ISPRM, Prague.

Mellinger, J., Nijboer, F., Pawelzik, H., Schalk, G., McFarland, D.J., Vaughan, T.M., Wolpaw, J.R., Birbaumer, N. and Kübler, A. (2004) P300 for communication: evidence from patients with amyotrophic lateral sclerosis (ALS). Biomed. Tech. (Berl.), 49: 71–74.

Middendorf, M., McMillan, G., Calhoun, G. and Jones, K.S. (2000) Brain–computer interfaces based on the steady-state visual-evoked response. IEEE Trans. Neural Syst. Rehabil. Eng., 8: 211–214.

Millan, J.d.R. and Mourino, J. (2003) Asynchronous BCI and local neural classifiers: an overview of the Adaptive Brain Interface project. IEEE Trans. Neural Syst. Rehabil. Eng., 11: 159–161.

Millan, J.d.R., Renkens, F., Mourino, J. and Gerstner, W. (2004a) Brain-actuated interaction. Artifi. Intell., 159: 241–259.

Millan, J.d.R., Renkens, F., Mourino, J. and Gerstner, W. (2004b) Noninvasive brain-actuated control of a mobile robot by human EEG. IEEE Trans. Biomed. Eng., 51: 1026–1033.

Musallam, S., Corneil, B.D., Greger, B., Scherberger, H. and Andersen, R.A. (2004) Cognitive control signals for neural prosthetics. Science, 305: 258–262.

Neumann, N. and Kotchoubey, B. (2004) Assessment of cognitive functions in severely paralyzed and severely brain-damaged patients: neuropsychological and electrophysiological methods. Brain Res. Protoc, 14(1): 25–36.

Neumann, N. and Kübler, A. (2003) Training locked-in patients: a challenge for the use of brain–computer interfaces. IEEE Trans. Neural Syst. Rehabil. Eng., 11: 169–172.

Neumann, N., Kübler, A., Kaiser, J., Hinterberger, T. and Birbaumer, N. (2003) Conscious perception of brain states: mental strategies for brain–computer communication. Neuropsychologia, 41: 1028–1036.

Neuper, C., Müller, G.R., Kübler, A., Birbaumer, N. and Pfurtscheller, G. (2003) Clinical application of an EEG-based brain–computer interface: a case study in a patient with severe motor impairment. Clin. Neurophysiol., 114: 399–409.

Nicolelis, M.A. (2003) Brain–machine interfaces to restore motor function and probe neural circuits. Nat. Rev. Neurosci., 4: 417–422.

Niedermeyer, E. (1998) The normal EEG of the waking adult. In: Niedermeyer E. and Lopes da Silva F.H. (Eds.), Electroencephalography: Basic Principles, Clinical Applications and Related Fields (4th ed.). Williams and Wilkins, Baltimore.

Obermaier, B., Müller, G. and Pfurtscheller, G. (2001) 'Virtual keyboard' controlled by spontaneous EEG activity. In: Dorffner G., Bischof H. and Hornik K. (Eds.) ICANN 2001, Lecture Notes in Computer Science, Vol. 2130. Springer, Heidelberg, pp. 636–641.

Paninski, L., Fellows, M.R., Hatsopoulos, N.G. and Donoghue, J.P. (2004) Spatiotemporal tuning of motor cortical neurons for hand position and velocity. J. Neurophysiol., 91: 515–532.

Perelmouter, J., Kotchoubey, B., Kübler, A., Taub, E. and Birbaumer, N. (1999) Language support program for thought-translation-devices. Automedica, 18: 67–84.

Pfurtscheller, G. (1999) EEG event-related desynchronization (ERD) and event related synchronization (ERS). In: Niedermeyer E. and Lopes da Silva F.H. (Eds.), Electroencephalography: Basic Principles, Clinical Applications and Related Fields (4th ed.). Williams and Wilkins, Baltimore, pp. 958–967.

Pfurtscheller, G., Flotzinger, D., Pregenzer, M., Wolpaw, J.R. and McFarland, D. (1996) EEG-based brain computer interface. Medi. Progr. Technol., 21: 111–121.

Pfurtscheller, G., Guger, C., Müller, G., Krausz, G. and Neuper, C. (2000a) Brain oscillations control hand orthosis in a tetraplegic. Neurosci. Lett., 292: 211–214.

Pfurtscheller, G., Müller, G.R., Pfurtscheller, J., Gerner, H.J. and Rupp, R. (2003) 'Thought' — control of functional electrical stimulation to restore hand grasp in a patient with tetraplegia. Neurosci. Lett., 351: 33–36.

Pfurtscheller, G., Neuper, C., Guger, C., Harkam, W., Ramoser, H., Schlogl, A., Obermaier, B. and Pregenzer, M. (2000b) Current trends in Graz brain–computer interface (BCI) research. IEEE Trans. Neural Syst. Rehabil. Eng., 8: 216–219.

Posse, S., Fitzgerald, D., Gao, K., Habel, U., Rosenberg, D., Moore, G.J. and Schneider, F. (2003) Real-time fMRI of temporolimbic regions detects amygdala activation during single-trial self-induced sadness. Neuroimage, 18: 760–768.

Schalk, G., McFarland, D.J., Hinterberger, T., Birbaumer, N. and Wolpaw, J.R. (2004) BCI2000: a general-purpose brain–computer interface (BCI) system. IEEE Trans. Biomed. Eng., 51: 1034–1043.

Scherer, R., Graimann, B., Huggins, J.E., Levine, S.P. and Pfurtscheller, G. (2003) Frequency component selection for an ECoG-based brain–computer interface. Biomed. Tech. (Berl.), 48: 31–36.

Scherer, R., Müller, G.R., Neuper, C., Graimann, B. and Pfurtscheller, G. (2004) An asynchronously controlled EEG-based virtual keyboard: improvement of the spelling rate. IEEE Trans. Biomed. Eng., 51: 979–984.

Schwartz, A.B. (2004). Direct cortical control of 3D neuroprosthetic devices. Paper presented at the 4th Forum of European Neuroscience, Lisbon, Portugal.

Serruya, M.D., Hatsopoulos, N.G., Paninski, L., Fellows, M.R. and Donoghue, J.P. (2002) Instant neural control of a movement signal. Nature, 416: 141–142.

Sutter, E.E. (1992) The brain response interface: communication through visually-induced electrical brain responses. J. Microcomput. Appl., 15: 31–45.

Taylor, D.M., Tillery, S.I. and Schwartz, A.B. (2002) Direct cortical control of 3D neuroprosthetic devices. Science, 296: 1829–1832.

Taylor, D.M., Tillery, S.I. and Schwartz, A.B. (2003) Information conveyed through brain-control: cursor versus robot. IEEE Trans. Neural Syst. Rehabil. Eng., 11: 195–199.

Vernon, D., Egner, T., Cooper, N., Compton, T., Neilands, C., Sheri, A. and Gruzelier, J. (2003) The effect of training distinct neurofeedback protocols on aspects of cognitive performance. Int. J. Psychophysiol., 47: 75–85.

Weiskopf, N., Mathiak, K., Bock, S.W., Scharnowski, F., Veit, R., Grodd, W., Goebel, R. and Birbaumer, N. (2004) Principles of a brain–computer interface (BCI) based on real-time functional magnetic resonance imaging (fMRI). IEEE Trans. Biomed. Eng., 51: 966–970.

Weiskopf, N., Scharnowski, F., Veit, R., Goebel, R., Birbaumer, N. and Mathiak, K. Self-regulation of local brain activity using real-time functional magnetic resonance imaging (fMRI) (in press). J. Physiol., Paris.

Weiskopf, N., Veit, R., Erb, M., Mathiak, K., Grodd, W., Goebel, R. and Birbaumer, N. (2003) Physiological self-regulation of regional brain activity using real-time functional magnetic resonance imaging (fMRI): methodology and exemplary data. Neuroimage, 19: 577–586.

Wolpaw, J.R., Birbaumer, N., McFarland, D.J., Pfurtscheller, G. and Vaughan, T.M. (2002) Brain–computer interfaces for communication and control. Clin. Neurophysiol., 113: 767–791.

Wolpaw, J.R. and McFarland, D.J. (2004) Control of a two-dimensional movement signal by a noninvasive brain-computer interface in humans. Proc Natl. Acad. Sci., USA, 101(51): 17849–17854.

Wolpaw, J.R., McFarland, D.J., Neat, G.W. and Forneris, C.A. (1991) An EEG-based brain–computer interface for cursor control. Electroencephalogr. Clin. Neurophysiol., 78: 252–259.

Wolpaw, J.R., McFarland, D.J., Vaughan, T.M. and Schalk, G. (2003) The Wadsworth center brain–computer interface (BCI) research and development program. IEEE Trans. Neural Syst. Rehabil. Eng., 11: 204–207.

Yoo, S.S., Fairneny, T., Chen, N.K., Choo, S.E., Panych, L.P., Park, H., Lee, S.Y. and Jolesz, F.A. (2004) Brain–computer interface using fMRI: spatial navigation by thoughts. Neuroreport, 15: 1591–1595.

Yoo, S.S. and Jolesz, F.A. (2002) Functional MRI for neurofeedback: feasibility study on a hand motor task. Neuroreport, 13: 1377–1381.

S. Laureys (Ed.)
Progress in Brain Research, Vol. 150
ISSN 0079-6123

CHAPTER 36

Neural plasticity and recovery of function

Nick S. Ward*

Wellcome Department of Imaging Neuroscience, Institute of Neurology, University College London, 12 Queen Square, London WC1N 3BG, UK

Abstract: Recovery of the function after stroke is a consequence of many factors including resolution of oedema and survival of the ischaemic penumbra. In addition there is a growing interest in the role of central nervous system (CNS) reorganization. Much of the evidence supporting this comes from animal models of focal brain injury, but non-invasive techniques such as functional magnetic resonance imaging, transcranial magnetic stimulation, electroencephalography and magnetoencephalography now allow the study of the working human brain. Using these techniques it is apparent that the motor system of the brain adapts to damage in a way that attempts to preserve motor function. This has been demonstrated after stroke, as part of the ageing process, and even after disruption of normal motor cortex with repetitive transcranial magnetic stimulation. The result of this reorganization is a new functional architecture, one which will vary from patient to patient depending on the anatomy of the damage, the biological age of the patient and lastly the chronicity of the lesion. The success of any given therapeutic intervention will depend on how well it interacts with this new functional architecture. Thus it is crucial that the study of novel therapeutic strategies for treating motor impairment after stroke take account of this. This review maps out the attempts to describe functionally relevant adaptive changes in the human brain following focal damage. A greater understanding of how these changes are related to the recovery process will allow not only the development of novel therapeutic techniques that are based on neurobiological principles and designed to minimize impairment in patients suffering from stroke, but also to target these therapies at the appropriate patients.

Introduction

Neurological damage accounts for nearly half of those requiring daily help for severe disabilities, and for the majority of those with complex disabilities (Wade and Hewer, 1987). The management of these patients relies on a multidisciplinary rehabilitative approach that attempts to facilitate restoration after, and adaptation to, impairments. There is little doubt that this overall approach is effective. For example, stroke unit care reduces death, institutionalized care and dependency at one year (Stroke Unit Trialists' Collaboration, 2000) and also in the long-term (Indredavik et al., 1999). However, disability secondary to residual impairment all too often remains. The study of whether, and particularly how, treatments can reduce impairments is thus crucial, and it is in this field that the clinical neurosciences can make a unique contribution.

*Corresponding author. Tel.: +44 (0) 207 833 7486;
Fax: +44 (0) 207 813 1420; E-mail: n.ward@fil.ion.ucl.ac.uk

Plasticity in the damaged brain

Despite the limited capacity of the central nervous system (CNS) to regenerate there is evidence that improvements in specific impairments do occur. Experiments in both animals and humans show that some regions in the normal adult brain, particularly

DOI: 10.1016/S0079-6123(05)50036-0

the cortex, have the capacity to change structure and consequently function in response to environmental change (Schallert et al., 2000). This reorganization at the systems level is often referred to as plasticity. In addition, work in animal models has clearly demonstrated that focal damage in the adult brain can lead to a number of molecular and cellular changes in both perilesional and distant brain regions, normally seen only in the developing brain (Cramer and Chopp, 2000). This suggests that the damaged brain is more able to change structure and function in response to afferent signals. In other words it is more "plastic". It is hypothesized that similar injury-induced changes occur in the human brain, and that targeted therapy interacts with these changes and thereby provides a means of reducing impairment in patients with focal brain damage via activity-dependent plastic change. These advances are clearly of great interest to clinicians and scientists alike. This review will concentrate on the current level of understanding of the mechanisms of recovery from motor impairment in human patients. Brain research in humans is largely performed at the systems level. The tools available include techniques such as functional magnetic resonance imaging (fMRI) and positron emission tomography (PET), which allow measurement of task-related brain activation with excellent spatial resolution; transcranial magnetic stimulation (TMS), a safe, non-invasive way to excite or inhibit the human cortex with high temporal resolution; and magneto- and electroencephalography (MEG and EEG) with even greater temporal resolution.

Cerebral reorganization in chronic stroke patients

What is the evidence that reorganization in the human brain mediates recovery from stroke? Until recently there has in fact been very little. Early functional imaging studies were performed in recovered chronic stroke patients. Compared to normal subjects, these patients demonstrated relative overactivations in a number of motor-related brain regions when performing a motor task compared to control subjects. In particular, overactivations were seen in dorsolateral premotor cortex (PMd), ventrolateral premotor cortex (PMv), supplementary motor area (SMA), cingulate motor areas (CMA), parietal cortex, insula cortex and cerebellar hemispheres (Chollet et al., 1991; Weiller et al., 1992, 1993; Cramer et al., 1997; Seitz et al., 1998). In addition, brain activation patterns were more likely to be bilateral than those in controls. Thus it was suggested that recruitment of these brain regions, particularly those in the unaffected hemisphere, might be responsible for recovery.

However, in order to make inferences about the mechanisms underlying functional recovery, it is necessary to study patients with different degrees of recovery. A more recent study has looked at chronic subcortical stroke patients with a range of late outcomes. Patients were asked to make repeated isometric hand grips. This allowed patients without return of fractionated finger movements to be studied, as hand grip can return relatively quickly. However, patients with no movement of the hand could not be studied with this paradigm. There were no differences between the activation maps of control subjects and patients without residual impairment. However, patients with more marked impairment showed relative overactivations bilaterally in a number of primary and secondary motor areas (Ward et al., 2003a). More formally, a negative correlation was found between the size of brain activation and outcome in a number of these brain regions, almost all of them part of the normal motor network.

This does not initially seem to support the notion that recruitment of activity in these regions facilitates recovery. However, the capacity for motor recovery after stroke is strongly influenced by the integrity of the fast direct motor output pathway from primary motor cortex (M1) to spinal cord motor neurons (Heald et al., 1993; Cruz et al., 1999; Pennisi et al., 1999). Thus it is likely that those patients with poorer outcome have greater disruption to this cortico-motoneuronal (CMN) pathway. The work of Strick (1988) suggests that in primates at least, the non-primary motor system is organized as a number of neural networks, or loops, involving premotor (both lateral and medial wall), parietal and subcortical regions (Strick, 1988). Each has its own direct projection to the spinal cord motor neurons, and also to the loops involving M1. The generation of an output to the musculature following disruption to the CMN pathway requires an increase in signals to spinal cord motor neurons via alternative pathways. The non-primary motor loops described by Strick provide the ideal substrate. The implication is that damage in one of these networks could be partially compensated for by the activity in another. However, the negative correlation can be explained by the fact that these secondary motor systems are unlikely to completely substitute for projections from M1 as the projections to spinal cord are less numerous and less efficient at exciting spinal cord motor neurons (Maier et al., 2002).

In attempting to reconcile these results with those from early studies it seems likely that patients in many previous studies may not have been fully recovered. Some patients found the finger tapping task often used more effortful to perform (Chollet et al., 1991) and results in those patients were similar to results in patients with residual motor deficit in a later study (Ward et al., 2003a).

The question of the functional relevance of these regions remains. Increased activity in ipsilesional (contralateral to the affected hand) PMd has been associated with therapy-induced improvement in both upper limb (Johansen-Berg et al., 2002a) and gait (Miyai et al., 2003) function. There is also evidence to suggest that disruption of ipsilesional PMd (Fridman et al., 2004) and contralesional (ipsilateral to the affected hand) PMd (Johansen-Berg et al., 2002b) using TMS impairs performance of a simple motor task in chronic stroke patients but not controls, suggesting that these regions are functionally useful. However, TMS to ipsilesional PMd is more disruptive in those patients with less impairment (Fridman et al., 2004) and TMS to contralesional PMd is more disruptive in patients with greater impairment (Johansen-Berg et al., 2002b).

Thus evidence is emerging to support the functional relevance of secondary motor area recruitment, but the degree to which a region is useful is likely to depend on a number of factors including the extent of damage to the normal motor network. As well as being recruited after stroke, there is evidence that some of these regions take on different functions. For example, ipsilesional PMd takes on an executive motor role, such that task-related blood oxygen level dependent (BOLD) signal increases linearly as a function of hand grip force in chronic stroke patients with significant impairment, but not in good recoverers or in controls (Ward et al., 2003a). Thus it is unlikely that the response to focal injury involves the simple substitution of one cortical region for another, as nodes within the remaining network may take on new roles.

Functional imaging studies have also reported shifts in ipsilesional M1 hand representation (Weiller et al., 1993). Ventral and posterior shifts tend to be more common, although the size of this shift does not appear to be related to the outcome (Calautti et al., 2003). However, a negative correlation between size of brain activation and outcome has been reported in a part of ipsilesional M1 ventral to the peak hand representation as well as in posterior M1 (Ward et al., 2003a). The former region is situated within brodmann area (BA) 4a, and the latter within BA 4p. Overall, these data suggest that there is a certain amount of remapping of hand representation in ipsilesional M1, even though undamaged, which may be the result of functionally relevant changes in both its afferent and efferent connections. In other words these shifts may facilitate access to undamaged portions of the CMN pathway, or alternatively may reflect changes in connections between M1 and secondary motor areas. For example, changes in somatotopic representation within non-primary motor regions might result in stronger connections with different (e.g., more ventral or posterior) regions of M1, which might again facilitate access to undamaged portions of the CMN pathway.

Once again we need to ask whether these changes are functionally relevant. The negative correlation between recruitment of parts of ipsilesional M1 and outcome, as with secondary motor areas, does not prove a causative link. Disruption of ipsilesional M1 function by TMS just prior to the motor response in a reaction time task impairs performance in chronic stroke patients, suggesting an intact ipsilesional M1 is functionally relevant (Werhahn et al., 2003). However, this effect is less marked in those with poorer functional recovery in whom ipsilesional M1 is probably less important than secondary motor areas.

A negative correlation between size of activation seen with fMRI and outcome for contralesional posterior M1 (BA4p) but not anterior M1 (BA4a) has also been reported (Ward et al., 2003a). Other studies have suggested that contralesional M1 is activated in some stroke patients (Weiller et al., 1993). However a direct role for contralesional M1 in recovery remains controversial. Anatomical studies suggest that both direct (CMN) and indirect (cortico-reticulospinal) pathways from ipsilateral M1 end in projections only to axial and proximal stabilizing muscles rather than hand muscles (Brinkman and Kuypers 1973; Carr et al., 1994). In normal subjects however, repetitive TMS (rTMS) to M1 results in errors in both complex and simple motor tasks using the ipsilateral hand (Chen et al., 1997), suggesting that ipsilateral M1 may play a role in planning and organization of normal hand movement. Furthermore, ipsilateral M1 seems to compensate for rTMS induced disruption to contralateral M1 (Strens et al., 2003).

Structural changes such as dendritic growth undoubtedly occur in contralesional motor regions in animal models, likely to be the result of an interaction between effects of the lesion and an increased reliance on the unaffected limb (Jones and Schallert, 1994). It has been suggested that any contralesional hemisphere changes have minimal effect on recovery of *skilled* movement in the affected limb (Whishaw and Metz, 2002; Schmidlin et al., 2004), but greater sensorimotor impairments in the affected limb were found when the dendritic growth in contralesional hemisphere was blocked (Jones and Schallert, 1994). There is also a greater capacity to learn with the unaffected forelimb in those animals with lesions compared to sham operated animals (Luke et al., 2004), suggesting that dendritic growth confers an increased potential for plastic change in the unaffected contralesional cortex. Further work needs to be done, but it appears that complex interactions between the lesion and subsequent changes in behavior are important. Clearly it is important to define what is meant by recovery, as behavioral improvement may be seen in an affected limb, without a return of highly skilled motor characteristics.

In studies involving human stroke patients TMS that disrupted local cortical function have failed to find any functional significance of increased contralesional M1 activation after stroke (Johansen-Berg et al., 2002b; Werhahn et al., 2003). More recently, it has been suggested that in chronic stroke patients contralesional M1 exerts an abnormally high degree of interhemispheric inhibitory drive toward ipsilesional M1 while generating a voluntary movement of the affected hand (Murase et al., 2004). Thus it is even possible that activity in contralesional M1 impairs rather than facilitates motor performance in some patients. Nevertheless TMS to the motor *hot spot* for hand muscles (corresponding to M1) may affect predominantly BA 4a, rather than BA 4p. So it remains plausible that in patients with significant deficit in whom a dependency on alternative motor projections has developed, other parts of contralesional M1 play an indirect role in recovery.

It is interesting to note that the relationship between motor recruitment pattern and degree of recovery (at the time of scanning) holds true at 10–14 days post stroke as well as in the chronic phase (Ward et al., 2004), suggesting that motor system adaptation occurs rapidly, and secondary motor areas are available to participate in the generation of a motor task very early after stroke.

The relationship between sensory and motor function after stroke

Many aspects of motor cortex function rely on the input from sensory cortex, including the acquisition of new motor skills (Pavlides et al., 1993), and clearly sensorimotor integration is crucial to optimal recovery after hemiparesis. Increases in the strength of MEG signal in response to median nerve stimulation were seen in M1 and S1, and corresponded to improvements in motor and sensory function respectively (Huang et al., 2004). Furthermore the absence of sensorimotor response to sensory stimulation as measured by MEG was associated with poorer outcome (Gallien et al., 2003). Similarly there seems to be a correlation between outcome at 5–6 weeks post stroke and signal change in SMA and ipsilesional inferior parietal cortex at 1–2 weeks (Loubinoux et al., 2003). A physiotherapist might consider this sensory input to be something that might therapeutically drive plastic change in potentially hyperexcitable cortex. It is interesting to speculate that such inputs will facilitate motor recovery only to the degree that the cerebral networks with which they interact are intact and can influence motor output pathways. In this way it is possible to see how tools such as fMRI and MEG might be useful in predicting a likely response to a given intervention.

Adaptation within the motor system

Are these changes particular to stroke, or do they represent a motor network that can adapt to alterations in its connections in a way that attempts to maintain performance? Rapid adaptability within the motor system can be demonstrated in other ways. For example, 30 min of 1 Hz rTMS to left M1 reduced the responsiveness of the stimulated region to inputs from premotor and medial motor regions. In addition, increased task-related recruitment of PMd of the nonstimulated hemisphere was noted, and an increased coupling between an inferomedial portion of left M1 (BA 4p) and (1) PMd, and (2) SMA was observed. As rTMS did not lead to impairments in motor performance, the changes noted were considered to be compensatory (Lee et al., 2003).

Adaptability in the motor system is also seen with normal ageing. Older subjects who maintain motor performance appear to recruit more of the motor system than younger subjects. These recruited regions included parts of deep central sulcus contralateral to the moving hand, close to BA 4p, as well contralateral caudal cingulate sulcus and ipsilateral PMd (Ward and Frackowiak, 2003). These results raise the question of whether the capacity for plastic change in the brain may be finite. After injury-induced cerebral reorganization in animal models, the capacity for subsequent adaptive change is reduced (Kolb et al., 1998). Thus age-related changes may in turn limit the capacity for further reorganization after injury. This clearly has implications for what we can expect from therapeutic techniques designed to promote cerebral reorganization after stroke in older subjects.

The motor system has the capacity for functional reorganization, and this reorganization can appear quickly. In the chronic phase of stroke at least, this results in a *new functional architecture*. The exact configuration of this new functional architecture is likely to depend on a number of factors, not least the exact anatomy of the damage, including not only which gray and white matter structures are damaged, but which hemisphere is affected (Zemke et al., 2003). In addition the biological age of the subject and the premorbid state of their brain will also influence the potential for plastic change, either lesion-induced or therapeutically driven.

The evolution of cerebral reorganization after stroke

How does this reorganized state evolve? Two early longitudinal studies with early and late time points demonstrated initial task related overactivations in motor related brain regions, followed by a reduction over time in patients said to recover fully (Marshall et al., 2000; Calautti et al., 2001). Others described similar changes over time, but no consistent relationship with recovery scores (Feydy et al., 2002). More recently, a longitudinal fMRI study of patients with infarcts not involving M1 found an initial overactivation in many primary and non-primary motor regions that was more extensive in those with greatest clinical deficit (Ward et al., 2003b). The subsequent longitudinal changes were predictably different in each patient but there were changes common to all. After the early overactivation, a negative correlation between size of activation and the recovery score at each session was observed in all patients in ipsilesional M1 (ventral BA4a, and BA4p), contralesional M1 (BA4p), as well as in anterior and posterior PMd bilaterally, contralesional PMv, and ipsilesional SMA, prefrontal cortex (superior frontal sulcus) and caudal cingulate sulcus. There were no consistent positive correlations seen for the group.

Driving functional reorganization

After damage to the motor system a new functional architecture emerges, and its configuration depends not only on the anatomy of the lesion and biological age of the subject, but also the time since the lesion. Cross-sectional and longitudinal correlation analyses find similarities in relationship between motor activation patterns and the degree of recovery. It seems clear that the motor system can reorganize, and that there are probably rules to this reorganization as already discussed. The question of what drives these changes however is unanswered.

Motor learning as a substrate for recovery?

It seems reasonable to suggest that in the face of motor impairment, a highly preserved cerebral system such as that subserving motor learning, will be engaged in an attempt to minimize the impairment. One model of motor learning suggests that during early motor learning movements are very attentionally demanding, and are encoded in terms of spatial coordinates within a frontoparietal network (Hikosaka et al., 2002). Once learning has occurred and the task has become automatic, movement is encoded in terms of a kinematic system of joints, muscles, limb trajectories etc., by a network involving primary motor cortex. Interaction between these parallel systems, and the transfer of reliance from one to the other, occurs in cerebellum and basal ganglia, but also via intracortical connections involving particularly premotor cortex and pre-SMA (Hikosaka et al., 2002).

The need to re-learn simple motor tasks after stroke is likely to engage these motor learning mechanisms, but the degree to which this can occur will depend on whether the necessary neural structures are intact. The patients in the longitudinal study had subcortical damage, and it is plausible that the longitudinal recovery related changes described are in part similar to those seen when normal subjects learn a motor sequence.

An important aspect of any motor learning model is the generation of an error signal, and it can be seen that attempted movements by hemiparetic patients will result in significant discrepancies between predicted and actual performance. The role of error signal generation in the damaged motor system is clearly of interest, particularly as it may diminish with chronicity of impairment (Ward et al., 2004).

Does the time since stroke matter?

Thus the "neural machinery" of motor learning may be one substrate of motor recovery after stroke. This principle can be applied to patients in the early or late stages after stroke. There are other factors besides the anatomy of the damage that are important when considering how the motor system might reorganize. It is likely that the time that has passed since the lesion plays a very important role in determining the role that a number of structures play in the recovery process. Evidence from animal models suggests that the lesioned brain has an increased capacity for plastic change, but only in the early phase after damage. For example, widespread areas of cortical hyperexcitability appear immediately after cerebral infarction in animal brains, changes which subside over subsequent months (Buchkremer-Ratzmann et al., 1996). These changes occur in regions structurally connected to the lesion in both hemispheres as a consequence of down-regulation of the $\alpha 1$ gamma-aminobutyric acid (GABA), a receptor subunit and a decrease in GABAergic inhibition (Neumann-Haefelin et al., 1998). $GABA_A$ receptor blockade, also resulting in reduced GABAergic inhibition, facilitates long-term potentiation (LTP) in primary motor cortex (Castro-Alamancos et al., 1995). Furthermore, it is easier to induce LTP in hyperexcitable cortex (Hagemann et al., 1998). Thus it seems that conditions favor the induction of LTP for at least some months after stroke. This finding is clearly of great interest to the clinician as LTP has long been considered a substrate of learning.

In humans, acute limb deafferentation leads to reduced levels of GABA within minutes (Levy et al., 2002), and it is tempting to think that the same thing may happen in areas of partially disconnected cortex after stroke. There is evidence of hyperexcitability in the contralesional motor cortex within 1 month of both cortical and subcortical stroke in humans with at least moderate recovery (Bütefisch et al., 2003). This hyperexcitability appears to tail off after 3–4 months (Shimizu et al., 2002; Delvaux et al., 2003) in keeping with the data from animals (Witte, 1998). Others have used paired pulse TMS in well-recovered stroke patients 20–42 days post stroke to demonstrate reduced intracortical inhibition in the ipsilesional, although not the contralesional, hemisphere. Inconsistencies across studies are probably due to differences in patient characteristics, particularly anatomical site of the lesions, chronicity of stroke, and degree of recovery at the time of study.

Lesion-induced cortical hyperexcitability and the underlying neurochemical changes are considered to be a potential substrate for activity dependent changes at synaptic level, which may in turn facilitate the systems-level cerebral reorganization thought to underlie reduction of impairment after stroke. However, the window of opportunity, if therapeutically useful, may last only for a limited time.

Increasing brain plasticity in chronic stroke patients

There has been recent interest in the theoretical possibility of increasing the potential for activity-driven functional reorganization in the chronic phase after stroke, once the early changes such as hyperexcitability have disappeared. This might allow therapeutic input, for example, targeted physical therapy, to have a greater effect. There is interest in the role of pharmacological treatment just prior to a physiotherapy session. Agents such as amphetamine (Martinsson et al., 2003) and L-dopa (Scheidtmann et al., 2001) have been used with promising results. In addition the manipulation of cortical hyperexcitability by rTMS as a means of "conditioning" the brain to be more responsive during therapy is an example of an interesting theoretically driven approach to the treatment of impairment (Uy et al., 2003; Bütefisch et al., 2004). Small-scale proof-of-principle studies suggest that these approaches are worth pursuing.

Conclusions

The motor system of the brain adapts to damage in a way that tries to preserve motor function. This has been demonstrated after stroke, as part of the ageing process, and even after rTMS to M1. The resulting reorganized state represents a new functional architecture that will vary from patient to patient depending on the site of the lesion, and the patient's age. The success of any given intervention will depend on how well it interacts with this new functional architecture, and so it is crucial that the study of novel therapeutic strategies for treating motor impairment after stroke take account of this. Furthermore, the chronicity of the stroke also influences the way the post-lesional motor system operates, and how it responds to environmental change or therapeutic input. Early after stroke a number of changes occur at the molecular, cellular and systems level which facilitate activity dependant functional reorganization. In the chronic stage, clinicians and scientists are developing novel strategies designed to increase the damaged brain's capacity to respond to therapeutic inputs. Although it is tempting to push on with trying out these therapies in all types of stroke patients, it is crucial that we continue in our efforts toward understanding the dynamic relationship between structure and function in the damaged brain, otherwise attempts at therapeutic interventions are likely to founder.

Acknowledgments

The author is supported by The Wellcome Trust, UK.

References

Brinkman, J. and Kuypers, H.G. (1973) Cerebral control of contralateral and ipsilateral arm, hand and finger movements in the split-brain rhesus monkey. Brain, 96: 653–674.

Buchkremer-Ratzmann, I., August, M., Hagemann, G. and Witte, O.W. (1996) Electrophysiological transcortical diaschisis after cortical photothrombosis in rat brain. Stroke, 27: 1105–1109.

Bütefisch, C.M., Netz, J., Wessling, M., Seitz, R.J. and Homberg, V. (2003) Remote changes in cortical excitability after stroke. Brain, 126: 470–481.

Bütefisch, C.M., Khurana, V., Kopylev, L. and Cohen, L.G. (2004) Enhancing encoding of a motor memory in the primary motor cortex by cortical stimulation. J. Neurophysiol., 91: 2110–2116.

Calautti, C., Leroy, F., Guincestre, J.Y. and Baron, J.C. (2001) Dynamics of motor network overactivation after striatocapsular stroke: a longitudinal PET study using a fixed-performance paradigm. Stroke, 32: 2534–2542.

Calautti, C., Leroy, F., Guincestre, J.Y. and Baron, J.C. (2003) Displacement of primary sensorimotor cortex activation after subcortical stroke: a longitudinal PET study with clinical correlation. Neuroimage, 19: 1650–1654.

Carr, L.J., Harrison, L.M. and Stephens, J.A. (1994) Evidence for bilateral innervation of certain homologous motoneurone pools in man. J. Physiol., 475: 217–227.

Castro-Alamancos, M.A., Donoghue, J.P. and Connors, B.W. (1995) Different forms of synaptic plasticity in somatosensory and motor areas of the neocortex. J. Neurosci., 15: 5324–5333.

Chen, R., Gerloff, C., Hallett, M. and Cohen, L.G. (1997) Involvement of the ipsilateral motor cortex in finger movements of different complexities. Ann. Neurol., 41: 247–254.

Chollet, F., DiPiero, V., Wise, R.J., Brooks, D.J., Dolan, R.J. and Frackowiak, R.S. (1991) The functional anatomy of motor recovery after stroke in humans: a study with positron emission tomography. Ann. Neurol., 29: 63–71.

Cramer, S.C., Nelles, G., Benson, R.R., Kaplan, J.D., Parker, R.A., Kwong, K.K., Kennedy, D.N., Finklestein, S.P. and Rosen, B.R. (1997) A functional MRI study of subjects recovered from hemiparetic stroke. Stroke, 28: 2518–2527.

Cramer, S.C. and Chopp, M. (2000) Recovery recapitulates ontogeny. Trends Neurosci., 23: 265–271.

Cruz, M.A., Tejada, J. and Diez, T.E. (1999) Motor hand recovery after stroke. Prognostic yield of early transcranial magnetic stimulation. Electromyogr. Clin. Neurophysiol., 39: 405–410.

Delvaux, V., Alagona, G., Gerard, P., De Pasqua, V., Pennisi, G. and de Noordhout, A.M. (2003) Post-stroke reorganization of hand motor area: a 1-year prospective follow-up with focal transcranial magnetic stimulation. Clin. Neurophysiol., 114: 1217–1225.

Feydy, A., Carlier, R., Roby-Brami, A., Bussel, B., Cazalis, F., Pierot, L., Burnod, Y. and Maier, M.A. (2002) Longitudinal study of motor recovery after stroke: recruitment and focusing of brain activation. Stroke, 33: 1610–1617.

Fridman, E.A., Hanakawa, T., Chung, M., Hummel, F., Leiguarda, R.C. and Cohen, L.G. (2004) Reorganization of the human ipsilesional premotor cortex after stroke. Brain, 127: 747–758.

Gallien, P., Aghulon, C., Durufle, A., Petrilli, S., de Crouy, A.C., Carsin, M. and Toulouse, P. (2003) Magnetoencephalography in stroke: a 1-year follow-up study. Eur. J. Neurol., 10: 373–382.

Hagemann, G., Redecker, C., Neumann-Haefelin, T., Freund, H.J. and Witte, O.W. (1998) Increased long-term potentiation in the surround of experimentally induced focal cortical infarction. Ann. Neurol., 44: 255–258.

Heald, A., Bates, D., Cartlidge, N.E., French, J.M. and Miller, S. (1993) Longitudinal study of central motor conduction time following stroke 2. Central motor conduction measured within 72 h after stroke as a predictor of functional outcome at 12 months. Brain, 116: 1371–1385.

Hikosaka, O., Nakamura, K., Sakai, K. and Nakahara, H. (2002) Central mechanisms of motor skill learning. Curr. Opin. Neurobiol., 12: 217–222.

Huang, M., Davis, L.E., Aine, C., Aine, C., Weisend, M., Harrington, D., Christner, R., Stephen, J., Edgar, J.C., Herman, M., Meyer, J., Paulson, K., Martin, K. and Lee, R.R. (2004) MEG response to median nerve stimulation correlates with recovery of sensory and motor function after stroke. Clin. Neurophysiol., 115: 820–833.

Indredavik, B., Bakke, F., Slordahl, S.A., Rokseth, R. and Haheim, L.L. (1999) Stroke unit treatment. 10-year follow-up. Stroke, 30: 1524–1527.

Johansen-Berg, H., Dawes, H., Guy, C., Smith, S.M., Wade, D.T. and Matthews, P.M. (2002a) Correlation between motor improvements and altered fMRI activity after rehabilitative therapy. Brain, 125: 2731–2742.

Johansen-Berg, H., Rushworth, M.F., Bogdanovic, M.D., Kischka, U., Wimalaratna, S. and Matthews, P.M. (2002b) The role of ipsilateral premotor cortex in hand movement after stroke. P. Natl. Acad. Sci. USA,, 99: 14518–14523.

Jones, T.A. and Schallert, T. (1994) Use-dependent growth of pyramidal neurons after neocortical damage. J. Neurosci., 14: 2140–2152.

Kolb, B., Forgie, M., Gibb, R., Gorny, G. and Rowntree, S. (1998) Age, experience and the changing brain. Neurosci. Biobehav. Rev., 22: 143–159.

Lee, L., Siebner, H.R., Rowe, J.B., Rizzo, V., Rothwell, J.C., Frackowiak, R.S. and Friston, K.J. (2003) Acute remapping within the motor system induced by low-frequency repetitive transcranial magnetic stimulation. J. Neurosci., 23: 5308–5318.

Levy, L.M., Ziemann, U., Chen, R. and Cohen, L.G. (2002) Rapid modulation of GABA in sensorimotor cortex induced by acute deafferentation. Ann. Neurol., 52: 755–761.

Loubinoux, I., Carel, C., Pariente, J., Dechaumont, S., Albucher, J.F., Marque, P., Manelfe, C. and Chollet, F. (2003) Correlation between cerebral reorganization and motor recovery after subcortical infarcts. Neuroimage, 20: 2166–2180.

Luke, L.M., Allred, R.P. and Jones, T.A. (2004) Unilateral ischemic sensorimotor cortical damage induces contralesional synaptogenesis and enhances skilled reaching with the ipsilateral forelimb in adult male rats. Synapse, 54: 187–199.

Maier, M.A., Armand, J., Kirkwood, P.A., Yang, H.W., Davis, J.N. and Lemon, R.N. (2002) Differences in the corticospinal projection from primary motor cortex and supplementary motor area to macaque upper limb motoneurons: an anatomical and electrophysiological study. Cereb. Cortex, 12: 281–296.

Marshall, R.S., Perera, G.M., Lazar, R.M., Krakauer, J.W., Constantine, R.C. and DeLaPaz, R.L. (2000) Evolution of cortical activation during recovery from corticospinal tract infarction. Stroke, 31: 656–661.

Martinsson, L., Hardemark, H.G. and Wahlgren, N.G. (2003) Amphetamines for improving stroke recovery: a systematic cochrane review. Stroke, 34: 2766.

Miyai, I., Yagura, H., Hatakenaka, M., Oda, I., Konishi, I. and Kubota, K. (2003) Longitudinal optical imaging study for locomotor recovery after stroke. Stroke, 34: 2866–2870.

Murase, N., Duque, J., Mazzocchio, R. and Cohen, L.G. (2004) Influence of interhemispheric interactions on motor function in chronic stroke. Ann. Neurol., 55: 400–409.

Neumann-Haefelin, T., Staiger, J.F., Redecker, C., Zilles, K., Fritschy, J.M., Mohler, H. and Witte, O.W. (1998) Immunohistochemical evidence for dysregulation of the GABAergic system ipsilateral to photochemically induced cortical infarcts in rats. Neuroscience, 87: 871–879.

Pavlides, C., Miyashita, E. and Asanuma, H. (1993) Projection from the sensory to the motor cortex is important in learning motor skills in the monkey. J. Neurophysiol., 70: 733–741.

Pennisi, G., Rapisarda, G., Bella, R., Calabrese, V., Maertens, D.N. and Delwaide, P.J. (1999) Absence of response to early transcranial magnetic stimulation in ischemic stroke patients: prognostic value for hand motor recovery. Stroke, 30: 2666–2670.

Schallert, T., Leasure, J.L. and Kolb, B. (2000) Experience-associated structural events, subependymal cellular proliferative activity, and functional recovery after injury to the central nervous system. J. Cereb. Blood Flow Metab., 20: 1513–1528.

Scheidtmann, K., Fries, W., Muller, F. and Koenig, E. (2001) Effect of levodopa in combination with physiotherapy on functional motor recovery after stroke: a prospective, randomized, double-blind study. Lancet, 358: 787–790.

Schmidlin, E., Wannier, T., Bloch, J. and Rouiller, E.M. (2004) Progressive plastic changes in the hand representation of the primary motor cortex parallel incomplete recovery from a unilateral section of the corticospinal tract at cervical level in monkeys. Behav. Brain Res., 1017: 172–183.

Seitz, R.J., Hoflich, P., Binkofski, F., Tellmann, L., Herzog, H. and Freund, H.J. (1998) Role of the premotor cortex in recovery from middle cerebral artery infarction. Arch. Neurol., 55: 1081–1088.

Shimizu, T., Hosaki, A., Hino, T., Sato, M., Komori, T., Hirai, S. and Rossini, P.M. (2002) Motor cortical disinhibition in the unaffected hemisphere after unilateral cortical stroke. Brain, 125: 1896–1907.

Strens, L.H., Fogelson, N., Shanahan, P., Rothwell, J.C. and Brown, P. (2003) The ipsilateral human motor cortex can functionally compensate for acute contralateral motor cortex dysfunction. Curr. Biol., 13: 1201–1205.

Strick, P.L. (1988) Anatomical organization of multiple motor areas in the frontal lobe: implications for recovery of function. Adv. Neurol., 47: 293–312.

Stroke Unit Trialists' Collaboration. (2000) Organised inpatient (stroke unit) care for stroke (Cochrane Review). In: The Cochrane Library, Issue 2. Oxford: Update Software.

Uy, J., Ridding, M.C., Hillier, S., Thompson, P.D. and Miles, T.S. (2003) Does induction of plastic change in motor cortex improve leg function after stroke? Neurology, 61: 982–984.

Wade, D.T. and Hewer, R.L. (1987) Epidemiology of some neurological diseases with special reference to work load on the NHS. Int. Rehabil. Med., 8: 129–137.

Ward, N.S., Brown, M.M., Thompson, A.J. and Frackowiak, R.S. (2003a) Neural correlates of outcome after stroke: a cross-sectional fMRI study. Brain, 126: 1430–1448.

Ward, N.S., Brown, M.M., Thompson, A.J. and Frackowiak, R.S. (2003b) Neural correlates of motor recovery after stroke: a longitudinal fMRI study. Brain, 126: 2476–2496.

Ward, N.S., Brown, M.M., Thompson, A.J. and Frackowiak, R.S. (2004) The influence of time after stroke on brain activations during a motor task. Ann. Neurol., 55: 829–834.

Ward, N.S. and Frackowiak, R.S. (2003) Age-related changes in the neural correlates of motor performance. Brain, 126: 873–888.

Weiller, C., Chollet, F., Friston, K.J., Wise, R.J. and Frackowiak, R.S. (1992) Functional reorganization of the brain in recovery from striatocapsular infarction in man. Ann. Neurol., 31: 463–472.

Weiller, C., Ramsay, S.C., Wise, R.J., Friston, K.J. and Frackowiak, R.S. (1993) Individual patterns of functional reorganization in the human cerebral cortex after capsular infarction. Ann. Neurol., 33: 181–189.

Werhahn, K.J., Conforto, A.B., Kadom, N., Hallett, M. and Cohen, L.G. (2003) Contribution of the ipsilateral motor cortex to recovery after chronic stroke. Ann. Neurol., 54: 464–472.

Whishaw, I.Q. and Metz, G.A. (2002) Absence of impairments or recovery mediated by the uncrossed pyramidal tract in the rat versus enduring deficits produced by the crossed pyramidal tract. Behav. Brain Res., 134: 323–336.

Witte, O.W. (1998) Lesion-induced plasticity as a potential mechanism for recovery and rehabilitative training. Curr. Opin. Neurol., 11: 655–662.

Zemke, A.C., Heagerty, P.J., Lee, C. and Cramer, S.C. (2003) Motor cortex organization after stroke is related to side of stroke and level of recovery. Stroke, 34: e23–e28.

S. Laureys (Ed.)
Progress in Brain Research, Vol. 150
ISSN 0079-6123

CHAPTER 37

Thirty years of the vegetative state: clinical, ethical and legal problems

Bryan Jennett*

Institute of Neurological Sciences, University of Glasgow, Glasgow, Scotland, UK

Abstract: The vegetative state is the rarest form of disability in patients now frequently rescued from life-threatening severe brain damage by resuscitation and intensive care. Many doctors have never seen such cases, yet it provokes great interest among professionals and the public because of the paradox of a person who is awake yet not aware. The commonest cause is head injury and it is more common in countries with a high incidence of severe head injury. The most consistent brain damage is in the subcortical white matter of the cerebral hemispheres and in the thalami; although the cerebral cortex is often severely damaged, it may be relatively spared. Diagnosis depends on prolonged expert observation to determine that there is no evidence of awareness in spite of a wide range of reflex responses, some of which may involve cortical activity. Functional imaging confirms that there is some residual cortical function in many vegetative patients. Mistaken diagnosis is less likely since the recent definition of clinical criteria for the vegetative state and for the minimally conscious state. Many patients recover consciousness and even regain independence after a month in a vegetative state after head injury, but few do so after non-traumatic insult. The longer the state persists the less likely the recovery, and eventually permanence can be declared. Patients can survive for many years in a vegetative state. Many consider that indefinite survival in a vegetative state is of no benefit to the patient and that there is no moral or legal obligation to continue life-sustaining treatment, including artificial nutrition and hydration. Ethical issues include how to respect the autonomy of the legally incompetent patient, and uphold the right to refuse unwanted treatment. Many cases have been brought to court in several North American, Northern European and some other jurisdictions where it has been ruled that it is legally permissible to withdraw life-sustaining treatment once a patient is declared permanently vegetative, and such withdrawal seems likely to be what that person would want done.

Introduction

The strange and harrowing sight of a person who is awake but unaware, with no evidence of a working mind, provokes intense debate among scientists, health professionals, philosophers, ethicists and lawyers. When Professor Plum of New York and I coined the term persistent vegetative state (PVS) in our Lancet article on April Fool's Day 1972, we were neither the first to describe the condition nor to propose a name. But the time was right to capture interest because resuscitation and intensive care techniques were by then saving the lives of increasing numbers of severely brain-damaged patients – many of whom were left permanently disabled. Vegetative survival is the most severe and rarest form of disability in such survivors but its curious features have resulted in its attracting a degree of interest out of all proportion to its relative rarity. Internet references to "vegetative state" on Google amount to 135,000 entries (January 2005).

*Corresponding author. Tel./Fax: +44 141 3345148;
E-mail: bryan.jennett@ntlworld.com

DOI: 10.1016/S0079-6123(05)50037-2

Frequency of occurrence

It is in fact quite difficult to determine exactly how many vegetative patients there are in a community. This is because such patients are so widely distributed in several different types of institution and medical specialties as well as at home. Surveys of prevalence also vary in their definition of how long a patient needs to have been vegetative to be counted, from 1 month to 3 or 6 months. Only some surveys include cases due to chronic, progressive conditions. It is much easier to calculate the annual incidence of new acute cases, based on follow-up after episodes of acute brain damage. The best estimates for cases vegetative 1 month after an acute insult are, for the US, 46 per million population (PMP), and 14 PMP for the UK — the higher rate in the US reflecting a three times greater frequency of severe head injuries there (Jennett, 2002). After 6 months these figures have fallen to 17 and 5 PMP, because by then many cases have either died or recovered and by a year the figures are even lower. Extrapolating from the US figures, with many assumptions, there might be in the UK at any one time at least 12–15 PMP vegetative for 6 months after an acute incident. This means that there are many doctors who have never seen such a patient, while few have experience of many such patients.

Causes and pathology

Head injury accounts for 50–70% of acute cases and because severe head injury is much less common in the UK than in North America, continental Europe or Australia so is the frequency of the vegetative state. The commonest non-traumatic cause is cardio-respiratory arrest resulting from disease or medical accidents (often under anesthesia). Other causes include intracranial hemorrhage or infection and hypoglycemia. The neuropathology of the underlying brain damage varies with the cause (Adams et al., 2000), and there are differences between those who are vegetative and those who are very severely disabled (Graham and Jennett, this volume). Note that some vegetative cases do not have severe damage in the cerebral cortex, the most consistent lesions being in the subcortical white matter and the thalami. This means that terms to describe this state that are based on supposed cortical pathology, such as neocortical necrosis or the apallic syndrome, are therefore inappropriate. The word vegetative describes observed behavior, which is the basis of the diagnosis. The Oxford English Dictionary defines "vegetative" as "an organic body capable of growth and development but devoid of sensation and thought." Some commentators, particularly patient advocates, have expressed concern that this term is demeaning and asked for an alternative to be found. Some have suggested "the wakeful unconscious state," but the vegetative state is now so widely used by many different disciplines and in the public domain that it seems unlikely to be replaced. It is, however, worth noting that the qualifier "persistent" is now seldom used, the preference being to refer to a vegetative state of a certain duration. But the new "P" word for "permanent" is now often used for patients whose state has persisted long enough for the chance of any recovery to have become vanishingly small. It is only then that the ethical and legal issues arise as withdrawal of treatment may then be considered.

Diagnosis

The diagnosis depends on expert observation of the patient's behavior over a sufficient period of time to discover whether there is any evidence of awareness. This may not always be easy because patients retain a variable range of reflex responses, some of which may mistakenly be interpreted as evidence of awareness, while limitations on the motor responses that are possible may impede reactions that might indicate awareness (Majerus et al., this volume). There have been a number of published criteria for diagnosis, mostly from the US, but the latest and most authoritative are those from the Royal College of Physicians (2003) — an update of its 1996 document. Suffice it to say that some patients during their eyes-open periods show facial grimacing and groaning in response to noxious stimuli. There may be transitory smiling or weeping but not in response to appropriate stimuli. Visual menace may evoke blinking. The head and eyes may turn fleetingly to a loud sound, a voice or a moving light, and a very few patients may have more sustained visual tracking. A variety of brain stem reflexes may persist — corneal, oculo-cephalic and oculo-vestibular. Claims have been made that as some of these responses seem likely to involve some cortical activity they are incompatible with the diagnosis of the vegetative state. However, recent functional imaging and electrophysiological studies in vegetative patients have shown considerable residual cortical function (Schiff, this volume; Owen et al., this volume). Sometimes these are linked to observed isolated fragments of higher level behavior (Schiff et al., 2002). These may reflect the survival of islands of cortex no longer part of a sufficient and coherent cortico-thalamo-cortical system to generate awareness, or it may be that cortical activity is limited to primary sensory and motor areas with connections to secondary and tertiary areas severely limited (Kotchoubey, this volume). Whatever the explanation, it is clearly no longer valid to require as a diagnostic criterion of the vegetative state that there is no evidence of any cortical activity.

The recent publication of an international consensus statement defining the minimally conscious state has listed particular behaviors that indicate minimal but definite evidence of awareness (Giacino et al., 2002; Giacino, this volume). This state may be the end-point of recovery from the vegetative state or a staging post on the way to further recovery. There have been a number of reports of the mistaken diagnosis of the vegetative state in patients in whom experts later detected evidence of awareness (Treesch et al., 1991; Childs et al., 1993; Andrews et al., 1996). It is likely that such mistakes will be fewer after the publication of authoritative criteria for both the vegetative and the minimally conscious states. Although functional brain imaging is a useful tool in understanding the mechanisms underlying these states (Hirsch, this volume), it is not a practical diagnostic aid in the wide range of locations where the diagnosis has to be made. In any event laboratory tests can detect only functional activities in the brain that might be associated with awareness; they do not directly detect awareness — the lack of which is the key to the diagnosis of the vegetative state.

Prognosis for recovery and survival

Many patients in VS a month after an acute insult regain consciousness and some become independent, the probability of recovery depending on the cause of the acute insult and on how long the patient has been vegetative. Prognosis is based on data from the Multi-Society Task Force (1994), which was endorsed by five major medical societies in the US. The 1 year outcome in 754 vegetative cases published in the English language was analyzed for children and adults, and for traumatic and non-traumatic cases, depending upon whether the vegetative state had persisted for 1, 3, or 6 months (Fig. 1). Outcomes considered were dead, still vegetative or conscious, and if so whether independent. While a quarter of traumatic cases vegetative at 1 month became independent, only 5% of non-traumatic cases did so. On the basis of these studies the conclusion was reached that permanence could be declared after 3 months in non-traumatic cases but only after a year in traumatic cases.

The Task Force recognized that there were occasional reports of limited recovery beyond the 3–12 month periods, and that declaring permanence indicated only reasonable medical certainty that further recovery was very unlikely. Analyzing a number of reports in the media claiming late recovery the Task Force discovered that some had clearly not been vegetative, while most of those who had been had clearly shown some signs of recovery before the time limit for permanence was reached. Of 11 medically verified late recoveries, 9 before the Task Force report, 8 were cases of anoxia of which in 3 the recovery was after less than 6 months in a vegetative state. Almost all alleged late recoveries have made very limited progress — to minimally conscious or only marginally better than that. It might have been expected that any new cases of late recovery since the Task Force report would have been published but only two have been. The Medical Ethics Committee of the British Medical Association (1992) stated that permanence should not be declared until a year in the

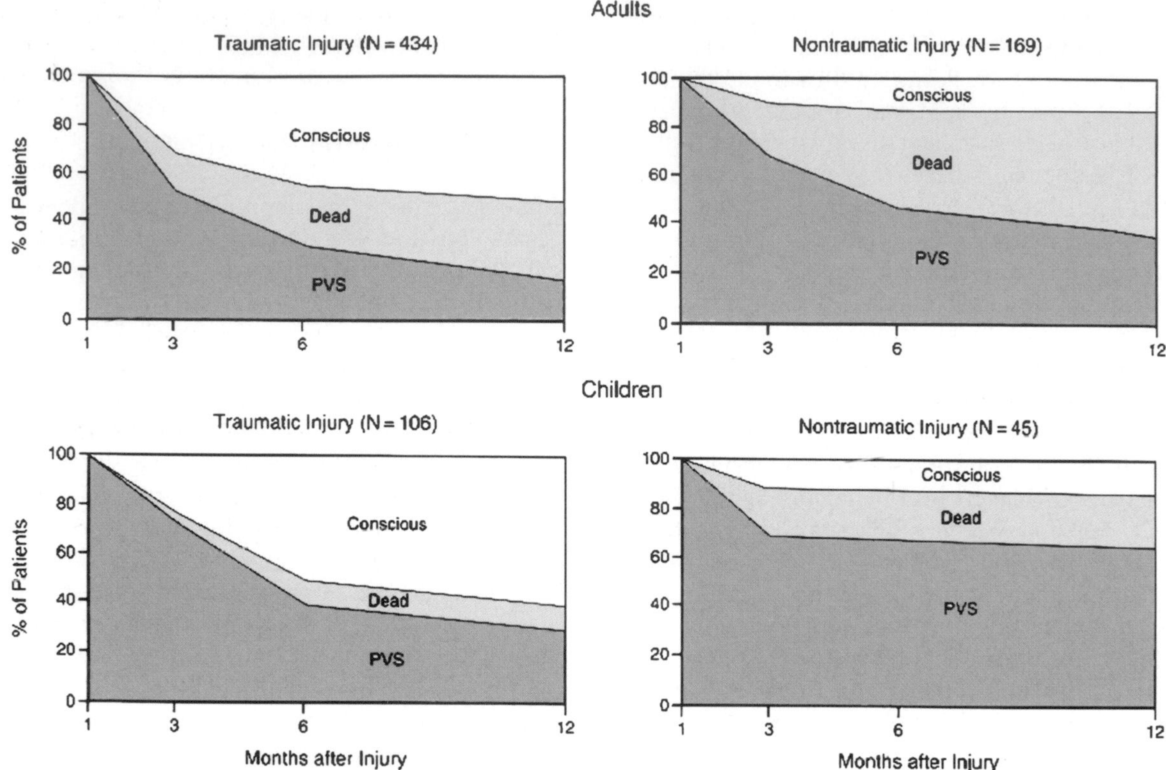

Fig. 1. Outcome for Patients in a Persistent Vegetative State (PVS) after a traumatic or Non-traumatic injury. Fifty-two percent of adults and 62% of children who are in a PVS 1 month after a traumatic injury recover consciousness within 1 year. The majority recover within the first 6 months; recovery after 6 months is unusual. In contrast, for patients in a PVS 1 month after a non-traumatic injury, recovery of consciousness is much less frequent (15% of adults and 13% of children) and is extremely unlikely after 3 months. From the Multi-Society Task Force on PVS (1994).

vegetative state, regardless of cause. But four years later the Royal College of Physicians (1996) declared that for non-traumatic cases 6 months was long enough. This was accepted for permanence by the Official Solicitor in his updated Practice Note of 1996 for lawyers in England and Wales who were seeking court approval for withdrawal of treatment from permanently vegetative patients.

Mean survival in the Task Force series was 2–5 years, a figure influenced by the number of cases vegetative for only a month. Those still alive after many months may have much longer survival. There have been many reports of survival for 10 years and some of 20, 30 and even 40 years. Duration of survival in some cases is determined by a decision to withhold antibiotics or life-saving interventions for complications having been taken once the condition was considered permanent. This is not the place to discuss the treatment of vegetative patients. Suffice it to say that no measures have been discovered that seem definitely to promote recovery. Nonetheless it is clearly important that every effort is made to provide maximal support in the form of nutrition, physiotherapy and treatment for complications as long as there remains the potential for a reasonable spontaneous recovery.

Attitudes to the permanent vegetative state

The next question is how various people regard indefinite survival in a permanent vegetative state. In an address to British doctors the Bishop of Durham (1972) said that, when considering vegetative patients, the principle of respect for life had to be qualified by an explicit concept of personality and the quality of life. When an anonymous questionnaire was given to 130 health professionals and lawyers at a San Francisco meeting on severe brain damage, and the following week to 130 British medical students early in their clinical course, nearly 90% of both groups answered that they considered that to be permanently vegetative was worse than death (Jennett, 1976). The US President's Commission for the Study of Ethical Problems in Medicine (1983) declared that there was no benefit in prolonging the life of a permanently vegetative patient. This body was made up of doctors, nurses, philosophers and lawyers. A similar view was taken later by the American Medical Association Council on Ethical and Judicial Affairs (1986) and by a Working Party of the British Medical Association (1988). This view was reiterated and expanded later by the Medical Ethics Committee of the British Medical Association (1992), prior to the first case going to the English High Court. Since then there have been a number of surveys asking people whether they would want life-sustaining treatment to be continued if they themselves were permanently vegetative. Surveys of doctors, nurses, outpatients and the public show that 80–99% would not want such continuing treatment. The views of those outside the health professionals are also reflected in the records of the many courts that have authorized withdrawal of treatment from such patients, mostly in the US and England. In the Cruzan case (1990) a US Supreme Court judge said, "It is improper to imply that continued existence and treatment in a permanent vegetative state is either beneficial or neutral. Not to terminate life support robs a patient of the very qualities protected by the right to avoid unwanted medical treatment. A degraded existence is perpetuated, his family's suffering is protracted and the memory he leaves behind becomes more and more distorted." Two years later a judge in the highest court in England said, "The pitiful and humiliating helplessness of the vegetative patients offends the values of human dignity and personal privacy that belong to every person. I cannot conceive what benefit his continued existence could be thought to give him" (Airedale NHS Trust versus Bland, 1993). There is, however, a minority of people who hold that life of any kind is sacred and who reject considerations of quality of life or of the patient's autonomy. This vitalist view is held by some conservative Roman Catholics, fundamental protestants and orthodox Jews, but it has been formally rejected by many secular and theological moral philosophers and organizations. Within the Catholic church, however, disagreement is particularly evident. While some have spoken out vehemently against treatment withdrawal, several of the most high-profile cases reaching the courts in the US and the UK have been from Catholic families with the support of their priests. In the Irish Supreme Court, a Catholic family had churchmen supporting their case for withdrawal of treatment, while other clergy opposed them. In the event, withdrawal was allowed by the trial judge and later supported by 4 of the 5 appeal judges in that predominantly Catholic country (Re a Ward of Court, 1995). However, in March 2004, the Pope, addressing an international meeting on the vegetative state, declared that artificial nutrition and hydration should not be withdrawn from Catholic patients in a permanent vegetative state (http://www.vegetativestate.org).

Ethical issues

The ethical debate revolves largely around the decision about continuing or withdrawing artificial nutrition and hydration once a patient is considered to be permanently vegetative. The question is what action is consistent with the ethical principles of proportionality in balancing the benefits and burdens of medical intervention, and on how best to respect the autonomy or self-determination of the patient. Until the last decade or so it was assumed that the patient's best interests should be determined by the doctor who would then act accordingly, but such paternalism is now increasingly challenged in many countries. In such places it is now accepted that all patients have a right to refuse treatment, even that which aims to save or prolong life. The question is how to ensure that the incompetent patient does not lose this right, usually by trying to discover how this patient would likely want to be treated in their present condition. Ideally this would be discovered from a living will or advance directive — but few people have such a document, especially younger patients. Both in the UK and in the US, there have been strong recommendations that more people should make such provision, particularly because of the possibility of vegetative survival. Recently, both countries have stressed the need for these declarations to be specific about the kind of treatment it is wished to refuse — in particular whether this is to include artificial nutrition and hydration. In practice reliance has usually to be put on the testimony of the family about the known beliefs and values of the patient, and in particular on recalling any verbal statements that might have been made about their wishes in the event of vegetative survival. Failing any definitive information, a doctor's decision to withdraw treatment could legitimately be based on what most reasonable people would want in these circumstances.

Some who oppose withdrawal claim that food and water are part of basic care to which every patient has a right. However, most doctors and philosophers as well as the courts have agreed that it constitutes medical treatment, and thereby differs from the symbolic significance of providing oral food and fluids to an infant or a conscious but helpless adult. As medical treatment it may therefore be withdrawn if considered by the treating doctor to be no longer of benefit to the patient. Moralists and lawyers agree that there is no difference between withholding and withdrawing treatment, although doctors and nurses may feel more reluctant to withdraw. If withdrawing were not allowed, however, doctors might be reluctant to embark on any trial intervention lest, if it proved unsuccessful, they would be committed to a prolonged period of futile treatment. There is an agreement also about the distinction between killing and letting die — were this not so, then any doctor who decided to abandon futile life-prolonging treatment of any kind from a hopelessly ill patient might be held to have killed that patient. There is also agreement that death after treatment withdrawal does not constitute euthanasia or suicide.

Legal issues

As to the legal aspects of withdrawal of nutritional support, the UK approach followed in the wake of a series of cases in the US. The landmark case for withdrawing life support was that of Karen Quinlan in 1976 who, unusually for a vegetative patient, was on a ventilator. The New Jersey appeal court ruled that her father could be appointed her guardian ad litem and could order withdrawal of her ventilator (*In re Quinlan*, 1976). In the event, she was weaned from the respirator and lived for another 10 years on tube feeding. Several different types of treatment withdrawal were subsequently authorized by State courts in the US, most for acutely ill patients with terminal conditions, some of whom were competent to refuse treatment. It was 10 years before a court approved withdrawal of artificial nutrition and hydration from a vegetative patient. Although some judges then ruled that all treatment withdrawal cases should come to court, doctors objected. It was not long before most judges agreed with the Quinlan judge that decisions about withdrawal should be made by the family and the physician alone, unless there was a dispute between them. All along the American courts have tended to give the family a more dominant role in decision-making than has subsequently been the case in the UK, where the decision is still seen as a medical one, albeit taken after consultation with the family. By the time the vegetative case of Cruzan came to US Supreme Court in 1990, there had been 27 vegetative cases through various State courts. Since the Bland case in 1992, when all 9 English judges (including 5 in the House of Lords) agreed to withdrawal of artificial nutrition and hydration (Airedale NHS Trust v Bland, 1993), there have been more than 25 cases in the English High Court (none of them refused). Single cases have been approved by courts in Ireland, Scotland, South Africa, New Zealand, Germany, and the Netherlands. In two cases in October 2000, the English High Court ruled that such withdrawal did not infringe the Human Rights Act, which had just become law, and had reflected the European Convention on Human Rights. Only in England, however, is it required to seek court approval before withdrawal of artificial nutrition and hydration from a vegetative patient, and this is the only condition for which court approval is required in England before withdrawing non-beneficial treatment. In its latest update of its guidance document on withdrawing treatment, the Medical Ethics Committee of the British Medical Association (2001) states that it hopes that in future vegetative cases would no longer inevitably require court review, where consensus exists. Some UK lawyers also do not believe that the legal situation there has yet been satisfactorily resolved (McLean, 2001). The American approach of limiting court intervention only to cases when there is an irresolvable dispute between involved parties depends largely on most hospitals in that country having in-house ethics committees always available to advise on difficult clinical decisions and to mediate where there are conflicting views. According to a newspaper report in 2004 hospitals in one NHS region in Scotland are soon to launch a pilot project to develop such committees.

References

Adams, J.H., Graham, D.I. and Jennett, B. (2000) The neuropathology of the vegetative state after an acute brain insult. Brain, 123: 1327–1338.

Airedale NHS Trust v. Bland (1993) 1 All ER 821.

American Medical Association Council on Ethical and Judicial Affairs. (1986) Statement on withholding or withdrawing life-prolonging treatment. JAMA, 256: 471.

Andrews, K., Murphy, L., Munday, R. and Littlewood, C. (1996) Misdiagnosis of the vegetative state: retrospective study in a rehabilitation unit. Br. Med. J., 313: 13–16.

Bishop of Durham. (1972) Moral problems facing the medical profession at the present time. The Bishoprick, 47: 48–61.

Childs, N.L., Mercer, W.N. and Childs, H.W. (1993) Accuracy of diagnosis of persistent vegetative state. Neurology, 43: 1465–1467.

Cruzan, v., Director, Missouri Dept of Health (1990) 110 S. Ct. 2841.

Giacino, J.T., Ashwal, S., Childs, N., Cranford, R., Jennett, B., Katz, D.I., Kelly, J.P., Rosenberg, J.H., Whyte, J., Zafonte, R.D. and Zasler, N.D. (2002) The minimally conscious state: definition and diagnostic criteria. Neurology, 58: 349–353.

In re Quinlan (1976) 70.NJ 10 355 A2d 647.

Jennett, B. (1976) Resource allocation for the severely brain damaged. Arch. Neurol, 33: 595–597.

Jennett, B. (2002) The vegetative state: medical facts, ethical, and legal dilemmas. In: Epidemiology Chap. 3. Cambridge University Press, Cambridge, UK, pp. 36–41.

McLean, S. (2001) Persistent vegetative state and the law. J. Neurol. Neurosur. Ps., 71(suppl 1): 126–127.

Medical Ethics Committee of the British Medical Association. (1992) Discussion Paper on the Treatment of Patients in Persistent Vegetative state. BMA, London.

Medical Ethics Committee of the British Medical Association. (2001) Withholding and Withdrawing Life-Prolonging Medical Treatment (2nd ed). BMJ Books, London.

Multi-Society Task Force on PVS. (1994) Medical aspects of the persistent vegetative state (part 2). N. Engl. J. Med., 330: 1572–1579.

President's Commission for the study of Ethical Problems in Medicine. (1983) Deciding to forego life-sustaining treatment. US Government Printing Office, Washington, DC.

Re a Ward of Court (1995) 2 ILRM 401 (Ir. Sup. Ct.)

Royal College of Physicians. (1996) Report on the permanent vegetative state. J. Roy. Coll. Physicians Lond., 30: 119–121.

Royal College of Physicians. (2003) The vegetative state: guidance on diagnosis and management. Clin. Med., 3: 249–254.

Schiff, N.D., Ribary, U., Moreno, D.R., Beattie, B., Kronberg, E., Blasberg, R., Giacino, J., McCagg, C., Fins, J.J., Llinas, R. and Plum, F. (2002) Residual cerebral activity and behavioural fragments can remain in the persistently vegetative brain. Brain, 125: 1210–1234.

Treesch, D.D., Sims, F.H., Duthie, E.H., Goldstein, M.D. and Lane, P.S. (1991) Clinical characteristics of patients in the persistent vegetative state. Arch. Intern. Med., 151: 930–932.

Working Party of the British Medical Association. (1988) The Euthanasia Report. BMA, London.

S. Laureys (Ed.)
Progress in Brain Research, Vol. 150
ISSN 0079-6123

CHAPTER 38

Assessing health-related quality of life after severe brain damage: potentials and limitations

Corinna Petersen and Monika Bullinger*

Centre of Psychosocial Medicine, Institute and Policlinics of Medical Psychology, University of Hamburg, Martinistr. 52, S35, 20246 Hamburg, Germany

Abstract: Assessment of health related quality of life (HRQoL) has been addressed in medicine since more that 20 years. However, little is known about the HRQoL of patients with traumatic brain injury (TBI). The main reason for this seems to be the challenge of attempting to assess HRQoL in the light of the diversity of consequences associated with the condition. Especially in severely brain-damaged patients with altered states of consciousness, patients self-report on own function and well-being seems at least difficult, if not impossible, and the question is whether other persons, like caregivers, might be able to truly judge the patients quality of life. A prerequisite to examining HRQoL in patients with TBI is the availability of appropriate measures. While generic measures are available, the development of condition-specific measures has only recently begun. This chapter reviews conceptual and methodological issues in current HRQoL research, applies these to the area of brain injury and critically summarizes challenges to HRQoL assessment in TBI in terms of potentials and limitations. It is concluded that in spite of recent promising approaches to HRQoL assessment in TBI, much has to be done to understand, measure and finally improve HRQoL in severely brain-damaged patients.

Introduction

The assessment of self-reported health-related quality of life (HRQoL) in traumatic brain injury (TBI) is an important issue, clinically as well as scientifically. It is widely acknowledged that TBI places high demands on health care services. TBI — as a consequence of a non-degenerative, non-congenital insult to the brain through an external mechanical force — can result in temporary or permanent impairment of cognitive, physical, psychological, behavioral, and/or social functions. Pattern and degree of these impairments depend greatly on the severity of the injury. The definition of severe TBI is in flux, among the accepted criteria are the Glasgow coma scale (GCS score of 8 or below) and extent of posttraumatic amnesia (PTA duration of more than 24 h). Clinical conditions encountered following TBI encompass vegetative state, coma, minimally consciousness state and locked-in syndrome. According to recent epidemiological studies the annual incidence of TBI is between 180 and 250 per 100,000 persons in the US (Bruns and Hauser, 2003, US data), while the reported annual incidence of "severe brain injury" is 8.5 per 100,000 persons (Masson et al., 2003, French data).

The topic of HRQoL in patients with TBI emerged only recently and very few data are available. Cognitive, but also physical, psychological, behavioral, and social problems after TBI are well known among practitioners and have been frequently addressed in clinical research. In most of these studies, however, the main focus was to describe the functional outcome of patients after

*Corresponding author. Tel.: +49/40/42803-6430;
Fax: +49/40/42803-4940; E-mail: bullinger@uke.uni-hamburg.de

DOI: 10.1016/S0079-6123(05)50038-4

severe brain injury rather than focusing on the patients' experiences. In HRQoL research, well-being and functioning have primarily been assessed in patient populations without cognitive impairments, and consequently few existing measures address mental functioning. Several instruments, however, have recently been developed for patients with Alzheimer's disease (see Ready and Ott, 2003 for a review). As a result of cognitive restrictions, most of these measures are available not as self, but as proxy-report (i.e., evaluated by others) rather than self-report (i.e., evaluated by the patients themselves). While patients with Alzheimer's disease are conscious, able to perceive, and fairly able to express themselves, the level of cognitive processing and understanding required for self-report of HRQoL is often lacking. Another population in which HRQoL has been studied are patients after stroke. Depending on the localization of the lesion, these patients may have less difficulty in perceiving or processing information, but may be more challenged in the expression of responses. Problems in assessing HRQoL also arise in children, especially in very young children. Since the ability to self-report is related to the developmental stage of the child, proxy reports of parents are frequently elicited (Eiser and Morse, 2001). These examples highlight a special challenge to HRQoL assessment in patients with TBI: the role of self-report. The question is whether TBI patients are able to process, i.e. perceive, understand, and respond adequately to questions regarding their well-being and function. The chapter will address these issues from the background of the HRQoL field, using recent studies of HRQoL in TBI patients as examples.

Health-related quality of life research

Over the past years, considerable progress has been made in measuring HRQoL and in acknowledging its importance as an outcome indicator in health care. One reason for this development was the World Health Organizations definition of "health" as "a state of complete physical, mental, and social well-being, and not merely the absence of disease or infirmity" (World Health Organization, 1948). In addition, the need for a "new medical model" in which the patient is in the foreground provided an impetus to HRQoL assessment (Engel, 1977). Since in medicine, quality of life (QoL) relates to the subjective perception of well-being and functioning of a patient with regards to his/her health state, the term HRQoL has been coined (e.g., Guyatt et al., 1996; Feldman et al., 2000). By now, the term is widely used to reflect the subjective representation of health (Baker and Intagliata, 1982; Spilker, 1990; Aaronson, 1992; Felce and Perry, 1995). HRQoL is defined as referring to the specific impact of an illness or injury, medical treatment, or health care policy on an individual's well-being and functioning. As a multidimensional construct it represents the physical, emotional, cognitive, social, and everyday life aspects of well-being and functioning associated with an individual's disease and/or its treatment (Eisen et al., 1979; Aaronson et al., 1991; Strand and Russell, 1997; Bullinger, 1997; Fig. 1).

HRQoL differs from other constructs, such as health status by its focus on the patients' subjective evaluation beyond the clinical assessments of physical or psychological status. To assess HRQoL, the self-report of the individual is crucial. Because patients are the experts for their own well-being and functioning, their self-report plays an important role in capturing their perception of the impact on HRQoL of health conditions and treatment regimes.

Bullinger, 1997). The epidemiological perspective focuses on HRQoL as a descriptor of well-being and function in the general population (e.g., in health surveys). From a clinical perspective, for which the evaluation of treatment effects is central, HRQoL is an outcome measure (e.g., in clinical

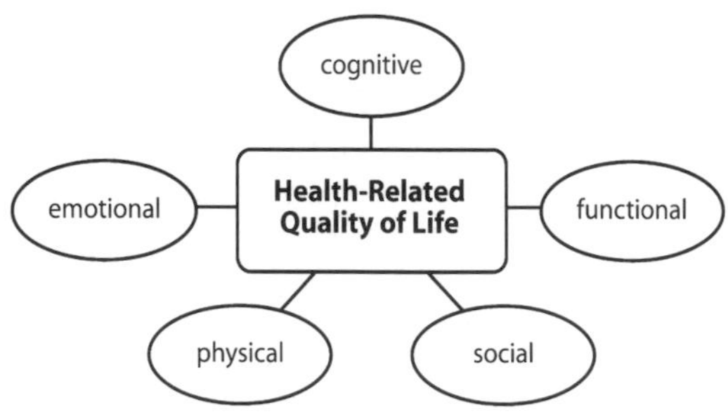

Fig. 1. Dimensions of health-related quality of life.

trials). The health economic perspective centers around the analysis of costs of care in which HRQoL denotes a potential benefit (e.g., in utility studies). The health systems perspective addresses HRQoL as an indicator of the quality of care (e.g., in quality assurance studies).

HRQoL research in medicine started as early as the 1970s, when concepts of HRQoL were discussed and established. From the 1980s, the development of assessment tools received most attention. From the 1990s, the application of the assessment tools began. Since 2000, the interpretation, clinical significance, and policy impact of HRQoL studies is increasingly being addressed (Symonds et al., 2002).

HRQoL assessment

According to the different objectives of HRQoL assessment, its measures vary in their format. As Guyatt et al. (1996) stated, measures can be classified across three dimensions: populations, scores, and respondents (see Table 1).

At the *population* level, measures can be divided into generic and condition-specific instruments. While generic instruments measures assess HRQoL across different health conditions (healthy versus ill), condition-specific measures do so with regard to specific clinical conditions (e.g., symptoms, diseases, or treatments). The disadvantage of using generic measures is that small changes in HRQoL of specific patient groups may not be detected in specific populations. On the other hand, while condition-specific instruments may provide clinically relevant information, comparison across health conditions is not possible (Bullinger, 1997).

As concerns *scores*, HRQoL measures can be further categorized into health profiles or single indicators, the former giving an overview over several HRQoL domains and the latter resulting in one single number. Finally, in terms of respondents, information can be obtained from the patients themselves (self-report) or from significant others, such as the spouse (proxy-report). This taxonomy has proven useful for the classification of HRQoL instruments based on their nature, scope, and applicability.

Table 1. Dimensions of HRQoL instruments

Dimension	Variation
Population	Generic versus condition-specific
Scores	Single indicator, profile or battery approach
Respondent	Self-report versus proxy-report

Quality of life measures in TBI

While several generic measures for HRQoL assessment exist, condition-specific measures for individuals with TBI are lacking. With respect to generic assessment, several HRQoL questionnaires have been used in patients after TBI. Examples are the sickness impact profile (Bergner et al., 1981; Damiano, 1996), the Nottingham Health Profile (Hunt et al., 1981; McEwen and McKenna, 1996), the WHOQOL (The WHOQOL Group, 1993, 1998), and the SF-36 health survey (Ware, 1996). However, while most of these generic measures have been carefully tested for psychometric properties (e.g., reliability, validity, and sensitivity) they fail to capture meaningful QoL issues for specific patient groups (such as individuals with TBI). Another challenge for researchers is to determine the relevance of HRQoL results and to disseminate this information to clinicians (Cella et al., 2002).

By comparison, the development of an HRQoL measure designed specifically for TBI started only very recently. Bullinger et al. (2002) summarized the available instruments to assess HRQoL in TBI patients. They identified several condition-specific HRQoL instruments especially developed for TBI patients, for example, the European brain injury questionnaire (EBIQ, Teasdale et al., 1997; Deloche et al., 2000) or the BICRO-39 (Powell et al., 1998) questionnaire. In addition, Berger et al. (1999) provided a literature review of HRQoL research in TBI patients in the last decade, and pointed out that although research had started to address HRQoL in TBI patients, standardized instruments had not been widely used; research had mainly focussed on physical and functional outcomes, and in most studies only certain dimensions of HRQoL were assessed.

By defining HRQoL as the subjective perspective of a patient with regard to health functioning and well-being, the use of functional outcome measures such as the functional independence measure (FIM) or the disability rating scale (DRS) in order to measure the physical/functional dimension of HRQoL clearly has its limitations. These ratings are useful to depict functional problems, yet do not assess the degree to which the patient is bothered or stressed because of these problems. It is therefore necessary to include standardized HRQoL instruments developed for this specific purpose. In the existing literature on TBI research, the Sickness Impact Profile (Bergner et al., 1981; Damiano, 1996) has been a frequently used HRQoL measure. For example, Powell et al. (2002) examined rehabilitation effectiveness in a non-randomized study using this measure. Levine et al. (2000) investigated the relationship between the sickness impact profile and performance scores of neuropsychological tests.

There is a need to assess TBI-specific HRQoL especially with regard to questions about outcome assessment in clinical trials. So far, validated and reliable condition-specific HRQoL for TBI patients have not been fully developed, but the development of those measures has started. Generic measures might not detect subtle changes in HRQoL, whereas condition-specific measures may provide clinically important information. To be considered as acceptable outcome measures, these questionnaires, either generic or condition-specific, have to comply with psychometric quality criteria including reliability, validity, sensitivity, and the ability to identify and differentiate changes over time due to treatment.

Given the many open questions in the assessment of HRQoL in TBI patients and the heterogeneity of the published studies, an international group of clinicians and QoL experts reviewed the literature and has given several recommendations for future research (Bullinger et al., 2002). With regard to the measurement time point, HRQoL assessment was recommended to take place not upon admission to the hospital, but in the early phase of rehabilitation (i.e., within 1 year after trauma) and in the post-rehabilitation phase. Self-report should be preferred to proxy-report. However, it was recognized that the family's view of the patient's HRQoL provides an additional, important source of information, particularly in the acute phase of treatment when the patient may not be fully capable of participating in the assessment of HRQoL.

The expert group recommended that the assessment of HRQoL includes both generic and condition-specific instruments. For the generic assessment, the SF-36 Health Survey (Ware and Sherbourne,1992), the EuroQol EQ5D (Kind, 1996), and the WHOQOL (The WHOQOL Group, 1993) were recommended. Although studies with condition-specific measures like the EBIQ (Teasdale et al., 1997; Deloche et al., 2000) are available, the group could not confidently recommend a condition-specific measure (Bullinger et al., 2002). Given the strengths and weaknesses of generic and condition-specific measures, the choice of measure should depend on the aim of the study. For international research, the necessity to have at hand a cross-nationally tested TBI-specific HRQoL questionnaire was addressed.

To better understand HRQoL in TBI, an European research group has formed to develop a new questionnaire for assessment of HRQoL in TBI (Von Steinbüchel et al., 2005). The Quality of Life in Brain Injury (QOLIBRI) group has collected data from 500 out of 1000 including adult TBI patients, confronting clinical data with QoL assessments such as the generic QoL questionnaires (SF-36 Health Survey) and the newly developed QOLIBRI questionnaire. The latter questionnaire is the result of a consensus discussion within the European research group, in which available questionnaires (such as the EBIQ questionnaire) as well as ideas and responses from focus groups were included. The aim of the QOLIBRI project is to offer a condition-specific questionnaire, which assesses the relevant dimensions of QoL in TBI research.

HRQoL research in severe traumatic brain injuries

HRQoL as a multidimensional construct has only been referred to in a few studies of TBI patients and study results were rather heterogeneous. According to Johnston and Miklos (2002) a number of papers make conclusions about HRQoL without actually having assessed it. In addition, most studies excluded patients with severe limitations, so that little is known about the HRQoL of the severely impaired patients. Dijkers (2004) reviewed HRQoL research in patients after TBI and found a substantial variation in HRQoL in patients after injury across studies. Importantly however, injury severity did not predict subjective well-being. The author called for further investigations to examine the feasibility of self-report in patients after TBI and pointed to the necessity to identify the relevant dimensions of HRQoL. Berger et al. (1999) reviewed 16 studies of patients with TBI, in which at least two dimensions of HRQoL were assessed. They found that in most studies HRQoL was either inferred from patients' clinically rated health status or was performed with measures not developed for HRQoL research (i.e., assessment other constructs such as social support). In addition, the heterogeneity of measures made comparisons across studies difficult. With respect to the potential predictors of HRQoL, Kalpakjian et al. (2004) examined HRQoL and psychosocial outcomes in adults after TBI living in community settings. They found social support, community integration, and positive affect to be associated with higher HRQoL scores. Kozlowski et al. (2002) explored HRQoL in patients and their relatives 3 years after severe TBI. Cognitive and behavioral problems as well as dependence on others in certain activities affected patients HRQoL. In this study authors compared self- and proxy-report and found that self-reported HRQoL was lower than ratings by proxies.

Brain injuries not only affect the dimensions of a patient's HRQoL but also the caregivers' HRQoL. Wedcliffe and Ross (2001) explored the psychosocial impact of TBI on the HRQoL of the spouse or partner of persons with severe brain injuries. Although most spouses and partners accepted their new roles and responsibilities, loneliness, and restriction in every day life activities were very frequently reported. Altered communication patterns, changes in family relations and financial strain were identified as major problems.

Steadman-Pare et al. (2001) examined the factors associated with HRQoL in persons with long-standing TBI. Perceived mental health, self-rated health, gender, participation in work and leisure, and the availability of emotional support were associated with better HRQoL many years after the trauma. According to a study conducted by Koskinen (1998), the strain on relatives was still high 10 years after injury, although it had decreased over the years.

Challenges of measuring HRQoL in TBI

TBI researchers have come to the conclusion that current generic measures underestimate the extent in which cognitive problems impact a patient's ability to report on HRQoL. Existing HRQoL assessment methods generally require a patient who is able to report on his or her perceptions of HRQoL: as prerequisite to assess HRQoL, the patient must not only be conscious and able to express himself or herself, but also have a sufficient cognitive capacity for understanding and replying to questions. This prerequisite poses a challenge to attempts measuring HRQoL in patients with severe TBI, which needs to be addressed theoretically, methodologically, and practically. On the basis of an information-processing model, functions to be present to elicit QoL information via patient self-report include:

- (i) the ability to take in information in terms of being aware of external stimuli (perceiving — visually or auditorially).
- (ii) processing of input, in terms of understanding and interpreting the input (cognitive processing).
- (iii) responding in terms of expressing results of the information processing as a message encoded via some chosen channel (answering — verbally or non-verbally).

Ideally, all functions should be present if a patient is asked for self-report. But also in patients lacking full function in one or all categories, self-report may be possible. Only if no function is present, the possibility to elicit self-report is critically compromised. Yet, even though self-report might be desired, there still is an obvious diversity of situations in which HRQoL assessment is difficult due to impairment or lack of function on different levels of the information processing chain. Nevertheless, self-report is important and in HRQoL assessment is considered vital, because patients are experts in their own well-being and function.

Is self-report of quality of life possible in severe TBI?

Generally, HRQoL information is elicited primarily through questions, either in written form (questionnaires) or interviews (spoken word). However, the type of questions used limits the range of possibilities for patients to perceive, process, and react. Furthermore, responding to HRQoL questions requires a high level of metacognitive processing, for example, in the comparison of one's actual stage with previous experiences of own standards, or with the situation of other persons, or with regard to own needs (as specified by models of QoL such as role function, social comparison, and need fulfillment). If it is the case that eliciting information about patients HRQoL requires full functioning in terms of perceiving, understanding, and answering, then the question arises whether it is possible at all to include the patient's view in some of the disorders associated with severe brain injury. The perceptions of other persons, such as caregivers, family, or medical staff have therefore a frequently used alternative. Nevertheless, the impetus to also assess the patient's perspective through self-report is increasing. Resorting to ask significant others — the so called proxies — about patient's HRQoL, which has been the method used for several neurological conditions is a possible solution, but, it does not reflect the patient's own account. Research on observer ratings in HRQoL research shows that they are often not representative of patients' own experiences, but have to be considered as additional sources of information. Although the caregiver perspective on patients' HRQoL as well as caregivers' perspectives on their own HRQoL are highly important clinical issues, they can never fully substitute for self-report in severely brain-damaged patients.

If one assumes that lack of consciousness precludes any onset of the information-processing chain, then assessing HRQoL by self-report in these cases is not possible. Options such as the observer rating still exist but will not represent the patient's experiential universe. If the patient is conscious, but has problems with adequately perceiving the question (e.g., visual dysfunction), it is necessary to identify other sources of questioning (e.g., by spoken rather than written word). The same is true for the mode of expression: spoken or written responses can be supplemented by other channels of information (e.g., lines, pictograms, and colors). Additionally, a computer or communication device may be used, as has been done with young children: HRQoL information was elicited by a computerized voice and responses were made easy by a touch screen or simple mouse clicks, thus eliminating the need to understand written language (Ravens-Sieberer et al., 2000). Even in patients with nearly completely absent motor output (e.g., locked-in patients) it is still possible to elicit information through innovative techniques such as EEG supported devices (see Kübler and Neumann, this volume).

HRQoL research in patients with aphasia or ataxia could also take innovative methods into account, however, in patients with agnosia or dementia the problems culminate. For the first group of patients, the question arises whether they can appropriately account for their individual experience. For the second group of patients the question is whether they can appropriately link the question to an answer.

Tackling the aforementioned problems of HRQoL assessment embarks onto a philosophical quandary: if HRQoL represents solely subjective experiences then only the individual can account for it and consequently there are no, right, wrong, or inappropriate answers. For example, if patients have anosognosia, his or her view in spite of the condition reflects his or her current level of experience, should be taken as such. Basically, the same is true for patients with dementia, although it is difficult to accept information about HRQoL which is based on impaired cognitive processing. This may put the validity and reliability of HRQoL information into question but could likely be resolved both at the subject level as at the instrument level. At the subject level, validity in terms of external criteria against which the appropriateness of individual QoL information can be judged, does not exist. However, on the measurement level, standardized instruments have been developed and tested for reliability and validity in various patient populations, making it possible to verify if the instrument is measuring the construct of HRQoL in an appropriate and psychometrically sound way.

For the conditions under study, especially those under the heading of severe TBI, the psychometric performance of scales has not yet been sufficiently studied. It is necessary to do so in order to understand the performance of standard measures as well as to construct measures that are able to grasp the individual experience in patients with these conditions.

Conclusions

A major issue in the assessment of QoL in patients with severe brain damage is the role of self-report. Owing to the fact that QoL has only very recently emerged in TBI research, the question whether and under which conditions patients can self-report on their QoL is of the utmost importance. Generally, and this is the conclusion reached by the European TBI consensus group, patients with a GCS score of 8 and below are not able to self-report. Self-report should also not be elicited in patients who have been admitted to an intensive care or hospital ward due to patients' reduced consciousness. However, in the course of the condition, self-report of QoL, using instruments such as the QOLIBRI measure, may be possible (Mellick et al., 2003). In such cases where the capacity for self-report evolves or vanishes, the use of proxy-report over time is recommended. Such proxy-report is possible also with generic instruments, such as the SF-36 health survey proxy version or the proxy version of the QOLIBRI questionnaire.

Self- and proxy-report should thus be considered as complementary sources of information and not as mutual replacements. In patients who cannot self-report on their QoL due to the severity or type of brain damage, proxy ratings can be helpful, however taking into account that if the patient could respond his or her judgment could differ significantly from that of the proxy.

It is clear that severe TBI, as well as other conditions such as non-traumatic or degenerative brain damage, or very young age, is one of the challenges in the field of QoL assessment. There is a strong need and desire to (i) elicit QoL information from the patient as much as possible, (ii) resort to proxy measures as the main source of information only if patient self-report is not possible, and (iii) use proxy measures as independent sources of information, taking into account that they do not necessary reflect the patients own view.

HRQoL research can contribute to the above mentioned challenges by identifying in specific patient populations: (i) which problem areas are relevant for the specific pathological entity, (ii) which functions in the chain of information processing (perceiving, processing, and answering) are present, absent, or impaired; and (iii) which innovative means to elicit information while taking into account the processing problems, can be utilized. With exception of patients who are unconscious, there are many more possibilities to obtain HRQoL information in patients with TBI. Yet, more research is needed in order to identify both novel techniques as well as their psychometric properties in order to gain valuable and clinically relevant information.

A new approach is to view HRQoL as a multifaceted construct to which all actors in a clinical situation can respond (patients, family, and caregivers). Rather than expecting that information from a single respondent will suffice, the focus should be on the pattern of responses from multiple sources. This "triangulation" of information has long been advocated for in psychiatry, however, experience has demonstrated that researchers and clinicians must be prepared to deal with the possibility of unexpected and conflicting results (e.g., the patient reporting excellent HRQoL while the family draws another picture, or viceversa). These potential conflicting reports are thought to be a consequence of the well known fact (as outlined in the theory of the social — and individual — constructions of reality) that different individuals' views of the world and themselves do not converge and that reflection about one's own actual state of well-being and function is a process in flux. As such, recognition that the patient's view is important should be maintained despite the fact that much work remains to be done in HRQoL assessment. Hopefully, work in this area will be encouraged through interdisciplinary cooperation with the aim to give voice to people with severe neurological handicap. These patients, as all persons, have feelings, thoughts, emotions, and wishes regarding their mental well-being and regarding their physical functioning they need to be able to express and share with others in order to maintain or improve their HRQoL.

References

Aaronson, N.K. (1992) Assessing the quality of life of patients in cancer clinical trials: common sense problems and common sense solutions. Eur. J. Cancer, 28: 1304–1307.

Aaronson, N.K., Meyerowitz, B.E., Bard, M., Bloom, J.R., Fawzy, F.I., Feldstein, M., Fink, D., Holland, J.C., Johnson, J.E., Lowman, J.T., Patterson, W.B. and Ware, J.E. (1991) Quality of life research in oncology: past achievements and future priorities. Cancer Nurs., 67: 839–843.

Baker, F. and Intagliata, J. (1982) Quality of life in the evaluation of community support systems. Eval. Program. Plann., 5: 69–79.

Bergner, M., Bobbit, R.A., Carter, W.B. and Gilson, B.S. (1981) The sickness impact profile: Development and final revision of a health status measure. Med. Care, 19: 780–805.

Berger, E., Leven, F., Pirente, N., Bouillon, B. and Neugebauer, E. (1999) Quality of life after TBI: A systematic review of literature. Restor. Neurol. Neurosci., 14: 93–102.

Bruns Jr., J. and Hauser, W.A. (2003) The epidemiology of TBI: a review. Epilepsia, 2: 2–10.

Bullinger, M. (1997) Health related quality of life and subjective health. Overview of the status of research for new evaluation criteria in medicine. Psychother. Psychosom. Med. Psychol., 47: 76–91.

Bullinger M. and The TBI Consensus Group. (2002) Quality of life in patients with TBI- basis issues, assessment and recommendations. Restor. Neurol. Neurosci., 20: 111–124.

Cella, D., Bullinger, M., Scott, C., Barofsky, I. and The Clinical Significance Consensus Meeting Group (2002) Group vs individual approaches to understanding the clinical significance of differences or changes in quality of life. Mayo Clin. Proc., 77: 384–392.

Damiano, A.M. (1996) The sickness impact profile. In: Spilker B. (Ed.), Quality of Life and Pharmaeconomics in Clinical Trials. Lippincott-Raven, Philadelphia, PA, pp. 347–354.

Deloche, G., Dellatolas, G. and Christensen, A. (2000) The European Brain Injury Questionnaire. Patients' and families' subjective evaluation of brain-injured patients' current and prior to injury difficulties. International Handbook of Neuropsychological Rehabilitation, Kluwer Academic Publishers, New York.

Dijkers, M.P. (2004) Quality of life after TBI: a review of research approaches and findings. Arch. Phys. Med. Rehabil., 85: 21–35.

Engel, G.L. (1977) The need for a new medical model: A challenge for biomedicine. Science, 196: 129–136.

Eisen, M., Ware, J.E., Donald, C.A. and Brook, R.H. (1979) Measuring components of children's health status. Med. Care, 17: 575–579.

Eiser, C. and Morse, R. (2001) Quality-of-life measures in chronic diseases of childhood. Health Tech. Assess., 5: 1–157.

Felce, D. and Perry, J. (1995) Quality of life: its definition and measurement. Res. Dev. Disabil., 16: 51–74.

Feldman, B.M., Grundland, B., McCullough, L. and Wright, V. (2000) Distinction of quality of life, health related quality of life, and health status in children referred for rheumatologic care. J. Rheumatol., 27: 226–233.

Guyatt, G.H., Jaeschke, R., Feeney, D.H. and Patrick, D.L. (1996) Measurement in clinical trials: Choosing the right approach. In: Spilker B. (Ed.), Quality of life and pharmacoeconomics in clinical trials. Lippincott-Raven, Philadelphia, PA, pp. 41–48.

Hunt, S.M., McEwen, J., McKenna, S.P., Williams, J. and Papp, E. (1981) The Nottingham health profile: Subjective health status and medical consultations. Soc. Sci. Med., 15: 221–229.

Johnston, M.V. and Miklos, C.S. (2002) Activity-related quality of life in rehabilitation and TBI. Arch. Phys. Med. Rehabil., 83: 26–38.

Kalpakjian, C.Z., Lam, C.S., Toussaint, L.L. and Merbitz, N.K. (2004) Describing quality of life and psychosocial outcomes after TBI. Am. J. Phys. Med. Rehabil., 83: 255–265.

Kind, P. (1996) The EuroQOL instrument: an index of health-related quality of life. In: Spilker B. (Ed.), Quality of Life and Pharmacoeconomics in Clinical Trials. Lippincott-Raven, Philadelphia, PA, pp. 191–203.

Koskinen, S. (1998) Quality of life 10 years after a very severe TBI (TBI): the perspective of the injured and the closest relative. Brain Injury, 12: 631–648.

Kozlowski, O., Pollez, B., Thevenon, A., Dhellemmes, P. and Rousseaux, M. (2002) Outcome and quality of life after three years in a cohort of patients with severe TBI. Ann. Readapt. Med. Phys., 45: 466–473.

Levine, B., Dawson, D., Boutet, I., Schwartz, M.L. and Stuss, D.T. (2000) Assessment of strategic self-regulation in TBI: its relationship to injury severity and psychosocial outcome. Neuropsychology, 14: 491–500.

Masson, F., Thicoipe, M., Mokni, T., Aye, P., Erny, P. and Dabadie, P. (2003) Epidemiology of traumatic comas: a prospective population-based study. Brain Injury, 17: 279–293.

McEwen, J. and McKenna, S.P. (1996) Nottingham health profile. In: Spilker B. (Ed.), Quality of Life and Pharmaeconomics in Clinical Trials. Lippincott-Raven, Philadelphia, PA, pp. 281–286.

Mellick, D., Gerhardt, K.A. and Whiteneck, G.G. (2003) Understanding outcomes based on the post-acute hospitalization pathways followed by persons with TBI. Brain Injury, 17(1): 55–71.

Powell, J.H., Beckers, K. and Greenwood, R.J. (1998) Measuring progress and outcome in community rehabilitation after brain injury with a new assessment instrument- the BICRO-39 scales. Arch. Phys. Med. Rehabil., 79: 1213–1225.

Powell, J.M., Temkin, N.R., Machamer, J.E. and Dikmen, S.S. (2002) Nonrandomized studies of rehabilitation for TBI: can they determine effectiveness? Arch. Phys. Med. Rehabil., 83(9): 1235–1244.

Ravens-Sieberer, U., Bettge, S., Heilmann, M. and Walleser, S. (2000) Assessment of quality of life in young children with a computer assisted touch screen program (CAT-SCREEN ©). Qual. Life Res., 9: 298.

Ready, R.E. and Ott, B.R. (2003) Quality of life measures for dementia. Health Qual. Life Outcomes, I: 11.

Spilker, B. (1990) Introduction. Quality of life assessments in clinical trials. In: Spilker B. (Ed.), Quality of Life and Pharmacoeconomics in Clinical Trials. Raven Press, New York, pp. 3–9.

Steadman-Pare, D., Colantonio, A., Ratcliff, G., Chase, S. and Vernich, L. (2001) Factors associated with perceived quality of life many years after TBI. J. Head Trauma Rehabil., 16: 330–342.

Strand, C.V. and Russell, A.S. (1997) WHO/ILAR taskforce on quality of life. J. Rheumatol., 24: 1630–1633.

Symonds, T., Berzon, R., Marquis, P., Rummans, T. and the Clinical Significance Consensus Meeting Group (2002) The clinical significance of quality of life results: practical considerations for specific audience. Mayo Clin. Proc., 77, 572–583.

Teasdale, T.W., Christensen, A., Willmes, K., Deloche, G., Braga, L., Stachowiak, F., Vendrell, J.M., Castro-Caldas, A., Laaksonen, R. and Leclercq, M. (1997) Subjective experience in brain-injured patients and their close relatives: A European brain injury questionnaire study. Brain Injury, 11: 543–563.

The WHOQOL Group. (1993) Study Protocol of the World Health Organization project to develop a quality of life assessment instrument (WHOQOL). Qual. Life Res., 2: 153–159.

The WHOQOL Group. (1998) The World Health Organization quality of life assessment (WHOQOL): Development and general psychometric properties. Soc. Sci. Med., 46: 569–1585.

Von Steinbüechel, N., Petersen, C., Bullinger, M. and the Qolibri Group. (2005) Assessment of health-related quality of life in persons after traumatic brain injury–development of the Qolibri, a specific measure. Acta. Neurochir., 93: 1–7.

Ware, J.E. (1996) The SF-36 health survey. In: Spilker B. (Ed.), Quality of Life and Pharmaeconomics in Clinical Trials. Lippincott-Raven, Philadelphia, PA, pp. 337–346.

Ware, J.E. and Sherbourne, C.D. (1992) The MOS 36-item short-form health survey (SF-36). I. Conceptual framework and item selection. Med. Care, 30: 473–483.

World Health Organization. (1948) Constitution of the World Health Organisation. Geneva.

Wedcliffe, T. and Ross, E. (2001) The psychological effects of TBI on the quality of life of a group of spouses/partners. S. Africa J. Commun. Disord., 48: 77–99.

S. Laureys (Ed.)
Progress in Brain Research, Vol. 150
ISSN 0079-6123

CHAPTER 39

Outcome and ethics in severe brain damage

Jean-Louis Vincent*

Department of Intensive Care, Erasme University Hospital, Free University of Brussels, B-1070 Brussels, Belgium

Abstract: The patient with severe brain damage represents a considerable ethical challenge for the medical team due to uncertainties in diagnosis and prognosis, the younger age of many of these patients, and the frequent acute nature of the disease, which allows little time for discussion of end-of-life issues with the patient. Surrogates are often relied on to fill in the gaps and provide their, not always reliable, interpretation of how they feel the patient would want to have been treated. The debate regarding the withdrawing/withholding of life-sustaining treatment is discussed but may not apply to many patients with severe brain damage who do not usually require invasive life support. However, withdrawal of artificial feeding and hydration is very relevant to such patients and is highly controversial. These issues are highly emotive and subjective, and individuals' views will depend on many factors including cultural background and religion. There are relatively few published data regarding ethical issues in the severely brain damaged patient and open discussion of the multiple facets of this difficult area must be encouraged.

Introduction

Many of the ethical decisions facing intensivists deal with end-of-life care in patients close to death. Patients with severe brain damage may not be dying, but their outcome is questionable; many will remain permanently vegetative and those who do recover will almost certainly never regain their pre-damage quality of life. The fundamental ethical principles of autonomy, non-maleficience, beneficence, and distributive justice apply equally to the patient with severe brain damage and to the dying, critically ill patient, but as patients with severe brain damage are often younger and previously healthy, decisions may have more profound consequences and be more emotive. In addition, the uncertainty regarding the long-term future of patients with severe brain damage and the very nature of these patients who often look so "normal" and are relatively independent of medical support, makes ethical decision-making especially difficult. Nevertheless, even if death is not imminent and a patient is permanently unconscious, withdrawing life-prolonging medical treatment may be considered an ethical decision (Council on Scientific Affairs and Council on Ethical and Judicial Affairs, 1990). There are relatively few data on the ethical issues surrounding patients with severe brain damage (also see with regard to brain death (Bernat this volume), vegetative state (Jennett this volume), and minimally conscious state (Fins this volume)). Here, we will discuss end-of-life issues from a more general intensive care viewpoint, for which more data are available, and which are generally applicable to the patient with severe brain damage.

Definitions

The need for a definition of death may at first glance seem rather superfluous — after all, is "dead" not a rather obvious state? But with modern medical technology able to maintain cardiac

*Corresponding author. Tel.: +32 2 555 3380;
Fax: +32 2 555 4555; E-mail: jlvincent@ulb.ac.be

DOI: 10.1016/S0079-6123(05)50039-6

and respiratory functions almost indefinitely, the situation is not quite so simple, and while for centuries death was diagnosed when cardiovascular and respiratory functions ceased, this definition is now not always adequate. The concept of "brain death" was introduced to deal with such patients and is now widely accepted internationally as distinguishing a dead patient from a living one. "Brain death" is defined as the absence of all brain functions and demonstrated by the presence of coma, apnea, and lack of brain stem reflexes, with exclusion of potentially confounding conditions such as severe hypothermia, severe electrolyte, or endocrine disturbance, or history of drug intoxication (Wijdicks, 2001). The precise methods needed to diagnose brain death and how many physicians of what specialties need to make the diagnosis vary between countries (Haupt and Rudolf, 1999; Wijdicks, 2002) and are influenced by cultural and religious factors. These factors are covered in more detail in the paper by Bernat in this issue: "The Concept and Practice of Brain Death".

Most patients with severe brain damage, however, do not meet the strict criteria for brain death and fall into a "gray" area, known as vegetative state, a term first used in 1972 by Jennett and Plum (1972). Vegetative state is a perturbing phenomenon, fundamentally different from cerebral death. Such patients are awake but unaware of themselves or their environment. They have no adapted, consistent, reproducible, or voluntary response to stimuli (visual, auditory, pain, etc.), but may sometimes make movements that seem coordinated, such as scratching, smiling, changing position, grasping, chewing, although they are transient and non-purposeful. Patients typically present sleep–wake cycles, they may blink when threatened, may even give the impression of following an object with the head or eyes, but the visual movements are never sustained. Patients may groan but do not speak. They may chew and even swallow food and drink, but in the majority of cases feeding must be provided enterally via a nasogastric tube or a gastrostomy tube. Patients in a vegetative state breathe spontaneously and do not need ventilatory support or even oxygen therapy. Persistent vegetative state is defined as a vegetative state remaining 1 month after acute traumatic or non-traumatic brain damage, while the vegetative state is said to be permanent (irreversible), if it lasts more than 3 months after non-traumatic brain damage, or 12 months after traumatic damage (The Multi-Society Task Force on PVS, 1994). The difficulties in measuring consciousness, along with case-reports of patients apparently recovering many years after a diagnosis of permanent vegetative state, make this a field fraught with uncertainty and littered with ethical and moral dilemmas.

Perhaps the issue needs to be turned around and a definition of life proposed such that death would simply mean "without life". But then, what is "life"? Merriam–Webster (http://www.m-w.com) defines life as "the quality that distinguishes a vital and functional being from a dead body", but what is this mysterious "quality"? The ability to rationalize, to interact with others, to conceive and conduct projects, to be aware of oneself and ones surroundings — all these concepts have been used to define "life" (Lizza, 1993; Randell, 2004). The problem remains, however, that whatever it is, the quality of "personhood" cannot be measured and thus remains a matter of philosophical debate (Kerridge et al., 2002; Randell, 2004). The question of whether a baby with anencephaly or a patient in persistent vegetative state is really "alive" is a huge ethical and moral problem. The loss of awareness in a person in a vegetative state, for example, even though brain functions may still exist at a lower level enabling vital organs to continue functioning, could be said to be a loss of human existence, and in losing their characteristic humanity, are such patients not, in essence, "dead"? (Fig. 1).

Intensive care unit decisions

Withholding versus withdrawing treatment: are they ethically identical?

Most deaths on the intensive care unit (ICU) are now preceded by a decision to withhold (refuse medical treatment that will not benefit the patient) or withdraw (reduce and remove medical treatment

Fig. 1. Futility of medical technology maintaining bodily functions in cases with irreversible loss of brain function (Reproduced with permission – copyright C. Laureys).

that is no longer of benefit) life-sustaining therapy (Buckley et al., 2004; Eidelman et al., 1998; Gajewska et al., 2004; Holzapfel et al., 2002; Prendergast et al., 1998; Sprung et al., 2003; Winter and Cohen, 1999). Most Western ethicists agree that there is no ethical difference between withholding and withdrawing therapy and this is supported by guidelines from various national bodies (American Thoracic Society, 1991; Ferdinande et al., 2001; British Medical Association, 2001); in both cases, the decision determines whether the treatment in question (e.g., vasopressor infusion, mechanical ventilation) will be administered in the immediate future and hence the outcome of both decisions is the same — whether the treatment is withheld or withdrawn, after the decision has been made, the patient will not be receiving it. Indeed, if withdrawal of therapy is seen as less acceptable than withholding, we may hesitate before starting or escalating therapy because we want to be sure that we do not start something that will turn out to be futile but cannot later be withdrawn. In the acute situation, even a short pause for thought could deprive a patient who would benefit from that therapy from their optimum chance of survival. If we know that we can withdraw a therapy at a later date, each patient can be given every chance of benefiting from that intervention. Importantly, this approach should not result in greater numbers of patients receiving prolonged futile care, or worse, falling into a permanent vegetative state — once the therapeutic "trial" has been shown to be of no effect, it must be withdrawn. Good communication with relatives (and patients when possible) is essential during this process to avoid raising false hopes and to ensure understanding that the "trial" will be stopped if there is no response.

Nevertheless, once a treatment has been started, the decision to stop can be more difficult than if it had not been started in the first place. This psychological and emotional "difference" between withholding versus withdrawing has been noted in several studies (Vincent, 1999; Giannini et al., 2003). In a questionnaire survey of 504 ICU physicians from 16 western European countries, 93% of respondents stated they sometimes withhold, but only 77% said that they sometimes withdraw therapy (Vincent, 1999). Religious and cultural differences can influence these decisions, and there are considerable international differences (Vincent, 1999; Sprung et al., 2003).

Withdrawal of feeding

In patients with severe brain damage, in a permanent vegetative state, the issue can be even more complex as such patients often require no sophisticated medical support and therapeutic withdrawal is therefore not applicable, although withholding treatment may be justifiable if a complication, such as an infection, presents. One key issue in such patients is the withdrawal of feeding. Some consider feeding to be a form of treatment and withdrawal of feeding would therefore be ethically acceptable as withdrawal of treatment in these patients with no hope of meaningful recovery. In a questionnaire survey of American neurologists and medical directors, 89% replied that they believed it ethical to withdraw artificial hydration and feeding from patients in persistent vegetative state (Payne et al., 1996). In a Belgian questionnaire, 56% of neurosurgeons, neurologists, and rehabilitation doctors felt withdrawing artificial feeding was sometimes appropriate (Dierickx et al., 1998), and in the UK the figure was 65% (Grubb et al., 1996). The General Medical Council (GMC) in the UK recently issued guidelines stating that "Where death is imminent and artificial hydration and/or nutrition are already in use, it may be appropriate to withdraw them if it is considered that the burdens outweigh the possible benefits to the patient" (http://www.gmc-uk.org/standards/whwd.htm). However, this is a highly controversial field and others would feel that nutritional support is non-medical everyday care and by stopping it and allowing death by malnutrition to become inevitable, the situation is akin to deliberately administering drugs to cause death, even if death occurs somewhat more slowly. The GMC guidelines have since been challenged and are currently undergoing judicial review (see e.g., Leslie Burke case; Dyer, 2004).

"Passive" versus "active"?

There have been several high profile media cases in which doctors have been charged with murder after deliberately administering drugs sufficient to kill the patient in order to relieve their suffering. Is there an ethical difference between the deliberate injection of large doses of opioids or sedatives to fasten death and withdrawing or withholding therapy, which will also allow death to occur more rapidly than if life-support is continued? Many feel that the deliberate injection of drugs is more "active" and hence less ethically acceptable than the more "passive" withholding/withdrawing. Nevertheless, in a questionnaire survey of European intensivists, Vincent (1999) noted that 40% of survey respondents reported that they sometimes inject fatal doses of drugs to speed death in patients with no hope of survival. In the Ethicus study, 6.5% of deaths were preceded by active shortening of the dying process (Sprung et al., 2003). In the questionnaire sent to American neurologists and medical directors, Payne et al. (1996) reported that 20% of respondents felt it would be ethical to speed death by lethal injection in patients in persistent vegetative state. When the prognosis is clearly established, which is not always easy, and the patient has no hope of surviving, the end result of withdrawing/withholding and deliberate drug administration is the same — the patient will die sooner than if life-sustaining treatments had been continued or the injection of opiate had not been given — but the cause of death is still the underlying illness. This whole area is fraught with ethical and legal "gray-zones". A key problem, both legally and ethically, is in determining the physician's intent in deliberately giving drugs to a dying patient. If the reasons are purely to relieve pain and suffering, deliberate injection of drugs can be considered ethical even if it inadvertently hastens death (the so-called rule of double effect), because it was not the doctor's intent to help the patients die, even though death may be foreseen as a result of the doctor's actions (Quill et al., 1997). However, patients with severe brain damage in a persistent vegetative state do not feel pain or discomfort and it is more difficult to apply the ethical double-effect escape clause.

The decision-makers

Surrogates

An important question in ethical decision-making is who should make the decision. The ICU situation is somewhat different from other medical specialties in that the intensivist will rarely have had time to get acquainted with the patient or to have developed any idea of their likely beliefs or preferences in terms of end-of-life care. In addition, the ICU patient will rarely be able to give any current expression of their wishes and family members are often the only persons who can assess what the patient would want. However, surrogates are not always an ideal reference as they may not necessarily have the same views and preferences as the patient, and it may be difficult for them to objectively separate how the patient would have wanted to be treated from the way in which they themselves would like the patient treated, particularly when faced with the fear and emotion of losing a loved one (Pochard et al., 2001). Ideally, the patient will have designated a preferred surrogate before the illness, but in many ICU patients the onset of illness is often too acute for this to be possible.

Advance directives

End-of-life decisions should be made in advance whenever possible. It may be difficult to broach the subject of death or vegetative state with patients and relatives, when all they really want to hear is that they or their loved one will recover, but these issues are best discussed openly and early to allow time for reflection and to avoid having to make an on-the-spot, impulsive judgment. Death is a natural process and although in many cases, particularly in the young, it may not seem fair in its choice of "victim", failing to raise and discuss the possibility of death early will ultimately benefit none. Issues of quality of life may also need to be discussed (see Bullinger, this volume), particularly in the patient with severe brain damage who may survive but likely with severe disability or permanently unconscious. The use of advance directives, "a description of one's future preferences for medical treatment made in anticipation of a time when one may not be able to express those preferences because of serious illness or damage" (Prendergast, 2001), has been promoted widely, particularly in the United States. The idea is not a bad one if it promotes early discussion of how a patient would feel if the worst scenario became a reality, but the presence of an advance directive should not preclude the possibility that patients may have changed their minds when faced with the certainty of their situation. Indeed, if asked, the healthy individual would normally say they would not want to be kept alive as a "vegetable", but the benchmark for what is considered as "keeping alive" may alter as one's health deteriorates. Exactly as teenagers visiting elderly relatives in a nursing home cannot imagine wanting to "end up like that", yet when that teenager reaches a certain age and mobility is reduced by illness, the idea of life in a nursing home may not seem so bad, so the healthy individual's opinions about what they want for their future may change when struck down by a serious, life-threatening illness (see Laureys, this volume, on "locked-in syndrome" patients). Moreover, advance directives may "penalize" those who do not sign one, making an advance directive more of an "opt-out" clause from society's assumptions as to how such patients should be treated. For example, if the assumption is that one would not want to have all support continued if one's brain were irreversibly damaged, if there were no advance directive to state otherwise, life-sustaining therapy would be withdrawn even if the patient would actually have wanted it to be continued. If the assumption were that society's preference would be to continue all therapy, unless an advance directive had been completed to say otherwise, life-sustaining therapy would be administered even when clearly of no benefit to the patient.

The ICU team

Intensive care medicine is very much a team effort and all members of the team, including nurses and other paramedical staff, should be able to be involved in discussions related to these decisions in addition to family members and the patient (whenever possible). Decisions should not be taken by the "paternalistic" physician alone, nor by the grief-stricken family, but should be consensus-built "shared" decisions (Carlet et al., 2004). While team involvement is essential, prolonging or postponing discussions to include absent members of the care team, or delegating the decision to a large ethical committee, can unnecessarily delay the decision and be distressing for family and patient. Importantly, any such discussions and decisions should be documented clearly in the patient notes, although documentation of end-of-life decisions and practice is often woefully inadequate (Heyland et al., 2003; Kirchhoff et al., 2004).

Family involvement and good communication

Although the physician is ultimately responsible for any end-of-life decisions, the family should be kept fully informed. However, despite widespread agreement that families should be involved in such decisions, frequently they are not. In a European questionnaire, only 49% of end-of-life decisions were made with full consultation among staff, family, and patient (Vincent, 1999). In France, the family was involved in decision-making in 44% of cases but simply informed in 13%, and in 11% of patients who underwent withdrawing or withholding, there was no patient or family involvement in the decision (Ferrand et al., 2001). Within Europe, doctors from southern European countries are less likely to involve the patient and the family than doctors from northern Europe (Vincent, 1999). In a Spanish study, the family was not informed in 28% of such decisions (Esteban et al., 2001). Even in the United States, one study reported that a third of surveyed doctors had continued life-sustaining therapies despite patient and surrogate wishes that it be discontinued, and more than 80% had sometimes unilaterally discontinued life-supporting treatment judged to be futile (Asch et al., 1995).

Considerable progress has been made in improving end-of-life care in recent years, particular with regard to pain and symptom control, but communication with the patient, or more often, the patient's family, remains poor (SUPPORT Principal Investigators, 1995; Malacrida et al., 1998; Baker et al., 2000), despite recent guidelines stressing its importance (Truog et al., 2001; Carlet et al., 2004). As shown in the SUPPORT study (1995), an interventional study aimed at improving end-of-life care in ICU patients, global attempts to improve communication are unlikely to be successful. These skills need to be incorporated already into medical school curricula and there needs to be an individual commitment to providing adequate information to patients and their families. Inadequate communication can be an avoidable source of conflict and distress between families and medical staff (Norton et al., 2003).

Practicalities

Despite the fact that most ICU deaths are now preceded by a decision to withdraw or withhold life-sustaining therapy, only recently have guidelines begun to be established regarding the practical approach in such situations (Truog et al., 2001), although the ethical issues involved have been well discussed. The aim for all involved, patient, family, and healthcare staff, is that the patient dies speedily, without pain or distress, in as dignified a manner as possible (Singer et al., 1999). Importantly, although by withdrawing life-sustaining therapy we may have defined the beginning of the last stage of the dying process, the patient is not yet dead, and care needs to be taken to ensure the rest of the dying process is as comfortable as possible. Death occurs at its own pace and must be allowed to unfold in its own way (Chapple, 1999) with assistance when needed to relieve pain, distress, or discomfort. Adequate pain relief and sedation is a crucial part of good end-of-life care, and care should be taken to avoid under- or over-prescribing. In many ICU patients, assessment of pain relies on the evaluation of the level of consciousness and awareness to certain stimuli, and measures of respiratory and hemodynamic status (Truog et al., 2001), although these are far from being perfect and determination of the degree of pain and discomfort being experienced by patients can be difficult. Opioids are widely used for their analgesic and sedative properties, benzodiazepines for their anti-anxiety and sedative effects. Neuroleptic drugs, such as haloperidol, may also be useful for patients with delirium. Doses of these drugs should be titrated to the individual patient's needs remembering that many patients will already have been receiving analgesics or sedatives and may have developed some tolerance. In line with the discussions above, all doses, including frequency and rationale, should be noted in the patient's charts to insure it is clear that the intention is to relieve suffering rather than to cause death; interestingly, several studies have noted that the doses of drugs used to hasten death may be no higher than the doses used to relieve pain (Sprung et al., 2003; Chan et al., 2004). Guidelines have been published by various groups to assist the intensivist in achieving adequate sedation and avoiding under- or over-dosing (Truog et al., 2001; Hawryluck et al., 2002). In addition to pain and anxiety, other symptoms commonly encountered at the end-of-life include dyspnea, dehydration, nausea, and vomiting, and the management of these is well discussed in the recent Society of Critical Care Medicine (SCCM) guidelines (Truog et al., 2001). Avoidance of dehydration has perhaps been overemphasized.

Terminal extubation versus terminal wean

There has been some debate about the "optimal" approach to ventilator withdrawal at the end-of-life. Essentially there are two ways in which mechanical ventilation can be withdrawn: "terminal extubation" and "terminal weaning". In the first, the endotracheal tube is withdrawn usually after bolus doses of sedatives and analgesics, in the second, terminal weaning, the ventilator settings of FiO_2 (fraction of inspired oxygen) and/or respiratory rate are gradually reduced so that hypoxemia and hypercarbia develop; this can be done relatively rapidly over a few minutes or much more slowly. An advantage of the latter approach is that patients do not develop any signs of acute respiratory distress, which can be uncomfortable for the patient and distressing for relatives to watch, when weaning is performed slowly with sedative and analgesic administration. Proponents of "terminal extubation" would say that it is a more rapid and "cleaner" way of withdrawing, allowing the patient to die without unnecessary tubes and equipment. In a study of 32 patients on a neurocritical care unit who underwent terminal extubation, the median duration of survival after extubation was 7.5 h; morphine or fentanyl was needed to relieve respiratory distress in 68% of cases (Mayer and Kossoff, 1999). In Faber-Langendoen's 1992 survey of 273 critical care physicians, 83 (33%) of 256 preferred terminal weaning while 32 (13%) preferred rapid extubation and the remainder reported that they used both methods, depending on patient's needs (Faber-Langendoen, 1994).

Conclusion

In addition to their skill as traditional curative physicians, intensivists are increasingly being called on to display skills in providing terminal care. The diagnostic and prognostic uncertainties in the patient with severe brain damage make this a particularly difficult ethical area. Time must be taken to discuss likely prognoses and outcomes clearly with patients and family early in the disease process to enable opinions and wishes to be expressed and understood, although the acuteness of the illness in severe brain damage may limit options for early discussion with the patient, and the family is often the only source of information regarding a patient's likely wishes. Consideration needs to be made for the moral and religious standpoints of patient, family, and also staff, as difficult ethical decisions are discussed and made. Legal considerations are also important and vary considerably between countries. Within Europe, for example, the legal situation regarding end-of-life issues is more "liberal" in Northern Europe, than in Southern European countries, where withdrawal of therapy can be seen as tantamount to murder. Key legal cases, for example, the Quinlan and Cruzan cases in the United States and the Bland case in the United Kingdom have set some precedents, and legislation related to the role of surrogates, withdrawal of nutrition, physician- and family-assisted suicide, etc., is now being reviewed and altered in many countries.

Recent years have begun to see a new openness in the discussion of end-of-life care in general and of the ethical concerns related to patients with severe brain damage, fuelled in part by individual cases that have created media interest. Nevertheless, these ethical dilemmas remain a difficult area of clinical medicine that requires great sensitivity and understanding.

References

American Thoracic Society. (1991) Withholding and withdrawing life-sustaining therapy. This Official Statement of the American Thoracic Society was adopted by the ATS Board of Directors, March 1991. Am. Rev. Respir. Dis., 144: 726–731.

Asch, D.A., Hansen-Flaschen, J. and Lanken, P.N. (1995) Decisions to limit or continue life-sustaining treatment by critical care physicians in the United States: conflicts between physicians' practices and patients' wishes. Am. J. Respir. Crit. Care Med., 151: 288–292.

Baker, R., Wu, A.W., Teno, J.M., Kreling, B., Damiano, A.M., Rubin, H.R., Roach, M.J., Wenger, N.S., Phillips, R.S., Desbiens, N.A., Connors, Jr., A.F., Knaus, W. and Lynn, J. (2000) Family satisfaction with end-of-life care in seriously ill hospitalized adults. J. Am. Geriatr. Soc., 48: S61–S69.

British Medical Association. (2001) Withholding and Withdrawing Life-Prolonging Medical Treatment: Guidance for Decision-Making. BMJ Books, Tavistock.

Buckley, T.A., Joynt, G.M., Tan, P.Y., Cheng, C.A. and Yap, F.H. (2004) Limitation of life support: frequency and practice in a Hong Kong intensive care unit. Crit. Care Med., 32: 415–420.

Carlet, J., Thijs, L.G., Antonelli, M., Cassell, J., Cox, P., Hill, N., Hinds, C., Pimentel, J.M., Reinhart, K. and Thompson, B.T. (2004) Challenges in End-of-life Care in the ICU Statement of the 5th International Consensus Conference in Critical Care, Brussels, Belgium, April 2003. Intensive Care Med., 30: 770–784.

Chan, J.D., Treece, P.D., Engelberg, R.A., Crowley, L., Rubenfeld, G.D., Steinberg, K.P. and Curtis, J.R. (2004) Narcotic and benzodiazepine use after withdrawal of life support: association with time to death? Chest, 126: 286–293.

Chapple, H.S. (1999) Changing the game in the intensive care unit: letting nature take its course. Crit. Care Nurse, 19: 25–34.

Council on Scientific Affairs and Council on Ethical and Judicial Affairs. (1990) Persistent vegetative state and the decision to withdraw or withhold life support. JAMA, 263: 426–430.

Dierickx, K., Schotsmans, P., Grubb, A., Walsh, P. and Lambe, N. (1998) Belgian doctors' attitudes on the management of patients in persistent vegetative state (PVS): ethical and regulatory aspects. Acta Neurochir. (Wien), 140: 481–489.

Dyer, C. (2004) GMC faces challenge over withdrawing treatment. BMJ, 328: 68.

Eidelman, L.A., Jakobson, D.J., Pizov, R., Geber, D., Leibovitz, L. and Sprung, C.L. (1998) Foregoing life-sustaining treatment in an Israeli ICU. Intensive Care Med., 24: 162–166.

Esteban, A., Gordo, F., Solsona, J.F., Alia, I., Caballero, J., Bouza, C., Alcala-Zamora, J., Cook, D.J., Sanchez, J.M., Abizanda, R., Miro, G., Fernandez Del Cabo, M.J., de Miguel, E., Santos, J.A. and Balerdi, B. (2001) Withdrawing and withholding life support in the intensive care unit: a Spanish prospective multi-centre observational study. Intensive Care Med., 27: 1744–1749.

Faber-Langendoen, K. (1994) The clinical management of dying patients receiving mechanical ventilation. A survey of physician practice. Chest, 106: 880–888.

Ferdinande, P., Berre, J., Colardyn, P., Damas, P., de Marre, F., Devlieger, H., Groenen, M., Grosjean, P., Installe, E., Lamy, M., Laurent, M., Lauwers, P., Lothaire, T., Reynaert, M., Roelandt, L., Slingeneyer de Goeswin, M. and Vincent, J.L. (2001) La fin de la vie en médecine intensive. Réanimation, 10: 340–341.

Ferrand, E., Robert, R., Ingrand, P. and Lemaire, F. (2001) Withholding and withdrawal of life support in intensive-care units in France: a prospective survey. French LATAREA Group. Lancet, 357: 9–14.

Gajewska, K., Schroeder, M., de Marre, F. and Vincent, J.L. (2004) Analysis of terminal events in 109 successive deaths in a Belgian intensive care unit. Intensive Care Med., 30: 1224–1227.

Giannini, A., Pessina, A. and Tacchi, E.M. (2003) End-of-life decisions in intensive care units: attitudes of physicians in an Italian urban setting. Intensive Care Med., 29: 1902–1910.

Grubb, A., Walsh, P., Lambe, N., Murrells, T. and Robinson, S. (1996) Survey of British clinicians' views on management of patients in persistent vegetative state. Lancet, 348: 35–40.

Haupt, W.F. and Rudolf, J. (1999) European brain death codes: a comparison of national guidelines. J. Neurol., 246: 432–437.

Hawryluck, L.A., Harvey, W.R., Lemieux-Charles, L. and Singer, P.A. (2002) Consensus guidelines on analgesia and sedation in dying intensive care unit patients. BMC Med. Ethics, 3: E3.

Heyland, D.K., Rocker, G.M., O'Callaghan, C.J., Dodek, P.M. and Cook, D.J. (2003) Dying in the ICU: perspectives of family members. Chest, 124: 392–397.

Holzapfel, L., Demingeon, G., Piralla, B., Biot, L. and Nallet, B. (2002) A four-step protocol for limitation of treatment in terminal care. An observational study in 475 intensive care unit patients. Intensive Care Med., 28: 1309–1315.

Jennett, B. and Plum, F. (1972) Persistent vegetative state after brain damage. A syndrome in search of a name. Lancet, 1: 734–737.

Kerridge, I.H., Saul, P., Lowe, M., McPhee, J. and Williams, D. (2002) Death, dying and donation: organ transplantation and the diagnosis of death. J. Med. Ethics,, 28: 89–94.

Kirchhoff, K.T., Anumandla, P.R., Foth, K.T., Lues, S.N. and Gilbertson-White, S.H. (2004) Documentation on withdrawal of life support in adult patients in the intensive care unit. Am. J. Crit. Care,, 13: 328–334.

Lizza, J.P. (1993) Persons and death: what's metaphysically wrong with our current statutory definition of death? J. Med. Philos., 18: 351–374.

Malacrida, R., Bettelini, C.M., Degrate, A., Martinez, M., Badia, F., Piazza, J., Vizzardi, N., Wullschleger, R. and Rapin, C.H. (1998) Reasons for dissatisfaction: a survey of relatives of intensive care patients who died. Crit. Care Med., 26: 1187–1193.

Mayer, S.A. and Kossoff, S.B. (1999) Withdrawal of life support in the neurological intensive care unit. Neurology, 52: 1602–1609.

Norton, S.A., Tilden, V.P., Tolle, S.W., Nelson, C.A. and Eggman, S.T. (2003) Life support withdrawal: communication and conflict. Am. J. Crit. Care,, 12: 548–555.

Payne, K., Taylor, R.M., Stocking, C. and Sachs, G.A. (1996) Physicians' attitudes about the care of patients in the persistent vegetative state: a national survey. Ann. Int. Med., 125: 104–110.

Pochard, F., Azoulay, E., Chevret, S., Lemaire, F., Hubert, P., Canoui, P., Grassin, M., Zittoun, R., Le Gall, J.R., Dhainaut, J.F. and Schlemmer, B. (2001) Symptoms of anxiety and depression in family members of intensive care unit patients: ethical hypothesis regarding decision-making capacity. Crit. Care Med., 29: 1893–1897.

Prendergast, T.J. (2001) Advance care planning: pitfalls, progress, promise. Crit. Care Med., 29: N34–N39.

Prendergast, T.J., Claessens, M.T. and Luce, J.M. (1998) A national survey of end-of-life care for critically ill patients. Am. J. Respir. Crit. Care Med., 158: 1163–1167.

Quill, T.E., Dresser, R. and Brock, D.W. (1997) The rule of double effect — a critique of its role in end-of-life decision making. New Engl. J. Med., 337: 1768–1771.

Randell, T.T. (2004) Medical and legal considerations of brain death. Acta Anaesthesiol. Scand., 48: 139–144.

Singer, P.A., Martin, D.K. and Kelner, M. (1999) Quality end-of-life care: patients' perspectives. JAMA, 281: 163–168.

Sprung, C.L., Cohen, S.L., Sjokvist, P., Baras, M., Bulow, H.H., Hovilehto, S., Ledoux, D., Lippert, A., Maia, P., Phelan, D., Schobersberger, W., Wennberg, E. and Woodcock, T. (2003) End-of-life practices in European intensive care units: the Ethicus Study. JAMA, 290: 790–797.

SUPPORT Principal Investigators. (1995) A controlled trial to improve care for seriously ill hospitalized patients. The Study to Understand Prognoses and Preferences for Outcomes and Risks of Treatments (SUPPORT). JAMA, 274: 1591–1598.

The Multi-Society Task Force on PVS. (1994) Medical aspects of the persistent vegetative state. New Engl. J. Med., 330: 1499–1508.

Truog, R.D., Cist, A.F., Brackett, S.E., Burns, J.P., Curley, M.A., Danis, M., DeVita, M.A., Rosenbaum, S.H., Rothenberg, D.M., Sprung, C.L., Webb, S.A., Wlody, G.S. and Hurford, W.E. (2001) Recommendations for end-of-life care in the intensive care unit: The Ethics Committee of the Society of Critical Care Medicine. Crit. Care Med., 29: 2332–2348.

Vincent, J.L. (1999) Forgoing life support in Western European intensive care units: The results of an ethical questionnaire. Crit. Care Med., 27: 1626–1633.

Wijdicks, E.F. (2001) The diagnosis of brain death. New Engl. J. Med., 344: 1215–1221.

Wijdicks, E.F. (2002) Brain death worldwide: accepted fact but no global consensus in diagnostic criteria. Neurology, 58: 20–25.

Winter, B. and Cohen, S. (1999) ABC of intensive care. Withdrawal of treatment. BMJ, 319: 306–308.

S. Laureys (Ed.)
Progress in Brain Research, Vol. 150
ISSN 0079-6123

CHAPTER 40

Clinical pragmatism and the care of brain damaged patients: toward a palliative neuroethics for disorders of consciousness

Joseph J. Fins*

Division of Medical Ethics, Weill Medical College of Cornell University, 435 East 70th Street, Suite 4-J, New York, NY 10021, USA

Abstract: Unraveling the mysteries of consciousness, lost and regained, and perhaps even intervening so as to prompt recovery are advances for which neither the clinical nor the lay community are prepared. These advances will shake existing expectations about severe brain damage and will find an unprepared clinical context, perhaps even one inhospitable to what should clearly be viewed as important advances. This could be the outcome of this line of inquiry, if this exceptionally imaginative research can continue at all. This work faces a restrictive research environment that has the potential to imperil it. Added to the complexity of the scientific challenges that must be overcome is the societal context in which these investigations must occur. Research on human consciousness goes to the heart of our humanity and asks us to grapple with fundamental questions about the self. Added to this is the regulatory complexity of research on subjects who may be unable to provide their own consent because of impaired decision-making capacity, itself a function of altered or impaired consciousness. These factors can lead to a restrictive view of research that can favor risk aversion over discovery. In this paper, I attempt to explain systematically some of these challenges. I suggest that some of the resistance might be tempered if we view the needs of patients with severe brain injury through the prism of palliative care and adopt that field's ethos and methods when caring for and conducting research on individuals with severe brain damage and disorders of consciousness. To make this argument I draw upon the American pragmatic tradition and utilize *clinical pragmatism*, a method of moral problem-solving that my colleagues and I have developed to address ethical challenges in clinical care and research.

Introduction

The world is not ready for the potential discoveries and innovations described in this collection. Unraveling the mysteries of consciousness, lost and regained, and perhaps even intervening so as to prompt recovery are possibilities for which neither the clinical nor the lay community are prepared. These developments will shake existing expectations about severe brain damage and will find an unprepared clinical context, perhaps even one inhospitable to what should clearly be viewed as important advances.

This could be the outcome of this line of inquiry, if this exceptionally imaginative research can continue at all. This work faces a restrictive research environment that has the potential to imperil it (Fins, 2003a,b). Added to the complexity of the scientific challenges that must be overcome is the societal context in which these investigations must occur. Research on human consciousness goes to

*Corresponding author. Tel.: +1 (212) 746 4246;
Fax: +1 (212) 746 8738; E-mail: jjfins@mail.med.cornell.edu

DOI: 10.1016/S0079-6123(05)50040-2

the heart of our humanity and asks us to grapple with fundamental questions about the self (Fins, 2004a). Added to this is the regulatory complexity of research on subjects who may be unable to provide their own consent because of impaired decision-making capacity, itself a function of altered or impaired consciousness. These factors can lead to a restrictive view of research that can favor risk aversion over discovery.

In this chapter, I attempt to explain systematically some of these challenges. I suggest that some of the resistance might be tempered if we view the needs of patients with severe brain injury through the prism of palliative care and adopt that field's ethos and methods when caring for and conducting research on individuals with severe brain injury and disorders of consciousness. To make this argument I draw upon the American pragmatic tradition and utilize *clinical pragmatism*, a method of moral problem-solving that my colleagues and I have developed to address ethical challenges in clinical care and research (Fins and Bacchetta, 1995; Fins, 1996; Fins et al., 1997; Fins, 1998; Miller et al., 1997) (Fig. 1).

John Dewey and clinical pragmatism

Clinical pragmatism has its philosophical roots in the American pragmatic tradition of Oliver Wendell Holmes Jr., William James, Sanders Peirce and the work of John Dewey (1859–1952) in particular (Menand, 2001; Dewey, 1988; Dewey, 1991a, b). Dewey was a leading American philosopher of the first half of the 20th century as well as a psychologist, democratic theorist and education reformer (Ryan, 1995; Hook, 1995; Miller et al., 1996; Fins, 1999a). Deweyan pragmatism is an appealing philosophical method through which to view current advances in neuroscience and assess the novel ethical challenges posed by this work.

Dewey had a real-world perspective. He was interested in the intersection of theory and practice and had a strong interest in using the scientific method as a means to address normative questions (Dewey, 1997, 1998). His analytical approach was inductive. He drew heavily upon empirical observations that would lead to intelligent interventions by better appreciating which ethical principles — autonomy, beneficence, nonmaleficence and justice — were applicable in a given case. But while principles matter in his pragmatism, Dewey was neither an essentialist nor a moral absolutist. Influenced by Darwin's work, he accepted the fact that advances in science and society called for an evolution in our thinking and mores. These advances were not to be feared, or categorically proscribed, but rather understood as part of the evolving history of humankind.

That Dewey accepted an evolving state of knowledge as normative makes his work especially relevant to the ethical assessment of scientific novelty. Although his pragmatism would apply prevailing ethical principles to established or settled questions, he acknowledged that in novel circumstances one cannot know in advance which ethical principles mattered or how competing ethical principles might be balanced and specified (Beauchamp and Childress, 1994). Such questions in the clinical and research context might be addressed through the application of clinical pragmatism. Building upon Dewey's pragmatism, clinical pragmatism does not view ethical principles as absolute or fixed truths independent of context. Principles, instead, are viewed as hypotheses that need to be understood and accepted in view of the consequence of their application in a particular narrative context. In these ethically contingent cases, moral judgment is better served by a disciplined process of analysis, which attends to the facts and particulars under consideration. In this way, we can delineate which principles are applicable and discern consequent ethical responsibilities and obligations.

To help in the balancing and specification of principles, we need to engage in a contextually situated analysis that Dewey called the process of *inquiry*. The purpose of inquiry is to be comprehensive and systematic, to consider all the material medical and narrative facts that would be necessary to make a reasonable judgment about a decision or course of action. This process of inquiry is analogous to the inductive reasoning of differential diagnosis used in clinical medicine. And like that process of diagnosis, pragmatic inquiry begins with the recognition of what Dewey described as

Fig. 1. Flemish renaissance artist Hieronymus Bosch's (1450–1515) "The Cure of Folly" or "The Stone Operation" (detail) showing removal of a stone from the brain, thought to cause madness. The stone has the form of a lotus, interpreted to symbolize spiritual awareness. Bosch's "psychosurgery" can be seen as an allegory of human stupidity or as depicting medical "quackery".

the problematic situation. A problematic situation is one that should prompt deeper reflection because ethical tensions are present but as yet unexplored or fully appreciated. It should be noted that recognizing problematic situations is easier said than done.

Although brewing ethical problems may become obvious over time, the key is to identify them as early as possible so that an explicit process of inquiry may be brought to bear upon the problem. This is especially true with considering the ethical dimensions of the management of patients with impaired consciousness where, as we shall see, prevailing biases prevent us from recognizing what should be viewed as compelling ethical obligations.

Inquiry can begin once the problematic situation has been made explicit. Although missed opportunities for inquiry are obvious in retrospect, the goal is to identify them prospectively so that a careful analysis can proceed. When this occurs, it is possible to collect the medical, narrative and contextual details necessary for the analytical process necessary to address the problem. This process is a series of related steps that can occur in a sequence or simultaneously. But in the clinical context, addressing ethical issues in practice, the process generally begins with an attempt to clarify the patient's diagnosis and prognosis and to ascertain what the patient and their family understand about these issues. We then seek to determine the patient's preferences, ability to make decisions and his values, religious beliefs, needs and desires. This mix of clinical and narrative information is then understood in the context of the broader family dynamics and how the patient's burden of illness affects the family and other intimates. And as we do this, we seek to determine which surrogate speaks for the patient if the patient has lost decision-making capacity.

From patient and family issues we next consider institutional and broader social determinants that may influence decisions about care or its provision. At the institutional level variables like the doctor–patient relationship and the continuity or discontinuity of care become important as does where care is being provided and their experience with disorders of consciousness, brain damage and rehabilitation. And finally, given the sociological complexity of disorders of consciousness, we need to be cognizant of how broader societal biases may inform the provision of care and the attitudes of surrogates or clinical staff.

It is only after this disciplined assessment of the clinical, narrative and broader contextual features of the case that we can formulate what might be termed *an ethics differential diagnosis* or our speculations about the moral considerations relevant to the case that might inform a negotiated and workable consensus about care plans with the patient and/or family. Reaching this consensus among all those who have a stake in the deliberations is also an important feature of Deweyan pragmatism and the method of clinical pragmatism. It helps foster an inclusive process and preempts unilateral judgments which may be ill-informed or morally absolutist.

It should be noted that like a differential diagnosis, these speculations may be in error and cannot be known a priori. Instead, these normative hypotheses emerge from the careful analysis of the relevant facts and with the humility of a scientist using the experimental method-seeking "truth." Sometimes this deliberative process will confirm an expected conclusion. On other occasions, this same process will help prevent a rush to moral judgment that might constrain the articulation of the range of morally reasonable outcomes.

In either case, the use of this inductive mode has instrumental value. It optimizes the deliberative process by fostering a principled and coherent mode of reasoning that remains contextually grounded and particularistic, an asset when considering the novel ethical challenges posed by the diagnosis and treatment of disorders of consciousness.

Finally, this deliberative process will engender new insights through the periodic review that should come after every case. In this way, we allow ourselves to take part in a process of experiential learning in which lessons borne of experience and being in the world make us more attentive to the next problematic situation, thus setting the stage for inquiry.

To summarize, clinical pragmatism as a method of moral problem-solving begins with the recognition of the problematic situation. This prompts a process of inquiry that includes the collection of medical, narrative and contextual data. This information leads to the articulation of an ethics differential diagnosis. These speculations in turn inform a negotiation with stakeholders about a plan of action with this consensus leading to an intervention. This process is completed with a periodic review that will foster experiential learning. (An outline of this approach is summarized in Table 1.)

Clinical pragmatism and disorders of consciousness

The problematic situation

The very clinical context of care encountered by patients with disorders of consciousness is a problematic situation that generally goes unrecognized. I have previously characterized this as a *neglect syndrome*, borrowing the diagnostic category from clinical neurology (Fins, 2003a). Although notable exceptions exist among highly dedicated neurologists, neurosurgeons and rehabilitation specialists, the vast majority of patients with disorders of consciousness are a population who remain out of our gaze.

Table 1. Clinical pragmatism and inquiry

I. Recognition of the problematic situation and the need for inquiry
II. Data collection: medical, narrative, contextual
1. Medical facts
2. Patient/surrogate preferences
3. Family dynamics
4. Institutional arrangements
5. Broader societal issues and norms
III. Interpretation (ethics differential diagnosis)
IV. Negotiation
V. Intervention
VI. Periodic review/experiential learning

Without careful consideration or an evidence base, it is taken for granted that patients who have sustained severe brain damage are beyond hope of any remediation. This leads to a clinical context marked by errors of omission and a sense that, ethically, nothing can or should be done for patients with catastrophic brain dysfunction. These prevailing sentiments have undermined the usual intellectual curiosity that marks other areas of practice. Surveys of neurologists' knowledge of diagnostic categories related to disorders of consciousness are rife with errors that would be intolerable elsewhere in practice (Childs et al., 1993), though they are tolerated here because of the sense that diagnostic clarity in the face of one form of futility or another just does not matter.

This is indeed paradoxical because while most in the clinical context are dismissive of these patients, questions related to consciousness are among the most fascinating topics in science and society. In an essay in the *New York Review of Books*, Oliver Sacks suggests that new approaches like MRI, PET and neural modeling, "... now make the quest for neural correlates of consciousness the most fundamental and exciting adventure in neuroscience today." (Sacks, 2004, p. 43). His review featured books written for a lay audience by distinguished scientists including Nobel Laureates Edelman and Crick as well as the latter's colleague Koch (Crick, 1994; Edelman, 2004; Koch, 2004). Despite the distinction of these commentators, and the wider audience to which they appeal, disinterest and lack of intellectual curiosity have marked the clinical context. Even diagnostic rigor is lacking (Fins and Plum, 2004). It is not uncommon for the same patient with impaired consciousness to be described as nearly brain dead, vegetative or comatose.

These practices can create a problematic situation that can go unrecognized because such attitudes become so pervasive so as to become the norm. The potential recovery of a brain-damaged patient goes unexamined because it is assumed that the intellectual consequences will be dire. Even survivors with intact cognitive abilities, like Jean Dominique Bauby — the author of *The Diving Bell and the Butterfly*, a memoir of the Locked-in-State (LIS) — comments, "But improved resuscitation techniques have prolonged the agony." (Bauby, 1997, p. 4). These perceptions prevail despite the fact that evidence-based outcomes demonstrate a wide variety of outcomes depending upon the etiology and anatomy of the injury, whether it was traumatic or anoxic, and demographic determinants like the patient's age and overall health status.

Although one would think that absence of careful diagnosis and informed prognostication would be an obvious deficit calling out for more precise assessment, confident pronouncements that there is "no hope for meaningful recovery" are taken for granted, unexamined and without the requisite skepticism that marks prudential medicine. Meaningful to whom? The neurologist? The patient? The family?

Recognition of the problematic situation requires that these questions are asked and that answers are forthcoming. It is also critical to appreciate that notions of the life that might be worth living can evolve as disability is confronted. Quality of life considerations can also be informed by perspective. The literature indicates clinicians forsaking the quality of life that their care will provide to chronically ill patients under their care, although these patients were satisfied with their quality of life (Uhlman et al., 1988; Uhlman and Pearlman, 1991).

Another unarticulated problem that informs the care provided to these patients are the norms that inform decisions about life-sustaining therapy, namely decisions about withholding and withdrawing care. Our views on do-not-resuscitate (DNR) orders and decisions to withdraw life-sustaining therapies like ventilators have grown out of our experience in intensive care units and the provision of acute care (Zussman, 1992). End-of-life decisions are informed by the patient's course with their level of acuity, indeed their viability, guiding choices to withhold or withdraw care.

These decisions can occur during a relatively short period of time once the patient's prognosis turns dire and the outcome becomes clear. In a retrospective chart review of end-of-life practice patterns of patients who had in-hospital deaths (Fins et al., 1999, 2000), we found that DNR orders were written — on median — 2 days after admission when the average length of stay of the patients studied was 16.9 days (median 9.0 days). It is also interesting to note that most DNR orders (64%) were agreed to by surrogates after the patient had lost decision-making capacity, suggesting surrogates may feel more willing to forgo resuscitation and pursue less aggressive care once consciousness has been lost. While these data retrospectively examined practice patterns of patients who had died, they do point to the relatively abridged timeframes that inform end-of-life decisions and the value placed on consciousness. Similar findings using the same analytical framework were noted in Australia (Middlewood et al., 2001).

This is the context of care for critically ill patients who have sustained brain damage, where clinical routines can become an excuse for unreflective practice. It is a setting with decisional constructs that operate in days and weeks, not months. And these contextual factors lead to a paradox that also informs the problematic situation: decisions about withholding and forgoing life-sustaining therapy are made prematurely for patients with disorders of consciousness because options for withdrawals are increasingly abridged as the patient moves from the acute to chronic stage of illness, even though these decisions may become more justifiable once the prognosis becomes clearer.

As has been noted elsewhere in this collection of essays, the course of recovery from the vegetative states can take from 3 to 12 months, depending upon whether the etiology of the insult was anoxic or traumatic (Kobylarz and Schiff, 2004; Jennett, this volume). Thus, to fully know whether recovery from the vegetative state is possible, a patient with traumatic brain injury might have to be observed for 12 months. By then he will have become *medically* stable and been extubated (Winslade, 1998). His only life-sustaining therapy might be his percutaneous gastrostomy tube and hence the paradox. While the legal and ethical norms clearly view a ventilator as an extraordinary measure that can be withdrawn within weeks of an injury, the ethical (and theological consensus) on decisions to withdraw artificial nutrition and hydration is less clear even after a patient is irreversibly in a permanent vegetative state.

These difficulties, coupled with the culture of the intensive care, lead to decisions to remove life-sustaining therapy, while it remains feasible, even though the dimensions of the patient's recovery remain unknown. This can lead to pronouncements about expected outcomes that may reflect personal biases and fall short on the evidence.

A better approach would be to be transparent with families and to explain to them that the patient's prognosis and hope for recovery of consciousness will become clearer within weeks and months. Although we should not compel families to continue to treat patients who are persistently but not permanently vegetative in order to develop greater prognostic clarity, it seems prudent and ethically within the norm of informed consent *and* refusal, to be intellectually honest about what can be predicted so early in the course of care. At the very least, clinicians need to be careful about making expedient claims that there is "no hope for meaningful recovery" when the evidence is lacking in order to "scientifically" justify decisions to forgo life-sustaining therapy.

Data collection

Clarifying the medical facts: brain damage and the challenge of diagnosis and prognosis

The first issue that will be seen as immediately different in the care of patients with disorders of consciousness is the fact that it may be difficult to clarify the medical facts. The diagnostic categories that inform practice in this area — the vegetative and minimally conscious states (Giacino et al.,

2002; Giacino, this volume) — remain *descriptive.* Diagnoses are just on the cusp of becoming more physiologic and precise utilizing advanced imaging techniques and sophisticated electroencephalography, but our state of knowledge remains rudimentary compared with other domains in medicine.

A colleague suggested that our state of knowledge is just a bit more evolved from the time when infectious diseases were termed "the fevers" (Barondess, 2003). A review of the history of medicine from Hippocrates and Galen to Osler and beyond indicates how etiology and prognostication evolved from rudimentary observations and theoretical speculations concerning fever (Wilson, 1997). Asserting that patients had a pattern of fevers carried less diagnostic and prognostic precision than linking these empirical manifestations of illness to an actual pathogen.

The state of affairs in neurological diagnosis, because of the complexity of the object of study, *is less evolved than other areas of medicine.* Families depend upon diagnostic clarity to make life-altering and family defining choices but even this basic element of decision-making may be lacking clarity in disorders of consciousness. We take some semblance of diagnostic clarity elsewhere in medicine for granted. Clinicians counseling families about disorders of consciousness need to be careful to be as precise as possible and to acknowledge the limitations of what is currently known and predictable.

Although progress has been made since Jennett and Plum described the vegetative state as "a syndrome in search of a name" (Jennett and Plum, 1972), descriptive brain states are only now being refined in subcategories such as the persistent and permanent vegetative states and the more recent minimally conscious state based on correlations between clinical observation neuropathological studies and neuroimaging (Hirsch et al., 2001; Jennett et al., 2001; Jennett, 2002; Laureys et al., 2000, 2002; Menon et al., 1998; Schiff and Plum, 1999; Schiff et al., 2002; Wilson et al., 2001).

But as noteworthy as this work is, these efforts are rudimentary and fundamentally still descriptive with patients with variable injuries and outcomes clumped together into broad inclusive categories. This imprecision can lead to disorders of consciousness of different etiologies being clinically indistinguishable although their causes and potential course are dramatically different, making prediction and prognostication so difficult. Such diagnostic questions are further confounded by diagnostic error. Even the rather straightforward diagnosis of brain death is prone to errors of omission in clinical assessment as a report from Los Angeles County General Hospital has shown (Wang et al., 2002). This suggests that errors are even more likely when the diagnostic questions become more complex and subtle, for example, distinguishing a patient with no motor activity in the LIS from one who is in the vegetative state (Laureys et al., this volume).

These issues create a challenging problem for clinical care, which may be unique to this area in medicine, namely, the difficulty of providing patients and families with a clear and meaningful diagnosis, which is an essential element of any ethical analysis about care. Families accustomed to diagnostic precision in other disciplines have similar expectations when a loved one has a disorder of consciousness. They expect precision and could easily mistake an MRI scan as meeting their expectations. Their expectations are heightened by popular articles on neuroethics, increasingly a mode of technological critique versus a scholarly discipline, that warn of imaging studies that might be able to read your mind and monitor your thoughts (The Economist, 2004; Farah and Wolpe, 2004). If this is the case, they might logically ask, why can't my child's neurologist tell me if he is going to come out of his coma?

And when we do return from the realm of science fiction to the less promising confines of the clinic, we appreciate how crude our categories are. Is there really a functional or meaningful difference between a patient with persistently vegetative state versus one who is on the lower end of the minimally conscious state (MCS)? Here families also encounter a profession and society that has yet to reach a clear consensus on emerging diagnostic domains, as demonstrated by the controversy over the status of MCS. Minimally *contentious* to some, there is the scientific critique about the use of a consensus panel to create such categories (Cranford, 1998; Shewmon, 2002) as well as the political concerns of advocates asserting disability rights and the right to die. Some disability advocates have been wary of MCS because they fear that a new diagnostic category could equate patients with higher functioning to those in the vegetative state and lead to the devaluation of their lives (Coleman, 2002). Conversely, right to die advocates fear that the new category may open Pandora's box by suggesting that patients with impaired consciousness might be candidates for treatment and more aggressive care (Cranford, 1998). If they are, these advocates worry, will this erode society's willingness to accept withdrawal of care?

Despite the forces that might conspire to rob patients of the most accurate diagnosis currently available, establishing the medical facts and the diagnosis is a critical and necessary step in making ethical choices about patient care. This becomes clear if we consider the exceptional case of Terry Wallis who was misdiagnosed, or perhaps simply ignored, while in the MCS (Schiff and Fins, 2003). Wallis was 19 in 1984 when he was as unrestrained passenger in a motor vehicle accident. He received acute medical care, survived and was discharged to a nursing home only months after his injury (Giacino, 2004). He carried the diagnosis of the persistent vegetative state. But on July 11, 2003, headlines described a "miracle awakening" while in "custodial care" in a nursing home. He spoke tentatively, single words at a time like "Mom" and "Pepsi". As the weeks passed, he spoke still haltingly but with greater fluency. In his mind he had never aged, and Reagan was still President. Although Wallis had been labeled as being in the persistent vegetative state, the behaviors noted by family members — evidence of episodic but unreliable demonstrations of awareness and consciousness — went unheeded by the nursing home staff. Their views about him were codified by his diagnosis on transfer. He had been and would always remain vegetative. The family had requested that a neurologist reassess him during this period and were refused, so certain or indifferent were the staff to the question of diagnosis. So Terry Wallis went 19 years without a neurological assessment. It is sad to say that while his recovery is noteworthy and clearly reportable, the diagnostic neglect of patients with brain injury once they have left the acute care setting and rehabilitation is not uncommon (Winslade, 1998).

Patients who fail to demonstrate medical necessity or demonstrate improvement with rehabilitative interventions are transferred to what is ungraciously described as "custodial care," where assessment is rare and incomplete. Given the exigencies of discharge planning, it is quite possible for a patient to be discharged while still in the persistent vegetative state and move to the minimally conscious state once transferred to a nursing home within the first year of traumatic injury. If this occurs in an inattentive or disinterested setting, crossing this diagnostic demarcation may go unnoticed to the detriment of the patient and the family.

Another pressure that makes the usually routine task of diagnosis so difficult is that these diagnoses have become too value-laden. If we recall the Terri Schiavo case in Florida, we are reminded of how the objective process of diagnosis can be eroded by ideology (Wallis, 2003).

Terri Schiavo had sustained anoxic injury in 1990, which left her in the permanent vegetative state. Her case has also attracted national media attention as an important right to die case. The case centered on a dispute over Ms. Schiavo's previous articulated wishes about end-of-life care as well as her medical diagnosis. Her husband, the surrogate decision-maker recognized by multiple courts, brought forth evidence that indicated that Ms. Schiavo would not have wanted to be sustained in the permanent vegetative state. Her parents confested these wishes and questioned Mr. Schiavo's motives. The Schindlers also questioned their daughter's diagnosis of the vegetative state (Charatan, 2003). Like many lay people, they had difficulty accepting an eyes-open state of "wakeful unresponsiveness," as one devoid of consciousness. Her ocular saccades were interpreted as meaningful tracking and evidence of awareness, despite expert testimony concluding she was in fact vegetative. Not satisfied, the family appealed to Governor Jeb Bush for intercession and he pushed legislation through the Florida legislature to reverse a court decision to remove her feeding tube. It was reported that Florida legislators viewed a videotape taken by the Schindlers to make their "diagnosis" (Reuters, 2003).

That these diagnoses are prone to becoming so value-laden has the potential to further complicate family counseling. So though each of us can place a different moral valuation on life in a vegetative state, we should not let these values turn the objective clinical evidence into something that it is not. Moral valuation should follow *upon* the clinical facts and *not transmute them* to meet political or ideological needs. Diagnostic clarity, or its distortion, has consequences beyond the patient. It also has implications for other patients and society in general (Fins and Plum, 2004).

Asserting by executive or legislative fiat, that Terri Schiavo was conscious when she was not obscures material differences between patients like Schiavo and Wallis (see Table 2). Such conflations have the potential to further the neglect of patients who retain residual elements of consciousness and who should be the object of intense study and clinical concern.

Patient and surrogate preferences

Once the medical facts have been clarified, it is time to turn to the narrative dimensions of the case. Most patients do not engage in advance planning, that is talking with family and friends about their wishes in the event of decisional incapacity (SUPPORT, 1995). Of those who do, even fewer envision a cognitively altered state while still a young person, the demographic most affected by traumatic brain injury. Nonetheless, strong views persist about cognitive impairment from Alzheimer's disease or following stroke later in life. In a study, we recently concluded using structured vignettes to assess patient and proxy views on end-of-life decisionmaking, respondents were averse to continued life support following a stroke that left the patient with "no hope for meaningful recovery" (Fins et al., 2005), a phrase as we have noted that is in the common prognostic discourse in the clinical setting. It is important to be cautious about these *background* views about impaired consciousness when they are analogized from the geriatric context and the setting of degenerative diseases to disorders of consciousness, which are traumatic, occurring in an otherwise healthy younger patient cohort. Although analogic reasoning is how we often analyze difficult choices, it is important to appreciate the salient differences between various types of cognitive impairment throughout the life cycle and make explicit these potential biases.

Parents of young adults considering their own views about cognitive impairment later in life or in their parents may erroneously generalize these preferences to a context that is biologically and developmentally quite different, at the risk of being too pessimistic about hopes for some modicum of recovery. It is worth remembering and counseling surrogates that recovery from brain damage is more variable than the inexorable decline that follows a diagnosis of Alzheimer's disease. And as the compelling account of the Central Park jogger's account of her experience shows, though statistically unusual, even patients who have a grim Glasgow Coma Scale of 4–5 have the potential to recover cognitive function, as Meli's account indicates (Meli, 2003).

In addition to pre-existing views of cognitive impairment analogized from other contexts, families will also bring their religious and cultural traditions to bear upon these deliberations. Life in the persistent vegetative state may look different to a family from a more fundamentalist religious tradition that holds that life itself is precious and does not sanction quality of life distinctions. Such a vitalist approach, that any life is worth preserving, is often accompanied by an obligation to provide what, in the Catholic tradition, is called ordinary care such as artificial nutrition and hydration (Beauchamp and Childress, 1994). As has been noted, this becomes a critical point because late withdrawals will not likely involve ventilator support but simple feeding tubes, the subject of a recent Papal discourse (John Paul II, 2004).

If surrogates are looking to forgo life-prolonging therapies, the process should be analogous to the process of an informed consent discussion, although in this context it would be an informed refusal of care. The practitioner should avoid value-laden statements that can engineer outcomes and preclude choice and provide surrogate decisionmakers with the best available evidence of the

Table 2. Diagnostic discernment: Tale of two Terries

	Terri Schiavo	Terry Wallis
Etiology, year	Anoxia, 1990	Traumatic, 1984
Diagnosis at 1 year	Permanent vegetative state	Minimally consciousness state
Assessment	Evaluated by court-appointed physicians	Not evaluated for years
Awareness	None even with deep-brain stimulation	Episodic and intrinsic
Current state (as of 2004)	Reflexive behavior only	Emergence from MCS
Prognosis	Static	Unclear with progressive improvement

patient's outcome as well as its time course and any burdens associated with continued treatment. The patient's ability to perceive pain and experience suffering should be discussed and efforts to provide symptomatic relief should be reviewed (Schiff, 1999).

Given the potential for confusion about the sequence of recovery, clinicians should carefully explain how a patient who is comatose, who neither dies nor regains consciousness, moves into the "wakeful unresponsiveness" of the vegetative state after a couple of weeks. Clinicians should appreciate that families may perceive that eye opening is an improvement over the eyes-closed state of coma. This misconstrual needs to be clarified and families made aware that the failure to recover consciousness and awareness coming out of coma is a negative prognostic sign. The move into the vegetative state, though, needs to be tempered with information that a persistent vegetative state is not yet permanent and that prospects for the MCS and subsequent emergence can remain for months depending upon the etiology of the injury. These discussions become even more complex if the patient recovers and is able to communicate preferences.

When the patient is able to participate in these decisions, they are confronted with the challenge of being an altered self. Paradoxically, an improvement in awareness and self awareness may also bring a greater appreciation of how devastating an injury has been and how much farther one has to travel to regain skills and the ability for independent living (Fins, 2000).

Is the goal to recapture a former self or approximate a reasonable facsimile? The literature is rife with narratives of brain injury survivors who provide first-hand accounts of this very private journey. Theirs are an admixture of physiology and psychology in which the nature of their injury determines *what* their deficits will be while their personal narratives will determine *who* they will now become. Claudia Osborn, a physician who sustained head trauma and lost much of her executive function characterized the psychological evolution necessitated by her injury:

> I desperately need a vision of an achievable future — one I wanted, not just one I could attain now that the me I knew no longer existed, I had to build another identity and move on, or wither and die. (Osborn, 1998, p. 180.)

The challenge of crafting a new future becomes even more formidable when the body has been paralyzed but the mind remains intact as in the LIS. In these cases it becomes especially important *to speak with the patient* using assistive devices or rudimentary efforts at blinking (Morris, 2004). Indeed, if we take self-determination seriously, we need to ensure that those trapped in a lifeless body are not robbed of the opportunity to direct their care by turning to surrogates for guidance. Contrary to the expectations held by the able-bodied and anecdotal reports (Powell and Lowenstein, 1996), a systematic assessment of the quality of life of patients in the LIS indicate that they can maintain a quality of life, enjoy social intercourse and that depression is not the norm (Leon-Carrion et al., 2002a; Laureys et al., this volume).

Finally, it is important to appreciate that views about an acceptable existence are plastic. They evolve over time, often accommodating an increasing burden of disability. Studies have shown that patients are willing to continue treatment that physicians think are burdensome. Physician's and patient's views about quality of life can be discordant.

These valuations can inform the clinicians' view of "appropriate" end-of-life decisions, even though patients may see things differently (Uhlmann et al., 1988; Uhlmann and Pearlman, 1991). Physicians must recognize these potential biases so as to enable neutral counseling about what might constitute proportionate care (Fins, 1992a).

Family dynamics

Thrown into this mix is the impact of care on the family. The aforementioned prognostic uncertainty mixed with the unclear diagnostic status of loved ones require an empathetic response and support. Having a family member with severe brain damage can lead to social isolation, or as one commentator put it, "of the loneliness of the long term care giver" (Levine, 1999). Relationships change with the patient when the person has changed but one's marital status has not. "For better or worse" takes on a new connotation when one spouse no longer recognizes the other or when a frontal head injury has led to disinhibition and dramatic changes in personality and temperament. When this occurs, love can turn to compassion and compassion might turn to resentment. Spouses can feel as trapped as the patient, burdened by loving vows that now seem unfillable and tethering.

It is important to appreciate the stress that families operate under when challenged by brain damage in a loved one. The patient's dependency, indeed fragility, becomes the epicenter of family life, altering roles and obligations. Though he wrote from the LIS, Bauby captures this dynamic, contrasting his children's vitalism with his dependency upon his wife, who herself, is delegated to a custodial role. His children too now shoulder a burden too large for their small frames:

> Hunched in my wheelchair, I watch my children surreptitiously as their mother pushes me down the hospital corridor. While I have become something of a zombie father, Théophile and Céleste are very much flesh and blood, energetic and noisy. I will never tire of seeing them walk along aside me, just walking, their confident expressions masking the unease weighing on their small shoulders. (Bauby, 1997, p. 69.)

Writing as a man in his early 40s, Bauby laments that he has become a zombie — a reanimated corpse — and what this state means for his wife and kids. From the other side of the sickbed, he describes his sense of abdication of family role, a sense of loss akin to bereavement. The only difference is that he mourns, even as life continues.

Institutional arrangements

Whatever the family dynamics, it will be played out within an institutional setting. For the most part, these institutions will not be geared to the chronic care needs of patients with severe head injury but rather to the provision of acute or what has been euphemistically called "custodial care" (Winslade, 1998). These contextual factors can distort decision-making and engineer outcomes that may not be in the patient or family's best interest.

We have already described how the time frame of end-of-life decisions are discordant with the pace of recovery from head injury. Similar economic pressures exist, at least in the North American context, to discharge patients from acute care if they do not demonstrate what has been termed "medical necessity" or more colloquially, show improvement. Although it is appreciated that recovery from brain damage can take months and that the trajectory of improvement may not be linear, the prevailing bureaucracy regulates the length of stay in ways that may truncate admissions and deprive patients of adequate diagnostic assessment and proper placement in appropriate rehabilitation settings.

The pace of discharge from hospital, and the sequestration of the acute and rehabilitative medical communities, can lead to distortions among the former about what may be achievable with continued treatment over time. Because acute-academic and rehabilitative care centers are generally geographically separate, acute care practitioners can have a distorted view about patient outcomes. They suffer from what has been described as a "bureaucratization of prognosis", viewing likely outcomes from their own perspective in the care continuum (Christakis, 1999; Fins, 2002). Understandably, they see those who they have "saved" as being damned to a horrific existence because these patients are no longer hospitalized if and when they later demonstrate cognitive improvement. This can lead to a sense of nihilism and a perception that aggressive efforts are futile.

At its worst, institutional perceptions about what might constitute futile care can become objectionable. Consider the conflicting goals of caring for patients with severe brain damage and the need to obtain organs for transplantation. The contemporary example of organ procurement practices has historical roots that also inform our utilitarian views about severely brain injured and our societal obligations. In 1968, the diagnosis of brain death was justified by the greater good that might come from organ harvesting (Beecher, 1968; Stevens, 1995). These powerful perspectives have expanded to those who are still not brain dead and today it is not uncommon for organ-procurement personnel to urge referrals of patients with Glasgow Coma scales of 3–5 for assessment as potential donors, even though these patients may yet harbor the potential for recovery (Meli, 2003). My goal is not to undermine laudable transplantation efforts but rather to illustrate how views about the viability and potential of patients with severe brain injury can be shaped by institutional practices that may go unexamined.

Societal issues and norms

All of the above factors are occurring against a broader societal backdrop as illustrated by the impact of cases like Schiavo and Wallis. I have addressed this broader context elsewhere (Fins, 2003a; Fins and Plum, 2004), suffice it to say here that our collective views about disorders of consciousness are informed by the tangled history of brain damage and the evolution of the right to die in modern medical ethics. American bioethics since the 1960s has been predicated upon the evolution of self-determination and autonomy. This right evolved as the right to be left alone and to have life-sustaining therapy withdrawn. The important right to die, to direct one's care at the end of life autonomously, was founded upon a number of landmark cases involving patients in the vegetative state, most notably Quinlan and Cruzan (Cantor, 2001; Cruzan v Director, 1990). This is critical because the test cases to establish this *negative* right were vegetative patients whom society concluded to be beyond hope and beyond care. Further treatment in such cases was paradigmatic of medical futility, (Cranford, 1994) simply geared at the preservation of vegetative functions and not, as the Quinlan court put it restoration of a "cognitive sapient state" (Annas, 1996).

By considering the potential for recovery in patients with severe brain injury, we are now asking society to intervene in a population that resembles those in whom the negative right to be left alone was first established. This has created a bit of cognitive dissonance — an oxymoron — seen so dramatically in the cases of Schiavo and Wallis, in which the laudable goals of *preserving the right to die* and *affirming the right to care* have come into conflict.

Interpretation (Ethics differential diagnosis): toward a palliative neuroethics

In considering how to reconcile these two conflicting obligations — preserving the hard-won right to direct care at the end of life and to care for those who may yet be helped — I would like to suggest the value of viewing these cases through the ethos of palliative care. The World Health Organization (WHO) defines palliative care as "...the active total care of patients whose disease is not responsive to curative treatment. Control of pain, of other symptoms, and of psychological, social and spiritual support is paramount. The goal of palliative care is the achievement of the best quality of life for patients and their families" (WHO, 1990, pp. 11–12). Conceptually, palliative care can accommodate the oxymoron of affirming the right to die and preserving the right to care in patients with disorders of consciousness. Clinically, it does not preclude active care but at the same time it attends to symptom management and quality of life issues when the patient has eluded cure.

Palliation is an especially good metaphor to describe interventional strategies being developed for patients with disorders of consciousness. If we consider the philological origins of palliation, we are reminded that to palliate means to cloak or disguise (Fins, 1992b). This is precisely what will occur if deep-brain stimulation can restore impaired consciousness or (Fins, 2004b) if brain–computer interfaces can help LIS communicate (Leon-Carrion, et al., 2002b; Kübler and Neumann, this volume). Neither of these interventions are curative but rather assistive devices that palliate by *cloaking* or placing a *veneer* over underlying disability which remains.[1]

The palliative care ethos also fits. As the WHO definition illustrates, palliative care views the entire family as the unit of care, an approach especially well suited to the needs of families whose lives have been touched by brain injury. But more fundamentally, palliative medicine is concerned with questions of meaning. It asks practitioners to consider difficult questions that are so central to patient care. What causes distress for the patient and family? What aspect of the self matters? Is the goal of care restoration or personal identity or a tolerable facsimile? Or is it the restoration of affect, memory and executive function? These are both empirical questions to be studied and ones that can be discussed with patients and families as they make individualized choices.

But at a fundamental level, these difficult questions raise an issue, which is central in the palliative care literature, namely the distinction between pain and suffering. Eric Cassell has written eloquently about this critical definitional issue. Pain is physiologic. Suffering is a threat to the self, an existential question (Cassell, 1991). At the most fundamental level, disorders of consciousness are about the endangered and altered self, raising the possibility, and the potential for suffering as the patient contemplates what has been lost, what remains and what still might be.

It is vitally critical that practitioners attend to this alteration of the self and appreciate that there are indeed psychological connections between the patient before and after injury. Although some philosophers, like Derek Parfit, maintain that there is a discontinuity between the former and current self (Parfit, 1987), this is more a theoretical argument than a pragmatic one as there is indeed continuity of memory and affect that links one stage of life to another. Though these linkages may be tenuous at the margins of consciousness, they do remain and inform relationships with intimates who look upon the injured patient and see what once was as much as what is. Turning again to that remarkable memoir of Jean-Dominique Bauby, we can see that he speaks to this notion of affirming the ties between who he was and what he has become. Conveying but the grandeur of his intellect and the depths of disability he tells us:

> But I see in the clothing a symbol of continuing life. And proof that I still want to be myself. If I must drool, I may as well drool on cashmere. (Bauby, 1997, p. 17.)

These sentiments remind us of the importance of appreciating that even as we distinguish differing brain states from another, there is a unique psychological element at each bedside, even when there are disorders of consciousness. Consciousness is about questions of meaning and although physiologically based, is psychological in its expression. Oliver Sacks captures this when he observes that, "... consciousness is always active and selective — charged with feelings and meanings uniquely our own, informing our choices and interfusing our perceptions." (Sacks, 2004, p. 44). Having said this, the connection between former and present selves requires the ability to interact with others and the environment, and to communicate even at the most rudimentary level. Central to the notion of a self is a self embedded in a web of relationships connected by the ability to interact and communicate. Here, consciousness is closely linked to speech, intentionality and the social context within which it is received (Searle, 2002). As Winslade has observed, "Being persons requires having a personality, being aware of our selves and our surroundings, and possessing human capacities, such as memory, emotions, and the ability to communicate and interact with other people." (Winslade, 1998, p. 121).

[1]If neuromodulation were to promote neural plasticity and regeneration of tissue, it would make this intervention a mosaic blend of the curative and palliative.

It is this ability to communicate and interact that the distinguished Spanish filmmaker, Pedro Almodovar, portrayed in his Academy award-winning film *Habla con Ella* (Almodovar, 2002). Properly translated in English as *Talk with Her*, it is a film about social isolation following brain injury and those who have lost consciousness can become re-engaged *with* others as they rejoin a human community marked by communication, words that share a common root (Fins and Plum, 2004).

Negotiation and intervention

In the absence of a particular patient narrative, it is difficult to script how to negotiate a plan of care with patients or their surrogates, but as a rule it is helpful to suggest plausible and achievable goals of care. Surrogates confronted with the spectre of a loved one with brain damage may want to precipitously withdraw care or cling to hope when there is none. These emotional responses may reflect more about their preconceptions about brain injury than the clinical reality that they face. To temper these responses and help ensure that surrogates are adequately informed, it is essential that surrogates understand the patient's diagnosis, prognosis and prospects for recovery. It is important to avoid misconstruals that might distort the surrogate's thinking and help ensure that they appreciate the likelihood and time course of recovery, its scope and the foreseeable burdens that might be imposed by ongoing care. This information needs to be conveyed empathically and with compassion, appreciating that most of us never consider the prospect of brain injury touching our families or close friends.

As I have suggested, a palliative approach may help frame the goals of care by acknowledging both the right to die and the right to care while seeking an optimal quality of life through the mitigation of the patient's symptom burden. Palliative care, however, should be carefully introduced into the discourse, because it is generally associated with end-of-life care and may itself be open to misconstrual by families and surrogates. To avoid this potential confusion, it may be best not to explicitly label care as "palliative". It would be more effective to be descriptive about the *elements* of palliative care including the right to withhold or withdraw life-sustaining therapies.

When offering palliation, or any other care strategy, it is important that the appropriate surrogate decisionmaker retains the ability to direct care. Ultimately, any decisionmaking authority resides in the patient's surrogate. This moral authority is based on what they know about the patient's preferences or values and their pre-existing relationship (Fins, 1999b). The practitioner's task is to help weave a consensus with the surrogate that takes account of the medical facts and the patient and surrogate's values and balances burdens and benefits.

Once a care plan has been agreed upon, it is helpful to suggest a *time trial* to see how and if the patient improves, leaving open the possibility that the goals of care can evolve as the situation changes. Time trials are also an important way to achieve a consensus and balance the power dynamics between clinician and family. They are essential to the *process of negotiation* and provide time for the surrogates to accommodate themselves to the sudden and often tragic reality of severe brain damage. Most importantly, they help safeguard the surrogate's authority and help them make decisions without being dictated to by the clinical team.

Periodic review

The final step in the process of clinical pragmatism is periodic review. This is a critical step because it allows for the reassessment of decisions made in particular cases and for modification of a course of action. As critically observed this process allows us to organize empirical observations to reform practice and public policy. Public policy that governs research on individuals with disorders of consciousness is one critical issue that calls for reassessment in the wake of this consideration of the clinical care of such patients.

It is beyond the scope of this paper to address fully the challenge of engaging in clinical research with subjects who have decisional incapacity (National Bioethics Advisory Commission, 1998; Fins and Miller, 2000). These subjects are rightly considered a vulnerable population and are subject to special protections because they are unable to provide their autonomous consent for enrollment in clinical trials. Although their next-of-kin or surrogates may consent to therapeutic procedures with demonstrated benefit, surrogates' ability to authorize enrollment in research is constrained when it has yet to demonstrate medical benefit, unless it was authorized prospectively by the patient before decisional incapacity. This severely limits the potential for phase I research in individuals who cannot provide consent (Michels, 1999; Miller and Fins, 1999).

But unlike other vulnerable populations, the disorder that precludes autonomous consent is the object of the intervention when trying to restore consciousness. A compelling argument emerges that clinical trials to restore consciousness would be ethically proportionate, even with the challenges posed by surrogate consent, when we appreciate this critical distinction and the burdens imposed on these patients and families (Fins, 2000).

While those with disorders of consciousness should be protected from harm, balancing and specifying the ethical principles of respect for persons, beneficence, and justice compel us to craft a responsible and responsive research ethic geared, for now, toward the pursuit of palliation (National Bioethics Advisory Commission, 1998). This becomes a fiduciary obligation grounded in a justice claim to meet the needs of patients so long misunderstood and historically neglected by society (Fins, 2003a).

Conclusion

The terra nova of neuroscience exploring disorders of consciousness is, to invoke an over-used phrase, paradigm breaking (Kuhn, 1996). This line of inquiry will challenge assumptions, stir up misconceptions and engender both unrestrained hopes and unsubstantiated fears. If we are to grapple with the promise and peril of this work, it is critical to engage in a deliberative process of inquiry that allows us to see all sides of the argument, identify the range of stakeholders who may be affected by clinical and scientific developments, and to reach a societal consensus on how these efforts will proceed. It is my contention that the process of inquiry provided by Deweyan pragmatism, and its derivative clinical pragmatism, can help structure these deliberations in the face of scientific novelty. An inductive approach, which is reminiscent of diagnostics, pragmatism can help constructively apply ethical principles to the context of care (Schmidt-Felzmann, 2003) and bring principled reasoning to complex ethical questions posed by cognitive neuroscience.

In this chapter, I have suggested that a pragmatic assessment of this emerging area of neuroscience might steer us to viewing this work through the prism of palliative care so as to acknowledge the malignancy of disorders of consciousness and outline an approach to mitigate this burden through clinical practice and scientific research. But perhaps most importantly, clinical pragmatism reminds us of the risks of absolutism in clinical practice, while sensitizing us to the moral underpinnings of medicine itself (Fins and Bacchetta, 1995; Thomasma, 1997). It is an approach, and a process, that argues for humility in the face of the unknown and suggests that moral certainty, founded upon uncertain and evolving facts, is tenuous at best. Acknowledging such contingencies, both scientific and normative, will be an important characteristic of the neuroethics that will comment upon and critique developments in the study of disorders of consciousness (Fins, 2004c). Although others may hope for broad categoricals about this work, I will content myself with its rich ambiguity.

Contingency, not moral certitude is the norm in science and society, and our normative approach to the world needs to reflect our evolving place in it. Although he could not imagine the course that the history of medicine would take in the century and half since he lectured students at Harvard Medical School, Professor Oliver Wendell Holmes, Sr. bequeathed advice about the nature of scientific inquiry that remains instructive. The physician father, of the famous jurist and American pragmatist, Oliver Wendell Holmes, Jr., cautioned a judicious approach to scientific knowledge:

> Science is the topography of ignorance. From a few elevated points we triangulate vast spaces, enclosing infinite unknown details. We cast the lead, and draw up a little sand from the abysses we shall never reach with our dredges. The best part of our knowledge is that which teaches us where knowledge leaves off and ignorance begins. Nothing more clearly separates a vulgar from a superior mind, than the confusion in the first between the little that it truly knows, on the one hand, and what it half knows and what it knows, on the other. That which is true of every subject is espccially true of the branch of knowledge which deals with living beings... (OW Holmes, 1862, p. 7).

Acknowledgements

The author thanks Dr. Steven Laureys, the Mind Brain Foundation and the organizers of the Satellite Symposium on Coma and Impaired Consciousness at The Eighth Annual Conference of the Association for the Scientific Study of Consciousness for the invitation to present an earlier version of this paper at the University of Antwerp, Belgium, on June 24, 2004.

Dr. Fins' work is supported by the Dana Brain Science Foundation and is a Project on Death in America Faculty Scholar of the Soros Open Society Institute.

References

Almodovar, P. (2002) Talk to Her. http://www.sonyclassics.com/talktoher/.

Annas, G.J. (1996) The "right to die" in America: sloganeering from Quinlan and Cruzan to Quill and Kevorkian. Duquesne Law Rev., 34(4): 875–897.

Barondess, J.A. (2003) Personal communication to Joseph J. Fins.

Bauby, J.-D. (1997) The Diving Bell and the Butterfly. Vintage International, New York.

Beauchamp, T.L. and Childress, J.F. (1994) Principles of Biomedical Ethics (4th ed). Oxford University Press, New York.

Beecher, H.K. (1968) Ethical problems created by the hopelessly unconscious patient. New England J. Med., 278(26): 1425–1430.

Cantor, N.L. (2001) Twenty-five years after Quinlan: a review of the jurisprudence of death and dying. J. Law and Med. Ethics, 29(2): 182–196.

Cassell, E.J. (1991) The Nature of Suffering and the Goals of Medicine. Oxford University Press, New York.

Charatan, F. (2003) Governor Jeb Bush intervenes in "right to die" case. Brit. Med. J., 327: 949.

Childs, N.L., Mercer, W.N. and Childs, H.W. (1993) Accuracy of diagnosis of persistent vegetative state. Neurology, 43: 1465–1467.

Christakis, N.A. (1999) Death Foretold: Prophecy and Prognosis in Medical Care. The University of Chicago Press, Chicago, p. 41.

Coleman, D. (2002) The minimally conscious state: definition and diagnostic criteria. Neurology, 58: 506–507.

Cranford, R.E. (1994) Medical futility: transforming a clinical concept into legal and social policies. J. Am. Geriatrics Soc., 42: 894–898.

Cranford, R.E. (1998) The vegetative and minimally conscious states: ethical implications. Geriatrics, 53: S70–S73.

Crick, F. (1994) The Astonishing Hypothesis: The Scientific Search for the Soul. Scribner, New York.

Cruzan v Director (1990) Missouri Department of Health, 110 S. Ct. 2841.

Dewey, J. (1988) The quest for certainty. The Later Works (Vol. 4). Southern Illinois University Press, Carbondale, p. 1929.

Dewey, J. (1991a) Theory of Valuation. The Later Works (Vol. 13). Southern Illinois University Press, Carbondale, pp. 1938–1939.

Dewey, J. (1991b) Logic: the theory of inquiry. The Later Works (Vol. 12). Southern Illinois University Press, Carbondale, p. 1938.

Dewey, J. (1997) Experience and Nature. Open Court Publishing Company, Chicago and LaSalle Illinois, pp. 134–137.

Dewey, J. (1998) The logic of judgments of practice. In: Hickman L.A. and Alexander T.M. (Eds.) The Essential Dewey: Ethics, Logic, Psychology, Vol. 2. Indiana University Press, Bloomington, pp. 236–271.

The Economist (2004) Inside the mind of the consumer. June 10, 2004.

Edelman, G.M. (2004) Wider than the Sky: The Phenomenal Gift of Consciousness. Yale University Press, New Haven.

Farah, M.J. and Wolpe, P.R. (2004) Monitoring and manipulating brain function, new neuroscience technologies and their ethical implications. Hastings Cent. Rep., 34(3): 35–45.

Fins, J.J. (1992a) The Patient Self-Determination Act and patient–physician collaboration in New York State. J. Med., 92(11): 489–493.

Fins, J.J. (1992b) Palliation in the age of chronic disease. Hastings Cent. Rep., 22(1): 41–42.

Fins, J.J. (1996) From indifference to goodness. J. Relig. Health, 35(3): 245–254.

Fins, J.J. (1998) Approximation and negotiation: clinical pragmatism and difference. Cambridge Quarterly of Healthcare Ethics, 7(1): 68–76.

Fins, J.J. (1999a) Klinischer Pragmatismus und Ethik-Konsultation (Clinical pragmatism and ethics case consultation). Das Parlament 49.Jahrgang/Nr. 23.4 Juni: 18.

Fins, J.J. (1999b) From contract to covenant in advance care planning. J. Law, Med. Ethics, 27: 46–51.

Fins, J.J. (2000) A proposed ethical framework for interventional cognitive neuroscience: a consideration of deep brain stimulation in impaired consciousness. Neurol. Res., 22: 273–278.

Fins, J.J. (2002) When the prognosis leads to indifference. J. Palliative Med., 5(4): 571–573.

Fins, J.J. (2003a) Constructing an ethical stereotaxy for severe brain injury: balancing risks, benefits and access. Nat. Rev. Neurosci., 4: 323–327.

Fins, J.J. (2003b) From psychosurgery to neuromodulation and palliation: history's lessons for the ethical conduct and regulation of neuropsychiatric research. Neurosurg. Clin. N. Am., 14(2): 303–319.

Fins, J.J. (2004a) Neuromodulation, free will and determinism: lessons from the psychosurgery debate. Clin. Neurosci. Res., 4(1–2): 113–118.

Fins, J.J. (2004b) Deep Brain Stimulation. In: (3rd ed) Post S.G. (Ed.) Encyclopedia of Bioethics, Vol. 2. MacMillan Reference, New York, pp. 629–634.

Fins, J.J. (2004c) "Important Human Issues" in technology that enhances the brain. Letter to the Editor re: Gray Matters. The Chronicle of Higher Education. August 6, 2004, p. B4.

Fins, J.J. and Bacchetta, M.D. (1995) Framing the physician-assisted suicide and voluntary active euthanasia debate: the role of deontology, consequentialism and clinical pragmatism. J. Am. Geriatrics Soc., 43(5): 563–568.

Fins, J.J. and Miller, F.G. (2000) Enrolling decisionally incapacitated subjects in neuropsychiaric research. CNS Spectrums, 5(10): 32–42.

Fins, J.J. and Plum, F. (2004) Neurological diagnosis is more than a state of mind: diagnostic clarity and impaired consciousness. Arch. Neurol., 61(9): 1354–1355.

Fins, J.J., Bacchetta, M.D. and Miller, F.G. (1997) Clinical pragmatism: a method of moral problem solving. Kennedy Inst. Ethics J., 7(2): 129–145.

Fins, J.J., Miller, F.G., Acres, C.A., Bacchetta, M.D., Huzzard, L.L. and Rapkin, B.D. (1999) End-of-life decision-making in the hospital: current practices and future prospects. J. Pain Symptom Manag., 17(1): 6–15.

Fins, J.J., Guest, R.S. and Acres, C.A. (2000) Gaining insight into the care of hospitalized dying patients: an interpretative narrative analysis. J. Pain Symptom Manag., 20(6): 399–407.

Fins, J.J., Maltby, B.S., Friedmann, E., Green, M., Norris, K., Adelman, R. and Byock, I. (2005) Contracts, covenants and advance care planning: an empirical study of the moral obligations of patient and proxy. J. Pain Symptom Manag., 29(1): 55–68.

Giacino, J.T. (2004) Personal communication with Joseph J. Fins.

Giacino, J.T., Ashwal, S., Childs, N., Cranford, R., Jennett, B., Katz, D.I., Kelly, J.P., Rosenberg, J.H., Whyte, J., Zafonte, R.D. and Zasler, N.D. (2002) The minimally conscious state: definition and diagnostic criteria. Neurology, 58(3): 349–353.

Hirsch, J., Kamal, A., Rodriguez-Moreno, D., Petrovich, N., Giacino, J., Plum, F. and Schiff, N.D. (2001) fMRI reveals intact cognitive systems in minimally conscious patients. Soc. Neurosci. 30th Annu. Meet. Abs., 529: 14.

Holmes, O.W. (1862) Border Lines of Knowledge in Some Provinces of Medical Knowledge. Ticknor and Fields, Boston.

Hook, S. (1995) John Dewey: An Intellectual Portrait. Prometheus Book, Amherst, New York.

Jennett, B. (2002) The Vegetative State. Cambridge University Press, Cambridge.

Jennett, B. and Plum, F. (1972) Persistent vegetative state after brain damage. A syndrome in search of a name. Lancet, 1(7753): 734–737.

Jennett, B., Adams, J.H., Murray, L.S. and Graham, D.I. (2001) Neuropathology in vegetative and severely disabled patients after head injury. Neurology, 56(4): 486–490.

John Paul II (2004) Speech to the Participants at the International Congress, "Life Sustaining Treatments and Vegetative State: Scientific Advances and Ethical Dilemmas." March 20, 2004.

Kobylarz, E.J. and Schiff, N.D. (2004) Functional imaging of severely brain injured patients: progress challenges, and limitations. Arch. Neurol., 61(9): 1357–1360.

Koch, C. (2004) The Quest for Consciousness: A Neurobiological Approach. Roberts & Co., Englewood, Colorado.

Kuhn, T.S. (1996) The Structure of Scientific Revolutions (3rd ed.). University of Chicago Press, Chicago.

Laureys, S., Faymonville, M.E., Moonen, G., Luxen, A. and Maquet, P. (2000) PET scanning and neuronal loss in acute vegetative state. Lancet, 355(9217): 1825–1826.

Laureys, S., Antoine, S., Boly, M., Elincx, S., Faymonville, M.E., Berre, J., Sadzot, B., Ferring, M., De Tiege, X., van Bogaert, P., Hansen, I., Damas, P., Mavroudakis, N., Lambermont, B., Del Fiore, G., Aerts, J., Degueldre, C., Phillips, C., Franck, G., Vincent, J.L., Lamy, M., Luxen, A., Moonen, G., Goldman, S. and Maquet, P. (2002) Brain function in the vegetative state. Acta Neurol. Belg., 102(4): 177–185.

Leon-Carrion, J., van Eeckhourt, P., Dominguez-Morales, M.D.R. and Santamaria, F.J.P. (2002a) Survey: the locked-in-syndrome: a syndrome looking for a therapy. Brain Injury, 16(7): 571–582.

Leon-Carrion, J., van Eeckhourt, P., Dominguez-Morales, M.D.R. and Santamaria, F.J.P. (2002b) Review of subject: the locked-in-syndrome: a syndrome looking for a therapy. Brain Injury, 16(7): 555–569.

Levine, C. (1999) The loneliness of the long-term care giver. New Eng. J. Med., 340(20): 1587–1590.

Meli, T. (2003) I am the Central Park Jogger. Scribner, New York, pp. 57–58.

Menand, L. (2001) The Metaphysical Club. Farrar, Straus and Giroux, New York.

Menon, D.K., Owen, A.M., Williams, E.J., Minhas, P.S., Allen, C.M., Boniface, S.I. and Pickard, J.D. (1998) Cortical processing in persistent vegetative state Wolfson Brain Imaging Centre Team. Lancet, 18 352(9123): 200.

Michels, R. (1999) Are research ethics bad for our mental health? New Engl. J. Med., 340(18): 1427–1430.

Middlewood, S., Gardner, G. and Gardner, A. (2001) Dying in the hospital: medical failure or natural outcome? J. Pain Symptom Manag., 22(6): 1035–1041.

Miller, F.G. and Fins, J.J. (1999) Protecting vulnerable research subjects without unduly constraining neuropsychiatric research. Arch. Gen. Psychiat., 56: 701–702.

Miller, F.G., Fins, J.J. and Bacchetta, M.D. (1996) Clinical pragmatism: John Dewey and clinical ethics. The J. Contemp. Health Law Policy, 13(27): 27–51.

Miller, F.G., Fletcher, J.C. and Fins, J.J. (1997) Clinical pragmatism: a case method of moral problem solving. In: Fletcher J.D., Lombardo P.A., Marshall M.F. and Miller F.G. (Eds.), Introduction to Clinical Ethics (2nd ed.). University Publishing Group, Frederick, Maryland.

Morris, K. (2004) Mind moves onscreen: brain–computer interface comes to trial. The Lancet Neurol., 3: 329.

National Bioethics Advisory Commission. (1998). Research Involving Persons with Mental Disorders that may Affect Decisionmaking Capacity. Rockville, MD.

Osborn, C.L. (1998) Over My Head. Andrews McNeal Publishing, Kansas City.

Parfit, D. (1987) Reasons and Persons. Clarendon Press–Oxford University Press, New York.

Powell, T. and Lowenstein, B. (1996) Refusing life-sustaining treatment after catastrophic injury: ethical implications. J. Law, Med. and Ethics, 24: 54–61.

Reuters (2003) Parents ask Florida government to save comatose daughter. New York Times, October 13, 2003.

Ryan, A. (1995) John Dewey and the High Tide of American Liberalism. W.W. Norton, New York.

Sacks, O. (2004) In the river of consciousness. The New York Review of Books, Januray 15: 41–44.

Schiff, N.D. (1999) Neurobiology, suffering, and unconscious brain states. J. Pain Symptom Manage., 17(4): 303–304.

Schiff, N.D. and Plum, F. (1999) Cortical function in the persistent vegetative state. Trends in Cognitive Sci., 3(2): 43–44.

Schiff, N.D., Ribary, U., Moreno, D.R., Beattie, B., Kronberg, E., Blasberg, R., Giacino, J., McCagg, C., Fins, J.J., Llinas, R. and Plum, F. (2002) Residual cerebral activity and behavioural fragments can remain in the persistently vegetative brain. Brain, 125: 1210–1234.

Schiff, N.D. and Fins, J.J. (2003) Hope for "comatose" patients. Cerebrum, 5(4): 7–24.

Schmidt-Felzmann, H. (2003) Pragmatic principles-methodological pragmatism in the principle-based approach to bioethics. J. Med. and Philos., 28(5–6): 581–596.

Searle, J.R. (2002) Consciousness and Language. Cambridge University Press, New York.

Shewmon, D.A. (2002) The minimally conscious state: definition and diagnostic criteria. Neurology, 58: 506–507.

Stevens, M.L. (1995) Redefining death in America, 1968. Caduceus, 1995 Winter, 11(3): 207–219.

The SUPPORT Investigators. (1995) A controlled trial to improve care for seriously ill hospitalized patients: the study to understand prognoses and preferences for outcomes and risks of treatment (SUPPORT). J. Am. Med. Assoc., 274: 1591–1598.

Thomasma, D.C. (1997) Antifoundationalism and the possibility of a moral philosophy of medicine. Theor. Med., 18: 127–143.

Uhlmann, R.F. and Pearlman, R.A. (1991) Perceived quality of life and preferences for life-sustaining treatment in older adults. Arch. Internal Med., 151: 495–497.

Uhlmann, R.F., Pearlman, R.A. and Cain, K.C. (1988) Physicians' and spouses' predictions of elderly patients' resuscitation preferences. J. Gerontol., 43: M115–M121.

Wallis, C. (2003) The twilight zone of consciousness. Time Mag., October 27, 2003. pp. 43–44.

Wang, M.Y., Wallace, P. and Gruen, J.P. (2002) Brain death documentation: analysis and issues. Neurosurgery, 51(3): 731–736.

Wilson, B.A., Gracey, F. and Bainbridge, K. (2001) Cognitive recovery from "persistent vegetative state": psychological and personal perspectives. Brain Injury, 15(12): 1083–1092.

Wilson, L.G. (1997) Fevers. In: Bynum W.F. and Porter R. (Eds.) Companion Encyclopedia of the History of Medicine, Vol. 1. Routledge, London and New York, pp. 382–411.

Winslade, W. (1998) Confronting Traumatic Brain Injury: Devastation, Hope and Healing. Yale University Press, New Haven.

World Health Organization. (1990) Cancer Pain Relief and Palliative Care. World Health Organization, Geneva, Switzerland, pp. 11–12.

Zussman, R. (1992) Intensive Care: Medical Ethics and the Medical Profession. University of Chicago, Chicago.

Subject Index

Colour

Plate

Section

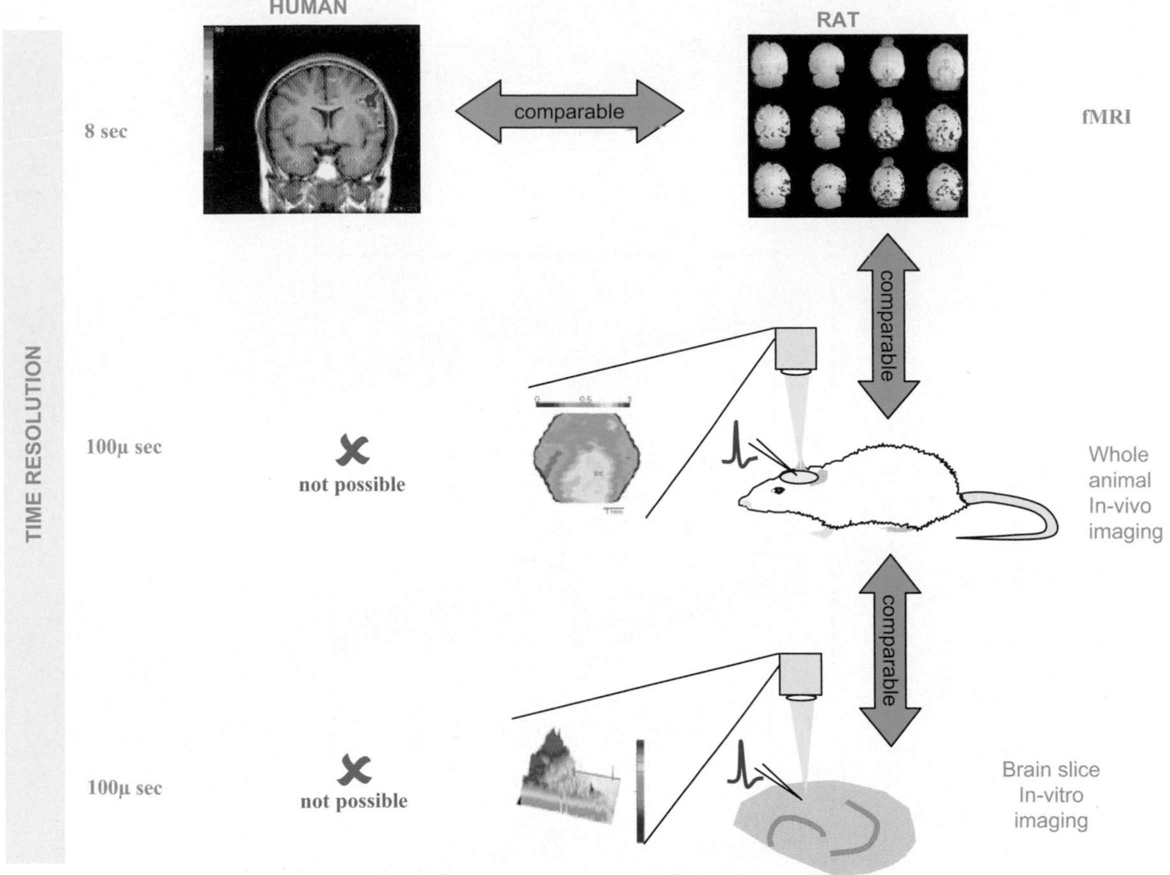

Plate 2.2. The different approaches of this study and how they inter-relate to compensate for the respective shortcomings of each. The limits of fMRI are primarily temporal, but by comparing human studies with those of lower mammals we can bridge the gap with invasive optical imaging techniques. Optical imaging provides the necessary temporal and spatial resolutions needed to investigate the characteristics of neural assemblies and the in vivo data can be compared with the commonly used in vitro brain slice preparation.

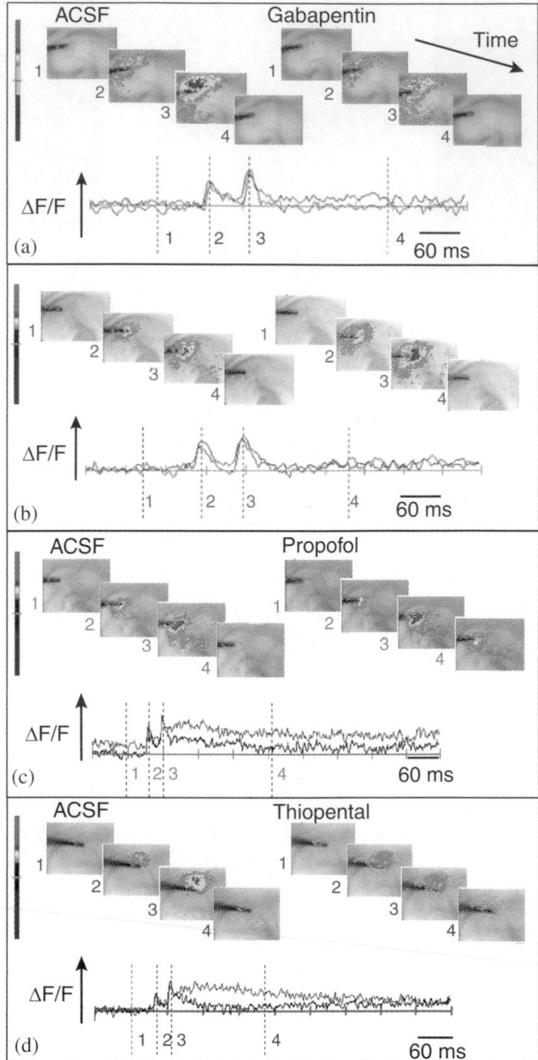

Plate 2.4. Analgesic and anesthetic effects on hippocampal activity. Optical measurements of analgesic and anesthetic action in the brain. Voltage-sensitive dye imaging of neuronal population activity in in vitro hippocampal slices of rat brain in response to paired-pulse electrical stimulation along the Schaffer Collaterals. The left panels show the responses under control conditions (in artificial cerebrospinal fluid, ACSF) and the right panels show treatment conditions. Both analgesics and anesthetics are linked with the actions of the inhibitory GABAergic system and so two compounds from each class were tested (analgesics: gabapentin and morphine) and (anesthetics: propofol and thiopental). Warmer colors show a higher degree of depolarizing excitation. The analgesic treatments do not significantly differ from their controls (a) and (b) but the anesthetics (c) and (d) cause a prolonged period of excitation that lasts for up to 500 ms after the second stimulating pulse. The optical recordings are also represented by traces below; black, control; red, treatment (Collins and Greenfield, *submitted*).

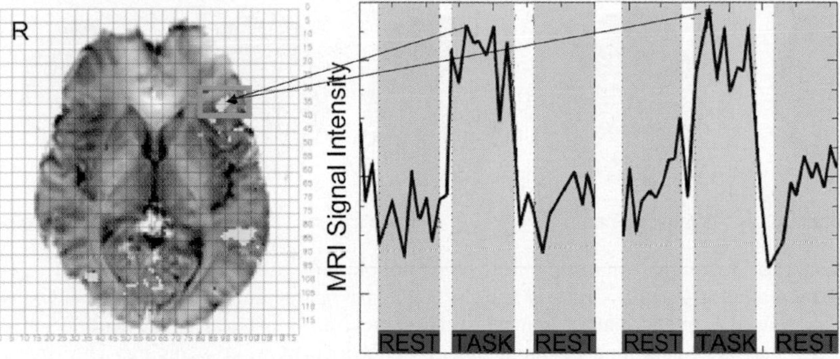

Plate 3.2. Cortical activity associated with naming visual objects. Signals illustrate the BOLD changes in MR susceptibility observed in response to naming visually presented objects in a healthy volunteer. Signals originate from a single voxel (1.5 × 1.5 × 4.5 mm) on two separate runs. Each run lasted 2 min 24 sec, during which 36 images were acquired, including 10 images for each of three epochs: initial resting baseline (*blue bar*), task (naming) (*pink bar*), and final resting baseline (*blue bar*). All voxels in the brain for which the statistical criteria were met (the average amplitude of the signal during the activity epoch was statistically different from the baseline signal) are indicated by either a *yellow*, *orange*, or *red* color superimposed on the $T2^*$-weighted image at the voxel address, and signify decreasing levels of statistical confidence. Arrows point to the source voxel, which is centered within a cluster of similar (*yellow*) voxels and located in the left hemisphere of the brain in the inferior frontal gyrus, putative Broca's area. This slice of brain also shows activity in left superior temporal gyrus, putative Wernicke's area, and in the right lateral occipital extra striate area (putative object processing area), and along the medial and posterior occipital regions (putative primary visual cortex).

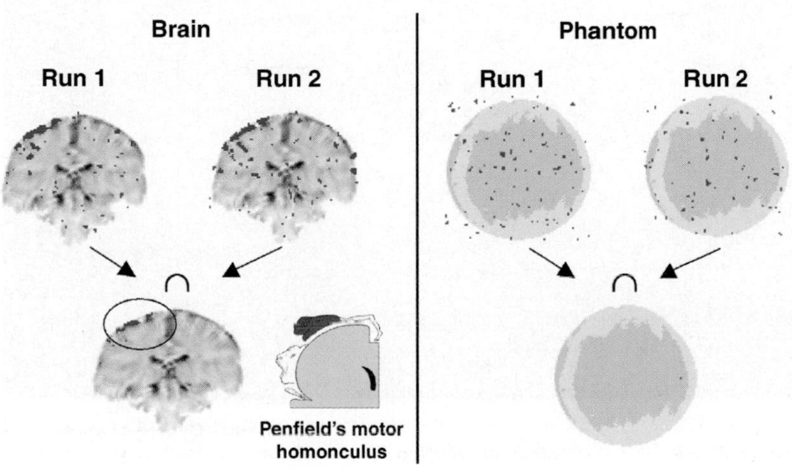

Plate 3.5. Double-pass method. Illustrations of voxels that meet statistical criteria at each stage of statistical decision for the brain (coronal slice) of a subject (left panel) and for the copper sulfate phantom simulating the brain (right panel). The stages of the analysis are outlined. Each average signal must be higher during the stimulation epoch than the initial baseline (stage 1) and the recovery baseline (stage 2) on two independent test-runs (runs 1 and 2). The task performed was finger-thumb tapping using the left hand. As would be expected, the right hemisphere of the precentral gyrus (motor strip) is activated (circled activity). The location of activity within the motor strip matches the known location of the hand in the motor strip in the Homonculus, as shown in the schematic on the right. The conjunction-across-2 runs is based on the principle that real signals can be differentiated from the noise by the probability of a repeat occurrence in the same place through multiple acquisitions. This double-pass acquisitions simulation (described above) is shown for a copper sulfate phantom control. Activity passing all conjunction criteria includes only three "active" voxels. This procedure is repeated many times, using both copper sulfate phantoms and also human brains under resting conditions, to empirically determine that the average rate of a false-positive signal is less than 1 part in 10,000, i.e., on average there is one voxel observed by chance for each brain slice.

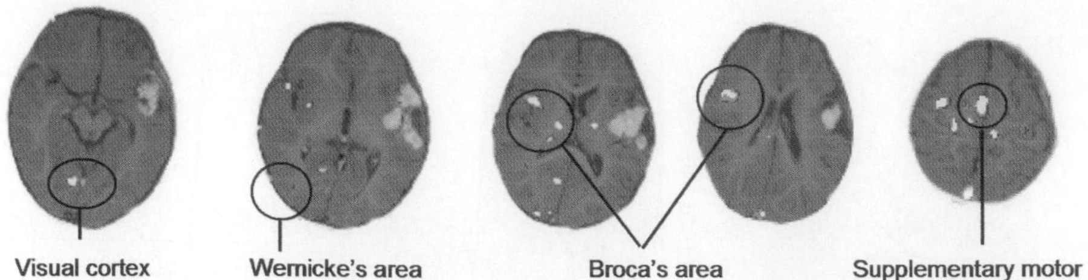

Plate 3.6. Sedated child "listening" to mother's narrative. Fifteen-month-old girl with an aggressive tumor in the left hemisphere (hyperintense region). In-plane resolution was 1.5 × 1.5 mm and the double-pass paradigm and analysis (see text) was employed. The patient was sedated with propofol and maintained at the lightest level that prevented movement of the head. A recorded passage of the mother talking to the patient was played through earphones worn by the patient while scanning. The circles indicate putative Wernicke's area, Broca's area, and supplementary motor area, consistent with the expected language network in preparation for the emerging language function.

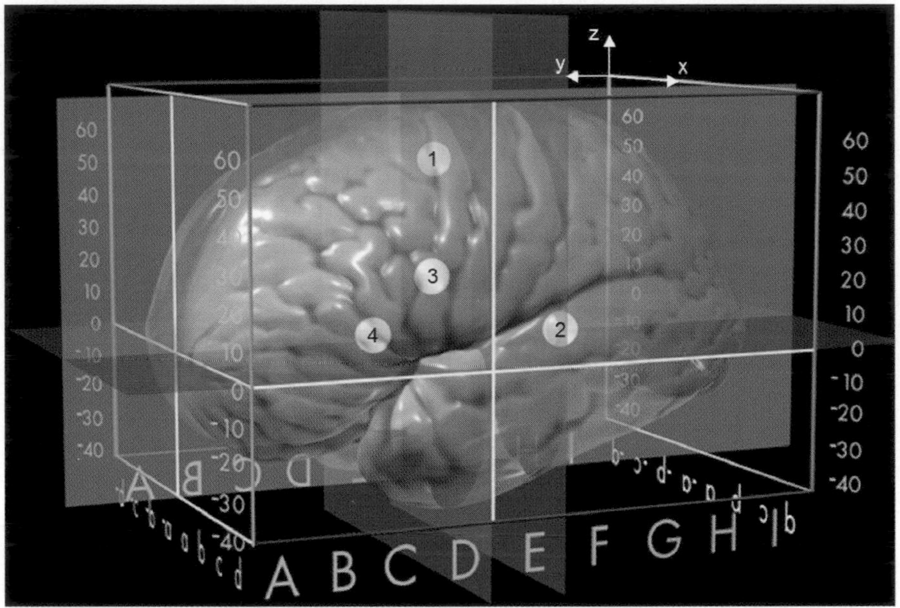

Common Name	Anatomical Region	Brodmann Area	Center of mass x	y	z
① Supplementary Motor	Medial Frontal Gyrus (GFd)	6	9	-6	53
② Wernicke's Area	Superior Temporal Gyrus (GTs)	22	57	-26	9
③ Broca's Area	Inferior Frontal Gyrus (Gfi)	44	49	10	25
④ Broca's Area	Inferior Frontal Gyrus (Gfi)	45	40	25	8

Plate 3.7. A fixed large-scale network for object naming. The network of areas which subserve object naming as determined by conjunction across three sensory modalities consists of medial frontal gyrus (GFd, BA 6), superior temporal gyrus (GTs, BA 22; putative Wernicke's area), and inferior frontal gyrus (Gfi, BA 44,45; putative Broca's area), in the left hemisphere. These areas are portrayed as colored circles in a three-dimensional glass brain based upon the Talairach and Tournoux Human Brain Atlas and stereotactic coordinate system. The table appearing below contains the average group coordinates (x, y, z) of the included regions (from Hirsch et al., 2001).

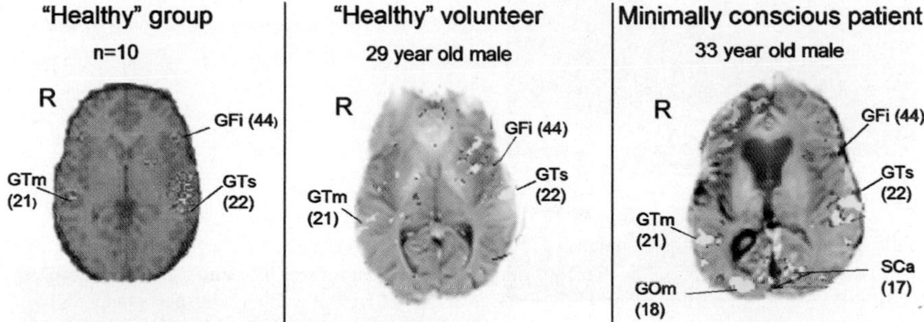

Plate 3.8. Passive listening to narratives presented as normal speech (forward) excluding responses elicited when the same speech was presented backwards. Results from the healthy volunteers are shown on one slice of an average brain (left panel), results from an age- and sex-matched healthy volunteer acquired at the same session and using the same narratives as the patient in an MCS are shown on a comparable slice (middle panel), and results from the MCS patient (see text) are shown at a similar slice level (right panel). Note the structural abnormalities in the anterior right hemisphere secondary to the injury responsible for the patient's altered state of consciousness. All three brains show activity in the left inferior frontal gyrus (GFi), Brodmann's area (BA) 44, putative Broca's area; left hemisphere temporal gyrus (GTs), BA 22, putative Wernicke's area; and middle temporal gyrus (GTm), BA21, bilaterally, putative auditory-cortex. Additionally, the MCS subject shows activity in the extrastriate visual cortex (middle occipital gyrus, GOm, BA18), and in the primary visual cortex (calcarine sulcus, SCa, BA17). The levels of statistical stringency and analysis procedures were identical for the patient and comparison healthy volunteer, suggesting that the responses elicited by the patient were (1) more robust as they activated a large volume of cortex (larger number of significant voxels) and (2) more distributed as the global pattern of activity included visual system responses. This was unexpected since there was no visual stimulation (only auditory), and the patient had his eyes closed. We cannot rule out the possibility that this patient was "visualizing" some representation of the narrative which was personalized for him.

Plate 5.2. Half-field tinted spectacles worn by A. Bompas.

Plate 5.6. Progressive change from red to blue is very difficult to notice if it occurs very slowly (10 s from Auvray and O'Regan, 2003).

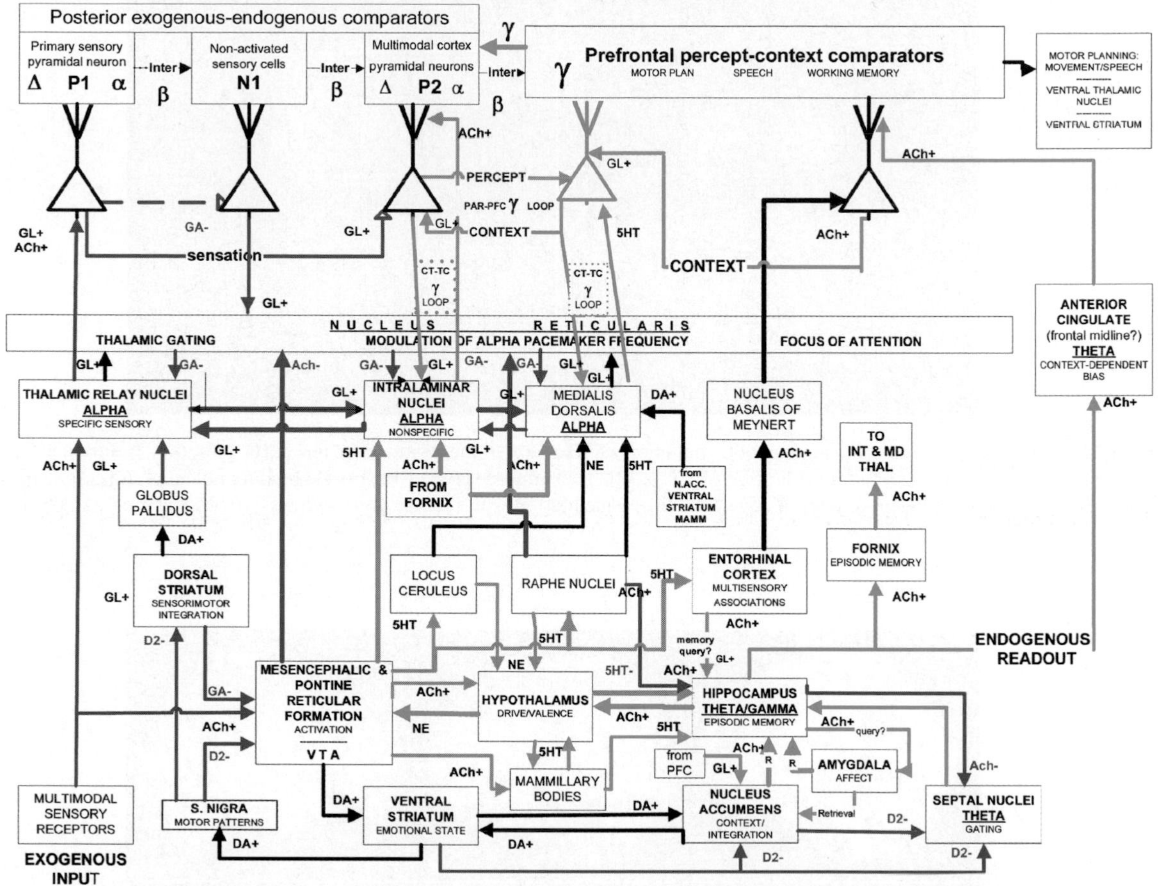

Plate 11.1. Oversimplified scheme of the neuroanatomical structures and neurotransmitters comprising the homeostatic system that regulates the electrical rhythms of the brain. The diagram assigns putative roles in the generation of EEG oscillations in the delta, theta, alpha, beta, and gamma frequency ranges and to components of event related potentials. Elements of the exogenous system processing sensory specific inputs are encoded as blue, elements performing nonsensory specific processing are encoded as gold, inhibitory influences are encoded as red, and endogenous processing of readout from the memory system is encoded as green. The behavioral functions that various regions are believed to contribute to are tentatively indicated for some of these brain structures (see text for details).

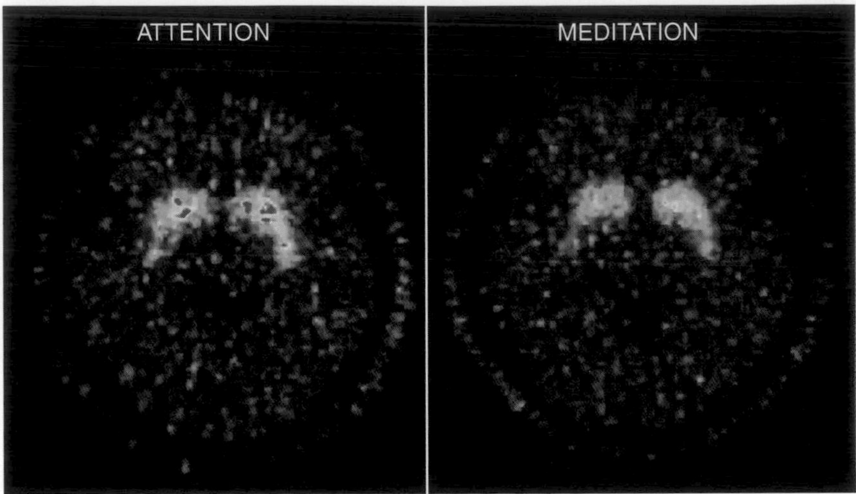

Plate 14.1. The [11C] — raclopride binding potential images at the level of the striatum for one participant (no. 8) during attention to speech (left panel), and Yoga Nidra meditation (right panel). The reduced [11C] — raclopride binding potential in ventral striatum is evidence of increased endogenous dopamine release during meditation with focused consciousness. (from Kjaer et al., 2002a, b).

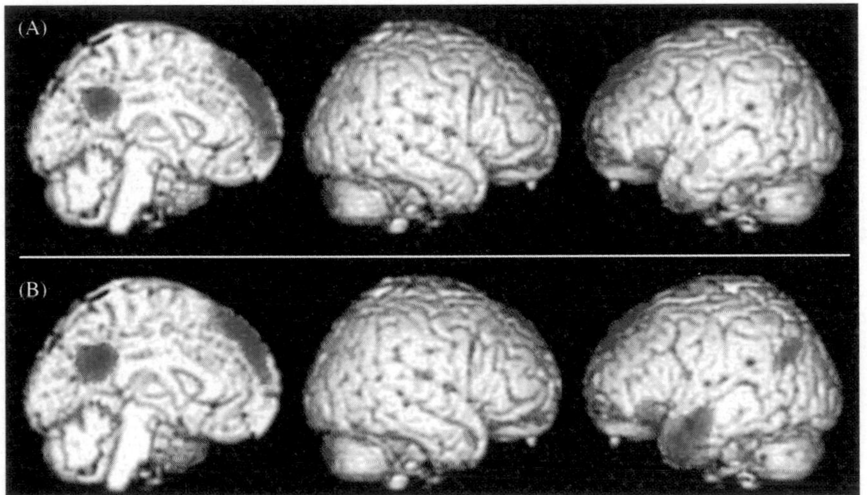

Plate 14.2. rCBF distribution in retrieval of previous judgment of mental characteristics, compared with control state. (A) Emergence of self-representation. Differential activity is noted in medial prefrontal and parietal/posterior cingulated regions, together with bilateral occipital and parietal regions, and the confluent left inferior prefrontal and temporal region ($P<0.05$, corrected for multiple comparisons). (B) Emergence of representation of Other (i.e., Danish Queen). Activation of nearly similar regions. Note that the relative contributions of the above regions are, however, different: For Self, activity is comparatively higher in right parietal region and lower in left lateral temporal region. (from Lou et al., 2004, with permission; Copyright 2004, by the National Academy of Sciences.)

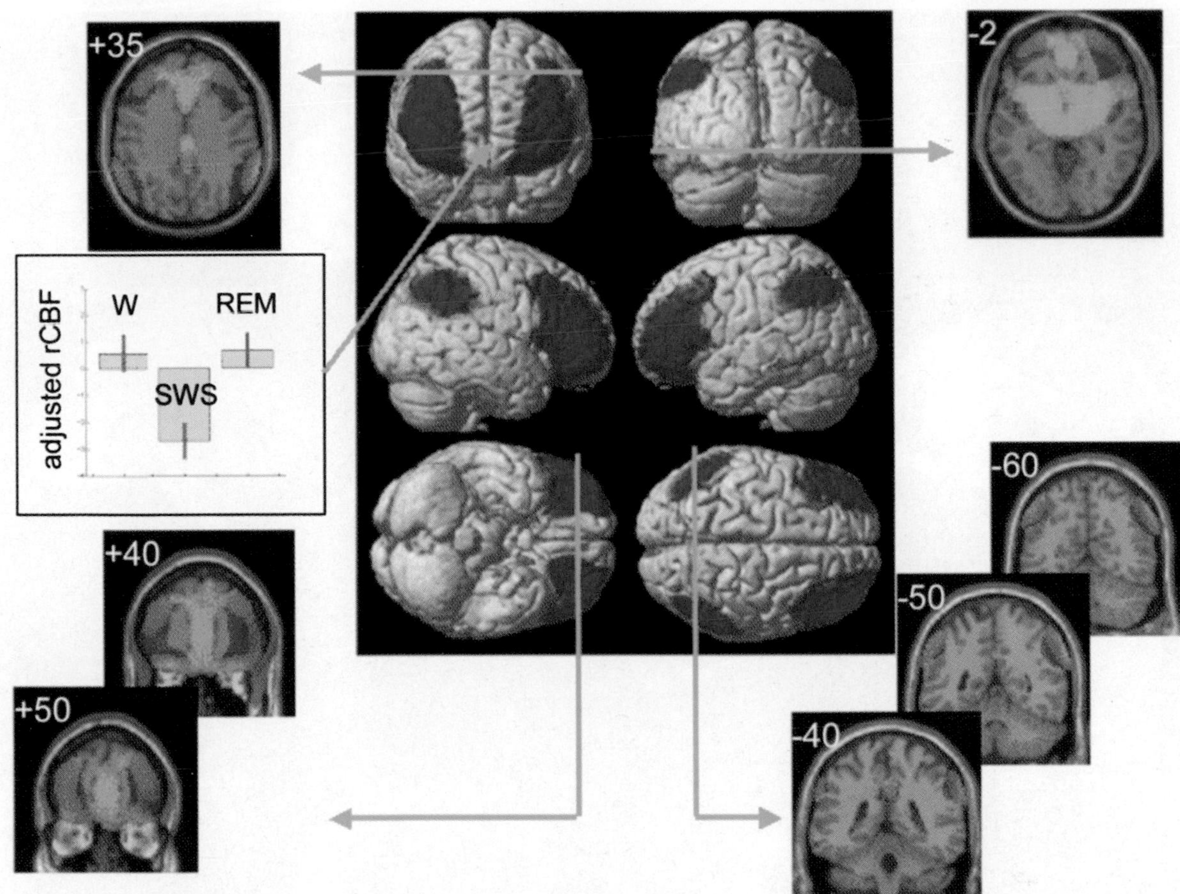

Plate 16.1. *Central panel*: Three-dimensional rendering of the brain areas that are less active during REM sleep than during wakefulness (in red). These areas involve the lateral part of the frontal cortex, the inferior lateral lobule, and medial aspects of the parietal cortex. Anterior and posterior views (first row), lateral views (second row) and bottom and top views (third row). *Side panels*: For comparison, the brain areas that are significantly less active during slow wave sleep (SWS) than during wakefulness are indicated in green. Orange areas indicate regions where regional cerebral blood flow is decreased in both SWS and REM sleep as compared to wakefulness. *Left and right upper panels*: Transverse sections 35 (left) and −2 (right) mm from the anterior-posterior commissural plane. The least active areas during REM sleep (as compared to wakefulness, in red) do not reach the medial frontal cortex. In contrast, the medial frontal cortex is one of the least active area during SWS (as compared to wakefulness, in green). *Left lower panel*: Frontal sections through the frontal cortex, at 40 and 50 mm from the anterior commissure. The least active areas during REM sleep (as compared to wakefulness, in red) involve only the inferior and middle frontal gyri but do not reach the superior frontal gyrus or the medial prefrontal cortex. The latter are significantly deactivated during SWS (green). *Right lower panel*: Frontal sections through the parietal cortex, at −40, −50, and −60 mm from the anterior commissure. The least active areas during REM sleep (as compared to wakefulness, in red) involve only the inferior parietal lobule but do not reach the intraparietal sulcus or the superior parietal cortex. They overlap (orange) with the least active area during SWS (green). *Inset*: The adjusted cerebral blood flow in the medial prefrontal area is similar during wakefulness and REM sleep and is decreased during SWS.

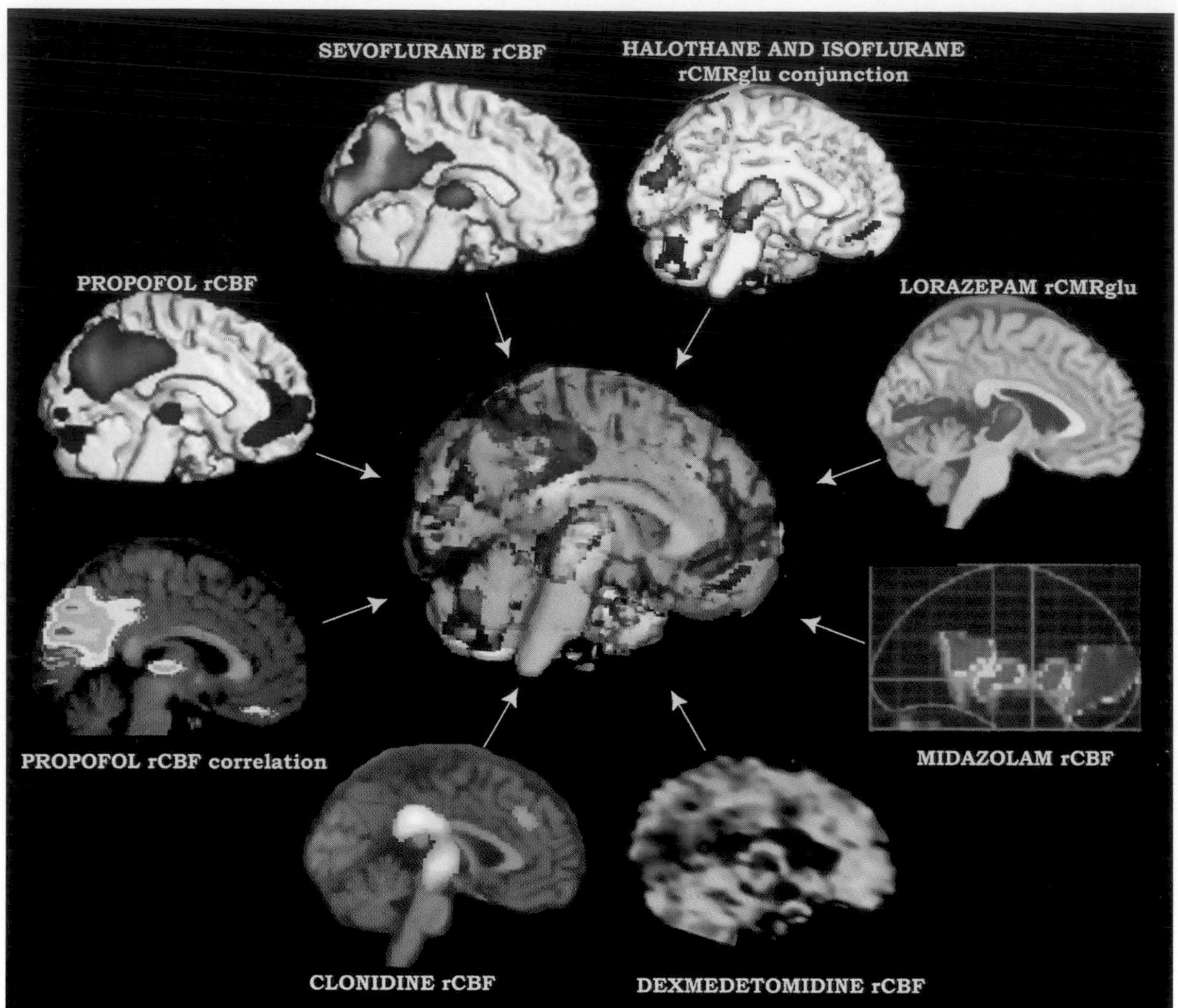

Plate 17.1. The regional effects of anesthetics on brain function are shown in humans who were given various anesthetic agents at doses that caused, or nearly caused, a loss of consciousness. The data are from seven different groups of investigators and encompass the study of eight different agents. Clockwise from the 1:00 O'clock position the agents studied were halothane and isoflurane (Alkire et al., 2000), lorazepam (Schreckenberger et al., 2004), midazolam (Veselis et al., 1997), dexmedetomidine (Prielipp et al., 2002), clonidine (Bonhomme et al., 2004), propofol (Fiset et al., 1999), propofol (Kaisti et al., 2002), and sevoflurane (Kaisti et al., 2002). The regional effects were measured using either blood flow- or glucose metabolism-based techniques. The images were reoriented, and resized to allow the direct overlapping effects between studies to be shown in the central image. The original color scales were used. Nevertheless, all images show regional decreases in activity caused by anesthesia compared to the awake state, except the propofol correlation image, which shows where increasing anesthetic dose correlates with decreasing blood flow. The figure identifies that the regional suppressive effects of anesthetics on the thalamus is a common finding that is also associated with anesthetic-induced unconsciousness.

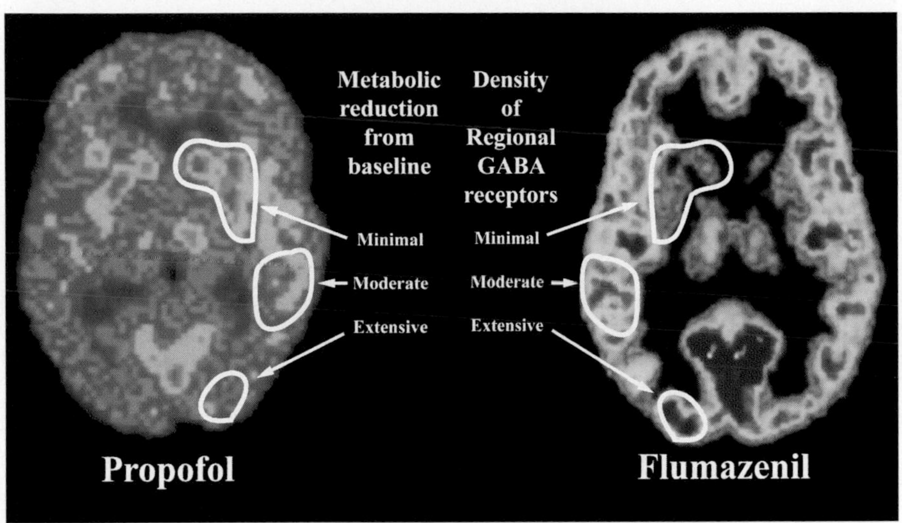

Plate 17.4. Regional cerebral metabolic suppressive effects of propofol anesthesia, on the left (Alkire et al., 1995), are compared with the regional distribution of ^{11}C-flumazenil binding (Roland, 1993). The figure suggests that regional cerebral metabolism is more depressed during propofol anesthesia (a presumed GABA agonist) in those brain regions with higher density of GABA receptors. Representative areas of metabolic reduction and receptor density are shown as regions-of-interest. A formal region-of-interest analysis examining this idea revealed that the magnitude of the metabolic suppression caused by propofol in various brain regions is highly correlated with the density of the GABA receptors in those regions (Alkire and Haier, 2001). Future imaging work with other ligands may reveal more directly how the regional metabolic effects of various anesthetics might be interpreted as a simple reflection of their underlying biochemical interactions.

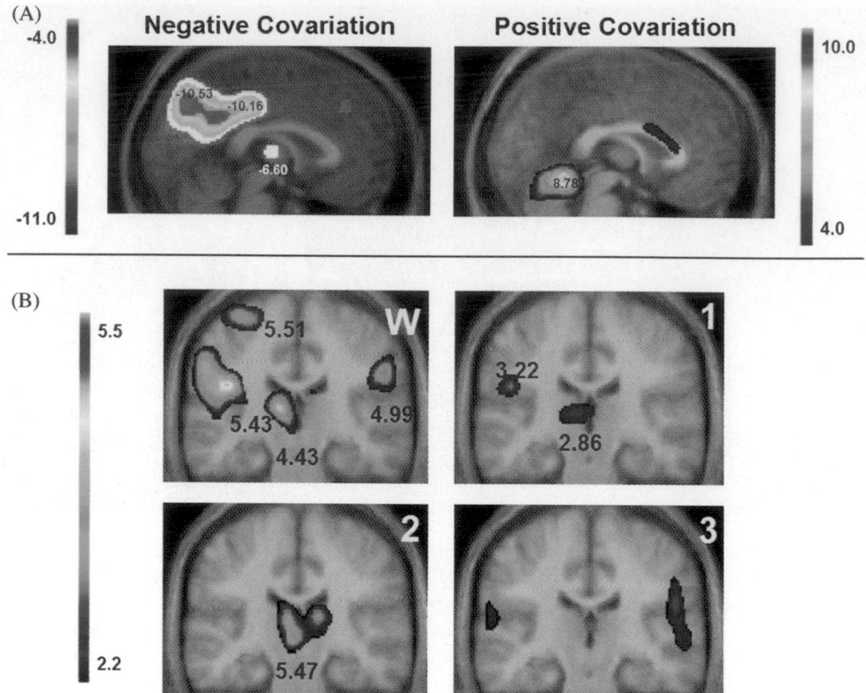

Plate 18.1. (A) Representative slices of t-statistic maps illustrating the negative covariation (left panel) and the positive covariation (right) between the measured plasma concentration of propofol and regional cerebral blood flow. The left color scale is plotted for t values ranging from −11.0 to −4.0 and the right color scale for t values ranging from 4.0 to 10.0. The maximal t value of each peak is plotted on these maps: the maximal t value of the negative covariation observed in the precuneus is −10.53, it is −10.16 in the posterior cingulate cortex, and −6.60 in the thalamus. The maximal t value of the positive covariation observed in the cerebellar vermis is 8.78. These results have been replicated in our three studies using H_2O (Dunnet et al., 1994). From Fiset et al. (1999).
(B) Representative slices of subtraction t-statistic maps showing the effect of somatosensory stimulation (vibration-on versus vibration-off) for each level of anesthesia. W: awake condition, propofol concentration, 0 μg/mL; 1: Level 1, propofol concentration, 0.5 μg/ml; 2: Level 2, propofol concentration, 1.5 μg/mL; 3: Level 3, propofol concentration, 3.5 μg/mL. The color scale is plotted for t values ranging from 2.2 to 5.5. The maximal t value of each peak is plotted on these maps: the t value of the thalamic peak is 4.43 at Level W, 2.86 at Level 1 and 5.47 at Level 2; the t value of the left primary somatosensory cortex peak is 5.51 at Level W, the t value of the left secondary somatosensory cortex peak is 5.43 at Level W and 3.22 at Level 1, the t value of the right secondary somatosensory cortex peak is 4.99 at Level W. The t values in colored regions visible in the statistical map at Level 3 are not significant. Those areas are extensions of non-significant peaks centered elsewhere, in the temporal lobes. From Bonhomme et al. (2001).

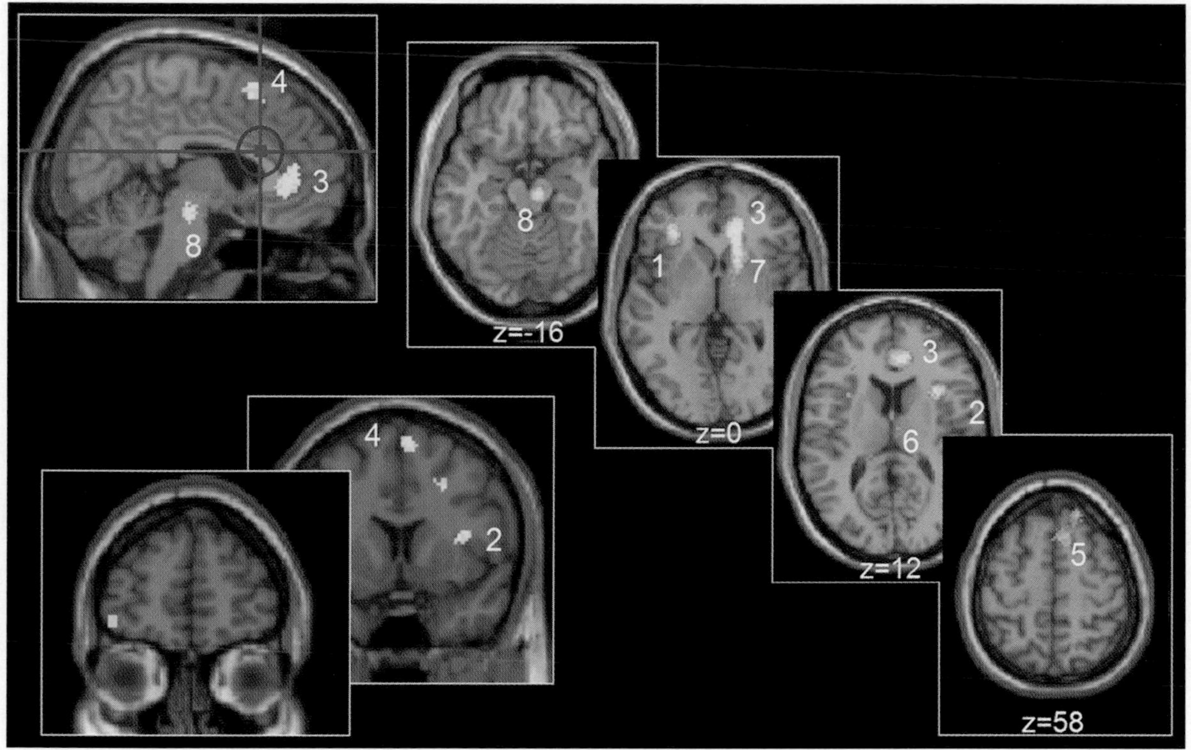

Plate 19.4. During hypnosis an increase in activity in midcingulate cortex increases activity in a wide cortical and subcortical neural network (part of the 'pain matrix'), much more so than is observed under control conditions (rest or distraction tasks). Regions that showed such hypnosis-related increased functional connectivity with midcingulate cortex (peak voxel marked by red crosshair in circle) are: left insula (1), right insula (2), perigenual cortex (3), pre-supplementary motor area (4), superior frontal gyrus (5), right thalamus (6), right caudate nucleus (7), and midbrain/brainstem (8). Adapted from Faymonville et al. (2003).

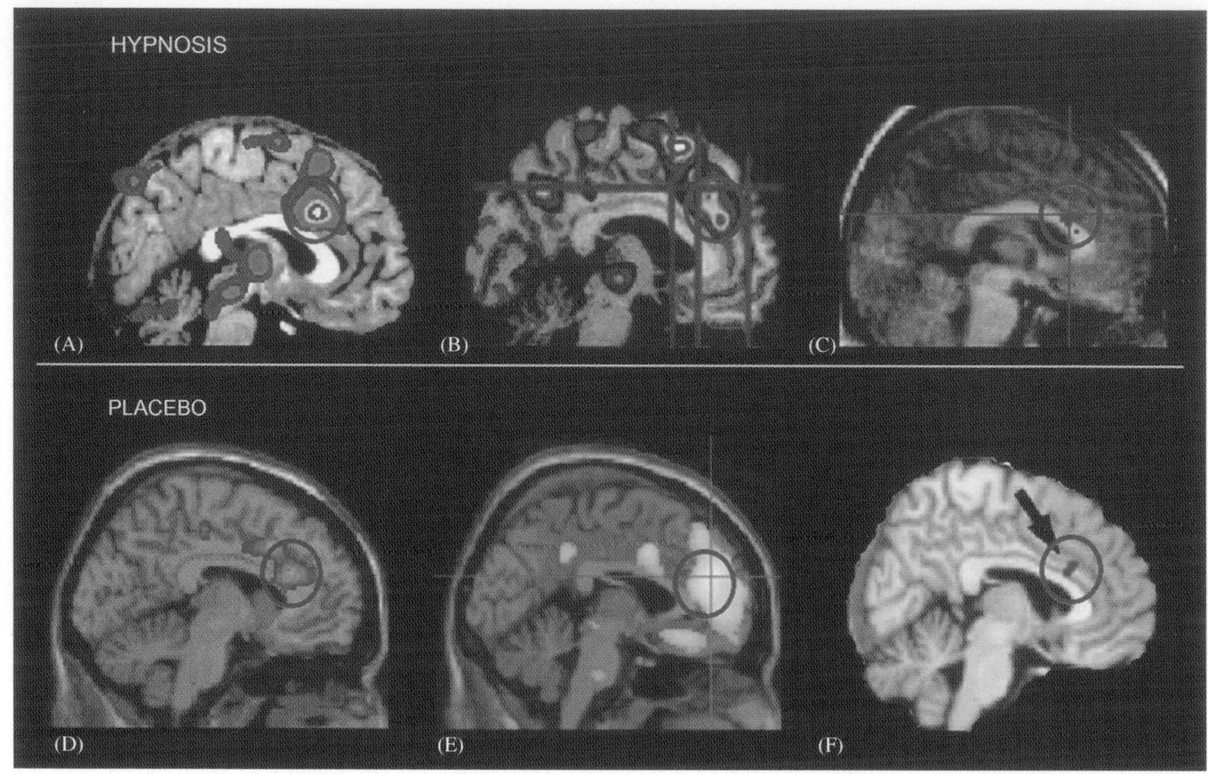

Plate 19.7. Both hypnosis- and placebo-induced modulation of pain perception seems to be mediated by the ACC (red circles). Note that mean peak coordinates from hypnosis- and placebo-induced pain modulation studies, using very different methodology, are within 5 mm from each other ($x = 2$, $y = 22$, $z = 28$ versus $x = 6$, $y = 17$, $z = 31$ mm, respectively). *Hypnosis*: (A) Suggestion-related changes in rCBF during pain perception: the image shows rCBF changes for the subtraction hypnosis-with-suggestion minus hypnosis-without-suggestion, both under painful stimulation (coordinates of peak voxel: $x = 7$, $y = 20$ and $z = 29$ mm). From Rainville et al. (1997). (B) Pain-related activity associated with hypnotic suggestions of high unpleasantness (painfully hot/high unpleasantness minus neutral/hypnosis control condition; peak voxel: $x = 0$, $y = 29$, $z = 34$ mm). From Rainville et al. (1999). (C) rCBF increases in proportion to pain sensation ratings, in the specific context of the hypnotic state (difference in pain ratings versus rCBF regression slopes between the hypnotic state and control conditions (i.e., non-hypnotic mental imagery and resting state; peak voxel: $x = 2$, $y = 18$, $z = 22$ mm). From Faymonville et al. (2000). *Placebo*: (D) rCBF increases with increased intestinal discomfort in placebo-treated patients with chronic abdominal pain (irritable bowel syndrome): a greater reduction in rCBF response to visceral stimulation from pre- to post-placebo was associated with greater self-reported symptom improvement (peak voxel: $x = 8$, $y = 19$, $z = 34$ mm). From Lieberman, M. D., Jarcho, J. M., Berman, S., Naliboff, B. D., Suyenobu, B. Y., Mandelkern, M. and Mayer, E. A. (2004). Neuroimage, 22: 447–455. (E) Activated areas in high placebo responders (healthy volunteers) during heat pain and opioid treatment minus heat pain only (peak voxel: $x = 3$, $y = 18$, $z = 34$ mm). From Petrovic et al. (2002). (F) Pain regions showing correlations between placebo effects in reported pain (control minus placebo) and placebo effects in neural pain (i.e., measured activity in pain related brain areas) (control minus placebo) (peak voxel: $x = 6$, $y = 14$, $z = 26$ mm). From Wager et al. (2004).

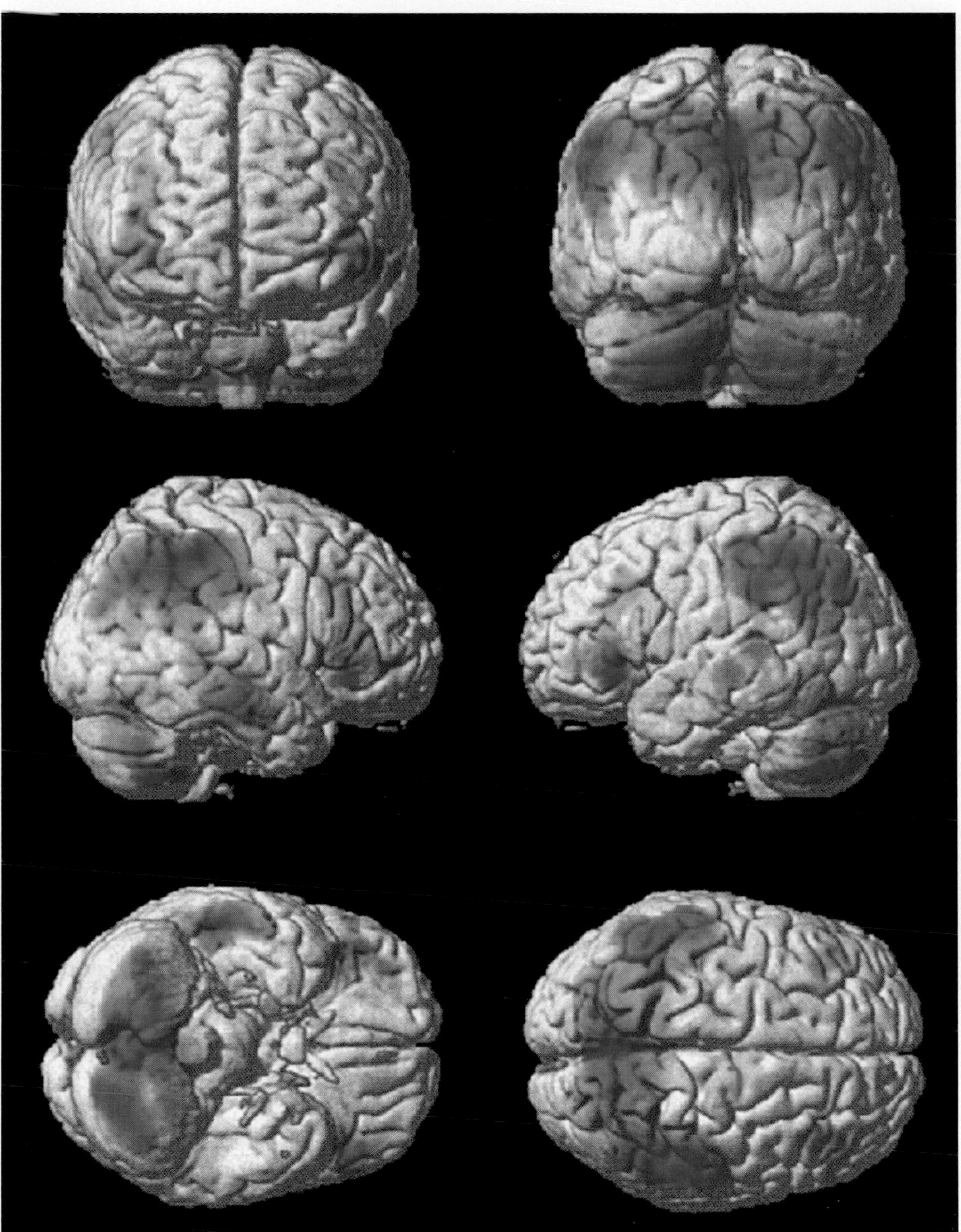

Plate 20.1. Fronto-parietal cortical involvement in generalized tonic-clonic seizures. Generalized tonic-clonic seizures were induced by electroconvulsive therapy for treatment of refractory depression (American Psychiatric Association, 2001) and cerebral blood flow (CBF) was imaged by single photon emission computed tomography (SPECT). Red represents relative increases in ictal compared to interictal CBF on SPECT scans, and green represents decreases. Despite the clinically generalized convulsions, focal relative signal increases are present in higher-order frontal and parietal association cortex, while many other brain regions are relatively spared. Ictal versus interictal SPECT images were analyzed with statistical parametric mapping. Reproduced with permission from Blumenfeld et al. (2003b)..

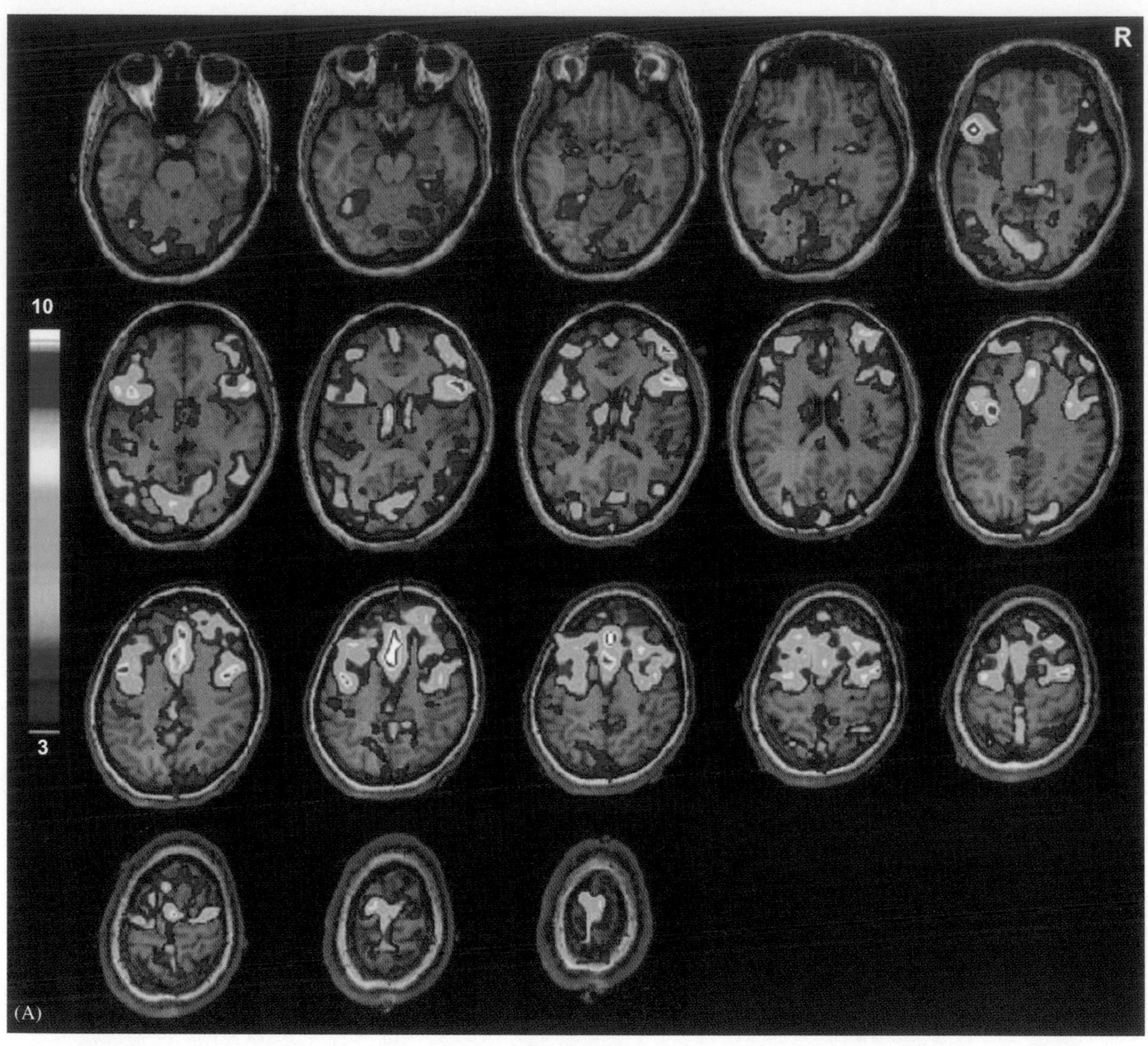

Plate 20.5. fMRI during human absence seizures. 18-year-old patient with onset of absence seizures at age 8. EEG and fMRI were recorded simultaneously. (A) fMRI activation. (B) fMRI deactivation. Focal regions of increased and decreased signal are seen bilaterally. Color bars show t scores. Reproduced with permission from Aghakhani et al. (2003).

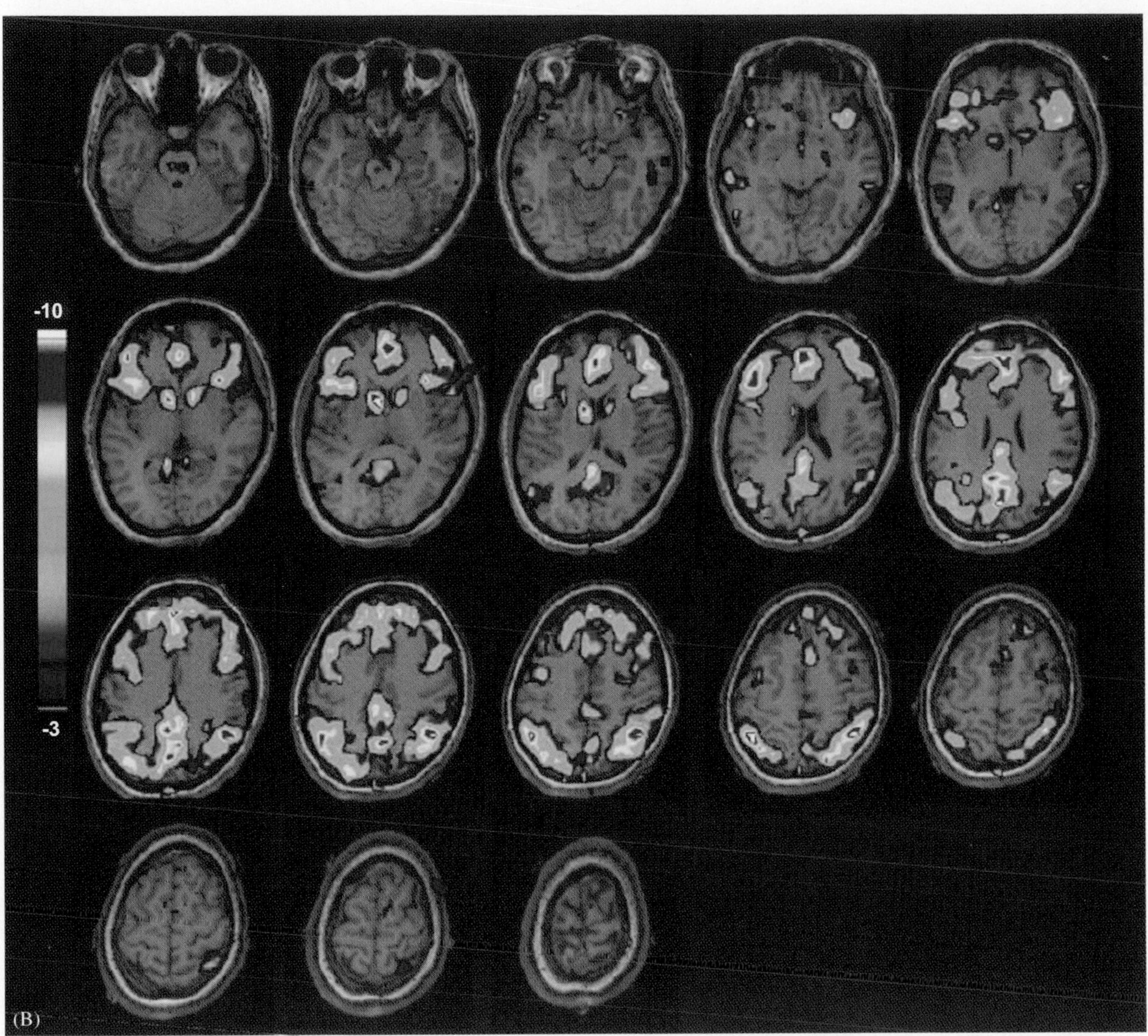

Plate 20.5. (*Continued*).

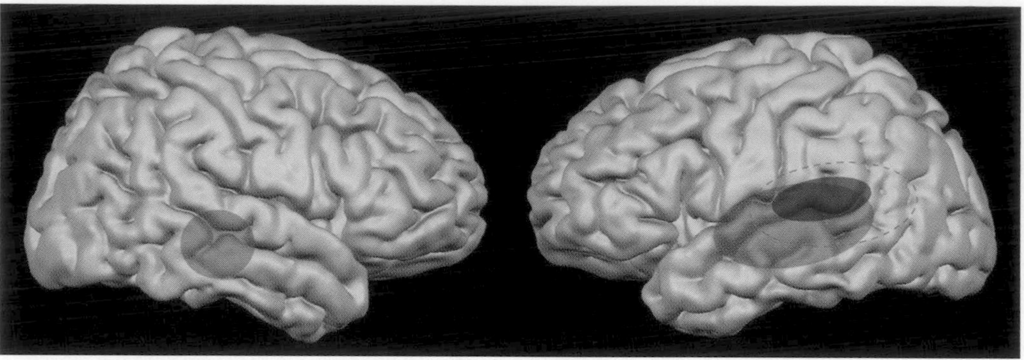

Plate 22.1. A schematic illustration of neural activity changes during auditory hallucinations and formal thought disorder in schizophrenia. Areas shaded in red represent activations during auditory hallucinations. The area shaded in blue represents reduced activation compared with normal language when positive formal thought disorder is present. The green, dotted line indicates the location of Wernicke's area (from Kircher et al., 2004b.).

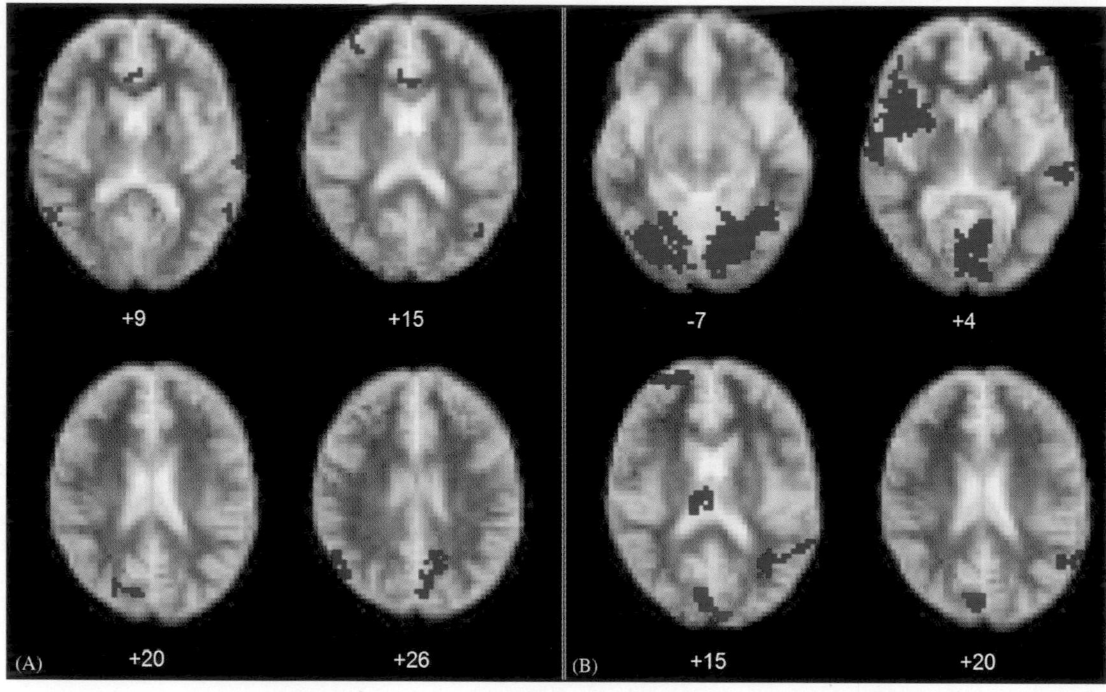

Plate 22.2. (A) Neural correlates of "negative" formal thought disorder in patients with schizophrenia. In this study, patients with schizophrenia speak continuously while their brain activity is measured with fMRI. The amount of negative formal thought disorder (e.g. impoverishment of thought, blocking, poverty of content of speech as measured with the Thought and Language Index, Liddle et al., 2002) per 20 s speech epoch was correlated with the BOLD effect. Red voxels indicate positive and blue voxels negative correlation ($p < 0.001$; from Kircher et al., 2003). Note that left precuneus, cuneus, medial frontal gyrus and right inferior parietal lobe are activated when negative formal thought disorder is prominent. (B) In healthy volunteers, short speech pauses during fluent speech correlate with activation in the superior temporal sulcus, between superior/middle temporal gyrus and parietal lobe (marked in red; $p < 0.001$). Note that this region, part of Wernicke's speech area, is underactivated during the production of formal thought disorder in schizophrenia. Activated voxels during overt speech are shown in blue. Talairach z coordinates are shown below the transverse sections. The left side of the image represents the right side of the brain (from Kircher et al., 2004a).

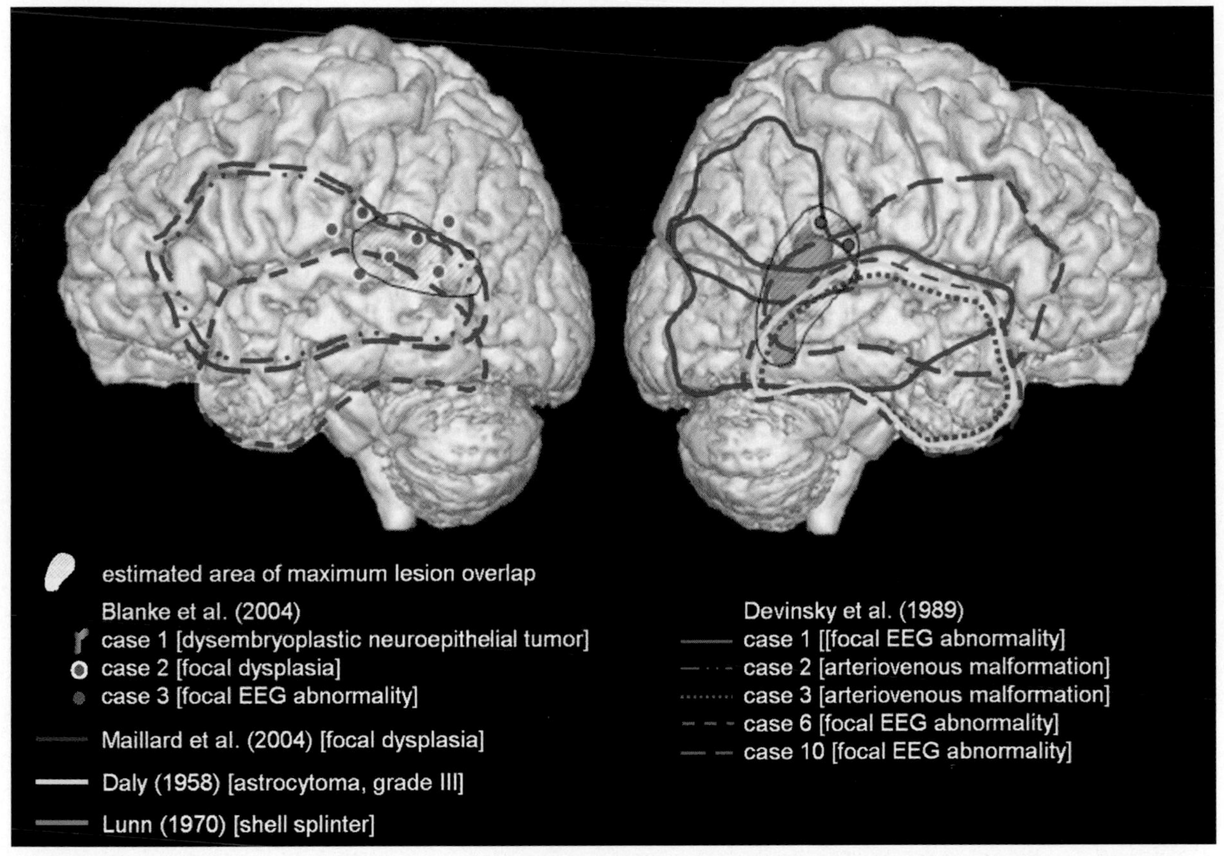

Plate 24.2. Lesion location in OBE patients. MRI-based lesion overlap analysis of all previously reported OBE patients due to focal brain damage. The data for all patients are drawn on the MRI of one of the patients from Blanke et al. (2004). Lesion data are plotted separately for the left (left figure) and right hemisphere (right figure). For each study a different color was chosen: Daly (1958; yellow; 1 patient), Lunn (1970; green; 1 patient), Devinsky et al. (1989; blue; 5 patients), Blanke et al. (2004; pink; 3 patients), and Maillard et al. (2004; red; 1 patient). For the three patients by Blanke et al. (2004) the pink dots surrounded by white represent the zone of seizure onset for case 2, the pink dots surrounded by black represent the zone of seizure onset for case 3, and the pink, shaded, area represents the area of lesion overlap for case 1. Lesion location is indicated by cortical area surrounded by a line for each patient. The lesion sites of all patients by Blanke et al. (2004) were transformed to Talaraich space and transformed to the coordinates of the MRI of one of the patients. Thus, when the lesion location (or the epileptic focus) was described as right temporal, we marked the whole temporal lobe for that patient, if the lesion was characterized as left fronto-temporal parts of the fronto-temporal lobe were marked. The two patients with bilateral EEG abnormalities from Devinsky et al. (1989; cases 6 and 10) have been plotted on the right and left hemisphere. The regions of maximal overlap are indicated by the hatched area for each hemisphere and were found on the right and left TPJ. Most of the OBEs were due to interference with the right TPJ. See Blanke et al. (2004) for further details.

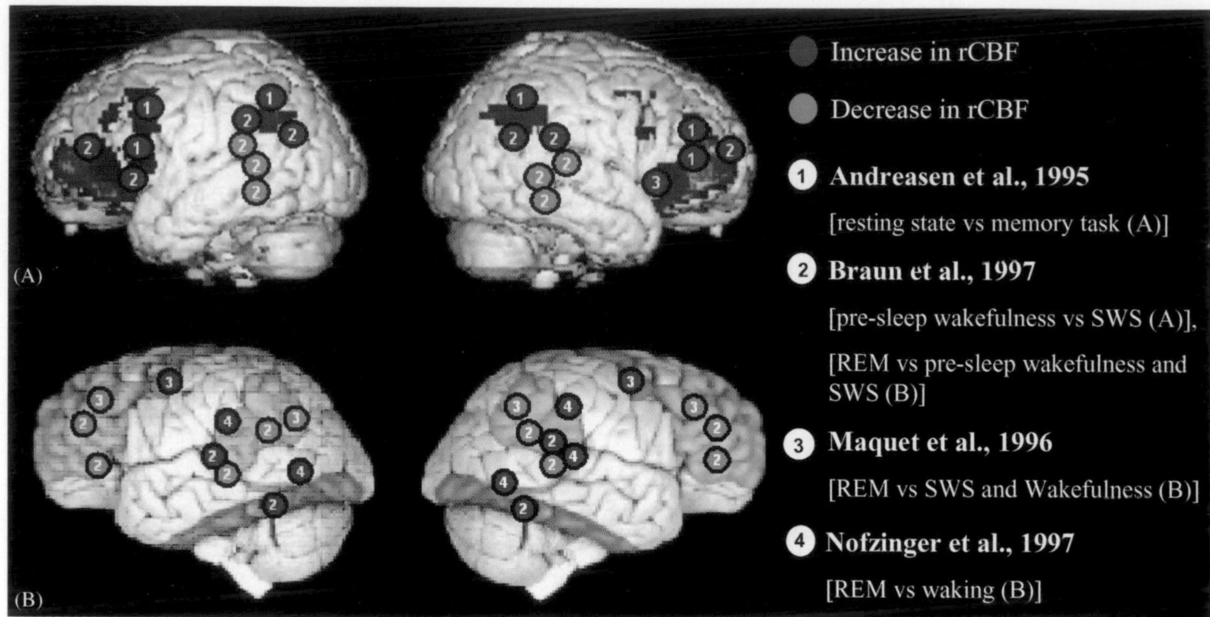

Plate 24.3. Regional cerebral blood flow during pre-sleep wakefulness, resting state, and during REM sleep. (A) The figure shows brain activation patterns during pre-sleep wakefulness and resting state compared to sleep (see text). The results of several studies are shown (1, Andreasen et al., 1995; 2, Braun et al., 1997). Estimated peak of activation (red) and deactivation (blue) are plotted on a figure reproduced with permission from Maquet (2000) showing activated brain areas during pre-sleep wakefulness as compared to sleep. The main voxels of activation on the lateral cortical surface are shown. Note the focalization on two lateral cortical areas: prefrontal cortex and TPJ. Whereas the prefrontal cortex is activated during pre-sleep wakefulness, the TPJ is characterized by an activation of the inferior parietal lobule and deactivation of the posterior parts of the superior, middle, and inferior temporal gyri. (B) Schematic representation of the relative increases and decreases in neural activity associated with REM sleep. The figure is modified and reproduced with permission from Schwartz and Maquet (2002). The results of several studies (2, Braun et al., 1997; 3, Maquet et al., 1996; 4, Nofzinger et al., 1997) are plotted. The main voxels of activation on the lateral cortical surface are shown. Red color indicates increases and blue color indicates decreases in rCBF. Brain deactivations focus on the lateral prefrontal cortex and the TPJ. Again, the TPJ shows patterns of activation and deactivation in close proximity. (See text for more details).

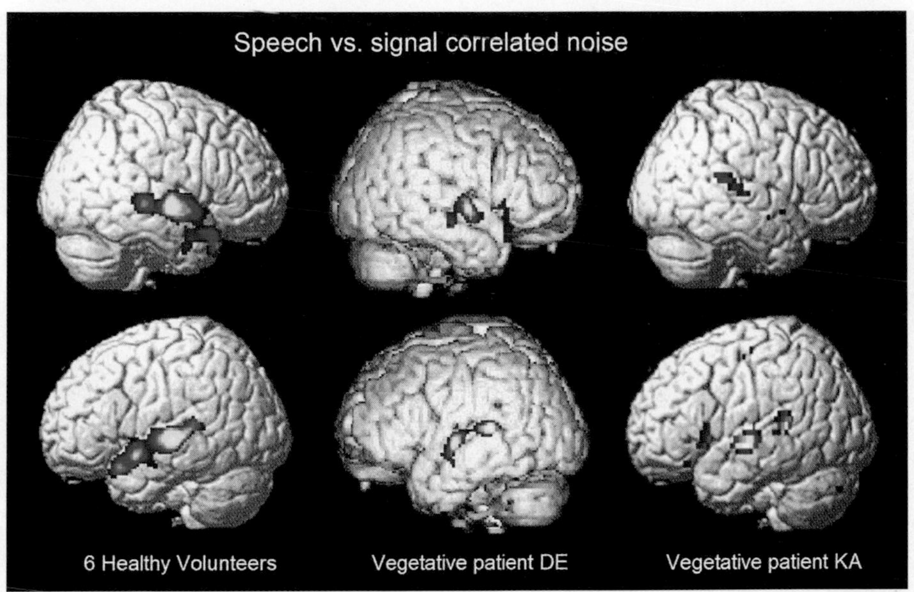

Plate 32.1. PET data illustrating the comparison of speech with signal correlated noise. On the left, control data (adapted from Mummery et al., 1999), and on the right data from two patients meeting the clinical criteria for vegetative state. All data were pre-processed and analyzed using Statistical Parametric Mapping software (SPM99, Wellcome Department of Imaging Neuroscience, London, UK).

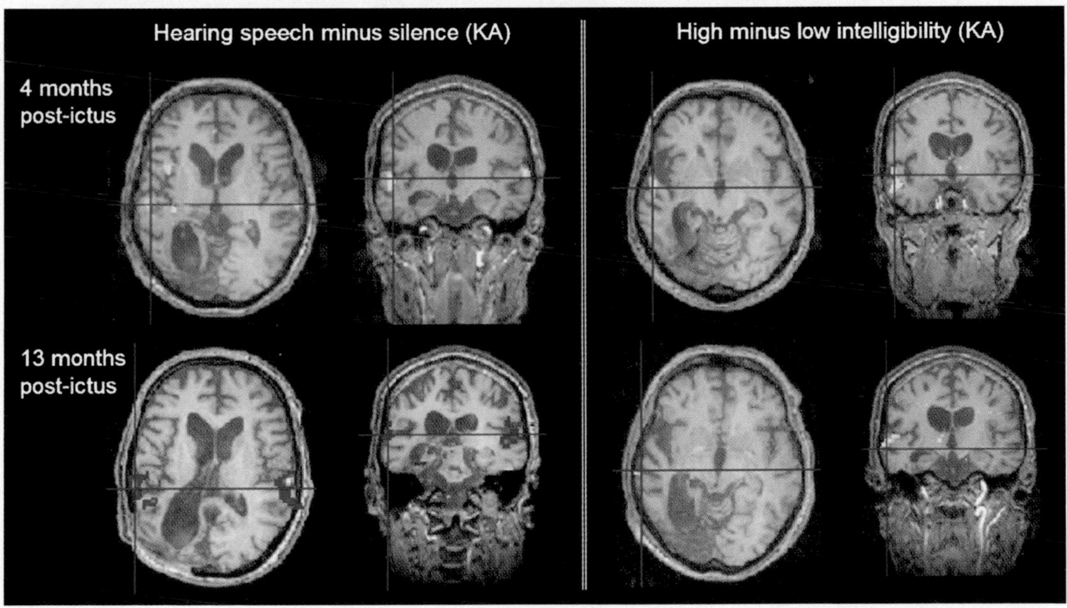

Plate 32.5. Activation data from the two PET sessions, 9 months apart. (i) Hearing speech minus silence during the first PET session (top left) reveals activity bilaterally in the superior temporal gyrus. (ii) Hearing speech minus silence during the second PET session (bottom left) reveals very similar activity bilaterally in the superior temporal gyrus. (iii) High intelligibility speech minus low intelligibility speech during the first PET session (top right) reveals activity predominantly in the left superior temporal gyrus. (iv) High intelligibility speech minus low intelligibility speech during the second PET session (bottom right) reveals very similar activity predominantly in the left superior temporal gyrus. All data were pre-processed and analyzed using Statistical Parametric Mapping software (SPM99, Wellcome Department of Imaging Neuroscience, London, UK).

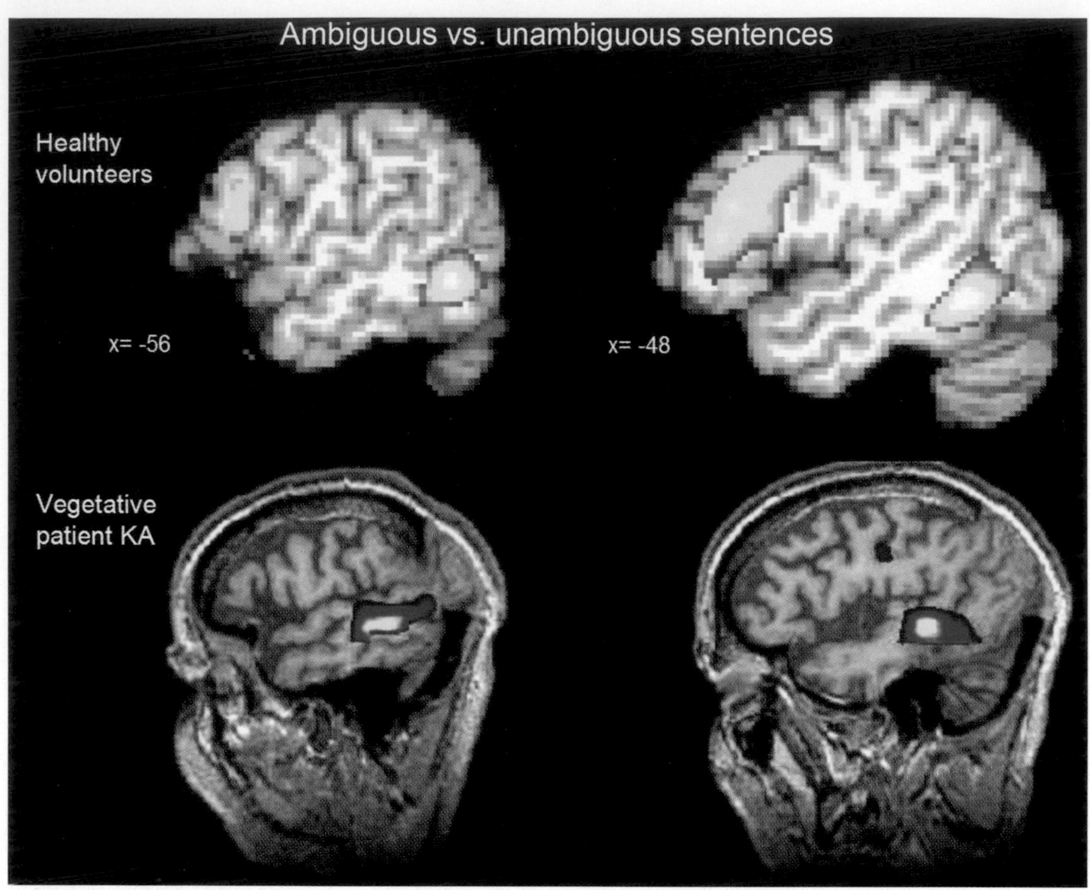

Plate 32.6. fMRI data for the ambiguous sentences versus unambiguous sentences comparison. Like controls (top; adapted from Rodd et al., 2005), the patient (bottom) exhibited significant signal intensity changes in the left posterior inferior temporal cortex, but (unlike controls) not in the inferior frontal gyrus. All data were pre-processed and analyzed using Statistical Parametric Mapping software (SPM99, Wellcome Department of Imaging Neuroscience, London, UK). Data presented are thresholded at $p<0.05$ (corrected).

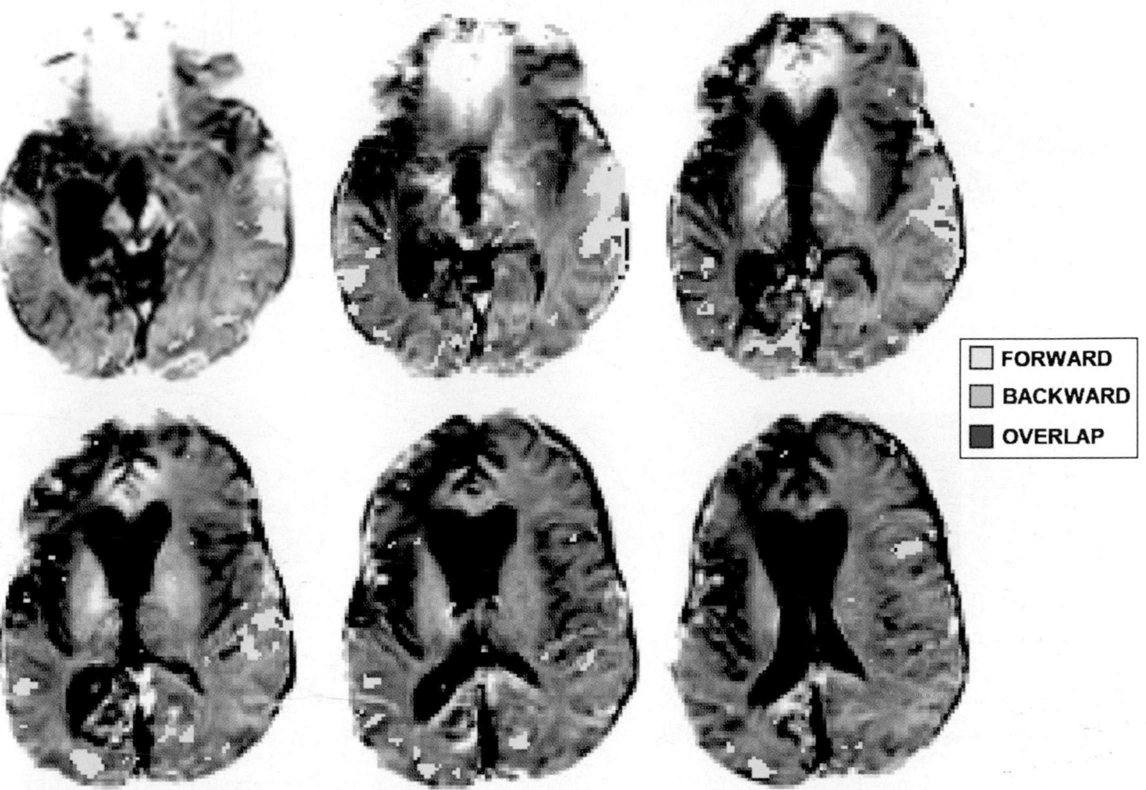

Plate 33.3. Functional magnetic resonance imaging activation patterns of BOLD signal in response to passive language presentations. Reproduced with permission from MIT Press.

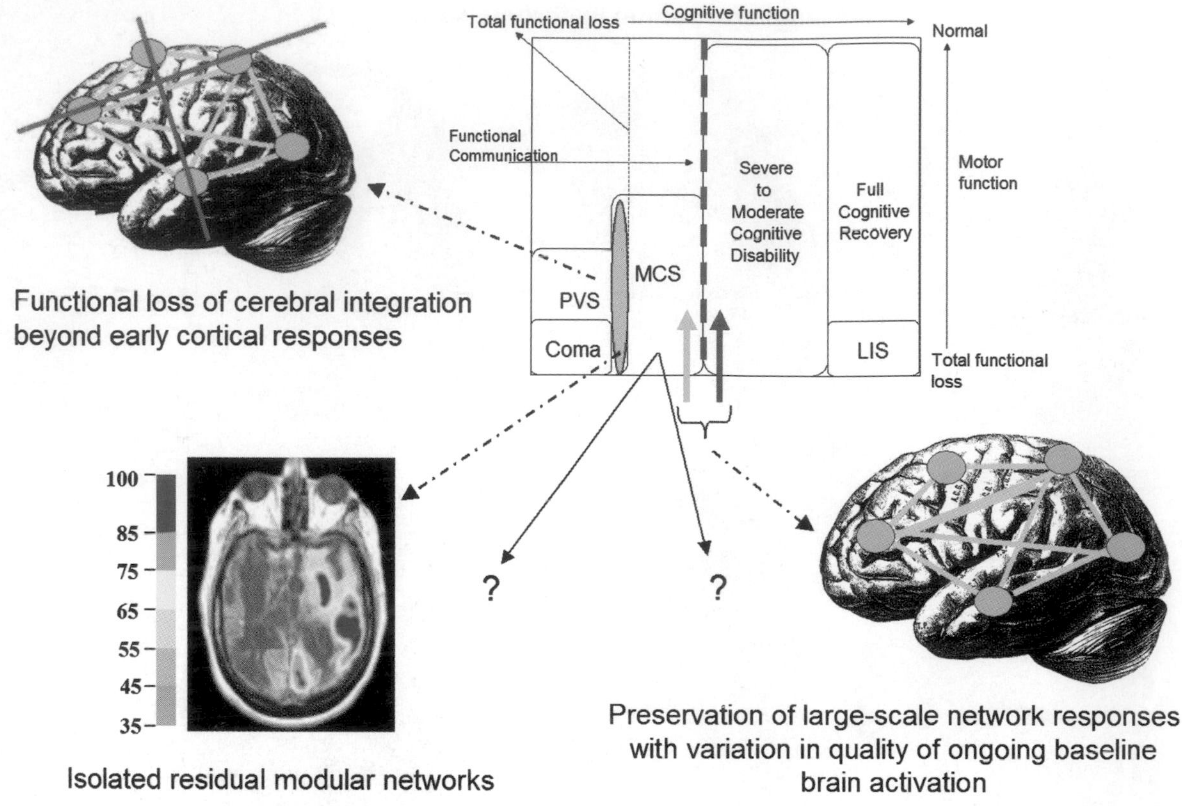

Plate 33.4. Mechanisms underlying functional levels across spectra of vegetative state and minimally conscious state patients. Co-registered FDG-PET and MRI image from patient in Fig. 2 with color scale indicating percentage of normal regional metabolic rates (from Schiff et al., 2002; see text for further discussion).

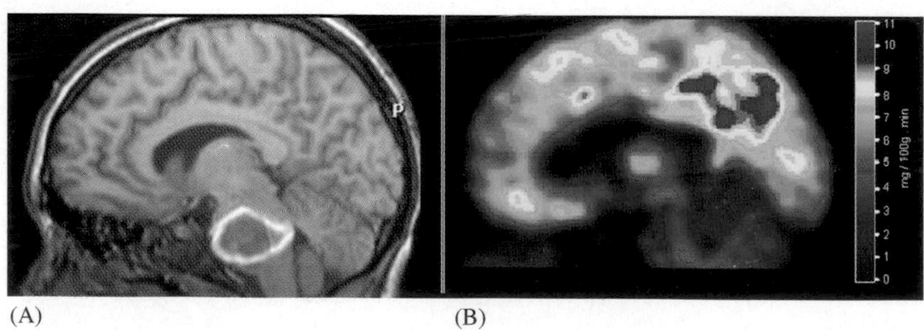

Plate 34.2. (A) Magnetic resonance image (sagittal section) showing a massive hemorrhage in the brainstem (circular hyperintensity) causing a locked-in syndrome in a 13 year-old girl. (B) ^{18}F-fluorodeoxyglucose — Positron emission tomography illustrating intact cerebral metabolism in the acute phase of the LIS when eye-coded communication was difficult due to fluctuating vigilance. The color scale shows the amount of glucose metabolized per 100 g of brain tissue per minute. Statistical analysis revealed that metabolism in the supratentorial gray matter was not significantly lower as compared to healthy controls (Adapted from Laureys et al., 2004a).